100 YEARS OF SUPERCONDUCTIVITY

100 YEARS OF SUPERCONDUCTIVITY

Edited by

Horst Rogalla and Peter H. Kes

CRC Press
Taylor & Francis Group
Boca Raton London New York

CRC Press is an imprint of the
Taylor & Francis Group, an **informa** business

A TAYLOR & FRANCIS BOOK

CRC Press
Taylor & Francis Group
6000 Broken Sound Parkway NW, Suite 300
Boca Raton, FL 33487-2742

Printed in the United States of America on acid-free paper
Version Date: 20110811

International Standard Book Number: 978-1-4398-4946-0 (Hardback)

1007081949

Library of Congress Cataloging-in-Publication Data

100 years of superconductivity / editors, Horst Rogalla, Peter H. Kes.
   p. cm.
  Includes bibliographical references and index.
  ISBN 978-1-4398-4946-0 (hardback)
  1. Superconductors--Research--History. 2. Superconductors--Industrial applications. I. Rogalla, H. (Horst) II. Kes, P. H. (Peter H.) III. Title: One hundred years of superconductivity.

QC611.96.A16 2011
537.6'23--dc23
                                         2011030774

Visit the Taylor & Francis Web site at
http://www.taylorandfrancis.com

and the CRC Press Web site at
http://www.crcpress.com

Contents

Preface   **xvii**

List of Authors   **xix**

Glossary   **xxv**

1 Early History   **1**
  Editor: Peter H. Kes
  1.1   The Discovery and Early History of Superconductivity
        Rudolf de Bruyn Ouboter, Dirk van Delft and Peter H. Kes   1
        1.1.1   The Real Story   1
        1.1.2   Elemental Superconductors 15
        1.1.3   Superconductive Alloys, the Spongy Road to Type II Superconductors. . . 24
  1.2   The Historical Context of Josephson's Discovery
        A. B. Pippard 29
  1.3   Further Reading 50

2 Theory   **51**
  Editor: Jan Zaanen
  2.1   Ginzburg–Landau Equations
        V. M. Vinokur and T. I. Baturina 51
        2.1.1   Introduction 51
        2.1.2   A History 52
        2.1.3   The Equation 53
        2.1.4   Consequences and Applications 58
        2.1.5   Conclusion: Strings, the Universe, and All the Rest... 64
  2.2   The Emergence of BCS
        David Pines 65
        2.2.1   Introduction 65
        2.2.2   A Welcoming Environment 65
        2.2.3   Starting Work with Bardeen 66
        2.2.4   Polaron Theory 67
        2.2.5   Effective Electron Interactions in Metals 67
        2.2.6   Cooper's Pair Instability 68
        2.2.7   Seattle and Stockholm 69
        2.2.8   The Schrieffer Wave-Function 70
        2.2.9   Deciphering, Teaching, and Applying BCS 70
        2.2.10 BCS Theory 71
  2.3   Theory of Superconductivity: From Phenomenology to Microscopic Theory
        Lev P. Gor'kov 72
        2.3.1   Introduction 72
        2.3.2   Early Days 73
        2.3.3   The 1950s 77

      2.3.4   Microscopic Theory 79

      2.3.5   Instead of Conclusion 91

  2.4   A Modern, but Way Too Short History of the Theory of Superconductivity at a High Temperature

      Jan Zaanen 92

      2.4.1   Introduction 92

      2.4.2   Pushing BCS to the Limits: The Spin Fluctuation Superglue 94

      2.4.3   The Legacy of Philip W. Anderson (I): Mottness 99

      2.4.4   The Legacy of Philip W. Anderson (II): Resonating Valence Bonds and Their Descendants 100

      2.4.5   Theory that Works: The Competing Orders of the Pseudogap Regime . . . 106

      2.4.6   Quantum Critical Metals and Superconductivity 110

  2.5   Intrinsic Josephson Effect in Layered Superconductors

      L. N. Bulaevskii and A.E. Koshelev 115

      2.5.1   Introduction: Layered Superconductors and Lawrence-Doniach Model . . . 115

      2.5.2   Vortex Matter in Layered Superconductors 118

      2.5.3   Josephson Plasma Resonance 122

      2.5.4   Radiation from Intrinsic Josephson Junctions 124

  2.6   Mixed State Properties

      Ernst Helmut Brandt 125

      2.6.1   Introduction 125

      2.6.2   Abrikosov's Periodic Vortex Lattice Near B_{c2} 126

      2.6.3   Distorted Vortex Lattice and Vacancy Near B_{c2} 127

      2.6.4   Nonlocal Elasticity of the Vortex Lattice 129

      2.6.5   Vortex Arrangements at Low Inductions 131

      2.6.6   Vortex Lattice Solution for All κ and $\bar{B}$ 132

      2.6.7   Vortex Motion, Pinning, and Thermal Depinning 135

      2.6.8   Anisotropic Superconductors 136

  2.7   Thermomagnetic Effects in the Mixed State

      R. P. Huebener 137

3   Experiments   **145**

  Editor: J. Mannhart

  3.1   Tunneling and the Josephson Effect

      John M. Rowell 145

      3.1.1   The Josephson Effect 153

  3.2   The Discovery of Fluxoid Quantization

      Dietrich Einzel 161

      3.2.1   Introduction 161

      3.2.2   Theoretical Foundations 162

      3.2.3   The Doll-Näbauer Experiment 163

      3.2.4   The Deaver-Fairbank Experiment 165

      3.2.5   The IBM Conference 1961 167

      3.2.6   Post-1961 168

      3.2.7   Summary and Conclusion 169

  3.3   The Search for the Pairing Symmetry in the High Temperature Superconductors

      Dale J. Van Harlingen 170

      3.3.1   Why the Symmetry Is Important 171

      3.3.2   Determining the Symmetry 171

      3.3.3   Samples 173

      3.3.4   Parity 174

        3.3.5   Penetration Depth 174
        3.3.6   ARPES 175
        3.3.7   Josephson Interferometry 177
        3.3.8   Beyond d-Wave Symmetry 182
   3.4 Half-Integer Flux Quantization in Unconventional Superconductors
        C. C. Tsuei and J. R. Kirtley 182
        3.4.1   The Design of the Tricrystal Experiments 183
        3.4.2   Direct Observation of the Half Flux Quantization 187
        3.4.3   Elucidation of the Nature of Half Flux Quantum Effect . . . 191
        3.4.4   Universality of the $d_{x^2-y^2}$ Pair State 192
        3.4.5   Large-Scale Arrays of the Half Flux Vortices 194
        3.4.6   Concluding Remarks 196
   3.5 Electric Field Effect Tuning of Superconductivity
        Jean-Marc Triscone and Marc Gabay 197
        3.5.1   Preamble 197
        3.5.2   Introduction 197
        3.5.3   Electric Field Effect in Complex Oxides 197
        3.5.4   Electric Field Effect in Superconductors: Physics and Lengthscales 199
        3.5.5   Thomas Fermi Screening and Interface Quality 201
        3.5.6   Field Effect Setups in a Nutshell 201
        3.5.7   What's New? 203
        3.5.8   Conclusions 206
   3.6 The Grain Boundary Problem of High-T_c Superconductors
        J. Mannhart and D. Dimos 206
   3.7 Overview of the Experimental History, Status and Prospects of HTS
        M. R. Beasley 213
        3.7.1   The Allure of High Temperature Superconductivity 213
        3.7.2   The Broad Sweep of History 213
        3.7.3   From Onnes to Bednorz and Mueller 215
        3.7.4   After Bednorz and Mueller 218
        3.7.5   Prospects for Higher Temperature Superconductors 219
   3.8 Vortex Matter in Anisotropic Superconductors
        Eli Zeldov 223
        3.8.1   Local Magnetization and Geometrical Barriers 223
        3.8.2   Vortex Lattice Melting 225
        3.8.3   Vortex Matter Phase Diagram 228
   3.9 Further Reading 231

4 Materials   **233**
   Editor: C. W. Chu
   4.1 Introduction
        C. W. Chu 233
   4.2 The Route to High Temperature Superconductivity in Transition Metal Oxides
        A. Bussmann-Holder and K. A. Müller 234
   4.3 Superconductivity above 10 K in Non-Cuprate Oxides
        David C. Johnston 239
        4.3.1   Introduction 239
        4.3.2 $LiTi_2O_4$ 240
        4.3.3 $Ba(Pb_{1-x}Bi_x)O_3$ 242
        4.3.4 $(Ba_{1-x}K_x)BiO_3$ 244
   4.4 Cuprates—Superconductors with a T_c up to 164 K

        C. W. Chu 244
        4.4.1 $RBa_2Cu_3O_7$ 245
        4.4.2 $Bi_2Sr_2Ca_{n-1}Cu_nO_{2n+4}$ 250
        4.4.3 $Tl_2Ba_2Ca_{n-1}Cu_nO_{2n+4}$ 252
        4.4.4 $HgBa_2Ca_{n-1}Cu_nO_{2n+3-\delta}$ 253
  4.5 Fe-Pnictides and -Chalcogenides
        Hideo Hosono 255
        4.5.1 Research Led to Discovery of Iron-Based Superconductors 255
        4.5.2 Discovery of Iron-based Superconductors 257
        4.5.3 Advances in Materials 260
        4.5.4 Current Status 262
        4.5.5 Perspective 263
  4.6 Superconductivity in MgB_2
        Takahiro Muranaka and Jun Akimitsu 265
        4.6.1 Introduction 265
        4.6.2 Crystal Structure of MgB_2 266
        4.6.3 Electronic Structure of MgB_2 268
        4.6.4 Two-gap Superconducting State of MgB_2 270
        4.6.5 Conclusion 271
  4.7 Fullerenes
        Kosmas Prassides and Matthew J. Rosseinsky 272
        4.7.1 Alkali Fulleride Superconductors 272
        4.7.2 Fullerides with Increased Interfullerene Separations 274
        4.7.3 Cs_3C_{60}—Fullerene Superconductivity Reborn 275
  4.8 Elemental Superconductors
        K. Shimizu 278
        4.8.1 Periodic Table for Superconducting Elements 279
        4.8.2 Experimental Technique 280
        4.8.3 Example of Pressure-Induced Superconductivity in Elements 280
        4.8.4 Summary: 3D Periodic Table for Superconducting Elements 282
  4.9 Heavy-Fermion Superconductivity
        F. Steglich 283
  4.10 Ruthenate Superconductor Sr_2RuO_4
        Yoshiteru Maeno 288
        4.10.1 Same Crystal Structure but Different Unconventional Superconductivity . . 290
        4.10.2 Novel Superconducting Phenomena 292
  4.11 Magnetic Superconductors (and Some Recollections of Professor Bernd T. Matthias)
        M. Brian Maple 293
        4.11.1 Introduction 293
        4.11.2 Localized Magnetic Moments in Conventional Superconductors . . . 294
        4.11.3 Coexistence of Superconductivity and Magnetic Order Involving the Same Set of Electrons 299
        4.11.4 Some Thoughts about High Temperature Superconductors 304
        4.11.5 Some Personal Recollections of Professor Bernd T. Matthias 305
  4.12 Further Reading 307

5 SQUIDs and Detectors   **311**
    Editor: A. I. Braginski
  5.1 Introduction
        John Clarke, Arnold Silver and A. I. Braginski 311

5.2 History and Device Fundamentals
    John Clarke and Arnold Silver 313
    5.2.1 Initial Discovery 313
    5.2.2 Thin-Film Josephson Tunnel-Junction Quantum Interferometer 315
    5.2.3 Point Contact Devices — dc SQUID 317
    5.2.4 The rf SQUID 318
    5.2.5 Linear SQUID Response and Single Flux Quantum Transitions 320
    5.2.6 R-SQUID, Oscillators and Detectors 320
    5.2.7 SLUGs at Cambridge 321
    5.2.8 SQUIDs at Berkeley 324
    5.2.9 The Square Washer SQUID 326
    5.2.10 The rf SQUID Revisited 327
    5.2.11 Today's SQUIDs 327
5.3 High-T_c SQUIDs
    A. I. Braginski 328
    5.3.1 Past and Present 328
    5.3.2 High-T_c SQUID Limits of Performance 330
5.4 Geophysical Applications of SQUIDs
    C. P. Foley 331
    5.4.1 Introduction 331
    5.4.2 Magnetic Measurements Used in Geophysical Prospecting 332
    5.4.3 Early SQUIDs in Geophysical Prospecting 333
    5.4.4 Transient ElectroMagnetics Using High-T_c SQUIDs 335
    5.4.5 Tensor Gradiometry 338
    5.4.6 Laboratory Systems 339
    5.4.7 Final Comments 340
5.5 Application to Nondestructive Evaluation of Materials and Structures
    A. I. Braginski 342
    5.5.1 Past and Present 342
    5.5.2 NDE Testing Methods Using SQUIDs 344
    5.5.3 Bridge Testing 344
    5.5.4 Airplane Testing 346
    5.5.5 Two Examples of SQUID NDE Now in the Field 346
    5.5.6 SQUID NDE Microscopy in Semiconductor Technology 348
    5.5.7 Concluding Remark 349
5.6 SQUIDs — from Laboratory Devices to Commercial Products
    Ronald E. Sager 349
    5.6.1 Early SQUIDs as Laboratory Devices 349
    5.6.2 First Commercialization of SQUIDs 350
    5.6.3 SHE — Early SQUID Instrumentation 351
    5.6.4 Quantum Design — Advanced SQUID Instruments 353
    5.6.5 The Market for SQUID-based Instruments Today 356
    5.6.6 Future Directions for SQUIDs in Laboratory Instrumentation 356
5.7 Electromagnetic and Particle Detection and Readout
    John Clarke, Kent Irwin and Peter Michelson 358
    5.7.1 DC SQUID Amplifiers 358
    5.7.2 Transition Edge Sensors 361
    5.7.3 Photon Energy Resolving Detectors 362
    5.7.4 X-Ray Astronomy and Materials Analysis 363
    5.7.5 Dark Energy and Cold Dark Matter 364
    5.7.6 Dark Energy: Searching for Galaxy Clusters 366

      5.7.7   Cold Dark Matter: WIMPs and Axions 368

      5.7.8   Gravitational Radiation Detectors 369

      5.7.9   Future Directions 371

  5.8 Concluding Remarks

      A. I. Braginski, John Clarke and Ronald E. Sager 371

  5.9 Further Reading 373

6 Qubits   **375**

  Editor: J. E. Mooij

  6.1 Qubits

      J. E. Mooij 375

      6.1.1   Introduction 375

      6.1.2   Early Experiments 376

      6.1.3   Qubit Types 379

  6.2 General Aspects

      J. E. Mooij 384

      6.2.1   (De)coherence 384

      6.2.2   Qubit Types, General Considerations 385

      6.2.3   Recent Results 386

      6.2.4   Future 387

7 Digital Electronics   **389**

  Editor: Shinya Hasuo

  7.1 Introduction

      Shinya Hasuo 389

  7.2 Operating Principles of Digital Circuits

      Theodore Van Duzer 390

      7.2.1   The Basic Switch 390

      7.2.2   Logic Circuits 392

      7.2.3   Memory 395

  7.3 Digital Electronics in Japan

      Shinya Hasuo 397

      7.3.1   The Dawn of Research on Digital Applications 398

      7.3.2   Brief History of Japanese Projects on Superconducting Digital Electronics . 399

      7.3.3   Progress in the 1980s 402

      7.3.4   Progress in the 1990s 403

      7.3.5   Progress in the 2000s 405

  7.4 Digital Electronics in the USA

      Fernand (Doc) Bedard 407

      7.4.1   Cryotrons 408

      7.4.2   Josephson Signal Processor (JSP) 409

      7.4.3   Superconductive Crossbar Switch 410

      7.4.4   Hybrid Technology Multi-Threaded Architecture (HTMT) 413

      7.4.5   Superconducting Technology Assessment (STA) 414

  7.5 Digital Electronics in Europe

      Horst Rogalla 415

      7.5.1   The Early Days 415

      7.5.2   European Projects 419

      7.5.3   Organization of Superconducting Electronics in Europe 422

      7.5.4   Road Ahead 423

  7.6 Integrated Circuit Fabrication Process

Mutsuo Hidaka 424
    7.6.1   Circuit Elements 424
    7.6.2   Integration of Circuit Elements 427
    7.6.3   Planarized Multilayer Process 428
    7.6.4   Toward Further Progress 430
  7.7 High-Speed Digital Circuits
       Akira Fujimaki 431
    7.7.1   Principle of the Rapid Single Flux Quantum Circuit 432
    7.7.2   LSI Design Technology 433
    7.7.3   Demonstration of RSFQ LSIs 436
    7.7.4   New Directions in SFQ Circuits 439
  7.8 History of Superconductor Analog-to-Digital Converters
       Oleg Mukhanov 440
    7.8.1   Superconductor Parallel-Type ADCs 442
    7.8.2   Superconductor Oversampling ADCs 446
    7.8.3   Superconductor Materials for ADC Implementation: LTS vs HTS 456
    7.8.4   Conclusions 457

8 Microwave Applications   **459**
  Editor: D. E. Oates
  8.1 Microwave Measurements of Fundamental Properties of Superconductors
       D.E. Oates 459
    8.1.1   Introduction 459
    8.1.2   Two-Fluid Model 460
    8.1.3   Microwave Measurement Techniques 462
    8.1.4   Early History 463
    8.1.5   Post World War II 463
    8.1.6   1960s and 1970s 464
    8.1.7   High-T_c Era 466
    8.1.8   Summary and Future Prospects 471
  8.2 Applications of Passive Microwave Filters and Devices in Communication and Related Systems
       R. B. Hammond, N. O. Fenzi and B. A. Willemsen 471
    8.2.1   Introduction 472
    8.2.2   HTS Filters for Cellular Telephone Base Station Receivers 473
    8.2.3   Other SC Passive Microwave Devices 483
    8.2.4   Summary and Looking Ahead 484
  8.3 Superconducting Quantum Electronics Enabling Astronomical Observations
       T. M. Klapwijk 484
    8.3.1   The Electrodynamics of Superconducting Films 485
    8.3.2   Photon-Assisted Tunneling with Superconductors 486
    8.3.3   Dynamic Pair Currents: Josephson Tunneling 487
    8.3.4   Passive Nonlinear Device: Quantum Mixing 488
    8.3.5   First Experimental Demonstration: Technology Base 489
    8.3.6   Submillimeter-Wave Astronomy 490
    8.3.7   From Proof-of-Principle to Demanding Use 491
    8.3.8   Nonequilibrium Mixers: Hot-Electron Bolometers 491
    8.3.9   Conclusions 492
  8.4 Microwave Cooling of Superconducting Quantum Systems
       W. D. Oliver 493
    8.4.1   Introduction 493

8.4.2   Superconducting Artificial Atoms   494
8.4.3   Microwave Cooling   496
8.4.4   Summary   499
8.5   Applications of Superconducting Microresonators
      Jonas Zmuidzinas   499
      8.5.1   Introduction   499
      8.5.2   Linear Electrodynamics: A Brief Review   499
      8.5.3   Superconducting Microresonators   501
      8.5.4   Superconducting Microresonator Detectors   505
      8.5.5   Parametric Amplifiers   510
      8.5.6   Other Applications   511
      8.5.7   Summary   511
8.6   Further Reading   511

9   Quantum Metrology   515
Editors: Richard E. Harris and Jürgen Niemeyer
9.1   Introduction
      Richard E. Harris and Jürgen Niemeyer   515
9.2   Josephson Voltage Standard — The Ultimate Precision
      Jaw-Shen Tsai and James E. Lukens   518
9.3   The First Josephson Voltage Standards
      Thomas J. Witt   522
9.4   First Josephson Junction Array Voltage Standard
      Tadashi Endo   525
      9.4.1   Introduction   525
      9.4.2   First Josephson Array   526
      9.4.3   Applications of the Multiple Josephson Junction   527
      9.4.4   Conclusion   528
9.5   How the DC Array Standards Were Developed
      Jürgen Niemeyer   528
      9.5.1   Introduction   528
      9.5.2   Single Junction Design   529
      9.5.3   Circuit Design and First Realization   530
      9.5.4   1 Volt Circuits   532
      9.5.5   10 Volt Circuits   533
9.6   Making the Josephson Voltage Standard Practical
      Clark A. Hamilton   535
9.7   Programmable Josephson Voltage Standards: from DC to AC
      Johannes Kohlmann   540
      9.7.1   Introduction   540
      9.7.2   Principles and Fundamentals of PJVS   540
      9.7.3   Overdamped Josephson Junctions for PJVS   540
      9.7.4   Realization of Series Arrays   543
      9.7.5   Applications   545
      9.7.6   Conclusions and Outlook   545
9.8   Quantum-Based Voltage Waveform Synthesis
      Samuel P. Benz   546
9.9   Superconductivity and the SI (Metric) System Based on Fundamental Constants
      Edwin Williams and Ian Robinson   553
9.10 Further Reading   557

10 Medical Applications **559**
Editor: Harold Weinstock
10.1 Introduction
Harold Weinstock 559
10.2 Medical Applications of Magnetoencephalography
*Cosimo Del Gratta, Stefania Della Penna, Vittorio Pizzella and Gian-Luca
        Romani* 562
    10.2.1 Introduction: The Origin of Magnetoencephalography 562
    10.2.2 Basics of MEG 564
    10.2.3 First MEG Studies 569
    10.2.4 A Step Ahead: Technological Developments 572
    10.2.5 Brain Mapping and Clinical Studies 577
    10.2.6 Recent Developments and Perspectives 580
10.3 MCG Instrumentation and Applications
Riccardo Fenici, Donatella Brisinda, Anna Rita Sorbo and Angela Venuti . 582
    10.3.1 A New Boost in the 1980s 582
    10.3.2 The History of High Resolution MCG 585
    10.3.3 From Magnetic Field Mapping to 3D Cardiac Source Localization 588
    10.3.4 From 3D localization to Magnetic Source Imaging 591
    10.3.5 To Shield or Not to Shield? The History of Unshielded MCG 592
    10.3.6 Clinical Perspectives 595
    10.3.7 Detection of Myocardial Ischemia and Viability (Rest and Stress MCG) . . 595
    10.3.8 3D Electro-anatomical Imaging of Arrhythmogenic Substrates and Pre-
           interventional Virtual Programming of Ablation Procedures 598
    10.3.9 Arrhythmogenic Risk Assessment 599
    10.3.10 Fetal Magnetocardiography 600
    10.3.11 Conclusions 601
10.4 MRI (Magnetic Resonance Imaging) Instrumentation and Applications
Jim Bray and Kathleen Amm 602
    10.4.1 Basic Principles of MRI 603
    10.4.2 History of MRI — A General Electric (GE) Perspective 604
10.5 Ultralow Field NMR and MRI
John Clarke 610
    10.5.1 Introduction 610
    10.5.2 Background 610
    10.5.3 Why ULF NMR and MRI? 611
    10.5.4 Ultralow Field NMR 611
    10.5.5 Ultralow Field MRI 612
    10.5.6 T_1-weighted Contrast Imaging 614
    10.5.7 T_1 Contrast in Ex Vivo Prostate Tissue 616
    10.5.8 Research in Other Institutions 617
    10.5.9 A Perspective 618
10.6 Superconductivity in Medical Accelerators for Cancer Therapy
Peter A. Zavodszky 619
10.7 Further Reading 625

11 Wires and Tapes **627**
Editor: David Larbalestier
11.1 The Long Road to High Current Density Superconducting Conductors
David Larbalestier 627
    11.1.1 Overview 627

11.1.2 Onnes's Vision for Superconducting Conductors and 10 T Magnets 628
11.1.3 Alloyed Superconductors, Type I and type II Transition and the Collective
       Failure to Understand It. 630
11.1.4 Theory and the Long Disconnect from Experiment 634
11.1.5 The 1960s: The Age of High Critical Current Densities and, Finally, High
       Field Magnets 635
11.1.6 Coexistence: High Temperature and Low Temperature Superconductors —
       1987 to Present 636
11.1.7 Superconducting Conductors beyond the 2011 Centennial 638
11.2 Nb-Ti — from Beginnings to Perfection
     Peter J. Lee and Bruce Strauss 643
11.2.1 The High Field Revolution in Retrospect 643
11.2.2 The High Field Revolution—The One that Was Noticed 643
11.2.3 First International Conference on High Magnetic Fields 644
11.2.4 God Save the Queen 645
11.2.5 The Slow Emergence of Nb-Ti 646
11.2.6 Rutherford CEGB/IMI Strand and the 1968 Brookhaven Summer School . 648
11.2.7 Filamentary Superconductors 649
11.2.8 After the 1968 Summer School 650
11.2.9 Making Multifilamentary Strand 651
11.2.10 The First Strain Measurements 652
11.2.11 The Birth of Nb-46.5wt%Ti 652
11.2.12 Manufacturing of Cable for the Fermilab Tevatron 655
11.2.13 Toward a Complete Description of Nb-Ti 656
11.2.14 Nb-Ti as a Commodity 660
11.3 History of Nb_3Sn and Related A15 Wires
     Kyoji Tachikawa and Peter J. Lee 661
11.3.1 Introduction 661
11.3.2 Chronological Progress in the Fabrication of Nb_3Sn and V_3Ga Wires 662
11.3.3 Bronze-Processed Nb_3Sn Wires 666
11.3.4 Internal Sn-Processed Nb_3Sn Wires 667
11.3.5 Conclusions and Future Outlook 669
11.4 Bi-Sr-Ca-Cu-O HTS Wire
     Martin W. Rupich and Eric E. Hellstrom 671
11.4.1 Introduction 671
11.4.2 2223 Wires 674
11.4.3 2212 Conductors 681
11.4.4 Conclusions 688
11.5 Coated Conductor: Second Generation HTS Wire
     A. P. Malozemoff and Y. Yamada 689
11.5.1 Texturing the Template 691
11.5.2 REBCO HTS Layer Deposition 694
11.5.3 Artificial Pinning 699
11.5.4 Drivers for 2G HTS Wire 700
11.6 The MgB_2 Conductor Story
     René Flükiger and Hiroaki Kumakura 702
11.6.1 Reasons for Excitement at the Unexpected Discovery of Superconductivity
       in MgB_2 702
11.6.2 The Effect of Carbon or Carbon-based Additives on the Transport Properties 703
11.6.3 Factors Influencing the Transport Properties of MgB_2 Conductors 704
11.6.4 The Fabrication Methods of MgB_2 Wires and Tapes 705

11.6.5 Densification Effects in MgB_2 Wires 708
11.6.6 Perspectives for Further Development of MgB_2 709
11.7 Further Reading 711

12 Large Scale Applications   **713**
   Editors: Peter Komarek, Bruce Strauss, and Steve St. Lorant
12.1 Introduction
   Steve St. Lorant 713
12.2 The History of Superconductivity in High Energy Physics
   Steve A. Gourlay and Lucio Rossi 716
   12.2.1 Introduction 716
   12.2.2 Accelerators 719
   12.2.3 Detectors 721
   12.2.4 Main Characteristics of Accelerator Magnets 723
   12.2.5 The Benefits of Superconductivity 724
   12.2.6 Early History 726
   12.2.7 The Large Hadron Collider 728
   12.2.8 Applications in Japan 731
   12.2.9 RF Superconductivity 732
   12.2.10 New Applications 735
12.3 Magnet Engineering—Study in Stability and Quench Protection
   Luca Bottura and Al McInturff 737
   12.3.1 Outline 737
   12.3.2 The Infancy of Superconducting Magnet Technology 737
   12.3.3 The Appeal of Nb_3Sn and the First Stumbling Steps 738
   12.3.4 Malleable Nb-Based Alloys: Technology Goes to Industry 739
   12.3.5 Training and Degradation, the Discovery of Stability 739
   12.3.6 Stabilization Strategies 740
   12.3.7 Cryogenic Stability 740
   12.3.8 Adiabatic Stability 741
   12.3.9 Dynamic Stability 742
   12.3.10 MultiFilamentary Wires and Twisting 743
   12.3.11 Cable Stability: The Invention of Rutherford Cable and CICC . . . 743
   12.3.12 Stability in the New Millennium 746
   12.3.13 Quench and Protection—a Burning Issue 747
   12.3.14 Hot Helium Bubbles 748
   12.3.15 Coda 749
   12.3.16 Appendix: Recollections and Reminiscences 749
12.4 The History of Fusion Magnet Development
   Jean-Luc Duchateau, Peter Komarek and Bernard Turck 753
   12.4.1 General Introduction 753
   12.4.2 The Large Coil Task (LCT) 753
   12.4.3 Early Work on Poloidal Field Coils for Tokamaks 755
   12.4.4 *Tore Supra* 755
   12.4.5 Tokamaks with Superconducting Magnet Systems in Recent Time . 759
   12.4.6 ITER as a Worldwide Collaboration Project in Thermonuclear Fusion . . . 760
   12.4.7 Superconducting Magnets for Stellarator-Type Fusion Devices 767
12.5 Electric Power Applications of Superconductivity
   William Hassenzahl and Osami Tsukamoto 769
   12.5.1 Introduction 769
   12.5.2 Fault-Current Limiters 772

12.5.3 SMES 777
12.5.4 Superconducting Rotating Machines 782
12.5.5 Superconducting Cables 788
12.5.6 Superconducting Transformers 792
12.5.7 Conventional Transformer Characteristics 794
12.6 Magnetic Separation
       Christopher Rey 797
12.6.1 Introduction 797
12.6.2 Principles of Magnetic Separation 797
12.6.3 Magnetic Separation Dynamics 799
12.6.4 Open Gradient Magnet Systems 800
12.6.5 Matrix/Filter Systems 800
12.6.6 Characteristics of High Gradient Magnetic Separation 801
12.6.7 Magnetic Separation Equipment 802
12.6.8 Applications of Magnetic Separation 806
12.6.9 Summary 811
12.7 Superconducting Induction Heating of Nonferrous Metals
       Niklas Magnusson and Larry Masur 811
12.7.1 Conventional Aluminum, Copper and Brass Induction Heating 811
12.7.2 AC Superconducting Induction Heating 812
12.7.3 DC Superconducting Induction Heating 812
12.7.4 Commercial Deployment 814
12.7.5 Simple Technical Design 815
12.7.6 Low Frequency Billet Heating 815
12.7.7 A Field Report: Experiences with Operation 817
12.8 Superconducting Magnets for NMR
       Gerhard Roth 817
12.8.1 Introduction 817
12.8.2 Stability 819
12.8.3 Homogeneity 820
12.8.4 Cryostat 822
12.8.5 Major NMR Magnet Development Steps 823
12.8.6 Nb_3Sn Technology 824
12.8.7 Subcooling Technology 826

Preface

Superconductivity came as a big surprise. But more amazing perhaps is that even a hundred years after its discovery this peculiar phenomenon continues to bring us new surprises. The research of superconductivity is characterized by times of relative quietness, interrupted by periods of exciting activities, often preceded by fundamental breakthroughs that later won Nobel Prizes in physics or chemistry. "Fundamental" can in this case both relate to an emerging theoretical insight and to a new class of materials displaying entirely unforeseen properties. In both cases, the prospect of new applications has been an important drive for commercial ambitions.

In the meantime, applications of superconductivity have found a place in science and industry; just like superconductivity in general, there were periods of quiet progress and stormy phases, but overall a steady progress took place, e.g., high-T_c superconductors were judged in the beginning as "never applicable in high magnetic fields" because of their grain boundary problem. In the meantime, they are essential building blocks for the creation of very high permanent fields: the result of excellent ideas and the cooperation between fundamental and applied scientists. Superconducting magnets for MRI can be found in any major hospital, and high-energy physics without superconducting magnets is practically unthinkable.

A similar breathtaking development took place in superconducting electronics: quantum-limited detectors, quantum-information processing, MEG and MCG, high-speed computing, and analog-to-digital and digital-to-analog converters with quantum precision, incorporating tens of thousands of Josephson junctions, are available. It seems to be a question only of time (and cooling) until they will enter the industrial product cycle.

In this centennial book on superconductivity many authors who themselves were responsible for the important steps forward, or were very close to those pioneers, report on the historical developments. The contributions represent their personal views and therefore this book does not provide the ultimate answers to all questions about superconductivity. Nonetheless it has become a very interesting collection of recollections and reviews of almost all the subdisciplines of superconductivity. The first three chapters concentrate on the interesting stories of the discovery and the succeeding gradual progress of theory and experiment. Much emphasis is given to the important developments in the early 1950s and 1960s. From then on superconductivity started to penetrate society and most applications today are based on the innovations of those years. But long before superconductivity could celebrate its centennial, a genuine revolution occurred with the discovery of the high temperature superconductors. A new episode began which is nicely described in a number of articles that bring this book up to the present time.

Originally it was the intention to present a small booklet (about 50 pages) to the attendants of the Superconductivity Centennial Conference taking place in the Netherlands in September 2011. This project quickly grew out to a book project of now more than 800 pages — primarily due to the excitement of scientists and engineers intending to show how far superconductivity has come after 100 years. Due to the limitation in space we had to restrict the number of contributions and many famous colleagues who would have deserved to take part in the project had to be left out. Even extending the size of the book to more than 800 pages did not help — superconductivity has become a vast field.

Being editors of the book was an exciting task, extremely enjoyable and nerve-wracking — hard deadlines melt in the sun of wonderful science and engineering! And preparing a book in LATEX was an experience on its own. The help of students of the University of Twente (M. Garcia, W.A.G. Vessies, S. Waanders) and of the CU in Boulder (P. Lippert) is appreciated. The work of Daan Boltje in Leiden, Troy Christensen in Houston, and Peter Lee and Dmytro Abraimov in Tallahassee was essential to finish chapters 1, 2, 4 and 11. Finally our thanks to all authors, who were too engaged to stick to the page limits, and to the section editors, who did a wonderful (and difficult) job to organize and edit their chapters and to keep the excitement of the authors in balance with the available space in the book.

Horst Rogalla and Peter H. Kes

List of Authors

Editors

Horst Rogalla
University of Twente
Enschede
The Netherlands

Peter H. Kes
Kamerlingh Onnes Laboratory
Leiden University
Leiden
The Netherlands

Chapter 1

Peter H. Kes (editor)
Kamerlingh Onnes Laboratory
Leiden University
Leiden
The Netherlands

Dirk van Delft
Boerhaave Museum
Leiden
The Netherlands

Rudolf de Bruyn Ouboter
Kamerlingh Onnes Laboratory
Leiden University
Leiden
The Netherlands

A. B. Pippard (deceased)
Cavendish Laboratory
Cambridge
England

Chapter 2

Jan Zaanen (editor)
Instituut-Lorentz for
  Theoretical Physics
Leiden University
Leiden
The Netherlands

David Pines
ICAM and Physics Department
U C Davis
Davis, CA
USA

A. E. Koshelev
Materials Science Division
Argonne National Laboratory
Argonne, IL
USA

V. M. Vinokur
Materials Science Division
Argonne National Laboratory
Argonne, IL
USA

Lev P. Gor'kov
NHMFL
Florida State University
Tallahassee, FL
USA

Ernst Helmut Brandt (deceased)[1]
Max Planck Institute for Metals
  Research
Stuttgart
Germany

T. I. Baturina
Materials Science Division
Argonne National Laboratory
Argonne, IL
USA

L. N. Bulaevskii
Los Alamos National Laboratory
Los Alamos, NM
USA

R. P. Huebener
Eberhard Karls Universität
Tübingen
Germany

---

[1] On September 1, 2011, Ernst Helmut Brandt passed away peacefully. We remember him as a fine colleague.

Chapter 3

J. Mannhart (editor)
Max Planck Institute for
    Solid State Research
Stuttgart
Germany

John M. Rowell
Arizona State University
Tempe, AZ
USA

Dietrich Einzel
Walther-Meissner Institut
Garching
Germany

Dale J. Van Harlingen
University of Illinois
    at Urbana-Champaign
Champaign, IL
USA

C. C. Tsuei
IBM Thomas J. Watson Research
    Center
Yorktown Heights, NY
USA

J. R. Kirtley
Center for Probing the Nanoscale
Stanford University
Stanford, CA
USA

Jean-Marc Triscone
University of Geneva
Geneva
Switzerland

Marc Gabay
Université Paris-Sud
Orsay
France

D. Dimos
Engineering Sciences Center
Sandia National Laboratories
Albuquerque, NM
USA

M. R. Beasley
Stanford University
Stanford, CA
USA

Eli Zeldov
Weizmann Institute of Science
Rehovot
Israel

Chapter 4

C. W. Chu (editor)
University of Houston
Houston, TX
and Lawrence Berkeley
    National Laboratory
Berkeley, CA
USA

A. Bussmann-Holder
Max-Planck Institut für
    Festkörperforschung
Stuttgart
Germany

K. A. Müller
Physik Institut der
    Universität Zürich
Zürich
Switzerland

David C. Johnston
Ames Laboratory and Depart-
    ment of Physics and
    Astronomy
Iowa State University
Ames, IA
USA

Hideo Hosono
Tokyo Institute of Technology
Tokyo
Japan

Takahiro Muranaka
Aoyama Gakuin University
Tokyo
Japan

Jun Akimitsu
Aoyama Gakuin University
Tokyo
Japan

Kosmas Prassides
Department of Chemistry
Durham University
Durham
UK

Matthew J. Rosseinsky
Department of Chemistry
University of Liverpool
Liverpool
UK

K. Shimizu
Osaka University
Osaka
Japan

F. Steglich
Max Planck Institute for
    Chemical Physics of Solids
    Dresden
Germany

Yoshiteru Maeno
Kyoto University
Kyoto
Japan

M. Brian Maple
Department of Physics
University of California
    at San Diego
La Jolla, CA
USA

Chapter 5

A. I. Braginski (editor)
Forschungszentrum Jülich
Jülich
Germany

C. P. Foley
Commonwealth Scientific and
Industrial Research
Organization (CSIRO)
Lindfield
Australia

Kent Irwin
National Institute of Standards
and Technology (NIST)
Boulder, CO
USA

John Clarke
University of California,
Berkeley and Lawrence
Berkeley National Laboratory
Berkeley, CA
USA

Ronald E. Sager
Quantum Design
San Diego, CA
USA

Peter Michelson
Stanford University
Stanford, CA
USA

Arnold Silver
Rancho Palos Verdes, CA
USA

Chapter 6

J.E. Mooij (editor)
Kavli Institute of Nanoscience
Delft University of Technology
The Netherlands

Chapter 7

Shinya Hasuo (editor)
ISTEC
Tokyo
Japan

Horst Rogalla
University of Twente
Enschede
The Netherlands

Akira Fujimaki
University of Nagoya
Nagoya
Japan

Theodore Van Duzer
University of California,
Berkeley
Berkeley, CA
USA

Mutsuo Hidaka
ISTEC
Tokyo
Japan

Oleg Mukhanov
HYPRES
Elmsford, NY
USA

Fernand (Doc) Bedard
National Security Agency (NSA)
Silver Spring, MD
USA

Chapter 8

D. E. Oates (editor)
MIT Lincoln Laboratory
Lexington, MA
USA

N. O. Fenzi
Superconductor Technologies
    Inc.
Santa Barbara, CA
USA

T. M. Klapwijk
Kavli Institute of Nanoscience
Delft University of Technology
Delft
The Netherlands

R. B. Hammond
Superconductor Technologies
    Inc.
Santa Barbara, CA
USA

B. A. Willemsen
Superconductor Technologies
    Inc.
Santa Barbara, CA
USA

W. D. Oliver
MIT Lincoln Laboratory
Lexington, MA
USA

Jonas Zmuidzinas
California Institute of
    Technology
Pasadena, CA
USA

Chapter 9

Richard E. Harris (editor)
National Institute of Standards
and Technology
Boulder, CO
USA

Jürgen Niemeyer (editor)
Physikalisch-Technische
Bundesanstalt
Braunschweig
Germany

Samuel P. Benz
National Institute of Standards
and Technology
Boulder, CO
USA

Jaw-Shen Tsai
RIKEN/NEC
Tsukuba
Japan

Tadashi Endo
MTA Japan Ltd.
Tokyo
Japan

Edwin Williams
National Institute of Standards
and Technology
Gaitherburg, MD
USA

James E. Lukens
SUNY
Stony Brook, NY
USA

Clark A. Hamilton
National Institute of Standards
and Technology
Boulder, CO
USA

Ian Robinson
National Physical Laboratory
Teddington
UK

Thomas J. Witt
Bureau International des Poids et
    Mesures
Sèvres
France

Johannes Kohlmann
Physikalisch-Technische
Bundesanstalt
Braunschweig
Germany

Chapter 10

Harold Weinstock (editor)
US Air Force Office of Scientific
    Research
Arlington, VA
USA

Cosimo Del Gratta
Gabriele D'Annunzio University
Chieti
Italy

Stefania Della Penna
Gabriele D'Annunzio University
Chieti
Italy

Vittorio Pizzella
Gabriele D'Annunzio University
Chieti
Italy

Gian-Luca Romani
Gabriele D'Annunzio University
Chieti
Italy

Riccardo Fenici
Catholic University of Rome
Rome
Italy

Donatella Brisinda
Catholic University of Rome
Rome
Italy

Anna Rita Sorbo
Catholic University of Rome
Rome
Italy

Angela Venuti
Catholic University of Rome
Rome
Italy

Jim Bray
GE Global Research
Albany, NY
USA

Kathleen Amm
GE Global Research
Albany, NY
USA

John Clarke
University of California at
   Berkeley and Lawrence
   Berkeley Laboratory
Berkeley, CA
USA

Peter A. Zavodszky
GE Global Research
Albany, NY
USA

Chapter 11

David Larbalestier (editor)
National High Magnetic Field
   Laboratory
Tallahassee, FL
USA

Peter J. Lee
National High Magnetic Field
   Laboratory
Tallahassee, FL
USA

Bruce Strauss
U. S. Department of Energy
Washington, DC
USA

Kyoji Tachikawa
Tokai University
Tokai
Japan

Martin W. Rupich
American Superconductor
   Corporation
Devens MA
USA

Eric E. Hellstrom
National High Magnetic Field
   Laboratory
Tallahassee, Florida
USA

A. P. Malozemoff
American Superconductor
   Corporation
Devens MA
USA

Y. Yamada
ISTEC
Tokyo
Japan

René Flükiger
University of Geneva
Geneva
Switzerland

Hiroaki Kumakura
National Institute for Materials
   Science
Tsukuba
Japan

Chapter 12

Peter Komarek (editor)
Karlsruhe Institute of
   Technology (KIT)
Karlsruhe
Germany

Bruce Strauss (editor)
U. S. Department of Energy
Washington, DC
USA

Steve St. Lorant (editor)
SLAC National Accelerator
   Laboratory
Menlo Park, CA
USA

Luca Bottura (editor)
CERN
Geneve
Switzerland

Al McInturff (editor)
Texas A&M University
College Station, TX
USA

Steve A. Gourlay
Lawrence Berkeley National
    Laboratory
Berkeley, CA
USA

Lucio Rossi
CERN
Geneve
Switzerland

Jean-Luc Duchateau
CEA
Cadarache
France

Bernard Turck
CEA
Cadarache
France

William Hassenzahl
Advanced Energy Analysis
Las Vegas, NV
USA

Osami Tsukamoto
Faculty of Engineering
Yokohama National University
Yokohama
Japan

Christopher M. Rey
Oak Ridge National Laboratory
Oak Ridge, TN
USA

Niklas Magnusson
SINTEF Energy Research
Trondheim
Norway

Larry Masur
Zenergy Power, Inc.
Burlingame, CA
USA

Gerhard Roth
Bruker Biospin
Rheinstetten
Germany

Glossary

Glossary of Frequently Used Acronyms, Symbols, Terms and Physical Constants

We list here the fundamental physical constants, symbols, terms and also acronyms appearing throughout this book. Some of the terms, necessary to understand this book, are briefly defined.

Fundamental Physical Constants

$c = 2.997925 \times 10^8$ m/s	velocity of light
$e = 1.6022 \times 10^{-19}$ C	electron charge
$h = 6.6261 \times 10^{-34}$ Js	Planck constant
N_A	Avogadro constant
$\hbar = h/2\pi = 1.0546 \times 10^{-34}$ Js	Planck constant
$\Phi_0 \equiv h/2e = 2.0678 \times 10^{-15}$ Vs	flux quantum
$k_B = 1.3807 \times 10^{-23}$ J/K	Boltzmann constant
$\epsilon_0 = 8.8542 \times 10^{-12}$ As/Vm	permittivity of vacuum
$\mu_0 = 4\pi \times 10^{-7}$ Vs/Am	permeability of vacuum

Symbols and Terms

A	vector potential (vectors are denoted by bold, Roman symbols)
B	magnetic induction, $\mathbf{B} = \text{curl}\,\mathbf{A}$
H	magnetic field
E	electric field
$B = \|\mathbf{B}\|$	magnetic induction (magnitude)
$H = \|\mathbf{H}\|$	magnetic field (magnitude)
$E = \|\mathbf{E}\|$	electric field (magnitude)
c	specific heat
H_c	thermodynamic critical field
H_{c1}	lower critical field, field at which flux penetrates into a type II superconductor
H_{c2}	upper critical field, field at which the normal state comes back. Between H_{c1} and H_{c2} the material is in the Shubnikov phase (mixed state)
λ	magnetic penetration depth
λ_L	London penetration depth
ξ_0	coherence length, size of a Cooper pair
ξ	Ginzburg-Landau coherence length, length scale describing variation of

	Cooper pair density
κ	Ginzburg-Landau parameter, $\kappa = \lambda/\xi$, parameter distinguishing between type I and type II superconductors
β_A	Abrikosov parameter, determines the structure of the Abrikosov vortex latt
a_0	lattice parameter of Abrikosov vortex lattice
σ	electric conductivity
ρ	electric resistivity
l	electron mean free path
F, f	Helmholz free energy, Helmholz free energy density
m	electron mass
n_s	density of superconducting electrons
$\mathbf{v}_s$	average velocity of superconducting electrons
$\mathbf{p}_s$	momentum of superconducting condensate
$\mathbf{v}_F, \mathbf{p}_F$	Fermi velocity, Fermi momentum
ϵ_F, E_F	Fermi energy
$N(E_F)$	density of electronic states at the Fermi energy
$\mathbf{j}, \mathbf{J}$	electric current density
Δ_χ	generalized phase difference over Josephson junction
γ	Sommerfeld constant of electron specific heat
$\Delta(T)$	temperature dependent energy gap
V	BCS parameter determining the strength of the electron-phonon interaction
ω_D	Debye frequency
ω_P	Josephson plasma frequency
γ	anisotropy parameter of a layered superconductor
J_0	interlayer Josephson critical current
λ_J	Josephson length
s	layer thickness in a layered superconductor
$\Phi_{\alpha\beta}$	elastic matrix of a vortex lattice
b	reduced magnetic induction, $b = H/H_{c2}$
c_{ii}	elastic moduli of a vortex lattice
c_{11}	uniaxial compression modulus
c_{44}	tilt modulus
c_{66}	shear modulus
η	friction coefficient of a moving vortex lattice
κ	heat conductivity
ν	Nernst coefficient
$\beta_{dc} \equiv 2LI_c/\Phi_c$	dc SQUID parameter
$\beta_{rf} \equiv 2\pi LI_c/\Phi_0$	rf SQUID parameter
C	electric capacitance, heat capacity
$\delta \equiv \varphi_1 - \varphi_2$	phase difference across a Josephson junction
E	energy

$\epsilon(f)$	energy resolution of a SQUID
f	frequency
Φ	magnetic flux
$\phi(r,t)$	phase of the function of state
I	electric current
I_B, I_b	bias current
I_0, I_c	critical current (at which, for given B and T the superconductor normalizes)
I_c	critical current
J_c	critical current density
J_{Cu}	current density in the copper
J_e	engineering critical current density
$J_{overall}$	overall current density
L	electric inductance
ω	angular frequency
ω_c	cutoff or critical frequency
$\Psi(\mathbf{r},t) = |\Psi(\mathbf{r},t)|\exp[i\phi(\mathbf{r},t)]$	macroscopic function of state
$\mathbf{r}$	space vector
R, r	electric resistance
R_n	normal state resistance of a Josephson junction
S_B	spectral density of magnetic field noise
$S_\Phi(f)$	spectral density of the flux noise
$S_I(f)$	spectral density of current noise
$S_V(f)$	spectral density of voltage noise
T_c	critical temperature (of transformation to superconducting state below T_c)
T_N	noise temperature of an electric device, circuit
t	time
τ	time constant
V, U	electric voltage
V_Φ	flux-to-voltage transfer coefficient (of a SQUID)
V_g	energy gap voltage

Acronyms

2212	$Bi_2Sr_2CaCu_2O_{8+X}$ (X denotes deviation from stoichiometry)
2223	$Bi_2Sr_2Ca_2Cu_3O_{10+X}$
1G	1st generation $Bi_2Sr_2Ca_2Cu_3O_{10+x}$ wire
2G	2nd generation $YBa_2Cu_3O_{7-x}$ coated wire
A15 compounds	a group of superconducting chemical compounds
AC, ac	alternating current
ACJVS, ac	Josephson voltage standard

ADC	analog to digital converter
ADR	all-digital receiver
AEC	alkaline earth cuprate phase in BSCCO system
APC	artificial pinning center
BCS	Bardeen-Cooper-Schriefer theory of superconductivity
Bi-2201	$Bi_2Sr_2CuO_{6+X}$
BSCCO	superconductors in the Bi-Sr-Ca-Cu-O system
BW	bandwidth
BZO	barium zirconate
CC	correction coils in a tokamak
CICC	cable-in-conduit conductor (cables of strand inside tube through which He coolant flows)
CDMC	old dark matter
CF	copper free phase in BSCCO system
CMB	cosmic microwave background (radiation)
CMS	compact muon solenoid
CPW	coplanar waveguide
CQOS	complementary quasione junction SQUID
CS	central solenoid, in a tokamak
CT-OP	controlled over-pressure processing—process used by Sumitomo Electric for 1G processing
CVD	chemical vapor deposition
CW	continuous wave
DC, dc	direct current
DE	dark energy
dipole	beam-bending electromagnet with two poles; the b_1 normal multipole coefficient in the expansion of the complex magnetic field
dodecapole	the b_6 normal multipole coefficient in the expansion of the complex magnetic field; a magnet with twelve poles
DVM	digital voltmeter
EB	electron beam
EMI	electromagnetic interference
EPR	electron spin paramagnetic resonance
FCL	fault current limiter
FEL	free electron laser
FET	field effect transistor
FWHM	full width half maximum
FPGA	field programmable gate array
GB	grain boundary (usually in high-temperature superconductor)
GLAG	Ginzburg-Landau-Abrikosov-Gorkov theory of superconductivity
G-M	Gifford-McMahon

GZO	gadolinium zirconate
HEMT	high electron mobility transistor
Hexed	filaments drawn through hexagonal die so that they can be stacked efficiently
HF, hf	high frequency
HTS	high-temperature superconductor (cuprate)
I	insulator
IBAD	ion beam-assisted deposition
IC	integrated (electric) circuit
ID	Inside diameter
IR	infrared
IT	internal tin process for making Nb_3Sn strand
JJ	Josephson junction
JNT	Johnson noise thermometry
JVS	Josephson voltage standard
LED	light emitting diode
LHe	liquid helium
LN_2	liquid nitrogen
LTS	low-temperature superconductor
MCG	magnetocardiography (biomagnetic imaging of heart fields/currents)
MEG	magnetoencephalography (biomagnetic imaging of brain fields/currents)
MFL	magnetic flux leakage NDE technique
MHD	magnetohydrodynamic(s)
MIITS	The energy balance defining the basic parameters for quench protection in a magnet, unit thereof
MOCVD	metal-organic chemical vapor deposition
MOD	metal-organic deposition
MPMS	magnetic property measuring system
MRI	magnetic resonance imaging
N	normal conductor
NDE	nondestructive evaluation (of materials and structures)
NMR	nuclear magnetic resonance
octupole	The b_4 normal multipole coefficient in the expansion of the complex magnetic field; a magnet with eight poles
OD	outside diameter
OFHC Cu	oxygen-free high conductivity copper
OP	overpressure processing — used for BSCCO conductors
OPIT	oxide powder in tube
PAIR	preanneal intermediate rolling — process to make 2212 conductor developed by NRIM
pancakes	planar magnetic coils
PECVD	plasma enhanced chemical vapor deposition

PF   poloidal-field coils, in a tokamak

pinning   effect used to inhibit movement of flux lines (vortices)

PJVS   programmable Josephson voltage standard

PLD   pulsed laser deposition

PMD   phase modulation-demodulation

PIT   powder in tube

quadrupole   The b_2 normal multipole coefficient in the expansion of the complex magnetic field; a magnet with four poles

QOS   quasi-one Junction SQUID

Quench   superconducting magnet enters non-superconducting state

QRQ   quarter rate quantizer

QVNS   quantized voltage noise source

RABiTS™   rolling assisted bi-axially textured substrate

REBCO   rare-earth cuprate (generic)

RF, rf   radio frequency

rms   root-mean-square

ROSAT   rotation-symmetric arragned tape-in-tape — conductor geometry for 2212 conductors developed by Hitachi

RRP™   rod restacked process, an IT Nb_3Sn strand design developed by OST

RRR   residual resistivity ratio

RSFQ   rapid single flux quantum

RSJ   resistively shunted junction (free of hysteresis Josephson tunnel junction)

S   superconductor

SASE   self-amplified stimulated emission

SC   superconducting; superconductor

SCFCL   superconducting fault current limiter

sextupole   The b_3 normal multipole coefficient in the expansion of the complex magnetic field; magnet with six poles

SF   self field

SFQ   single flux quantum

SINIS   superconductor-insulator-normal-insulator-superconductor

SIS   superconductor-insulator-superconductor (tunnel junction, trilayer)

SLUG   superconducting low-inductance undulatory galvanometer

SMES   superconducting magnetic energy storage

SNS   superconductor–normal conductor–superconductor (proximity junction, trilayer)

SNR   signal-to-noise ratio

SQUID   superconducting quantum interference device

SCCO/Ag   bismuth strontium calcium copper oxide superconductor. The Ag indicates the substrate.

SRF   superconducting radio frequency

STO $SrTiO_3$ (strontium titanate)

SZE   Sunyaev-Zel'dovich effect

TEM	transmission electron microscopy
TEM	transient electromagnetics (geophysical exploration method)
TES	transition-edge sensor (superconducting detector of energy change)
TF	toroidal (magnetic) field in a tokamak
TDC	time to digital converter
V/F	voltage to frequency
WIMP	weakly interacting massive particle
YBCO	$YBa_2Cu_3O_{7-x}$
YSZ	yttria stabilized zirconia

Institution, Experiment and Instrument Acronyms

AI	Atomics International
AIST	National Institute of Advanced Industrial Science and Technology, Japan
AMSC	American Superconductor Corporation
ANL	Argonne National Laboratory
ASC	Applied Superconductivity Conference
BIPM	Bureau International des Poids et Mesures, Paris, France
BNL	Brookhaven National Laboratory
BOC	British Oxygen Company
BTL	Bell Telephone Laboratories
CEA	Commissariat à l'Energie Atomique et aux Energies Alternatives, France
CEBAF	Colliding Electron Beam Facility
CEGB	Central Electricity Generating Board (UK)
CERN	European Organization for Nuclear Research
CESR	Cornell electron positron storage ring
CGPM	General Conference on Weights and Measures
CSIRO	Commonwealth Scientific and Industrial Research Organization, Australia
DESY	Deutsches Elektronen-Synchrotron
DOE	US Department of Energy
EAST	Experimental Advanced Superconducting Tokamak: Chinese experimental tokamak
EPAC	European Particle Accelerator Conference
ESCAR	Experimental Superconducting Electron Ring, at LBL
ETL	Electrotechnical Laboratory, Japan; former name of AIST
FLUXONICS	European network for superconducting electronics
FNAL	Fermilab (Fermi National Accelerator Laboratory)
FSU	Florida State University
FZ-Karlsruhe	Forschungs zentrum Karlsruhe, Germany. Renamed KIT, Karlsruhe Institut für Technologie
GE	General Electric Corporation
GSI	Helmholtzzentrum (Gesellschaft) für Schwerionenforschung, Darmstadt

Harwell	United Kingdom Atomic Energy Authority (UKAEA) Research Establishment at Harwell
HEP	high energy physics
HERA	Hadron Electron Ring Accelerator, at DESY, Deutches Elektronen-Synchrotron
IEEE	Institute of Electrical and Electronics Engineers
IGC	Intermagnetics General Corporation.
IMI	Imperial Metal Industries
IPHT	Institute für Photonishe Technologien, Germany
IPK	International Prototype of the Kilogram
IPP	Max Planck Institute for Plasmaphysics, "Institut für Plasmaphysik"
IREE	Institute for Radio Electronics and Engineering
ISABELLE	Intersecting Storage Accelerator + "belle" at BNL
ISTEC	International Superconductivity Technology Center (Tokyo)
ITER	International Thermonuclear Experimental Reactor
IUPAP	International Union of Pure and Applied Physics
J-PARC	Japan Proton Accelerator Research Complex
JAERI	Japan Atomic Energy Research Institute
JINR	Joint Institute for Nuclear Research, Dubna, Russia
JLAB	Thomas Jefferson National Accelerator Facility (TJNAF)
JT-60SA	Experimental tokamak program preceding ITER, in Japan
KAERI	Korea Atomic Energy Research Institute
KAIST	Korea Advanced Institute of Science and Technology
KEK	High Energy Accelerator Research Organization, Japan
KERI	Korea Electro-Technology Research Institute
KSTAR	Korea Superconducting Tokamak Advanced Research
LAMPF	Los Alamos Meson Physics Facility
LASL	Los Alamos Scientific Laboratory, now LANL, ("National")
LBNL/LBL	Lawrence Berkeley National Laboratory
LEP	large electron positron collider, CERN
LHC	large hadron collider, CERN
LHD	large helical device superconducting: stellarator in operation in Japan
MCA	Magnetic Corporation of America
METAS	Federal Office of Metrology, Switzerland
MIT	Massachusetts Institute of Technology
MSU	Michigan State University (location of National Superconducting Cyclotron Laboratory, NSCL)
MSU	Moscow State University
NAL	National Accelerator Laboratory (founded 1967), renamed Fermilab in 1974
NBS	National Bureau of Standards, USA; former name of NIST
NIST	National Institute of Standards and Technology, USA
NMI	National Measurement Institute

NML	Former name of national measurement institute in Australia
NPL	National Physical Laboratory, Great Britain
NRIM/NIMS	National Research Institute for Metals, now National Institute for Materials Science (Japan)
NST/NKT	Nordic Superconductor Technologies A/S is a subsidiary of Denmark's NKT Holding A/S
OI	Oxford Instruments
ORNL	Oak Ridge National Laboratory
PTB	Physikalisch-Technische Bundesanstalt
RCA	Radio Corporation of America
RHIC	Relativistic heavy ion collider (at BNL)
RMI	Formerly Reactive Metals, Inc., now RMI Titanium Company Extrusion Plant
Rutherford	Rutherford Appleton Laboratory (RAL) — when not referring to physicist Ernest Rutherford
SI	International system of units
SIS	Schwerionen Synchroton at GSI, Darmstadt, Germany
SNS	spallation neutron source
SSC	superconducting super collider
SST1	Experimental tokamak in India with superconducting components
SEI	Sumitoma Electric Industries
Tevatron	Adopted name for the Energy Saver/Doubler hadron collider at FNAL
Tore Supra	Combination of "torus" and "superconductor," a French tokamak
TOSKA	Usually understood to be a composite of "tokamak," "supraleiter" and "Karlsruhe," Germany, a test facility
TRIAM	Tokamak at the Research Institute for Applied Mechanics (RIAM) at Kyushu University, Japan
UNK	Accelerator at JINR, Dubna, Russia. Acronym unknown
UTSI	University of Tennessee Space Institute
W7-X	Wendelstein 7-X: superconducting stellarator under construction in Germany
WDG	Wire Development Group — research collaborations organized by American Superconductor Corp. that has worked on 1G and 2G wire

1

Early History

Editor: Peter H. Kes

1.1 The Discovery and Early History of Superconductivity
   Rudolf de Bruyn Ouboter, Dirk van Delft and Peter H. Kes 1
1.2 The Historical Context of Josephson's Discovery
   A. B. Pippard 29
   Acknowledgments 50
1.3 Further Reading 50

1.1   The Discovery and Early History of Superconductivity

Rudolf de Bruyn Ouboter, Dirk van Delft and Peter H. Kes

1.1.1   The Real Story

On July 10, 1908, in his laboratory at Leiden University, the great Dutch physicist Heike Kamerlingh Onnes (1853–1926) experienced the most glorious moment of his career[1]. That day, after 25 years of hard work and perseverance, of building up from scratch a cryogenic laboratory and organizing superb technical support to run it, he liquefied helium, opening up an entire new research field of low temperature physics. In a triumphant report to the Royal Dutch Academy of Arts and Sciences (KNAW) this historical fact is documented in great detail[2,3]. Therefore it is remarkable that reliable details about his serendipitous discovery of superconductivity three years later are hard to come by. Lack of information has led to speculations about the discovery, in particular about the doubtful role played by a sleepy "blue boy"[4], and about the possible disappearance of Kamerlingh Onnes's laboratory notebooks. Enough reason, then, to have a close look at the Kamerlingh Onnes archive, stored at Boerhaave Museum in Leiden, to see whether any new clues could be found about the discovery of superconductivity — that most important consequence of the ability to reach liquid-helium temperatures.

Of course, it is roughly known when the first two experiments were carried out. Kamerlingh Onnes's two earliest reports to the KNAW about zero resistance and "supraconductivity," as this

---

[1]   Dirk van Delft, *Freezing Physics. Heike Kamerlingh Onnes and the Quest for Cold*, KNAW, Amsterdam (2008)

[2]   H. Kamerlingh Onnes, *Proc.* 11 (1909) 168, *Comm.* 108 (July 1908)

[3]   The regular articles in *Communications from the Physical Laboratory at the University of Leiden* (Comm.) were reprinted in the English-language version of the *Proceedings of the KNAW* (Proc.) and are available from the KNAW at http://www.dwc.knaw.nl/english/academy/digital-library/

[4]   J. de Nobel, *Physics Today*, (Sept. 1996) 40–42

phenomenon was initially called, are dated 28 April[5] and 27 May 1911[6]. According to the archive's inventory, two notebooks (numbers 56 and 57) should cover the period 1909–1912. But on the cover of number 56 is written "1909–1910," and 57 begins with entry dated 26 October 1911. So it does indeed seem as if a crucial notebook is missing. This would explain why so many speculations started to circulate! Another obscuring factor is Kamerlingh Onnes's terrible handwriting. He wrote his lab notes, in pencil, in small household notebooks. They are very hard to read. After a few desperate hours trying, one tends to give up. And that is a pity because, the cover notwithstanding, notebook 56 does indeed announce the 1911 discovery of superconductivity (see Figure 1.1). The entry reads: "De meting van temperatuur is gelukt. Kwik nagenoeg nul. Herhaald met goud. (The temperature measurement was successful. [The resistance of] Mercury practically zero. Repeated with gold)". That looks very much like the discovery of superconductivity.

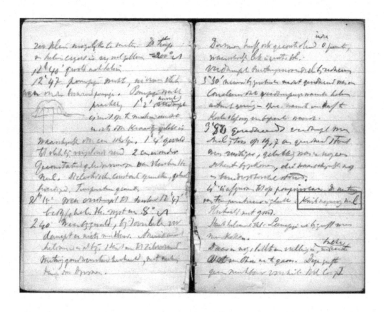

FIGURE 1.1:   A crucial page from the entry for 8 April 1911 in Kamerlingh Onnes's notebook 56. The highlighted sentence *Kwik nagenoeg nul* means "Mercury['s resistance is] practically zero [at 3.0 K]" announcing the first observation of superconductivity. The sketch of the functioning stirrer is seen on the left page. (Archive of the Boerhaave Museum, Leiden.)

When Kamerlingh Onnes took lab notes, he always started by writing down the date. In this case: the day was April 8, but he did not write the year! He dated the second experiment with mercury May 23, again without giving the year. It gets worse: Between those dates, he and Albert Perrier, a visitor from Lausanne, performed an entirely different experiment on the paramagnetism of liquid and solid oxygen. For that experiment the entry reads 19 May 1910. The year was specified this time, but the wrong one! It should have been 1911.

Why did Kamerlingh Onnes make that mistake? It is probably because an extensive series of similar experiments with Perrier had been carried out at the end of 1909 and during the first few months of 1910. In any case, that little slip of the pencil has led many astray. It is the most likely

---

[5]   H. Kamerlingh Onnes, *Proc.* 13 (1911) 1274, *Comm.* 120b (Apr. 1911)

[6]   H. Kamerlingh Onnes, *Proc.* 14 (1911) 113, *Comm.* 122b (May 1911)

reason that researchers exploring the archives were disappointed and, until now, did not look more closely at the notes. Had they made the effort, they would have found the excitement over the first successful transfer of liquid helium to a separate cryostat, the exact dates of the first resistivity experiments on a superconducting material, who was involved, and what their roles were. The notes also reveal that some nice stories about those events will always remain nice, but will never become true.

"Mercury practically zero": with this note a new field of physics was born. But probably at that moment Kamerlingh Onnes was simply thinking how right he had been to choose mercury. Zero resistance was what he expected to find in extremely pure metals at liquid-helium temperatures[7]. After he liquefied hydrogen in February 1906, he started a program to investigate the resistance of metals at low temperatures. There was a practical reason —thermometry— but he also had a purely scientific interest.

One of the issues in those days was the question what would happen to the resistance of a metal as its temperature approaches absolute zero[8]. It was accepted that electrons were responsible for the electric conductance and that the resistance was due to the scattering of electrons by the ions of the metal crystal. Would the scattering amplitude decrease fast enough with falling temperature to yield zero resistance at zero temperature? Or would the mobility of the electrons also diminish at lower temperature, thus resulting in zero *conductivity* at absolute zero? If nature would follow the latter prescription — put forward by Lord Kelvin in 1902[9] — the resistance of a pure metal would first fall with decreasing temperature, go through a minimum, and finally go up to infinity at absolute zero.

In the earliest investigations at liquid hydrogen temperatures in Leiden, Kamerlingh Onnes and his assistant Jacob Clay studied the resistance R versus temperature T in very thin gold and platinum wires[10]. Before July 1908 the lowest available temperature was 14 K, at which solid hydrogen sublimates under reduced pressure. That was low enough to observe that the almost linear decrease of R with T at higher temperatures starts to level off to an almost constant value. In one of his KNAW reports, Kamerlingh Onnes even mentioned a trace of a minimum in the $R(T)$ plot which indicates that he originally believed in Kelvin's model. But the effect could equally well be due to the measuring accuracy.

The almost linear $R(T)$ behavior of Pt above 14 K made that metal suitable as a secondary thermometer. It was much more convenient than the cumbersome helium gas thermometer. Only one calibration point at the temperature of liquid nitrogen was sufficient to provide a sufficiently accurate and reproducible temperature scale. For many decades these Pt resistors were fabricated in the Leiden Physics Laboratory, and provided an extra source of income. But a disadvantage of these temperature standards was the rather large size: 10 cm long and about 1 cm wide.

The calibration point was needed because, according to Matthiessen's rule, the resistance depended on the chemical and physical purity of the material. For instance, Kamerlingh Onnes showed that the resistance increase due to adding small admixtures of silver to the purest available gold was temperature independent and proportional to the concentration of added silver. So, improving purity would yield metal wires of very low resistance that could serve as secondary thermometers at temperatures far below 14 K.

Those very low temperatures came within reach after the successful liquefaction of helium in July 1908. But the race for absolute zero must have taken a great deal of Kamerlingh Onnes's energy, because in the fall of 1908, after visiting the First International Congress of Refrigeration in Paris, he collapsed. It took until June 1909 before he could resume his experiments. In his KNAW report he

---

[7]   R. de Bruyn Ouboter, *IEEE Transactions on Magnetics*, MAG-23 (1987) 355

[8]   Per Fridtjof Dahl, *Superconductivity, its historical roots and development from mercury to the ceramics oxides*, AIP, New York (1992) 13–49

[9]   Lord Kelvin, *Phil. Mag.* 3 (1902) 257

[10]   H. Kamerlingh Onnes and J. Clay, *Proc.* 9 (1906) 213, *Comm.* 95d (June 1906)

reminded his audience how close to failure he had been in July 1908, because at the time he only had a very provisional, too high estimate of the critical pressure of helium from the law of corresponding states. Therefore he repeated the helium circulation at a much lower pressure and produced as much as 60 mL "without serious difficulties". In a further attempt to solidify helium the pressure was reduced to 2.2 mm. At such low pressures the thermo-molecular corrections are substantial and instead of "2.5 K, perhaps 2 K" one can conclude in retrospect that Kamerlingh Onnes already had reached about 1.4 K. In a few subsequent experiments the helium gas thermometer was improved and the margin of errors of this instrument was so greatly reduced that the temperatures in the famous paper with the $R(T)$ plot of mercury[11] were reliable within 0.1 K.

The next important requirement was transferring helium from the liquefier, which lacked adequate space for experiments, to a separate cryostat. Considering the technology of those days this step was a real challenge. Thanks to the notebooks, we can follow quite closely the strategy followed by Kamerlingh Onnes, his technical manager of the cryogenic laboratory, Gerrit Jan Flim, and his master glassblower, Oskar Kesselring.

The first entry about the liquid-helium experiments in notebook 56 is dated Saturday 12 March 1910. It describes the first attempt to transfer helium to a separate cryostat. An extensive report can be found in the proceedings of the "Kältecongress, Wien, October 1910." It is accompanied by a few beautiful drawings made by Flim, for example, the one showing the setup of the experiment displayed in Figure 1.2. The only content of the cryostat consisted of a double-walled container in which an even smaller one was connected to an impressive battery of pumps. "The plan is transfer, then decrease pressure, then condense in inner glass, then pump with Burckhardt [pump to a pressure of] 1/4 mm [Hg], then with Siemens pump [to] 0.1 mm." Because there was nothing but glass inside the container, the experiment worked well and a new low-temperature record was registered: roughly 1.1 K. The goal of the next experiment, four months later, was to continue measuring $R(T)$ for the Pt resistor that had previously been calibrated down to 14 K. But the experiment failed because the extra heat capacity of the built-in resistor caused violent boiling and rapid evaporation of the freshly transferred liquid helium. So it was decided to drastically change the transfer system. And that would take another nine months.

Meanwhile, interest in the low temperature behavior of solids was growing rapidly. Specific heat experiments carried out in Berlin and Leiden exhibited unexpected decreases with descending temperatures. Einstein gave a qualitative explanation by modeling the energy distribution of the atomic degrees of freedom in terms of "Planck vibrators" (phonons)[12]. For the first time, the quantum world was connected to low temperature phenomena. Inspired by Einstein and a publication of Riecke, Kamerlingh Onnes worked out his own model for scattering of electrons by Planck vibrators. It described the decrease in resistance with diminishing temperatures reasonably well[13]. He probably wanted to check his model at lower temperatures and did not want to wait until the new liquid-transfer system was ready. Therefore he decided to expand the original liquefier so that it could house a platinum resistor. Thus, on 2 December 1910, he made the first measurement of $R(T)$ for a metal at liquid helium temperatures. Cornelis Dorsman assisted with the temperature measurements and student Gilles Holst operated the Wheatstone bridge with the galvanometer. Because of its sensitivity to vibrations, the galvanometer had been placed in a room far away from the cryogenic laboratory with all its heavy, thumping pumps.

The outcome of the experiment was striking. It is shown in Figure 1.3a together with some older data of Au resistors[13]. The resistance of Pt became constant below 4.25 K. There was no longer any doubt that Lord Kelvin's theory was wrong, the electron mobility did not freeze out near absolute zero. The resistance had fallen to a residual value that presumably depended on the impurity of the

---

[11] H. Kamerlingh Onnes, *Proc.* 14 (1912) 685, *Comm.* 124c (Nov. 1911)

[12] A. Einstein, *Annalen der Physik* 22 (1907) 180-190

[13] H. Kamerlingh Onnes, *Proc.* 13 (1911) 1107, *Comm.* 119b (Febr. 1911)

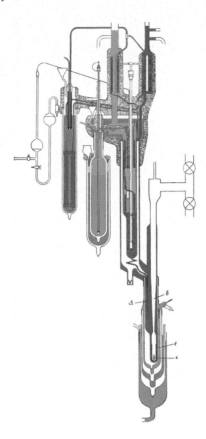

FIGURE 1.2: Set-up used in the first attempt (March 12, 1910) to transfer helium from the liquefier to a separate cryostat in which experiments could be done. The far right shows a drawing of the
cryostat consisting of several containers of glass fitting inside one another. Most of the containers
are in the form of a thermos bottle or Dewar (a double-walled glass container, pumped vacuum and
usually covered with silver coatings on the inside walls; for visibility the coatings were not applied
here). The outer vessel contained alcohol at 30–40°C to prevent condensation of water vapor on
the glass. The outer Dewar contained liquid air, the middle one liquid hydrogen, and the inner Dewar *c* was supposed to be filled with liquid helium coming from the liquefier on the left, entering
through the double-walled tube *d*. The bottom part of insert *b* was double walled and could be filled
with liquid helium by condensation of helium gas. Subsequently, it could be evacuated to isolate
the liquid helium inside *b* from the liquid helium in the cryostat. By pumping on the helium in *b* a
new low temperature record was obtained. The left part of the drawing shows the liquefier with the
Joule-Thompson valve clearly visible[2]. [The original drawing (like the ones in Figure 1.4 and 1.5)
was published in the Communications from the Physic Laboratory of the University of Leiden (referred to as Commun.). Colors have been added to indicate various cryogenic fluids: alcohol (pink),
liquid air (purple), liquid and gaseous hydrogen (dark and light green respectively), and liquid and
gaseous helium (dark and light red, respectively).]

wires. Those impurities could be either intrinsic or caused by the drawing process for the fabrication of thinner and longer wires to get higher resistances. Based on his model, Kamerlingh Onnes
concluded that resistors of sufficiently pure platinum or gold should become zero at liquid-helium
temperatures.

On 21 December 1910 and 27 January 1911 two more experiments were started up in the same

Kamerlingh Onnes adapted Riecke's theory by replacing the thermal energy $k_B T$ by $E_T = 3R\beta\nu/[\exp(\beta\nu/T) - 1]$, where k_B is Boltzmann's constant, R the gas constant, ν the frequency of Planck vibrators, and $\beta = h/k_B$ with h Planck's constant. The resistance ratio then could be written as $R(T)/R_0 = \sqrt{TE_T/T_0E_0}$, where '0' refers to 0 °C. The parameter $\beta\nu$ depended on the material properties related to the elastic constants of the vibrators, for instance the melting temperature. With this model he computed the $R(T)/R_0$ dependence of Pt, Ag, Au, and Pb and found down to hydrogen temperatures reasonable agreement with the data of the purest wires. At 4.3 K, however, the model yielded immeasurably low values for all those metals.

In a communication to the first Solvay Conference Kamerlingh Onnes also reported the first data for mercury (see Figure 1.3b). Down to a temperature of 14 K there was reasonable agreement with the predictions of his model (see below). We now can reproduce why he was so eager to continue his investigations with mercury. As he later also explained there were two reasons: by using his model he could compute $R(T)/R_0$ at a few relevant temperatures, say 20 K, 14 K, 4.3 K, and 3.0 K. If we repeat his exercise with the value of $\beta\nu$ he chose for Hg[a], namely $\beta\nu = 30$, we obtain 5.0×10^{-2}, 2.7×10^{-2}, 0.13×10^{-2}, and 0.024×10^{-2}, respectively. The two latter values were within the sensitivity range of his equipment and the resistance ratio would nicely fall off to zero at the lowest temperatures. Furthermore, the Leiden laboratory had a lot of experience with the purification of mercury by distillation, and it would not be contaminated by the necessity of drawing a thin wire. (The liquid mercury in a capillary simply freezes at 234 K [−39°C].)

---

[a] H. Kamerlingh Onnes, *Proc.* 13 (1911) 1093, *Comm.* 119b (Febr. 1911)

set-up, but they failed because of some technical difficulties. Again Flim was in charge of the helium liquefier and Kamerlingh Onnes took notes in his little notebook. Dorsman and Holst were present to assist with the experiments. Dorsman left in the summer of 1911, but Holst stayed until the end of 1913. He apparently enjoyed helping Kamerlingh Onnes with his research, rather than working on his thesis which was more like a routine project and much less exciting for a young researcher with ambitions.[1,14]

The discovery

At the beginning of April 1911, the new cryostat was ready for its first cooldown. It was a masterpiece of technical design, demonstrating the amazing levels of glassblowing skill and fine mechanical construction. An extensive description with detailed drawings, reproduced in Figure 1.4, appeared in June 1911[15]. The transfer tube had been replaced by a double-walled, vacuum-pumped glass siphon, externally cooled by a counter flow of liquid air forced through a copper capillary coil wound around it. The liquefier and the cryostat could be separated from each other simply by closing a valve operated from above. Another important contraption, necessary to establish a well-defined

---

[14] Holst received his doctorate in Zürich in September 1914. In *Commun.*144d an abstract of his thesis (on the thermodynamic properties of ammonia and methylchloride) is given. At the end he writes (translated from German): "At the end of this work it is my pleasant duty to thank Professor H. Kamerlingh Onnes for his kindness to put all the equipment necessary for this research at my disposal. The time, during which I had the honor to be his assistant and more in particular the years during which I had the pleasure to assist him with his own research, will always remain a period of my life of which I will only be able to think with a feeling of deep gratitude".

[15] H. Kamerlingh Onnes, *Proc.* 14 (1911) 204, *Comm.* 123a (June 1911)

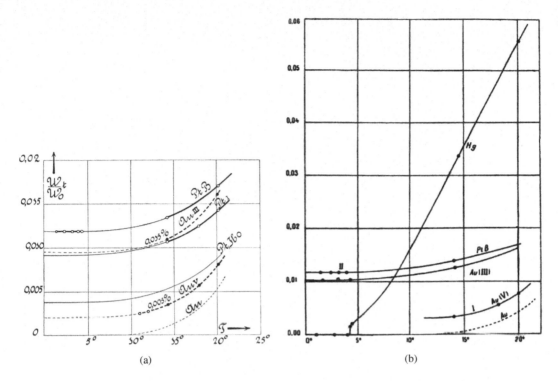

(a)   (b)

FIGURE 1.3: Resistance ratios of some metals versus temperature T (Kelvin). Left panel (a): Several platinum and gold resistors of various purities measured at different hydrogen temperatures. Pt-B was the first resistor ever to be cooled to helium temperatures in the experiment of 2 December 1910. The constant resistance below 4.3 K contradicted Kelvin's model for conductance[9]; the electrons did not freeze onto the ion lattice at absolute zero. The remaining resistance was due to scattering of the electrons on impurities. By making the metal wires purer, both chemically and physically (by annealing out the lattice disorder), the resistance was shifted downward over a constant value, demonstrating that Matthiessen's rule is valid down to the lowest temperatures. Right panel (b): The resistance ratio of Pt and Au compared to that of mercury (Hg). I denotes the temperature range of liquid hydrogen, II that of liquid helium[13].

temperature at the site of the measuring devices, was a stirrer connected to a magnet at the top of the cryostat that could be moved up and down by a motor. The action of valve and stirrer could be directly followed through uncoated strips in the silvered vacuum glasses. Kamerlingh Onnes wrote "Niveau (van helium) vlak onder bovenrand pompje (= de roerder). Pompje werkt prachtig. [Level (of helium) just below upper rim little pump (= the stirrer). Pump works splendidly]". In his notebook next to a little sketch visible in Figure 1.1 it shows how "It is peculiarly charming to see this little pump ejecting the light liquid over its upper edge when the level of the liquid helium sinks a little below it." The main purpose of the experiment was to test the transfer of liquid helium to the experimental cryostat. But according to notebook 56, inside the cryostat were installed: "1^e transparent helium gas thermometer [...], 2^e resistor gold, 3^e thermo-element AuAg, 4^e resistor mercury, 5^e resistor constantin, 6^e conductivity liquid helium and dielectric constant". So experiments could be done—just in case the helium transfer worked.

The mercury resistor was constructed by connecting seven U-shaped glass capillaries in series, each containing a small Hg reservoir to prevent the wire from breaking during cooldown. In Figure 1.5 one can see a similar design used in an experiment later that year. The electrical connections were

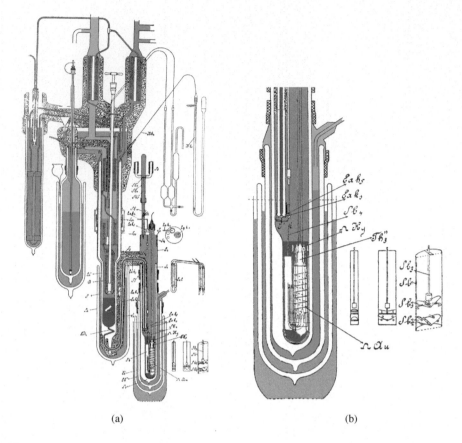

(a)　　　　　　　　　　　　　　　(b)

FIGURE 1.4: Set-up in which Heike Kamerlingh Onnes and coworkers carried out the 8 April 1911 experiment that first revealed superconductivity (same color scheme as in Figure 1.2). Left panel (a): On the left the liquefier with extended Dewar is schematically displayed. The liquid helium could be transferred through the double-walled vacuum-pumped siphon that could be closed with a valve (***Ea k*1**) at the end inside the cryostat. Right panel (b): Blow-up of the lower part of the cryostat. Handwritten by Gerrit Flim are labels for the mercury and gold resistors (Ω Hg and Ω Au), the gas thermometer (***Th*3**), components at the end (***Ea kn***) of the transfer tube from the helium liquefier, and parts of the liquid helium stirrer (***Sb***), which is also shown enlarged in several cross sections at the right[11].

made by four platinum feedthroughs with thin copper wires leading to the measuring equipment outside the cryostat. Kamerlingh Onnes had followed Gilles Holst's suggestion to distill the mercury by using liquid nitrogen. Apparently that worked well, but the Hg wire was *"yet not as pure as our former mercury"*.

To learn what happened on 8 April 1911, we just have to follow the notes in notebook 56. The experiment started at 7 a.m., and Kamerlingh Onnes joined in when they began to circulate helium at 11:20 a.m. The resistance of the mercury indicated the falling temperature. After 30 minutes, the gold resistor was at $-140°C$, and soon after noon the gas thermometer denoted 5 K. The valve also worked "very sensitively". Half an hour later, enough helium had been transferred to test the functioning of the stirrer (see above) and to measure the very small evaporation heat of helium.

The team established that the liquid helium did not conduct electricity, and they determined its dielectric constant. Holst made precise measurements of the resistances of Hg and Au and the temperature was determined by Dorsman. Then they started to reduce the vapor pressure. The helium

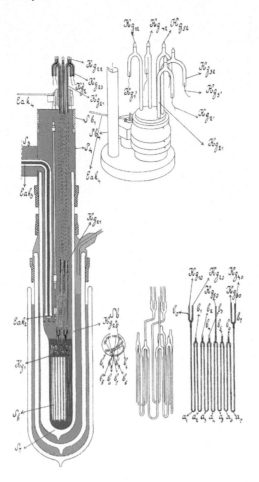

FIGURE 1.5: Cryostat with mercury resistor and mercury leads for the 26 October 1911 experiment (same color scheme as in Figure 1.2): seven U-shaped glass capillaries in series (inner diameter 0.07 mm), each with a mercury reservoir at the top and contact leads also made of glass capillaries filled with mercury. External contacts were made through Pt wires (denoted by Hgxx) shown in the top right drawing[11].

evaporated very quickly due to its small evaporation heat, they measured the specific heat, and stopped at a vapor pressure of 197 mmHg, corresponding to about 3 K.

Exactly at 4 p.m., says the notebook, the resistances of gold and mercury were determined again. The latter was in the historic entry "practically zero." The notebook further records that the helium level stood quite still ("*Staat helemaal stil*").

The experiments continued into the late afternoon. At the end of the day, Kamerlingh Onnes finished with an intriguing notebook entry: "Dorsman [who had measured the temperatures] really had to hurry to make the observations." The temperature had been surprisingly hard to control. "It should be remarked that just before the lowest temperature [about 1.8 K] was reached, the boiling suddenly stopped and was replaced by evaporation in which the liquid visibly shrank. So, a remarkably strong evaporation at the surface." Without realizing it, the Leiden team had also observed the superfluid transitions of liquid helium at 2.2 K. Two different quantum transitions had been seen for the first time, in one lab on one and the same day!

Three weeks later, Kamerlingh Onnes reported the results at the April meeting of the KNAW[5]. For the resistance of ultrapure mercury, he first told the audience, his model had yielded three predictions: (1) at 4.3 K the resistance should be much smaller than at 14 K, but still measurable with his equipment; (2) it should not yet be independent of temperature; and (3) at very low temperatures it should become zero within the limits of experimental accuracy. Those predictions, Kamerlingh Onnes concluded, had been completely confirmed by experiment. A look at Figure 1.3b clarifies why he came to that conclusion with only three data points (3.0 K, 4.3 K, and 14 K) at his disposal.

For the next experiment, on 23 May[4], the voltage resolution had been improved to about 30 nV. The $R(T)/R_0$ at 3 K turned out to be less than 10^{-7}! This value did not change anymore when T was lowered to 1.5 K. (The normalizing parameter R_0 was the calculated resistance of solid mercury extrapolated to 0 °C.) The notebook tells us that Ohm's law was checked at 4.3 K (supposedly both for the Au-III and Hg resistors). Then at 3:05pm:

> "Now proceeded to resistance [measurements] at low temperatures".[...] Everything is very regular now, no notion of violent boiling. All regarding installation had been disconnected from electrical wiring in building. It was feared that previous time this could have been of influence. Remains now [...] to suspect paper and paraffin. Evaporation seems decreased at these low temperatures. Now explored point between 190 mmHg and 760 mmHg [3.0 K and 4.3 K] because found resistance zero at just chosen point 3.2 K. At 4.00 [K] not yet anything to notice of rising resistance. At 4.05[K] not yet either. At 4.12 [K] resistance begins to appear."

That entry contradicts the oft-told anecdote about the key role of a "blue boy" — an apprentice from the instrumentmaker's school Kamerlingh Onnes had founded. (The appellation refers to the blue lab coats the boys wore. As the story goes, the blue boy's sleepy inattention that afternoon had let the helium boil, thus raising the mercury above its 4.2 K transition temperature and signaling the new state — by its reversion to the normal conductivity — with a dramatic swing of the galvanometer.

The experiment was done with increasing rather than decreasing temperatures because that way the temperature changes slowly and the measurements could be done under more controlled conditions. Kamerlingh Onnes reported to the KNAW[11] that slightly above 4.2 K the resistance was still found to be only $10^{-5}R_0$, but within the next 0.1 K it had increased by a factor of almost 400. Such a fast increase was much more than his model could account for; see above and Figure 1.3b. He used the remainder of the paper to explain how useful this vanishing of the electrical resistance could be. It is interesting that the day before Kamerlingh Onnes submitted that report, he wrote in his notebook that the team had checked whether "evacuating the apparatus influenced the connections of the wires, as caused by deformation of the top [*of the cryostat*]. It is not the case". Thus they ruled out inadvertent short circuits as the cause for the vanishing resistance.

That entry reveals how puzzled they were with the experimental results. Notebook 57 starts on 26 October 1911, "In helium apparatus mercury resistor with mercury contact leads (separate drawing made)". This drawing, reproduced in Figure 1.5 and published in *Commun.*124c (30 December 1911), shows the team had spent the whole summer replacing the platinum feed-throughs and copper leads by mercury wires in glass capillaries that went all the way through the cryostat's cap. And they investigated how the new setup could be cooled down in a controlled fashion without breaking the mercury wires or the glass.

That was quite a challenge. In retrospect, the effort turned out to be a waist of time, but it was motivated by the important question of how small the resistance actually was. To improve voltage resolution, they sought to minimize the thermoelectric effect in the voltage leads. The idea was to do that by using the same material for both sample and leads and bringing the connections with Pt outside the cryostat where they can be thermally isolated from temperature variations. It didn't work, because the transition from solid to liquid mercury actually turned out to be the source of a considerable thermoelectric voltage of about 0.5 mV.

Still, the October experiment produced the historic plot, shown in Figure 1.6, of the abrupt appearance of mercury's resistance at 4.20 K. The part of the plot above the transition temperature is of particular interest because it shows a gradual increase with temperature beyond the jump. To obtain those data, the Leiden team had to go beyond the normal boiling point of helium. They did that by closing the helium inlet valve so that the vapor pressure could rise and thus raise the boiling temperature. That the resistance above the jump depended on temperature showed that the electrons at 4.2 K were still predominantly scattered by phonons (Planck vibrators). From the sudden jump it was clear that a totally new and unexpected phenomenon had been discovered, which Kamerlingh Onnes called "supraconductivity".

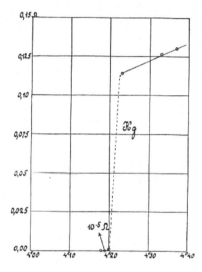

FIGURE 1.6: Historic plot of resistance (Ω) versus temperature (K) for mercury from the 26 October 1911 experiment shows the superconducting transition at 4.20 K. Within 0.02 K, the resistance jumps from unmeasurably small (less than 10^{-5} Ω) to 0.1 Ω.[11]

Just one week later he reported his discovery in Brussels to the elite of the physics world at the very first of the historic Solvay Conferences. In the discussion "M. Langevin asked if other properties of the substance displayed similar sudden changes, as would be the case if mercury underwent a structural modification at 4.20 K". Apparently, Langevin was thinking of a phase transition, although he might have had a structural phase transition in mind. This was certainly Kamerlingh Onnes's interpretation, for "... should there exist such a new modification, it would differ from ordinary mercury at higher temperatures chiefly by the property that the frequency of the vibrators in the new state has become greater, and therefore the conductivity rises to the extremely large value exhibited below 4.19 K". So he still saw a way to stick to his model, but he also announced new experiments to investigate Langevin's suggestion.

Further on in notebook 57, we notice preparations in March 1912 for specific-heat experiments near the transition of mercury. Holst was put in charge. But there turned out to be experimental difficulties that could not satisfactorily be resolved. Eventually, Holst and Kamerlingh Onnes published a paper together on the specific heat and thermal conduction of Hg, but the accuracy of the measurements at helium temperatures was not sufficient to reveal any features at the transition temperature[16]. A month later they also tried to determine the thermo power of a mercury-platinum contact at temperatures around the transition. But again, experimental problems were in the way of

---

[16] H. Kamerlingh Onnes and G. Holst, *Proc.* 17 (1914) 760, *Comm.* 142c (June 1914)

success.

Other Materials, Other Experiments

A notebook entry dated 20 June 1912 is very interesting: "Discussed with Holst ... alloying mercury with gold and Cd. Decided [to use] very small concentrations." A few days later the experiment had already been carried out. The results were not published until the following March: "To my surprise the resistance (of the mercury with admixtures) disappeared in the same way as with pure mercury; much of the time spent on the preparation of pure mercury ... might therefore have been saved .... Even with the amalgam that is used for the backing of mirrors, the resistance was found [to be] 0 at helium temperatures. Later, December 1912, it was found that if it disappears suddenly, as with the pure mercury, it happens at a higher temperature." In fact, the mercury period was closed with that experiment of December 1912. Not only did the Leiden team observe superconductivity in amalgam was observed, but they also discovered that lead and tin were superconductors with transition temperatures near 6 K and 3.8 K, respectively. These discoveries came as something of a relief. Since then the experiments were continued with these materials without worrying about laboratory problems peculiar to mercury: double distillation, broken threads, and very cumbersome cool down procedures.

The other notes stored in the Archives of the Boerhaave Museum concerning superconductivity are dealing with the persistent current experiments carried out in the spring and summer of 1914; see below. In the mean time Kamerlingh Onnes had published several papers on superconductivity[17]. They appeared early 1913 and gave a very open account of all the difficulties encountered with the mercury experiments. The destruction of superconductivity by a strong current surpassing a certain threshold value was a new issue. That depairing current or critical current, as it is called now, increased with diminishing temperature.

An experiment on 17 January 1914 revealed the destructive effect of a magnetic field on superconductivity. For Pb, the threshold value at 4.25 K was only 600 Gauss (60 mT)[18]. That must have been a great disappointment for Kamerlingh Onnes because on several occasions, for instance at the Third International Congress of Refrigeration in Washington-Chicago, he had dreamt aloud about coils made of superconducting material that could produce magnetic fields as high as a 100,000 Gauss (10 T). He finished the February 1914 paper with a nice twist: "An unforeseen difficulty is now found in our way, but this is well counterbalanced by the discovery of the curious property which is the cause of it". Nowadays the 8-tesla bending magnets of the Large Hadron Collider at CERN approach that dream, but they do so with a niobium-titanium alloy whose critical magnetic field far exceeds those of the superconductors known on Kamerlingh Onnes's time.

Kamerlingh Onnes next concentrated on the question of how small the "microresidual" resistance actually was in the superconducting state. He designed an experiment to measure the decay time of a magnetically induced current in a closed superconducting loop. He used the small lead coil which was made for the experiments about the magnetic field effect. To close the coil in itself the two ends were soldered together. And after lowering the temperature in presence of an external magnetic field a current could be induced in the superconducting closed loop by removing the magnet. From the decay time of the current and the self-induction of the coil the micoresidual resistance R could be determined. With the upper bound for R being extremely small, it was to be expected that the decay time would be very large. For the Pb coil it was expected that the decay time would be more than 24 hours[19].

To probe the decay of the current circulating in the closed loop after the induction magnet had been removed, he used a compass needle placed close to the cryostat and precisely to its east. To

---

[17] H. Kamerlingh Onnes, Proc. 16 (1913) 113, Comm. 133a,b,c (Febr. 1913); Proc. 16 (1913) 113, Comm. 133d (May 1913)

[18] H. Kamerlingh Onnes, Proc. 16 (1914) 987, Comm. 139f (Febr. 1914)

[19] H. Kamerlingh Onnes, Proc. 17 (1914) 12, 278, 514, Comm. 140b (April 1914), 140c (May 1914), 141b (June 1914)

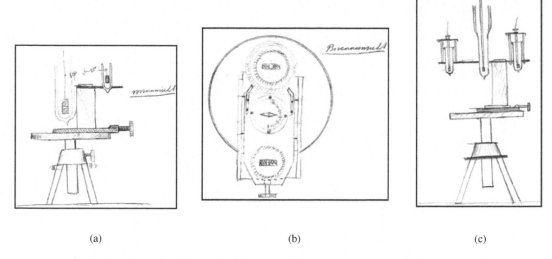

(a)   (b)   (c)

FIGURE 1.7: Original drawing by Gerrit Jan Flim showing the setup for the persistent-current experiments of 19 (a and b) and 29 May 1914. Left (a): front view (vooraanzicht) showing the lead coil in the helium cryostat and the copper compensation coil in the liquid air Dewar (actually, during the experiment, the lead and copper compensation coils were on the same height as the compass needle). Center (b): top view (bovenaanzicht) showing also the compass needle in the middle pointing north demonstrating good compensation of the fields from the Pb coil and that from the copper coil. Right (c): side view of the symmetric compensation set up with two copper coils in liquid air used in the experiment of 29 May 1914 (Archive of the Boerhaave Musuem, Leiden).

improve the sensitivity, the earth field was compensated and to calibrate the supercurrent in the loop, he positioned an almost identical copper coil on the other side of the compass needle which had to compensate the effect of the superconducting coil. In Figure 1.7a, a sketch of the setup made by Flim is reproduced. Kamerlingh Onnes reported his first results to the KNAW on 24 April 1914: "During an hour the current [0.6 A] was observed not to decrease perceptibly. [...] A coil cooled in liquid helium and provided with current at Leyden, might, if kept immersed in liquid helium, be conveyed to a considerable distance and there be used to demonstrate the permanent-magnetic action of a superconductor carrying a current. I would have liked to show the phenomenon in this meeting in the same way as I brought liquid hydrogen here in 1906. But the appliances at my disposal do not yet allow the transportation of liquid helium"[19].

Two decades later, such a travelling show had become possible. In 1932 six years after Kamerlingh Onnes's death, Flim flew to London with a portable Dewar containing a lead ring immersed in liquid helium and carrying a persistent current of 200 A. He made the trip to demonstrate this most sensational effect of superconductivity at the traditional Friday evening lectures of the Royal Institution. That was the same grand venue at which James Dewar had demonstrated the liquefaction of hydrogen in 1899.

The analogies expressed by the title of the May 1914 paper are worth mentioning: "The imitation of an Ampére molecular current or of a permanent magnet by means of a supraconductor". The latter effect is at the basis of superconducting levitation that we can demonstrate today on a human scale.

The precision of the first experiment, estimated to be about 10%, was further improved in two

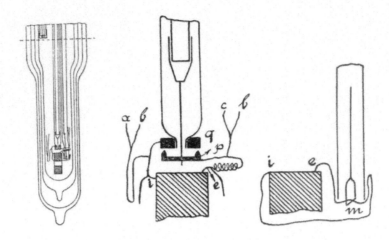

FIGURE 1.8: Design for the experiment on 20 June 1914 with cutting mechanism (right), the superconducting key (mechanical persistent mode switch) (center), and the cryostat with insert (left)[19]. The hatched element is part of the lead coil. All wires, as well as the ring q and the plate with protrusions p, are made of lead.

subsequent runs. The latter with two compensation coils of copper, both immersed in liquid air; see Figure 1.7b. It turned out that the change of the current was less than 1% per hour. That result raised the decay time to more than 4 days and diminished the upper limit of the resistance ratio to about 2×10^{-11}. In fact, that result was somewhat disappointing because it turned out to be only a factor of two lower than the ratio determined by means of the galvanometer. However, on spectators the impact of this direct observation of a persistent current was so much greater than all earlier reports, that this experiment of 1914 may be considered as the ultimate proof that superconductivity was indeed an entirely novel phenomenon.

The excitement spread quickly and widely. In Berlin Max Planck heard about it and wrote to Kamerlingh Onnes to express his amazement[20]. Paul Ehrenfest, who had witnessed the experiment himself, told Lorentz that he was flabbergasted. "I attended a fascinating experiment at the laboratory. ... Unsettling, to see the effect of this 'permanent' current on a magnetic needle. It is almost palpable, the way the ring of electrons goes round and round and round in the wire, slowly and virtually without friction"[21]. Popular news media covered the spectacular phenomenon too. The Dutch newspaper *NRC* (*Nieuwe Rotterdamsche Courant*) invited Keesom to write an article that appeared in the issues of Friday and Saturday 10 and 11 July 1914[22]. On 30 June 1914 Le Courrier de la Presse had reported "Une sensationelle communication a été fait hier à l'Academie des Sciences: un courant électrique, une fois amorcé, peut durer indéfiniment ...."[22]

Next Kamerlingh Onnes wished to supply a conclusive proof that the magnetic moment of the coil is really caused by a current. The essential elements for that experiment are displayed in Figure 1.8 on the right, reproduced from Commun. 141b[19]. This time the closed loop was constructed such that it could be cut from outside. Upon charging the coil and cutting the connection the current could escape only through a ballistic galvanometer which was connected to the coil on both sides of the

---

[20] Planck to Kamerlingh Onnes, 10 March 1915, Museum Boerhaave, Kamerlingh Onnes archive

[21] Ehrenfest to Lorentz, 11 April 1914, Noord-Hollands Archief, Lorentz archive, inv. nr. 20

[22] Museum Boerhaave, Kamerlingh Onnes archive

intersect. It worked and "the current does really flow through the coil" was the firm conclusion.

The same paper described two more experiments, both inspired by suggestions of colleagues in Leiden (Kuenen and Ehrenfest). Kuenen's suggestion would nowadays be recognized as a mechanical persistent mode switch. The first such device is displayed in the center of Figure 1.8. It consisted of a lead ring mounted on a glass tube and a lead plate that could be pulled up against the ring providing a superconducting contact through three small protrusions on the plate pointing towards the ring. Ring and plate were connected through lead wires to the coil yielding a superconducting loop once the contact was made. In addition to the wires of the ballistic galvanometer (a and c), two more wires (b and d) were attached to the coil and an external current source. In this way the following experiment could be done: with open key and disconnected galvanometer an external current could be sent through the coil. Upon closing the key by applying sufficient pressure, the superconducting loop could be established, so that the ballistic galvanometer could be connected and the current source disconnected without disturbing the current in the loop. Finally, when the key was opened, the current produced a throw of the galvanometer. All this could be followed by monitoring the position of the compass needle next to the cryostat. And it all worked convincingly well[19].

Ehrenfest came up with the idea that the experiment could be equally well performed with a lead ring. It worked perfectly. A current of 320 A was registered in a ring with a rectangular cross section of 3.0×3.5 mm^2. Kamerlingh Onnes concluded correctly that the current density was the important quantity, but the value he computed, 30 A/mm^2, was far too low. As we know today, the current is concentrated in a thin surface layer with thickness λ, the London penetration depth, which was introduced 20 years later, as will be discussed in Section 1.1.2.

1.1.2 Elemental Superconductors

The outbreak of the Great War in 1914 resulted in a period of silence which lasted until the mid 1920s. The only progress made was Silsbee's explanation[23] of the linear relation between the (critical) threshold current of a tin wire and its diameter[17]. Silsbee proposed that superconductivity would be disturbed when the magnetic field at the surface produced by the current through the wire equals the critical value $\mu_0 H_c$ at the temperature considered, where μ_0 is the magnetic permeability of free space. New experiments in Leiden, first published in the thesis of Tuyn (July 1924), confirmed Silsbee's conjecture[24]. At about the same time an overview of $R(H, T)$ data of the then-known superconductors was published[25]. That data showed, as can be seen in Figure 1.9, that the temperature dependence of the critical field follows a generic behavior that is well approximated by the parabola $H_c(T) = H_c(0)(1 - (T/T_c)^2)$. That critical field line thus separates a normal from a superconducting state and the resemblance with a thermodynamic phase diagram must have been compelling[26]. Nevertheless it took more then a decade before that step was made.[8]

It was believed that the description of superconductivity had to rely on the theory of perfect conductivity—not unreasonable, because new experiments on the persistent current by Tuyn had reduced the upper limit of $R(4.2K)/R_0$ to 10^{-13}. That believe culminated in the theory of Becker, Heller and Sauter[27]. By combining Faraday's law, curl $\mathbf{E} = -\partial \mathbf{B}/\partial t$, Ohm's law, $\mathbf{E} = \mathbf{J}/\sigma$, and the conductivity $\sigma \to \infty$, it followed that $\partial \mathbf{B}/\partial t = 0$ so that, once in a state of perfect conductivity, the magnetic induction $\mathbf{B}$ could not change anymore. A change of the applied magnetic field would therefore give rise to induced screening currents in the perfect conductor which would create a magnetic field that exactly compensated the applied field change in the interior. Those currents should

[23] F.B. Silsbee, *J. Wash. Acad.* Sci. 6 (1916) 597

[24] W. Tuyn, H. Kamerlingh Onnes, Franklin Institute, Journal 201 (1926) 379, Comm. 174a (1926); W.J. de Haas, J. Voogd, Proc. 14 (1911) 113, Comm. 212a (1931)

[25] W.J. de Haas, G.J. Sizoo, H. Kamerlingh Onnes, Proc. 29 (1926) 250, Comm. 180d (1926)

[26] W.H. Keesom, *Rapp. et Disc. 4^e Conseil Solvay* 288 (1924)

[27] R. Becker, G. Heller, F. Sauter, *Z. Phys.* 85 (1933) 772; G.L. de Haas-Lorentz, *Physica* 5 (1925) 384 (in Dutch)

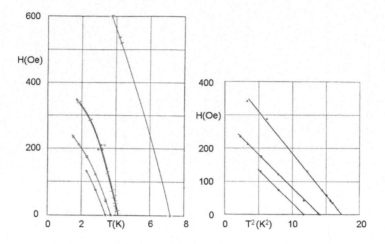

FIGURE 1.9:   Left panel: the threshold curves of lead, mercury, tin, and indium (high to low T_c) in a linear plot of H_c versus T. Right panel: same H_c data (except for lead) plotted versus T^2 revealing the seemingly good quadratic temperature dependence $H_c(T) \approx H_c(0)(1-(T/T_c)^2)$. [24,25]

flow in a small, but finite layer at the surface with thickness $\lambda = (m/\mu_0 n e^2)^{1/2}$. For most known superconductors this layer was estimated to be about 10^{-4} mm (-e, m, and n respectively denote charge, mass, and density of the (relevant) conduction electrons). Interestingly, a few years earlier, the daughter of Lorentz had also obtained that result by equating the kinetic energy of the screening currents with the magnetic energy.[26] As has been schematically illustrated in Figure 1.10a and 1.10b, the state of a perfect conductor thus depends on its history. The perfect conductor therefore could not represent a thermodynamic equilibrium phase.

Thermodynamics

The first evidence that a superconductor was more then a perfect conductor and instead represented a state of thermodynamic equilibrium was provided by Kok and Keesom[28] in specific heat experiments carried out in zero applied field. They had improved the sensitivity of earlier experiments of Keesom and Van den Ende so that for the first time a clear jump in the electronic heat capacity of tin at the critical temperature T_c could be observed; see Figure 1.11. In addition, they showed there was no latent heat at the transition and that below T_c the specific heat roughly varied as T^3. The absence of a latent heat at T_c implied a phase transition of the second kind. The possibility of such a new kind of phase transition was introduced by Keesom[29] and Ehrenfest[30] for the theoretical analysis of the lambda transition that was just discovered in liquid helium. A student of Ehrenfest, Rutgers, used this analogy to derive a relation between the jump in the specific heat Δc and the derivative of the critical field at T_c[31]

$$\Delta c = (c_s - c_n)_{T=T_c} = \mu_0 T_c \left(\frac{\partial H_c}{\partial T} \right)^2_{T=T_c}. \tag{1.1}$$

[28] W.H. Keesom, J.A. Kok, Proc. 35 (1932) 743, *Comm.* 221e (1932) and *Physica* 1 (1934) 175

[29] W.H. Keesom, *Commun.* Suppl. 75a (1933)

[30] P. Ehrenfest, *Commun.* Suppl. 75b (1933)

[31] A.J. Rutgers, *Physica* 1 (1934) 1055

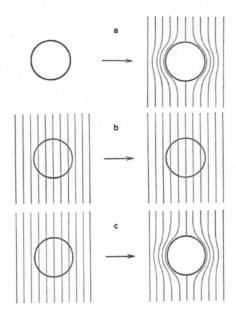

FIGURE 1.10: Difference between perfect conductor (a and b) and perfect diamagnetic (c). From left to right: a) sample cooled in zero field to T below T_c, then a magnetic field is applied. Magnetic induction B remains zero. b) Sample brought into magnetic field while in the normal state, then cooled to T below T_c. B doesn't change. c) Perfect diamagnetic brought into magnetic field $H \ll H_c(0)$ while in the normal state, then cooled through $T_c(H)$. At the transition to the superconducting state the magnetic field is spontaneously expelled.

That prediction agreed beautifully with the available experimental results for tin, and later for thallium as well[32]. Rutgers was a rather prudent person and did not want to publish his equation before it was certain that the transition of a superconductor was indeed a reversible transition between thermodynamic equilibrium phases. Cornelis Gorter, a student of De Haas, was more aware of the urgency of the developments and quoted Rutgers's findings in a pioneering thermodynamic analysis which he published in the spring of 1933[32]. To the regret of for instance Fritz and Heinz London[33], Gorter published his paper in the *Archives of the Teyler Museum in Haarlem*, an institution by which he was employed at the time[34]. This guaranteed fast publication, but was also rather inaccessible. A later publication in *Physica* with Casimir as coauthor[35] included most of the contents of the paper in the archive.

In the first place, Gorter emphasized that the transition from the normal to the superconducting state in an applied magnetic field could be interpreted only in a simple way, if a configuration was chosen in which demagnetization effects would not play a role, for instance, a long wire with the axis parallel to the applied field. Only then, the destruction of superconductivity would occur sharply, in a well-defined way. Such a sharp transition was indeed observed in the experiments of De Haas and Voogd[36] on single crystals in the "longitudinal" configuration with much larger dimensions in the direction of the field than in the transversal directions. Moreover, the experimental data of Kok and Keesom[28] restricted him in his thermodynamic analysis to consider only those closed

[32] C.J. Gorter, *Arch. Mus. Teyler* 7 (1933) 378

[33] G. Rickhayzen, *Theory of Superconductivity*, Wiley, New York, 1964, Chapter 1

[34] C.J. Gorter, *Rev. Mod. Phys.* 36 (1964) 3

[35] C.J. Gorter, H. B.G. Casimir, *Physica* 1 (1934) 306

[36] W.J. de Haas, J. Voogd, *Proc.* 34 (1931) 51, 192, *Comm.* 212c and 214c (1931)

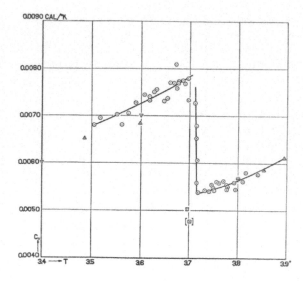

FIGURE 1.11:   Temperature dependence of the specific heat of tin measured in ambient magnetic field; the first clear evidence of a jump in c at the superconducting transition[28].

rectangular loops in the $H - T$ diagram which contained the zero-field transition between normal and superconducting states. Following the loop in clockwise direction starting above T_c and going along $H = 0$, that automatically resulted in a superconducting state in which the induction $\mathbf{B} = 0$. That, according to Fritz London, "came very close to predicting the Meissner effect[37]."

Under these conditions the validity of Rutgers's equation was equivalent to the statement that the second law of thermodynamics ($\mathrm{d}Q/T = \mathrm{d}S$) would apply for the transition in a magnetic field and that the transition would be reversible. Finally, Gorter concluded that in transversal field configurations (i.e., non-zero demagnetization) superconductivity would be destroyed in some parts of the specimen, while it could persist in other parts. Later on, this sequence of normal and superconductive layers was coined the "intermediate state".

Within the above assumptions Gorter could compute the difference in free energy between the superconductor in zero field and that in the normal state by integrating the specific heat data of Keesom and Kok down from a temperature above T_c. From that calculation he obtained the field at which superconductivity would become thermodynamically unstable avoiding the discussion about the expected irreversibility of the field induced transition to the normal state[34]. The good agreement with the measured data justified his approach from which the famous expression followed for the condensation free energy per unit volume, $\Delta f = f_s - f_n = -\mu_0 H_c^2 (T)/2$.

Diamagnetics

At the end of the 1920s, new very-low-temperature facilities outside Leiden were founded and with that also new ideas for experiments emerged. At the Physikalisch-Technische Reichsanstalt (PTR) in Berlin Walther Meissner started to investigate the nature of the current distribution in superconducting cylinders in an applied transversal magnetic field[38]. Assisted by Ochsenfeld he wanted to detect the changes in the magnetic field strength between two parallel, cylindrical su-

[37] F. London, *Superfluids*, vol. I, Dover, New York (1961) §2; 1st edition, John Wiley, New York (1950)

[38] W. Meissner, R. Ochsenfeld, *Naturwissenschaften* 21 (1933) 787

perconductors as they passed the transition to the superconducting state. In fact, Max von Laue, who was consultant at the PTR, had proposed to do the experiments and also assisted by calculating the expected field distribution outside the cylinders. Figure 1.12 shows the experimental set-up. A tiny search coil *b*, 1 mm wide and 15 mm long, probed the magnetic field strength between the mono-crystalline tin cylinders which were connected in series carrying a current *J*. The search coil was connected to a ballistic galvanometer and by flipping it over 180° the local field could be determined. When cooled through the superconducting transition a clear jump in the measured field strength was detected, indicating that the magnetic flux which had penetrated the cylinders in the normal state was entirely expelled from them at the transition to the superconducting state. On the other hand, no change was seen when the search coil was placed inside a hollow superconductor. They repeated the experiment by investigating several long, hollow, thick-walled cylinders in different external field configurations and each time confirmed that for pure superconductors the magnetic field distribution corresponded to zero induction in the bulk of the superconductor independent of the initial conditions (see Figure 1.10c). It implied that inside a superconductor an external magnetic field had to be completely compensated by the field generated by (super) currents in a layer at the surface.

The Meissner-Ochsenfeld effect came as a complete surprise and one can indeed state that $\mathbf{B} = 0$ signified a turning point in the history of superconductivity. It meant that a superconductor is not only a perfect conductor, it is above all a perfect diamagnetic. The theoretical implications are illuminated in Figure 1.10c: the magnetic flux should be spontaneously expelled from the pure superconductor. However, in practice, small impurities or inhomogeneitics of the material can be sufficient to suppress the appearance of the perfect Meissner effect. In such "non-ideal" conditions frozen-in flux had been found. The shape of the specimen and the field configuration without demagnetizing turned out to be of importance; both are needed for the establishment of the pure Meissner effect. Moreover, being a purely magnetic effect, it was hard to recognize by relying only on resistivity experiments as was done by De Haas and collaborators. That probably was the reason why that fundamental phenomenon remained hidden for so many years.

Zero Induction

Immediately after the appearance of Meissner's article its implications were picked up by Gorter[39]. He wrote in a short letter to *Nature*: "A few weeks ago, Meissner and Ochsenfeld published a series of very interesting observations on the establishment of

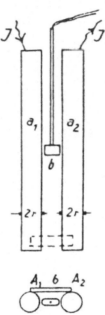

FIGURE 1.12: The Meissner-Ochsenfeld experimental setup (top and side view)[38]. The pickup coil b measured the magnetic field distribution between two monocrystalline tin cylinders. In that first experiment the cylinders were connected in series and a current *I* was applied (adapted from W. Meissner and F. Heidenreich, *Phys. Z.* 37 (1936) 451).

---

[39] C.J. Gorter, *Nature* 132 (1933) 931

superconductivity in a constant external field. Their results seem to indicate that in a superconductor B always equals 0." This point was indeed the missing link in his thermodynamic analysis earlier that year. The great importance of the Meissner-Ochsenfeld paper was immediately clear, but it also initiated a lot of scientific debates, since Gorter's interpretation was less obvious and not immediately accepted. While reading in Dahl's extensive history on superconductivity[8] about the vivid discussions at the small conferences and meetings organized in 1933–1934, one gets the impression of surprisingly great difficulties in distinguishing between thermodynamic equilibrium ((i.e.) the Meissner state) and metastable equilibrium (i.e., flux conservation due to persistent currents). That caused a lot of confusion about the phenomena observed in the many beautiful experiments of Meissner and others, for instance, that no effect was seen in the detected magnetic field inside the hole of a hollow superconductor when the external field was changed.

Equally remarkable was the doubt about the relation between the jump in the specific heat at T_c and a phase transition of the second kind. To elucidate the characteristics of a second order phase transition, Gorter in 1934 introduced a model[40] to illuminate the reasoning leading to the Keesom-Ehrenfest relations[29,30]. It goes as follows: suppose a vessel contains a small amount of liquid and its saturated vapor. The vessel is opaque so that the experimenter does not know what is happening inside, but he starts adding heat and measures the heat capacity. That consists of small contributions of the liquid and its saturated vapor, and a large contribution due to the evaporation of the liquid. If at T_c the evaporation of the last droplet takes place, it is clear that the heat capacity of the vessel will drop sharply to a lower value. The model also illustrates the notion of a free energy of condensation: when the temperature is lowered through T_c a gradual internal condensation takes place which finishes at $T = 0$. The amount of condensation energy follows from the area below the $\Delta c(T)$ curve.

The dispute about the interpretation of the Meissner-Ochsenfeld effect was eventually settled by a convincing experiment by Stark and Steiner.[41] They fabricated a hollow cylinder of superconducting material and wound a pick-up coil with the windings going through the hole and connected that to a ballistic galvanometer. The magnetic field caused by a current flowing along the length of the superconductor was thus embraced by the coil. Upon switching on the current in the normal state a ballistic deflection was observed which was equal but opposite to the signal that was detected when the superconductor was cooled through its transition temperature, proving unambiguously that both the field and the current were expelled from the bulk of the superconducting cylinder.

The thermodynamics of superconductors was further confirmed by experiments of Kok, Van Laer and Keesom[42]. Simple expressions for the differences in entropy and specific heat along the threshold curve were obtained by substituting the quadratic formula for $H_c(T)$. For instance, the apparent T^3 behavior of the electronic specific heat in the superconducting state followed as a natural result. Nevertheless, Kok also noted that considerable deviations from the parabolic $H_c(T)$ curve could be observed for lead and mercury indicating that the quadratic behavior was not a law of nature[42].

Two Electron Fluids

Gorter and Casimir[43] extended their thermodynamic treatment in a subsequent paper in which they introduced a phenomenological two-fluid model which was based on two general assumptions:

a) The superconductor possesses an ordered internal state, the condensed state, and its free energy F_c is characterized by an internal order parameter $x_s(T)$ that represents the fraction of (superconducting) electrons condensed in the superconducting state; $x_s(T)$ varies from zero at T_c to unity at $T = 0$.

[40] W.H. Keesom, *Helium*, Elsevier, Amsterdam (1942) Section 5.32
[41] J. Stark, K. Steiner, *Phys. Z.* 18 (1937) 277
[42] J.A. Kok, Nature 134 (1934) 532; *Physica* 1 (1934) 1103; W.H. Keesom, P.H. van Laer, *Physica* 5 (1938) 193
[43] C.J. Gorter, H.B.G. Casimir, *Phys. Z.* 35 (1934) 963

b) Because the superfluid fraction of condensed electrons x_s is completely ordered, it possesses no entropy. Therefore, the entropy of the system is entirely due to the non-condensed normal electrons indicated by a fraction $1 - x_s$ with free energy F_{nc}.

The two interpenetrating fluids were not supposed to be independent, which was taken into account by assuming that the thermodynamic functions of the normal fluid were proportional to some power of the normal fraction, $(1 - x_s)^r$, in stead of being simply linearly proportional. With this arbitrary choice, the free energy took the form $F = x_s F_c(T) + (1 - x_s)^r F_{nc}(T)$ and by substituting $r = 1/2$ a satisfactory fit to the experimental results was obtained, especially the T^3 dependence of the specific heat. The function $F_c(T)$ was taken independent of T to assure that the superconducting condensate had no entropy. By minimizing F with respect to x_s the equilibrium value $x_s(T) = 1 - (T/T_c)^4$ was obtained, in addition to simple expressions for the thermodynamic quantities such as entropy and specific heat.

The theory of second order phase transitions was soon extended fundamentally by Landau in 1937.[44] It also played an important role in the phenomenological Ginzburg-Landau theory of superconductivity[45] (see the theory chapter in this book). In his famous paper of 1941 on the two-fluid description of liquid helium[46] Landau stressed that the fundamental property of that model is the simultaneous existence of two motions, a superfluid flow and a normal flow, rather then two species of particles, one with a "super" and the other with a "normal" label. And that principle should also hold for the corresponding two-fluid picture of superconductivity. In analogy to Kapitza's experiment which showed a heat deficiency for superfluid helium streaming out of a superleak, Landau concluded that the supercurrent could not transfer heat. That phenomenon was supported by the absence of thermo-electric effects in superconductors[47].

London Equations

The Meissner effect proved that a superconductor in a magnetic field was in a single, thermodynamically stable state, at least, in absence of holes inside the material in which magnetic flux could be trapped. In order to describe the electromagnetic behavior of a pure superconductor within the Maxwell theory, one could try to replace Ohm's law by a relation that would hold for the superconducting condensate. Such an equation should give complete diamagnetism, $\mathbf{B} = 0$, for a bulk superconductor in a longitudinal magnetic field, as well as for a superconducting wire with a transport current going through it. The brothers Fritz and Heinz London were the first to make this very important step[48]. They proposed to introduce the relation (known as the 2nd London equation)[49]

$$\text{curl } \Lambda \mathbf{j}_s = -\mathbf{B} \qquad (1.2)$$

where $\Lambda = m/n_s e^2$ was a positive constant characteristic of the superconductor. By taking the curl on both sides and eliminating j_s by applying curl $\mathbf{B} = \mu_0 \mathbf{j}_s$ (ignoring displacement currents) they obtained $\Lambda \mu_0^{-1} \text{curl curl } \mathbf{B} = -\mathbf{B}$. Since div $\mathbf{B} = 0$ that yielded

$$\Lambda \mu_0^{-1} \Delta \mathbf{B} = \mathbf{B}, \qquad (1.3)$$

where Δ denotes the Laplace operator. The prefactor on the left defines a typical length scale, the London penetration depth, $\lambda_L = (\Lambda/\mu_0)^{1/2} = (m/\mu_0 n_s e^2)^{1/2}$. The above screening equation explained

[44] L.D. Landau, *Phys. Z. Soviet Union* 11 (1937) 26 and 545

[45] V.L. Ginzburg, L.D. Landau, *Zh. Eksperim. i Teor. Fiz.* 20 (1950) 1064

[46] L.D. Landau, *J. Phys. U.S.S.R.* 5 (1941) 71

[47] W.H. Keesom, C.J. Matthijs, Physica 5 (1938) 437 and earlier work quoted in that paper.

[48] F. London, H. London, *Proc. Roy. Soc.* A149 (1935) 71; *Physica* 2 (1935) 341

[49] F. London, *Proc. Roy. Soc.* A152 (1935) 24

the Meissner effect since it gave a local induction inside the superconductor that decays exponentially from its value at the surface $\mu_0 \mathbf{H}_e = \mathbf{B}_e$ to zero over a length of order λ_L which at zero temperature is about 10^{-4} mm. Also the supercurrent density was confined to the same thin surface layer. Historically, (1.2) was justified by considering the condensate of superconducting electrons and to apply an electric field E which would accelerate the superfluid according to $m \partial \mathbf{v}_s / \partial t = -e\mathbf{E}$. Together with $\mathbf{j}_s = -n_s e \mathbf{v}_s$, that yielded the first London equation

$$\frac{\partial \Lambda \mathbf{j}_s}{\partial t} = \mathbf{E}. \tag{1.4}$$

That equation alone leads to a screening equation in $\partial \mathbf{B}/\partial t$ with exponential decaying spatial solution in $\mathbf{B} - \mathbf{B}_0$, where $\mathbf{B}_0$ is the induction frozen in at the transition to the superconducting state.[27] In his first paper Gorter[32] could get the thermodynamics right by choosing $\mathbf{B}_0 = 0$, but not the electrodynamics. That required the invention of the 2nd London equation (1.2). By introducing the vector potential $\mathbf{A}$, defined by $\mathbf{B} = \text{curl } \mathbf{A}$ and $\mathbf{E} = -\partial \mathbf{A}/\partial t - \text{grad } \phi$, (1.2) resulted in

$$\Lambda \mathbf{j}_s = \mathbf{A} \tag{1.5}$$

for a simply connected superconductor applying the appropriate (London) gauge. Finally, by substituting the empirical Gorter-Casimir temperature dependence for the density of superconducting electrons, $n_s(T)/n = x_s(T) = 1 - (T/T_c)^4$, it followed that the penetration depth very near T_c should diverge as $\lambda_L(T) = (\lambda_L(0)/2)/(1 - (T/T_c))^{1/2}$, expressing the loss of screening capability with decreasing density of superconducting electrons.

Penetration Depth

Being very small, the penetration depth could not be easily measured. The first to succeed was Shoenberg in Cambridge in 1939.[50] By using very fine-grained colloidal mercury and in virtue of the strong increase of λ_L near T_c he could determine the change in the magnetic susceptibility due to the temperature dependence of λ_L. The comparison with the theoretical temperature dependence displayed in Figure 1.13 turned out to be very good. However, the non-uniform grain size did not allow to accurately determine the value of $\lambda_L(0)$. Using better experimental techniques, Shoenberg and collaborators could eventually, in 1947, measure $\lambda_L(0)$ of a 20 μm thin mercury cylinder and got 7.6×10^{-2} μm, quite close to the estimated theoretical value.

Subsequently, Pippard[51] realized a tenfold increase of sensitivity by using microwave techniques that became available after World War II. He could therefore investigate the effect of the electron mean free path on the penetration depth by alloying tin with indium[52].

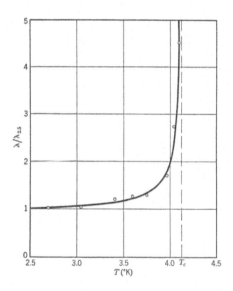

FIGURE 1.13: The temperature dependence of the penetration depth as determined by D. Shoenberg from measurements of the magnetic susceptibility of very fine grained colloidal mercury[50].

[50] D. Shoenberg, *Nature* 43 (1939) 433; *Proc. Roy. Soc.* A175 (1940) 49

[51] A.B. Pippard, *Proc. Roy. Soc.* A191 (1947) 747

[52] A.B. Pippard, *Proc. Roy. Soc.* A216 (1953) 547

It turned out that the simple local relation between $\mathbf{j}_s$ and $\mathbf{A}$ of (1.5) had to be amended for nonlocal effects according to:

$$j_s(r) = \int_V K((r - r')/\xi)A(r')dr'. \tag{1.6}$$

That is, the current density in a certain point r depends on the vector potential in a sphere around this point. The radius of the sphere is determined by the length scale over which the kernel $K(\mathbf{r}, \mathbf{r}')$ decays exponentially from unity in the center to $1/e$. In a superconductor this length scale, *the coherence length* ξ, represents the distance in which the amount of order of the superconducting condensate can change. In a pure superconductor this length scale is about equal to the coherence length ξ_0 introduced by Pippard. A few years later it was recognized that ξ_0 could be interpreted as the size of a Cooper pair. Using the Heisenberg uncertainty principle Pippard could estimate $\xi_0 \approx \hbar v_F / k_B T_c$, where v_F is the Fermi velocity. In 1953 he could explain with that formalism the deviations from the London theory that occurred when the coherence length was larger than the penetration depth. That happened to be the case in many pure elemental superconductors. Also the effect of changing the mean free path l by alloying could be qualitatively understood. Tin alloyed with a small amount of indium turned out to have a much greater penetration depth, without changing the thermodynamic properties like the critical temperature. Pippard proposed that the decrease of the mean free path would reduce the coherence length due to the extra scattering of electrons by the impurity ions and argued that $\xi^{-1} \approx \xi_0^{-1} + l^{-1}$.

Quantum Connection

Already in 1935 London had discussed a possible, quantum mechanical justification of the two London equations, referring to ideas of Bloch, Frenkel, and Landau[53]. In 1948 he came back to this issue[54]. The essential feature of superfluidity and superconductivity is, according to London, an internal condensation of a macroscopic number of particles in a single macroscopic ground state described by a macroscopic wave function. He further suggested that this state is characterized as "a kind of condensed state in momentum space which implies long-range order of the [average] momentum vector $\mathbf{p}_s = m\mathbf{v}_s - e\mathbf{A}$ in ordinary space [...][54]."According to a theorem of Bloch (see, for a discussion, Reference 53, section 2.6), the wave function should be rigid upon applying a magnetic field, which for a singly connected isolated superconductor implied $\mathbf{p}_s = 0$, or $m\mathbf{v}_s = e\mathbf{A}$. By taking the curl or the time derivative of both sides, (1.2), (1.4), and (1.5) could be derived.

London also showed that $\mathbf{p}_s = \hbar \operatorname{grad} \varphi$, where φ is the uniquely determined phase of the macroscopic wave function. He realized that this relation had very interesting physical consequences for multiply connected superconductors as, for instance, a hollow cylinder, because the macroscopic wave function should be single valued in any point. Consequently, the phase φ should change by 2π with each complete turn around the hole. If the circulation of $\mathbf{p}_s$ was computed along a path around the hole, but inside the superconductor so that $\mathbf{v}_s = 0$, it immediately followed that the magnetic flux Φ enclosed by the path is quantized. As is well known, London[37,54] predicted the flux quantum to be $\Phi_0 = h/e$, whereas it followed from the microscopic theory[55] that one should expect $h/2e$ because the superconducting electrons are bound in Cooper pairs[56]. That prediction was confirmed in the beautiful experiments of Deaver and Fairbank[57] and of Doll and Näbauer[58] in 1961 (see Section 3.2 in Chapter 3).

[53] L. Hoddeson, E Braun, J. Teichmann, S. Weart (ed.), *Out of the Crystal Maze*, Oxford University Press, New York (1992)

[54] F. London, *Phys. Rev.* 74 (1948) 562

[55] J. Bardeen, L.N. Cooper, J.R. Schrieffer, *Phys. Rev.* 108 (1957) 1175

[56] L.N. Cooper, *Phys. Rev.* 104 (1956) 1189

[57] B.S. Deaver, W.M. Fairbank, *Phys. Rev. Lett.* 7 (1961) 43

[58] M. Doll, M. Näbauer, *Phys. Rev. Lett.* 7 (1961) 51; *Z. Physik* 169 (1962) 526

Crucial Developments

In the beginning of the 1950s a few crucial experiments helped to clarify the true mechanism of superconductivity. Those results had a major impact on the eventual development of the microscopic theory by Bardeen, Cooper and Schrieffer[55]. Improved experimental techniques enabled Maxwell and Reynolds et al. to measure independently a small shift in the critical temperature as a function of the isotope mass of the mercury lattice[59]; the lighter the mass of the isotopes the higher was T_c. That isotope effect indicated that the electron-phonon interaction played a crucial role and confirmed the conjecture made in 1950 by Fröhlich[60]. Those experiments were soon repeated on tin isotopes and demonstrated the predicted square root dependence on the isotope mass M, $T_c \sim M^{-1/2}$ with high precision[61].

A second series of high sensitivity experiments revealed the true behavior of the electronic specific heat c_{es} at very low temperatures. The T^3 dependence observed in the 1930s that so nicely fitted in the phenomenological two-fluid model, turned out to transpose at the lower temperatures into an exponential temperature dependence

$$c_{es} = \gamma T_c \, a \, \exp(-bT_c/T), \tag{1.7}$$

where γT is the electronic specific heat in the normal state, and a and b are constants of order unity. Within a few years (1953–1956) the exponential temperature dependence was found in several elements, first in vanadium by Corak et al.[62] and later also in niobium and tin. It confirmed the existence of an energy gap Δ of order $k_B T_c$ in the electron excitation spectrum at the Fermi level between the superconducting ground state and the normal quasi-particle excitations. Fritz and Heinz London had already suggested the existence of such a gap in 1935[48] and experimental indications were first provided by Daunt and Mendelssohn in 1946.[63] Measurements of electromagnetic absorption in the far-infrared on thin superconducting films by Glover and Tinkham[64] definitively established the existence of an energy gap. Not only c_{es}, but also the entropy of the electrons goes exponentially to zero as the growing ratio between energy gap and temperature reduces the number of electronic excitations at low temperatures.

1.1.3 Superconductive Alloys, the Spongy Road to Type II Superconductors.

The hope of utilizing superconductivity began to rise again when De Haas and coworkers in Leiden started to investigate the behavior of superconducting alloys in a magnetic field. They found that superconductivity could survive in significantly stronger magnetic fields than were known for elemental superconductors. The most striking example was the eutectic of the lead-bismuth alloy ($Pb_{65}Bi_{35}$) with a threshold field of 1.4 T at 4.2 K, and even 2.1 T at 1.9 K.[65] On the other hand, when the critical current of a 0.3 mm diameter wire made of the same alloy was measured at 4.2 K a self field of only 30 mT was observed for a current of 23 A, and in an applied transverse field of 55 mT the wire could carry a current of only 11 A.[66] That was the more disappointing because the threshold field was expected to be 1.6 T. Clearly, the limiting behavior was not well understood.

[59] E. Maxwell, *Phys. Rev.* 78 (1950) 477; C.A. Reynolds et al., Phys. Rev. 78 (1950) 487

[60] H. Frölich, *Phys. Rev* 79 (1950) 845; see also H. Frölich, Proc. Roy. Soc. A215 (1952) 291; J. Bardeen, D. Pines, Phys. Rev. 99 (1955) 1140

[61] W.D. Allen, R.H. Dawton, J.M. Lock, A.B. Pippard, and D. Shoenberg, *Nature* 166 (1950) 1071, and W.D. Allen, R.H. Dawton, M. Bär, K. Mendelssohn, and J.L. Olsen, *Nature* 166 (1950) 1072

[62] W.S. Corak, B.B. Goodman, C.B. Satterthwaite, and A. Wexler, *Phys. Rev.* 96 (1952) 1442

[63] J.G. Daunt and K. Mendelssohn, *Proc. Roy. Soc.* A185 (1946) 225

[64] R.E. Glover, M. Tinkham, *Phys. Rev.* 104 (1956) 844; 108 (1957) 243

[65] W.J. de Haas, J. Voogd, *Proc.* 33 (1930) 262 and 34 (1931) 56, *Comm.* 208b (1930) and 214b (1931)

[66] W.H. Keesom, *Physica* 2 (1935) 35

In a quite different experimental approach De Haas and Casimir-Jonker[67] studied the penetration of a transverse magnetic field in Bi_5Tl_3 and $Pb_{35}Tl_{65}$ samples. They prepared cylindrical rods and along the central axis was a narrow cavity which contained a bismuth wire with current and voltage leads. The resistance of bismuth is very sensitive to magnetic fields and a change of its resistance therefore revealed the penetration of field. We reproduced their results in Figure 1.14. The data points denote the fields at which penetration was first detected. They noted that on both sides of that critical field line the sample remained superconducting which was quite remarkable. They also depicted a shaded region in which the "resistance was gradually coming back rising to its value in the normal state." Moreover, when the field was switched off, the resistance of the bismuth wire did not come back to its original value, indicating trapped magnetic induction in the cavity. Also Keeley, Mendelssohn, and Moore in Oxford had found magnetic hysteresis when pure lead was alloyed with only 4% bismuth[68].

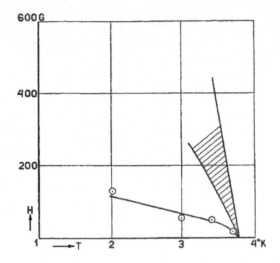

FIGURE 1.14: De Haas and Casimir-Jonker using the bismuth-wire technique, demonstrated the penetration of a transverse magnetic field into a sample of Bi_5Tl_3.[67] The data points denote the fields at which penetration was first detected. On both sides of the line through the data the sample remained superconducting. The shaded area depicts the region in which the resistance was gradually rising to its normal state value.

Today, we know that these features are typical for type II superconductors with weak flux pinning and that—taking into account demagnetization—the lower line in the figure denotes half the lower critical field $H_{c1}(T)$ and the highest line represents the upper critical field $H_{c2}(T)$. A simple estimate from these data tells us that the Ginzburg-Landau parameter κ (see Chapter 2) of the Pb_5Tl_3 alloy must have been at least 3.0. Groups in Oxford[68], Toronto, and Kharkov reported similar behavior for alloys. Most prominent were the results of Shubnikov and coworkers at the Ukrainian Physical-Technical Institute in Kharkov[69,70].

Shubnikov was an authority in growing single crystals from the melt. This was the reason that De Haas invited him to work at his laboratory in Leiden. In 1926 he moved from Leningrad to Leiden where he set up the growth facility for the very pure bismuth single crystals in which he found the magnetoresistance oscillations, now known as the Shubnikov-De Haas effect. The bismuth wires used by Casimir-Jonker to probe the field penetration were actually made by Shubnikov. In 1930 he moved to Kharkov where he built up his own low temperature laboratory.

Two Types of Superconductors

Shubnikov must have been an excellent experimentalist. He developed a technique to measure the magnetic moment by rapidly withdrawing the sample from a coil surrounding it, which resulted

---

[67] W.J. de Haas, J.M. Casimir-Jonker, *Nature* 135 (1935) 30; Physica 2 (1935) 935

[68] T.C. Keeley, K. Menselssohn, J.R. Moore, *Nature* 134 (1934) 773

[69] J.N. Ryabinin, L.V. Shubnikov, *Phys. Z. Sowjet Union* 7 (1935) 122; Nature 135 (1935) 581.

[70] See especially the more extensive review co-authored by V.I. Khotkevich and Yu.D. Shepelev in the memorial issue of the *Ukr. J. Phys.* 53 (2008) 42-52 with the English translation of the paper that originally appeared in *Zh. Eksper. Teor. Fiz.* 7 (1937) 221

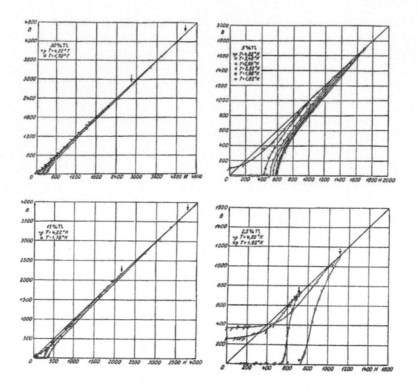

FIGURE 1.15: The $B(H)$ curves measured by Shubnikov and Ryabinin[69,70] at several temperatures for single crystals made of $Pb_{1-x}Tl_x$ alloys with various compositions (only the results for $x = 0.3, 0.15, 0.05$, and 0.025 are depicted here). Below the lower critical field H_{c1} the samples behaved as for a pure superconductor (no flux penetration, in accord with the Meissner effect, $B = 0$). Above H_{c1} flux started to penetrate until eventually at and above the upper critical filed H_{c2} (denoted by the arrows) $B = H$. The symbols with a dash represent the data obtained in decreasing field. The observed hysteresis was most pronounced at low fields and for the lowest Tl concentrations.

in a deflection of the connected ballistic galvanometer. This method worked very well as he earlier demonstrated by obtaining the $B(H)$ curve of a poly-crystalline rod of Pb, thereby convincingly confirming the Meissner effect[71]. Directly measuring the magnetization gave substantially more information about the magnetic behavior than the techniques used by De Haas. This can be seen in Figure 1.15 where some $B(H)$ curves for increasing and decreasing fields at several temperatures are plotted for single crystals of PbTl alloys of various compositions[69,70]. We immediately recognize the generic features of type II superconductors with κ values ranging between 1 and 5. Some hysteresis in the decreasing field curves due to weak flux pinning is also observed. The $H - T$ phase diagram resulting from one of these experiments (for a $PbTl_2$ single crystal) is given in Figure 1.16. The lower curve displays the field obtained from critical currents measured in wires of the same alloy in absence of an applied field. The dependence on the wire diameter showed that the superconductivity was destroyed by the field generated by the current at the surface of the wire. The value is low compared to the thermodynamic critical field and in retrospect we know that it actually should be closer to H_{c1}. It is amazing that such beautiful data were so long ignored. Although they were

---

[71] J.N. Ryabinin, L.V. Shubnikov, *Nature* 134 (1934) 286

extensively discussed in Shoenberg's book[72], it took 22 years before the theory that fully explained their meaning was published[73].

In one of his papers Shubnikov mentioned the results of specific heat measurements in a $Pb_{65}Bi_{35}$ alloy published earlier[74]. There was no appreciable jump at T_c, which is remarkable since the slope of the threshold curve was so large. He concluded that the specific heat and the magnetic measurements were not in agreement with Gorter's theory[32]. Probably, Gorter wouldn't have denied this conclusion because he realized that the difference between dirty-alloy and pure-elemental superconductors must have had a deeper reason which he tried to understand on thermodynamic grounds[75], rather than blaming it on concentration variations and the "freezing in of flux" in "rings formed by regions of higher threshold value" as in Mendelssohn's sponge model[76]. Quoting Per Dahl (reference 8, p.219): "This tenacious, albeit in the long run counterproductive model, was destined to exert an authoritative influence out of proportion to its worth for over two decades [...] distracting investigators from the final formulation of the theoretical underpinning of high-field superconductivity and discouraging concomitant progress in realizing the potentially enormous current densities that were then within sight." It is indeed strange that the sponge model remained so influential for over two decades, because one would expect that any of the typical length scales of superconductivity should play a role in comparison to the length scales of the material inhomogeneities that give rise to the hypothetical, multiply-connected superconducting sponge structure.

Gorter proposed the possibility of a fine partition in superconducting and normal regions (s- and n-regions) and argued that "there exists a minimum size k for the superconductor" (today we would identify k with the coherence length ξ. To avoid confusion we further will use ξ), and that the condensation energy "would decrease considerably as soon as the dimensions of the s-regions would become of the order of ξ[75]." Secondly, if the dimensions of the superconductive regions would become of the order of the penetration depth λ, the energy related to the screening will decrease. If now $\xi > \lambda$, there will be no tendency to form very small superconductive regions, and as soon as $H > H_c$ every trace of superconductivity will vanish. This appears to be on the whole in agreement with the behavior of single crystals of very pure metals. If, however, $\xi < \lambda$, small superconducting regions will be possible even at fields which are stronger than H_c. Gorter even estimated a value for the upper critical field, namely $(\lambda/\xi)H_c$. Above this value the small superconductive "blades or drops [...] will

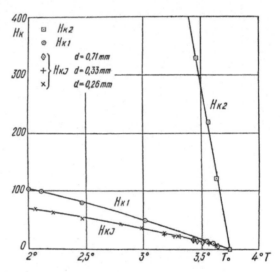

FIGURE 1.16: Phase diagram of $PbTl_2$ showing the lower and upper critical field data (H_{k1} and H_{k2} in Shubnikov's notation[70]) as well as the data determined from the critical current measurements on wires of the same material with different diameters, H_{kJ}.

[72] D. Shoenberg, *Superconductivity* (University Press, Cambridge, 1st edition 1938, 2nd edition 1952)

[73] A.A. Abrikosov, *Zh. Eksperim. I Teor. Fiz.*32 (1957) 1442 [*Soviet Phys. -JETP* 5 (1957) 1174]

[74] L.V. Shubnikov, W.J. Khotkevich, *Phys. Z. Sowjet Union* 6 (1934) 605

[75] C.J. Gorter, *Physica* 2 (1935) 449

[76] K. Mendelssohn, *Proc. Roy. Soc.* A152 (1935) 34

become impossible [...] and superconductivity will vanish entirely." As we now know, this was very close to Abrikosov's prediction[73]. However, we should as well note here that Gorter concluded by saying that "these remarks [...] do not even offer a suggestion why λ should be especially large or ξ especially small for an alloy[75]."

Surface Energy

Almost at the same time Heinz London in Oxford[77] showed that the threshold field H_t in a thin superconducting layer parallel to the field and of thickness $d < \lambda$ is larger than H_c of the bulk, namely $H_t = (\lambda/d)H_c$. Measuring H_t would thus provide a way to determine the penetration depth. From that result London says "one is tempted to conclude that every superconductor in a magnetic field above H_c should split up into a great number of thin, superconducting laminae or fibres separated from each other by thin normal conducting regions, as then the superconductivity could persist at these higher fields." He referred to Gorter's paper, but ignored the possible role of a second length scale that comes into play at the interface between superconducting and normal regions, thus always obtaining a negative interface surface energy.

The question of the surface energy was taken up again in 1951 by Pippard[78,51] who now explicitly introduced the coherence length expressing the range in which the gradual transition takes place from the ordered superconducting state to the disordered normal state, in contrast to the abrupt jump at the interface that was tacitly assumed in earlier discussions. (He apparently missed Gorter's paper, but corrected that later in a very interesting historical overview which has been reprinted as the next contribution in this centennial book.) Pippard continued by discussing the relation between the surface energy of an n-s interface and the purity of the material, coming to the conclusion that "in very impure metals and alloys [the surface energy] will probably be negative, the range of order being considerably less than the penetration depth. It is thus understandable why 'super-cooling' is observed only in pure metals and why there is in alloys a persistence of superconducting regions in fields greater than critical."

The ultimate theoretical derivation of the $n-s$ interface energy was eventually given by Ginzburg and Landau[45]. The intention of their paper was to describe the properties of the intermediate state and therefore they limited their attention to situations in which the surface energy was positive. But with their paper they paved the way for the final attack and the theoretical prediction of two types of superconductors by Abrikosov[73]. He introduced referring to the experimental results of Ryabinin and Shubnikov[69,70], the distinction between type I and type II superconductors depending on the sign of the $n-s$ interface energy. The type II superconductors are characterized by a mixed state consisting of superconducting regions encircling normal filaments containing one quantum of magnetic flux which form a triangular flux line lattice. When two years later Lev Gorkov showed that the Ginzburg-Landau equations could be derived from the microscopic BCS theory[79] and how to deal with mean free path effects,the GLAG (Ginzburg-Landau-Abrikosov-Gorkov) theory was born (see Chapter 2, Theory) providing the tools to explain the properties of all sorts of superconductors from first principles.

It should be mentioned though, that before 1961 the GLAG theory did not receive much attention from the scientific community in the Western world. That changed drastically after the discovery by Kuntzler et al.[80] of very high critical fields in Nb_3Sn in combination with high critical currents: 1 kA/mm^2 at 8.8 T. Eventually, Goodman bridged the gap when in 1962 he drew attention to the GLAG theory and demonstrated its ability to account quantitatively for the behavior of Pb-Tl al-

[77] H. London, *Proc. Roy. Soc.* A152 (1935) 650

[78] A.B. Pippard, *Proc. Camb. Phil. Soc.* 47 (1951) 617, Proc. Roy. Soc. A216 (1953) 547

[79] L.P. Gorkov, *Zh. Eksperim. I Teor. Fiz.* 37 (1959) 1407 [*Soviet Phys. JETP* 10 (1960) 998]

[80] J.E. Kunzler, E. Buehler, F.S.L. Hsu, and J.H. Wernick, *Phys. Rev. Lett.* 6 (1961) 89

loys.[81] Finally, any doubt about the exact magnetic structure of type II superconductors evaporated when the decoration experiments of Essmann and Träuble[82] in 1967 visualized the triangular lattice structure.

1.2   The Historical Context of Josephson's Discovery

A. B. Pippard

This contribution by the late Sir Brian Pippard is reprinted with kind permission from Springer Science and Business Media. It was originally published in the *Proceedings of a NATO Advanced Study Institute on Small-Scale Applications of Superconductivity* held in Gardone Riviera (Lago di Garda), Italy, September 1–10 (1976): *Superconductor Applications: SQUIDs and Machines,* Editors Brian B. Schwartz and Simon Foner (Plenum Press, New York & London, 1977), Chapter 1, p. 1-20.

---

[81] B.B. Goodman, *IBM J. Research Develop.* 6 (1962) 63
[82] U. Essmann, H. Träuble, Phys. Lett. 24A (1967) 526

THE HISTORICAL CONTEXT OF JOSEPHSON'S DISCOVERY

A. B. Pippard

Cavendish Laboratory

Cambridge, England CB3 OHE

In an introductory talk like this you might expect a complete history of superconductivity from 1911 onward. I was not a physicist in 1911; I never met Kamerlingh Onnes and I cannot tell you how it was then, nor am I going to try. I am going to start considerably later, about the time of the B.C.S. theory. I am not a historian, and when one starts to think about writing history, one realizes that historians may not be skilled in physics, but they certainly have expertise of their own. It is quite difficult to both write history and be involved in discovery. The thing which normally saves historians from being completely bogged down in detail is that they let enough time elapse between the events and the writing for almost everything to be forgotten; then it is possible to reconstruct the bits you care about and let the trivial details fade into insignificance. Because we are not talking about that sort of time lapse here it is very hard to separate the details from the important things. We feel the urge to find out what people were really thinking as the discoveries were made, but this is an extremely hard task.

Most of you probably have read Watson's book "The Double Helix" and his account of the discovery of the form of the DNA

Note: This paper is based on the opening talk presented by Sir Brian Pippard to the participants at the NATO Advanced Study Institute on Small-Scale Superconducting Devices. The talk was recorded and it represents essentially the transcript. The editors have requested that Professor Pippard allow us to maintain the informal nature of his talk in order to give the reader the flavor of his remarks. We encouraged only slight editing for clarity.

molecule. Watson was basing his story on copious notes and letters which he wrote at the time and was able to do a magnificent job of reconstruction. I was in Cambridge at the time that this work on the double helix was going on in the Cavendish laboratory. I can assure you that the flavor which comes through in Watson's book is very close to what I remember it to have been. To tell an equally good story of superconductivity would require someone with almost perfect recall and also with very good notes to make sure that his memory is not falsified by subsequent interpretations. I cannot promise you that this is what I can manage in this paper. The double helix is the only example that I know which gets at all close to the feel of events. One of the troubles one has in reconstructing important scientific events is that anyone working on a difficult problem goes through a great many fanciful interpretations. A lot of nonsense goes on in his mind until the moment when suddenly the right idea appears, and that usually comes through as a moment of enlightenment - sudden enough to drive out completely any feeling for what it was like before, so that almost immediately it is impossible to reconstruct the process of thought which led to that inspiration. Already history has begun to be blurred. One can try to go back and plot out the main lines which led to the new insight. However, once having seen the truth it is impossible to give an exact description of what it is like not to be able to see the truth. All I promise then is that I have done my best to find out what happened, but I cannot guarantee that any of the details are right.

I was, as far as the Josephson effect was concerned, an observer of the events but a very poor observer. One reason is that you could hardly find two people whose minds worked on more different lines than Brian Josephson and myself, so that I was never able to understand what he was talking about. The other thing is that one is not told at the time that something important is happening. It is only afterwards - years afterwards - that people will say, "You ought to have taken more notice of what was happening, because it was obviously important." All I can say is that it was not obviously important at the time - it was obviously interesting, but that is quite another matter. If something is interesting but you don't actually understand it, you can hardly recall it very closely. Apart from that, I was interested in other things at the time. It was only a few months since I had commissioned a high magnetic field laboratory. I was concerned with the behavior of electrons in high magnetic fields and had no deep interest in superconductivity at that time, apart from certain outstanding problems that I had a few students working on. My mind was going on other things, so I didn't take all that much notice. This is why I cannot present you with the evidence of a notebook in which I say "On January the third, Josephson told me that he had solved the Hamiltonian."

Well, where shall we start? I think the best thing is to go
back a little into history. It is very easy to overestimate the
importance of the Josephson effect when the major part of this
NATO Institute is devoted to its applications. I am not saying
that the Josephson effect is not important, but it is not the only
important thing that happened in superconductivity about that time,
and I think it is wise to get some perspective on it. How one
gets perspective depends on taste, and one way (a bit curious
perhaps) is through numerical research. I have studied Physics
Abstracts and counted the number of papers published year by
year on superconductivity and compared their total with the total
number of papers published. This (as illustrated in Fig. 1)
shows how the subject of superconductivity was far from dying
as a result of B.C.S. theory. Normalization to the total number
of papers is to allow for the overall five-fold expansion in the
number of papers published in physics annually in this period.
Far from superconductivity coming to an end in 1957, the B.C.S.
theory represents the moment when it took off both in terms of
the proportion of physicists actually publishing papers and (even
more striking) in absolute numbers; in ten years there is a
twelve-fold increase in the number of published papers, from 66
to 830. Some of this represents the rise of letter journals.
(It may not represent more work done but just shorter papers

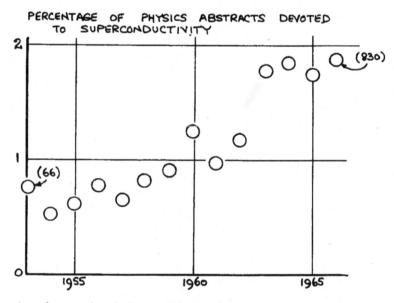

Fig. 1   A graph of the number of papers on superconductivity
in Physics Abstracts from 1953 to 1966.

being published, with still less in each paper even than before.) The data in Fig. 1 stops in 1965-66 since it was quite clear looking at later Physics Abstracts that the curve was still increasing and my patience was exhausted. There is an enormous amount of applications work in superconductivity nowadays. Although 1957 was the year of the B.C.S. theory, this rise in papers is not associated just with B.C.S. nor with the Josephson effect which was discovered in 1962. Some of the bump in 1960 is perhaps the result of Giaever's discovery of tunneling, but there is a general rise of interest at this time in all areas of superconductivity. Another very important factor, of course, was the emergence of technical superconductivity; high-field alloys and devices must account for a great deal of the increase. Before we leave the simple numerology of the subject, there was a conference at Colgate on the science of superconductivity in August 1963, which was published in the Reviews of Modern Physics. The published proceedings ran to 330 pages. Although the conference took place one and a half years after Josephson, only 40 pages were devoted to tunneling - just over 10% - whereas 130 pages, more than one-third of the work, was concerned with high-field superconductors and with transition metal alloys. Nowadays, technical applications of superconductivity are separately listed in Physics Abstracts and they overwhelm, by quite a large factor, the papers published on small-scale devices. Let us not then be too parochial when we are talking about small-scale devices. We are not the only pebble on the beach, there is quite a large rock standing quite close by.

Bearing in mind that many factors were responsible for the large increase of papers in superconductivity, I have traced historically four rather closely related fields, and have listed in Table I, against the year, some of the milestones. I hope nobody will be offended at being left out, for it is not meant to be anything like a complete list, but it includes some of the more significant developments. Let us start with the third column, dealing with type II superconductors which are possibly less intimately related to the main theme than the others. I start in 1935, when there was already quite a lot of experimental work by Mendelssohn, Shubnikov and others showing that alloy superconductors behave quite differently from pure superconductors. In 1935, Gorter specifically suggests that it is a short mean-free-path effect which is responsible for the peculiar properties of alloy superconductors. He doesn't know why, however. His ideas are in contrast to Mendelssohn's that alloys are inhomogeneous and form a superconducting sponge which can trap the flux. The war came along and the next significant event is the Ginsburg and Landau theory in 1950. I have put my own name there in 1951. These two developments are quite independent (it takes time for Russian papers to filter through). I revived Gorter's idea (without knowing he had had it already) in a more specific

Table I: Milestones in Superconductivity

	From Energy Gap to Tunneling	Weak Links	Type II Superconductors	Fundamental Theory
1935			Gorter	
1946	Ginsburg Daunt and Mendelssohn			
1950			Ginsburg and Landau	Frölich Maxwell
51			Pippard	
52				
53	Goodman			
54				
55				
56	Glover and Tinkham			Cooper
57			Abrikosov	Bardeen, Cooper Schrieffer
58		Meissner		Anderson
59			Gorkov	
1960	Giaever Nicol, Shapiro and Smith	Parmenter		
61			Kunzler, Buehler, Hsu & Wernick [Goodman]	
62	Cohen, Phillips and Falicov			
63	Josephson Anderson & Rowell Shapiro			
64	Zimmerman and Silver	De Gennes		

REFERENCES FOR TABLE I

FROM ENERGY GAP TO TUNNELING

1946   Ginsberg: Superconductivity (book; in Russian).
Daunt & Mendelssohn: Proc. Roy. Soc. A185, 225 (1946).
First suggestions of energy gap to account for London
equations (G), and absence of Thomson heat (D & M).

1953   Goodman: Proc. Phys. Soc. A66, 217 (1953).
Experimental evidence for energy gap from thermal
conductivity.

1956   Glover and Tinkham: Phys. Rev. 104, 844 (1956).
Absorption of far infrared reveals gap directly.

1960   Giaever: Phys. Rev. Lett. 5, 147 (1960).
Discovery of N-S tunneling through oxide layer.
Nicol, Shapiro & Smith: Phys. Rev. Lett. 5, 461 (1960).
Giaever: Phys. Rev. Lett. 5, 464 (1960).
S-S tunneling; explicit exhibition of zero-voltage tunneling
current, discussed as metallic bridge (G) or uncommented
upon (N, S & S).

1962   Cohen, Phillips and Falicov: Phys. Rev. Lett. 8, 316 (1962).
Tunneling Hamiltonian.
Josephson: Phys. Lett. 1, 251 (1962).
Prediction of pair tunneling through oxide layer separating
superconductors.

1963   Anderson & Rowell: Phys. Rev. Lett. 10, 230 (1963).
Observation and recognition of Josephson tunneling.
Shapiro: Phys. Rev. Lett. 11, 80 (1963).
Stepped characteristic with microwave irradiation.

1964   Zimmermann and Silver: Phys. Lett. 10, 47 (1964).
Double-junction device as forerunner of fabricated
interferometers.

WEAK LINKS

1958   Meissner: Phys. Rev. 109, 686 (1958).
Evidence of supercurrents through normal metals not
caused by superconducting bridges.
Parmenter: Phys. Rev. 118, 1173 (1958).
Application of BCS theory to NS sandwiches.

1964   De Gennes: Rev. Mod. Phys. 36, 225 (1964).
Refinement of earlier theories of proximity effect.
Anderson and Dayem: Phys. Rev. Lett. 13, 195 (1964).
Supercurrents through narrow bridges.

REFERENCES FOR TABLE I (CONTINUED)

TYPE II SUPERCONDUCTORS

1935   Gorter: Physica 2, 449 (1935).
        Difference between dirty and pure superconductors tenta-
        tively related to mean free path.

1950   Ginsburg & Landau: Zh. Eksperim. i Teor. Fiz. 20,
        1064 (1950).
        Phenomenological theory.

1951   Pippard: Proc. Camb. Phil. Soc. 47, 617 (1951).
        Coherence length shortened by collisions, leading to
        negative surface energy in impure superconductors.

1957   Abrikosov: Soviet Physics J.E.T.P. 5, 1174 (1957).
        Flux-lattice solution of Ginsburg-Landau equations.

1959   Gorkov: Soviet Physics J.E.T.P. 9, 1364 (1959).
        Derivation of Ginsburg-Landau equations from
        microscopic theory.

1961   Kunzler, Buehler, Hsu and Wernick: Phys. Rev. Lett. 6,
        89 (1961).
        Discovery of very high critical field in Nb_3Sn.

        [Goodman: see IBM Journal 6, 63 (1962).
        Draws attention of Western world to Abrikosov (1957).]

FUNDAMENTAL THEORY

1950   Fröhlich: Phys. Rev. 79, 845 (1950).
        Electron-phonon interaction proposed as cause of
        electronic phase change.
        Maxwell: Phys. Rev. 78, 477 (1950).
        Reynolds,   Serin, Wright and Nesbitt: Phys. Rev. 78
        487 (1950).)
        Isotope effect strongly supporting Fröhlich's proposal.

1956   Cooper: Phys. Rev. 104, 1189 (1956).
        Pairing of electron states lowers energy.
        Bardeen, Cooper and Schrieffer: Phys. Rev. 108,
        1175 (1956).

1958   Anderson: Phys. Rev. 112, 1900 (1958).
        Pseudo-spin formulation of BCS theory.

form and tried to give some reason why mean-free-paths, when they are short, can lead to negative surface energies and the possibility of high field superconductivity. Mine was essentially a qualitative idea, in contrast to Ginsburg and Landau who, as we all know, produced a phenomenological theory in which for the first time we get the famous concept of a wavefunction ψ describing the collective state of the superconductor. The (imaginary) gradient of the wavefunction $\nabla\psi$ is responsible for the current, in accordance with normal quantum mechanical principles. The gradient of the wavefunction squared, $|\nabla\psi|^2$, enters the Ginsburg-Landau equation in a way which represents in energy terms the difficulty of confining the wave function. You can't change the spatial distribution of particles without affecting their kinetic energy.

Here I am going to touch on personal history, because I think it illustrates how physics is not always as simple as it ought to be. We all know now that the Ginsburg-Landau theory has been established microscopically. Why in 1950 did it take such a long time for the Ginsburg-Landau theory to be recognized? Why, bearing in mind the extreme lucidity and clarity of their paper, didn't everybody accept it? Why, I ask in particular, did I not accept it, since I soon came to know it perfectly well? Well, there is quite a good reason for this. It wasn't merely the dislike of someone else's ideas, though we all suffer from that. It was that in the Ginsburg-Landau theory the parameter $\varkappa$, whose size controls whether the superconductor is type I or type II, is determined by the penetration depth and by certain other parameters such as the transition temperature. The penetration depth in London theory, which the Ginsburg-Landau theory incorporates, is fixed by the number of superconducting electrons and their mass. In other words, the penetration depth is a fundamental parameter according to London. What made me skeptical, at that time, was that in the early 1950's we knew that the penetration depth was changed by scattering. When the mean-free path was made shorter, the penetration depth increased, as could be explained easily by a non-local equation. One didn't have to infer that the number of superconducting electrons changes because of scattering. But Ginsburg and Landau implied that when you alloy a superconductor, making the mean-free-path shorter, the penetration depth increases and $\varkappa$ changes because the fundamental parameters which go into the theory change. I found that quite unacceptable. I couldn't understand how the number of superconducting electrons could be altered by alloying without, shall we say, altering the transition temperature. Of course all this was cleared up satisfactorily later by Gorkov, but for some years there was a feeling that the Ginsburg-Landau theory was altogether too arbitrary. You read it now, post-1957, and see how beautifully it ties onto the B.C.S. theory. Believe me, however, it's quite different when you haven't got a theory of superconductivity and chaps come out suddenly saying, "Let's write down a wavefunction ψ which will behave just like an ordinary wavefunction behaves in quantum mechanics, except that we will fudge up the boundary conditions because it is not convenient to have normal boundary conditions ...". You can say, "Well, Ginsburg and Landau are very clever chaps and they know much more than I do about it, but it doesn't necessarily mean they are right.". And it took quite a long time for Ginsburg and Landau to be accepted. It was only, I think, when Gorkov produced from the Green's function treatment of the B.C.S. theory an explicit demonstration of how the microscopic parameters could be interpreted, that the Ginsburg-Landau theory fell into place. So in the early 1950's there was a certain amount of conflict which wasn't helped, incidentally, by the fact that Ginsburg kept on writing small papers in which he said it would be much better if we interpreted the electronic charge as not being exactly e, but e times a small numerical factor which might be as large as 2! He didn't say it was exactly 2; instead he wanted to introduce a fudge factor of (say) 1.6, and Landau kept on telling him he couldn't just put in arbitrary numbers, and muttered darkly about gauge invariance going wrong if you did. So even Landau wasn't supporting Ginsburg, and the Soviet Union seemed to be falling apart at that time.

The same thing happened when Abrikosov developed the Ginsburg-Landau theory to produce the theory of vortices in type II superconductors. As he told the story at the time he received the Simon prize, Abrikosov was discouraged by Landau who loathed the theory and would have nothing at all to do with it for some years, so that it wasn't published till 1957. In this now classic paper Abrikosov shows how vortices (flux lines) can appear in the theory, which nobody took any notice of. I don't know when the Russians began to think seriously of it, but in the West we owe a debt to Goodman. He is on my list because at the IBM Conference in 1961 he drew attention to the existence of Abrikosov's paper. He said it would explain high field superconductors and flux penetration far better than any ideas that he (Goodman) and I had been developing. So there is a gap of about 7 years between Abrikosov's solution of the equations and the idea of type II superconductivity really becoming respectable. Meanwhile Kunzler and his colleagues at Bell Labs had found that Nb3Sn has a critical field higher than 80 kG. This discovery represents the sudden recognition of the technical importance of superconductivity, and opens up the development of high-field superconducting magnets. At this point a new story begins, which is not relevant to my theme.

I thought it worthwhile going into this matter in some detail simply because, although it was high-field superconductivity that

acted as the prime stimulus to the development and application of
Ginsburg-Landau, yet once Gorkov had firmly based the G-L
equations, they were soon recognized as the natural way of discussing weak links and many other things. In making applications of superconductivity one hopes to be allowed to forget fundamental theory and rely on the simpler G-L phenomenological
theory.

Turning to the fundamental theory, I propose to take a
moment to talk a little more about the Russian connection. This
is a personal recollection of an incident I found instructive. In
the early summer of 1957 I went to Moscow for the first and only
time. Up to then, ever since the War, there had been extremely
poor contact between Russia and the Western world. This was
one of the first moments when it was possible for scientists
working on superconductivity in Russia and the West to talk, and
we all found it extremely revealing to discover each other's
misapprehensions about work on the other side. In the course of
a few days spent in Moscow, I spent a lot of time arguing fiercely
with Landau, Ginsburg and Lifschitz. They are excellent people
to argue with; there's no one like a Russian for having a fight
with - no holds barred, intellectually speaking - and we all had a
good time. But I was seriously assailed by Landau particularly,
who would have nothing to do with non-local electrodynamics.
The London equation was good enough for superconductivity, and
it was a monstrous thing to throw this beautiful London equation
out and replace it by an ugly non-local equation which was quite
unnecessary. I did my best to explain why it was necessary, but
Landau was not a man to be easily convinced once his mind was
made up. Eventually I promised that I would do an experiment
to show, once and for all, that the penetration depth varied with
the direction of current flow in a non-tensorial manner. This,
they said, was the only thing which would persuade them that the
London equations were wrong. And so we parted.

Round about the end of the same year, I was rather aston-
ished to receive one morning in the mail a letter from the BBC
Monitor Service. They listen to foreign broadcasts and hand on
the information to anyone interested. They wrote to say that a
few days before Professor Lifschitz had been giving a general
talk on science, and had mentioned particularly the pleasure the
Russian physicists had in talking with people from the West. He
went on to say "... and if Dr. Pippard should happen to be
listening to this broadcoast, I would like him to know that in the
discussions which took place earlier in the summer, it was he
who was entirely right." Now I hadn't done any of the promised
experiments, so this was a surprise, and I was gratified to find
that these great scientists could change their minds so readily.
Of course what had happened was quite simple. Cooper's letter
on electron pairs had reached Russia in the Physical Review and

immediately Bogoliubov had worked out his theory of superconductivity* (Bogoliubov is not post-B.C.S., but contemporary with B.C.S., having a powerful formalism absolutely ready from his earlier work on superfluid helium). As soon as the essential idea of Cooper pairs was put to him he lost no time in developing his extremely tidy and mathematically satisfying formulation of superconductivity. It was this theory that convinced Landau and Ginsburg - they really didn't care any more about experimental evidence; what they wanted was a nice tidy theory that looked convincing, and they would then happily accept its consequences, even non-local electrodynamics. This is a lesson then on how to persuade your critics. Don't give them what they think they need, give them what you know they need (if you possibly can).

Brian Schwartz remarked in introducing me that when the B.C.S. theory came out, it was obviously correct and it seemed as though there was little more theory to be done. Now this raises an interesting historical point. He is quite right in saying that the majority of people working on superconductivity, once they realized what was in the B.C.S. paper, recognized its worth. It took a month or two, but not much more than that, for the majority to appreciate that this was, if not the right answer, so near that it really was a fundamental breakthrough. But when I say the majority, I don't mean all. I must exclude from the consensus a considerable proportion of the leading theoretical physicists of the world who disliked the theory very much! This was somewhat similar to what happened in 1950, when Fröhlich proposed a theory in which the electron-phonon interaction was responsible for superconductivity. It was supported by experimental evidence from the isotope effect discovered by Maxwell at NBS and the Rutgers group. Most of us working in the field then were convinced that the isotope effect substantiated the idea of electron-phonon interaction, but the reaction from many leading theorists was hostile. Why was there this adverse reaction to Fröhlich and to Bardeen, Cooper and Schrieffer? To some degree the criticism of Fröhlich's paper was justified; he quite correctly traced superconductivity to the electron-phonon interaction, but the detailed model he constructed is not right. But that wasn't what those who disagreed objected to. The real trouble lay in the fact that Fröhlich's electron-phonon interaction is something they had almost all thought of themselves and rejected for what seemed very good reasons. Fröhlich's criterion was that in superconductors the electron-phonon interaction is strong enough to cause electron-electron attraction. Earlier theorists who had discovered this attraction had, I think,

---

*This may not be correct. B.C.S. published a Letter in Physical Review early in 1957, and it was perhaps this, and not Cooper's Letter, that did the trick. At all events, Bogoliubov's paper was submitted to Il Nuovo Cimento before the main B.C.S. paper was published.

concluded that in these circumstances the lattice is unstable and a crystal modification occurs. Perhaps they were right in general, and superconductivity is the last resort of a metal which cannot find a better way of eliminating the embarrassing interactions between the electron and lattice. Anyone who had travelled that road and decided it led nowhere was understandably irritated by the suggestion that he had missed discovering the theory of superconductivity; and it was easy enough for the expert to discover flaws in the working out that enabled him to overlook any possible merits. I think by the time B.C.S. appeared, seven years later, the iron had entered into the soul of these eminent men. They had already been had once on the electron-phonon interaction, and when B.C.S. showed that this very mechanism which they had rejected was in fact capable of leading to the right theory, their response was distinctly ill-natured. For two or three years after the B.C.S. theory, many of the leading theoreticians were saying, "It's all very well, but the theory is not manifestly gauge invariant." I don't know quite what those words stand for, but what they really meant was that they weren't having the theory at any price. It didn't make the slightest difference, of course; everybody else had accepted it.

The more one had been involved in superconductivity, the less happily one accepted B.C.S. I confess I muttered away for six months or so after B.C.S. before I was finally convinced that muttering did no good, and it was better to join the majority. But my disinclination in no way reflected a private disappointment at having missed getting the answer first - I could never have hit on anything so clever as B.C.S. My reaction sprang from regret for the end of an era, when superconductivity as an unsolved mystery posed the sort of problem that keeps an experimental physicist happy. From now on the subject was basically different. I need not pursue the history of fundamental theory any further, but must note that my inclusion of Anderson's pseudo-spin formulation of the theory is not intended to imply that it was, in itself, a major advance; it did, however, provide the framework which inspired Josephson's first thoughts about tunneling.

Let's go on to tunneling, and approach it through the idea of an energy gap. Now we are in deep water. Who first thought there was an energy gap in superconductors? Mendelssohn tells us that this is an idea which was being bandied around at Oxford before the War,[*] but he and Daunt never had time to write it up

---

[*]Since delivering the talk I have looked at some of the pre-War literature, especially the abortive models of superconductivity developed by, among others, Slater and Welker. It is clear that the energy gap was part of the mental furniture of the time, even if its precise role was ill-defined.

until after the War, in 1946, as Table I records.   Ginsburg
published a book on superconductivity in 1946, in Russian of
course, which seems to have passed out of recollection.   I suspect the ideas in the book go back several years, before 1946.
Ginsburg is quite explicit that the London equations can come out
of an energy gap model, and he stands out as the one who has the
most cogent physical reasoning behind why he thinks there ought
to be a gap.

The gap was then lost sight of for a few years, until
Goodman revived it, to interpret his measurements on thermal
conductivity well below the transition temperature.   He found
that at very low temperatures the electronic conductivity was
exponentially dependent on temperature, and suggested an energy
gap.   Now you might expect him to have referred back to Daunt
and Mendelssohn, but he doesn't; instead, he refers back to
Koppe who isn't even in Table I.   This is because it is difficult
to give fair credit to an incorrect theory, even when it provides
helpful clues.   I refer to Heisenberg's theory of superconductivity which, like so many theories in the pre-BCS days, was
based on the wrong mechanism and worked out wrongly.   Nevertheless, we in Cambridge had cause to remember it with gratitude,
because when Heisenberg went wrong he did not cease to be
enlightening.   In fact, in my own experience, two things came
out of Heisenberg's theory.   One is that Goodman knew about it
because Heisenberg had lectured in Cambridge.   We had discussed his theory in great detail in Cambridge, and had not forgotten it even though we were unconvinced.   Goodman knew an
energy gap came out of the Heisenberg theory, as developed by
Koppe, and took it over as possibly the only correct thing to
emerge from this work.

The other valuable outcome that I recollect took place when,
in the early 1950's, I found compelling evidence for the non-local
character of the supercurrent equation (about which I had the
argument with Landau).   I needed to formulate the non-local
theory, and remembered how Heisenberg had explained the supercurrent as arising by a curious take-over mechanism from the
normal current.   So if the normal current obeyed a local
equation, so did the supercurrent, and that gave the London
equation; but if the normal current was non-local, as in the
anomalous skin effect (my own baby), so too was the supercurrent.   So I wrote down what the Heisenberg theory would have
given in the non-local case, and (hurray!) it fitted the experiments.   Here again, although the reasoning behind Heisenberg's
theory is totally wrong, it provided the incentive and the
formalism which was needed.   This is something which crops up
whenever one traces the development of ideas; almost always
they start wrong and finish up right.   There can be very few
creative scientists who do not remember the debt they owe to

other people's bad ideas which set their mind working in a new direction. The debt is greater still when we take another's good idea and improve it in a way he could not have imagined, and the apportioning of credit can then be difficult. I should not have broached this rather awkward topic but that it plays a part in the Josephson story, and I feel moved to try to resolve some of the tensions. But this comes later - let's get back to the energy gap.

Between 1953 and 1960 the energy gap gained acceptance steadily, and the matter was clinched by Giaever's tunneling experiments. He first found tunneling between a normal metal and a superconductor, and then a few months later between super-conductors. His second paper jostles that of Nicol, Shapiro and Smith in Physical Review Letters, and both of them produce evidence of zero-voltage currents. Giaever remarks on this explicitly and says it is due to a superconducting bridge. Nicol, Shapiro and Smith publish a beautiful oscillograph showing the current at zero voltage, but they do not draw attention to it. Here is another instance of researchers having the data to make an important discovery, but unable to break away from their traditional notions. We had to wait until 1962 for Brian Josephson to take a good result and make it even better.

Since we have at last reached the central theme, I'll go into a little more detail of this particular discovery. Josephson started as a research student under my direction in October 1961. The reason for this choice is worth recounting. He had had a brilliant undergraduate career, first in mathematics and then in physics. At that point he felt that his understanding of practical matters was deficient, and he would therefore do a thesis in experimental physics in order to balance up his expertise and try to compensate for what was easiest for him - the mathematical formulation of problems. But for that, he would have started as a theoretical research student under John Ziman, probably, and I don't know where he would have gone - not into superconductivity at any rate. If you enjoy arid speculation you might try to guess if, and when, the Josephson effects would have been discovered; and you might also try guessing what it is we don't know now that he might have discovered in different circumstances. But the simple fact is that he came to me to do an experimental thesis. I suggested to him that he should study the variation of penetration depth with magnetic field in superconductors, as measured by microwaves. This is a problem which still remains unsolved. Josephson studied this and obtained experimental results, but did not add anything to the theoretical understanding. He did a perfectly satisfactory thesis, ultimately, but was rather side-tracked in the middle by his independent theoretical work. I think it must have been early in 1962 that he began this, and not long afterwards came to tell me about it. In retrospect I feel no

shame at not understanding him then.   I don't know how many
people here have read his 1962 paper in Physics Letters.   If you
have, read it again, and remember that the paper is the result of
systematic pressure on my part and Phil Anderson's to get the
ideas into an intelligible form.   I still find it a very difficult
paper, but the first versions were really in a class of their own,
since he was a new research student with almost no experience
of technical writing.   I hope you will be sympathetic to my
reaction, which was to tell him that I just didn't have a clue what
he was talking about, and he had better go and talk to
Phil Anderson, who was in Cambridge as a visitor at the time.
Phil is a much more clever man than I am, of course, and he did
understand, though I think he too had to work for it.   A few
months later John Bardeen was very doubtful about Josephson's
ideas, and he was far from being alone in this.   I think this
shows how hard it is for someone with a new idea to visualize
what other peoples' difficulties will be in accepting it.   It was so
clear to Josephson in his particular way of looking at it, that he
could not understand why, for example, I should fail to see the
point.   He remarks in his Nobel lecture, "In 1961 Pippard had
considered the possibility that a Cooper pair could tunnel through
an insulating barrier such as that which Giaever used, but argued
that the probability of two electrons tunneling simultaneously
would be very small so that any effects would be unobservable.
This plausible argument is now known not to be valid.   However,
in view of it, I turned my attention to a different possibility, that
the normal currents might be affected by phase difference."
Let's look at this point.   If a single electron only has a probability of 1 in 10^{10} of getting through, the probability for two
electrons to tunnel simultaneously is 1 in 10^{20} - as near
impossible as no matter.   It is quite easy therefore to conclude
that no current flows in pairs.   I put this argument to
Brian Josephson early on, and I am quite sure that this criticism
of his theory was in the minds of almost everybody who was
thinking about pair tunneling.   I asked, "How can the pairs get
through?" and he explained that the wave-functions of the electrons
in the pair are phase-coherent, so that you have to add the
amplitudes before you square.   The electrons do not tunnel
independently, but more like a single particle, and the probability
of a pair going through is comparable to the probability for a
single electron.   It is like interference in optics with phase-
coherent waves mixing, and if he had only said that in his early
publication, he would have had no difficulty in carrying the world
with him.   But can you find it said in his 1962 paper?   Of course
you can't; because he didn't see the existence of a difficulty.   So
for some months a lot of people (especially those who had been
thinking about tunneling before) thought he might well be talking
nonsense because his orders of magnitude for the effect were
completely astray.

While I am trying to correct the record, let me make reference to Phil Anderson's account of the early days of the Josephson paper. In one paragraph he manages to create two obscurities, one important and one not important. He says that "we were all - Josephson, Pippard and myself as well as various other people who habitually sat at the Mond tea table and participated in the discussions of the next few weeks - very much puzzled by the meaning of the fact that the current depends on the phase (this is the famous formula $J = J_1 \sin \phi$ relation).
I think," he continues, "that it was residual uneasiness on this score that caused the two Brians (Pippard and Josephson) to decide to send the paper to Physics Letters, which was just then starting publication, rather than to Physical Review Letters."
I disregard the insult that I don't mind publishing wrong papers, if they are published obscurely; in fact the reason for the choice of journal is quite simple - Physical Review Letters has a page charge which is charged in dollars. At that time exchange regulations in England made it very difficult to get dollars. Therefore it was not published in Physical Review Letters, but in Volume 1 of the new European journal, Physics Letters. This is a trivial point; more important is Anderson's statement about the famous $J = J_1 \sin \phi$ relation. Here I have some difficulty because my recollection on this is quite at variance with what Anderson says. The interesting thing is that $J = J_1 \sin \phi$ is not to be found in Josephson's original paper, nor is it to be found in the fellowship thesis which he wrote later in the year. I believe that the equation was derived by Phil Anderson either at the time of these discussions or within the following months; he quotes it openly in his lectures given in the spring of 1963. There is no question but that Phil has been generous to a fault in what he has written about Josephson. He did an enormous amount to help Josephson get his ideas across, and some of the suggestions which are associated with Josephson are really Phil's. Although Josephson may have been aware, at the back of his mind, of the simple sine relationship, I believe it was Anderson who recognized its importance as a statement of the tunneling law, and also the related expression for the coupling energy of two superconductors as a cosine of the phase. I cannot find these formulae in anything Josephson wrote at that time. By 1963, though, in his article for the Colgate Conference, he quotes the sine relation in a way that leaves one uncertain whether he means it is actually in his original paper or implicit in it (which it certainly is). I'm sure the ambiguity is unintentional. This is not the only example of undue modesty on Anderson's part; he deserves great credit for recognizing two obstacles in the way of detecting Josephson tunneling: (1) The deleterious effect of stray magnetic fields such as the Earth's, and (2) much more subtle, room temperature noise getting down the leads. When Phil suggested they should be eliminated, John Rowell was able to exhibit a clear demonstration of Josephson tunneling.

I must try to make a few remarks about Josephson's own approach to the problem. I can't say much of value because, as I have already indicated, his habit of thought was (and remains) alien to mine. But I can paraphrase his own account. Anderson is on the scene from the start, but his role in this case is unconscious. Being on leave from Bell Labs and spending a year in Cambridge, he gave a course of lectures on solid state physics. At the end he devoted some time to the question of broken symmetry which was intriguing him at that time. Let me explain broken symmetry briefly. The lattice of a crystal has certain symmetry properties, say cubic symmetry, described by a set of rotations that transform it into itself. It is not isotropic, for any other rotation leaves it looking different. However, if you write down the Hamiltonian describing the particles that make up the block of solid, the Hamiltonian itself contains only isotropic central forces. There are no symmetry elements corresponding to the symmetry of the solid crystal which results as the ground state solution. If the particles had been non-interacting, the ground state would have been isotropic, the lowest state of a quantum gas. As you start increasing the interaction between the particles there will come a time when there will precipitate out, as the lowest state, a new ground state which contains certain symmetry properties not present in the original Hamiltonian - a gas-to-solid phase transition has occurred. In a sense the ground state is still isotropic, for the solid lattice can be oriented in any direction, and an isotropic wave-function can be constructed as the superposition of all the degenerate states describing orientations. But this is physically unrealistic - what we observe is one particular orientation at a time.

Now this phenomenon, which is widespread and not confined to solid lattices, has excited the interest of many physicists, including Anderson, and has been a source of inspiration for general ideas about how physics works. He was talking about it in his lectures and used his pseudo-spin formulation of B.C.S. theory to show how the phase transition into the superconducting state could also be seen as a symmetry transition. This intrigued Josephson very much because when a symmetry is changed a new parameter enters, which in general we may refer to as the order-parameter. In superconductors the energy gap plays the role of order-parameter, and as Josephson pondered this in the light of Anderson's general discussion, he recognized that the energy gap alone was an incomplete specification - a phase must be associated with it; moreover (and all this is still consistent with general broken-symmetry theory) this phase was a variable which was non-commutative with respect to the number of electrons in the sample. The Heisenberg uncertainty principle applies to phase and number - if you know exactly how many electrons you have in the superconductor you can say nothing about the phase of the wavefunction. This is the case for two pieces of superconductor quite separate from one another so that you can count the number of particles in each.   On the other hand, Josephson saw if you take a single piece of superconductor and imagine dividing it into two, it is easy to transfer particles from one half to the other and correspondingly, you can have complete knowledge of the phase.   Therefore you have a superconducting junction between one half and the other, because the Ginsburg-Landau equations tell one how the supercurrent is related to phase.   Thus Josephson recognized early on that the uncertainty relation of phase and number distinguished clearly between superconductors which were separate and superconductors which were joined together.   He then asked himself what happened in the intermediate case when they are weakly joined together.   I find it particularly interesting to see the weak link appearing as the primary concept, with the oxide barrier coming later as a way of realizing it, and not the other way round - a theory of oxide barriers later being generalized to other sorts of weak link. Josephson recognized that if it was possible to transfer electrons, though with difficulty, from one superconductor to another; you could know something about the number and something about the phase.   And since electron number is closely related to the current between the two, and so also is phase, through the G-L theory, he was on the way to describing the properties of a weak link.   But, as he points out in the passage I quoted, he was also convinced by the argument against pair tunneling.

At precisely the right moment Cohen, Falicov and Phillips provided the formalism he needed, with their tunneling Hamiltonian, and first he looked at it to see what effect a knowledge of phase would have on the normal current.   Not finding what he knew he needed there, he concentrated on the supercurrent and found pair tunneling as one element of the solution.   It may be worth recording that a complete analysis of the tunneling Hamiltonian is an extremely formidable undertaking; Cohen, Falicov and Phillips tackled the simpler case of superconductor-normal tunneling, but were daunted by the superconductor-superconductor case.   The thorough way they had approached the formulation of the problem prevented them in the end from seeing the wood for the trees.   Josephson's advantage lay in his having started from a very general conception of what he wanted, so that he could seize on those features of the tunneling theory that appealed to his physical sensibilities.   This is why I lay stress on the idea of weak links rather than tunneling.

Now weak links, though not so named, had been around for some time.   In 1958 Meissner published a lot of careful work on the junction of two superconductors with a thin normal film between them.   He provided a clear demonstration of the proximity effect (the normal metal between two superconductors

carrying pairs and enabling supercurrents to pass from one superconductor to another). Two years before Giaever, and four
before Josephson, then, he was talking about superconductors
being joined together by weak non-superconducting links. But
his conception of what was going on in the weak links is not
radical like Josephson's; nor is that of Parmenter who, following
Meissner's results, wrote a very solid and systematic account of
the proximity effect based on the B.C.S. theory. There were
many of us at that time wondering, more or less deeply, about
the proximity effect, and we agreed that it was perfectly possible
for this superconductor to infect that normal metal with superconducting pairs so that the supercurrent could pass from one
side to the other. But our thought was strictly limited to the
idea of the normal metal becoming a sort of dilute superconductor,
passing current according to G-L, that is, proportional to the
gradient of the wavefunction just as in a normal superconductor.
It never entered any of our minds that the current could be a
periodic function of the phase difference, $\sin \phi$; the essential
idea of a weak link was missing. Since Josephson, we
recognize the weak link as one in which the behavior of the
electrons between the two superconductors is determined by what
is going on in the superconductors themselves, and not by the
local state of affairs at each point between. And the difference
is crucial - it permits the current to be periodic in the phase
difference - but it took several years for many of us to see just
how significant the difference is in practical terms.

I could really stop here, but perhaps should make a comment
on the last entries in Table I, marking the beginning of the device
era. I have put Zimmerman and Silver here as the key
contributors and that may not be fair. I know other people,
Mercereau and others, whose names should also be mentioned.
But Zimmerman and Silver published the first description I can
find of a device with two Josephson junctions separated by a
macroscopic gap. What they did was to take a strip and a
V-shaped wire, both of Nb, and they bent the wire over the strip
so that it contacted the two sides and left a gap in between. They
then showed interference fringes in the critical current as a
function of magnetic flux through the loop. I think this is the
progenitor of the macroscopic devices based on the Josephson
effect.

Well, to sum up, I have tried to show some of the windings
in the interconnected research paths leading to the Josephson
effects, to bring out the mixture of logicality and illogicality,
inspiration and desperation, and how we get things wrong. And
of course this is still the way things are done. John Clarke will
tell you that his discovery of the Slug was the result of imperfectly
formulated ideas providing the inspiration for an elegant and
useful end-product, and almost everyone who believes himself

to have made a significant discovery will admit that the first
inklings had their origin in something read or heard, and very
likely misunderstood.   The final result is none the worse because
it was reached by stumbling - it is only our pride which is hurt
when we fail to measure up to that perfection of progress that the
great men of the past always seemed to achieve.   Or did they?
If we wish to boast of our achievements, let us not point to the
unerring pursuit of truth by a logically faultless thinking-machine,
but to the even more astonishing way in which truth can be caused
to emerge from the toils of error and stupidity.

Acknowledgments

We would like to acknowledge Bertram Schwarzschild, editor of *Physics Today*, for the editorial improvements of Section 1.1.1 and Louwrens van Dellen for his assistance with the addition of colors to the historic drawings.

1.3   Further Reading

1. Dirk van Delft, *Freezing Physics. Heike Kamerlingh Onnes and the Quest for Cold*, KNAW, Amsterdam (2008). The biography of Heike Kamerlingh Onnes.

2. Per Fridtjof Dahl, *Superconductivity, Its Historical Roots and Development from Mercury to the Ceramics Oxides*, AIP, New York (1992). An excellent book on the history of superconductivity.

3. L. Hoddeson, E Braun, J. Teichmann, S. Weart (ed.), *Out of the Crystal Maze*, Oxford University Press, New York (1992). An extensive overview of the history of condensed matter physics.

Textbooks with an extensive overview about the history of superconductivity:

4. F. London, *Superfluids*, vol. I, Dover, New York (1961).

5. D. Shoenberg, *Superconductivity*, University Press, Cambridge (1952).

6. G. Rickhayzen, *Theory of Superconductivity*, Wiley, New York (1964).

Proceedings of the Kamerlingh Onnes Symposium

7. H. Kamerlingh Onnes Symposium on the Origins of Applied Superconductivity—75th Anniversary of the Discovery of Superconductivity at the 1986 Applied Superconductivity Conference, September 28–October 3, 1986, Baltimore, Maryland, *IEEE Transactions on Magnetics*, MAG-23 (1987) 354–415.

2

Theory

Editor: Jan Zaanen

2.1 Ginzburg–Landau Equations
    V. M. Vinokur and T. I. Baturina 51
2.2 The Emergence of BCS
    David Pines 65
2.3 Theory of Superconductivity: From Phenomenology to Microscopic Theory
    Lev P. Gor'kov 72
2.4 A Modern, but Way Too Short History of the Theory of Superconductivity at a High Temperature
    Jan Zaanen 92
2.5 Intrinsic Josephson Effect in Layered Superconductors
    L. N. Bulaevskii and A.E. Koshelev 115
2.6 Mixed State Properties
    Ernst Helmut Brandt 125
2.7 Thermomagnetic Effects in the Mixed State
    R. P. Huebener 137
    Acknowledgments 144

2.1   Ginzburg–Landau Equations

V. M. Vinokur and T. I. Baturina

> *I dedicate this contribution to Vitaly Ginzburg, in grateful memory of his wonderful personality, his kind encouragement, and generous help.*
>
> *V. M. Vinokur*

2.1.1   Introduction

In 1950 Vitaly Ginzburg and Lev Landau introduced in their seminal work[1] (which will be referred to as **GL** hereafter) the equations which have been known ever since as Ginzburg-Landau (GL) equations and which became among the most universal equations of modern physics. Ginzburg-Landau equations either in their original or modified form describe a wealth of phenomena and systems including superconductivity and superfluidity, nonlinear waves in active media, pat-

---

[1]   V. L. Ginzburg and L. D. Landau, *Zh. Eksp. Teor. Fiz.* 20 (1950) 1064

tern formation and liquid crystals and supersymmetric conformal field theories[2]. Ginzburg-Landau equations were one of the first nonlinear theories to demonstrate solutions in the form of topological singularities. Historically, GL theory was an extension of the Landau theory of second-order phase transitions[3] onto the quantum phenomenon of superconductivity. It was based on the idea that the normal metal-superconducting state transition is, in the absence of a magnetic field, a thermodynamic second-order transition. An order parameter Ψ of the GL theory is an averaged wave function of superconducting electrons. Because of its (comparative) simplicity and physical transparency, GL theory has become one of the most universal and powerful tools in studies of superconductivity. In what follows, we briefly review the history and the content of GL theory and discuss its most standard applications.

2.1.2 A History

The discovery of superconductivity marked the beginning of a new era in physics, and it is no wonder that the brightest minds in physics strived to understand and reveal its nature and underlying mechanisms. The collegial work and intense scientific exchange, including strong yet forgiving and mutually supportive critique, brought a deserved reward to the scientific community: two theories that are justly considered to be among the top scientific achievements of the second part of 20th century. The first of these was the Ginzburg-Landau theory, which became a foundation for the phenomenological description of superconductivity. The second of these major achievements, which came 7 years later, was the microscopic theory of John Bardeen, Leon Cooper, and Robert Schrieffer[4]. Explosive theoretical efforts were triggered by Walther Meissner and Robert Ochsenfeld's experimental breakthrough[5]: 22 years after the discovery of superconductivity they demonstrated that superconductors expel magnetic fields irrespective of the route by which the superconducting state was achieved, either by cooling the sample in the applied magnetic field, below some magnetic field-dependent critical temperature T_c or by switching on the field after the cooled sample turned superconducting. The magnetic field B in the superconducting sample remains zero, as long as the applied magnetic field does not exceed some temperature-dependent field $H_c(T)$. This discovery was of primary importance, since it allowed us to consider the equality of magnetic induction to zero in the bulk of superconductor as a characteristic of the superconducting state, which hereafter acquired a status of a thermodynamic state of matter. In 1934, Cornelis ("Cor") Gorter and Hendrik ("Henk") Casimir provided strong evidence for the fact that superconductivity is an equilibrium phenomenon[6]. A year later the brothers Fritz and Heinz London published equations describing the linear electrodynamics of superconductors that explained the Meissner-Ochsenfeld effect[7]. They argued that the conventional Ohm's law $\mathbf{j} = \sigma \mathbf{E}$ should be replaced in superconductors simply by Newton's law of motion $\partial \mathbf{j}/\partial t = (n_s e^2/m)\mathbf{E}$, where n_s is the density of the superconducting electrons, since the motion of the superconducting charges must be dissipationless. Making use of Maxwell equation $\mathrm{curl}\,\mathbf{E} = (1/c)(\partial \mathbf{B}/\partial t)$, one arrives at the relation $(\partial/\partial t)[\mathrm{curl}\,\mathbf{j} + (n_s e^2/(mc))\mathbf{B}] = 0$, meaning that the quantity in the brackets is conserved. As in the bulk of a superconductor must be $\mathbf{j} = 0$ and

[2] Y. Kuramoto, *Chemical Oscillations, Waves and Turbulence*, Springer Series in Synergetics, Springer, Berlin (1984); M. C. Cross and P. C. Hohenberg, Rev. Mod. Phys. 65 (1993) 851; A. C. Newell, T. Passot, and J. Lega, Annu. Rev. Fluid Mech. 25 (1993) 399; T. M. Bohr, H. Jensen, G. Paladin, and A. Vulpiani, *Dynamical Systems Approach to Turbulence*, Cambridge University, New York (1998); G. Dangelmayr and L. Kramer, "Mathematical Approaches to Pattern Formation", in *Evolution of Spontaneous Structures in Dissipative Continuous Systems*, edited by F. H. Busse and S. C. Muller, Springer, New York, (1998) p. 1; L. M. Pismen, *Vortices in Nonlinear Fields*, Oxford University/Clarendon Press, Oxford/New York (1999); R. Dijkgraaf and Ed. Witten, Nucl. Phys. B 342 (1990) 486-522

[3] L. D. Landau, *Sov. Phys. JETP* 7 (1937) 19; ibid. 7 (1937) 627

[4] J. Bardeen, L. N. Cooper, and J. R. Schrieffer, *Phys. Rev.* 106 (1957) 162

[5] W. Meissner and R. Ochsenfeld, *Naturwiss.* 21 (1933) 787

[6] C. J. Gorter and H. Casimir, *Zeitschrift für Technische Physik* 15 (1934) 539

[7] F. London and H. London, *Proc. Roy. Soc.* A149 (1935) 71

$\mathbf{B} = 0$, the quantity in the brackets is zero and

$$\operatorname{curl}\mathbf{j} + \frac{n_s e^2}{mc}\mathbf{B} = 0. \tag{2.1}$$

This equality is now known as the London equation. It implies, in particular, that the external magnetic field (and current) can penetrate only a thin surface layer of a superconductor and that the corresponding penetration depth, now called the London penetration depth, is

$$\lambda = \left(\frac{mc^2}{4\pi n_s e^2}\right)^{1/2}. \tag{2.2}$$

This explained the Meissner-Ochsenfeld effect and became the basis of the electrodynamics of superconductors.

Around the same time that these discoveries were being made, Landau endeavored to construct a phenomenological theory of superconductivity. He was motivated by the ongoing debate on whether superconductivity could be interpreted in terms of the Drude formula for electronic conductivity, $\sigma = ne^2\tau/m$, where n is the charge carrier density, e is the electron charge, τ is the scattering time, i.e., the time between the two successive scattering events, and m is the electron mass. The infinite conductivity could be viewed then as a result of indefinitely growing τ. Remarkably, even before the discovery of Meissner and Ochsenfeld, Landau considered it highly implausible that the scattering time jumped to infinity at some finite temperature. Thus, ruling out the possibility of superconductivity due to infinite conductivity, Landau sought a theory yielding an equilibrium state with finite current. He wrote the free energy of a superconductor in the form of an expansion in even powers of the current[8]: $F(\mathbf{j}) = F(\mathbf{j} = 0) + (1/2)a\mathbf{j}^2 + (1/4)b\mathbf{j}^4$, and the equilibrium current $\langle\mathbf{j}\rangle$ is given by the minimization of the free energy (no odd terms occurred in current $\mathbf{j}$ in $F(\mathbf{j})$ since the energy should not depend on the current direction). Landau proposed that $b > 0$ and $a \propto (T - T_c)$, to make a transition into a state with the nonzero equilibrium current, $|\mathbf{j}| > 0$, below T_c, continuous.

The proposed free energy possesses immediately recognizable features of Landau's theory of second-order phase transitions. Landau knew, however, that the temperature behavior of the current on approach to T_c, $|\langle\mathbf{j}\rangle| \propto (T_c - T)^{1/2}$ disagreed with the experimental observations. His approach was inspired by the theory of ferromagnetism and already used the much more general concept of an order parameter that discriminates between different states of matter. The same idea of expansion with respect to the order parameter appeared in Landau's subsequent work on ferromagnetism[9], and came to full fruition in his celebrated theory of the second-order phase transition[3]. It is truly remarkable to consider how these ideas, originally conceived in 1933, reappeared 17 years later in the GL theory.

The tragedy of World War II interrupted the great enterprise of superconductivity, so that the next great discovery had to wait until 1948. In that year, Fritz London proposed that superconductivity is a macroscopic coherent quantum state[10]. This was the final stone that laid the foundation on which Ginzburg and Landau constructed their theory.

2.1.3 The Equation

Functional and Equation

Ginzburg and Landau start their remarkable paper of 1950 by stating that London's existing phenomenological theory, despite successfully describing several aspects of the electrodynamics of

---

[8] L. D. Landau, *Physikalische Zeitschrift der Sowjetunion* 4 (1933) 43

[9] L. D. Landau, *Physikalische Zeitschrift der Sowjetunion* 4 (1933) 675

[10] F. London, *Phys. Rev.* 74 (1948) 562

superconductors, cannot help determine the surface energy at the normal-superconducting phase interface. Moreover, they assert, his theory fails to describe the destruction of superconductivity by currents and magnetic fields. They thus wrote their paper to correct these inadequacies, but, in practice, the GL theory they introduced far outpaced this modest claim. Ginzburg and Landau's starting key observation was that, in the absence of a magnetic field, the transition into a superconducting state at the critical temperature T_c is a phase transition of the second order. They postulated the existence of some order parameter in the theory of this transition analogous with other second-order phase transitions, which is nonzero in the ordered phase and becomes zero in the nonordered one.

They defined the order parameter Ψ that is finite in a superconducting state and zero above T_c and postulated that Ψ plays the role of some "effective" wave function, a complex function to allow for supercurrent flow. It is important to remember, however, that Ψ is not the true wave function of the electrons in the metal but rather some averaged quantity. In a two-fluid model it must be related to the number of superconducting electrons. Thus its absolute square could be identified with the density of superconducting electrons $n_s = |\Psi|^2$, as it was done originally in **GL**. However, comparison with the microscopic theory showed later (see below) that the order parameter should instead be normalized to the number of Cooper pairs so that

$$|\Psi|^2 = n_s/2. \tag{2.3}$$

Near the second-order phase transition, the order parameter is small and its spatial variation is always slow. Constructing the free energy within the framework of the general theory of second-order phase transitions and considering first the case where Ψ is constant and there is no magnetic field, we realize that the free energy density should depend only on the density of superconducting electrons and can be written as an expansion in powers of $|\Psi|^2$:

$$F_s = F_n + \alpha|\Psi|^2 + \frac{1}{2}\beta|\Psi|^4, \tag{2.4}$$

where $\alpha = a(T - T_c)$, and $\beta > 0$. Such a presentation assumes that the free energy is analytic near the transition at $T = T_c$. Remembering that we are close to the second-order phase transition, we must prevent a too rapid spatial variation of the order parameter. Introducing the dependence of the order parameter at one point upon its values at others, we write:

$$\Psi(\mathbf{r}) = \int d\mathbf{r} K(\mathbf{r} - \mathbf{r}')\Psi(\mathbf{r}'). \tag{2.5}$$

Expanding near the point $\mathbf{r}$ in the integral, we obtain the contribution from the nonlocality as (the term proportional to $|\Psi|^2$ can be included into the α-term):

$$\int d\mathbf{r} K(\mathbf{r} - \mathbf{r}')\frac{1}{2}C(\mathbf{r}\nabla)^2\Psi(\mathbf{r}) \simeq \tilde{C}|\nabla\Psi(\mathbf{r})|^2, \tag{2.6}$$

where the constant represents the integral of the nonlocal kernel.

To ensure gauge-invariance in the presence of an applied magnetic field $\mathbf{H}_a$, we have to replace ∇ by $\nabla - i(e^\star \mathbf{A}/\hbar c)$, where the vector potential $\mathbf{A}$ is related to the magnetic induction via $\mathbf{B} = \text{curl}\mathbf{A}$.

The final free energy density can be written in the form:

$$F_s = F_n + \alpha|\Psi|^2 + \frac{1}{2}\beta|\Psi|^4 + \frac{1}{2m^\star}\left|\left(-i\hbar\nabla - \frac{e^\star \mathbf{A}}{c}\right)\Psi\right|^2 + \frac{1}{8\pi}(\mathbf{B}(\mathbf{r}) - \mathbf{H}_a)^2. \tag{2.7}$$

Minimizing the total free energy $\mathcal{F} = \int d\mathbf{r} F_s$ with respect to Ψ and $\mathbf{A}$, we arrive at (choosing the gauge where $\text{div}\mathbf{A} = 0$)

$$\frac{1}{2m^\star}\left(-i\nabla - \frac{e^\star}{c}\mathbf{A}\right)^2\Psi + \alpha\Psi + \beta\Psi|\Psi|^2 = 0, \tag{2.8}$$

$$\Delta \mathbf{A} = -\frac{4\pi}{c} \mathbf{j}_s, \tag{2.9}$$

$$\mathbf{j}_s = -\frac{i\hbar}{2m^\star}(\Psi^\star \nabla \Psi - \Psi \nabla \Psi^\star) - \frac{(e^\star)^2}{m^\star c}|\Psi|^2 \mathbf{A}. \tag{2.10}$$

If no additional conditions on Ψ at the surface of a superconductor are assumed then the so-called "natural" boundary condition taken in the original **GL** paper reads

$$\mathbf{n} \cdot \left(-i\hbar \nabla - \frac{e^\star \mathbf{A}}{c}\right)\Psi = 0, \tag{2.11}$$

where the subscript $\mathbf{n}$ is the vector normal to the surface. This condition only holds literally at the superconductor-insulator interface. The general condition, which is valid at the superconductor-normal metal, is obtained by replacing zero in the rhs of Eq. (2.11) by $-(1/d)\Psi$, where d has the dimensionality of length[11].

Characteristic lengths and surface energy

Two fundamental lengths appear in the GL equations. One of them defines the spatial scale of variations of currents and magnetic fields. In the zero field, the order parameter is constant over space,

$$|\Psi|^2 \equiv |\Psi_0|^2 = -\frac{\alpha}{\beta}. \tag{2.12}$$

Then the expression for the current assumes the form

$$\mathbf{j}_s = -\frac{(e^\star)^2}{m^\star c}|\Psi|^2 \mathbf{A}, \tag{2.13}$$

i.e., coincides with the London equation giving the London penetration depth, which is the characteristic length of the field and current variations as

$$\lambda^2 = \frac{m^\star c^2}{4\pi(e^\star)^2|\Psi_0|^2}. \tag{2.14}$$

The other characteristic length defining the spatial scale of the variation of the order parameter is the so-called *coherence length* and follows from the GL equation

$$\xi^2 = \frac{\hbar^2}{2m^\star|\alpha|}. \tag{2.15}$$

Both scales ξ and λ diverge proportionally to $\sqrt{T_c - T}$ as $T \to T_c$. The ratio of the two lengths, the so-called Ginzburg-Landau parameter,

$$\kappa = \frac{\lambda}{\xi}, \tag{2.16}$$

is temperature independent and is the fundamental characteristic of a superconducting material. It is related to the experimentally measurable thermodynamic critical magnetic field, H_c, above which the superconducting state is destroyed as

$$\kappa = \frac{2\sqrt{2}e}{\hbar c}H_c\lambda^2, \tag{2.17}$$

---

[11] P. G. De Gennes, *Rev. Mod. Phys.* 36 (1964) 225; R. O. Zaitsev, *Sov. Phys. JETP* 21 (1965) 1178

where $-e$ is the electron charge ($e^\star = 2e$).

Introducing the material parameter κ reveals the inherent beauty and elegance of the GL theory. After the proper normalization, $\Psi \to \Psi/\Psi_0$, $H \to H/(H_c \sqrt{2})$, $\mathbf{r} \to \mathbf{r}/\lambda$, where $\Psi_0^2 = -\alpha/\beta$, the GL equations acquire a dimensionless form:

$$[-(\mathrm{i}/\kappa)\nabla - \mathbf{A}]^2 \Psi - \Psi + |\Psi|^2 \Psi = 0, \qquad (2.18)$$

$$\mathbf{n}[(\mathrm{i}/\kappa)\nabla - \mathbf{A}]\Psi|_{\text{boundary}} = 0, \qquad (2.19)$$

$$\operatorname{curl}\operatorname{curl}\mathbf{A} = -(\mathrm{i}/2\kappa)(\Psi^\star \nabla \Psi - \Psi \nabla \Psi^\star) - |\Psi|^2 \mathbf{A}, \qquad (2.20)$$

i.e., the GL equations depend upon the unique material parameter κ.

This characteristic is connected with another fundamental property of superconductivity: the surface energy of the interface separating superconducting and normal phases. Solving GL equations in the case $\xi \ll \lambda$ (i.e., with $\kappa \ll 1$), we can find for the surface tension

$$\sigma_{\text{ns}} = (H_c^2/8\pi)(4\sqrt{2}\lambda/3\kappa). \qquad (2.21)$$

As long as the relation $\kappa < 1/\sqrt{2}$ holds, the surface energy at the interface separating superconductor and normal metal remains *positive*. Now superconductors with $\kappa < 1/\sqrt{2}$ are called type I superconductors. At $\kappa = 1/\sqrt{2}$, the surface tension becomes zero and turns negative at $\kappa > 1/\sqrt{2}$.

In the original **GL** paper the GL equations contained the electron charge e and mass m. Explaining replacement $-\mathrm{i}\hbar\nabla \to -\mathrm{i}\hbar\nabla - (e/c)\mathbf{A}$ the **GL** stated "... e is a charge and there is no reason to consider it different from the charge of the electron." The same concerns the mass m. As we now know it is the Cooper pair charge $e^\star = 2e$ and mass $m^\star = 2m$ that should enter the GL equations. Lev Gorkov established this in his paper[12], where the GL equations were derived from the Bardeen, Cooper, and Schrieffer (BCS) microscopic theory of superconductivity. The dramatic competition of opinions on the nature of the effective charge as well as on the possible properties of superconductors with a large κ is described in Vitaly Ginzburg's memoir[13] and in a recent essay by Eugene Maksimov[14]. After analyzing experimental data, Ginzburg concluded that results similar to the experimental ones could be achieved by choosing $e^\star = (2 \div 3)e$. Landau objected to the idea of an effective charge different from the charge of an electron, saying that the effective charge depended on the properties of the sample and possibly on the coordinate that would have violated the gauge invariance of the GL equations. Remarkably, both ideas have their merits: Ginzburg's original idea that the true value of $e^\star$ can be inferred from the data on the London penetration depth advocated in his works[15] eventually appeared to be incorrect, at least in the literal sense, since both $e^\star = e$ and $e^\star = 2e$ produce the same value of the penetration depth $\lambda = \sqrt{m^\star c^2/(4\pi|\Psi_0|^2(e^\star)^2)}$, provided $m^\star = 2m$ and the correct normalization $|\Psi_0|^2 = n_s/2$ are chosen. However, his guess, although based on the inaccurate experimental data, turned out to be right: the correct value $e^\star = 2e$ followed from the experimental data on magnetic flux quantization (see below). At the same time, Landau's physical insight and intuition proved correct as well, since no other physical quantity besides the electron charge enters the effective charge $e^\star$.

Another trace of internal debate in **GL** can be seen around the question of the relevant values of κ. **GL** says that since available experiments suggest that usually $\kappa \ll 1$, the solution of the GL equations in the limit $\kappa \to \infty$ is of no interest and will not be covered. Yet later in the text, the question of the values of $\kappa \geq 1/\sqrt{2}$ is discussed again. The authors remark that at $\kappa \geq 1/\sqrt{2}$ some

[12] L. P. Gorkov, JETP 36, (1959) 1918; ibid, 37 (1959) 1407

[13] V. L. Ginzburg, Sverprovodimost' and Sverhtekuchest' [Superconductivity and Superfluidity (What is accomplished and what is not)] in the book *O sverhprovodimosti i o sverhtekuchesti*, Moscow, FM (2006) [About superconductivity and superfluidity]

[14] E. Maksimov, *Physics Usp.* 53 (2010) 8470

[15] V. L. Ginzburg, *Uspehi Fiz. Nauk* 48 (1952) 26; V. L. Ginzburg, ZhETF 29 (1955) 748

instability of the normal phase may emerge and streaks of the superconducting phase may appear, i.e., the solution with $\Psi \neq 0$ may arise due to a negative energy of the superconductor-normal metal interface. This passage of **GL** contains the equation

$$\frac{d^2\Psi}{dx^2} = -\kappa^2(1 - x^2H^2)\Psi, \tag{2.22}$$

which is nothing but the Schrödinger equation for the harmonic oscillator and has solutions for Ψ going to zero at $x = \pm\infty$ provided $\kappa = 2H(n + 1/2)$, $n = 0, 1, 2, \dots$. However, as **GL** further states, "The character of the solutions appearing at $\kappa > \kappa_0[= 1/\sqrt{2}]$ was not investigated, since there is no necessity in doing that, so far as from the experimental data although tentative and processed with the aid of relation (2.22)[16] it follows that $\kappa \ll 1$." It is noteworthy that V. L. Ginzburg was discussing the issue of $\kappa > 1/\sqrt{2}$ in his reviews published in 1950 and later in 1952[17]. In 1950, he noticed that, just as the investigation of the original GL equations at $\kappa > 1/\sqrt{2}$ predicted, a peculiar instability of the normal metal phase arises. In 1952, Ginzburg argued that at negative values of the surface energy, the destruction of superconductivity at $H > H_c$ would not occur; rather, the sample would break up into alternating normal-superconducting layers.

The question of what happens at $\kappa > 1/\sqrt{2}$ was investigated by Alexei Abrikosov[18], who introduced the concept of type II superconductivity. In 1957 Abrikosov published a paper where he showed that below the certain field $H_{c2} = \sqrt{2}\kappa H_c$ the magnetic field penetrates the superconductor in a shape of flux filaments carrying one quantum of magnetic flux $\Phi_0 = \pi\hbar c/e$ and surrounded by circulating currents[19] and comprising a regular square lattice. These formations are known now as Abrikosov vortices and the pattern they make is called the Abrikosov lattice.

The field H_{c2} is called the *upper critical field* and is the maximal field at which superconductivity can survive. There is one more characteristic field, H_{c1}, at which the existence of vortices in the bulk of a superconductor becomes thermodynamically advantageous and it is called the *lower critical field*. Thus a magnetic field penetrates a type II superconductor without destroying it in the interval $H_{c1} < H < H_{c2}$. It is interesting that although Abrikosov seemed to start from the idea sketched in **GL**—namely that the negative surface energy of the superconductor-normal metal interface implies that the normal phase becomes unstable with respect to the formation of nuclei of the superconducting phase as soon as the magnetic field is decreased down to H_{c2}—what he actually found was rather the solution corresponding to the nucleation of the normal seeds within the superconducting phase.

But all this happened later. The original **GL** paper remained restricted to small κ. The Ginzburg-Landau equations were solved for the case of thin superconducting films in the parallel magnetic field, and the critical fields and currents destroying superconductivity were found. In the relatively thick film with thickness $d > \lambda$, the critical field appeared to be $\tilde{H}_c \approx H_c(1 + (\lambda/d))$. The situation was more interesting in thin films with $d \ll \lambda$, namely, a thin film was found to withstand much higher magnetic fields than a bulk sample: $\tilde{H}_c \approx H_c2\sqrt{6}(\lambda/d)$. These results demonstrated an excellent agreement with the experiment and marked an unprecedented success of the GL theory in describing phenomenology of superconductors.

---

[16] Eq. (2.22) from the GL paper reads: $\kappa^2 = 4.64 \times 10^{14} H_c^2 \lambda^4$, where λ is measured in centimeters and H_c is measured in oersteds, and $e^\star \equiv e = 4.8 \times 10^{-10}$.

[17] V. L. Ginzburg, *Usp. Fiz. Nauk* 42 (1950) 169; ibid 42 (1950) 333; ibid 48 (1952) 26

[18] A. A. Abrikosov, *Doklady Akademii Nauk SSSR* 86 (1952) 489, this work was submitted to "Doklady" by L. D. Landau.

[19] A. A. Abrikosov, *Sov. Phys. JETP* 5 (1957) 1174

2.1.4 Consequences and Applications

Flux quantization

An important step toward the GL theory was the recognition of the quantum nature of the order parameter in superconductors. Building on this concept, in 1950 F. London predicted the quantization of the magnetic flux piercing through the hole or cavity in a superconductor[20]. At that time, no one knew that the elemental charge carriers in superconductors are Cooper pairs, so London's assumption produced double the true magnetic flux quantum. Only in 1961 did Bascom Deaver and William Fairbank[21] and Robert Doll and Martin Näbauer[22] deliver convincing experimental evidence for electron pairing. They observed that the flux embraced by a superconducting ring varies in steps of half the size proposed by London. Lars Onsager[23] proved that this is a direct consequence of pairing so that $e^\star = 2e$. The flux quantization can be seen from the GL equations with the proper choice ($e^\star = 2e$, etc.) of parameters following Bardeen's line of reasoning[24]. Writing $\Psi = |\Psi| \exp(i\chi)$,

$$\mathbf{j} = \frac{\hbar e}{m}|\Psi|^2 \left[\nabla\chi - \frac{2e}{\hbar c}\mathbf{A} \right]. \tag{2.23}$$

Integrating this equation along the path holding the magnetic field,

$$\oint d\mathbf{l} \frac{\mathbf{j}}{|\Psi|^2} = \frac{\hbar e}{m}\left[\oint \nabla\chi - \frac{2e}{\hbar c}\oint d\mathbf{l}\mathbf{A} \right]. \tag{2.24}$$

The current is zero in the bulk of the superconductor; thus, setting the contour enclosing the hole within the body of the material, we find $\mathbf{j} = 0$ along this path. The second integral gives the flux captured by the contour: $\oint \mathbf{A}d\mathbf{l} = \int \mathbf{B}d\mathbf{S} \equiv \Phi$. Now we have to remember that in order for the Ψ be single valued, the phase must change by integral multiples of 2π in making a complete circuit, i.e., it has to change over $2\pi n$, n being an integer,

$$\oint \nabla\chi d\mathbf{l} = 2\pi n, \tag{2.25}$$

leading to

$$\Phi = n\Phi_0, \qquad \Phi_0 = \pi\hbar c/e = 2.07 \cdot 10^{-7} \, \mathrm{Oe \cdot cm^2}. \tag{2.26}$$

Flux quantization is the macroscopic analogue of the quantization of angular momentum in atoms, and is a fundamental concept governing the vast number of phenomena related to magnetic field-penetrating superconducting systems. This phenomenon stemming from the necessity for the phase to change over 2π when making a closed contour around the cavity containing a magnetic field was first demonstrated experimentally as the periodic depression of T_c in the thin-walled superconducting cylinder as a function of the trapped magnetic flux with a period of Φ_0[25].

Type II superconductors

The most spectacular application of the GL equations has become the prediction of type II superconductors in which $\kappa \geq 1/\sqrt{2}$. Verifying the predictions of the GL theory, Nikolay Zavaritskii undertook measurements of the critical field of thin films. As Alexei Abrikosov recalls in his Nobel lecture, the GL theory for films of superconductors with $\kappa < 1/\sqrt{2}$ fitted perfectly with experimentation up to the change of the nature of the transition: first order at larger thickness and second order

[20] F. London, *Superfluids*, Vol. 1, p. 152, Wiley, New York (1950)

[21] B. S. Deaver, Jr. and W. M. Fairbank, *Phys. Rev. Lett.* 7 (1961) 43

[22] R. Doll and M. Näbauer, *Phys. Rev. Lett.* 7 (1961) 51

[23] L. Onsager, *Phys. Rev. Lett.* 7 (1961) 50

[24] J. Bardeen, *Phys. Rev. Lett.* 7 (1961) 162

[25] W. A. Little and R. D. Parks, *Phys. Rev. Lett.* **9** (1962) 9

in very thin films. However, after Zavaritskii, on the advice of Alexander Shalnikov, changed the technique of the film preparation to ensure film homogeneity[26], the GL theory for small κ ceased to describe the experiment. In particular, Zavaritskii now saw the second-order transition even in thick films. During their discussions of the experiment, Abrikosov and Zavaritskii could not accept that the GL theory might be incorrect even for a split second. Ultimately, Abrikosov decided to explore the consequences of the possibility of κ being larger than $1/\sqrt{2}$ where the normal metal-superconductor surface energy becomes negative. The resulting theory nicely described the experimental data; in particular, it showed that the superconducting transition became of second order for any thickness. Abrikosov derived the general conclusion that superconductors with $\kappa > 1/\sqrt{2}$ comprise a distinct class of superconductors[19], which is called now type II superconductors.

Not only films but also the bulk type II superconductors exhibit a second-order transition in a magnetic field. Indeed, if a superconductor were split into thin alternating layers of the normal and superconducting phases, the critical field of the thin superconducting layer would be about $\kappa H_c > H_c$, provided $\kappa > 1$. Thus for these layers it would be advantageous to maintain superconductivity at higher fields. Type I superconductors would not split because of the surface energy needed to form a normal metal-superconductor interface, but for type II superconductors forming such an interface would lower the energy. Thus in a type II superconductor the transition into the normal state should occur gradually via the incremental expulsion of the superconducting regions until even the thinnest possible SC layer would cease to exist. This implies that the transition broadens into a finite interval of magnetic fields within which the superconducting and normal phase can coexist and the magnetic field can partially penetrate the superconductor. The upper boundary of such a mixed state is called the *upper* critical field and can be estimated as $H_{c2} \sim H_c \lambda/\xi$. In order to estimate the *lower* critical field, H_{c1}, we notice that the tiniest nucleus of the normal phase should have the linear size ξ across the direction of the magnetic field, but the field penetrating along the normal filament is screened on the distance λ and thus creates circular persistent currents. The circular motion of Cooper pairs is quantized (v_s is the velocity of the electrons comprising the pair): $\oint (2m) v_s d\mathbf{l} = 2m v_s \cdot 2\pi r = 2\pi n\hbar$. Thus $v_s = \hbar/2mr$. The kinetic energy due to these rotating pairs per unit length of the filament is $\varepsilon_0 = (\pi n_s \hbar^2/4m) \ln \kappa$, the magnetic moment created by rotating pairs $M = \pi e \hbar n_s \lambda^2/4m$. Then H_{c1} is estimated from the condition $\varepsilon_0 - HM = 0$: $H_{c1} = H_c \ln \kappa/(\sqrt{2}\kappa)$. The above estimate is based on the fact that the magnetic field penetrates type II superconductors in the form of *quantum vortices*. The existence of quantum vortices was established by Lars Onsager[27] and Richard Feynman[28] in the context of superfluid helium. A theory of the formation of the vortex lattice in type II superconductors near T_c became a triumph of the GL theory and won its creator Alexei Abrikosov the 2003 Nobel Prize.

The generally glorious story of the discovery of type II superconductivity contains a sad chapter. As mentioned above, during the writing of the **GL** Landau insisted that all the experimentally known superconductors possess $\kappa \ll 1/\sqrt{2}$. At the same time it is hard to imagine that he did not know the work of his friend Lev Shubnikov who first observed and studied the mixed state in superconducting alloys[29]. Both Landau and Shubnikov were in Kharkov during the time this work was completed. At the height of the Great Purge in 1937, the NKVD launched the "Ukrainian Physics and Technology Institute Affair," and Lev Shubnikov, along with several colleagues, was arrested. He was executed

---

[26] In 1938 Shalnikov developed an original technique for deposition of extremely thin films [down to a single atom layer (!!)] on a glass substrate at cryogenic temperatures [see A. Shalnikov, *Nature* 142 (1938) 74; A. I. Shalnikov, *Zh. Eksp. Teor. Fiz.* 10 (1940) 630]. In these works, which laid the foundation of the physics of two- and quasi-two-dimensional systems, Shalnikov, in particular, discovered a huge increase of the magnetic critical field in freshly deposited films of tin, and emphasized that all the film properties (except for the sharpness of the transition) are very reminiscent of those of superconducting alloys.

[27] L. Onsager, *Nuovo Cim. Suppl.* 6 (1949) 249

[28] R. P. Feynman: in *Progress in Low Temperature Physics*, ed. by C. J. Gorter, North-Holland, Amsterdam (1955) Vol. 1, Ch. 11

[29] L. V. Shubnikov, V. I. Khotkevich, Yu. D. Shepelev, and Yu. N. Ryabinin, *Zh. Eksp. Teor. Fiz.* 7 (1937) 221

by Stalin's regime on November 10, 1937. Landau was investigated as having been associated with this UPTI Affair in Kharkov as well, but he managed to leave for Moscow. Still, he was arrested on April 27, 1938 and held in an NKVD prison until April 29, 1939, when he was released after Pyotr Kapitsa wrote a letter to Stalin, personally vouching for Landau's behavior.

It is telling that, in spite of his opposition against including the discussion of superconductors with $\kappa > 1/\sqrt{2}$ into the **GL** text in 1950, Landau submitted Abrikosov's work introducing type II superconductors after Zavaritskii obtained his independent data on the thallium films[30]. According to Ginzburg's review of 1952, Landau was well aware that the case $\kappa > 1/\sqrt{2}$ can be realized in superconducting alloys. Yet Abrikosov's 1952 paper, submitted by Landau, did not contain any reference to Shubnikov's work. This is not surprising, given the situation in the Soviet Union at the time: mentioning the name of someone executed as an "enemy of people" even in private, much less in print, could cost a person his or her life. After Shubnikov was posthumously rehabilitated in 1956, Abrikosov's 1957 paper included a detailed comparison of his theoretical results with Shubnikov's data. Even in 1957 such an inclusion was a daring and courageous act. To honor Shubnikov's memory, we now often refer to the mixed state of type II superconductors as the *Shubnikov phase*.

Abrikosov vortex lattice

When applying the GL equations it is convenient to introduce dimensionless variables: $\Psi \to \Psi/\Psi_0$, $\mathbf{B} \to \mathbf{B}/H_c\sqrt{2}$, $\mathbf{r} \to \mathbf{r}/\lambda$, $\Psi_0^2 = -\alpha/\beta$. In these variables the GL equations contain only one parameter κ characterizing the material involved. In the vicinity of H_{c2}, where the first superconducting nucleus appears, $\Psi \to 0$ and choosing the gauge $A_y = Hx$, the GL equations read:

$$\frac{1}{\kappa^2}\frac{d^2\Psi}{dx^2} - (1 - A^2)\Psi = 0, \tag{2.27}$$

$$\frac{d^2A}{dx^2} = 0. \tag{2.28}$$

Finding $A = Hx$ from Eq. (2.28) and plugging it into Eq. (2.27), one obtains

$$\frac{d^2\Psi}{dx^2} + \kappa^2 H^2 x^2 \Psi = \kappa^2\Psi, \tag{2.29}$$

where the solution $\Psi \to 0$ as $x \to \pm\infty$ is sought. This equation looks formally like the quantum harmonic oscillator equation. The latter has solutions of the required kind provided $\varepsilon = \hbar\omega(n + 1/2)$. Comparing coefficients, one finds that this condition transforms into $\kappa^2 = 2\kappa H(n + 1/2)$. Hence, the highest field at which superconducting nuclei are still possible is $H_{c2} = \kappa$, or returning to physical units,

$$H_{c2} = H_c\kappa\sqrt{2}. \tag{2.30}$$

The Ψ-function corresponding to the nucleus located at some arbitrary point of the sample (which is taken into account by the gauge of the vector potential) is $\Psi \propto \exp[iky - \kappa^2(x - k\kappa^{-2})/2]$. A general solution is a linear combination of Ψ-functions of nuclei centered periodically:

$$\Psi = \sum_{-\infty}^{\infty} C_n \exp[ikny - \kappa^2(x - kn\kappa^{-2})/2]. \tag{2.31}$$

To fix the geometry of the periodic array one has to include nonlinear terms into the consideration. Abrikosov defined the quantity $\beta_A = \overline{|\Psi|^4}/(\overline{|\Psi|^2})^2$, which has to become minimal in a true configuration, where the minimum free energy is achieved. Originally, Abrikosov concluded that

---

[30] N. V. Zavaritskii, *Dokl. Akad. Nauk SSSR* 86 (1952) 501

it is the square array that realizes the lowest free energy. Only later[31] were Abrikosov's original numerical calculations refined and it was shown that it is the triangular array for which $\beta_A = 1.16$ possesses the smallest free energy, while for the square configuration $\beta_A = 1.18$. Analyzing the solution of the GL equations for the square array Abrikosov showed that it can be written in the form $\Psi = \text{const} \cdot \exp(-\kappa^2 x^2/2) \vartheta_3(1; i\kappa(x+iy)\sqrt{2\pi})$, where $\vartheta_3(x; y) \equiv \sum_{-\infty}^{\infty} \exp(-\pi x n^2 + 2\pi i n y)$ is the theta-function. Investigation of this formula reveals that there are points at which $\Psi = 0$ and that these points comprise a square lattice with the period $a_0 = \xi_0 \sqrt{2\pi}$. In the vicinity of zeros the order parameter can be written as $\Psi \propto r\exp(i\chi)$. The polar angle coincides with the phase of the order parameter, which therefore changes over 2π upon closing the contour around the zero; r is the distance from the zero. Accordingly the Cooper pair currents encircling the zeros are $j = (\hbar e/m)|\Psi|^2/r$, which, recalling that $j = e n_s v_s$, and $|\Psi|^2 = n_s/2$, transforms into $v_s = \hbar/(2mr)$, which is exactly the expression for quantum vortices. Thus Abrikosov's solution of the GL equation describes the square lattice of quantum vortices. The magnetic field penetrating a superconductor achieves its maximal value of about twice the external field H at the locations of the vortex cores, and each vortex (i.e., every elementary cell of the lattice) carries exactly one magnetic flux quantum Φ_0 (see above). From his solution Abrikosov determined the magnetization M of the superconductor near H_{c2}:

$$-4\pi M = (H_{c2} - H)/[(2\kappa^2 - 1)\beta_A], \tag{2.32}$$

which agreed perfectly with the experiment. The period of the square Abrikosov lattice increases upon departing down from the upper critical field as

$$a_0 = (\Phi_0/H)^{1/2}. \tag{2.33}$$

Remarkably, the concept of the Abrikosov lattice applies in the whole interval $H_{c1} < H < H_{c2}$ and the formula (2.33) holds until very close to H_{c1}. When vortices are far from each other, every vortex can be approximately viewed as almost independent and can be treated as an elastic line with the linear tension $\varepsilon_0 \approx [(\Phi_0)^2/(4\pi\lambda)^2]\ln \kappa$. The physics of Abrikosov vortices based on the phenomenological treatment of the mixed state of type II superconductors as the ensemble of interacting elastic lines which can be driven by the applied current and are subject to thermal and quantum fluctuations and quenched disorder became an independent discipline.

Vortex science came to fruition after the discovery of high temperature superconductors (HTS)[32]. Because of the layered nature (i.e., high anisotropy) and short coherence length of high-T_c cuprates, the Ginzburg number is much larger than that in conventional superconductors, typically, $Gi \approx 10^{-2}$, giving rise to the enhanced role of thermal fluctuations. In particular, the Abrikosov lattice can melt, and a considerable part of the phase diagram of the Shubnikov state appears to be a vortex liquid[33]. The competition between the basic characteristic energies, the elastic energy, and energies of fluctuations and disorder results in a rich vortex phase diagram[34]. A wealth of dynamic vortex behaviors is described in the review[35].

Especially interesting is the low-temperature disorder-dominated vortex dynamics. It was found in the context of the study of motion of dislocations and domain walls in materials with disorder that thermally activated dynamics of elastic objects in a random environment at *small* applied forces becomes highly nonlinear[36]. The velocity becomes a non-analytical function of the driving force f

$$v \propto \exp[-\text{const}(T)/f^\mu], \tag{2.34}$$

---

[31] W. H. Kleiner, L. M. Roth, and S. H. Autler, *Phys. Rev.* A 133 (1964) 1226

[32] J. G. Bednorz and K. A. Müller, Zs. *Physik.* B 64 (1986) 189

[33] D. R. Nelson, *Phys. Rev. Lett.* 60 (1988) 1973

[34] V. Vinokur, B. Khaykovich, E. Zeldov, M. Konczykowski, R. A. Doyle, and P. H. Kes, *Physica* C 295 (1998) 209

[35] G. Blatter, M. V. Feigel'man, V. B. Geshkenbein, A. I. Larkin, and V. M. Vinokur, Rev. *Mod. Phys.* 66 (1994) 1125

[36] L. B. Ioffe and V. M. Vinokur, *J. Phys.* C 20 (1987) 6149

where the exponent μ is defined by the dimensionality of the driven object and the space into which it is immersed. The reason for this highly nonlinear response is that the energy barriers controlling the motion diverge as $1/f^{\mu}$ as $f \rightarrow 0$, which is a characteristic of the glassy dynamic behavior. Application of these ideas to the vortex ensemble, which can be viewed as an array of the interacting elastic lines, brought up the concept of the *vortex glass* which forms out of the vortex lattice in the presence of disorder[37]. The characteristic property of a glass is a highly nonlinear response to the applied small dc current. The measured current-voltage (*I-V*) characteristic in the vortex glass state follows the law $V \propto \exp(-\mathrm{const}/j^{\mu})$, i.e., becomes extremely non-Ohmic at small currents. This kind of vortex motion is now called *vortex creep*. Because of their experimental accessibility, Abrikosov vortices became a unique laboratory for studying general properties of glasses (see the comprehensive review)[38].

To conclude this section we want to stress another aspect of the beauty of Abrikosov's solution, which demonstrated that being a nonlinear equation the GL equation contains a novel physical object, topological excitations. This discovery of the vortex lattice made an enormous impact on contemporary physics.

Josephson effect

In 1962 Brian Josephson predicted[39] that if two superconductors are separated by a very thin insulating layer, so that electron tunnelling is possible between the two superconductors, then the superconducting current can flow across this layer in the absence of an external bias:

$$I_{\mathrm{s}} = I_{\mathrm{c}} \sin \Delta \chi, \qquad (2.35)$$

where $\Delta \chi$ is the phase difference between the superconducting banks of the junction, and the critical current I_{c} is the maximal supercurrent which this junction can support. He further found that in the presence of a voltage V across the junction the phase difference between the banks evolves as

$$\frac{d(\Delta \chi)}{dt} = \frac{2eV}{\hbar}, \qquad (2.36)$$

so that the ac current with the frequency $\omega = 2eV/\hbar$ would flow. These two predictions are now known as the stationary and non-stationary Josephson effects and they were confirmed in numerous experiments. Moreover, they have become the basis of many technological applications of super-conductivity. In general, the Josephson junction can be comprised of two strong superconductors, separated by an insulating layer, or of two superconductors coupled through a thin normal metal where weak superconductivity is induced due to the so-called proximity effect. Alternatively, this can be a narrow and short constriction where superconductivity is suppressed bridging two bulky pieces of the same superconducting material. All these versions of Josephson junctions are generi-cally called "weak links."

The Josephson effect results from tunnelling of the wave function, Ψ, of the condensate across the tunnelling barrier with Ψ being expected to follow the GL equations. The power of an approach based on the GL equations is remarkably illustrated by the quantitative derivation of the expression for the Josephson current across the short, $L \ll \xi$, weak link in the form of a superconducting bridge connecting two massive superconductors[40]. As the passing current is squeezed into the narrow bridge, it can exceed the critical current there and the bridge turns into a normal "wire" connecting the two bulk superconductors. Since the Josephson current is low, its self-generated magnetic

[37] M. P. A. Fisher, *Phys. Rev. Lett.* 62 (1989) 1415; M. V. Feigelman, V. B. Geshkenbein, A. I. Larkin, and V. M. Vinokur, *Phys. Rev. Lett.* 63 (1989) 2303

[38] T. Nattermann and S. Scheidl, *Advances in Physics* 49 (2000) 607

[39] D. D. Josephson, *Phys. Lett.* 1 (1962) 251; *Adv. Phys.* 14 (1965) 419

[40] L. G. Aslamazov and A. I. Larkin, *Fiz. Tverd. Tela* (Leningrad) 10 (1968) 1104 [*Sov. Phys. Solid State* 10 (1968) 875]

field can be neglected and the one-dimensional GL equation (in dimensionless units) reads:

$$-(1/\kappa^2)\Delta\Psi - \Psi + |\Psi|^2\Psi = 0. \tag{2.37}$$

If the length of the bridge $L \ll \xi$, the first term is of the order $(\xi/L)^2\Psi$ and dominates over the other terms there. Thus this equation reduces to

$$\Delta\Psi = 0. \tag{2.38}$$

Far from the bridge $|\Psi| = |\Psi_0|$, where Ψ_0 is the order parameter in the bulk of a superconductor. We then take $\Psi_{1,2} \equiv |\Psi_0|\exp(i\chi_{1,2})$ at superconductors 1 and 2 connected by the bridge. Then, the solution to equation (2.38) is sought in the form $\Psi = |\Psi_0|\{f(\mathbf{r})\exp(i\chi_1) + [1 - f(\mathbf{r})]\exp(i\chi_2)]\}$. Plugging this into the expression for the superconducting current, one finds

$$I = (e\hbar/m)|\Psi_0|^2(\nabla f)\sin(\chi_1 - \chi_2), \tag{2.39}$$

which is nothing but the Josephson current $I_s = I_c\sin(\chi_1 - \chi_2)$, with $I_c = (e\hbar/m)\Psi_0^2(T)\nabla f$ where $\nabla f \simeq (\mathcal{A}/L)$, with $\mathcal{A}$ being the cross-sectional area of the bridge.

Time-Dependent Ginzburg-Landau Equation and Fluctuations

The Ginzburg-Landau equations proved to be so powerful and universal in describing equilibrium superconductivity that their generalization to nonstationary and relaxation processes seemed only natural. The standard recipe of constructing the relaxation dynamic equation, provided that the free energy functional $\mathcal{F}[\Psi]$ is known, is to write the relaxation of the slightly disturbed order parameter toward its equilibrium value as:

$$-\gamma\frac{\partial\Psi}{\partial t} = \frac{\delta\mathcal{F}}{\delta\Psi^\star}, \tag{2.40}$$

where $\mathcal{F}$ is the GL free energy functional and γ is the friction coefficient. The equation for the dynamics of the order parameter in superconductors should be gauge invariant. The gauge invariance implies that the scalar electric potential has to be added to the time derivative and the equation acquires the form:

$$-\gamma\left(\frac{\partial}{\partial t} + 2ie\phi\right)\Psi = \frac{\delta\mathcal{F}}{\delta\Psi^\star}. \tag{2.41}$$

In the nonequilibrium case the current comprises both supercurrent and current due to the motion of normal quasiparticles under the action of the electric field. Making use of Ohm's law for the normal component we thus arrive at

$$\mathbf{j} = \mathbf{j}_s + \mathbf{j}_n, \tag{2.42}$$

$$\equiv -\frac{i\hbar}{2m^\star}(\Psi^\star\nabla\Psi - \Psi\nabla\Psi^\star) - \frac{(e^\star)^2}{m^\star c}|\Psi|^2\mathbf{A} - \sigma_n\left(\nabla\phi + \frac{1}{c}\frac{\partial\mathbf{A}}{\partial t}\right), \tag{2.43}$$

where σ_n is the normal conductivity. We can determine γ by noticing that the right-hand side of Eq. (2.41) contains the term $-(\hbar^2/4m)\Delta\Psi$ and it thus resembles the diffusion equation, where the diffusion coefficient is $D = \hbar^2/(4m\gamma)$. On the other hand, in disordered superconductors the coefficient $\pi D\hbar/(8T_c)$ substitutes the expression $\hbar^2/(4ma)$ that appears in GL equations for clean superconductors. Writing down $\hbar^2/(4m) = \pi D\hbar a/(8T_c) = D\gamma$, we find the friction coefficient $\gamma = \pi\hbar a/(8T_c)$. The time-dependent GL equations (TDGL) are valid, strictly speaking, only at very small deviations from equilibrium where quasiparticle excitations remain in equilibrium with the thermostat. Thus, to remain absolutely rigorous, we must remember that these restrictions usually hold in the so-called gapless superconductors. However, TDGL proved to give a rather good description of the dynamics of the order parameter on a qualitative level, capturing many essential features of nonstationary processes in superconductors. Interested readers are referred to the book by Nikolay Kopnin[41]. Here we only mention that TDGL appeared to be an especially productive and powerful tool for studying dynamics of superconducting fluctuations.

Speaking of fluctuations, we recall that as a mean field theory the stationary GL equations leave fluctuations out. However, it turns out that in the case of conventional or low-T_c materials, the GL approach appears pretty accurate. This is because in conventional superconductors, fluctuations are tiny except in the immediate neighborhood of the transition temperature. The role of fluctuations is quantified by the so-called Ginzburg number: $Gi \simeq [T_c/H_c^2(0)\epsilon\xi_0^3]^2/2 \simeq (T_c/\varepsilon_F)^4$, where ε_F is the Fermi energy, and ϵ is the anisotropy parameter[42]. Thus the fluctuation corrections become relevant only in the vicinity of T_c, so that the condition of the validity of the GL equations in clean superconductors is $(T_c - T)/T_c \gg (T_c/\varepsilon_F)^4$. The ratio T_c/ε_F is very small in the conventional, low-T_c materials, yielding that Gi is less than or about 10^{-12}, ensuring that the GL approach gives a good quantitative description of equilibrium properties of superconductors until very close proximity to T_c. In disordered superconductors where the mean free path $\ell < \xi$, the effective coherence length $(\xi\ell)^{1/2}$, and the above condition transforms into $(T_c - T)/T_c \gg (T_c/\varepsilon_F)(\hbar/(\varepsilon_F\tau))^3$, where $\tau \sim \ell/v_F$ is the scattering time, thus the fluctuation domain around T_c appreciably widens. This implies that disorder significantly increases the fluctuation domain and the role of fluctuations. The role of fluctuations is enhanced as well in high-T_c materials, where due to high anisotropy (i.e. small $\epsilon \ll 1$ and small coherence lengths) the Ginzburg number increases up to 10^{-2} or even more. To reveal the role of fluctuations one has to take into account the non-equilibrium short-living Cooper pairs which appear "outside" the Cooper condensate. The lifetime of the fluctuation Cooper pair is of the order of the inverse coupling energy and is given by $\tau_{GL} = \pi\hbar/[8k_B(T - T_c)]$. One of the important phenomena where the TDGL model reveals its full power is the so-called paraconductivity: the additional conductivity due to creation of fluctuation Cooper pairs above but very close to T_c. These short-living fluctuation Cooper pairs cause flickering short circuits all over the superconducting sample, thus creating the additional channel for current[43]. Using TDGL, one can find the contribution of fluctuations to the conductivity deriving the correction to the order parameter Ψ, caused by the weak applied electric field via the perturbation theory. An especially interesting result appears for the two-dimensional films with thickness $d \ll \xi$:

$$\delta\sigma_{fl} = \frac{e^2}{16\hbar d} \frac{1}{\ln(T/T_c)}. \tag{2.44}$$

Unfortunately, we cannot discuss here all the beauty and wealth of the physics of superconducting fluctuations. We refer the reader to Anatoly Larkin and Anderi Varlamov's book[44].

To conclude we note that as the GL functional emerged as an expansion in powers of the order parameter, the low-temperature bound for the applicability of the GL approach is $(T_c - T)/T_c \ll 1$. However, the GL equations often work fairly well (within a few percent accuracy) even at temperatures well below T_c.

2.1.5   Conclusion: Strings, the Universe, and All the Rest...

The significance and impact of GL equations go well beyond superconductivity. The GL equations and the GL-based approach and philosophy became instrumental to new developments in particle physics and quantum field theories. More specifically, in particle physics any quantum field

[41] N. Kopnin, *Theory of Nonequilibrium Superconductivity*, Clarendon Press, Oxford (2001)

[42] V. L. Ginzburg, *Soviet Solid State Physics* 2 (1960) 61

[43] L. G. Aslamazov and A. I. Larkin, *Phys. Lett.* A 26 (1968) 238

[44] A. Larkin and A. Varlamov, *Theory of Fluctuations in Superconductors* (revised edition), Oxford University Press, New York (2009)

theory with a unique classical vacuum state and a potential energy with a degenerate critical point is called a Landau-Ginzburg theory. The GL-based descriptions became ubiquitous in the study of two-dimensional conformal field theories especially in the context of supersymmetric theories. The latter find numerous applications in string theory and its mathematical manifestations such as, for example, mirror symmetry. One of the earliest approaches to mirror symmetry was the so-called Landau-Ginzburg/Calabi-Yau correspondence. Since we cannot cover these appealing topics within our restricted format, we refer the interested reader to the fundamental works[45]. The time-dependent GL equations open a door to a whole world of nonlinear waves in active environments, chemical reactions, spatio-temporal chaos, spiral waves, pattern formation, and many other beautiful phenomena[46]. The study of the mathematical properties of the Ginzburg-Landau functional has become a dynamically developing branch of mathematics[47].

We have to stop here, but the Book of Life of the Ginzburg-Landau equations, one of the most universal approaches in modern physics, is not finished; there are more and more beautiful chapters still to be written.

2.2 The Emergence of BCS

David Pines

2.2.1 Introduction

Major contributions in theoretical physics do not emerge in a vacuum. They usually are inspired by experiment, require near-total immersion in the relevant theoretical and experimental literature, and are often preceded by years of preliminary work, including the publication of failed or incomplete attempts at producing a theory, while recognition of their success does not always come instantly. The 1957 theory of John Bardeen, Leon Cooper, and Robert Schrieffer, known as BCS, which provided, 46 years after the discovery of superconductivity, a highly successful microscopic theory of its emergent quantum behavior, fits well into this general paradigm. In this highly personal perspective on their work I address these issues, writing not as a historian, but as a colleague and sometime participant in the events that led up to the theory, and some that followed.

I begin with John Bardeen's 1951 move from Bell Laboratories to the Physics Department of the University of Illinois in Urbana-Champaign, and focus on the 5-year period between the summer of 1952, when I arrived to work with him on a postdoctoral research appointment, and the summer of 1957, when BCS sent off their historic paper for publication in the *Physical Review*.

2.2.2 A Welcoming Environment

For John Bardeen the move from Bell Laboratories to a joint appointment in the Physics and Electrical Engineering Departments at the University of Illinois could not have been better. In Urbana he was free to pursue his twin passions, searching for a theory of superconductivity in Physics, with early support for his research from the U. S. Army Office of Ordinance Research, and establishing a laboratory on semiconductors in the EE Department.

---

[45] R. Dijkgraaf and Ed. Witten, *Nucl. Phys.* B 342 (1990) 486; C. Vafa, *Mod. Phys. Lett.* A 6 (1991) 337; Ed. Witten, *Nucl. Phys.* B 403 (1993) 159

[46] See the comprehensive review by Igor Aranson and Lorenz Kramer in Rev. Mod. Phys. 74 (2002) 99

[47] F. Bethuel, H. Brezis, F. Helein. *Ginzburg-Landau vortices*, Birkhäuser (1994); E. Sandier, S. Serfaty, *Vortices in the magnetic Ginzburg-Landau model*, Birkhäuser (2007)

In Physics he found a department that was already famous for its mix of excellence and collegiality, a mix that can be directly attributed to its head, Wheeler Loomis, who had built a first-rate department whose members saw themselves as part of a community, not a collection of independent fiefdoms. Faculty in different sub-fields genuinely enjoyed one another's company as they met regularly over lunch or dinner, attended one another's seminars, and joined together in the weekly colloquia. New arrivals were welcomed, and even the most junior of these soon felt as much a part of the department as did the most senior professors.

Bardeen's friend from his Princeton days, Fred Seitz, was instrumental in attracting him to Urbana. On coming to Urbana 2 years earlier, Seitz had brought along three colleagues whose specialty was the then new sub-field of solid state physics, and with Bardeen's arrival, Urbana's leading role in the field [a position it continues to hold, as solid state has morphed into condensed matter] was firmly established.

2.2.3 Starting Work with Bardeen

When I arrived in Urbana in early July 1952 as the department's first research assistant professor (the position was invented to recognize my "advanced" postdoctoral status, since I had received my Ph.D. 2 years earlier) I was given a desk in the corner of Bardeen's office. It was from this vantage point that I had the privilege and the pleasure of working with Bardeen for the next 2 1/2 years, lunching with him almost daily, and being able to observe and absorb his approach to research on what was then the major challenge in theoretical physics, developing a microscopic understanding of superconductivity.

Two years earlier, the experiments of Serin, Maxwell, and their collaborators had provided an important clue to its physical origin. The dependence of the superconducting transition temperature on the mass of the ions in the lattice in which electrons moved showed that their coupling to phonons, quantized lattice vibrations, must play an essential role. However, the initial attempts by Bardeen and Fröhlich to develop a theory based on the modification in the self energy of an electron brought about by that coupling had failed, and there appeared to be no unique path forward.

Bardeen's approach was to pursue, more or less simultaneously, a number of different possible avenues in the hope that one of these might pay off. Looking back, I can identify at least four of these:

- learning from experiment–following closely the latest experimental results and trying to develop a phenomenological theory that tied these together, a process that culminated in his 1956 *Handbook of Physics* review article[48]

- following and absorbing the latest theoretical developments in many-body physics and field theory, such as the multiple scattering theory for strong nuclear interactions being developed by Keith Brueckner and his collaborators, and the collective description of electron interactions I was working on with David Bohm

- exploring matrix approaches to the problem of a few electrons excited above the Fermi surface

- asking whether a strong-coupling approach to the coupled electron-phonon problem might provide a further essential clue.

As it turned out, each played a role in setting the stage for the microscopic theory Bardeen developed 5 years later with my successor as his postdoc, Leon Cooper, and his graduate student, Bob Schrieffer.

---

[48] J. Bardeen, *Encyclopedia of Physics*, Springer-Verlag, Berlin (1956), Vol. 15, p 274

I begin with the last of these four approaches, because that was the *raison-d'être* for the problem Bardeen suggested I study on my arrival—developing a theory of polarons, single electrons moving in polar crystals that are strongly coupled to the optical lattice vibrations found there.

2.2.4 Polaron Theory

Bardeen suggested that I familiarize myself with earlier work on polarons by reading the seminal papers of Solomon Pekar and Herbert Fröhlich, who had studied the limit in which polarons were weakly coupled to lattice vibrations, and Lev Landau, who had shown that a polaron could become self-trapped by very strong coupling to lattice vibrations.

Not long after I started work on the problem, I encountered Tsung-Dao [T.D.] Lee in the hallway, and we started exchanging information on our summer research plans. Lee was spending that summer in Urbana, before moving to Columbia University, and we had met earlier in Princeton. Within minutes we realized that we could connect his planned work on the coupling of pions to nucleons, using a method developed by Sin-itro Tomanaga, to my fledgling effort on polaron theory, and we soon were able to send off a letter to *Physical Review* reporting on our intermediate coupling theory of the polaron self-energy.

When, a few weeks later, I described the progress T.D. and I had made on polarons to another close friend in the Physics Department, Francis Low, a quantum field theorist who had also arrived in the Physics Department that summer, he proposed what became known as the LLP wave function[49] as a much simpler way to express our results. In it, one sees clearly that the virtual phonons in the cloud of excitations moving with the polaron are emitted successively into the same momentum state:

$$\Psi_{\text{LLP}} \sim \exp[\sum_k f(k)[a_k^* + a_{-k}]]\Psi_0,$$

where Ψ_0 was the ground state wavefunction, the operators a_k^* and a_k act to create or destroy phonons, and $f(k)$ describes the phonon state. [In the language of Feynman diagrams, this corresponds to summing all the "rainbow" diagrams, in which each new virtual phonon is emitted successively into the same quantum state.]

Our work represented the first non-trivial extrapolation of approaches developed in particle physics to a problem in solid-state physics. Looking back, there were perhaps three reasons it came about. First, I was part of an emerging generation of theorists who grew up acquainted with the newly developing methods in quantum field theory due to Feynman, Schwinger, and Tomanaga. Second, T.D. was in Urbana that summer because Wheeler Loomis had realized that summer appointments made it possible to attract great young theorists to Urbana [C.N. (Frank) Yang was his immediate predecessor in that visiting position]. Third, the world of theoretical physics was still quite small, so T.D., Francis, and I frequently talked to colleagues in different sub-fields about our current research. It is perhaps worth noting that the summers of 1952 and 1953 in Urbana were not only remarkable for their record-setting heat records but also for the concentration of young theorists there, as T.D. came back for a second summer to work with John Bardeen on superconductivity and me on polarons, while the three of us were joined by Murray Gell-Mann who came down from Chicago to work with Francis, and briefly with me.

2.2.5 Effective Electron Interactions in Metals

After working on polarons, I returned to working, at a distance, with my thesis supervisor, David Bohm, on writing up the results of my 1950 Ph.D. thesis for publication,[50] and soon found my-

[49] T. D. Lee, F. E. Low, D. Pines, *Phys. Rev.* 90 (1955) 297

[50] D. Bohm and D. Pines, *Phys. Rev.* 92 (1953) 609; D. Pines, *Phys. Rev.* 92 (1953) 626

self trying, with limited success, to generalize our collective description of electron interaction to the more general problem of electrons coupled to each other and to phonons. Within the random phase approximation [RPA] which Bohm and I had used to treat electron-electron interaction, I could obtain the modification in the phonon dispersion relation brought about by their coupling to electrons, but was unsuccessful in arriving at a self-consistent treatment of the net effective electron interaction, and so was not in a position to answer the key question: how does the strong repulsive Coulomb interaction between electrons modify the electron–phonon interaction that gives rise to the apparently weak attractive phonon-induced interaction between electrons that Fröhlich[51] had conjectured in 1952 might be the way phonons make possible a transition to the superconducting state, and what is the net interaction that results when this Coulomb repulsion is included?

As I was describing my lack of progress to Bardeen one morning in late 1953, he suggested that I consider adding an explicit phonon contribution to the canonical transformation I was using to go from my introduction of an arbitrary supplementary field to describe collective motion to coordinates that could describe the collective modes of the coupled electron–electron–phonon system. I tried putting that in, and everything worked. We then joined forces and were soon able to obtain a self-consistent account, within the RPA, of the way electron interactions modified their coupling to ions and to one another and how the combined ionic and electronic Coulomb interactions gave rise to two collective modes, sound waves and plasma oscillations, and a net effective electron interaction.[52] The resulting phonon dispersion relation turned out to be just that Bardeen had calculated 17 years earlier in his seminal 1937 paper on lattice vibrations in simple metals,[53] while the effective electron–electron interaction turned out to have properties that were similar to those Fröhlich had found, but with full account now having been taken of all the interactions in play.

We found that for pairs of electrons whose energies differed from each other by less than a typical phonon energy, the attractive interaction brought about by phonon exchange could win out over the repulsion coming from the now present and properly screened Coulomb interaction. For these electrons, the net effective electron–electron interaction would be attractive; for larger energy differences, the repulsive screened Coulomb interaction would win out and their net interaction would be repulsive. Since a full account of the Coulomb interaction had now been taken into account, we closed our paper with the prophetic line, "The equations we have presented here should provide a good basis for development of an adequate theory."

As Nozières[54] subsequently pointed out, our results for the net effective frequency-dependent electron interaction at low frequencies, V_{eff}, could be put in an especially simple form if one neglected the periodicity of the ionic lattice, the so-called "jellium" model for a metal:

$$V_{\text{eff}}(q,\omega) = [4\pi e^2/q^2 \epsilon(q,0)][1 + \omega_q^2/(\omega^2 - \omega_q^2)],$$

where $\epsilon(q,0)$ is the static dielectric constant, and the second term, proportional to the square of the frequency of the phonon being exchanged, is the phonon-induced contribution to the net effective interaction. The screened Coulomb interaction sets the overall scale for the strength of the interaction, and it is evident that the net interaction will always be attractive for those electrons of momentum p and $p+q$ near the Fermi surface whose energy difference, $\omega = \epsilon_{p+q} - \epsilon_p$, is below the natural resonance frequency, ω_q, of the phonon that is being virtually exchanged.

2.2.6   Cooper's Pair Instability

When I left Urbana for Princeton in January 1955, Bardeen sought a successor who was familiar with quantum field theory. He asked the advice of Frank Yang at the Institute for Advanced Study,

---

[51] H. Fröhlich, *Proc. Royal Soc.* [London] A215 (1952) 291

[52] J. Bardeen and D. Pines, *Phys. Rev.* 99 (1955) 1140

[53] J. Bardeen, *Phys. Rev.* 52 (1937) 688

[54] P. Nozières, private communication to the author (1957)

which by then had a large postdoctoral program, and Frank recommended Leon Cooper. As Cooper tells the story,[55] when he arrived in Urbana that fall, he knew nothing of superconductivity, so his first months were devoted to a crash course in its fundamentals, reading David Shoenberg's excellent monograph, absorbing Bardeen's work-in-progress on his *Handbook* review on superconductivity. John encouraged him to look at the behavior of a few electrons excited above the Fermi surface, and within a year he made a key discovery as he studied the behavior of two electrons of opposite spin and momentum whose interaction would be attractive if they were close to the Fermi surface—a simplified version of the Bardeen–Pines attractive interaction. In a calculation[56] that allowed for the multiple scattering of the pair above the Fermi surface, he showed that such an attraction could give rise to a bound state. If his approach could be extended to an actual many-electron problem, it would correspond to the energy gap in the elementary excitation spectrum of superconductors that Bardeen had shown could provide a phenomenological explanation for the recent experiments on the specific heat and the penetration of magnetic fields into superconductors.

Cooper's work served to confirm the role for phonons in superconductivity that Fröhlich had conjectured, and that Bardeen and I had demonstrated could occur even in the presence of strong Coulomb repulsion—that a net attractive interaction produced by the phonon-induced electron interaction was the gateway for superconductivity. But it did not provide a path forward to explaining how that attraction gave rise to the signature hallmark of superconductivity—the emergence of a single quantum state to describe superfluid flow, whose rigidity London had eloquently argued could explain perfect diamagnetism.

2.2.7 Seattle and Stockholm

During the following year, superconductivity was front and center at a number of international conferences. At the September 1956 international conference on theoretical physics held in Seattle, John Blatt outlined the approach he, Butler, and Schafroth were making to superconductivity, in which they assumed that it resulted from the Bose–Einstein condensation of pre-formed pairs of electrons. Because there was no experimental evidence for such pre-formed pairs, their proposal was received with some skepticism by those present. At that same meeting, Richard Feynman, in the course of a lecture on his seminal work on liquid helium, made it clear that he had turned his attention to superconductors, but had no theory to present as yet.

Bob Schrieffer, by then in his third year of graduate work with John Bardeen in Urbana, with a planned thesis on superconductivity, was one of the principal note-takers on the Seattle lectures. He returned home worrying that not only had he yet to make significant progress, but also that in Feynman he and Bardeen had a formidable competitor. I can testify first-hand to Feynman's immersion in the problem. During a 2-week visit I made to Caltech in December 1956, almost every day over lunch he would try out another idea on me. For example, he proposed that since so few superconductors were monovalent, it must be the combination of multivalent lattice structure and vibrations that made the latter superconduct, and suggested an approach that I told him was pretty much identical to the one that Bardeen had first tried, unsuccessfully, to work through in 1937.

Additional competition for Schrieffer and Bardeen had emerged earlier that fall, when Bardeen received word that he had been awarded the 1956 Nobel prize for his discovery of the transistor. This meant he had to set aside his current work on superconductivity to prepare his Nobel lecture and go to Stockholm in December 1956 to receive the prize. What made Bardeen so different from most Nobel laureates was that instead of basking in the Nobel-induced limelight, he could scarcely wait to get home to resume his research.

---

[55] L.N. Cooper, in *Bardeen, Cooper, and Schrieffer 50 Years*, World Scientific Pub., Singapore (2010)

[56] L.N. Cooper, *Phys. Rev.* 104 (1956) 1189

2.2.8   The Schrieffer Wave-Function

The following month [January 1957], an International Conference on the Many-Body Problem was held at the Stevens Institute in Hoboken. Bob Schrieffer was among the attendees. As he describes in detail in his 1964 book, *The Theory of Superconductivity*,[57] he had been working with Bardeen to find a variational wavefunction that might describe the ground and excited states of a superconductor. While riding a New York City subway train following the Hoboken meeting, he had his "Aha moment". He realized that he could generate a candidate wave function by assuming that the key physics was the formation of a macroscopically occupied coherent quantum state made up of pairs of electrons (of opposite spin and momentum) and then adapting the Lee–Low–Pines intermediate-coupling ground-state wave function to describe it. In his variational approach, the LLP phonon field was replaced by the coherent pair field of the condensate, $b_k^* = c_{k\uparrow}^* c_{-k\downarrow}^*$, where c and c^* are the annihilation and creation operators for a single quasiparticle. When Schrieffer applied it to the model Hamiltonan used by Cooper, he was able to calculate the reduction in the ground state energy brought about by his hypothesized pairing condensate, and obtain an energy gap for excitations from that new ground state.

Schrieffer's proposed gateway to emergent superconducting behavior was a quite remarkable coherent state of matter; the pairs in the condensate need not be physically located close to one another, while their condensation is not the Bose condensation of pairs that pre-form above the superconducting transition temperature in the normal state. Instead the pairs condense only below the superconducting transition temperature, and the typical distance between them can be some hundreds of times larger than the typical spacing between particles.

On Schrieffer's return to Urbana, he showed his results to Cooper and then to Bardeen. Bardeen quickly recognized that this was the correct basis for finding a solution to superconductivity, and with great excitement, he, Schrieffer, and Cooper began to work out its consequences for the ground and excited states of a superconductor. Within some 2 weeks, they had developed the microscopic theory that soon became known as BCS, and on February 18, 1957 sent a brief account of their results for publication in *Physical Review Letters*.[58]

2.2.9   Deciphering, Teaching, and Applying BCS

In early 1955 I left Urbana for Princeton, and it was there, some 2 years later, that I received a brief letter from John reporting that he thought superconductivity had been solved, and enclosing a dittoed copy of their not-yet-published PRL. I shared the news with Elihu Abrahams, who was newly arrived at Rutgers, and my prize graduate student, Philippe Nozières. Filled with excitement, we decided to see if we could flesh out the details of what BCS had done. Since the details provided in their manuscript were scant, doing so proved to be non-trivial, but after three intensive days in our living room on Clover Lane we succeeded sufficiently well that I was able to teach it to my class later that spring.

In the course of these lectures I did some simple calculations showing how the effective interaction that John and I had derived, and that formed the starting point for BCS, led in a natural way to the famous empirical rules for the occurrence of superconductivity that had been proposed by Bernd Matthias.[59] I showed these to John when he came to Princeton to give what may have been his first colloquium on BCS; he encouraged me to publish these as what turned out to be a first application of their theory — showing that the requirement that the net interaction be attractive enabled one to distinguish between metals that would superconduct, and those that would not.

---

[57] J.R. Schrieffer, *The Theory of Superconductivity*, W.A. Benjamin, New York (1964)

[58] J. Bardeen, L. N. Cooper, J. R. Schrieffer, *Phys. Rev.* 106 (1957) 162

[59] B. Matthias, in *Progress in Low Temperature Physics*, ed. C.J. Gorter, North-Holland, Amsterdam (1957) vol. 2

What I did in this paper[60] was first to see how well one could do with a "minimalist" approach to calculating the average effective interaction V, which had to be attractive to bring about superconductivity, and the product $N(0)V$ that determined the superconducting transition temperature, T_c, in the then-fledgling BCS theory. I took the repulsive part to be a screened Coulomb interaction and considered separately the contributions to the attractive part coming from normal processes (in which the interacting electron momentum differences were less than a typical phonon wave vector, k_D and the Umklappprocesses (U processes) in which their momentum difference was $q + Q > k_D$, where Q is a reciprocal lattice vector. For the latter, I further assumed that the momentum of the phonon being exchanged was always k_D, which underestimated the contribution of U processes to V. I was thus able to obtain a simple analytic expression for $N(0)V$ that depended only on the effective ion charge, Z^*, and r_s, the dimensionless interelectron spacing.

Within this simple model, the net attraction coming from N processes alone was not strong enough to overcome the screened Coulomb repulsion, so that, for example, jellium would never superconduct. When U processes were included, the model turned out to do quite well at distinguishing between normal and superconducting elements in the periodic table and even had predictive power, in that it predicted the superconductivity of Mo, W, Y, Sc, and Pa, and, with minor modification, could explain the empirical rules developed by Matthias.

2.2.10 BCS Theory

As soon as their manuscript describing their initial results was sent off for publication, Bardeen, Cooper, and Schrieffer set about developing the theory in detail; less than 5 months later, they were able to send the long paper describing their theory to the *Physical Review*, which received it on July 8, 1957.[61] Their results included the microscopic description of the two fluids that characterize superconducting behavior: the superfluid, a single macroscopic quantum state, formed by the condensation of pairs whose average spacing is large compared to the inter-electron spacing, that flows without resistance and acts to screen out external magnetic fields; and a normal fluid that is made up of the "pair-breaking" elementary excitations that a finite amount of energy—the energy gap-is required to excite. These quasiparticles scatter against each other and impurities much as normal electrons do. In using Schrieffer's wave function to calculate various properties of the superconducting state, Bardeen, Cooper, and Schrieffer were guided at every stage by the phenomenological description Bardeen had enunciated 2 years earlier, with what quickly became recognized as remarkable success. Remarkably, their new theory was also able to explain the quite surprising results of a measurement that had just been completed in Urbana by Charlie Slichter and his student, Chuck Hebel, on the change in the nuclear spin-lattice relaxation rate when a material becomes superconducting.

So how was it that Bardeen and his young collaborators were able to solve the riddle of superconductivity? The solution had, after all, eluded their many distinguished theoretical colleagues who were working on the problem at that time—notably Feynman, Landau, Fröhlich, Ginzburg, Blatt, and Schafroth. The answer is, in part, to be found in Bardeen's emphasis on understanding the experimental facts and developing a phenomenological description of these while simultaneously pursuing a number of different theoretical scenarios with his younger colleagues. Equally important was his total dedication to cracking the problem, and the encouragement, support, and freedom to pursue their own ideas that he gave to his younger colleagues, who played such key roles in the development of the theory.

The story of BCS is now a highpoint in the history of physics in the twentieth century. The theory, for which its authors received the 1972 Nobel Prize in Physics, not only explained all existing

---

[60] D. Pines, *Phys. Rev.* 109 (1958) 280

[61] J. Bardeen, L. Cooper, J.R. Schrieffer, *Phys. Rev.* 108 (1957) 1175

experiments on superconductors, but made a number of predictions that were subsequently verified, It quickly had an impact on other fields of physics. In the summer of 1957 the key BCS idea—that a net attractive interaction between fermions (particles of intrinsic spin $\frac{1}{2}$) would always lead to a pairing state that was macroscopically occupied—was applied to atomic nuclei by Aage Bohr, Ben Mottelson, and the author,[62] and soon thereafter was taken into the realm of particle theory by Yoichiro Nambu and. and Gianni Jona-Lasinio.[63] Indeed, within 2 years it had become so clear that the BCS theory was successful that David Shoenberg, in his introductory remarks at a 1959 superconductivity conference at Cambridge, was led to make his classic remark, "Let us see to what extent experiment can explain the theoretical facts."

In preparing this perspective I have drawn upon my memoir of John Bardeen prepared for the American Philosophical Society,[64] on a chapter I wrote for the online text, Physics for the Twenty-first Century,[65] and on a talk given as part of the UIUC celebration of the 50th anniversary of BCS,[66] a written version of which recently appeared,[67] so the reader should not be surprised by any similarities in wording that may have resulted from this process.

2.3 Theory of Superconductivity: From Phenomenology to Microscopic Theory

Lev P. Gor'kov

2.3.1 Introduction

The foundations of the microscopic theory of SC were laid by Bardeen, Cooper, and Schrieffer (BCS) in 1957[68,69,70]. In the next few years the theory underwent significant developments. Of special importance was the period between the end of 1957 and the early 1960s during which, basically, the theory received its current accomplished form. In the author's opinion, the history of those years is not yet well known. The comprehension of profundity of ideas of the modern SC theory supposes knowledge of their origination. The modern reader, more often than not, is not familiar with the problems that puzzled physicists half a century ago. For that reason, in what follows, the author had to begin with a brief review of some early experimental findings and trace the progress of theoretical ideas before the BCS era. The BCS theory did not emerge from nothing. Some important concepts formulated earlier on phenomenological grounds continue to preserve their significance even today. In addition, it is also important to keep in mind that the superconductors that are being investigated in our days differ in many aspects from pure elemental metals that were studied, say, between the mid-1930s and mid-1950s. For instance, a number of new superconductors reveal no gap in the energy spectrum, whereas deriving such a gap was one of the main triumphs of the BCS theory[70]. To understand the logic that guided BCS, it is therefore necessary to return back to the ideas, hypotheses, and sometimes, even the prejudices of those days.

---

[62] D. Pines in *Proc. of the Rehovoth Conf. on Nuclear Structure*, Interscience Press (1957), p 26; A. Bohr, B. Mottelson, and D. Pines, *Phys. Rev.* 110 (1958) 936

[63] Y. Nambu and G. Jona-Lasinio, *Phys. Rev.* 122 (1961) 345; ibid 124 (1961) 246

[64] D. Pines, *Proc. Amer. Phil. Soc.* 153 (2009) 288

[65] D. Pines, in *Physics for the 21st Century*, http://www.learner.org/courses/physics/unit/text.html?unit=8&secNum=0 (2010)

[66] http://www.conferences.uiuc.edu/bcs50/video.html

[67] D. Pines, in *Bardeen, Cooper, and Schrieffer 50 Years*, World Scientific Pub., Singapore (2010)

[68] L. N. Cooper, Phys. Rev. 104 (1956) 1189

[69] J. Bardeen, L. N. Cooper, J.R. Schrieffer, *Phys. Rev.* 106 (1957) 162

[70] J. Bardeen, L. N. Cooper, J.R. Schrieffer, *Phys. Rev.* 108 (1957) 1175

2.3.2 Early Days

When in 1911 Kamerlingh Onnes discovered[71] the disappearance of resistivity below 4.15 K in his mercury samples, no one would have expected any help from theory. At that time even the discovery of the electron in 1897 by J. J. Thompson probably was considered to be "a recent discovery". The creation of quantum mechanics was yet at its very beginning, and it would not be earlier than 1928 that A. Sommerfeld applied the ideas of quantum mechanics for the first time to the thermodynamic properties of a free electron gas. Actually, a meaningful theoretical discussion became possible only after Meissner and Ochsenfeld[72] in 1933 proved the reversible character of the first-order transition in a magnetic field. The expulsion of a magnetic flux from the interior of a superconductor established the superconducting state as a *new thermodynamic phase* of a metal. Even then, in spite of the rapid accumulation of experimental data, another quarter of a century elapsed between the date of this pivotal experiment and the publication of the BCS theory in 1957. The difficulties that the microscopic theory had to overcome are briefly addressed below.

Phase Diagram in Magnetic Field

The Meissner effect proves that infinite conductivity, as such, falls short of uniquely identifying the superconducting state. Indeed, if an *ideal* conductor were placed into the magnetic field, the emerging state would depend on whether the material was cooled below T_c *before* or *after* switching on the magnetic field. In Reference 72 it was shown that the magnetic field is *always expelled from the bulk* independent of pre-history. The historical experiments[72] were performed on cylindrical tin samples with the external field H parallel to the cylinder axis to minimize the demagnetization coefficient. The induction B inside the cylinder was shown to be identically to zero at temperatures below some $T_c(H)$, i.e., $\mathbf{B} = \mathbf{H} + 4\pi\mathbf{M} \equiv 0$. The cylinder thus acquires a magnetic moment (per unit volume):

$$\mathbf{M} = -(1/4\pi)\mathbf{H} \tag{2.45}$$

The superconductor below $T_c(H)$ is an *ideal diamagnetic*, with magnetic susceptibility $\chi = -(1/4\pi)$. The interaction of the magnetic moment $\mathbf{M}$ with the external field $-(1/2)\mathbf{M} \cdot \mathbf{H}$ reduces the gain in free energy at the transition into the superconducting state $\Delta F = F_n - F_s$. With increasing field the superconductor undergoes a first-order transition back into the normal state, in agreement with Kamerlingh Onnes's experiments in 1914[73]. The field at which this happens is called the *thermodynamic* critical field H_c. It is defined by:

$$\Delta F = F_n - F_s \equiv \frac{H_c^2}{8\pi}. \tag{2.46}$$

That result establishes the important interrelation between the thermal and magnetic properties of a superconductor (along the transition line) first analyzed by Rutgers[74]. The popular two-fluid models of those days (e.g., the one by Gorter and Casimir[75]) depicted the superconducting phase as constituted of two independent "liquids": the "condensate" that appears below T_c and bears no entropy, and a dissipative component possessing normal properties. The fraction of electrons in the condensate was termed "the order parameter".

[71] H. Kamerlingh Onnes, *Commun. Phys. Lab. Leiden*, 120b, 122b, 124c (1911)

[72] W. Meissner, R. Ochsenfeld, *Naturwiss.* 21 (1933) 787

[73] H. Kamerlingh Onnes, *Commun. Phys. Lab. Leiden*, 139f (1914)

[74] A. J. Rutgers, *Physica* 1 (1934) 1055

[75] C.J Gorter and H.B. Casimir, *Phys. Zs.* 35 (1934) 963, *Physica* 1 (1934) 306

Intermediate State

A superconductor of an arbitrary shape would reveal a more complicated behavior in magnetic fields. Indeed, the applied external field being distorted by the field expelled from the sample is not homogeneous outside the superconductor. In particular, its value varies along the surface. With increasing field it first reaches the critical value H_c locally at a certain surface point. At such point the superconducting phase should return to the normal state, but in its vicinity the field still is below H_c. As a compromise, the sample gradually goes over into the so-called *"intermediate state"* by forming alternating normal and superconducting layers (R. E. Peierls,[76] F. London,[77] 1936).

Box 1. To illustrate, consider a superconducting slab placed in a field $H < H_c$, perpendicular to its plane. The total magnetic flux must be equal on both sides of the slab. To ensure this, the slab breaks into normal (N) and superconducting (S) layers; the inductance is equal to H_c in the N-layer and is zero in the S-layer. If the total area of the slab is S, its fraction occupied by the normal phase must be xS, where $x = (H/H_c)$. Note that the relative widths of the layers remain unfixed so far.

The theory of the intermediate state was elaborated by Landau in 1937. In Landau's theory[78] the pattern of the S–N layers is uniquely defined by adding the positive *surface energy* $\sigma > 0$ at the interface to the total energy balance.

Penetration Depth

The essence of the Meissner effect is that the non-dissipative (*diamagnetic*) currents at the surface shield the external field. These currents and the field should gradually attenuate into the sample and the question arises about their microscopic description. The electrodynamics of superconductors in weak magnetic fields was first addressed by F. and H. London[79] in 1935. They arrived at the expression for the local electric current:

$$\mathbf{j}(r) = -\frac{n_s e^2}{mc}\mathbf{A}(r). \tag{2.47}$$

Solving the Maxwell equation, $\mathrm{curl}\,\mathbf{B}(r) = \frac{4\pi}{c}\mathbf{j}(r)$ for the half-space, $z > 0$ the field distribution would be given by $B(z) = H(0)\exp(-z/\lambda_L)$. The parameter $\lambda_L = (mc^2/4\pi n_s e^2)^{1/2}$ is known in the literature as the London penetration depth; n_s stands for the "number of superconducting electrons" equal by order of magnitude to the density of electrons in a metal. The estimate leads to $\lambda_L \sim 10^{-6}$ cm for such elemental metals as Al or In. Note that this scale is much larger than the atomic scale $\sim 10^{-8}$ cm. At the time, expression (2.47) for the current caused much concern. In fact, the current in (2.47) is proportional to the vector potential $\mathbf{A}(r)$ but that is not a physical variable on its own. The magnetic field $\mathbf{B}(\mathbf{r})$ expressed through $\mathbf{A}(r)$ as $\mathbf{B}(\mathbf{r}) = \nabla \times \mathbf{A}(\mathbf{r})$, is the physical variable and the transformation $\mathbf{A}(\mathbf{r}) \Rightarrow \mathbf{A}(\mathbf{r}) + \nabla\varphi(\mathbf{r})$, called the *gauge transformation*, should not change the physical results. Consequently, in the London theory the conservation of current $\mathrm{div}\,\mathbf{j}(\mathbf{r}) = 0$ would impose severe restrictions on the choice of $\mathbf{A}(r)$: $\mathrm{div}\,\mathbf{A}(r) = 0$. F. London[80] argued that the current

---

[76] R. E. Peierls, *Proc. Roy. Soc.* A155 (1936) 613

[77] F. London, *Physica* 3 (1936) 450

[78] L. D. Landau, *Phys. Z.* (USSR) 11 (1937) 129

[79] H. London, F. London, *Proc. Roy. Soc.* A149 (1935) 71; *Physica* 2 (1935) 341

[80] F. London, *Superfluids*, vol.1, Wiley, New York (1950)

(2.47) is sustained by "...the rigidity of the collective wave function of electrons in the field" on a "coherence" distance which is at least of the order of the penetration depth.

Surface Energy

At the first-order transition at the critical field H_c the normal (N) and the superconducting (S) phases are in equilibrium with each other. What can be said about the *microscopic structure of the interface*? For usual first-order phase transitions, such as the liquid-gas transition, the boundary between phases is stabilized by the surface tension $\sigma > 0$. Commonly, the transition layer between two phases is of atomic thickness, but that may be different for the N–S boundary where the penetration depth alone is fairly large. Indeed, consider an N–S interface placed at $z = 0$; the superconducting and normal phases each occupying the half-spaces $z > 0$ and $z < 0$, respectively. The magnetic field which is $H = H_c$ inside the normal phase will attenuate from the boundary into the superconducting phase over a distance $\sim \lambda_L$. In addition, the scale describing the variation of the superconductivity parameter n_s is denoted by ξ_0. The local free energy $F_s(z)$ will *increase* as n_s gradually *decreases* approaching $z = 0$. If $\xi_0 > \lambda_L F_s(z)$ for $z > 0$ will be larger than F_s in the bulk on a distance of order $\sim \xi_0$, even though the field does not penetrate there and the existence of this effectively normal layer is not justified by the presence of any field of order H_c. The loss in the superconducting free energy in the transitional area near $z = 0$ is the reason for the positive surface energy[80]. The latter is usually presented as $\sigma = d(H_c^2/8\pi)$ (per surface unit), where the effective "interface width" d is given by $d \approx \xi_0 - \lambda_L$. The value of σ can be obtained from the experiments on the field distribution in the intermediate state[78]. Data were accumulating from the late 1940s to the mid-1950s and revealed that the width d is surprisingly large (for instance $d \approx 10^{-4}$ cm for tin). Thus, there are *two independent* spatial scales that characterize a superconductor: the penetration depth $\lambda_L \sim 10^{-6}$ cm and another, bigger scale $\xi_0 \sim 10^{-4}$ cm related to the internal structure of the superconducting phase itself.

Two Spatial Scales

One may wonder why in the old days the existence of the second scale ξ_0 attracted so much attention. The quantum nature of the SC phenomenon being unquestionable at the time, experiments on the penetration depth and on the interface structure were the only ones allowing a glimpse of the quantum properties of the superconducting electrons. With the London penetration depth accounting for such a local feature as n_s, it was intuitively perceived that ξ_0 must somehow characterize the *wave functions* of electrons in the new phase. Common expectations at the time were that SC emerges as the result of short-range (screened) electron–electron interactions. The transition temperature would then determine an energy scale rather than a spatial scale. The existence of another, larger spatial scale seemed to contradict such a perception. Meanwhile, in 1953 B. Pippard[81] found that the penetration depth for tin samples alloyed with indium ($< 3\%$) increased with the increase of the indium concentration. From that result Pippard concluded that the connection between the vector potential $\mathbf{A}(r)$ and the current $\mathbf{j}(r)$ must bear a non-local character, $\mathbf{j}(\mathbf{r}) = -\int Q(\mathbf{r} - \mathbf{r}')\mathbf{A}(\mathbf{r}')d\mathbf{r}'$ expressing that it is sensitive to the incoherence caused by the scattering of electrons on defects. The *non-locality* of the current again points to the existence of an additional scale, called the "coherence length" by Pippard. It does not only control the spatial variation of n_s, but it determines the minimal area inside which, for SC to exist, the wave functions of electrons must preserve quantum coherence. The non-locality of the electrodynamics in elemental superconductors was a remarkable finding, all the more because Pippard's result for $Q(\mathbf{R})$ was closely reproduced later by BCS[70]. Pippard's concept of the "coherence length" was initially met with skepticism, above all on the part of the Russian community. The presumptions were that SC in a sense should be akin to superfluidity of liquid helium in which the phenomenon is due to short-range interactions between helium

---

[81] A. B. Pippard, *Proc. Roy. Soc.* A216 (1953) 547

atoms. As Lev Landau would also point out, experimentally the temperature of the superconducting transition was not sensitive to defects, neither to geometrical factors such as the thickness of thin films. (This enigma was later resolved in the framework of the microscopic theory by Abrikosov and Gor'kov[82,83], and P.W. Anderson)[84].

Superconductivity and Superfluidity

Superfluidity was discovered by Kapitza in 1938 as the capability of liquid helium at temperatures below $T_\lambda = 2.19$ K to flow along a narrow capillary without viscosity. In this regard, superfluidity and SC seemed to have much in common. After the creation of the theory of superfluidity by L. D. Landau (1940–41)[85], some of his ideas were extended to SC. The most fundamental principle introduced by Landau in *Statistical Physics* was the notion of *quasiparticles* (*qps*). The concept maintains that any macroscopic system, whatever the interactions between its particles, at low enough temperatures may be represented by a gas of excitations (*qps*) with an energy spectrum $\epsilon(\mathbf{p})$. The latter, generally speaking, has little in common with the spectrum of non-interacting particles and should be determined empirically. According to the Landau theory the non-zero viscosity in flowing liquid helium appears only when the velocity of the flow exceeds a certain threshold value v_{cr} above which the spontaneous generation of excitations inside the liquid signals the beginning of dissipation:

$$v_{cr} = \min\{\epsilon(\mathbf{p})/p\}. \tag{2.48}$$

Box 2. Consider a liquid *He II* at rest. At $T = 0$ there are no excitation in the system. Assume then that helium begins to move with respect to the capillary walls with a constant velocity v. In the reference system that moves *with* helium the wave function of an excitation with momentum $\mathbf{p}$ and energy $\epsilon(\mathbf{p})$ is proportional to $\psi(\mathbf{r},t) \propto \exp[-i\epsilon(\mathbf{p})t + i\mathbf{p}\cdot\mathbf{r}]$. The substitution $\mathbf{r} \Rightarrow \mathbf{r} - \mathbf{v}t$ transforms this into the wave function of the same excitation in *the capillary reference frame*: $\psi(\mathbf{r},t) \propto \exp[-i(\epsilon(\mathbf{p}) + \mathbf{p}\cdot\mathbf{v})t + i\mathbf{p}\cdot\mathbf{r}]$. This is the familiar Galilean transformation for the energy of *qps*: $\epsilon(\mathbf{p}) \Rightarrow \epsilon(\mathbf{p}) + \mathbf{p}\cdot\mathbf{v}$. Thus, the cost of the excitation in the reference system at rest is smaller if $\mathbf{p}\cdot\mathbf{v} = -pv < 0$. Reaching the threshold $\epsilon(\mathbf{p}) - pv = 0$ marks the beginning of the spontaneous creation of excitations in helium, i.e., the beginning of dissipation.

The energy spectrum of helium starts with the linear, acoustic *qps* branch: $\epsilon(p) \approx sp$ with s the sound velocity. In reality, in Eq. (2.48) $v_{cr} < s$ and is determined by the so-called "*roton minimum*" in the helium spectrum[85]. The experimental fact that large enough currents (in thin films), as well as a large enough magnetic field destroy SC[73], seemed to support the Landau critical velocity concept. In a metal the electronic excitations are the Fermi excitations with spectrum: $\epsilon(\mathbf{p}) = v_F(p - p_F)$ (p_F and v_F stand for the Fermi momentum and the Fermi velocity, respectively). Obviously, such a spectrum cannot sustain non-dissipative currents. However, assume that the excitations in the superconducting state somehow were changed to acquire an energy gap Δ near the Fermi surface. The Landau criterion (2.48) would then give for the critical velocity $v_{cr} = \Delta/p_F$. Experimental evidences in favor of a gap in the spectrum began to accumulate rapidly before 1957. For instance, the

---

[82] A. A. Abrikosov, L. P. Gor'kov, *Sov. Phys. JETP* 8 (1959) 1090

[83] A. A. Abrikosov, L. P. Gor'kov, *Sov. Phys. JETP* 9 (1959) 220

[84] P. W. Anderson, *J. Phys. Chem. Solids* 11 (1959) 26

[85] L. D. Landau and E. M. Lifshitz, *Course of Theoretical Physics*: vol. 9, *Statistical Physics*, part 2 (1998); ibid. vol.6, *Fluid Mechanics*, Butterworth-Heinemann, Oxford (1999)

exponential behavior of the specific heat at very low temperatures

$$c_p(T) \propto A\gamma T_c \exp(-\alpha T_c/T) \qquad (2.49)$$

(with $\alpha \approx 1.50$) was observed[86] in vanadium.

2.3.3 The 1950s

By the mid-1950s the experimental literature on SC was immense. The thermodynamics, temperature dependencies of magnetic properties, the nonlinear behavior in strong magnetic fields, experiments on colloids, thin films, and other finite-size effects, and so on and so forth; everything was investigated down to the last detail for about a quarter of a century between 1933 and 1957. The early, purely empirical, two-fluid model[75] fit the data for the specific heat of a superconductor at not-too-low a temperature rather well, but failed to reproduce the exponential law (2.49) for the specific heat at the lowest T.

Ginzburg–Landau Phenomenology

One of the major achievements of the 1950s was the Ginzburg-Landau (GL) theory[87]. For convenience, we give first a concise summary of the GL phenomenology. The microscopic derivation[88,89] is discussed later.

The GL theory was the extension of the Landau theory of second-order phase transitions[90] modified to incorporate the charge and the current interaction with the magnetic field. The order parameter was chosen as a *complex* "wave function" $\Psi(r)$; the gradient term was then presented in the form $(\frac{1}{2m})|(-i\nabla - \frac{e}{c}\mathbf{A}(r))\Psi(r)|^2$, to ensure the gauge invariance of the theory. The phenomenology agreed well with known experimental facts, at least qualitatively, but its range of applicability, by definition, was limited by a narrow vicinity of T_c. (See for more details the preceding article by V. Vinokur.) Note that one of the main motivations of the GL paper was the calculation of the surface energy at the N–S interface. As it turned out, all results critically depended on one and only one dimensionless parameter which can be expressed through the experimentally observed quantities as

$$\kappa = (\frac{\sqrt{2}e}{\hbar c})H_c \lambda^2. \qquad (2.50)$$

If its value is small, $\kappa \ll 1$, the theory indeed discloses *two distinct* spatial scales: the penetration depth, λ, and the second one, which may be called "the coherence length", $\xi = (\lambda/\kappa) \gg \lambda$. In agreement with the speculations above (see "Two spatial scales"), the surface tension in this case is positive, $\sigma > 0$. The two GL scales diverge near T_c: $\lambda, \xi \propto 1/\sqrt{1 - T/T_c}$.

It turned out[87] that the surface energy changes sign at $\kappa = 1/\sqrt{2}$. A negative surface energy would make the N–S boundary unstable. No materials with such a property were known to the authors "at that time" (1950), so GL did not consider the case of negative $\sigma < 0$. (It is interesting, though, to find in the footnote to the Abrikosov paper[91]: "The suggestion that κ may be greater than $1/\sqrt{2}$ for an alloy was first made by L. D. Landau". Actually, such alloys had already been experimentally studied by Lev Shubnikov's group in 1937 (see below).

The physics of superconductors with negative surface energy $\sigma < 0$ was addressed in 1957 by Abrikosov[91] in the framework of the same phenomenological GL equations. He found that the

[86] W.S. Corak , B. B. Goodman, C.B. Satterthwaite, A. Wexler, *Phys. Rev.* 102 (1956) 656

[87] V. L. Ginzburg, L.D. Landau, *Zh. Exp. Teor. Fiz.* 20 (1950) 1064 [in Russian]

[88] L. P. Gor'kov, *Sov. Phys. JETP* 9 (1959) 1364

[89] L. P. Gor'kov, *Sov. Phys. JETP* 10 (1960) 998

[90] e.g. see in: L. D. Landau and E. M. Lifshitz, Course of Theoretical Physics: vol. 9, *Statistical Physics*, part 1, Butterworth-Heinemann, Oxford (1998)

[91] A. A. Abrikosov, *Sov. Phys. JETP* 5 (1957) 1174

magnetic properties of superconductors with a GL $\kappa > 1/\sqrt{2}$ were very unusual. Abrikosov dubbed such superconductors type II, to distinguish them from the known superconductors at that time forming the type I group. Like type I, type II superconductors show the Meissner effect, but only in a weak enough applied field. With field increase magnetic flux gradually penetrates into the bulk, first as a lattice of "vortices"—non-linear formations carrying a trapped "quantum of flux", $\Phi_0 = ch/2e$ (note the double charge in the denominator, $2e$, as follows from the microscopic theory[88,89]). Each individual vortex line possesses a "normal" core of order ξ in size. The GL order parameter $\Psi(r)$ inside the core decreases to zero at the center line. The magnetic field of a single vortex extends outward from the center to distances of order λ.

There are the two critical fields in type II superconductors, $H_{c1} < H_{c2}$. At H_{c1} the magnetic flux for the first time penetrates into the bulk as a rarified periodic vortex lattice; the diamagnetic moment decreases: $|M| < H/4\pi$. With further field increase, the lattice of vortices becomes denser and the vortices start to overlap, initially by the "clouds" of their trapped magnetic fields, then by the cores, so that the vortices gradually lose their individuality. At last, at H_{c2} the superconducting order parameter entirely disappears. Within the interval, $H_{c1} < H < H_{c2}$ the superconductor is in the so-called "mixed" state[91]. An order of magnitude estimate[91] gives for the fields $H_{c1} \sim H_c/\kappa$; $H_{c2} \sim \kappa H_c$. Consequently, for $\kappa \gg 1$, the field range for realization of the mixed phase may be very large.

Historically, two different critical fields were first seen by the group of Lev Shubnikov (Kharkov, 1936-37) in their experiments on single-phased samples of *Pb-Tl* and *Pb-In* alloys[92]. In the literature the mixed state is therefore also called the "Shubnikov phase". Type II materials often are superconducting alloys. However, many of the recently discovered *pure* superconducting materials also belong to the type II superconductors.

Eve of Microscopic Theory

As was repeatedly noted, the first superconductors were elemental metals or diluted alloys with the temperature of transition in the range 1–10 K. The properties of the recently discovered superconductors may in many respects differ from the properties of superconductors investigated several decades ago. However, in 1957 the BCS theory had to explain SC as it manifested itself in these "old" (or "conventional") superconductors. Among the key issues was the mechanism responsible for the formation of a gap in the excitation spectrum revealed below T_c, for instance, in the activation dependence of Eq. (2.49) for the low temperature specific heat data. The *isotope effect* was discovered in 1950. Reynolds et al.[93] and Maxwell[94] found that the transition temperature, T_c, of a metal depends on M, the mass of its isotopes: $T_c \propto (M)^{-1/2}$. The observation of the isotope effect gave the first clear clue to the SC mechanism: the dependence of the transition temperature on the isotopic mass of the element signifies the involvement of the lattice.

Electrons in a metal interact via screened repulsive Coulomb forces and scatter on the quantized vibrations of the lattice: the phonons. It turned out that, besides the direct Coulomb interaction, the virtual exchange by phonons between electrons generates an additional electron–electron (e–e) interaction. Fröhlich[95] in 1952 and later Bardeen and Pines[96] (also taking the Coulomb screening into account) derived the Hamiltonian for such phonon-mediated interaction, which we write in the

[92] L. V. Shubnikov, V. I. Kotkevich, Yu. D. Shepelev, J. N. Ryabinin, *Zh. Exp. Teor. Fiz.* 7 (1937) 221 [in Russian]

[93] C. A. Reynolds, B. Serin, W. H. Wright, L. B. Nesbitt, *Phys. Rev.* 78 (1950) 487

[94] E. Maxwell, *Phys. Rev.* 78 (1950) 477

[95] H. Fröhlich, *Proc. Roy. Soc.* A215 (1952) 291

[96] J. Bardeen, D. Pines, *Phys. Rev.* 99 (1955) 1140

following form:

$$\hat{H}_{\text{e-ph}} = \frac{1}{2} \sum_{k,k',q;\sigma,\sigma'} \frac{g^2(q)\omega_0^2(q)}{(\epsilon(k)-\epsilon(k-q))^2 - \omega_0^2(q)} \hat{c}_{k,\sigma}^+ \hat{c}_{k',\sigma'}^+ \hat{c}_{k'+q,\sigma'} \hat{c}_{k-q,\sigma}. \tag{2.51}$$

$(\hat{c}_{k,\sigma}^+, \hat{c}_{k',\sigma'}$ are, correspondingly, the creation and the annihilation operators for electrons in the second quantization; the notations in (2.51) slightly differ from those in Refs. 95, 96.) The remarkable feature of this Hamiltonian is the negative sign of the interaction (2.51) for low energy electrons: $|\epsilon(k)|, |\epsilon(k-q)| < \omega_0(q)$, thus providing a mechanism for the e–e attraction. Note that in (2.51) the electrons typically exchange momenta $|q| \sim p_F$: the latter is of atomic order. Therefore, *space-wise*, the effective radius of the phonon-mediated interaction has atomic order as well. To save on notations, rewrite (2.51) as:

$$\hat{H}_{\text{int}} = \frac{1}{2} \sum_{k,k',q';\sigma,\sigma'} V_q \hat{c}_{k,\sigma}^+ \hat{c}_{k',\sigma'}^+ \hat{c}_{k'+q,\sigma'} \hat{c}_{k-q,\sigma}. \tag{2.52}$$

Notice the *retired* character of the interaction, $\hat{H}_{\text{e-ph}}$, Eq. (2.51): atoms of the lattice move slowly as compared to electrons ($\omega_0 \ll E_F$). The group velocity of phonons, $d\omega_0(k)/dk \sim s$, is significantly lower than the electronic Fermi velocity, v_F.

Cooper Phenomenon

The analogy to superfluidity suggests that SC perhaps should have something to do with Bose statistics. As noted above, the superfluid flow of electrons cannot set in as long as their spectrum preserves the metallic form $\epsilon(\mathbf{p}) = v_F(p - p_F)$. One can speculate that an attractive interaction, such as discussed above, may facilitate the formation of bound states of electrons. The existence of a bound state would be consistent with the experiments[86] that disclose an energy gap in the low temperature specific heat data. Here lies the fundamental difficulty, though. Indeed, while SC is a widespread phenomenon over the Mendeleev Chart (Periodic System), the transition temperatures of the elemental metals, $T_c \sim 1 - 10$ K, are low in comparison with the typical value $E_F \sim 1$ eV, for the Fermi energy, or $\omega_D \sim 200 - 400$ K, for the phonon frequencies. One could speculate that for one reason or another, the phonon-mediated e–e attraction should be weak. However, such a guess would contradict the well-known theorem of quantum mechanics that in three dimensions two particles may form a bound state only if the attractive interactions between them would be *strong enough*.

The latter difficulty was surmounted in 1956 by L. Cooper[68] who demonstrated that for two electrons near the Fermi level interacting *on the background of the Fermi sea* the quantum mechanical problem reduces to a one-dimensional one. Therefore such electrons *will* form a bound state at any *arbitrary weak attraction*. Thereby, at low enough temperatures and in the presence of attractive forces between electrons the Fermi liquid becomes *absolutely unstable* with respect to the formation of bound pairs.

2.3.4 Microscopic Theory

The BCS theory[69,70] was built on the idea of the Cooper instability. In the emerging ground state the electronic and hole-like states from above and below the Fermi level are intermingled. Excitations in the superconducting phase acquire a gap in the spectrum by an order of magnitude equal to $|\Delta E|$ in Eq. (2.56). (Note in passing that the exponential dependence in (2.56) on the interaction strength would agree with the observed smallness of T_c even if real interactions were not so small: $\tilde{g} = VN(E_F) \leq 1$.)

Box 3. Write down the Schrödinger equation in momentum space for the wave function of two electrons with *opposite* momentum, $\phi(\mathbf{p}, -\mathbf{p}) \equiv \phi(\mathbf{p})$:

$$\{p^2/m - E\}\phi(\mathbf{p}) + \sum_{p'} \tilde{V}_{p-p'}\phi(\mathbf{p}') = 0.$$

With the notation $\chi(\mathbf{p}) = \{p^2/m - E\}\phi(\mathbf{p})$ the equation acquires the integral form

$$\chi(\mathbf{p}) = -\sum_{p'} \tilde{V}_{p-p'}\chi(\mathbf{p}')\{2\epsilon(p') - \Delta E\}^{-1}. \tag{2.53}$$

(The energy, $\epsilon(p) = v_F(p - p_F)$, counts from the chemical potential, $\mu = p_F^2/2m$; $\Delta E = E - p_F^2/m$). In Eq. (2.53) $\tilde{V}_{p-p'}$ stands for $V_{p-p'}$ from $\hat{H}_{\text{int}}$ modified to account for the changes in the matrix elements in the scattering processes that involve as intermediate states the *occupied* states below the Fermi sea, $\epsilon(p') < 0$. Cooper[68] merely confined himself by integrating in (2.53) over energy $\epsilon(p') > 0$. A more accurate analysis[97] leads to the equation:

$$\chi(\mathbf{p}) = -\sum_{p'} V_{p-p'} \, sign\epsilon(p')\{2\epsilon(p') - \Delta E\}^{-1}. \tag{2.54}$$

The interaction in Eq. (2.54) can be simplified even further by assuming that $V_{p-p'}$ is non-zero only inside a narrow energy strip near the Fermi surface:

$$V_{p-p'} = -V \; ; \text{at} \; |\epsilon(p)|, |\epsilon(p')| < \omega_D. \tag{2.55}$$

Choosing $\chi(\mathbf{p}) \cong \chi(p_F) = $ constant (i.e., choosing for s-wave pairing!), it is now easy to perform the one-dimensional integration over $\epsilon(p') = v_F(p' - p_F)$ in (2.54). Omitting the trivial solution $\chi(\mathbf{p}) = 0$, one obtains from (2.54) the relation: $1 = VN(E_F)\ln(\omega_D/|\Delta E|)$, with $N(E_F)$ the density of states at E_F, from which it follows:

$$|\Delta E| = \omega_D \exp(-1/VN(E_F)). \tag{2.56}$$

So, the solution for the bound state does indeed exist at any arbitrary weak attractive interaction:

$$\tilde{g} = VN(E_F) \ll 1. \tag{2.57}$$

The BCS Method

In Refs.[69,70] the interaction Hamiltonian (2.52) was reduced to the form of Eq. (2.55). In the spirit of Ref.[96], it was assumed that the screened Coulomb repulsion would not change the general attractive character of the interaction; its contribution was merely included in the same constant, $-V$. Technically, BCS used the variation method. Let $\hat{b}_k^* = \hat{c}_{k\uparrow}^* \hat{c}_{-k\downarrow}^*$ be the product of the two operators for the creation of a pair of electrons with opposite momenta and spins. In the new ground state the trial function

$$\Phi = \prod_k [(1 - h_k)^{1/2} + h_k^{1/2}\hat{b}_k^*]\Phi_0 \tag{2.58}$$

---

[97] A. A. Abrikosov, L. P. Gor'kov, I. E. Dzyaloshinskii, *Methods of the Quantum Field Theory in Statistical Physics*, Dover, New York (1963)

is a superposition of states formed of pairs created as the result of the Cooper instability; h_k are variation parameters. Φ_0 in (2.58) is the wave function of the normal metal "vacuum". (The energy spectrum of excitations is calculated on the manifold of trial functions

$$\Phi = \hat{c}_{p,\sigma}^* \prod_{kk \neq p} [(1-h_k)^{1/2} + h_k^{1/2} \hat{b}_k^*] \Phi_0.$$

(Since the BCS paper is well known, we will not dwell on lengthy calculations[70].) Eventually, BCS constructed a *qps* spectrum with an energy gap; built up the thermodynamics of the superconducting phase at all temperatures below T_c; and found the matrix elements for many essential physical processes, thus making it possible to calculate thermal, transport, and electromagnetic properties in the superconducting state. Experimental laboratories all over the world immediately set out to verify the theoretical predictions. (One of the first triumphs of the theory was the explanation for the famous Hebel-Slichter peak in the NMR near T_c[98].) The second spatial scale ξ_0 naturally emerges from the BCS physics as the "size" of the Cooper pair and is given by $\xi_0 = \hbar v_F / \pi \Delta_0$ (here Δ_0 is the energy gap in the spectrum at zero temperature). Regarding the Meissner effect and to ease the comparison with Pippard's result[81], the expression for the current should be presented in the form[70]:

$$\mathbf{j(r)} = -\frac{3}{4\pi c \Lambda \xi_0} \int \frac{\mathbf{R}[\mathbf{R} \times \mathbf{A}(\mathbf{r}')] J(R,T) d\mathbf{r}'}{R^4}. \tag{2.59}$$

The kernel $J(R,T)$ turned out to be numerically very close to its form in[81] $J(R,T) \Rightarrow \exp(-R/\xi_0)$ with ($\mathbf{R} = \mathbf{r} - \mathbf{r}'$, $\Lambda = m/ne^2$). If $\mathbf{A}(\mathbf{r}')$ varies slowly on distances $\sim \xi_0$ ($\xi_0 \ll \lambda_L$), the current will acquire the local London form of Eq. (2.47). The opposite case is known in the literature as *the Pippard limit*. It applies to most elemental superconductors. (As shown below, the question whether a superconductor belongs to type I or type II, is closely related to the character of its electrodynamics in weak fields.) Notwithstanding all these impressive achievements, the BCS variation method was rather awkward. The attention of theorists also concentrated on expression (2.59) for the Meissner current which is linear in the vector potential, $A(r)$, raising thereby concerns about the gauge invariance of the new theory. In addition, BCS considered only clean superconductors. Extending the variational method[70] to spatially inhomogeneous problems and to alloys, as well as the generalization of the method beyond the weak coupling phonon model seemed to be difficult, if not impossible. Last but not least: any second-order phase transition is associated with the breaking *of a certain symmetry in the system* (Landau[90]). There was no discussion at all of the superconducting order parameter in Refs.[68-70] except a rather naïve comment[69] in the spirit of the two-liquid theory: *"An order parameter, which might be taken as the fraction of electrons above the Fermi surface in virtual pair states, comes in a natural way"*.

Remarks on Science in the U.S.S.R.

Many of the key results of the later modern microscopic theory are based on ideas initiated by Russian theorists. Actually, two new powerful methods were developed even before the main BCS paper[70] appeared in the December issue of *Physical Review* (1957). Those two were Bogolyubov's method of the canonical transformation[99a,b], and the Quantum Field Theory (QFT) approach to SC developed by Gor'kov[100].

The exact sciences in Russia were traditionally top level. In the field of low temperature physics, for instance, come to mind names of such experimentalists as P. L. Kapitza, A. I. Shal'nikov, M.

---

[98] L. Hebel, C.P. Slichter, *Phys. Rev.*107 (1957) 901

[99] a) N. N. Bogolyubov, *Sov. Phys. JETP* 7 (1958) 41, 51; b) N. N. Bogolyubov, V. V. Tolmachev, D. V. Shirkov, *A New Method in the Theory of Superconductivity* (1958) (English translation: Consulting Bureau, Inc, New York, 1959).

[100] L. P. Gor'kov, *Sov. Phys. JETP* 7 (1958) 505

Khaikin, Yu. V. Sharvin (Moscow); L. V. Shubnikov and his group (Kharkov); and of the theorists: L. D. Landau, N. N. Bogolyubov, V. L. Ginzburg, the two brothers E. M. and I. M. Lifshitz, to name just a few. After the second World War II (WWII), the scientific community in the U.S.S.R found itself in a state of isolation from the West. Besides difficulties with buying any modern equipment, the shortage of foreign currency led to cuts in subscriptions to many Western scientific journals. The *Physical Review* and other journals used to arrive irregularly and with a considerable delay. *Sov. Phys. JETP*, the English translation of the Russian *ZhETF*, was initiated in 1955. Typically, after submission of the original manuscript to the Russian editors, it would take about 1 year for the English version to appear in *Sov. Phys. JETP*. In the absence of all personal contacts and with no facilities for publishing preprints, this often led to loss of priority. Restrictions on personal contacts were imposed by politicians on both sides of the notorious "Iron Curtain". In the context of the BCS story, it is worthwhile to quote John Bardeen[101]: "Although the preprint of our paper, submitted in June [1957], had wide circulation in the West, we were not allowed to send copies behind the Iron Curtain". Meanwhile, scientific life in the country was very intense. The Russian traditions of scientific schools, the high level of public education in the U.S.S.R, and the influx of fresh blood after WWII, young people being attracted by the high prestige of science in the country made it possible to preserve outstanding scientific standards, above all, in physics and mathematics. Speaking of the theory of superfluidity and SC more specifically, there were two leading theory groups with traditional interests in the area: the Landau School and the group led by Bogolyubov. At the time, N. N. Bogolyubov was the head of the Theory Department at the Moscow Steklov Mathematical Institute and he also held posts in Ukraine. In 1956 he became one of the organizers of the Joint Institute for Nuclear Research (JINR) in Dubna and Director of the Theory Laboratory there. JINR was designed as the international institution open to scientists from the countries belonging to the Soviet bloc. Restrictions on contacts with the West were more relaxed at JINR. The areas of interest of N. N. Bogolyubov and his collaborators, besides statistical physics, included high energy physics and non-linear mechanics. Characteristic of his scientific style was the profound interest in the mathematical aspects of a problem. From 1939 Lev Landau was the head of a small theory group at the Kapitza Institute of the Soviet Academy in Moscow. No one from the Landau group, let alone Landau himself, was permitted to go abroad in those years. Landau was one of the last scientists equally strong in practically every branch of theoretical physics of the day. He expected the same from his disciples. Anyone who wanted to have scientific contact with Landau had to pass the famous exams of the Landau Theoretical Minimum. Broad interests and a good education were obligatory. Diagrammatic methods in quantum electrodynamics were elaborated at the beginning of the 1950s by R. Feynman, F. Dyson, and G. C. Wick. The field continued to be active even in the mid-1950s, and many Russians worked in that area. (For instance, the subject of the author's Ph.D. thesis (1956) was quantum electrodynamics of scalar particles). Extending QFT methods to the physics of metals (at *zero temperature*), seemed to a large extent straightforward, with the occupied Fermi sea accepting the role of the "vacuum". In 1957–1958 the diagrammatic approach[102] to solving many-body problems of statistical physics in Fermi systems ($T = 0$) was already popular.

Bogolyubov's Canonical Transformation

Bogolyubov first applied[99a] his method to the Fröhlich Hamiltonian for the electron–phonon (*e-ph*) interaction (at $T = 0$):

$$\hat{H}_{\mathrm{Fr}} = \sum_{k,s} E(k)\hat{a}_{k,s}^{+}\hat{a}_{k,s} + \sum_{q} \omega(q)\hat{b}_{q}^{+}\hat{b}_{q} + \hat{H}' \quad ; \quad \hat{H}' = \sum_{k,q=k'-k,s} g\left\{\frac{\omega(q)}{2V}\right\}^{1/2} \hat{a}_{k,s}^{+}\hat{a}_{k',s}\hat{b}_{q}^{+} + h.c.$$

[101] J. Bardeen, Talk at LT XV, *J. de Phys. Colloque* C6, vol. 35 (1978) C6-1368

[102] V. M. Galitski, A. B. Migdal, *Sov. Phys. JETP* 7 (1958) 96

(Note the units, in which Plank's constant $\hbar \equiv 1$.) Actually, in Ref. 99a the phonons were integrated out by going over to the second-order approximation in **g**; thereby Bogolyubov worked with the same interaction Hamiltonian (2.55) as BCS[70]. The Bogolyubov transformation expresses the normal state operators, $\hat{a}^+_{k,s}$, $\hat{a}_{k,s}$, through a set of new operators, $\hat{\alpha}_{k,i}$, $\hat{\alpha}^+_{k,i}$, for qps in the superconducting state:

$$\hat{a}_{k,1/2} = u_k \hat{\alpha}_{k,0} + v_k \hat{\alpha}^+_{k,1}$$
$$\hat{a}_{-k,-1/2} = u_k \hat{\alpha}_{k,1} - v_k \hat{\alpha}^+_{k,0}. \tag{2.60}$$

The coefficients, u_k, v_k (with the unitary condition, $u_k^2 + v_k^2 = 1$) should then be determined with the help of what Bogolyubov called the principle of compensation of "dangerous diagrams", formulated as follows. At the transition to new operators of Eq. (2.60), there appear in the Hamiltonian, among others, non-diagonal terms of the following specific form: $(\hat{\alpha}^+_{k,1}\hat{\alpha}^+_{k,0} + h.c.)$. Such combinations emerge both from the free particle Hamiltonian and from the *e–ph* interaction. These terms have matrix elements to the intermediate states with (I quote): *"the energy denominator $1/2E(k')$ ...dangerous for integrating...; Thus, in the choice of the canonical transformation, it must be kept in mind that it is necessary to guarantee the mutual compensation of the diagrams which lead to virtual creation from the vacuum of pairs of particles with opposite moments and spins"*[99a]. The language was far from being transparent. Bogolyubov of course alluded to the logarithmic divergences in the Cooper channel. In practice, the Bogolyubov recipe for forbidding the "creation of 'dangerous' pairs" was reduced to the stipulation that in the transformed Hamiltonian the sum of all terms proportional to the product of the new operators $\hat{\alpha}^+_{k,1}\hat{\alpha}^+_{k,0}$ must be equal to zero. (At $T = 0$. For the generalization to finite temperatures, see Ref. 99b.) In whole, the derivation[99a] of the new excitation spectrum was simpler than in Ref. 70. In addition, once the spectrum is found, the matrix elements in the superconducting state for various physical processes can be automatically expressed in terms of Bogolyubov qps after the straightforward substitution of the operators (2.60) into the corresponding Hamiltonian for a given process. The Bogolyubov method shared some shortcomings with the BCS method. In particular, the order parameter driving the transition into the superconducting state remained unidentified.

Role of Coulomb Repulsion The effect of the repulsive Coulomb interaction on the value of T_c was investigated by Tyablikov and Tolmachev[103]. The retardation in the *e–ph* interactions, mentioned above, means the slowness of the ionic motion: the lattice distortions diffuse typically over a time $\sim \hbar/\omega_0$. At the exchange of a phonon the two electrons move fast and therefore they stay most of the time far from each other at a distance of order $\sim \hbar v_F/\omega_0$, which considerably exceeds the atomic Thomas–Fermi radius. This picture explains qualitatively why the phonon-mediated attraction seems to dominate over the screened Coulomb repulsion along the whole Mendeleev Chart. In Ref. 103 the problem was modelled using the simple "two energy scales" generalization of Eq. (2.55):

$$V_{p-p'} = -V + V_{Coul} \quad ;\text{at } |\epsilon(p)|, |\epsilon(p')| < \omega_D$$
$$V_{p-p'} = V_{Coul} \quad ;\text{at } \omega_D < |\epsilon(p)|, |\epsilon(p')| < E_F. \tag{2.61}$$

It was found that the renormalized constant, V_{eff} that determines the "actual" value of T_c in Eq. (2.56) changes to:

$$V_{eff} = -V + \frac{V_{Coul}}{1 + V_{Coul}N(E_F)\ln(E_F/\omega_D)}. \tag{2.62}$$

To the extent that $\ln(E_F/\omega_D)$ can be large, Eq. (2.62) points to a considerable reduction of the Coulomb repulsion impact on T_c. In the Russian folklore Eq. (2.62) was known as the "Tolmachev

---

[103] S. V. Tyablikov, V. V. Tolmachev, *Sov. Phys. JETP* **7** (1958) 867

logarithm". In 1962 Morel and Anderson[104] found basically the same result (2.62) for V_{eff}. The dependence of V_{eff} on ω_D in Eq. (2.62) changes the dependence of T_c on the atomic mass, thus changing the value of the *isotope effect*[104].

First Landau Seminar on Superconductivity

The 1956 paper by Cooper went unnoticed by Landau and his group at the Kapitza Institute. Curiously, the first direct information about such a pivotal breakthrough in studying SC actually came from N. N.Bogolyubov. In the fall of 1957 rumors spread in Moscow that Bogolyubov had created *the* theory of superconductivity. Landau invited Bogolyubov to give a talk at his seminar at the Kapitza Institute. The Landau Seminar was an important meeting place for Russian theorists. Bogolyubov started the presentation with the exposition of his principle of compensation of "dangerous diagrams". He was immediately interrupted by Landau who first wanted to hear about the underlying physics. The controversy took a rather long time. After the seminar break Bogolyubov reproduced in several lines on the blackboard the calculations of the Cooper paper[68]. The significance of the concept of the Fermi sea instability in the presence of any weak attractive e–e attraction was immediately recognized and highly appreciated by Landau. At the time I was a junior scientist in the Landau theory group. While listening to the Bogolyubov's talk, it crossed my mind that the pairing instability should create a macroscopic number of pairs, thus leading to the appearance of a kind of bosonic degrees of freedom in the system and I set myself to this question later that very day.

QFT Methods at Zero Temperature

The advantage of the diagrammatic methods lies in the automatism of calculations. This review of course is not intended to describe the quantum field theoretical (QFT) methods in detail apart from a few definitions. The main object studied by the QFT method is the so-called Green function. For the Fermi systems it is defined as $G_{\alpha\beta}(x,x') = -i < \hat{T}_t(\hat{\psi}_\alpha(x)\hat{\psi}_\beta^+(x')) >$. ($T = 0$; the average, $< \ldots >$ is taken over the exact ground state.)

The superconducting state cannot be described by the function $G_{\alpha\beta}(x,x')$ alone. The Cooper instability stabilizes itself by creating a macroscopic number of pairs in the new ground state. To single out these *bosonic* degrees of freedom, consider the average of the product of four $\hat{\psi}$-operators and decouple them in the following way[100]:

$$\langle T_t(\hat{\psi}_\alpha(x_1)\hat{\psi}_\beta(x_2)\hat{\psi}_\gamma^+(x_3)\hat{\psi}_\delta^+(x_4))\rangle = -\langle T_t(\hat{\psi}_\alpha(x_1)\hat{\psi}_\gamma^+(x_3))\rangle\langle T_t(\hat{\psi}_\beta(x_2)\hat{\psi}_\delta^+(x_4))\rangle +$$
$$\langle T_t(\hat{\psi}_\alpha(x_1)\hat{\psi}_\delta^+(x_4))\rangle\langle T_t(\hat{\psi}_\beta(x_2)\hat{\psi}_\gamma^+(x_4))\rangle + \qquad (2.63)$$
$$\langle N|T_t(\hat{\psi}_\alpha(x_1)\hat{\psi}_\beta(x_2))|N+2\rangle\langle N+2|T_t(\hat{\psi}_\gamma^+(x_3)\hat{\psi}_\delta^+(x_4))|N\rangle.$$

(N in (2.63) is the total number of particles.) The first two terms are present even in the normal state and could be omitted, at least in the weak coupling limit. The third term is proportional to the density of pairs and is absent in the normal state. The two *anomalous functions*, $F_{\alpha\beta}(x-x')$ and $F_{\alpha\beta}^+(x-x')$, in Eq. (2.63)

$$\langle N|T_t(\hat{\psi}_\alpha(x)\hat{\psi}_\beta(x'))|N+2\rangle = \exp(-2i\mu t)F_{\alpha\beta}(x-x')$$
$$\langle N+2|T_t(\hat{\psi}_\alpha^+(x))\hat{\psi}_\beta^+(x')|N\rangle = \exp(2i\mu t)F_{\alpha\beta}^+(x-x'), \qquad (2.63')$$

define the new symmetry in the superconducting state. (In the Josephson pre-factors μ stands for the chemical potential.) One may notice the similarity to the "c-numbers" $< N-1|\hat{a}_0|N >$ and $<$

---

[104] P. Morel, and P. W. Anderson, *Phys. Rev.* 125 (1962) 1263

Box 4. The operators $\hat{\psi}_\alpha(x)$ and $\hat{\psi}_\alpha^+(x)$ are the *exact* fermionic field operators related to the annihilation and creation operators in the second quantization formalism, $\hat{\psi}_\alpha(\mathbf{r}) = V^{-1/2} \sum_{k,\sigma} \hat{c}_{k\sigma} s_{\sigma\alpha} \exp(i\mathbf{k}\cdot\mathbf{r})$, $\hat{\psi}_\alpha^+(\mathbf{r}) = V^{-1/2} \sum_{k,\sigma} \hat{c}_{k\sigma} s_{\sigma\alpha} \exp(-i\mathbf{k}\cdot\mathbf{r})$, as:

$$\hat{\psi}_\alpha(x) = e^{-i\hat{H}t}\hat{\psi}_\alpha(\mathbf{r})e^{i\hat{H}t} \; ; \; \hat{\psi}_\alpha^+(x) = e^{-i\hat{H}t}\hat{\psi}_\alpha^+(\mathbf{r})e^{i\hat{H}t}.$$

Here $x = (\mathbf{r}, t)$, $\hat{H}$ is the total Hamiltonian of the system, and $\hat{T}_t$ is the well-known Feynman time-ordering operator. A variety of important physical quantities can be expressed with the help of the Green function: the electrical currents, the electron density, the magnetization, etc. For a homogeneous system the Green function may be presented in the Fourier representation: $G_{\alpha\beta}(x - x') = \int G(p)\exp(ipr)d^4p/(2\pi)^4$; $r = x - x'$, $pr = \mathbf{p}\cdot\mathbf{r} - \omega t$. In the Landau Fermi liquid theory the Green function, $G(p)$, contains the singular term $G(p) \approx z(\omega - \epsilon(\mathbf{p}) + i\delta\,\mathrm{sign}\omega)^{-1}$. The pole, $\omega = \epsilon(\mathbf{p})$, determines the spectrum of the elementary excitation in the system. The imaginary part $i\delta\,\mathrm{sign}\omega$ stands for the attenuation, which is small near the Fermi surface. One sees that the Green function $G_{\alpha\beta}(x - x')$ contains the essential information about the Fermi liquid.

$N + 1|\hat{a}_0^+|N>$ introduced by Bogolyubov in the problem of the weakly interacting Bose gas at zero temperature[105]. The closed system of equations for the normal and the anomalous Green functions at $T = 0$ was derived in coordinate space[100]. Thereby, the gauge-invariance of the new equations was made self-evident. Note that to the extent that all relevant energy scales pertinent to the problem are small, $\epsilon(p)$, $T_c \ll \omega_D$, the Hamiltonian Eq. (2.52), (2.55) in *coordinate* space corresponds to the local interaction:

$$\hat{H}_{int} = -\frac{1}{2}V \int \hat{\psi}_\alpha^+(r)(\hat{\psi}_\beta^+(r)\hat{\psi}_\beta(r))\hat{\psi}_\alpha(r)d^3r \tag{2.64}$$

The anomalous functions are anti-symmetric in the spin indices: $F_{\alpha\beta}(x,x') = -\sigma_{\alpha\beta}^y F(x,x')$, as expected in the case of "s-wave" (singlet) pairing, which is the only possibility with the local BCS Hamiltonian Eq. (2.52), (2.64). From the main results[100], we here mention only that the gap, $|\Delta|$, in the energy spectrum, $\epsilon(p) = \pm\sqrt{v_F^2(p - p_F)^2 + |\Delta|^2}$ is equal to the absolute value of *the order parameter* defined as:

$$\Delta = VF(x = x'). \tag{2.65}$$

In turn, the function $F(x,x')$ can be expressed through Δ from the set of equations for $G(x,x')$ and $F(x,x')$, $F^+(x,x')$. The order parameter, Eq. (2.65), is thereby determined self-consistently, without invoking either Bogolyubov's "compensation principle," or the use of the BCS probe functions. (For the thermodynamic version of the Gor'kov equations see below.)

Extension of QFT Approach of Finite Temperatures

The success of the QFT method[100] at zero temperature intensified the search for an analogous technique at finite temperatures. An attempt to develop a QFT scheme workable in quantum statistics at $T \neq 0$ was already made in 1955 by Matsubara[106]. Broadly speaking, Matsubara studied the *perturbation expansion* of the partition function in the *Grand canonical ensemble*. Instead of the familiar field operators $\hat{\psi}_\alpha(x)$ and $\hat{\psi}_\alpha^+(x)$ (see Box 4), Matsubara defined the new operators $\hat{\bar{\psi}}_\alpha(x) = e^{-\hat{H}\tau}\hat{\psi}_\alpha(\mathbf{r})e^{\hat{H}\tau}$ and $\hat{\bar{\psi}}_\alpha^+(x) = e^{-\hat{H}\tau}\hat{\psi}_\alpha^+(\mathbf{r})e^{\hat{H}\tau}$, where τ is a certain *"imaginary time"*, vary-

[105] N. N. Bogolyubov, *J. Phys. USSR* 11 (1947) 23–32
[106] T. Matsubara, *Progr. Theor. Phys.* 14 (1955) 351

ing inside the interval $0 < \tau < 1/T$ (T is the temperature). The formal correspondence with the methods at $T = 0$ becomes complete after introducing the new, thermodynamic Green function as $\overline{G}_{\alpha\beta}(x, x') = - \ll T(\hat{\overline{\psi}}_{\alpha}(x)\hat{\overline{\psi}}_{\beta}^{+}(x')) \gg$ (here $\ll \ldots \gg$ means average over the *Grand canonical ensemble*). The fundamental drawback of Reference 106 was that, since the variation of τ was restricted to the *finite* interval $\{0, 1/T\}$, the expansion of $\overline{G}_{\alpha\beta}(x, x')$ over the "time" variable in the Fourier *integral* was not possible anymore. The way out of that difficulty was found by Abrikosov et al.[107] by expanding $\overline{G}_{\alpha\beta}(x_1, x_2)$ over $\tau = \tau_1 - \tau_2$ in a *Fourier series*:

$$\overline{G}(\tau - \tau'; r) = T \sum_{n} \overline{G}(\omega_n; r)\exp(-i\omega_n(\tau - \tau')).$$

Here $\omega_n = \pi T n$; n is even for a Bose system and odd in the Fermi case. (The function $\overline{G}(\omega_n; r)$ admits the analytic continuation from the "imaginary" axis, $z = i\omega_n$ onto the real frequency axis, $z \equiv \omega$; this allows the application of the method to non-stationary processes.) Apart from minor changes in the numeric coefficients, the diagrammatic rules of the thermodynamic technique are in a one-to-one correspondence with the rules for the zero temperature technique, while the only difference is that the integrations over the frequency variable must be replaced by summations over its discrete values. Before its publication, the method[107] was successfully applied to the theory of superconducting alloys[83].

Defects and Impurities

One of the most challenging problems faced by the new microscopic theory of SC was how to include superconducting alloys into the theoretical framework. This was achieved in 1958 by Abrikosov and Gor'kov in two publications[82,83]. In metals and semiconductors scattering of electrons on impurities plays an important role in the electronic transport and other kinetic processes. In the normal metallic phase scattering on defects is routinely treated by means of the Boltzmann kinetic equation. However, it turns out that in the superconducting state impurities may affect the equilibrium properties as well. To address this issue, Abrikosov and Gor'kov[82] first needed to develop what is now known as the "cross technique": the diagrammatic expansion of the Green functions over the strength of the impurity potential. For superconductors, the technique amounts to the generalization of the Boltzmann equation to calculations of the *thermodynamic* characteristics. Properties of alloys in the new method are studied with the help of Green functions averaged over the positions of the impurity atoms.

Gor'kov Equations at Finite Temperature

For completeness, we write down the pair of Gor'kov equations[88] in the thermodynamic technique[107]:

$$\{i\omega_n + \frac{1}{2m}(\nabla - i\frac{e}{c}\mathbf{A}(r))^2 - \mu\}\overline{G}(\omega_n; r, r') + \Delta(r)\overline{F}^{+}(\omega_n; r, r') = \delta(r - r')$$

$$\{i\omega_n - \frac{1}{2m}(\nabla + i\frac{e}{c}\mathbf{A}(r))^2 - \mu\}\overline{F}^{+}(\omega_n; r, r') + \Delta^{*}(r)\overline{G}(\omega_n; r, r') = 0. \tag{2.66}$$

The self-consistency condition is:

$$\Delta(r) = VT \sum_{n} \overline{F}(\omega_n; r, r) = V\overline{F}(x = x'). \tag{2.66'}$$

The gauge-invariance of (2.66)) is self-evident: upon the substitution of $\mathbf{A}(\mathbf{r}) \Rightarrow \mathbf{A}(\mathbf{r}) + \nabla\varphi$ the gap parameter transforms according to $\Delta(r) \Rightarrow \Delta(r)\exp[i(2e/\hbar c)\varphi(r)]$. As above, the order parameter (2.66') is defined self-consistently with the help of the whole set of Eqs. (2.66).

---

[107] A. A. Abrikosov, L. P. Gor'kov, I. E. Dzyaloshinskii, *Sov. Phys. JETP* 9 (1959) 636; E. S. Fradkin, ibid. (1959) 912

Energy Spectrum in External Fields

In 1959 Bogolyubov discussed whether it was possible to extend his principle of compensation of the "dangerous diagram" to the case of a superconductor in an external magnetic field[108]. Indeed, without external fields the term "dangerous diagrams"[99 a,b] was relevant to the creation of electronic pairs with *exactly opposite* directions of momenta and spin. That definition loses its meaning in the case of spatially varying fields. Bogolyubov provided the formal proof that the canonical transformation (2.60) can indeed be constructed even in the latter case but with coefficients $u_\lambda(r), v_\lambda(r)$ that will depend on the spatial coordinates. (The far from simple argumentation[108] was additionally complicated by using, instead of the Feynman Green functions, the hierarchy of the Bogolyubov–Born–Green–Yvon equations, notorious for their poor control of accuracy.) Actually, the equations for $u_\lambda(r), v_\lambda(r)$ can be obtained just in a few lines, if one remembers the very fact that below T_c the superconducting order parameter is non-zero *by definition*. In 1965 de Gennes simplified the Bogolyubov discussion[108] by substituting Gor'kov's parameters $\hat{\Delta}(r), \hat{\Delta}^*(r)$ directly into the interaction Hamiltonian (2.64):

$$\hat{H}_{\text{int}} = -\frac{1}{2}V\int \hat{\psi}_\alpha^+(r)(\hat{\psi}_\beta^+(r)\hat{\psi}_\beta(r))\hat{\psi}_\alpha(r)d^3r \Rightarrow \int d^3r\{\Delta(r)\hat{\psi}_\uparrow^+(r)\hat{\psi}_\downarrow^+ + \Delta(r)\hat{\psi}_\downarrow(r)\hat{\psi}_\uparrow(r)\}.$$

Together with the kinetic energy part the total Hamiltonian becomes then quadratic in the operators $\hat{\psi}, \hat{\psi}^+$ and may be diagonalized.

Box 5. To determine the spectrum of *qps* in a *clean* superconductor in static fields, one must consider the operator in the L.H.S. of Eq. (2.66):

$$\hat{L}(i\omega_n) = \begin{pmatrix} i\omega_n + \left(\frac{1}{2m}\right)(\nabla - i\frac{e}{c}\mathbf{A}(r))^2 + \mu & \Delta(r) \\ \Delta^*(r) & i\omega_n - \left(\frac{1}{2m}\right)(\nabla + i\frac{e}{c}\mathbf{A}(r))^2 - \mu \end{pmatrix}.$$

After the substitution, $i\omega_n \Rightarrow E$ the eigenvalues, E_λ together with the corresponding eigenfunctions could be determined by solving the system of equations:

$$\hat{L}(E_\lambda)\begin{bmatrix} u_\lambda(r) \\ v_\lambda(r) \end{bmatrix} = 0. \tag{2.67}$$

In turn, the matrix of the Green functions is expressible through $u_\lambda(r), v_\lambda(r)$ as

$$\hat{\bar{G}}(\omega_n; r, r') = \begin{pmatrix} \bar{G}(\omega_n; r, r') & \bar{F}(\omega_n; r, r') \\ \bar{F}^+(\omega_n; r, r') & -\bar{G}(-\omega_n; r, r') \end{pmatrix}$$

$$= \sum_\lambda \frac{1}{i\omega_n - E_\lambda}\begin{pmatrix} u_\lambda(r) \\ v_\lambda(r) \end{pmatrix} \otimes (u_\lambda^*(r'), v_\lambda^*(r')).$$

The equations (2.67) for a superconductor in a magnetic field are known in the literature as the Bogolyubov–de Gennes equations[109].

---

[108] N. N. Bogolyubov, *Uspekhi Fiz. Nauk.* 67 (1959) 549

[109] P. G. de Gennes, *Superconductivity of Metals and Alloys*, W.A. Benjamin, New York (1966)

Ginzburg–Landau Equations and Microscopic Theory

The GL theory comprises the essential part of the modern microscopic theory of SC. Indeed, solving the differential GL equations looks much simpler than dealing with the machinery of the full theory. The GL equations were derived[88] from the BCS theory at the beginning of 1959. As the *limiting case* of the *microscopic* theory, the GL theory is quantitatively correct only near T_c but the main results are qualitatively meaningful in the whole temperature interval below T_c. The derivation starts by expanding the anomalous function $\overline{F}(\omega_n; r, r')$ in Eq. (2.66') in powers of $\Delta(r)$, $\Delta^*(r)$ (and its gradients), and $\mathbf{A}(r)$: all of them are small near T_c. After a not-too-lengthy calculation[88], one obtains:

$$\left\{ \frac{1}{2m}(\nabla - i\frac{2e}{c}\mathbf{A}(r))^2 + \frac{1}{\lambda}\left[\frac{T_c - T}{T_c} - \frac{2}{n}|\Psi(r)|^2 \right] \right\} \Psi(r) = 0 \qquad (2.68)$$

$$\mathbf{j}(r) = -\frac{i(2e)}{2m}(\Psi^*\nabla\Psi - \Psi\nabla\Psi^*) - \frac{(2e)^2}{mc}\mathbf{A}|\Psi|^2. \qquad (2.69)$$

$\Psi(r)$ is proportional to the *wave function of the Cooper pair*: $\Psi(r) = \Delta(r)\sqrt{7\varsigma(3)n}/4\pi T_c$. ($\lambda = 7\varsigma(3)E_F/12(\pi T_c)^2$; n is the electronic density.) (The expression for the current (2.69) was obtained from the expansion of the normal Green function $\overline{G}(\omega_n; r, r')$ in powers of Δ and $\mathbf{A}(r)$.) Note the double charge of the Cooper pair, $2e$ in the Eq. (2.68) and (2.69). The GL parameter κ from Eq. (2.50) for a pure superconductor can be presented in the form[88]:

$$\kappa \approx 0.96\lambda_L/\xi_0. \qquad (2.70)$$

(Here $\lambda_L^{-2} = 4\pi e^2 n/mc^2$; the parameter, $\xi_0 = 0.18\hbar v_F/k_B T_c$ was defined in Ref. 70). It follows from (2.70) that the parameter κ not only classifies whether a superconductor belongs to type I or type II; its value also describes *the type of electrodynamics*, which may be either local ($\kappa \gg 1$, the London type) or strongly non-local ($\kappa \ll 1$, the Pippard limit). Since the ratio between λ_L and ξ_0 is not fixed by the microscopic theory, even a clean superconductor may be type II. One well-known example[88] is elemental *niobium*, Nb, for which κ is slightly above the border line, $1/\sqrt{2}$. A number of superconductors discovered more recently also belong to type II and among them, the so-called high T_c-cuprates and the group of intermetallic compounds known as *heavy fermions*. It is useful to single out the dependence of κ on the basic characteristics of a material. For an isotropic three-dimensional superconductor one finds $\kappa \propto T_c m^{3/2}/n^{5/6}$. In new superconductors that have a pronounced layered structure, the 2D-limit would give for $\kappa \propto T_c m^{3/2} n$. So, larger masses, m higher T_c and low densities, n favor type II superconductivity. Qualitatively, this agrees with the experimental trend. Among conventional materials, type II SC is usually realized in alloys. The GL equations for alloys were obtained by Gor'kov[89] for the full range of concentrations. Not to overcrowd the text, consider only the case of the so-called "dirty" alloys ($l \ll \xi_0$). It turns out that in this limit the dependence on T_c drops out and κ can be expressed through parameters of the normal phase only:

$$\kappa = 0.065 ec\gamma^{1/2}/\sigma k_B,$$

(γ and σ stand for the linear coefficient of the electronic specific heat and σ for conductivity, respectively.) Together with the GL and Abrikosov's phenomenology, the above results are known as the GLAG theory (after Ginzburg, Landau, Abrikosov, and Gor'kov).

Thermodynamics and Electrodynamics of Alloys

The independence of the transition temperature on alloying initially seemed to be in conflict with Pippard's concept of the "coherence length," because experimentally the latter varied with the concentration of the alloy[81] . The microscopic theory sorted out the difference. It turned out that the *thermodynamics* of conventional superconductors is indeed weakly affected by the presence of defects. This was proved for the first time in References 82 and 83. Abrikosov and Gor'kov calculated the Green functions averaged over the positions of the impurities (in the isotropic model) and found that in coordinate space the Green functions, including the anomalous function $F(\omega_n; \mathbf{r}, \mathbf{r}')$, in the presence of impurities are merely multiplied by an exponential factor

$$F(\omega_n; t - t', R) \Rightarrow F(\omega_n; t - t', R)\exp(-R/l), \qquad (2.71)$$

where $R = |\mathbf{r} - \mathbf{r}'|$ and l is the mean free path. From the definition of the order parameter (2.66') it directly follows that the gap in the spectrum and, hence all of the thermodynamics of an alloy, remains the same as in the pure material. (This assertion is known in the Western literature as the "Anderson Theorem"[84]). Note though, that this statement is only correct for s-wave pairing, while neglecting a possible anisotropy of the Fermi surface. The *electrodynamics* of superconductors, on the contrary, is extremely sensitive to alloying[82]. In particular, with alloying, the non-local Pippard electrodynamics of a pure elemental metal rapidly evolves into the local London electrodynamics of Eq. (2.47) (n_s becomes a function of the transport mean free path, l_{tr}). In the limit of so-called "dirty alloys", $l \ll \xi_0$, one has[82]:

$$\lambda_L = \left(\frac{c}{2\pi}\right) \sqrt{\frac{1}{\sigma\Delta(T)\tanh(\Delta(T)/2T)}} \qquad (2.72)$$

Pippard's "coherence length", hence, is the length that defines *the current–current correlations*. The most fundamental length scale is the BCS coherence length $\xi_0 = \hbar v_F/\pi\Delta_0$, the effective size of a Cooper pair at zero temperature.

Gapless Superconductors

The electronic spectrum with a gap was justly considered the triumphal achievement of BCS[70]. In particular, the finite gap was in accord with Landau's criterion (2.48). (Criterion (2.48) is applicable to ordinary alloys due to the gapped density of states (DOS), as well as to clean materials.) In 1960 Abrikosov and Gor'kov[110] showed that scattering of electrons on *paramagnetic* impurities (atoms or ions possessing a non-zero spin, **S**) destroys SC. The transition temperature $T_c(x)$ decreases with increasing concentration x of the impurities and becomes zero at a certain critical concentration, x_{cr}. As to the gap in the DOS, it decreases as well. In itself, that result was not too surprising: scattering on the spin-dependent potential of a paramagnetic impurity, $V(p - p') \Rightarrow V(p - p') + \tilde{V}(p - p') \times (\hat{\sigma} * \hat{\mathbf{S}})$ causes the misalignment of spins of the Cooper pairs. It came as a surprise, though, that there exists a narrow interval of concentrations below x_{cr} inside which the energy gap in the DOS *is exactly zero*, although the superconducting order parameter remains finite. In other words, SC may persist in spite of a *zero gap in the DOS*[110]. Thereby, it was shown for the first time that the capability of a superconductor to carry persistent currents is controlled by the non-zero superconducting *order parameter* Δ rather than the non-zero gap in the energy spectrum. Hence, the Landau criterion of Eq. (2.48) may be violated. Interactions with an external spin **S** break the time reversal invariance. The time reversal symmetry is also broken by the Pauli term, $\mu_B \sigma \cdot \mathbf{B}$ describing the interaction of the electronic spins with the external magnetic field[111]. The physical effects that *destroy* the Cooper pairing are called "pair-breaking mechanisms". The superconducting gap may also be absent in the case of the so-called "unconventional" superconductors.

Up to now we have discussed only the BCS "s-wave" singlet pairing. Consider, however, a more general, momentum-dependent interaction, $V_{k-k'}$ in Eqs. (2.52) and (2.53): it was already recognized in 1957 that, in principle, for the isotropic Fermi gas the Cooper pair also may form with

[110] A. A. Abrikosov, L. P. Gor'kov, *Sov. Phys. JETP* 12 (1961) 1243

[111] K. Maki and T. Tsuneto, *Progr. Theor. Phys.* 31 (1964) 945

non-zero angular momentum, $l \neq 0$, provided that the pairing were the strongest in such a channel. (A discussion of p-wave pairing ($l = 1$) as in superfluid ^{3}He lies beyond the scope of this chapter.) In *crystalline* materials the permitted symmetries of the superconducting order parameters depend on the symmetry of the lattice point group itself[112]. However, in certain cases the non-trivial "gap" parameter may possess zeros along the symmetry lines or at the nodal points. Such "gapless" energy spectrum should manifest itself in a T^3- (or T^2) -*power law* dependence of the specific heat at low temperatures. The so-called "d-wave" pairing was indeed observed experimentally in certain new superconductors with tetragonal symmetry of the lattice. Note that any unconventional pairing is very sensitive to the presence of ordinary defects.

Phonon Mechanism

Most of the exact results of the new theory do not go beyond the scope of the weak coupling scheme discussed above. The BCS-like theory was not generalized yet to the case of strong enough e–e interactions in real materials. One remarkable exception is the "strong coupling" version of the *phonon mechanism* for which the theory can be extended to include materials with T_c values of the order of typical phonon frequencies, $T_c \sim \omega_D$. One known example of a "strong coupling" phonon SC is lead, Pb. In lead with its $T_c = 7.2$K and a Debye temperature $\omega_D = 96$K, deviations from the weak coupling scheme become quite sizable. For a normal metal Migdal[113] was able to show that the problem of strong e–ph interaction can be solved *rigorously*. He analyzed corrections to the so-called "self energy part" ($T = 0$):

$$\sum(\omega; p) = \int D(\omega - \omega'; p - p')G(\omega'; p')\frac{d^4 p'}{(2\pi)^4} \qquad (2.73)$$

where $D(\omega; k) = g^2 \frac{\omega_0^2(k)}{\omega^2 - \omega_0^2(k)}$ is the phonon Green function multiplied by the square of the e–ph interaction constant. Technically, Migdal had proven that within the adiabatic approximation, $\omega_D \ll E_F$, all corrections to (2.72) can be neglected. Eliashberg[114] extended Migdal's arguments to the phonon-mediated SC. In addition to the normal self-energy part $\sum(\omega_n; p)$, Eliashberg introduced the *anomalous* self-energy parts:

$$\Delta(\omega_n; p) = T \sum_m \int D(\omega_n - \omega_m; p - p')F(\omega_m; p')\frac{d^3 p'}{(2\pi)^3} \qquad (2.74)$$

(Note the thermodynamic frequency axis, $\omega \rightarrow i\omega_n$, in Eq. (2.74).) The complete set of equations for the Green functions is obtained by substituting Eqs. (2.73) and (2.74) into the generalized Gor'kov Eqs. (2.66). McMillan and Rowel[115] devised a procedure of "inverting" the Eliashberg "gap" equations (2.74) that allows determining the product of the DOS for phonons and the coupling constant from the tunneling experiments. The analysis[115] confirmed the phonon mechanism of SC in lead and using the available neutron data the participating phonon modes could be identified. To the author calculating T_c of real materials does not look like a promising or even sensible task. The abundance of the parameters for any specific material that must be known experimentally before the calculation of T_c could start to look sufficiently enough realistic, makes the whole problem rather dubious. Of interest for the theory is the interpolation by Allen and Dynes[116] undertaken as an attempt to establish the upper limit imposed on the value of T_c in the case of the phonon mechanisms. Note in

[112] G. E. Volovik and L. P. Gor'kov, JETP Lett. 39 (1984) 674; Sov. Phys. JETP 61(1985) 843; P. W. Anderson, Phys. Rev. B30 (1984) 4000

[113] A. B. Migdal, *Sov. Phys. JETP* 7 (1958) 996

[114] G. M. Eliashberg, *Sov. Phys. JETP* 11 (1960) 696; 12 (1961) 1000

[115] W. L. McMillan and J. M. Rowel, *Phys. Rev. Lett.* 14 (1965) 108

[116] P. B. Allen and R. C. Dynes, *Phys. Rev.* B12 (1975) 905

passing that the above analysis of the "strong-coupling" phonon SC was obviously beyond reach of the BCS variational method or of Bogolyubov's canonical transformation method.

Eilenberger Equations

The making of the microscopic theory of SC was essentially accomplished sometime in 1961–62. This summary, however, would be incomplete without mentioning the subsequent important simplification of the Gor'kov equations that was formulated[117,118] in 1968 with the help of the so-called "quasi-classical" approximation. In this new method the Gor'kov equations are reduced to a set of non-linear equations for the "integrated" Green functions:

$$\int \hat{\bar{G}}(\omega_n; \mathbf{p}, \mathbf{p}+\mathbf{k}; \xi) \frac{d\xi}{i\pi} \Rightarrow \begin{pmatrix} g(\omega_n; \mathbf{p}, \mathbf{k}) & f(\omega_n; \mathbf{p}, \mathbf{k}) \\ f^{+}(\omega_n; \mathbf{p}, \mathbf{k}) & \bar{g}(-\omega_n; \mathbf{p}, \mathbf{k}) \end{pmatrix}. \tag{2.75}$$

(The vector $\mathbf{p}$ lies on the Fermi surface.) In the momentum representation the vector $\mathbf{k}$ stands for the dependence on the spatial coordinates. Being converted into coordinate space, the equations for the integrated functions (2.75) are local. In the limit of a short mean free path ("dirty" alloys!) the dependence on the direction of the momentum $\mathbf{p}$ drops out, leading to the Usadel equations[119]. In the presence of defects the new equations, known as the Ellenberger equations,[117] bear a formal resemblance to the kinetic equation and are rather convenient, especially for the use of numerical methods.

2.3.5 Instead of Conclusion

The microscopic theory of SC is one of the most beautiful recent theoretical achievements. Its fundamental ideas are used nowadays in many areas of modern quantum physics far beyond the field of condensed matter, from high energy physics to astrophysics. It also stimulated new experimental studies of SC which were elevated to higher levels. The SC phenomena are exceedingly rich and the current literature on SC is so immense that the notes given here are not comprehensive.

To date, a few new classes of superconductors have been discovered (cuprates, Heavy Fermions, organic superconductors, and others) for which the BCS-like physics seems to be insufficient (see Zaanen's contribution in this Chapter) There are two sides to the problem. On the one hand, SC in new materials can be due to non-conventional and yet unknown mechanism; the strong-coupling approach may be necessary for its rigorous treatment. While this would be beyond the existing theory, the pertinent physics may still be properly discussed on the basis of the weak-coupling scheme, at least qualitatively. On the other hand, strong experimental evidence exists that difficulties could already begin in the normal phase because of the violation of the Fermi liquid approach in the material under study. Although not much can yet be said in such cases, the basic concepts of the current microscopic theory concerning the appearance of the symmetry of the order parameter below T_c and probably of the double charge $2e$ originating from a pairing instability do remain in force.

[117] G. Eilenberger, *Z. Phys.* 214 (1968) 195

[118] A. I. Larkin and Yu. N. Ovchinnikov, *Sov. Phys. JETP* 28 (1969) 1200

[119] K. D. Usadel, *Phys. Rev. Lett.* 25 (1970) 507

2.4   A Modern, but Way Too Short History of the Theory of Superconductivity at a High Temperature

Jan Zaanen

2.4.1   Introduction

One cannot write the history of a war when it is still raging and this certainly applies to the task we are facing dealing with the theory of superconductivity in the post-BCS era. The BCS theory was of course a monumental achievement that deserves to be counted among the greatest triumphs in physics of the twentieth century. With the discovery of high-T_c superconductivity in the cuprates in 1986[120] a consensus emerged immediately that something else was at work other than the classic (i.e., phonon-driven) BCS mechanism. A quarter of a century later this subject is still contentious, perhaps even more so than at any instance in the past. It is plainly impossible to write an "objective" history of any kind and this piece has no pretense to do so. I thought it would be useful instead to write some sort of eyewitness account, serving two potential readerships: The first are the newcomers who wonder about the prehistory of some standard notions that appear as established and controversial at the same time. The other readers right now are no more than a potentiality. Imagine that the theory of high-T_c superconductivity will turn truly glorious at some point in the future. The era that I am discussing here will then have a similar status to the "dark ages" that preceded the discovery of BCS theory as described elsewhere in this volume. I can imagine that an eyewitness account like this one might be entertaining for the professional historians who want to chronicle this triumph of 21st century physics.

I am aware that this exposition is far from complete—there are quite a number of ideas that are very interesting in themselves, and they might be on the long run more consequential than the matters I discuss here. However, given the length restrictions I do not have the space to give a truly comprehensive overview of everything that has been explored. I have to restrict myself to a sketch of the history behind the notions that are right now on the foreground. Similarly, I do not claim to be thorough with regard to distributing credit to the various individuals involved. Although in some instances I will draw attention to contributions that I find are underexposed, this material is surely not intended as background material for award committees.

This eyewitness account is also colored with regard to the subjects that are discussed. Whether we like it or not, humanity is a religious species and belief systems play a big role when science is still in the making. One must be aware of this. In physics I like to call myself a "denomination tourist". I have familiarized myself with various belief systems to a degree that I felt myself to be a believer, only then to hop at that point to another system. In the course of time I have settled into a faith that is right now quite popular: the "agnostic denomination in high-T_c superconductivit". Its dogma is as follows: "The research in unconventional superconductivity is flourishing. It is however owned by the experimentalists. Energized by the high-T_c puzzle, the experimentation in quantum matter leaped forward in the last twenty years, due to a variety of highly innovative instrumental developments. These new experimental facts have however just deepened the mystery for the theorists. The agnostic zeal is anchored in the conviction that these fantastic new data are begging for an explanation in terms of mathematical equations of a beauty and elegance that will measure up to the products of Einstein and Dirac. Right now we have not the faintest idea what these equations are but that should not keep us away from attempting to find them".

Surely this has had an impact on the organization of this text. I will discuss matters in a sequence

---

[120] K. A. Müller and G. Bednorz, *Z. Phys.* B64 (1986) 189

that is in historical order (from old to new), while at the same time the sense of the awareness of dealing with a mystery (the agnostic dogma) is increasing.

The point of departure is the highly influential "denomination" that rests on the believe that the theoretical framework that emerged in the 1950s in essence still suffices to explain the physics of the cuprates and so forth. The normal state is a Fermi-liquid that is perhaps a bit obscured by perturbative processes. However, the essence of BCS theory is still fully intact and to explain the high-T_c superconductivity one needs a "superglue" that has more punch than the traditional phonons. This superglue is than supposed to be embodied by spin fluctuations. Since this "50s paradigm" has been around for a long time it has turned into a framework that pretends to be quite qualitative. This "holy trinity" (Local Density Approximation (LDA) band structure, Random Phase Approximation (RPA) spin fluctuation superglue, Eliasberg equations for superconductivity) is right now in the midst of a revival in the context of superconductivity in the iron pnictides. It surely should be taken quite seriously but more than anywhere else the danger of religious illusions is around the corner. In this regard I perceive it as beneficial to be aware of the long and contentious history of the spin fluctuation superglue idea. This is the subject of Section 2.4.2.

Although it has been a bit on the retreat recently, the next influential school can be called "RVB, and related phenomena". This refers to a highly imaginative set of ideas that emerged in the wild early years of high-T_c superconductivity under the inspired intellectual leadership of Phil Anderson. Regardless of whether it has anything to say about the real life of high-T_c superconductors these ideas deserve attention if only because they developed into a quite interesting and innovative branch of theoretical physics.

The spin fluctuation theories are rooted in the perturbative schemes of the fermiology that was established in the 1950s. The Resonating Valence Bond (RVB) ideas eventually rest on the recognition that doped Mott insulators have to be different. This story starts with the last skirmishes of a war that is much older: the understanding of the Mott-insulating state itself. As I will discuss in Section 2.4.3, this pursuit started a long time ago. By the late 1980s a rather detailed and complete understanding of the real life Mott insulators was established as a by-product of the high-T_c research. Section 2.4.4 deals with the RVB school of thought. This started from an appealing idea of Phil Anderson's that evolved during the wild early years of high-T_c into something of a mathematical science by itself. In contrast to the 1950s spin fluctuations, these "gauge theories of strongly correlated fermions" rest conceptually on the discoveries in the 1970s in high energy physics. Although it still lacks both convincing experimental support and real mathematical control, these theories surely serve well the purpose of confusing the high energy theorists. At stake is the meaning of the gauge principle. In fundamental physics it is a primordial principle. The "RVB" gauge theories insist that the gauge principle is governed by collectivity, telling how large numbers of "un-gauged" microscopic degrees of freedom co-operate in unison to generate "gauged" collective behaviors.

Keeping the historical order, in Section 2.4.5 I focus on the subject that has been dominating the field during the last 10 years or so: the physics of the underdoped "pseudogap regime". This has been so much on the foreground because it turned out that there was much to discover using the modern experimental techniques. It is now clear that there is a collection of exotic competing orders at work, including stripes, quantum liquid crystals, and spontaneous diamagnetic currents. Apparently theoretical physics as we know it does provide a grip on order, because theorists have made a big difference in this particular pursuit.

In the last section I discuss what may be the beginning of the history of the future of high-T_c superconductivity. This revolves around the idea that the quantum physics of the mysterious superconductors is in the grip of the overarching symmetry principle of scale invariance. In analogy with the classical critical state, this "quantum criticality" is believed to be caused by a phase transition happening at zero temperature, driven by quantum fluctuations. Given the increased understanding of the competing orders at work in the pseudogap phase this notion is at present becoming quite popular in the cuprates. In the context of the heavy fermions, however, it has been obvious for a long time that their superconductivity is closely related to quantum phase transitions and at stake is whether these share the essential physics with the cuprates. This is the stronghold for the agnostic theorists because nowhere else is it so obvious that we are lacking mathematical machinery that works as in the context of quantum critical "fermionic" metals. I end this chronicle on an optimistic note by expressing my hope that the string theorists might actually have such machinery in the offering.

2.4.2   Pushing BCS to the Limits: The Spin Fluctuation Superglue

The BCS theory is a remarkable achievement. A constant factor in the debate during the last 25 years has been whether it is at all possible to get around the essence of BCS to explain superconductivity at a high temperature. Its central wheel is the Cooper instability which in turn rests on the physics of nearly free fermions. A first condition is that the normal state closely approaches the Fermi-liquid fixed point near the superconducting transition temperature. A second condition is that the net residual quasiparticle interactions turn attractive in the appropriate pairing channel. At this instance the Fermi-surface exerts its power, by singling out the pairing channel as uniquely singular. Feeding a generic fermion system with a very strong attractive interaction that overwhelms the Fermi energy would just cause a clumping of the particles since in (nearly) classical systems there is no special stability associated with two particle bound states[121]. The remainder of BCS is simple. Cooper pairs are bosons and the Cooper pairs Bose condense at the moment they form, of course modulo the possibility of "dangerous" thermal-order parameter fluctuations.

The question is whether the normal state is sufficiently like a Fermi liquid for the BCS mechanism to apply directly. This has been dividing the minds more than anything else—the "radical" theories discussed in the next sections depart from the assertion that the normal state is something other than a Fermi liquid. Assuming the governance of the Fermi liquid, the only issue is then to explain the origin of the attractive interactions. This is far from trivial since electrons actually repel one another rather strongly on the microscopic scale through the Coulomb interactions. Much of the hard work of diminishing these bare interactions can be ascribed to the magic of the Fermi liquid. The ubiquitous nature of the quasiparticle gas is much more of a mystery than is often realized. The particular limiting case where on the microscopic scale the interactions are substantially smaller than the bare Fermi energy is nowadays well understood, although this involves the fanciful "Polchinski-Shankar" functional renormalization group. By principle one has to keep track of an infinity of coupling constants "spanning up" the Fermi surface, when integrating out short distance degrees of freedom. In the renormalization process, one finds that besides the "harmless" marginality of the Landau F functions, one has only to watch a possible (marginal) relevancy in the Cooper channel.[122] The bottom line is that with the exception of the BCS superconductivity the Fermi liquid is generically (i.e., away from special nesting conditions) stable as an infrared fixed point.

The problem is, however, that the systems of interest in the present era are invariably the kind where the microscopic interaction scale exceeds the bare kinetic energy by order(s) of magnitude. One can no longer rely on the perturbative renormalization group and it is just an experimental fact that Fermi liquids are nevertheless ubiquitous. A classic example is ^{3}He. At microscopic distances this is like a quantum version of a dense, highly correlated van der Waals liquid. However, upon cooling it down to milli-kelvins one finds a long wavelength physics corresponding to an impeccable Fermi gas of quasiparticles. The miracle is that the "hard helium balls" have become effectively completely transparent relative to one another, only communicating via the Pauli principle, while the effect of interactions is that these "quasi helium atoms" are three times heavier than the real atoms. I learned from Robert Schrieffer that in the 1960s and 1970s theorists tried hard to explain

---

[121] Although it is usually pushed under the rug, this problem haunts ideas involving local pairing, like the bipolarons.

[122] For a fanciful recent discussion see S. A. Parameswaran, R. Shankar, and S. L. Sondhi, arXiv:1008.2492

this wonder, but got anywhere.[123] Fermi liquids are unreasonably resilient and this is perhaps the best reason to take the "conservative" theories of high-T_c superconductivity seriously.

The next question becomes: how to turn the residual weak quasiparticle repulsions of the Fermi-liquid into serious attractive interactions? The classic BCS answer is, of course, phonons. The exchange of bosons that are not governed by the gauge principle will generically give rise to an induced attractive interaction. The phonons are just perfect: they are external to the electron system since they originate in the lattice, their interactions with the quasiparticles emerge naturally, and they have the added benefit of causing retarded interactions. Phonon energies are tiny compared to the Fermi energy of conventional metals and the associated "Migdal parameter" can be exploited to simplify the perturbation theory into a virtually exact "resummed" Migdal–Eliashberg theory.

This control over the classic phonon-driven BCS theory explains the sociological earthquake that rumbled through the physics community following the Bednorz–Müller discovery of superconductivity at 34 K in $La_{2-x}Ba_xCuO_4$ in 1986. Back then it was believed that phonons could not be responsible for a T_c exceeding 30 K and a consensus emerged immediately that something new was at work. In hindsight this is actually a bit embarrassing. It is a number game and numbers are dangerous. The strongest claim rested on stability considerations[124] that were subsequently challenged by theorists of Ginzburg's high-T_c superconductivity group at the Lebedev Institute[125]. In fact, in 2001 superconductivity at 40 K was discovered in the simple MgB_2 system. In no time a consensus emerged that the superconductivity is phonon driven, with a number of material details (high-phonon frequencies, multi-bands) conspiring to optimize the phonon mechanism, while the phonons responsible for the superconductivity are unrelated to the factors rendering the crystal to be stable. This development gave also credibility to the claim that such phonon-driven superconductivity can be quantitatively studied using LDA type band structure theory. Computing T_cs from first principles is very demanding and has only become possible recently because of the increase of computer power. MgB_2 and a number of other cases have lent credibility to the case that these numbers can be trusted. This supports the "computational" case that 40 K need not be the limit of conventional phonon-driven superconductivity[126].

Despite these caveats I am not aware of any expert who believes that conventional phonons can be responsible for the superconductivity in cuprates or iron pnictides. Perhaps the best reason is that the extreme electron–phonon coupling required to explain a T_c of 150 K should somehow reveal itself in band structure calculations, while these just indicate quite moderate couplings. This is not to say that phonons do not play any role. In fact, there are striking experimental indications that especially at very low doping the electron–phonon coupling is so strong that some variety of small polaron physics is at work. This appears to invoke polar couplings that are ignored in the LDA calculations where implicitly metallic screening is assumed. It cannot be excluded that even at optimal doping such a "poor metal" polaronic physics is at work and claims that bipolarons are behind the superconductivity cannot be dismissed beforehand.[127]

The mainstream thinks differently. Staying within the "BCS paradigm" the quest to explain the high-T_cs turns into the issue whether one can identify non-phonon stuff that can take the role of the phonons, mediating very strong attractive interactions. The only option on the table is to look for "glue" that emerges from the interacting electron system itself. There are quite a number of options

---

[123] Remarkably, as a very recent development it appears that the quantum-fields gravity holographic dualities of string theory shed a new light on this old problem: the Fermi liquid is related to the gravitational stability of special black holes with "fermion hair"; see Section 2.4.5.

[124] M. L. Cohen and P. W. Anderson, in *Superconductivity in d- and f-band Metals*, ed. D.H. Douglass, AIP, New York (1972) 17

[125] O.V. Dolgov, D.A. Kirzhnits and E.G. Maksimov, *Rev. Mod. Phys.* 53 (1981) 81

[126] See, e.g., Marvin Cohen's talk at the 2009 KITP superconductivity conference:
http://online.kitp.ucsb.edu/online/highertc09/cohen/

[127] A. S. Alexandrov and N. F. Mott, *Polarons and Bipolarons*, World Scientific, Singapore (1995)

here, like the excitonic mechanism first proposed by Little, emphasizing special units in the crystal structure characterized by a high electronic polarizability.[128] Another old idea is the Kohn–Luttinger overscreening effect in 2D systems[129] and the closely related idea of the plasmon taking the role of glue. However, by far the most popular view is spin fluctuations. The basic idea dates back to Berk and Schrieffer in the 1960s. The aimed to explain why metals that are close to a transition to an itinerant ferromagnetic state usually do not superconduct[130]. They introduced a particular approximate way to compute matter that forms the conceptual core of the myriad improved spin fluctuation theories that have been developed since then.

A first hurdle is that in the electron system at microscopic distances there is no such thing as a system of spins that has a separate existence of electrons—the electrons just carry around the spins themselves. The "spin glue" has therefore itself to be a highly cooperative phenomenon emerging at "large" distances in the interacting electron system. Upon approaching a transition to a magnetic state in an itinerant metal one does however expect precursors of the magnetism in the paramagnetic state. These are in the form of short-lived magnetic fluctuations that become increasingly better defined upon approaching the magnetic transition where they turn into real magnons—the paramagnons. Using time-dependent mean field theory (RPA) the paramagnons can be computed perturbatively assuming weak interactions. Assuming in addition residual interactions between these spin fluctuations and the conduction electrons, their propagators can just be fed into the Migdal-Eliasberg formalism to take the role of phonons.

The particular benefit of this approach is that it is quite predictive with regard to the pairing symmetries. For the case of ferromagnetic fluctuations, Schrieffer and Berk showed that these generically cause repulsions in the s-wave channel explaining why nearly ferromagnetic metals usually do not superconduct. In the early 1970s superfluidity in fermionic ^{3}He was discovered. It was already known that the normal state is characterized by a strongly enhanced magnetic susceptibility indicating that the helium Fermi-liquid is on the verge of becoming ferromagnetic. Given that there are no phonons around, the idea that this superfluidity had dealings with paramagnons was immediately realized. Dealing with phonons, the induced attractive interaction surely favors a spin-singlet order parameter, while in good metals the effective coupling is localized in space and thereby is momentum independent, favoring s-wave superconductivity. However, ferromagnetic fluctuations favor spin-triplet pairing. This in turn implies that the pairs need to be in an orbital state with uneven angular momentum, having the added benefit that the real space pair wave function has to have a node at the origin. The fermions in the pair thereby avoid each other automatically such that the hard core microscopic repulsions are circumvented. The bottom line is that although there might be a net repulsion between the quasiparticles, this might well turn out to sum up to an attractive interaction in a higher angular momentum channel.

The superfluidity in ^{3}He is of the triplet variety and a discussion doing justice to the richness of this subject deserves a separate chapter in a book like this.[131] With regard to the microscopic pairing mechanism, a worthwhile lesson is the use of the Berk-Schrieffer logic by Anderson and Brinkman to get with great effect the details of the A-phase right.[132] The primary message is that ^{3}He demonstrates that an interacting fermion system as dominated by microscopic repulsive interactions can eventually turn into a superconductor, with the pairing mechanism having undoubtedly dealings with the spin fluctuations. However, the Anderson-Brinkman result just shows that the spin-fluctuations contribute to the attractive interactions, stabilizing a particular superconducting state, but it does not prove that the RPA-style spin fluctuations do all the pairing work explaining in

[128] This idea revived in the context of pnictides: see G.A. Sawatzky et al., *Eur. Phys. Lett.* 86 (2009) 174409

[129] W. Kohn and J. M. Luttinger, *Phys. Rev. Lett.* 15 (1965) 524

[130] N. F. Berk and J. R. Schrieffer, *Phys. Rev. Lett.* 17 (1966) 433; an important follow-up is S. Doniach and S. Engelsberg, *Phys. Rev. Lett.* 17 (1966) 750, focusing on the renormalization of the Fermi liquid.

[131] The classic, hard-to-improve treatise is A. J. Leggett, *Rev. Mod. Phys.* 47 (1995) 331

[132] P. W. Anderson and W. F. Brinkman, *Phys.Rev.Lett.* 30 (1973) 1108

detail why T_c is what it is. For instance, much attention was paid as well to the role of the attractive tail of the microscopic Van der Waals interaction. Another example of this lack of microscopic understanding is the question whether the mass enhancements of the ^{3}He Fermi liquid can be accounted for by the coupling to the ferromagnetic spin fluctuations or whether they are related to the proximity of the crystallization transition[133]. It appears that the quest to explain the nature of the residual attractive force responsible for the pairing in ^{3}He ended in a stalemate. Many factors can contribute and it is just a number game. Given that there are no controlled mathematical methods that tell us how to connect the strongly interacting microscopy to the weakly interacting infrared, it is in fact plainly impossible to address this subtle quantitative matter with the available theoretical technologies. With an eye on the remainder, one should be aware that ^{3}He is, in a way, a best case scenario for glue ideas since the normal state is a well-developed Fermi-liquid, with a superfluidity that is clearly in the weak coupling limit, while complications like (lattice) Umklapp and phonons are absent.

The notion that the superconductivity of electron systems in solids could be caused by intrinsic electronic interaction effects got on the main stage for the first time by the discovery of the heavy fermion superconductors in the late 1970s and 1980s. The gigantic mass renormalizations in these systems leave no doubt that electron-electron interaction effects are dominant. In addition it gradually became clear that the superconducting order parameters in these systems are unconventional by default while magnetism is usually around the corner. Over the last twenty years, a consensus emerged that in these systems the superconductivity is closely related to the presence of a quantum phase transition where at zero temperature a magnetically ordered state gets destroyed by quantum fluctuations. In section 2.4.6 I take up this theme in detail, including a discussion of the Hertz-type theories that are closely related to the spin fluctuations.

The spin fluctuation idea directly acquired prominence in the early history of cuprate high-T_c superconductivity. Obviously an agent different from phonons causing the superconductivity was at work, while in the frenzy of the late 1980s it became clear in no time that cuprate superconductivity emerged from doping an antiferromagnet. In addition, both inelastic neutron scattering and NMR experiments showed that the metals at superconducting doping concentrations show the signatures of pronounced spin fluctuations. The spin-fluctuation-glue idea was lying on the shelf and it was predictably taken up by part of the community. The most visible advocates were Douglas Scalapino and David Pines, representing somewhat complementary views that in 2010 are still discernible. Pines embodied the "bottom-up" phenomenological approach, analyzing in great detail experimental information on the spin fluctuations, aiming for a quantitative phenomenological description of the superconductivity. Scalapino approached it from a more theoretical angle, trying to cast the (incomplete) information from quantum Monte Carlo computations and so forth in a diagrammatic framework tailored according to the Berk-Schrieffer framework and improvements thereof such as the "fluctuation-exchange" (FLEX) perturbative theory.

In fact, the spin fluctuation advocates faced a heavy battle in the early history of high-T_c superconductivity. The predictive power with regard to the pairing symmetry was their stronghold. In the presence of Umklapp associated with the scattering of the electrons against the crystal lattice, the pairing symmetry story becomes more interesting and specific. The lattice warps the electron dispersions and the Fermi-surface shape with the ramification that on the level of RPA calculations the spin fluctuations tend to get quite structured in momentum space. The cuprate fermi-surface computed by LDA bandstructure (and confirmed by photoemission measurements) shows "nesting" features that promote a strong enhancement of spin fluctuations at large momenta. This in turn implies a strong momentum dependence of the pairing interactions mediated by these fluctuations that strongly favor a $d_{x^2-y^2}$ superconducting order parameter. In the 1987–1990 era however, it was

---

[133] D. Vollhardt, *Rev. Mod. Phys.* 56 (1984) 99; this in turn rested on the discovery by Maurice Rice and Bill Brinkman of the description of the Mott transition in terms of Gutzwiller projected wave functions: W.F. Brinkman and T.M. Rice, *Phys. Rev.* B2 (1970) 4302. These ideas were in turn inspirational in the development of the RVB-type theories of section 2.4.4.

taken for granted by the community at large that cuprate superconductivity was s-wave. In hindsight, these strong beliefs are not easy to explain on rational grounds: the data were poor quality and they appear to be rooted instead in a conservative sociological reflex—even d-wave was too risky. After a hard fight that lasted for a couple of years, Scalapino and Pines *et al.* were proven right with their d-wave claim by the phase-sensitive measurements of Tsuie, Kirtley, van Harlingen and coworkers as discussed in Chapter 3. This is remarkable and I find it respectable that this success has motivated a stubborn attitude in the spin fluctuation community; there has to be something to the notion given that it got the d-wave right. However, after it became clear that the pairing symmetry is d-wave, in no time it turned out that this is also the natural ground state in, for instance, the RVB context (section 2.4.4), and even invoked the spontaneous current phases of Varma, as discussed in Section 2.4.5. Although d-wave superconductivity is rather unnatural dealing with conventional phonons, the consensus has emerged that it is just too generic dealing with electronic mechanisms for it to be regarded as conclusive evidence for any specific mechanism.

At this moment in time the idea of a spin-fluctuation "superglue" is very much alive—counting heads in the community it might be the most influential hypothesis altogether. Part of this appeal is surely associated with the fact that the hypothesis poses a crisp challenge to the experimentalists; the superglue is like a holy grail, and one just has to dig with the modern experimental machinery of condensed matter physics into the electron matter to isolate it. The fact of the matter however, is that after 20 years of concerted effort in the cuprates, the "evidence" is as indirect as it has been all along. The state of the art is perhaps best represented by a tour de force that was published in 2009.[134] Empirical information obtained from direct measurements of the spin fluctuations by neutron scattering and the electron self-energy effects picked up in photoemission is combined with the results from numerical calculations, arriving at the claim that the spin-fluctuation glue can fully account for the superconductivity. But this work also reveals the weaknesses of this "paradigm". The evidence is highly quantitative, while it departs from oversimplified toy models where one nevertheless loses mathematical control because of the intermediate strength of the couplings involved. Equally convincing cases have appeared arguing that the self-energy kinks in photoemission are entirely due to rather weak electron-phonon coupling. The magnetic resonance that dominates the spin fluctuation spectrum in optimally doped cuprates can be interpreted rather as an effect than as the cause of the superconductivity. Perhaps the most devastating counter evidence available at present for such an interpretation is the staggering "electronic complexity" of the underdoped pseudogap regime as discussed in section 2.4.5. The simple fermiology "RPA view" on this competing order appears as a gross oversimplification of this reality.

The advent of the pnictide superconductors appeared initially as a shot in the arm for the spin-fluctuation idea. The LDA calculations indicate nesting properties of the Fermi-surface pockets that are beneficial for the RPA paramagnons, while they are consistent with the "$(\pi, 0)$" magnetism found in the parent compounds. The case was made vigorously that since the parent systems are not Mott insulating it has to be that the iron superconductors are quite itinerant and typical candidates for spin-fluctuation superconductivity. Last but not least, a peculiar "$s_\pm$" pairing symmetry was predicted based on the LDA-RPA-Eliashberg "trinity".[135] It is now widely believed that this symmetry is indeed realized in at least some of these systems, although definitive evidence is still missing.

Much of the early experimental activity in this young field was aimed at trying to find evidence for the spin-fluctuation superglue. Remarkably, when the moment seemed fit to cry victory the building started to crumble. Data appeared indicating that unexpected "nematic" phenomena are happening in the underdoped pnictides that seem to parallel the pseudogap complexities of the cuprates. In addition, the pnictide metals at optimal doping have weird properties such as a linear resistivity that

---

[134] T. Dahm et al., *Nature Physics* 5 (2009) 217

[135] This appears to have been a close race, with winners: I.I. Mazin et al., *Phys. Rev. Lett.* 101 (2008) 057003, being cited already 500 times.

again suggests that these have uncanny commonalities with the mystery stuff of the cuprates. This is a very fresh development and it is much too early to say where it will land. However, I do expect that it is beneficial for the newcomers in the pnictide field to be aware of the very long and contentious history of the spin-fluctuation glue idea.

2.4.3 The Legacy of Philip W. Anderson (I): Mottness

When in some future the real history of T_c can be written, the odds are high that Philip Anderson will have a legendary role to play. The influential Andersonian tradition in high-T_c superconductivity revolves around the assertion that the ground rules of metal physics as laid down in the 1950s do not apply to the electrons in cuprates, because high-T_c superconductivity emerges from a Mott insulator that is doped. This is in stark contrast to the spin-fluctuation school of thought as discussed in the previous section. It departs from the idea that the normal state is in essence a Fermi-liquid. Although shaken hard by the powerful glue bosons, the quasiparticles forming a Fermi-gas are eventually ruling the waves, and it is just a matter of sorting out the Feynman graphs governing the perturbations of the quasiparticle gas. When high-T_c was discovered in 1986 this was taken as a self-evident truism by nearly everybody. However, Anderson published one of the first papers ever on the subject and it is still the most influential paper in the history of the subject introducing the "resonating valence bond" (RVB) idea[136] that is the focus point of the next section. This section is preliminary: the success story of nailing down the specifics of the parent cuprate Mott insulator in the late 1980s.

Mott insulators are in fact much simpler than band insulators; they are just the incarnation of rush hour traffic in the world of electrons. The Coulomb repulsion dominates with the effect that electrons have to avoid each other. When only one electron fits every unit cell, while there are as many unit cells as electrons, all "electron traffic comes to a complete standstill". Doping has the effect of making some room for motions—it is the hbar version of stop-and-go traffic. The first fact that was settled in high-T_c superconductivity was the Mott insulating nature of the undoped cuprate parents.

This matter actually jump-started my own career. I was quite lucky having George Sawatzky as thesis supervisor. In the early 1980s photoemission was still quite young, and George had the vision to use this technique to focus on the classic problem of the electronic structure of insulating transition metal salts. This was also the era that the local density approximation (LDA) band structure description of the electronic structure was at the height of its success, having demonstrated that it delivered a remarkably accurate description of the band structures of conventional metals and insulators. Unaware of its contentious history, the claim was made by band structure theorists in 1984 that straightforward spin-dependent LDA did correctly describe the electronic structures of 3d transition metal oxides. This triggered a ferocious response by Anderson. I have a vivid recollection of reading a short paper Anderson wrote that never got published, written with a literary quality that is rare in the physics literature: "Williams *et al.* arrive at the laughable contention that cobalt oxide is a metal because their band structure calculation is telling so." (P. W. Anderson, private communication)

Apparently this response was rooted in developments in the 1950s. Slater might be viewed as the father of modern band structure theory. With his $X - \alpha$ method he had outlined the idea of effective exchange-correlation mean-field potentials and one can view the later work of Kohn and so forth as an a-posteriori formal justification of the basic notion. Apparently Slater believed that his mean-field potentials were the last word and it is rumored that Slater gave the young Anderson a very hard time.

In this era Mott and van Vleck, Anderson's intellectual fathers, had discovered the basics of

---

[136] P. W. Anderson, *Science* 235 (1987) 1196

what we now understand as "strongly correlated electron physics". This revolved initially around the notion that the electron physics in transition metal salts is dominated by strong, local electron repulsions. Although Hubbard can be credited as the first who wrote down the Hubbard model explicitly in 1964, Mott introduced the notion of a Mott insulator in 1956, while Van Vleck already had in the 1940s a clear notion of the local Coulomb interaction (U) competing with the hopping t[137]. Anderson himself has a big stake in this early development in the form of his 1959 work on the superexchange mechanism for the spin interactions in Mott insulators[138], as well as his Anderson impurity problem (awarded with the Nobel prize), representing landmarks of the physics of strongly correlated electron systems.

In the period when the band structure/fermiology view was turning into the mainstream the solid state physics community was not particularly receptive to the Mott-insulator bad news. When this old struggle revived in 1984, in Groningen we were making good progress, finding interesting signals of the "Mottness" at work in transition metal salts in the form of satellites and so forth in various electron spectroscopies. Together with Jim Allen, Sawatzky realized that this new information could be used to make a difference in the old debate and this culminated in the "Zaanen-Sawatzky-Allen" theory for the electronic structure of Mott-insulating salts. We explained how the essence of the toy models designed by Anderson and Hubbard could be combined with the complexities of the electronic structure of the real salts. In the first half of the 1980s nobody seemed interested, but this changed drastically when high-T_c superconductivity was discovered. Cuprates were an important part of the ZSA ploy as archetypical "charge transfer insulators" where the holes of the p-type cuprates are actually formed from oxygen states. My thesis work was, to a degree, rediscovered by others: Emery with his "Emery model" and Zhang and Rice with their "singlet" (Sawatzky's idea of starting with $3d$ impurities). Rather quickly this converged into a mainstream view, insisting that one should be able to get away with the minimal $t - J$ model in the doped systems, much in the spirit of Anderson's RVB philosophy. The marked exception was Chandra Varma who was convinced that "chemistry matters" and that one has to worry about holes being in oxygen. After a lonesome but resilient search of some twenty years this triumphed recently, in the form of Varma's intra-unit cell spontaneous currents, discussed in Section 2.4.5.

In the years that followed, a massive effort ensued aimed at understanding the parent Mott insulators. The experimental and theoretical technologies were just laying ready to be used to clarify completely their physics. It appears that this resulted in a seemingly complete understanding of Mott-insulators. We have learned how to modify LDA so that it can cope quantitatively with the real-life Mott insulators through the LDA+U and LDA+DMFT functionals. A detailed understanding has been achieved of the physics of an isolated carrier moving through the antiferromagnet resting on linear spin-wave theory and the self-consistent Born approximation, with the recent addition that in the insulators the polar electron-phonon couplings are actually quite important. The study of Mott-insulators has turned in the meantime into a form of standard science in material-science laboratories, but unfortunately this has not made much of a difference with regard to the still highly mysterious physics of Mott-insulators turning into metals and superconductors when doped.

2.4.4 The Legacy of Philip W. Anderson (II): Resonating Valence Bonds and Their Descendants

Upon learning in 1986 that high-T_c superconductivity did occur in a doped copper-oxide, Anderson directly arrived at the hypothesis that the phenomenon should be rooted in the "strangeness" of the electron states subjected to the "Mottness". His RVB paper heralded a remarkably intense and

---

[137] According to Phil Anderson, Van Vleck used to call it the UT model, a typical side reference to the University Theater, Harvard Square (private communication)

[138] P. W. Anderson, *Phys. Rev.* 115 (1959) 2

creative period in condensed matter theory. The "spins and holes" ideas that are still at the center of the thinking regarding unconventional mechanisms were laid down at a high pace in 1987–1988, under the intellectual leadership of Anderson. The emphasis is here on "ideas" which actually amount to inspired guesswork, not disciplined by controlled mathematics, while until the present day empirical evidence for "RVB-ish" physics is quite thin. With the exception of the notion that Mottness is changing the ground rules, my impression is that even Anderson will admit that the problem of high-T_c superconductivity is still wide open. However, regardless whether they apply literally to cuprate superconductors, the ideas that emerged in 1987–1988 are embodying a very interesting theoretical framework that deserves to be realized by nature somewhere.

The essence of the Mott-insulator is that the electrons, that should form a high density degenerate Fermi-gas, turn instead into a system of spins due to the domination of the local Coulomb repulsions. This is simple to understand, but the consequences are remarkable. The free Fermi-gas governed by Fermi-Dirac statistics is rooted in the fact that electrons are indistinguishable fermions. Due to the interactions the electrons localize, thereby becoming distinguishable particles, and all that is left are their spins that live as distinguishable degrees of freedom in tensor product space. Upon doping, the moving holes release "here and there" the indistinguishable nature of the electrons. In the doped Mott insulator the electrons are most of the time distinguishable spins, but once in a while when a hole passes by they remember that they are actually fermions. At stake is that "Mottness" interferes with the fundamental rules of quantum statistics. This is a rather recent insight—I learned it from Zheng-Yu Weng—and it appears that it is not quite appreciated yet by the community at large. Whatever it means, I find it myself a particularly helpful point of departure to appreciate why RVB and so forth make sense.[139]

The point of departure of Anderson's seminal RVB paper is the idea that the pairing of electrons is already imprinted in the spin system formed in the Mott-insulator. The common phenomenon in these Heisenberg spin systems is that the spins order in simple antiferromagnets. After the dust settled in 1987/1988 this turned out also to be the case in the high-T_c parents. However, in two dimensions and dealing with small $S = 1/2$ spins there is a potentiality that the quantum fluctuations become so strong that some quantum spin liquid is formed. Anderson envisaged such a liquid, actually inspired by old ideas of Pauling. Configurations from pairs of spins forming singlets ("valence bonds") are formed, and the spin liquid corresponds with coherent superpositions ("resonating") of all possible tilings of the lattice with such singlet configurations. Upon doping, the electrons forming the singlet pairs become mobile, and the valence bonds turn into Cooper pairs that condense in a superconductor.

The idea is simple and appealing but the hard part is to demonstrate that the doped spin system is actually dominated by these pair-wise singlet configurations. After all these years definitive evidence for such a state is still missing. There is no controlled mathematical or computational procedure available for the doped Mott-insulator problem, as rooted in the statistical troubles alluded to in the above ("fermion signs"). At the same time, there is no experimental machinery available that can measure the RVB "entanglement" directly and the evidence that is quoted to support RVB-type physics is invariably indirect—the pseudogaps, d-wave superconductivity and so forth can be explained as well within e.g., the conventional spin fluctuation view.

The "idea revolution" that ensued in 1987–1988 was driven by the desire to pack mathematical meat on the conceptual bones of Anderson's RVB idea. Anderson took a first step in this direction in his 1987 *Science* paper by introducing the idea of a particular kind of variational wave function. This dates back to the helium era when Jastrow introduced the idea of ansatz wave functions that are derived from the free Fermi/Bose gas, but where one wires in that the particles avoid each other

---

[139] See J. Zaanen and B. Overbosch, *arXiv*:0911.4070. See also P. Phillips. *Rev. Mod. Phys. Coll.* 82 (2010) 1719 for a complimentary perspective on Mottness.

locally due to strong local repulsions.[140] Gutzwiller suggested a special form of such ansatz wave functions, amounting just to a free fermion ground state where the double occupied states are projected out.[141] These were applied in 1970 by Maurice Rice and Bill Brinkman, in the context of the Mott metal-insulator problem (W. F. Brinkman, T. M. Rice, *Phys. Rev.* B2 (1970). Anderson pointed out in his RVB paper that projecting out double occupancy from a free BCS ground state wave function (instead of the Fermi gas) yields a state which is just like his envisaged RVB superconductor. To do this well one has to evaluate matters numerically, and this variational Monte-Carlo approach turned mainstream with hundreds of papers published.

Despite strong claims, this numerical activity has not been decisive up to the present day. The Gutzwiller style variational approaches suffer intrinsically from the "garbage in-garbage out" problem. One wires in through the ansatz a particular prejudice regarding the physics and it is hard to judge from the outcomes whether this was a good idea, given the well-established wisdom that ground state energies and so forth are not very sensitive measures of the correctness of the wave function. In fact, the failure of the computational methods at large can be viewed as the best evidence for the profundity of the doped Mott insulator problem. All numerical methods characterized by a real measure of control show that they eventually lose this control. Quantum Monte Carlo and the closely related high temperature expansions crash before temperature is low enough for matters to become truly significant because of the sign problem. The density matrix renormalization group fails in two (and higher) dimensions by its inability to keep track of the quantum entanglements. Dynamical mean field theory (DMFT) is based on the uncontrolled adhoc assumption that the consequences of electron correlation effects are purely local and in this regard it is not that much different from a primitive Gutzwiller ansatz. Now the most promising approach is perhaps the "dynamical cluster approximation" as designed by Jarrell which can be viewed as either a finite cluster Quantum Monte Carlo with cleverly designed effective medium boundary conditions or a finite range extension of the DMFT. At the least this approach has delivered a real surprise in the form of a phase separation type quantum phase transition, happening in the Hubbard model at intermediate coupling.[142]

The theoretical pursuit triggered by the RVB idea eventually settled in what is called the "gauge theories of strongly correlated electron systems". Several motifs that were combined into a new view on the collective behaviors that, at least in principle, can exist in strongly interacting quantum matter. Perhaps the most important motifs is the notion of quantum number fractionalization: the idea that the strongly interacting, highly collective state of matter might carry excitations that behave like weakly interacting particles/elementary excitations carrying, however, "fractions" of the quantum numbers of the elementary particles. In the context of the RVB school of thought this turned specifically into the notion that the electron quasiparticle of Fermi-liquid theory "falls apart" in particles that carry separately spin 1/2 ("spinons") and electrical charge ("holons"). The fractionalization "phenomenon" was first identified in the 1970s by Jackiw and Rebby in the mathematical study of field theories in one-space dimensions (Thirring, Sine-Gordon models) while it acquired great fame in condensed matter physics when it was rediscovered by Su, Schrieffer and Heeger (SSH) in the simple, physical setting of the excitations of polyacetylene.[143]

This is just a dimerized (double bond-single bond) chain of carbon atoms and SSH recognized that a kink/domain wall in this dimerization pattern carries an electronic bound state that is formed from half a conduction band and half a valence band state. The consequence is that it corresponds with a pure S=1/2 excitation carrying no charge ("spinon"). When one binds a carrier to this soliton it turns into a spin-less excitation carrying the electron charge ("holon"). Kivelson, Rokshar and

[140] R. Jastrow, *Phys. Rev.* 98 (1955) 1479

[141] M.C. Gutzwiller, *Phys. Rev. Lett.* 10 (1963) 159

[142] See E. Khatami et al., *Phys. Rev.* B81 (2010) 201101

[143] See the classic review: A.J. Heeger, S. Kivelson, J.R. Schrieffer and W.P. Su, *Rev. Mod. Phys.* 60 (1988) 781

Sethna recognized that a "short range" RVB state formed from nearest-neighbor singlet pairs (described by "quantum dimer" models) is for the purpose of fractionalization just a two dimensional extension of the SSH physics.[144] As long as the dimers form a liquid it is simple to see that isolated spins and holes are sharply defined topological excitations that carry either spin 1/2 or charge. Fradkin, Kivelson and others soon thereafter managed to link the quantum dimer models to gauge theory.[145] They showed that it is dual to a version of compact quantum electrodynamics in two space dimensions, linking the fractionalization to the confinement-deconfinement phenomenon, as known from, e.g., quantum chromodynamics.

However, compared to the high energy physics context, the perspective is warped in a way that continues to confuse high energy theorists. The lore in fundamental physics is to consider the quarks as the fundamental particles and, due to the anti-screening property of the gluon fields, these are confined at low energy in baryons. The rules precisely reverse in the quantum dimer model: the electron is the fundamental object, but it exists only in the confining state of the theory. The spinons and the holons have the status of the quarks, but they appear as highly collective excitations rooted in the global properties of the "topologically ordered" RVB liquid. These become real only when the ground state/vacuum of the gauge theory is deconfining. Therefore, deconfinement is a profound emergence phenomenon while the confining state is almost "no-brainer" microscopic "chemistry" surviving up to the longest distances. This interesting motif is shared by all theories dealing with the "emergent gauge principle" in condensed matter physics.

The quantum dimer problem on the square lattice actually died in a rather interesting way. Compact pure gauge electrodynamics is always confining in $2 + 1D$. In the "dual" quantum dimer model context this translates into the fact that at least on a square lattice the ground states always correspond with valence bond crystals where the valence bonds just stack in a regular manner breaking the lattice translations. This has the consequence that spinons and holons are confined since they are tied together by domain wall "strings" in the valence bond crystals. Much later Moessner and Sondhi discovered a loophole, by their demonstration that the dimer model of the two dimensional triangular lattice does support a deconfining state.[146] This is perhaps still the best proof of principle that fractionalization as linked to the notion of deconfinement in gauge theories does make sense, at least theoretically.

Another, at first glance, unrelated way that the gauge principle enters was recognized by Baskaran and Anderson, now directly associated with the very nature of Mottness itself. This relates directly to the "stay at home" principle that emerges at low energy in the Mott-insulator. The Mott-localization implies that in fact the *local* number operator is sharply quantized: in the Mott-insulator one can count precisely the (integer) number of electrons within a finite localization volume. Baskaran and Anderson argued that one can work with free fermions as long as one couples these to a compact $U(1)$ gauge field that in turn imposes the local conservation of charge.[147] Affleck and Marston took this one step further by noticing that the Heisenberg-type spin Hamiltonian is in turn also invariant under charge conjugation (it does not matter whether one describes spins in terms of electrons or holes), showing that together with the "stay at home" gauge the overall symmetry of the gauge fields in the Mott insulator is SU(2).[148]

The final crucial motif that was required for the mathematical underpinning of the gauge theories for doped Mott-insulators is the idea of the large N limit. N refers here to the number of degrees of freedom in a field theory. The notion that one can organize matter in terms of a perturbation expansion, where $1/N$ acts as the small quantity, dates back to the hay days of non-abelian Yang-Mills theory in the 1970s. The Dutch theorist 't Hooft realized that the diagrammatic perturbation theory

[144] S.A. Kivelson, D.S. Rokhsar and J.P. Sethna, *Phys. Rev.* B35 (1987) 8865

[145] See Eduardo Fradkin's book *Field theories of Condensed Matter Systems*, Addison-Wesley (1991)

[146] R. Moessner and S.L. Sondhi, *Phys. Rev. Lett.* 86 (2001) 1881; see also A. Kitaev's honeycomb model, *Ann. Phys.* 321, (2006) 2

[147] G. Baskaran and P.W. Anderson, *Phys. Rev.* B37 (1988) 580

[148] I. Affleck et al., *Phys. Rev.* B38 (1988) 745

of the pure gauge $SU(N)$ Yang-Mills theory drastically simplifies in the limit that $N \to \infty$, since one has to consider only the so-called planar diagrams. When the gauge coupling is strong (the interesting and physically relevant case) one still has to re-sum an infinity of diagrams, but these now form dense two dimensional "nets" in spacetime that are topologically trivial (no handles). This motif evolved in a crucial conceptual pillar underpinning the gauge-gravity dualities of modern string theory; these planar diagram networks can be viewed as string worldsheets that are topologically trivial, with the ramification that the large N limit of the gauge theory behaves like string theory at small string couplings.

The way that large N enters in the context of strongly interacting electron systems is actually very different, obviously so because there is no such thing as fundamental non-abelian gauge bosons at the microscopic scale. This involved a leap of imagination by Ramakrishnan and Sur,[149] who showed that the large degeneracy of the strongly interacting 4f atomic states in the valence-fluctuating/heavy fermion intermetallics could be exploited for the purpose of a large N expansion. It soon became clear that the Kondo and Anderson impurity problems drastically simplify in the limit of high orbital degeneracy. A next crucial step was taken by S. Barnes, followed soon thereafter by Newns and Read with the discovery that "slave" mean field theories control the physics at large N.[150]

This construction can be considered as the central wheel in the "modern" formulation of the emergent gauge principle in strongly interacting fermion matter. It demonstrates the potentiality of gauge symmetry and dynamics as a controlling principle for the emergent highly collective physics at long distances and times, in a manner that is completely different from the way the gauge principle is understood "as coming from god" in high energy physics. At first sight the construction seems absurd. One starts out with, for instance, an Anderson lattice model describing a lattice of strongly interacting f-shell electrons hybridizing with a non-interacting valence band, such that one of the two low energy valence states of the f-shell has a large orbital degeneracy N. One then asserts that the f-electron field operators can be written as products of creation/annihilation operators of two particles, one carrying the spin ("spinon") and the other the charge ("holon") of this electron. The electron is a fermion and therefore one has to choose how to divide the quantum statistics over the slave particles. One can either associate a bosonic statistic to the holon such that the spinons are fermions ("slave fermions"), or the other way around ("slave bosons"), while pending the choice one ends up with quite different theories. Surely these slave particles are unreal; they correspond to redundant (gauge) degrees of freedom that can be removed by imposing gauge invariance invoking explicit gauge fields. But then one discovers that in the large N limit the local constraints that tie the spinons and holons together turn into simple global chemical potentials, which is the same statement that the gauge couplings disappear. The consequence is that the effective gauge theory is in a deconfining regime such that the spinons and holons become as real as quarks above the confinement transition!

One is not done yet because the resulting theory in terms of "physical" spinons and holons has no free part: it contains only two particle (spinon-spinon, spinon-holon) interaction terms. But these can be handled in terms of mean-field theory, introducing single boson (Bose condensation) and bilinear fermion vacuum expectation values. In the original context of Anderson lattice models this was used to explain the emergence of the heavy electron bands. However, soon after the RVB ideas of Anderson had sunk in, a variety of theorists realized more or less independently that this large N slave theory route seemed quite promising with regard to explaining high-T_c superconductivity when applied to the appropriate large N generalizations of the $t - J$ model. In case the fermion statistics is attached to the spinons these mean field theories are quite like the Gutzwiller-type theories for

[149] T.V. Ramakrishnan and K. Sur, *Phys. Rev.* B26 (1982) 1798

[150] See D.M. Newns and N. Read, *Adv. Phys.* 36 (1987) 799 for an account of this early era of slave theories. Piers Coleman deserves special mention for his work on the impurity models and is apparently the father of the idea of slave particles: *Phys. Rev.* B 29 (1984) 3035

RVB states, while they also include the fractionalized excitations suggested by the quantum dimer theories. The spin vacuum is built up from pair singlets with a hard-wired d-wave type coherence, described in terms of a BCS-like mean field structure at "infinite" coupling. The doped holes turn into holons, that Bose condense directly giving rise to a "holon superfluid" with a superfluid density that increases with the number of doped carriers. At the same time, the spinon BCS state has a T_c that decreases with doping. From gauge invariance it is easy to understand that one has a physical superconductivity only when both holons and spinons are condensed, explaining the doping dependence of the superconductivity. Above all, it becomes trivial to explain why T_c is high because the scale is set by the large superexchange interaction J. The underdoped state at higher temperatures where the spinons are condensed, but the holons are still normal is suggestive of the pseudogap regime, while it can also be argued that the (overdoped) state where the spinons form a metal while the holons are condensed should eventually correspond with a Fermi-liquid.

The benefit of slave theories in the large N limit surely lies in the sense of mathematical control while the physical implications are remarkable.[151] The headache is, however, that N in the physical case is only two (spin degeneracy) and it is highly questionable whether the large N limit has anything to say about the real system. For any finite N the gauge couplings in the bare theory jump to infinite since Maxwell terms for the gauge fields are nonexistent. This would mean that one lands in a strongly confining regime, where the spinons and holons bind in the original electrons such that nothing is achieved. But one can then argue that short distance perturbative corrections involving the spinons and holons will generate dynamically a gauge kinetic energy. This might then stabilize a deconfining state, but the next problem is that one has to find reasons why, in the two space dimensions of the cuprates, the monopoles associated with the compactness of the gauge field can be kept out of the vacuum. This gave rise to much debate over the last twenty years and a consensus emerged that, in principle, even in two dimensions the deconfining state has a chance to exist.

Up to the present day, the slave particle gauge theory is highly contentious as an explanation for superconductivity at a high temperature, and not only because of the lack of real mathematical control in the physically relevant small N regime. The greater problem is that despite many attempts this theory has failed to produce striking insights and predictions for experiment. It offers suggestive explanations for a number of observations, but these can also be explained in a multitude of different ways (such as the superconducting dome, the pseudogap, and so forth). In order to be taken seriously, a theory that departs as radically from established wisdom as these gauge theories, it should produce "smoking gun" predictions that can be tested by experiment. After more than twenty years, such hard empirical evidence is still not available.

Another matter is that this development in high-T_c context has inspired a revival of the field of quantum magnetism. Pure spin systems tend to order "classically" in the form of (anti)ferromagnets. When the microscopic spins have a small magnitude (like the $S = 1/2$ spins of the cuprates) the quantum fluctuations around this classical order can be quite serious. The challenge is to find out whether quantum spin systems exist that do not order in a conventional magnet but instead have a "quantum spin liquid" ground state. This is a natural arena of the slave particle gauge theories discussed in the above paragraphs. There are a number of indications that the slave theories are landing on the real axis in this context. An early success is the spinon theory in bosonic incarnation by Arovas and Auerbach.[152] This representation is especially suited to track the physics of quantum spin systems that fail "at the last minute" to turn into classical order; it turned out to be very closely related to the renormalization group for the quantum nonlinear sigma model discussed in the last section. The next success was the large N slave fermion spinon theory in the CP_N incarnation by Sachdev and Read[153], predicting the valence bond crystals in two space dimensions. It appears

---

[151] For a recent in-depth review see P.A. Lee, N. Nagaosa and X.-G. Wen, *Rev. Mod. Phys.* 78 (2006) 17

[152] D.P. Arovas and A. Auerbach, *Phys. Rev.* B38 (1988) 361

[153] N. Read and S. Sachdev, *Phys. Rev.* B42 (1990) 4568

that in the meantime this state is now well established in the $J_1 - J_2$ frustrated Heisenberg spin systems[154], while the case is compelling that this valence bond "crystallization" is playing an important role in the stripe physics discussed in the next section. The key question, however, is whether *compressible* quantum spin liquids exists, systems that show no manifest classical order while their excitation spectrum is gapless. Believing the fermion slave theories, such states should be generic, corresponding with states where the fermions form effective Fermi-liquid systems characterized either by large Fermi-surfaces or states that look like unconventional superconductors with massless nodal excitations. In fact, in this context the most fanciful mathematical considerations were formulated, in the form of the classification of spin liquids in terms of the projective symmetry groups by X.-G. Wen.[155] This work exposes more than anything else the unusual nature of the symmetry at work in the "slave worlds". The essence of Wen's construction is the recognition that the gauge volume of the slave theories is not spanned by separate gauge fields but instead by gauge equivalent slave mean-field ground states. Wen derives the precise symmetry algebras for this "projective symmetry", and shows how to derive the elementary excitations in Goldstone theorem style, coming up with a zoo of compressible "fermion like" spin liquids.

Quite recently an organic system was discovered in Japan where the empirical evidence is quite serious for the existence of such a "spinon Fermi-liquid". This system forms a geometrically frustrated two dimensional lattice of organic molecules containing one electron per site, forming a Mott insulator that is very close to the transition to a metal. After being applied to a small pressure, it undergoes a first-order transition into a metallic state that suggestively turns into an unconventional superconductor at low temperatures. The insulating magnet does not show any hint of long-range order down to the lowest temperatures while several properties are quite suggestive of a "spinon" Fermi-liquid type of physics.[156]

2.4.5   Theory that Works: The Competing Orders of the Pseudogap Regime

In the study of cuprates in the last decade or so the emphasis shifted to the investigation of the underdoped pseudogap phase. The reason is that high-T_c superconductivity is a very empirical branch of physics and there is much to explore in the pseudogap phase using the modern equipment of condensed matter physics. This has developed into a kind of embarrassment of riches. It is now clear that a competing order is at work, in tune with the idea of a quantum phase transition at optimal doping (see next section), but the evidence points actually at a whole collection of exotic ordering phenomena. Not so long ago there was much disagreement among the different subcommunities focused on particular orders, but a consensus emerged that these are all real. The present state of affairs was anticipated a number of years ago by Bob Laughlin, who described this situation in his popular book[157] as the "dark side of emergence", arguing that the strongly interacting electron matter of the cuprates is complex to a degree that approaches biology. The ramification would then be that theoretical physics is sidetracked dealing with such complexity. Although Laughlin's prophecy regarding the pseudogap phase turned out to be quite accurate, I do find his view overly pessimistic. In the first place, when one heats up the pseudogap "stuff" one reenters the "marginal Fermi-liquid" normal state at some higher temperature. As discussed in the next section, this is by itself a monument of simplicity that deserves to be explained by equations of an Einstein-Dirac quality. The significance of the pseudogap "complexity" then lies in the fact that apparently the simple quantum critical metal of the cuprates carries the *potentiality* to "fall apart" in a variety of exotic ordered states. Andy Mackenzie gives it a wonderful engineering twist: the "stuff" that

---

[154] This got much attention in the entertaining context of "deconfined quantum criticality", T. Senthil et al., *Science* 303 (2004) 5663

[155] This was introduced in: X. G. Wen, *Phys. Rev.* B65 (2002) 165113

[156] See P. A. Lee, *Science* 321 (2008) 1306

[157] R. B. Laughlin, *A Different Universe: Reinventing Physics from the Bottom Down*, Basic Books (2005)

makes up the quantum critical metals of the next section corresponds to construction material that enables building states of quantum matter that cannot be constructed from stable states like the Fermi liquid.

Laughlin gets it really wrong with regard to the role of theorists. Although my guild has failed miserably with regard to impacting the understanding of the disordered quantum liquid states, theory has been remarkably successful in guiding the experimentalists in discovering the exotic orders at work in the pseudogap phase. Apparently the mathematical machinery that is lying on the theoretical physics shelf is particularly geared towards understanding order in the sense of symmetry breaking.

Among the phenomena that have been identified in the pseudogap phase, only the notion of low superfluid density, strongly phase-fluctuating superconductivity was understood in principle before 1987. The naive version of this is the idea of "preformed pairs" but the reality in cuprates is better captured by Laughlin's metaphorical "gossamer superconductivity".[158] The phase fluctuations of the superconductor are caused by the struggle for domination of the superconductivity with the competing states. The proximate Mott-insulator that hangs like a big thundercloud over the pseudogap regime is in this regard a major factor (the "gossamer" idea).

The other "exotic" pseudogap orders that have been identified in the meantime were invariably not known in 1987. These were first discovered theoretically, motivating experimentalists to look for them in less obvious corners. These gained general acceptance only after much controversy: testimony for the novelty involved in the discoveries. In historical order this is about the stripes, the quantum liquid crystal orders and the intra-unit cell spontaneous diamagnetic currents.

The stripes were the first to be discovered and they are, at present, the best established among the exotic pseudogap orders. They came into existence on a some Friday afternoon in early October 1987, when I was fooling around with computer code that produced a very unexpected output.[159] This landed on the real axis by the observation of static stripes with neutron scattering in the so-called 214 LTT cuprates by Tranquada et al. in 1995,[160] while it took until this century for it to become fashionable. At another occasion I will give a detailed account of this long history. The bottom line is that I eventually got myself lavishly rewarded for my resilience, but I also learned to be quite cynical regarding the agility and open mindedness of the relevant part of the physics community.

What are the stripes about? When it started in 1987 it was perhaps at first a symbol for the idea that the electron matter in cuprates could be more organized than in conventional metals. The resistance against the idea that the electron stuff could be "inhomogeneous" is just rooted in the way hard condensed matter physicists are trained. Much emphasis is put on the physics of Fermi and Bose gasses that make much sense dealing with the kinetic energy-dominated electron matter in conventional metals. However, dealing with doped Mott-insulator, rush hour traffic is a more useful metaphor: in which the insulator is like the complete traffic jam, and stripes are like the dynamical patterns that occur when the electron traffic enters the stop-and-go regime of the lightly

---

[158] Given that the concept of phase fluctuating superconductivity is quite conventional, the literature is large. It starts with several groups in the late 1980s that are largely ignored because they were ahead of the fashion. At least counting citations, V.J. Emery and S.A. Kivelson, *Nature* 374 (1995) 434 has been perhaps most influential. Laughlin stresses the proximity to the Mott-insulator as a perspective to understand the "thin superconductor": R.B. Laughlin, *Phil. Mag.* B86 (2006) 1165.

[159] For the record, this work was presented for the first time in a seminar in December 1987 at the ETH Zürich, and in a large oral session at the Interlaken M2S meeting in February 1988. The paper was submitted in March 1988 to *Physical Review Letters* and after 1.5 years it was transferred to *Physical Review* B while the original submission date was removed: J. Zaanen and O. Gunnarsson, *Phys. Rev.* B 40 (1989) 7391. I kept the files with referee reports—it is a hilarious read. A couple of months after submission I got an e-mail by Machida explaining that he had seen our preprint while they had independently discovered the same motif. All other papers in this early era came much later. Especially, the late Heinz Schulz approached me after my Interlaken talk explaining that he found it very interesting, planning to take up the theme by himself. Apparently, back in Paris he had forgotten where he got the idea.

[160] J.M. Tranquada et al., Nature 375 (1995) 561

doped insulator.[161] Early on, there were more theorists playing around with the inhomogeneity motif, including Lev Gorkov, as well as Vic Emery and Steve Kivelson who were pursuing in the late 1980s the idea of frustrated phase separation droplets.[162]

Stripes are, however, a quite specific form of organization that is directly tied to the proximity of the Mott-insulator. The carriers form lines ("rivers of charge") in the two dimensional cuprate planes, separated by Mott-insulating "stripes" that support antiferromagnetic order, with the specialty that the "rivers of charge" have a double role as domain walls in the antiferromagnetism. From the point of view of symmetry breaking, these rivers correspond just with a conventional electron charge density wave coexisting with an incommensurate colinear antiferromagnetic order ("spin density wave"). However, the microscopic physics is quite special for the doped Mott-insulator. The main contribution of Olle Gunnarsson to the first paper was his vision that the stripes should be viewed as the electronic version of a *discommensuration lattice*—we borrowed the designation "stripes" from the naming of such states as they occur in classical incommensurate systems. Hubbard and t-J models are just crude toy models, introduced to capture some essence of the physics. Viewed generally, Mott-insulators are electron crystals that are commensurate with the underlying ionic lattice, and the Mott gap is nothing else than a "commensurate pinning energy" causing a gap in the acoustic phonon spectrum of the electron crystal. Upon doping, the lattice constant of the electron crystal becomes incommensurate relative to the ion lattice, and, since the pinning is strong the mismatch gets concentrated in small areas (discommensurations, the charge stripes) while elsewhere the electron crystal stays commensurate (the Mott-insulating domains). A difference with the classical incommensurate systems is the spin degeneracy. This is lifted by the short-range quantum fluctuations turning the spin system into an incommensurate antiferromagnet. My 1987 computation was rooted in Hartree-Fock ("classical saddle point", "old fashioned mean field theory", whatever), and dealing with classical orders (like antiferromagnets, electron crystals) the power of this method is at least in qualitative regards remarkable. It got the stripes quite right.

Stripes are now well known to be ubiquitous in doped Mott-insulators. However, they usually (nickelates, manganites) submit to the Hartree-Fock stability rule that they should keep the doped Mott-insulator insulating. The cuprates are in fact the exceptions since they can be made metallic. Static stripes seem exceptional in cuprates: they occur at the crystal surfaces (the scanning tunneling microscopy measurements) while only in very special (214-LTT) cuprates they seem to exist in the bulk, where they are very bad for superconductivity. So much is clear that in the cuprates stripes are shaken hard by quantum fluctuations. The density matrix renormalization group computations by Steve White and Doug Scalapino in the mid 1990s were quite revealing in this regard.[163] Different from the "deeply classical" Hartree-Fock stripes, the stripes in the t-J model seem to be governed by a RVB-like organization of the spin system, such that they can be viewed in essence as a crystallization of valence bond pairs.[164] Quantum physics also seems to rule the stripes in the long time limit. Resting on the information obtained on the dynamical spin fluctuations by inelastic neutron scattering, the case has been forcefully made[165] that the quantum liquid in underdoped cuprates looks like a stripe state that has fallen prey to what the quantum field theorists call "dynamical mass generation". These dynamical stripes might behave as static stripes at short distances but going to longer times the *collective* quantum fluctuations associated with the order increase, to get out of hand at a relatively long time, beyond which one is dealing with a uniform quantum liquid.

---

[161] See J. Zaanen, *Science* 315 (2007) 1372 for a further elaboration of this metaphor.

[162] V.J. Emery and S.A. Kivelson, *Physica* C209 (1993) 597

[163] S.R. White and D.J. Scalapino, *Phys. Rev. Lett.* 80 (1998) 1272

[164] As first suggested by S. Sachdev and N. Read, *Int. J. Mod. Phys* B 5 (1991) 219: this appears to be quite consistent with the Cu-O-Cu "valence bond" units comprising the stripes seen by the STS "Mott maps" by the Davis group.

[165] Our Leiden group was in this regard (too) far ahead of the time, investing much energy in the problem of fluctuating stripes in the early 1990s: See J. Zaanen, M.L. Horbach and W. van Saarloos, Phys. Rev. B53 (1996) 8671 (1996) and J. Zaanen et al. Phil. Mag. B 81 (2001)1485. See also S.A. Kivelson et al., *Rev. Mod. Phys.* 75 (2003) 1201 and for a recent review see: M. Vojta, *Adv. Phys.* 58 (2009) 699

It has proven difficult to get a handle on these dynamical stripes. However, inspired by this question a next great idea was born. Static stripes should first be seen as an electron crystal. Resting on the analogy of the "partial" melting of classical crystals due to thermal fluctuations, Steve Kivelson, Eduardo Fradkin and Vic Emery introduced in 1997 the idea[166] that, given the severe quantum fluctuations, the stripe crystal in the cuprates could perhaps quantum melt step by step. This would then yield a hierarchy of quantum liquid crystal states: the quantum smectic (crystal in one direction, quantum liquid in the others) and the quantum nematic (a quantum liquid breaking spontaneously rotational symmetry).

I was immediately enchanted by the idea. Realizing that this was potentially highly fertile territory, I made it into a main focus of my own research, but as far as I am aware this was for a long time the only substantive effort outside the Kivelson-Fradkin group. The lack of interest in this obviously interesting pursuit was quite frustrating, but the show was eventually saved again by the experimentalists. Right now the quantum liquid crystals are turning into a fashionable research area. In the cuprates the first evidence was produced by the inelastic neutron scattering work by the group of Keimer, followed by the very recent work of Tranquada et al. on the "sliding phase" physics in the LTT cuprates, and especially the Nernst effect anisotropy discovered by Taillefer's group and the peculiar "hidden" orientational symmetry breaking in the STM stripes of Davis et al. This is supplemented by the particularly neat quantum nematic found at the metamagnetic quantum critical point in a ruthenate by the Mackenzie group, as well as the very recent evidence for the famous hidden order in URu_2Si_2 to be of this kind. Perhaps the most spectacular evidence was presented in 2010, in the form of the extreme transport anisotropies in detwinned underdoped pnictides by the Fisher group.

I find the greatest star among the exotic pseudogap orders to be the spontaneous diamagnetic current phase that started with the theoretical work by Chandra Varma.[167] This opinion is motivated by the fact that this type of order is completely detached from anything that was known before 1987. The essence of the phenomenon is that below the pseudogap temperature diamagnetic currents start to flow spontaneously around plaquettes in the copper-oxide lattice.

The idea that something of this kind can happen dates back to the wild early years of high-T_c superconductivity, where it arose in the context of the gauge theories discussed in Section 2.4.4. In their study of the slave mean field theories controlled by gauge symmetry, Ian Affleck and Brad Marston[168] observed that besides the spinon d-wave superconductor there was another stable mean field state. In this "staggered flux phase" diamagnetic currents occur spontaneously, forming a staggered pattern on the plaquettes of the square lattice. In the context of the gauge theories this flux phase has a very special status: in Wen's modern projected symmetry language, the d-wave superconductor and the flux phase span up the gauge volume in the sense that one can *locally* rotate one in the other without physical ramifications for the spin liquid. Upon doping, this gauge invariance is lifted and the flux phase turns into a real phase that takes the role of competitor of the d-wave superconductor. A main result of the work by Patrick Lee and coworkers on the $SU(2)$ formulation is the prediction that one should find collective modes in the d-wave superconductor that correspond with exciting infinitesimal flux phase modes.[169]

This idea was recycled in 1999 by Chetan Nayak[170] showing that staggered flux phases have a generic significance also in a more conventional (Hartree-Fock) setting. It is well known that under certain fine tuning conditions one can store the particle-particle Cooper pairs and particle-hole charge density wave in a single (global) $SU(2)$ multiplet (Anderson's pseudo-spins), where the superconductivity is associated with the XY plane and the CDW with the z-direction of the pseudo-

[166] S.A. Kivelson, E. Fradkin, and V.J. Emery, *Nature* 393 (1997) 550

[167] C. M. Varma, *Phys. Rev. Lett.* 83 (1999) 3538

[168] I. Affleck and J.B. Marston, *Phys. Rev.* B37 (1988) 3774

[169] P.A. Lee and N. Nagaosa, *Phys. Rev.* B68 (2003) 024516

[170] C. Nayak, *Phys. Rev.* B62 (2000) 4880; this was quite effectively advertised by S. Chakravarty et al., *Phys. Rev.* B63 (2001) 094503

spin internal space. In this way one can show that the $SU(2)$ partner of a s-wave superconductor is a conventional charge density wave. Nayak observed that a d-wave superconductor "rotates" in a similar guise into a staggered flux phase. This led to the name "d-density wave (d-DW)," and it gave further impetus to the idea that, as such, it would be a natural candidate for the "hidden" competing pseudogap order. However, despite a concerted experimental effort this idea failed to generate any empirical support.

Varma actually departed from a quite different perspective. From the very beginning of the cuprate odyssey he had been in the grip of the idea that one has to look in more detail into the chemistry of the cuprate planes, taking a quite dissident stance in a climate dominated by the divinity of the simple t-J model. Varma departed from the wisdom that the holes are physically located on the oxygen ions, while he observed that in the presence of "excitonic" Cu-O nearest neighbor Coulomb interactions, one can run into instabilities when doping increases. Much later, he realized that such microscopic physics opens up the possibility for d-DW type currents to develop inside the Cu-O unit cell, forming patterns involving the various O-Cu-O intra-unit cell triangular plaquettes. The specialty of such a phase is that it breaks only time reversal symmetry, with the ramification that in an experiment one has to know where to look for it in order to find it. The first indications for the presence of such a phase in the pseudogap regime were derived from ARPES measurements, but the data analysis turned out to be a contentious affair. This landed on its feet in 2005 when Philippe Bourges et al. found direct evidence for such intra-unit cell "antiferromagnetism" breaking time reversal symmetry using rather straightforward spin-polarized neutron scattering. Around the same time complimentary evidence was found by Kapitulnik et al. employing highly sensitive optical dichroism measurements.

I was among the first to see this paper—as "Mr. Stripes" I had perhaps something to lose—but I was immediately convinced that Varma's currents are real. Given the persuasiveness of the evidence, and the fact that this was a genuine new type of ordering phenomenon, I took it for granted that this would make headlines by the next day. But reality was quite different. Again the brick wall of hard-headed conservatism was erected. The authors had a hard time geting this published and initially it was largely ignored. As with the stripes and the nematics, the neutron scattering community once again demonstrated its open mindedness. In a concerted effort, with a particularly prominent role of Martin Greven with his superb mercury crystals, much additional evidence was generated. The latest evidence is by Philippe Bourges et al., showing that in the 214 stripe phase the stripes not only live together with nematic order (according to Davis et al.), but also with a short-ranged Varma current order, making the case that the pseudogap phase deserves a name that was recently conceived by Uchida: the monster.

To summarize, although the pseudogap regime has still its mysterious sides (such as the normal state "Fermi-arcs", and the "two gap" behavior), the theorists can at the least claim that they have played a decisive role in uncovering the various exotic ordering phenomena that are taking place. Another matter is what this history reveals regarding the sociology of this community. As I implicitly emphasized, to my perception the community operates in this regard in a less than optimal way. Scientific skepticism is a great good, and perhaps it is just part of human nature that scientific communities have to be conservative. However, an atmosphere a bit more receptive toward exciting new scientific potentialities would have undoubtedly rendered the above research enterprise a more efficient affair.

2.4.6 Quantum Critical Metals and Superconductivity

The last major theme in the context of theories of unconventional superconductivity is the school of thought that revolves around the idea that the physics of the cuprates, heavy fermion superconductors, and perhaps the pnictides and organics revolves around a quantum phase transition. The emphasis is on "quantum": the transition is happening at zero temperature and driven by quantum fluctuations. In a way, this general idea has many fathers. It is also a prevalent theme elsewhere in physics; I recently learned that in string theory the notion of what is known in condensed matter as quantum criticality is among the fundamental organizing principles in this seemingly very different area of physics. In fact, I will end this history reporting on a very recent attempt to mobilize the powerful mathematics of string theory to corner the physics of high-T_c superconductivity.

Symmetry is the best friend of the theoretical physicists and the critical state (quantum or not) is in the grip of one of the most powerful symmetries that have been identified: scale invariance. This came to flourish in the classical realms in the 1970-1980s by the discovery of Wilson's renormalization group theory, which is in essence the theory that exploits the power of the scale invariance emerging at the classical continuous phase transitions. In a way it is remarkable that it took so long before it was realized how to exploit its full powers in the quantum realms. In high energy physics this is likely due to the preoccupation with QCD. This is about the "marginal" case with its running coupling constants, as echoed in condensed matter physics in the form of the attention to the Kondo impurity problem and the non-abelian "Haldane" spin chains subjected to the phenomenon of dynamical mass generation. It appears that in string theory, conformal field theory (also beyond the $1 + 1D$ worldsheet context) came fully into focus when supersymmetric field theories were explored. Due to the non-renormalization theorems coming with supersymmetry, these are characterized by strongly interacting unstable fixed points (in the statistical physics jargon) that span up planes in the space of coupling constants.

In condensed matter physics the idea of quantum phase transitions entered the mainstream in 1988, via the seminal work of Sudip Chakravarty, David Nelson and Bert Halperin (CNH).[171] Bert Halperin told me that this work was initially motivated by his conviction that Phil Anderson got it all wrong with his claim that the magnetism in the Mott-insulating La_2CuO_4 had to do with resonating valence bonds. Bert got it right: the quantum magnetism in this system is of the quasiclassical kind, controlled by the collective quantum precessions of the Néel order parameter. Compared to many other antiferromagnets, this $S = 1/2$ two-dimensional Heisenberg spin system is not that far from the point where these precessions get completely out of hand, quantum melting the Néel order in a continuous quantum phase transition. This physics is captured by the O(3) quantum nonlinear sigma model field theory which amounts to a "classical" Heisenberg system that is living in the $2+1$ dimensional spacetime of the Euclidean (thermal) path integral. Sudip Chakravarty was then on sabbatical in Harvard, and it appears that due to his high-energy physics training he was pivotal in the realization that this becomes really interesting when one asks what the proximity to the quantum phase transition means for the *finite temperature* physics. As far as I am aware, they were the first to realize that by raising temperature one re-enters the quantum critical "wedge" where the low-energy physics is controlled by the zero temperature quantum physical scale invariance. They demonstrated how the spin stiffness controlling the thermal order parameter fluctuations is renormalized by the quantum fluctuations, as was soon thereafter spectacularly confirmed experimentally.

In the same year the Bell Labs theorists Varma and Littlewood with coworkers had independently constructed the marginal Fermi liquid phenomenology for the metallic states of cuprates at optimal doping.[172] This is actually not much of a mathematical theory, but it was quite effective in highlight-

---

[171] S. Chakravarty, D.R. Nelson and B.I. Halperin, *Phys. Rev.* B39 (1989) 2344; In this same era the notion of "quantum critical hydrodynamics" was also fully realized in the context of DC transport phenomena, involving both (fractional) quantum Hall plateau transitions (S.L. Sondhi et al., *Rev. Mod. Phys.* 69 (1987) 315) and the superconductor-insulator transition in amorphous systems (M.P.A. Fisher et al., *Phys. Rev.* B40 (1989) 546; ibid., *Phys. Rev. Lett.* 64 (1990) 587)

[172] C. M. Varma at al., *Phys. Rev. Lett.* 63 (1989) 1996. Undoubtedly Phil Anderson also played a key role in attracting attention to the strangeness of the normal state metal. With his "tomographic Luttinger liquid" (see P. F. Anderson, The Theory of Superconductivity in the high T_c cuprates, Princeton University Press Princeton (1997)) and his more recent "hidden Gutzwiller quasiparticle gas" (*Nature Physics* 4 (2008) 208) he was (and is) searching for a truly non perturbative formulation for a state that behaves temporally in a critical way while it is still "organized around a Fermi surface". This contrasts strongly with the perturbative (Hertz-Millis like) attitude of the marginal Fermi liquid. At least on the phenomenological level his constructions are remarkably similar to the very recent results coming from the Anti-de-Sitter/Conformal Field Theory correspondence. Historically the marginal Fermi liquid had perhaps more impact for no other reason than its lack of pretense, to be no more than a way to memorize the strangeness in the experimental data.

ing and unifying the strange behaviors of these normal state metals in a single phenomenological (or "heuristic", "mnemonic") framework. Forced by the experimental circumstances, Varma et al. imported the "Planckian dissipation" (energy/temperature scaling) as a crucial building block of the construction, among others giving a heuristic explanation for the famous linear-in-temperature resistivity of the optimally doped metals.

Subir Sachdev discovered soon afterward that the classical, relaxation hydrodynamics inside the quantum critical wedge is highly unusual[173]. Much later I named it "Planckian dissipation"[174]: it is a dissipative state, controlled by Planck's constant through a "universal" relaxation time $\tau_\hbar \simeq \hbar/(k_B T)$. In my own reference frame I became fully aware of the meaning of quantum criticality-proper at the Aspen winter conference in 1992 where Sachdev presented analysis of the NMR spin-spin and spin-lattice relaxation times in superconducting, underdoped cuprates employing the Planckian dissipation.[175] The implication of this work was that at least the spin fluctuations of the optimally doped "marginal Fermi liquid" was in the grip of the quantum scale invariance. This was an important moment in my career: Since then, I have been a devotee of the quantum critical state, in fact as a goal by itself regardless its relevance towards superconductivity and so forth. During the 1990s Sachdev developed the quantum critical hydrodynamics theme further and this culminated in his quantum phase transition book. [176]

The theme was picked up in the high-T_c community because the relevancy of the notion towards the then widely accepted marginal Fermi liquid phenomenology was rather obvious. A difficulty was, however, that for quantum criticality to control the physics at optimal doping, one needs a quantum phase transition, and for this one needs some form of order that comes to an end in the best superconductors. Loram and Tallon early on came up with thermodynamical evidence for such a quantum critical point, but this was then highly controversial. As described in the previous section, in the last few years this has changed drastically with the solid empirical evidence for the "monster" competing orders that are simultaneously at work in the underdoped regime. It remains to be seen whether a single order parameter is responsible for the quantum critical regime in the sense of a "conventional" isolated quantum unstable fixed point. A very recent idea is that the competing orders of the underdoped regime are more an effect than a cause, reflecting instead the collapse of the Mott-ness at optimal doping, which is the real driving force behind the quantum criticality at optimal doping.[177] Altogether, the idea that the cuprate superconductivity is rooted in the quantum critical nature of the optimally doped metal is, at present, again very popular, but more work needs to be done to nail down how it precisely works. The empirical situation is in this regard more transparent in the development that came next: the heavy fermions.

This unfolded in the early 1990s with the discovery of a phletora of quantum phase transitions in the heavy fermion intermetallics, as triggered by the seminal work of the Lonzarich group at Cambridge.[178] These quantum phase transitions involve rather straightforward magnetic order that is disappearing. However, on both sides of such quantum critical points (QPT) one finds Fermi liquids

[173] S. Sachdev and J. Ye, *Phys. Rev. Lett.* 69 (1992) 2411

[174] J. Zaanen, *Nature* 430 (2004) 513; embarasingly, I called such a state maximally viscous not realizing that viscosity is proportional to the relaxation time. The Planckian dissipator is therefore as close to a perfect (un-viscous) fluid as possible, as was realized in the same period in the context of string theory and the quark-gluon plasma.

[175] A.V. Chubukov and S. Sachdev, *Phys. Rev. Lett.* 71 (1993) 169

[176] S. Sachdev, *Quantum Phase Transitions*, Cambridge Univ. Press (1999)

[177] As argued by, for instance, Philip Phillips, the Hubbard projections might come to an end for increasing doping when U is finite (*Rev. Mod. Phys.* 82 (2010) 1607; for closely related ideas in the heavy fermion context see C. Pepin, *Phys. Rev. Lett.* 98 (2007) 206401). As we argued (*arXiv*:0911.4070), the big Fermi surface in the overdoped regime might be taken as evidence that the Hubbard projections have disappeared in the overdoped regime (in the spirit of the considerations in Anderson's book) while the Mott-ness dominated underdoped regime falls prey to competing orders because of the altered "Weng quantum statistics". The ramification would be that the quantum criticality is rooted in a quantum-statistics clash between the different statistics of the Mott-ness and Fermi liquid regimes.

[178] See, e.g., J. Zaanen, *Science* 319 (2008) 1205 and references therein.

that are characterized by a quasiparticle mass that tends to diverge at the critical point. The fermions are in one or the other way "on the critical surface" and perhaps the most dramatic empirical fact is that one invariably finds that a "last minute" instability sets in, centered at and "shielding" the quantum critical point. Usually this corresponds with an unconventional superconductor that takes over from a normal state showing the traits of a quantum critical metal like a linear resistivity. Since these systems involve the localized f electrons of lanthanides or actinides the microscopic scales are much smaller than in the cuprates. All along it has been realized that by "boosting" up their microscopic energy cutoff, the physics of the heavy fermion systems would be strikingly similar to that of the cuprates, adding credibility to the idea that also in the latter case the superconductivity is eventually rooted in the quantum criticality itself.

As discussed in Section 2.4.2, the idea that spin fluctuations are responsible for the cuprate super-conductivity has been around since the very beginning. Catalyzed by the discoveries in the heavy fermion systems, this merged with the quantum criticality idea, actually falling back on a theoretical interpretation of quantum criticality that was introduced many years before CNH entered the stage. This dates back to the development of paramagnon theory discussed in the second section. Surely, in the context of itinerant magnetism paramagnons become significant when one gets close to the transition to ordered magnetism. In 1976 John Hertz reformulated this in the modern thermal field theory language, respecting the essence of quantum criticality, but eventually ending up with a story that is operationally similar to the RPA treatment (the Moriya paramagnon theory).[179] It starts out with the weakly interacting Fermi gas, the Hubbard-Stratonovich order parameter fields are introduced, and it is asserted that the fermions can be integrated out on the RPA level, forming a heat bath damping the order parameter. This lifts the effective dimensionality of Euclidean spacetime felt by the order parameter dynamics such that it lands at or above its upper critical dimension, implying that the fixed point is Gaussian. In the modern era it was realized that these order parameter fluctuations "backreact" on the fermions in the form of causing a singular interaction in the pairing channel.[180] This in effect boils down to a fancy version of the spin fluctuation "super-glue" idea, where now the spin fluctuations appear in a quantum critical incarnation causing the superconducting dome surrounding the quantum critical point.

This has turned into a quite popular view in both the heavy fermion and cuprate communities. It could even be correct, but the problem is that the construction rests on a manifestly uncontrolled approximation scheme. The fundamental problem is the fermion signs. Regardless the representation, fermions are not governed by an effective Boltzmannian physics in Euclidean spacetime due to the destructive workings of the Fermi-Dirac statistics rendering the quantum partition sum to be one with mathematically ill-behaved "negative probabilities". In dealing with a fermionic critical state, one can therefore a priori not fall back on the "hidden" classical criticality, which is the secret behind the rigor of the CNH/Sachdev treatment of the bosonic problem. In the Hertz treatment this "fermion nightmare" was worked under the rug in the beginning, by the assumption that the fermions can just be integrated out. The flaw is that the integration of the fermions requires them to be massive but the fermion sector is massless, and therefore the Fermi surface itself is somehow part of the zero temperature critical state. In fact, it has become clear more recently that in some heavy fermion systems the Fermi surfaces reconstruct drastically at the quantum critical point, demonstrating directly that the fermionic degrees of freedom are "active" part of the criticality.[181]

With regard to the key question of why T_c is as high as it is, we have arrived at the present day. Although one still finds plenty of strong opinions regarding the truth of spin fluctuations or the di-

---

[179] J. Hertz, *Phys. Rev.* B14 (1976) 1165, with some corrections by A. Millis, *Phys. Rev.* B48 (1993) 7183

[180] For a review see Ar. Abanov, A.V. Chubukov and J. Schmalian, *Adv. Phys.* 52 (2003) 119

[181] See the recent review: H. von Löhneisen et al., *Rev. Mod. Phys.* 79 (2007) 1015

vinity of RVB in the community, the "agnost view" to which I referred in the beginning has become steadily more influential in the community. At least my own perspective has changed considerably in the last quarter century, from the desire to just explain why T_c is high to the conviction that the cuprates have been instrumental in making visible a very deep and general problem in fundamental physics at large. The modern code word is "quantum matter", the question regarding the general nature of matter as governed by the fundamental laws of quantum physics and the general principles of emergence. Remarkably, it appears that physics is entering an era where the empirical condensed matter and the mathematical high energy communities are merging their activities for a head-on attack on this, in a way, very new problem.

Some twenty years ago there was an implicit but widespread belief that we knew what matter is. However, high-T_c superconductivity and related phenomena have played the role of the Michelson-Morley experiment, opening our eyes for the fact that we actually have no clue and that there is plenty of room for surprises. I already alluded a number of times to the "fermion signs", the "negative probabilities" that spoil the mathematics that we can handle, which also show up in various guises in problems that are not governed by fermion statistics (such as frustrated spins, bosons when time reversal is broken, any non-equilibrium quantum-field problem). The bottom line is that these "quantum signs" are ubiquitous and in their presence we have no mathematical tools for the description of matter that really works. The exception is the incompressible states (like the fractional quantum Hall states, topological insulators) where topology is helping us out, but for compressible states it is just guess work: we just know from experiment that Fermi liquids and the "Hartree Fock states" (including those of Section 2.4.5) are part of nature's portfolio of forms of matter.

The problem of fermionic quantum criticality is rather interesting from this viewpoint. Scale invariance is a powerful symmetry and it should somehow make the general sign problem easier. However, departing from the 50 year old established mathematical technology, one runs immediately into a deep problem: to have perturbative control one needs the fact that the Fermi liquid has a scale (the Fermi energy). What to think of a fermionic matter without Fermi energy, and how to think about a Fermi surface when there are no quasiparticles? In the Hertz approach these questions are worked in the rug since these constructions are departing implicitly from the stability of the Fermi liquid. We obviously need mathematical weaponry to address these issues, which is not lying on a shelf somewhere.

With the reservation that what follows is very fresh, it might be that, in this regard, 2010 will end in the history books as the year that it started to happen. During the last forty years, string theorists have been working hard developing some very new mathematical machines. Although their performance as mere mathematical constructs is breath taking, it has been quite frustrating that these string-machines do not seem to have an obvious application to explain observable phenomena in nature. But this is now changing: it appears that the most powerful machine of string theory (the Anti-de-Sitter/Conformal Field Theory correspondence) is tailored to compute the properties of states of sign-full quantum matter. The way this works is at first sight outrageous, but it becomes very beautiful when one get used to the idea. It departs from the notion that, in a special sense, black holes behave as material objects when combined with quantum physics. They carry entropy and so forth, and in their simplest incarnation they just behave as black body radiators giving off thermalized Gaussian fields. The modern string theoretical version has turned this into a "generalized particle wave" duality, insisting in a surprisingly precise mathematical way that such black hole worlds in $d + 1$ dimensional space times are in a dual sense equivalent to strongly interacting forms of quantum matter living in a boring flat d-dimensional space-time—the stuff that one finds in the earthly material science and cold atom labs!

Right now the string theorists are sorting out collections of colorful "hairy black holes" that by themselves represent a leap forward in quantum gravity. These in turn render a description of quantum matter that extends way beyond the fifties' paradigm. It started with the recognition in 2007 by Sachdev, Son and coworkers[182] that the correspondence encodes the "Planckian dissipation" hydrodynamics associated with the finite temperature quantum critical state in the condensed matter systems. Since last year this development sped up, by the discovery that Fermiliquids emerge "out of the blue" from a strongly interacting critical ultraviolet, while their stability is encoded by a quantum-mechanical "Dirac hair" hanging some distance way from the black hole horizon.[183] However, such "Dirac hair" can be "shaved away" and the resulting "extremal" Reissner-Nordstrom black-hole codes for an emergent quantum critical fermion state. This is breathtaking: it describes an infrared that has a sole temporal scale invariance, while the "matching" with the bare ultraviolet fermion generates a singularity structure in momentum space that corresponds with the generalization of the Fermi-surface to the non-Fermi-liquid realms[184]. This is quite like the marginal Fermi liquid, which was introduced on mere empirical grounds!

Last but not least, the "hairy black hole" pursuit started in 2008 by the "H^3" discovery[185] of the holographic encoding of the superconductor in terms of "scalar hair". This tells us a fascinating story of a "glueless" BCS-like superconductivity, where the superconducting instability is driven by a kind of "perfect quantum frustration" of the non-superconducting quantum critical state (of the "bald" extremal kind), as manifested by the fact that the "naked" quantum critical point would carry a zero temperature entropy that is "eaten" by the superconductor. This resonates with the idea that has been around, especially in the heavy fermion community, for quite some time: were it not for the intervention of the "last minute" superconductivity, one would have to deal with a Fermi liquid acquiring an infinite quasiparticle mass right at the critical point, representing of course such a quantum frustration. The end of the present history is that right now the first attempts are undertaken to nail down this quantum frustration in the experimental laboratories.[186]

I might be overly optimistic but I have to admit that the developments discussed in the last paragraphs have been a source of inspiration to do my best to formulate the present "chronicles of the dark ages of quantum matter". In case the black hole dreams land on the real axis, the professional historians will be quite curious regarding the hairy details of the prehistory of this scientific revolution. When this fails, it doesn't matter since at some point in the future a very young physicist will likely scribble on a black board equations of a quality that in no time a general consensus will emerge that the mystery of high-T_c superconductivity is solved. Given our present understanding of the depth of the mystery there is no doubt in my mind that the insights gained from these equations should be at least as good as the black holes!

2.5 Intrinsic Josephson Effect in Layered Superconductors

L. N. Bulaevskii and A.E. Koshelev

2.5.1 Introduction: Layered Superconductors and Lawrence-Doniach Model

The first known superconductors (Hg, Pb, Sn, Nb, . . .) had nearly cubic crystal structures. For that reason, the free energy functional proposed by Ginzburg and Landau (GL) in 1950 [see elsewhere

[182] C.P. Herzog et al., *Phys. Rev.* D75 (2007) 085020, see also J. Zaanen, *Nature* 448 (2007) 1000 for a light perspective.

[183] M. Cubrovic, J. Zaanen and K. Schalm, *Science* 325 (2009) 439

[184] H. Liu, J. McGreevy and D. Vegh, *arXiv*:0903.2477, T. Faulkner et al., *Science* 329 (2010) 1043; see also S. Sachdev, *arXiv*:1006.3794 for inspirational ideas of how this relates to the matters discussed in section 2.4.4.

[185] S.A. Hartnoll, C. Herzog and G. Horowitz, *Phys. Rev. Lett.* 101 (2008) 031601

[186] See J. Zaanen, *Nature* 462 (2009) 15

in this chapter] to describe properties of these materials was written for an isotropic case,

$$F_s\{\Psi(\mathbf{r})\} = \int d\mathbf{r}\left\{a(T)|\Psi|^2 + \frac{1}{2}b|\Psi|^4 + \frac{\hbar^2}{4m}\left|\left[-i\nabla - \frac{2\pi}{\Phi_0}\mathbf{A}(\mathbf{r})\right]\Psi\right|^2 + \frac{\mathbf{B}^2}{8\pi}\right\}. \tag{2.76}$$

Here $a(T) = \alpha(T - T_c)$, $\mathbf{B} = \nabla\times\mathbf{A}$ is the magnetic induction, $\mathbf{A}$ is the vector potential, m is the electron effective mass, and Φ_0 is the flux quantum. A first attempt to describe *anisotropic* super-conductors was made by Ginzburg even before the GL functional was proposed, in 1944 [187]. He introduced an anisotropic form of the supercurrent energy in the framework of the London theory. This corresponds to replacement of the effective mass by the effective mass tensor in the GL functional,

$$\frac{\hbar^2}{4m}\left|\left(-i\nabla - \frac{2\pi}{\Phi_0}\mathbf{A}\right)\Psi\right|^2 \Rightarrow \frac{\hbar^2}{4}(m^{-1})_{ij}\left(i\nabla_i - \frac{2\pi}{\Phi_0}A_i\right)\Psi^*\left(-i\nabla_j - \frac{2\pi}{\Phi_0}A_j\right)\Psi. \tag{2.77}$$

Here the effective mass tensor is the same as that in the electron spectrum, $\epsilon(\mathbf{k}) = (\hbar^2/2)(m^{-1})_{ij}k_ik_j$, in crystals with elliptical Fermi surface. Such a functional is valid at temperatures close to T_c and, when the superconducting correlation lengths $\xi_i = \hbar/(4m_{ii}\alpha)^{1/2}$ are much bigger than interatomic distances as the atomic structure, is ignored in this phenomenological approach.

Superconductors with significant anisotropy like the dichalcogenides of transitions metals ($NbSe_2$, TaS_2) were synthesized in the 1960s[188]. In spite of the layered crystal structure and the considerable effective mass anisotropy[189], $\gamma^2 = m_c/m_{ab} \lesssim 50$, the c-axis coherence length was estimated as $\xi_c \approx 65$ Å, which is still much larger than the interlayer spacing s. Hence, the anisotropic GL model is adequate for dichalcogenides. As was realized in the 1980s, though the anisotropic GL functional is a simple generalization of the isotropic one, it describes the behavior in magnetic fields qualitatively different from that inherent to isotropic crystals. The reason is that while the energy of supercurrents may be rescaled, $(x,y)/\xi_{ab} \to (x',y')$, $z/\xi_c \to z'$, to the isotropic form, such a transformation does not rescale the magnetic field energy $\mathbf{B}^2/(8\pi)$. As a consequence, the magnetic moment induced by an applied magnetic field tilted with respect to the principal crystal axes deviates from the direction of the magnetic induction leading to a finite torque acting on the crystal[190].

The next important step in the study of anisotropic superconductors was made when dichalco-genide crystals were intercalated by large organic molecules[191]. These molecules are placed be-tween the layers significantly increasing the interlayer spacing. As a result, tunneling of electrons between the layers is strongly reduced. For example, intercalation of $2H$-TaS_2 by pyridin increases anisotropy of the conductivity and of the effective mass from 50 to 10^5. As a consequence, the c-axis coherence length ξ_c drops below the layer spacing. The anisotropic GL model does not describe such a situation. In this case, it is natural to think about interlayer electron motion as a tunneling process with an interlayer kinetic energy smaller than the superconducting condensation energy. Lawrence and Doniach, invoking an analogy with superconductor-insulator-superconductor structures, where electrons tunnel through the insulating layer, proposed a model of Josephson coupling between su-perconducting layers (LD model)[192]. The model uses a discrete variable n, the layer index, instead of a continuous variable z to describe the dependence of the order parameter, $\Psi_n(\mathbf{r})$ in the direction perpendicular to the layers. Here $\mathbf{r} = (x,y)$ are the coordinates in the layers. The LD functional has

[187] V. L. Ginzburg, Zh. Eksp. *Teor. Fiz.* 14 (1944) 134

[188] J. Wilson and A. Yoffe, *Adv. Phys.* 18 (1969) 193

[189] We use the subscript "*ab*" for parameters in the layer plane *ab* (x,y), while "*c*" defines those along the c-axis (z-axis), perpendicular to the layers.

[190] V. G. Kogan and J. R. Clem, *Phys. Rev.* B 24 (1981) 2497; V. G. Kogan, *Phys. Rev.* B 38 (1988) 7049

[191] F. Gamble, F. DiSalvo, R. Klemm, and T. Geballe, *Science* 168 (1970) 568

[192] W. E. Lawrence and S. Doniach, in *Proc. 12th Int. Conf. Low Temp. Phys.*, Kyoto, Japan, edited by E. Kanda, p. 361, 1970

the form

$$F_s\{\Psi(\mathbf{r})\} = \int d\mathbf{r} \left\{ s \sum_n \left[a(T)|\Psi_n|^2 + \frac{1}{2}b|\Psi_n|^4 + \frac{\hbar^2}{4m_{ab}} \left| \left(-i\nabla - \frac{2\pi}{\Phi_0}\mathbf{A}(\mathbf{r}) \right)\Psi_n \right|^2 \right. \right.$$

$$\left. \left. + \eta \left| \Psi_n - \Psi_{n+1}e^{-i\chi_n} \right|^2 \right] + \int dz \frac{\mathbf{B}^2}{8\pi} \right\}, \text{ with } \chi_n = \frac{2\pi}{\Phi_0} \int_{ns}^{(n+1)s} dz A_z(\mathbf{r}). \tag{2.78}$$

The parameter $\eta = 16t^2/\epsilon_F$ was derived using the "tunneling Hamiltonian" method. This functional corresponds to the tight-binding electron spectrum $\epsilon(\mathbf{k}) = (\hbar^2/2m_{ab})[k_x^2 + k_y^2] + 2t[1 - \cos(k_z s)]$, where t is the interlayer transfer integral. For such a spectrum at $t \ll \epsilon_F$, the Fermi surface is open and has the shape of a corrugated cylinder, while $m_c = \epsilon_F/t^2 s^2$. The LD functional has been derived from the BCS Hamiltonian with a tight-binding electron spectrum[193] in the manner similar to Gor'kov's derivation of the GL functional (2.76). In this derivation the dependence of the superconducting order parameter on the coordinate z across the layers was written in a discrete representation with the Wannier functions localized on the layers n, neglecting the functions' overlap for adjacent layers. Then the expansion in the small parameter t/T_c gives the LD functional (2.78).

The LD model is characterized by the dimensional crossover at a temperature $T^* = T_c (1 - \xi_c^2/s^2)$. Close to T_c, in the temperature interval (T^*, T_c) one has $\xi_c(T) \gg s$ and the LD model is equivalent to the anisotropic GL one with the anisotropy $\gamma^2 = m_c/m_{ab}$. Below T^* the LD model exhibits new properties. On the one hand, the vector-potential components A_x, A_y, describing the magnetic field perpendicular to the layers, enter the LD functional in the same way as in the isotropic GL case. The behavior of a superconductor in such magnetic fields is not expected to be very different from the isotropic case. Indeed, expressions for the $H_{c2,c}$ and $H_{c1,c}$ remain the same below T^*, but a vortex in the z-direction is not a continuous Abrikosov vortex but rather a discrete stack of two-dimensional (2d) vortices (pancakes) with screening currents localized mainly in the superconducting layers[194]. On the other hand, the dependence of the LD functional on the component A_z, describing the fields B_x, B_y and E_z, is oscillating, and the behavior in such fields should be of the Josephson type. The properties in parallel fields change drastically, as the weak supercurrents induced by such fields cannot destroy the superconducting pairing inside the layers. Thus $H_{c2,ab}$ is bound only by the spin paramagnetic effect [193,195], while vortices in parallel fields become Josephson vortices without normal core.[193] Lawrence and Doniach noticed that the behavior in a perpendicular electric field is of the Josephson type. The frequency of the Josephson plasma oscillations, $\omega_{p,c}$, induced by such a field, depends on the average phase difference $\Delta\varphi$ between the layers,

$$\omega_{p,c}^2 = (eJ_0/\hbar s\epsilon_c)\cos(\Delta\varphi), \quad J_0 = c\Phi_0/(8\pi^2 s\lambda_c^2), \tag{2.79}$$

where λ_{ab} and λ_c are the London penetration depths, ϵ_c is the c-axis dielectric constant, and J_0 is the Josephson interlayer critical current. At $T = 0$ and $\Delta\varphi = 0$ one obtains expression for the plasma frequency in the normal state, $\omega_{p,c}^2 = 4\pi Ne^2/(m_c\epsilon_c)$, where N is the electron density. The phase difference $\Delta\varphi$ may be induced by the interlayer supercurrent, for example. As was found later, Eq. (2.79) leads to the dependence of the Josephson plasma frequency on the perpendicular magnetic field because $\Delta\varphi$ is induced by misalignment of the pancake vortices along the c-axis due to thermal fluctuations and pinning (see below).

Josephson effects in intercalated compounds have never been demonstrated experimentally, probably because nonuniform intercalation resulted in shorts between the layers. The situation changed dramatically after the discovery of Bi- and Tl-based layered high-temperature cuprate superconductors, such as $Bi_2Sr_2CaCu_2O_{8+\delta}$ (Bi-2212). In these crystals thick insulating layers strongly suppress

[193] L. N. Bulaevskii, *Zh. Eksp. Teor. Fiz.* 64 (1973) 2241 [*Sov. Phys.-JETP* 37 (1973) 1133]

[194] J. R. Clem, *Phys. Rev.* B 43 (1991) 7837

[195] R. A. Klemm, M. R. Beasley, and A. Luther, *J. Low Temp. Phys.* 16 (1974) 607

electron tunneling between the superconducting CuO_2 layers. In Bi-2212 an anisotropy $\gamma \approx 500$ has been evaluated in optimally doped crystals and up to 1000 in the underdoped ones. Also, these materials have a very short in-plane coherence length $\xi_{ab} \approx 2nm$. By use of the anisotropic GL theory one estimates $\xi_c = \xi_{ab}/\gamma \ll s$, except very close to T_c . The conclusion made in 1992 was that the anisotropic GL model is inadequate for Bi-2212, and it should be replaced by the Josephson LD model which accounts for the layered structure along the c-axis[196,198].

To check the validity of the LD model in Bi- and Tl-based superconductors, Bulaevskii, Clem, and Glazman[196] proposed to study the oscillations in the dependence of the c-axis critical current on the magnetic field H parallel to the layers and perpendicular to the crystal side with the length W smaller than the Josephson length $\lambda_J = \gamma s$. In this case the interlayer critical current, J_c, is not limited by vortex depinning but solely by the Josephson nature of interlayer coupling, leading to Fraunhofer-like oscillatory dependence on H, $J_c(H) = J_0[\sin(\pi H W s/\Phi_0)]/(\pi H W s/\Phi_0)$ with a periodicity of the oscillations Φ_0/Ws. Such oscillations were indeed observed in tiny Bi-2212 crystals in 2006 [197].

In the same year, 1992, Kleiner et al.[198] proved experimentally that Bi-2212 crystal consists of a stack of intrinsic Josephson junctions. They measured the I-V characteristics for an interlayer current and observed many discrete branches corresponding to the transition from the superconducting to resistive state when the current exceeds J_0 for a given junction. They also observed strong hysteresis for each single junction at descending bias current; i.e. the junctions remain resistive at currents well below J_0 on descending branch. The authors concluded that Bi-2212 single crystals are naturally grown stacks of underdamped Josephson junctions. This message accelerated the following experimental studies of the c-axis properties of layered high-temperature superconductors and resulted in a deeper understanding of the rich physics described by the LD model.

The LD equations[192] describe the static properties of intrinsic Josephson junctions (IJJ). To study IJJ dynamics, the time-dependent equations were needed. The direct extension of the LD equations was achieved[199] by use of the Maxwell equations with currents following from the LD functional (2.78),

$$\mathbf{j}_{ab} = -\frac{c}{4\pi}\frac{\delta F_s}{\delta \mathbf{A}(\mathbf{r})} + \sigma_{ab}\mathbf{E}_{xy}, \quad j_{z,n} = -\frac{c}{4\pi}\frac{\delta F_s}{\delta A_{z,n}(\mathbf{r})} + \sigma_c E_z, \tag{2.80}$$

where $j_{z,n}$ is the current between layers n and $n+1$, while $\mathbf{A}_{z,n}(\mathbf{r})$ is the average vector potential there, σ_{ab} and σ_c are the quasiparticle conductivities, while $(\mathbf{E}_{xy}, E_z)$ is the electric field. A year earlier Sakai et al.[200] derived equations for multilayer consisting of superconducting slabs with a thickness t characterized by the bulk penetration length λ separated by insulating layers of a thickness d. In principle, at $t \ll d$ this "multilayer" model is equivalent to the LD-based dynamic model. However, the parameters, t, d, and λ do not have direct physical meaning and have to be related to the LD parameters, s, λ_{ab}, and λ_c, accessible experimentally.

2.5.2 Vortex Matter in Layered Superconductors

External magnetic fields generate vortices in layered superconductors. The vortex matter in Josephson coupled layered superconductors displays an amazingly rich spectrum of physical phenomena since the effects of parallel and perpendicular fields are drastically different and two very different vortex species exist in layered superconductors. A magnetic field applied along the layers creates Josephson vortices without the normal cores, see Figure 2.1a. A magnetic field applied perpendicular to the layers creates 2d vortices piercing the layers, named "pancake vortices" by

[196] L. N. Bulaevskii, J. R. Clem, and L. I. Glazman, *Phys. Rev.* B 46 (1992) 350

[197] S. Kim, S. Urayama, H. B. Wang, S. Kawakami, K. Inomata, M. Nagoa, K. S. Yun, Y. Takano, K. Lee, and T. Hatano, J. Phys. *Chem. Solids* 57 (2006) 438

[198] R. Kleiner, F. Steinmeyer, G. Kunkel, and P. Müller, *Phys. Rev. Lett.* 68 (1992) 2394

[199] L. N. Bulaevskii, et al., *Phys. Rev.* B 50 (1994) 12831

[200] S. Sakai, P. Bodin, and N. F. Pedersen, *J. Applied Physics* 73 (1993) 2411

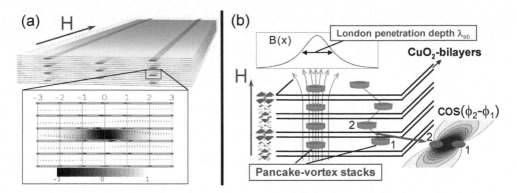

FIGURE 2.1: **(a)** Josephson vortex lattice in magnetic field applied parallel to the layers and the structure of an isolated Josephson vortex. The arrows represent the current distribution. The gray-level codes the cosine of the interlayer phase difference. The horizontal scale is in units of the Josephson length. **(b)** Aligned and misaligned pancake-vortex stacks in magnetic field applied perpendicular to the superconducting layers. Phase perturbation caused by two misaligned pancake vortices is also illustrated.

Clem[194].

Josephson vortices. The concept of a vortex in a long superconducting tunneling junction was introduced by Josephson in 1965 [201]. The key feature of this object is the existence of a single length scale, the Josephson length, within which the phase difference between the superconducting banks changes by 2π and magnetic flux is concentrated. In their seminal paper, Lawrence and Doniach assumed that the flux line in a layered superconductor directed parallel to the layers has exactly the same structure[192]. Later, a more careful analysis[193] revealed substantial differences. The vortex center is located in the insulating region between two superconducting layers so that there is no region of suppressed order parameter (see Figure 2.1a). There is indeed a length scale $\lambda_J = \gamma s$, within which the phase difference between adjacent layers changes by 2π. This length scale determines the nonlinear core size of the vortex and usually is also called the Josephson length. However, in contrast to the vortex in a tunnel junction, the magnetic field decays over a much larger length scale, λ_c, along the layer direction and at the scale λ_{ab} in the perpendicular direction. In Bi-2212 λ_J, λ_{ab} and λ_c are 0.8, 0.25 and 130 μm, respectively. Away from the nonlinear core the "layeredness" is not essential and the vortex has the same structure as in three-dimensional anisotropic materials.

At finite concentrations, the Josephson vortices form a triangular lattice stretched along the layer direction (see Figure 2.1a). The ratio of the lattice constants along and perpendicular to the layers is given by the anisotropy factor γ. The distance between the vortices decreases inversely proportional to the square root of the magnetic field. In fields of the order $B_{cr} = \Phi_0/2\pi\gamma s^2$ the in-plane intervortex distance is of the order of the nonlinear core size λ_J. For Abrikosov vortices, the overlap of normal cores marks the destruction of superconductivity. In the case of Josephson vortices which do not have normal cores, this field just corresponds to a crossover to a new dense-vortex-lattice regime[202]. In that regime the Josephson vortices homogeneously fill all layers. This state is characterized by rapid oscillations of the Josephson current in the direction of the layers and by very weak modulation of the in-plane supercurrent. As the magnetic field applied along the layers does not destroy superconductivity via the orbital effect, the dense-lattice regime may survive up to extremely high fields, only limited by the paramagnetic effect.

---

[201] B. D. Josephson, *Adv. Phys.* 14 (1965) 419

[202] L. Bulaevskii and J. R. Clem, *Phys. Rev.* B 44 (1991) 10234

Pancake vortices. The perpendicular applied magnetic field determines the density of pancakes which is the same in all layers because the energy of an extra pancake in a layer diverges logarithmically with crystal size, as was shown by Efetov in 1979 [203]. This implies that at low temperature and low magnetic fields the pancakes must form stacks, (Figure 2.1b). The stacks are aligned by the relatively small line-tension energy which consists of two contributions: a magnetic and a Josephson coupling. The magnetic coupling appears because pancake currents generate a magnetic field which extends over many layers, leading to a long-range interaction between pancakes. In contrast, the Josephson interaction is mostly of the nearest-neighbor type. The local Josephson coupling energy between layers n and $n+1$, $E_J = (\Phi_0 J_0/2\pi c)\cos\varphi_{n,n+1}$, is determined by the phase difference between these layers $\varphi_{n,n+1}$. The pancakes in a straight stack do not induce phase difference. However, if pancakes in the stack are positioned at different coordinates $\mathbf{R}_n \neq \mathbf{R}_{n+1}$, a phase difference is induced which has a very simple form at points $\mathbf{r}$ within a distance λ_J from the stack,

$$\Delta\varphi_{n,n+1}(\mathbf{r}) = f(\mathbf{r} - \mathbf{R}_n) - f(\mathbf{r} - \mathbf{R}_{n+1}), \quad f(\mathbf{r}) = \tan^{-1}(y/x). \tag{2.81}$$

Thus, misaligned pancakes increase the phase difference between the layers and reduce the Josephson coupling[204]. The relative strength of Josephson and magnetic couplings is determined by the ratio λ_J/λ_{ab}. For most layered superconductors $\lambda_J/\lambda_{ab} \ll 1$, meaning that the Josephson contribution is typically stronger than the magnetic one. Only in Bi-2212, due to its extreme anisotropy, are these contributions of the same order. Due to the weakness of the interlayer coupling, the pancake-vortex stacks are very flexible with respect to thermal fluctuations and pinning. Disorder of the pancake lattice results in a reduction of the Josephson coupling and the Josephson critical current densities which become proportional to the averaged cosine of the phase differences, $\langle\cos\varphi_{n,n+1}(\mathbf{r})\rangle$.

At finite densities and low temperatures the pancake-vortex stacks form a triangular lattice, just as the Abrikosov vortices. This lattice is subject to strong thermal fluctuations and melts at fields much smaller than the upper critical fields. The nature of fluctuations and melting change when the inter-vortex distance becomes of the order of the Josephson length, λ_J, corresponding to the field $B_J = \Phi_0/\lambda_J^2$ [204,205]. At fields $B < B_J$, the vortex lattice fluctuates as an ensemble of interacting elastic lines, while at $B > B_J$ the lattice behaves like a stack of weakly interacting 2d lattices. Correspondingly, at small fields the lattice melts into a liquid of elastic lines and at high fields it melts into nearly independent 2d liquids in the layers. The first theoretical predictions for the location of a melting transition line using the Lindemann criterion were given in Brandt[206].

Experimentally, the vortex-lattice melting was reported in 1992 for high-quality $YBa_2Cu_3O_7$ (Y-123) crystals with $\xi_c \sim s$. It was seen as a sharp drop of the resistivity at a well-defined field-dependent temperature [207]. Since the resistivity is not a thermodynamic quantity, this "melting" interpretation of the resistivity jump was generally accepted only after the thermodynamic signatures of the first-order transition were found several years later in the specific heat and the magnetization, exactly at the location of the resistivity jump. In a parallel development, at approximately the same time, the vanishing of crystalline order of the vortex matter in Bi-2212 at a well-defined line in the field-temperature phase diagram was directly demonstrated by neutron scattering[208]. Very soon after that magnetization jumps were found at this line [209]. The last skeptics were convinced after the melting transition was directly imaged by scanning Hall probes [210]. Due to the much higher anisotropy, the melting fields in Bi-2212 are about 100 times lower than in Y-123, typically in the

[203] K. B. Efetov, Zh. Eksp. *Teor. Fiz.* 76 (1979) 1781 [*Sov. Phys.-JETP* 49 (1979) 905

[204] L. I. Glazman and A. E. Koshelev, *Phys. Rev.* B 43 (1991) 2835

[205] D. S. Fisher, M. P. A. Fisher, and D. A. Huse, *Phys. Rev.* B 43 (1991) 130

[206] E. H. Brandt, *Phys. Rev. Lett.* 63 (1989) 1106; A. Houghton, R. A. Pelcovits, and A. Sudbø, *Phys. Rev.* B 40 (1989) 6763

[207] H. Safar et al., Phys. Rev. Lett 69 (1992) 824; W. K. Kwok et al., Phys. Rev. Lett. 69 (1992) 3370

[208] R. Cubitt et al., *Nature* 365 (1993) 407

[209] H. Pastoriza et al., Phys. Rev. Lett. 72 (1994) 2951; E. Zeldov et al., *Nature* 375 (1995) 373

[210] A. Oral et al., *Phys. Rev. Lett.* 80 (1998) 3610

range 20-200 Gauss.

Real materials always contain disorder caused by defects in the crystal lattice. The crystalline disorder creates a random potential for the vortices which strongly influences the "vortex matter" phase diagram. Early magnetization measurements of the highly anisotropic cuprates revealed the so-called "second peak" effect, a sharp increase of the irreversible magnetization, caused by an increase of effective vortex-lattice pinning, above a well-defined field in the range 200-400 gauss [211]. Further research demonstrated that at higher temperatures the "second peak" line smoothly merges with the melting line[208]. This led to the conclusion that the "second peak" field B_{sp} marks the destruction of the vortex lattice by disorder. In real Bi-2212 crystals at high fields, an ordered vortex lattice does not exist and a quasi-2d melting regime at fields $B > B_J$ has never been observed. Pinning results in a suppression of the Josephson coupling in addition to the suppression caused by thermal fluctuations. Both effects lead to a magnetic-field induced suppression of the c-axis critical current.

Lattice structures in tilted magnetic fields. Layered superconductors in a magnetic field tilted with respect to the layers have an amazingly rich phase diagram due to the complex interplay between various interactions. In moderately anisotropic superconductors, $1 < \gamma < \lambda_{ab}/s$, the vortex lattice consists of tilted vortex lines. The direction of the vortices differs from that of the applied magnetic field, and the interaction of the tilted vortices is modified in comparison with the standard repulsion in isotropic superconductors. For some directions of the applied field, the vortex interaction becomes attractive, leading to the formation of chains[212]. In addition, there is an angular range within which the tilted line is unstable[213]. As a consequence, in some range of tilting angles the lattice has to be composed of vortex rows tilted at different angles. In superconductors with very high anisotropy $\gamma > 2\lambda_{ab}/s$ tilted magnetic fields create two interpenetrating sub-lattices: a sub-lattice of Josephson vortices and a sub-lattice of pancake stacks. This very unusual state of vortex matter is known as crossing or combined lattices [214]. The idea that highly anisotropic superconductors are almost independently responding to the H_{xy} and H_z field components has been proposed by Kes et al.[215].

When the magnetic field is applied at a very small angle with respect to the layers the Josephson vortices remain locked along the layer direction up to a critical lock-in angle[214,216]. Above this angle, the penetration mechanism of the c-axis magnetic flux depends on the anisotropy. For small anisotropy, $\gamma < 2\lambda_{ab}/s$, the flux penetration starts with the formation of kinks on the Josephson vortices leading to the tilted kinked chains. For large anisotropy, $\gamma > 2\lambda_{ab}/s$, the pancake stacks are formed[214]. Due to the attractive interaction between pancake-vortex stacks and Josephson vortices[217], the stacks are located on Josephson vortices forming crossing chains. An interesting behavior was found experimentally in Bi-2212 at temperatures close to T_c [218]. It was observed that the penetration starts with the formation of kinks, but at some critical c-axis field the density of pancake vortices jumps and aligned pancake stacks are formed on Josephson vortices. Such behavior was predicted theoretically for intermediate anisotropies $\gamma \approx 2\lambda_{ab}/s$ [219]. Upon a further increase of the magnetic field, the phase-separated state is formed which is composed of the pancake-stack chains located on Josephson vortices separated by several rows of the triangular lattice with larger intervor-

---

[211] V. N. Kopylov, I. F. Shegolev, and T. G. Togonidze, *Physica* C 162-164 (1989) 1143; N. Chikumoto, et al., *Physica* C 185–189 (1991) 2201

[212] A. I. Buzdin and A. Y. Simonov, Pisma Zh. *Eksp. Teor. Fiz.* 51 (1990) 168 [*Sov. Phys.-JETP Lett.* 51 (1990) 191]

[213] E. Sardella and M. A. Moore, *Phys. Rev.* B 48 (1993) 9664

[214] L. N. Bulaevskii, M. Ledvij, and V. G. Kogan, *Phys. Rev.* B 46 (1992) 366

[215] P. H. Kes, J. Aarts, V. M. Vinokur, and C. J. van der Beek, *Phys. Rev. Lett.* 64 (1990) 1063

[216] D. Feinberg, *Physica* C 194 (1992) 126

[217] A. E. Koshelev, *Phys. Rev. Lett.* 83 (1999) 187

[218] A. Grigorenko, S. Bending, T. Tamegai, S. Ooi, and M. Henini, *Nature* 414 (2001) 728

[219] A. E. Koshelev, *Phys. Rev.* B. 68 (2003) 094520

tex separation. Such a state was observed in Bi-2212 by Bitter decorations in 1991 [220]. Since then various unusual lattice structures were visualized in a number of cuprates in small fields by several techniques, see review[221] and references therein.

2.5.3   Josephson Plasma Resonance

The excitation spectrum of superconductors consists of the quasiparticle branch and collective modes corresponding to phase oscillations. The latter branch was considered first by Bogoliubov for the Bose-Einstein condensate of neutral gases[222]. He found that the dispersion of collective mode, $\omega(\mathbf{k})$, is linear in the momentum k, in agreement with the Goldstone theorem stating that in a degenerate ground state with broken symmetry the collective mode should be gapless. For superconductors, within the BCS framework and neglecting the long-range Coulomb interaction, Bogoliubov found for the collective mode dispersion $\omega(\mathbf{k}) \approx (v_F/3)k$, similar to that in neutral gases[223]. That statement was revised very soon by Anderson who, accounting for the long-range Coulomb interaction, has proved that the collective mode in bulk superconductors acquires a gap and that "collective modes are essentially unperturbed plasma oscillations in the charge case"[224]. Indeed, in isotropic superconductors the plasma frequency lies well above the superconducting gap, and thus superconductivity has negligible effect on the plasma mode.

The situation is quite different in Josephson-coupled layered superconductors, where the frequency of the uniform collective mode, Eq. (2.79), is well below the superconducting gap. This frequency is the same in the superconducting and normal states, as Anderson concluded. However, the relaxation rate of this mode is drastically reduced in the superconducting state. In the normal state the c-axis plasma oscillations are carried by tunneling normal quasiparticles which are strongly scattered by impurities. As a result, the plasma relaxation rate, $2\pi\sigma_c/\epsilon_c$, is comparable to or larger than the c-axis plasma frequency, so that a sharp plasma resonance is absent. In the superconducting state, plasma oscillations are carried by the tunneling Cooper pairs for which impurity scattering is ineffective. Such oscillations are accompanied by quasiparticle tunneling, but the quasiparticle concentration is low. In s-wave superconductors at temperatures well below T_c, the quasiparticle density is exponentially small, while in the d-wave cuprates quasiparticles occupy only a small portion of the Fermi surface near the gap nodes. In Bi-2212 the c-axis conductivity, σ_c, drops from 35 to 2 $(k\Omega\,cm)^{-1}$ as the temperature drops from 80 to 4 K.[225] Correspondingly, the quasiparticle relaxation rate drops from 17.5 to 1 GHz. Thus in the superconducting state below 80 K, the frequency of the c-axis plasma mode is higher than the relaxation rate and a sharp resonance is observable.

The effect of Josephson plasma resonance (JPR) on the optical properties of $La_{2-x}Sr_xCuO_4$ was first observed in 1992 by Tamasaku et al.[226] That superconductor is characterized by T_c in the interval 27-34 K depending on x and by a conductivity anisotropy above 500 in the normal state. Studying the infrared reflectivity spectrum above and below T_c for $\mathbf{E} \parallel \mathbf{c}$, these authors have seen a dramatic change at T_c from a featureless spectrum above T_c to one with a sharp reflection edge below T_c in the interval (600-1500) GHz depending on x. Two years later, Tachiki et al.[227] attributed the sharp edge to the JPR frequency and stressed the importance of the plasma damping reduction in the superconducting state which makes the JPR observation possible.

A next natural step would be the observation of JPR in Bi-2212 where the resonance is expected

[220] C. A. Bolle et al., *Phys. Rev. Lett.* 66 (1991) 112

[221] S. J. Bending and M. J. W. Dodgson, *J. Phys.: Condens. Matt.* 17 (2005) R955

[222] N. N. Bogoliubov, *J. Phys.* (Moscow) 11 (1947) 23

[223] N. N. Bogoliubov, *Zh. Eksp. Teor. Fiz.* 34 (1958) 41 [*Sov. Phys. JETP* 34 (1958) 41]

[224] P. W. Anderson, *Phys. Rev.* B 112 (1958) 1900

[225] Y. I. Latyshev, A. E. Koshelev, and L. N. Bulaevskii, *Phys. Rev.* B 68 (2003) 134504

[226] K. Tamasaku, Y. Nakamura, and S. Uchida, *Phys. Rev. Lett.* 69 (1992) 1455

[227] M. Tachiki, T. Koyama, and S. Takahashi, *Phys. Rev.* B 50 (1994)

at much lower frequencies because of higher anisotropy. However, similar experimental studies in reflectivity or absorption were not done in Bi-2212 at that time because tunable radiation sources in the 30-200 GHz frequency range were not available. Nevertheless, JPR in Bi-2212 was observed in 1994 with a different technique. Interestingly, JPR in Bi-2212 was not recognized at first because it appeared as a resonance in the absorption dependence on the magnetic field at fixed frequency rather than as an absorption peak in the frequency dependence: Tsui et al. recorded narrow resonances in the magnetoabsorption of microwave radiation [in the range 30-50 GHz] in the vortex state with $\mathbf{H} \parallel$ c using fixed-frequency sources and sweeping the magnetic field[228]. The origin of these resonances was not identified at that time.

Soon after this observation, it was concluded theoretically[229] that, since $\omega_{p,c}^2$ is proportional to the Josephson critical current, Eq. (2.79), $\omega_{p,c}$ should be strongly reduced by the magnetic field applied along the c axis. Such a reduction of the Josephson coupling by disordered pancake vortices had been considered earlier [204]. The misaligned pancakes introduce a phase difference $\Delta\varphi$ (Eq. (2.81)), which plays the same role as the phase difference introduced by Lawrence and Doniach in Eq. (2.79). As the magnetic field increases, so does the number of pancakes and $\omega_{p,c}$ drops with H.

Subsequent measurements[230] proved that the magnetoabsorption peak is indeed caused by the Josephson plasma resonance and they also revealed the dependence $\omega_{p,c}^2(H) \propto H^{-\mu}$ with $\mu \approx 0.7 - 0.8$ below the irreversibility line in the (H, T) plane. The resonance line was found to be asymmetric and almost universal. Those properties were qualitatively explained by the model of JPR affected by pancakes that are displaced from straight stacks due to random pinning and thermal fluctuations[231]. At low temperatures the scale of the field dependence of the plasma frequency is given by the field at which the distance between the pancake stacks becomes of the order of the layer-to-layer pancake wandering length, r_w, $r_w^2 = \langle (\mathbf{R}_{n+1} - \mathbf{R}_n)^2 \rangle$. This scale coincides with the "second peak" field, $B_{sp} = 200\text{-}400\text{G}$. Surprisingly, the JPR was also observed deeply in the pancake-liquid state[231] where interlayer phase coherence is expected to be destroyed by pancake thermal motion. This apparent paradox was resolved in Koshelev[232], where it was demonstrated that the JPR actually probes the short-time coherence, which is preserved in the liquid state. In this regime the JPR frequency can be evaluated using a high-temperature expansion with respect to the Josephson coupling leading to a remarkably simple result $\omega_{p,c}^2 \approx 2\pi s J_0^2 \Phi_0 / (\epsilon_c T B_z)$ which agrees well with experiment. Moreover, in the liquid state the JPR frequency is directly related to the pancake-density correlation function, which may be extracted from the dependence of $\omega_{p,c}$ on the component of $\mathbf{H}$ parallel to the layers at a fixed c-axis component[233]. Being sensitive to the pancake configurations, JPR is a powerful tool to study the vortex matter. JPR in different vortex states was explored in Shibauchi et al. and Gaifullin [234] and in many subsequent papers. In particular, jumps of the Josephson energy across the vortex phase transitions have been measured.

Finally, Gaifullin, et al.[235] studied JPR in zero magnetic field by measuring the frequency dependence of the absorption. They found that the JPR frequency drops with temperature proportional to $1/\lambda_c(T)$, decreasing from 125 to 40 GHz as the temperature increases from 4 to 80 K in agreement with Eq. (2.79) at $\Delta\varphi = 0$ and with the strong-resonance condition $\omega_{p,c}(T) \gg 2\pi\sigma_c(T)/\epsilon_c$.

The concept of the JPR with strong dependence of the resonance frequency on the magnetic field $\mathbf{H} \parallel \mathbf{c}$ is at present firmly established. Obviously, such a dependence on the very low scale is absent in the normal state. This strong field dependence is another remarkable property, in addition to the

[228] O. K. C. Tsui et al., *Phys. Rev. Lett.* 73 (1994) 724

[229] L. N. Bulaevskii, M. P. Maley, and M. Tachiki, *Phys. Rev. Lett.* 74 (1995) 801

[230] Y. Matsuda et al., *Phys. Rev. Lett.* 75 (1995) 4512; O. K. C. Tsui et al., Phys. Rev. Lett. 76 (1996) 819

[231] L. N. Bulaevskii et al., *Phys. Rev. B* 54 (1996) 7521

[232] A. E. Koshelev, *Phys. Rev. Lett.* 77 (1996) 3901

[233] A. E. Koshelev, L. N. Bulaevskii, and M. P. Maley, *Phys. Rev. B* 62, (2000) 14403

[234] T. Shibauchi et al., *Phys. Rev. Lett.* 83 (1999) 1010; M. B. Gaifullin et al., *Phys. Rev. Lett.* 84 (2000) 2945

[235] M. B. Gaifullin et al., *Phys. Rev. Lett.* 83 (1999) 392

strong suppression of damping, which makes the c-axis plasma resonance in the superconducting state distinct from that in the normal state.

2.5.4   Radiation from Intrinsic Josephson Junctions

Josephson junctions (JJs) as a possible source of electromagnetic radiation have been discussed for a long time since the discovery of the AC Josephson effect. A DC voltage V across a JJ in the resistive state leads to an oscillating current with the frequency $\omega_J = 2eV/\hbar$, which should generate electromagnetic waves. To bring a JJ into the resistive state, the bias DC current J has to exceed the Josephson critical current J_0 and the largest part of the DC current is carried by quasiparticles at $J > J_0$. The emission from a single JJ[236] turns out to be very weak, of the order $\sim (10^{-4} - 10^{-6})$ μW, due to the small size of the radiating area proportional to the junction thickness as compared to the radiation wave length. Since then, the JJ arrays have been developed as sources of coherent radiation with the goal to achieve high radiation power proportional to the number of junctions squared [237]. For an array of 500 Nb junctions a radiation power of up to 10 μW was reported at discrete frequencies below 0.4 THz [238]. Since the typical size of an artificial junction is about 10 μm, it is possible to fit at most several hundred of them within the radiation wave length. For arrays of larger sizes, JJs have to be distributed over several wavelengths and that complicates their synchronization. On the other hand, layered superconductors present natural stacks of the intrinsic Josephson junctions (IJJs). Due to the atomic thickness of an individual junction, it is possible to have up to 10^5 IJJs on the scale of the wavelength in the terahertz frequency range.

In 1995 Koyama and Tachiki[239] proposed IJJs in Bi-2212 as a radiation source at $\sim$(0.1 - 1) THz. They suggested using the resistive flux-flow state of Josephson vortices induced by the magnetic field parallel to the layers. The magnetic field suppresses the critical current and makes it possible to reach a state with the necessary interlayer voltage and frequency. In addition, interaction between the Josephson vortices in different layers helps frequency synchronization of the IJJ oscillations. However, the Josephson vortex lattice typically has a triangular structure whose motion generates Josephson oscillations with π-phase shifts between neighboring layers. Unlike the uniform plasma mode, such oscillations practically do not couple to the electromagnetic field in the outside space.

Since then different approaches have been proposed which do not utilize a DC magnetic field[240,241]. In zero magnetic fields the resistive state in the IJJ stacks can be accessed by applying a current exceeding the critical current J_0. At such current the junction voltages jump from zero to very high values, generating enormous heat. To reach the target voltage (frequency), the current has to be reduced. Due to strong hysteresis, the junctions remain in the resistive states down to currents much smaller than J_0 and a wide range of voltages can be swept. It is known from transport measurements that on the current-descending branch the resistive state can be preserved down to a voltage corresponding to ~ 0.15 THz in optimally doped Bi-2212. The c-axis-uniform resistive state can be achieved if the coupling between IJJs synchronizes their oscillations in frequency and phase. The electromagnetic field induced by the Josephson oscillations provides such a coupling because all the junctions contribute to the electromagnetic field inside and/or outside the crystal. The same mechanism synchronizes radiation from individual atoms in lasers. To be effective, the synchronization field inducing stimulated radiation should be sufficiently strong.

---

[236] I. K. Yanson, V. M. Svistunov, and I. M. Dmitrenko, *Zh. Eksp. Teor. Fiz.* 48 (1965) 976 [*Sov. Phys.-JETP* 21 (1965) 650]; D. N. Langenberg et al., *Phys. Rev. Lett.* 15 (1965) 294

[237] A. K. Jain, K. K. Likharev, J. E. Lukens, and J. E. Sauvageau, *Phys. Rep.* 109 (1984) 309; J. Lukens, p. 135, *in "Superconducting Devices"*, Acad. Press, Boston, 1990, ed. by S. T. Ruggiero and D. A. Rudman

[238] S. Han, B. Bi, W. Zhang, and J. Lukens, *Appl. Phys. Lett.* 64 (1994) 1424

[239] T. Koyama and M. Tachiki, *Sol. St. Comm.* 96 (1995) 367

[240] L. N. Bulaevskii and A. E. Koshelev, *Phys. Rev. Lett.* 99 (2007) 057002

[241] L. Ozyuzer et al., *Science* 318 (2007) 1291

The internal electromagnetic modes inside the crystal (working as a cavity) were proposed and utilized to synchronize electromagnetic radiation from IJJs. The frequency of synchronization resonance is set by the lateral mesa size, and for the resonance frequency in the terahertz range the width has to be rather large ($\gtrsim 40\,\mu m$). Part of the electromagnetic energy stored in the resonance mode is emitted outside. The recent experimental demonstration of this mechanism[241] is an important step toward practical Josephson terahertz sources. In these experiments the radiation was generated by mesas fabricated on the top of Bi-2212 crystals; the mesas were 1 μm high corresponding to 640 junctions, 300 μm in length and 40-100 μm in width. The radiation frequencies were inversely proportional to the mesa width and varied in the range of 0.3-0.8 THz. The typical total radiation power was 0.5 μW. It was observed that the power scales as the the square of the number of the resistive junctions in the mesa, characteristic of coherent radiation. A disadvantage of the resonance mechanism is a limited tunability, as the resonance frequency is fixed, roughly, by the mesa size.

Another possible way to synchronization is to use interactions between the IJJs via the generated external radiation[240]. In this case, for efficient coupling to the radiation field, a crystal should have a plate-like geometry with the c-axis along the plate, a large number of junctions, up to 10^5 (the radiation power is proportional to N^2), a thickness $> Ns/\epsilon_c$ (but sufficiently small to provide effective cooling), and the length should be larger than the radiation wavelength. Such a crystal would be a powerful frequency-tunable source with a considerable power conversion efficiency. This design is technologically challenging and it has not yet been implemented.

The mechanism of radiation from the IJJs directly into the space outside of the crystal is quite different from that in usual radiating systems. The latter consist of an antenna loaded by a known AC current source; the electromagnetic fields on the antenna surface and the radiating fields depend on the antenna geometry and on properties of the surrounding space. In extended Josephson junctions neither AC voltage nor the AC current are known in advance. Both are determined by equations for the phase differences $\varphi_n(\mathbf{r})$, defining the fields inside the crystal, by the Maxwell equations outside of the crystal and by the boundary conditions that match inside and outside fields. Thus, both the radiation (external fields) and the source (internal fields) are determined self consistently. In this way, the radiation power has been evaluated for a design proposed by Bulaevskii [240].

The prospects for utilization of Josephson coupled layered superconductors as a radiation source at frequencies between microwave and far infrared are a powerful stimulus for new studies of Josephson phenomena. Early hopes to use artificial junctions for this purpose, in fact, did not materialize. Superconductors with IJJ are more promising, but many technological challenges, such as the growth of Bi-2212 crystals with the needed geometry and orientation, are yet to be met.

2.6 Mixed State Properties

Ernst Helmut Brandt

2.6.1 Introduction

Using Ginzburg-Landau (GL) theory,[242] Landau's thesis student Alexei Abrikosov predicted that superconductors with a GL parameter $\kappa > 1/\sqrt{2}$ may contain a lattice of vortices of supercurrent, or flux lines (fluxons) with quantized magnetic flux $\Phi_0 = h/2e = 2\cdot10^{-15}$ Tm^2. Abrikosov had linearized the GL equations with respect to a small order parameter $|\psi|^2$ and discovered a solution

---

[242] V. L. Ginzburg and L. D. Landau, *Zh. Exp. Teor.* Fiz. 20 (1950) 1064 [Engl. Transl. *Men of Physics: L. D. Landau*, ed. D. ter Haar, Pergamon, New York (1965) vol. 1 pp 138-167]

$\psi(x, y)$ possessing a regular lattice of zero lines. This lattice solution appears when the applied magnetic field B_a (along z) is decreased below the upper critical field $B_{c2} = \Phi_0/(2\pi\xi^2)$, where $\xi = \lambda/\kappa$ is the GL coherence length. At $B_a = B_{c2}$ one has the average induction $\bar{B} = B_{c2}$. With decreasing B_a, the induction decreases and reaches $\bar{B} = 0$ at the lower critical field $B_a = B_{c1} = \Phi_0(\ln\kappa + \alpha)/(4\pi\lambda^2)$ with $\alpha(\kappa) \approx 0.5$ for $\kappa \gg 1$ (see below), $\lambda = \kappa\xi$ is the GL magnetic penetration depth. At the same time the vortex lattice spacing $a \approx (\Phi_0/\bar{B})^{1/2}$ increases and diverges when $\bar{B} \to 0$. Abrikosov tells[243] that he had obtained this vortex solution already in 1953 but Landau didn't like it, stating that there are no line-like singularities in electrodynamics. Only when Feynman[244] had published his paper on vortices in superfluid Helium, did Landau agree and Abrikosov could publish his solution in 1957[245]. For his prediction of the vortex lattice Abrikosov 50 years later in 2003 received the Nobel Prize in Physics together with Vitalii Ginzburg and Anthony Leggett.

After first evidence of the triangular vortex lattice in superconducting Niobium by small-angle neutron scattering in Saclay[246], Träuble and Essmann in Stuttgart succeeded[247,248] to observe the vortex lattice directly by decorating the surface of a superconductor with iron microcrystallites ("magnetic smoke"). At that time I joined this research group headed by A. Seeger and wrote my thesis on the theory of defects in the vortex lattice[249]. Parts 1 and 2 deal with low inductions $\bar{B} \ll B_{c2}$ when London theory may be used and the vortices interact with each other pairwise, similar to 2D atomic lattices. Parts 3 and 4 consider high inductions $\bar{B} \approx B_{c2}$, where the shape of the GL solutions $\psi(x, y)$ and $B(x, y)$ may be obtained from linearized GL theory, while the nonlinear GL terms determine the amplitudes of this ψ and B. My thesis extended Abrikosov's theory of periodic vortex lattices to non-periodic vortex arrangements (see below). Such distorted-lattice solutions are required to calculate the elastic energy of the vortex lattice and the energy of lattice defects like vacancies and dislocations. They are also helpful to visualize where the solutions of the linearized GL theory apply and how they have to be modified at lower inductions.

2.6.2 Abrikosov's Periodic Vortex Lattice Near B_{c2}

In the usual reduced units (length λ, induction $\sqrt{2}B_c$, energy density B_c^2/μ_0, where $B_c = B_{c2}/\sqrt{2}\kappa$ is the thermodynamic critical field) the spatially averaged free energy density F of the GL theory referred to the Meissner state ($\psi = 1$, $\mathbf{B} = 0$) within the superconductor reads

$$F = \left\langle \frac{(1 - |\psi|^2)^2}{2} + \left| \left(\frac{\nabla}{i\kappa} - \mathbf{A} \right)\psi \right|^2 + \mathbf{B}^2 \right\rangle. \tag{2.82}$$

Here $\psi(\mathbf{r}) = f \exp(i\varphi)$ is the complex GL function, $\mathbf{B}(\mathbf{r}) = \nabla \times \mathbf{A}$ the magnetic induction, $\mathbf{A}(\mathbf{r})$ the vector potential, and $\langle \ldots \rangle = (1/V) \int_V d^3r \ldots$ means spatial averaging over the superconductor with volume V. Introducing the super velocity $\mathbf{Q}(\mathbf{r}) = \mathbf{A} - \nabla\varphi/\kappa$ and the magnitude $f(\mathbf{r}) = |\psi|$ one may write F as a functional of the real and gauge-invariant functions f or $f^2 = \omega$ and $\mathbf{Q}$,

$$F = \left\langle \frac{(1 - f^2)^2}{2} + \frac{(\nabla f)^2}{\kappa^2} + f^2 Q^2 + (\nabla \times \mathbf{Q})^2 \right\rangle. \tag{2.83}$$

In the presence of vortices $\mathbf{Q}(\mathbf{r})$ has to be chosen such that $\nabla \times \mathbf{Q}$ has the appropriate singularities along the vortex cores, where f vanishes. By minimizing this F with respect to ψ, $\mathbf{A}$ or f, $\mathbf{Q}$, one

[243] A. A. Abrikosov, *Physics Today* 26 (Jan 1973) 56

[244] R. Feynman, *Progress in Low Temperature Physics*, editor C. J. Gorter, Vol. 1, North Holland, Amsterdam (1955), p. 36

[245] A. A. Abrikosov, *Zh. Exp. Teor. Fiz.* 32 (1957) 1442, [Sov. Phys.-JETP 5 (1957) 1174]

[246] D. Cribier, B. Jacrot, L. Madhav Rao, and B. Farnoux, *Progress in Low Temperature Physics*, editor C. J. Gorter, Vol. 5, North Holland, Amsterdam (1967), p. 161

[247] U. Essmann and H. Träuble, *Phys. Lett.* A 24 (1967) 526; Phys. Status Solidi 32 (1969) 337

[248] H. Träuble and U. Essmann, *J. Appl. Phys.* 39 (1969) 4052

[249] E. H. Brandt, *Phys. Status Solidi* 35 (1969) 1027 ; 36 (1969) 371 ; 36 (1969) 381 ; 36 (1969) 393

obtains the GL equations together with the appropriate boundary conditions. For superconducting films one has to add the energy of the magnetic stray field outside the film, which makes the perpendicular component B_z of $\mathbf{B}$ continuous at the film surface (see below).

The two GL equations are obtained by minimization of F (2.82) with respect to ψ and $\mathbf{A}$, $\delta F/\delta \psi = 0$ and $\delta F/\delta \mathbf{A} = 0$, yielding

$$(\nabla/i - \kappa\mathbf{A})^2\psi = \kappa(1 - |\psi|^2)\psi, \quad \nabla \times [\nabla \times \mathbf{A}] = |\psi|^2\mathbf{Q}. \tag{2.84}$$

With B_a and $\bar{B}$ chosen along the z axis and in the gauge $A_x = -\bar{B}y + \tilde{A}_x(x,y)$, $A_y = \tilde{A}_y(x,y)$ ($\tilde{A}_x$, $\tilde{A}_y$ are terms of higher order) the linearized first GL equation, obtained by omitting the term $|\psi|^2\psi$ in (2.84a), has the general solution

$$\psi(x,y) = \exp(-\kappa\bar{B}y^2/2)\, g(x,y), \quad \frac{\partial g}{\partial x} + i\frac{\partial g}{\partial y} = 0. \tag{2.85}$$

This means $g(x,y) = g(z)$, $z = x + iy$, can be any analytical function. For a periodic solution satisfying $|\psi|^2(\mathbf{r} + \mathbf{R}_{mn}) = |\psi|^2(\mathbf{r})$, $\mathbf{r} = (x,y)$, with real and reciprocal lattice vectors

$$\mathbf{R}_{mn} = (mx_1 + nx_2; ny_2), \quad \mathbf{K}_{mn} = (2\pi/x_1 y_2)(my_2; -mx_2 + nx_1), \tag{2.86}$$

($m,n = 0,\pm1,\pm2,\ldots$; triangular lattice: $x_1 = a$, $x_2 = x_1/2$, $y_2 = x_1\sqrt{3}/2$; square lattice: $x_1 = y_2 = a$, $x_2 = 0$) and with a zero at $\mathbf{r} = 0$, one obtains for $g(z)$ the theta function ϑ_1 defined as[250]

$$\vartheta_1(z,\tau) = 2\sum_{n=0}^{\infty}(-)^n \exp[\,i\pi\tau(n + \tfrac{1}{2})^2\,]\sin(2n + 1)z. \tag{2.87}$$

Thus, the periodic Abrikosov solution with zeros at the $\mathbf{r} = \mathbf{R}_{mn}$ may be written as

$$\psi_A(x,y) = \exp\left(-\frac{\pi y^2}{x_1 y_2}\right)\vartheta_1\left(\frac{\pi}{x_1}(x + iy), \frac{x_2 + iy_2}{x_1}\right). \tag{2.88}$$

This solution has the mean induction $\bar{B} = \Phi_0/(x_1 y_2)$, normalized order parameter $\langle|\psi_A|^2\rangle = 1$, and the Fourier series of $|\psi_A|^2 = \omega_A(x,y)$,

$$\omega_A(\mathbf{r}) = \sum_{\mathbf{K}_{mn}}(-)^{mn+m+n}\exp\left(-\frac{K_{mn}^2 x_1 y_2}{8\pi}\right)e^{i\mathbf{K}\mathbf{r}}. \tag{2.89}$$

From the zero $\omega_A(0,0) = 0$ it follows that the sum over all Fourier coefficients in (2.89) is zero for all lattice symmetries. (Abrikosov[245] chose a different position for the zero $\omega_A = 0$ and thus obtained a different theta function ϑ_3.[250])

2.6.3 Distorted Vortex Lattice and Vacancy Near B_{c2}

The GL solution for a distorted vortex lattice near B_{c2} is obtained as follows. Assume that each of the straight and parallel vortex lines is displaced from its ideal lattice positions $\mathbf{R}_{mn} = \mathbf{R}_\nu = (X_\nu, Y_\nu)$ by displacements $\mathbf{s}_\nu = (s_{\nu x}, s_{\nu y})$, $\mathbf{r}_\nu = (x_\nu, y_\nu) = \mathbf{R}_\nu + \mathbf{s}_\nu$, such that the displacement field itself is periodic with a super lattice N times larger than the vortex lattice, but with same symmetry, $\mathbf{s}(\mathbf{r} + N\mathbf{R}_{mn}) = \mathbf{s}(\mathbf{r})$. Where needed we use a continuous displacement field $\mathbf{s}(\mathbf{r})$ defined such that it has the same Fourier transform as the discrete $\mathbf{s}_\nu = \mathbf{s}(\mathbf{R}_\nu)$. A distorted triangular vortex lattice with spacing $x_1 = a$ then has the solution (2.85),

$$\psi(x,y) = c_1 \exp\left(-\frac{2\pi y^2}{\sqrt{3}a^2}\right)\vartheta_1\left(\frac{\pi}{a}z,\tau\right)\prod_\nu \frac{\vartheta_1[(\pi/Na)(z - z_\nu - s_\nu),\tau]}{\vartheta_1[(\pi/Na)(z - z_\nu),\tau]} \tag{2.90}$$

[250]E. T. Whittaker and G. N. Watson, *Modern Analysis*, Cambridge Unversity Press, Cambridge (1927)

with $z = x + iy$, $z_\nu = X_\nu + iY_\nu$, $\tau = (1 + i\sqrt{3})/2$, $s_\nu = s_{x\nu} + is_{y\nu}$, the product is over one super cell, and $c_1 \approx 1$ is a normalization constant. When all $s_\nu = 0$, the product in (2.90) is unity, $\prod_\nu = 1$, thus the first two factors in (2.90) are the ideal lattice solution with $c_1 = 1$, cf. Eq. (2.88). Each factor of the product shifts one zero from $\mathbf{r} = \mathbf{R}_\nu$ to $\mathbf{r} = \mathbf{R}_\nu + \mathbf{s}_\nu$. The absolute value of $|\psi|^2 = \omega$ of this GL function may also be expressed in terms of the Fourier series $\omega_A(\mathbf{r})$ (2.89),

$$\omega(\mathbf{r}) = c_1^2\, \omega_A(\mathbf{r}) \prod_\nu \frac{\omega_A[(\mathbf{r} - \mathbf{R}_\nu - \mathbf{s}_\nu)/N]}{\omega_A[(\mathbf{r} - \mathbf{R}_\nu)/N]}. \tag{2.91}$$

In the limit of infinite super cell, $N \to \infty$, one may use $\vartheta_1(z/N, \tau) \propto z/N$ for $|z|/N \ll 1$, thus one may replace the function ϑ_1 by its argument since all the constant factors cancel or combine to a normalization factor that follows from numerics. One then obtains simply

$$\omega(\mathbf{r}) = c_1^2\, \omega_A(\mathbf{r}) \prod_\nu \frac{|\mathbf{r} - \mathbf{R}_\nu - \mathbf{s}_\nu|^2}{|\mathbf{r} - \mathbf{R}_\nu|^2}. \tag{2.92}$$

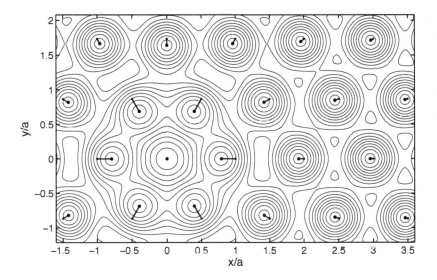

FIGURE 2.2: Contour lines of the order parameter $\omega(x, y)$ (2.93) of a vortex lattice with one vacancy at $x = y = 0$ and complete relaxation of the vortex lattice. The vortex displacements are indicated by short bold lines.

Removing the central vortex at $\mathbf{R}_\nu = 0$ adds a factor $1/r^2$ to the linearized solution $\omega(\mathbf{r})$, (2.92). Obviously, if the other vortices are not allowed to relax, this solution at large distances vanishes as $1/r^2$; it cannot be normalized and its energy is infinite. However, if the relaxation of the other vortices is chosen appropriately it will minimize the defect energy and make it finite. This can be seen from the solution

$$\omega(\mathbf{r}) = c_1^2\, \frac{\omega_A(\mathbf{r})}{r^2} \left| \frac{h(z)}{h(0)} \right|^2, \quad h(z) = \prod_{\nu \neq 0} \left(1 - \frac{s_\nu}{z - z_\nu} \right). \tag{2.93}$$

The constant factor $|1/h(0)|^2$ was added to force convergence of the infinite product. The solution for a super lattice of vacancies positioned at the $N\mathbf{R}_\nu$, is given by expression (2.91) divided by $\omega(\mathbf{r}/N)$ that removes the zeros at positions $N\mathbf{R}_\nu$.

The energy of both the ideally periodic and the distorted vortex lattices is calculated via the Abrikosov parameter $\beta = \langle|\psi|^4\rangle/\langle|\psi|^2\rangle^2 = \langle\omega^2\rangle/\langle\omega\rangle^2 \geq 1$. This β enters the free energy of the linearized GL theory (referred to the normal state), that has to be minimized when $\bar{B}$ is held constant,

$$F = \frac{\bar{B}^2}{2\mu_0} - \frac{(B_{c2} - \bar{B})^2}{2\mu_0[1 + (2\kappa^2 - 1)\beta]} \tag{2.94}$$

and the free enthalpy that has to be minimized when B_a is held constant,

$$G = F - \frac{\bar{B}B_a}{\mu_0} = -\frac{(B_{c2} - B_a)^2}{2\mu_0(2\kappa^2 - 1)\beta}. \tag{2.95}$$

The elastic energy of the distorted vortex lattice is the product of the derivatives $\partial F/\partial\beta$ or $\partial G/\partial\beta$ times the change of $\beta\{\psi\}$ times the volume, with the limit of infinite volume taken. This means that all elastic energies and energies of structural defects near B_{c2} vanish as $(B_{c2} - \bar{B})^2 \propto (B_{c2} - B_a)^2$. This is true also for the shear modulus c_{66} of the vortex lattice, which can be obtained using Abrikosov's periodic lattice solution.[249,251]

The defect energy of the vortex vacancy is finite only if the vortices relax (shift towards the removed vortex, Figure 2.2) such that at large distances the vortex displacements are $\mathbf{s}_v = -\mathbf{R}_v/(2\pi n R_v^2)$ with $n = \bar{B}/\Phi_0 = 1/(x_1 y_2)$ the vortex density. If the radial displacements were chosen smaller (larger) than that, the order parameter (2.93) would vanish (diverge) at large distances r. But with the correct displacements that minimize β and thus the defect energy, the amplitude of the oscillating order parameter stays almost constant, even near the vacancy. Note that the continuous field $\mathbf{s} = -\mathbf{r}/(2\pi n r^2)$ satisfies $\nabla\cdot\mathbf{s}(\mathbf{r}) = 0$ and thus describes a pure shear deformation. More precisely, one has $\nabla\cdot\mathbf{s}(\mathbf{r}) = (1/n)\delta_2(\mathbf{r})$ (δ_2 is the 2D delta function), i.e., this displacement field "remembers" that one vortex cell area was removed. Further interesting properties of structural defects in vortex lattices and other two- or three-dimensional soft lattices are discussed in Brandt [251d].

2.6.4 Nonlocal Elasticity of the Vortex Lattice

As shown with the vacancy example, the distorted-lattice solution (2.93) of the linearized GL equations yields finite energies of lattice defects only if the vortex position is allowed to relax appropriately. This unphysical divergence of defect energies of the vortex lattice is removed when the influence of the nonlinear GL terms on the solutions $\omega(x, y)$ and $B(x, y)$ is accounted for. This calculation was performed in a series of four papers[252]: Parts 1 and 2 deal with the linear elastic energy of the vortex lattice at low and high inductions $\bar{B}$. Parts 3 and 4 derive the GL solutions for the distorted vortex lattice when the vortex lines are straight and parallel or arbitrarily curved. The essential result is that the long-ranging modulation factors like $(1 - 2s/r)$ in the linearized version of solution (2.93) become exponentially damped over a new length $\xi' = 1/k_\psi = \xi/\sqrt{2(1-b)}$ with $b = \bar{B}/B_{c2}$. As $b \to 1$; this screening length becomes infinite and the linearized solution (2.93) is recovered. At $b < 1$, the distorted-lattice solution (2.92) should be replaced by

$$\omega(\mathbf{r}) = \omega_A(\mathbf{r})\left[1 + \sum_v \mathbf{s}_v \nabla K_0(|\mathbf{r} - \mathbf{R}_v|k_\psi)\right]^2 + O(s^2), \tag{2.96}$$

where $K_0(x)$ is a modified Bessel function with limits $K_0(x) \approx -\ln x$ ($x \ll 1$), $K_0(x) \approx (\pi/2x)^{1/2}e^{-x}$ ($x \gg 1$). This generalized expression up to terms linear in the vortex shifts $\mathbf{s}_v$ reproduces the linearized solution (2.92) when $k_\psi \to 0$, but it does not possess the correct zeros at $\mathbf{r}_v = \mathbf{R}_v + \mathbf{s}_v$. This

[251] R. Labusch, *Phys. Status Solidi* 32 (1969) 439; E. H. Brandt, *Phys. Status Solidi* b77 (1976) 551; M. A. Moore, *Phys. Rev.* B39 (1989) 136; E. H. Brandt, *Phys. Rev.* B56 (1997) 9071

[252] E. H. Brandt, *J. Low. Temp. Phys.* 26 (1977) 709 ; 26 (1977) 735 ; 28 (1977) 263 ; 28 (1977) 291

may be corrected by replacing in (2.96) the periodic order parameter $\omega_A(\mathbf{r})$ by the "phase modulated" $\omega_A[\mathbf{r} - \mathbf{s}(\mathbf{r})]$ and cutting the infinity of K_0 off. The resulting solution is still exact up to linear terms in $\mathbf{s}_v$ since $\nabla\omega_A(\mathbf{r})$ vanishes at the $\mathbf{R}_v$ and thus the expansion of $\omega_A[\mathbf{r} - \mathbf{s}(\mathbf{r})]$ contains no linear term.

In a similar way, the solution for the induction $B(x, y)$ of the linearized GL theory is modified by the nonlinear terms to give for periodic displacements $\mathbf{s}(\mathbf{r}) \propto \exp(i\mathbf{kr})$

$$B(\mathbf{r}) = \bar{B} + B_{c2}\frac{\langle\omega\rangle - \omega(\mathbf{r})}{2\kappa^2} \rightarrow B_0[\mathbf{r} - \mathbf{s}(\mathbf{r})] - \frac{\bar{B}\,\nabla\mathbf{s}(\mathbf{r})}{1 + k^2/k_h^2} + O(s^2) \tag{2.97}$$

with $k_h^2 = 1/\lambda'^2 = \langle\omega\rangle/\lambda^2 \approx (1 - b)/\lambda^2$ and $B_0(x, y)$ the ideal periodic solution for $\mathbf{s} \equiv 0$. In deriving (2.97) all terms containing k_ψ have canceled. The generalization to non-periodic displacement fields is obtained by Fourier transform. For extension to curved vortices see[252,253].

The distorted-lattice solution up to terms linear in the displacements $\mathbf{s}$ can be used to calculate the linear elastic energy of the vortex lattice, $F_{\text{elast}} = F\{\mathbf{s}_v\} - F\{\mathbf{s}_v \equiv 0\}$, referred to the perfect lattice (the equilibrium state). The most general expression quadratic in the 2D displacements $\mathbf{s}_v(z)$, or in their Fourier transforms $\tilde{\mathbf{s}}(\mathbf{k}) = (\tilde{s}_x, \tilde{s}_y, 0) = \sum_v \int dz\, \mathbf{s}_v(z)\exp(-i\mathbf{kR}_v)$ is

$$F_{\text{elast}} = \frac{1}{2}\int_{\text{BZ}} \frac{d^3k}{8\pi^3 n}\, \tilde{s}_\alpha(\mathbf{k})\Phi_{\alpha\beta}(\mathbf{k})\tilde{s}_\beta(-\mathbf{k}), \tag{2.98}$$

where the sum over the indices $\alpha, \beta = (x, y)$ is taken. The 2×2 matrix $\Phi_{\alpha\beta}$ is the elastic matrix. This expression applies for both an elastic continuum and for a lattice. For a lattice, $\Phi_{\alpha\beta}$ is periodic, $\Phi_{\alpha\beta}(\mathbf{k} + \mathbf{K}) = \Phi_{\alpha\beta}(\mathbf{k})$, and thus the integral should be restricted to the first Brillouin zone (BZ). The BZ for the triangular lattice is a hexagon, and for the square lattice a square. Where required, the BZ may be approximated by a circle with radius $k_B = (2b)^{1/2}/\xi$, $b = \bar{B}/B_{c2}$, and area $\pi k_B^2 = 4\pi^2 n$, $n = \bar{B}/\Phi_0$. For a uniaxial elastic continuum the elastic matrix $\Phi_{\alpha\beta}(k_x, k_y, k_z)$ has the form

$$n\Phi_{\alpha\beta}(\mathbf{k}) = (c_{11} - c_{66})k_\alpha k_\beta + \delta_{\alpha\beta}[(k_x^2 + k_y^2)c_{66} + k_z^2 c_{44}]. \tag{2.99}$$

In it the coefficients are the elastic moduli: $c_{11} - c_{66}$ the isotropic compression modulus, c_{11} the uniaxial compression modulus, c_{66} the shear modulus, and c_{44} the tilt modulus. The elastic moduli of the vortex lattice are obtained by deriving the elastic energy, e.g., from GL theory and comparing it at small $k_x^2 + k_y^2 \ll k_B^2$ with the continuum limit (2.99). This yields

$$c_{11}(k) = \frac{\bar{B}^2}{\mu_0}\frac{\partial B_a}{\partial \bar{B}}\frac{1}{(1 + k^2/k_h^2)(1 + k^2/k_\psi^2)} + c_{66} \tag{2.100}$$

$$c_{66} = \frac{\bar{B}B_{c2}}{8\kappa^2\mu_0}(1 - b)^2\frac{(2\kappa^2 - 1)2\kappa^2}{[2\kappa^2 - 1 + 1/\beta_A]^2}(1 - 0.3b) \tag{2.101}$$

$$c_{44}(k) = \frac{\bar{B}^2}{\mu_0}\frac{1}{1 + k^2/k_h^2} + \frac{\bar{B}(B_a - \bar{B})}{\mu_0}. \tag{2.102}$$

These expressions are exact at large reduced induction $b = \bar{B}/B_{c2} \rightarrow 1$ and for all κ, but they are written such that they reduce to the correct values also in the limit of small induction $\bar{B} \ll B_{c2}$. In c_{66}, $\beta_A = 1.1596$ is the Abrikosov parameter of the triangular lattice (the square lattice is unstable and thus has negative c_{66}); the third factor reduces to 1 for $2\kappa^2 \gg 1$ and to $(2\kappa^2 - 1)\beta_A^2 \rightarrow 0$ for $\kappa \rightarrow 1/\sqrt{2}$, which means the shear stiffness of the vortex lattice is zero in superconductors with $\kappa = 0.71$; the factor $1 - 0.3b$ interpolates between the correct limits at $b \rightarrow 1$ and $b \rightarrow 0$. In particular, for $b \ll 1$ and $2\kappa^2 \gg 1$, (2.101) reproduces the London result $c_{66} = \bar{B}B_{c2}/(8\kappa^2\mu_0)$.

---

[253] E. H. Brandt, *Rep. Prog. Phys.* 58 (1995) 1465

An interesting result is the dependence of c_{11} (2.100) and c_{44} (2.102) on $k = |\mathbf{k}|$, which means the elasticity of the vortex lattice is **nonlocal**. In the limit of uniform stress, $k \to 0$, these expressions reproduce the known values of the compression and tilt moduli obtained by thermodynamics, $c_{11} - c_{66} = (\bar{B}^2/\mu_0)\partial B_a/\partial \bar{B}$, $c_{44} = \bar{B}B_a/\mu_0$. However, when the wavelength of a periodic compression or tilt decreases, i.e., the wave vector k increases, these moduli decrease. This means, the vortex lattice is softer for short-wavelengths compression and tilt than it is for long wavelengths. The two characteristic lengths or wave vectors are $k_h = 1/\lambda' \approx \sqrt{1-b}/\lambda$ and $k_\psi = 1/\xi' = \sqrt{2(1-b)}/\xi$.

This dispersion or elastic nonlocality means, e.g., that a point force exerted by a small pinning center on the vortex lattice deforms the vortex on which it acts not like plugging a string but more, causing a sharp cusp since a local deformation costs little energy. If the interaction of the vortices with the pinning center is via the order parameter $|\psi|^2$ or via the gradient term in the GL functional, then this interaction itself is nonlocal, smeared over the length $\xi' = 1/k_\psi$. In the expressions for the elastic force and the elastic energy, there is thus a factor $1 + k^2/k_\psi^2$ in the numerator that compensates the same factor in the denominator originating from $c_{11}(k)$, (2.100). Therefore, the factor $1/(1 + k^2/k_\psi^2)$ in c_{11} has no physical meaning in pinning problems since near B_{c2} where ξ' can be larger than the vortex spacing a, it is not possible to exert a pinning force on one single zero of the order parameter but only on an area with radius ξ' containing several such zeros. The nonlocality factor $1/(1 + k^2/k_h^2)$ in c_{44}, however, is important in pinning theories since it strongly enhances the elastic deformations caused by small pins acting on the vortex cores.

The correct, nonlocal elasticity thus effectively softens the vortex lattice and leads to large, pinning-caused distortions and disorder of the vortex lattice. Furthermore, the thermal fluctuations of the vortex lattice are strongly enhanced by this nonlocal elasticity. In both cases the lattice softening is caused mainly by the dispersion of $c_{44}(k)$, while the dispersion and reduction of $c_{11}(k)$ is not so important since the shear modulus c_{66} is typically much smaller than $c_{11}(k)$ and the shear modes of the elastic deformation thus dominate over the compressional modes.

2.6.5 Vortex Arrangements at Low Inductions

At low inductions $b < 0.2$ and not too small $\kappa > 2$, the GL theory for arbitrary 3D arrangements of vortices reduces to the London theory, which may be expressed by the energy functional

$$F\{\mathbf{B}\} = \frac{\mu_0}{2} \int d^3 r [B^2 + \lambda^2 (\nabla \times \mathbf{B})^2]. \tag{2.103}$$

Here λ is the London penetration depth equal to the GL magnetic penetration depth. Minimizing $F\{\mathbf{B}\}$ with respect to the induction $\mathbf{B}(\mathbf{r})$ using $\nabla \mathbf{B} = 0$, and adding appropriate singularities along the positions $\mathbf{r}_\nu(z)$ of the vortex cores, one obtains the modified London equation[253],

$$(-\lambda^2 \nabla^2 + 1)\mathbf{B}(\mathbf{r}) = \Phi_0 \sum_\nu \int d\mathbf{r}_\nu \delta_3(\mathbf{r} - \mathbf{r}_\nu) \tag{2.104}$$

with δ_3 the 3D delta function. From this one obtains the energy of an arbitrary arrangement of straight or curved vortices,

$$F\{\mathbf{r}_\nu(z)\} = \frac{\Phi_0^2}{8\pi\lambda^2\mu_0} \sum_\mu \sum_\nu \int d\mathbf{r}_\mu \int d\mathbf{r}_\nu \frac{\exp(-r_{\mu\nu}/\lambda)}{r_{\mu\nu}}. \tag{2.105}$$

In this double sum the terms $\mu \neq \nu$ describe the pairwise interaction of the vortex line elements $d\mathbf{r}_\mu$, $d\mathbf{r}_\nu$ over the distance $r_{\mu\nu} = |\mathbf{r}_\mu - \mathbf{r}_\nu|$. The term $\mu = \nu$ is the self-energy of the μth vortex line, which depends on the shape of this vortex. In it an inner cutoff is needed, obtained, e.g., by putting $r_{\mu\mu}^2(z, z') = |\mathbf{r}_\mu(z) - \mathbf{r}_\mu(z')|^2 + r_c^2$ with $r_c \approx \xi$ the vortex core radius, to avoid divergence when in the integral the parameters equal, $z = z'$.

From the GL nonlocal elastic energy (2.98)- (2.102) one may construct an effective interaction potential between vortex line elements such that the full nonlocal linear elastic energy is reproduced at small displacements[254]. This interaction at low $b \ll 1$ reproduces the London interaction for arbitrary vortex arrangements, and it is an approximate GL interaction valid at all b and κ,

$$F\{\mathbf{r}_\nu(z)\} = \frac{\Phi_0^2}{8\pi\lambda'^2\mu_0} \sum_\mu \sum_\nu \left[\int d\mathbf{r}_\mu \int d\mathbf{r}_\nu \frac{\exp(-r_{\mu\nu}/\lambda')}{r_{\mu\nu}} - \int |d\mathbf{r}_\mu| \int |d\mathbf{r}_\nu| \frac{\exp(-r_{\mu\nu}/\xi')}{r_{\mu\nu}} \right] \quad (2.106)$$

with $r_{\mu\nu} = |\mathbf{r}_\mu - \mathbf{r}_\nu|$. For $b \ll 1$ the first term in (2.106) reproduces the magnetic repulsion of London vortices with $\lambda' = \lambda/\sqrt{1-b} \approx \lambda$; this magnetic interaction is vectorial due to the product $d\mathbf{r}_\mu \cdot d\mathbf{r}_\nu$ containing the cosine of the angle between two line elements. The second term of shorter range $\xi' = \xi/\sqrt{2(1-b)}$ may be interpreted as an attraction caused by the overlap of the vortex cores, the regions where the order parameter is reduced: two overlapping cores require less (positive) condensation energy than two separated cores, thus the cores attract. This attraction has scalar character, hence the product $|d\mathbf{r}_\mu||d\mathbf{r}_\nu|$. The attractive second term in (2.106) removes the logarithmic divergence of the magnetic repulsion at zero distance, since both terms have the same singularity but of opposite sign. The total potential (2.106) is thus a smooth function that at $r_{\mu\nu} = 0$ starts with a finite value and then decreases monotonically to zero with increasing distance $r_{\mu\nu} \to \infty$.

For straight parallel vortex lines the general 3D energy expression (2.106) simplifies to the sum of the vortex self-energies, $F_{\text{self}} = \Phi_0 B_{c1}/\mu_0$ per unit length and per vortex, and the interaction energy F_{int} of all vortices per unit length,

$$F_{\text{int}}\{\mathbf{r}_\nu\} = \frac{\Phi_0^2}{2\pi\lambda'^2\mu_0} \sum_\mu \sum_{\nu>\mu} \left[K_0\left(\frac{|\mathbf{r}_\mu - \mathbf{r}_\nu|}{\lambda'}\right) - K_0\left(\frac{|\mathbf{r}_\mu - \mathbf{r}_\nu|}{\xi'}\right) \right]. \quad (2.107)$$

Here $K_0(x)$ is a modified Bessel function (see (2.96)). The effective 2D interaction potential in (2.107) is a smooth, monotonically decreasing function with a finite value at $r_{\mu\nu} = 0$ since the two logarithmic singularities of the K_0 functions cancel. The first term in (2.107) is the magnetic repulsion of the straight vortices, and the second term is an attraction due to gain in condensation energy when the vortex cores overlap.

2.6.6 Vortex Lattice Solution for All κ and $\bar{B}$

Abrikosov's solution method for the periodic vortex lattice starts from the linearized GL theory and is thus valid only at large inductions $\bar{B}$ near the upper critical field B_{c2}. First numerical solutions for all $\bar{B}$ and κ were obtained by the "circular cell method"[255] that approximates the hexagonal Wigner-Seitz cell of the triangular vortex lattice by a circle and solves a cylindrically symmetric problem. The periodic solution in the entire ranges of reduced induction $0 < b = \bar{B}/B_{c2} < 1$ and GL parameter $1/\sqrt{2} \le \kappa < \infty$ may be obtained for bulk and film superconductors by the numerical method[256]: It minimizes the free energy functional F (2.83), with respect to the real and periodic functions $\omega(x,y) = f^2 = |\psi|^2$ (order parameter) and $\mathbf{Q}(x,y) = \mathbf{A} - \nabla\varphi/\kappa$ (negative super velocity) or $\hat{\mathbf{z}}B(x,y) = \nabla \times \mathbf{Q}$ (induction) using periodic trial functions with a finite number of Fourier coefficients $a_\mathbf{K}$ and $b_\mathbf{K}$,

$$\omega(\mathbf{r}) = \sum_\mathbf{K} a_\mathbf{K}(1-\cos\mathbf{K}\mathbf{r}), \quad B(\mathbf{r}) = \bar{B} + \sum_\mathbf{K} b_\mathbf{K}\cos\mathbf{K}\mathbf{r}, \quad \mathbf{Q}(\mathbf{r}) = \mathbf{Q}_A(\mathbf{r}) + \sum_\mathbf{K} b_\mathbf{K}\frac{\hat{\mathbf{z}}\times\mathbf{K}}{K^2}\sin\mathbf{K}\mathbf{r}, \quad (2.108)$$

[254] E. H. Brandt, *Phys. Rev.* B34 (1986) 6514

[255] D. Ihle, *Phys. Status Solidi* b47 (1971) 423

[256] E. H. Brandt, *Phys. Status Solidi* b51 (1972) 354; *Phys. Rev. Lett.* 78 (1997) 2208; *Phys. Rev.* B68 (2003) 054506; Phys. Rev. B71 (2005) 014521

with $\mathbf{r} = (x, y)$ and $\mathbf{K} = (K_x, K_y)$ from (2.86). In all sums the term $\mathbf{K} = 0$ is excluded. $\mathbf{Q}_A(x, y)$ is the super velocity of the Abrikosov B_{c2} solution $\omega_A(x, y)$, (2.89). One has

$$\nabla \times \mathbf{Q}_A = \left[\bar{B} - \Phi_0 \sum_{\mathbf{R}} \delta_2(\mathbf{r} - \mathbf{R}) \right] \hat{\mathbf{z}}, \quad \mathbf{Q}_A(\mathbf{r}) = \frac{\nabla \omega_A \times \hat{\mathbf{z}}}{2 \kappa \omega_A}. \tag{2.109}$$

$\mathbf{Q}_A$ is the velocity field of a lattice of ideal vortex lines but with zero average rotation. Close to each vortex center one has $\mathbf{Q}_A(\mathbf{r}) \approx \mathbf{r}' \times \hat{\mathbf{z}}/(2\kappa r'^2)$ and $\omega(\mathbf{r}) \propto r'^2$ with $\mathbf{r}' = \mathbf{r} - \mathbf{R}$.

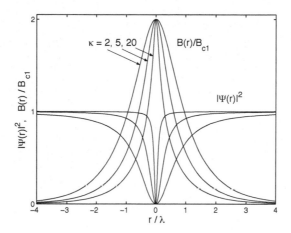

FIGURE 2.3: The magnetic field $B(r)$ and order parameter $|\psi(r)|^2$ of an isolated vortex line calculated from the Ginzburg-Landau theory for GL parameters $\kappa = 2$, 5, and 20. For such large κ the field in the vortex center is twice the applied equilibrium field, $B(0) \approx 2B_{c1} = 2B_a$.

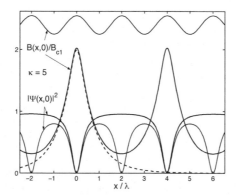

FIGURE 2.4: Two profiles of the magnetic field $B(x, y)$ and order parameter $|\psi(x, y)|^2$ along the x axis (nearest neighbor direction) for triangular vortex lattices with lattice spacings $a = 4\lambda$ ($b = 0.073$, bold lines) and $a = 2\lambda$ ($b = 0.018$, thin lines). The dashed line shows the magnetic field of the isolated flux line from Figure 2.3. From Ginzburg-Landau theory for $\kappa = 5$.

Minimization of F is achieved by iterating the two GL equations $\delta F/\delta \omega = 0$ and $\delta F/\delta \mathbf{Q} = 0$ written in appropriate form. Figures 2.3 and 2.4 show some resulting profiles $B(x, 0)$, $|\psi(x, 0)|^2$. The

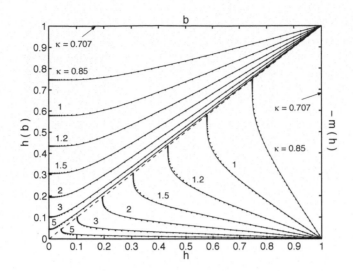

FIGURE 2.5:   The magnetization curves of the triangular vortex lattice (solid lines, numerical result) coinciding within line thickness with those of the square lattice. Shown are $h = B_a/B_{c2}$ versus $b = \bar{B}/B_{c2}$ (upper left triangle) and $-m = h - b$ versus h (lower right triangle). The dots show the fit (2.112), good for $\kappa \leq 20$.

numerics also confirms Clem's isolated vortex model as an excellent approximation[257,258],

$$|\psi(r)|^2 \approx 1/(1 + 2\xi^2/r^2), \quad B(r) \approx (\Phi_0/2\pi\lambda^2)K_0[(r^2 + 2\xi^2)^{1/2}/\lambda]. \tag{2.110}$$

The reversible magnetization $M = B - B_a$ and the equilibrium field $B_a = \mu_0 \partial F/\partial \bar{B}$ (the applied field) are easily computed from Doria's virial theorem[259], which in our reduced units reads $B_a = \langle f^2 - f^4 + 2B(x,y)^2\rangle/2\langle B\rangle$. In this way we find the lower critical field, $B_{c1}(\kappa) = \lim_{\bar{B}\to 0} B_a(\bar{B},\kappa)$,

$$B_{c1}(\kappa) = \frac{\Phi_0}{4\pi\lambda^2}[\ln\kappa + \alpha(\kappa)], \quad h_{c1} = \frac{B_{c1}}{B_{c2}} = \frac{\ln\kappa + \alpha(\kappa)}{2\kappa^2},$$

$$\alpha(\kappa) = \alpha_\infty + \exp[-c_0 - c_1\ln\kappa - c_2(\ln\kappa)^2] \pm \epsilon \tag{2.111}$$

with $\alpha_\infty = 0.49693$, $c_0 = 0.41477$, $c_1 = 0.775$, $c_2 = 0.1303$, and $\epsilon \leq 0.00076$. This expression yields at $\kappa = 1/\sqrt{2}$ the correct value $h_{c1} = 1$ and for $\kappa \gg 1$ it has the limit $\alpha = 0.49693$. A simpler expression for $\alpha(\kappa)$, yielding an h_{c1} with error still less than 1% and with the correct limits at $\kappa = 1/\sqrt{2}$ and $\kappa \gg 1$, is $\alpha(\kappa) = 0.5 + (1 + \ln 2)/(2\kappa - \sqrt{2} + 2)$. The resulting magnetization curves $M = \bar{B} - B_a$ are shown in Figure 2.5. They are well fitted by

$$h(b,\kappa) = \frac{B_a}{B_{c2}} \approx h_{c1} + \frac{c_1 b^3}{1 + c_2 b + c_3 b^2}, \quad c_1 = (1 - h_{c1})^3/(h_{c1} - p)^2,$$

$$c_2 = (1 - 3h_{c1} + 2p)/(h_{c1} - p), \quad c_3 = 1 + (1 - h_{c1})(1 - 2h_{c1} + p)/(h_{c1} - p)^2 \tag{2.112}$$

with h_{c1} from Eq. (2.111) and $p = -dm/db|_{b=1} = 1/[(2\kappa^2 - 1)\beta_A + 1]$ ($m = b - h = M/B_{c2}$), $\beta_A = 1.15960$ (1.18034) for the triangular (square) vortex lattice. This $h(b)$ satisfies the exact relations:

[257] J. R. Clem, *J. Low Temp Phys.* 18 (1975) 427

[258] A. Yaouanc, P. Dalmas de Réotier, and E. H. Brandt, *Phys. Rev.* B55 (1997) 11107

[259] M. M. Doria, J. E. Gubernatis, and D. Rainer, *Phys. Rev.* B39 (1989) 9573; ibd. 41 (1991) 6335; see also: U. Klein and B. Pöttinger, *Phys. Rev.* B44 (1991) 7704

$h(0) = h_{c1}, h'(0) = h''(0) = h''(1) = 0, h(1) = 1, h'(1) = 1 - p(\kappa)$. The $M(B_a)$ from the fit (2.112) applies for not too large $\kappa \lesssim 10 \ldots 20$. For larger κ, better fits are given in Brandt [256c], where it is also shown that the often used "logarithmic law at $B_{c1} \ll B_a \ll B_{c2}$" for the magnetization $M(B_a) = \bar{B} - B_a$ has very limited range of validity. For extension of these vortex lattice computations to superconducting films of arbitrary thickness see Brandt [256d] and Figure 2.6.

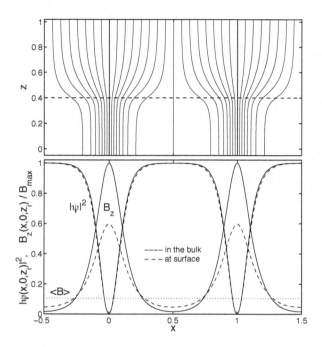

FIGURE 2.6: Magnetic field lines (top) and profiles of order parameter $|\psi(,0,z_i)^2|$ and of induction $B(x,0,z_i)$ for a superconducting film or half space for $\kappa = 1.4$ and $b = 0.04$. Length unit is the vortex spacing. The surface is at $z = 0.4$.

2.6.7   Vortex Motion, Pinning, and Thermal Depinning

Under the action of an applied electric current, the vortices can move and dissipate energy. This undesired vortex drift is suppressed by introducing material inhomogeneities that pin the vortices, e.g., structural defects or precipitates of a different material. Vortex pinning is crucial for the application of superconductors as loss-free conductors in cables and magnet coils, but also in electronic devices where it reduces losses and suppresses noise originating from jumping vortices.

The force density on the vortex lattice is $\mathbf{F}_v = \mathbf{j} \times \mathbf{B}$, where $\mathbf{j}$ is the current density and $\mathbf{B}$ the magnetic induction averaged over a few vortices. In the absence of pinning, the drift velocity v is assumed to be proportional to $\mathbf{F}_v$, and the electric field caused by the vortex motion is $\mathbf{E} = \mathbf{B} \times \mathbf{v} = \rho_{ff}\mathbf{j}$. This defines the flux-flow resistivity ρ_{ff}. Pinning stops the vortices[260], but finite temperature T can activate the vortices above a current-dependent barrier[253,261], e.g., $U(j) = U_0 \ln(j_c/j)$. Inserting this into an Arrhenius law $E = E_0 \exp(-U/kT)$, one obtains $E(j) = E_0(j/j_c)^n$ with $n = U_0/kT \gg 1$. This

[260] A. M. Campbell and J. E. Evetts, *Adv. Phys.* 21 (1972) 199

[261] G. Blatter et al., *Rev. Mod. Phys.* 66 (1994) 1125

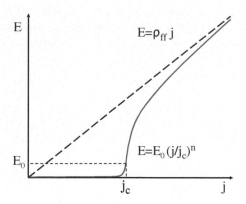

FIGURE 2.7:   The electric field E versus current density j in a superconductor with thermally activated depinning of vortices. Schematic.

defines the critical current density j_c via a voltage criterion, typically chosen as $E_0 = 0.1\,\mu V/mm$. At larger current density j this power law has to go over to the linear flux-flow law as shown in Figure 2.7. For more $E(j)$ models, e.g., the vortex-glass picture and for $H - T$ phase diagrams of vortex lattices with thermal fluctuations and with pinning see[261,262,263,264] and references therein.

In the limit of $n \to \infty$, the power law $E(j) = E_0(j/j_c)^n$ reproduces the critical state model for flux pinning, which means that the magnitude j everywhere in the superconductor is either zero or j_c. In general, $j_c(B)$ depends on the induction B. If this dependence is disregarded one has the Bean critical state model[262]. Using any such nonlinear $E(j,B)$ model law, one may compute the dynamics of induction and current density in a superconductor with thermally activated depinning for any time-dependent applied magnetic field H_a or transport current I_a. For large creep exponent $n \gg 1$ one obtains the static limit. Such computations have been done for long slabs and cylinders in parallel field H_a meaning absence of demagnetization effects[262]. For strong demagnetization effects, namely thin and thick strips and discs, and rectangular plates and films in a perpendicular or tilted H_a, computations were done, e.g., by inversion of a matrix[265]. In these continuum computations the magnetic penetration depth λ may be chosen to be zero (ideal screening) or finite.

2.6.8   Anisotropic Superconductors

Some superconducting materials are anisotropic, in particular high-T_c ceramics like YBCO. They possess three different coherence lengths ξ_a, ξ_b, ξ_c, and three magnetic penetration depths λ_a, λ_b, λ_c, describing the screening of supercurrents flowing along the three principal axes (a, b, c) of the crystal. The corresponding anisotropic GL theory is obtained by multiplying in Eq. (2.82) the gradient term by a tensor. If the coherence lengths are sufficiently small, one may use the anisotropic London theory (AL). In many cases one may assume uniaxial symmetry by defining $\lambda_a \approx \lambda_b \approx \lambda_{ab} = (\lambda_a \lambda_b)^{1/2}$ (and similarly for ξ_{ab}) and a GL parameter $\kappa = \lambda_{ab}/\xi_{ab}$ and anisotropy $\Gamma = \lambda_c/\lambda_{ab} = \xi_c/\xi_{ab}$; see Brandt[253] and Blatter et al.[261] and references therein. For YBCO one has $\kappa \approx 50$, $\Gamma \approx 7$, and $\xi_a/\xi_b \approx 1.3$.

[262] T. Nattermann and S. Scheidl, *Adv. Phys.* 49 (2000) 607

[263] G. P. Mikitik and E. H. Brandt, *Phys. Rev.* B64 (2001) 185414; *Phys. Rev.* B68 (2003) 054509

[264] J. Kierfeld and V. Vinokur, *Phys. Rev.* B61 (2000) 14928

[265] E. H. Brandt, *Phys. Rev.* B64 (2001) 024505

From GL theory, properties of the vortex lattice were calculated, e.g., by Kogan and Clem[266]. The nonlocal elasticity of the vortex lattice within AL theory and the anisotropic vortex interaction are given in Brandt [253]. The thermal fluctuations of the vortices $\langle u \rangle^2$ are enhanced by the anisotropy factor Γ when the vortices are along the c axis. The line energy J and line tension P of a vortex are highly anisotropic[253]. In general, J and P do not coincide. One has $P(\alpha) = J(\alpha) + \delta^2 J / \delta \alpha^2$ where the angle α defines the vortex orientation in the plane of tilt. For vortices along the c axis ($\alpha = 0$) these two terms almost compensate; one then has $J(\alpha) = J(0)(\cos^2 \alpha + \Gamma^{-2} \sin^2 \alpha)^{1/2}$ and thus $P(0) = J(0)/\Gamma^2$. For vortices in the ab plane one has $P(\pi/2) = J(0)/\Gamma$ and for tilt out off the ab plane one has $P_{\mathrm{out}}(\pi/2) = J(\pi/2)\Gamma^2 = P(0)\Gamma^3$. For tilt within the ab plane the vortices do not feel the anisotropy and thus $P_{\mathrm{in}}(\pi/2) = J(\pi/2)$.

2.7   Thermomagnetic Effects in the Mixed State

R. P. Huebener

Not long after A.A. Abrikosov had introduced the concept of type II superconductivity and the existence of the magnetic flux-line lattice in 1957, the motion of magnetic flux lines under the influence of external forces had been discussed and experimentally observed. In 1962 C. J. Gorter and P. W. Anderson proposed the phenomenon of current-induced motion of magnetic flux lines and the concept of flux-flow resistance in type II superconductors. Previously, current-induced magnetic domain motion in the intermediate state of type I superconductors had been discussed already by D. Shoenberg and C. J. Gorter.[267]

The phenomenon of flux-flow resistance arises because of the Lorentz force $\mathbf{j} \times \mathbf{\Phi_0}$ acting on a (single-quantum) magnetic flux line. Here $\mathbf{j}$ is the applied electric current density and $\mathbf{\Phi_0}$ the magnetic flux quantum, oriented along the direction of the magnetic field. We note that the Lorentz force is directed perpendicular to the vectors $\mathbf{j}$ and $\mathbf{\Phi_0}$ (see Figure 2.8a). The resulting flux motion is governed by the force equation

$$\mathbf{j} \times \mathbf{\Phi_0} - \eta \mathbf{v}_\phi = 0 \tag{2.113}$$

where η is a damping coefficient and $\mathbf{v}_\phi$ the flux-flow velocity. In Eq. (2.113) the Lorentz force $\mathbf{j} \times \mathbf{\Phi_0}$ and the damping force $\eta \mathbf{v}_\phi$ are given per unit length of flux line. During the flux-flow process magnetic flux lines are nucleated at the sample edge and are annihilated at the opposite edge. In this discussion we have simplified the situation by ignoring flux pinning and a small component of the driving force leading to the Hall effect. In the simplified case of Eq. (2.113) flux motion occurs exactly perpendicular to $\mathbf{j}$ resulting in a longitudinal (resistive) electric field $\mathbf{E}$ given by

$$\mathbf{E} = -\mathrm{grad}\, V = -\mathbf{v}_\phi \times \mathbf{B} \tag{2.114}$$

where V is the voltage and $\mathbf{B}$ the magnetic flux density. $\mathbf{B}$ can be expressed as

$$\mathbf{B} = n\mathbf{\Phi_0} \tag{2.115}$$

where n is the (two-dimensional) density of magnetic flux lines.

The motion of the magnetic flux lines across the superconductor induced by the Lorentz force is accompanied by the transport of entropy, since within the normal core of a vortex line the entropy density is larger than in the surrounding superconducting phase. This difference in entropy density

[266] V. G. Kogan and J. R. Clem, *Phys. Rev.* B24 (1981) 2497

[267] R. P. Huebener, *Magnetic Flux Structures in Superconductors*, 2nd edition, Springer, Berlin (2001)

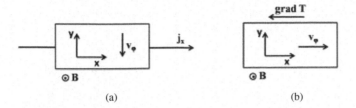

(a)   (b)

FIGURE 2.8: Geometry of the flux motion in the case of an applied electric current density j in x-direction (*a*) or temperature gradient in x-direction (*b*). The magnetic flux density B is oriented in the z-direction.

is expressed in terms of the transport entropy S_ϕ per unit length of flux line. The resulting heat current density $\mathbf{U}$ is

$$\mathbf{U} = nTS_\phi \mathbf{v}_\phi \tag{2.116}$$

where T is the temperature. During this process, heat energy is absorbed from the sample at the edge where the flux lines are nucleated and is delivered back to the sample at the opposite edge. Hence, a temperature gradient is established along the direction of flux motion. In the stationary state the heat current density is compensated by regular heat conduction. Assuming the geometry shown in Figure 2.8a, $\mathbf{U}$ is given by the y-component U_y, and we have

$$U_y = nTS_\phi v_{\phi y} = -\kappa(\partial T/\partial y) \tag{2.117}$$

where κ is the heat conductivity. Inserting (2.114) and (2.115), we find

$$|\partial T/\partial y| = \frac{TS_\phi}{\kappa\phi_0}|\partial V/\partial x|. \tag{2.118}$$

The appearance of a transverse temperature gradient during electric current flow in a perpendicular magnetic field is referred to as the Ettingshausen effect. The current-induced resistive state due to the self-field of the current in the absence of an applied magnetic field represents an interesting special case. Assuming a flat superconducting strip carrying a sufficiently large electric current, magnetic flux lines of opposite sign are nucleated at the opposite edges of the strip and then move to the center of the strip in opposite direction. At the center they annihilate each other. In this case in the two halves of the strip a temperature gradient develops in opposite direction, with the maximum of the temperature appearing in the center.

In addition to the Lorentz force of an applied electric current, there exists another force acting on magnetic flux lines: the thermal force due to a temperature gradient. In this case we deal with the general phenomenon of thermal diffusion. This phenomenon always appears in the case of a mixture of particles exposed to a temperature gradient. A famous example are dust particles in the air, experiencing thermal diffusion in the temperature gradient between heating pipes and walls in a house and they are driven to the wall. Another example is the separation column based on thermal diffusion and invented by Klaus Clusius, utilized for isotope separation.[268]

In the case of the mixed state, in a temperature gradient the magnetic flux lines experience the thermal force $-S_\phi \text{grad } T$, resulting in the motion of flux lines down the temperature gradient (see Figure 2.8b). This process is governed by the force equation

$$-S_\phi \text{grad } T - \eta\mathbf{v}_\phi = 0 \tag{2.119}$$

[268] R. P. Huebener, *Electrons in Action*, WILEY-VCH, Weinheim (2005)

where ηv_ϕ represents the damping force also appearing in Eq. (2.113). Here we have ignored again flux pinning and a small force component leading to flux-line motion transverse to grad T. With the geometry shown in Figure 2.8b, together with Eq. (2.114), we have

$$|\partial V/\partial y| = (-S_\phi)\frac{B}{\eta}|\partial T/\partial x|. \tag{2.120}$$

The appearance of a transverse electric field due to a longitudinal temperature gradient in a perpendicular magnetic field is referred to as the Nernst effect.

At this point we add some historic remarks.[269] In 1885, in his third year as a student, Walther Nernst went to the Karl-Franzens University in Graz, intending to study theoretical physics, in particular with Ludwig Boltzmann. In Graz, Nernst developed a close friendship and collaboration with Albert von Ettingshausen, who at the time, in addition to Ludwig Boltzmann, occupied the position of Professor at the Physics Institute. Motivated by Boltzmann, Nernst and von Ettingshausen studied experimentally the galvanomagnetic and thermomagnetic phenomena, and they discovered new effects, which were subsequently named after them. Based on these experiments in Graz, in 1887 Nernst submitted a thesis at the Royal Julius-Maximilians University of Würzburg, where he obtained his PhD under Friedrich Kohlrausch. The title of his dissertation was "On the electromotoric forces generated by the magnetism within metal plates through which a heat current is flowing".

It is interesting that after his PhD, Nernst left the subject area of his thesis and turned to the new field of Physical Chemistry, where he became one of its founders. As he explained much later, "... the group of the thermomagnetic and galvanomagnetic phenomena ... should find a satisfactory theoretical explanation only after the theory of the metallic conduction would have been developed still further."[269] The latter development of the theory had to wait until the 1930s and later, when the quantum theory was successfully applied to the electrons in solids. We see that Nernst possessed deep visionary insight into the overall status of science. (This also motivated him much later, when in 1911 he took the initiative and organized the famous First Solvay Congress dealing with the spectacular developments in quantum physics.)

The Ettingshausen and the Nernst effect are intimately connected with each other due to the Onsager reciprocity relations of irreversible thermodynamics. With the geometry of Figure 2.8a, the Ettingshausen coefficient ϵ is defined as[270]

$$\epsilon \equiv \frac{1}{B\kappa}\frac{U_y}{j_x}. \tag{2.121}$$

Using (2.113) through (2.115) and (2.117), one finds

$$\epsilon = \frac{TS_\phi}{\kappa\eta}. \tag{2.122}$$

With the geometry of Figure 2.8b, the Nernst coefficient ν is defined as[270]

$$\partial V/\partial y \equiv -\nu B(\partial T/\partial x). \tag{2.123}$$

Using (2.120) one finds

$$\nu = S_\phi/\eta \tag{2.124}$$

and finally

$$\epsilon\kappa = T\nu. \tag{2.125}$$

[269] H.-G. Bartel and R. P. Huebener, *Walther Nernst — Pioneer of Physics and of Chemistry*, World Scientific, Singapore (2007)

[270] S. R. De Groot, *Thermodynamics of Irreversible Processes*, North-Holland, Amsterdam (1951)

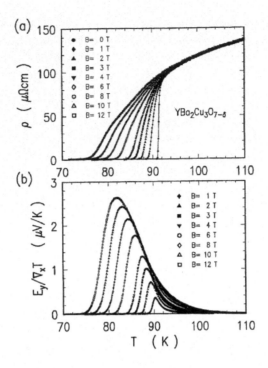

FIGURE 2.9:   (*a*) Resistivity and (*b*) normalized Nernst electric field versus temperature of an epitaxial c-axis oriented film of $YBa_2Cu_3O_{7-\delta}$ for different magnetic fields oriented in the c-direction.

Equation (2.125) is referred to as the Bridgman relation of irreversible thermodynamics (representing one entry of the Onsager reciprocity scheme).

The Ettingshausen and the Nernst effect in the mixed state have been observed experimentally in a series of type II superconductors, including epitaxial thin films of the cuprate high-temperature superconductors (see Figure 2.9). A summary of the literature can be found in Huebener [267,271]. These thermomagnetic measurements had been extended in some cases also to the intermediate state of type I superconductors, where the motion of multiquantum magnetic flux tubes or of normal domains causes the corresponding phenomena.

In the case of measurements of the Nernst effect, the application of a large temperature gradient to the sample can present complications because of the rapidly changing properties of superconductors with temperature. Such problems can be avoided in the case of a geometry where the sample dimension in the direction of the temperature gradient is kept small, say, about 100 μm. Then a large gradient can be achieved even for a small temperature difference between the hot and the cold side of the sample.[272] From thermodynamic arguments, in the case of low temperatures and small magnetic fields, for a type II superconductor one finds for the transport entropy per unit length of flux line[273]

$$S_\phi/\Phi_0 = -\frac{1}{4\pi}\partial H_{c1}/\partial T \tag{2.126}$$

where H_{c1} is the lower critical magnetic field. In the case of (2.126) it is assumed that the flux lines are well separated from each other. From (2.126) we see that in the limit $T \to 0$ the transport entropy

[271] R. P. Huebener, *Supercond. Sci. Technol.* 8 (1995) 189

[272] H.-C. Ri, R. Gross, F. Gollnik, A. Beck, R. P. Huebener, P. Wagner, and H. Adrian, *Phys. Rev.* B 50 (1994) 3312

[273] M. J. Stephen, *Phys. Rev. Lett.* 16 (1966) 801

vanishes as expected from the third law of thermodynamics. On the other hand, as one approaches T_c, the transport entropy is reduced again to zero since the flux lines overlap more and more. Usually, the experimental observation of the thermomagnetic effects due to the motion of magnetic flux lines is restricted to temperatures not far below T_c because of flux pinning preventing flux motion at low temperatures. In the high-field regime close to the upper critical field H_{c2}, the time-dependent Ginzburg-Landau theory has been used to calculate the temperature dependent function $S_\phi(T)$ by Maki and others. A summary can be found in Huebener [267]. In (2.113) and (2.119) we had ignored the (small) force component leading to a finite Hall angle θ of the flux-line motion. This force component $f_m = -\alpha(\mathbf{v}_\phi \times \mathbf{n})$ is referred to as the Magnus force, and (2.113) is extended to

$$\mathbf{j} \times \mathbf{\Phi}_0 - \eta \mathbf{v}_\phi - \alpha(\mathbf{v}_\phi \times \mathbf{n}) = 0. \tag{2.127}$$

A similar extension applies to (2.119). Here $\mathbf{n}$ is a unit vector in the magnetic field direction. In the case shown in Figure 2.8a, the Hall angle θ is given by [267]

$$\tan\theta = \frac{v_{\phi x}}{v_{\phi y}} = -\frac{\alpha}{\eta}. \tag{2.128}$$

(We note that the flux-line velocity $\mathbf{v}_\phi$ has a small component $v_{\phi x}$ in the x-direction.) Because of the Magnus force, the heat current density $\mathbf{U}$ has a small component in the x-direction. This leads to the Peltier effect. Similarly, if (2.119) is extended including the Magnus force, in the case shown in Figure 2.8b a small component $v_{\phi y}$ appears perpendicular to the direction of the temperature gradient. This leads to an electric field in the x-direction due to (2.114), i. e., to the Seebeck effect. For the geometry shown in Figure 2.8a, the Peltier coefficient Π is defined as

$$\Pi = \frac{U_x}{j_x} \tag{2.129}$$

and for the geometry shown in Figure 2.8b, the Seebeck coefficient S is defined as

$$S = \frac{E_x}{(\partial T/\partial x)}. \tag{2.130}$$

Both coefficients are connected with each other because of the Thomson relation

$$\Pi = TS. \tag{2.131}$$

We see that both the Peltier and the Seebeck effect result from the finite Hall angle appearing during the motion of the flux-line lattice experiencing the Lorentz force or the thermal force, respectively. Further below we discuss how the quasiparticles still existing in the mixed state also contribute significantly to both of these thermoelectric phenomena.

In general, the magnitude of the Ettingshausen effect and the Nernst effect in the mixed state of a superconductor is much larger than in the normal state of the material. Therefore, the measurements of these effects do not require a special subtraction procedure for eliminating a normal-state background. This aspect is particularly interesting in the temperature range near and above T_c, where fluctuation effects become dominant. A discussion of this subject can be found in Ri et al. [272]

Recently, the Nernst effect due to flux flow in type II superconductors attracted special interest because of the following reason. Ong and coworkers have reported the unexpected observation of an unusually large Nernst voltage 50–100 K above T_c in a series of underdoped $La_{2-x}Sr_xCuO_4$ crystals.[274,275,276] This temperature range coincides with that of the pseudogap discovered in the

[274] Z. A. Xu, N. P. Ong, Y. Wang, T. Kakeshita, and S. Uchida, *Nature* (London) 406 (2000) 486

[275] Y. Wang, Z. A. Xu, T. Kakeshita, S. Uchida, S. Ono, Y. Ando, and N. P. Ong, *Phys. Rev.* B 64 (2001) 224519

[276] Y. Wang, N. P. Ong, Z. A. Xu, T. Kakeshita, S. Uchida, D. A. Bonn, R. Liang, and W. N. Hardy, *Phys. Rev. Lett.* 88 (2002) 257003

underdoped cuprate high-temperature superconductors.[277] It has been suggested that the anomalous Nernst voltage in the pseudogap region is due to the thermal diffusion of vortex-like excitations. Here the underlying idea assumes that the pseudogap is due to (preformed) Cooper pairs without long-range phase coherence ("pairing without condensation"). Of course, the anomalous Nernst voltage being clearly absent in the normal state represents a crucial point in the argument. Experimental evidence supporting this concept of preformed pairs continues to be found.[278,279] However, this issue requires more work for a final clarification. The Nernst effect in underdoped $La_{2-x}Sr_xCuO_4$ crystals in the vicinity of the superconductor-insulator transition has been studied by Capan et al.[280]

Finally, we turn to the role of the quasiparticles in the thermomagnetic phenomena in the mixed state. (A previous discussion can be found in Huebener [267,271].) Starting with the Seebeck effect and assuming the temperature gradient is oriented along the x-direction, in a normal conductor the electric charges q experience the thermal force $f_{th} = -S^*(\partial T/\partial x)$ and the electrostatic force $f_{el} = qE_x$. Here S^* is the transport entropy of the charge carriers. The electric field E_x is generated by space charges accumulated due to their thermal diffusion. Under stationary conditions, we have $f_{th} + f_{el} = 0$, yielding for the normal-state Seebeck coefficient (2.130)

$$S_n = \frac{E_x}{(\partial T/\partial x)} = \frac{S^*}{q}. \tag{2.132}$$

(We note that the Seebeck coefficient is the transport entropy per charge.) The electric current density j_n associated with the quasiparticle thermal diffusion process is

$$j_n = -E_x/\rho_x = -(S_n/\rho_n)(\partial T/\partial x) \tag{2.133}$$

where ρ_n is the normal-state electric resistivity. This diffusion current is compensated by the drift current of density $E_x/\rho_n = -j_n$.

These concepts have been extended to the superconducting state by Ginzburg[281] and were discussed by him again recently.[282,283] Noting that in the superconducting state electric fields such as generated by space charges cannot exist, Ginzburg showed that in a zero magnetic field the thermal diffusion current density j_n of (2.133) is locally compensated by a counter flowing supercurrent density j_s: $j_n + j_s = 0$. This phenomenon is the exact analog of the counterflow in superfluid helium, leading to the fountain effect. Below T_c the quasiparticle concentration becomes rapidly smaller with decreasing temperature, and the counter flow eventually vanishes. From a simple two-fluid model we expect $j_n = -j_s \sim (T/T_c)^4$. For the classical superconductors the counterflow effect predicted by Ginzburg[281] has been observed in a series of experiments.[284] Such an experimental demonstration has also been reported for a polycrystalline YBaCuO film.[285]

In the mixed state we have an inhomogeneous situation, and the counterflow model of Ginzburg must be extended.[286,287] The magnetic field is assumed to be oriented along the z-direction. As

[277] T. Timusk and B. Statt, *Rep. Prog. Phys.* 62 (1999) 61

[278] A. Kanigel, U. Chatterjee, M. Randeria, M. R. Norman, G. Koren, K. Kadowaki, and J. C. Campuzano, *Phys. Rev. Lett.* 101 (2008) 137002

[279] O. Yuli, I. Asulin, Y. Kalcheim, G. Koren, and O. Millo, *Phys. Rev. Lett.* 103 (2009) 197003

[280] C. Capan, K. Behnia, J. Hinderer, A. G. M. Jansen, W. Lang, C. Marcenat, C. Marin, and J. Flouquet, *Phys. Rev. Lett.* 88 (2002) 056601

[281] V. L. Ginzburg, *Zh. Eksp. Teor. Fiz.* 14 (1944) 177 [Engl. Trans. *J. Phys.* USSR 8 (1944) 148]

[282] V. L. Ginzburg, *JETP Lett.* 49 (1989) 58

[283] V. L. Ginzburg, *On Superconductivity and Superfluidity*, Springer, Berlin (2009)

[284] D. J. Van Harlingen, *Physica* B 109-10 (1982) 1710

[285] A. V. Ustinov, M. Hartmann, and R. P. Huebener, *Europhys. Lett.* 13 (1990) 175

[286] R. P. Huebener, A. V. Ustinov, and V. K. Kaplunenko, *Phys. Rev.* B 42 (1990) 4831

[287] H.-C. Ri, F. Kober, A. Beck, L. Alff, R. Gross, and R. P. Huebener, *Phys. Rev.* B 47 (1993) 12312

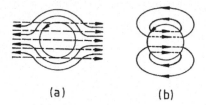

(a)   (b)

FIGURE 2.10:   Extended counterflow model of Ginzburg. In a temperature gradient oriented hor-
izontally from right to left the counterflow of j_n (dashed lines) and j_s (solid lines) near a flux line
is shown. (*a*) T close to T_c: The quasiparticle concentration is high everywhere. (*b*) $T \ll T_c$: Away
from the flux line the quasiparticle concentration is negligible.

discussed also in Huebener [267,271] and illustrated in Figure 2.10, the pattern of the counterflow
supercurrent depends on the temperature. Not much below T_c the quasiparticle concentration is relatively high everywhere, and far from the core of the flux lines the two-fluid counterflow concept
of Ginzburg is valid. However, close to the flux lines the supercurrent redistributes itself, flowing
around the flux line, whereas the normal current does not. On the other hand, in the temperature
range $T \ll T_c$, where the quasiparticle concentration away from the vortex core vanishes, the quasiparticle thermal diffusion current is restricted to the vortex-core region and is compensated again
by the superfluid backflow. In both cases shown in Figure 2.10, the supercurrent backflow generates
a driving force acting on the flux lines and causing vortex motion in the y-direction. According
to (2.114) a longitudinal electric field E_x develops, and it is this mechanism, which generates the
dominant component of the Seebeck effect in the mixed state (see Figure 2.11).

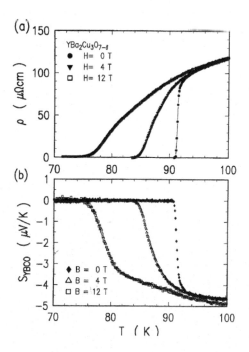

FIGURE 2.11:   (*a*) Resistivity and (*b*) Seebeck coefficient versus temperature of an epitaxial c-
axis oriented film of $YBa_2Cu_3O_{7-\delta}$ for different magnetic fields oriented in the c-direction.

By introducing the flux-flow resistivity ρ_{fl} in the mixed state through the relation

$$E_x = \rho_{fl} j_s \qquad (2.134)$$

inserting $j_s = -j_n$ from (2.133), and using (2.130), we obtain for the Seebeck coefficient $S_m(T)$ in the mixed state

$$S_m(T) = (\rho_{fl}(T)/\rho_n(T))S_n(T). \qquad (2.135)$$

This result has also been obtained by the time-dependent Ginzburg-Landau theory. It has been well confirmed experimentally. References to the literature can be found in Huebener [267,271].

Similar to the Seebeck effect, the Peltier effect associated with the quasiparticle current in a flux-flow electric field must be expected because of the Thomson relation (2.131). However, it appears that such measurements have not yet been reported. Furthermore, a small Hall component is connected with the quasiparticle motion discussed above. Details can be found in Ri et al.[272,287].

Acknowledgments

Ginzburg–Landau Equations

We are delighted to thank our friends and colleagues who shared with us their expertise and knowledge: Igor Aranson, Leonid Berlyand, Andreas Glatz, Alex Koshelev, Eugene Maksimov, who also made his manuscript available to us prior to publication, Nikita Nekrasov, Nikolay Shchelka-chev, and Andrei Varlamov. We are grateful to Maya Vinokour for critical reading of the manuscript. This work was supported by the U.S. Department of Energy Office of Science through the contract DE-AC02-06CH11357.

Theory of Superconductivity: From Phenomenology to Microscopic Theory

The work was supported by NHMFL through the NSF Cooperative agreement No. DMR-0654118 and the State of Florida.

Intrinsic Josephson Effect in Layered Superconductors

We thank V. Kogan, A. Saxena, I. Martin, C. Batista and P. Kes for helpful comments. The work of L.N.B. was carried out under the auspices of the National Security Administration of the U.S. Department of Energy at Los Alamos National Laboratory under Contract No. DE-AC52-06NA25396 and supported by LANL/LDRD Program. A.E.K. is supported by U Chicago Argonne, LLC, under contract No. DE-AC02-06CH11357.

3

Experiments

Editor: J. Mannhart

3.1   Tunneling and the Josephson Effect
    John M. Rowell 145
3.2   The Discovery of Fluxoid Quantization
    Dietrich Einzel 161
3.3   The Search for the Pairing Symmetry in the High Temperature Superconductors
    Dale J. Van Harlingen 170
3.4   Half-Integer Flux Quantization in Unconventional Superconductors
    C. C. Tsuei and J. R. Kirtley 182
3.5   Electric Field Effect Tuning of Superconductivity
    Jean-Marc Triscone and Marc Gabay 197
3.6   The Grain Boundary Problem of High-T_c Superconductors
    J. Mannhart and D. Dimos 206
3.7   Overview of the Experimental History, Status and Prospects of HTS
    M. R. Beasley 213
    Acknowledgments 222
3.8   Vortex Matter in Anisotropic Superconductors
    Eli Zeldov 223
    Acknowledgments 231
3.9   Further Reading 231

3.1   Tunneling and the Josephson Effect

John M. Rowell

In his Nobel lecture in 1973, Ivar Giaever quoted from his laboratory notebook of May 2, 1960, as follows: "Friday April 22, I performed the following experiment aimed at measuring the forbidden gap in a superconductor"[1]. The experiment was a success, and so began 50 years of a field of research, applications and technology that includes superconducting tunneling spectroscopy, the Josephson Effect, SQUIDS, biomagnetism, low field MRI, far infrared detectors, quantum computing and single flux quantum logic. The work began somewhat earlier: after Esaki's demonstration of electron tunneling in semiconductor diodes, Fisher and Giaever at General Electric began a program to see if tunneling could be observed in a very different system, namely between two metals separated by a thin insulator. After some unsuccessful experiments with other insulators between

---

[1]   Ivar Giaever, *Rev. Mod. Phys.* 46 (1974) 245

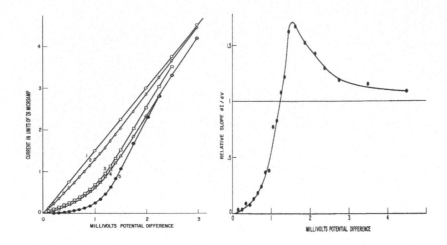

FIGURE 3.1:   The first observation of the energy gap of a superconductor as revealed in the current-voltage and conductance-voltage characteristics of an Al/AlOx/Pb tunnel junction, by Ivar Giaever.

the metals, they chose a structure of crossed thin films of aluminum-aluminum oxide-aluminum, with the aluminum films being evaporated and the oxide being grown by exposure to air. Essentially the same structure or device, with the aluminum films replaced by niobium, is used today in all the SQUIDS, detectors and digital circuits. Initially, they observed an exponential dependence of current (I) on the voltage (V) at high voltages (0.2 to 1.4 V), as expected for tunneling, but a number of their GE colleagues suggested other explanations of their results. Luckily, Giaever attended a lecture about superconductivity and the BCS theory in his Ph.D. studies (he received his Ph.D. in 1964) and the energy gap was mentioned. Intuitively, he believed that tunneling into a superconductor should reveal the gap, as it would not be possible to tunnel into a region having no electron states. Ignoring the doubts of experts, he tried the experiment with a junction of aluminum-aluminum oxide-lead, and when the temperature was reduced below the T_c of Pb (7.2 K), the gap was clearly revealed as a reduced current at low voltages, as shown in Figure 3.1. Giaever further realized that the derivative plot, dI/dV versus V, resembled the density of states expected from BCS, when thermal kT effects were allowed for. These results were submitted to *Physical Review Letters* in July 1960 and published six weeks later[2]. Giaever's Nobel lecture[1] is a most entertaining account of the history of the GE experiments. In this chapter, I will give a personal account of some events related to tunneling between superconductors in the period from 1960 to about 1975. As the science has been adequately described previously, I will try to relate aspects of the history that have not appeared in publications and reviews.

The Giaever experiment was elegant and simple; current to the junction was provided from a battery and potentiometer, and I and V were read from an ammeter and voltmeter and plotted by hand. However, by January 1961, Giaever and Megerle had described using an X-Y recorder, so reading from the two meters and plotting I-V must have quickly become too tedious. The films were made in a glass bell jar that probably achieved pressures in the 10^{-6} Torr range, while the junctions were immersed in liquid helium in a glass dewar, temperatures below 4.2 K being provided by a rotary pump. The contrast with today's omnipresent PPMS measurements and stainless steel MBE systems, and their relative costs, is striking. In my biased view, much has been lost now that X-

---

[2]   Ivar Giaever, *Phys. Rev. Lett.* 5 (1960) 147

Y recorders have been replaced by digital meters and computers. The continuous trace of the pen on the 11×15 inch sheet of paper was quite unforgiving, noise was obvious or even audible from vibration of the pen (especially from ground loops), so were "spikes" caused by interference or instabilities in the line voltage, and so was drift when the same derivative trace (say) was recorded a number of times and did not reproduce exactly. Recently I enthusiastically described the advantages of X-Y recorders to a group of graduate students, until it became obvious that none of them had any idea what I was talking about. (The Moseley Company, formed in 1951 to manufacture the first X-Y recorders, was purchased by HP in 1958, their first acquisition.)

When Al/AlOx/Pb junctions were cooled below the T_c of Al (1.2 K), a small energy gap opened in the Al, and the I-V characteristics exhibited a negative resistance region. This was shown by the second group to enter the tunneling field, Nicol, Shapiro and Smith at Arthur D. Little, in a letter submitted in October 1960 [3], and by Giaever in a letter submitted 4 days later. I have always believed that the best way to determine the energy gap of any superconductor with a higher T_c is to note exactly where in voltage the Al gap opens just as T_c of about 1.2 K is reached.

Interest in tunneling between superconductors grew extremely slowly, compared, for example to events after the discovery of HTS. I often feel that today's younger scientists cannot comprehend how small the physics research enterprise was in the 1960s. For a year or two after Giaever's publication, the number of groups studying superconducting tunneling could be counted easily on one hand. In January 1961, Burstein, Langenberg and Taylor[4] at the University of Pennsylvania proposed that NIS and SIS junctions could be used as detectors of microwave and sub-millimeter-wave radiation, in that the photons would "help" the quasiparticles across the barrier when the voltage bias was less than the gap, or the sum of the gaps. This proposal is the basis for today's SIS detectors used on many telescopes. Note that this effect has nothing to do with the Josephson Effect, and was proposed before that effect was even known. The interaction they proposed was demonstrated by Dayem and Martin[5] of Bell Labs in early 1962, and was seen as a series of steps in I-V both just below and above the rise in current at the sum gap.

During the time of Giaever's discovery of superconducting tunneling, I was carrying out my thesis research at Oxford. The strong interest there in low temperature physics resulted in his work being noticed and discussed, so I was made aware of it by a colleague before I joined Bell Labs on April 12 of 1961. During the first week there, I was interviewed by a number of department heads, including Ted Geballe, Alan Chynoweth, Bill Boyle and those reporting to Rudi Kompfner (including the group of Dayem and Martin), who had recruited me from Oxford. I decided to join Alan's department, I suppose because I felt more comfortable reporting to an English boss. This choice of department, however, had important consequences, as Alan, with Ralph Logan and Don Thomas, had earlier used electronics built by Don with Jack Klein[6] that took both the first and second derivatives of I-V characteristics, and displayed them on an X-Y recorder. From their studies of silicon tunnel diodes, they had published results showing phonon emission as strong features in both the dI/dV and d^2I/dV^2 versus voltage characteristics. As my thesis work at Oxford was on impurity band conduction in germanium, it would have been easy to carry on in semiconductor research, which was the dominant topic in the department and indeed in the research area at Bell. It was, after all, only 14 years after the invention of the transistor, and previous heads of the department had been Shockley and Pearson. However, hoping that similar "phonon emission" effects might be seen in normal metal and superconductor tunnel junctions using the Thomas/Klein electronics, Alan and I agreed that I should begin such a research program. A diffusion-pumped evaporator with a glass Bell jar, similar to that used by Giaever, was available, I soon built a circuit to record I-V

[3] James Nicol, Sidney Shapiro, and Paul H. Smith, *Phys. Rev. Lett.* 5 (1960) 461

[4] E. Burstein, D. N. Langenberg, and B. N. Taylor, *Phys. Rev. Lett.* 6 (1961) 92

[5] A. H. Dayem and R. J. Martin, *Phys. Rev. Lett.* 8 (1962) 246

[6] D. E. Thomas and J. M. Klein, *Rev. Sc.. Instr.* 34 (1963) 920

characteristics on the first X-Y recorder I had ever seen, and a glass dewar for measurements at 4.2 K and below was ordered.

Initially, I deliberately tried to make tunnel junctions without copying the Al/AlO_x process already described by Giaever. My attempts to use dipping into diluted varnish were a failure, while Langmuir films made by Harold Schonhorn had some limited success. In his Nobel lecture, Giaever described similar initial ideas. These experiments, however, did allow me to make small capacitors on bismuth, which were studied with George Smith and Gene Baraff to determine the g factor of electrons and holes. Looking back recently at my earliest laboratory notebooks, I was surprised by how much time was spent in trying to make junctions on bulk pieces of Nb_3Sn and V_3Si, and later on V_3Si films made by Hauser. Interest in these A15 materials was high at Bell, as the first measurement of high critical currents at the highest magnetic fields then available (8.8 T) had just been made by Kunzler and his colleagues[7]. One Langmuir film junction did reveal a poorly defined gap on V_3Si, but it is possible that the surface also had a layer of oxide. Another topic taking many pages of my notebooks was trying to make cold tunneling cathodes, following work reported by Philco. Even 27 years after the disintegration of the Bell System, the notebooks of Bell Labs employees can still be studied at a storage facility in New Jersey, which is owned by today's AT&T.

Perhaps as a concession to the interest in semiconductors in the department, my first successful junctions were made on 12/8/61 by evaporating lead onto a heavily doped Si surface, and I observed the energy gap, "estimated as 1.5 mV at 4.2 K", using the Schottky barrier as the tunneling insulator. On 12/18/61, the derivatives $dI/dV - V$ and $d^2I/dV^2 - V$ were taken with Jack Klein using the equipment mentioned earlier. The d^2I/dV^2 sensitivity of this system could be increased by sweeping the voltage more rapidly, but at the expense of losing voltage resolution. We observed a series of dips and peaks extending up to about 15 mV; in hindsight this was my first rather crude step into tunneling spectroscopy. In analogy with the studies of semiconductor tunnel diodes performed in the department earlier, Alan and I thought that these features might be due to phonon emission processes, and that sums and harmonics of the Pb phonons were responsible for the features above about 10 mV, which was the highest phonon energy in Pb indicated by neutron scattering. The same type of silicon junction was also made with a drop of mercury, and revealed the smaller gap of Hg. In fact it was not necessary to evaporate Pb on Si; I found it could simply be pressed onto the surface.

Due to their nonlinear and asymmetrical I-V characteristics, the Schottky barriers were not ideal. On February 22, 1962 I wrote in my notebook of a decision to make Al/AlOx/Pb junctions by Giaever's method: "Try one shot only to see if structures suitable for dI/dV and d^2I/dV^2 appear — if not, get from Ali Dayem?" Ali Dayem had already made such junctions successfully. February 22 was Washington's birthday, a holiday at Bell Labs, but one I must have felt it was inappropriate to observe. The first junctions made that day behaved exactly as expected, and the derivative equipment was used on the next day. Quoting again from my notebook: "Jack ran off $d^2I/dV^2 \times V$ — dramatic kinks and structure out to about 15 mV either side of zero bias. Magnetic field only alters magnitude of structure but not the position of any peak. Pumped down - structure becomes even more dramatic but no change in position. At about 1.7 K > 1500 gauss required to destroy gap. . . . .$dI/dV \times V$ also taken, looks like - ". I had drawn a rough sketch of the result there, and the remainder of the page explained the features in terms of phonon emission, as Alan and I had discussed earlier.

A month later, I attended my first March APS meeting, where Giaever presented his latest results, including the density of states of Pb measured at 0.3 K using a derivative technique with a modulation signal to measure $dI/dV - V$. (Lock-in amplifiers were invented in 1961 by Dicke, so were just becoming available commercially from Princeton Applied Research.) This was the first time I had seen Giaever's derivative data, although it was submitted to *Phys. Rev.* in November 1961 and pub-

---

[7]   J. E. Kunzler, E. Buehler, F. S. L. Hsu, and J. H. Wernick, *Phys. Rev. Lett.* 6 (1961) H515

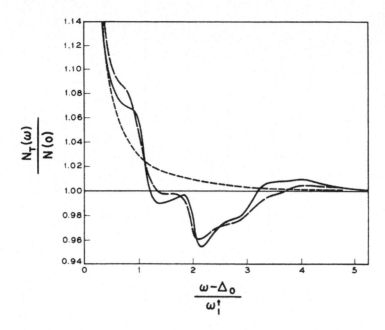

FIGURE 3.2: A comparison of the density of states of superconducting Pb calculated using a realistic model of the phonon spectrum by Schrieffer, Scalapino and Wilkins, with the measurement made by me with Phil Anderson and Don Thomas.

lished in May of 62 [8] — there was no Internet "prepublication" of preprints in those days. Because of the use of 0.3 K and better electronics, their data showed more precision than my results, which prompted me to decide to adopt the lock-in technique. With Alan and Jim Phillips, I submitted our dI/dV and d^2I/dV^2 results for publication in June[9]. This PRL suggested that the series of features in d^2I/dV^2 was due to sums and harmonics of a single phonon frequency appearing in the density of electron states through the gap equation, which was incorrect.

Although it was clear that the features Giaever and we had seen in the derivative plots were related to phonons, exactly how that took place remained unexplained for a while. For me, interest increased markedly and quickly when in the late summer of 1962 Phil Anderson returned from a sabbatical year in Cambridge. Given his recent work with Morel, he was very interested in my derivative data; in fact I believe he had been made aware of it while still in Cambridge. He was also aware that Schrieffer, with Culler, Fried and Huff, and with Scalapino and Wilkins, had begun to make solutions of the Eliashberg equations using somewhat realistic models of the superconductor's phonon density. So he arranged a meeting with the University of Pennsylvania group, and we went to Philadelphia for a discussion with them. After examining my derivative data, they decided that a model spectrum of Pb should have two gaussian peaks where the density of states dropped most sharply, i.e., where $d^2I/dV^2 - V$ had dips (with energy being the voltage measured from the gap). A short while later, they visited us at Bell, and the results of their calculation, showing remarkable similarity to the $dI/dV - V$ data (Figure 3.2), led to the publication of the letters by them[10] and by Phil and myself[11]. A key player in the improvement of my derivative measurements was Don Thomas, who was earlier responsible for the equipment used to study semiconductor diodes. It was

[8] Giaever, H. R. Hart, Jr., and K. Megerle, *Phys. Rev.* 126 (1962) 941

[9] J. M. Rowell, A. G. Chynoweth and J. C. Phillips, *Phys. Rev. Lett.* 9 (1962) 59

[10] J. R. Schrieffer, D. J. Scalapino, and J. W. Wilkins, *Phys. Rev. Lett.* 10 (1963) 336

[11] J. M. Rowell, P. W. Anderson, and D. E. Thomas, *Phys. Rev. Lett.* 10 (1963) 334

one of the strengths of Bell Labs that people like Don, an expert in the design and assembly of electronics in the days when much less equipment was for sale commercially, were available for help and collaboration. He made possible my measurements of $dI/dV - V$ and d^2I/dV^2 using the lock-in, then later designed and built the bridge circuit that greatly increased the stability of the dI/dV measurements, allowing reproducibility over long time periods to a part in 10^5. This was essential, in that dI/dV had to be measured with the junction superconducting, and then normal, which could take two hours or so.

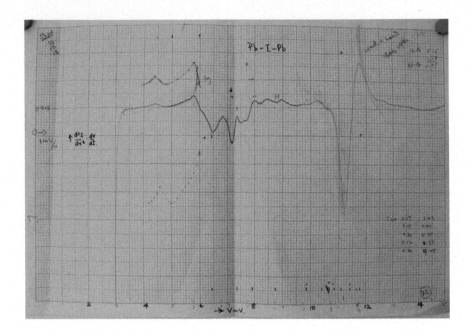

FIGURE 3.3: A photograph of an X-Y recorder trace I took late in 1962 of $d^2I/dV^2 - V$ for a Pb/PbO$_x$/Pb junction.

Don's second derivative circuit, when used with Al/AlO$_x$/Pb junctions, and even more so with Pb/PbO$_x$/Pb ones, revealed a fascinating fine structure, showing that very detailed features of the Pb phonon spectrum were contained in the electronic density of states. Perhaps it is not unfair to say that these d^2I/dV^2 traces represented the first high resolution tunneling spectroscopy of superconductors. Of all the measurements I made, including the first of the Josephson Effect, it was always the spectroscopy of Pb revealed in $d^2I/dV^2 - V$ (of course, on an X-Y recorder) that I found most satisfying. The initial d^2I/dV^2 traces published[12] with Phil Anderson and Don were a direct tracing made by the art department from an X-Y recorder sheet. A photograph of one such original trace in shown in Figure 3.3. The much weaker structures from an Al/AlO$_x$/Sn junction were also included in that letter, which, as a PRL, accompanied the work of the Penn group.

In addition to the use of improved derivative equipment, almost coincident advances were made in the late 1950s and early 1960s in both the understanding of the electron-phonon interaction by, for example, Migdal, Eliashberg, Nambu, Morel and Anderson, and Swihart, and of tunneling into strong coupling superconductors by Cohen, Falicov and Phillips. Schrieffer, with his collaborators at TRW and his colleagues at Penn, showed that structure in the density of states, and hence in the tunneling derivatives, arose from the complex and energy-dependent gap parameter, and not from

---

[12] J. M. Rowell, P. W. Anderson, and D. E. Thomas, *Phys. Rev. Lett.* 10 (1963) 334

emission processes (see chapters 10 and 11 of Parks book). However, a final and major step in the development of tunneling spectroscopy was needed, in that the calculations of Schrieffer, Scalapino and Wilkins only closely resembled our data, and did not contain any of the detail revealed by d^2I/dV^2. This final step was provided by Bill McMillan, a student of John Bardeen at Illinois, who joined the Bell theory group as a post-doc in 1964, soon after the publication of the pair of letters by the Penn group and ourselves. Almost immediately, I believe prompted by Phil, Bill became interested in whether an exact phonon spectrum (more strictly, the phonon density weighted with the coupling constant α^2) could be derived from my tunneling data. The "McMillan program", as it came to be known, made a solution of the gap equation with a model α^2F spectrum, say the one used by the Penn group, and then calculated the density of states. The key step was then to adjust the input spectrum over a number of iterations, usually 5 to 8, until the calculated density of states agreed with the measurement in all the exact details, including the fine structure in $d^2I/dV^2 - V$. The α^2F for Pb, derived in this way[12], is shown in Figure 3.4. As Phil once said about the work represented by this figure "The fat lady had sung".

The need for the best possible data had prompted changes in my lab and measurement system, as one day Bill watched as I took the $dI/dV - V$ traces for an Al-I-Pb junction with the Pb superconducting and then normal. At high voltages, say 30 mV, the two traces should have overlapped, but they did not. Drift in the lock-in was a serious problem, made worse for me because Bill had observed it! Earlier the "Klein machine", which was first used to measure semiconductor tunnel diodes and then to show both the first and second derivatives of the Si/Pb and the $Al/AlO_x/Pb$ junctions, had been replaced by my first lock-in amplifier and the second derivative circuit built by Don Thomas. The first derivative circuit followed soon afterwards. Further changes were made about the time that Ron Parks asked Bill and me to write a chapter for his book on superconductivity, which eventually turned out to be our only publication on the details of tunneling spectroscopy except for *Physical Review Letters*. I often wonder, if we had not been asked to write that chapter, whether I would ever have had the incentive to improve the measurement capabilities to the extent that I did. The drift problem in the dI/dV traces turned out to be impossible to solve with the direct measurement, and I thought about a Wheatstone bridge, which was made useful in a cir-

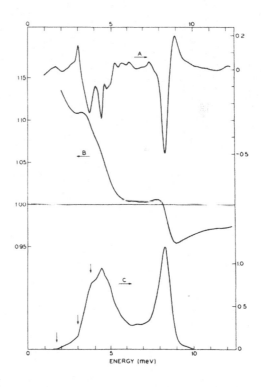

FIGURE 3.4: From the density of states of Pb (B), including the details revealed $d^2I/dV^2 - V$ (A), is shown the phonon spectrum of Pb (weighted by the coupling parameter $\alpha^2(w)$, as derived by Bill McMillan from my data.

cuit again designed and built by Don Thomas. This circuit provided the stability of 1 part in 10^5 over the couple of hours required for the two measurements. The final measurement system used for both derivatives employed two HP voltmeters to measure the output of the lock-in and the voltage from the junction, both using a 10 second integration time per data point to reduce noise. Thus the voltage sweep was slow, say 30 or 40 minutes, and was derived from a battery and potentiometer that was driven by a DC motor with an O-ring as a belt drive.

With a large Edwards pump, reaching temperatures in ^{4}He below 1 K became routine, although initially a dewar was built based on Russian work, in which a small insert dewar, inside a larger one at 4.2 K, was pumped to near 0.8 K. The need for a shielded room became very clear one evening when I was measuring the I-V characteristics of a junction in the nonlinear gap region. Suddenly the I-V changed and became linear, and then over time moved smoothly between the two extremes a number of times. As the dewar still had ample liquid helium, the only explanation was that someone with an rf-furnace was powering an induction coil to melt samples, which I had mistakenly assumed would not happen in the evening. This experience justified the expense of a shielded room and its installation, and it also meant that I acquired a 50% larger "3 bay" lab, rather than the usual "2 bay". As far as I know, it was the first shielded room to be installed at Bell Labs. The walls between the labs at Murray Hill were metal panels that could be moved and "3 bays" meant the lab included 3 windows. After the invention of the transistor, industrial research expanded in many American companies, with the hope that similar inventions would result. One day in the Westinghouse Lab in Pittsburgh, I noticed that the wall panels and "bays" were identical to those in Murray Hill. I did not like to ask if they had increased the probability of inventing the next transistor.

Today, digital recording into a computer is routine in all laboratories, but was not in the 1960s. However, we needed the data in digital form for Bill's inversion program. The output of the two voltmeters was recorded on punched paper tape, and then transferred to cards in the only computer facility at Murray Hill. There was one dI/dV and one V value per card. Bill added his software cards to the front of the stack, handed this into the facility, and by the next day received the results on large sheets of paper. Later, as this became more routine, I used his program in the same way. Today, the same tasks can be done readily on a personal computer. In addition to the inversion program (the "McMillan program"), Bill wrote a number of other programs, less well known, to handle the data. They are described, tersely, in the Parks' chapter. For example, as we preferred to use $Pb/PbO_x/Pb$ junctions, the first step was to "unfold" the data to give the density of states in a single Pb film. As well as normalizing the data taken in the superconducting state to that in the normal state, he merged the normalized $dI/dV - V$ and the $d^2I/dV^2 - V$ data, so that if the dI/dV data was then used to digitally generate d^2I/dV^2, it exactly reproduced all the fine detail while the overall magnitude of the dI/dV features were unchanged. Also, any phonon emission processes[13] in the normal state would appear shifted in voltage by the gap (or sum gap) in the superconducting state, so this was taken into account in the normalization.

With improvements in the derivative equipment, it became possible to use Bill's program in studies of the weaker phonon features in superconductors with lower Tc, such as Sn, In and Tl [14]. Lawrence Shen studied junctions on both Nb and Ta, using foils rather than films. The vacuum systems we had at the time were not adequate to make clean films of transition metals, so Lawrence melted foils in a good vacuum and made junctions on the melted surface. The Ta junctions had good characteristics that allowed spectroscopy measurements[15]. At the time, Bernd Matthias was not convinced that transition metals were BCS superconductors (shades of HTS?), as he had classified superconductors into various types, so when Lawrence obtained $\alpha^2 F$ for Ta, and a publication of the data was in the release procedure, I was asked to visit the Executive Director, Al Clogston, in his office — a rare event. He asked if I was sure of the data, given "reservations" in some quarters. I assured him I was, and heard nothing more on the issue.

As an example of our informal working interactions, on most mornings I would drink coffee with Bill in his office, usually Phil would join us and sometimes Bill Brinkman and Joel Appelbaum. One day there was discussion of impurity phonon modes due to light or heavy ions in a lattice; I do not

[13] J. M. Rowell, W. L. Mcmillan, and W. L. Feldmann, *Phys. Rev.* 180 (1968) 658

[14] J. M. Rowell, W. L. McMilland, and W. L. Feldmann, *Phys. Rev.* B3 (1971) 4065
J. M. Rowell, W. L. McMillan, and R. C. Dynes (1979). This tabulation of the results of tunneling spectroscopy of Hg, In, Sn, Ta, Tl, Pb, Pb allots with Tl and Bi, and In Tl alloys was not published, but was made available on request.

[15] L. Y. L. Shen, *Phys. Rev. Lett.* 24 (1970) 1104

remember the context now. Back in the lab, I consulted the lists of superconductors due to Roberts and decided that In in Pb should show such a mode in the phonon spectrum, as my derivative traces soon showed[16].

Beyond the study of single superconductors, my interest (prompted by Doug Scalapino) turned to the proximity effect, which had become a popular topic in bilayers and later in multilayers (superlattices). The question was simple: if a junction was made so that tunneling was into the Ag film of a sandwich of say, Ag/Pb, and the Ag was not too thick, would the pairing induced in the Ag by the Pb appear in the density of electron states, and if so, how? I made measurements on Ag/Pb by tunneling into the Ag and Pb sides of bilayers with the same film thicknesses, and observed the same minimum energy gap in both the Ag and Pb. At higher energies, the results were much more complex than I expected and were the topic of ongoing coffee discussions with Bill and Phil, who explained them in terms of Andreev scattering. Suddenly we saw the publication of Tomasch[17], who had carried out experiments for tunneling into very thick Pb films, and in less than two months Bill and Phil quickly submitted their PRL[18] explaining the interference patterns he had reported as virtual bound states. Bill and I submitted our studies of tunneling into the Ag side of Ag/Pb sandwiches[19] two weeks later. Seven years afterwards, by making the clean Zn of a Zn/Pb sandwich very thick, I managed to see real bound states in the Zn at energies below the Pb gap[20]. Watching those appear on the recorder was rewarding — experiments that worked as designed always were. During a year in Cambridge ("sabbaticals" did not officially exist at Bell), Bill continued to consider the proximity effect. His model of the SN interface[21] in terms of a Josephson junction has always struck me as an illustration of the remarkably simple and clear approach he took to problems.

Post-docs at Bell were sometimes visitors from abroad for a year; the position was also used to make a hire when no permanent slot was currently available. Adrian Wyatt joined me as a visitor, and discovered zero bias anomalies, which were explained by Joel Appelbaum in terms of the Kondo effect. Bob Dynes was initially my post-doc for a short while; I'm not sure now whether that was because no permanent position was available, or because a post doc was considered more appropriate for someone from McMaster University! Bell could be an elitist place. Bob and I worked extensively together as close friends; the study of changes in T_c and the phonon spectrum with the e/a ratio in TlPbBi alloys[22] was one project I found particularly satisfying, maybe because it encroached on the "Matthias rules" of e/a ratios.

3.1.1 The Josephson Effect

Writing this chapter in this way, with the Josephson Effect as a separate story, is completely misleading. There were days when a set of junctions, say Pb/PbO$_x$/Pb, were made in the morning, and in the afternoon either derivative measurements would be made, or traces of the Josephson current I_c would be taken, depending simply on the junction resistance we had obtained. However, for the sake of clarity, I will describe the Josephson Effect as if it happened quite separately, even though it did not. Even the stimulus for the two studies was the same, in that when Phil returned from Cambridge with his interest in the derivative data, he had also just left his interactions with Brian Josephson, who had submitted his *Physics Letter*[23] on June 8, 1962.

My reaction to Phil's enthusiasm to search for Josephson's Effect was probably to say that I had

---

[16] L. Y. L. Shen, *Phys. Rev. Lett.* 24 (1970) 1104

[17] W. J. Tomasch, *Phys. Rev. Lett.* 15 (1965) 672

[18] W. L. McMillan and P. W. Anderson, *Phys. Rev. Lett.* 16 (1966) 85

[19] J. M. Rowell and W. L. McMillan, *Phys. Rev. Lett.* 16 (1966) 453

[20] J. M. Rowell, *Phys. Rev. Lett.* 30 (1972) 167

[21] W. L. McMillan, *Phys. Rev.* 175 (1968) 537

[22] R. C. Dynes and J. M. Rowell, *Phys. Rev.* B 11 (1974) 1884

[23] B. D. Josephson, *Phys. Lett.* 1 (1962) 25

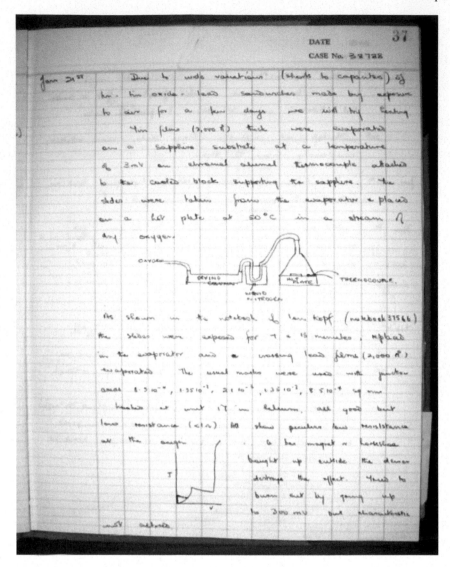

FIGURE 3.5: A photograph of my notebook page for January 21, 1963, in which I describe the way Len Kopf and I prepared Sn/SnO$_x$/Pb junctions, and the first observation of the Josephson Effect and its magnetic field sensitivity.

never seen any supercurrents in any of my Al/AlOx/Pb junctions. On August 10, 1962 my first notebook entry related to the Effect briefly mentions the need for "$H = 0$ space" and for a mu metal can. Two weeks later, my wife and I left for a 6,200 mile tour of the National Parks in the west of both America and Canada. We had our priorities. My first attempt to make an observation of a Josephson current was to cool an Al/Pb junction well below the 0.8 K that I could reach in my own lab. Only one ^{3}He system existed at that time in Bell Labs, that used by Ernie Corenzwit with Ted Geballe and Bernd Matthias to search for new superconductors. With Ted's help, measurements were made on January 3, 1963, on an Al/Pb junction in that system, but they did not reveal any supercurrent either. There is no mention in my notebook of the junction needing to be of lower resistance than I typically made for phonon studies, and the noise level in Ted's lab was probably higher than in mine. Again, the history is not recorded, but Phil obviously believed that I should have seen a Josephson current, so began to ask why I had not. He had earlier considered the same question in Cambridge when Josephson had failed to observe his own effect. This led him to the realization that the junctions had a coupling energy proportional to I_c, and that noise and thermal kT fluctuations could prohibit observation of the effect. Experimentally, this translated into the need to make junctions with lower resistances, but the time needed to change shadow masks between the Al and Pb film depositions prohibited that for Al/Pb junctions. So Len Kopf (my assistant at the time) and I began to experiment with the oxidation of other types of base electrodes, first Sn in Sn/SnO$_x$/Pb junctions, and later Pb in Pb/PbO$_x$/Pb ones. (I prefer the convention that the film deposited first is mentioned first.) On January 21, 1963, my notebook (Figure 3.5) reads "Looked at unit 17 in helium, all good but low resistance ($< 1\ \Omega$). All show peculiar low resistance at the origin. A bar magnet or horseshoe brought up outside the dewar destroys the effect". After taking a few X-Y recorder traces, I walked down to Phil's office and we returned together to the lab. The most convincing moment was when Phil, holding the bar magnet that was usually on the metal wall of the lab, retreated back towards the lab door, swinging the magnet from side to side. I had set the current just below its maximum, and the pen of the recorder jumped back and forth between zero voltage to the sum gap. This behavior was so novel, there seemed no doubt we were seeing the Josephson Effect. Strangely, in the first publication[24], we simply noted that the supercurrent was reduced in very small fields and was not observed in a field of 20 gauss. We did not mention the unusual way in which we changed the field, and instead made arguments against shorts by comparing the resistance of a metallic short with the junction resistance below the gap. Later that afternoon, I made the field measurement more "quantitative". The glass dewar was mounted into a "unit bench"; I had cut holes for it in both the wooden bench top and the shelf below. The shelf was roughly at the level of the sample in the dewar, so by placing the magnet on the shelf, and moving it in steps towards the dewar, I took a series of I-V plots. The critical current I_c depended on the magnet position in a Fraunhofer-like way. I made a plot of $I_c(H)$ on the X-Y recorder paper, which unfortunately did not survive. In hindsight, I regret that in the first publication, we did not include a figure entitled "Critical junction current versus magnet position on the bench". In a way, it was unfair

FIGURE 3.6: A photograph of five Al/AlO$_x$/Cu junctions made with the masks that were used for Josephson junctions. The copper was used simply to give a film of a different color for the photograph.

not to do so, in that arguments that our data were due to shorts persisted for a while, and we were already convinced that those were not present.

A photograph of my notebook entries on January 21 is reproduced in Figure 3.5. The top half of the page describes the new method Len Kopf and I tried to control the oxide growth on the lower Sn film, by heating it to 50 °C for a short time. The lower half describes and sketches the results. It is interesting that this notebook entry must have been made before I went to see Phil, as it does not mention just how sensitive the supercurrent was to the field. On the next page, written on the next day, I mention the arguments Phil made against shorts. The experiment was repeated with a

[24] P. W. Anderson and J. M. Rowell, *Phys. Rev. Lett.* 10 (1963) 230

different junction the next day and junctions of this kind were described in our PRL[25]. According to *Physical Review*, the submission date was January 11, but that must be incorrect, as the manuscript was mailed from Bell on February 8.

Clearly I needed a more precise magnetic field, so I designed a solenoid to be immersed in the helium and within the mu-metal shield, to apply known fields in the plane of the junction. I also ordered much better shadow masks from a supplier, so that different junction widths could be measured at the same time — the notion of the Josephson penetration length had developed by then. The company, using etching of a multilayer metal strip, was able to produce slots in the mask as narrow as 0.04 mm. A photograph of junctions made with those masks is shown in Figure 3.6. I also decided to use $Pb/PbO_x/Pb$ junctions in the next experiments, in order to see the supercurrent at all temperatures below 7.2 K, instead of just below the T_c of tin (3.7 K) as in the case of the earlier $Sn/SnO_x/Pb$ junctions.

I have chosen to show (Figure 3.7) a page from my notebook from May 21, 1963, not because it was the day when I obtained the best data, but because it illustrates the point I made earlier, that Len made junctions in the morning, but we were never quite sure what might surprise us in the afternoon! In this case, the junctions were far from ideal, but displayed, in the subgap voltage region, a most interesting series of features, which I had never observed before. I called them subharmonics, because they were — although much more recent work has pointed out that they are not exactly so. Reading my notebook, it seems clear that I was much more interested that day in these new features than in the Josephson Effect — after all, by then we were convinced that was real! I did, however, make a sketch of the Fraunhofer pattern, and noted that for this non-ideal junction, there was always a residual current. At the time, I considered subharmonic structure to be different from multiparticle tunneling, which had been described by Adkins and by Taylor and Burstein. Certainly experimentally they appeared quite distinct, as I noted when I sketched the shape of the feature at say $2\Delta/2$, showing it was a decrease in conductance rather than an increase, and as I emphasized in a publication[26]. Years later Arnold showed that they were both caused by Andreev scattering, but in junctions with very different transmission probabilities.

The following day, much better junctions were made, which had no unusual subgap currents. To obtain $I_c(H)$, I traced the I-V characteristics a number of times on the X-Y recorder at each field setting in the solenoid, and noted the maximum value. I never did set up a system to record $I_c(H)$ automatically, although later other people did. In the PRL submitted on July 24, 1963, I chose to plot I_c on a log scale (Figure 3.8 is the hand-drawn plot I gave to the art department), as I was so impressed by the fact that the current at the first minimum was so small compared to the maximum at zero field[27].

A superconductivity conference was held at Colgate University in the summer of 1963, and I have three recollections of it. One is of waiting in the breakfast line behind Kurt Mendelssohn, while he tried to teach the person serving how to make a cup of tea, by pouring boiling water on the tea bag, and not by serving him lukewarm water with a bag to be added later. After my talk, Brian Pippard commented that my $I_c(H)$ pattern for the Pb junction implied a very high uniformity of the oxide thickness. I replied that it seemed to be unique to Pb oxide junctions and, as far as I know, it still is. Another comment, I believe also by Pippard, was more important; in an evening rump session he pointed out that as I had seen the single slit Fraunhofer pattern in $I_c(H)$, it should be possible to also see the double slit interference pattern. I thought (briefly!) about this, but could not immediately see an easy way to make the two junctions, and so went back primarily to phonon spectroscopy. The two junctions were made some 3 months later by the Ford group of Jaklevic,

[25] P. W. Anderson and J. M. Rowell, *Phys. Rev. Lett.* 10 (1963) 230

[26] J. M. Rowell, *Phys. Rev*, 172 (1968) 393

[27] J. M. Rowell, *Phys. Rev. Lett.* 11 (1963) 200

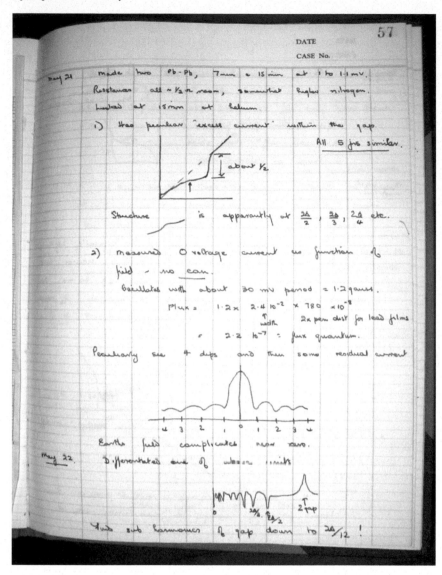

FIGURE 3.7: A photograph of my notebook page for May 21, 1963, describing a Pb/PbO$_x$/Pb junction in which I observed "subharmonic structures" and the Fraunhofer pattern of the Josephson current versus field.

Lambe, Silver and Mercerau[28], who painted a stripe of formvar varnish down the middle of a single Sn/SnOx/Sn junction, and so was born the SQUID, or at least, one version of it. See the chapter by John Clarke and Arnold Silver. Some readers might wonder why I did not single-mindedly pursue the double junction experiment; in answer I can only point out that, while the best $I_c(H)$ data was being observed in May, the PRL showing phonons in d^2I/dV^2 had been submitted in March with the work of the Penn group, and some months later Bill McMillan arrived at Bell. The honest answer is that I always found phonon spectroscopy much more enjoyable as an experiment than a Josephson current measurement.

---

[28] R. C. Jaklevic, John Lambe, A. H. Silver, and J. E. Mercereau, Phys. *Rev. Lett.* 12 (1964) 159

A by now well-known discussion between John Bardeen and Brian Josephson took place at the Low Temperature Conference LT8 in September 1962, and later at a small workshop in London. The details have been described by McDonald in a delightful article in *Physics Today*[29]. The bottom line was that Bardeen could not understand that pairs could tunnel through the barrier of the junction with the same probability as single electrons, and Brian could not understand how John could not understand it! Ted Geballe recalls that "at the workshop, after Josephson presented his idea and Bardeen gave his arguments why it wouldn't work, Brian, this quiet-spoken grad student, then got up and said, 'well I believe it will work and if nobody else does I am going into the laboratory to find it myself', or words to that effect".

Another key feature of the Josephson Effect, that the junctions would generate an AC signal at a frequency proportional to the voltage, was confirmed by Sidney Shapiro at A.D. Little, who showed the constant voltage steps induced by incident microwaves[30]. Parker, Taylor and Langenberg at Penn demonstrated that this could become

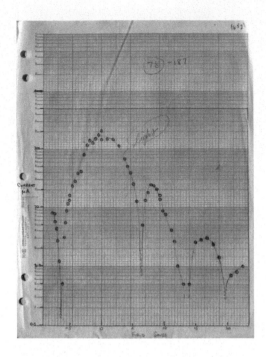

FIGURE 3.8: The field dependence of the Josephson current in a Pb/PbOx/Pb junction. This photograph is of the data I plotted by hand, and gave to the graphics department to prepare the figure in reference 26.

the basis for a standard volt, and that has since happened in standards labs around the world.

For 18 months after Josephson's startling prediction, his effect was confined to tunnel junctions. Bardeen's disagreement with him concerned the tunneling probability of electrons and pairs through a thin insulator free of pinholes. This changed in an unusual way in February, 1964, when Ron Parks and Jack Mochel[31] reported studies of narrow films of tin deposited on a fiber of glue. They explained the results in terms of arrays of flux vortices fitting in the width of the film, but in my notebook of 2/3/64, I wrote "Phil suggests that the structures of Parks — narrow superconducting constrictions — should (in low *H*) be weak coupling regions between bulk suprs so should exhibit Josephson effects — in fact he considers Parks' results are Josephson measurements." Phil had realized that the Josephson Effect could be generalized to all "weak links" between two superconductors. I experimented with Al and Sn films deposited on fibers, but Ali Dayem made a much better defined weak link, being a narrow and short neck in a film made by evaporation through a mask. That such a link will carry a supercurrent was obvious, but the results of applying microwaves were spectacular; in fact the constant voltage steps in I-V were more dramatic than in any tunnel junction to date[32]. As the years went by, this history of junctions and links has become blurred. Some have commented that work before 1962 on what would now be called weak links — contacts between crossed wires — should be regarded as the first observation of the Josephson Effect. That comment ignores the history; after all, I doubt that Bardeen would have argued that a supercurrent would not

[29] Donald G. McDonald, *Physics Today* (2001) 46

[30] Sidney Shapiro, *Phys. Rev. Lett.* 11 (1963) 80

[31] R. D. Parks and J. M. Mochel, *Phys. Rev. Lett.* 11 (1963) 354

[32] P. W. Anderson and A. H. Dayem, *Phys. Rev. Lett.* 13 (1964) 195

flow between two Nb wires that were in contact with each other.

I will not enter here into any further discussion of Phil's role in the expansion of the Josephson Effect; clearly I was not an impartial observer as I had daily interactions with Phil while his understanding of many further issues and implications grew beyond the content of the initial *Physics Letter* written by Brian. For those interested in the history, I recommend reading the references listed in Further Reading.

Having ignored for almost 50 years the comments that "others" first observed the Josephson Effect, maybe it is time to respond. Giaever[1] and Shapiro[30] both made Al/AlO$_x$/Pb junctions exhibiting supercurrents, either a Josephson current or from a metallic short. All they had to do was reach for a bar magnet. Neither of them did so. All my Al/Pb junctions had high resistances and hence small coupling energies, and showed no supercurrents, and theirs were made in a similar way. Possibly their labs were quieter, but maybe they were not.

When the second letter with the $I_c(H)$ figure was ready to be submitted to PRL, it was circulated within the Labs for comments, and also sent to the Patent Department, as were all manuscripts being considered for publication. I assumed this would be a formality, but it was decided that a patent should be applied for, and an attorney was assigned to the case. The claims suggested that Josephson junctions could be used as digital devices, being switched between two states (zero voltage and 2Δ) by applying a field (rather similar to the earlier cryotron) or by adding two currents to exceed I_c. An application was eventually filed on 1/17/64 and became the first Josephson patent when it was granted on October 25, 1966, number 3281609. Clearly it was one of many patents that did not make any money for Bell Labs, but possibly it was used in cross licensing agreements.

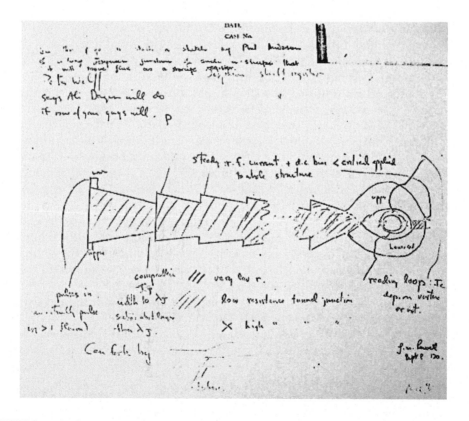

FIGURE 3.9: A photograph of Phil Anderson's proposal that single flux quanta could be manipulated in a "flux shuttle".

A more interesting incident concerning patents occurred at the time when there was a strong interest within Bell in magnetic bubbles as memory devices. In a discussion with Art Hebard, the notion of using flux quanta in superconducting films as an alternative memory arose, and we arranged a meeting with a patent attorney, who bluntly told us to go away, as the Bobeck patent covered the use of any magnetic entity in every possible material. Somewhat later, we mentioned this to Phil, who commented that fluxoids in a film were not a good choice anyway, in that the movement of a normal core in a superconductor would be slow. He pointed out that Josephson vortices in a long junction would be a much faster device. A day or so later, he showed me a page of notes, on which he had sketched his design of the flux shuttle. Of all my communications with Phil, this page, reproduced in Figure 3.9, has always been my favorite — perhaps because it shows Phil as an inventor, and probably because the page does not have a single equation! Note that at the top left of the page, Phil has written "Peter Wolff says Ali Dayem will do if none of you guys will". Persuasion has many forms! This was the first description of a single flux quantum device using Josephson junctions, 15 years before RSFQ was described by Mukhanov, Semenov and Likharev in 1987. The first demonstration of such a device was by Fulton, Dynes and Anderson[33], and a patent application was filed by them on July 11, 1972.

For some years, applications of the Josephson Effect suffered from the lack of a reliable materials technology, although considerable progress was made at IBM using PbInAu alloys, and at Sperry using barriers of SiNb with Nb electrodes. Today's solution was discovered by accident. While tunneling spectroscopy had been successful on Ta, as mentioned earlier, junctions made on Nb in the same way had never been quite good enough for Bill McMillan's program to converge with a reasonable value of μ^*. I believed that the problems with Nb were that the tunnel barrier formed by thermal oxidation contained multiple oxides, one of which (NbO) was thought to be superconducting, and that oxygen penetrating into the Nb surface degraded the T_c by 1 K per atomic percent. So I had given up trying to make ideal junctions on Nb. However, when Jochen Geerk spent a year's leave from Karlsruhe working with me, he was making Nb/Al multilayers, with the individual layers of Nb and Al being as thin as 20 Å. He was preparing junctions by thermal oxidation of the layer that was deposited last, either Nb or Al, covered by a second electrode. One day he made the comment that when he made the junctions on the Nb surface, they were bad ("niobium-like" was his phrase), but on the Al surface, they were good ("aluminum-like"). This immediately implied that the final 20 Å (say) of Al were completely covering the Nb of the layer beneath, and hence that a thick film of Nb, if it was covered with 20 Å of Al, would result in good junctions. This was indeed shown to be true in a phonon spectroscopy study of Nb/Al-AlO$_x$/Ag junctions[34]. Michael Gurvitch used the same barrier process but replaced the top electrode by Nb[35]; initially I recommended that he use Al/Nb to isolate the Nb from the barrier oxide, but that proved to be unnecessary. If I had read the literature more carefully earlier, maybe I would have noticed the use of Al/AlO$_x$ in 1961 by Adkins[36], or the thicker Al used in the same way at IBM[37]. Other relevant factors in this history are the great improvement in commercially available stainless steel vacuum systems in the 1960s, and the invention of the magnetron sputter gun in 1968, both of which meant that in 1981 much cleaner films of Nb could be made than in 1963 and 1964.

This is an appropriate place to end my story of the early days of tunneling and the Josephson Effect. Tunneling became a much more versatile tool with the invention of the Scanning Tunneling Microscope, which has allowed, for example, atomic resolution of the "gap features" on cleaved surfaces of some cuprate superconductors. I have been surprised that tunneling spectroscopy has

---

[33] T. A. Fulton. R. C. Dynes, and P. W. Anderson. *Proc. IEEE* 61 (1973) 28

[34] J. Geerk. M. Gurvitch, D. B. McWhan, and J. M. Rowell, *Physica* 109 & 110B (1982) 1775; J. M. Rowell, M. Gurvitch, and J. Geerk, *Phys. Rev.* 175 (1981) 537

[35] M. Gurvitch, M. A. Washington. and H. A. Huggins, *Appl. Phys. Lett.* 42 (1983) 472

[36] C. J. Adkins, *Phil. Mag.* 8 (1963) 1051

[37] R. B. Laibowitz and A. F. Mayadas, *Appl. Phys. Lett.* 20 (1972) 254

been more useful in studies of how non-uniform these features are across the surface, than in revealing the coupling mechanism, although Jochen Geerk and others, using more conventional junctions, showed phonon features in the density of states of YBCO. The applications of both Giaever junctions and of the Josephson Effect are described in subsequent chapters, particularly in SQUIDS, detectors, and in RSFQ circuits.

I have mentioned many collaborators in this chapter, but they were more than that; they were and are my friends. Without them, my career would have been not only less successful, it would have been much less rewarding and enjoyable. It was my great pleasure to work with them all.

I would also like to thank George Kupczak at the AT&T facility in Warren, NJ for kindly providing access to my old lab notebooks, and the facility to study them and copy some of the pages.

3.2   The Discovery of Fluxoid Quantization

Dietrich Einzel

3.2.1   Introduction

Since the seminal theoretical work by John Bardeen, Leon Cooper and Robert Schrieffer, (BCS)[38], we understand the phenomenon of superconductivity in terms of a condensation of fermion pairs in **k**-space. Therefore quantization phenomena (like that of the fluxoid) are seen to happen on a macroscopic scale, which means that not individual fermion pairs but rather the whole paircondensate is involved. Hence we start from a condensate of (possibly composite) particles with mass $m = km_0$ and charge $q = ke$ (m_0 and e are the mass and the charge of the electron, respectively). The parameter k (1 for bosons and 2 for fermion pairs) is retained in order to see how the pair character of the superconducting charge carriers can be traced back during the calculations. Such a condensate can be described in terms of macroscopic pseudo-bosonic wave function $\Psi(\mathbf{r}, t) = a(\mathbf{r}, t) \exp[i\varphi(\mathbf{r}, t)]$. The Ψ-function represents a two-component *order parameter* (or *pair field*), characterized by what is referred to as *off-diagonal long-range order* (ODLRO)[39] and, equally importantly, by *macroscopic phase coherence*. It can be represented either by the set of two functions $\{\Psi, \Psi^*\}$ or, alternatively, by $\{a, \varphi\}$, the latter being the more physical representation. An interpretation of $n^s(\mathbf{r}, t) = ka^2(\mathbf{r}, t)$ as *condensate* or *superfluid density* is then obvious.

Two early phenomenological theoretical treatments, based on Ψ, are the local electrodynamic theory of superconductivity by Fritz and Heinz London[40] and the celebrated stationary thermodynamic (magnetostatic) Ψ-theory by Ginzburg and Landau[41].

Both theories predict the screening of magnetic fields from the superconductor's bulk volume, characterized by the screening length or magnetic (London) penetration depth

$$\lambda_L^2 = (km_0)c^2 / 4\pi(n^s/k)(ke)^2 = m_0 c^2 / 4\pi n^s e^2$$

which is governed by the condensate density n^s and from which the pair charge parameter k is clearly seen to drop out. The superfluid density n^s characterizes also the general form of the supercurrent density $\mathbf{j}^s = (n^s e^2 / m_0 c)[(\Phi_0/2\pi)\nabla\varphi - \mathbf{A}]$, with $\mathbf{A}$ the vector potential. It has turned out that one of the few physical quantities that explicitly depends on the pair charge parameter k is the so-called fluxoid

[38] J. Bardeen, L. N. Cooper and J. R. Schrieffer, *Phys. Rev.* 106 (1957) 162

[39] C. N. Yang, *Rev. Mod. Phys.* 34 (1962) 694

[40] F. London and H. London, *Proc. Roy. Soc.* (London), A 145 (1935)

[41] V. L. Ginzburg and L. L. Landau, *Zh. Eksp. Theor. Fiz.* 20 (1950) 1064

quantum $\Phi_0 = hc/ke$. The experimental detection of Φ_0, and even more so, the related macroscopic quantization phenomenon, turned out to be rather nontrivial and tricky and therefore had to be wait for until four more years after the publication of the BCS theory in 1957.

This article is motivated by the historical fact that precisely 50 years ago there were two experiments which not only proved the quantization of the fluxoid but also showed that its explicit form displays a pair charge rather than the ordinary elementary charge. They therefore strongly supported the pairing hypothesis put forward by BCS in 1957. The one experiment was carried out by Robert Doll and Martin Näbauer[42,43] in Herrsching at the Ammersee in Bavaria/Germany where the Low Temperature Institute under the leadership of Walther Meißner was located at that time. The other experiment was carried out by Bascom S. Deaver, Jr. at Stanford University in California/USA, in connection with his Ph.D. work, proposed and supervised by William M. Fairbank[44,45].

3.2.2   Theoretical Foundations

Let us now aim at an up-to-date theoretical description of the phenomenon of fluxoid quantization. In what follows, we use (certain extensions of) London's phenomenological theory of superconductivity[46]. For a detailed description of the Doll-Näbauer experiment using the Ginzburg-Landau theory, the reader is referred to the articles of Douglass[47] and Doll and Einzel[48], the latter written on the occasion of Vitaly Ginzburg's 90th birthday.

We consider a hollow superconducting cylinder of length L, inner radius R, and thickness d, oriented in the $\hat{z}$-direction, to which an external field H^{ext} is applied. In terms of cylindrical coordinates $\{r, \theta, z\}$, one has the following field configuration: $H_z(r)$ outside $[H_z(R+d) = H^{\text{ext}}]$ and inside $[H_z(r) = H_i = \Phi_i/\pi R^2; 0 \leq r \leq R]$ the cylinder, with H_i and Φ_i the field and flux, respectively, trapped in the cylinder hole. Using the condition that the pair field Ψ, representing a quantum-mechanical wave function, must be single-valued, one may derive a quantization condition for a quantity Φ', which has been called by London the *fluxoid*[46]: $\Phi' = \Phi_i + \Phi_{\mathbf{j}} = n\Phi_0$. Here $\Phi_0 = hc/ke$ denotes the *fluxoid quantum*. The difference between the ordinary flux Φ_i and the fluxoid is given by a contribution from the supercurrent j_θ^s in the cylinder wall $\Phi_{\mathbf{j}} = (4\pi/c)2\pi R\lambda_L^2 j_\theta^s$, flowing in the azimuthal direction, with $j_\theta^s = (c/4\pi)(H_i - H^{\text{ext}})/d = $ const, if $\lambda_L > d$ is assumed. Only the final result for the magnetic flux Φ_i in the presence of a finite supercurrent j_θ^s in the cylinder wall can be given here for lack of space. For details, the reader is referred to Eq. (15) of Doll and Einzel [48], in which $f_0 \to 1$ has to be taken in the London limit:

$$\Phi_i = \frac{n\Phi_0 + (2\lambda_L^2/Rd)\Phi^{\text{ext}}}{1 + 2\lambda_L^2/Rd} \quad ; \quad \Phi_0 = \frac{hc}{ke}.$$

This equation nicely explains the difference between the flux Φ_i and the fluxoid $n\Phi_0$ as originating from geometry effects $\propto 2\lambda_L^2/Rd$. With London's early prediction in mind, and influenced by the pairing hypothesis by BCS, the correct form for the fluxoid quantum was given in 1961 by Byers and Yang[49] in a publication adjacent to that of Deaver and Fairbank[44]. If $\lambda_L > d$, the geometry parameter causes a relevant correction, leading to a geometry dependence of the flux, which was

[42] R. Doll and M. Näbauer, *Phys. Rev. Lett.* 7 (1961) 51

[43] R. Doll and M. Näbauer, Z. *Phys.* 169 (1962) 526

[44] B. S. Deaver and W. M. Fairbank, *Phys. Rev. Lett.* 7 (1961) 43

[45] B. S. Deaver Jr., Ph.D. Thesis, Stanford University (1962)

[46] F. London, *Superfluids*, Wiley, New York, (1950) 152

[47] D. H. Douglass, Jr., *Phys. Rev.* 132 (1963) 513

[48] R. Doll and D. Einzel, *J. Superconductivity and Novel Magnetism* 19 (2006) 173

[49] N. Byers and C. N. Yang, *Phys. Rev. Lett.* 7 (1961) 46

first discussed by Keller and Zumino[50] and Bardeen [51]. The natural question arises, which quantum number n is realized by nature at a given value for the external field H^{ext}. The answer can be found in Doll and Einzel [48]: the transition between the quantum numbers n and $n+1$ is to be expected for external fields $H_n^{\text{ext}} = (n + \frac{1}{2})H_0$, with H_0 denoting the field corresponding to the fluxoid quantum Φ_0. This short theoretical review should serve as a motivation to understand why the physics behind the two experimental discoveries is so fascinating.

In their initial publication[3.2.1], Bardeen, Cooper and Schrieffer gave the result (c.f. Eq. (5.46) on page 1195 of ref. [3.2.1]) $\mathbf{j}^s = -(n^s/k)[(ke)^2/(km_0)c]\mathbf{A} = -(n^s e^2/m_0 c)\mathbf{A}$ for the supercurrent density. This result is, however, independent of the pair charge k, and it clearly lacks the (k-dependent!) phase gradient term. Therefore this expression did not allow for the prediction of flux quantization. This is at least part of the cause for many speculations, having taken place still in the year 1961 w.r.t. to the form of the fluxoid quantum. The only available source of theoretical information world-wide at this time was therefore Fritz London's proposal in a footnote in the book[46], in which he claimed that the fluxoid should be quantized in units of $\Phi_0 = hc/e$.

According to Deaver et al.[52] and Careri[53], however, in 1959, Lars Onsager mentioned in a private conversation with William Fairbank the possibility that the fluxoid quantum could be of the form $hc/2e$.

Around 1961, the time was ultimately ripe for an experimental verification of the precise form of the fluxoid quantum. In the following sections, we review, historically, what motivated the experimentalists separately in Herrsching/Bavaria and Stanford/California and what led to their independent discoveries.

3.2.3 The Doll-Näbauer Experiment

This section is based to a large extent on a detailed personal interview of the author with Robert Doll, which took place on September 12, 2007 at the Walther-Meißner Institute (WMI) in Garching. Since Robert Doll still comes to work at the WMI on a regular (almost daily) basis, there have been many additional occasions of getting profound answers to history-related questions.

Robert Doll was born in Munich on January 16, 1923. In 1946 he started studying physics at the Technische Hochschule (TH) in Munich. In 1949 Doll began his diploma thesis with Professor Walther Meißner, chair of Technical Physics of the TH Munich. Walther Meißner (born on December 16, 1882 in Berlin) can certainly be regarded as a pioneer of low temperature physics in Germany. He was renowned in the physics community at that time for various scientific and social activities and achievements[54]. The diploma work ended in 1953. In the same year, Walther Meißner offered Robert Doll the position of a permanent research assistant with the possibility to write a doctoral thesis. The topic of the doctoral thesis was the measurement of the "Gyromagnetic effect in superconductors". Doll eventually received his Ph.D. from the TH in Munich in 1958.

In the beginning of 1960, Doll, his colleague Martin Näbauer (born November 1, 1919 in Karlsruhe, Germany, who worked at Walther Meißner's institute in Herrsching since January 12, 1951), and Gerhard Schubert, a visiting Professor from the University of Mainz, were inspired by a footnote in the book of Fritz London[46], as well as by a comprehensive treatment of multiply-connected superconductors in a book written by Max von Laue[55], to develop the idea of measuring the fluxoid quantum in a hollow superconducting Pb cylinder. They believed this fluxoid to be of the form

[50] J. B. Keller and B. Zumino, *Phys. Rev. Lett.* 7 (1961) 164

[51] J. Bardeen, *Phys. Rev. Lett.* 7 (1961) 162

[52] B. S. Deaver in *Near Zero – New Frontiers in Physics*, J. D. Fairbank, B. S. Deaver, C. W. F. Everitt, P. F. Michelson, Eds., W. H. Freeman and Co., New York (1988) 260

[53] Giorgi Careri, *Phys. Perspect.* 2 (2000) 204

[54] R. Groß and D. Einzel, *Akademie Aktuell* 02/2009, p. 47

[55] Max von Laue, *Die Theorie der Supraleitung*, Springer-Verlag, Berlin, Göttingen, 1947

$\Phi_0 = hc/e$ according to London [46], which was even cited in a review article by Bardeen and Schrieffer as late as 1961 [56]. Meißner was, at that time, definitely not involved in this decision making process. In order to inform Meißner about their intentions, Doll and Näbauer wrote a very careful and detailed development proposal, which they showed to Meißner. Meißner agreed after a certain period of hesitation and signaled that he would not like to be involved with this project in the near future.

The experimental apparatus (i.e., the suspension of the torsional oscillator) for the measurement of the quantizaton effect existed already, having been used before to measure the gyromagnetic effect of small superconducting spheres in connection with Doll's doctoral thesis. The obvious plan was to use it to measure the torque exerted on a hollow superconducting Pb cylinder connected to it, which could directly be translated into the value of the trapped flux. So only the hollow Pb cylinder had to be newly constructed, and the sensitivity of the oscillator had to be estimated carefully. Major difficulties arose in connection with the fabrication of the Pb hollow cylinder, particularly with the determination of its inner and outer diameters. This was the state of affairs around the middle of 1960. In fact, the experiment clearly proceeded to its success in less than one year. The experimental procedure contained the following different steps:

1. Application of a certain value of an external field H^{ext} parallel to the axis of the Pb cylinder.
2. Cooling the sample in the field through the (field-dependent) transition temperature $T_c(H^{\text{ext}})$.
3. Complete removal of the external field.
4. Cooling the sample with the trapped flux down further to a final temperature of about $0.6T_c$.
5. Application of a small field B_M perpendicular to the axis of the hollow cylinder in order to produce a torque.
6. Measurement of the torque and derivation from it the step height $\propto \Phi_i$.
7. Warming above T_c to start a new measurement.

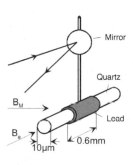

FIGURE 3.10: The torsional oscillator arrangement in the experiment by Doll and Näbauer[42].

The first run did not lead to any exploitable result since the external field increments were too large. This fact can be traced back to the fluxoid quantum proposed by London being a factor of 2 too large. In early 1961, in a second run, Doll and Näbauer reduced the increments for the steps in the externally applied field dramatically, beginning at zero field. At a field H^{ext} of 0.1 Oe, they realized that there was a trapped flux inside their cylinder, accompanied by a sharp step in the observed torque. This way, by April 1961, they had discovered the first step in the fluxoid quantization profile. At this time, Doll and Näbauer could, however, not answer the question of why their measured flux quantum differed from the London proposal by a factor of 1.6...1.7. Doll's comment was *"Es hat halt nicht gestimmt! (It just didn't agree!)"*.

But the precise value of the fluxoid quantum $\Phi_0 = hc/ke$ can be deduced from the range of external fields ΔH^{ext}, for which no torque was measured. It must *not* be determined from the step height, into which complicated factors due to the properties of the torsional oscillator enter. Taking the *outer* radius $a = R + d$ for the computation of the trapped flux $\Phi_i = \Delta H^{\text{ext}} \pi a^2$, which is meaningful in the limit $\lambda_L > d$ just below T_c, when the flux gets frozen in, Doll and Näbauer obtained the result $\Phi_0 = (0.49 \pm 20\%) hc/e \approx hc/2e$, surprisingly close to the result predicted by the BCS theory[42,43].

When it became clear that Doll and Näbauer had discovered the quantization effect, various things

---

[56] J. Bardeen and J. R. Schrieffer, *Prog. Low Temp. Phys.*, Vol. III (1961) 170

happened. *First*, there was the plan that Martin Näbauer should participate at the IBM conference in Yorktown Heights in June 1961 in order to give a seminar on the quantization effect. *Second*, a publication for *The Physical Review Letters* was planned. *Third*, when the success of this experiment became visible, Walther Meißner, having adopted a neutral position before, changed his mind and all of a sudden requested to be accepted as a coauthor on the publication. This was the situation in Herrsching right before the beginning of the IBM conference in Yorktown Heights, in June 1961. At this stage, it appears to be meaningful to split the scene and investigate what happened in a completely uncorrelated manner, on the other side of the earth, namely at Stanford University.

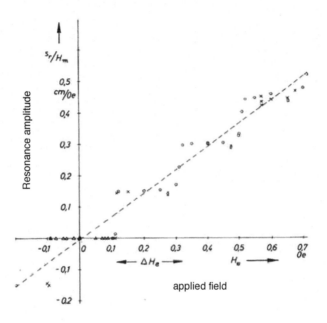

FIGURE 3.11: The fluxoid quantization data by Doll and Näbauer.

3.2.4 The Deaver-Fairbank Experiment

This section is based to a large extent on the contribution "Fluxoid quantization: experiments and perspectives" by Bascom S. Deaver[52] in the book *Near Zero – New Frontiers of Physics*, which contains contributions to a conference held at Stanford University in March 23–25, 1982 on the occasion of the retirement of Professor William Martin Fairbank. A second source is an article by Bascom Deaver, written on the occasion of the retirement of Professor William Little[57].

Bascom S. Deaver was born on August 16, 1930 in Macon, Georgia, USA. He received his undergraduate B.S. degree from the Georgia Institute of Technology in 1952 and his masters degree at the Washington University in St. Louis in 1954. Between 1954 and 1957, he was in the U. S. Air Force at the Air Force Special Weapons Center, Kirtland Air Force Base, in New Mexico as a

---

[57] Bascom S. Deaver in *From High-Temperature Superconductivity to Microminiature Refrigeration*, B. Cabrera, H. Gutfreund, and V. Kresin, Eds., Plenum Press (New York) (1995) 243

physicist and commissioned lieutenant. In spring 1958, Deaver, while working as a part-time graduate student at the Stanford Research Institute, Menlo Park, CA, consulted his friend and Stanford Professor George Pake about what kind of research he should pursue. Pake strongly recommended working for Professor William Martin Fairbank, who would be arriving at Stanford University in Fall 1959. Like Walther Meißner, William M. Fairbank (born on February 14, 1917 in Minneapolis, MN) can be viewed as a pioneer of low temperature physics. Around 1959, he could already look back to a very productive academic career, the details of which can be studied in Deaver [52]. In Fall 1959, William Fairbank joined the faculty at Stanford University and became the Max von Stein Professor of Physics. It was actually Felix Bloch who persuaded Fairbank to a change from Duke to Stanford. He stayed there until his retirement in 1985, and as an emeritus Professor until his sudden death in 1989. Following Professor Pake's advice, Deaver applied for a doctoral thesis at Fairbanks' research group. Among the experimental topics that William Fairbank proposed to Deaver was the measurement of the fluxoid, proposed by Fritz London in 1950. Deaver found this idea *enormously appealing* and, without great hesitation, asked Fairbank to be allowed to work on it in Fall 1959.

Fairbank's first experimental concept was the use of a thousand identical loops encircled by a thousand small pickup coils. Later he changed the idea and decided to use a *hollow cylinder*. Eventually, a tiny *tin* cylinder (13 μm i.d., $L = 1$ cm) was vibrated along its symmetry axis at 100 Hz with an amplitude of 1 mm. Two pickup coils measured the magnetic flux in the cylinder at various values of the applied external field. The output voltage from the pickup coils was calibrated by cooling the cylinder from its normal state to well below the superconducting transition temperature in zero applied field, so that no flux was trapped, and then measuring the voltage as a function of applied field as the cylinder was vibrated. Assuming that the cylinder was completely diamagnetic, knowing the applied field and the measured outside diameter of the cylinder and thus the area, so the ejected flux was just the applied field times the outside area of the cylinder, the voltage was calibrated directly in flux (Gauss cm^2).

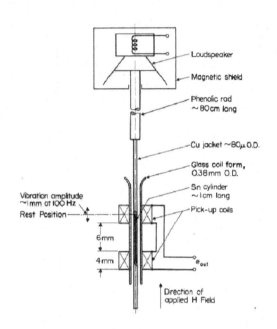

FIGURE 3.12:   The experimental arrangement by Deaver and Fairbank[52].

During the progress of Deaver's work, he heard of three other endeavors to search for quantized flux, all of them unsuccessful. Therefore it is quite remarkable that Deaver and Fairbank did not become aware of the fourth endeavor, namely the ongoing Doll-Näbauer experiment at Walther Meißner's institute in Herrsching, and vice versa. On May 3, 1961 Deaver could report the first definitive signature of quantized flux. He was at first surprised, though, that the measured flux was so different from the value London predicted. However, Fairbank quickly recalled Onsager's comment about the possibility of $hc/2e$. In this way, they found agreement with $hc/2e$ within about 20%.

This was the situation in the middle of June 1961, when both experimental groups, represented by Martin Näbauer (Herrsching) and Bill Little (Stanford), intended to attend the IBM conference in Yorktown Heights. Bill Little had taken the task, besides presenting his own talk, to show the data points of Deaver and Fairbank, which he had received only a few hours before he left for the conference.

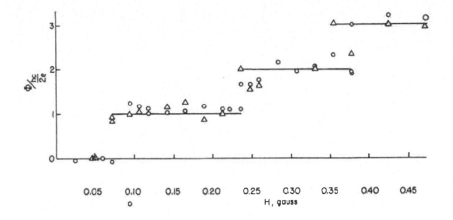

FIGURE 3.13: The fluxoid quantization data by Deaver and Fairbank.

3.2.5 The IBM Conference 1961

The IBM Conference of Fundamental Research in Superconductivity took place in Yorktown Heights, New York, in June 1961 [58]. The conference was part of the dedication ceremonies for the new Thomas J. Watson Research Laboratory.

At the IBM conference Martin Näbauer had prepared a talk entitled "Experimente zur Quantisierung des magnetischen Flusses in Supraleitern". Näbauer recalled[59], that on the evening before his presentation, Professor Little visited him in the hotel and made inquiries about both the experimental apparatus and the results, which Näbauer readily provided in all details. The following dialogue is reported by Robert Doll: Little: "Is the flux quantized?" ("Ist der Fluss quantisiert?"). Näbauer: "Sure!" ("Ja freilich ist er quantisiert!").

On June 15, 1961, Näbauer presented his talk, followed by Bill Little's contribution "Kapitza resistance of metals in the normal and superconducting states". During the discussion of the latter, Bill Little showed the data of Deaver and Fairbank.

Bill Little recalled[60] that prior to the beginning of the IBM conference, he had a lively discussion with Brian Pippard over drinks. On this occasion, he showed to him Deaver's and Fairbank's results. Pippard was adamant that the experiments were *wrong* and that he could prove this fact using a gauge-invariance argument. On June 15, Näbauer presented his and Doll's data on the flux quantization and Little presented the data of Deaver and Fairbank. A heated discussion developed which culminated in the conviction that the factor $k = 2$ is ultimately needed, in order to understand *both* of the experimental results. Immediately after these presentations, according to Little, Phil Anderson spoke up and claimed that the factor of $k = 2$ was an *obvious consequence of the BCS pairing hypothesis*. Brian Pippard did not rebut further.

It became immediately clear to Näbauer that the submission of the PRL paper was overdue. So he wrote a telegram back to Herrsching, asking what was going on with the paper. Clearly, the time delay was due to a series of discussions still going on between Doll and Meißner about Meißner's coauthorship. When Näbauer's telegram arrived, however, Meißner had eventually recognized that his coauthorship was not such a good idea and the paper was submitted to PRL on June 19, 1961. Three days earlier, on June 16, Deaver and Fairbank had submitted their paper to PRL. The long and

[58] IBM J. Res. Dev., Vol. 6, No. 1 (January 1962)

[59] G. Möllenstedt, *Physikalische Blätter* 43 (1987) 60

[60] W. A. Little, private conversation, 2008

short of it, the paper by Doll and Näbauer appeared in the same volume of PRL as that of Deaver and Fairbank. Extended versions of the publications of the two groups can be found in Doll and Naebauer [43] and Deaver [45].

Bill Little recalled, that neither Deaver nor Fairbank nor he knew anything about Robert Doll's and Martin Näbauer's activities before the IBM conference. "It came as a big surprise and some relief that both parties had recognized the factor of two." Bascom Deaver commented the "$2e$" problem as follows: "However obvious this last point seems now, it was surprisingly difficult to understand at that time".

3.2.6 Post-1961

In 1962 Doll and Näbauer received awards from both the Bavarian Academy of Sciences as well as the Academy of Sciences in Göttingen for the discovery of the fluxoid quantization.

In the same year, Martin Näbauer passed away completely unexpectedly on September 10.

In 1986, the award "Bene Merenti" in silver was given to Robert Doll for his "*Contribution to the experimental discovery of the quantum nature of the magnetic flux in superconductors*" at the "Solemn Annual Meeting" of the Bavarian Academy by its President Professor Arnulf Schlüter.

Known already all over the world for his quantum flux experiment, Bascom Deaver began his career as a Professor at the University of Virginia (UVA) in 1965. There he continued his basic research on superconductivity, and went on to explore its applications. He and his students did significant research on flux quantization and developed superconducting magnetometers, which have had important applications. As a whole, Bascom Deaver's research has spanned the entire range from the most basic physics to practical applications. For more than 40 years on the teaching side of his university activities, Deaver was responsible for a tremendous success of the undergraduate program at UVA for which he has worked tirelessly to ensure that it became first rate. Under his direction, two undergraduate concentrations (computational physics and optics) have been developed, and a new B.A. program has been launched in order to expose the students to the intellectual beauty of physics without sophisticated mathematics.

In 1967 the Commission of Low Temperature Research moved from Herrsching into a new building on the university campus in Garching (about 20 km north of Munich), which was renamed Walther-Meissner-Institut (WMI) in 1982. Robert Doll worked there until his retirement as an Academic Director. His name does not stand only for experimental skills, but also for the development of numerous technical high precision devices, in which he took part significantly. These include the valve-free expansion reciprocating engine for the pre-cooling of helium prior to liquefaction and the maintenance-free submerged rotary pump for the filling of liquefied helium into the transport dewars. Robert Doll retired in 1988. He decided to stay at the WMI, though, as a permanent guest, dealing with tasks requiring precision engineering via optics and interferometry, with problems in astrophysics and even theoretical physics. His particularly strong affinity to the mathematical and esthetic beauty of the Ginzburg-Landau theory was given proof of in a publication, written on the occasion of Vitaly Ginzburg's 90th birthday, in which he used it for a comprehensive analysis of his experimental data on fluxoid quantization[48].

On November 15, 1974, Walther Meißner died at the age of 91 in Munich. Fifteen years later, on October 1, 1989, William M. Fairbank suffered a heart attack during his daily jogging in Palo Alto and died at the age of 72.

On January 16, 2008, Robert Doll could celebrate his 85th birthday in the best of health. On this occasion, an article of appreciation was written for the *Journal of the Bavarian Academy of Sciences* by two of his friends[61]. In May, 2010 Deaver officially retired and became professor emeritus. He continued though to have his office and research lab and can be found at the department regularly

---

[61] D. Einzel and R. Hackl, *Akademie Aktuell* 01/2008, p. 44

working on various projects. Bascom Deaver could celebrate his 80th birthday on August 16, 2010 in the best of health.

FIGURE 3.14: Recent portraits of Bascom S. Deaver (left) and Robert Doll.

3.2.7 Summary and Conclusion

We have provided a short historical review of the discovery of the fluxoid quantization by Doll/ Näbauer in Bavaria and Deaver/Fairbank in California. A remarkable coincidence turned out to exist with respect to a couple of completely nontrivial ingredients for the success of both groups. These were the basic idea, which could be traced back to London's footnote in London[46], the starting time around 1959–1960, the duration of the experiments from the construction to the final answers in the first half of 1961, the conference on superconductivity, where the two groups presented their results and, finally, the physics journal (*The Physical Review Letters*) where the respective results were eventually brought to the attention of the world-wide physics community.

From the completely uncorrelated experimental activities of Doll and Näbauer and Deaver and Fairbank, one may conclude that progress in science may, in some cases, happen on colliding paths. In the first place, there exists a certain reservoir of ideas, which is shared by scientists all over the world. It may happen that the same idea is taken seriously by scientists at different places in the world, creating a—certainly unintended—competitive situation. In the end one might feel happy if the results coincide. In such a case there is no need for the elimination of competition. What counts in the end is the quality of the scientific publication.

It is an exceptional fact in the history of science, that an important discovery is made independently by two experimental groups, which worked at fairly remote places, but came to the same conclusion almost at the same time, namely that (i) the fluxoid is quantized and (ii) the fluxoid quantum is $\Phi_0 = hc/2e$. The discovery of quantized flux is often cited, however, sometimes by omitting the Doll-Näbauer experiment on the American side and the Deaver-Fairbank experiment on the European side. Given the importance of this discovery and the courage and ingenuity of the experimentalists, we believe that both parties deserve the same credit for their work and should always be cited together.

The impact of the fluxoid quantization is seen to exceed largely the mere fundamental understanding of superconductivity. It now lies most notably in the high precision measuring technique and the information technique. Superconducting quantum interferometers, for example, can detect magnetic fields, which are by a factor of 10^{-11} smaller than the earth's magnetic field. Application areas include medicine, minimally invasive material testing, sensors in geo-prospection and quantum information processing, to name only a few.

What should remain at the end of this article, is the awareness, that the fluxoid quantization in its significance as a quantum phenomenon, seen on a macroscopic scale, detected for the first time fifty years ago, can be added, together with its theoretical interpretation, to the set of most exciting discoveries of the last century.

3.3   The Search for the Pairing Symmetry in the High Temperature Superconductors

Dale J. Van Harlingen

This article is dedicated to the memory of my colleague Donald Ginsberg—I loved him for his friendship, his wit, and his crystals that made possible the experiments described here.

One of the most exciting chapters in the history of superconductivity took place in the years between 1990 and 1995 when the issue of the pairing symmetry of the high temperature cuprates stimulated an intense and often contentious debate. During this period, theorists and experimentalists worked worldwide to find ways to predict or determine the pairing symmetry. Few topics in superconductivity have attracted such attention—for many years I think it is fair to say that it was the single most discussed question in condensed matter physics.

We now of course have accepted that the cuprates have d-wave symmetry of the form $d_{x^2-y^2}$, although we still do not have a microscopic model for that state or even a consensus on the pairing mechanism. A vast number of experiments have now confirmed the d-wave symmetry; in most, this conclusion now seems obvious and necessary to explain the observed data. However, in the early 1990s, this conclusion seemed far from obvious and the path by which we determined it is an interesting story of innovation and discovery. It is also a story that has not ended since the verification of unconventional pairing symmetry in the cuprates has opened the door to the discovery of many new materials that are confirmed or suspected to exhibit exotic order parameter symmetry, even beyond d-wave into odd parity and complex states.

Here, I describe some of the early experiments that first hinted at and ultimately confirmed the unconventional nature of the pairing state in the cuprates. This presentation is not at all intended as a comprehensive review of what transpired in that exciting period. Indeed, I could not write such a review even if I wanted since there were hundreds of scientists worldwide addressing this problem and hundreds of papers that contributed to the literature on this subject. Instead, I will present it from my perspective, just one scientist from one place who followed this debate and had the opportunity to make a contribution to settling it.

It is also fair to say that the one place in which I work is not an ordinary place in the history of superconductivity. The University of Illinois at Urbana-Champaign has a long and storied legacy in this field. It has been the home to many researchers who have made important contributions to the understanding of both classical and unconventional superconductors. Of course, the most famous of those were the trio of theorists—John Bardeen, his postdoc Leon Cooper, and his graduate student Robert Schrieffer—who presented the microscopic theory of conventional superconductivity in their legendary BCS paper. This paper, one of the most significant scholarly achievements of the last century, essentially solved the problem of superconductivity in a comprehensive and elegant way 46 years after its discovery and has had enormous impact in many fields: condensed matter physics, astrophysics, nuclear physics, high energy physics, and atomic-molecular-optical (AMO) physics. In fact, after the BCS paper, many considered the problem of superconductivity solved and well understood, and the emphasis in the field had turned more to applications that exploited the extraordinary properties of superconductors and superconducting coherence. The discovery of the high temperature superconductors in the mid-1980s changed all that and I think now all of us in the field would agree that superconductivity remains complicated and challenging and utterly fascinating.

It is interesting to note that the BCS paper dwells very little on the issue of the pairing symmetry. The phonon-propagated electron-electron interactions responsible for the Cooper pairing of electrons give rise very directly to an order parameter and energy gap that is (at least very nearly) isotropic in momentum space. We now call that s-wave, although there was no need at the time for BCS to identify it as such. This has been confirmed by numerous measurements, most elegantly by quasiparticle tunneling spectroscopy pioneered by Ivar Giaever at General Electric in the early 1960s. The only significant attention to symmetry has consisted of some careful mapping of relatively small anisotropies in the energy gap arising from band structure effects.

The situation changed immediately after the discovery of the high temperature superconductors. The high transition temperature, low dimensionality, and unusual properties seemed unlikely to be explained by the BCS model that had worked well for virtually all known superconductors up to that time. Both the phonon mechanism and the order parameter symmetry were questioned, although the presence of Cooper pairing was quite quickly established by a series of flux quantization experiments. I think we all knew that we had a family of superconductors that was unconventional in symmetry and mechanism, although it was not yet clear what it was and what caused it. Only the first of these questions has been fully answered; the second remains one of the greatest unsolved problems in condensed matter physics, awaiting the next "BCS" breakthrough.

3.3.1 Why the Symmetry Is Important

It is perhaps appropriate to diverge and address why knowing the symmetry is so important and why determining it occupied so much attention in the period under consideration. One primary reason is that it provides a clue to the mechanism. Indeed, a microscopic model leads directly to knowledge of the order parameter symmetry, as it did for the BCS model. However, the inverse is certainly not true–knowing the symmetry does not uniquely specify the mechanism and we now appreciate that there are many different ways to get d-wave symmetry. Nonetheless, the symmetry does severely constrain the possible microscopic approaches and guides further investigations. The symmetry also determines the behavior of the superconducting material, dictating the number and distribution of excitations that dominate the temperature dependence of all thermodynamic, electromagnetic, thermal, and electronic properties. This is particularly evident in comparing a fully-gapped state as in s-wave, in which electronic properties exhibit exponential temperature dependences, with one that has nodes as in d-wave and gives rise to power law dependences from the excitations present at all temperatures. Finally, the order parameter anisotropies affect possible applications of unconventional superconductors—the excess of quasiparticles excited at the nodes can create detrimental effects such as high frequency losses, but in turn the phase anisotropies can enable novel superconductor architectures that are just beginning to be explored.

3.3.2 Determining the Symmetry

In thinking about how one can go about determining the pairing symmetry of any superconductor, it is useful to look at a roadmap of candidate pairing states such as in Figure 3.15.

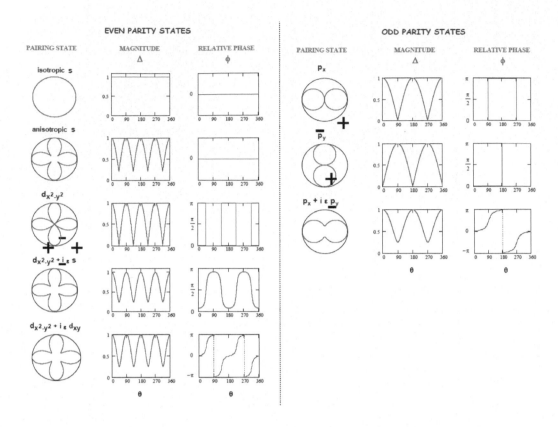

FIGURE 3.15: Magnitude and phase of the superconducting order parameter for candidate pairing symmetries. This chart provides a roadmap for determining the pairing symmetry of unconventional superconductors.

Figure 3.15 shows the variation of the magnitude and phase of the order parameter as a function of k-space direction for a series of possible two-dimensional pairing states. Included are both even and odd orbital pairing states corresponding to single and triplet spin pairing respectively, single component order parameters with s, p, and d symmetry, and two-component order parameters that have complex symmetries that break time-reversal symmetry. By looking at this chart, it is possible to map out strategies for determining the pairing symmetry. Most experiments probe the symmetry by looking at the magnitude of the order parameter, in most cases by measuring the number and distribution of quasiparticles excited above the energy gap. These measurements can determine if the energy gap is isotropic or varies with k-space angle, and in particular if there are nodes in the energy gap. The complication is that magnitude measurements are generally model dependent and can be confused by impurities that in some cases can mask and in other cases mimic the presence of nodes. For example, in the two leading candidates for the cuprates, anisotropic s-wave and d-wave, the only discernible difference is whether the nodes are fully formed or not. In contrast, a measurement of the phase anisotropy provides a more direct way to distinguish different candidate states and ultimately can determine the symmetry unambiguously.

At the risk of over-simplifying a very active and intense period of research, I will highlight just four experiments that from my perspective contributed key pieces to completing the puzzle of the pairing symmetry:

1. NMR experiments at the University of Illinois at Urbana-Champaign that established the

singlet pairing and even symmetry of the order parameter.

2. Microwave resonator penetration depth measurements at the University of British Columbia that first showed that there were nodes in the order parameter.

3. Angle-resolved photo-emission spectroscopy (ARPES) experiments at Stanford and at the University of Illinois at Chicago that identified the location of the nodes.

4. Josephson interferometry experiments at the University of Illinois at Urbana-Champaign that demonstrated the phase anisotropy of the order parameter and confirmed the d-wave symmetry.

It is interesting that the last three of these measurements pioneered novel experimental techniques that were developed specifically for studying the high temperature superconductors and played a key role in the determination of the symmetry of that class of materials and subsequent unconventional superconductors.

The story of the pairing symmetry starts before any of these experiments, in fact immediately after the discovery of the high temperature superconductors with the suggestion first by Doug Scalapino and co-workers that the symmetry could be d-wave. This hypothesis was based on pairing by antiferromagnetic spin fluctuations. However, early estimates did not seem to give sufficient pairing strength to explain the high transition temperatures observed so this idea did not gain significant traction at the time with most theorists. A notable exception was David Pines at Illinois, who had independently concluded that d-wave was the most likely symmetry based on NMR experiments from the group of his experimental colleague Charlie Slichter. Their conviction stemmed from the temperature dependence of the spin susceptibility and motivated the spin fluctuation model that attributed the pairing to magnetic antiferromagnetic fluctuation in the copper oxide planes. Via extensive numerical calculations, Pines and co-workers concluded that the spin pairing was sufficient to explain the cuprate T_c. The key to these calculations was the need to include all excitations—as Pines described it at the time, "there is no such thing as a free lunch". For many years, Pines became perhaps the most vocal advocate for d-wave pairing, insisting that it had to describe the cuprates and keeping a list of experiments that supported d-wave. At the time, this list was close to zero, but, as we will see, grew rapidly over a period of several years. Pines also gave a very simple and compelling physical picture for why the symmetry would be d-wave, arguing that the order parameter would have to vanish along the diagonals in order to avoid the strong onsite repulsion. Although the spin fluctuation model likely does not fully apply to the cuprates, David Pines' role in pushing the d-wave picture was crucial to motivating experiments to test the pairing symmetry—we in Urbana certainly were driven to resolve this interesting and challenging question by Pines' passionate stance.

3.3.3 Samples

Before summarizing some of the important experiments that eventually confirmed the d-wave symmetry, it is important to highlight the crucial importance of good samples to this success. In the early days of the HTSC research explosion, poor quality and poorly characterized samples led to a flood of inconsistent, questionable, and simply inaccurate results that confused the field. As both the samples and the recognition of the importance of samples improved, there was corresponding progress toward understanding the intrinsic properties of the cuprates. Work in many groups worldwide in the growth of thin films and single crystals contributed to this progress and it is not possible to acknowledge them all. However, it is appropriate to celebrate the excellent single crystals from two particular groups that enabled the seminal experiments that I will describe here: Don Ginsberg's group at the University of Illinois and Ruixing Liang's group at the University of British Columbia.

From a wider perspective, it is important never to forget what we learned from the quest to determine the pairing symmetry—the critical importance of good samples in the study of complex and strongly-correlated materials.

3.3.4 Parity

The NMR experiments of Charlie Slichter and co-workers at Illinois had already provided a significant clue to the pairing symmetry. The Knight shift in YBCO single crystals grown by Ginsberg shows a distinct change at T_c, indicating that the spin pairing was singlet and restricting the possible symmetries to odd parity states such as s-wave or d-wave.[62] Further work motivated the spin fluctuation model of Pines and gave early clues for d-wave pairing.

3.3.5 Penetration Depth

One of the most important probes of the pairing symmetry turned out to be measurements of the penetration depth, the parameter that characterizes the spatial extent of magnetic fields into a superconducting sample. The penetration depth λ depends on the superfluid density, the number of Cooper pairs available to screen applied magnetic fields. At the lowest temperatures, the penetration depth drops proportionally to the number of excitations and so the temperature dependence of λ serves as a test of the energy gap structure. In particular, fully-gapped s-wave superconductors are expected to exhibit an exponential variation in $\lambda(T)$, whereas nodes give rise to linear behavior in the simplest models.

Early measurements of penetration depth in the high temperature cuprates, both in bulk materials and thin films, did not reveal a linear variation, exhibiting instead a much more rapid function, and this was generally interpreted as evidence for a fully-gapped state as in conventional s-wave BCS superconductors. However, in 1991, Nigel Goldenfeld and collaborators carried out a re-examination of the penetration depth data, showing that it was consistent with a quadratic power law dependence at low temperatures and proposing several explanations for this functional form that allowed unconventional symmetry.[63] This work slowed the growing consensus that the cuprates were s-wave and motivated further experiments.

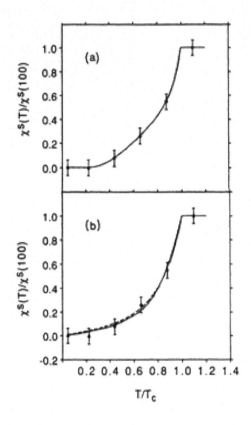

FIGURE 3.16: Knight shift measurements in YBCO showing the drop in spin susceptibility that indicates a spin-singlet state. (a) Fit to s-wave. (b) Fit to anisotropic s-wave (solid line) and d-wave (dashed line) models. From Annett et al.[63].

---

[62] S. E. Barrett et al., *Phys. Rev.* B 41 (1990) 6283

[63] James Annett et al., *Phys. Rev.* B 43 (1991) 2778

The most convincing of these were the penetration depth measurements performed by the group of Walter Hardy and Doug Bonn at the University of British Columbia on YBCO single crystals grown by their colleague Ruixing Liang. Motivated by the suggestion of nodes in the energy gap, Hardy designed a novel superconductor cavity resonator experiment while on a trip to Paris in the summer of 1992. During that fall, the experiment came together and achieved a remarkable resolution of 0.2 Å in changes in the penetration depth, recorded as a frequency shift in the resonator. These gave a clear indication of a linear variation of the penetration depth at low temperatures down to at least 4 K, as shown in Figure 3.18, in

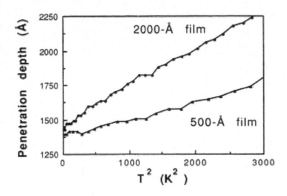

FIGURE 3.17: The low temperature penetration depth showing a quadratic variation. From Annett et al.[63].

sharp contrast to the exponential variation observed in conventional BCS superconductors[64]. After careful checks for spurious sources of the linear dependence, which according to Walter, "just wouldn't go away", they published the results that gave the first strong evidence for nodes in the energy gap. This linear dependence was also consistent with their microwave conductivity measurements on the cuprates.

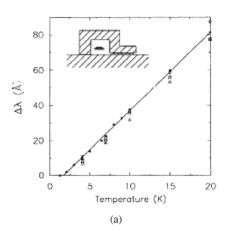

(a)

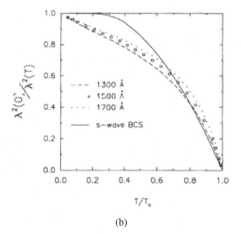

(b)

FIGURE 3.18: (a) The change in the low temperature penetration depth of YBCO at low temperatures as measured by the resonator technique. (b) Temperature dependence of the $\lambda^2(0)/\lambda^2(T)$, proportional to the superfluid density, for different estimates of $\lambda(0)$ and compared to the BCS form. From Hardy et al.[64].

3.3.6 ARPES

Many of the measurement techniques that were employed to study the cuprates were probes that had been developed to characterize the conventional s-wave superconductors. However, several new approaches emerged that were particularly well suited to look at the cuprates. One of those

---

[64] W. N. Hardy et al., *Phys. Rev. Lett.* 70 (1993) 3999

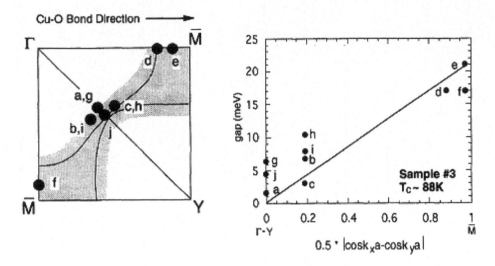

FIGURE 3.19: The ARPES measurements of the Stanford group first showed evidence for the strong gap anisotropy and identified the nodal directions. From Shen et al. [65].

was ARPES — angle resolved photo-emission spectroscopy. Although this was not a totally new technique, a combination of the larger energy scale of the cuprates coupled with advances in the technique, data acquisition, and analysis of ARPES signals enabled it to evolve as a key probe of the momentum and energy distribution of the quasiparticles in the cuprates.

Z. X. Shen and collaborators at Stanford were one of the first groups to develop the capabilities of ARPES for the study of the superconducting state in the cuprates. By using crystals of BSCCO, fabricated by his colleague Aharon Kapitulnik, that could be cleaved to reveal a pristine surface, Shen and his group found evidence for a strong anisotropy in the energy gap of BSCCO samples[65]. As shown in Figure 3.19, the gap in the (110) directions is an order of magnitude less than that along the (100)-directions. This was not an easy experiment. Even with the larger energy gap scale, it was necessary for them to implement a novel multi-channel spectrophotometer detection scheme that gave an order of magnitude improvement in efficiency. The data shown above took as long as 10 hours to acquire, during which the spectra changed with time as the surface aged, significantly complicating the analysis, but in the end the evidence for the gap anisotropy became apparent.

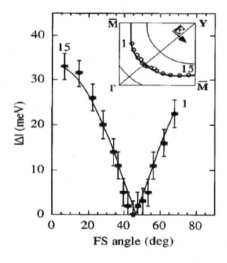

FIGURE 3.20: Variation of the energy gap in BSCCO by ARPES measurements by the UIC group. From Ding et al. [66].

[65] Z.X. Shen et al., *Phys. Rev. Lett.* 70 (1993) 1553

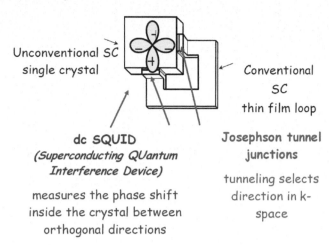

Unconventional SC single crystal

Conventional SC thin film loop

dc SQUID
(Superconducting QUantum Interference Device)

measures the phase shift inside the crystal between orthogonal directions

Josephson tunnel junctions

tunneling selects direction in k-space

FIGURE 3.21: The corner SQUID experiment, the prototype for phase sensitive measurements of the pairing symmetry of unconventional superconductors.

To demonstrate how rapidly the technique of ARPES has advanced, just a few years later the gap was mapped directly by J.-C.Campuzano and co-workers at UIC, giving a clear map of the energy gap variation in the nodal direction[66].

Since those early experiments, ARPES has evolved into one of the standard tools for probing excitations in unconventional superconductors. It continues to be limited by energy resolution, making it difficult to probe exotic superconductors with small energy gaps, and surface properties, requiring materials such as BSCCO that are easily cleaved.

3.3.7 Josephson Interferometry

In Urbana, the issue of the pairing symmetry was a topic of much discussion, motivated primarily by David Pines who was convinced that the order parameter had to be d-wave. My own involvement in this problem started over the winter holidays in December of 1991 when Don Ginsberg approached to tell me about an idea that Tony Leggett had devised to determine the pairing symmetry. Don and I had adjacent labs and it was not uncommon for us to talk about experiments after the students had disappeared for the holidays and the two us had time to be in the lab. The essential idea was to connect the orthogonal faces of a cuprate single crystal with a conventional superconductor to form a multiply-connected loop and detect the phase difference by interferometry, the way in which phase differences are always measured in quantum systems. For several hours we discussed this idea and designed an experiment, the corner SQUID experiment, that we believed could make an unambiguous determination of the phase anisotropy and hence the pairing symmetry.

The corner SQUID experiment, shown in Figure 3.21, is a relatively simple experiment. The crystal to be studied is mounted on a substrate using glue that both holds the crystal and provides a smooth ramp from the substrate to the edge of the crystal. Josephson junctions between the crystal and a thin film conventional s-wave counterelectrode are fabricated on the orthogonal faces of the crystal and probe the order parameter in a direction perpendicular to the tunneling surface. The nature of the tunneling is not critical as long as it forms a Josephson junction—in most cases, we use a normal metal barrier in order to contact the unconventional superconductor and form SNS (superconductor-normal-superconductor) junctions. Connecting the two junctions forms a dc SQUID (superconducting quantum interference device).

---

[66] H. Ding et al., *Phys. Rev.* B 54 (1996) R9678

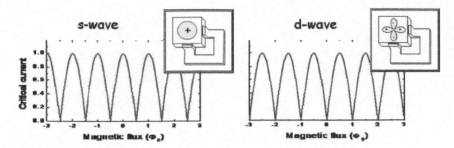

FIGURE 3.22: Modulation of the critical current across the SQUID for s-wave and d-wave symmetry.

This device is the quantum analog of a two-slit optical interference experiment–just as the intensity of light through the slits is a periodic function of the angle of the outgoing beam, the critical current across the SQUID (the maximum current before the device develops a finite voltage) modulates with a magnetic field applied through the junction, as shown in Figure 3.22.

If the crystal is s-wave, the orthogonal directions probe the same order parameter sign and the critical current has a maximum at zero applied magnetic field. In contrast, in a d-wave junction the signs probed by the two edges have opposite signs and the critical current vanishes at zero field (or is at least strongly suppressed if the junctions are not of equal size); this suppression is accompanied by a circulating current in the SQUID to maintain flux quantization. The elegance of this experiment is that it is sensitive only to the phase anisotropy and relatively independent of other effects so that an unambiguous determination can be made. So although we did not know the answer to the pairing symmetry, we were quite confident as 1992 began that this technique would give us that answer.

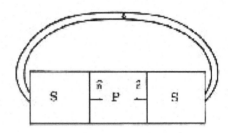

FIGURE 3.23: The circuit proposed to test for spontaneous currents generated by p-wave pairing in the heavy fermion superconductors. From Geshkenbein [67].

It turned out that this scheme had previously been proposed in a beautiful paper by Vadim Geshkenbein a few years earlier as a possible way to test for p-wave symmetry of the heavy fermion superconductors[67]. The layout is shown in Figure 3.23.

Interestingly, that experiment was finally done 17 years later by Ying Liu and collaborators at Penn State, indeed giving evidence for odd symmetry pairing in the ruthenate superconductor Sr_2RuO_4. However, as pointed out by Tony Leggett, there is a distinct and significant difference between the implementation of this approach for even *vs* odd symmetry. In an even symmetry state such as s-wave or d-wave, there is a direct correspondence between the real space tunneling direction and the lobe of the order parameter being probed. This is not the case for odd symmetry states since lobes of both signs are probed. In fact, the supercurrent should vanish in the absence of spin-orbit coupling or some other perturbation that breaks the parity symmetry.

The task of actually doing the experiment was taken on by David Wollman, a remarkable graduate student in my group who had extraordinary experimental creativity and skill. Wollman became the hands and eyes of the experiment, succeeding in fabricating up to 10 Josephson junctions on one

_____
[67] V. B. Geshkenbein, et al., *Phys. Rev.* B 36 (1987) 235

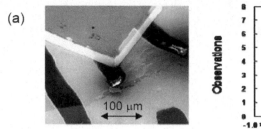
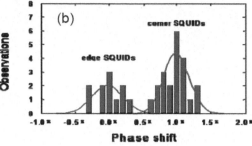

FIGURE 3.24: (a) SEM micrograph of a corner SQUID showing the edge junctions on orthogonal faces. (b) the location of the peak in the critical current for a series of edge and corner SQUIDs, showing evidence for the sign change (phase shift of π) between orthogonal directions. From Wollman et al. [68].

of Ginsberg's single crystals that were roughly the size of a grain of pepper, and taking most of the measurements, as shown in Figure 3.24(a). The phase shift between orthogonal directions in a corner SQUID for a number of samples, compared to that of edge SQUIDs in which both junctions are on the same face and should sample the same phase are shown in Figure 3.24(b). The phase shift of π is evident and gave evidence for the d-wave state[68]. However, also evident is a distribution of phase values that results from trapped magnetic flux near the SQUID loop that shifts the critical current pattern. These shifts were not large enough to obscure the sign change in the order parameter, but made it difficult to quantify the shift with any precision.

To improve the experiment, we adopted an alternative form of it in which we eliminated the loop, forming a single corner junction. This configuration is equally sensitive to the phase difference across the corner of the crystal, yielding the standard Fraunhofer diffraction pattern familiar from single slit optics for an s-wave crystal with a uniform phase and a modified form with a minimum at zero field for d-wave crystal in which the two halves of the junction probe lobes of opposite sign, as shown below in Figure 3.25. The advantage of this approach is that there is virtually no area for trapped flux to couple into the junction so the phase difference can be directly detected. For a corner junction as shown, the critical current shows the precipitous drop at zero field characteristic of a phase change of π, only failing to go to zero at the dip due to the areas of the junctions on the two faces being unequal[69]. This gives unambiguous evidence of the d-wave symmetry. The symmetry of the peaks in the data puts a strong limit of less than 1% on the existence of any out-of-phase component in the order parameter—such a term would cause a notable difference of peak heights.

We first presented our data at the March 1993 meeting. Since we had not submitted an abstract on this experiment, I gave the talk in a session on "Order Parameter Symmetry" in which a talk had been withdrawn and the chair of the session kindly allowed us to show our results. After the publication of our results later that year, combined with the growing list of experiments that were consistent with unconventional superconductivity, the case for d-wave symmetry began to grow rapidly. At the winter meeting in Aspen in January 1994, at which the pairing symmetry of the cuprates was widely debated, a proposed vote to record who favored s-wave and who favored d-wave in the cuprates was aborted when Mac Beasley stated that "if the Illinois phase-sensitive experiments are correct, the issue is already decided". There was no vote taken.

[68] D. A. Wollman et al., *Phys. Rev. Lett.* 71 (1993) 2134
[69] D. A. Wollman et al., *Phys. Rev. Lett.* 74 (1995) 797

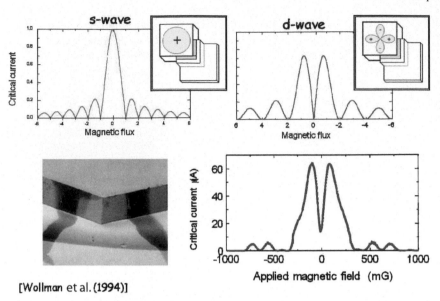

[Wollman et al. (1994)]

FIGURE 3.25: Predicted critical current modulation patterns for s-wave and d-wave symmetry in the single junction Josephson interferometry experiment and the results from the corner junction shown, clear evidence for the d-wave pairing. From Wollman et al. [69].

Within a year, the validity of the Josephson interferometry experiments was confirmed, in large part by alternative ways to carry out phase-sensitive interference experiments. Of particular note are two experiments that directly measured the spontaneous circulating current that arises in the d-wave state (for sufficiently large SQUID inductance): a measurement by Fred Wellstood and co-workers at Maryland, and the beautiful tricrystal experiments of the IBM group headed by John Kirtley and Chang Tsuei that will be described in the next sections. These experiments gave further evidence for the d-wave state in a very visual and clear way. The experiments are also notable because they were carried out on thin films which are micro-twinned, a result of the orthorhombic crystal structure of YBCO. When we began the corner SQUID and corner junction experiments,

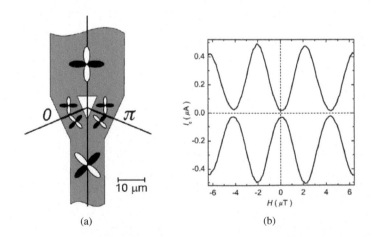

(a)   (b)

FIGURE 3.26: (a) Design of the thin film π-SQUID and (b) flux modulation showing minimum critical current at zero applied magnetic field. From Schulz et al.[70].

it was not known if the signs of the order parameter lobes would align globally or follow the a-b axes of the twins. By performing the experiments on both twinned and untwined crystals, we determined that, in Tony Leggett's words, "the order parameter aligns north-south instead of a-b". This phenomenon, presumably the result of the excessive energy cost required to form domain walls of opposite sign, allows the Josephson interferometry measurements to be done on thin films, but, even more importantly, allows the possibility of thin film superconductor devices that utilize the functionality of the sign change. One example is the π-SQUID demonstrated by Jochen Mannhart and Hans Hilgenkamp that shows the shifted SQUID pattern originally proposed for the corner SQUID in a precise and dramatic fashion[70].

I am often asked when we actually knew in our group that the cuprates were d-wave. There is always a time in an experiment when you become convinced of what nature is telling you, often well before you have gathered enough "publication quality" data to make the case to others and before you have explored all of the potential concerns. For me, that occurred in December 1992. Each year, my group hosts a holiday "cookie" party for the Condensed Matter group in Urbana at which it was traditional for some of my students to be taking data during the party in order to showcase our laboratory. In that year, David Wollman was taking corner junction data that was showing the clear signature of the precipitous dip at zero field, a striking and unusual behavior. When asked about what it was by some of my colleagues, I remember answering "that's because it's d-wave", the first time I ever told anyone that and I think the moment when I really became certain myself. That was almost exactly one year after Don and I sat and launched the experiment. It was a remarkable year and a remarkable quest in the story of high temperature superconductivity. As we now know, the determination of the microscopic mechanism has proven to be a much longer and more challenging task.

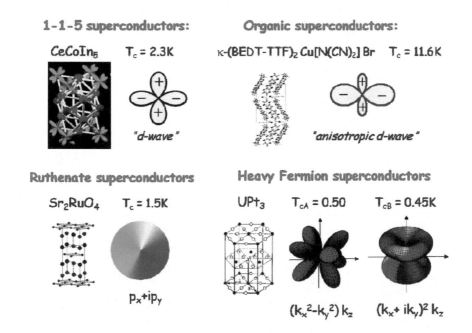

FIGURE 3.27: The growing family of unconventional superconductors.

[70] R. R. Schulz et al., *Appl. Phys. Lett.* 76 (2000) 912

3.3.8 Beyond d-Wave Symmetry

Since the confirmation of d-wave symmetry in the cuprates, a large number of superconductors has been discovered that are suspected to exhibit unconventional superconductivity. Figure 3.27 shows just a few of the materials that are believed to be unconventional.

Includeed are the so-called 115-heavy fermion superconductors such as CeCoIn5 that are believed to be d-wave like the cuprates, and the two-dimensional organic superconductors that are believed to be anisotropic d-wave. Even more exotic are materials that exhibit complex order parameters that break time-reversal symmetry. In the ruthenate superconductor Sr_2RuO_4, substantial evidence points to a complex order parameter of the form px+ipy, the superconducting analog of the order parameter in the A-phase of superfluid He^3. Perhaps the most exotic superconducting pairing state is that of the heavy fermion superconductor UPt_3, a three-dimensional material that exhibits two superconducting phases. Below 550 mK, the superconductor has a real order parameter with point nodes along the z-axis, a line node in the basal plane, and d-wave-like line nodes along the (110)-directions, as shown in Figure 3.27. At the lower transition ~500 mK, an imaginary component onsets which lifts the d-wave nodes and creates a chiral order parameter of the form $(k_x + ik_y)^2 k_z$. This order parameter form has recently been confirmed by Josephson interferometry by the Illinois group. Ironically, UPt_3 was the first superconductor that was believed to be unconventional, having been discovered a year or two before the discovery of the high temperature superconductors.

3.4   Half-Integer Flux Quantization in Unconventional Superconductors

C. C. Tsuei and J. R. Kirtley

Integer flux quantization is one of the most striking demonstrations of the macroscopic phase coherence of the charge-carrier pair wavefunction in superconductors. Based on the BCS theory, the ensemble of Cooper pairs can be described by a single complex wavefunction of electron wavevector:

$$\Psi(\mathbf{k}) = |\Psi(\mathbf{k})|e^{i\varphi(\mathbf{k})} \tag{3.1}$$

The amplitude $|\Psi(\mathbf{k})|$ determines properties like T_c, and the size of the BCS energy gap, $\Delta(\mathbf{k})$, which is a measure of pairing strength. The well-known hallmarks of superconductivity, such as zero resistance and the Meissner effect, are all derived from the phase $\varphi(\mathbf{k})$ and quantum phase coherence established among all the Cooper pairs on a global scale. The phenomenon of integer flux quantization is a consequence of the single-valueness of the pair wavefunction, or equivalently of the phase winding in multiples of 2π around any singly-connected path in a superconductor. The magnetic flux Φ threading through a superconducting loop, hence quantized in units of Φ_0 ($\Phi_0 = h/2e = 2.07 \cdot 10^{-15}$Wb):

$$\Phi = n\Phi_0 \quad n = 0, 1, 2, \dots \tag{3.2}$$

One can artificially incorporate a phase-shift of π in a superconducting loop, a so called "π-loop". The magnetic flux states in a π-loop are governed by the following expression:

$$\Phi = (n + 1/2)\Phi_0 \quad n = 0, 1, 2, \dots \tag{3.3}$$

The ground state, $\Phi = \pm\Phi_0/2$, represents a pair of time-reversed degenerate states spontaneously generated in the absence of an external field, corresponding to supercurrent flowing clockwise and counter-clockwise in the loop. As a function of the loop configuration, the presence and absence of the *half-integer flux quantum effect*[71] has been employed as the basis of phase-sensitive tests of

---

[71] C.C. Tsuei and J.R. Kirtley, *Rev. Mod. Phys.* 72 (2000) 969

pairing symmetry.

In this article, we give a historical and personal account of the advent of our tricrystal phase-sensitive symmetry experiments using a scanning SQUID microscope for establishing d-wave pairing symmetry in the cuprate superconductors. On the occasion of commemorating the 100th anniversary of the discovery of superconductivity by Kamerlingh Onnes, we are honored to present this article as part of the Jubilee Celebration.

3.4.1 The Design of the Tricrystal Experiments

Historical Remarks

The tricrystal experiments were conceived in the early Spring of 1993. At the five-day 1993 APS March meeting in Seattle, there was furious debate over the pairing symmetry in the high temperature superconductors. The root of the problem was that conventional techniques such as quasiparticle tunneling, NMR, ARPES, penetration depth, etc. can provide information only about the magnitude of the pair wavefunction, but not its phase. At this meeting Dale van Harlingen presented his exciting results using SQUID interference measurements in YBCO/Pb SQUIDs and junctions to support *d*-wave pairing symmetry (see in Section 3.3 by van Harlingen). These preliminary experiments were highly controversial, so much so that the need for alternative phase-sensitive tests of pairing symmetry was clear to many attending the meeting. During the flight back to New York, Chang Tsuei (CT) thought very hard about how to do a definitive phase-sensitive pairing symmetry experiment. By the time he landed at the airport in New York, he had formulated a basic tricrystal configuration for testing *d*-wave symmetry in cuprate superconductors. The crystallographic orientations of the three crystals were deliberately arranged to create a phase shift of π in a superconducting loop, interrupted by three grain-boundary weak links around the tricrystal meeting point. The tricrystal idea was a natural reflection of CT's experience in studying critical current density in bicrystal YBCO films[72,73] and in dc SQUIDs, made with two grain-boundary Josephson junctions, which resulted in an US Patent in 1988 [74]. Shortly after the APS March meeting, CT was asked to present his plan for the tricrystal experiment at a group meeting, called by Mark Ketchen, a senior manager of the Physical Sciences Department of the T.J. Watson Research Center. According to Ketchen's Lab Notebook, the meeting was named "S-wave d-wave superconductivity". Although the response to CT's presentation was quite positive and supportive, there were also strong reservations concerning whether it was feasible to make the proposed tricrystal $SrTiO_3$ (STO) substrates (needed for growing epitaxial cuprate films) with the three precisely-oriented grains, and three atomic-scale sharp grain boundaries between them. At that time, high-quality bicrystal STO substrates were commercially available for growing (001) tilt grain boundary junctions. No one had any experience or expertise of synthesizing the needed tricrystal substrates. CT ignored the skeptics and chose to work with Sinkosha in Tokyo, a bicrystal STO manufacturer. Sinkosha was willing to give it a try, with no guarantee of the quality of their final product. It turned out that the tricrystal substrates we received from Sinkosha were of truly exceptionally high quality. The grain boundaries in the substrates were so sharp that they could not be inspected without the aid of an optical microscope equipped with polarized light. Furthermore, all other specifications were met as well.

With this initial technical hurdle overcome, the tricrystal pairing symmetry experiment[75] was underway in earnest. As one would say: The rest is just history!

[72] P. Chaudhari, J. Mannhart, D. Dimos, C.C. Tsuei, J. Chi, M.M. Oprysko, and M. Scheuermann, *Phys. Rev. Lett.* 60 (1988) 1653

[73] J. Manhart, P. Chaudhari, D. Dimos, C.C. Tsuei, and T.R. McGuire, *Phys. Rev. Lett.* 61 (1988) 2476

[74] P. Chaudhari, C.C. Chi, J. Mannhart, and C.C. Tsuei "Grain Boundary Junction Devices Using High Tc Superconductors", US Patent 5162298.

[75] C.C. Tsuei, J.R. Kirtley, C.C. Chi, Lock See Yu-Jahnes, A. Gupta, T. Shaw, J.Z. Sun, and M.B. Ketchen, *Phys. Rev. Lett.* 73 (1994) 593

Key Considerations of the Tricrystal Design

Phase-sensitive pairing symmetry tests are based on two macroscopic phase coherence phenomena: flux quantization and Josephson pair tunneling. Theoretical suggestions for such a probe of unconventional gap symmetry had been reported in the literature many years before. Geschkenbein, Barone and Larkin suggested half-flux quantization could result from an order parameter with nodes in polycrystalline heavy fermion superconductors in 1987 [76]. It is interesting to note that this important theoretical work was not extended to its application in cuprate high-temperature superconductors, although it was published soon after the discovery of Bednorz and Mueller in 1986. Sigrist and Rice argued that the observed paramagnetic shielding effects in granular cuprate superconductors could be due to sign changes in a d-wave pair wavefunction and in 1992 proposed a SQUID interference experiment to verify such a nodal gap structure [77]. This theoretical proposal was exactly realized with a Pb-YBCO SQUID by D. Wollman et al. in 1993 [78].

The first tricrystal symmetry experiments were initiated at the IBM T.J. Watson Research Center in 1993. Key considerations that went into their design were:

Tricrystal geometry designed with Sigrist-Rice and maximum disorder formula

For testing for $d_{x^2-y^2}$-wave pairing symmetry, the sign of I_s is determined by the Sigrist-Rice formula (i.e. in the clean limit where the junction interface is assumed perfectly smooth) [77]

$$I_s^{ij} = \left[|I_c^{ij}| \cos 2\theta_i \cos 2\theta_j \right] \sin \left(\Delta\varphi_{ij} \right) \tag{3.4}$$

where $\Delta\varphi_{ij}$ is the phase difference between grains i and j, and θ_i and θ_j are the angles of the crystallographic axes (100) in the grains i and j with respect to the junction interface GB_{ij}.

To take into account the microstructural disorder effects arising from defects such as impurities, micro-facets, etc., on tunneling normal to the junction interface, a maximum disorder (angular deviations up to a maximum of $\pi/4$ in a d-wave configuration) formula is derived for I_s [75]:

$$I_s^{ij} = |I_c^{ij}| \cos 2 \left(\theta_i + \theta_j \right) \sin \left(\Delta\varphi_{ij} \right). \tag{3.5}$$

With the combined constraints imposed by (3.4) and (3.5), the 3-grain-boundary-junction ring shown in Figure 3.29a was designed to be a π-ring if the cuprate under test is a superconductor with $d_{x^2-y^2}$-wave pairing symmetry. The misorientation angles α_{12}, α_{31} and the angle between the grain-boundary planes (β) were chosen to ensure that the net sign of the supercurrent in the 3-junction ring is negative as required by the d-wave pair state [71,75]. The design point of the first tricrystal experiment corresponds the solid dot in the light yellow regions of Figures 3.29b and 3.29c.

---

[76] V.B. Geshkenbein, A.I. Larkin, and A. Barone, *Phys. Rev.* B 36 (1987) 235

[77] M. Sigrist and T. M. Rice, J. *Phys. Soc. Japan*, 61 (1992) 4283

[78] D.A. Wollman, D.J. Van Harlingen, W.C. Lee, D. M. Ginsberg, and A. J. Leggett, *Phys. Rev. Lett.* 71 (1993) 2134

Phase-Sensitive Tests of Pairing Symmetry

The basic idea of phase-sensitive pairing symmetry tests of unconventional (non-simple s-wave) superconductors had existed in the literature for some time[76,77]. Due to the nature of unconventional pairing symmetry such as in the $d_{x^2-y^2}$-wave pair state, one can configure a Josephson junction (between i and j, of which one is an unconventional superconductor) that is characterized by a negative supercurrent I_s:

$$I_s = -|I_c|\sin(\Delta\varphi_{ij}) = |I_c|\sin(\Delta\varphi_{ij}+\pi). \tag{3.6}$$

This remarkable effect of negative pair tunneling can be understood as an intrinsic phase shift of π added to the phase difference of the junction. In a superconducting singly-connected loop containing such junctions, the phase shift is gauge-invariant and the single-valueness of the pair wavefunction $\Psi(\mathbf{k})$ demands that phase can vary only in multiples of 2π in going around a closed contour, leading to the phenomenon of flux quantization,

$$\Phi + I_s L + \frac{\Phi_0}{2\pi}\Sigma_{ij}\Delta\varphi_{ij} = n\Phi_0, \tag{3.7}$$

where the self-inductance is L, and n is an integer. One needs to count the number of sign changes in the loop to distinguish a π-loop (half-integer flux quantization, see (3.3)) and 0-loop (standard integer flux quantization, see (3.2)). The odd-numbered sign changes cost energy in the Josephson coupling across the junctions. This excess energy is reduced by the spontaneous generation of circulating currents, which result in a Josephson vortex with half of the flux quantum Φ_0,

$$U(\Psi,\Psi_a) = \frac{\Phi_0^2}{2\pi}\left\{\left(\frac{\Phi+\Phi_a}{\Phi_0}\right)^2 - \frac{LI_c}{\pi\Phi_0}\cos\left[\left(\frac{2\pi\Phi}{\Phi_0}+\theta\right)\right]\right\}; \ \theta = 0,\pi. \tag{3.8}$$

The applied flux Φ_a is zero in the ground state. The doubly degenerate half flux quantum states, as shown in Figure 3.28, correspond to the supercurrent circulating in the loop clockwise and counter-clockwise respectively. As a function of the loop geometry, the observation or non-observation of the half flux quantum effect can be used to probe the phase of the order parameter, $\Delta(\mathbf{k})$ (i.e. $\Psi(\mathbf{k})$).

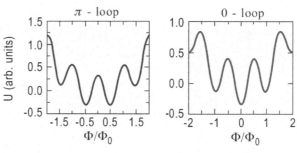

FIGURE 3.28: Free energy of superconducting π-loop (left), 0-loop (right).

Based on flux quantization and energy considerations, such a π-ring in the ground state should exhibit spontaneously generated magnetization of a half flux quantum, regardless of whether the grain boundaries are in the clean or dirty limit. This makes the tricrystal experiment a yes − or − no test of d-wave pair symmetry in terms of observing the half flux quantum or not.

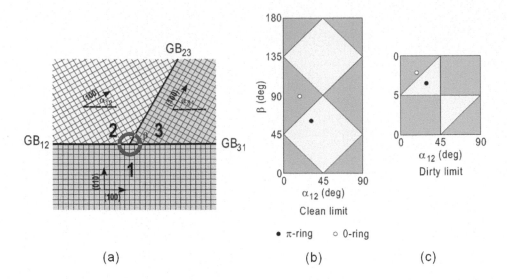

(a)   (b)   (c)

FIGURE 3.29: (a) Tricrystal design parameters, regions of π- and 0-rings in the clean limit (b) and the dirty limit (c).

Symmetry-Independent Mechanisms for π-Phase-Shift:

Several theoretical studies have predicted that magnetic interactions (including the spin-flip scattering due to impurities[79] at the junction can induce a π-phase-shift[79,80] in the tunnel barrier region, hence the name π-junction. In particular, such π-junctions have been demonstrated in various superconductor/ferromagnet/superconductor (SFS) thin-film structures through the exchange-field induced spatial oscillation in the order parameter in the junction barrier[81], (for a review of recent developments in SFS junctions and their potential in device applications, see Hilgenkamp[80]).

A superconducting loop containing an odd number of such π-junctions can also exhibit the half flux quantum effect. It is important to differentiate between the magnetically driven π-junction and the π-loop derived from the d-wave pairing symmetry observed in the original tricrystal experiment[75]. Using the criteria of the Sigrist-Rice and maximum disorder formula, (3.4) and (3.5), a second tricrystal experiment was designed by CT to rule out any symmetry-independent mechanism for the observed π-phase-shift effect. The design point of this tricrystal geometry is shown in Figures 3.29b and 3.29c as an open circle. Since the magnetic impurity effect would not be sensitive to such small variation in the tricrystal configuration as shown in Figures 3.29b and 3.29c, a half flux quantum would be observed in both tricrystal experiments if the spin-flip mechanism was in operation, while only in the first tricrystal experiment for the case of d-wave pairing symmetry.

Experimental Verification of the Tricrystal Design Parameters

X-ray diffraction measurements were carried to ensure a single-phase, high-quality c-axis epitaxial film growth of cuprate films, e.g., YBCO on the tricrystal STO substrate with the geometry depicted in Figure 3.30a. In-plane scanning x-ray diffraction and electron backscattering measurements demonstrated strong in-plane alignment in the tricrystal cuprate film with the misorientation

[79] L.N. Bulaevskii, V.V. Kuzii, A.A. Sobyanin, *JETP Lett.* 25 (1977) 290

[80] Hans Hilgenkamp, *Supercond. Sci. Technol.* 21 (2008) 034011

[81] V.V. Ryazanov, V.A. Oboznov, A.Yu. Rusanov, A.V. Veretennikov, A.A. Golubov, and J. Aarts, *Phys. Rev. Lett.* 86 (2001) 2427

angles at each grain boundary within 4° of the intended design angle. The IV characteristics of each grain-boundary junction in the tricrystal ring were shown to be those of a Josephson weak-link. The misorientation angles of the tricrystal configuration in Figure 3.30a were designed to be identical to avoid any extrinsic effect due to the difference between the three junctions. The IV measurements indicated the Ic values of the three boundary junctions were in agreement within 20%.

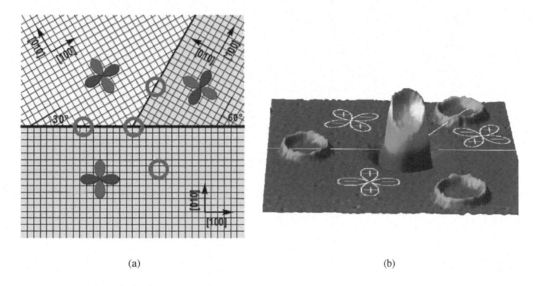

(a)   (b)

FIGURE 3.30: (a) Schematic of the d-wave testing tricrystal geometry and (b) SSM image of YBCO rings at 4.2 K and nominal zero field.

3.4.2   Direct Observation of the Half Flux Quantization

In February 1994 a c-axis-oriented epitaxial YBCO film was deposited on the d-wave-testing STO substrate and micro-patterned into four rings as shown in Figure 3.30a. The 3-junction ring at the center was surrounded by 3 0-rings (with an even number of grain-boundary junctions), which were designed to serve as the controls of the experiment. The basic design parameters of all the rings were thoroughly checked and we were finally ready to do the symmetry test. It was a great fortune that John Kirtley (JK) joined the tricrystal experiment with his scanning SQUID microscope (SSM); see Figure 3.4.2 and box below, "Development of a high-resolution Scanning SQUID Microscope".

There are several advantages of using SSM over the SQUID interference technique for determining the amount of magnetic flux threading through a superconducting loop. First, it provides a direct imaging of the magnetic flux state in the ring without relying on interpreting the measured SQUID characteristics[78]. Due to the small coupling between the SSM's pickup coil and the cuprate ring, the magnetic flux ground state of the rings can be probed precisely and noninvasively. With our SSM, any spurious flux trapped on the ring can be detected to a level of 10^{-3} Φ_0. Figure 3.30b shows a SSM image, taken at 4.2 K and nominal zero field, of four YBCO rings on a tricrystal STO substrate with the configuration depicted in Figure 3.30a. We were all very excited that there was flux in the 3-junction ring at the tricrystal point but not in the 3 control rings, as expected if YBCO were a d-wave superconductor. However, we needed to be sure that the magnitude of the flux through the 3-junction was exactly a half flux quantum, $\Phi_0/2$.

Hence, a quantitative determination of the flux state of the rings became the crucial part of the tricrystal experiment. We eventually used a number of flux calibration techniques. The most straight-forward one was direct calculation of the flux through the pickup loop due to currents in the rings. The flux Φ_p through the pickup loop can be written as $\Phi_p = \Phi_r M/L$, where Φ_r is the flux in the ring, M is the mutual inductance between the pickup loop and the ring, and L is the self-inductance of the ring. Calculating M is straightforward, but estimating L is more difficult since field penetration into the superconducting body of the ring must be carefully accounted for. Nevertheless, Mark Ketchen had previous experience with this type of problem, and his initial estimates for L were within 10% of the real value. CT vividly recalled that Mark called CT at home around one o'clock in the morning to discuss with him the L value he obtained. They concluded that it was probably a half flux quantum threading through the 3-junction ring. John C.C. Chi and Jonathan Sun also provided their calculated L values using different approximations. All these calculations agreed within 5%, and very convincingly led us to the conclusion that it was a $\Phi_0/2 = h/4e$ flux trapped in the 3-junction ring!

We nicknamed the second calibration technique the "oil drop" method: we cooled the rings repeatedly in various fields and measured the difference between the SQUID signal when it was centered above a ring vs that away from a ring. For all of the rings these differences were integer multiples of a value corresponding to a flux of Φ_0, but the differences were offset by half of this value for the 3-junction ring relative to that for the control rings—exactly what one would expect if the 3-junction ring was experiencing the half-flux quantum effect while the control rings had conventional flux quantization. A third calibration technique was to apply an external field to the sample, while monitoring the flux through the SQUID with the pickup loop centered above one of the rings. Over a certain range of fields the SQUID flux jumped in discrete steps as the applied field was increased, each step corresponding to an increase in flux of Φ_0 through the ring. A fourth calibration technique was to run cross sections through the center of one of the rings as a function of externally applied field B_{ext}. The SQUID signal was the same inside the ring as outside the ring when the externally applied flux $\Phi_{ext} = B_{ext}A$, where A is the effective area of the ring. The SQUID signal difference vs B_{ext} characteristics crossed zero at B_{ext} values corresponding to integer multiples of Φ_0, while the 3-junction ring had zero crossings offset by $\Phi_0/2$, within experimental errors of a few percent. We called this last method "titration" because the mechanism for detecting the zero crossing was not important, just as in a chemical titration. All four methods agreed, making it clear that the 3-junction rings in the first tricrystal geometry were exhibiting the half flux quantum effect.

It was both tedious and quite challenging to pattern rings lithographically from the expitaxial cuprate films with the 3-junction ring centered on the tricrystal point. The patterned ring configuration was essential for a straightforward flux determination, without the need of modeling, in the initial tricrystal experiments. Based on the discussion earlier in this paper, a Josephson vortex with $\Phi_0/2$ of total flux is spontaneously generated at the tricrystal point for a d-wave superconductor epitaxially grown on a tricrystal substrate with the appropriate geometry as expected from the design. If such a sample is cooled in a zero field, only the half flux quantum Josephson vortex should be present. Upon cooling in a finite field, Abrikosov vortices in the bulk and integer Josephson vortices along the grain boundaries should also be observed. Careful integration of the flux, taking into account the SSM-and-sample geometrical configuration, showed that the vortex at the tricrystal point had half of the flux of the other single vortices (see Figure 3.31) [82]. The technique of using unpatterned tricrystal epitaxial cuprate films is especially well adapted for testing pairing symmetry in various cuprate superconductors other than YBCO.

---

[82] J.R. Kirtley, C.C. Tsuei, M. Rupp, J.Z. Sun, Lock See Yu-Jahnes, A. Gupta, and M.B. Ketchen, *Phys. Rev. Lett.* 76 (1996) 1336

Development of a High-Resolution Scanning SQUID Microscope

The development of the scanning SQUID microscopes used for pairing symmetry tests at York-town Heights began in early 1992. At first this effort had nothing to do with pairing symmetry tests; tunneling measurements in the high-T_c cuprates often show a linear dependence of the dynamic conductance dI/dV on voltage. There have been a number of possible explanations for this so-called "linear conduction background", but Doug Scalapino and JK proposed that it could be due to inelastic tunneling mediated by a broad, flat distribution of excitations in the tunneling region. A review of the literature found that linear conductance backgrounds were common, including in Cr-CrO_x-Pb planar tunnel junctions. Al-AlO_x-CrO_x-Pb junctions had large linear conduction backgrounds, with the largest effects occurring for a few angstroms of Cr evaporated on an aluminum film before oxidation. Since a possible inelastic excitation causing the linear conduction background is spin fluctuations, it would be interesting to compare the magnetic properties of these films with the size of the linear conductance background. However, detecting magnetism in such thin films is difficult.

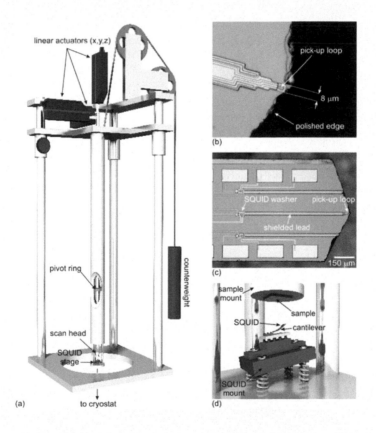

Scanning microscope

A conventional SQUID magnetometer is not sensitive enough for this application, but a scanning SQUID system, in which the SQUID is much closer to the local dipole moments, could work.

Twenty years previously IBM researchers had built a SQUID microscope[a] for imaging trapped vortices in superconducting circuitry as part of IBM's Josephson computer effort. This system no longer existed, so one was built from scratch, with Mark Ketchen designing and overseeing the fabrication of the SQUIDs, and JK building and running the microscope. JK converted an existing point contact tunneling probe[b], which used a mechanical differential thread and spring mechanism for the approach, as well as a long piezoelectric tube for the scanner, into a scanning SQUID microscope. We first used SQUIDs connected to small superconducting pickup loops through superconducting wire bonds, then SQUIDs with pickup loops integrated into them through strip line leads, and finally SQUIDs with pickup loops integrated through coaxial sheaths. It soon became apparent that piezo tube scanning was not a good choice for a SQUID microscope, both because the largest scan range attainable was too small, but also because the large voltages associated with piezo scanning caused arcing, which destroyed the SQUIDs. JK replaced the differential thread and spring approach of the original microscope with a lever connected at one end to a 3-axis optical stage and linear actuators at room temperature, with the sample mounted on the other end. At first this was intended to be the coarse positioning mechanism, while still using piezoelectrics for the fine scanning. However, the lever mechanism worked so well that the piezotube scanner was removed. The resulting microscope[c] could scan an area about 400 microns on a side. It was never used for measuring the magnetic properties of AlO_x-CrO_x films, in part because of the success of its application to the tricrystal experiments.

---

[a] F.P. Rogers, "A device for experimental observation of flux vortices trapped in superconducting thin films", Master's thesis, Massachusetts Institute of Technology, Cambridge, Massachusetts (1983).

[b] A.P. Fein, J.R. Kirtley, and R.M. Feenstra, *Rev. Sci. Instrum.* 58 (1987) 1806.

[c] J.R. Kirtley, M.B. Ketchen, K.G. Stawiasz, J.Z. Sun, W.J. Gallagher, S.H. Blanton, and S.J. Wind, *Appl. Phys. Lett.* 66 (1995) 1138.

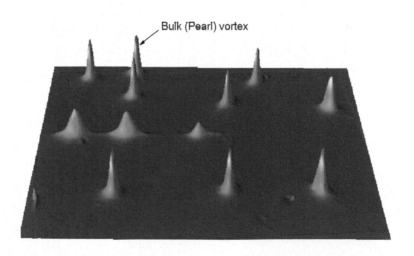

FIGURE 3.31:   SSM picture of unpatterned d-wave tricrystal – YBCO.

3.4.3 Elucidation of the Nature of Half Flux Quantum Effect

After the first tricrystal experiment, a series of phase-sensitive symmetry experiments were carried out to elucidate the nature of the observed half flux quantum effect.

Symmetry-Independent Mechanism for π-Phase-Shift

Right after we published the results of the first tricrystal experiment, there were several suggestions to us that the half flux quantum effect we reported might be due to the impurity and defects at the grain boundary of our tunnel junctions through the spin flip scattering or electronic correlation effect. In response to these criticisms, CT designed a new tricrystal experiment to demonstrate that by a small variation in the tricrystal configuration (the design point in the 0-loop regime as shown in Figure 3.29b) we could turn **on** and **off** the half flux quantum effect in accordance with the d-wave pairing symmetry assumption ((3.4) and (3.5)). As shown in Figure 3.32b, the SSM image of the new 3-junction ring and the controls exhibited no flux in their ground state[83]. This would not happen if any symmetry-independent mechanism was in operation.

Bilayer Effect in Cuprate Superconductors:

In the early stage of determining d-wave pairing symmetry in cuprates, all phase-sensitive experiments were done with YBCO whose crystal structure is characterized by two CuO_2 planes per unit cell. It was suggested that the two Cu-O layers can carry two s-wave order parameters but with opposite signs and mimic the observed $d_{x^2-y^2}$-wave gap function[84,85]. This issue was resolved by a d-wave pairing symmetry test with single-layer tetragonal $Tl_2Ba_2CuO_{6+\delta}$ (Tl2201) superconductors[86]. The observation of spontaneously generated half flux quantum at the 3-junction ring centered at the tricrystal meeting point is convincing evidence to rule out the bilayer effect. The half flux quantum effect was also observed in the tricrystal blanket film of Tl2201.

Tetracrystal Pairing Symmetry Experiment:

All the phase-sensitive pairing symmetry experiments with YBCO cannot distinguish between pure d-wave pairing and a d-wave dominant $d + s$ mixed pair state. The order parameter should transform in accordance with the symmetry operations of the relevant crystal point group. In an orthorhombic superconductor such as YBCO, s-wave and d-wave spin-singlet pairings correspond to the same irreducible representation A_{1g} of the point group C_{2v}. A $d + s$ mixed pair state is expected. For a tetragonal single-layer Tl2201 cuprate superconductor with point group C_{4v}, s-wave and d-wave pairings correspond to the irreducible representation A_{1g} and B_{1g}, respectively. An admixture of s-wave and d-wave pairings is not allowed. To test these notions a tetracrystal experiment with a $\pi/4$-rotated wedge geometry was designed and implemented. The interpretation of this experiment depends only on symmetry arguments and is therefore independent of any model for the Josephson pair tunneling current (i.e., (3.4) and (3.5)). The observation of the half flux quantum effect in the tetracrystal experiment represents strong and model-independent evidence for pure and $d_{x^2-y^2}$-wave pairing state in cuprates with tetragonal crystal structure[87]. This work also served as

[83] J.R. Kirtley, C.C. Tsuei, J.Z. Sun, C.C. Chi, Lock-see Yu-Jahnes, A. Gupta, M. Rupp, and M.B. Ketchen, *Nature* 373 (1995) 225

[84] D.Z. Liu, K. Levine, J. Malay, *Phys. Rev.* B 51, 8680 (1995).

[85] A.I. Liechtenstein, I.I. Mazin, O.K. Andersen, *Phys. Rev. Lett.* 74 (1995) 2303

[86] C.C. Tsuei, J.R. Kirtley, M. Rupp, J.Z. Sun, A. Gupta, M.B. Ketchen, C.A. Wang, Z. F. Ren, J.H. Wang, M. Bhushan, *Science* 271 (1996) 329

[87] C.C. Tsuei, J.R. Kirtley, Z.F. Ren, J.H. Wang, H. Raffy, and Z.Z. Li, *Nature* 387 (1997) 481

the basis of the later development for all d-wave dc SQUIDs[88] and related superconducting device applications[89,90,91].

3.4.4 Universality of the $d_{x^2-y^2}$ Pair State

In the past, numerous theoretical studies suggested that the stability of the *d*-wave pair state in various cuprates could be significantly affected by the details in the band structure and pairing potential. Especially the competing nature of the *s*-wave and *d*-wave pairing channels was emphasized. To clarify these issues, several phase-sensitive tricrystal experiments were carried out.

Doping Effect on Pairing Symmetry in Hole-Doped Cuprates

High-temperature superconductivity in cuprates is achieved by doping the Mott insulators with charge carriers, electrons and holes. A definitive determination of doping effect on pairing symmetry in cuprate superconductors is important in the study of high-T_c mechanism and quantum critical phenomena. We have done systematically a series of tricrystal experiments to study the doping dependence of gap symmetry in a variety of hole-doped cuprate superconductors. By an appropriate heat treatment of the $Bi_2Sr_2CaCu_2O_{6+\delta}$ system, tricrystal experiments were carried out at the optimum doping and to scan the pairing symmetry over a wide range of doping

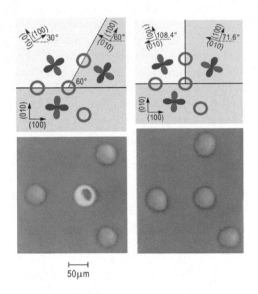

FIGURE 3.32: Tricrystal geometrical configurations and SSM images of rings: (left) designed to be a π-ring at the tricrystal point for a *d*-wave superconductor, (right) 0-ring designed to rule out any symmetry-independent mechanism for the half flux quantum effect.

covering both the over- and under-doped regimes. Also included in the study were $La_{2-x}Sr_xCuO_4$, a historically and fundamentally important high-T_c superconducting system, Tl2201-system, optimally doped YBCO and Ca-doped YBCO system. Our results indicate that the *d*-wave pairing persists through the entire phase diagram[92].

Electron-Doped Cuprates

It turned out to be quite difficult to do a tricrystal experiment with the electron-doped cuprate superconductors $Nd_{1.85}Ce_{0.15}CuO_{4-y}$ and $Pr_{1.85}Ce_{0.15}CuO_{4-y}$. This is mainly because the critical

[88] R.R. Schulz, B. Chesca, B. Goetz, C.W. Schneider, A. Schmehl, H. Bielefeldt, H. Hilgenkamp, J. Mannhart, and C.C. Tsuei, Appl. *Phys. Lett.* 76 (2000) 912

[89] T. Ortlepp, Ariando, O. Mielke, C.J.M. Verwijs, K.F.K. Foo, H. Rogalla, F.H. Uhlmann, H. Hilgenkamp, *Science* 312 (2006) 1495

[90] L.B. Ioffe, V.B. Geshkenbein, M.V. Feilgel'man, A.L. Fauchere, and G. Blatter, *Nature* 398 (1999) 679

[91] Thilo Bauch, Tobias Lindström, Francesco Tafuri, Giacomo Rotoli, Per Delsing, Tord Claeson, and Floriana Lombardi, *Science* 311 (2006) 57

[92] C.C. Tsuei, J.R. Kirtley, G. Hammerl, J. Mannhart, H. Raffy, Z.Z. Li, *Phys. Rev. Lett.* 93 (2004) 187004

current density J_c of the grain boundary with 30° misorientation angle (as shown in Figure 3.29a) is five orders of magnitude smaller than that of YBCO. As a result, the spatial extension of half flux quantum signal, as measured by $2\lambda_J \sim 100$ μm (the Josephson penetration depth $\lambda_J \propto J_c^{-1/2}$), is much broader than in the YBCO case. Despite the weak SSM signal at the tricrystal point, we managed to demonstrate and use the presence and absence of half flux quantum effect in a series of phase-sensitive tricrystal experiments[93] designed to be π- and 0-loops at the tricrystal point for a d-wave superconductor (see Figure 3.33). These results represent strong evidence that the electron-doped cuprates are d-wave superconductors as their hole-doped counterparts. In addition, there was convincing evidence that the time reversal symmetry is also conserved in both election-doped superconductors. It is remarkable that, although their normal-state properties are quite different from those of their hole-doped counterparts, the d-wave pairing dominates in all cuprates.

Temperature Dependence of the Half Flux Quantum Effect

Although the tricrystal experiments were widely accepted to provide conclusive evidence for d-wave superconductivity in the cuprates at liquid helium temperatures, there was speculation that the pairing symmetry could change with temperature, with the inclusion of a real or imaginary subdominant component at either low or high temperatures. We therefore built a scanning SQUID microscope with the capability of varying the sample temperature over a large range. Temperatures down to 0.5 K were achieved by condensing ^{3}He in the sample space and pumping on it; temperatures up to 100 K were attained by evacuating the sample space, providing a high thermal conduc-

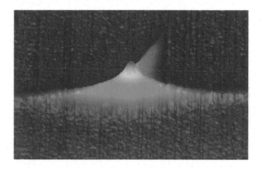

FIGURE 3.33: Tricrystal pairing symmetry tests of electron-doped cuprates.

tivity link between the SQUID sensor and the ^{4}He bath, and heating the sample. Measurements of a YBCO tricrystal sample using this microscope showed that the half flux quantum vortex had, to within experimental error of a few percent, $\Phi_0/2$ total flux over a temperature range from 0.5 K to within a few degrees of $T_c = 90$ K [94]. This indicated that there was no change in the dominant d-wave symmetry, and little if any imaginary component in the order parameter, over the entire temperature range.

In brief summary, the results of a systematic study using tricrystal experiments have demonstrated that the d-wave pair state in cuprates is robust against large variations in temperature up to T_c and doping covering both the under- and over-doped regimes. This finding underscores the important role of strong onsite Coulomb interaction, a universal characteristic of all cuprates, in favoring the d-wave over the s-wave pairing channel. When the strong correlation condition is relaxed, as presumably happened in the recently discovered Fe-based pnictides and tellurides, the s-wave pairing can prevail. In fact, due to the multi-orbital nature of these Fe-based high-temperature superconductors, a nodeless s-wave order parameter with sign reversal is also possible [95,96].

[93] C.C. Tsuei and J.R. Kirtley, *Phys. Rev. Lett.* 85 (2000) 182

[94] J.R. Kirtley, C.C. Tsuei, and K.A. Moler, *Science* 285 (1999) 1373

[95] I.I. Mazin, D.J. Singh, M.D. Johannes, and M.H. Du, *Phys. Rev. Lett.* 101 (2008) 057003

[96] C.-T. Chen, C.C. Tsuei, M.B. Ketchen, Z.A. Ren, and Z.X. Zhao, *Nature Physics* 6 (2010) 260

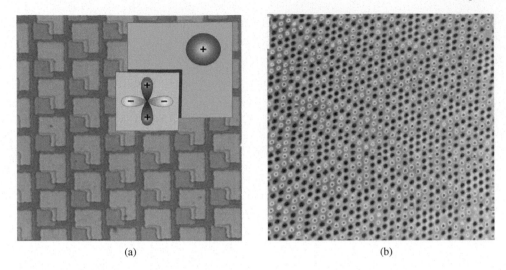

(a)   (b)

FIGURE 3.34:   Two-dimensional triangular π-arrays, (a) SEM micrograph, (b) SSM micrograph of the flux state in a triangular lattice.

3.4.5   Large-Scale Arrays of the Half Flux Vortices

The tricrystal and tetracrystal experiments described so far were all characterized by one single half flux quantum in a π-ring per sample. To gain insight in the gap structure of YBCO, for example, and for paving the road to future device applications in quantum computing and so on, one needs the capability of fabricating large-scale integration of arrays of π-loops on one single wafer. Two of such efforts are presented below.

Two-Dimensional Arrays of π-Loops

A large-scale integration of superconducting π-loops on a single wafer was made possible due to the development of a process, by the H. Hilgenkamp and D. Blank groups of the University of Twente in the Netherlands, for making high quality ramp-edge Josephson junctions between YBCO and Nb.[97]. This process used an in-situ etch and regrowth of the YBCO after photolithographic definition of the ramp edge that allowed for junction interfaces for making Josephson junctions with high critical current density and high reproducibility. In collaboration with the Twente group, fabrication of a triangular lattice of 25000 π-loops on a single chip, of size 5 mm $\times$ 10 mm, was demonstrated. We were able to use the SSM to image and to study the ordering and manipulation of the half flux quantum states in such large arrays[98] (see Figures 3.34a and 3.34b). This work opens the door to fundamental studies, including phase transitions in geometrically frustrated and non-frustrated model systems (2D-Ising models), and possible superconducting device applications[89].

Angle-Resolved Determination of Gap Anisotropy in YBCO

After doing all the tricrystal and tetracrystal experiments described above, there was one important unresolved issue in pairing symmetry. Group theory says that a $d + s$ mixed pair state in YBCO is unavoidable due to its orthorhombic crystal symmetry C_{2v}. However, the size of the gap anisotropy was a highly controversial topic for many years. Early attempts to measure the in-plane

[97] H.-J.H. Smilde, H. Hilgenkamp, G. Gijnders, H. Rogalla, and D.H.A. Blank, *Appl. Phys. Lett.* 80 (2002) 4579

[98] Hans Hilgenkamp, Ariando, Henk-Jan H. Smilde, Dave H. A. Blank, Guus Rujnders, Horst Rogalla, John R. Kirtley, and Chang C. Tsuei. *Nature* 422 (2003) 50

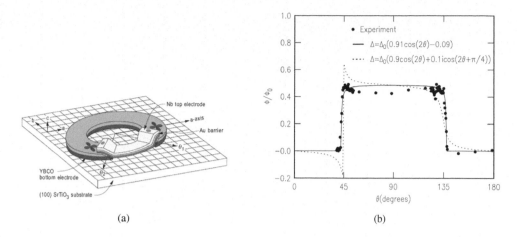

(a)   (b)

FIGURE 3.35: Angle-resolved phase-sensitive determination of the gap in YBCO: (a) Schematic of the 2-junction ring, (b) angular dependence of integrated flux through the rings.

momentum dependence of the order parameter in YBCO and other cuprates using tunneling measurements had limited success. Even the phase-sensitive experiments, for example the interference using a Pb-YBCO corner or single Josephson junction, can only conclude qualitatively that $d > s$, i.e. the s-wave component in the s + d pairing admixture is less than 50%. This is due to the fact that these experiments were based only on a sign change in order parameter between the orthogonal a and b faces of a YBCO single crystal, and therefore do not provide any quantitative information about the d/s ratio. A golden opportunity for solving this problem arose when Hans Hilgenkamp visited IBM in Yorktown after he attended the APS March 2004 meeting in Montreal. It was at a meeting of HH, JK and CT at the Cafeteria of the Watson Research Center, when CT presented his proposal for making a series of 2-junction Nb/YBCO loops to map out quantitatively the in-plane gap anisotropy in YBCO. The proposed experiment demanded the fabrication of many high quality 2-junction rings with the second ramp-edge junction angle varying, from ring to ring, in increments of a few degrees and less. Furthermore, the critical currents of all the ramp-edge junctions should be large enough for a viable study of the magnetic flux state in about seventy 2-junction rings. Even with the prior experience of fabricating large-scale π-arrays[98], this experiment represented a tremendous challenge to HH and his colleagues. Several months later, we were pleased to receive a set of beautiful arrays of the 2-junction YBCO/Nb rings (see Figure 3.35). One of the ramp-edge junction angles relative to the a-axis direction was held fixed at -22.5°. The second junction angle was varied in 5° intervals between $-107.5°$ and 242.5°. There were a total of 72 2-junction rings on the same chip!! HH and his team took great care optimizing the ring design to ensure that the rings would have the required uniformity (among all the rings on the same chip) and large I_cL products of each individual ring for observing well-defined flux quanta.

JK analyzed his SSM results on the first sample and we concluded that the data indicated the nodal transition occurred at an angle deviating from 45°, the angular position for the pure $d_{x^2-y^2}$ gap. While we were enjoying the promising data from the first sample, it was clear that we needed a second sample with the second junction angle to vary in 0.5° intervals in the nodal regions around 45° and 135°. However, our project suffered a "phase-shift" because a team member, Ariando, at Twente graduated with his Ph.D. At that point, a new graduate student, C.J.M. Verwijs joined the team. HH and CT worked together to get him to phase-in quickly and he managed to produce "super chips" that allowed us to fill in the data points in the nodal transition region with angular resolution in the second angle better than 0.5°! See Figure 3.35b and Figure 3.36. The combined results of the two samples were fit by JK with a functional form $\Delta(\theta) = \Delta_0(0.91\cos(2\theta) - 0.09)$ suggesting that

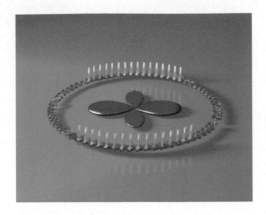

FIGURE 3.36:   A polar plot of SQUID microscope images of 72 rings with different geometries.

the gap along the b-axis of YBCO is at least 20% larger than that along the a-axis direction[99]. The results also showed that any imaginary component to the order parameter, if present, must be quite small.

3.4.6   Concluding Remarks

In summary, building on the results obtained with conventional gap-amplitude measurements (quasiparticle tunneling spectroscopy, NMR, ARPES, and penetration-depth studies), the development of phase-sensitive techniques has finally settled the decade-long d-wave versus s-wave debate in favor of an order parameter with $d_{x^2-y^2}$ symmetry in both hole- and electron-doped cuprate superconductors. In this article, we recall our experience of doing the scanning SQUID microscope phase sensitive experiments, using the half flux quantum effect as the definitive signature for establishing d-wave pairing symmetry. Through a series of tricrystal experiments, we have demonstrated that the d-wave pair state is robust against a wide range temperature and doping variations and time-reversal symmetry breaking.

The journey to d-wave has been very exciting and rewarding for us. Along the way, we were supported and encouraged by many colleagues. Without their contributions, the work we described here would have been impossible. Here, we wish to thank a few of them: Ariando, C. C. Chi, W. J. Gallagher, H. Hilgenkamp, R. P. Huebener, M. B. Ketchen, D. H. Lee, Z. Z. Li, J. Mannhart, K. A. Moler, D. M. Newns, H. Raffy, Z. F. Ren, J.Z. Sun, G. Trafas, C. J. M. Verwijs, M. B. Walker, J. H. Wang and S. K. Yip.

We would like to end our article with Maurice Rice's comment in *AIP Research Highlights* of the 1994: " ⋯. These experiments are an important milestone on the way to a complete microscopic theory of this spectacular phenomenon. Only when a reliable formula for T_c with predictive power has been obtained can one say the journey is ended".

---

[99] J.R. Kirtley, C.C. Tsuei, Ariando, C.J.M. Verwijs, S. Harkema, H. Hilgenkamp, *Nature Physics* 2 (2006) 190

3.5 Electric Field Effect Tuning of Superconductivity

Jean-Marc Triscone and Marc Gabay

3.5.1 Preamble

Serendipitously, the search for superconducting materials in the last 50 years has taken us down a winding road not unlike that followed by some of our nations' space programs. The essence of this quest can indeed be spelled out using a slight variation on the celebrated words spoken by Neil Armstrong, the US astronaut, as he was about to step foot on the moon on July 20th, 1969: "one small step for T_c, one giant leap for superconductivity". To boot, from the time when Kamerlingh Onnes made his momentous discovery up until the mid-1970s, when the superconducting critical temperature reached the 22–23 K mark, most of the drive towards reaching the highest possible T_c focused primarily on anything but oxides. At that time, doped $SrTiO_3$ was a known oxide superconductor, but its measly (<1 K) transition temperature discounted it as a role model for high T_c superconductivity. As we all know, oxides made a come back with a vengeance a decade later, and, since the mid-1990s, copper oxides hold the record with a T_c on the order of 164 K. If we have learned our lessons, now might be the right time, once again, to go against the tide and take another close look at these rather low T_c superconductors, in the hopes of uncovering new clues allowing us, perhaps, to break the 164 K barrier. As we will see in this short article, the electric field effect approach has become a very promising technique to study and possibly discover new superconductors.

3.5.2 Introduction

The nuts and bolts of today's electronics consist mainly of semiconductor materials and field effect transistors. A field effect transistor allows an electric field control of the conductivity in a semiconductor channel defining the 0 and 1 of our binary logic. In such a device, as the electric field is applied across a dielectric such as SiO_2 or these days a "high-k" Hf oxide, the charge carrier density is modulated in the semiconductor channel and so is the electrical resistance, which is measured. On general grounds, it is very tempting to apply such a successful concept to other materials, for instance to superconductors. There, one can imagine tuning superconductivity with an electric field, eventually realizing a superconducting switch. An electric field control of superconductivity would open a wide avenue of possibilities but, as we will see, such control turns out to be difficult and it is only recently that major advances have been achieved, opening the way to exciting new experiments and possibly to novel devices and circuits. The scope of this short article is a far cry from exhaustively reviewing all the important body of work which has been accomplished over the years in this area of research. Rather it aims at discussing and explaining the ideas and challenges behind field effect experiments in superconductors, at briefly overviewing some of the exciting new techniques that have been used and systems that have been investigated and at giving some perspectives on the potential of this approach.

3.5.3 Electric Field Effect in Complex Oxides

Before addressing the issue of electric field effects on superconductivity, let us first discuss interesting field effects in complex materials such as correlated oxide systems[100]. A key difference between such oxides and semiconductors is that, for the former class of systems, electronic correla-

---

[100] C.H. Ahn, J.-M. Triscone, and J. Mannhart, *Nature* 424 (2003) 1015

tions often lead to complex phase diagrams, to competing phases and thus to a particular sensitivity of the ground state to external parameters such as pressure and magnetic field. In many of these materials, ranging from high T_c superconductors to manganites, a key control parameter is the carrier density in the material, which is usually controlled by chemical doping[101]. Figure 3.37 illustrates this point and shows that, upon changing the carrier density in several interesting correlated electron systems, a series of transitions between ground states is observed.

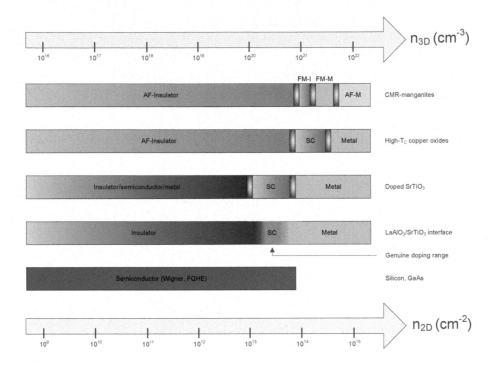

FIGURE 3.37: Adapted from Triscone and Mannhart [100]. Illustration of the zero-temperature behavior of various correlated materials as a function of the volume and sheet charge densities (n_{3D}, n_{2D} respectively). Silicon is shown as a reference. For high-T_c superconductors and for colossal magnetoresistive (CMR) manganites, figures pertain to $YBa_2Cu_3O_7$ and $(La,Sr)MnO_3$ respectively. AF, antiferromagnetic; FM, ferromagnetic; I, insulator; M, metal; SC, superconductor; FQHE, fractional quantum Hall effect; Wigner, Wigner crystal. For the $LaAlO_3/SrTiO_3$ system, the genuine doping level is indicated as well as the boundaries between the different phases inferred form field effect experiments [A. D. Caviglia, S. Gariglio, N. Reyren, D. Jaccard, T. Schneider, M. Gabay, S. Thiel, G. Hammerl, J. Mannhart, J.-M. Triscone, Nature 465, 624 (2008)]. There are some uncertainties on the exact locations of these boundaries. The thick bottom arrow indicates the sheet carrier density (n_{3D} multiplied by an assumed 1 nm film thickness).

Tuning a material across quantum phase transitions and controlling its state with an electric field are obviously exciting perspectives. Unfortunately, one obvious difficulty in carrying out such experiments is readily seen in Figure 3.37. Most of the transitions are occurring for high metallic ($3D$) carrier densities. Since screening of the electric field occurs on the scale of the Thomas-Fermi

---

[101] M. Imada, A. Fujimori and Y. Tokura, *Rev. Mod. Phys.* 70 (1998) 1039

screening length (we will come back to this point), which is very short for a metal (<1 angstrom), the effect will be substantial only for very thin films of the considered materials. Additionally, if one multiplies the $3D$ carrier density by the thickness of one unit cell of the material of interest (say 1 nm), one obtains the sheet carrier densities indicated by the bottom thick arrow in Figure 3.37. Most of the interesting physics occurs at high sheet carrier densities, typically $10^{14} - 10^{15}$ cm^{-2}. Such experiments hence require very thin films (ideally one or a couple of unit cells thick) and very high polarizations, which sets challenging technical constraints when using field effect devices.

In this short article we will not review the history of the field but direct the reader to the few existing review articles[100,102,103]. As we will see, although these experiments are very demanding, much progress has been achieved in recent years.

3.5.4   Electric Field Effect in Superconductors: Physics and Lengthscales

Now, let us go back to the influence of an electric field E on superconductivity. Such experiments aim at modifying and controlling T_c by field tuning the system carrier density. In conventional BCS superconductors, the mean field critical temperature is given by $k_B T_c = 0.18 h\omega_D \exp(-1/N(0)V)$. T_c is thus related to the density of states at the Fermi level ($N(0)$), the Debye frequency (ω_D) and the pairing interaction between the electrons (V). A change in the carrier density will not result in a large change in ω_D nor in the electron-phonon interaction. As the Fermi level is also only marginally changed by an electrostatic field in a metallic sample, one does not expect large field effects in conventional metallic superconductors as confirmed by the early experiments of Glover and Sherill[104]. The situation is quite different in unconventional superconductors such as high T_c cuprates or doped SrTiO$_3$ for which T_c is experimentally found to depend markedly on the doping level (see for instance the review by Jochen Mannhart[105] and references therein). Figure 3.38 displays the well-known "generic" temperature-doping phase diagram of high T_c superconductors with the under-doped, optimally doped and over-doped regimes[106].

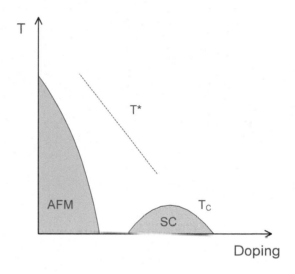

FIGURE   3.38: Generic   temperature-doping phase diagram for high T_c superconductors. The undoped material is an antiferromagnetic insulator. Upon doping, the Neel temperature goes down. In some doping range, superconductivity develops. The dashed line marks the onset of the pseudogap regime at a characteristic temperature T*.

In high T_c compounds, the physics controlling T_c changes across the phase diagram. In the under-

[102] C. H. Ahn, A. Bhattacharya, M. Di Ventra, J. N. Eckstein, C. D. Frisbie, M. E. Gershenson, A. M. Goldman, I. H. Inoue, J. Mannhart, A. J. Millis, A. F. Morpurgo, D. Natelson, J.-M. Triscone, *Rev. Mod. Phys.* 78 (2006) 1185

[103] J. Mannhart, *Superconductor Sci. Tech.* 9 (1996) 49

[104] R. E. Glover and M. D. Sherrill, *Phys. Rev. Lett.* 5 (1960) 248

[105] J. Mannhart, *Mod. Phys. Lett.* B6 (1992) 555

[106] H. Alloul, J. Bobroff, M. Gabay and P.J. Hirschfeld, *Rev. Mod. Phys.* 82 (2009) 45

doped regime, the observed "Uemura scaling" [107,108,109] of the critical temperature with $\frac{1}{\lambda^2}$ (λ is the London penetration depth) is interpreted in some scenarios as a signature of a phase fluctuation driven transition (pairing would occur at higher temperatures)[110,111]. In the over-doped regime, the transition appears to be of the more standard type, when both pairing and phase coherence develop at T_c. The anisotropy of the system is also markedly changing as the doping level is modified. As noticed by Jochen Mannhart and coworkers[112], these unconventional materials, having naturally low carrier densities and critical temperatures linked to the doping level, offer a very unique opportunity for field effect experiments. Of great interest is the under-doped region of these compounds. There, not only is the carrier density low but, as mentioned above, T_c may be controlled by classical and quantum phase fluctuations and can be substantially different from the mean field T_c. Inasmuch as the field effect is able to change the superfluid density without affecting the background disorder, this regime seems ideal to explore the physics, close to the superconducting-nonsuperconducting boundary. Some studies have indeed dealt with the approach to the quantum critical point in the under-doped regime of $NdBa_2Cu_3O_7$ [113].

It is important to realize that several lengthscales are at play in such experiments. One is the Thomas-Fermi screening length $d_{TF} = \sqrt{\frac{2\varepsilon E_F}{3ne^2}}$. Applying an electrostatic field E perpendicularly to the surface or the interface of the material (c-direction) causes an inhomogeneous modulation of the carrier density in that direction which accompanies the screening of E, and presents an extremum near the boundary. For an electron fluid of density n and Fermi energy E_F, d_{TF} depends sensitively on the dielectric constant ε. In some of these oxides—for instance in $SrTiO_3$—ε shows a strong T and field variation and this leads to a 2D to 3D crossover upon increasing E [114].

In the superconducting state, $\lambda = \left(\frac{m}{2\mu_0 n_S e^2}\right)^{\frac{1}{2}}$ describes the magnetic screening of the medium, such that, when the superfluid density n_S vanishes at the transition, $\lambda \to \infty$. The coherence length $\xi = \frac{\hbar v_F}{\pi \Delta}$ (Δ is the gap and v_F the Fermi velocity) controls the characteristic distance over which the superconducting order parameter amplitude varies (for instance it gives an estimate of the size of a vortex core in conventional superconductors). When an electric field is applied along c, ξ plays the role of a cut-off for the variations of the carrier density[115,116]. The mean field T_c then changes by a relative amount $\frac{\delta T_c}{T_c} \sim (\delta n\, d_{TF}/n\xi)^2$ (δn is the field induced variation of (3D) density over the distance d_{TF}). For a film geometry, a Kosterlitz-Thouless transition is ideally expected at $T_{KT} = \frac{d}{2\lambda^2(T_{KT})}(\frac{\Phi_0}{4\pi})^2$ whenever ξ in the c direction is much larger than the film thickness d. d then plays the role of an effective coherence length along c. In a field effect experiment, a change in T_{KT} by an amount $\frac{\delta T_{KT}}{T_{KT}} \sim (\delta n\, d_{TF}/nd)$ is then predicted. This expression shows that a large carrier density change in the top unit cell will lead to the same change in T_c as a small change across the entire thickness of the film.

In high T_c compounds, we have argued that in the under-doped regime, T_c signals the establishment of phase coherence and thus is controlled by the penetration depth λ, while in the over-doped regime, T_c is controlled by the pairing energy and is linked to the superconducting coherence length

[107] Y. Uemura et al, *Phys. Rev. Lett.* 62 (1989) 2317

[108] Y. Uemura et al, *Phys. Rev. Lett.* 66 (1991) 2665

[109] Y. Uemura, *J. Phys. Condens. Matter* 16 (2004) 4515

[110] V. Emery and S. Kivelson, *Phys. Rev. Lett.* 74 (1995) 3253

[111] V. Emery and S. Kivelson, *Nature* 374 (1995) 434

[112] J. Mannhart, J. G. Bednorz, K. A. Müller, D. G. and Schlom, *Zeitschrift für Physik B Condens. Matter* 83 (1991) 307

[113] D. Matthey, S. Gariglio, C. H. Ahn, J.-M. Triscone, *Physica C* 372 (2002) 583

[114] O. Copie, V. Garcia, C. Bodefeld, C. Carretero, M. Bibes, G. Herranz, E. Jacquet, J.-L. Maurice, B. Vinter, S. Fusil, K. Bouzehouane, H. Jaffres and A. Barthelemy, *Phys. Rev. Lett.* 102 (2009) 216804

[115] B. Ya. Shapiro, *Sov. Phys. JETP* 61 (1985) 998

[116] B. Ya. Shapiro, *Physics Letters* A197 (1995) 361

of the system. These considerations should be kept in mind when designing a field effect experiment since one would expect variations of order $\frac{\delta T_{KT}}{T_{KT}}$ in the former case and $\frac{\delta T_c}{T_c}$ in the latter case.

An additional consideration comes into play in the case of layered materials (e.g. in high T_c). If ξ in the c direction is less than or of order of a unit cell, there will only be a weak (Josephson and magnetic) coupling between unit cell layers parallel to the interface. In this limit, one has essentially decoupled parallel channels with their own T_c's. The lengthscale over which the carrier density is modified is then the thickness of the modified T_c channel. If the coherence length is large, the proximity effect will "average" the pairing interaction and a unique T_c will be found over a depth typically of the order of ξ along c. One thus expects quite different behaviors for a field effect experiment in a very anisotropic high T_c film or, for instance, in a doped $SrTiO_3$ film for which the coherence length is large (typically 100 nm).

3.5.5 Thomas Fermi Screening and Interface Quality

As mentioned above, a material will be sensitive to the field effect over a screening length which is usually fairly short in the case of superconductors. This implies that the structural and chemical control of the interface is a key issue as well as the ability to grow ultrathin films. For oxide systems, impressive advances in the growth and control of complex heterostructures have been achieved, see for instance[117], and have been reviewed recently[118]. In short, an atomic layer-by-layer control is possible with abrupt interfaces between different compounds. Intermixing–interdiffusion at the atomic level is however obviously difficult to avoid. Also, the structure of the interface and possible reconstructions have to be considered. For interfaces between a film and a liquid electrolyte (discussed below), film-electrolyte interactions may affect the intrinsic properties of the material. Another issue to bear in mind is the potential impact of the electric field on the electronic properties (an example with the $LaAlO_3$-$SrTiO_3$ (LAO-STO) system will be given below) and on possible ionic motion at interfaces. Although one should always be cautious about possible pitfalls, we will see that several recent experiments demonstrate efficient and impressive field effect control and tuning of superconductivity.

3.5.6 Field Effect Setups in a Nutshell

As illustrated in Figure 3.39, there are experimentally several different possibilities to realize field effect experiments.

Figure 3.39(a) shows a "standard" field effect device geometry. On a substrate, a superconducting film is grown—the channel—and is capped by a gate dielectric (for example SiO_2 or $SrTiO_3$). Source (**S**), drain (**D**), and gate (**G**) electrodes allow a voltage to be applied across the dielectric (between **G** and **S** for instance) and transport properties to be probed (between **S** and **D**; in general a four-point geometry is used). Another option is illustrated in Figure 3.39(b) where the substrate itself is used as a gate dielectric. In this back-gate geometry, the advantage is to have a thick (low leakage) substrate (with a high dielectric constant at low T, if $SrTiO_3$ is used). Large voltages are, however, necessary, the substrate being typically $0.1 - 0.5$ mm thick. In general, with a SiO_2 gate dielectric, polarizations on the order of 2×10^{13} ecm^{-2} can be reached whereas polarizations close to 10^{14} ecm^{-2} can be obtained with $SrTiO_3$ (note that the nonlinear $SrTiO_3$ dielectric constant has to be taken into account). 10^{14} ecm^{-2} corresponds to 16 μC cm^{-2}. In Figure 3.39(c), the gate dielectric is replaced by a ferroelectric. Switching of the ferroelectric polarization (P_r) induces a change of the sheet carrier density in the electrodes of $2P_r$. Since P_r can be of the order of 100 μC cm^{-2} in $BiFeO_3$, for instance, the change in polarization could be, in principle, of order 10^{15} ecm^{-2} (in prac-

[117] A. Ohtomo, D. A. Muller, J. L. Grazul, and H. Y. Hwang, *Nature* 419 (2002) 378

[118] D. G. Schlom, L.-Q. Chen, X. Pan, A. Schmehl, and M. A. Zurbuchen, *J. Amer. Ceramic Soc.* 91 (2008) 2429

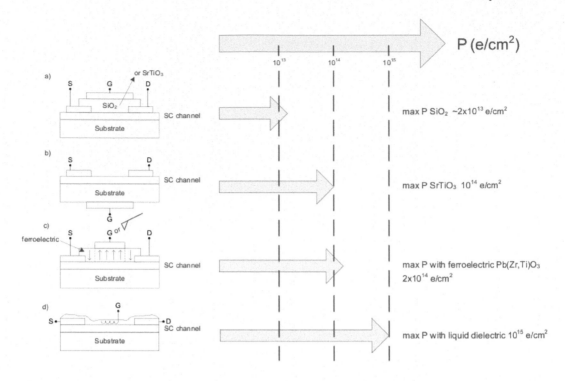

FIGURE 3.39:   Selected field effect devices.

tice, the experimentally measured value is a few 10^{14} ecm^{-2}). A ferroelectric field effect transistor is a nonvolatile device allowing switching between two states. Since the ferroelectric domain structure can be controlled at the nanoscale using an atomic force microscopy approach[119,120], it is possible to replace the metallic **G** gate by a conducting atomic force microscope tip (Figure 3.39(c)). Such an approach can lead to the design of electronic nanofeatures and circuits[121]. One should however note here that continuous tuning of the carrier density is more difficult with ferroelectrics. Finally, Figure 3.39(d) shows the liquid gate dielectric approach that has recently been followed very successfully[122,123]. There an electrolyte is used, such that when a voltage is applied, the solvated ions move towards an electrode according to their polarity. They form a charged ionic layer on the electrode surface, effectively resembling a capacitor with sub-nanometer gap. The sheet charge density that will accumulate in the electrodes can reach the amazing value of a few 10^{15} cm^{-2} [122,123]. Drawbacks of this new approach, however, are, for one, the difficulty to vary the induced carrier density at low T (the liquid freezes typically between 220 and 250 K) since a change in polarization is possible only in the liquid phase and, for two, possible chemical reactions between the liquid and the channel. We have just discussed various possible devices and device geometries allowing field effect experiments to be performed and indicated the magnitude of the polarization changes that can be expected in each case. As one often looks for the largest effects, not only the maximum available

[119] C. H. Ahn, T. Tybell, L. Antognazza, K. Char, R. H. Hammond, M. R. Beasley, O. Fischer, J.-M. and Triscone, *Science* 276 (1997) 1100

[120] P. Paruch, T. Tybell and J.-M. Triscone, *App. Phy. Lett.* 79 (2001) 530

[121] C. H. Ahn, K. M. Rabe, and J.-M. Triscone, *Science* 303 (2004) 488

[122] J. T. Ye, S. Inoue, K. Kobayashi, Y. Kasahara, H. T. Yuan, H. Shimotani and Y. Iwasa, *Nat. Mater.* 9 (2009) 125

[123] K. Ueno, S. Nakamura, H. Shimotani, A. Ohtomo, N. Kimura, T. Nojima, H. Aoki, Y. Iwasa, and M. Kawasaki, *Nat. Mater.* 7 (2008) 855

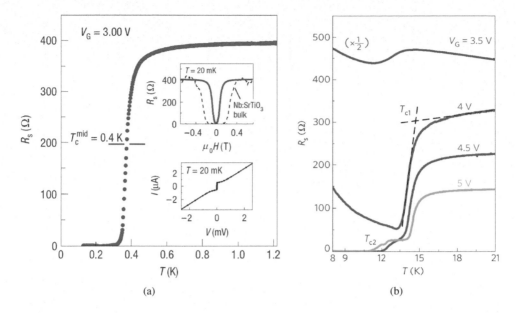

FIGURE 3.40: (a) Superconducting properties of the $SrTiO_3$ channel subject to a gate bias voltage $V_G= 3$ V. Main panel, temperature dependence of R_s, the sheet resistance. The dashed line marks the mid-point of the transition. Insets show : top, the sheet resistance versus magnetic field at and bottom, a current-voltage curve at 20 mK. (b) Gate-voltage dependence of the superconducting transition temperatures in ZrNCl.Temperature dependence of the sheet resistance R_s showing superconducting transitions at different gate voltages, V_G.

polarization matters but also the "parent" channel system. Should one start from a low or from a high doping state? From the insulating state? Owing to the limitation posed by the dielectric breakdown issue, most of the experiments, to date, are based on a (super-)conducting channel, the field effect increasing or decreasing T_c. Ideally, starting from the insulating state seems a better idea since it should allow an off-on switching of superconductivity. This approach is unfortunately dauntingly challenging and it is only recently that the teams of Kawasaki and Iwasa succeeded in switching on superconductivity in an insulating $SrTiO_3$ crystal[123].

3.5.7 What's New?

In this section, we do not attempt to review all the landmarks pertaining to field effect tuning of superconductivity. Many attempts to tune and control T_c have ben reported in the literature. Instead, we will focus on a few recent experiments underscoring the impressive progress that has been recently achieved in this area.

Figure 3.40a shows the resistance versus temperature of a $SrTiO_3$ crystal subjected to a 3 V gate bias voltage[123]. The crystal is initially undoped. Using the electrical double layer transistor concept illustrated in Figure 3.39(d), the insulating surface of the crystal is doped with a sheet carrier density of around $10^{14}cm^{-2}$. Analysis of the data suggests a thickness of the superconducting layer of 5–15 nm and $2D$ superconductivity[123]. In this experiment, the use of the electrolyte insulator played a key role in order to avoid dielectric breakdown. This is the first demonstration of electric field induced superconductivity in an insulator. Soon after this discovery, the same approach was followed with the layered nitride compound ZrNCl, which is insulating in the absence of chemical doping. Figure 3.40b shows the resistance versus temperature for different applied gate voltages[122]. As can be seen, a T_c of about 15 K has been attained in this material. Such an amazing technique, allowing changes in sheet carrier densities of the order of a few 10^{15}cm^{-2}, certainly underscores the potential of the field gating technique which, applied to many other insulating compounds, could hopefully lead to the discovery of new superconductors with possibly higher T_cs.

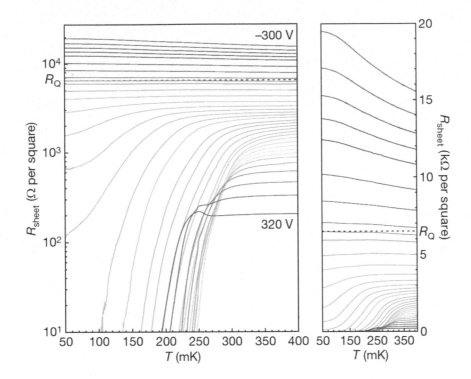

FIGURE 3.41: Field effect modulation of the transport properties of the LaAlO$_3$-SrTiO$_3$ interface. (Left) Measured sheet resistance as a function of temperature for gate voltages varying in 20 V steps between −300 V and 320 V, plotted on a semi-logarithmic scale. (Right) The same data plotted on a linear resistance scale.

Finally, let us turn to recent field effect experiments performed on the 2D electron gas found at the interface between LaAlO$_3$ and SrTiO$_3$ (LAO-STO). In 2004, Othomo and Hwang discovered that the interface between these two band insulators is conducting[124]. This discovery generated a considerable amount of work, part of the effort devoted to an attempt to understand the origin of this conduction and the properties of this electron gas[125,126]. For this particular system, the 2D electron gas is naturally sandwiched between two insulators. The sheet carrier densities are found to be typically in the range $2 - 8 \cdot 10^{13}$ cm^{-2} (a relatively low doping; see Figure 3.37). The thickness

[124] A. Ohtomo and H. Y. Hwang, *Nature* 472 (2004) 423

[125] H. Y. Hwang, *Mater. Res. Soc. Bull.* 31 (2006) 28

[126] J. Mannhart, D. H. A. Blank, H. Y. Hwang, A. J. Millis and J.-M. Triscone, *Mater. Res. Soc. Bull.* 33 (2008) 1027

of the electron gas is on the order of a few nanometers[127,114,128,129,130]. Details of the properties of this system and how they can be influenced by growth parameters can be found in several reviews[125,126]. This interface system seems to be ideal for field effect experiments. With the help of a back gate geometry (Figure 3.39(b)), tuning of the superconducting properties and on-off switching of superconductivity could be achieved, as illustrated in Figure 3.41. The temperature versus gate voltage phase diagram was established and, as shown in Figure 3.42, it resembles that of high T_c superconductors (Figure 3.38), albeit with rescaled doping levels and T_cs.

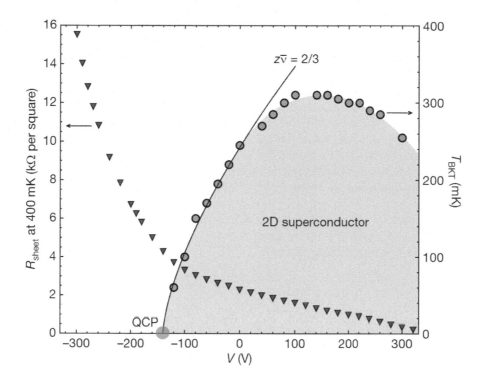

FIGURE 3.42: Electronic phase diagram of the LaAlO$_3$-SrTiO$_3$ interface. Critical temperature T_{BKT} (right axis, blue dots) versus gate voltage, revealing the superconducting region of the phase diagram. The solid line describes the approach to the quantum critical point using the scaling relation $T_{BKT} \propto (V - V_c)^{z\bar{\nu}}$, with $z\bar{\nu} = 2/3$. Normal state resistance, measured at 400 mK (left axis, red triangles) as a function of gate voltage.

An outstanding issue is whether the similarities observed between the high T_c and the LAO-STO phase diagram originate in the same physics; for instance, phase fluctuations might control the establishment of phase coherence in both systems, in the under-doped regime. In the LAO-STO heterostructure, where doping cannot be changed in an obvious way, field effect experiments have

[127] M. Basletic, J.-L. Maurice, C. Carretero, G. Herranz, O. Copie, M. Bibes, E. Jacquet, K. Bouzehouane, S. Fusil and A. Barthelemy, *Nat. Mater.* 7 (2008) 621

[128] A. Dubroka, M. Rössle, K. W. Kim, V. K. Malik, L. Schultz, S. Thiel, C. W. Schneider, J. Mannhart, G. Herranz, O. Copie, M. Bibes, A Barthélémy and C. Bernhard, *Phys. Rev. Lett.* 104 (2010) 156807

[129] N. Reyren, S. Gariglio, A. D. Caviglia, D. Jaccard, T. Schneider and J.-M. Triscone, *Appl. Phys. Lett.* 94 (2009) 112506

[130] M. Sing, G. Berner, K. Goß, A. Müller, A. Ruff, A. Wetscherek, S. Thiel, J. Mannhart, S. A. Pauli, C. W. Schneider, P. R. Willmott, F. Schäfers and R. Claessen, *Phys. Rev. Lett.* 102 (2009) 176805

played a key role in obtaining the phase diagram of this system. Magnetotransport studies have also revealed the presence of a strong spin orbit coupling stemming from the breaking of inversion symmetry. Its strength, which depends on the doping level, develops around the gate voltage where superconductivity appears at "zero" T. The connection between superconductivity and spin-orbit coupling as well as the possible influence of a strong spin splitting of the Fermi surface on the superconducting order parameter are issues that still have to be addressed. These experiments also show that the electric field may not only change the system carrier density but may also deeply modify the electronic structure at the interface.

3.5.8 Conclusions

In this short article, we have discussed field effect experiments in superconductors. We gave a brief outline of various possible ways to perform field effect experiments and the related challenges. We then attempted to highlight the influence of a change in carrier density on the evolution of the system ground state and superconducting critical temperature. The change in T_c will depend on the dimensionality of the system and on the physics controlling the superconducting transition. We then focused on recent advances in the techniques that have been used and discussed recent achievements that show that superconductivity can be efficiently controlled by an electric field. Continuous tuning of T_c, on-off switching of superconductivity and promoting superconductivity in insulating materials are among the recent amazing breakthroughs. The electrical double layer transistor concept—with huge induced carrier density changes—shows that the quest for an electrostatic modulation/control/induction of superconductivity is an exciting area of research which will undoubtedly continue to develop. It is our hope that many new discoveries and novel superconductors can be found using this appealing approach.

3.6   The Grain Boundary Problem of High-T_c Superconductors

J. Mannhart and D. Dimos

In their first years, the high-T_c superconductors created overwhelming excitement plus a few reserved reactions. While the enthusiasm about these materials was abounding, the understanding grew that the cuprates are exceedingly difficult to explore scientifically and that it would be a formidable challenge to fabricate useful high-T_c products, if ever possible[131]. The reasons for these difficulties are rooted in the fundamental physics of the high-T_c superconductors. Their low carrier density, their small coherence lengths, and the correlated character of their electron systems render them sensitive to defects, to give just one example.

The easy-to-fabricate and commonly used polycrystalline samples behaved like irregular networks of Josephson junctions. Already early in 1987, the Birmingham group used them to provide evidence that in Y-Ba-Cu-O samples the holes form Cooper-pairs[132]. Also, first exploratory SQUIDs which operated at 77 K were fabricated by using bulk, granular Y-Ba-Cu-O [133]. The performance of such SQUIDs was not reproducible, though, and even after optimization, their $1/f$ noise at 77 K exceeded the noise of He-cooled Nb-SQUIDs by some six orders of magnitude. All attempts to fabricate Josephson junctions in controlled ways yielded frustrating results.

---

[131] A.M. Wolsky et al., *Scient. Amer.*, p. 45 (February 1989)

[132] C. Gough et al., *Nature* 326 (1987) 855

[133] M.S. Colclough et al., *Nature* 328 (1987) 47

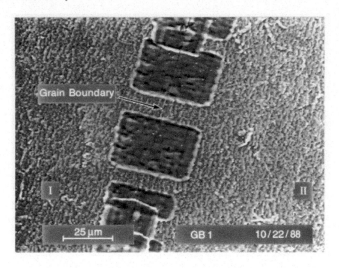

FIGURE 3.43: Photograph of the first bicrystal SQUID[140].

First YBa$_2$Cu$_3$O$_{7-x}$-wires, fabricated in pioneering work at Bell Labs, had critical current densities of $J_c \sim 200$ A/cm^2 at 77 K [134], several orders of magnitude below practical values, so that it was rightfully questioned whether these materials could ever be useful. This notion changed when the first epitaxial YBa$_2$Cu$_3$O$_{7-x}$ films were grown and found to support at 77 K supercurrent densities exceeding 10^5 A/cm^2 [135]. As suggested by the large current densities of the epitaxial films, grain boundaries and the large anisotropy of the cuprates were obvious possible culprits for the small critical currents of the wires. Yet, in low-T_c superconductors, grain boundaries enhance the critical currents by pinning flux lines, and into the 1990s for YBa$_2$Cu$_3$O$_{7-x}$ diverse large angle boundaries were reported to be strongly linked[136]. Also, intragrain weak links such as twin boundaries and oxygen defects were proposed to cause the low current densities[137].

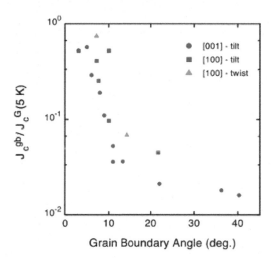

FIGURE 3.44: Original measurements of the grain boundary critical current density at 4.2 K divided by the grain critical current densities as a function of grain boundary angles[141].

In the fall of 1987, a group at the IBM Research Division in Yorktown Heights, NY, around Praveen Chaudhari, Chang Tsuei and the authors, set out to measure directly the properties of single, well-defined grain boundaries by using some of the first epitaxial YBCO films. In the beginning,

[134] S. Jin et al., *Appl. Phys. Lett.* 51 (1987) 203

[135] P. Chaudhari et al., *Phys. Rev. Lett.* 58 (1987) 2684

[136] see, *e.g.*, S.E. Babcock et al., *Nature* 347 (1990) 167

[137] see, *e.g.*, G. Deutscher and K.A. Müller, *Phys. Rev. Lett.* 59 (1987) 1745, D.C. Larbalestier et al., *Physica* C 153-155 (1988) 1580

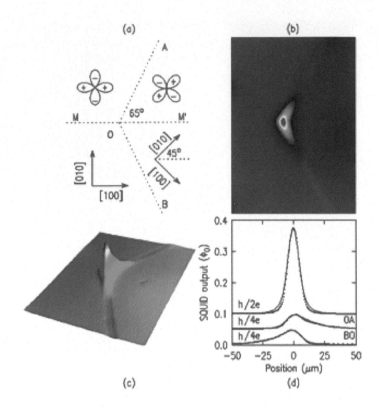

FIGURE 3.45:   Scanning SQUID investigation of a tetracrystalline $Tl_2Ba_2CuO_{6+\delta}$ film: (a) tetracrystal geometry; (b) scanning SQUID microscope image (4.2 K); (c) three-dimensional rendering of the data of (b); (d) cross sections through a bulk Abrikosov vortex and through the half-vortex of (b) and (c), along the directions indicated in (a); dots: experimental data; lines: modeling, assuming the Abrikosov vortex has $h/2e$ flux trapped in it, and the vortex at the tetracrystal point has $h/4e$ flux (figure courtesy of C.C. Tsuei and J.R. Kirtley, *Rev. Mod. Phys.* 72, 969 (2000).

the group included John (C.C.) Chi, David Clarke, Modest Oprysko, Mike Scheuermann, and Tom Shaw. At the IBM Zurich Research Lab, high-T_c superconductivity had been discovered in 1986[138], and in 1987 substantial efforts were underway at all three IBM research labs to explore these new materials. Enthusiasm, excitement, pride, and competitiveness were tangible. We enjoyed great freedom in choosing our field of work, and the company cared that we could focus on science.

Two goals were in our minds. First, see if the grain boundaries behave as weak links. If so, use the standard Josephson junction techniques, e.g tunnel spectroscopy, to find the superconducting gap and identify the pairing interaction. Second, in case the boundaries did behave as weak links, understand why and how, and use that understanding to enhance the critical currents.

Praveen was IBM vice president at that time, but nevertheless managed to participate actively in research. Following his ideas to use large-grained, polycrystalline and, later, bicrystalline $SrTiO_3$ substrates, we implemented the bicrystal technique. After developing a couple of approaches to fabricate these substrates, bicrystal substrates were used to grow $YBa_2Cu_3O_{7-x}$ films, applying the e-beam evaporation technique Praveen and colleagues had just developed[135]. Since we had learned how to handle the precious, but still barely reproducible, ex-situ annealed cuprate films, these were

---

[138] J.G. Bednorz and K.A. Müller, Z. *Phys.* B 64 (1986) 189

then patterned by writing with a laser beam in a process we had devised[139]. For illustration, Figure 3.43 shows a micrograph of the first bicrystal SQUID[140]. We were concerned about questions that are nonexistent today: Does photoresist or acetone affect the J_c? Do chemical problems or second phases at the grain boundaries affect their electronic properties? Do the grain boundaries develop superconducting shorts if mistreated?

Our experiments showed that the grain boundaries indeed are weak links and, while the grains exhibited high critical current densities, high-angle boundaries are excellent Josephson junctions with single boundaries being preferable for devices. The critical current density of the boundaries increases drastically with grain alignment along all directions (Figure 3.44)[141]. This behavior of the boundaries, now understood to arise from several fundamental causes[142], allowed the reproducible fabrication of high quality Josephson junctions and pointed a way to the manufacturing of high-T_c conductors by grain alignment.

Our results were questioned, for example because polycrystals had critical current densities that seemed incompatible with the bicrystal J_c. Polycrystalline superconductors, however, comprise complex grain boundary networks, the critical current density of which may be smaller or higher than that of the grain boundaries. Indeed, the use of grain boundaries with large effective areas was found to provide another principle route to high-T_c wires.[143,144,145]

In a great, joint effort, Zdravko Ivanov, Per-Åke Nilsson, Doug Winkler, José Alarco, and Tord Claeson at Chalmers University together with Evgueni Stepantsov and Alex Tzalenchuk from the Institute of Crystallography of the USSR Academy of Sciences in Moscow, who studied $YBa_2Cu_3O_{7-x}$ junc-

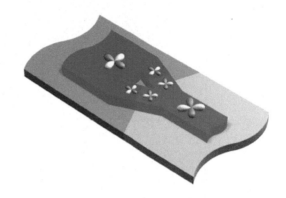

FIGURE 3.46: Drawing of a π-SQUID. In the device, a tetracrystal is used to fabricate a dc-SQUID, one junction of which is biased with a phase-shift of π [151].

tions on Y-ZrO$_2$ bicrystals, found that the drop of the grain boundary J_c with boundary angle is amazingly exponential[146]. At the same time, also at IBM, the progress continued and important discoveries were made. For example, Rudolf Gross, Arunava Gupta and Gad Koren, using the now available in-situ grown, laser ablated films managed to lower substantially the SQUID noise[147]. A team around Masashi Kawasaki and Ettore Sarnelli discovered that the grain boundary properties they investigated are universal for the high-T_c cuprates, for they observed them in the bismuth-, neodymium-, and thallium-based superconductors[148].

This surprising behavior of the high-T_c grain boundaries was put to use to an extent that nobody could have imagined. Here, we describe only the most important ones.

---

[139] J. Mannhart et al., *Appl. Phys. Lett.* 52 (1988) 1271

[140] J.M. Hagerhorst et al., *MRS Proceedings* 129 (1989) 347

[141] D. Dimos et al., *Phys. Rev.* B 41 (1990) 4038

[142] S. Graser et al., *Nat. Phys.* 6 (2010) 609

[143] J. Mannhart and C.C. Tsuei, Z. *Phys.* B 77 (1989) 53

[144] J. Mannhart in *"Earlier and Recent Aspects of Superconductivity"*, edited by J.G. Bednorz and K.A. Müller, Springer Series in Solid State Physics 90 (1990) 208

[145] L.N. Bulaevskii et al., *Phys. Rev.* B 45 (1992) 2545

[146] Z.G. Ivanov et al., *Appl. Phys. Lett.* 59 (1991) 3030

[147] R. Gross et al., *Appl. Phys. Lett.* 57 (1990) 727

[148] M. Kawasaki et al., *Appl. Phys. Lett.* 62 (1993) 417

FIGURE 3.47:   Photograph of a Cessna carrying a cryostat with a bicrystal-based gradiometer (Tristan Technologies, model 701G) in its tail boom. Such systems are used for detection of mineral resources and unexploded ordnance (photo courtesy of Tristan Technologies).

As bicrystalline substrates became commercially available, the bicrystal technology was quickly used to reliably fabricate high-quality Josephson junctions. To avoid restrictions imposed by the existence of only one grain boundary per bicrystal, the biepitaxial technique was successfully developed by Kookrin Char and coworkers at Conductus. With biepitaxy, well-defined grain boundaries can be generated at any location on the chip[149].

Grain boundary Josephson junctions turned out to be highly useful to elucidate the fundamental properties of the high-T_c superconductors, for example by tunnel spectroscopy. Chang Tsuei and John Kirtley performed elegant experiments[150] by using tricrystals, and later even tetracrystals, to explore the order parameter symmetry in the high-T_c cuprates. The tricrystals were designed such that superconducting rings patterned around the tricrystal points generate half a magnetic flux quantum $h/4e$, if and only if the superconducting order parameter has a $d_{x^2-y^2}$ symmetry (see the articles of D. van Harlingen in Section 3.3 and of J.R. Kirtley and C.C. Tsuei in Section 3.4 of this chapter). Imaging the magnetic flux density generated by the superconducting rings or even by plain films covering tricrystal or tetracrystal points with scanning SQUID microscopy, the group indeed found the half flux quanta in all p-doped and n-doped compounds they investigated. Figure 3.45 shows the magnetic flux generated at the tetracrystal point of a $Tl_2Ba_2CuO_{6+\delta}$ film; it clearly has a value of $h/4e$.

The π-phase shifts across the grain boundaries induced by the order parameter symmetry can be used for Josephson electronic circuits that combine standard Josephson junctions and junctions biased with an additional phase change of π [151]. Figure 3.46 shows in comparison a tetracrystalline, standard dc-SQUID and one with such a π-phase shift. Replacing the grain boundaries

[149] K. Char et al., *Appl. Phys. Lett.* 59 (1991) 733

[150] C.C. Tsuei et al., *Phys. Rev. Lett.* 73 (1994) 593

[151] R. Schulz et al., *Appl. Phys. Lett.* 76 (2000) 912

$YBa_2Cu_3O_{7-x}$-Nb junctions, chips containing thousands of π-junctions have been built by now[152].

Many SQUIDs and other Josephson devices based on grain-boundary Josephson junctions were developed and partly brought to market. Figure 3.47 shows an example. The dewar at the end of the tail boom of the Cessna carries bicrystal Josephson junctions forming a HTS SQUID gradiometer to locate mineral deposits or unexploded ordnance by mapping magnetic anomalies.

While it became clear that high-T_c Josephson junctions could be easily fabricated by using large-angle grain boundaries with well-defined angles, three rather different approaches were developed to fabricate high-T_c wires with big critical currents.

The straightforward approach is to avoid grain boundaries altogether by growing single-crystal-type bulk materials using melt texturing pioneered by Sungho Jin and his collaborators at Bell Labs[153] and by Masato Murakami and colleagues at IS-TEC[154]. These melt-textured samples can trap enormous magnetic flux densities, as used, for example, in magnetic bearings for flywheels or as "permanent" magnets in magnetic separators.

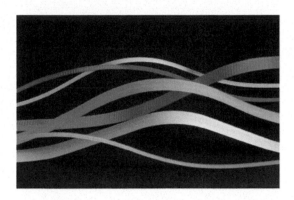

FIGURE 3.48: Photograph of coated conductors fabricated by the IBAD-process using Hastelloy tapes. The tapes with widths up to 12 mm carry more than 200 A/cm at 77 K. Their redish or white color is given by the electric insulation (photo courtesy of SuperPower, Inc).

The second approach is based on a decisive breakthrough achieved in 1989 by Klaus Heine and his colleagues at the Vakuumschmelze in Hanau (Germany), who partially melted BSCCO (2212) powder in a silver tube[155]. The polycrystalline wires they obtained had critical current densities of 1200 A/cm² at 77 K, and at 4.2 K and 26 T more than 10^4 A/cm². For the first time high-T_c wires became a practical technology. This "powder-in-tube" technology was further developed by several groups and companies to fabricate high-T_c wires, the so-called first generation high-T_c conductors (see article by M. Rupich and E. Hellstrom). In these wires, the BSCCO grains are partially aligned in the vicinity to the rolled silver. The critical current is enhanced by the brick-wall type stacking of the platelet-shaped grains, which provides an enormous effective grain boundary area.

The most astounding progress in the fabrication of cost-effective high-T_c wires was achieved by tackling the problem head-on: fabricating wires in which all these submillimeter-sized grains are aligned in all directions with a precision of very few degrees. Essentially, such a conductor exhibits behavior approaching that of a single crystal that is a mile long. In this technology, the superconductors are epitaxially grown as films on substrates coated with buffer layers. The alignment is done by orienting the grains of the buffer layers during growth, for example by ion bombardment. Alternatively, the buffer layers and superconductors can be deposited epitaxially on substrates with grains aligned by a rolling process. These two processes are named IBAD (ion beam assisted deposition) and RABiTS (rolling-assisted biaxially-textured substrates); another variant, the alignment by growth on tilted substrates is named ISD (inclined substrate deposition).

The IBAD technique originated from a group around James Harper and Jerome Cuomo at

[152] H. Hilgenkamp et al., *Nature* 422 (2003) 50

[153] S. Jin et al., *Appl. Phys. Lett.* 54 (1989) 584

[154] M. Tomita and M. Murakami, *Nature* 421 (2003) 519

[155] K. Heine et al., *Appl. Phys. Lett.* 55 (1989) 2441

the IBM T.J. Watson Research Center who used it to grow niobium films[156]. In seminal work, the potential of IBAD to grow aligned high-T_c superconductors was recognized by the group around Yasuhiro Iijima at Fujikura Ltd., who, by aligning zirconia layers, oriented the grains of $YBa_2Cu_3O_{7-x}$ films grown epitaxially on the zirconia buffer[157]. Recognizing the possibility to align very thin (~10 nm) MgO layers, the Stanford group greatly improved the speed of the process[158], which in particular at Los Alamos was further optimized (see also article by A. Malozemoff and Y. Yamada). More or less simultaneously, the RABiTS technique was pioneered at Oak Ridge National Labs, initiated by a group around Amit Goyal and David Norton [159]. These techniques are presented in detail in the article by A. Malozemoff and Y. Yamada. Several companies such as American Superconductor and SuperPower joined in, and coated conductors with engineering cross sections of 5 mm × 0.2 mm that carry more than 100 A at 77 K are now on the market (see Figure 3.48). These supercurrents exceed 100 times the electrical current of the equivalent copper wire. Present and future applications of these so-called second generation of high-T_c wires include cables, generators, motors, and fault current limiters (Figure 3.49) (see also article by B. Hassenzahl and O. Tsukamoto).

FIGURE 3.49: Photograph of a 3-phase, 114 MVA cable that uses coated conductors (photo courtesy of SuperPower, Inc.).

The adventure of exploring and optimizing grain-boundary interfaces in the cuprates led to substantial progress in the scientific understanding of the high-T_c superconductors as well as in their practical use in devices, wires, and cables. These technologies are spreading to be applied to non-superconducting materials to understand, use, and optimize grain boundaries, or to cost effectively grow single-crystal-like films over large areas, for example for cheap solar-cell fabrication. It is amazing how closely, and yet unforeseeably, important applications are interlinked with apparently arcane scientific questions.

[156] L. See et al., *Appl. Phys. Lett.* 47 (1985) 932

[157] Y. Iijima et al., *Appl. Phys. Lett.* 60 (1992) 769

[158] C.P. Wang et al., *Appl. Phys. Lett.* 71 (1997) 2955

[159] A. Goyal et al., *Appl. Phys. Lett.* 69 (1996) 1795

3.7 Overview of the Experimental History, Status and Prospects of HTS

M. R. Beasley

3.7.1 The Allure of High Temperature Superconductivity

Superconductivity is a spectacular phenomenon. It is even more spectacular when one realizes that it is macroscopic quantum phenomenon — quantum mechanics on a real-world-sized scale. Add to this the prospects for the applications of superconductivity, and one can understand the allure of high temperature superconductivity. Can this wonder of cryogenic temperatures exist at much higher temperatures? Astrophysicists tell us that superconductivity likely exists in neutron stars, so any theoretical limit in T_c in nature is very high indeed. But this does not count. The allure is for superconductivity at the temperatures of the earthly world in which we live — so-called room temperature superconductivity. Scientifically, the question is whether superconductivity can exist in conventional matter (natural or synthetic) at room temperature.

And there is another issue that has only recently entered the discussion. Will the properties of such a very high temperature superconductor be useful? This might seem self evident, but, in fact, as we shall discuss later, it is an interesting question. Also, to be fair, from an engineering point of view, one does not need room temperature operation to make a very large difference in the prospect of widespread use of superconductivity. Still, scientists, being human, like alluring goals.

So, how does one find higher temperature superconductors? How has it happened in the past, and how might it be done in the future? And what have we learned along the way? To address these questions, it is helpful to sketch the history of the discovery of high temperature superconductors.

3.7.2 The Broad Sweep of History

No discussion of high temperature superconductivity is complete without the canonical plot of the evolution of T_c upward over the course of time. The one here is from a recent US Department of Energy report on *Basic Research Needs for Superconductivity*, published in 2006 [160]. It is easy to be inspired by the inexorable rise in T_c over the years. In particular, there has been much activity in the past couple of decades. Indeed, the new Fe pnictide superconductors do not even appear on this plot. They were discovered in 2008. They now reach $T_c's$ as high as 55 K.

Actually, there has always been high temperature superconductivity. It is only a matter of what material was the record setter at any given point in time. Examination of this plot shows that in the era of elemental superconductors, Nb was the champion. In the present era, it is the cuprate superconductors. But the plot teaches us much more. It shows that there are clear classes of superconductors in which researchers sought higher temperature superconductivity. There were the elemental superconductors, transition metal alloys (not shown in the figure), transition metal compounds (usually based on Nb) and finally oxides along with other contemporary materials such as MgB_2 and doped C_{60}. Of course the important heroes in this story are those who found the first member of a given class. The pantheon of these is shown in Table 3.1.

The desire for higher temperature superconductors is, however, not the only motivation for seeking new superconductors. Establishing the range of existence of superconductivity among materials types is obviously an important scientific question, and this purely scientific motivation drives the search for new superconductors to this day. Some of the most interesting entries in this category

---

[160] http://www.sc.doe.gov/bes/reports/files/SC_rpt.pdf

TABLE 3.1:   The Pantheon of Heroes

K. Onnes	Hg
W. Meissner	NbC/NbB
J. Hulm and G. Hardy	V_3Si
A. Sleight	BaPbBiO
J. Bednorz and A. Mueller	LaBaCuO
R. Haddon and A. Hebard	K_3C_{60}
J. Akimitsu	MgB_2
H. Hosono	LaOFP

appear in the lower part of Figure 3.50. They are discussed elsewhere in this volume and surely deserve a pantheon of heroes of their own. They give us new superconductors where new physics is found. But in this Section, our focus is on high T_c superconductivity for its own sake. Still, some of these emerging materials have strikingly steep trajectories in Figure 3.50, and they may yet surprise us.

Of course the members of our high T_c pantheon are not the only ones who make singular contributions. One has to admire those researchers who extend these seminal discoveries to their maximal potential. Here we have to acknowledge Gavaller for finding Nb_3Ge, Wu and Chu for finding YBaCuO and Chu for finding the current reigning champion HgBaCaCuO with a transition temperature of 164 K under pressure.

But without question, the most dramatic feature in Figure 3.50 is the sharp rise in T_c that followed the discovery of high T_c superconductivity in LaBaCuO by Bednorz and Mueller. This was without

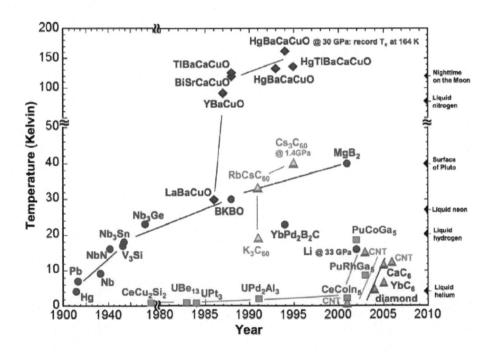

FIGURE 3.50:   Evolution of high temeprature superconductivity over time. (Figure courtesy of the US DoE.)

question a landmark event. All of our lives changed. The focus on where to look for higher temperature superconductors radically shifted. The theoretical thinking radically shifted as well. Before Bednorz and Mueller, for all high T_c materials known at the time (with the exception of BaPbBiO), the pairing mechanism had been convincingly shown to be the electron-phonon interaction, and the pairing symmetry of the s-wave singlet variety. We now have s-, p-, d-wave pairing symmetries, not to mention even more exotic ones, and except for MgB_2 (an electron-phonon interaction superconductor) there is no consensus on the pairing mechanisms. Historically then, there are clearly two epochs: *From Onnes to Bednorz and Mueller*, and *After Bednorz and Mueller*.

3.7.3 From Onnes to Bednorz and Mueller

It is now the stuff of legend that Kamerlingh Onnes discovered superconductivity in a slender wire of Hg in 1911. After Onnes' discovery, it was natural to explore the existence of superconductivity in the elements, with Nb emerging a couple of decades later as the champion in terms of transition temperature. But this search was not, to the knowledge of this author, motivated by seeking higher T_cs per se. It was just good systematic science. Not until the 1950s and 1960s did a search for new superconductors motivated specifically at achieving higher transition temperatures really emerge. Why was this so? Two historical threads intertwined.

First, theory provided a way to think about the issue. The newly available BCS theory of superconductivity and the Eliashberg theory of the electron-phonon mechanism, taken together, provided a complete theory of superconductivity in the sense that they explained the phenomenon of superconductivity (in terms of Cooper pairs that locked their individual quantum phases to form a coherent macroscopic quantum state) and provided an understanding of how the attractive electron-phonon interaction could work to form pairs in the presence of the very strong coulomb repulsion between electrons. The key idea in the latter case was the appreciation that the electron-phonon interaction is retarded in time. One electron polarizes the lattice, leaving an attractive potential for a second electron that lasts long enough for the second electron to come by after the first has left, avoiding thereby their mutual coulomb repulsion. BCS simply postulated an attractive interaction of unspecified origin, although clearly they had the electron-phonon interaction in mind. In the modern understanding, we now know that the BCS theory is very general and applies for any interaction in the weak coupling limit. In its simplest form, this work culminated in the famous equation

$$T_c = \omega_c e^{-1/(\lambda - \mu^*)} \tag{3.9}$$

where

$$\lambda = N(0)V \tag{3.10}$$

is the dimensionaless attractive interaction parameter in BCS theory, $N(0)$ is the density of states at the Fermi level and V is the attractive interaction. The parameter μ^* is the Coulomb repulsion renomalized by retardation. Finally, ω_c sets the energy scale around the Fermi energy over which the attractive interaction is effective. In the Eliashberg theory λ can be calculated as an integral over the electron-phonon interaction spectral function (famously denoted as $\alpha^2 F(\omega)$) and ω_c is a measure of the extent of the interaction spectral function as a function of frequency.

Thus for the first time one could consider how to optimize the electron-phonon interaction in fairly concrete terms. Of course, doing accurate calculations of the relevant materials parameters for real materials is quite another matter. Nonetheless, there was specific guidance: increase ω_c and λ, assuming of course that they are independent, which they are not. Still having some basis for optimization was useful. Moreover, it was possible to think in terms of other mechanisms of superconductivity that might provide a dynamic polarization that could provide an attractive interaction between the electrons.

The second thread that contributed to the focus on finding higher T_c superconductors involved real materials. As the author understands the story, in the early 1950s before BCS theory, Bernd

Matthias and John Hulm, both new immigrants to the US, were young researchers at the University of Chicago when Enrico Fermi asked them why don't you go look for new superconductors. They did. And later, after Matthias went to Bell Labs, Ted Geballe joined the group, and this triumvirate dominated the scene in the search for new and higher temperature superconductors in that era. Eschewing theory in colorful rhetoric—and in Matthias' case, wonderful theatre—as unable to predict the T_c of real materials (while of course thinking about its content), they led the systematic search for superconductivity in the d- and f-band transition metals, alloys and compounds. An informal conference on *D- and F-band Superconductivity* emerged during this era. Over the years, it has evolved into an international conference on *Materials and Mechanisms of Superconductivity* (M^2S) that is arguably the most important conference on superconductivity today.

On the high T_c front, in the culmination of this era, Gavaller (a protogee of Hulm) teased out a $T_c = 23$ K from Nb_3Ge, a record that stood for some time as Figure 3.50 shows. This era also provided two superconductors, a Nb-Ti alloy and Nb_3Sn, that are the present mainstays of the large-scale applications of superconductivity. In addition, during this era, Mathias articulated his famous rules for the existence of superconductivity based on the accumulated experience at the time. They are shown in Table 3.2 below.

TABLE 3.2: Matthias' Rules for
the Existence of Superconductivity

---

1. Seek high symmetry
2. Seek peaks in the density of states
3. Stay away from oxygen
4. Stay away from magnetism
5. Stay away from insulators

These rules were a reflection of hard work and solid empirical thinking, and they were useful and important in their day. They embody the wisdom of an era. The rule about peaks in the density of states had theoretical support from the BCS relation $\lambda = N(0)V$. But viewed from a contemporary perspective, these rules clearly did not anticipate Bednorz and Mueller. The cuprate superconductors have low symmetry. They are oxides. And they are derived from a parent material that is an antiferromagnetic insulator. Clearly, suggesting where high temperature superconductors will be found is a trecherous business. We will return to the rules emerging in the modern era in the concluding section of this chapter.

But Matthias, Hulm and Geballe did something perhaps even more important. They laid the foundations (at least in the US) for what we now call materials physics. The high-energy physics community chastised them as doing "schmutz" physics. But they persevered, and now the search for new materials with novel properties is central to condensed matter physics. It is firmly appreciated that new phenomena are found in new materials. New materials are the leading edge. A respect for materials was historically stronger in Europe, Japan and China, and it is probably no accident that so many of the major new classes of high temperature superconductors were discovered and/or advanced in those countries. The unique position in science of the US after World War II is now clearly over.

In any event, due to these developments in the 1950s, the stage was set for a search for higher temperature superconductors. There was a theory that provided guidance and a very talented materials community eager to try. Many ideas were put forth in addition to the electron-phonon mechanism. Perhaps the most influential was to use electronic polarization to provide the attractive pairing interaction as opposed to the lattice polarization. Examples are excitonic and plasmonic superconductivity. In these so-called electronic mechanisms, the energy scale of the interactions (essentially ω_c in the BCS equation for T_c) is very large, suggesting high Tc might follow. On the other hand, such a high energy scale makes retardation less effective in reducing the direct coulomb interaction, at least for s-wave superconductors.

The most important proponents of this point of view were Little and Ginzburg. Little proposed the use of conducting organic molecules with highly polarizable side chains. Ginzburg imagined the use of quasi-two-dimensional materials with alternating metallic and polarizable layers. While the mechanism has never been convincingly demonstrated in a real material, there can be no doubt that these ideas stimulated research in organic materials and a variety of layered compounds, which in turn stimulated interest in reduced dimensional superconductors, quite independent of the magnitude of T_c. Other ideas that got play were bipolaron superconductivity in the presence of a strong electron-phonon interaction. These ideas stimulated serious theoretical work. Interested readers might want to examine the books by Ginzburg[161] and by Salje[162].

At the same time, however, the powerful technique of superconducting tunneling spectroscopy, which yields the critically important spectral weight of the interaction (i.e., $\alpha^2 F(\omega)$), was being used to demonstrate, one by one, that the mechanism of superconductivity in all the high T_c superconductors of the time was the electron-phonon interaction, with one possible exception. The exception was the bismuthate class of superconductors discovered by Sleight and pushed to a higher T_c by Cava. These superconductors were different and did not naturally fall into any of the convenient categories of the time. They were the first high T_c superconducting oxides, in manifest violation of Matthias' Rules. The parent compound $BaBiO_3$ is a charge density-wave insulator due (in chemical language) to charge disproportionation of the Bi ($B^{4+} \rightarrow B^{3+} + B^{5+}$), in which in some sense a preformed pair exists on every other Bi site. In physics nomenclature this is a negative-U material in the Andersonian sense, and to this day the relative roles of charge disproportionation and the associated screening by the lattice in producing the negative U (and the superconductivity) of this material are matters of debate. Put another way, it is not clear whether the mechanism operating here is an electronic interaction, the electron-phonon interaction, or both.

Unfortunately, this exception not withstanding, the ubiquity of the electron-phonon mechanism in the superconductors of the time and the lack of progress in finding new higher temperature superconductors took the wind out of the sails of the various searches for higher temperature superconductors. Funding agencies lost interest. Thus wound down an exciting era in the search for high temperature superconductivity.

One post script to this era is in order. Later, in the early 1980s, as part of the study of disorder in metals (weak localization), it became clear that the coulomb interaction increases in the presence of disorder. The physics is simple. Diffusing electrons do not screen as well as ballistic electrons. The consequences are a reduction of the density of states at the Fermi level and an increase in μ^*, and hence a reduction in T_c. Disorder was not included in the Eliashberg theory, and it makes an important difference. As established experimentally, the reduction in T_c due to disorder can be substantial. For example, in the case of the A15 compounds (V_3Si, Nb_3Sn etc), which are clean superconductors in their parent state, T_c universally decreases significantly as the residual resistivity is increased (e.g., by radiation damage). Carrying this logic to the transition metal alloys, their "intrinsic" (disorder free) T_cs are certainly much higher than what is observed. This might be academic, but it surely demonstrates that disorder is detrimental to high T_c, at least for BCS-Eliashberg superconductors.

[161] V.L. Ginzburg and D.A. Kirzhnits (Editors), *High-Temperature Superconductivity*, Consultants Bureau, New York and London, 1982

[162] E. K. H. Salje, A. S. Alexandrov, and W. Y. Liang (Editors), *Polarons and Bipolarons in High-T_c Superconductors and Related Materials*, Cambridge University Press, 1955.

Indeed, strong disorder depresses T_c to zero, leading to a superconductor/insulator transition.

3.7.4   After Bednorz and Mueller

As already noted, after Bednorz and Mueller, everything changed. Their discovery not only ushered in the era of very high temperature cuprate superconductivity, it also focused attention on highly correlated materials generally. Then, out of nowhere came doped C_{60}, MgB_2 and the Fe pnictides. So much has been discovered, debated (sometimes heatedly) and speculated about these materials that it is hard to know what to say. With the exception of MgB_2, there is no definitve understanding of the mechanism of the superconductivity in any of these material classes. On the other hand, each has lessons relevant to the search for higher temperature superconductors.

MgB_2 has the clearest lessons. The mechanism of its superconductivity is the electron-phonon interaction. In fact, an excellent theoretical account of this material using the BCS/Eliashberg theoretical framework was established within a year of its discovery. This success, in and of itself, is an important lesson. The potential of stiff covalent bonds for superconductivity is another. It would be foolish in light of these lessons to discount the potential of the electron-phonon interaction for higher temperature superconductors, and possibly the use of theory to provide some explicit guidance. In addition, perhaps the most important lesson is that checking for superconductivity even in known classes of materials is important. It is literally true that MgB_2 was an "on the shelf" high temperature superconductor.

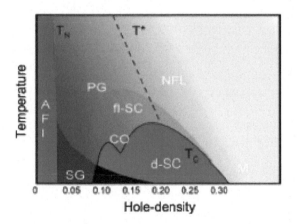

FIGURE 3.51:   Schematic phase diagram of the cuprate superconductors. The various regions at low temperature are antiferromagnetic insulator (AFI), spin glass (SG), charge order (CO), d-wave superconductor (d-SC) and normal metal Fermi liquid (M), and at higher temperatures a pseudogap phase(PG), fluctuation superconductivity (fl-SC) and non-Fermi liquid (NFL). (Figure courtesy of the US DoE.)

The cuprate supercondutors are surely the archetype of a highly correlated superconductor. The undoped parent material is an antiferromagnetic Mott (more precisely, a charge transfer) insulator (AFI). Doping leads to the high temperature superconducting phase, but that is not the whole story by any means. For low doping the normal state properties of the cuprates are not those of a classic Fermi liquid. Only at very high doping is Fermi liquid behavor thought to arise. Clearly, there must be crossover from a Fermi liquid to an antiferromagnetic insulator as doping is reduced. The optimal doping for superconductivity arises in the crossover regime. These various regions are shown in the schematic phase diagram of the cuprate superconductors in Figure 3.51.

How to think about what is going on in the cuprates then becomes a matter of where one chooses to start. One can start in the Fermi liquid regime at high doping and ask how a Fermi liquid transforms into an AF insulator in the presence of strong correlation as the doping is reduced. Experimentally, this transformation is rich in new physics (e.g., arcs of Fermi surface and charge ordering in the form of stripes). In any event, in this view, superconductivity arises due to exchange of AF spin fluctuations, which are the precursor of the AFI end point. At least at high doping, one can think in terms of the AF equivalent of Eliashberg interaction function $\alpha^2 F(\omega)$, but where the attraction is in the spin channel as opposed to the charge channel as in the electron-phonon interaction or the negative-U model. The direct coulomb repulsion is avoided by virtue of the d-wave nature of the superconductivity (i.e., a node at the origin of the internal Cooper pair wave function) if not retardation.

Alternatively, one can start from the Mott insulator side and think in terms of Cooper pairs as latent in the AF insulator that are freed up by doping in a way that may require a fundamentally new theory. These are very deep theoretical questions, and their ultimate resolution in the form of *the* (not *a*) theory of superconductivity in the cuprates will surely be instructive in how to seek higher temperature superconductivity in this class of materials or its relatives. In the case of the Fe pnictides, the parent compound is an antiferromagnet metal (AFM). The degree of correlation is clearly less, and one starts from an Fermi liquid picture, as in the over-doped cuprates. On the other hand, the multiple orbitals in the Fe pnictides could conveivably play some role. Time will tell. All this said, even without a complete theory of either of these materials, the basic lesson is clear: proximity to a competing AF phase is favorable for high T_c superconductivity.

But there are other lessons as well. All these materials (including MgB_2) have layered structures, which to varying degrees makes them quasi-two-dimensional. Reduced dimensionality cuts in two directions, however. For example, in the case of the cuprates, some would argue that reduced dimensionality is necessary in order to enable the large quantum fluctuations necessary to weaken the AFI order and permit superconductivity to emerge upon doping. Others would say that in reduced dimensions, the cost of enhanced phase fluctuations (discussed below) may become too large and prohibit very high temperature superconductivity. Also, all these superconductors are charge transfer materials in the sense that the carriers that become superconducting are donated into the superconducting regions from other parts of the unit cell (e.g., the so-called blocking layers in the cuprates). Charge transfer of this sort avoids disorder in the regions of the superconductivity, much as modulation doping avoids disorder in the conducting layers of semiconductor hetrostructures. In any event, avoiding disorder is good for high T_c superconductivity.

Doped C_{60} would seem to be a similar story to MgB_2, but the situation became more complicated with the discovery that Cs_3C_{60} shows signs of being highly correlated. The phase diagram of Cs_3C_{60} as a function of doping is very similar to that of the cuprates, in particular the existence of a Mott insulator at low doping. Thus, the role of strong correlation in this material remains uncertain. Nonetheless, the lessons of doped C_{60} appear to be in accord with those of MgB_2 and the highly correlated cuprates.

3.7.5   Prospects for Higher Temperature Superconductors

One does not need to be a particularly astute reader of the history of the discovery of higher temperature superconductors to see that serendipity plays a major role. It is simply a fact that, for example, the superconductivity of MgB_2 and the Fe pnictides were discovered in materials being studied for other reasons. And, in what must be one of the supreme ironies in the history of the discovery of high temperature superconductivity, in their conscious search for high temperature superconductivity in the cuprates, Bednorz and Mueller were, as far as this author understands the story, motivated by the possibility of a strong electron-phonon interaction, not strong correlation per se, in these materials. Of course, this in no way detracts from the profundity of their discovery

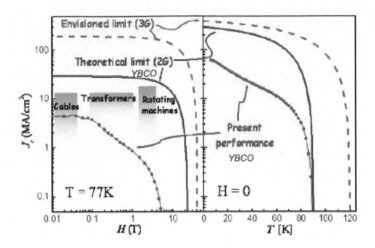

FIGURE 3.52:   Limitations of YBCO. (Figure courtesy of the US DoE.)

or their courage to look in a new place. But it certainly shows that researchers should examine new and old classes of materials as widely as possible for the existence of superconductivity. We should also certainly try to understand those superconductors that just will not fit comfortably into the existing conceptual framework. Some would argue that this should include even those reports of trace superconductivity at very high temperatures that are not clearly nonsense.

But these are collective strategies that apply to the community as a whole. Each individual has to have a more specific approach. The very best seekers of new and/or higher T_c superconductors use their hard-earned intuition to motivate looking in particular classes of materials (e.g., Bednorz and Mueller). This wisdom is deep and difficult to state concisely. Interested readers should study the thoughts of these researchers. One useful place to start might be some of the recorded lectures at the program on *The Physics of Higher Temperature Superconductors* held at the Kavli Institute of Theoretical Physics at the University of California at Santa Barbara in the summer of 2009. These can be found on line [163].

At the same time, interest in searching for better and higher temperature superconductors is growing presently. Why is this? First and foremost, as concluded in the DoE report mentioned at the beginning of this Section[160], there is a need for higher temperature superconductors beyond the cuprates, if one is to have electric power applications of superconductivity operating above liquid nitrogen temperature. The situation is summarized in Figure 3.52 taken from that report. The point is that the performance of YBCO, which has the best in-field properties of any cuprate superconductor, is approaching its fundamental limit in critical current density J_c. This fundamental limit of J_c is essentially that current density at which the kinetic energy density of the current exceeds the condensation energy of the superconductor. Thus the limitation is fundamental and unavoidable. And, more importantly, these limiting values of J_c are marginal at best to reach the regimes of performance needed for the various applications of interest when operating at liquid nitrogen temperature, much less at higher temperatures. The report concludes that a fundamentally new "third generation" (3G) superconductor is needed. No known supercondutor can do the job.

This striking conclusion raises a very important basic question. Why is it that these wonderful high T_c superconductors do not have higher fundamental limits of J_c? Simple scaling from the behavior of low T_c superconductors indicates that they should. The answer lies in the low carrier

---

[163] http://online.kitp.ucsb.edu/online/highertc09/

densities in the cuprate superconductors. The extrapolation from low T_c supercondutors fails because it assumes similarly high pair densities. The cuprate superconductors have low pair densities for several reasons but mainly because the normal carrier density is low to start with. They are, after all, doped Mott insulators. These observations raise another basic question. Does the very high pairing interaction in the cuprates (or possible all highly correlated superconductors) necessarily require low carrier density? This is a deep question that clearly needs to be addressed.

Moreover, current thinking suggests that T_c itself can be affected when pair density is low. The argument is actually very simple and, again, based on thermodynamic reasoning. As the absolute temperature increases, so too do thermal fluctuations of the superconducting pair wave function. The pair wave function, being a complex number, has both an amplitude and a phase, and both degrees of freedom can fluctuate. The problem at low pair density arises due to phase fluctuations, even when the amplitude (the pair density) is robust against fluctuations. Thus, it can happen that phase fluctuations destroy the coherence of the macroscopic quantum state (and hence zero resistance) even if the pair density is finite[164]. Thus superconductivity is governed by two temperature scales: the temperature T_P at which Cooper pairs form and the temperature T_ϕ at which the phases of these pairs order. T_P is a measure of the pairing interactions, T_ϕ is determined by the phase stiffness of the pair wavefunction, which in turn depends on the pair density. Obviously, if $T_\phi > T_P$, superconductivity arises at T_P, as in the BCS theory. If $T_\phi < T_P$, superconductivity arises at T_ϕ, and one has preformed pairs above this temperature.

In short,

$$T_c = \min[T_P, T_\phi]. \tag{3.11}$$

Being more precise, T_ϕ is estimated as the temperature at which thermal phase fluctuations (proportional to T and inversely proportional to the phase stiffness) are sufficient to cause an rms phase difference of π across the size of a Cooper pair. This condition leads to the equation

$$T_\phi \approx \frac{1}{2} \frac{1}{\Lambda_K} \frac{\xi_{ab}}{\gamma} = \frac{1}{2} \frac{\hbar^2 n_s^*}{m^*} \frac{\xi_{ab}}{\gamma} \tag{3.12}$$

where $1/\Lambda_K = \hbar^2 n_s^*/m^*$ is the inverse phase stiffness, n_s^* the pair density and $\gamma = M^*/m^*$ is the Ginzburg–Landau mass anisotropy. Clearly, large T_ϕ requires large n_s^* and γ as close as possible to unity (i.e., an isotropic superconductor)[164].

Do these phase fluctuations prohibit room temperature superconductivity? No, fortunately not, but one will need low anisotropy and increased superfluid density (compared with YBCO) to avoid a limitation due to phase fluctuations. This conclusion obviously has implications for where one needs to look for very much higher T_c superconductors. Presently, these considerations are not widely appreciated in the superconducting materials community.

Also contemporary thinking by some theorists suggests that T_c is not a monotonic function of the interaction parameter λ. Rather, as the interaction strength increases, T_c initially increases as predicted by BCS but eventually rolls over and decreases, provided there is no interceding instability to another competing phase. The physics operating here is that, as the interaction strength increases, the pairs become small ($\xi \propto 1/T_P$ theoretically and pretty close to this empirically) and eventually localize. By the laws of quantum mechanics, localized pairs are subject to large quantum phase fluctuations. To be clear, these two arguments involving fluctuations are not rigorous as they stand, but they are based on very general reasoning, and, at a minimum, they need to be a central issue in the dialog about how to reach higher temperature superconductors.

So what specific guidance can one give in the search for higher temperature superconductors? One school of thought says, as in the past, ignore all this theory stuff; just use your intuition and look in new places. At a minium, this author would advocate that the old strategy of seeking the

---

[164] E. W. Carlson, V. J. Emery, S. A. Kivelson, and D. Orgad, Concepts in high temperature superconductivity, *arXiv*:cond-mat/0206217 v1 12 Jun 2002

highest interaction strength (λ) and the highest energy scale (ω_c) be updated to, say, seek the optimal interaction strength and the highest energy scale, consistent with large phase stiffness. Still another school of thought says, at least for the electron-phonon interaction, we may be entering the era of materials design. Their point is that the BCS/Elisashberg framework generalized to explicitly include a dependence on wave number over the Fermi surface, combined with modern electronic structure calculations, are now good enough to give useful specific guidance, at least in non-highly correlated materials—an exciting claim to be sure. Surely we all hope that someday it will be true. Maybe that day is close at hand.

The author has asked many people what they would suggest in the way of specific guidance. A tabulation of the responses is shown in Table 3.3.

TABLE 3.3:   Guidance in the Search for Higher Temperature Superconductors

---

    1. Strong correlation is good
    2. Electronic interactions are good (charge and spin channels)
    3. Proximity to a competing phase (AFI, AFM, CDW) is good
    4. Reduced dimensionality is good
    5. Reduced dimensionality is bad
    6. Low carrier density is good
    7. Low carrier density is bad
    8. Optimize the electron-phonon interaction
    9. Covalent bonds are good
    10. Small mass and high spring constants are good
    11. Disorder is bad

Yes, they are varied and contradictory. There is much to do. There is no known reason that would prohibit very much higher temperature superconductivity. Let us admire those willing to search. Let us also appreciate that on its 100th birthday our field is alive and well.

Acknowledgments

The author would like to thank T. H. Geballe for his advice in writing this section. It was written under the support of the US Air Force Office of Scientific Research.

3.8 Vortex Matter in Anisotropic Superconductors

Eli Zeldov

Vortices threading type-II superconductors in presence of magnetic field are remarkable extended particles that interact with each other and with the underlying material disorder forming a substance that is generally referred to as vortex matter[165]. A striking property of vortex matter is that the various energy scales can be readily tuned over a wide range of parameters thus providing a unique system for investigation of a broad range of fundamental questions in statistical mechanics and condensed matter physics[166]. Since the discovery of high-T_c superconductors (HTS) it has been recognized that the complexity of vortex matter results in a very rich phase diagram and in various unique equilibrium structures, much of which are still unresolved[167]. Even less understood is the response of the vortex lattice to applied forces, i.e., the dynamics of vortex matter. It is expected that a moving lattice should display a dynamic phase diagram that is even more intriguing and diverse than the already rich static phase diagram. In addition, the most relevant aspect of superconductors for technological applications—their capability to carry loss-free currents—is determined entirely by vortex dynamics.

Our level of knowledge of vortex matter in superconductors has experienced a revolution in recent years. Because of the elevated temperatures and high anisotropy of HTS, the role of thermal fluctuations are significantly enhanced and the mean-field picture of an ordered solid vortex lattice made of parallel straight vortices that was used in the context of conventional superconductors became vastly insufficient. A multitude of new phenomena were discovered both experimentally and theoretically, most of them arising from the interplay of the enhanced fluctuations, different types of point and correlated disorder, layered structure of the HTS, symmetry of the order parameter, and surface and geometrical effects. Clearly, it is impossible to provide a meaningful overview of the field in such a limited space. I have therefore chosen to describe three intriguing aspects of the static properties of the vortex matter which I have encountered personally. These phenomena are just a few examples of the new insight that emerged from development of novel experimental tools that provide local information in contrast to the commonly utilized global tools. They provide a historic perspective of the advent of local measurements and their impact on our current understanding of vortex matter in HTS.

3.8.1 Local Magnetization and Geometrical Barriers

Magnetization is one of the most commonly studied properties of superconductors that provides information on both the static and dynamic behavior of vortices. Of particular interest are the magnetic hysteresis and the irreversibility line on the field-temperature (H-T) diagram that demarcates the hysteretic magnetization behavior at low fields and temperatures from the reversible magnetization above the line. The common source of magnetic hysteresis is vortex pinning due to material disorder, giving rise to a finite critical current below which vortices are immobile. In this case the magnetic hysteresis of a sample is described by the Bean model according to which upon increasing magnetic field vortices gradually penetrate from the sample edges forming steep slopes like sand dunes moving into the sample. The gradient of the vortex density is determined by the critical current with a maximum in vortex density at the edges and minimum in the center. On decreasing

[165] M. Tinkham, *Introduction to Superconductivity* Mc Graw Hill, New York, 1996

[166] G. Blatter, M. V. Feigelman, V. B. Geshkenbein, A. I. Larkin, V. M. Vinokur, *Rev. Mod. Phys.* 66 (1994) 1125

[167] T. Giamarchi and S. Bhattacharya, *High Magnetic Fields: Applications in Condensed Matter Phys., Spectroscopy* (Springer, 2002), p. 314.

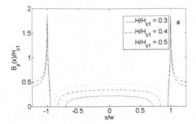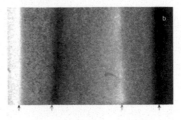

FIGURE 3.53: (a) Calculated field profile $B_z(x)$ across the width $-W < x < W$ of an infinitely long superconducting strip of thickness d ($d/W = 0.1$) for three values of increasing applied field H. The vortex dome in the center of the sample is the result of a geometrical barrier and it grows with increasing H. (b) Differential magneto-optical image of a segment of a long BSCCO crystal strip at $H = 6.8$ Oe and $T = 75$ K, showing the vortex dome between the inner arrows[171]. The outer arrows mark the edges of the crystal of width $2W = 420$ μm. The differential image is obtained by subtracting images with positive and negative periodic transport current of 10 mA applied along the strip. The bright and dark shades of the dome edges result from the differential imaging of the small periodic right and left shift of the dome caused by the applied current.

magnetic field vortices resemble a sand pile with highest vortex density in the center and lowest near the edges. This mechanism causes magnetic hysteresis which is a convenient contactless measure of the critical current of a superconductor that is of central importance for applications.

The magnetic hysteresis is commonly studied by global measurements that determine the total magnetic moment of the sample. In early 1990s we developed with M. Konczykowski arrays of very sensitive microscopic Hall sensors using a two-dimensional electron gas in GaAs. Our early studies of BSCCO crystals revealed a very surprising phenomenon in which vortices seemed to "appear" mysteriously in the center of the sample forming a vortex puddle with a dome-shaped density surrounded by a vortex free region[168]. With increasing field the dome grew from the sample center outwards. Similar observations were obtained using magneto-optical imaging[169]. These puzzling findings were counterintuitive in view of the Bean model according to which a vortex front should gradually penetrate from the edges towards the center upon increasing the field. Having studied extensively the expected current and field distributions in platelet samples in perpendicular fields within the Bean model with John Clem[170], we soon gained the following qualitative understanding of the phenomenon. Bulk pinning in the anisotropic HTS is very weak at elevated temperatures and therefore one would expect a reversible magnetization in a bulk sample. The situation, however, is very different in a thin sample in perpendicular field. If the sample has an elliptical cross-section it can be shown that the magnetic induction B inside the sample will be uniform at any applied field. This result can be understood as following. A test vortex that resides inside the sample experiences two forces. The first force arises from the fact that in elliptical sample the length of the vortex and hence its total line energy is position dependent which results in an outward force that is proportional to the derivative of vortex elongation. The second force is due to the Meissner shielding currents that circulate along the surface and push the vortex towards the sample center. It turns out that for elliptical shape these two forces cancel out exactly, resulting in position-independent vortex energy and hence in a constant B throughout the sample. In platelet crystals with rectangular cross-section the first force is absent and thus the vortices that penetrate through the edges are rapidly driven towards the center forming a dome-shaped vortex puddle as shown in Figure 3.53a. This dome grows

---

[168] E. Zeldov, A. I. Larkin, V. B. Geshkenbein, M. Konczykowski, D. Majer, B. Khaykovich, V. M. Vinokur, and H. Shtrikman, *Phys. Rev. Lett.* 73 (1994) 1428

[169] M.V. Indenbom, H. Kronmüller, T.W. Li, P.H. Kes and A.A. Menovsky, *Physica* C 222 (1994) 203

[170] E. Zeldov, J. R. Clem, M. McElfresh, and M. Darwin, *Phys. Rev.* B 49 (1994) 9802

as the applied field is increased giving the impression that the vortices appear in the sample center and expand outwards.

Having some understanding of the underplaying mechanism, our next challenge was to find a proper solution. In the early 1990s, the Physics Department at the Weizmann Institute had established a special program which hosted a large group of leading Soviet physicists from the Landau Institute for extended periods of time. I discussed this puzzle with Anatoly I. Larkin and his former student Dima Geshkenbein, who were part of the Landau-Weizmann program. In the first couple of meetings, which we had to conduct with my very limited Russian at that time, most of the interaction was with Geshkenbein, while Larkin observed quietly. Then, in a following meeting Larkin all of a sudden started to talk and right away came up with an ingenious insight: since the currents are present in the vortex-free regions and there is no current in the vortex-filled region, B and J are mutually exclusive, and hence they have to be the imaginary and real parts of a complex function. He then immediately guessed that this complex function has to be of the form $\sqrt{(b^2 - x^2)/(W^2 - x^2)}$, where b is half-width of the dome and W is half-width of the sample. By proper integration of this trial function I was amazed to find out that Larkin's guess was an exact solution and fully described the current and field distributions. This vivid encounter with Larkin's genius left an extraordinary impression on me. We coined the phenomenon geometrical barrier, since the vortices have to overcome an extended potential barrier of geometrical origin, and expanded this approach to take a proper account of sample edges and of bulk pinning[168]. Figure 3.53a shows the calculated field profiles $B_z(x)$ at various values of increasing applied field including more detailed treatment of the sample edges in absence of bulk pinning.

Figure 3.53b shows an experimental manifestation of the geometrical barrier and of the vortex dome in a BSCCO crystal using differential magneto-optical imaging[171]. The vortices in the dome are trapped by the Meissner currents and cannot leave the sample. Upon decreasing the field the dome expands while preserving the total flux trapped in it. Vortices can leave the sample only when the edges of the dome reach the sample edges. As a result, the geometrical barrier gives rise to magnetic hysteresis even in the absence of any bulk pinning. Vortex penetration starts at field $H_p \approx H_{c1} \sqrt{d/W}$ and the hysteresis persists up to fields of the order of H_{c1}. The effect of the geometrical barrier can be further significantly enhanced if a Bean Livingston surface barrier is also present, giving rise to a combined edge barrier that can persist to much higher fields[172,173].

3.8.2 Vortex Lattice Melting

Following the discovery of HTS an extensive discussion of the possible melting of the vortex lattice emerged. In the context of conventional low T_c SC this question was of no real significance since the melting should occur essentially at the disappearance of superconductivity at H_{c2}. In HTS, in contrast, theoretical studies suggested that the highly enhanced thermal fluctuations may lead to melting of the lattice well below H_{c2} [166,167,174]. In addition to the great scientific interest, this possibility had a major practical implication since the vortex liquid cannot support any critical current. It was predicted that in disordered systems the melting may occur through a second order transition, whereas in clean systems with an ordered lattice the melting should occur through a first-order phase transition (FOT) [165,166,167,174,175,176,177]. The issue became a vigorously debated topic with numerous controversial experimental studies. Neutron scattering experiments showed disappearance of

[171] Y. Segev, I. Gutman, S. Goldberg, Y. Myasoedov, E. Zeldov, E. H. Brandt, G. P. Mikitik, and T. Sasagawa, unpublished.

[172] E. H. Brandt, *Phys. Rev.* B 60 (1999) 11939

[173] J. Clem, *J. Superconductivity and Novel Magnet.* 21 (2008) 343

[174] D. R. Nelson, *Phys. Rev. Lett.* 60 (1988) 1973

[175] M. P. A. Fisher, *Phys. Rev. Lett.* 62 (1989) 1415

[176] E. H. Brandt, *Rep. Prog. Phys.* 58 (1995) 1465

[177] T. Nattermann and S. Scheidl, *Adv. Phys.* 49 (2000) 607

Bragg peaks[178], which indicated loss of long range order of the lattice, and transport measurements in YBCO single crystals revealed a sharp drop in resistivity consistent with freezing of the lattice[179,180]. None of the experiments, however, provided a thermodynamic signature of a FOT.

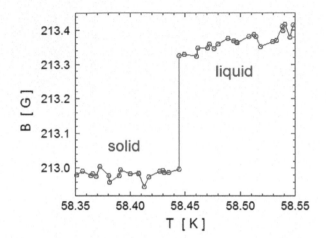

FIGURE 3.54: Discontinuous step in the local vortex density and induction B across a first-order melting transition in a BSCCO crystal. The data were measured by an array of $10 \times 10\ \mu m^2$ Hall sensors on sweeping temperature at a constant applied field of 240 Oe [184]. The transition occurs at slightly different temperature at the location of each sensor. The vortex density in the liquid is higher than in the solid. Vortex liquid thus behaves like water that expands upon freezing into ice when cooled at a constant pressure.

All the initial searches for the melting transition were based on global measurements. It turned out that having local information was essential for the detection of the FOT. At that time we used arrays of Hall sensors to investigate the so-called second magnetization peak in BSCCO crystals[181]. This large peak in hysteretic magnetization is visible at low temperatures at which the critical current of the sample appears to increase at some characteristic field B_{sp} that is almost temperature independent. With increasing temperature the peak shrinks and seems to disappear at some intermediate temperature. We noticed that at higher temperatures a small step appeared in the data in place of the second magnetization peak. This step was hardly visible but seemed to be reproducible and hence we decided to study it more carefully. We were very fortunate to have very high quality BSCCO crystals grown by N. Motohira in the lab of K. Kishio and K. Kitazawa in University of Tokyo[182], which turned out to be the cleanest and most uniform crystals I have ever encountered. We improved the sensitivity of our Hall probes and carried out measurements of the local induction B using very dense temperature and field sweeps. A small but an extremely sharp step ΔB became evident (see Figure 3.54), the location of which traced a well-defined $B_m(T)$ line on the H-T phase diagram extending from the region of the second magnetization peak to T_c. Moreover, this step

[178] R. Cubitt, E. M. Forgan, G. Yang, S. L. Lee, D. McK. Paul, H. A. Mook, M. Yethiraj, P. H. Kes, T. W. Li, A. A. Menovsky, Z. Tarnawski, and K. Mortensen, *Nature* 365 (1993) 407

[179] H. Safar, P. L. Gammel, D. A. Huse, D. J. Bishop, J. P. Rice, and D. M. Ginsberg, *Phys. Rev. Lett.* 69 (1992) 824

[180] W. K. Kwok, S. Fleshler, U. Welp, V. M. Vinokur, J. Downey, G. W. Crabtree, and M. M. Miller, *Phys. Rev. Lett.* 69, 3370 (1992).

[181] B. Khaykovich, E. Zeldov, D. Majer, T. W. Li, P. H. Kes, and M. Konczykowski, *Phys. Rev. Lett.* 76, 2555 (1996).

[182] N. Motohira, K. Kuwahara, T. Hasegawa, K. Kishio, and K. Kitazawa, *J. Ceram. Soc. Jpn. Int. Ed.* 97 (1989) 994

occurred at slightly different field or temperature at different places across the sample as revealed by the different Hall sensors in the array. We had numerous debates of this step with Geshkenbein, Larkin, and Vinokur considering its different possible origins. Shortly prior to a workshop organized by M. Konczykowski in the Ecole Polytechnique in 1994, a paper by the group of Paco de la Cruz had been published[183] which showed evidence of such a step in global magnetization measurements at high temperatures and was interpreted as a FOT of the lattice. Following further exciting discussions at the workshop, I became convinced that the sharp step that we observed is a direct thermodynamic observation of the first-order melting of the vortex lattice[184]. Locally this transition is very sharp; however, at each location it occurs at a somewhat different field or temperature due to disorder and geometrical barriers, and therefore is very hard to be detected by global magnetization measurements. In YBCO crystals the melting transition extends to much higher fields due to lower anisotropy, and therefore the broadening of the transition is relatively smaller. As a result, subsequent high sensitivity studies of magnetization and specific heat succeeded to resolve the FOT also in YBCO crystals by global measurements[185,186].

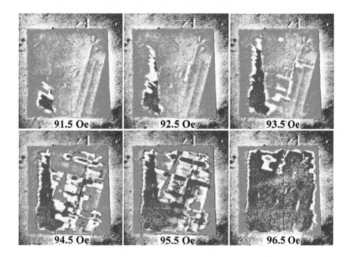

FIGURE 3.55: Vortex lattice melting process in BSCCO crystal of 1.1×1.2 mm^2 as revealed by differential magneto-optical imaging upon increasing field at constant $T = 70$ K [187]. At low fields the entire crystal is in the solid phase (brown). At 91.5 Oe an irregular liquid droplet (blue) is formed in the lower left corner. On the right hand side three parallel defects are visible along one of the crystallographic directions. With increasing field complicated melting patterns are obtained with numerous solid and liquid domains that have preferential orientation either parallel or perpendicular to the crystallographic directions. At 96.5 Oe a large liquid region is present in the center, with a few solid islands in the top part. The presence of the liquid phase in the sample center surrounded by vortex solid is the result of the geometrical-barrier vortex dome profile. The full movie is available at http://www.weizmann.ac.il/condmat/superc/.

The arrays of Hall probes showed that there is a nontrivial coexistence of the solid and liquid phases in the sample over a wide range of fields and temperatures. We therefore were very interested

[183] H. Pastoriza, M. F. Goffman, A. Arribére, and F. de la Cruz, *Phys. Rev. Lett.* 72 (1994) 2951

[184] E. Zeldov, D. Majer, M. Konczykowski, V. B. Geshkenbein, V. M. Vinokur, and H. Shtrikman, *Nature* 375 (1995) 373

[185] U. Welp, J. A. Fendrich, W. K. Kwok, G. W. Crabtree, and B. W. Veal, *Phys. Rev. Lett.* 76 (1996) 4809

[186] A. Schilling, R. A. Fisher, N. E. Phillips, U. Welp, D. Dasgupta, W. K. Kwok, G. W. Crabtree, *Nature* 382 (1996) 791

in visualization of this melting process on a microscopic scale. For this purpose we developed differential magneto-optical (DMO) imaging[187]. In conventional magneto-optics (MO) typical sensitivity is of the order of a few Gauss which is insufficient to distinguish the difference in the induction between the solid and liquid phases ΔB that is typically of the order of 0.1 G. The sensitivity of the MO is mainly limited by the inhomogeneities in the MO indicators and the shot noise in the CCD cameras. Both these limitations can be levitated by periodically modulating either the applied field by a small δH or the temperature by δT and averaging over many corresponding differential images. If vortex solid and liquid regions coexist in the sample, increasing the temperature by a small δT will slightly expand the liquid domains. In the regions where the solid-liquid interface has shifted, the local induction increases by ΔB while essentially no change occurs in the rest of the sample. As a result in DMO image the solid-liquid interface will appear bright as shown in Figure 3.55. By taking a series of DMO images as a function of temperature or field a movie of the entire melting process is thus obtained[187]. We found that the quenched material disorder results in a complicated melting temperature T_m landscape[188] that causes formation of intricate melting patterns and coexistence of solid and liquid domains as exemplified in Figure 3.55. As a result the sharp local FOT becomes significantly broadened and hard to be resolved in global measurements.

3.8.3 Vortex Matter Phase Diagram

The discovery of the FOT established that the vortex matter in HTS displays at least two thermodynamic phases. Additional studies showed that the phase below the transition has a finite shear modulus and some degree of long-range order as reflected by the Bragg peaks in neutron scattering and magnetic decoration experiments, and therefore is a vortex solid. According to Larkin and Ovchinikov, however, quenched disorder should always destroy the long-range order of the lattice[189]. A conceptual breakthrough was obtained when the theory of Bragg glass was developed [190,191], according to which at sufficiently weak disorder no dislocations are formed and as a result the lattice displays algebraic quasi-long-range translational order. This theory was consistent with the experimental observation of the Bragg peaks and the existence of the first-order melting, at which dislocations proliferate causing loss of order. In weakly anisotropic HTS like YBCO the liquid phase is believed to constitute a highly entangled state of vortex lines dominated by thermal fluctuations. In materials with high anisotropy like BSCCO, there is experimental evidence for simultaneous melting and decoupling at the FOT, namely in the high temperature phase the vortex lines decompose into a gas-like state of uncorrelated vortex pancakes in the individual CuO_2 planes[192].

This understanding, however, applied only to the high temperature part of the vortex matter phase diagram. The experimental difficulty is that at low temperatures vortex matter is dominated by strong pinning due to quenched disorder. As a result, the vortex matter is usually far away from the equilibrium state and magnetization measurements show highly hysteretic behavior which completely masks the thermodynamic properties of the lattice. Consequently the FOT was found to terminate at some intermediate temperatures[184]. The resolution to this experimental limitation came with the introduction of vortex shaking by an in-plane ac field that was originally applied to YBCO crys-

[187] A. Soibel, E. Zeldov, M. Rappaport, Y. Myasoedov, T. Tamegai, S. Ooi, M. Konczykowski, and V. B. Geshkenbein, *Nature* 406 (2000) 282

[188] A. Soibel, Y. Myasoedov, M. L. Rappaport, T. Tamegai, S. S. Banerjee, and E. Zeldov, *Phys. Rev. Lett.* 87 (2001) 167001

[189] A. I. Larkin and Y. N. Ovchinnikov, *J. Low Temp. Phys.* 34 (1979) 409

[190] T. Nattermann, *Phys. Rev. Lett.* 64 (1990) 2454

[191] T. Giamarchi and P. Le Doussal, *Phys. Rev. Lett.* 72 (1994) 1530

[192] D. T. Fuchs, R. A. Doyle, E. Zeldov, D. Majer, W. S. Seow, T. Tamegai, S. Ooi, R. Drost, M. Konczykowski, and P. H. Kes, *Phys. Rev.* B 55 (1997) R6156

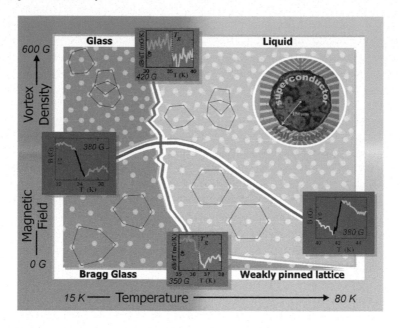

FIGURE 3.56: Schematic vortex matter phase diagram in BSCCO as revealed by local magnetization measurements in presence of vortex shaking by in-plane ac field[197]. Four thermodynamic phases are suggested to be formed by two intersecting phase transition lines. The red line is a single first-order transition $B_m(T)$ that gradually changes its character from thermally driven melting at high temperatures to disorder-driven transition displaying an inverse melting behavior at low temperatures. A positive step ΔB in the local induction occurs upon increasing temperature across the thermal melting (right inset) and a negative step across the inverse melting (left inset). The green line is apparently a second-order transition T_g characterized by a step-like change in the slope dB/dT (top and bottom insets). The top-right phase is vortex liquid that undergoes a glass transition into a glassy phase upon decreasing temperature. The two blue phases are entangled with high concentration of dislocations. The two brown phases are vortex solid with no dislocations. The low-temperature phase is apparently strongly pinned Bragg glass. The high-temperature region is weakly pinned and may display a higher extent of order. The top-right inset shows a disk-shaped BSCCO crystal residing on an array of microscopic Hall sensors.

tals at high temperatures[193,194]. We found that in highly anisotropic materials like BSCCO vortex shaking is very effective in equilibrating the vortex lattice at low temperatures, thus opening a new avenue for the investigation of thermodynamic phase diagram in regions that were previously inaccessible[195]. Addition of an in-plane field in highly anisotropic HTS results in formation of crossing lattices in which stacks of pancake vortices (PVs) coexist with a lattice of Josephson vortices (JVs) that reside between the CuO$_2$ planes[196]. The JVs are weakly pinned and therefore remain highly mobile even at low temperatures. An ac in-plane field thus results in periodic motion of JVs that repeatedly intersect the stacks of PVs at random places. The circulating currents of the JVs exert

[193] M. Willemin, A. Schilling, H. Keller, C. Rossel, J. Hofer, U. Welp, W. K. Kwok, R.J. Olsson, and G. W. Crabtree, *Phys. Rev. Lett.* 81 (1998) 4236

[194] E. H. Brandt and G. P. Mikitik, *Phys. Rev. Lett.* 89 (2002) 027002

[195] N. Avraham, B. Khaykovich, Y. Myasoedov, M. Rappaport, H. Shtrikman, D. E. Feldman, T. Tamegai, P. H. Kes, M. Li, M. Konczykowski, K. van der Beek, and E. Zeldov, *Nature* 411 (2001) 451

[196] A. E. Koshelev, *Phys. Rev. Lett.* 83 (1999) 187

a local force on the PVs at the intersections thus causing a random local shaking of the PVs that results in gradual equilibration of the vortex lattice.

By combining local magnetization measurements with vortex shaking we discovered that the FOT extends to much lower temperatures and traces the location of the second magnetization peak. Moreover, in this region the vortex matter displays a unique phenomenon of inverse melting in which the lattice crystallizes upon increasing rather than decreasing temperature[195,197]. The red curve in Figure 3.56 shows the full FOT line $B_m(T)$ in BSCCO which displays a maximum at intermediate temperatures followed by a change of slope in the inverse melting region. At high temperatures the solid vortex lattice is destroyed by thermal fluctuations, whereas at low temperatures the lattice is destroyed by pinning due to quenched disorder which becomes dominant over the elastic energy as the field is increased[198]. As a result, the lattice transforms into an amorphous entangled state with a high concentration of dislocations. The FOT is thus a single line that gradually changes its character from predominantly thermally driven melting to predominantly point disorder driven transition[199]. The inverse melting behavior results from the fact that with increasing temperature the vortex pinning is reduced by thermal fluctuations resulting in increasing $B_m(T)$ [200,201,202].

By increasing the temperature at intermediate fields, the FOT line can be crossed twice. In this interesting case, a negative ΔB step is observed on the first crossing in the inverse melting part (left inset in Figure 3.56), followed by a positive ΔB step in the region of thermally driven FOT (right inset). Moreover, between the two FOT steps, we found a sharp kink in $B(T)$ which indicates a possible existence of a second-order transition (SOT) line as shown by the green curve in Figure 3.56. Along this line a sharp change in the slope dB/dT is revealed in local measurements in regions both below and above the FOT as shown by the top and bottom insets[197]. The existence of a SOT at elevated fields implies that instead of a single vortex liquid phase above the FOT line, two distinct phases are present: the vortex liquid and glassy phase separated by a glass transition line. The continuation of the SOT line into the vortex solid region below the FOT line, however, is much more surprising. This entire region of the phase diagram is believed to be the Bragg glass phase[167,191,202]. The existence of the SOT line implies that this region is subdivided into two distinct thermodynamic phases[203]. The low-temperature phase displays strong pinning and hence is presumably the Bragg glass phase. In the high-temperature part the pinning is extremely weak due to enhanced thermal fluctuations which could lead to a solid phase with partially recovered long-range order. Figure 3.56 with the proposed four thermodynamic phases calls for new experimental and theoretical studies in order to comprehend the evermore intriguing nature of the vortex matter in high-temperature superconductors.

[197] H. Beidenkopf, N. Avraham, Y. Myasoedov, H. Shtrikman, E. Zeldov, B. Rosenstein, E. H. Brandt, and T. Tamegai, *Phys. Rev. Lett.* 95 (2005) 257004

[198] V. Vinokur, B. Khaykovich, E. Zeldov, M. Konczykowski, R. A. Doyle, and P. H. Kes, *Physica* C 295 (1998) 209

[199] P. Olsson and S. Teitel, *Phys. Rev. Lett.* 87 (2001) 137001

[200] D. Ertas and D. R. Nelson, *Physica* C 79, 272 (1996).

[201] H. Beidenkopf, T. Verdene, Y. Myasoedov, H. Shtrikman, E. Zeldov, B. Rosenstein, D. Li, and T. Tamegai, *Phys. Rev. Lett.* 98 (2007) 167004

[202] G. P. Mikitik and E. H. Brandt, *Phys. Rev.* B 68 (2003) 054509

[203] I.D. Li, B. Rosenstein, and V. Vinokur, J. *Superconductivity and Novel Magnetism* 19 (2006) 369

Acknowledgments

The Discovery of Fluxoid Quantization

Useful and stimulating comments on this manuscript by Werner Biberacher, B. S. Chandrasekhar, Rudolf Groß and Rudi Hackl are gratefully acknowledged. I am particularly indebted to Bascom S. Deaver Jr., Robert Doll, and Bill Little for their continuous and encouraging support of this manuscript.

The Grain Boundary Problem of High-T_c Superconductors

The authors gratefully acknowledge discussions with V. Mathias and support by G. Hammerl, A. Herrnberger, and C. Hughes in writing this article.

3.9   Further Reading

Tunneling and the Josephson Effect

1. "The Discovery of Electron Tunneling into Superconductors", Roland W. Schmitt, *Physics Today* (1961) 38.

2. "Coupled Superconductors", B. D. Josephson, *Rev. Mod. Phys.* (1964) 216.

3. Nobel Lecture, B. D. Josephson, *Science* 184 (1974) 527.

4. "The Historical Context of Josephson's Discovery", A. B. Pippard, A talk presented to a NATO Advanced Study Institute on Small-Scale Superconducting Devices.

5. "How Josephson discovered his effect", Philip W. Anderson, *Physics Today* (1970) 23.

6. Oral History Transcript - Philip Anderson (1999), http://www.aip.org/history/ohilist/23362_-1.html

The Search for the Pairing Symmetry in the High Temperature Superconductors

1. D. J. Van Harlingen. "Phase-sensitive tests of the symmetry of the pairing state in the high-temperature superconductors – Evidence for $d_{x^2-y^2}$ symmetry". *Rev. Mod. Phys.* 67, 515 (1995).

2. C. C. Tsuei and J. R. Kirtley "Pairing symmetry in cuprate superconductors". *Rev. Mod. Phys.* 72, 969 (2000).

The Grain Boundary Problem of High-T_c Superconductors

1. S.E. Babcock, High-Temperature Superconductors from the Grain Boundary Perspective, *MRS Bulletin*, August 1992, pp. 20.

2. P. Chaudhari and J. Mannhart, High-Tc Grain Boundaries, *Phys. Today*, Nov. 2001, pp. 48.

3. J. Hilgenkamp and J. Mannhart, Grain Boundaries in High-Tc Superconductors, *Rev. Mod. Phys.* 74, 485 (2002).

4. C.C. Tsuei and J.R. Kirtley, Pairing Symmetry in Cuprate Superconductors, *Rev. Mod. Phys.* 72, 969 (2000).

4

Materials

Editor: C. W. Chu

4.1 Introduction
   C. W. Chu 233
4.2 The Route to High Temperature Superconductivity in Transition Metal Oxides
   A. Bussmann-Holder and K. A. Müller 234
4.3 Superconductivity above 10 K in Non-Cuprate Oxides
   David C. Johnston 239
4.4 Cuprates—Superconductors with a T_c up to 164 K
   C. W. Chu 244
4.5 Fe-Pnictides and -Chalcogenides
   Hideo Hosono 255
4.6 Superconductivity in MgB_2
   Takahiro Muranaka and Jun Akimitsu 265
4.7 Fullerenes
   Kosmas Prassides and Matthew J. Rosseinsky 272
4.8 Elemental Superconductors
   K. Shimizu 278
4.9 Heavy-Fermion Superconductivity
   F. Steglich 283
4.10 Ruthenate Superconductor Sr_2RuO_4
   Yoshiteru Maeno 288
4.11 Magnetic Superconductors (and Some Recollections of Professor Bernd T. Matthias)
   M. Brian Maple 293
   Acknowledgments 306
4.12 Further Reading 307

4.1 Introduction

C. W. Chu

Materials form the foundation of most scientific and technological endeavors. This is particularly true for superconductivity research and development. The discovery and perfection of a new class of superconductors invariably leads to the development of new physics and/or new technology. Examples abound. For instance, today's powerful magnetic resonance imaging technology for medical diagnoses and the omnipotent accelerator technology for exploring the deep secrets of the universe would not have been possible without the perfection of Nb-Ti and Nb_3Sn in the 1960s and 1970s, although these are superconductors with low T_c. It was the study on the transition metal dichalco-genides for possible high T_c in the 1970s that led to the discovery of charge density waves, which have become an important subfield of physics. The discovery of high temperature superconductivity in 1986 has created an exciting subfield of physics and promised a new vista in science and technology.

Many superconducting material systems have been found since the discovery of superconductivity 100 years ago. The selection of superconductor systems for this Chapter is indeed a challenge. However, the search for novel superconducting materials with higher T_c has been the major driving force in sustaining superconductivity research for decades and the discovery of high temperature superconductivity (HTS) in 1986 has been considered to be a major advance in modern science. The material systems chosen therefore focus on high T_c and possible mechanisms for high T_c.

Because of the limited space, this chapter consists of 10 articles:

4.2 "The route to High Temperature Superconductivity in Transition metal Oxides" by Annette Bussmann-Holder and K. Alex Müller—beginning of the epoch of cuprate HTS;

4.3 "Superconductivity above 10 K in Non-cuprate Oxides" by David C. Johnston—the dawn of cuprate HTS;

4.4 "Cuprates—Superconductors with a T_c up to 164 K" by Paul C. W. Chu—beginning of the era of liquid nitrogen HTS;

4.5 "Fe-pnictides and Chalcogenides" by Hideo Hosono—beginning of the era of iron-based HTS;

4.6 "Superconductivity in MgB_2" by Takahiro Muranaka and Jun Akimitsu—expecting the unexpected;

4.7 "Fullerenes" by Kosmas Prassides and Matthew J. Rosseinsky—a possible alternate route to high T_c;

4.8 "Elemental Superconductors" by Kasuya Shimizu—high T_c via high pressures;

4.9 "Heavy Fermion Superconductivity" by Frank Steglich—a novel superconducting state and a possible road to understanding HTS;

4.10 "Ruthenate Superconductor Sr_2RuO_4" by Yoshiteru Maeno—a new superconducting state; and

4.11 "Magnetic Superconductors (and Some Recollections of Professor Bernd T. Matthias)" by M. Brian Maple—competition and cooperation between antagonists.

4.2 The Route to High Temperature Superconductivity in Transition Metal Oxides

A. Bussmann-Holder and K. A. Müller

The discovery of high temperature superconductivity in cuprates was possible only through an intimate knowledge of perovskite oxides which have been synthesized and characterized for decades at the IBM in the Zürich laboratory. Especially $SrTiO_3$ and $LaAlO_3$ have been in the focus at IBM as was presented in Volume 1 of a series.[1] Probably for the first time, detailed microscopic investigations of the local properties of these compounds have been obtained by studying by means of

---

[1]   K. A. Müller and T. N. Kool, *Properties of perovskites and other oxides*, World Scientific, May 2010.

EPR the surroundings of transition metal impurities in these materials. These experiments enabled the identification of the order parameter of the structural instability observed in these oxides. However, the idea to search for superconductivity came later, motivated by theoretical considerations that metallic hydrogen could become superconducting at high temperatures. Since $SrTiO_3$ is an insulator it was thought that the implantation of hydrogen would render it metallic and eventually also superconducting. This approach failed since the carrier density remained always too small. In sequence it was then tried to achieve a metallic state in oxide perovskites by varying their composition which was insofar promising as reduced $SrTiO_3$ exhibits superconductivity at 0.3 K.[2] Furthermore it was subsequently shown[3] that T_c can be enhanced to 1.2 K by doping $SrTiO_3$ with Nb. In spite of the fact that T_c was far below values achieved in A15 compounds, a remarkable observation was connected with the Nb-doped perovskite, namely for the first time long before predicted two-gap superconductivity was realized here.

In spite of the rather disappointing low transition temperatures achieved in $SrTiO_3$ another observation, namely the discovery of T_c enhancements in granular Al as compared to crystalline Al, kept the interest in this field alive. Opposite to crystalline Al, the small metallic grains in the non-crystalline Al, are surrounded by amorphous Al_2O_3. These small grains couple by Josephson junctions and enhance T_c by a factor of up to three, however, still remaining on the side of low T_c materials.[4,5]

Early predictions by Matthias to search for high T_c superconductivity in simple metallic or at most binary compounds have been followed for many years but unfortunately could not exceed values of $T_c > 23$ K in Nb_3Sn in the early 1970s. This fact encouraged to search for other than intermetallic compound as new superconductors with special focus on oxides, actually definitely excluded from the Matthias considerations. Since in these days it was undoubtedly accepted that the BCS theory is best suited to explain superconductivity, a simple inspection of the BCS T_c defining equation shows the material limits: $k_B T_c = 1.13 \hbar \omega_D \exp\{-1/[N(E_F)V_{ph}]\}$. In oxides the density of states at the Fermi level $N(E_F)$ is low, $N(E_F) = 4 \times 10^{21}/cm^3$; on the other hand the electron-phonon interaction V_{ph} is rather strong. Thus it could be an interesting route to enhance T_c by increasing the carrier density via appropriate doping, or by strengthening the electron-phonon interaction. Especially, it was suggested that mixed valency compounds could support an increasing $N(E_F)$, whereas polaron formation was shown to enhance V_{ph}. Chakraverty[6] was one of the first to calculate the phase diagram of a polaronic superconductor which has many features in common with the one of cuprate superconductors. By defining $\lambda = N(E_F)V_{ph}$, the phase diagram (Figure 4.1) shows metallic properties for small values of λ, whereas it is insulating bipolaronic for λ large. In the intermediate λ regime superconductivity is realized which may exhibit exceedingly large values of T_c. This quite intriguing idea left the question open regarding in which systems such a scenario could be realized.

At this stage a new concept came into play, namely the idea of the Jahn-Teller polaron, as suggested by Höck et al.[7] and discussed in more detail below. The basic ingredient of the Jahn-Teller effect is based on the instability of orbitally degenerate electronic states towards a lattice distortion whereby the degeneracy is lifted. Here a competition between electron localization and their kinetic energy sets in, which is dominated by the Jahn-Teller energy E_{JT} versus the bandwidth W. For $E_{JT} \gg W$ localization takes place and the electron is trapped in a phonon cloud. In the opposite case $E_{JT} \ll W$ the electron travels almost unaffectedly through the lattice experiencing only small disturbances from it. The interesting case $E_{JT} \approx W$ combines the limiting cases since here the electron travels through the lattice with its own displacement field. This is the Jahn-Teller polaron.

2 J. F. Schooley, W. R. Hosler, and M. L. Cohen, *Phys. Rev. Lett.* 12 (1964) 474

3 G. Binnig, A. Baratoff, H. E. Hoenig, and J. G. Bednorz, *Phys. Rev. Lett.* 45 (1980) 1352

4 G. Deutscher and S. A. Dodds, *Phys. Rev. B* 16 (1977) 3936

5 K. A. Müller, M. Pomerantz, C. M. Knoedler, and D. Abraham, *Phys. Rev. Lett.* 45 (1980) 832

6 B. K. Chakraverty, J.Phys.Lett. 40 (1979) L99; *J. Phys.* 42 (1981) 1351

7 K.-H. Höck, H. Nickisch, and H. Thomas, *Helv. Phys. Acta* 56 (1983) 237

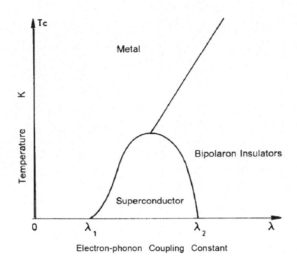

FIGURE 4.1: Phase diagram as a function of electron-phonon coupling λ (©The Nobel Foundation, 1987).

Regarding oxides, the knowledge of these established that many of them contain transition metal ions with partially filled e_g orbitals which are known to act as Jahn-Teller centers. Of special interest are in this respect Ni^{3+}, Fe^{4+} or Cu^{2+} which were then considered to be possible candidates for high temperature superconductivity.

The search started in 1983 with the La-Ni-O systems. $LaNiO_3$ is metallic and the Jahn-Teller energy is smaller than the bandwidth. In order to reduce the bandwidth, Al was substituted for Ni where for small concentrations an increase in the resistivity sets in and a semiconductor-type behavior is obtained for large concentrations turning into a localization regime with decreasing temperature. Superconductivity could not be realized by this approach. Another option to reduce the bandwidth is strain which can be achieved via replacement of the La^{3+} ion by the smaller Y^{3+} ion. However, the results were comparable to the previous replacement and remained unsuccessful with respect to superconductivity.

Two years later improvement in the experimental situation enabled the final breakthrough. Since the experiments with Ni^{3+} did not reveal the expected results, Cu^{2+} was considered as another candidate for the Jahn-Teller polaron effect. By partially replacing the Jahn-Teller ion Ni^{3+} by the-non-Jahn-Teller ion Cu^{3+} an increase in the sample resistivity was achieved, however, preserving its metallic character down to 4 K. Shortly afterwards a report on the La-Ba-Cu-O system appeared which showed that this compound is metallic between 300 and 100 °C.[8] The special property of this system is the mixed valency of Cu which is present as Cu^{2+} and Cu^{3+}. Accordingly, it is possible to tune the Cu valency continuously by changing the La/Ba ratio. When cooling these samples a metallic like resistivity decrease was obtained followed by an increase at lower temperature which is a signature of localization. This behavior continued to 30 K when a sudden drop in resistivity set in around 11 K. These data could be reproduced on several samples confirming the same temperature behavior (Figure 4.2).

By varying the La/Ba ratio and the thermal treatment, the drop could be shifted to temperatures as high as 35 K, much higher than the one observed in Nb_3Ge at 25 K. The origin of the resistivity drop remained open since the Meissner-Ochsenfeld effect had not been demonstrated. Nevertheless, the first paper with the carefully selected title "Possible high T_c superconductivity in the La-Ba-

---

[8]   C. Michel and B. Raveau, *Chim. Min.* 21 (1984) 407

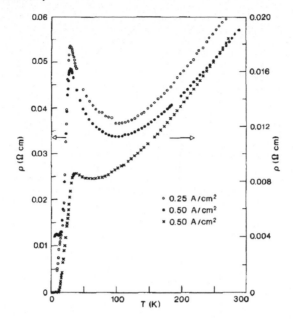

FIGURE 4.2: Temperature dependence of the resistivity in $Ba_xLa_{5-x}Cu_5O_{5(3-y)}$ for samples with x(Ba)= 1 (upper curves, left scale) and x(Ba)= 0.75 (lower curve, right scale). The first two cases also show the influence of current density [after Bednorz and Müller [9]]. (With kind permission from Springer Science Business Media: "Possible high T_c superconductivity in he Ba-La-Cu-O system," J. G. Bednorz and K. A. Müller, Z. Phys. B: Condens. Matter 64 (1986) 189–193, Fig. 1.)

Cu-O system" was submitted for publication.[9] In addition, it turned out that the such synthesized samples contained two different phases and it had to be verified which of them corresponded to the superconducting one. By systematically changing the composition and measuring electrical and lattice properties, the tentative assignment was made that the Ba containing La_2CuO_4 was responsible for superconductivity. Interestingly, a correlation between Ba doping, structural distortion and superconductivity was observed, namely, that the orthorhombic distortion decreases with increasing Ba content to make the structure more tetragonal with the maximum T_c almost coinciding with the tetragonal to orthorhombic transition.

In September 1986, it was possible to carry out magnetic measurements where first a compound with low Ba content was measured. In this system, metallic conductivity was observed down to 100 K followed by a transition to localization. The susceptibility was Pauli-like positive, temperature independent, changing to Curie-Weiss behavior at low temperatures. Samples exhibiting a resistivity drop underwent a transition from paramagnetic to diamagnetic, characteristic of superconductivity-related shielding currents. The diamagnetic transition always started slightly below the resistivity drop, evidencing percolative superconductivity. The final proof for superconductivity, namely the Meissner-Ochsenfeld effect, had thus been demonstrated and the superconducting phase identified (Figure 4.3[10,11]).

In the following, the magnetic characterization of the samples was continued and evidence for a glass state discovered.[12] More important, however, was the idea to replace La not only by Ba

[9] J. G. Bednorz and K. A. Müller, Z. *Phys. B* 64 (1986) 189

[10] J. G. Bednorz, M. Takashige, and K. A. Müller, *Europhys. Lett.* 3 (1987) 379

[11] J. G. Bednorz and K. A. Müller, Nobel Lecture, *Angew. Chemie* 100 (1987) 5

[12] K. A. Müller, M. Takashige, and J. G. Bednorz, *Phys. Rev. Lett.* 58 (1987) 1143

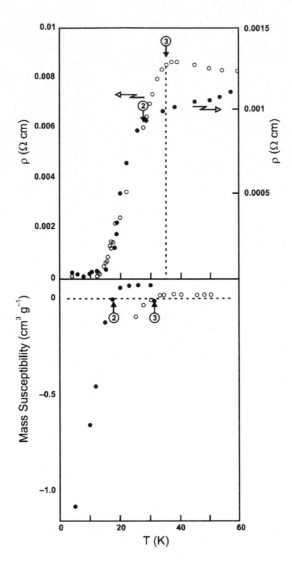

FIGURE 4.3: Low-temperature resistivity and susceptibility of (La-Ba)-Cu-O samples 2 (solid circles) and 3 (open circles). Arrows indicate the onset of the resistivity and the paramagnetic to diamagnetic transition, respectively (© The Nobel Foundation, 1987).

but to try substitutions with Sr and Ca. Sr substitutions induced superconductivity at even higher temperatures than Ba with a maximum onset of 40 K.[13] It is worth mentioning that the radius of Sr is almost identical to the one of the La ion and consequently this ion fits better into the structure than Ba.

Even though the above results created an enormous positive international response, the breakthrough was the confirmation of the data by other groups. At the end of 1986, a Japanese group reproduced the data[14] and the community, which had been skeptical before, became very attentive.

[13] J. G. Bednorz, K. A. Müller, and M. Takashige, *Science* 236 (1987) 73

[14] H. Takagi, S. Uchida, K. Kitazawa, and S. Tanaka, *Jpn. J. Appl. Phys.* 26 (1987) L123; S. Uchida, H. Takagi, K. Kitazawa, and S. Tanaka, ibid. (1987) L151

4.3 Superconductivity above 10 K in Non-Cuprate Oxides

David C. Johnston

Beginning in 1973, several non-cuprate transition metal and non-transition metal oxides were discovered with superconducting transition temperatures between 10 and 30 K. Retrospectives about these discoveries are given.

4.3.1 Introduction

The quest for high superconducting transition temperatures T_c led to a slow increase in the maximum observed T_c from 4.2 K in Hg in 1911, which was the discovery of superconductivity itself by Onnes, to 22–23 K in thin films of the cubic A-15 structure compound Nb_3Ge as reported by Gavaler et al. in 1973 and 1974.[15] Oxides and non-transition metal compounds were not seriously considered by most researchers as contenders for high T_c. Superconducting oxides such as NbO and doped $SrTiO_3$ exhibited $T_c \lesssim 1$ K. The highest T_c oxide, hexagonal rubidium tungsten bronze Rb_xWO_3, showed $T_c \leq 6.6$ K.[16] Until 1986, "high T_c" was considered to be any T_c above 10 K.

Beginning in 1973, a series of discoveries of $T_c > 10$ K in oxides and non-transition metal compounds was made that changed our view of their potential for high T_c. A maximum T_c onset of 13.7 K was discovered by the author and coworkers in 1973 for $LiTi_2O_4$ with the cubic spinel structure,[17] and T_c up to about 13 K was found by Sleight et al. for the non-transition metal oxide $Ba(Pb_{1-x}Bi_x)O_3$ with the cubic perovskite structure in 1975.[18] These T_cs were the highest for oxide and/or non-transition metal compounds until the discovery by Bednorz and Müller in 1986 of superconductivity up to about 30 K in the layered copper oxide compound $La_{2-x}Ba_xCuO_4$ that contains Cu square lattice layers,[19,20] as discussed in other articles within this chapter by Alex Müller and by Paul Chu. The current maximum T_c of 164 K for this class of compounds was obtained under pressure by Gao et al. in 1994 and is currently also the record T_c for any material.[21] This T_c was reached by applying 31 GPa pressure to a $Hg_{1-x}Pb_xBa_2Ca_2Cu_3O_{8+\delta}$ sample that had a zero-pressure T_c of 134 K.[21] Superconductivity at temperatures up to about 30 K was found in the non-transition-metal cubic perovskite oxide compound $(Ba_{1-x}K_x)BiO_3$ by Mattheiss et al.[22] and Cava et al.[23] in 1988. In related developments, bulk high T_cs were discovered by Tanigaki et al. in non-transition-metal alkali metal-doped A_xC_{60} Buckyballs up to about 33 K in 1991,[24] and in MgB_2 at 39 K by Nagamatsu et al.[25] in 2001 as reviewed in a separate article in this chapter by Jun Akimitsu and Takahiro Muranaka. In 2008, the Fe-containing tetragonal compound $LaFeAsO_{1-x}F_x$ was discov-

[15] J. R. Gavaler, *Appl. Phys. Lett.* 23 (1973) 480; J. R. Gavaler, M. A. Janocko, and C. K. Jones, *J. Appl. Phys.* 45 (1974) 3009

[16] J. P. Remeika, T. H. Geballe, B. T. Matthias, A. S. Cooper, G. W. Hull, and E. M. Kelly, *Phys. Lett.* 24A (1967) 565

[17] D. C. Johnston, H. Prakash, W. H. Zachariasen, and R. Viswanathan, *Mater. Res. Bull.* 8 (1973) 777

[18] A. W. Sleight, J. L. Gillson, and P. E. Bierstedt, *Solid State Commun.* 17 (1975) 27

[19] J. G. Bednorz and K. A. Müller, *Z. Phys. B* 64 (1986) 189

[20] For reviews, see D. C. Johnston, *Handbook of Magnetic Materials*, Ch. 1, Vol. 10, ed. K. H. J. Buschow (Amsterdam: Elsevier, 1997), pp. 1–137; M. A. Kastner, R. J. Birgeneau, G. Shirane, and Y. Endoh, *Rev. Mod. Phys.* 70 (1998) 897

[21] L. Gao, Y. Y. Xue, F. Chen, Q. Xiong, R. L. Meng, D. Ramirez, C. W. Chu, J. H. Eggert, and H. K. Mao, *Phys. Rev. B* 50 (1994) 4260

[22] L. F. Mattheiss, E. M. Gyorgy, and D. W. Johnson, Jr., *Phys. Rev. B* 37 (1988) 3745

[23] R. J. Cava, B. Batlogg, J. J. Krajewski, R. Farrow, L. W. Rupp, Jr., A. E. White, K. Short, W. F. Peck, and T. Kometani, *Nature* 332 (1988) 814

[24] K. Tanigaki, T. W. Ebbesen, S. Saito, J. Mizuki, J. S. Tsai, Y. Kubo, and S. Kuroshima, *Nature* 352 (1991) 222

[25] J. Nagamatsu, N. Nakagawa, T. Muranaka, Y. Zenitani, and J. Akimitsu, *Nature* 410 (2001) 63

ered by Kamihara et al. to have a high $T_c = 26$ K,[26] as discussed by Hideo Hosono in another article in this chapter. Members of this general class of materials have crystal structures containing iron square lattice layers with a maximum T_c of 56 K,[27] which coincidentally(?) is the same transition metal sublattice structure as the Cu atoms have in the layered cuprate high-T_c superconductors.

Herein are presented retrospectives of the discoveries of $T_c > 10$ K in the non-cuprate oxides $LiTi_2O_4$, $Ba(Pb_{1-x}Bi_x)O_3$, and $(Ba_{1-x}K_x)BiO_3$, with an emphasis on $LiTi_2O_4$ because of the author's familiarity with this compound. Lack of space precludes discussions of their normal or superconducting state physical properties other than their T_cs and of their mechanism(s) for superconductivity.

4.3.2 $LiTi_2O_4$

The face-centered-cubic normal spinel crystal structure of $LiTi_2O_4$ is shown in Figure 4.4 (space group $Fd\bar{3}m$, No. 227). The structure prototype is the mineral spinel with the composition $MgAl_2O_4$. The structure consists of a nearly cubic-close-packed array of O atoms layered along the [111] directions of the unit cell, with the Mg atoms in tetrahedral interstices and Al atoms in octahedral interstices between adjacent O layers. The cation sublattices, shown separately in the bottom panel of Figure 4.4, are both geometrically frustrated for antiferromagnetic ordering due to the triangular connectivity of the respective nearest neighbors. Most $3d$ transition metal spinel structure compounds become distorted at low temperatures, which can partially relieve the frustration. There are very few transition metal oxide spinel compounds that remain metallic and cubic at low temperatures. The intermediate-valent spinel compound $LiTi_2O_4$ is one of these and LiV_2O_4 is another, where the formal oxidation state of both the Ti ($d^{0.5}$) and V ($d^{1.5}$) is +3.5. This nonintegral valence state together with the undistorted crystal structure require both compounds to be metals to low temperatures.

The discovery of superconductivity in $LiTi_2O_4$ came about in the following way. At about the middle of the period (1969–1975) when I was a graduate student at the University of California at San Diego, my research advisor, Bernd T. Matthias, asked me to make a sample of $Li_xTi_{1.1}S_2$, which he and coworkers at Bell Laboratories, Murray Hill, NJ, had recently (1972) reported to become superconducting at a high temperature of 10–13 K for $0.1 < x \leq 0.3$.[28] Lithium is a very reactive alkali metal, so I asked a chemistry professor to make some powder of $Li_xTi_{1.1}S_2$ that I could process further. His postdoc Hari Prakash, using wet chemical methods, provided me with a batch of amorphous $Li_xTi_{1.1}S_2$ which, in retrospect, probably had a high surface area. In order to make a crystalline sample, I wanted to melt it in an arc-furnace (like a commercial arc-welder, but in an inert argon atmosphere). This was certainly not an optimum way to crystallize the material! However, even so, I wanted to do this experiment, so I first poured some of the powder out of the bottle in air in order to make a pressed pellet, but the powder started to smoke! Therefore I made the pellet as fast as possible and loaded it into the arc-furnace. When I tried to melt the pellet, the sample made a lot of smoke that filled up the arc-furnace because of the volatility of Li and S, but it also made a melted round ball that was a beautiful dark blue color (metals are usually silvery and don't usually have colors; pure bulk elemental Os metal is light blue, though). Then I measured pieces of the blue ball for superconductivity using ac susceptibility measurements and found that the sample became superconducting at 11 K. This seemed to confirm Bernd's findings. But all was not as it seemed. My powder x-ray diffraction pattern of the sample contained lines that I could index as Li_2TiO_3, and I subsequently found that the intensity of these lines correlated with the volume fraction of superconductivity in similarly prepared samples.

---

[26] Y. Kamihara, T. Watanabe, M. Hirano, and H. Hosono, *J. Am. Chem. Soc.* 130 (2008) 3296

[27] For a comprehensive critical review, see D. C. Johnston, *Adv. Phys.* 59 (2010) 803

[28] H. E. Barz, A. S. Cooper, E. Corenzwit, M. Marezio, B. T. Matthias, and P. H. Schmidt, *Science* 175 (1972) 884

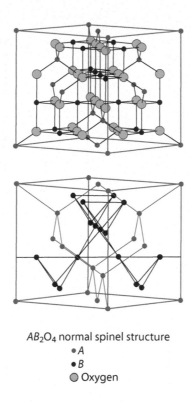

AB_2O_4 normal spinel structure
- A
- B
- Oxygen

FIGURE 4.4: (top panel) Face-centered-cubic crystal structure of a normal spinel oxide compound in which the A atoms occupy only tetrahedral sites and the B atoms only the octahedral sites between O layers that are close-packed along the [111] directions. The unit cell edges are shown. The unit cell contains eight formula units. (Reprinted figure with permission from S. Kondo, D. C. Johnston, and L. L. Miller, Phys. Rev. 59 (1999) 2609. Copyright 1999 by the American Physical Society. http://link.aps.org/doi/10.1103/PhysRevB.59.2609.) (bottom panel) Cation sublattice of the normal spinel structure. The B sublattice consists of corner-sharing tetrahedra.

Then, as luck would have it, Bernd had a famous crystallographer friend, Willie Zachariasen, who happened to be visiting Bernd's lab at UCSD at the time. I pointed out to him the set of x-ray diffraction peaks that correlated with the volume fraction of superconductivity. After a couple of days analyzing the peaks, he figured out that the composition of the superconducting material in my samples was close to $LiTi_2O_4$ with a face-centered-cubic (fcc) normal-spinel structure. It turned out that the strongest lines in the pattern were at about the same positions as those for the different phase Li_2TiO_3 noted above. Undoubtedly Hari did provide me with pure $Li_xTi_{1.1}S_2$, but when I exposed it to air, it reacted with the air and oxidized before I could make it into a pellet. Then when I subsequently arc-melted it in pure Ar, part of the sample turned into $LiTi_2O_4$ which was superconducting at 11 K. Furthermore, I later discovered that the metallic spinel structure compound $LiTi_2O_4$ transforms to a semiconducting Ramsdellite structure compound with the same composition upon heating above 950 °C, and that the conversion of the structure to the spinel structure on cooling to below 950 °C is a slow process. Apparently when the melted sample cooled in the arc-furnace, at least some of it converted to the low-temperature spinel crystal structure. The discovery of superconductivity in $LiTi_2O_4$ was thus pure serendipity.

I subsequently learned how to synthesize $LiTi_2O_4$ in a rational way and further confirmed that the superconductivity arose from $LiTi_2O_4$ with the spinel structure at temperatures up to 13.7 K. R. Viswanathan at UCSD carried out heat capacity measurements on the initial samples and demonstrated that $LiTi_2O_4$ is a bulk superconductor. The first announcement of "High-Temperature Superconductivity in the Li-Ti-O Ternary System" was in 1973.[17] I did my Ph.D. Thesis on the follow-up synthesis, structure, and properties of polycrystalline samples of $LiTi_2O_4$ and of the solid solution $Li_{1+x}Ti_{2-x}O_4$.[29,30,31] Some of these studies were carried out in collaboration with Bill McCallum, Carlos Luengo, Brian Maple, Robert Shelton, and Hermann Adrian who were all also at UCSD at the time.

Before I worked on $LiTi_2O_4$, it was known that there exists a complete solid solution of spinel-structure compounds from $LiTi_2O_4$ to $Li(Li_{1/3}Ti_{5/3})O_4$, which can be written $Li_{1+x}Ti_{2-x}O_4$ with $0 \leq x \leq 1/3$.[29] In this solid solution, some Li substitutes for the Ti on the octahedral sites of the structure. At the same time, the formal oxidation state of the Ti changes from +3.5 to +4.0. The compound with $x = 1/3$ is a colorless nonmagnetic insulator. Therefore there had to be a metal-to-insulator transition with increasing x. As part of my thesis work, I investigated this issue and found a metal-insulator transition at $x \approx 0.10$ with T_c being nearly constant for $0 \leq x \leq 0.1$. This suggested that if a metal-insulator transition occurred in a specific materials system, the potential for high T_c superconductivity might be greatest for the metallic compound in that system with a composition closest to the metal-insulator boundary. I made a compilation of oxide and chalcogenide compounds and found that this correlation indeed held about 80% of the time. It was subsequently found to be applicable to the cuprate high-T_c superconductors. This correlation is also mentioned in Arthur Sleight's commentary below. A summary of the literature on $LiTi_2O_4$ up to 1999 was given by Moshopoulou.[32]

Before leaving the topic of $LiTi_2O_4$, one might think that the very similar metallic normal spinel structure compound LiV_2O_4, which has the same beautiful dark blue color, would also be a superconductor. However, in 1973, it was already known that LiV_2O_4 was not superconducting above 4 K, and that this compound, despite being metallic, showed a Curie-Weiss temperature dependence to the magnetic susceptibility with a Curie constant indicating that each V atom had a spin $S = 1/2$ with a spectroscopic splitting factor $g \sim 2$. Since my advisor Bernd Matthias was mainly interested in discovering new superconductors at UCSD, I did not pursue the properties of LiV_2O_4 at that time. It was not until 1997 that my group at Iowa State University discovered that LiV_2O_4 is a very rare example of a d-electron heavy fermion compound, in collaboration with Clayton Swenson and Ferdinando Borsa's and Alan Goldman's groups at Iowa State and with the groups of James Jorgensen, Brian Maple, and Yasutomo Uemura.[33,34]

4.3.3 $Ba(Pb_{1-x}Bi_x)O_3$

This compound has a primitive cubic perovskite crystal structure (space group $Pm\bar{3}m$, No. 221), named after the prototype mineral perovskite, $CaTiO_3$. The structure is shown in Figure 4.5. The structure of an ABO_3 cubic perovskite compound consists of corner-sharing BO_6 octahedra with A atoms in the tunnels between the octahedra. The structure is very susceptible to lattice distortions, sometimes by cooling the cubic structure from room temperature, and many insulating compounds with distorted variants of this structure are ferroelectric.

[29] D. C. Johnston, *J. Low Temp. Phys.* 25 (1976) 145

[30] R. W. McCallum, D. C. Johnston, C. A. Luengo, and M. B. Maple, *J. Low Temp. Phys.* 25 (1976) 177

[31] R. N. Shelton, D. C. Johnston, and H. Adrian, *Solid State Commun.* 20 (1976) 1077

[32] E. G. Moshopoulou, *J. Am. Ceram. Soc.* 82 (1999) 3317

[33] S. Kondo, D. C. Johnston, C. A. Swenson, F. Borsa, A. V. Mahajan, L. L. Miller, T. Gu, A. I. Goldman, M. B. Maple, D. A. Gajewski, E. J. Freeman, N. R. Dilley, R. P. Dickey, J. Merrin, K. Kojima, G. M. Luke, Y. J. Uemura, O. Chmaissem, and J. D. Jorgensen, *Phys. Rev. Lett.* 78 (1997) 3729

[34] For a review, see D. C. Johnston, *Physica* B 281–282 (2000) 21

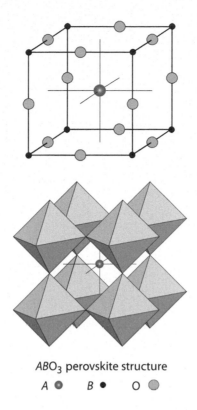

ABO_3 perovskite structure

A ● B ● O ○

FIGURE 4.5: (top panel) Cubic perovskite crystal structure of an ABO_3 compound. The unit cell edges are shown. The unit cell contains one formula unit. (bottom panel) Figure showing the corner-sharing BO_6 octahedra with the A atoms in between. The centers of the eight BO_6 octrahedra shown are each occupied by B atoms and are at the corners of the unit cell shown in the top panel.

The following account of the discovery of superconductivity in the $Ba(Pb_{1-x}Bi_x)O_3$ system at temperatures up to 13 K in 1975 was provided by Arthur Sleight, showing that serendipity was again at work:

> I had just discovered a first-order metal-insulator [MI] transition in $Tl_2Ru_2O_7$ and a second-order metal-insulator transition in $Cd_2Os_2O_7$. I was wondering whether there could be a metal-insulator transition in an oxide without a transition metal. $BaBiO_3$ was known, but it had not been well characterized. The diffraction data that existed at that time suggested that there might be only one type of Bi present. If this were true, $BaBiO_3$ should be metallic with a half-filled $6s$ conduction band. I was thinking that as a function of temperature there might be a transition from $BaBi^{4+}O_3$ (metal) to $Ba_2Bi^{3+}Bi^{5+}O_6$ (insulator). I decided crystals were needed, and none have ever been reported. I grew some crystals. At first, I thought they might be metallic because they were a gold color with a metallic luster. However, electrical measurements showed that the crystals were semiconducting with a rather high resistivity. I knew that $BaPbO_3$ was a very good conductor. So I thought there must be some sort of metal-insulator transition in the $BaPbO_3$–$BaBiO_3$ solid solution, which had never been investigated. I first made 50/50 ($BaPb_{1/2}Bi_{1/2}O_3$). It was not metallic. So I then made $BaPb_{0.75}Bi_{0.25}O_3$, and it was metallic and superconducting. ... I wanted to be very sure of things before we published. I grew crystals of $BaPb_{0.75}Bi_{0.25}O_3$ and they were superconducting. We were not set up for magnetic measurements appropriate for superconductors. But I gave

a sample to a guy who measured magnetic susceptibility. He said it must be a super-conductor because the sample jumped out of its holder when the temperature of zero resistivity was reached. I was confident of our results, and I did not want our paper to be reviewed by skeptics. So I sent it to a journal where an Editor would communicate it.[18] I then tried $BaBiO_3$–$KBiO_3$ at that time, but only once and saw no superconductivity.

I had previously worked on W and Re oxides superconductors, and I did appreciate at that time that superconductivity frequently occurred in compositions that were close to a phase boundary between metallic and insulating properties. The actual value of T_c in the $Ba(Bi,Pb)O_3$ system was a surprise to me. So I was not really looking for a new superconductor, but my search for a MI transition based on $BaBiO_3$ made this discovery inevitable.

4.3.4 $(Ba_{1-x}K_x)BiO_3$

Superconductivity was discovered in 1988 at temperatures up to around 22 K in the system $(Ba_{1-x}K_x)BiO_3$ by Mattheiss et al.[22] using a specific focused approach. Several factors led to this discovery.[22] First, the superconductivity at temperatures up to 13 K in the related perovskite system $Ba(Pb_{1-x}Bi_x)O_3$ described above was well known. Second, according to Mattheiss et al., based on previous work, "These results suggest that it should be possible to suppress the ordering waves [that result in insulating behavior] and extend the metallic regime in $BaPb_{1-x}Bi_xO_3$ closer to the half-filled band condition ($BaBiO_3$) where the electron-phonon interaction is a maximum, by leaving the conducting Bi-O complex intact and instead doping substitutionally at the inactive Ba donor sites. This situation, which is analogous to that in $La_{2-x}(Ba,Sr)_xCuO_4$, is expected to produce marginal stability and enhanced T_cs. In earlier studies, a combination of K and Pb dop-ing in a $Ba_{0.9}K_{0.1}Pb_{0.75}Bi_{0.25}O_3$ sample has produced similar critical temperatures ($T_c \sim 12$ K) but sharper transitions than those observed in Pb-doped $BaPb_{0.75}Bi_{0.25}O_3$ samples. In the present in-vestigation we have carried out dc magnetization measurements on K-doped $Ba_{0.9}K_xBiO_3$ samples with $x \approx 0.2$".[22] (Reprinted excerpt with permission from L. F. Mattheiss, E. M. Gyorgy, and D. W. Johnson, Jr., Phys. Rev. B 37 (1988) 3745. Copyright 1988 by the American Physical Society. http://link.aps.org/doi/10.1103/PhysRevB.37.3745.) This investigation resulted in superconducting onset temperatures up to about 22 K as noted above.

Cava and coworkers subsequently refined the synthesis procedures and obtained a supercon-ducting onset temperature of 29.8 K in a single-phase cubic perovskite structure sample of $Ba_{0.6}K_{0.4}BiO_3$, thus firmly confirming the source of the superconductivity as the compound with the cubic perovskite structure.[23] The onset temperature was determined by Cava et al. from a mag-netization measurement of the Meissner effect upon cooling the sample in an applied magnetic field of 19 Oe. We note that direct Meissner effect measurements such as this on other materials are of-ten not informative or conclusive due to flux trapping effects that prevent the magnetic flux from escaping from the sample upon cooling below T_c.

4.4 Cuprates—Superconductors with a T_c up to 164 K

C. W. Chu

The important role of cuprates in the development of high temperature superconductivity (HTS) as we know it today can never be overstated. To date, a superconducting transition temperature (T_c) above the liquid nitrogen boiling point can be found only in cuprates. It was the seminal discovery of the first cuprate high temperature superconductor (HTS), the Ba-doped La_2CuO_4 (214), with a T_c of 35 K in 1986 by Alex Müller and George Bednorz,[35] who were awarded the Nobel Prize in 1987, that ushered in the era of cuprate high temperature superconductivity. Before 1986, the highest T_c was 23 K observed in Nb_3Ge and the study of the then HTS was mainly on intermetallic compounds. In the ensuing 24 years since the discovery of 214, more than 200 cuprates have been found to superconduct at temperatures up to the current T_c-record of 134 K at ambient pressure[36] and 164 K at 30 GPa.[37] They all display a layer structure and can be represented by a generic formula $A_mE_2R_{n-1}Cu_nO_{2n+m+2}$, where A = Bi, Tl, Pb, Hg or Cu; E = Ca, Sr or Ba; R = Ca, Y or rare-earth; m = 0, 1 or 2 and n = 1, 2, ... The generic formula can be rewritten as $A_mE_2R_{n-1}Cu_nO_{2n+m+2}$ = $[(EO)(AO)_m(EO)] + [(CuO_2)R_{n-1}(CuO_2)_{n-1}]$, which consists of two substructures: the active block of $[(CuO_2)R_{n-1}(CuO_2)_{n-1}]$ and the charge reservoir block of $[(EO)(AO)_m(EO)]$. The space group for compounds with m = 2 or 0 is I4/mmm but changes to P4/mmm when m = 1. The active block comprises n square-planar (CuO_2) layers interleaved by $(n-1)$ R layers and the charge reservoir block contains m (AO) layers bracketed by 2 (EO) layers, which are relatively inert chemically. Superconducting current flows mainly in the (CuO_2) layers and doping takes place in the charge reservoir block, which transfers charges without introducing defects into the (CuO_2) layers (similar to modulation-doping in semiconducting superlattices). All cuprate HTSs can simply be designated as $Am2(n-1)n$ or just $0(n-1)n$-E when the two (AO) layers are absent.[38]

The history of the discovery of the 214 superconductor has been described by A. Bussmann-Holder and K. A. Müller in section 4.2 in this chapter. I shall recall briefly the discoveries of only four main cuprate HTS families that have played significant roles in the development of HTS science and technology, while leaving other families derived from these four to literature for more serious readers. In chronological order, they are:

- $RBa_2Cu_3O_7$ [RBCO, 123, or Cu1212 with R = Y or rare-earth (La, Nd, Sm, Eu, Gd, Tb,Dy, Ho, Er, Tm, Yb, or Lu)];

- $Bi_2Sr_2Ca_{n-1}Cu_nO_{2n+4}$, where n = 1, 2, 3, ... [BSCCO or Bi22$(n-1)n$];

- $Tl_2Ba_2Ca_{n-1}Cu_nO_{2n+4}$, where n = 1, 2, 3, ... [TBCCO or Tl22$(n-1)n$]; and

- $HgBa_2Ca_{n-1}Cu_nO_{2n+3-\delta}$, where n = 1, 2, 3, ... [HBCCO or Hg12$(n-1)n$].

4.4.1 $RBa_2Cu_3O_7$

The first cuprate family that displays a T_c in the 90s K was discovered by Paul C. W. Chu, Maw-Kuen Wu, and colleagues in their respective groups at the University of Houston and the University of Alabama at Huntsville in January 1987.[39] The discovery represents a giant advancement in modern science and has drastically changed the psyche of superconductivity research, heralding the new era of HTS. In spite of the many cuprate HTSs subsequently discovered, RBCO remains the most desirable HTS material for applications to date due to its physical robustness and superior superconducting behavior in high magnetic field.

With RBCO being the first HTS family to bring down the liquid nitrogen temperature barrier of 77 K and my heavy personal involvement, it is rather natural to indulge myself in a little more space in this article to recall selected events before and during its discovery, although some of these have been presented previously. At the same time, I also remember that R. W. Emerson once said,

---

[35] J. G. Bednorz and K. A. Müller, *Z. Phys. B* 64 (1986) 189

[36] A. Schilling et al., *Nature* 363 (1993) 56

[37] L. Gao et al., *Phys. Rev. B* 50 (1994) 4260(R)

[38] See for example, C. W. Chu, *AAPPS Bulletin* 18 (2008) 9

[39] M. K. Wu et al., *Phys. Rev. Lett.* 58 (1987) 908

"there is no history; there is only biography." This is particularly true when the events are recounted by a person who himself is a player and the line between history and biography can be blurred. Incompleteness of the account is inevitable. For this I have to apologize.

Until his untimely death in 1980, Bernd Matthias was an undisputed leader in the search of novel high T_c superconductors. Matthias was an Edisonian and was deeply skeptical about theories. As a former student of his, I have been continuously influenced, for better or worse, by his style of doing physics and his taste in selecting problems, especially in the pursuit of high temperature superconductivity. I had learned to pay attention to materials that are usually considered too mundane by many physicists and not to be intimidated by theoretical predictions that are often treated as Sacred Writ by many experimentalists. Matthias' unceasing enthusiasm and optimism about the future of superconductivity with higher T_c were contagious. I have dreamed of a superconductor of a practically high T_c ever since I became a graduate student of his in the 1960s.

In the 1970s and 1980s, one of the major concerns for higher T_c was structural instabilities often observed in superconductors with a relatively high T_c. To determine the correlation of lattice instabilities with superconductivity, I developed an ac calorimetric high pressure technique to vary and detect simultaneously the structural and superconducting transitions without introducing any chemical complexity. We found that lattice instabilities do affect the T_c of the A15 superconductors, but only slightly, by no more than a few tenths of one degree.[40] The observation gave me the confidence that superconductivity at higher T_c might be achievable. During the same period of time, I also studied compounds with unexpectedly high T_c, trying to see if a novel superconducting mechanism was in operation. They included the ~ 13 K oxide superconductors, $Ba(Pb_{1-x}Bi_x)O_3$ and $Li_{1+x}Ti_{2-x}O_4$, the precursors of HTSs.

On September 1, 1986, I took up the so-called rotator position at the National Science Foundation (NSF) as a co-director with Joe Trivisonno of the Solid State Physics Program, while serving simultaneously as the Director of the NASA Space Vacuum Epitaxy Center (SVEC) established by me at the University of Houston (UH) only on August 1, 1986. Due to the flextime rule of NSF, and the sympathetic management of NSF and UH, I experienced one of the most productive and exciting years in my life as a physicist. I traveled frequently between Washington, D.C., and Houston to review the NSF grant proposals and to supervise my research group. I took the first trip home on November 7, 1986, to prepare for the formal inaugural meeting of SVEC. On Saturday, November 8, I came to my office early and saw the paper by Bednorz and Müller, which was left on my desk by Ruling Meng, a visiting scientist in my group. I read it with fascination and excitement. I foresaw that a new chapter in HTS was about to unfold and decided immediately to duplicate their results as quickly as possible.

It should be noted that the report by Bednorz and Müller was initially met with skepticism and did not attract too much attention except by a few, since oxides are often insulators and not even metallic, let alone superconducting at high temperature. In addition, their decisive magnetic data showing the Meissner effect did not reach the U.S. until late January 1987. Little action was taken after the news broke except by a lucky few. My group was among those lucky few who took the report seriously due to our previous experience mentioned above. On November 21, we observed reproducibly a resistance transition at 30.2 K in the first sample (LBCO) cut from the mixed-phase $LaBa_4Cu_5O_{15}$ disk following the stoichiometry of Müller and Bednorz. Unfortunately, a day or two later, we failed to detect any resistance drop in the other four samples from the same disk. I often wondered what would have happened had we not observed superconductivity in the first four samples. Fortunately, more positive results were observed by us later. I briefly mentioned our results on December 4 at the 1986 MRS Fall Meeting in Boston after I finished my scheduled talk on $Ba(Pb_{1-x}Bi_x)O_3$. At the end of the Q&A, Koichi Kitazawa of Tokyo walked up to the podium and showed his magnetic results that clearly showed the Meissner effect in LBCO, removing any remaining doubt concerning

---

[40] C. W. Chu and V. Diatschenko, *Phys. Rev. Lett.* 41 (1978) 572 and references therein

the nature of the transition. The genie of high temperature superconductivity in LBCO was finally out of the bottle.

After reproducing the results of Müller and Bednorz, we subjected the LBCO mixed-phase samples to high pressure, focusing only on the superconducting phase in an attempt to reveal the cause for such an unexpectedly high T_c in oxides. The T_c was unexpectedly raised to 40 K and then to 52 K at a rate more than ten times that of the intermetallic superconductors. The observation of a T_c higher than 40 K shattered the then-theoretically predicted T_c-limit of 30s K and raised serious questions about the theoretical T_c-predictions of the time. The unusually large positive pressure effect on T_c observed further suggested to me that higher T_c might be achievable through chemical pressures by replacing elements in LBCO with those of smaller ionic radii and the same valences, such as Ba by Sr or Ca and La by the nonmagnetic Y or Lu. The Ba/Sr replacement to raise T_c was quickly confirmed, but the Ba/Ca substitution was unfortunately found to lower the T_c. At the urging of our dean at UH, Roy Weinstein, and the university legal counsel, Scott Chaffin, I filed the patent application with the US Patent Office on January 12, 1987, while ordering the nonmagnetic oxides of Y and Lu for the replacements conjectured.[41]

At the 1986 Fall MRS Meeting in Boston, after my presentation, Kitazawa also told me that his group had identified La214 as the superconducting phase in the original LBCO mixed-phase samples. It was natural for me to decide to make pure 214-phase samples, preferably single crystals, and to examine the origin of the 35 K T_c before contemplating the next step to raise T_c. We tried but failed to grow La214 single crystals, following the destruction of two of my three expensive crystal-growing Pt-crucibles. "A kick of luck," as Müller put it, clearly played a role in many scientific discoveries (K. A. Müller personal communication). Subsequently, I decided to turn our attention to stabilizing the high temperature anomalous resistance drops, indicative of superconductivity. They were detected sporadically in the multiphase LBCO samples above 70 K as early as November 25, 1986, but not in the pure 214 ones. Regardless of its unstable nature, I showed the ∼ 70 K preliminary data to my former student M. K. Wu, who was then with the University of Alabama at Huntsville, at the MRS meeting in Boston and successfully convinced him to join our search. On January 12, 1987, we observed a large diamagnetic shift or Meissner signal up to ∼ 96 K in one of our mixed-phase LBCO samples, representing the first definitive superconducting sign detected above the liquid nitrogen temperature of 77 K. It is interesting to note that the XRD-data taken at the time already displayed the La123 phase but this was recognized only later after the structure of Y123 was determined. Unfortunately, the sample degraded and the diamagnetic signal disappeared the following day. No effort of ours in the ensuing two weeks succeeded to reproduce and stabilize this high temperature superconducting signal. I decided to write up details of the experiment and let other better equipped groups stabilize and identify the high temperature superconducting phase. No sooner than half of the paper was drafted, Wu called from Alabama in the afternoon of January 29, 1987, with the exciting news that a resistive drop indicative of a superconducting transition above 77 K was detected in the mixed-phase samples $Y_{1.2}Ba_{0.8}CuO_4$ (YBCO). We were all ecstatic. On January 30, Wu brought to Houston his samples, in which a resistive transition was immediately reproduced and the Meissner effect observed. Stable superconductivity at ∼ 93 K was finally achieved as shown in Figure 4.6, nearly tripling the T_c of La214. The excitement preempted my desire to complete the manuscript on the unstable 90 K LBCO, but it was left as a footnote in the YBCO paper. It took me less than 24 hours to complete two drafts entitled, "Superconductivity at 93 K in a New Mixed Phase Y-Ba-Cu-O Compound System at Ambient Pressure" and "High Pressure Study on the New Y-Ba-Cu-O Compound System." The latter was done to demonstrate that the 90 K YBCO belonged to a different phase from the 35 K 214. After being circulated among my collaborators for comment, the manuscripts were FedExed to the *Physical Review Letters* on February 5, 1987. Both were accepted on February 11 and appeared back-to-back in the March 2, 1987 issue. Contrary to

---

[41] C. W. Chu, patent application, January 12, 1987

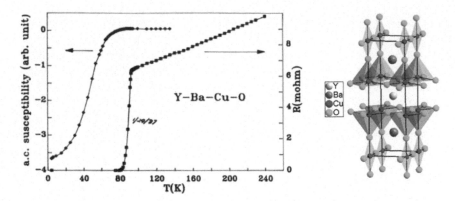

FIGURE 4.6: YBCO temperature dependence of resistance[39] and structure. (Reprinted figure with permission from M. K. Wu, J. R. Ashburn, C. J. Torng, P. H. Hor, R. L. Meng, L. Gao, Z. J. Huang, Y. Q. Wang and C. W. Chu, Phys Rev Lett. 58 (1987) 908. Copyright 1987 by the American Physical Society. http://link.aps.org/doi/10.1103/PhysRevLett.58.908.)

the initial hesitation toward 214, the 90 K YBCO results were reproduced rapidly worldwide after the news broke. The world of HTS was on fire and YBCO heralded the modern era of HTS.

I still vividly remember the extraordinary emotion I felt when I wrote the second sentence in our first YBCO paper,[39] "To obtain a superconducting state reaching beyond the technological and psychological temperature barrier of 77 K, the liquid nitrogen boiling point, would be one of the greatest triumphs of scientific endeavor of this kind," knowing that we had achieved it. Being the first to claim superconductivity with a T_c above 77 K placed upon us tremendous psychological burden. In spite of my confidence in our results, the thought—"could it be too good to be true?"—did enter my mind occasionally after submitting the manuscripts. A mistake of this magnitude would have meant the abrupt end of my career in superconductivity. More than once, I dropped by the offices of Meng and Pei-Herng Hor, my then student, and asked, "could there be phenomena other than superconductivity that can account for our observations? Please think and think hard!" Even in the cover letter to *PRL* editor Myron Strongin, I confirmed the earlier verbal agreement with him over the phone that "information in these manuscripts should not be released nor experimentally tested prior to their formal publications by the journal." To do so was to let us have more time to crack the mystery and to correct possible mistakes.

The next challenge was to isolate and identify the 93 K superconducting phase from the greenish-looking mixed phase YBCO samples and to determine its composition and structure. After failing to resolve the problems with colleagues at Houston, I sought help from Dave Mao and Bob Hazen at the Geophysical Lab in Washington, D.C., knowing that they had expertise in determining the structures and compositions of tiny crystals from rocks, and they graciously agreed. I brought them a sample the third week of February 1987. They started working feverishly on the sample. Within a week, they determined the compositions and symmetries of the black and green phases closely intertwined in the small grains of the crushed sample. At the same time, we determined that the black phase was superconducting by correlating the sample color with the superconducting volume in samples of different compositions and synthesized under different conditions in Houston. Through continued and instant exchanges of information between Hazen and me, an almost pure YBCO sample was obtained in Houston on March 1 and was brought to Hazen and Mao on March 5. Within a few days they solved the structure of the first liquid nitrogen HTS as $YBa_2Cu_3O_{7-\delta} = (BaO)(CuO_{1-\delta})(BaO)(CuO_2)(Y)(CuO_2)$ as shown in Figure 4.6, except for the oxygen vacancy

FIGURE 4.7: The Woodstock of Physics, March 18, 1987. (Photo courtesy Phillip Schewe, American Institute of Physics.)

locations.[42] The more than one CuO_2 layer per unit cell was proposed to be responsible for the 90 K-T_c.

We asked if Y can be completely replaced by nonmagnetic Lu,[41] with the same T_c and if other superconductors of similar or higher T_c could be found. We decided to determine the role of Y in YBCO. With the structure information in hand, we quickly found through partial magnetic substitution that Y in YBCO is electronically isolated from the superconducting component of the compound and serves only as a stabilizer to hold the crystal structure together. The information led us to the subsequent discovery of the series of $RBa_2Cu_3O_7$ ($RBCO$, R123 or Cu1212) with a T_c varying between 93 and 100 K where $R = Y$ and all rare-earth elements except Ce and Pr, expediently using the reduced atmosphere method. The results appeared in an article in the May 1987 issue of *PRL* entitled, "Superconductivity above 90 K in the Square-Planar Compound System $ABa_2Cu_3O_{6+x}$ with $R = $ Y, La, Nd, Sm, Eu, Gd, Ho, Er, and Lu."[43] Two related structures were later found by others, i.e. $YBa_2Cu_4O_8$ and $Y_2Ba_4Cu_7O_{15}$ with a T_c around 80 K.

The discovery broke the liquid nitrogen temperature barrier of 77 K for superconductivity. It has brought superconductivity applications a giant step closer to reality and at the same time poses serious challenges to physicists concerning the origin of HTS. The excitement at the time was amply demonstrated by the Special Panel Discussion on Novel Materials and High Temperature Superconductivity organized by the American Physical Society (APS) on March 18, 1987, at the New York Hilton (Figure 4.7) and by the Federal Conference on High Temperature Superconductivity Science and Technology in July 1987 in Washington, D.C., attended by President Reagan and his cabinet members. The special APS discussion panel was initiated by me and hastily organized by APS. It started at 7:30 p.m. with short presentations by five panelists: Alex Müller of the IBM Zurich Lab; Shoji Tanaka of the University of Tokyo; Paul Chu of the University of Houston; Bertrum Batlogg of Bell Labs; and Zhongxian Zhao of the Physics Institute of the Chinese Academy of Sciences, followed by short contributions and discussions that lasted until the wee hours of the next morning. The 1,200-seat meeting room was packed with more than 2,000 people. Many more who could not

---

[42] R. M. Hazen et al., *Phys. Rev. B* 35 (1987) 7238

[43] P. H. Hor et al., *Phys. Rev. Lett.* 58 (1987) 1891

get into the room watched on TV screens outside the room to witness this exciting event, called the Woodstock of Physics by the late Bell Labs physicist Mike Schluter, referring to the legendary 1969 Woodstock Music and Art Festival in upstate New York. In spite of later discoveries of many other cuprate HTS families, some of which have higher T_c, YBCO remains the most desirable material for HTS science and technology due to its superior sample quality, current carrying capacity in the presence of high magnetic fields, and physical robustness in thin-film form. A YBCO puck was selected as an entry for the White House's National Millennium Time Capsule in 2000 (Figure 4.8), which was created in the spirit of "honor the past—imagine the future" to contain discoveries and achievements in all areas by Americans over the previous 100 years considered to be significant. It will be opened in 2100 to communicate to future generations about US accomplishments and visions made in the 20th Century.

FIGURE 4.8: Closing ceremony of the National Millennium Time Capsule that included a YBCO puck. (Courtesy White House Millennium Council.)

4.4.2 $Bi_2Sr_2Ca_{n-1}Cu_nO_{2n+4}$

The first cuprate family that displays a T_c up to 110 K at ambient and ~ 135 at ~ 35 GPa was discovered without rare-earth element by Hirosh Maeda et al. of the National Research Institute for Metal in January 1988.[44] The discovery suggested that HTS had a broader material base than originally thought and that more HTSs with a higher T_c were possible. BSCCO has been the material used for the first generation HTS-wires and for spectroscopic studies, due to its graphitic-like behavior.

Euphoria permeated the field following the announcement of superconductivity above 90 K in YBCO and RBCO, and the sky seemed to be the only limit to T_c. As 1987 drew to an end, an impatient physicist at an industrial lab told the *Wall Street Journal* that the accumulated person-hours devoted to cuprate-HTSs in 1987 had exceeded all previous effort devoted to low temperature superconductors in the preceding 75 years since the discovery of superconductivity and that any

---

[44] H. Maeda et al., *Jpn. J. Appl. Phys.* 27 (1988) L209

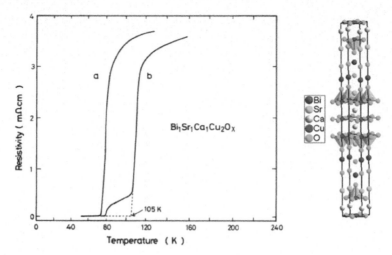

FIGURE 4.9: BSCCO temperature dependence of resistance[44] and structure. (Reprinted with permission from H. Maeda et al., Japanese Journal of Applied Physics 27 (1988) L209. Copyright 1988 by Japanese Society of Applied Physics.)

cuprate with a T_c above 90s K should have been found. He went on to propose that one should search outside cuprates for superconductors with higher T_c. The fallacy of prophecy based on past statistics immediately faced the truth in the first week of January 1988 when Maeda et al. reported superconductivity above 100 K in their mixed-phase samples of Bi-Sr-Cu-O that later appeared in an article in the February 20 issue of *Japanese Journal of Applied Physics* entitled "A New High-T_c Oxide Superconductor without a Rare Earth Element."[44]

In their attempt to broaden the material base for HTSs, Maeda et al. examined elements in the VB group of the periodic table, such as Bi, which is trivalent and has a similar ionic radius to those of the rare earth elements, along the lines of 214 and *R*BCO. They found superconductivity in samples of nominal $Bi_1Sr_1Ca_1Cu_2O_x$ above 105 K, as shown in Figure 4.9. The resistivity data suggested that the sample might consist of multiple phases. Indeed, shortly afterward, three members of the Bi22(n-1)n family with $n = 1$, 2, and 3 were isolated and the structures determined by several groups, including the Carnegie Institute Geophysical Lab and our own. The layer stacking sequence of the highest T_c member of the family is $Bi_2Sr_2Ca_2Cu_3O_{10} = (SrO)(BiO)(BiO)(SrO)(CuO_2)(Ca)(CuO_2)(Ca)(CuO_2)$, as also shown in Figure 4.9. Layer structural modulation appears in the (BiO) double layers for all n. The maximum T_cs for members of the homologous series increase with n and are ~ 22, ~ 80, and ~ 110 K for $n = 1$, 2, and 3, respectively. The T_c of Bi2223 was reported to reach 135 K without sign of saturation at ~ 35 GPa by Chen et al. at the Carnegie Geophysical Lab in August 2010. However, for $n > 3$, T_c starts to drop and the drop is attributed to the combined influence of electrostatic shielding and proximity effects of the (CuO_2) layers. Aided by the experience on Y123, it took less than two days for Hazen et al. of Carnegie Geophysical Lab and Veblen et al. of Johns Hopkins to crack the structure of BSCCO after receiving the samples from us, in contrast to the more than 10 days taken by Hazen et al. to do the same for Y123 in 1987. From their structures, it was immediately evident that the weak van der Waal force between the (BiO) double layers in BSCCO is responsible for the graphitic-behavior of the compound, making cleaving the sample in vacuum easy for spectroscopic studies and mechanical rolling possible for aligning the (CuO_2) layers in the first generation HTS-wire processing. Unfortunately, the softness and structural modulation nature of the compound make thin-film synthesis of BSCCO for devices a challenge.

It is interesting to note that in the summer of 1987 Akimutsu et al. of Aoyama-Gakuin University and Raveau et al. of University of Caen reported an 8 K superconducting transition in the Bi-Sr-Cu-O, which was later found to be associated with the $n = 1$ member of the BSCCO family. This important piece of information was lost at the time due to the mad rush for superconductors with a higher T_c than that of YBCO. I even marked their preprints with "exciting, more study needed!" in August 1987, but took no immediate action. Should one have paid greater attention to these early results, the discovery of the 110 K superconducting BSCCO family without rare-earth could have been advanced by half a year. A lesson from this episode is never to get trapped in the turbulence of excitement presented by fashionable ideas.

4.4.3 $Tl_2Ba_2Ca_{n-1}Cu_nO_{2n+4}$

The second cuprate family without rare earth that displays a T_c up to 125 K at ambient and 131 K at 7 GPa was soon discovered by Zhengzhi Sheng and Allen Hermann of the University of Arkansas in February 1988.[45] The discovery appeared to justify the early optimism after RBCO and BSCCO that more high HTSs would be on their way to be found. The physical softness and complex synthesis procedure of TBCCO make its application difficult, in spite of the early optimism regarding its possible use for thin film devices due to its higher T_c and greater stability.

After examining the role of R in R123, Sheng and Hermann of the University of Arkansas started to replace the trivalent R in R123 with the trivalent Tl with an ionic radius similar to R. They found superconductivity up to 90 K in their mixed phase samples of Tl-Ba-Cu-O with a nominal composition of $Tl_2Ba_2Cu_3O_{8+x}$ in early January 1988, about the same time as Maeda et al. announced their discovery of superconductivity up to 110 K in BSCCO. Sheng and Hermann soon partially replaced Ba with Ca and observed superconductivity above 120 K in their multiphase samples of nominal $Tl_2Ca_{1.5}BaCu_3O_{8.5+x}$ during the second week of February 1988 as shown in Figure 4.10. The results appeared in an article in the March 10, 1988, issue of *Nature* entitled, "Bulk superconductivity at 120 K in the Tl-Ca/Ba-Cu-O system."[45] The $n = 2$ and 3 members of the Tl22$(n-1)n$ homologous series were quickly identified in these samples by Hazen et al. of the Carnegie Geophysical Lab and Torardi et al. of Du Pont Research Lab only two days after the announcement of the 120 K superconductivity. Parkin et al. of the IBM Almaden Lab obtained pure-phase samples of Tl2223 and achieved a T_c of 125 K, which was the record-T_c at ambient pressure until April 1993. The T_c of Tl2223 was raised to 131 K by Berkeley et al. at the Naval Research Lab by pressures up to 7 GPa in September 1992. The homologous series Tl22$(n-1)n$ display the layered stacking sequence, e.g., (BaO)(TlO)(TlO)(BaO)(CuO$_2$)(Ca)(CuO$_2$)(Ca)(CuO$_2$) for $Tl_2Ba_2Ca_2Cu_3O_{10}$ with $n = 3$ as shown in Figure 4.10. Similar to Bi22$(n-1)n$, the T_c of Tl22$(n-1)n$ increases with n up to 3 and decreases when $n > 3$. The maximum T_cs are 90, 110, and 125 K, respectively, for $n = 1$, 2, and 3. Depending on the oxygen content, the T_c can be varied by more than 10 K. The structure of Tl22$(n-1)n$ is similar to that of Bi22$(n-1)n$. In spite of the gross similarity between Tl22$(n-1)n$ and Bi22$(n-1)n$, there exists no structural modulation along the double (TlO) layers in contrast to that along the double (BiO) layers, suggesting that such a structural anomaly does not play a role in their high T_c as was initially thought. The single (TlO)-layer TBCCO homologous series Tl12$(n-1)n$ was later discovered. The layer stacking sequence is similar to that for Tl22$(n-1)n$ except that the double (TlO) layers in Tl22$(n-1)n$ are replaced by the single (TlO) layer with an I4/mmm symmetry. The T_c of members of this series is, in general, lower than that for the corresponding members of Tl22$(n-1)n$. It was found to increase continuously with n up to $n = 4$ before it decreases, i.e., $T_c = 50$, 82, 110, and 120 K for $n = 1$, 2, 3, and 4, respectively. The reasons for the lower T_cs and for the continuous increase of T_c to $n = 4$ could be interesting for understanding the inner atomic structure influence on T_c, but unfortunately remain unknown.

The news of the TBCCO discovery caught many in the field and in the media by surprise, for such

---

[45] Z. Z. Sheng and A. M. Hermann, *Nature* 332 (1988) 138

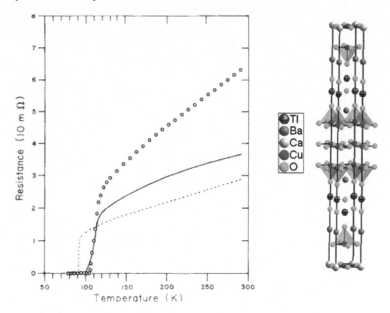

FIGURE 4.10: TBCCO temperature dependence of resistance[45] and structure. (Reprinted by permission from Macmillan Publishers Ltd: Nature 332 (1988) 138, copyright 1988.)

an exciting discovery came from an institute not known for superconductivity study. Sheng brought his first 120-K sample to me later to confirm his discovery. This demonstrated that HTS is a level playing field—there is a chance for everyone with vision and willingness to try.

4.4.4 $HgBa_2Ca_{n-1}Cu_nO_{2n+3-\delta}$

$HgBa_2Ca_{n-1}Cu_nO_{2n+3-\delta}$, where n = 1, 2, 3, ... [HBCCO or Hg12$(n-1)n$] with a T_c up to 134 K at ambient and 164 K at ~ 30 GPa was discovered by A. Schilling et al. of ETH in mid-April 1993.[36] A T_c of 134 K is above the temperature in the cargo bay of the Space Shuttle when orbiting Earth opposite the Sun and above the boiling point of liquid natural gas (LNG) on earth. The former makes HBCCO a possible material for HTS devices operable on the Space Shuttle without liquid cryogen and the latter may enable the development of a combined HTS/LNG system for the efficient delivery of electrical and chemical energies simultaneously.

In late 1989, the T_c of cuprates appeared to have stagnated at 125 K since the Spring of 1988. A prominent chemist speculated that the T_c of cuprates could not exceed 160 K based on his physical chemistry arguments. However, $HgBa_2Ca_2Cu_3O_{9-\delta}$ was discovered in 1993 to display a T_c at ~ 133 K by Schilling et al. of ETH at Zurich as shown in Figure 4.11 and the results appeared in an article in the May 6, 1993, issue of *Nature* entitled "Superconductivity above 130 K in the Hg-Ba-Ca-Cu-O system." The homologous series of HBCCO, Hg12$(n-1)n$, was quickly identified. The highest T_c member with $n = 3$ of the series displays the layer stacking sequence $HgBa_2Ca_2Cu_3O_{9-\delta}$ = (BaO)(HgO$_{1-\delta}$)(BaO)(CuO$_2$)(Ca)(CuO$_2$)(Ca)(CuO$_2$), as shown in Figure 4.11. When Hg1223 is subjected to high pressures, its T_c rises to 164 K at ~ 30 GPa before exhibiting any sign of saturation as shown in Figure 4.12, clearly surpassing the predicted 160 K T_c-limit made in 1989 for cuprate HTSs. A T_c = 134 K and 164 K at ambient and high pressure, respectively, remain today's records.

It is interesting to note that attempts were made as early as in 1991 to substitute the linearly coordinated Hg^{+2} for the similarly coordinated Cu^{+2} in the (CuO)-chain-layer of R123. Compounds of $HgBa_2EuCu_2O_x$ with a structure similar to Tl1212 were made but found not superconducting by

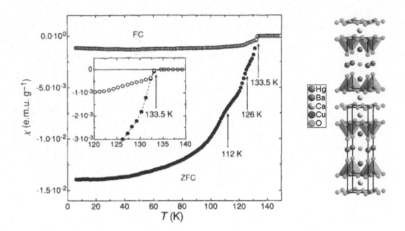

FIGURE 4.11: HBCCO temperature dependence of the magnetic susceptibility[36] and structure. (Reprinted by permission from Macmillan Publishers Ltd: Nature 363 (1993) 56, copyright 1993.)

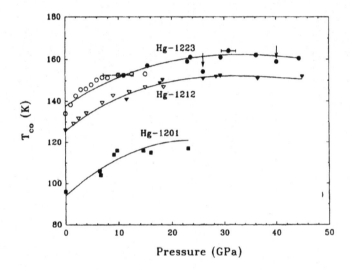

FIGURE 4.12: Pressure dependence of the T_c of Hg1201, 1212, and 1223. Figure reprinted with permission from L. Gao, Y. Y. Xue, F. Chen, Q. Xiong, R. L. Meng, D. Ramirez, C. W. Chu, J. H. Eggert, and H. K. Mao, Phys. Rev. B 50 (1994) 4260. Copyright 1994 by the American Physical Society. http://link.aps.org/doi/10.1103/PhysRevB.50.4260.)

Antipov of Moscow State University. That same year, we detected a small superconducting signal at 94 K in one of our $HgBa_2EuCu_2O_x$ samples due to an impurity phase, without recognizing that it was the Hg1201. The whole story of HBCCO did not start to unfold until September 1992 when Antipov of Moscow State University and Marezio of CNRS at Grenoble successfully synthesized Hg1201 with a $T_c = 94$ K. Knowing that increasing the number of (CuO_2) layers per cell will lead to an increase of T_c, A. Schilling et al. of ETH added Ca to the compound to achieve $n = 2$ and 3 for $Hg12(n-1)n$ and reported an enhanced T_c up to 133 K. After overcoming issues with the complex chemistry of HBCCO, we and several other groups isolated the different phases and showed the maximum T_c s to be 97, 128, and 134 K, for $n = 1$, 2, and 3, respectively.

Although $Hg12(n-1)n$ has a layer structure similar to $Tl12(n-1)n$, there exists a subtle difference, presumably arising from the linear oxygen coordination of Hg^{+2} ions in HBCCO as reflected in the relatively short Hg-O bond length along the c-axis and the large number of voids in the $HgO_{1-\delta}$ layer. Higher T_c was therefore expected in HBCCO under pressures. Experiments on optimally doped pure Hg1201, 1212, and 1223 were carried out under pressures. We, in collaboration with Mao et al. at the Carnegie Geophysics Lab, found that the T_c of these members grows with pressure in parallel and peaks at $\sim$ 118 K in Hg1201 at $\sim$ 24 GPa, at $\sim$ 154 K in Hg1212 at $\sim$ 29 GPa, and at 164 K in Hg1223 at $\sim$ 30 GPa, as shown in Figure 4.12. The observation strongly suggests that superconductivity of members of the $Hg12(n-1)n$ homologous series arises from a common origin. However, the unusually large pressure-induced T-enhancement, even in their optimally doped states, cannot be accounted for by models commonly used to explain the pressure effect on T_c for other HTSs. A modified rigid band model has been proposed for the observations.

4.5 Fe-Pnictides and -Chalcogenides

Hideo Hosono

4.5.1 Research Led to Discovery of Iron-Based Superconductors

My background is ceramic science, especially exploration of transparent electro-active oxide materials such as wide-gap transparent crystalline/amorphous semiconductors. The motivation of this research is a desire of "We wish to convert transparent insulating oxide composed of abundant elements to conducting materials". The top 10 elements in our territorial crust are oxygen, silicon, aluminum, iron, calcium, sodium, potassium, magnesium, hydrogen and titanium. These elements exist in the form of stable oxides and all the oxides except iron are optically transparent (the band gap > 3 eV) but electrically insulating. Thus, these oxides have been utilized as the main ingredients for traditional ceramics such as cements, glasses, and porcelains but have not been regarded as a platform of electro-active materials like semiconductors. Although many insulating materials have been successfully converted to conducting materials by innovative doping methods, these transparent oxides such as alumina and silica still remained typical insulators as described in college textbooks. I started exploration of electro-active functions in transparent oxide in 1993, aiming at cultivation of transparent oxide electronics. At this point, we had two major concrete targets to be realized, i.e., p-type transparent oxide semiconductor (p-TOS) and transparent amorphous oxide semiconductor (TAOS) with large electron mobility. We thought that the absence of p-TOS was the major obstacle to establish transparent oxide electronics because pn-junction is the major origin of the active functions of semiconductors. As for the latter, I expected that TAOS would open a new world in amorphous semiconductor with which I was fascinated as a graduate student.

My research proposal described above was approved in 1999 as an ERATO (Exploratory Research for Advanced Science and Technology) project, the most prestigious project for objective fundamental research, and "Hosono Transparent Electro-active Materials (TEAM) project" had started under the sponsorship of the Japan Science and Technology Agency (JST) from October, 1999.

Figure 4.13 summarizes the track of our research to date. We started the research with exploiting p-TOS and TAOS on the basis of our material design concepts based on a simple consideration of chemical bonding of oxides. The former and the latter led to the realization of transparent pn-junction for UV-LED and transparent, high performance TFTs which can be fabricated on plastic substrates, respectively. In particular, the latter has become attractive as the universal backplane to drive next generation flat panel displays (FPDs) and various prototype FPDs have been fabricated for practical application.

In the course of this research, we discovered two classes of new superconductors. One is a heavily

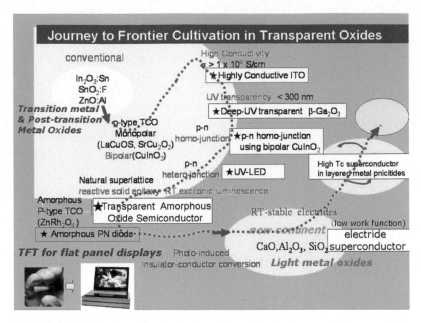

FIGURE 4.13: Our research progress in transparent oxide semiconductors and discovery of two types of new superconductors.

electron-doped $12CaO \cdot 7Al_2O_3$ (abbreviated as $C12A7:e^-$), a first room temperature stable electride. The other is an iron (or nickel) oxy-pnictide which is the focus of this section. Superconductivity in $C12A7:e^-$ was found as a consequence of a strong will toward realization of superconductive cement (C12A7 is a constituent of commercial alumina cements). The driving force of this subject was a great surprise received by discovery of high temperature cuprates in 1986. At that time, I was an assistant professor in the department of inorganic materials, Nagoya Institute of Technology, and was shocked by news that high T_c was realized from insulating oxides based on Cu^{2+}-based insulating oxides by doping. So many active researchers with different backgrounds joined this research in Japan and performed concentrated research. This devotion led to the establishment of new research areas in materials science and condensed matter physics. However, I did not participate in this research except provide some help for the laboratory to which I belonged. The reason is quite simple: I had no thought to compete with those already well engaged in the research, to my regret. I was concentrating on elucidation of the mechanism for photosensitivity of dopant-free $CaO\text{-}Al_2O_3$ glass, which I discovered in 1985. Thus, the successful conversion of insulating C12A7, which is the crystalline phase corresponding to the $CaO\text{-}Al_2O_3$ glass, to transparent semiconductor and transparent metal by injecting electrons into the 3-dimensionally connected cages with the expectation that electrons could tunnel through mono-oxide layers (Figure 4.14) was due to our own approach and the eventual conversion of C12A7 to a superconducting state was a goal of this study.

On the other hand, the iron (nickel) pnictide superconductors were discovered in the course of systematic survey of $LaTmPnO$ (Tm: 3d transition metal, Pn = P, As, and Sb) with a layered structure which is the same as that of p-type transparent semiconductor $LaCuChO$ (Ch = chalcogen). Although we had a faint hope that a new superconductor might be found (as most researchers would when they start to examine transition metal compounds) no concrete approach or chemical intuition on the emergence of superconductivity in this material system stayed in our group including me.

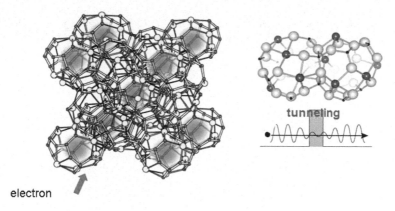

electron

FIGURE 4.14: C12A7 electride. $12CaO \cdot 7Al_2O_3$ (C12A7) is an insulator with 3-dimensionally connected cages with 0.44 nm in diameter. Since the cage is positively charged, O^{2-} ions are entrapped randomly as counter anions in 2 out of 12 cages constituting a unit cell. When these O^{2-} are extracted by chemical reduction, electrons are injected in place of O^{2-} to keep electroneutrality. The resulting material containing electrons of 2×10^{21} cm^{-3} instead of the counter anion O^{2-} in the cages is called C12A7 electride because crystals in which electrons serves as anions are called electrides. We synthesized C12A7 electride, finding it exhibits a metallic conduction in 2004. This high electronic conductivity comes from electron tunneling through the thin cage wall. We examined resistivity at low temperature with an expectation of emergence of T_c from 2006 (and eventually realized in 2007!).

4.5.2 Discovery of Iron-based Superconductors

From *p*-type Transparent Semiconductor to Magnetic Semiconductors

A layered compound composed of alternatively stacked active layers is an attractive platform for exploration of wide gap semiconductors and/or functions due to the following reasons:

1. The carrier doping layer can be separated spatially from the carrier transport layer, which may enhance the carrier mobility through suppression of impurity scattering.

2. Electrons and/or holes are confined in the active layers which serve as wells, stabilizing excitons even at room temperature.

3. Magnetic interaction tends to be weakened because of intrinsic magnetic instability in 2-dimensions.

4. Band gap is increased due to suppression of 3-dimensional band dispersion.

We chose a Cu^+-based mixed oxyanion-layered materials for the candidate for *p*-type transparent oxide semiconductors (*p*-TOS). In general, a hole at the valence band top tends to localize at oxygen $2p$ orbitals primarily constituting the VBM, due to its strong ionic nature of oxides. For *p*-type TCOs, how to delocalize the positive hole at the VBM of oxides works as a strategy. We focused a Cu^+ with $[Ar](3d^{10})$ configuration which has $3d$ electron levels comparable to those of O $2p$, leading to the discovery of *p*-TOS, $CuAlO_2$ with a layered structure (delafossite). Further improved performance was obtained in a mixed oxychalcogenide LaCu*Ch*O (*Ch*: chalcogenide). We found the unique optoelectronic properties for epitaxial thin films of LaCu*Ch*O, i.e., large hole mobility, RT-stable exciton, a large 3rd order optical nonlinearity, 2D-electronic structure and blue light-emitting diode utilized PN-junction and efficient excitonic emission.

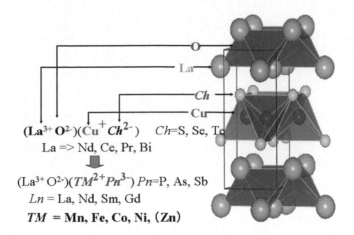

$(La^{3+}O^{2-})(Cu^{+}Ch^{2-})$ Ch=S, Se, Te
La => Nd, Ce, Pr, Bi

$(La^{3+}O^{2-})(TM^{2+}Pn^{3-})$ Pn=P, As, Sb
Ln = La, Nd, Sm, Gd
TM = Mn, Fe, Co, Ni, (Zn)

FIGURE 4.15: From *p*-type transparent semiconductor LaCuO*Ch* (*Ch* = chalcogenide) to La*TmPn*O (*Tm* = +2 charged state 3*d* transition metal cation, *Pn* = pnictgen anion) as candidate for magnetic semiconductors. Both materials belong to ZrCuSiAs-type crystal structure composed of insulating LaO layer with a positive charge and Cu*Ch* or *TmPn* with a negative charge.

Intriguing properties of a transition metal primarily come from the freedom combined with spin and charge of 3*d*-electrons. Although a Cu⁺ with a closed 3*d*-shell structure is appropriate for keeping optical transparency (a large gap), no such properties can be expected for its compounds. We thus determined to start the exploration of new magnetic semiconductors in the La*TmPn*O system (where *Tm* is a 3*d* transition cation with +2 charge state) with the same crystal structure as LaCu*Ch*O as shown in Figure 4.15.

Electromagnetic properties of La*TmPn*O

The binary compounds of transition metal pnicides exhibit a diversity of electromagnetic properties by changing the combination of *Tm* and *Pn* elements. However, few studies on electromagnetic properties of layered compounds La*TmPn*O had been performed, notwithstanding that the existence of many compounds were already reported. We tried to synthesize the bulk compounds La*TmPn*O (*Tm*: 3*d* transition metal, *Pn*: P,As,Sb) by solid state reactions in evacuated silica glass tubes and measured electromagnetic properties on the resulting materials. For device application, thin film fabrication by pulsed laser deposition method was examined.

Figure 4.16 summarizes the electromagnetic properties elucidated for LaFeAs*Pn*O. The properties drastically vary with the 3*d* electron number in the *Tm*: an anti-ferromagnetic semiconductor with Néel temperature (T_N) > 400 K for the Mn system, a Pauli parametal with a superconducting critical temperature (T_c) at ~ 4 K for the Fe, a ferro-magnetic metal with a Curie temperature of 60 K for the Co, a Pauli parametal with T_c = 3 K for the Ni, and a diamagnetic semiconductor for the Zn. It is of interest to note that metal-superconducting transition is observed for the Tm^{2+} with an even number of 3*d* electrons, i.e., Fe^{2+} with $3d^6$ and Ni^{2+} with $3d^8$. The optical properties of La*TmPn*O differ from those of binary *TmPn*, but the magnetic properties are similar to each other, AFM for the Mn and FeAs, FM for the Co, and Pauli Para for the Fe and Ni. Noteworthy is the striking difference between FeP and FeAs. FeP is a Pauli parametal, while FeAs is an AF metal.

La Fe*Pn*O: Striking Difference between LaFePO and LaFeAsO

Figure 4.17 shows ρ-T curves between LaFePO and LaFeAsO$_{1-x}$F$_x$. The resistivity of the former simply decreases with temperature and suddenly drops to zero around 4 K, while the latter has higher

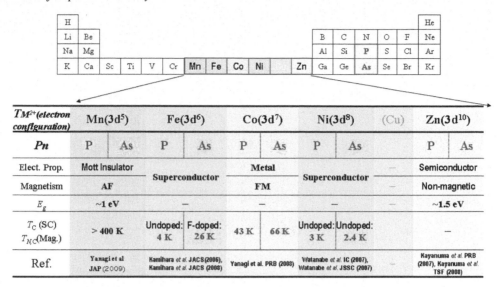

FIGURE 4.16: Electrical and magnetic properties of La$TmPn$O.

resistivity by an order of magnitude than the former but does not change monotonically with temperature, showing a distinct break around 150 K and ρ increases with T. When electrons are doped via partial replacement of the oxygen sites by fluorine, this break disappears and instead the bulk superconductivity of $T_{c(onset)}$ of 32 K emerges. This break in the ρ-T curve is not seen for LaFePnO and LaNiPnO with T_c = 3 K. After we submitted a communication reporting LaFeAsO$_{1-x}$F$_x$ with $T_{c(mid-point)}$ = 26 K, we concentrated on examining the superconductivity in the compounds with multi Fe(Ni)Pn layers following a drastic jump of T_c in cuprates by discovery of YBCO with bilayers of CuO$_2$. The first material we chose was BaNi$_2$P$_2$ with 122 structures. We found a T_c but it remained $\sim$4 K. Next we chose BaFe$_2$As$_2$ and tried to electron dope it by replacing oxygen sites by fluorine but the result was totally unsuccessful (it took almost 2 years to succeed in synthesizing superconducting Sr122 compounds by electron-doping via aliovalent substitution of alkaline earth ion with a help of high pressure). At this time, a paper reporting Ba$_{1-x}$K$_x$Fe$_2$As$_2$ with a maximum T_c = 38 K by the Johrendt group of Munich (M. Rotter, M, Tegel, and O. Johrendt, Phys. Rev. Lett. 101 (2008) 107006.) was posted on the cond-mat preprint server. It is evident that BaFe$_2$As$_2$, the parent compound of T_c = 38 K, exhibits a break at $\sim$140 K like LaFeAsO but BaNi$_2$P$_2$ has no such anomaly as LaFePO and LaNiAsO (T_c = 3 K). So, it is natural to postulate that the presence of a break in ρ-T curve is requisite for the parent material for high T_c, but the invalidity of this idea will be shown later.

What happens around this break temperature? First, we suspected the crystallographic transition and measured powder XRD by using SR-ring as a function of temperature. As a consequence, a phase transition from tetragonal to orthorhombic (not monoclinic from the extinction rule) symmetry was clearly observed at $\sim$155 K. Measurements of heat capacity and the internal field at the Fe-site monitored by Mossbauer spectroscopy revealed that crystallographic transition and antimagnetic ordering occur at $\sim$150 K and $\sim$140 K, respectively. The most decisive information on the magnetic structure was elucidated by neutron diffraction. The spin configuration and the magnitude of local moment at Fe ion at the AFM state in the parent compound were determined by a group at Oak Ridge National Lab (ORNL). Figure 4.18 shows electronic phase diagrams for the 1111-type and 122-type iron-based compounds. The superconductivity emerges by introducing carriers to the parent compounds which are antiferromagnetic metals. The major differences between these 2-systems are the presence of an overlapped region between AF phase and superconducting phase

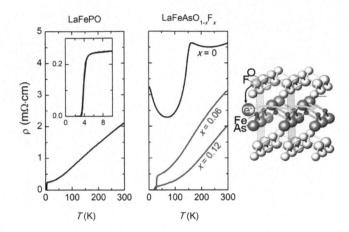

FIGURE 4.17: Temperature dependence of resistivity in LaFePO and $LaFeAsO_{1-x}F_x$. Right figure shows electron-doping to FeAs-layers by replacement of oxygen with fluorine.

and the convergence of structural transition and magnetic transition temperature for the 122 system.

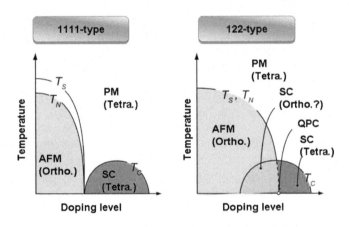

FIGURE 4.18: Phase diagrams for 1111 and 122 Iron pnictide superconductors. T_s: crystallographic transition temperature; T_N: magnetic transition (Néel) temperature; QCP: quantum critical point (phase transition at 0 K).

4.5.3 Advances in Materials

When we discovered the high T_c in $LaFeAsO_{1-x}F_x$ in October 2007, the target was switched from $LaFeAsO_{1-x}F_x$ and we found a rise in T_c to 43 K under a pressure of $\sim$ 4 GPa. This temperature exceeded the T_c of MgB_2 and is next-to-the highest T_c for cuprates. We submitted this paper on February 26, 2008, and started to synthesize the $LuFeAsO_{1-x}F_x$ at the same time. It was evident that high pressure effectively works to raise the T_c in this system. We thought if La ion with the largest radius among rare-earth ions could be replaced by nonmagnetic rare-earth ions with the smallest radius, the highest T_c would have been realized. This trial was totally unsuccessful as well. The compound LuFeAsO itself was not obtained (even to date). During the struggle with synthesis of

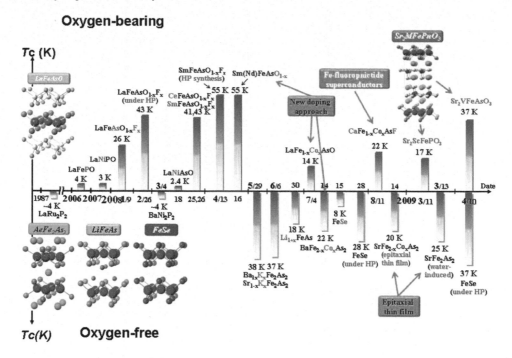

FIGURE 4.19: Progress in iron(nickel)-based superconductors. The ordinate is the date the paper was accepted by the journal or posted on cond-mat preprint server.

LuFeAsO, higher T_cs were reported on a cond-mat-server by several Chinese groups at USTC and IOP in late March and early April 2008. They obtained the high T_c by replacing La ion with Ce, Pr, or Sm ion with an open $4f$-shell. These reports shocked us because we intentionally avoided the use of magnetic rare-earth ions based on the experimental results (presented at the 15th Conference on Intercalation at Korea in May 2007) that superconductivity disappears when the La ion in LaFePO ($T_c = 4$ K) was replaced by Ce. That is, we misunderstood without careful checking whether or not the $4f/5d$ orbitals of rare-earth ions participate to the Fermi level which are primarily composed of Fe $3d$ orbitals.

Figure 4.19 summarizes the progress in Fe(Ni)-based superconductors viewed from material aspects. The abscissa is the date received in a journal or posted on the preprint server. The T_c reached a maximum of 56 K in a short period by replacing La ion in LaFeAsO with an appropriate rare-earth ion. Two types of new materials were reported in June 2008: one is $Ba_{1-x}K_xFe_2As_2$, 122-type by Johrendt's group, and $FeSe_{1-x}$, 11-type by Wu's group of Taiwan. Subsequently, Na(Li)FeAs, 111-type was reported by Chu's group of Houston and Jin's group of IOP, and in March 2009, Shimoyama's group of Tokyo found T_c in FePn compounds with perovskite-like blocking layers. These 5 types of compounds summarized in Figure 4.20 have a common structural unit, a square lattice of Fe^{2+} ions (Figure 4.21). The Fermilevel is primarily composed of Fe $3d$ orbitals and the contribution of pnicogen/chalcogen orbitals is ~10% at the Fermilevel. This situation is rather different from high T_c cuprates in which both orbitals of Cu $3d$ and O $2p$ participate to the Fermi-level.

In June 2008, a totally unique phenomenon was posted by a group of ORNL. They found the superconductivity is induced by partially replacing the Fe-sites in the 1111 compounds by Co ions. Similar results were observed in the 122 phases. This result strikingly differs from that in high T_c cuprates in which partial substitution of Cu sites by other ions diminishes the T_c. The group led by Clearfield at Iowa found that T_c in Fe-substituted 122 compounds with $3d/4d$ transition metal could be scaled by the excess number of electrons per Fe ion.

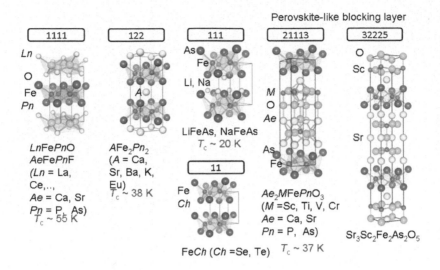

FIGURE 4.20:   Crystal structure of 5-representaive type parent materials.

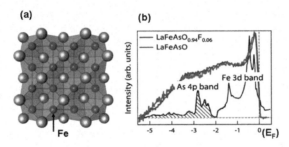

FIGURE 4.21:   Common structural unit in 5-types parent materials and contribution of Fe $3d$ and As $4p$ bands (calculated) to the Fermi level of LaFeAsO. The observed density of state measured by photoemission is shown.

4.5.4   Current Status

What is The primary Factor Controlling T_c?

The striking difference in T_c between LaFePO and LaFeAsO systems or $BaFe_2As_2$ and $BaNi_2As_2$ systems implied that materials with higher T_N exhibit high T_c. However, this idea is discounted in Figure 4.22 showing the anti-correlation between T_c and T_N for each material system. If this anti-correlation is valid, highest T_c should be derived from the parent material with higher T_N, but in fact it is reversed. Thus, this supposition is invalid.

Lee found a good correlation between T_c and the bond angle of $Pn(Ch)$-Fe-$Pn(Ch)$, i.e., high T_c is observed as this bond angle approach to that (109.28′) of the regular tetrahedron. Kuroki *et al.* proposed a height (h_{Pn}) of pnictgen (chalcogen) from the iron plane as a measure of T_c on the basis of calculation on spin-fluctuation-mediated superconductivity using five $3d$ orbitals of Fe. Since the energy level of a band derived from Fe $3d_{x^2-y^2}$ is most sensitive to h_{Pn}, the shape of the Fermi surface is rather changed with h_{Pn} as illustrated in Figure 4.23. The γ-pocket at (π,π) hidden for low h_{Pn} appears as h_{Pn} is increased. As a consequence, nesting becomes satisfactory and T_c is increased as h_{Pn}. Figure 4.24 shows a plot of T_c *vs.* h_{Pn}. Aside from some exceptions, such as the 11 system, a good correlation is observed as a whole.

Comparison with Cuprates and MgB$_2$

Table 4.1 summarizes the features of iron pnictide along with cuprates and MgB$_2$, (the representative of BCS superconductors). Three unique properties are evident for iron pnictides: robustness to impurity doping, very high upper critical magnetic fields, and low crystallographic anisotropy in physical performances. These features are favorable for application to superconducting wires.

TABLE 4.1: Comparison among 3 representative superconductors.

	Fe-pnictides	MgB$_2$	Cuprates
Parent Material	(bad) metal ($T_N \sim 150$ K)	metal	Mott Insulator ($T_N \sim 400$ K)
Fermi Level	$3d$ 5-bands	2-bands	$3d$ single band
Max T_c	56 K	40 K	~ 140 K
Impurity	robust	sensitive	sensitive
Sc gap symmetry	sign inverted s-wave(?)	s-wave	d-wave
$H_c^2(0)$	100–200 T >	~ 40 T	~ 100 T
γ	2–4 (122)	~ 3.5	5–7 (YBCO); 50–90 (Bi system)
J_c	?		

4.5.5 Perspective

Iron, a representative magnetic element, was believed to be the last constituent for emergence of superconductivity because long-range magnetic ordering competes with the formation of Cooper pair required for superconductivity. However, once LaFeAs(O,F) with $T_c = 26$ K was discovered, many iron-pnictide (chalcogenide) superconducting materials have been found and the maximum T_c reached 56 K, which is next to the high T_c cuprates exceeding MgB$_2$. I think there are two significances in discovery of iron-based superconductors. First, we realized that a magnetic element is not an enemy but a powerful friend to realize high T_c superconductors. Second it provides a large opportunity to find new high T_c materials because there exist several hundreds of layered compounds containing square lattice of transition metal cations taking tetrahedral coordination with

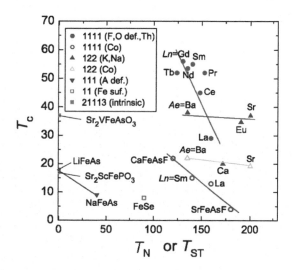

FIGURE 4.22: T_c *vs.* magnetic (T_N)/structural (T_s) transition temperature.

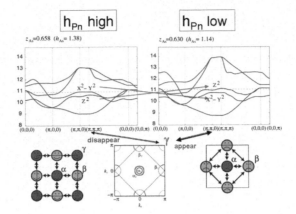

FIGURE 4.23: Change in Fe 3*d* energy level and Fermi surface with the height of pnictgen from the iron plane (h_{Pn}). When h_{Pn} increased, nesting becomes satisfactory via appearance of γ-pocket. This result well explains the striking difference in T_c between LaFePO and LaFeAsO$_{1-x}$F$_x$.

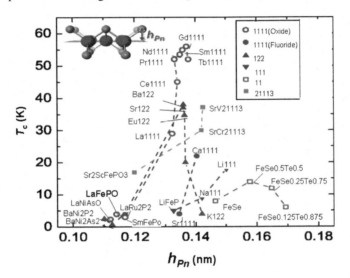

FIGURE 4.24: T_c vs. pnictgen height from iron plane. (Reprinted with permission from H. Hosono et al. Bulletin of the Physical Society of Japan 64 (2009) 807. Copyright 2009 by the Physical Society of Japan.)

non-oxide anions. We expect materials with higher T_c and/or novel class of superconductors to be hidden among these. For our purposes, the crystal structure of 122 is the same as that of a representative heavy fermion superconductor CeCu$_2$T$_2$ (T = Si,Ge). One may expect some clue to bridge these two superconducting systems to be found.

What we cannot forget is the historical fact that most groundbreaking materials, including high T_c superconductors, have been discovered by serendipity in the course of concentrated explorative effort. I am anticipating new material functions to be discovered as a result of concentrated material explorative with the help of theoretical modeling and advanced characterization. Iron is the most important element "to the material progress of civilization". I hope iron will serve same role in the history of superconductivity. *Strike while the iron is hot.* I think this saying is still true for superconductivity research.

4.6 Superconductivity in MgB_2

Takahiro Muranaka and Jun Akimitsu

4.6.1 Introduction

After the discovery of superconductivity in Hg in 1911, a number of superconductors among metal elements and intermetallic compounds have been discovered (Figure 4.25). In the early stage of superconductive material investigations, A15-type superconductors,[46] such as Nb_3Sn, V_3Ga, $Nb_3(Al,Ge)$,[47] and Nb_3Ge,[48] were found. At the same time, AlB_2-type structure had already been recognized as candidate material for "high-T_c" superconductors, mainly by B. T. Matthias in the early 1950s (B. T. Malthias and J. K. Hulm, *Phys. Rev* 87 (1952) 799). The superconduc-

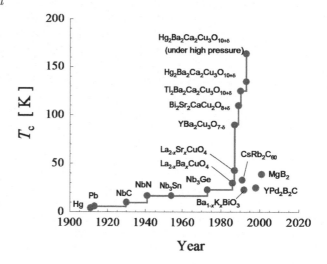

FIGURE 4.25: Chronology of T_c (until 2001).

tors discovered in this stage are called "BCS superconductors" because their behavior can be well explained within the framework of the BCS theory. However, the discovery of Cu-oxide superconductors in 1986[49] required a theoretical interpretation with a new key concept. The important point is that T_c rose to 138 K at ambient pressure,[50] which is far above the highest T_c record of 23 K in Nb_3Ge[48] (Figure 4.25). With the high-T_c record being broken one after another in Cu-oxides, one of the most challenging questions in superconductivity became "how much will T_c increase in non-Cu-oxide superconductors?"

Up to 2001, the highest T_c in this class was 33 K in an electron-doped $Cs_xRb_yC_{60}$[51] and the next highest T_c was 30 K in $Ba_{1-x}K_xBiO_3$.[52] Against this background, we reported superconductivity at 39 K in MgB_2 in 2001,[53] which is the highest T_c among intermetallic superconductors. Although MgB_2 is a well-known, popular material, its high-T_c superconductivity (T_c = 39 K, Figure 4.26) had been hidden until our discovery. MgB_2 opened a new frontier into the physical properties and applications of intermetallic superconductors. Because the limit of T_c in metallic superconductors had been believed to be about 30 K in the framework of the BCS theory, the discovery of unexpectedly high-T_c superconductivity in this simple binary intermetallic compound triggered enormous interest by the world. In the short period since the discovery of its high-T_c superconductivity, a large

[46] For example, G. Hardy and H. Franz, Phys.Rev. 89 (1953) 884; B. T. Matthias, *Phys. Rev.* 92 (1953) 874

[47] B. T. Matthias, T. H. Geballe, K. Andres, E. Corenzwit, G. W. Hull, and J.P. Maita, *Science* 156 (1967) 645

[48] J. R. Gavaler, *Appl. Phys. Lett.* 23 (1973) 480

[49] J. G. Bednorz and K. A. Müller, *Z. Phys. B* 64 (1986) 189

[50] S. N. Putilin, E. V. Antipov, A. M. Abakumov, M. G. Rozova, K. A. Lokshin, D. A. Pavlov, A. M. Balagurov, D. V. Sheptyakov, and M. Marezio, *Physica* C 338 (2001) 52

[51] K. Tanigaki, T. W. Ebbesen, S. Saito, J. Mizuki, J. S. Tsai, Y. Kubo, and S. Kuroshima, *Nature* 352 (1991) 222

[52] R. J. Cava, B. Batlogg, J. J. Krajewski, R. C. Farrow, L. W. Rupp Jr, A. E. White, K. T. Short, W. F. Peck Jr., and T. V. Kometani, *Nature* 332 (1988) 814

[53] J. Nagamatsu, N. Nakagawa, T. Muranaka, Y. Zenitani, and J. Akimitsu, *Nature* 410 (2001) 63

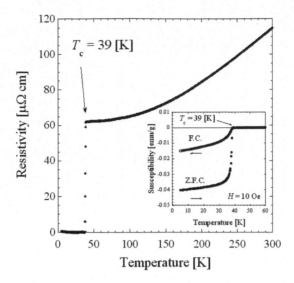

FIGURE 4.26: Superconductivity in MgB_2. (Reprinted by permission from Macmillan Publishers Ltd: Nature 410 (2001) 63, copyright 2001.)

number of experimental and theoretical works have been performed,[54,55] and the interpretation of this superconductivity focuses on the metallic nature of the 2D layer formed by B atoms. Moreover, MgB_2 is a particularly attractive material for its multiple superconducting gaps caused by the characteristic electronic structure derived from 2D σ-bands and 3D π-bands. Although multiple-gap superconductivity had been proposed theoretically by Suhl et al.[56] and since then, discussed in relation to other materials, for example, Nb-doped $SrTiO_3$ by Binnig et al.,[57] MgB_2 is the first material containing intrinsic multiple gaps, and many basic elements of multiple-gap superconductivity have been investigated from various experimental and theoretical aspects.

The enthusiasm caused by the discovery of superconductivity in MgB_2[53] has led to a search for a new high-T_c material in a similar system containing light elements, B and C. Thus, we concentrated our interest on searching for a new material with high-T_c, particularly in transition metal boride and carbide, and lanthanide carbide system.

4.6.2 Crystal Structure of MgB_2

MgB_2 has the AlB_2-type structure with $a = 3.08$ Å and $c = 3.51$ Å (hexagonal, space group: $P6/mmm$).[58] In this structure, the characteristic 2D honeycomb layers formed by boron atoms are sandwiched by the triangular metal layers, like intercalated graphite, as shown in Figure 4.27. Each Mg atom is at the center of a hexagonal prism of boron atoms at a distance of 2.5 Å. Each boron atom is surrounded by three other boron atoms, forming an equilateral triangle at a distance of $a/\sqrt{3} \sim 1.78$ Å, while the Mg-Mg distance in the plane is equal to the lattice constant a. Because lattice constants a and c in the AlB_2-type structure are in the range of 3.0–3.2 Å and 3.0–4.0 Å, respectively, MgB_2 has the middle lattice constant among this type of structure. The crystal struc-

[54] For review, C. Buzea and T. Yamashita, *Supercond. Sci. Technol.* 14 (2001) R115

[55] *Physica* C 385(1-2) (2003)

[56] H. Suhl, B. T. Matthias, and L. R. Walker, *Phys. Rev. Lett.* 3 (1959) 552

[57] G. Binnig, A. Baratoff, H. E. Hoenig, and J. G. Bednorz, *Phys. Rev. Lett.* 45 (1980) 1352

[58] M. E. Jones and R. E. Marsh, *JACSAT* 76 (1954) 1434

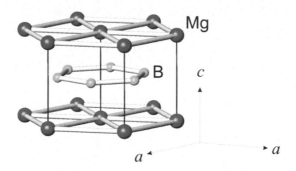

FIGURE 4.27: Crystal structure of MgB_2.

ture of MgB_2 has been investigated by high-resolution neutron powder diffraction[59,60,61] and x-ray diffraction.[62,63,64,65,66,67,68] These experiments mostly focused on the temperature dependence of lattice constants and the lattice compression versus external pressure. The lattice constants, a and c, observed by several research groups are basically in good agreement; 3.085–3.090 Å and 3.520–3.529 Å, respectively, at room temperature. The tiny differences may reflect some uncontrollable displacement of the B atom from the ideal position and/or a slight defect at the Mg site. MgB_2 remains a hexagonal unit to the lowest temperature of 2 K or the highest pressure of 40 GPa, and no sign of structural transition was observed. The results reported by Jorgensen et al.,[59] Margadonna et al.,[60] and Oikawa et al.[61] are basically in good agreement with each other in terms of lattice thermal expansion. The anisotropy between a- and c-axis corresponds to a difference in the bond strength; the B-B bonds are more rigid than the Mg-B bonds. Jorgensen et al. also pointed out that there is a small hump in the variation of the lattice constant along the a-axis at around T_c ($\Delta a/a \approx 1.6 \times 10^{-5}$), while no hump exists along the c-axis.

They speculated that these behaviors are related to the change of B-B bonding for superconductivity. Margadonna et al. observed a slope discontinuity in the expansion along the c-axis just above T_c ($\Delta c/c \approx 4.0 \times 10^{-5}$), but not along the a-axis. However, Oikawa et al. did not observe any anomaly in the temperature dependence of either lattice constant or the unit cell volume around T_c. The difference in these experimental results might be ascribed to the sample preparation process, which affects the stoichiometry, and structural properties such as imperfect stacking.

[59] J. D. Jorgensen, D. G. Hinks, and S. Short, *Phys. Rev. B* 63 (2001) 224522

[60] S. Margadonna, T. Muranaka, K. Prassides, I. Maurin, K. Brigatti, R. M. Ibberson, M. Arai, M. Takata, and J. Akimitsu, *J. Phys.: Condens. Matt.* 13 (2001) L795

[61] K. Oikawa, T. Kamiyama, T. Mochiku, H. Takeya, M. Furuyama, S. Kamisawa, M. Arai, and K. Kadowaki, *J. Phys. Soc. Jpn.* 71 (2002) 2741

[62] K. Prassides, Y. Iwasa, T. Ito, D. H. Chi, K. Uehara, E. Nishibori, M. Takata, M. Sakata, Y. Ohishi, O. Shimomura, T. Muranaka, and J. Akimitsu, *Phys. Rev. B* 64 (2001) 012509

[63] E. Nishibori, M. Takata, M. Sakata, H. Tanaka, T. Muranaka, and J. Akimitsu, *J. Phys. Soc. Jpn.* 70 (2001) 2252

[64] S. Lee, H. Mori, T. Masui, Y. Eltsev, A. Yamamoto, and S. Tajima, *J. Phys. Soc. Jpn.* 70 (2001) 2255

[65] T. Vogt, G. Schneider, J. A. Hriljac, G. Yang, and J. S. Abell, *Phys. Rev. B* 63 (2001) 220505(R)

[66] A. F. Goncharov, V. V. Struzhkin, E. Gregoryanz, H. K. Mao, R. J. Hemley, G. Lapertot, S. L. Bud'ko, P. C. Canfield, and I. I. Mazin, cond-mat/0106258 (2001)

[67] P. Bordet, M. Mezouar, M. Núñez-Regueiro, M. Monteverde, M. D. Núñez-Regueiro, N. Rogado, K. A. Regan, M. A. Hayward, T. He, S. M. Loureiro, and R. J. Cava, *Phys. Rev. B* 64 (2001) 172502

[68] S. I. Schlachter, W. H. Fietz, K. Grube, and W. Goldacker, *Advances in Cryogenic Engineering: Proceedings of the International Cryogenic Materials Conference*—ICMC 48 (2002) 809, cond-mat/0107205 (2001)

4.6.3   Electronic Structure of MgB$_2$

Despite its crystal structure being similar to that of a graphite intercalated compound, MgB$_2$ has a qualitatively different and uncommon structure of the conducting states. The band structure has been calculated by several groups after the discovery of superconductivity.[69,70,71,72] The band structure of MgB$_2$ is similar to that of graphite, and is formed by three bonding σ bands (in-plane $sp_x p_y$ hybridization) and two π bands (bonding and antibonding; p_z hybridization). MgB$_2$ has two imperfectly filled σ bands, and these σ bands correspond to the sp^2-hybrid bonding within the 2D honeycomb layer. The holes of σ bands along the ΓA line localized within the 2D boron layer manifest 2D properties.

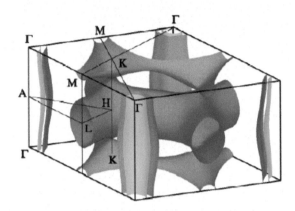

FIGURE 4.28:   Fermi surface of MgB$_2$. Green and blue cylinders (hole-like) come from the bonding $p_{x,y}$ bands, the blue tubular network (hole-like) from the bonding p_z bands, and the red (electron-like) tubular network from the antibonding p_z band.

In contrast, the electrons and holes in 3D π bands are delocalized. The 2D σ bands and 3D π bands at E_F contribute equally to the total density of states, while 2D σ bands have a strong interaction with longitudinal vibrations within the 2D boron layer. Moreover, because k_z dispersion of σ bands is weak, two cylindrical sheets appear around the ΓA line (see Figure 4.28). On the other hand, π bands form two tubular networks: an antibonding electron-type and a bonding hole-type sheet. These two sheets touch at one point on the KH line. An investigation of the charge density distribution would give a better understanding of how the superconductivity is related to the electronic and crystal structures of MgB$_2$. Precise x-ray structure analysis by Nishibori et al. (in a polycrystalline sample),[63] Lee et al. (in a single crystal),[64] and Mori et al. (in a single crystal; Mg$_{1-x}$B$_2$ $x = 0.045$)[73] yielded accurate charge densities in MgB$_2$. The charge density obtained at room temperature revealed a strong B-B covalent bonding feature. On the other hand, there was no bond electron between Mg and B atoms, and Mg atoms were found to be fully ionized and in the divalent state. Nishibori et al. also reported that these characteristic density features were preserved in the charge density obtained at 15 K and were consistent with the calculated band structures indicating a two-band model.[70] Moreover, they examined the valence of the atoms by accumulating

[69] J. Kortus, I. I. Mazin, K. D. Belashchenko, V. P. Antropov, and L. L. Boyer, *Phys. Rev. Lett.* 86 (2001) 4656

[70] J. M. An and W. E. Pickett, *Phys. Rev. Lett.* 86 (2001) 4366

[71] G. Satta, G. Profeta, F. Bernardini, A. Continenza, and S. Massidda, *Phys. Rev. B* 64 (2001) 104507

[72] K. D. Belashchenko, M. van Schilfgaarde, and V. P. Antropov, *Phys. Rev. B* 64 (2001) 092503

[73] H. Mori, S. Lee, A. Yamamoto, S. Tajima, and S. Sato, *Phys. Rev. B* 65 (2002) 092507

the number of electrons around a certain atom in the MEM density (see Figure 4.29). The numbers of electrons at room temperature and 15 K were estimated to be $10.0(1)e$ and $10.0(1)e$ around the Mg atom and $9.9(1)e$ and $10.9(1)e$ around the boron 2D sheets, respectively. The values for Mg atoms are very close to the number of electrons around Mg^{2+} ions, so Mg atoms are fully ionized in the MgB_2 crystal at whole temperatures. On the other hand, the total numbers of electrons around the boron 2D sheet show significant difference, which can be attributed to the valence of the whole boron 2D sheet changing from neutral to monovalent at 15 K. These results suggest that the electrons transfer from π bands (p_z orbitals) to in-plane σ bands (p_{xy} orbitals) at 15 K. However, these results do not agree with the valence electron distribution at room temperature determined by Wu et al. using synchrotron x-ray and electron diffraction techniques.[74] They reported that two electrons from each Mg atom moved to the B plane. Therefore, the boron layer had the same number of valence electrons, and these electrons were mainly located in the $p_x p_y$ orbitals between neighboring boron atoms. This disagreement is not yet resolved.

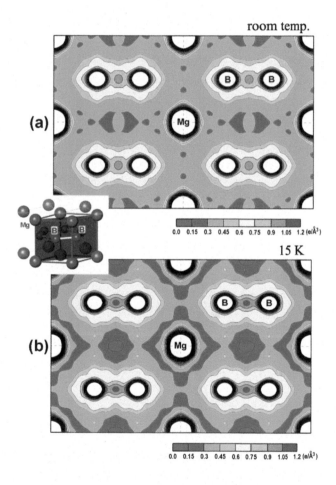

FIGURE 4.29: The (110) sections of the MEM charge density of MgB_2 at room temperature (a) and 15 K (b).

[74] L. Wu, Y. Zhu, T. Vogt, H. Su, and J. W. Davenport, *Phys. Rev. B* 69 (2004) 064501

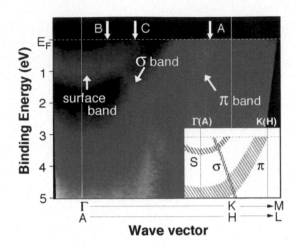

FIGURE 4.30:   Band structure near E_F along ΓKM (AHL) at 45 K obtained by ARPES.

4.6.4   Two-gap Superconducting State of MgB$_2$

In an early stage after the discovery of MgB$_2$, the superconducting gap size Δ of 2–5.9 meV ($2\Delta/k_B T_c$ = 1.2–3.5) was reported, as determined by spectroscopic measurements assuming an isotropic s-wave symmetry gap.[75,76,77] On the other hand, later experimental reports showed clear signatures for two-gap superconductivity. Theoretically, in MgB$_2$, superconducting gaps with significantly different magnitudes were expected to open in the π and σ bands, as a result of the strong k dependence of electron-phonon coupling.[78,79] PES of a high-density polycrystalline sample[80] and a single crystal[81,82,83] yielded experimental evidence of two-gap superconductivity and indicated the band structure of MgB$_2$ along ΓKM (AHL), in studies by Uchiyama et al. and Souma et al. (see Figure 4.30), and along ΓΣM in a study by Tsuda et al. The observed band structure showed considerable agreement with the results of band calculations.[69,70] However, Uchiyama et al. and Souma et al. observed different experimental results with theoretical calculations. The difference was that a small electron-like pocket was observed around the Γ(A) point, and they ascribed the observed band to a surface state. Tsuda et al. and Souma et al. reported Δπ and Δσ to be 2.2 ± 0.4 and 5.5 ± 0.4 meV[81] and 1.5 ± 0.5 and 6.5 ± 0.5 meV[82] with s-wave symmetry. In the surface band observed by Souma et al., the gap size was close to Δσ (6.0 ± 0.5 meV). Tsuda et al. reported that both gaps Δπ and Δσ closed at same temperature (see Figure 4.31). These behaviors indicate the existence of strong interband pairing interaction in MgB$_2$. Moreover, other experimental reports in specific

75 A. D. Caplin, Y. Bugoslavsky, L. F. Cohen, L. Cowey, J. Driscoll, J. Moore, and G. K. Perkins, *Supercond. Sci. Technol.* 16 (2003) 176

76 G. Karapetrov, M. Iavarone, W. K. Kwok, G. W. Crabtree, and D. G. Hinks, *Phys. Rev. Lett.* 86 (2001) 4374

77 T. Takahashi, T. Sato, S. Souma, T. Muranaka, and J. Akimitsu, *Phys. Rev. Lett.* 86 (2001) 4915

78 A. Y. Liu, I. I. Mazin, and J. Kortus, *Phys. Rev. Lett.* 87 (2001) 087005

79 H. J. Choi, D. Roundy, H. Sum, M. L. Cohen, and S. G. Louie, *Nature* 418 (2002) 758

80 S. Tsuda, T. Yokoya, T. Kiss, Y. Takano, K. Togano, H. Kito, H. Ihara, and S. Shin, *Phys. Rev. Lett.* 87 (2001) 177006

81 S. Tsuda, T. Yokoya, Y. Takano, H. Kito, A. Matsushita, F. Yin, J. Itoh, H. Harima, and S. Shin, *Phys. Rev. Lett.* 91 (2003) 127001

82 S. Souma, Y. Machida, T. Sato, T. Takahashi, H. Matsui, S.-C. Wang, H. Ding, A. Kaminski, J. C. Campuzano, S. Sasaki, and K. Kadowaki, *Nature* 423 (2003) 65

83 H. Uchiyama, K. M. Shen, S. Lee, A. Damascelli, D. H. Lu, D. L. Feng, Z.-X. Shen, and S. Tajima, *Phys. Rev. Lett.* 88 (2002) 157002

heat and tunneling spectroscopy, using polycrystalline samples or single crystals[84,85,86,87,88,89] presented clear signatures for two-gap superconductivity and offered the opportunity for comparison with theoretical predictions. Thus, MgB_2 is the first example of this model and opens up the possibility of interesting new phenomena.

4.6.5 Conclusion

Since the discovery of superconductivity at 39 K in MgB_2, of which T_c has the highest record among the intermetallic superconductors, a huge number of theoretical and experimental investigations have been reported in a short period because of its unexpectedly high-T_c. In the framework of BCS theory, the theoretical interpretation of this superconductivity focuses on the metallic nature of a 2D layer formed by boron atoms and also on strong electron-phonon coupling caused mainly by the E_{2g} phonon mode (B in-plane bond stretching), which is supported by experimental results. The results have become consistent, except for small inconsistencies which are probably due to some disorder of the boron atom and/or a slight defect at the Mg site. In particular, the two-gap (on 2D σ-band and 3D π-band) nature in MgB_2 has been confirmed from the results of several spectroscopic measurements, and MgB_2 is now recognized as the first material to be found that containing intrinsic multigaps.

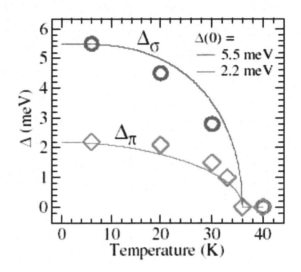

FIGURE 4.31: Temperature dependence of superconducting gap $\Delta(T)$. Open circles and diamonds show the gap values on the σ and π bands.

The entire scenario of superconductivity in MgB_2 has essentially been clarified.

MgB_2 has opened up a new frontier of investigation of the physical properties of intermetallic superconductors. Now, one of the most exciting questions is "whether MgB_2 is merely a special example in compounds including p-electron elements." Because the "BCS limit" has been broken by the appearance of MgB_2, the discovery of new higher T_c non Cu-oxides is strongly expected. Unidentified exotic high-T_c superconductors may yet be waiting to be found.

[84] F. Bouquet, R. A. Fisher, N. E. Phillips, D. G. Hinks, and J. D. Jorgensen, *Phys. Rev. Lett.* 87 (2001) 047001

[85] F. Bouquet, Y. Wang, R. A. Fisher, D. G. Hinks, J. D. Jorgensen, A. Junod, and N. E. Philips, *Europhys. Lett.* 56 (2001) 856

[86] Y. Wang, T. Plackowski, and A. Junod, *Physica* C 355 (2001) 179

[87] P. Szabó, P. Samuely, J. Kačmarčík, T. Klein, J. Marcus, D. Fruchart, S. Miraglia, C. Marcenat, and A. G. M. Jansen, *Phys. Rev. Lett.* 87 (2001) 137005

[88] R. S. Gonnelli, D. Daghero, G. A. Ummarino, V. A. Stepanov, J. Jun, S. M. Kazakov, and J. Karpinski, *Phys. Rev. Lett.* 89 (2002) 247004

[89] F. Bouquet, Y. Wang, I. Sheikin, T. Plackowski, A. Junod, S. Lee, and S. Tajima, *Phys. Rev. Lett.* 89 (2002) 257001

4.7 Fullerenes

Kosmas Prassides and Matthew J. Rosseinsky

Following the discovery of C_{60} (a quasi-spherical molecule with dimensions of ~ 1 nm) in 1985, the subsequent isolation and preparation of bulk crystalline samples of fullerenes—a set of hollow, closed-cage molecules consisting purely of carbon—from arc-processed carbon in 1990 sparked a remarkable interdisciplinary research activity, encompassing diverse fields of chemistry, physics and materials science. Fullerene solids act as "molecular sponges" with almost all atoms or molecules that come into contact with them, readily trapped in the large interfullerene voids and resulting in molecular adduct formation. Very importantly such intercalation reactions can be accompanied by charge transfer to the fullerene units, thereby leading to salt or donor-acceptor adduct formation. The early research activity quickly culminated in the synthesis of alkali metal derivatives of C_{60} with stoichiometry A_3C_{60} (A = alkali metal) which were found to be superconducting with considerably enhanced superconducting transition temperatures, T_c, when compared with their graphite analogues ($T_c = 0.15$ K)[90]—T_c in graphite intercalated compounds has also been raised subsequently to 11.5 K for C_6Ca.[91]

There are two important features of the molecular and electronic structures of C_{60} that endow its alkali intercalated derivatives with their remarkable properties. The nearly spherical symmetry of the molecule leads to an extremely simple high-symmetry crystal structure in the solid state, namely face-centered-cubic (fcc) with a lattice constant of 14.157 Å at room temperature.[92] At the same time, the weak intermolecular van der Waals forces between the molecules in the crystal make it an excellent host for intercalation chemistry. Of particular importance in this fcc C_{60} packing motif is the presence of two types of unoccupied interstitial holes with high symmetry: the smaller tetrahedral (two per C_{60} unit with a radius of 1.12 Å) and the larger octahedral (one per C_{60} unit with radius of 2.06 Å) sites. Occupation of these holes by spherical moieties of appropriate size allows retention of the fcc crystal symmetry with unit cell dimensions tunable by the radii of the dopants. At the same time, the high electron affinity of C_{60} endows it with an extensive reduction chemistry and easy access to oxidation states ranging from -1 to -6 by reaction with electron donors such as the alkali metals. This high electron affinity can be ascribed to the 12 five-membered rings present in all members of the fullerene family and required to close the curved structures.

4.7.1 Alkali Fulleride Superconductors

Metallic behavior and superconductivity were first observed in potassium-doped C_{60} films[93] with C_{60}^{3-} charge corresponding to half-filling of the t_{1u} band and thus full occupancy of the tetrahedral and octahedral interstitial sites in the fcc lattice.[94] This was quickly followed by the isolation of bulk alkali fulleride phases[95] with stoichiometries K_3C_{60} and Rb_3C_{60} which showed the onset of

[90] N. B. Hannay, T. H. Geballe, B. T. Matthias, K. Andres, P. Schmidt, and D. MacNair, *Phys. Rev. Lett.* 14 (1965) 225

[91] T. E. Weller, M. Ellerby, S. S. Saxena, R. P. Smith, and N. T. Skipper, Nat. Phys. 1 (2005) 39

[92] W. I. F. David, R. M. Ibberson, J. C. Matthewman, K. Prassides, T. J. S. Dennis, J. P. Hare, H. W. Kroto, R. Taylor, and D. R. M. Walton, *Nature* 353 (1991) 147

[93] R. C. Haddon, A. F. Hebard, M. J. Rosseinsky, D. W. Murphy, S. J. Duclos, K. B. Lyons, B. Miller, J. M. Rosamilia, R. M. Fleming, A. R. Kortan, S. H. Glarum, A. V. Makhija, A. J. Muller, R. H. Eick, S. M. Zahurak, R. Tycko, G. Dabbagh, and F. A. Thiel, *Nature* 350 (1991) 320

[94] A. F. Hebard, M. J. Rosseinsky, R. C. Haddon, D. W. Murphy, S. H. Glarum, T. T. M. Palstra, A. P. Ramirez, and A. R. Kortan, *Nature* 350 (1991) 600

[95] K. Holczer, O. Klein, S. M. Huang, R. B. Kaner, K. J. Fu, R. L. Whetten, and F. Diederich, *Science* 252 (1991) 1154

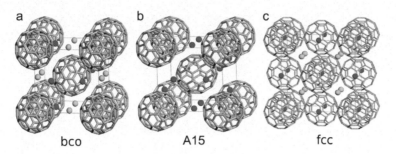

a   b   c

bco   A15   fcc

FIGURE 4.32: Crystal structures of A_3C_{60} (A = alkali metal) fullerides. (a) Body-centered orthorhombic structure of Cs_3C_{60} with partial (75%) occupancy of the indicated cation sites. (b) Primitive cubic A15 structure of Cs_3C_{60} based on body-centered cubic anion packing. (c) Face-centered cubic structure of A_3C_{60} (A = K, Rb, Cs). In (a) and (b), one unique orientation of the C_{60}^{3-} anions is present, whereas in (c) two orientations related by 90° rotation about [100] occur in a disordered manner (merohedral disorder)—only one of these orientations is shown for clarity. (Reprinted by permission from Macmillan Publishers Ltd: Nature Materials 7 (2008) 367, copyright 2008.)

superconductivity at 19 and 29 K,[94,96] respectively. The superconducting compositions adopted expanded face-centered-cubic (fcc) structures,[97] in which the three cations occupy the available octahedral and tetrahedral interstitial sites in the fcc structure of solid C_{60} (Figure 4.32c). Charge transfer between the alkali metals and the C_{60} molecules is essentially complete and the conduction band of C_{60}, which arises from its lowest unoccupied molecular orbital (LUMO) of t_{1u} symmetry, is half filled.

Although purely electronic models[98] were also proposed to account for the origin of superconductivity, conventional electron-phonon coupling mechanisms in which pair binding is dominated by intramolecular on-ball vibrational modes[99] quickly became the dominant way of explaining the properties of these systems. Within a BCS-derived model, the observed T_cs could be understood in terms of (i) a high average phonon frequency (notably though of comparable magnitude to the Fermi energy), resulting from the light carbon mass and the large force constants associated with the intramolecular modes; (ii) a moderately large electron-phonon coupling constant with contributions from both radial and tangential C_{60} vibrational modes;[100] and (iii) a high density-of-states at the Fermi level arising from the weak intermolecular interactions and resulting narrow t_{1u} bandwidth. This appeared to firmly place the T_cs of the alkali fulleride superconductors near the upper limit expected for conventional phonon-mediated superconductors. Additional support for such an interpretation also came from the experimental evidence derived from the relationship between lattice dimensions (i.e., interfullerene spacing) and superconducting transition temperatures[101] in A_3C_{60} both at ambient and at high pressures—as the size of the intercalant increases (or applied pressure decreases), T_c increases monotonically reaching a maximum at 33 K for the most expanded

[96] M. J. Rosseinsky, A. P. Ramirez, S. H. Glarum, D. W. Murphy, R. C. Haddon, A. F. Hebard, T. T. M. Palstra, A. R. Kortan, S. M. Zahur, and A. V. Makhija, *Phys. Rev. Lett.* 66 (1991) 2830

[97] P. W. Stephens, L. Mihaly, P. L. Lee, R. L. Whetten, S. M. Huang, R. B. Kaner, F. Diederich, and K. Holczer, *Nature* 351 (1991) 632

[98] S. Chakravarty, M. P. Gelfand, and S. Kivelson, *Science* 254 (1991) 970

[99] M. Schluter, M. Lannoo, M. Needles, and G. A. Baraff, *Phys. Rev. Lett.* 68 (1992) 526

[100] K. Prassides, J. Tomkinson, C. Christides, M. J. Rosseinsky, D. W. Murphy, and R. C. Haddon, *Nature* 354 (1991) 462

[101] R. M. Fleming, A. P. Ramirez, M. J. Rosseinsky, D. W. Murphy, R. C. Haddon, S. M. Zahurak, and A. V. Makhija, *Nature* 352 (1991) 787

experimentally available fulleride, $RbCs_2C_{60}$.[102] This is consistent with T_c being modulated by the density-of-states at the Fermi level—as the interfullerene spacing increases, the overlap between the molecules decreases; this leads to a reduced bandwidth and, for a fixed band filling, to an increased density of states.

However, although such a consensus on the superconductivity mechanism rapidly developed and was almost universally accepted, there were a number of disturbing features of their electronic structure parameters that did not fit into the established phenomenology. These were related to the narrow t_{1u}-derived conduction band in A_3C_{60} superconductors and therefore to any role played by electronic correlations in determining the normal and superconducting state properties. Namely, in the fcc fullerides, the t_{1u} orbital degeneracy is retained in the solid because of the cubic symmetry and typical values of the conduction bandwidth, W, are on the order of 0.5 eV. Estimates of the on-site Coulomb repulsion, U, for the C_{60} molecule are on the order of 3 eV.[103] This value is reduced significantly in the solid to ~ 1 eV but still results in a (U/W) ratio much larger than 1 whence the t_{1u} band should split into two and, as a result, a transition to a Mott-Hubbard insulating state should be observed. In order to resolve this apparent paradox, it was originally proposed that the metallic and superconducting behavior of the fullerides could arise through doping induced by non-stoichiometry of the materials, namely the true composition should be $A_{3-\delta}C_{60}$.[103] However, no evidence for such deviations in stoichiometry was found as the A_3C_{60} fullerides are strictly line phases. To account for the observed metallic behavior with integer occupancy of the t_{1u} orbitals it was shown theoretically that when the orbital degeneracy of the LUMO states is taken into account, as a result of the additional hopping channels possible for the carriers, the boundary of the metal-insulator transition shifts to much higher values of the (U/W) ratio; namely, the critical ratio $(U/W)_c$ scales with the square root of the orbital degeneracy.[104] In the fullerides, the cubic symmetry retains the triple t_{1u} orbital degeneracy and therefore metallic behavior survives—this physical picture is especially favorable for the fcc structural variant where the frustrated topology of the underlying lattice can further stabilize the highly correlated metallic state to even higher values of (U/W), $(U/W)_c \sim 2.3$.

4.7.2   Fullerides with Increased Interfullerene Separations

An important issue that arises is therefore the search for new materials with large lattice parameters in order to address the issue of whether these will be superconducting with increased T_c or the further band narrowing at such interfullerene distances will lead to electron localization and a transition to a Mott insulating state. While U is a molecular quantity and does not vary across the A_3C_{60} series, W depends sensitively on the interfullerene separation and (U/W) increases with increasing lattice constant. As a result, for large enough interball separations, (U/W) may exceed the critical value of 2.3 predicted by theory for the transition to an insulating state. An excellent method to synthesize such systems is to use as structural spacers alkali ions solvated with neutral molecules, such as ammonia[105] and methylamine.[106] The large effective radii for the resulting species, which reside in the interstitial sites of the fullerene structure have increased effective radii while maintaining the extent of charge transfer. This strategy successfully led to the isolation of fullerides such as $(NH_3)K_3C_{60}$ and $(CH_3NH_2)K_3C_{60}$ with interfullerene separations exceeding those of $RbCs_2C_{60}$ but the crystal symmetry was lowered to orthorhombic (Figure 4.33) and superconductivity was suppressed. Instead the electronic ground state of these expanded fullerides was authenticated as

[102] K. Tanigaki, T. W. Ebbesen, S. Saito, J. Mizuki, J. S. Tsai, Y. Kubo, and S. Kuroshima, *Nature* 352 (1991) 222

[103] R. W. Lof, M. A. VanVeenendaal, B. Koopmans, H. T. Jonkman, and G. A. Sawatzky, *Phys. Rev. Lett.* 68 (1992) 3924

[104] E. Koch, O. Gunnarsson, and R. M. Martin, *Phys. Rev. Lett.* 83 (1999) 620

[105] M. J. Rosseinsky, D. W. Murphy, R. M. Fleming, and O. Zhou, *Nature* 364 (1993) 425

[106] A. Y. Ganin, Y. Takabayashi, C. A. Bridges, Y. Z. Khimyak, S. Margadonna, K. Prassides, and M. J. Rosseinsky, *J. Am. Chem. Soc.* 128 (2006) 14784

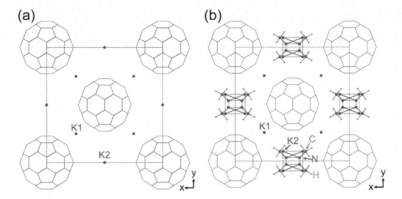

FIGURE 4.33: Basal-plane projection of the structures of (a) K_3C_{60} and (b) $(CH_3NH_2)K_3C_{60}$. Four K^+-NH_2-CH_3 units are disordered over the corners of two interpenetrating rectangles per octahedral site. The unit cell volume of K_3C_{60} expands by ~7% upon CH_3NH_2 co-intercalation (Y. Takabayashi et al., Chem. Commun. (2007) 870. Reproduced by permission of The Royal Society of Chemistry (RSC). http://pubs.rcs.org/en/content/ArticleLanding/2007/CC/B614596E).

that of a $S = 1/2$ long-range-ordered antiferromagnetic insulator,[107] providing for the first time an intriguing commonality with the phenomenology in organic and high-T_c superconductors. The suppression of metallic behavior can be understood by the crystal symmetry lowering which lifts the degeneracy of the t_{1u} orbitals, removes the effects of frustration and decreases the critical value of the (U/W) ratio for the transition to the AF Mott insulating state.

4.7.3 Cs$_3$C$_{60}$—Fullerene Superconductivity Reborn

Despite the insight provided by the expanded ammoniated/aminated alkali fullerides through the suppression of the metallic state and the appearance of magnetic interactions, there has been no definitive experimental evidence for a non-BCS origin for superconductivity in fullerides, where correlation or orbital degeneracy would play a role—the lower symmetry of the ammoniated materials lifts the degeneracy required to suppress the metal-insulator transition in the cubic systems. The established fulleride chemistry had chronically failed to deliver new materials and the physical picture of fullerene superconductivity had remained unaltered since 1992. An ideal material for understanding the interactions producing superconductivity in these structurally and chemically complex correlated electron systems would allow the isolation of the influence of purely electronic factors without the complications of structural transitions, while maintaining the site symmetry required for orbital degeneracy in all the potentially competing electronic ground states.

The Cs_xC_{60} phase field remained poorly understood around the critical $x = 3$ composition. Cs_3C_{60} had been a key target since the discovery of superconductivity in these materials, but direct synthesis by thermal combination of Cs and C_{60} in gas-solid or solid-solid reactions has been unsuccessful. This is ascribed to competition from the stable Cs_1C_{60} and Cs_4C_{60} phases. Trace superconductivity (shielding fraction < 0.1%) at 40 K under pressure was reported in 1995 in multiphase samples with nominal composition Cs_3C_{60}.[108] Despite numerous attempts by many groups worldwide, this remained unconfirmed and the structure and composition of the material responsible for superconductivity unidentified. Thus the possibility of enhancing fulleride superconductivity and under-

---

[107] K. Prassides, S. Margadonna, D. Arcon, A. Lappas, H. Shimoda, and Y. Iwasa, *J. Am. Chem. Soc.* 121 (1999) 11227

[108] T. T. M. Palstra, O. Zhou, Y. Iwasa, P. E. Sulewski, R. M. Fleming, and B. R. Zegarski, *Solid State Commun.* 93 (1995) 327

standing the structures and properties of these archetypal molecular superconductors close to the Mott-Hubbard metal-insulator (M-I) transition had remained elusive.

The development of new synthetic ideas based on low-temperature solution routes has now allowed removal of this bottleneck by using methylamine and/or ammonia as solvents to prepare solvated precursors with a suitable Cs distribution and structure to nucleate the bulk formation of two cubic polymorphs of superconducting Cs_3C_{60}.[109,110,111] Besides fcc Cs_3C_{60}, which is isostructural with all other A_3C_{60} superconductors (Figure 4.32c), a less-dense A15-structured Cs_3C_{60} polymorph (Figure 4.32b) based uniquely among fullerides on body-centered-cubic packing can be also synthesized. All of the known fcc A_3C_{60} fullerides are superconducting at ambient pressure but both expanded Cs_3C_{60} polymorphs are not. They are instead $S = 1/2$ antiferromagnetic insulators at ambient pressure but with distinctly different ordering temperatures—the magnetic ordering occurs at an order of magnitude lower temperature in the magnetically frustrated fcc polymorph ($T_N \sim 2.2$ K) than in the bipartite bcc-based A15 packing ($T_N \sim 46$ K). Thus, unlike in the necessarily less expanded systems with smaller alkali metals, the electron correlation effects represented by U dominate despite the retained orbital degeneracy in the cubic symmetry of both Cs_3C_{60} polymorphs, providing the first direct evidence that correlation directly competes with superconductivity in cubic A_3C_{60} materials.

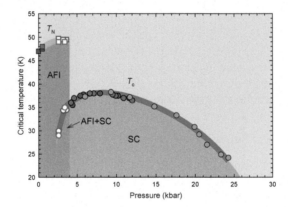

FIGURE 4.34: Electronic phase diagram of A15 Cs_3C_{60} showing the evolution of the Néel temperature T_N (squares) and the superconducting transition temperature T_c (circles), and thus the isosymmetric transition from the ambient-pressure AFI state to the high-pressure superconducting state with change in pressure. Different symbol colors represent data obtained for different sample batches. (Reprinted with permission from Y. Takabayashi et al., Science 323 (2009) 1585. Copyright 2009 by the American Association for the Advancement of Science.)

Both Cs_3C_{60} polymorphs become superconducting without crystal structure change on the application of moderate applied pressures. Superconductivity emerges out of the insulating state with a rapidly increasing T_c and reaches a maximum (35 K fcc Cs_3C_{60}, 38 K A15 Cs_3C_{60}—this is the highest T_c observed in a bulk molecular material) at $\sim$ 7 kbar before decreasing upon further pressurization (Figure 4.34). The resulting maximum in T_c is strikingly reminiscent of the T_c (hole

[109] A. Y. Ganin, Y. Takabayashi, Y. Z. Khimyak, S. Margadonna, A. Tamai, M. J. Rosseinsky, and K. Prassides, *Nature Mater.* 7 (2008) 367

[110] Y. Takabayashi, A. Y. Ganin, P. Jeglič, D. Arčon, T. Takano, Y. Iwasa, Y. Ohishi, M. Takata, N. Takeshita, K. Prassides, and M. J. Rosseinsky, *Science* 323 (2009) 1585

[111] A. Y. Ganin, Y. Takabayashi, P. Jeglič, D. Arčon, A. Potočnik, P. J. Baker, Y. Ohishi, M. T. McDonald, M. D. Tzirakis, A. McLennan, G. R. Darling, M. Takata, M. J. Rosseinsky, and K. Prassides, *Nature* 466 (2010) 221

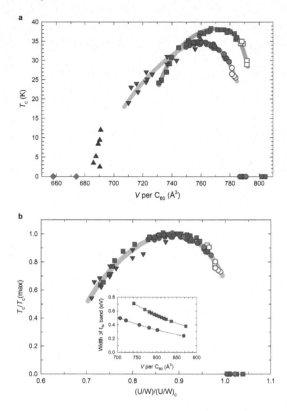

FIGURE 4.35: (a) Superconducting transition temperature, T_c, as a function of volume occupied per fulleride anion, V, at low temperature in the two sphere packings of A_3C_{60} superconductors. The red/green/blue circles and pink squares correspond to the bulk $T_c(V)$ behavior observed in fcc- and A15-structured Cs_3C_{60}, respectively. Open symbols represent data at pressures where trace superconductivity is observed and where in the A15 phase superconductivity coexists with anti-ferromagnetism. The yellow rhombi, dark blue triangles and brown inverted triangles correspond to the ambient pressure T_c of fcc C_{60}^{3-} anion packings with Li_2CsC_{60}, Pa-3 symmetry, and Fm-3m symmetry, respectively. (b) Normalized superconducting transition temperature, $T_c/T_c(\text{max})$, as a function of the ratio (U/W) divided by the critical value $(U/W)_c$ required to produce localization in the A_3C_{60} fulleride structures with fcc- and bcc-sphere packings. Inset: dependence of the t_{1u} conduction bandwidth on volume occupied per fulleride anion, V, for fcc-sphere (red circles) and bcc-sphere (pink squares) packings, as determined by electronic structure calculations. (Reprinted by permission from Macmillan Publishers Ltd: Nature 466 (2010) 221, copyright 2010.)

density) behavior of the high-T_c copper oxides as they are chemically doped to cross the Mott-Hubbard insulator-metal transition—in the present case, however, the complicating site disorder associated with crystal-chemical substitution in the oxides is not introduced. The absence of any structural change associated with the pressure-induced insulator-metal transition supports its purely electronic origin. The observed maximum in the dependence of T_c on P is consistent with theoretical treatments that explicitly take into account the orbital degeneracy and the repulsion between the electrons as well as the classical electron-phonon coupling.[112] These effects are not seen in the less expanded fcc A_3C_{60} systems, which are too far from the metal-insulator transition for differences from the conventional BCS predictions of the dependence of T_c on $N(E_F)$ to become apparent.

[112] M. Capone, M. Fabrizio, C. Castellani, and E. Tosatti, *Rev. Mod. Phys.* 81 (2009) 943

The different sphere packings in the fcc and A15 bcc-based Cs_3C_{60} polymorphs produce distinct dome-like $T_c(V)$ relationships, where V is the cubic unit cell volume per C_{60}^{3-} unit (Figure 4.35a). Superconductivity emerges out of the insulating state in fcc Cs_3C_{60} at a critical value of $V_c \sim 786.0$ Å^3 per C_{60}^{3-}, smaller than that for the less-dense A15 Cs_3C_{60} (at ~ 796.6 Å^3 per C_{60}^{3-}). Observation of the packing densities at which the metal-insulator transition takes place experimentally allows estimation of the bare bandwidth, W_c, at which the metallic state becomes unstable in the two lattice topologies (~ 0.35 eV for fcc *vs* ~ 0.55 eV for bcc-based lattices). The frustrated fcc packing thus stabilizes strongly correlated metallic states to much smaller W [much larger (U/W) ratios] than does the bipartite bcc packing.

As W_c is controlled by the specific cubic lattice packing, it can be used to scale T_c in the two polymorphs by replacing the lattice-specific parameter, V, with $(U/W)/(U/W)_c = (W_c/W)$. This permits the scaling of the two distinct $T_c(V)$ relationships onto a single dome-shaped lattice-packing-independent universal curve for all currently known C_{60}^{3-} superconductors (Figure 4.35b), suggesting that it is the separation from the competing localized electron ground state that determines T_c in fullerides. Electron correlations are also important in high-T_c superconductors like the copper oxides, where the packing of the electronically active ions is essentially identical (two-dimensional square) in all known superconducting families. Cs_3C_{60} is unique, as the same electronically active unit can be packed in two distinct lattice motifs to reveal that T_c scales in a lattice-independent manner with proximity to the correlation-driven metal-Mott insulator transition. This poses the synthetic chemical challenge of identifying new methods to tune the separation and interaction of orbitally degenerate molecules near the metal-insulator-superconductor transition to both enhance T_c and further understand the structural control of electronic behavior required to access higher T_c . The newly demonstrated commonality with the copper oxide and iron chalcogenide/pnictide systems of the key role of electron correlations shows the importance of understanding these systems for the development of high T_c superconductivity in new classes of materials.

4.8   Elemental Superconductors

K. Shimizu

Which metals become superconductors? The answer remains incompletely understood and the mechanism of superconductivity is still unclear. The treasure hunting started just after the discovery of superconductivity in Hg (mercury) in 1911 and we are still searching for new superconductors and meet them accidentally. Figure 4.36 shows a timeline of the development of superconducting temperature in elements with plots of the highest-T_c in each element from Hg. Nb (niobium) has the highest ever found in elements at ambient (atmospheric, 1013 hPa) pressure. After La (lanthanum) many elements that show higher T_c than Nb are plotted on the timeline. They were observed under pressures. There are 30 superconducting elements at ambient pressure, but that number increases to 53 if pressure is applied. Pressure can change the

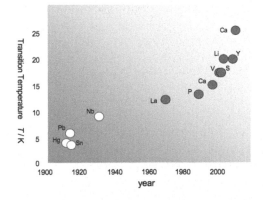

FIGURE 4.36:   The timeline of the development of superconducting temperature of elements. Superconducting temperatures measured at ambient (white circles) and high pressure (gray circles).

structural distance of atoms in elements and their superconducting properties. Actually, it has been found that non-superconducting elements and compounds show superconductivity under high-pressure conditions. The number of compounds formed from combinations of elements is infinite, but hints for the best selection of elements for high-T_c compounds should be found through a deep investigation of elements. In this section, superconducting elements and pressure effect on the superconductivity are reviewed.

4.8.1 Periodic Table for Superconducting Elements

Figure 4.37 shows a periodic table of superconducting elements. Elements that are superconducting at ambient pressure are colored in pink and the number is 30. Superconducting elements under pressure are in red and non-superconducting elements are in white. Non-colored (white) elements are categorized into four groups: non-metal element, noble metal, magnetic metal, and alkaline metal. Each group seems to have a reason why its members are not superconducting and it is important to investigate these reasons to understand the mechanism of superconductivity. Elements in Figure 4.37 with symbols written in white are typical for their groups and were found to be superconducting after 1996. No elements in the noble metal group have been found to be superconducting as their transition temperatures are expected to be inaccessibly low, in the micro-Kelvin range.

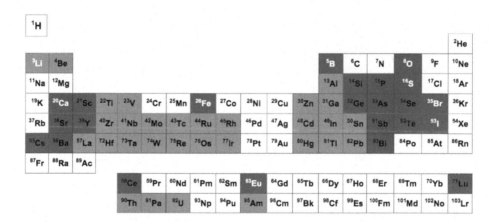

FIGURE 4.37: The periodic table of superconducting elements. Pink: superconducting at ambient pressure (30 elements); red: superconducting under pressure (23 elements); white: non-superconducting (50 elements).

Generally "superconductivity" is known as a very rare phenomenon that can be seen only in limited compounds. However, almost half of the elements in the periodic table can become superconducting. Thus, one may suppose that "superconductivity" is a rather common phenomenon in materials and is caused through a variety of mechanisms. When we look at the hypothesis that all elements can be superconducting, searching for superconductivity in simple elements is a shortcut not only for understanding the superconducting mechanism but also for investigating the unrevealed possibilities of materials.

To study the hidden possibilities of elements or compounds, "pressure" is one of the most effective and powerful tools. For example, although most elements and compounds show a negative pressure dependence on the superconducting temperature, some elements show positive pressure dependence and are found to keep the positive dependence. How high can the superconducting temperature be raised by pressure? As shown in Figure 4.36, the current highest T_c of elements is ~ 25 K in Ca (calcium) at high pressure of ~160 GPa.[113] Moreover, the slope of the T_c curve remains positive.

4.8.2   Experimental Technique

To study superconductivity under high pressure, a pressure cell was designed for low-temperature experiments of resistance and magnetization measurement. Figure 4.38 shows a photograph and the schematic drawing of a typical high-pressure device, the diamond anvil cell (DAC), which can be used for generating up to 250 GPa at low temperature down to 30 mK. The DAC consists of a pair of diamond anvils and is made from a nonmagnetic Be-Cu alloy.

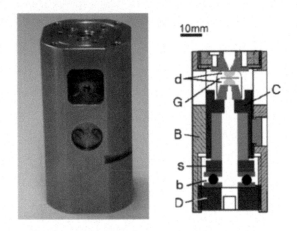

FIGURE 4.38:   Photograph and schematic drawing of a diamond anvil cell (DAC), a typical pressure generating device. B: main body; C: piston with lower diamond; D: loading nut; G: metal gasket; b: ball bearing; d: diamond; s: plastic ring.

The size of the sample volume for experiments at 100 GPa is typically as small as 10^{-8} cm^3, which corresponds to dimensions of about $30\,\mu$m$^2 \times 10\,\mu$m. For ac four-terminal electrical resistance measurements, a lithographic technique for fabricating a fine electrical network on the pressure surface of the diamond anvils is employed for experiments above 100 GPa. The schematic drawing of the sample chamber and setup of the electrodes is shown in Figure 4.39. In addition to the electrical resistance measurements, a sensitive magnetization measurement technique using a SQUID magnetometer is employed for Meissner signal detection. The technical details are described in elsewhere.[114]

4.8.3   Example of Pressure-Induced Superconductivity in Elements

To provide an example of pressure-induced superconductivity, experimental results leading to the discovery of superconductivity in O (oxygen), in the non-metal elements group, are reviewed. Oxygen forms diatomic molecules (O_2) and is a gas at ambient conditions. Gas can be liquefied and then solidified by cooling and pressurizing. Liquid oxygen was introduced into a small sample

[113] T. Yabuuchi, T. Matsuoka, Y. Nakamoto, and K. Shimizu, *J. Phys. Soc. Jpn.* 75 (2006) 083703

[114] K. Shimizu, K. Amaya, and N. Suzuki, *J. Phys. Soc. Jpn.* 74 (2005) 1345

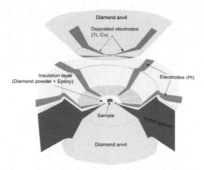

FIGURE 4.39: The schematic drawing of the sample chamber and setup of the electrodes. In this drawing, the gasket is cut and shown in a 1/4 (90°) section. The sample is set into a chamber on the gasket and squeezed by a pair of diamonds; contact is made with the four deposited electrodes on the upper diamond surface.

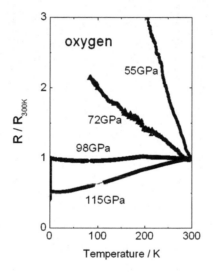

FIGURE 4.40: Normalized resistance (R/R_{300K}) as a function of temperature at various pressures. The slope of the curves changed by applying pressure and became positive at ~ 100 GPa, which indicates the metallization of oxygen. The abrupt drop in resistance near 0 K is the onset of superconductivity.

chamber in a pressure device, a DAC, inside a cryostat. Oxygen becomes solid crystal under pressure of around 5 GPa.

The occurrence of metallization was suggested by an appearance of light reflection and an abrupt drop of the resistance by an order of $7 \sim 8$ up to 100 GPa. Proof of the metallization, the temperature dependence of resistance, was measured, and the change from semiconductive to metallic behavior was found at ~100 GPa.[115] Onset of superconductivity was also observed at the pressure of onset of metallization at a temperature of 0.6 K. The superconductivity was confirmed through a separate experiment by detecting the Meissner effect. Figure 4.40 shows the transition of oxygen from semiconductor to superconducting metal measured by temperature dependence of resistance. The

---

[115] K. Shimizu, K. Suhara, M. Ikumo, M. I. Eremets, and K. Amaya, *Nature* 393 (1998) 767

observed superconducting temperature of oxygen (0.6 K) is found to be rather low in comparison to those of other elements in the VIb group like S (sulfur), Se (selenium), and Te (tellurium). Structural experiments show that oxygen remains in the diatomic molecular structure even in the metallic state. It was revealed that oxygen become superconductive in its molecular metallic state.

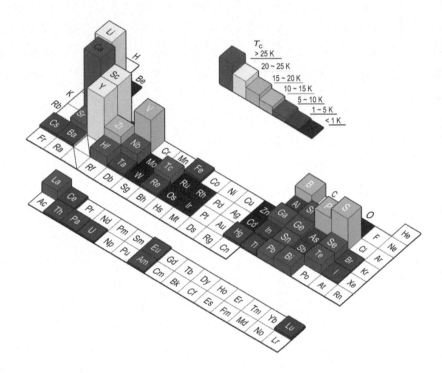

FIGURE 4.41: Three-dimensional periodic table for superconducting elements. The height of the column represents the highest (maximum) superconducting transition temperature observed as of 2010.

4.8.4 Summary: 3D Periodic Table for Superconducting Elements

Figure 4.41 shows the relative superconducting transition temperatures of elements on the periodic table. Column height is proportional to the highest (maximum) superconducting temperature ever observed in each element. Most of the maximum T_cs that have been recorded at ambient pressure, but at least some of the elements show rather higher temperatures under pressure. Additionally, some elements show increasing T_c under pressure. According to Figure 4.41 one can say that high-T_c elements are located at the horizontal edge of the periodic table. The number of superconducting elements is 53 at this moment, which means that almost half of all elements can become superconducting. Moreover, elements adjacent to superconducting ones in the table can also be expected to become superconducting because they have similar properties. Thus, can all elements be expected to become superconducting?

At the unexplored high-pressure frontier, unrevealed high-T_c phases can be expected. Even simple elements are typical candidates to be explored for the possibility of superconductivity and still have hidden capacity for superconducting properties. Recent theoretical investigations of high-pressure structure and superconducting transition temperature using first-principle calculations is another of the most powerful tools in such an exploration.

One of the ultimate goals of superconductivity in elements is the realization of a metallic hydrogen room temperature superconductor. The expected pressure needed is 450 GPa, which has not yet been achieved with hydrogen. Studying superconductivity in simple elements at high pressures will accelerate the rate of development of superconductivity in its next century.

4.9 Heavy-Fermion Superconductivity

F. Steglich

The 1979 discovery[116] of heavy-fermion superconductivity below $T_c = 0.6$ K in tetragonal $CeCu_2Si_2$ was not completely accidental, as it followed the earlier discoveries[117,118] of superfluidity in ^{3}He and heavy-fermion phenomena in $CeAl_3$. Given the phenomenological similarity of the low-temperature Landau Fermi-liquid (LFL) phase of the latter compound with (charge-neutral) ^{3}He in its normal-fluid state,[119] it became obvious to ask whether systems like $CeAl_3$ could possibly show unconventional superconductivity (SC), similar to the superfluid phases of ^{3}He. In fact, when the first indications of SC were found in 1978 on low-quality polycrystals of $CeCu_2Si_2$, it was the anticipation of an analog to superfluid ^{3}He which called for the preparation of cleaner sample material and, eventually, led to the discovery of bulk SC in $CeCu_2Si_2$ (see Figure 4.42).

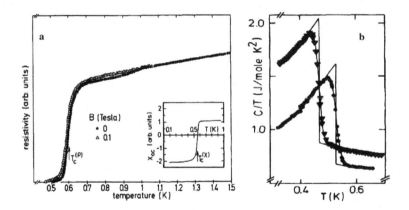

FIGURE 4.42: Low-temperature properties of $CeCu_2Si_2$ indicating a bulk superconducting transition: (a) resistivity (main part) and low-field ac susceptibility (inset); (b) electronic specific heat as C/T *vs.* T for two equally prepared polycrystals. From Steglich [116]. (Reprinted figure with permission from F. Steglich, J. Aarts, C. D. Bredl, W. Lieke, D. Meschede, W. Franz and H. Sch afer, *Phys. Rev. Lett.* 43 (1979) 1892. Copyright 1979 by the American Physical Society. http://link.aps.org/doi/10.1103/PhysRevLett.43.1892.)

Given the antagonistic nature of SC and magnetism, this discovery came as a big surprise. While

[116] F. Steglich et al., *Phys. Rev. Lett.* 43 (1979) 1892

[117] D. D. Osheroff, R. C. Richardson, and D. M. Lee, *Phys. Rev. Lett.* 28 (1972) 885

[118] K. Andres, J. E. Graebner, and H.R. Ott, *Phys. Rev. Lett.* 35 (1975) 1779

[119] See, e.g., D. Vollhardt and P. Wölfle, *The Superfluid Phases of ^{3}He* (London: Taylor and Francis, 1990)

all then-known (BCS) superconductors lose their SC when doped with, typically, less than 1at% of magnetic impurities, in $CeCu_2Si_2$ 100at% of magnetic Ce^{3+} ions are prerequisite to generate SC: The nonmagnetic reference compound $LaCu_2Si_2$ is not a superconductor, and doping $CeCu_2Si_2$ with a low concentration of nonmagnetic dopants suppresses SC completely.[120]

Like $CeAl_3$, the "Kondo-lattice system" $CeCu_2Si_2$ adopts a nonmagnetic state well below its Kondo temperature $T_K \approx 15$ K (cf. inset of Figure 4.42a). Here, a huge electronic ($4f$-derived) specific heat, $C(T)$, emerges as a result of the Kondo effect. As $T \to 0$, the Sommerfeld coefficient, $\gamma = C/T$, is of order 1 J/K^2mol (see Figure 4.42b) which is huge compared to about 1 mJ/K^2mol for ordinary metals like Cu. Most interestingly, the jump $\Delta C/T_c$ at T_c is of the same gigantic order as $\gamma(T_c)$. This led to the conclusion that Cooper pairs are formed by quasiparticles with extremely large effective mass $m^* = (100–1000)$ (m_e: bare electron mass). The Fermi velocity of these "heavy fermions" (HFs) is correspondingly small, of the order of the sound velocity only. Therefore, the electron-phonon coupling is not retarded and not apt to mediate Cooper pairing. Consequently, given the similarity to ^{3}He, magnetically driven pairing was proposed for $CeCu_2Si_2$ and its subsequently discovered U-based counterparts, like UBe_{13} ($T_c \approx 0.9$ K)[121] and UPt_3 ($T_c \approx 0.5$ K),[122] already by the mid-1980s.

The early studies on $CeCu_2Si_2$ were severely plagued by the fact that physical properties were strongly varying from one sample to another (cf. Figure 4.42b). It took several years before these "sample dependences" could be understood as being due to a rather complex chemical Ce-Cu-Si phase diagram. The latter displays a narrow homogeneity range of $CeCu_2Si_2$ which prevents Cu-Si exchange by more than 1% and contains an antiferromagnetic (AF) instability close to the exact 1:2:2 stoichiometry. This enables one to prepare *homogeneous* $CeCu_2Si_2$ single crystals either with a tiny deficit of Cu showing AF order ("A-type" samples), or with a slight Cu excess being superconducting ("S type"). In the vicinity of the stoichiometric composition, SC and AF order compete with each other without microscopic coexistence ("A/S type").

By the end of the 1990s, about ten HF superconductors were known. Meanwhile, this number has increased to more than 35. Interestingly, the vast majority of the recently discovered HF superconductors belong to two distinct groups of intermetallic compounds, i.e., (i) systems of $Ce_nT_mIn_{3n+2m}$ type (T: transition metal), like $CeCoIn_5$,[123] and (ii) non-centrosymmetric superconductors, like $CePt_3Si$.[124] The former systems are quasi-two-dimensional (2D) variants of the cubic superconductor $CeIn_3$ ($T_c \approx 0.2$ K).[125] They are formed by stacking alternating layers of $CeIn_3$ and TIn_2 sequently along the tetragonal c-axis. As theoretically predicted, the so reduced dimensionality causes an increase of T_c, which amounts to more than one order of magnitude ($T_c = 2.3$ K for $CeCoIn_5$). A further substantial T_c-enhancement was achieved by the same group when replacing Ce (with well-localized $4f$ shell) by Pu (whose $5f$ shell is spatially more extended), $PuCoGa_5$ with $T_c = 18.5$ K presently being the record holder.[126] Its Rh-homologue as well as $NpPd_5Al_2$ also show strongly enhanced T_c values.[127,128]

In HF metals, SC often occurs in the vicinity of some type of instability. For example, $CePd_2Si_2$ exhibits in its temperature-pressure phase diagram a narrow dome of SC ($T_{cmax} \approx 0.4$ K) centered around the critical pressure $p_c \approx 2.8$ GPa at which AF order becomes suppressed (Figure 4.43a).

[120] H. Spille, U. Rauchschwalbe, and F. Steglich, *Helv. Phys. Acta* 56 (1983) 165

[121] H. R. Ott et al., *Phys. Rev. Lett.* 50 (1983) 1595

[122] G. R. Stewart et al., *Phys. Rev. Lett.* 52 (1984) 679

[123] C. Petrovic et al., *J. Phys.: Condens. Matter* 13 (2001) L337

[124] E. Bauer et al., *Phys. Rev. Lett.* 92 (2004) 027003

[125] N. D. Mathur et al., *Nature* 394 (1998) 39

[126] J. L. Sarrao et al., *Nature* 420 (2002) 297

[127] F. Wastin et al., *J. Phys.: Condens. Matter* 15 (2003) 52279

[128] H. Aoki et al., *J. Phys. Soc. Jpn.* 76 (2007) 063701

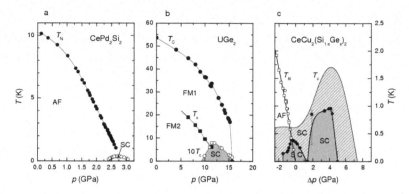

FIGURE 4.43: Superconductivity near magnetic and valence instabilities in the temperature-pressure phase diagrams of (a) $CePd_2Si_2$,[125] (b) UGe_2,[129] and (c) $CeCu_2(Si_{1-x}Ge_x)_2$.[132] T_N, T_C, T_c: critical temperatures of AF, FM and superconducting order, respectively. T_x: transition temperature between two FM states. a: (Reprinted by permission from Macmillan Publishers Ltd: Nature 394 (1998) 39, copyright 1998.) b: (Reprinted with permission from N. Tateiwa et al., "Pressure induced superconductivity in a ferromagnet UGe_2," J. Phys.: Condens. Matter 13 (2001) L17–L24. Copyright 2001 by IOP. http://dx.doi.org/10.1088/0953-8984/13/1/103) c: (Reprinted with permission from H. Q. Yuan et al., Science 302 (2003) 2104. Copyright 2003 by the American Association for the Advancement of Science.)

UGe_2,[129,129B] like a few other U-based compounds,[130,131] shows SC within a regime of ferromagnetic (FM) order (see Figure 4.43b). Pure $CeCu_2Si_2$[132] as well as $CeCu_2Ge_2$[133] are superconductors in wide pressure ranges. Upon doping $CeCu_2Si_2$ with 10 at% Ge, two distinct superconducting domes emerge (Figure 4.43c). The low-p dome is centered at $p_c \approx 0$ where AF order disappears continuously, resembling the phase diagram of $CePd_2Si_2$ (cf. Figure 4.43a). SC under the second dome, which occurs near a weak valence instability at $p \approx 5.5$ GPa, is believed to result from almost critical valence fluctuations.

While most of the Ce-based HF superconductors exhibit an anomalous low-T normal state, often referred to as a non-Fermi-liquid (NFL) state, several of their U-based counterparts are moderately heavy LFLs below the critical temperature of AF (in the case of URu_2Si_2: "hidden"[134]) order. This LFL phase apparently coexists with AF/hidden order and becomes unstable against the formation of HFSC below T_c.

Commonly, HF superconductors have a highly anisotropic, even-parity ($S = 0$) order parameter: Non-exponential T-dependences, i.e., often simple power laws, of the specific heat and related quantities seem to rule out s-wave symmetry and make d-wave pairing very likely. As an empirical rule, this type of HFSC is suppressed by potential scattering resulting in a mean free path of the charge carriers, l, which is shorter than the superconducting coherence length, ζ. A small number of HF superconductors, however, are prime candidates for odd-parity ($S = 1$) pairing.[135,136] Inter-

[129] N. Tateiwa et al., *J. Phys.: Condens. Matter* 13 (2001) L17

[129B] S. S. Saxena et al., *Nature* 406 (2000) 587

[130] F. Lévy et al., *Science* 309 (2005) 1343

[131] N.T. Hung et al., *Phys. Rev. Lett.* 99 (2007) 067006

[132] H. Q. Yuan et al., *Science* 302 (2003) 2104

[133] A. T. Holmes, D. Jaccard, and K. Miyake, *Phys. Rev. B* 69 (2004) 024508

[134] C. R. Wiebe et al., *Nature Phys.* 3 (2007) 96

[135] H. Tou et al., *Phys. Rev. Lett.* 80 (1998) 3129

[136] K. Ishida et al., *Phys. Rev. Lett.* 89 (2002) 37002

estingly, in each of these cases SC coexists with either AF (e.g., UPt$_3$) or FM (e.g., UGe$_2$) order. Recently, non-centrosymmetric HF superconductors have attracted much theoretical interest. This derives from the fact that the lack of inversion symmetry allows for a mixing of even- and odd-parity pair states, the degree of mixing depending on the strength of the antisymmetric spin-orbit coupling.[137,138]

UPt$_3$ is probably the best studied HF superconductor. It exhibits a temperature-magnetic-field phase diagram which displays several superconducting phases (Figure 4.44a[139]) similar to the different superfluid phases of ^{3}He. Multiphase diagrams were also discovered,[140,141] for both U$_{1-x}$Th$_x$Be$_{13}$ (Figure 4.44b[142]) and PrOs$_4$Sb$_{12}$. The latter compound is unique in that here, quadrupolar fluctuations are believed to mediate the Cooper pairing.

a   b

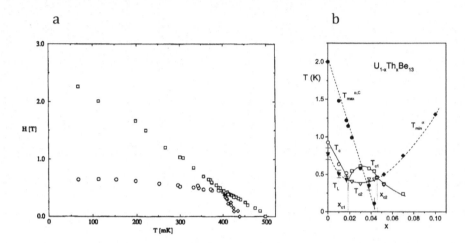

FIGURE 4.44: (a) Magnetic field-temperature phase diagram of UPt$_3$ for the field perpendicular to the c-axis.[139] (b) Temperature-concentration phase diagram of U$_{1-x}$Th$_x$Be$_{13}$. Solid (dotted) lines: second-order phase transitions[140] (crossovers).[142] (a: "The order parameter for the superconducting phases of UPt$_3$," J. A. Sauls, Advances in Physics 43 (1994) 113, reprinted by permission of the publisher (Taylor & Francis Group, http://www.informaworld.com).) (b: With kind permission from Springer Science+Business Media: "Magnetic Field Response in (UTh) Be$_{13}$," F. Kromer, N. Oeschler, T. Tayama, K. Tenya, T. Cichorek, M. Lang, F. Steglich, J. S. Kim and G. R. Stewart, Journal of Low Temperature Physics 126 (2002) 815–833. Fig. 1.)

The first experimental evidence for magnetically driven SC came from inelastic neutron-scattering (INS) results on the weak antiferromagnet UPt$_3$ ($T_N \approx 5$ K), where the magnetic scattering intensity was found to drop upon cooling to below $T_c \approx 0.5$ K.[143] A combined analysis of tunneling and INS results revealed that in UPd$_2$Al$_3$ strong-coupling SC ($T_c \approx 2$ K), which microscopically coexists with local-moment AF order ($T_N = 14.3$ K, $\mu_{ord} = 0.85\mu_B/U$), is mediated by the acoustic magnon at the AF ordering wave vector ("magnetic exciton").[144] Guided by the theoretical

[137] See, e.g., L. P. Gor'kov and R. I. Rashba, *Phys. Rev. Lett.* 87 (2001) 037004

[138] D. F. Agterberg et al. *Physica B* 378–380 (2006) 351

[139] See, e.g., J. A. Sauls, *Adv. Phys.* 43 (1994) 113

[140] H. R. Ott et al., *Phys. Rev. B* 31 (1985) 1651

[141] M. B. Maple et al., *J. Phys. Soc. Jpn.* 71(Supplement) (2002) 23

[142] F. Kromer et al., *J. Low Temp. Phys.* 126 (2002) 815

[143] G. Aeppli et al., *Phys. Rev. Lett.* 63 (1989) 676

[144] N. K. Sato et al., *Nature* 410 (2001) 340

prediction of HFSC being mediated by AF spin fluctuations, several Ce-based HF antiferromagnets could be turned to superconductors under applied pressure. Early examples were $CeCu_2Ge_2$[145] and $CeRh_2Si_2$[146] for both of which it was found that (i) AF order disappears abruptly at a critical pressure p_c and (ii) heavy LFL behavior develops at sufficiently low temperatures for pressures $p \gtrsim p_c$. The majority of the p-induced Ce-based superconductors, however, show pronounced NFL, phenomena near the critical pressure, up to sometimes surprisingly high temperatures. For such a material, p_c marks a quantum critical point (QCP) at which the Néel temperature vanishes smoothly.

Since QCPs in proximity to HFSC usually occur under pressure, it is experimentally very difficult to probe their magnetic excitation spectrum. $CeCu_2Si_2$ is ideally suited for such a study because here, SC forms in the vicinity of an AF QCP at ambient (low) pressure. Neutron diffractometry showed AF order in this compound to be of spin-density wave (SDW) type, with a small ordered moment (0.1 μ_B/Ce) and an incommensurate propagation vector $\mathbf{Q}_{AF}$. The latter could be identified as a nesting wave vector of the renormalized Fermi surface.[147] Pronounced NFL behavior in the thermodynamic and transport properties observed in the field-driven low-T normal state of S-type $CeCu_2Si_2$ single crystals[148] strongly suggest that the QCP in $CeCu_2Si_2$ is of the 3D SDW type.

Recent INS spectra on single crystals of S-type $CeCu_2Si_2$ clearly revealed spin excitations which are localized in $\mathbf{Q}$-space at the incommensurate ordering wave vector $\mathbf{Q}_{AF}$. As seen in Figure 4.45a, they appear broadly distributed in energy transfer and are well described by a Lorentzian.[149] Out of this quasielastic excitation spectrum found in the normal state, an inelastic line at 0.2 meV (as $T \to 0$) emerges in the superconducting state. The missing spectral weight at low energies indicating a spin-excitation gap is recovered at the gap edge, thereby constituting an inelastic line. Because its temperature dependence coincides with the scaled gap function of a d-wave BCS superconductor, this spin gap at $\mathbf{Q}_{AF}$ was related to the superconducting gap. Upon increasing temperature in the normal state, the magnetic response weakens in intensity and broadens substantially; i.e., there is a considerable slowing down of the response upon cooling. As $T \to 0$, the width of the quasielastic line remains finite, because the S-type $CeCu_2Si_2$ single crystal is located on the paramagnetic side of the low-p phase diagram (cf. Figure 4.43c). The slowing down of the response, which was found to be nearly isotropic, therefore confirms the close vicinity of this crystal to a 3D SDW QCP.

Figure 4.45b displays the momentum dependence of the magnetic response near $\mathbf{Q}_{AF}$ in the superconducting state, recorded at different energy transfers $\hbar\omega$. The single peak due to the AF excitations at 0.2 meV splits upon increasing $\hbar\omega$ into two sub-peaks which move further away from each other, with a strong decrease in intensity. These data show convincingly that the nearly quantum-critical 3D SDW fluctuations behave as an *overdamped, dispersive* mode. As inferred from the inset of Figure 4.45b, the positions of the two sub-peaks for different $\hbar\omega$ yield a linear dispersion relation. The mode velocity as read off the slope of this dispersion relation, $v_m \approx 680$ m/s, turns out to be substantially smaller than the already very small Fermi velocity of the heavy charge carriers, $v_F^* \approx 8700$ m/s. This means that the coupling between the HFs and the 3D quantum-critical SDW fluctuations is retarded, and the latter may be capable of acting as glue for SC. This is supported by a huge reduction in exchange energy when going from the normal to the superconducting state, which exceeds the superconducting condensation energy by a factor of more than 20.

$CeCu_2Si_2$ is the first superconductor for which almost quantum-critical AF spin fluctuations could be identified as a major driving force for SC. However, this scenario may well prove to be more relevant, e.g., for other p-induced HF superconductors like $CePd_2Si_2$ as well as for the unconventional high-T_c cuprates, like $YBa_2Cu_3O_{6.6}$.[150] For the electron-doped Fe-pnictide superconductors

[145] D. Jaccard, K. Behnia, and J. Sierro, *Phys. Lett. A* 163 (1992) 475

[146] R. Movshovich et al., *Phys. Rev. B* 35 (1996) 8241

[147] O. Stockert et al., *Phys. Rev. Lett.* 92 (2004) 136401

[148] P. Gegenwart et al., *Phys. Rev. Lett.* 81 (1998) 1501

[149] O. Stockert et al., *Nature Phys.* 7 (2011) 119

[150] T. Dahm et al., *Nature Phys.* 5 (2009) 217

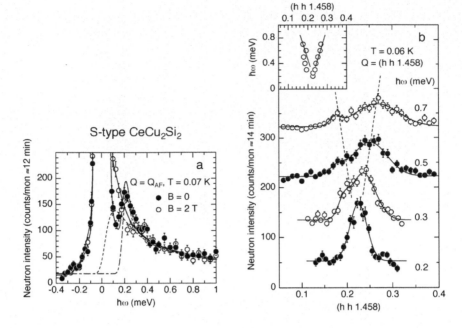

FIGURE 4.45: Inelastic neutron-scattering results for S-type $CeCu_2Si_2$. (a) Magnetic response at $T = 70$ mK at the AF wave vector $\mathbf{Q}_{AF}$ as a function of energy transfer in the superconducting ($B = 0$) and normal states ($B = 2$ T). (b) Dispersion and damping of the magnetic response at $T = 60$ mK. From Stockert et al.[149]. (Reprinted by permission from Macmillan Publishers Ltd: Nature Physics 7 (2011) 119, copyright 2011.)

crystallizing in the same $ThCr_2Si_2$ structure as $CeCu_2Si_2$, $Ba(Fe_{1-x}Co_x)_2As_2$[151] being an example, temperature-charge carrier (or hydrostatic pressure) phase diagrams were established which are very similar to the T-p phase diagram of $CePd_2Si_2$ (see Figure 4.43a). This suggests that nearly quantum-critical SDW fluctuations may essentially contribute to the Cooper pairing in these new "high-T_c" superconductors as well.

To what extent a Mott transition, like in the cuprates, or an "orbital-selective" Mott transition in HF metals, perhaps existing in p-induced $CeRhIn_5$[152,153] and β-$YbAlB_4$,[154] may be involved in the formation of unconventional SC remains a challenging open question.

4.10   Ruthenate Superconductor Sr_2RuO_4

Yoshiteru Maeno

With the announcement of the discovery of revolutionary high-temperature superconductivity in 1986 by J. G. Bednorz and K. A. Müller,[155] immense worldwide efforts immediately surged to

[151] J.-H. Chu et al., *Phys. Rev. B* 79 (2009) 014506

[152] H. Shishido et al., *J. Phys. Soc. Jpn.* 74 (2005) 1103

[153] T. Park et al., *Nature* 440 (2006) 65

[154] S. Nakatsuji et al., *Nature Phys.* 4 (2008) 603

[155] J. G. Bednorz and K. A. Müller, *Z. Phys. B: Cond. Matter* 64 (1986) 189

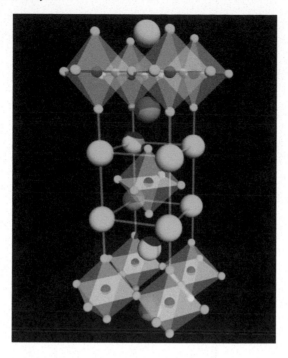

FIGURE 4.46: Crystal structure of Sr_2RuO_4. The CuO_2 plane of the original Bednorz-Müller cuprate superconductor is replaced by the RuO_2 plane. Blue: Sr; red: Ru; green: O. Figure by K. Deguchi.

search for other copper-oxide (cuprate) superconductors with higher transition temperatures T_c. At the same time, searches for similar superconductors without copper began. However, it was eight years later when the first oxide superconductor in the same layered structure as the high-T_c cuprates was discovered (Figure 4.46).[156] It is the ruthenium-oxide (ruthenate) superconductor Sr_2RuO_4, and, in spite of its low T_c of 1.5 K, this superconductor stimulated extensive research efforts with the possibility of firmly establishing the "spin-triplet" pairing state.

In nearly all the known superconductors, conduction electrons are paired in the "spin-singlet" state and become superfluids carrying electric charge. An electron as an elementary particle is characterized by its mass, charge $-e$ (e is called the elementary charge), and the spin $s = 1/2$. When they form pairs in the superconducting state, each pair becomes spinless. In contrast, it is in principle possible that pairs with the total spin of $S = 1$, called the spin-triplet state, form. In fact it is firmly established that the superfluidity of an isotope of helium, ^{3}He, below 3 mK is carried by atomic pairs of total spin one. In a spin-triplet superconductor, active spins of electrons, as well as their charges, are expected to exhibit superfluidity.

The author spent one year at IBM Zurich laboratory as an assistant to J.G. Bednorz. The research topic he chose there was the search for superconductivity in ruthenates. Soon after the discovery of high-T_c cuprate, Bednorz was extending his search for new superconductors to other oxide materials including ruthenates. In fact it was exactly on the day of the Nobel prize ceremony in December 1987 when a summer student in Bednorz' lab synthesized a powder sample of Sr_2RuO_4, a metallic oxide with layered crystal structure. Y. Maeno and later F. Lichtenberg continued to investigate both

---

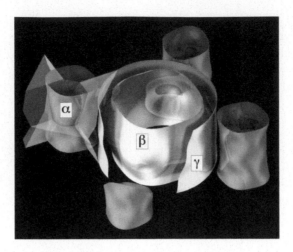

FIGURE 4.47:   Three conduction bands (Fermi surfaces) of Sr_2RuO_4 determined experimentally from quantum oscillations. (Figure created by Christoph Bergemann, copyright by AIP.)

cubic and layered ruthenates in search of superconductivity in Zurich. Because of such a strong influence of the cuprate superconductivity based on the parent insulating material with divalent copper ions (Cu^{2+}) with odd number of electrons ($3d^9$ configuration), the main focus was on materials with odd number of electrons and also the search was extended only down to 4.2 K, the boiling point of liquid ^{4}He. In 1994 Maeno and coworkers at Hiroshima University found a strong sign of superconducting transition using polycrystalline Sr_2RuO_4 below 1.5 K and, using single crystals grown by Lichtenberg, the Hiroshima group immediately obtained firm evidence for its superconductivity.

The availability of high-quality single crystals and the relative simplicity of the electronic structure of Sr_2RuO_4 promoted a large number of experimental as well as theoretical studies. Sr_2RuO_4 is now established as one of the archetypal unconventional superconductors, along with heavy fermions, organics, cuprates, iron pnictides, etc. In particular, it is considered as the strongest candidate of the spin-triplet superconductors, compatible with the strongest candidate of the odd-parity, pseudo-spin-triplet superconductor UPt_3.

4.10.1   Same Crystal Structure but Different Unconventional Superconductivity

Despite its structural similarity to high-T_c cuprates, with the CuO_2 plane replaced by the RuO_2 plane, Sr_2RuO_4 has a number of properties contrasting to those of the cuprates, especially the symmetry of the Cooper pairs. First the parent compound itself is metallic. The electronic structure is derived from Ru^{4+} state with four $4d$ electrons distributed in three bands as shown in Figure 4.47, in contrast to the single-band nature of the cuprates. The observed quantum oscillations clarified the nature of the three bands with the effective masses enhanced from the bare electron mass as much as 15 times due to interactions among electrons. Such strong electron correlations, in fact, led to the Mott insulating state simply by replacing Sr with isovalent Ca in Ca_2RuO_4. Thus Sr_2RuO_4 is indeed on the verge of a Mott insulator, somewhat similar to the cuprate superconductors. The superconducting state is extremely sensitive to the presence of even nonmagnetic impurities. These results suggest that the superconductivity occurs in the presence of strong Coulomb repulsion among the electrons and is "unconventional," namely with the pairing state different from the conventional spin-singlet s-wave state, which would be robust against nonmagnetic scattering.

More direct evidence for the exotic pairing state comes from the measurements using microscopic probes. First, the measurements of the electron spin susceptibility through nuclear magnetic resonance (NMR) decisively show that the electron pairs maintain active spins in the superconducting state, i.e., they are spin-triplet pairs.[157] It should be emphasized that the NMR Knight shift in Sr_2RuO_4 has a large magnitude and can be clearly separated into spin and orbital contributions of the susceptibility. Also both Ru and oxygen nuclei exhibit clear invariance of the spin susceptibilities in the superconducting state. Secondly, the spin rotation of positive muons injected into the crystal indicates that a local magnetic field emerges only in the superconducting state.[158] Such behavior is possible if the superconductivity involves symmetry breaking with respect to time reversal, in addition to the symmetry breaking of gauge, which is the hallmark of the superconducting state in general. From these results, along with a number of other strong evidence such as polarized neutron scattering and magneto-optic Kerr effect, the superconducting state of Sr_2RuO_4 is characterized by spin triplet with spontaneously broken time-reversal symmetry (TRS) as depicted in Figure 4.48. The broken TRS is attributed to the orbital moment of the Cooper pairs aligning in one direction, $L_z = +1$ or -1, and thus called the "chiral" p-wave state. Such a state is a two-dimensional analogue of the A-phase of superfluid ^{3}He with three-dimensional character.[159]

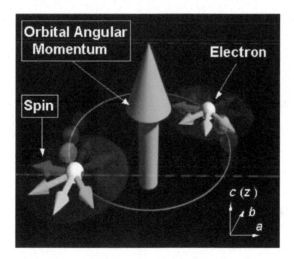

FIGURE 4.48: Spins and angular momentum of the spin-triplet Cooper pair deduced for Sr_2RuO_4. Image by K. Deguchi.

With the availability of detailed electronic structure of all three bands involved in the conduction, the microscopic mechanism of pairing has been extensively developed. Since strong ferromagnetic fluctuation is absent, it has been strongly suggested that the Coulomb repulsion among the electrons itself is important for the pairing with anisotropic energy gap.

---

[157] K. Ishida, H. Mukuda, Y. Kitaoka, K. Asayama, Y. Mori, Z. Q. Mao, and Y. Maeno, *Nature* 396 (1998) 658

[158] G. M. Luke, Y. Fudamoto, K. M. Kojima, M. I. Larkin, J. Merrin, B. Nachumi, Y. J. Uemura, Y. Maeno, Z. Q. Mao, Y. Mori, H. Nakamura, and M. Sigrist, *Nature* 394 (1998) 558

[159] T. M. Rice and M. Sigrist, *J. Phys. Cond. Matter* 7 (1995) L643

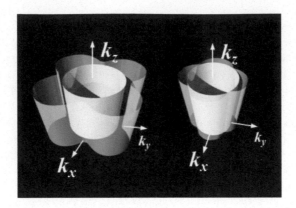

FIGURE 4.49: Superconducting gap deduced from the specific heat under magnetic field. The size and anisotropy of the gap is different for each band reflecting its orbital character. Image by K. Deguchi.

4.10.2 Novel Superconducting Phenomena

The anisotropic superconducting energy gap shown in Figure 4.49 has been deduced from the specific heat under oriented magnetic fields. It shows a multi-gap superconducting state: the gap on the main band originating from the two-dimensional electronic state has a deep minimum in the direction of the zone-boundary as required from the odd parity of the superconducting wave function. The gap on the other two bands, originating from quasi-one-dimensional electronic states with antiferromagnetic nesting vector, has a much smaller magnitude and moreover nearly closes in the direction relevant to the magnetic fluctuation mode. Such a gap structure is more complicated than a simple spin-triplet p-wave state with broken TRS, but can be described within such symmetry.

Although a large majority of the data are explained by an existing p-wave scenario, there are at present some unresolved issues concerning the exact superconducting symmetry of Sr_2RuO_4. First, although the upper critical field H_{c2} is 20 times higher for the field direction parallel to the RuO_2 plane than for the perpendicular direction, it exhibits a limiting behavior within a few degrees from the exact parallel direction. Such suppression of H_{c2} in a similar magnitude has also been observed for UPt_3. In Sr_2RuO_4, the limiting in H_{c2} is accompanied with an emergence of the first-order like transition in a very narrow range of the H-T phase diagram near H_{c2} at low temperatures. Another outstanding issue is the absence of the spontaneous current around the edge of a Sr_2RuO_4 sample, expected from the broken TRS. Such an edge current has been predicted for spin-triplet superfluid ^3He-A phase, but has not been observed either.

In order to demonstrate definitively the realization of spin-triplet superconductivity, it is important to find phenomena which cannot be explained unless spin-triplet or odd-parity is assumed. The following exemplifies the current efforts in this direction.

There are a variety of superconducting phases related to Sr_2RuO_4. By adding excess Ru metal in the crystal growth, one can obtain "eutectic" crystals in which micron-sized Ru metal platelets are periodically embedded in the matrix of Sr_2RuO_4. Surprisingly, the T_c doubles to 3 K in the interfacial region of Sr_2RuO_4 ($T_c = 1.5$ K if pure) and Ru ($T_c = 0.5$ K). The interface is atomically clean; it forms a self-organized epitaxial junction. In fact, conventional s-wave superconductivity induced in Ru interferes strongly with the surrounding superconductivity of Sr_2RuO_4. Such quantum interference can be explained naturally only by the odd-parity wave function of Sr_2RuO_4. The mechanism of enhancement of T_c of Sr_2RuO_4 near the interface is likely due to slight distortion of the RuO_2 square lattice. It should be noted that the Mott insulator Ca_2RuO_4 becomes metallic under pressure, and superconductivity was recently discovered at pressures above 10 GPa.[160]

A truly striking phenomenon expected for the spin-triplet pairing state is the half-quantum vortices (HQV), carrying half of the flux quantum $\Phi_0 = h/2e$. Quantization of magnetic flux in a superconductor originates from the single-valuedness of the wave function around a singularity, the vortex core. In a spin triplet state, the novel HQV may be realized if the spin state has a winding of phase π around a singularity with the orbital phase change of π. Since it is the orbital phase that couples with the external magnetic field through the vector potential, this state is accompanied with $\Phi_0/2$. The coupling of spin with the magnetic field may be used to stabilize the HQV state to overcome additional energy associated with the spin current. The emergence of HQV is predicted for superfluid ^{3}He in a restricted geometry where the Cooper-pair orbital moment is pinned, but so far this has not been observed. In both systems, investigations are underway by taking advantage of the microfabrication techniques not available in the past. In fact, using micron-sized rings of Sr_2RuO_4, the first observation of a phenomenon intimately related to HQV has been reported very recently.[161]

As described above, Sr_2RuO_4 has established its important role as an archetypal unconventional superconductor in the vicinity of Mott insulating phase. In spite of the layered perovskite structure common to cuprate high-T_c superconductors, its superconductivity is characterized by multiband gaps and most probably spin-triplet p-wave symmetry with time-reversal symmetry breaking. High-quality single crystals and peculiar eutectic systems allow detailed and extensive investigations towards complete characterization of its superconducting symmetry. The detailed information of the correlated electronic states obtained from quantum oscillations, angle resolved photoemission spectroscopy (ARPES), band-structure calculations, and other techniques enable us to develop microscopic mechanism of superconductivity based on the actual multi-bands. The methods and concepts introduced to unveil the exotic superconductivity of Sr_2RuO_4 have already been adopted to the investigation of many other unconventional superconductors.

4.11 Magnetic Superconductors (and Some Recollections of Professor Bernd T. Matthias)

M. Brian Maple

4.11.1 Introduction

During the first half-century that followed the discovery of superconductivity by H. Kammerlingh Onnes in 1911, superconductivity and magnetism were generally regarded as being mutually exclusive. Within the past half-century, research on the interplay between superconductivity and magnetism has flourished and revealed a rich variety of extraordinary phenomena in novel materials that arise from superconducting-magnetic interactions. In this short paper, we briefly survey some of the noteworthy developments on this subject from a personal perspective. A more complete description of this rich field of research can be found in review articles that summarize various aspects of this subject.

The author also includes some personal recollections of Professor Bernd T. Matthias who was his Ph.D. thesis advisor and mentor, and later, faculty colleague in the Department of Physics at the University of California, San Diego (UCSD). Matthias was one of the pioneers and a major figure

---

[160] P. L. Alireza, F. Nakamura, S. K. Goh, Y. Maeno, S. Nakatsuji, Y. T. C. Ko, M. Sutherland, S. Julian, and G. G. Lonzarich, *J. Phys. Cond. Matter* 22 (2010) 052202

[161] J. Jang, D. G. Ferguson, V. Vakaryuk, R. Budakian, S. B. Chung, P. M. Goldbart, and Y. Maeno, *Science* 311 (2011) 186

in superconducting materials research and performed the first experiments on the interrelation of superconductivity and magnetism.

Vitaly Ginzburg[162] first addressed the question of whether superconductivity and magnetism could coexist microscopically in 1957. Bernd Matthias, Harry Suhl, and Ernie Corenzwit[163] carried out the first experimental studies of this problem in 1958. In the early experiments of Matthias and coworkers, various rare earth elements with a partially-filled $4f$-electron shell and corresponding magnetic moment were dissolved into superconductors such as La and YOs_2. As a result of the short-range or "glassy" types of magnetic order encountered in these alloy systems, it was not possible to address the question of the coexistence of superconductivity and magnetic order in a definitive way. However, these early experiments revealed the basic form of the interaction between superconducting electrons and localized magnetic moments of the rare earth solutes and the adverse effect of paramagnetic impurities on superconductivity. They also stimulated the development by Philip Anderson and Harry Suhl of some interesting theoretical predictions such as the formation of a "cryptoferromagnetic" state in superconductors containing localized magnetic moments.[164] It became possible to study the interaction between superconductivity and long-range magnetic order and test some of these predictions in the 1970s when several families of superconducting ternary compounds containing an ordered sublattice of rare earth ions were discovered.

This article is divided into three parts. In the first part, we consider situations in which the magnetic behavior is confined to ions that are embedded in a conventional BCS superconductor where the superconducting electrons form pairs in which the electrons have opposite momenta and spins. In the second part, we discuss examples in which the same set of electrons (quasiparticles that are derived from the admixture of itinerant and localized d- or f-electrons) take part in both superconductivity, which is generally unconventional in nature, and magnetic order. In the third part, we make a few remarks about magnetically mediated superconducting electron pairing in high temperature superconductors. The final part consists of some personal recollections of Professor Bernd T. Matthias.

4.11.2   Localized Magnetic Moments in Conventional Superconductors

Superconducting-Magnetic Interactions

In a conventional superconductor, the superconductivity involves electron pairs (Cooper pairs) in which the electrons in each pair have opposite momenta and spins ($\mathbf{p} \uparrow, -\mathbf{p} \downarrow$). An applied magnetic field $\mathbf{H}$ or magnetic moment μ of an ion in a superconductor can interact with the superconducting electrons in two ways: via the Zeeman interaction of $\mathbf{H}$ or μ with the conduction electron spin $\mathbf{s}$ (of the form $\mathbf{s} \cdot \mathbf{H}$) and the electromagnetic interaction of the vector potential $\mathbf{A}$ associated with $\mathbf{H}$ or μ with the momentum $\mathbf{p}$ of the electrons (the $\mathbf{p} \cdot \mathbf{A}$ term in the one-electron Hamiltonian). Both of these interactions raise the energy of one member of a Cooper pair and lower the energy of the other. Such "pair breaking" interactions are very destructive for superconductivity and generally lead to a rapid suppression of the superconducting critical temperature T_c.

When transition metal, rare earth, or actinide ions with partially-filled d- or f-electron shells are introduced into a conventional superconductor, extraordinary phenomena emerge that can be traced to the exchange and electromagnetic interactions described above. We distinguish two cases — one in which the ions that carry the magnetic moments occupy random sites and do not undergo magnetic order (paramagnetic impurities in superconductors), and another in which the ions occupy an ordered sublattice and exhibit long-range magnetic order (magnetically ordered superconductors).

[162] V. L. Ginzburg, *Sov. Phys. JETP* 4 (1957) 153

[163] B. T. Matthias, H. Suhl, and E. Corenzwit, *Phys. Rev. Lett.* 1 (1958) 92

[164] P. W. Anderson and H. Suhl, *Phys. Rev.* 116 (1951) 898

Paramagnetic Impurities in Superconductors

The striking phenomena that have been observed in superconductors containing paramagnetic impurities include a rapid suppression of T_c with paramagnetic impurity concentration, "gapless" superconductivity, and "reentrant" superconductivity associated with the Kondo effect.[165] These phenomena are produced by the spin-dependent exchange interaction between the conduction electrons and the paramagnetic impurity ions, which has the form

$$\mathcal{H}_{ex} = -2\mathcal{J}\mathbf{S}\cdot\mathbf{s}, \tag{4.1}$$

where $\mathcal{J}$ is the exchange interaction parameter, $\mathbf{S}$ is the spin of the paramagnetic impurity ion, and $\mathbf{s}$ is the spin of the conduction electron. For rare earth (R) ions, it is more appropriate to project $\mathbf{S}$ onto the total angular momentum $\mathbf{J}$ of the Hund's rule ground state so that Eq. 4.1 takes the form

$$\mathcal{H}_{ex} = -2\mathcal{J}(g_J - 1)\mathbf{J}\cdot\mathbf{s}, \tag{4.2}$$

where g_J is the Landé g-factor. The exchange interaction was implicated in the experiments carried out in 1958 by Matthias, Suhl, and Corenzwit[163] who observed a rapid and nearly linear depression of the T_c of La with R impurity concentration at a rate that correlated with the R ion's spin rather than its effective magnetic moment. The depression of T_c was actually found to scale reasonably well with the de Gennes factor $\mathcal{G}_J = (g_J - 1)^2 J(J + 1)$ of the R impurity ion which follows from Eq. 4.2, except for Ce, which is discussed below. These pioneering experiments inspired the author who was then a graduate student in the Matthias group at UCSD to perform similar measurements of the depression of T_c of the superconducting compound LaAl$_2$ by R impurities.[166] A plot of the initial rate at which R solutes depress T_c of the superconductors La and LaAl$_2$ *vs.* R solute, respectively, is shown in Figure 4.50. In both cases, the depression of T_c scales reasonably well with $\mathcal{G}_J$ except for Ce, which is anomalously large for both La and LaAl$_2$ host superconductors.

The exchange interaction parameter $\mathcal{J}$ characterizes the strength and sign of the interaction; parallel alignment of the conduction electron and impurity spin is favored when $\mathcal{J} > 0$, whereas antiparallel alignment is favored when $\mathcal{J} < 0$. The case $\mathcal{J} > 0$ leads to behavior that is consistent with the "pair breaking" theory of Abrikosov and Gor'kov (AG).[167] According to the AG theory, both T_c and the energy gap Δ are rapidly suppressed as "universal functions" of the concentration x of the paramagnetic impurities wherein T_c vanishes at a critical concentration x_c and Δ vanishes when $x = 0.91x_c$. In the concentration region between $0.91x_c$ and x_c, superconductivity persists without an energy gap. The form of the T_c vs. x curve predicted by AG has been shown to be consistent with experiments on La$_{1-x}$Gd$_x$Al$_2$ alloys carried out by the author[168] in 1968 and the phenomenon of gapless superconductivity was observed experimentally by means of electron tunneling experiments on Pb$_{1-x}$Gd$_x$ quenched films by Woolf and Reif[169] in 1965 and specific heat measurements on La$_{1-x}$Gd$_x$ alloys by Finnemore et al.[170] in 1965. The initial rate of the depression of T_c with x according to the AG theory is given by the expression

$$(dT_c/dx)_{x=0} = -[\pi^2 N(E_F)/3k_B]\mathcal{J}^2(g_J - 1)^2 J(J + 1), \tag{4.3}$$

and increases rapidly with the magnitude of $\mathcal{J}$ and scales with $\mathcal{G}_J$. The deviations of the rate of depression of T_c from $\mathcal{G}_J$ scaling have been attributed to variations in $|\mathcal{J}|$ with R solute and splitting of the R ion's Hund's rule ground state multiplet by the crystalline electric field.

[165] M. B. Maple, in *Magnetism: A Treatise on Modern Theory and Materials*, ed. H. Suhl (New York: Academic Press, 1973) Vol. V, Chapter 10; and references cited therein

[166] M. B. Maple, *Appl. Phys.* 9 (1976) 179, and references cited therein

[167] A. A. Abrikosov and L. P. Gor'kov, *Sov. Phys. JETP* 12 (1961) 1243

[168] M. B. Maple, *Phys. Lett.* 26A (1968) 513

[169] M. A. Woolf and F. Reif, *Phys. Rev. A* 137 (1965) 557

[170] D. K. Finnemore, D. L. Johnson, J. E. Ostenson, F. H. Spedding, and B. J. Beaudry, *Phys. Rev. A* 137 (1965) 550

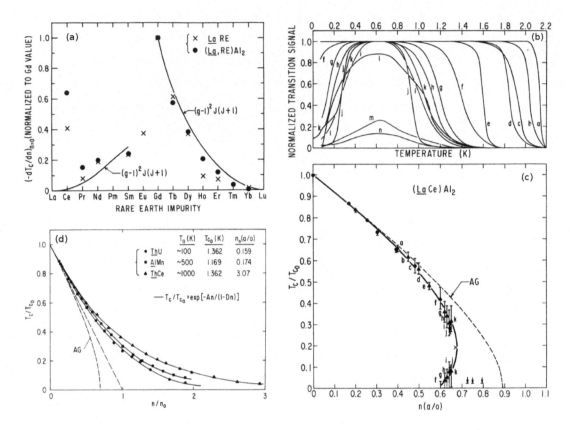

FIGURE 4.50: (a) Initial rate of depression of the superconducting critical temperature T_c of $La_{1-x}R_x$ and $La_{1-x}R_xAl_2$ alloys *vs.* R impurity (normalized to the Gd value). The solid line is the de Gennes factor $G_J = (g_J - 1)^2 J(J + 1)$, normalized to the Gd value. The values of $(-dT_c/dx)_{x=0}$ for Gd impurities are 5.3 and 3.79 K/at.% Gd substitution in La for $La_{1-x}R_x$ and $La_{1-x}R_xAl_2$, respectively.[165] (b) Normalized χ_{ac} transition signal *vs.* temperature and (c) T_c/T_{c0} *vs.* Ce impurity concentration n for the $La_{1-x}Ce_xAl_2$ system.[172] (d) Reduced transition temperature T_c/T_{c0} *vs.* reduced concentration n/n_0 for the systems $Th_{1-x}U_x$, $Th_{1-x}Ce_x$, and $Al_{1-x}Mn_x$. The values of n_0 were chosen to normalize the initial slope $[d(T_c/T_{c0})/d(n/n_0)]_{n=0}$ to the value -1.[166] (a: This figure was published in "Paramagnetic Impurities in Superconductors," M. B. Maple, Ch. 10 in Magnetism: A Treatise on Modern Theory and Materials, H. Suhl, Ed., Vol. V, 1973, pp. 289–325, Fig. 5. Copyright Academic Press, Inc. (1973).) (b: This figure was published in "The re-entrant superconducting-normal phase boundary of the Kondo system (La, Ce)A12," M. B. Maple, W. A. Fertig, A. C. Mota, L. E. DeLong, D. Wohlleben and R. Fitzgerald, Solid State Commun. 11 (1972) 829–834, Fig. 1a. Copyright Elsevier (1972).) (c: This figure was published in "The re-entrant superconducting-normal phase boundary of the Kondo system (LA, Ce)Al₂," M. B. Maple, W. A. Fertig, A. C. Mota, L. E. DeLong, D. Wohlleben and R. Fitzgerald, Solid State Commun. 11 (1972) 829–834, Fig. 1b. Copyright Elsevier (1972).) (d: With kind permission from Springer Science+Business Media: "Superconductivity: A probe of the magnetic state of local moments in metals," M. B. Maple, Appl. Phys. 9 (1976) 179–204, Fig. 14.)

The Kondo effect occurs in a nonmagnetic metallic host containing small concentrations of paramagnetic impurities when the exchange interaction parameter $\mathcal{J} < 0$. The negative exchange interaction is generated by hybridization of the localized d- or f-electron states of the paramagnetic impurity ions and conduction electron states and has the form $\mathcal{J} \sim -\langle V_{kl}^2 \rangle / \varepsilon_l$ where V_{kl} is the matrix element that admixes the localized d- or f-electron states and conduction electron states, $\varepsilon_l = E_F - E_l$ is the localized d- or f-electron binding energy, E_F is the Fermi energy, and E_l is the energy of the centroid of the localized d- or f-electron state. A many body singlet ground state, in which the spin of each paramagnetic impurity ion is screened by antiferromagnetically aligned conduction electron spins, gradually forms as the temperature is lowered through the Kondo temperature $T_K = T_F \exp(-1/N(E_F)|\mathcal{J}|)$, where T_F is the Fermi temperature. At temperatures $T \gg T_K$, the system behaves magnetically; the magnetic susceptibility χ conforms to a Curie-Weiss law $\chi(T) = N\mu_{\text{eff}}^2/3k_B(T - \theta)$, where N is the number of paramagnetic impurities, μ_{eff} is the effective magnetic moment, and θ is the Curie-Weiss temperature, which is negative and has a magnitude $|\theta| \sim 3T_K$, and the electrical resistivity ρ diverges logarithmically with decreasing temperature; i.e., $\rho(T) \sim -ln(T)$. At temperatures $T \ll T_K$, the system behaves as a nonmagnetic local Fermi liquid: $\gamma = C(T)/T \sim \chi(T) \sim$ constant and $\rho(T) \sim 1 - (T/T_K)^2$.

When the metallic host is a superconductor with a critical temperature T_{c0}, the interplay between superconductivity and the Kondo effect produces some remarkable effects due to the competition between singlet spin pairing of electrons in the superconducting state with characteristic energy $k_B T_{c0}$ and the formation of the Kondo many-body singlet state involving the conduction electrons and each paramagnetic impurity ion with characteristic energy $k_B T_K$. The initial depression of T_c with impurity concentration x, $(-dT_c/dx)_{x=0}$, exhibits a pronounced maximum as a function of T_K/T_{c0} at $T_K/T_{c0} \sim 1$, the T_c vs. x curve has negative curvature and is reentrant for $T_K/T_{c0} \ll 1$, while the T_c vs. x curve has positive curvature with a nearly exponential shape for $T_K/T_{c0} \gg 1$. The systems $La_{1-x}Ce_x$ and $La_{1-x}Ce_xAl_2$ are both superconducting Kondo systems with $T_K/T_{c0} \ll 1$, which accounts for the anomalous depression of T_c of La and LaAl$_2$ by Ce impurities shown in Figure 4.50. The system $La_{1-x}Ce_xAl_2$ ($T_{c0} = 3.3$ K; $T_K \approx 0.1$ K) is particularly interesting because it exhibits the phenomenon of "reentrant superconductivity," in which superconductivity vanishes below a second critical temperature T_{c2} that is lower than the critical temperature T_{c1} at which it first appears, and which occurs in the limit $T_K/T_{c0} \ll 1$. This extraordinary phenomenon was observed in the system $La_{1-x}Ce_xAl_2$ (shown to exhibit a Kondo effect by the author and Zachary Fisk) in experiments carried out at the University of Cologne by Riblet and Winzer[171] and UCSD by the author and his coworkers.[172] Müller-Hartmann and Zittartz[173] and Ludwig and Zuckermann[174] predicted the phenomenon of reentrant superconductivity due to the Kondo effect on theoretical grounds. The author's mentor, Bernd Matthias, was very intrigued by the reentrant superconductivity exhibited by $La_{1-x}Ce_xAl_2$, but did not think it could be caused by the Kondo effect, since he did not believe in it!

Examples of systems that have exponential-like T_c vs. x curves include $Th_{1-x}U_x$, $Th_{1-x}Ce_x$, and $Al_{1-x}Mn_x$ for which $T_K/T_{c0} \gg 1$.[166] Through the application of an external pressure, or alloying the superconducting host metal with a nonmagnetic element, it is possible to vary the ratio T_K/T_{c0} systematically from $T_K/T_{c0} \ll 1$ to $T_K/T_{c0} \gg 1$ and observe the evolution of the T_c vs. x curves from negative curvature with reentrant behavior to positive curvature with exponential-like shapes. This behavior has been observed experimentally when the $La_{1-x}Ce_x$ system is subjected to pressure and when the La host is alloyed with Th [i.e., $(La_{1-y}Th_y)_{1-x}Ce_x$], both of which result in an increase of T_K/T_{c0}.[166] The specific heat jump ΔC at T_c is also predicted to exhibit strong deviations from the

[171] G. Riblet and K. Winzer, *Solid State Commun.* 9 (1971) 1663

[172] M. B. Maple, W. A. Fertig, A. C. Mota, L. E. DeLong, D. Wohlleben, and R. Fitzgerald, *Solid State Commun.* 11 (1972) 829

[173] E. Müller-Hartmann and J. Zittartz, *Phys. Rev. Lett.* 26 (1971) 428

[174] A. Ludwig and M. J. Zuckermann, *J. Phys. F* (1971) 516

AG theory that vary with the ratio T_K/T_{c0}. These predictions are in good agreement with experiment except in the extreme limit $T_K/T_{c0} \gg 1$, where the theory predicts that ΔC should approach the AG value, indicative of pair breaking, whereas experiment yields values of ΔC that conform to the BCS law of corresponding states, $\Delta C/\Delta C_0 = T_c/T_{c0}$. The relatively strong and nearly exponential depression of T_c with transition metal, rare earth, and actinide solute concentration and the BCS behavior of the specific heat jump ΔC at T_c are consistent with pair weakening, primarily due to the strong Coulomb repulsion between two electrons when they scatter into the nonmagnetic d- or f-electron resonant states. For this reason, it was surprising when heavy fermion compounds, some of which have γ values as large as several J/mol K^2, were later found to exhibit superconductivity. This discrepancy and several other issues such as the possible existence of a transition back to the superconducting state at $T_{c3} < T_{c2} < T_{c1}$ in the limit $T_K/T_{c0} \ll 1$ in superconducting-Kondo systems remain to be resolved. Shown in Figure 4.50 are plots of the reentrant T_c vs. x curve of the $La_{1-x}Ce_xAl_2$ system ($T_K/T_{c0} \ll 1$) and the nearly exponential T_c vs. x curves of the $Th_{1-x}U_x$, $Th_{1-x}Ce_x$, and $Al_{1-x}Mn_x$ systems ($T_K/T_{c0} \gg 1$).

Magnetically Ordered Superconductors

In the mid-1970s, certain ternary compounds containing R ions that occupy an ordered sublattice were found to exhibit long-range magnetic order in the superconducting state.[175,176] The series of isostructural ternary rare earth compounds that have been investigated most extensively in connection with the interaction between superconductivity and long-range magnetic order include the rhombohedral rare earth (R) molybdenum chalcogenides RMo_6S_8 and RMo_6Se_8, and the tetragonal rare earth rhodium borides RRh_4B_4. Subsequent investigations of other systems such as RRh_xSn_y and RNi_2B_2C have also yielded significant information about the interaction of superconductivity and long-range magnetic order.[177] The weak exchange interaction between the conduction electron spins and R magnetic moments in these compounds accounts for the persistence of superconductivity, even in the presence of relatively large concentrations of R ions. The long-range magnetic ordering occurs via the "RKKY" mechanism, an indirect interaction between R magnetic moments that is mediated by the conduction electrons through the exchange interaction, and the dipolar interaction.

Antiferromagnetic Superconductors

The first observations of the coexistence of superconductivity and long-range antiferromagnetic (AFM) order were made independently at UCSD on certain RMo_6Se_8 and RRh_4B_4 compounds and at the University of Geneva on various RMo_6S_8 compounds.[175,176] The occurrence of AFM order in the superconducting state was initially inferred from features in the specific heat and the upper critical field at the Néel temperature T_N and then confirmed by means of neutron diffraction experiments. Other systems in which the coexistence of superconductivity and AFM order has been investigated include RNi_2B_2C compounds[177] and high T_c cuprates containing R ions such as the $RBa_2Cu_3O_{7-\delta}$ compounds.[178]

Ferromagnetic Superconductors

In 1977, reentrant superconductivity due to the onset of long-range ferromagnetic (FM) order

[175] *Superconductivity in Ternary Compounds I, Topics in Current Physics*, eds. Ø. Fischer and M. B. Maple (Berlin; Heidelberg; New York: Springer-Verlag, 1982) Vol. 32; and references cited therein

[176] *Superconductivity in Ternary Compounds II, Topics in Current Physics*, eds. M. B. Maple and Ø. Fischer (Berlin; Heidelberg; New York: Springer-Verlag, 1982) Vol. 34; and references cited therein

[177] P. C. Canfield, P. L. Gammel, and D. J. Bishop, *Physics Today* 51 (1998) 40

[178] *Physical Properties of High Temperature Superconductors I-V*, ed. D. M. Ginsberg (New Jersey: World Scientific, 1989-1996)

was observed in $ErRh_4B_4$ at UCSD[179] and in $HoMo_6S_8$ at the University of Geneva.[180] These two materials, which become superconducting at a critical temperature T_{c1}, lose their superconductivity at a lower critical temperature $T_{c2} < \theta_C < T_{c1}$, where θ_C is the Curie temperature. Thermal hysteresis in various physical properties and a spike-shaped feature in the specific heat near T_{c2} reveal a first-order transition from the superconducting to the FM normal state at T_{c2}. Neutron diffraction measurements have confirmed that $ErRh_4B_4$ and $HoMo_6S_8$ have FM ground states, while small-angle neutron-scattering studies have revealed the existence of a sinusoidally-modulated magnetic state with a wavelength of the order of ~ 100 Å that coexists with superconductivity in a narrow temperature interval above T_{c2}.[181,182] The neutron scattering studies on a single crystal specimen of $ErRh_4B_4$ have shown that the sinusoidally modulated magnetic state in this compound is a linearly polarized state with a propagation vector that lies in the *a-c* plane and is oriented at an angle of 45° with respect to one of the *a*-axes while the polarization is parallel to the other *a*-axis.[183] Hysteresis in the scattering intensity in $ErRh_4B_4$ was ascribed to the nucleation of normal FM domains within the paramagnetic superconducting regions above T_{c2}. Thus, the regions within which superconductivity and the sinusoidally modulated magnetic state coexist appear to be interspersed with normal FM domains to form a spatially inhomogeneous state. Shown in Figure 4.51 are plots of the electrical resistance R and ac magnetic susceptibility χ_{ac} *vs. T* for $ErRh_4B_4$, revealing the destruction of superconductivity at T_{c2} due to the onset of FM order.

Experiments on various pseudoternary R compounds, such as the RRh_4B_4 and RMo_6X_8 ($X = S$, Se) compounds, involving the interplay between superconductivity and competing types of magnetic moment anisotropy and/or magnetic order, have yielded some rich and complex temperature-composition phase diagrams. An example is the T *vs. x* phase diagram of the $Er_{1-x}Ho_xRh_4B_4$ system shown in Figure 4.51.[184]

4.11.3 Coexistence of Superconductivity and Magnetic Order Involving the Same Set of Electrons

Both *d*- and *f*-electron systems have been observed to exhibit the coexistence of superconductivity and magnetic order, in which both phenomena involve the same set of electrons. Many examples of this behavior are found in the class of "heavy fermion" *f*-electron materials and the new iron arsenide *d*-electron compounds.

Superconducting and Magnetic Heavy Fermion Compounds

More than 20 "heavy fermion" or "heavy electron" superconductors have been found to date; most are based on Ce and U, one on Pr, one on Yb, and, possibly, even one on Pu! The values of T_c of these compounds are rather low, usually less than 2 K. In an appreciable fraction of these compounds, superconductivity coexists with AFM order, with a T_c that is smaller than the Néel temperature T_N. In some compounds, the AFM state is reminiscent of an SDW and forms a gap over a portion of the Fermi surface, while the superconductivity forms a gap over the remainder of the Fermi surface at lower temperatures.

At temperatures well below T_c, the superconducting properties generally have power-law *T*-

[179] W. A. Fertig, D. C. Johnston, L. E. DeLong, R. W. McCallum, M. B. Maple, and B. T. Matthias, *Phys. Rev. Lett.* 38 (1977) 987

[180] M. Ishikawa and Ø. Fischer, *Solid State Commun.* 23 (1977) 137

[181] D. L. Moncton, D. R. McWhan, P. H. Schmidt, G. Shirane, W. Thomlinson, M. B. Maple, H. B. MacKay, L. D. Woolf, Z. Fisk, and D. C. Johnston, *Phys. Rev. Lett.* 45 (1980) 20

[182] J. W. Lynn, G. Shirane, W. Thomlinson, R. N. Shelton, and D. L. Moncton, *Phys. Rev. B* 24 (1981) 3817

[183] S. K. Sinha, G. W. Crabtree, D. G. Hinks, and H. A. Mook, *Phys. Rev. Lett.* 41 (1982) 950

[184] M. B. Maple, *J. Magn. Magn. Mater.* 31-34 (1983) 479

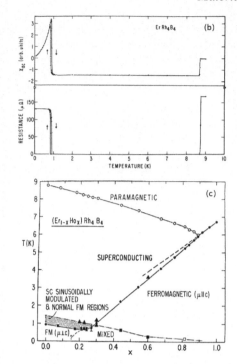

FIGURE 4.51: (a) Bernd Matthias and the author performing low temperature measurements on $ErRh_4B_4$ in the author's laboratory at UCSD in 1977. (Photograph taken by A. C. Lawson.) (b) Plot of the electrical resistivity and ac magnetic susceptibility vs. temperature of $ErRh_4B_4$ showing the transition back to the normal state at T_{c2} due to the onset of ferromagnetic order at $\theta_C \approx$ 1.5 K.[176] (c) Phase diagram of the $Er_{1-x}Ho_xRh_4B_4$ system, showing the temperatures T and Ho concentrations x where superconductivity and ferromagnetism are found. In a limited range of T and x, superconductivity and the sinusoidally modulated magnetic state coexist microscopically.[184] (a: Reproduced with permission from A. C. Lawson.) (b: With kind permission from Springer Science+Business Media: "Superconductivity, Magnetism, and Their Mutual Interaction in Ternary Rare Earth Rhodium Borides and Some Ternary Rare Earth Transition Metal Stannides," M. B. Maple, H. C. Hamaker, and L. D. Woolf, Ch. 4 in Superconductivity in Ternary Compounds II, Superconductivity and Magnetism, Vol. 34, eds. M. B. Maple and Ø. Fischer (Berlin; Heidelberg; New York: Springer-Verlag, 1982), p. 109, Figure 4.7.) (c: This figure was published in "Experiments on Magnetically Ordered Superconductors," M. B. Maple, J. Magn. Magn. Mater. 31–34 (1983) 479–483, Fig. 2. Copyright Elsevier (1983).)

dependences, T^n, where n is an integer, rather than exponential T-dependences, $exp(-\Delta/T)$, where Δ is the energy gap, expected for a conventional BCS superconductor. These properties include the specific heat, thermal conductivity, ultrasonic attenuation and nuclear-spin-lattice relaxation rate. The power-law T-dependences provide evidence for anisotropic superconductivity, in which the energy gap vanishes at points or lines on the Fermi surface. In a conventional BCS superconductor, the energy gap is isotropic, or only weakly varying over the Fermi surface. There are strong indications that the superconducting electron pairing in these materials is mediated by magnetic dipole or, perhaps, even electric quadrupole excitations, rather than phonons, and that the pairs are in p- or d-wave angular momentum states. In some materials, such as UPt_3, UBe_{13}, and $PrOs_4Sb_{12}$, there is evidence for the existence of multiple superconducting phases with different order-parameter symmetries.[185,186]

The "heavy fermion" or "heavy electron" compounds derive their name from their enormous electronic-specific heat coefficients, which can reach values as high as several J/mol K^2, corresponding to quasiparticle effective masses as large as several hundred times the mass of the free electron. The underlying physics of the heavy fermion state is believed to involve the Kondo effect, generalized to the case of a lattice of ions (Kondo lattice), or valence fluctuations, both of which can be traced to the hybridization between the localized f-electron and conduction electron states. As noted above, the Kondo effect is based on the exchange interaction and corresponds to the situation where the valence, or, equivalently, the average occupation number n_{av} of the f-electron shell is nearly integral so the magnetic moment is well defined. Valence, or interconfiguration, fluctuations are appropriate when the valence, or n_{av}, is nonintegral. The valence of rare earth compounds can be estimated from measurements of various properties such as the lattice parameter, Mössbauer isomer shift, and photoemission spectra. The terminology "valence fluctuations," which was coined by Dieter Wohlleben and the author,[187] refers to temporal fluctuations of the f-electron shell between two configurations, one with n electrons (f^n) and the other with $n-1$ electrons (f^{n-1}) plus a conduction electron. The conclusion that the admixture of the two f-electron configurations is temporal, rather than spatial, was based on magnetic susceptibility measurements on the high-pressure "collapsed" phases of elemental Ce and the compound SmS, α-Ce and the "gold" phase of SmS, which have intermediate valences of ~ 3.7 and ~ 2.7, respectively. The measurements revealed Pauli-like paramagnetic behavior indicative of a nonmagnetic ground state, which implies that the admixture of the two valence states is temporal. In this situation, each f-electron shell emits and absorbs an electron and "fluctuates" between two states with integral occupation number with a frequency $f \sim k_B T_0/h$, where T_0 is the characteristic temperature separating high-T "magnetic behavior" from low-T "nonmagnetic behavior." Phenomenologically, T_0 plays the same role as the Kondo temperature T_K.

A few theoreticians criticized our use of the term "valence fluctuations" because they thought we were ignoring the quantum mechanical nature of the nonmagnetic ground state of rare earth compounds with "intermediate valence," although this was not our intent. Nevertheless, the name "valence fluctuations" caught on and is widely used today. When we mentioned our work on valence fluctuations in rare earth compounds to the distinguished magnetician Peter Wohlfarth, renowned for his theoretical work on itinerant ferromagnetism, he exclaimed: "Transition metals are for men, rare earths are for boys"! It seems fair to say that the extraordinary phenomena found in rare earth materials during the past three decades have demonstrated that rare earths are for men, too! In so far as Kondo and valence fluctuation physics is responsible for the heavy fermion state in rare earth and actinide compounds, it plays a key role in the unconventional superconductivity exhibited by these extraordinary materials. Both magnetic and valence fluctuations have been implicated as possible pairing mechanisms for the superconductivity of heavy fermion f-electron materials.[188,189,190]

Superconductivity and Non-Fermi Liquid Behavior Near Quantum Critical Points

In many of the f-electron compounds, superconductivity occurs in the vicinity of a critical pressure P_c at which a magnetic transition is suppressed towards 0 K. For example, both CeIn$_3$ and CePd$_2$Si$_2$ exhibit AFM order with a Néel temperature $T_N \approx 10$ K at atmospheric pressure. Upon application of pressure, T_N of both compounds decreases smoothly towards 0 K at $P_c \approx 28$ kbar,

[185] R. H. Heffner and M. R. Norman, *Comm. Condens. Matter Phys.* 17 (1996) 361

[186] D. L. Cox and M. B. Maple, *Phys. Today* 48 (1995) 32

[187] M. B. Maple and D. Wohlleben, in *AIP Conf. Proc.* (No. 18) on "Magnetism and Magnetic Materials — 1973", eds. C. D. Graham, Jr., J. L Rhyne (1974) 447, and references cited therein

[188] Y. Onishi and K. Miyake, *J. Phys. Soc. Jpn.* 69 (2000) 3955

[189] D. Jaccard, K. Behnia and J. Sierro, *Phys. Lett. A* 163 (1992) 475

[190] H. Q. Yuan, F. M. Grosche, M. Deppe, C. Geibel, G. Sparn, and F. Steglich, *Science* 302 (2003) 2104

in the vicinity of which a dome-shaped region of superconductivity is observed.[191] In contrast, the compounds URhGe, UCoGe, UGe_2,[192] and UIr exhibit weak FM order that coexists with superconductivity, although the latter two compounds become superconducting only under pressure. A particularly intriguing aspect of these studies is the possibility that these weak ferromagnets exhibit triplet spin superconductivity as predicted, for example, by Fay and Appel.[193] Another superconductor of current interest is $PrOs_4Sb_{12}$,[194] the only known example of a Pr-based heavy fermion superconductor. This compound has a nonmagnetic ground state with an effective mass $m^* \approx 50$ m_e and a superconducting critical temperature $T_c = 1.86$ K, but exhibits a magnetic field-induced antiferroquadrupolar ordered phase in fields between 4 T and 15 T. The unconventional superconductivity breaks time reversal symmetry according to μSR measurements, and there is evidence for nodes in the energy gap and, possibly, multiple superconducting phases. This has led to the speculation that electric quadrupole fluctuations may be responsible for superconductivity in this compound. Examples of $T - P$ phase diagrams for $CePd_2Si_2$ in which superconductivity occurs near an AFM QCP,[195] and UGe_2, where superconductivity occurs within the FM phase,[196] are shown in Figure 4.52.

Another interesting characteristic of many heavy fermion materials is non-Fermi liquid (NFL) behavior in their normal state properties in the vicinity of pressures or chemical compositions where magnetic transitions are suppressed to 0 K, such as the critical pressures mentioned in the previous examples. In general, these quantum critical points occur where some control parameter, such as pressure, chemical composition, or magnetic field, suppresses a phase transition towards 0 K. NFL behavior is distinguished by properties that have T-dependences at low temperatures that do not conform to Landau's Fermi liquid theory. There are good indications that the quasiparticle picture of electronic interactions breaks down in these regimes, which is currently a topic of intense research. The NFL behavior displayed by heavy fermion compounds relates them to a larger class of materials that exhibit NFL physics, and suggests that a general common mechanism lies at its core.

Superconducting and Magnetic Iron-Based Compounds

Recently, the coexistence of SDW-like AFM and superconductivity has been observed in the new iron-based high T_c superconductors such as $SmFeAsO_{1-x}F_x$ and $BaFe_{2-x}Co_xAs_2$ where the AFM order involves the d-electrons of Fe.[197] In these systems, the SDW-like AFM order can be traced to nesting of the Fermi surface. It also appears that the superconductivity may arise through pairing of electrons on different Fermi surface sheets by spin fluctuations where the order parameter changes sign (so-called $d+$ pairing) in order to render the pairing interaction attractive.[198] In contrast, in the high T_c cuprate superconductors, in which the superconductivity emerges from an AFM Mott insulating phase with increasing carrier concentration, the AFM and superconductivity do not occur in the same carrier concentration range. A schematic representation of the generic T vs. dopant

[191] N. D. Mathur, F. M. Grosche, S. R. Julian, I. R. Walker, D. M. Freye, R. K. W. Haselwimmer, and G. G. Lonzarich, *Nature* 394 (1998) 39

[192] S. S. Saxena, P. Agarwal, K. Ahilan, F. M. Grosche, R. K. W. Haselwimmer, M. J. Steiner, E. Pugh, I. R. Walker, S. R. Julian, P. Monthoux, G. G. Lonzarich, A. Huxley, I. Sheikin, D. Braithwaite, and J. Flouquet, *Nature* 406 (2000) 587

[193] D. Fay and J. Appel, *Phys. Rev. B* 16 (1997) 2325

[194] M. B. Maple, N. A. Frederick, P.-C. Ho, W. M. Yuhasz, and T. Yanagisawa, *J. Super. Nov. Magn.* 19 (2006) 299; and references cited therein

[195] N. P. Butch, M. C. de Andrade, and M. B. Maple, *Am. J. Phys.* 76 (2008) 106; and references cited therein

[196] M. B. Maple, R. E. Baumbach, N. P. Butch, J. J. Hamlin, and M. Janoschek, *J. Low Temp. Phys.* 161 (2010) 4; and references cited therein

[197] D. C. Johnston, *Adv. in Phys.* 59 (2010) 803; and references cited therein

[198] I. Mazin, *Nature* 464 (2010) 183

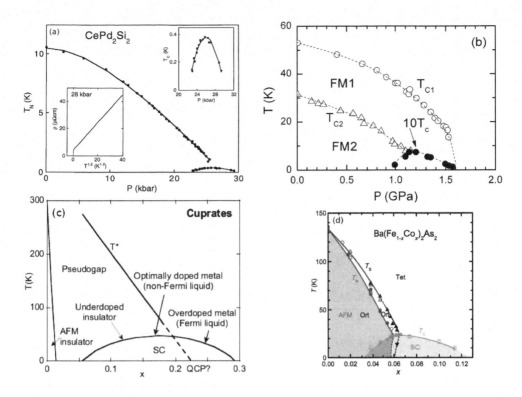

FIGURE 4.52: (a) Suppression of the Néel temperature T_N and onset of superconductivity in $CePd_2Si_2$ with applied pressure. The superconducting phase boundary has a dome shape, magnified in the upper right inset. The lower left inset shows the unusual $T^{1.2}$ dependence of the low T electrical resistivity at 28 kbar. (Figure from Butch et al.[195] based on data from Mather et al.[191].) (b) Temperature-pressure phase diagram of UGe_2. At ambient pressure, UGe_2 is a ferromagnet, but under applied pressure it becomes superconducting with a maximum transition temperature of about 0.7 K around 12 kbar. Superconductivity and magnetic order in UGe_2 disappear at the same pressure. (Figure from Maple et al.[196] based on data from Saxena et al.[192].) (c) Generic temperature-dopant concentration (T-x) phase diagram for cuprate superconductors (schematic). The line denoted T^* represents the crossover into the pseudogap state.[199] (d) T vs. x phase diagram of $BaFe_{2-x}Co_xAs_2$ showing the structural, SDW, and superconducting transitions. Superconductivity coexists with both the orthorhombic phase and the AFM phase below $x \approx 0.06$.[200] (a: With kind permission from American Institute of Physics, "Resource Letter Scy-3: Superconductivity," N. P. Butch, M. C. de Andrade, and M. B. Maple, American Journal of Physics 76 (2008) 106, Fig. 7.) (b: With permission from Springer Science+Business Media: "Non-Fermi Liquid Regimes and Superconductivity in the Low Temperature Phase Diagrams of Strongly Correlated d-and f-Electron Materials," M. B. Maple, R. E. Baumbach, N. P. Butch, J. J. Hamlin and M. Janoschek, Journal of Low Temperature Physics 161 (2010) 4–5, Fig. 15a) (c: With kind permission from Springer Science+Business Media: "Unconventional Superconductivity in Novel Materials," M. B. Maple, E. D. Bauer, V. S. Zapf and J. Wosnitza, Ch. 13 in Superconductivity, K. H. Bennemann and J. B. Ketterson, Eds., Vol. 1, 2008, p. 639, Fig. 13.104.) (d: Reprinted figure with permission from S. Nandi, M. G. Kim, A. Kreyssig, R. M. Fernandes, D. K. Pratt, A. Thaler, N. Ni, S. L. Bud'ko, P. C. Canfield, J. Schmalian, R. J. McQueeney and A. I. Goldman, Phys. Rev. Lett. 104 (2010) 057006. Copyright 2010 by the American Physical Society. http://link.aps.org/doi/10.1103/PhysRevLett.104.057006.)

concentration phase diagram of cuprate high T_c superconductors[199] and the T vs. x phase diagram of the $BaFe_{2-x}Co_xAs_2$ system[200] are shown in Figure 4.52.

4.11.4 Some Thoughts about High Temperature Superconductors

It is interesting that superconductivity with the highest values of T_c (> 40 K) is found in two classes of correlated electron materials: the cuprates, where the maximum value of T_c reaches 130 K at atmospheric pressure and 160 K at several hundred kbar in $HgBa_2Ca_2CuO_8$,[201,202] and the iron arsenides, where T_c reaches 55 K in $SmFeO_{1-x}F_xAs$.[197] These compounds have rich phase diagrams, which contain insulating, metallic, spin and charge ordered, and superconducting regions, which apparently arise from the delicate interplay between competing interactions. Upon changing the charge carrier concentration, superconductivity emerges from an AFM Mott insulating phase in the cuprates and an SDW phase in the iron arsenides. Many researchers believe that the pairing of superconducting electrons in both classes of these novel materials is mediated by spin fluctuations. Thus, while magnetism suppresses superconductivity in many of the situations discussed here, it may play the key role in producing high T_c superconductivity! Both classes of materials have layered crystal structures and anisotropic properties (although the anisotropy in the iron arsenides is much smaller than in the cuprates). Thus, the current strategy for searching for high temperature superconductivity embraced by the author and his research group is to investigate strongly correlated electron materials, which contain elements with partially-filled d- and f-electron shells, and have low symmetry. In view of the fact that superconductivity and other spin and charge ordered phases are often found in proximity to one another in such systems, a reasonable prescription for searching for high temperature superconductivity might be to look for something else! Superconductivity may be lurking nearby!

The spectacularly high values of T_c found in the cuprates following the initial discoveries in 1986 shocked the scientific community and ignited an explosion of research on high T_c superconductivity on a worldwide scale. Some of the surprise was based on the belief that oxides were poor prospects for high T_c superconductivity, according to a set of empirical "rules" attributed to Bernd Matthias. These "rules" were based on investigations by Matthias and his coworkers of a vast number of superconducting elements, and binary and pseudobinary alloys. Used in conjunction with the periodic table, they served as a guide in the search for superconductors with high T_c. According to W. E. Pickett,[203] Matthias' "loosely formulated" set of "rules" can be succinctly summarized as follows:

1. Transition metals are better than simple metals;

2. There are favorable valence electron per atom ratios [$N(E_F)$];

3. High symmetry is best, especially cubic;

4. Stay away from oxygen;

5. Stay away from magnetism; and

6. Stay away from insulating phases.

---

[199] M. B. Maple, E. D. Bauer, V. S. Zapf, and J. Wosnitza, in *Superconductivity*, Vol. 1, eds. K. H. Benneman, J. B. Ketterson (Berlin: Springer-Verlag, 2008) Ch. 13, p. 639–762

[200] S. Nandi, M. G. Kim, A. Kreyssig, R. M. Fernandes, D. K. Pratt, A. Thaler, N. Ni, S. L. Bud'ko, P. C. Canfield, J. Schmalian, R. J. McQueeney, and A. I. Goldman, *Phys. Rev. Lett.* 104 (2010) 057006

[201] C. W. Chu, L. Gao, F. Chen, Z. J. Huang, R. L. Meng, and Y. Y. Xue, *Nature* 365 (1993) 323

[202] M. Nuñez-Regueiro, J.-L. Tholence, E. V. Antipov, J.-J. Capponi, and M. Marezio, *Science* 262 (1993) 97

[203] W. E. Pickett, *Physica B* 296 (2001) 268

As a result of Matthias' public chiding of theorists about the inability of theory to predict where to find high T_c superconductors, another "rule" appears to have been added: (7) Stay away from theorists! Rule (2) is based on the observation that T_c exhibits a peak at an average valence electron count of 5 and 7 in a number of transition metal systems. Rule (3) is illustrated by the cubic A15 compounds (e.g., Nb_3Sn, Nb_3Ge), which had the highest values of T_c known prior to the discovery of the high T_c cuprate superconductors in 1986.

Matthias' "rules" are clearly violated by the high T_c cuprate and iron arsenide superconductors. These compounds have layered structures and contain oxygen, and the superconductivity occurs in the vicinity of an insulating phase (in the case of the cuprates) and magnetism (AFM or SDWs). It is noteworthy that Matthias actually proposed a magnetic pairing mechanism[204] based on experiments in which substitution of Fe into Ti was found to raise the T_c of Ti much more rapidly than expected from his "rules." Actually, the Fe additions stabilized another (nonmagnetic) phase of Ti with a higher value of T_c. While the interpretation of this experiment was incorrect, such a magnetic pairing mechanism as envisaged by Matthias is believed by many researchers to be responsible for the high T_c superconductivity in the cuprates and iron arsenides, as well as the unconventional superconductivity in certain heavy fermion, organic and weakly ferromagnetic compounds. Interestingly, during the last several years of his life, Matthias was quite interested in oxide superconductors which were then known to have T_c values as high as ~ 12 K for $Ba(Bi,Pb)O_3$ and $LiTi_2O_4$.

4.11.5   Some Personal Recollections of Professor Bernd T. Matthias

Bernd Matthias was one of the founding faculty members of the Physics Department of UCSD (then called UCLJ, for University of California, La Jolla), which was established in 1960. He was a leading practitioner in the discovery of superconductivity, ferroelectricity, and ferromagnetism in new materials, and he synthesized hundreds of superconducting and ferroelectric compounds. He had a powerful personality and was very charismatic. Working with Matthias wasn't always easy, but it was always interesting. One of the most important things we learned from him was the excitement of doing research and thrill of making a discovery. Matthias traveled extensively, between UCSD, Bell Laboratories and Los Alamos National Laboratory to do research, and to places throughout the world to lecture at conferences, universities, and research institutes. He had many students, postdocs, and collaborators at UCSD, Bell Laboratories and Los Alamos, many of whom are listed in the Biographical Memoir written by his long-time colleagues and friends, T. H. Geballe and J. K. Hulm, for the National Academy of Sciences.[205] In the early years at UCSD, Matthias apparently had an agreement that he wouldn't have to teach regular courses, although in later years he taught some large classes to undergraduate students on special topics such as "metaphysics," which were very popular. From time to time he would come into the lab after a trip with an exciting new project that he would set us to work on and then, shortly thereafter, leave on another trip. Upon return he would ask us what we had done while he was away. We would report on all of the progress we had made on the project he had assigned us, to which he would sometimes exclaim: "Why are you working on that? You should work on this other project. It is far more interesting!" This illustrates the penchant he had for discovering new materials with superior properties or displaying new types of behavior, and then moving on to another challenge, leaving the details for other researchers to work out. In order to encourage us to exploit opportunities he thought might be found within a new class of materials, he would often say: "Even a blind chicken can find a corn."

While Matthias publicly railed against theorists, he did so to emphasize the fact that theory had not been useful in predicting where to find high T_c superconductors (although it was very successful in accounting for the pairing mechanism and physical properties of conventional superconductors)

[204] B. T. Matthias, *Physics Today* 16 (1963) 23

[205] T. H. Geballe and J. K. Hulm, *Biographical Memoir of Bernd Theodor Matthias* (National Academy Press, 1996)

and for effect. In fact, he interacted with a number of distinguished theorists, many of whom were also his friends, such as Phil Anderson, John Bardeen, Al Clogston, Herbert Fröhlich, Walter Kohn, Lu Sham, Harry Suhl, Peter Wolfe, and others. For example, Bardeen and Fröhlich used to visit his laboratory at UCSD in the 1960s and 1970s. As a young student in Matthias' research group, I was assigned to talk with Bardeen during his visits. This was initially rather intimidating because of Bardeen's extraordinary scientific accomplishments and stature as a twice-honored Nobel Laureate, and because it took him a long time to respond when you would ask him a question, which was, at first, a bit unnerving. However, he was kind and personable, and it was an honor to be able to discuss physics with him. In fact, the first talk I gave about my own research was in a seminar series at UCSD organized by Bardeen and Matthias during one of Bardeen's visits. It would have been interesting to see how Matthias would have reacted to the discovery of high T_c superconductivity in the cuprates and what role he would have played in the frenetic days that immediately followed, had he lived long enough to witness it.

Acknowledgments

Superconductivity above 10 K in Non-Cuprate Oxides

I am grateful to Arthur Sleight and Leonard Mattheiss for communicating to me historical aspects of their discoveries of superconductivity in the non-transition metal perovskites. Work at the Ames Laboratory was supported by the Department of Energy-Basic Energy Sciences under Contract No. DE-AC02-07CH11358.

Cuprates—Superconductors with a T_c up to 164 K

For the work at Houston I would like to acknowledge the hard and dedicated work of my former students and research staff and helpful discussions with colleagues both in and outside Houston. The support of the sympathetic management and colleagues at NSF and UH that made possible my travel between Washington, D.C., and Houston to tend to my NSF responsibilities and to supervise my group research deserves special mention. For instance, during the critical period between January 29 and March 18, 1987, I was able to spend 33 days out of 50 to work in my lab at Houston. Without the generous help from Joe Trivisonno, my co-director colleague of the Solid State Physics Program at NSF, our HTS work would have been extremely difficult, if not impossible. Support from NSF, NASA, AFOSR, DoE, the State of Texas through the Texas Center for Superconductivity at the University of Houston, the T. L. L. Temple Foundation, the John J. and Rebecca Moores Endowment, and Cullen Foundation over the years is greatly appreciated.

Magnetic Superconductors

The author would like to thank James Hamlin for assistance in preparing this manuscript and Ryan Baumbach, Marc Janoschek, and Diego Zocco for critical comments. The support of the U. S. Department of Energy, National Science Foundation, and the Air Force Office of Scientific Research for research relevant to this article is gratefully acknowledged.

4.12   Further Reading

Cuprates

1. Randy Simon and Andrew Smith, *Superconductors: Conquering Technology's New Frontier* (New York: Plenum, 1988).

2. Gianfranco Vidali, *Superconductivity: The Next Revolution?* (Cambridge; New York: Cambridge University Press, 1993).

3. Robert M. Hazen, *The Breakthrough: The Race for the Superconductor* (New York: Summit, 1988).

4. Gerald Holton, Hasok Chang, and Edward Jurkowitz, "How a Scientific Discovery Is Made: A Case History," *American Scientist* 84 (1996) 364.

5. Helga Nowotny and Ulrike Felt, *After the Breakthrough: The Emergence of High-Temperature Superconductivity as a Research Field* (Cambridge; New York: Cambridge University Press, 1997).

6. Paul C. W. Chu, "High-Temperature Superconductors," *Scientific American* 273(3) (1995) 162.

7. Paul C. W. Chu, "A Possible Path to RTS," *AAPPS Bulletin* 18 (2008) 9.

Fe-Pnictides and -Chalcogenides

1. Reviews of Hosono group research on transparent oxide semiconductors

    H. Hosono, Chapter 10 and 13 in *Handbook of Transparent Conductors*, ed. D. Ginley, H. Hosono, and D. Paine (Springer, 2010).

2. Review articles on iron-based superconductors

    H. H. Wen, *Adv. Mat.* 20 (2008) 3764.

    K. Ishida, Y. Nakai, and H. Hosono, *J. Phys. Soc. Jpn.* 78 (2009) 062001.

    C. Day, *Phys. Today* 62(8) (2009) 36.

    M. Jacoby, *Chemical and Engineering News*, October 20, 2008.

    P. M. Aswathy et al., *Supercon.Sci.Technol.* 23 (2010) 073001.

    I. I. Mazin, *Nature* 464 (2010) 183.

    D. C. Johnston, *Adv. Phys.* 54 (2010) 803.

3. Special journal issues

    J. Phys. Soc. Jpn. 77(Supplement C) (2008).

    New. J. Phys. 11(February) (2009).

    J. Supercond. Nov. Magn. 22(6) (2009).

    Physica C 469 (2009).

Fullerenes

1. O. Gunnarsson, *Alkali-Doped Fullerides: Narrow-Band Solids with Unusual Properties* (World Scientific, 2004).

2. N. Manini and E. Tosatti, *Jahn-Teller and Coulomb correlations in fullerene ions and compounds* (Lambert Academic Publishing, 2010).

Elemental Superconductors

1. C. Buzea and K. Robbie, "Assembling the puzzle of superconducting elements: a review," *Supercond. Sci. Technol.* 18 (2005) R1.

Heavy-Fermion Superconductivity

1. Heavy-Fermion Physics

    An excellent introduction into the physics of heavy-fermion metals is *The Kondo Problem to Heavy Fermions* by A. C. Hewson (Cambridge: Cambridge University Press, 1993).

    A recent review on heavy fermions is given in "Heavy fermions—Electrons at the edge of magnetism" by P. Coleman in *Handbook of Magnetism and Advanced Magnetic Materials*, ed. by H. Kronmüller and S. Parkin, Vol. 1: Fundamentals and Theory (Wiley, 2007), p. 95.

2. Heavy-Fermion Superconductivity

    A very good review on unconventional Cooper pairings is given in "Superconductivity without phonons" by P. Monthoux, D. Pines, and G. G. Lonzarich, *Nature* 450 (2000) 1177.

    An excellent overview on the status of heavy-fermion superconductivity at the beginning of 2009 is given in "Superconducting phases of f-electron compounds" by C. Pfleiderer, *Rev. Mod. Phys.* 81 (2009) 1551.

3. Quantum Phase Transitions

    An excellent entry into quantum phase transitions is *Quantum Phase Transitions* by S. Sachdev (New York: Cambridge University Press, 1999).

    A very recent review on heavy-fermion quantum criticality is given by Q. Si and F. Steglich, *Science* 329 (2010) 1161.

Ruthenate Superconductor Sr_2RuO_4

1. A review on Sr_2RuO_4 with an introduction to the physics of spin-triplet superconductivity can be found in "The superconductivity of Sr_2RuO_4 and the physics of spin-triplet pairing" by A. P. Mackenzie and Y. Maeno, *Rev. Mod. Phys.* 75 (2003) 657–712.

2. An introduction to general science readers is "The Intriguing Superconductivity of Strontium Ruthenate" by Y. Maeno, T. M. Rice, and M. Sigrist, *Physics Today* 54 (2001) 42–47. The cover of this January 2001 issue of Physics Today features Sr_2RuO_4.

3. Insightful discussions on profound problems covering superconductors and superfluids, as well as the author's assessment on "exotic" superconductors (Ch. 8), can be found in *Quantum Liquids* (Oxford University Press, 2006) by A. J. Leggett.

Magnetic Superconductors

1. A. C. Hewson, *The Kondo Problem to Heavy Fermions* (Cambridge: Cambridge University Press, 1997). Comprehensive treatment of the impurity Kondo problem, with an overview of how it relates to heavy fermion compounds.

2. M. B. Maple, "Coexistence of Superconductivity and Magnetism," in *Advances in Superconductivity, NATO ASI Series, Series B Physics*, eds. B. Deaver and J. Ruvalds (New York; London: Plenum, 1983), Vol. 100, pp. 279–346. Thorough review of magnetic superconductors including much experimental data.

3. Ø. Fischer, "Magnetic Superconductors," in *Ferromagnetic Materials*, eds. K. H. J. Buschow and E. P. Wohlfarth (New York: Elsevier, 1990), Vol. 5, Chap. 6, pp. 465–550. Comprehensive review of magnetically ordered superconductors.

4. M. B. Maple, "Novel types of superconductivity in f-electron systems," *Phys. Today* 39 (1986) 72. Review of novel superconductivity just prior to the discovery of high T_c cuprates.

5. A. Huxley, E. Ressouche, B. Grenier, D. Aoki, J. Flouquet, and C. Pfleiderer, "The co-existence of superconductivity and ferromagnetism in actinide compounds," *J. Phys.: Condens. Mat.* 15 (2003) S1945. Discusses properties of UGe_2 and URhGe.

6. A. Gasparini, Y. K. Huang, N. T. Huy, J. C. P. Klaasse, T. Naka, E. Slooten, and A. de Visser, "The Superconducting Ferromagnet UCoGe," *J. Low Temp. Phys.* 161 (2010) 134. Review of the properties of the superconducting ferromagnet UCoGe.

7. M. Sigrist and K. Ueda, "Phenomenological theory of unconventional superconductivity," *Rev. Mod. Phys.* 63 (1991) 239. Theoretical discussion of a generalized theory for unconventional superconductors.

8. Y. Onuki, R. Settai, K. Sugiyama, T. Takeuchi, T. C. Kobayashi, Y. Haga, and E. Yamamoto, "Recent advances in the magnetism and superconductivity of heavy fermion systems," *J. Phys. Soc. Jpn.* 73 (2004) 769. Overview of current understanding of several heavy fermion superconductors, including discussion of quadrupolar interactions.

9. G. R. Stewart, "Heavy-fermion systems," *Rev. Mod. Phys.* 56 (1984) 755. Overview of the properties of heavy fermion compounds known at the time.

5

SQUIDs and Detectors

Editor: A. I. Braginski

Forschungszentrum Jülich GmbH, Jülich, Germany

5.1 Introduction
     John Clarke, Arnold Silver and A. I. Braginski 311
5.2 History and Device Fundamentals
     John Clarke and Arnold Silver 313
5.3 High-T_c SQUIDs
     A. I. Braginski 328
5.4 Geophysical Applications of SQUIDs
     C. P. Foley 331
5.5 Application to Nondestructive Evaluation of Materials and Structures
     A. I. Braginski 342
5.6 SQUIDs — from Laboratory Devices to Commercial Products
     Ronald E. Sager 349
5.7 Electromagnetic and Particle Detection and Readout
     John Clarke, Kent Irwin and Peter Michelson 358
5.8 Concluding Remarks
     A. I. Braginski, John Clarke and Ronald E. Sager 371
     Acknowledgments 372
5.9 Further Reading 373

5.1 Introduction

John Clarke, Arnold Silver and A. I. Braginski

The Superconducting QUantum Interference Device (SQUID) combines the phenomena of flux quantization and Josephson tunneling. First predicted by F. London[1], flux quantization was observed experimentally by Deaver and Fairbank[2] and Doll and Näbauer[3] in 1961. They showed that the flux contained in a closed superconducting loop is quantized in units of the flux quantum $\Phi_0 \equiv h/2e \approx 2.07 \times 10^{-15}$ Wb, where, $h \equiv 2\pi\hbar$ is Planck's constant, and $2e$ is the charge on the Cooper pair—the cornerstone of the Bardeen-Cooper-Schrieffer (BCS) theory of superconductivity[4]. Flux quantization originates in the fact that the macroscopic wave function $\Psi(\vec{r},t) = \Psi_0(\vec{r},t)exp[i\phi(\vec{r},t)]$ must be single-valued in going once around a superconducting loop. In the absence of applied fields

---

[1] F. London *Superfluids*, Wiley, New York (1950)

[2] B.S. Deaver and W.M. Fairbank, *Phys. Rev. Lett.* 7 (1961) 43

[3] R.Doll and M. Näbauer, *Phys. Rev. Lett.* 7 (1961) 51

[4] J. Bardeen et al., *Phys. Rev.* 108 (1957) 1175

or currents, the phase $\phi(\vec{r}, t)$ takes the same value throughout the superconductor for all Cooper pairs. In the case of a loop threaded by a magnetic flux, however, the phase around the loop changes by $2\pi \cdot n$, where n is the number of enclosed flux quanta. In the year preceding the observation of flux quantization, Giaever[5] demonstrated the tunneling of single electrons between a superconductor (S) and a normal metal (N) separated by a thin insulating (I) layer. Subsequently, he observed the tunneling of single electrons through SIS junctions[6]. Single particle tunneling between superconductors was explained in terms of a tunneling Hamiltonian by Cohen, Falicov and Phillips[7].

In 1962, Brian Josephson (Figure 5.1) carried through this Hamiltonian to higher order to predict the tunneling of Cooper pairs through a barrier separating two superconductors[8]. He showed that the supercurrent I flowing through a junction is given by $I = I_0 \sin\delta$ where $\delta \equiv \phi_1 - \phi_2$ is the difference between the phases ϕ_1 and ϕ_2 of the condensates in the two superconducting electrodes and I_0 is the critical current. Furthermore, in the presence of a voltage V between the electrodes, δ evolves with time t according to $d\delta/dt = 2eV/\hbar = 2\pi V/\Phi_0$. These equations predict that, as the current through a junction is increased from zero, the voltage across the junction remains zero until the current exceeds the critical current, at which point a voltage appears.

Just one year later, Anderson and Rowell[9] made the first observation of the dc Josephson effect, using a thin-film, Sn-SnOx-Pb junction cooled to 1.5 K in liquid helium. The observations of flux quantization and Josephson tunneling set the scene for the invention of the SQUID.

SQUIDs are a class of superconductor electronics devices consisting of one or more Josephson junctions that are shunted by small superconducting inductors to form a multiply connected circuit. There are three classes of SQUIDs:

FIGURE 5.1: Brian Josephson discussing Giaever's work on single electron tunneling; circa 1965 (Courtesy Brian Josephson).

- dc SQUIDs contain two junctions and have a dc current-voltage characteristic,

- rf SQUIDs contain a single junction and have only rf response, and

- R-SQUIDs are rf SQUIDs with a small resistor in series with the inductance and junction, also called relaxation oscillator SQUIDs.

All types were invented nearly 50 years ago and have remained at the core of active electronic devices, circuits and systems based on superconductivity since shortly after their discovery. In Section 5.2, Arnold Silver and John Clarke, respectively, describe the early discoveries and developments at The Ford Motor Company in the U.S. and at Cambridge University in the U.K., which culminated in today's SQUID devices deployed in applications ranging from fundamental physics to a variety of sensors, amplifiers and devices for signal processing, digital computing and, recently, for quantum computation and communication.

---

[5]   I. Giaever, *Phys. Rev. Lett.* 5 (1960) 147

[6]   I. Giaever, *Phys. Rev. Lett.* 5 (1960) 464

[7]   M.H. Cohen et al., *Phys. Rev. Lett.* 8 (1962) 316

[8]   B.D. Josephson, *Phys. Lett.* 1 (1962) 251

[9]   P.W. Anderson and J.M. Rowell, *Phys. Rev. Lett.* 10 (1963) 230

With the advent of high-T_c cuprates, high-T_c Josephson junctions and SQUIDs were also developed, and eventually found applications. In Section 5.3, Alex Braginski outlines milestones in their developments and characterizes their current status. Developments leading to the successful use of SQUIDs as sensors for geophysical exploration of minerals are described in Section 5.4 by Cathy Foley, while in Section 5.5 Alex Braginski discusses attempts to use SQUID sensors for nondestructive evaluation of materials and structures. Section 5.6 of this chapter is written by Ron Sager. He presents the saga of the development of the world's most successful industrial SQUID instruments, now widely used in laboratory characterization of materials. Finally, in Section 5.7, John Clarke, Kent Irwin and Peter Michelson address the research leading to SQUID amplifiers and transition edge detectors of radiation and particles. These now find rapidly growing applications in various areas of science, such as astronomy and cosmology, as well as in analysis of materials. Concluding remarks are offered in Section 5.8.

Many other applications of SQUIDs as sensors are not included here, but are covered by or mentioned in other chapters. For example, Chapter 6 describes ongoing research in quantum computing with SQUIDs as qubits (quantum bits), while Chapter 10 includes the use of SQUID sensors in medical research and diagnostics. SQUIDs are widely used for the sensing and imaging of biological magnetic fields, such as those emitted by the brain and heart of humans and animals.

Overall, our chapter does not strive for comprehensive coverage of SQUIDs and their various uses. Rather, the co-authors present their personal perspectives on representative examples of SQUID research and development leading to selected practical applications. "Further Reading" offers a short list of textbooks devoted to a more systematic presentation of the subject and provides references to recently published conference proceedings presenting the current status of SQUIDs, detectors, and their applications.

5.2   History and Device Fundamentals

John Clarke and Arnold Silver

5.2.1   Initial Discovery

Arnold Silver

The first observations of macroscopic quantum interference in superconductors occurred at the Ford Motor Company Scientific Laboratory during 1963. Subsequent conceptual understanding, experimental verification, and expansion of those observations and concepts would eventually produce SQUIDs.

Based on prior developments and ongoing research in other institutions, SQUIDs would have been invented and developed somewhere, although under another name. However, the series of discoveries and inventions at the Ford labs occurred by an unusual sequence of events. It combined the collaborative efforts of Robert Jaklevic, John Lambe, James Mercereau, James Zimmerman, and myself (Figure 5.2), five physicists with differing technical backgrounds and interests. Here, I summarize those events and results. Since, after nearly 50 years, few records aside from publications are available, I rely on my memory to fill in the narrative.

The Ford work started in 1963, as the first publication date in February 1964 indicates. John Lambe and his collaborators first observed an anomalous periodic microwave response during electron-spin paramagnetic resonance (EPR) studies of doped Si at liquid helium (LHe) temperatures. These experiments were carried out at X-band ($\sim$ 10 GHz) in a high field electromagnet, an unlikely environment for SQUID research. That the measuring equipment was a very low-power super-heterodyne X-band receiver was essential to observing the very low-power signals. His observations could be characterized as follows: signals were periodic in magnetic field and microwave amplitude; they appeared as the temperature was lowered below the boiling point of LHe (4.2 K); they disappeared eventually as the temperature was lowered further; they disappeared as the magnetic field was increased significantly; and the signal was identified as arising from the Si samples.

FIGURE 5.2:   Photograph of the research team at Ford in 1964. Standing, left to right, J. Lambe, J. Zimmerman, R. Jaklevic, J. Mercereau. Seated, A. Silver. Courtesy of Ford Research Labs (Previously printed in *Physics Today*).

At the time, I was doing nuclear magnetic resonance (NMR) research in the same laboratory and was exposed to and intrigued by Lambe's unexplained discovery. Although the exact sequence of the events is uncertain, I joined him to help resolve the dilemma.

Critical to our success was that the Ford Motor Company Scientific Laboratory afforded most staff members the freedom to conduct self-motivated research. At one point, our new department manager asked if we knew what we were doing. We said probably not, but were determined to find out. That was that and we continued on.

Although we had only a general knowledge of superconductivity, we were inclined to attribute the effects to superconductivity of In contacts on the Si samples. The signals appeared at about the critical temperature of In (3.4 K) as we pumped on LHe. I should note here that the In contacts had been scratched in the process of characterizing the Si. Since direct observation of flux quantization in superconductors had recently been reported, we thought this was also relevant. We set out to understand the relevant literature, particularly F. London's theory of flux quantization[1].

The next task was to design experiments to replicate the microwave effects. Jaklevic, who had thin-film vacuum-deposition equipment, was engaged to deposit thin film superconductors such as In, Sn, and Pb on microscope slides. Small samples that fitted into the microwave cavity were cut; no signals were observed. However, when we damaged the films, either by scratching or, in the case of Pb, heating in air to corrode the Pb, we began to get familiar signals.

Jim Mercereau had joined the Laboratory from the California Institute of Technology, and was interested in flux quantization in superconductors. He informed us of Josephson's recent publication[8], P.W. Anderson's interpretation[10], and the Anderson and Rowell experiment[9]. These papers were certainly intriguing but not fully understood at that time. We were particularly interested in Anderson's physical description in terms of a macroscopic quantum wavefunction representing the Cooper pairs and its associated phase as being responsible for the behavior of the Josephson current in an applied magnetic field. Relating the magnetic field periodicity to the flux quantum by "carving" multiply connected thin films with microbridges was not very successful because of strong demagnetizing effects. Since there was no way we could relate the physics of thin film microbridges to Josephson tunnel junctions, we did not publish the microwave experiments until later[11].

5.2.2 Thin-Film Josephson Tunnel-Junction Quantum Interferometer

Arnold Silver

At that time, Jaklevic, Lambe, Mercereau, and I decided to try to produce the analog of a two-slit optical interferometer by using two Josephson tunnel junctions (JJ). The motivation was to demonstrate phase control of the superconducting current. Bob Jaklevic had the lead role here, first producing Josephson junctions, and then the two-junction thin-film interferometer shown in Figure 5.3. Since the magnetic field was in the plane of the films, we would be able to make quantitative measurements. Jaklevic's fabrication was successful and in late 1963 we began testing the two-junction device. Jaklevic called after lunch to say that he had a device cooled and was looking at the characteristic static current-voltage $(I - V)$ curve of the JJ on an oscilloscope as it varied rapidly under a changing magnetic field. We assembled quickly and discussed what the next step should be. Mercereau suggested we take data and we began recording the critical current (I_c), read from the oscilloscope trace, as a function of the magnet current. The device was situated in a completely unshielded double glass cryostat in common use at that time, with liquid nitrogen (LN2) in the outer dewar and LHe in the inner dewar. Anything that moved in the lab or the adjacent hallway caused the critical current to change. The magnetic field period was so small that it soon became clear that the noise in the local magnetic field would render data, taken at the speed of manual reading and recording, almost meaningless. We needed to record the data on a time scale short compared with the changes in the ambient magnetic field. We had no equipment that could record the critical current directly.

At this point, I thought that I could assemble an automated recording system in short order. Jaklevic's junctions were of sufficiently high quality that when we displayed the voltage vs time, we observed a nearly square waveform switching between $V = V_g$ (the energy gap voltage) and $V = 0$. The fraction of the period where $V = V_g$ depended linearly on the critical current. From my NMR experience, I knew we could record this signal vs magnetic field using a synchronous detector, essentially one part of a lock-in amplifier. I collected a synchronous detector and X-Y recorder from my lab and found the dynamic range insufficient to record large and small I_c on the same trace. I remedied this by operating the synchronous detector out-of-phase with the current waveform. After reconvening, we made X-Y recordings of I_c vs magnetic field for two devices that Jaklevic had available. Figure 5.4 shows the two-junction interference patterns superimposed on the Fraunhofer-like dependence of the junction I_c. By then, the afternoon had passed and we went out to celebrate with pizza and wine!

The next task was to publish the work as soon as possible. We were concerned that Anderson and Rowell at Bell Labs were trying the same experiment with substantial lead time. Fortunately, this was not the case. We prepared the manuscript and submitted it to the *Physical Review Letters* as

---

[10] P.W Anderson, *Lectures on the Many-Body Problem*, Vol.II, edited by E.R. Caianiello (Academic Press, New York, 1964)

[11] J. Lambe et al., *Phys. Lett.* 11 (1964) 16

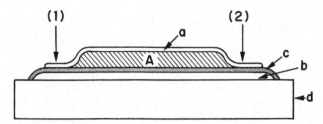

FIGURE 5.3: Cross-section of the first two-junction, thin-film, interferometer. (1) and (2) are Josephson junctions. The superconducting films are labeled a and b; c is the oxide barrier, d - the quartz substrate. Cross-section of the deposited insulating film is labeled A - it defines the loop area. Reprinted by permission from ref.12 (©1964 American Physical Society).

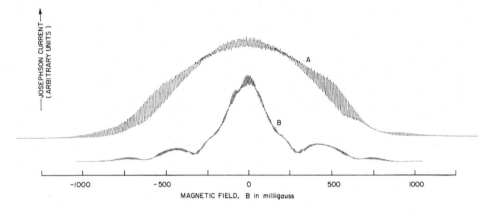

FIGURE 5.4: The first interference patterns illustrating both the single junction and two-junction interference in devices A and B. Reprinted by permission from ref.13 (©1964 American Physical Society).

rapidly as possible[12]. While we recognized the magnetic sensitivity of the technology, our focus was on the physics of long-range phase coherence. Meanwhile we planned and successfully performed two new experiments[13] with thin film JJs: First, the observation of quantum interference due to the magnetic vector potential in a magnetic field-free region, the Aharonov-Bohm (A-B) effect. Although we used the same measurement techniques as for the first interferometer experiments, the results were noisier. The A-B experiment presented a challenge to fabricate and insert a small solenoid into the thin film interferometer (see Figure 6 in Jaklevic et al.[13]). The solenoid current leads acted as an antenna for rf noise. In addition, the A-B experiment does not show the Fraunhofer suppression of I_c because there is no magnetic field at the junctions, although there is a vector potential.

Second, we measured the de Broglie wavelength of the supercurrent. This involved injecting a current into the superconducting film between the two JJs and measuring the phase shift as a function of the supercurrent.

---

[12] R.C. Jaklevic et al., *Phys. Rev. Lett.* 12 (1964) 159

[13] R.C. Jaklevic et al., *Phys .Rev.* 140 (1965) A1628

5.2.3 Point Contact Devices — dc SQUID

Arnold Silver

As our experiments were proceeding, Jim Zimmerman joined the activity. This marked the beginning of what would become SQUIDs. Fabricating tunnel junction devices was a slow, low-yield process. Zimmerman tried some other approaches to producing the "junctions" needed for future experiments and devices. He crossed two Nb ribbons and applied pressure from a hanging weight (Figure 5.5). A dab of epoxy permanently fastened the two ribbons together. This produced two weak-links with a small opening between them, matching the topology of the two-junction interferometers. These weak-links exhibited single-valued $I - V$ characteristics, which modulated as a magnetic field was applied to the aperture. From this we observed the periodic interference behavior determined by the size of the small aperture between the ribbons. There was no single-contact Fraunhofer effect. The single-valued $I - V$ curves permitted direct measurements with a low-noise amplifier without the need for a synchronous detector[14]. Since all our previous experiments were with type I superconductors, a new feature was the observation of macroscopic quantum interference in type II superconductors.

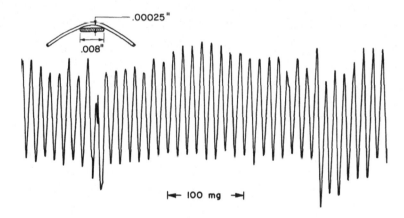

FIGURE 5.5: Cross-section of crossed Nb ribbons (upper left) and the recorded quantum interference pattern[14]. Reprinted by permission of Elsevier.

Eventually the team dissolved. Lambe and Jaklevic turned to other activities and Mercereau established an effort at Ford in California to pursue applications of quantum interferometers. Zimmerman and I continued to seek better methods of adjusting the critical current of the contacts. Devices made with a sharp point raised on a flat Nb surface and pressed against a second flat Nb surface were successful and provided impetus for an adjustable point-contact device. Jim fashioned a 000-120 (0.66 mm) Nb screw and tapped a corresponding hole in a Nb block. This block was pressed against a second Nb block and the screw adjusted to make contact with the second block. In this way, we had a mechanically adjustable contact. By using two screws with a hole drilled between them, we fashioned what became for a considerable time the dc-SQUID prototype (Figure 5.6). The adjustable-screw point-contact was our device of choice during the remainder of the 1960s. While others adopted this approach using 0-80 (1.19 mm) screws, we used 000-120 screws because Jim had Nb wire whose diameter was very nearly that size. Therefore, we did not have to machine the

---

[14] J.E. Zimmerman and A.H. Silver, *Phys. Lett.* 10 (1964) 47

screw, only run the wire through the die. At one point, our machine shop fabricated a device that did not work; we found they did not make it of Nb because Nb is difficult to machine; eventually our machine shop refused to do it. Consequently, we acquired a small milling machine and fabricated Nb devices in our lab.

Continuously repeating "superconducting quantum interference device" became tiresome and Zimmerman suggested we shorten the name to SQUID, which was not used outside Ford for some time. Consequently, when Jim and I authored a chapter in Academic Press' *Applied Superconductivity* in 1969-70, we avoided the name. By the time the book was published in 1975, the name SQUID had been widely adopted.

As the development of the SQUID structure was proceeding, we continued both to improve the measurement techniques and to develop a satisfactory model to understand and design SQUIDs. Our thinking shifted from an interferometer controlling a quantum wave to the device measuring the magnetic flux in units of the flux quantum, Φ_0.

Two concerns occupied a great deal of our time: what was the current-phase relation for the point contacts and how could we express the circulating currents and magnetic flux in

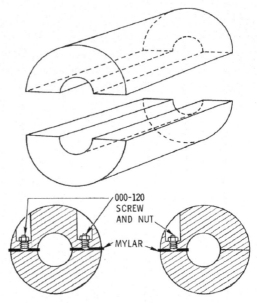

FIGURE 5.6: Drawing of typical point-contact SQUIDs machined from bulk Nb.

terms of the independent magnetic field or external current? There was no a priori reason to accept the Josephson current-phase relation for point contacts. We tried to use a linear (London-like) relation with strong depairing. This did not yield the continuous mathematical solution that we desired; the Josephson relations certainly did. We adopted the Josephson model and developed equations describing the magnetic flux and the circulating supercurrent, I_J, in the SQUID ring as a function of applied currents and magnetic fields.

At first, we adjusted I_c at room temperature and then cooled and tested the devices. Jim devised wrenches with long handles that enabled us to adjust the contacts in situ at LHe temperatures. Using a point-contact SQUID, we measured the magnetic field (London moment) produced by a rotating superconductor. This SQUID had a hole about 10 mm in diameter, and was rotated at a fixed frequency in LHe as we measured the shift of the interference pattern due to the induced magnetic field[15]. To do this experiment we moved our setup to a remote nonmagnetic building on the Oakland County campus of the University of Michigan. We could correlate occasional shifts in the interference pattern with the motion of vehicles that were visible several hundred yards away, illustrating the extreme sensitivity of SQUIDs.

5.2.4   The rf SQUID

Arnold Silver

The RF SQUID was invented in another case of serendipity. In addition to improving the device structure, we sought continuously to improve the measurements. Measurements of point contact

---

[15] J.E. Zimmerman and J.E. Mercereau, *Phys. Rev. Lett.* 14 (1965) 887

devices were initially performed with low frequency transformers and dc amplifiers connected to an oscilloscope or X-Y recorder. These transformers and amplifiers did not have very wide bandwidths and the amplifiers were noisy. I thought that we could improve both by using rf instrumentation that was available from my NMR research. We used an rf source to supply the current and a low noise rf amplifier as the detector. A simple *LC*-resonant circuit connecting the SQUID to the amplifier achieved large voltage gain, large signal bandwidth, and limited the noise bandwidth. An integrated diode functioned as a homodyne detector to produce a base-band signal.

This approach worked well and was adopted. We occasionally observed that the rf signature changed and became independent of one of the contacts; this was the single-junction ring that became the rf SQUID[16]. This produced new results that we would not have discovered had we continued with traditional dc electronics. In addition to the $V(I_c)$ *vs* magnetic field response, there is also an rf-current-dependent response, which we called the rf $V - I$ curve. This is illustrated in Figure 5.7. We developed a simplified magnetic state equation for the rf SQUID (Φ *vs* Φ_x): $\Phi + LI_c \sin\delta = \Phi_x + n\Phi_0$. Here, Φ and Φ_x are the enclosed and external fluxes, δ is the phase difference across the junction and n is an integer. We derived the magnetic states of the rf SQUID in terms of one universal parameter that characterizes the SQUID, $\beta_{rf} \equiv 2\pi LI_c/\Phi_0$. Inserting β_{rf} into the expression above yields a frequently used equation in terms of the junction phase, $\delta + \beta_{rf}\sin\delta = \delta_x$.

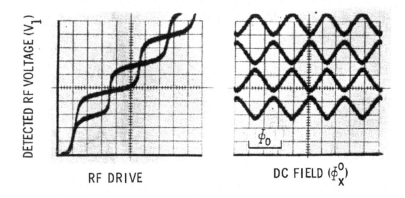

FIGURE 5.7: Typical response of an rf-SQUID. At left, "rf $V - I$ curve" at zero and half-flux quantum applied magnetic flux. At right, magnetic field dependence of rf voltage for fixed values of rf drive. Reprinted by permission from Silver and Zimmerman *Applied Superconductivity* vol. 1, New York: Academic Press 1975.

For $\beta_{rf} < 1$, the magnetic states are single valued, the SQUID can proceed continuously from one state to the next, and can be set stably at $\Phi = \Phi_x = \Phi_0/2$ and $I_J = 0$. This condition cannot exist for dc SQUIDs no matter the value of LI_c/Φ_0. This appeared to be an unusual situation and we spent considerable time trying to understand its significance, without success. It is now important for qubit research. Further, we reduced the wiring between room temperature and the SQUID at LHe to one rf cable that carried the rf drive, reflected signal and the flux generating static and low frequency current. We frequently injected the flux-generating current into the SQUID inductance instead of using a second coil.

[16] A.H. Silver and J.E. Zimmerman, Phys.Rev.Lett. 15 (1965) 888

5.2.5   Linear SQUID Response and Single Flux Quantum Transitions

Arnold Silver

While the SQUID was demonstrably a very sensitive detector, its response was nonlinear and periodic. We conceived and developed a "flux-locked loop" that linearized the response without sacrificing sensitivity, but reduced signal bandwidth. Again, this was a result of prior experience using modulation techniques to stabilize magnetic fields, and the availability of suitable equipment. Once we had a linear magnetometer, we used one SQUID to observe the magnetic states and single flux quantum transitions of a second SQUID[17]. We confirmed the behavior of $\beta_{rf} > 1$ (hysteretic in flux) and $\beta_{rf} < 1$ (nonhysteretic in flux) SQUIDS, and attempted to measure the current-phase relation of the point contacts.

5.2.6   R-SQUID, Oscillators and Detectors

Arnold Silver

Once we settled on the rf SQUID as the preferred device, we eliminated the second point contact and tried several other methods to close the ring. We found that a SQUID with a small resistance r in the loop (R-SQUID) behaved dynamically in much the same way as the rf SQUID. The rf $V-I$ curves were similar except for the absence of flux modulation at low frequencies. While the SQUID has perfect long-term memory, the R-SQUID has an L/r decay time. A second difference is that the junction in the R-SQUID can have a static voltage V_0 developed by a static current through r. Then the junction oscillates at the Josephson frequency given by the relation $V_0/f = \Phi_0 \approx 2.07\ \mu V/GHz$.

The physics is that the junction pulses single flux quanta into the SQUID L, and the flux escapes with a time constant L/r. The pulses are non-sinusoidal and have high harmonic content until the frequency approaches the plasma frequency. We used r between 10 and 50 $\mu\Omega$. These oscillations could be absorbed to excite resonant rf and microwave structures that were closely coupled to the SQUID loop. Embedding an R-SQUID in an X-band cavity in the cryogenic probe, we directly observed the cavity resonances[18]. We measured the standing wave resonance of the coaxial cable connecting the SQUID to the rf electronics. This experiment exhibited additional nonlinear properties of the SQUID. We observed the zero-field nuclear magnetic resonance of Co^{60} by placing finely powdered cobalt in the SQUID inductance and scanning the Josephson frequency across the Co^{60} resonance frequency. The latter two experiments were performed in the usual rf SQUID mode using our 27 MHz measuring system. The resonances were observed in a Josephson frequency scan with additional responses at $f_{res} \pm 27$ MHz, representing parametric idlers. In effect, the SQUID was operating as a parametric amplifier. Decades later, such parametric interactions were used to produce very low-noise microwave amplifiers[19].

Most experiments were performed with the SQUID devices sealed in a cylindrical stainless-steel tube. The stainless-steel tube was evacuated and refilled with a small amount of He heat-exchange gas. Thus, the devices were slightly isolated from the boiling LHe. We noticed, in the absence of any static current, a very-low-frequency oscillation whose linewidth was linearly dependent on the temperature. We attributed the voltage to a thermal emf from a temperature gradient across the metal resistor in the R-SQUID and the linewidth to Nyquist noise at LHe temperatures. This work continued jointly with R. Kamper at NBS in Boulder in an investigation of a primary temperature

[17]   A.H. Silver and J.E. Zimmerman, *Phys. Rev.* 157 (1967) 317

[18]   J.E. Zimmerman, et al., *Appl. Phys. Lett.* 9 (1966) 353

[19]   A. Smith, et al., *IEEE Trans. Magn.* 21 (1985) 1022

standard[20].

In 1968, Jim left the Scientific Lab and joined Ford's Aeronutronics subsidiary in California to work on applications of SQUIDs. From there he moved to the National Bureau of Standards in Boulder, CO in 1970. I left Ford in 1969 to join The Aerospace Corporation in California.

5.2.7 SLUGs at Cambridge

John Clarke

I became a research student in the Royal Society Mond Laboratory, then part of the Cavendish Laboratory at the University of Cambridge, in October 1964. The Mond was presided over—with great kindness—by David Shoenberg. My thesis supervisor was Brian Pippard (Figure 5.8). He gave me a project to investigate the electrical resistance of the superconductor-normal metal (SN) interface, a topic that later became of considerable interest. Brian suggested I measure the resistance of SNS sandwiches in which the normal metal was too thick to sustain a supercurrent. Since the voltages would be tiny (~ 1 pV), Brian asked me to look into improving the cryogenic galvanometer he and George Pullan[21] had previously developed. This device involved a mirror and magnet suspended by a quartz fiber inside a vacuum can, surrounded by liquid helium, at the center of a Helmholtz coil through which one passed the current to be measured.

My thinking about ways to upgrade the galvanometer was brought to an abrupt end, however, by a Mond seminar given by Brian Josephson. He described the experiments demonstrating flux quantization, his own theory of pair tunneling, and the observation of quantum interference at Ford earlier that year[13]. To me, as a new research student, it was utterly fascinating, and I remember thinking how I would like to work on these new ideas. My opportunity came much sooner than I could possibly have expected! The very next day, Brian Pippard bounced into my laboratory with a big smile on his face. "John, how would you like a voltmeter with a resolution of $2 \cdot 10^{-15}$ V in 1 second?" he asked. I somewhat cautiously admitted that this did seem like an interesting possibility. Brian excitedly explained that he had dreamed up his new concept of a voltmeter the previous evening after hearing Brian Josephson's seminar. He sketched on the blackboard a voltage source V in series with a resistance R and an inductance L that was

FIGURE 5.8: Brian Pippard (ca 1973; courtesy Cavendish Laboratory).

perfectly coupled to a SQUID, also with inductance L. The current through the resistor and inductor required to generate the flux quantum in the SQUID was simply Φ_0/L, and the corresponding static voltage V was thus $(\Phi_0/L)R$. Setting the time constant $\tau = L/R = 1$ s, Brian pointed out that the resolution was $2 \cdot 10^{-15}$V. His initial idea was to make a digital voltmeter, each digit being a flux quantum.

Needless to say, I was thrilled—not least because my enthusiasm for dangling a tiny magnet and mirror from a quartz fiber inside a vacuum can had already begun to wane. I talked to John Adkins, who had already made thin-film Josephson junctions in the Mond. He was not encouraging about the longevity of such devices, and I looked around for better alternatives. I came across an article

---

[20] A.H. Silver, et al., *Appl. Phys. Lett.* 11 (1967) 209

[21] A.B. Pippard and G.T. Pullan, *Proc. Camb. Phil. Soc.* 48 (1952) 188

by Zimmerman-Silver[14], which seemed like a good place to start. I made a similar device with the additional twist that I could adjust the vertical force on the bent wire (top of Figure 5.5) from the top of the cryostat. I remember rolling out a piece of niobium wire to make the narrow piece of sheet. In those days, one took one's glass cryostat to the helium liquefier for Paul Booth to fill it. After returning my dewar safely to my lab, I connected the simple current-voltage measurement system I had built—and my SQUID worked! I spent much of that day adjusting the force on the wire, and discovered that I could vary the critical current quite readily. Sometimes I observed oscillations in critical current when I varied the applied fields and sometimes I did not. I made several variants of this design, and they all more or less worked, but I was not entirely convinced that the device was sufficiently stable to use routinely as a voltmeter.

An important part of one's life in the Cavendish was—and still is—coffee at 11 a.m. and tea at 4 p.m. The Mond students and staff always sat at the same table. One tea-time, I discussed how I was looking for a Josephson junction technology that did not require thin films yet was mechanically stable. Paul Wraight—with whom I shared my lab—suddenly looked at me and said something like "How about a blob of solder on a piece of niobium wire? Solder is a superconductor and you keep telling me that niobium has a surface oxide layer." We rushed back to our lab, where fortunately I still had some liquid helium left in my dewar from the experiment I had been running earlier in the day. I made two devices consisting of a blob of lead-tin solder melted onto a short length of niobium wire, attached some leads and lowered the devices into the helium bath. They both worked! The critical currents were roughly 1 mA. Paul and I were thrilled. The next morning, Brian Pippard wandered into our lab to see how things were going, and Paul and I proudly showed him one of our new gadgets. Brian contemplated it thoughtfully for a while, and then—with a smile—said, "It looks as though a slug crawled through the window overnight and expired on your desk!"

I tried hard to make a SQUID by freezing a solder blob on a piece of niobium wire doubled back on itself so that there were two junctions. I applied a magnetic field to the loop of wire sticking out of the solder blob. This never worked, in retrospect because the inductance of the loop was too high. One day, I decided instead to pass a current along the wire, and immediately saw oscillations in the critical current when I changed the current. It did not take me very long to discover that the loop was irrelevant—I needed to pass the wire through the solder only once, and apply a current to the wire [Figure 5.9(a)]. How did it work? Apparently, there were typically just two or three dominant junctions between the wire and the solder. The current in the wire generated a magnetic field in the penetration depths of the wire and solder, so that the area of the "SQUID" was given by the sum of the penetration depths times the separation of the junctions[22]. The periodicity in current ranged randomly from about 0.2 to 1 mA. However unlikely, the majority of these devices showed interference between two or three junctions, and they generally survived at least scores of thermal cyclings. It is interesting to note that, as I increased the current through the niobium wire, I could generally observe thousands of oscillations in critical current with no evident diminution in amplitude. This lack of a discernible Fraunhofer pattern [Figure 5.9(b)] suggested that the junctions were essentially points.

The name for this new device had already been unwittingly provided by Brian Pippard, and we simply had to work out what it stood for. And so the "Superconducting Low-inductance Undulatory Galvanometer" (SLUG) was born.

Brian Pippard's original concept was a digital voltmeter, but I soon realized that one could readily measure changes in flux that were much less than one flux quantum. I applied a static current and a sinusoidal current through the SLUG [leads "I" in Figure 5.9(a)] so that voltage pulses appeared across the voltage leads (leads "V"). The area under these pulses depended on the critical current of the SLUG, enabling me to determine the critical current, and hence the current in the niobium wire [leads "I_B" in Figure 5.9(a)] to about ± 1 μA, using simple electronics. Once I had this worked

---

[22] John Clarke, *Phil. Mag.* 13 (1966) 115

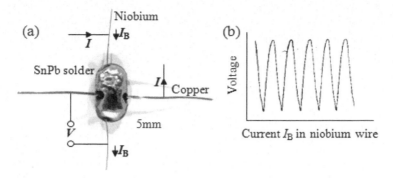

FIGURE 5.9: The SLUG. (a) Photograph showing attachment of current (I), voltage (V) and flux bias (I_B) leads. (b) V *vs* I_B.

out, I immediately put the SLUG to use as a voltmeter. A postdoctoral fellow in the Mond, Steve Lipson, was interested in measuring the resistance of a single crystal of copper (to be used for measurements of magnetoresistance) that he estimated to be about $10^{-8}\Omega$. We made a voltmeter by connecting the Cu block in series with a manganin wire, with a measured resistance of about $10^{-5}\Omega$, and the niobium wire through the SLUG. The idea was to make a potentiometer, adjusting the currents in the two resistors to produce zero current in the SLUG. After Steve and I cooled down our potentiometer, I was chastened to observe that the output from the SLUG was horribly noisy. Further inspection showed that the "noise" was in fact due to currents induced in the loop by 50-Hz magnetic fields. For the next run, Steve soldered a lead-foil box that enclosed the circuit, and our pickup problem was solved. This was an invaluable lesson for me: if you have a sensitive magnetometer or voltmeter, you have to protect it from the real world, which is a very noisy place.

I used the SLUG as a voltmeter for the rest of my graduate career (Figure 5.10), mostly to investigate SNS Josephson junctions. With a typical circuit resistance of 10^{-8} Ω, I could measure a voltage of about 10^{-14} V in a second. I built a feedback circuit incorporating an amplifier and an integrator that fed current back into the manganin resistor to maintain zero current in the Nb wire of the SLUG. Other members of the Mond subsequently used the SLUG, notably Eric Rumbo for measuring thermoelectric voltages and Pippard, Shepherd and Tindall for measuring the resistance of the SN interface—my original thesis topic.

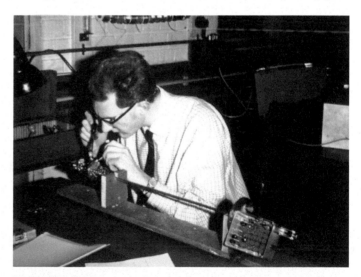

FIGURE 5.10: John Clarke makes a SLUG circa 1966. In those days, research students wore ties! (Courtesy Gordon Donaldson)

In addition to making a femtovoltmeter, I played with various other ideas with SLUGs. One was a magnetometer, based on the idea of a superconducting flux transformer[23] (Figure 5.11). I inserted the two ends of the niobium wire of the SLUG into a blob of molten solder to make a superconducting contact, forming a superconducting loop, typically 50 mm in diameter. When I applied a magnetic field to the loop, the induced current was detected by the SLUG, which thus became a magnetometer. I made this null-balancing by connecting the output from my feedback loop to a 1000-turn, copper-wire coil that was tightly coupled to the niobium loop. I remember showing the magnetometer to Brian Pippard who stared at it thoughtfully for a while, and then said something like, "You know, John, if you were to put a twist in the loop, you would have a gradiometer." As usual, Brian was ahead of his time, but I did not need a gradiometer at that point in my life and did not then pursue the idea. In an attempt to improve the current-sensitivity of the SLUG, I coupled a 10-turn, niobium wire coil to the niobium-wire loop[23]. The voltage signal was connected to the ends of the 10-turn coil, in series with a resistor. Because of the relatively poor coupling between the 10-turn coil and the SLUG loop, as I recall, this arrangement improved the current sensitivity by a factor of only 2 or 3, but at least it worked.

As an extension of these ideas, I connected the voltmeter with the 1000-turn input coil to the output of a second SLUG, biased into the voltage state (Figure 5.11). Because setting up the bias current of the input SQUID produced a substantial flux in the superconducting loop of the magnetometer, I devised a thermal switch—a length of bifilar-wound manganin wire wrapped around a region of the niobium loop—and surrounded it with a blob of epoxy to provide thermal insulation from the helium bath. By passing a current through the heater, I could drive the niobium wire into the normal state, releasing the flux trapped in the loop. After juggling the four bias currents for a while, I actually made these cascaded SLUGs work[23].

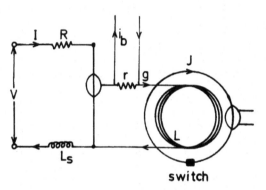

FIGURE 5.11:   Two-SLUG amplifier. The niobium wire of the right-hand SLUG is connected into a superconducting loop, forming a magnetometer. A change in the current I in the left-hand SLUG induces a current into the 1000-turn coil, thus generating a current J in the magnetometer loop and finally an output voltage from the right-hand SLUG.

Another idea was an inductor-resistor relaxation oscillator[23]. SLUGs with higher values of critical current usually had hysteretic current-voltage characteristics. I connected a resistor in series with an inductor across the current-voltage leads of a SLUG, and applied a bias current to produce square wave oscillations. I had visions of using this SLUG oscillator to read out a second SLUG, but never actually attempted to do this.

5.2.8   SQUIDs at Berkeley

John Clarke

I moved to the Department of Physics at the University of California, Berkeley in January 1968. During the early 1970s, there was a general shift towards using the rf SQUID, including devices made of thin films deposited on cylindrical substrates. In 1973, a new postdoctoral scholar, Wolf Goubau, and a new graduate student, Mark Ketchen, and I decided to make dc SQUIDs borrow-

[23] John Clarke, in *Proc. Symposium on the Physics of Superconducting Devices*, University of Virginia, Charlottesville, April 28-29, 1967, Bascom S. Deaver, Jr. and William S. Goree, eds., page D1.

ing three ideas from the rf SQUID: a cylindrical geometry (which gives a large area for a low inductance), a tank circuit readout and thin films. At about the same time, Paul Hansma's group, at the University of California, Santa Barbara, had achieved good reproducibility and longevity with Nb-NbOx-Pb tunnel junctions[24], and we decided to adopt his technique. Figure 5.12 shows the geometry of our SQUID, grown on a 3-mm-diameter quartz tube[25]. We used shadow masks to pattern the films, which had a minimum linewidth of 75 μm. The slit in the PbIn cylinder was scribed with a razor blade. Subsequently, we submerged the SQUID in a solution of Duco cement (our favorite insulator!) dissolved in acetone and deposited a PbIn film over the slit and the various metal strips to reduce their inductances. This process sounds extraordinarily primitive by today's standards, but we could cool these devices to liquid helium temperature many times with no degradation.

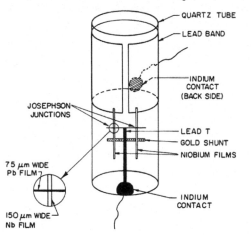

QUARTZ TUBE

LEAD BAND

INDIUM CONTACT (BACK SIDE)

JOSEPHSON JUNCTIONS

LEAD T
GOLD SHUNT
NIOBIUM FILMS

75 μm WIDE Pb FILM

INDIUM CONTACT

150 μm WIDE Nb FILM

FIGURE 5.12: Configuration of cylindrical dc SQUID.

We tested our SQUIDs using 100-kHz flux modulation and a homemade flux-locked loop. We boosted the 100-kHz voltage across the SQUID with a cold resonant circuit, giving a substantially enhanced signal into the preamplifier, and a correspondingly reduced flux noise. The white noise in one device, $3.5 \cdot 10^{-5} \, \Phi_0 \, \text{Hz}^{-1/2}$, was a record at the time, and yielded a magnetic field noise of about 10 fTHz$^{-1/2}$. We used our cylindrical SQUIDs in a variety of experiments. For example, Wolf, Tom Gamble and I used them for magnetotellurics, a geophysical surveying technique[26].

During the time we were working on our experiments, Claudia Tesche and I developed the theory for the dc SQUID[27]. Claudia spent roughly a year carrying out the computer simulations including the effects of white noise, which arises from Nyquist noise generated in the resistive shunts required to eliminate hysteresis on the current-voltage characteristics of the tunnel junctions[28]. We worked out the parameters required to minimize the noise energy of the SQUID, $\epsilon(f) = S_\Phi(f)/2L$; here $S_\Phi(f)$ is the spectral density of the flux noise and L is the inductance of the SQUID loop. At 4.2 K and for the optimized parameters $\beta_{\text{dc}} \equiv 2LI_0/\Phi_0 \approx 1$ and $\beta_c \equiv 2\pi I_0 R^2 C/\Phi_0 \leq 1$, the essential results are $S_\Phi(f) \approx 16k_BTL^2/R$ and $\epsilon(f) \approx 9k_BTL/R \approx 16k_BT(LC)^{1/2}$, where R and C are the shunt resistance and self capacitance of each junction. These results make it clear that "smaller is better" and that $S_\Phi(f)$ and $\epsilon(f)$ decrease linearly with temperature. This theory has proved to be a good predictor of experimental results over a wide range of SQUID parameters and temperature. For a typical SQUID at 4.2 K, one finds $S_\Phi^{1/2}(f) \approx 10^{-6} \, \Phi_0\text{Hz}^{-1/2}$ and $\epsilon(f) \approx 10^{-32} \, \text{JHz}^{-1} \approx 100\hbar$. When the temperature of the SQUID is lowered to a few tens of millikelvin, the noise energy is typically on the order of $\hbar$ at frequencies f above the $1/f$ noise. Claudia and I also calculated the noise current circulating around the SQUID loop and its partial correlation with the voltage noise[29]. These terms turn out to be crucial to the optimal design of SQUID amplifiers (see Section 5.7).

[24] C.M. Falco et al., *Phys. Rev. B* 10 (1974) 1865

[25] John Clarke et al., *J. Low Temp. Phys.* 25 (1976) 99

[26] T.D. Gamble et al., *Geophys.* 44 (1979) 53

[27] C.D. Tesche and J. Clarke, *J. Low Temp. Phys.* 29 (1977) 301

[28] The theory for the non-hysteretic characteristics of resistively shunted junctions (RSJ) was independently developed in D. E. McCumber, J. Appl.Phys. 39 (1968) 3113 and W. C. Stewart, *Appl. Phys. Lett.* 12 (1968) 277.

[29] C.D. Tesche and J. Clarke, *J. Low Temp. Phys.* 37 (1979) 397

5.2.9   The Square Washer SQUID

John Clarke

I think the cylindrical SQUID played a role in turning the attention of the community back from the rf SQUID to the dc SQUID despite the fact that, to my knowledge, no other group adopted this design. One reason may have been the realization that higher performance required tunnel junctions with smaller areas. Nobody was excited about using photolithography—having just been imported from the semiconductor industry—on a cylindrical surface! After graduating from Berkeley, Mark Ketchen moved to IBM, Yorktown Heights, where he and Jeffrey Jaycox developed the square washer SQUID[30]. Mark described this device as "The result of putting your thumb on one end of the cylindrical SQUID and squashing it flat, transforming it into a washer with a spiral input coil". The SQUID itself is a thin-film square washer, typically 1 mm across, with a hole in the middle and a slit that runs to the outer edge where two tunnel junctions are grown (Figure 5.13). The junctions are completed with an upper film that connects them, thus closing the SQUID loop. An insulating layer is deposited over the square washer, followed by a thin-film, spiral coil. A current passed through the coil generates a magnetic flux that the washer focuses into the hole, giving efficient coupling between the coil and the washer. However, these SQUIDs had initially one drawback: they were made from a PbInAu alloy, and were not particularly robust. Fortunately, in 1980 John Rowell and coworkers[31] developed the niobium-based junction technology that is universally used today. One first deposits a Nb film on a silicon wafer, followed immediately by an Al film a few nanometers thick. The Al film is subsequently oxidized in a controlled oxygen pressure for a prescribed time. The gas is pumped out and the upper Nb electrode is deposited. Subsequently, the "trilayer" is patterned to form Josephson junctions. The technology for the Nb-based, square washer dc SQUID–the workhorse of today's SQUID applications–was essentially in place by the early 1980s.

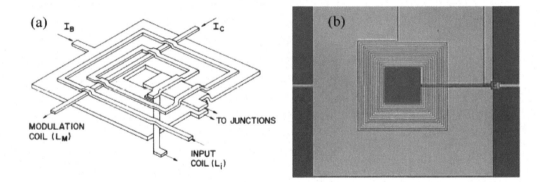

FIGURE 5.13:   Square washer dc SQUID. (a) Schematic of original design. (Reproduced by permission from Reference 30.) (b) Today's typical practical device. The two resistively shunted Josephson tunnel junctions are at the right-hand edge, one on each side of the slit.

---

[30] M.B. Ketchen and J.M. Jaycox, *Appl. Phys. Lett.* 40 (1982) 736

[31] J.M. Rowell et al., *Phys. Rev. B* 24 (1981) 2278

5.2.10   The rf SQUID Revisited

John Clarke

The theory of the rf SQUID was worked out by Kurkijarvi[32]. In the hysteretic or dissipative mode ($\beta_{rf} > 1$), the applied rf flux causes the SQUID to make transitions between quantum states and to dissipate energy at a rate that is periodic in Φ_x. This periodic dissipation in turn modulates the tank circuit Q, so that when it is driven on resonance with a current of constant amplitude, the rf voltage is periodic in Φ_x (Figure 5.7). A detailed analysis shows that the SQUID is optimized when $k^2 Q \approx 1$, where k is the coupling coefficient between the SQUID loop and the inductor of the resonant circuit. The intrinsic noise energy is $\epsilon(f) \approx L I_0^2 (2\pi k_B T / I_0 \Phi_0)^{4/3} / 2\omega_{rf}$. Since the noise energy scales as $1/\omega_{rf}$, there is a strong motivation to use a microwave driving frequency. For rf SQUIDs at 4.2 K coupled to a room-temperature preamplifier, however, extrinsic noise contributions—notably preamplifier noise and loss in the line coupling the SQUID and preamplifier—may far exceed the intrinsic noise.

Subsequently, it was realized that the noise energy can be much lower in the nonhysteretic or nondissipative regime $\beta_{rf} < 1$ in which the SQUID remains in the zero voltage state[33]. The SQUID behaves as a flux-sensitive, nonlinear inductor since the inductance of a Josephson junction for $I < I_0$ is given by $\Phi_0/2\pi(I_0^2 - I^2)^{1/2}$. Thus, the circulating current produced by an applied flux induces a change in the SQUID inductance. When the tank circuit is driven off-resonance at constant amplitude and frequency, a flux change in the SQUID changes the resonant frequency, so that the rf voltage is periodic in the applied flux. In the limit $I_0\Phi_0/2\pi \gg k_B T$, the intrinsic noise energy is $\epsilon(f) \approx 3k_B T/\beta_{rf}^2 \omega_c$ when the drive frequency is set equal to the SQUID cutoff frequency, ω_c[33]. In principle, at low temperatures the nondissipative SQUID can rival the noise energy of the dc SQUID, but in practice at the required high frequencies its noise is dominated by the post-amplifier.

5.2.11   Today's SQUIDs

John Clarke

For most practical applications at and below He^4 temperatures, the dc SQUID is dominant. The square washer dc SQUID, involving Nb trilayer tunnel junctions and typically made in batches of several hundred on silicon wafers that are diced to produce individual devices, is virtually indestructible. At 4.2 K, its white noise is typically $10^{-6}\Phi_0$ Hz$^{-1/2}$, with a $1/f$ knee of roughly 1 Hz. In most applications, the SQUID is inductively coupled to a superconducting flux transformer, to form a magnetometer—with a magnetic field noise of around 1 fTHz$^{-1/2}$—or a first- or second-order gradiometer. As the temperature of the dc SQUID is lowered, its performance can approach the quantum limit[34]. Its simplicity of use—requiring only low frequency readout techniques—is very appealing, especially in applications requiring large numbers of SQUIDs, for example magnetoencephalography (Chapter 10). There is, however, one situation in which rf SQUIDs have an advantage: high-T_c, nondissipative rf SQUIDs at 77 K can be operated with a higher loop inductance—and thus a larger area—than dc SQUIDs, giving them a lower magnetic field noise[35]. Finally, both rf[33] and dc SQUIDs can be used as parametric amplifiers, and can achieve the quantum limit at low temperatures[36].

[32] J. Kurkijarvi *J. Appl. Phys.* 44 (1973) 3729

[33] K.K. Likharev, *Dynamics of Josephson Junctions and Circuits* (Gordon Breach, New York, 1986)

[34] M. Mück et al., *Appl. Phys. Lett.* 78 (2001) 967

[35] Y. Zhang et al., *IEEE Trans. Appl. Supercond.* 3 (1993) 2465

[36] M. Hatridge et al., *Phys. Rev. B* 83 (2011) 134501.

5.3   High-T_c SQUIDs

A. I. Braginski

5.3.1   Past and Present

After the news of high-temperature superconductivity in lanthanum-barium cuprate spread across the superconductivity community worldwide, the early weeks and months in 1987 witnessed incredibly hectic activity by every group capable of synthesizing one of the new materials in polycrystalline bulk or thin film form (Chapter 4). The scouting went in all possible directions and high-T_c SQUIDs were naturally included. It was not immediately obvious to all of us that higher-angle cuprate grain boundaries (GBs) act as Josephson weak links, so the first bulk and thin-film SQUIDs[37] were fabricated from polycrystalline yttrium cuprate $YBa_2Cu_3O_{7-\delta}$ (YBCO). It exhibited an incredibly high $1/f$ noise due to the penetration and motion of vortices at grain boundaries. As I recall, it was the late Masaki Suenaga who first pointed to the weak link properties of GBs in his late night talk at the "Woodstock of Physics", at the American Physical Society March Meeting of 1987. From then onwards the more fruitful efforts were directed towards synthesizing epitaxial, nearly single-crystalline thin films of YBCO and other rare earth cuprates (REBCO), from which controlled weak links and SQUID structures could be patterned and fabricated.

Glossary of High-T_c Junction Terms

- A bicrystal epitaxial substrate consists of two crystals of the same material, but different crystalline orientation, that are fused together. The grain boundary between the two crystals is reproduced in the film growing epitaxially on the polished and undamaged substrate surface.

- A step-edge junction is formed when an epitaxial film grows over a sharp step etched by a photolithographic method in the undamaged single-crystalline surface of a substrate. A change in the crystalline orientation at any sharp edge nucleates a grain boundary there.

- An edge or ramp junction, either low- or high-T_c, involves a relatively shallow edge or wedge patterned in a superconducting thin film electrode using photolithographic techniques. A thin barrier layer and superconducting counter-electrode are deposited on top and patterned such that a layered narrow bridge structure exists over the inclined edge (ramp) area.

The research on epitaxial films and on Josephson properties of grain boundaries in epitaxial YBCO thin films on bicrystalline $SrTiO_3$ (STO) substrates was initiated at IBM Yorktown Heights under the leadership of the late Praveen Chaudhari[38,39]. This seminal work on the critical current I_c dependence upon the grain boundary angle (Chapter 3) laid the foundation of what is still today the most frequently used and reliable high-T_c Josephson junction technology for SQUIDs operating at

---

[37] R. H. Koch, et al., *Appl. Phys. Lett.* 51 (1987) 200

[38] P. Chaudhari et al., *Phys. Rev. Lett.* 58 (1987) 2684

[39] D. Dimos et al., *Phys. Rev. Lett.* 61 (1988) 219

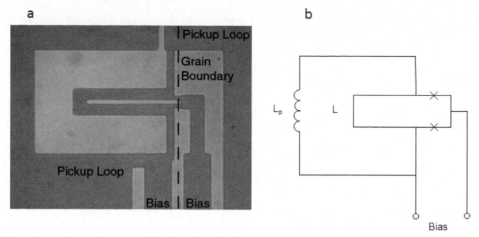

FIGURE 5.14: Directly-coupled high-T_c dc SQUID magnetometer: (a) photograph of layout on bicrystal substrate; (b) schematic diagram, where the two junctions are represented by crosses (reproduced from *The SQUID Handbook*, Figure 5.15, with permission).

or near the liquid nitrogen (LN2) temperature of 77 K. The dc SQUID loop is patterned such that two symmetric microbridges cross the GB. The typical range of angles is from 24° to 36° with a likely optimum around 30°. Many approaches to fabricating SQUIDs without using bicrystal substrates were demonstrated in the past, but only two withstood the test of time and are occasionally used in some applications: step-edge junctions, especially those on MgO crystal substrates which involve only one [100] tilt grain boundary[40] and edge or ramp junctions with PrBCO barriers[41]. All high-T_c junctions are internally shunted and their current-voltage characteristics roughly conform to the RSJ model. Therefore, high-T_c SQUIDs do not require shunting resistors (see the box "Glossary of High-T_c Junction Terms").

During the 1990s, epitaxial thin film and junction fabrication technologies reached a level of maturity sufficient to fabricate single-layer patterned structures, usually of YBCO, more or less reproducibly. On bicrystal substrates, the most typical and successful single-layer design of a magnetometer involves a small slit-like loop directly coupled to a much larger pickup coil, both shown in Figure 5.14[42]. This whole structure is patterned on a large, typically 10×10 mm^2 substrate. Of course, there is a large mismatch between the input coil and SQUID loop inductance, and the effective coupling coefficient is rather low. When a larger pickup coil is needed for magnetic field sensitivity, a single-layer flux transformer is patterned on a suitably large substrate with its input coil being a replica of the SQUID's pickup loop. For coupling, the two are brought into close contact by clamping them together in the so-called flip-chip configuration.

Multilayered flux transformers, analogous to low-T_c structures, which offer a much better coupling between the transformer input coil and the SQUID loop, are rarely used. It is still rather difficult to fabricate high-quality multilayers and multiturn coils; the yield is so low that the cost becomes too high. In special cases such transformers are made separately from the SQUID chip and used in the flip-chip configuration—allowing one to select and match the best performing SQUID and transformer chips. High-T_c SQUID magnetometers with planar single- and multilayer flux transformers are commercially available from several sources[43]. Flux transformers with pickup

---

[40] C.P. Foley et al., *IEEE Trans. Appl. Supercond.* 9 (1999) 4281

[41] Gao, J. et al., *Physica* C 171 (1990) 126

[42] An early design of this type: D. Koelle et al., *Appl. Phys Lett.* 63 (1993) 3630

[43] For example, STAR Cryoelectronics www.starcryo.com, SUPRACON http://www.supracon.com/, and MAGNICON http://www.magnicon.com/

coils wound of high-T_c wire do not exist because superconducting contacts between such wires and films cannot be made by known techniques.

As in early SQUID days, high-T_c rf SQUIDs were initially believed to be advantageous because only one junction had to be fabricated. However, the bicrystal GB topology favored the dc SQUID design with two symmetrical microbridges. Therefore, the majority of high-T_c devices in use today are dc SQUIDs. Early thin-film high-T_c rf SQUIDs usually employed step-edge junctions so that a single junction could be fabricated at the desired location on the substrate. For magnetometers, a large superconducting washer with a small hole has been used as a flux focuser to enhance the field sensitivity. This was the direction we pursued at FZJ[44], where the high-T_c SQUID development was led by my collaborator Yi Zhang. Today, the most successful rf SQUID design employs a single crystal, usually $SrTiO_3$, substrate as a dielectric microwave resonator instead of a patterned or wire-wound tank circuit[45]. The large flux-focusing washer covers the substrate and is flip-chip mated with the much smaller SQUID chip. Such SQUIDs are also commercially available[46]. Figure 5.15 shows schematically the components of the device.

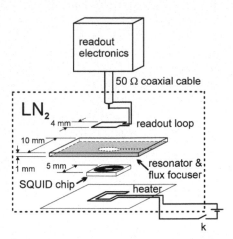

FIGURE 5.15: Schematic drawing of the substrate resonator SQUID. Dotted line surrounds the components immersed in liquid nitrogen. Adapted from Zhang et al.[45], with permission.

5.3.2   High-T_c SQUID Limits of Performance

The main advantage of the high-T_c SQUID is that it can operate at or near 77 K rather than at LHe temperatures, so cooling is much easier and less expensive. Furthermore, the liquid nitrogen cryostat walls are thinner than those of LHe cryostats allowing the SQUID pickup loop or coil to be positioned closer to the measured object thus partly compensating for the lower sensitivity due to higher noise. Indeed, the higher operation temperature sets the intrinsic limit of performance. As shown in Section 5.2.10, the thermal noise energy in the junction resistance scales with temperature over a wide range of frequencies. Furthermore, the $1/f$ flicker noise caused by critical current fluctuations, and by the motion of vortices in thin films and junctions, is dramatically higher than in low-T_c SQUIDs. The flux and energy resolution at a given loop inductance L of high-T_c SQUIDs are correspondingly limited. When L is on the order of 100 pH, a value typical for magnetometers with coupled input circuits, the lower limit imposed by thermal (white) noise is not much below 10 $\mu\Phi_0$ Hz$^{-1/2}$, an order of magnitude higher than that of low-T_c SQUIDs at 4.2 K. For single-layer magnetometers on a typical substrate (10×10 mm^2), the white magnetic field noise floor is at best $S_B^{1/2} = 40$ to 50 fTHz$^{-1/2}$ for both dc and rf SQUIDs. With flip-chip multilayer flux transformer, the best attained[47] is $S_B^{1/2} \leq 10$ fTHz$^{-1/2}$, while commercially available devices have $S_B^{1/2} \approx 15$ fTHz$^{-1/2}$.

At frequencies below 1 kHz the $1/f$ noise dominates and limits the device sensitivity. The component of that noise caused by critical current fluctuations can be removed by the use of ac bias (bias

---

[44] FZJ is the German acronym for the Jülich Research Center (Forschungszentrum Jülich).

[45] Y. Zhang et al. (2002), *Physica* C 372-276, part I (2002) 282

[46] JSQuid, see www.jsquid.com

[47] E. Dantsker et al., *Appl. Phys. Lett.* 67 (1995) 725

reversal), but the flux noise caused by vortex motion cannot. Vortices readily enter the less than perfect YBCO epitaxial films, even in very weak magnetic fields, B_0, like those generated by screening currents in the SQUID loop, and hop around due to the higher thermal energy and relatively weak flux pinning. Consequently, the $1/f$ noise measured with bias reversal strongly increases with B_0[48]. A partial remedy is to pattern high-T_c device structures with very narrow film lines of width w of only a few micrometers, because the vortex entry field is inversely proportional to the square of w: $B_{\text{entry}} \approx \Phi_0/w^2$.

The intrinsic limit of high-T_c SQUID performance in the presence of large thermal fluctuations was evaluated theoretically for dc and rf SQUIDs both analytically and by extensive simulations. The analytical work was done by Boris Chesca[49], the simulations by Reinhold Kleiner et al.[50]. For dc SQUIDs these simulations predicted thermal noise several times lower than experimentally measured. The consequence of lower flux and field sensitivity and dramatically higher low-frequency device noise is that after a decade of development the interest in high-T_c SQUIDs subsided and their use today is rather limited compared to that of low-T_c devices. Although the absence of LHe should facilitate portability and mobility, motion in external magnetic fields (i.e., the Earth's magnetic field) results in flux penetration and vortex motion, creating noise and hysteresis much higher than in mobile LHe-cooled devices. Although available commercially, at least in small quantities, high-T_c devices are also more expensive than their low-T_c counterparts due to low fabrication yield and limited reproducibility.

5.4 Geophysical Applications of SQUIDs

C. P. Foley

5.4.1 Introduction

Detection of variations or "anomalies" in the Earth's magnetic field was identified as an obvious application for SQUIDs soon after their invention. These anomalies could be due to the subsurface presence of bodies that are magnetic (such as iron ore, kimberlite pipes that contain diamonds, submarines or bomb casings), bodies that are electrically conducting (such as silver, gold or nickel sulphide), or bodies such as gas and oil that cause a local change in the Earth's magnetic field. Although initial trials were successful and early commercialization was attempted, by the late 1980s it became clear that SQUIDs would not be adopted for these applications, largely because of the need for LHe cryogenics, related cost and limited LHe availability in remote areas. However, the discovery of high critical temperature, T_c, superconductors in 1986 gave SQUIDs for geophysical prospection a second chance.

At this time, in 1988, I joined a team of researchers led by John Macfarlane at the Australian Commonwealth Scientific and Industrial Research Organization (CSIRO) to work on the emerging high-T_c material, YBCO. I had previously developed indium nitride thin films which I used for simple semiconducting photodiodes. Based on this background, I was charged with developing Josephson junctions and then SQUIDs from this new granular ceramic material.

From early 1967 on, CSIRO has made many successful contributions to superconducting electronic metrology and cryogenics and we were hopeful to build on these. Our HTS team, in collabora-

[48] A.H. Miklich et al. (1994), *Appl. Phys. Lett.* 64 (1994) 3494.

[49] B. Chesca (1998), *J. Low Temp. Phys.* 112 (1998) 165-196, and J.LowTemp.Phys. 110 (1998) 963.

[50] R. Kleiner et al. (2007), *J. Low Temp. Phys.* 149 (2007) 230 and 261 (2 parts).

tion with Australian companies, BHP[51], Nucleus and AWA, had secured a \$1M generic government-industry research and development grant to develop HTS SQUIDs, with the first application being nondestructive testing of steel. In 1990, we demonstrated the detection of a 1 mm wide slot in a piece of steel using our first YBCO thin film dc SQUID which had a poor flux noise level of $3.8 \cdot 10^{-4}$ $\Phi_0 \mathrm{Hz}^{-1/2}$ and was housed in a crude Styrofoam dewar. The device was powered by homemade SQUID drive electronics designed and fabricated by David Dart, the BHP company staff member who worked with us in our laboratory at Lindfield. After this first "success" in non-destructive evaluation, BHP management recommended that we turn our efforts to the development of SQUIDs for mineral exploration. This challenge required the detection of magnetic fields of an ore body with a magnetic field strength less than 2 pT in the presence of the Earth's magnetic field — seven orders of magnitude stronger. Detection was to occur in remote locations (where there is limited mains frequency noise), on the ground, in the air or under water when the system is either stationary or in motion. This ambitious goal ignited our enthusiasm for SQUIDs for geophysical applications, in spite of their lapse in this field during the early 1980s. I'll describe our successful work after introducing briefly the magnetic measurements used in geophysical prospecting, and equally briefly addressing the early geophysical measurements performed using low-T_c SQUIDs.

5.4.2   Magnetic Measurements Used in Geophysical Prospecting

Three types of magnetic measurements are made in geophysical prospecting: scalar measurements usually using vapor magnetometers (this provides the total magnetic field, B, intensity values — TMI), vector measurements which give the three axes of the magnetic field which can be directly measured by flux gates or SQUIDs, B_x, B_y and B_z, or their time differentials, $\partial B_i(t)/\partial t$, measured by induction coils with the vector components obtained by the integration of the measurements, and, thirdly, the gradient measurements, which measure the spatial variation of the magnetic field leading to the nine components of the Earth's magnetic field tensor.

Magnetotellurics (MT), a "passive" magnetic exploration method, determines the distribution of electrical conductivity in the Earth's subsurface by measuring the natural electromagnetic field variations (i.e., of both the electric and magnetic fields) caused by solar radiation/particles or atmospheric sources, such as lightning discharges that induce electric currents in the subsurface. By modeling and inversion (i.e., by solving the magnetic inverse problem[52]), an image of the distribution of electrical conductivity is obtained. Active variations of MT include Audio-MT and the underwater Controlled Source Electromagnetics (CSEM) where electromagnetic field variations are induced by audio or very-low-frequency electromagnetic waves from a transmitter.

Transient ElectroMagnetics (TEM) is an "active" exploration technique used to prospect for conducting ore bodies such as nickel, gold and silver and can be used with either an airborne or ground-based system. This technique evolved from the active MT methods. A transmitter coil is mounted either around the wing tips of an aircraft or placed on the Earth's surface. Such an antenna emits a short magnetic pulse which induces in the ground electrical eddy currents that decay over time. A sensitive magnetometer measures the associated decaying "secondary" magnetic field. Once the decay is completed, the process is repeated many times, so that the measured decay signal can be averaged to improve the signal-to-noise ratio (SNR) of the measurement. The measured decaying magnetic field data can be processed to derive the electrical conductivity of the ground as a function of depth. In this way, a TEM survey is used to generate three-dimensional (3-D) conductivity maps. Where conducting anomalies occur, the decay of the secondary magnetic field will be slower than in

---

[51] BHP Billiton is a global leader in the resources industry and was formed from a merger between BHP and Billiton in 2001.

[52] The magnetic inverse problem or "inversion" addresses the description of a magnetic source (such as the current distribution within a body) from measurements of its external magnetic field at various locations. The three-dimensional inverse problem is ill-posed, i.e., has no unique solution, and thus requires additional boundary conditions to be known.

the less-well conducting surroundings. This is manifest in the later stages of the decay curve where the SNR is poorest so high-sensitivity magnetometers are required.

Magnetic Tensor Gradiometry is a "passive" method that measures the magnetic field tensor. In orthogonal x, y, z coordinates, the tensor components characterize the magnetic field rate of change in one coordinate direction, say x, determined in the three different (x, y, z) directions. The combination of all of these components provides important information about the target characteristics and location that cannot be obtained from total or vector field measurements. Whereas a magnetic field vector has only three components, the gradient requires nine separate quantities. These nine elements form a square matrix representing the magnetic field in a so-called "tensor form". Of these, five are unique. Gradient measurements are relatively insensitive to sensor orientation. This is because gradients arise largely from anomalous (local) sources while the background gradient arising from the Earth's core is nearly negligible. This contrasts with the magnetic field vector, which is dominated by the background field. Gradient measurements are, therefore, most appropriate for airborne applications. Another advantage is they obviate the need for base stations and corrections for diurnal Earth's field variations. They also greatly reduce the need for regional corrections, which are required in TMI surveys because of deeper crustal fields that are not of exploration interest, or the normal (quasi-) latitudinal intensity variation of the global field.

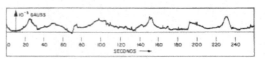

Fig. 11. Recording of fluctuations in vertical component of earth's field.

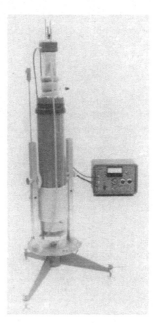

FIGURE 5.16: Earth's microfluctuations (top graph) measured using a SQUID magnetometer setup shown; taken from the original publication[53], with permission.

5.4.3 Early SQUIDs in Geophysical Prospecting

Soon after the invention of the SQUID, in 1967, Forgacs and Warnick from the Ford Motor Company measured microfluctuations of the Earth's magnetic field[53] with the Silver-Zimmerman SQUID using the setup shown in Figure 5.16. This was the first-ever geomagnetic measurement using a SQUID. They used a single SQUID housed in a cryostat cooled by liquid helium contained within a nitrogen-cooled jacket; the cryogen hold-time was over 24 hours. The assembly could be rotated and tilted for the calibration required to determine the field components. During the survey, the probe assembly was protected from wind gusts by a tent to reduce vibrations—a method still commonly used. The magnetometer electronics and recorder were separated by approximately 32 m of cable from the probe to minimize noise from the drive electronics. This first survey was followed

---

[53] R.L. Forgacs and A. Warnick, *IEEE Trans. Instrum. Meas.* IM-15 (1966) 113

by the unsuccessful attempt by Jim Zimmerman to measure changes in the Earth's magnetic field due to perturbation of ionospheric current (equatorial electrojet) during a solar eclipse. His latter attempt (1978) to measure currents in the Alaskan oil pipeline induced by the polar electrojet was also unsuccessful.

The first commercial geophysical vector magnetometer for MT, with three SQUIDs measuring B_x, B_y, B_z, was a few years later introduced by the SHE Corporation (see Section 5.6). Of all the early geophysical uses of SQUIDs, the most enduring impact on geophysical prospecting was John Clarke's (Figure 5.17) research on MT conducted with Goubau and Gamble of the Earth Science Division, Lawrence Berkeley Laboratory, Berkeley. In 1975, their aim was to use a SQUID magnetometer with a sensitivity of $0.01 \text{ pTHz}^{-1/2}$ to reduce the bias errors that were plaguing MT measurements[54]. After several field experiments using the conventional single station scheme to measure the two orthogonal horizontal components of the electric and magnetic fields, they found that the significantly improved instrument noise did not improve the MT data quality. Rather, the noise was inherent in the measurement technique. Therefore, in 1979, Clarke, Gamble and Goubau introduced a second remote magnetometer to "lock-in detect" the naturally occurring plane wave magnetic and electric signals at the MT site. This remote reference technique substantially reduced the bias error of the MT measurements and provided reliable confidence limits on them. The technique makes surveying possible in regions of high cultural magnetic and electric noise. It is now standard practice in MT and magnetic gradiometry, and it is usually performed with nonsuperconducting systems

FIGURE 5.17:   John Clarke transfers liquid helium into a fiberglass cryostat containing three orthogonal cylindrical dc SQUIDs that form a three-axis LTS magnetometer. The cryostat is below ground to eliminate wind-induced motion. Circa 1982 (courtesy John Clarke).

such as flux gates and vapor magnetometers. It is an excellent example of Harold Weinstock's oft-repeated axiom: "Never use a SQUID when a similar, less expensive, technology will do the job. Use a SQUID only when nothing else (less expensive) will satisfy your need."

Momentum in the use of SQUIDs for a range of geophysical applications led Harold Weinstock and Bill Overton to organize a workshop "SQUID Applications to Geophysics" held at Los Alamos Scientific Laboratory, New Mexico in June 1980. Every paper presented at that workshop was a benchmark in innovative geophysical measurement. However, soon after, other measuring instruments, such as vapor magnetometers, dominated the field. This was because they were cheaper and did not require the management of cryogenics and LHe in remote areas.

Also, SQUID electronics at that time did not have sufficiently high slew rates necessary to track

---

[54]   J. Clarke et al., *Geophys.Prospect.* 31 (1983) 149

the SQUID output in unshielded operation. SQUID use in geophysics was practically abandoned until the discovery of high-T_c materials and demonstration of practical high-T_c junctions and SQUIDs.

5.4.4 Transient ElectroMagnetics Using High-T_c SQUIDs

I now return to our work with high-T_c SQUIDs. By changing the research direction to mineral exploration, we could capitalize on the CSIRO's TEM system (see 5.4.2), SIROTEM, that had been developed by Ken McCracken in a sister division. This system used induction coils as the magnetic receiver. In the 1980s, the TEM method became popular and has been credited with the discovery and delineation of commercially viable ore bodies. Consequently, many TEM transceiver systems from a range of suppliers appeared on the market.

The TEM method has two major advantages: (1) excellent transmitter signal rejection, because the measurements start after the transmitter has pulsed, and (2) a broadband frequency/time spectrum due to the pulses having a spread Fourier transform. The latter provides depth discrimination because lower frequencies penetrate deeper into a conducting body than higher ones and therefore appear later in the decay of the secondary signal.

The disadvantages of TEM are that the measurement SNR decays rapidly in time with the signal (at a constant noise level), so the most interesting region from the point of view of resolving conductivity anomalies and looking deeper is compromised by poor SNR. Induction coil TEM receivers have additional limitations. First, the time derivative $\partial B(t)/\partial t$ decays faster than B, so the measurement may not detect the high quality ore bodies which have long decay times. Second, the SNR of an induction coil is proportional to frequency, so at low frequencies, i.e., at the longest decay times, the sensor performance is poor. Not only does the signal decay with time but the detector's sensitivity does as well, compounding the SNR problem. Another disadvantage of measuring $\partial B(t)/\partial t$ and integrating is that the initial integration point is unknown and different initial boundary conditions can lead to nonunique solutions to the location of the ore.

SQUIDs and fluxgate magnetometers offer a direct measurement of the magnetic field that removes all of these disadvantages. Optically pumped total field magnetometers are also sensitive enough, but their bandwidth is too narrow for active systems. For the deepest conductivity anomalies, SQUIDs are the only suitable sensors, although, in the early 2000s, there was a largely unsuccessful effort to promote fluxgates as the preferred receiver for TEM.

In 1991, our task became to develop a SQUID receiver that was inexpensive to fabricate, had low noise and could operate without a magnetic shield. Our original high-T_c SQUIDs fabricated in 1989 used Josephson junctions that were dependent on the natural formation of grain boundaries and constrictions in the SQUID loop. However, by that time epitaxial films on STO crystal substrates and controlled bicrystal grain-boundary junctions had been already reported (see Section 5.3). My colleague, Nick Savvides, soon developed epitaxial YBCO films on MgO substrates, chosen because they are significantly cheaper than STO, but still offer a reasonable lattice match. As bicrystal substrates were expensive and difficult to source, I was inspired by the original step-edge junction paper[55] to develop step-edge junctions on MgO substrates. I initially fabricated junctions which had multiple grain boundaries at the top and bottom of the step edge, which contributed to device noise. I wondered if it were possible to have a step with only one junction and a gentle return path allowing a single junction to be placed anywhere on the substrate. With the assistance of an undergraduate student, Simon Lam, I varied the angle of the ion beam to the substrate coated with a photoresist mask to align with the crystal lattice and successfully created a single grain boundary junction on MgO[40]. Our substrate selection was fortuitous as YBCO growth on MgO over a step-edge is different than that on other substrate materials often used (such as STO). We had better control of the junction formation leading to a single engineered grain boundary. Indeed, our resulting junctions did

---

[55] K. P. Daly et al., *Appl. Phys. Lett*, 64 (1990) 228

not contain any other grain boundaries cutting the device and their critical current was relatively stable in changing magnetic fields. Our best junctions, which operated unshielded in motion, required YBCO films to have a specific morphology to minimize flux movement, and a high film critical current to achieve this. These junctions are less than 2 μm wide. Over twenty years, our research group optimized our junction performance and increased both their yield and reproducibility.

The next step was to place our junction into a SQUID. We made a significant decision in 1991 to use rf SQUIDs for our TEM receiver. We thought that there were several advantages: a single junction (only one to get right), inductive coupling to the device and no need for wire bonding which can be fragile. Furthermore, the rf drive created automatic bias reversal which reduced the low-frequency device noise. It was necessary to fabricate a SQUID that could operate in varying magnetic field and remain in lock during the TEM pulse current time. We set a specification of 1 pTHz$^{-1/2}$ noise at 30 Hz and designed a 100×100 μm^2 hole, 2 mm square simple washer SQUID with a 125 pH inductance. In 1991, I also gave birth to my second child and fabricated devices late at night returning to the laboratory after the babies had been put to bed to make sure there were devices ready for testing the next day. As we used three SQUIDs to measure B_x, B_y and B_z, we needed three similar rf SQUIDs to assemble a sensor head capable of measuring the three orthogonal components of the magnetic field.

Rex Binks, our talented technical officer, recommended that we use glass cryostats to reduce the boil-off times. These cryostats were initially fabricated by the CSIRO in-house glass blower. The SQUIDs were inductively coupled to a small pancake coil of rf tuned circuit located adjacent to the washer and operated at about 180 MHz. Our mark one version used the traditional dip probe. We used Quantum Design VHF SQUID Amplifier Model 2000 and SQUID Control Unit Model 2010. David Dart developed the custom electronics to provide a synchronous reset unit allowing us to reset the Quantum Design amplifier synchronously with the pulsed output of the SIROTEM unit. Furthermore, David modified the electronics to achieve a higher linearity. We also introduced a novel feature of a separate large feedback coil to provide a uniform magnetic field environment for the SQUID to eliminate noise and nonlinearities arising from flux movement within the superconducting film. The system had a magnetic field sensitivity of 1-2 pTHz$^{-1/2}$ with a bandwidth of 100 kHz and slew rates of 300 mTs^{-1}. We undertook several experiments using a variety of rf shielding materials to arrive at a shield design which gave an acceptable self response[56] in the TEM survey.

FIGURE 5.18: The first field trial in Cooper Pedy.

We performed our first outdoor trial on the grounds of CSIRO in mid-1992. This was within sight of some TV broadcast towers, and proved to be an important test site as it set the standard for rf screening that was sufficient to keep out the rf while not creating self-field effects that reduced the

---

[56] Self response is the interference between SQUIDs and adjacent electronic components or metal shielding used to screen rf radiation. This response can be mitigated by appropriate design of shielding and increased distances between interacting components to provide satisfactory isolation. Whenever the rf screening is placed around a cryostat or located in the vicinity of SQUID sensors, it may increase noise due to eddy currents induced in the rf screen by the external changing fields.

system sensitivity. Over time, we optimized the rf shielding using homemade cables, shielded glass dewar and carefully anchored connectors. The first system affectionately called "The Rocket" was trialed at Cooper Pedy (a well-known opal mining town in South Australia) in December 1992. Figure 5.18 shows the historical photo taken during this trial. The trial led to three ground-based field trials at Murray Bridge, Cannington and Mt Isa, lasting until August 1995. Our most significant trial was that at Cannington where a large silver lode was known to be present. BHP was gearing up to undertake bore hole digging, the next stage for determination of the extent and quality of the ore deposit. This SQUID trial collected data that clearly indicated a double lode rather than a single one that had been suggested by the induction coil receiver measurements. Mike Asten, the geophysicist leading our trial for BHP, thus recommended a change in the location of the first bore hole. This advice led to the first core passing through the center of the $1B silver reserve. The BHP Chief Scientist at the time acknowledged informally that the Cannington Silver Mine came online 18 months earlier than initially planned because of our SQUID TEM survey.

After this success, BHP was keen to trial airborne SQUID magnetometers. When airborne, SQUIDs are under constant motion while data is recorded, unlike in the stationary ground-based system. The ground-based systems can be degaussed between stations to expel the trapped flux and reset the system. Moving an operating magnetometer in the Earth's field is the most challenging of tasks. It is also necessary to have a light and compact cryostat and electronics that is suitable to fit into a "bird" that is towed behind the aircraft. Rex Binks used a KMART 1 liter thermos flask, modified with rf screening and a metal cap. Rex invented also a housing that attached a module containing the three SQUIDs to the bottom of the dewar with twisted pair cables brought through the metal cap. Removing the dip probe increased the liquid nitrogen hold time to 13 hours and we achieved a very compact system. The liquid sloshing and the gas bubble noise were removed by filling the cryostat with cotton wool before charging with the cryogen. BHP developed a novel mechanical suspension system that successfully removed the motion noise above 20 Hz. Chris Lewis and Wayne Murray developed in-house rf electronics that operated at 300 MHz. The tank circuit had a low Q at the cost of sensitivity to provide a robust system that did not require retuning during the flight. Rex also developed a nonmetal, mechanical pressure valve to maintain the pressure above the liquid nitrogen at one atmosphere thus preventing changes in the nitrogen boiling point with the airborne altitude. The system operation was disrupted by flux jumps as at that stage we had not yet introduced a chip-based heater to allow resetting the SQUID when the flight reversed direction (this caused a field change of up to 100 μT). All these somewhat "small" innovations were essential to achieving a three-axis SQUID magnetometer that was effectively airborne. Figure 5.19 shows the "bird" (left photo) and the miniaturized cryostat with electronics (right photo). We undertook five airborne trials between September 1994 and November 1997. This work was confidential and we were unable to publish until 1999[57].

Over the next 14 years, we worked with several different mining and mineral exploration companies to further develop the CSIRO SQUID TEM receiver which we called LANDTEM®. After working with Falconbridge P/L, a Canadian mining company, on the recommendation of Ken Witherly, an ex-BHP Chief Geophysicist, we developed an auto-tuned robust LANDTEM system that is now licensed to a manufacturing company, Outer Rim Developments. Such systems have been since manufactured, are commercially available and extensively used in many countries. Their use assisted in the discovery and delineation of many billions of dollars worth of mines[58]. Figure 5.20 shows a photo of the commercial system at a test location. Over the years, and in addition to my colleagues mentioned above, leading technical contributions to this system were made by the late Graeme Sloggett, Jia Du and Emma Mitchell. Keith Leslie was instrumental in leading the system integration and commercial adoption.

[57] C.P. Foley et al., *IEEE Trans. Appl. Supercond.*, 9 (1999) 3786

[58] C.P. Foley, K.E. Leslie and R.A. Binks, *First Break*, 25 (2007) 73

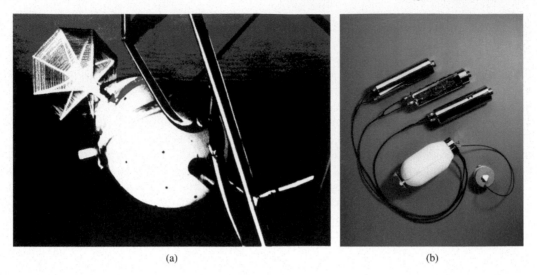

(a)   (b)

FIGURE 5.19: The CSIRO high-T_c SQUID system in flight as it is released from a fixed wing aircraft on a tow rope (left photo). Right photo shows the light-weight, three axis, SQUID system with the miniature cryostat, electronics and pressure regulator.

The high-T_c SQUID TEM for ground-based applications has been so successful that there are currently Japanese, German, US and Australian systems on the market. Furthermore, the desire to improve sensitivity led to the development of two low-T_c systems exclusively used by two different major mining companies, Anglo American and BHP Billiton. It is now common for a mining company to send out a press release announcing a SQUID-based exploration survey with a view to increasing their share price. However the airborne TEM system has not been commercially successful so far, although there are renewed efforts underway with new systems being developed by CSIRO and IPHT (See the box "High Competition").

FIGURE 5.20: The CSIRO commercial LANDTEM system.

5.4.5 Tensor Gradiometry

It was a natural progression to develop a tensor gradiometer to survey localized disturbances in the dc Earth's magnetic field resulting from the presence of concentrations of ferromagnetic material. The main applications of magnetic tensor gradiometry include military sensing of submarines, unexploded ordnance, non-invasive archaeology and geomagnetic surveying.

However, I will consider only geophysical applications here. During the early stages of mineral exploration, it is mandatory to carry out passive magnetic surveys. The acquisition and processing of magnetic surveys using TMI data from airborne cesium vapor magnetometers have improved remarkably over the past two decades. It is now possible to create high quality magnetic images, similar to topographical maps, from survey data. Nevertheless, due to timeconstraints on surveys

High Competition

Competition was so "hot" for the use of HTS SQUIDs in airborne TEM that, in 1997, a major mining multinational secretly funded two competing research groups to develop an airborne SQUID magnetometer which was tested in Timmins in Canada over a two-week period. In order to prevent the two groups knowing about each other's involvement, the company did not have a researcher from either research group to assist with the trials. The trials were hampered by a pin-out problem, flux jumps and suspension system failures. However five days of flying were achieved and data were collected that showed a factor-of-two improvement in the late time channels in SNR compared to the coil systems flown at the same time. This trial was undertaken during a period of downturn in the minerals industry due to low commodities prices, and a ten-times improvement in SNR was required from this trial if further development were to continue. Thus the support for further development was not forthcoming. At that time, the exploration industry underwent a major restructuring and it became clear that the price point of $27 per kilometer of survey must be attained to be economically viable. More recently, the IPHT Jena group has developed an airborne AeroTEM III (low-T_c) which has been trialed slung under a helicopter in 2008 with effective surveying results still elusive.

and the lack of appropriate technologies, much magnetic information that could assist in mineral discovery remained inaccessible.

Our colleagues at the CSIRO Division of Exploration and Mining, Phil Schmidt and Dave Clark, published a paper that set the specification for gradiometry for mineral exploration and outlined the opportunities a magnetic tensor gradiometer survey would provide[59]. The sensitivity requirement of a tensor gradiometer is 0.01 nTm^{-1} gradient sensitivity for detection of a vertical contact between two paramagnetic rock units such as a mafic and a felsic gneiss, which contain no magnetite or pyrrhotite with a susceptibility contrast of about $6 \cdot 10^{-4}$ SI units at 100 meter range. The scale of the motion noise problem in this application can be comprehended when one considers that a SQUID magnetometer with 5.6 fTHz$^{-1/2}$ noise floor, edgeways on to a 56 μT Earth field needs only about a 10^{-10} degree rotation in one second before the motion is seen on the output. Luckily, a gradiometer with the same sensor noise floor (at the two points of measurement along the baseline) that is balanced to the 1 part in 10^6 level has a residual response of 5.6 nTHz$^{-1/2}$ and an allowable angular motion of about 10^{-4} degrees, still a formidable requirement.

For geophysical use, IPHT Jena has been most successful in airborne tensor gradiometry of minerals for Anglo American Company (see the box "Airborne Full Tensor Magnetic Gradiometry" below). The best results were obtained when using low-T_c gradiometer systems slung under a helicopter. Fixed-wing aircraft trials have also been performed[60]. The helicopter trials used a towed platform which carried the tensor gradiometer constructed of planar gradiometers that achieved a gradient resolution after software balancing of better than 10 pTm^{-1}Hz^{-1} for frequencies above 0.1 Hz. The reports on the fixed-wing gradiometer trial have not been formally published but some conference presentations have revealed a full tensor gradient measurement with software compensation using tri-axial (triple) SQUID magnetometers providing data for the gradiometer compensation.

5.4.6 Laboratory Systems

The Superconducting Rock Magnetometer, a laboratory-based instrument for paleomagnetics, is a very successful product of long-term development of SQUIDs for geophysics. This specialized

---

[59] P. Schmit and D. Clark, *Preview,* April, 26, 2000

[60] H-G. Meyer, et al., *Phys. Stat. Sol.,* (c), 5 (2005) 1504

low-T_c instrument, manufactured by 2G Enterprises, has sold widely over the years. 2G Enterprises was founded in October 1981 as a result of a marketing agreement between the late Bill Goree of William S. Goree, Inc. and Bill Goodman of Applied Physics Systems, Inc. They focused on high performance superconducting systems for paleomagnetics. 2G received the first two contracts for such systems in 1981–1982. Figure 5.21 shows the two founders with their instrument. 2G has since installed and upgraded over 100 LHe-cooled systems and is currently offering their third generation "liquid helium-free" systems.

The magnetometer has a horizontal room-temperature access and is aimed specifically at determining the magnetic moment—along three axes—of rock core samples up to 0.12 m in diameter and 1.5 m in length. With the aid of cryo-cooled thermal radiation shields, the system can run for a remarkable 1000 days between liquid helium refills. Thus, the cryogenics is virtually invisible to the user. This instrument has become the standard rock magnetometer of the geophysics community. One significant application has been

FIGURE 5.21: Bill Goree (left) and Bill Goodman (right) with a 2G Rock Magnetometer (courtesy of 2G Enterprises).

to measure the magnetic moment of sedimentary cores taken from the ocean basins to study the polarity reversal of the Earth's field over geologic time.

5.4.7 Final Comments

SQUID systems operating in real environments to measure the magnetic fields and their gradients originating from remote mineral sources remain technically challenging, particularly if the sensors are in motion. In applications where the SQUID environment can be controlled, such as rock magnetometers, SQUIDs are now well established at the high-sensitivity end of this fairly small market.

For field applications, where the sensor's magnetic environment cannot be controlled, industry acceptance has grown significantly since 2005 for SQUID ground-based TEM, as witnessed by a specialist workshop at the Society for Exploration Geophysics conference in New Orleans in 2006 and the fact that five companies are offering HTS SQUID TEM receivers for sale. Indeed, mining companies are announcing their plans for SQUID surveys with the intention to boost their share price. The growing acceptance has resulted from strong user requirements for higher sensitivity and better ore body discrimination than current conventional systems can offer.

Several SQUID research systems for geophysical survey applications other than TEM have been successfully tested in operation. Although these tests demonstrated that SQUIDs can be used in such applications, the only established first-choice technology is thus far the ground-based TEM. The success of this relatively small application, which has big indirect economic impact, has been significant and I am proud to have been involved.

The IPHT team with their LTS magnetic tensor gradiometer exclusively used by Anglo American has shown that even complex technology and cooling to liquid helium temperatures can be adopted by industry when the system clearly demonstrates economic and technical benefits. The market for geophysical SQUID systems can be expected to remain fairly limited; their economic impact is and will remain indirect—via the value of discovered ores. Market growth could occur if SQUID military

Airborne Full Tensor Magnetic Gradiometry
(by Ronny Stolz, IPHT-Jena)

The development of our full tensor magnetic gradient systems (FTMG) started in 1997 with a request from the German Naval Research Center for a system to localize magnetic dipoles. Our team at the Institute for Physical High Technology (IPHT) in Jena, Germany included Slava Zakosarenko, Marco Schulz, Andreas Chwala, Hans-Georg Meyer and me. For sensors we used our planar-type, long-baseline first-order low-T_c gradiometers with noise as low as 15 $fTm^{-1}Hz^{-1/2}$. Because of their good intrinsic balance, i.e., the suppression of the homogeneous Earth's magnetic field by a factor higher than 10,000, they could be freely rotated in the Earth's magnetic field without compromising the dynamic range of the SQUID electronics. Patience was necessary in those early days because only one sensor could be patterned on a 3-inch-long substrate; it took more than five weeks to produce such a device.

During his visit in 2001, Eddie Koestlin, chief geoscientist at Anglo American at that time, inspired us to develop a FTMG system with similar sensors for geophysical exploration of magnetic ore bodies like iron, kimberlites and so on. We demonstrated to Eddie the performance of our dual channel system against that of a total field gradiometer, which was already tested in 1999 in South Africa, and in motion.

We performed the first airborne tests (a tow bird below a helicopter and a stinger mounted on an aircraft) in 2002, together with Anglo American and Fugro Airborne systems. The big breakthrough occurred in the production-like tests during 2006, when the FTMG system was almost three months in operation and flew more than 22,000 line kilometers in different areas in Southern Africa without faults. Even a rough takeoff with a tumbling-down bird evaporating quite a bit of liquid helium,

FIGURE 5.22: Evolution of field tests of the IPHT-magnetic tensor gradiometer, 2001 to 2009. 2001 — Ronny carries the cryostat with sensors for an early ground-based test; 2002 — cryostat mounted on airplane stinger; 2006 — the system pod to be towed by helicopter; 2009 — the system "bird" towed by helicopter.

so precious in the southern hemisphere, couldn't do any harm to the system. Figure 5.22 shows pictorially the evolution of FTMG field tests.

The development is still ongoing. We aim for a noise floor an order of magnitude better than that of conventional airborne gradiometers—via better electronic components and software tools for post processing, as well as decreased motion. Nevertheless, our present system is already a very important tool for magnetic exploration as it measures a new physical quantity in the exploration process–the magnetic gradient tensor—allowing us to invert better the magnetic potential to reliably pinpoint magnetic sources which might become the mines for tomorrow. Many parts of our developed FTMG technology are used in other systems such as SQUID systems for ground-based TEM, the archaeological scanner, or the THz safety scanning camera.

applications, such as magnetic anomaly detection, for example of unexploded ordnance, would come into broad use. The system cost could then decrease through the economies of scale.

5.5   Application to Nondestructive Evaluation of Materials and Structures

A. I. Braginski

5.5.1   Past and Present

Nondestructive evaluation (NDE) of materials and structures or objects in industry, construction, transportation and other domains is a vast field of great economic and societal relevance. Public safety critically depends on it. The purpose is to detect cracks, material fatigue, impurity inclusions and other defects not accessible to visual inspection. Here, I address only the opportunities offered to NDE by SQUIDs used as sensitive magnetic flux and field detectors.

In the early 1980s, Harold Weinstock and Marty Nisenoff in the U.S., and independently Gordon Donaldson and collaborators in Scotland, U.K., were the first to demonstrate the usefulness of SQUIDs for NDE[61,62]. Harold's personal reminiscences are reproduced here (see box on p. 343). Obviously, they both used low-T_c SQUIDs then available.

Many NDE applications require portability, use in motion and/or in the presence of relatively strong magnetic fields. Furthermore, testing must be economical, i.e., inexpensive. All these requirements are significant handicaps for any SQUID sensor. Nevertheless, with the advent of high-T_c SQUIDs, many of us believed that once they reached a certain level of performance, reliability and cost they would become ideal detectors for difficult NDE tasks.

In early 1989, I joined the newly created Jülich group led by the late Christoph Heiden[63] and became responsible for all high-T_c R&D programs. Until 1991, the main part of our efforts was devoted to establishing a base for technology of low- and high-T_c materials and devices for superconducting electronics. The low-T_c SQUID work was pursued by a young and talented scientist, Michael Mück, who joined Christoph when he temporarily moved from Giessen to Jülich, while bulk and thin film high-T_c rf SQUIDs were worked on by a doctoral student from China, Yi Zhang. Yi soon graduated, but remained in the group and successfully continued his work. By early 1992, Yi had produced the (then) most sensitive epitaxial thin film rf SQUIDs in the world, based on step-edge junctions, and through Christoph's initiatives we became involved in evaluating their use for NDE. Our initial work was on highway bridge structures built from steel-reinforced concrete, and since 1993 we also worked on NDE of airplanes. Both initiatives were pursued in joint projects with, respectively, a bridge supervision authority, a major air carrier, a manufacturer of conventional eddy-current devices for airplane NDE and also a manufacturer of airplane turbine (jet) engines. After Christoph returned to Giessen in 1993, we continued to collaborate with his group there, especially in airplane testing.

Our projects were technically very successful; one was implemented (under our license) in aircraft engine manufacturing. Nevertheless, a decade later, it became clear that high-T_c SQUIDs had not been accepted in NDE of airplanes and bridges, mainly because of cost, necessity of cooling, re-

---

[61]   H. Weinstock and M. Nisenoff, in *SQUID '85: Proc. 3rd Intl. Conf. on SQUIDs*, deGruyter, Berlin, (1985) 853

[62]   R.J.P. Bain et al., ibid., (1985) 841

[63]   In 1988, the late Prof. C. Heiden (1935-2000) from the University of Giessen, Germany, was invited by the Research Center Jülich, Germany, then known under the acronym KFA, to organize and head a new Institute devoted to superconductivity and its electronic applications.

First NDE Ideas and Tests (by H. Weinstock, AFOSR, US)

As a faculty member at IIT in Chicago, I went on sabbatical in 1972–1973 to the Naval Research Lab (NRL) specifically to use a second-derivative SQUID gradiometer to detect magnetic signatures of cryocoolers being built under contract to the Office of Naval Research. When I discovered these cryocoolers were unlikely to arrive that year, 2 potential applications came to mind.

The first was to see if this sensitive gradiometer could be used to find holes in a buried pipeline by passing a current through a pipe and seeking magnetic field anomalies due to nonuniform current distribution. Since that was not something I could do directly, I thought it possible to simulate it. Marty Nisenoff, my host at NRL, produced a 5-cm diameter pipe about 70 cm long. With no elegance whatsoever, I drilled and sawed a couple of ragged holes several centimeters apart. When the pipe was found to be ferromagnetic, I decided to use a low frequency AC signal and an oscilloscope to monitor the height of the AC signal in the output of the SQUID's room-temperature electronics.

This arrangement allowed me to move the pipe beneath the SQUID dewar to simulate hunting for a leak and locating the pipe. It soon became clear that the SQUID gradiometer could detect the presence of the pipe when it was directly under the dewar, and when one of the holes was directly under it. It also was possible to rotate the pipe around its own axis to detect an anomalous signal when one of the holes was on the bottom while the SQUID dewar was above the pipe. It was possible to see this signal grow as the pipe was rotated about its own axis. Another remarkable thing about these crude measurements was that we obtained these definitive results despite the fact that the signal came from a distance several times greater than the baseline of the second derivative magnetic gradiometer.

The second attempt at nondestructive evaluation evolved from an idea I had that if we placed the tail of the dewar near a piece of steel clamped in a tensile testing machine, we might see changes in the ambient magnetic field near the center of the steel strip being tested. Despite some false starts, we did see a strong signal from the steel specimen, so strong that we had to switch to the lowest level of amplification. As stress increased, the magnetic field changed as well, with a linear relationship between the applied stress and the measured magnetic field. At a stress about 2/3 of the elastic limit, the magnetic field started to decrease. Beyond the elastic limit, the slope of the stress-strain curve flattened, indicative of plastic flow.

I learned later that at a stress about 2/3 of the elastic limit, one reaches the "endurance limit", indicative of microscopic particle movement that can be thought of as a precursor to the elastic limit. We also saw magnetic field oscillations as a function of stress in the plastic regime, and discovered these were analogous to stress-strain oscillations known as Lüders bands. Since the signals obtained were so large, these last observations were verified using a fluxgate magnetometer by others at NRL. Thus, while a SQUID gradiometer was not needed to see this effect, it might never have been discovered using conventional magnetometry.

liability in the field and relatively complicated handling requirements. Also comprehensive studies by others on SQUID NDE in application to power plant component testing concluded that conventional sensors suffice[64]. Today, either high- or low-T_c devices are only occasionally used to solve

[64] Studies at Siemens and KWU, performed in collaboration with IPHT-Jena, Germany, 1997-2003. G. Daalmans — private communication (2010)

especially tough NDE problems.

In the following I will reminisce about our past efforts and give examples of NDE applications which apparently survived the test of time, are currently in use or still hold promise for the future.

5.5.2 NDE Testing Methods Using SQUIDs

Testing using magnetic field sensors usually involves scanning the sensor over the tested object, at some distance from it, and is based on one or another of the following techniques:

- Detection of (weak) magnetic moments, e.g., of ferromagnetic impurities dispersed in a non-magnetic matrix, or of magnetization anomalies in nominally homogeneous ferromagnetic objects.

- Magnetic flux leakage technique (MFL), where the tested ferromagnetic object is magnetized, usually by a permanent magnet, and the induced stray field is then measured above the surface. This method was used in the pioneering work of Donaldson's group and later by us for testing of bridge structures.

- Eddy current technique of testing metallic nonmagnetic objects, semiconductors and laminated carbon-fiber-reinforced plastic (CFRP). Here, eddy currents are induced in the object below its surface and local anomalies in the distribution of the resulting magnetic field can be measured above the surface. Very low excitation frequency is required to avoid limitation by skin depth and to probe deep defects. Such excitation is practical when using SQUID detectors because the SQUID senses the magnetic field vector $\vec{B}$ rather than its time derivative $\partial\vec{B}/\partial t$ as is the case of induction (sensing) coils, making the SQUID's response frequency independent. The eddy-current technique with induction coil detectors is typically used for testing airplane parts and fuselage. Alternative magnetic response excitation methods by light, heat, etc., have been also used in other applications. Pulsed eddy current excitation permits conductivity tomography and flaw depth determination (in analogy to the TEM method of geomagnetic exploration).

In all such techniques the fault causes a local magnetic anomaly or deviation from the expected field distribution, which is detectable. Although theoretical calculations of such field distributions have been performed, practical testing usually involves empirical comparison with a fault-free reference sample. What SQUIDs bring to the game in comparison with conventional detectors (e.g., fluxgate or magnetoresistive magnetometers and induction coils) is the much higher sensitivity and signal-to-noise ratio when measuring field changes, the wider dynamic range and ability to operate in the presence of static fields. As noted above, in eddy-current testing much lower excitation frequency can be used, thus expanding the practical depth range. However, the penalty for SQUID use is the more sophisticated equipment, more difficult handling and maintenance, and, above all, the need for cryogenics. Consequently, the much higher investment and operational cost of the SQUID system can be justified only when no other sensor can do the job, as might be the case when searching for faults or impurities located very deep beneath the surface of the object to be tested.

5.5.3 Bridge Testing

In steel-reinforced concrete structures the pre-stressed steel tendons are located deep below the surface of a concrete bridge or roof. When water penetrates the tendon beam duct, tendon rupture due to corrosion cracks may have catastrophic consequences.

In May 1991, at Christoph Heiden's initiative, Yi Zhang's high-T_c rf SQUID, not yet at its ultimate sensitivity, was demonstrated at the German industrial fair "Sensor". At that show, a representative of an industrial construction company became interested in using our SQUID for locating

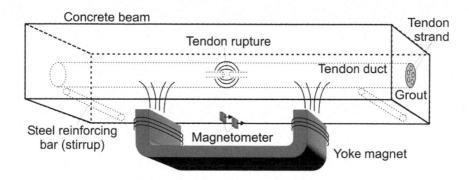

FIGURE 5.23: Principle of magnetic flux leakage detection of ruptures in tendon strands of pre-stressed concrete-reinforcing beams (from The SQUID Handbook, Figure 13.3, with permission).

reinforcing steel beams in thick concrete structures, which eventually resulted in our joint work with the bridge inspection authority of the German Land Baden-Württemberg (FMPA). Using the MFL method, a probe carriage containing a permanent yoke magnet and SQUID sensors was moved on rails over the surface to magnetize the local area and record the the magnetic stray field versus position. The principle of such detection is shown in Figure 5.23. In this application we eventually used an array of high-T_c dc SQUIDs with ramp junctions, which were much less sensitive to static magnetic fields than step-edge junctions in our rf SQUIDs[65].

FIGURE 5.24: Portable rail support frame with the magnetizing and sensing carriage during a bridge test (1999).

The bridge projects, started by Herbert Bousack and later led by H. Jochen Krause, were technically very successful and resulted also in significant progress in NDE signal analysis: new methods were developed for the separation of signals of the nonpre-stressed steel reinforcements from the signals of ruptures in pre-stressed steel tendons[66]. The culminating success consisted of finding dangerous flaws in two freeway bridges while automobile traffic continued (Figure 5.24).

The test results, confirmed by opening the bridge deck, resulted in decommissioning, demolition and replacement of these bridges. However, our success was somewhat self-defeating in the sense that the much improved signal analysis eliminated the absolute need for a sensor having a high signal-to-noise ratio in this application; SQUIDs were soon replaced by the conventional magnetometers used today. The sensitivity offered by SQUIDs would be necessary only if mandatory periodical bridge inspections were instituted to detect minute changes in the tendons' magnetic signature.

---

[65] These junctions and dc SQUIDs were developed by Ulrich Poppe and Michael Faley at another Jülich Institute, and are still available commercially.

[66] G. Sawade et al., in J.H. Edwards, B. Gasper, P. Flewitt, B. Tomkins, P. Stanley and A. McLarty (eds.) *Proc. 4th Conf. on Engineering Structural Integrity Assessment*, Cambridge, U.K., (1998) 353

5.5.4   Airplane Testing

In eddy-current testing, the very low excitation frequency can be an essential advantage when using a SQUID to test for flaws deep beneath the surface. This was the motivation for our involvement in eddy-current airplane testing also initiated by Christoph Heiden. The project included testing of both the fuselage and airplane wheels, which we were developing together with the Lufthansa air carrier and Rohmann GmbH, a German company manufacturing conventional eddy-current test devices ("Elotest"), which use induction coil sensors. Soon thereafter Christoph's group active at Giessen University joined the project. The work involved all aspects of testing with SQUIDs, starting with the development of appropriate excitation coils, which assured differential self-referencing excitation[67], and eventually including signal analysis and the implementation of a commercial Joule-Thomson cryocooler suitable for cooling the SQUID to ≈ 77 K. In all airplane tests we used our rf SQUIDs capable of operating in most unshielded environments due to the elaborate rf shielding perfected over the years (one can find more on rf shielding in Section 5.4).

After our Giessen colleagues first demonstrated the detection of deep-lying cracks in airplane wheels using Yi Zhang's SQUID[68], the joint work concentrated on developing an automated prototype system for testing such wheels in airport facilities with a high level of electromagnetic noise. System tests in the wheel testing facility at the Lufthansa base of the Frankfurt international airport were successfully performed: even the smallest specified flaws of 10% of the wheel rim's cross-section were detected among the much stronger signals of the wheel's keys, an achievement which was then impossible with sensing coils. Nevertheless, such systems were never implemented in practice, mainly because of concerns over high-T_c SQUID sensor's limited reliability and high cost.

The only airplane testing using our high-T_c SQUIDs that found its way into practice was the aircraft turbine engine blade inspection for ferrous inclusions[69]. A three-SQUID second-order gradiometer operating without magnetic shielding was used for that purpose. However, here also the SQUIDs were eventually replaced by conventional magnetometers, in spite of a sensitivity loss by a factor of 20. Overall, our technically successful projects contributed to the further development of NDE methods, but had otherwise no lasting impact on the NDE field. However, in the following I highlight a few successful examples of SQUID NDE that are currently used in practice or expected to be of great value soon.

5.5.5   Two Examples of SQUID NDE Now in the Field

A tough NDE problem, which is solved by the use of low-T_c SQUID, is the eddy-current testing of high-purity niobium sheets and aluminum wires at the manufacturer's facility. In the box "SQUID NDE at Heraeus," you will find a succinct reminiscence of Michael Mück (University of Giessen) on how this testing became a reality at a prominent specialty metals company, W.C. Heraeus.

To Michael's reminiscence I would like to add that quality control of superconducting striated high-T_c conductors for future electric ac power applications seems also to necessitate SQUID gradiometer sensor arrays to detect minute defects in striated conductor filaments[70].

Another example of NDE in the field that requires SQUID sensitivity is the detection of magnetic contaminants in commercial products, such as cheese. After the Product Liability Law was promulgated in Japan in 1995, the Japanese food industry started implementing new methods of quality control to avoid lawsuits. As a result of such a lawsuit, a major Japanese dairy company purchased and implemented in their production line a high-T_c SQUID system capable of detecting, by the MFL

---

[67] Self referencing compares signals from two spatially separated areas excited by a differential exciter, such as the double-D coil.

[68] M. von Kreutzbruck et al., *IEEE Trans. Appl. Supercond.* 7 (1997) 3279

[69] Y. Tavrin et al., *IEEE Trans. Appl. Supercond.* 9 (1999) 3809.

[70] T. Hato et al., *Physica C*, 469 (2009) 1630

SQUID NDE at Heraeus (by M. Mück)

At Giessen University, we were approached by David Lupton of W. C. Heraeus, Hanau, who was interested in testing Russian platinum coins for ferromagnetic inclusions. The amount of such inclusions could hint to the origin of the platinum metal of which the coins were made. As only a few of these coins survive today, a chemical analysis, which might have damaged the coin, was out of the question. Instead, a simple measurement of the static field produced by the coins using our SQUID microscope revealed abundant ferromagnetic inclusions in the coin, and from this their place of origin could be traced. That was back in 1999. Details about this work were published in the journal *Platinum Metals Review*. These measurements in turn raised the interest of Friedhold Schölz of Heraeus, and Waldemar Singer of DESY, Hamburg, who were working on superconducting cavities for particle accelerators. Such cavities are made from extremely pure niobium sheets; inclusions of foreign material in such sheets, even extremely small ones, will dramatically reduce the critical current density of the cavities and thus the obtainable acceleration voltage. As forming the cavities is quite expensive, Schölz and Singer were interested in a nondestructive testing method which could reliably find even the smallest inclusions in a rather short time. We quickly determined that when using eddy-current testing, the required field sensitivity of the SQUID should be better than 50 fTHz$^{-1/2}$, a sensitivity which can reliably be obtained only from a niobium (low-T_c) SQUID.

The potential customers quickly agreed that the additional few Euros per day for the liquid helium were not a major obstacle as the major cost of testing is in the labor. Subsequently, we developed two SQUID systems in Giessen, and made two more for Heraeus, two for WSK Meß-und Datentechnik, Hanau, and one for DESY, all designed for testing flat metal sheets. Subsequently, we tested several hundred sheets and compared the results to those obtained with a conventional eddy-current system; this proved the much higher detection sensitivity of the SQUID system. For the projected (and partly funded) tera-electron-volt accelerator TESLA, several hundred thousand niobium sheets will have to be tested[a]. The usefulness of the SQUID systems triggered their application to many other testing purposes at Heraeus. They are now regularly testing products, such as noble-metal sheets or sintered sputtering targets, with the SQUID system. Recently, we developed for them a new SQUID system which is used for testing 1-mm-dia. aluminum wires which they use to make 25-micron-diameter aluminum bond wires. When drawing the aluminum wire to smaller diameters, foreign inclusions (e.g., of aluminum oxide), which are larger than a few microns, will cause the wire to break. With a specially-designed helium cryostat that allows us to pull the wire at room temperature through superconducting pick-up loops connected to a SQUID, we have been able to detect minute inclusions of aluminum oxide or silicide, which were only a few microns in size. This eddy-current test allows now for discarding faulty wires even before drawing them to smaller diameters, and saves both labor and cost.

[a] That would require a higher number of test systems.

method, minute magnetic inclusions in cheese blocks at a depth up to 15 cm, which is impossible for conventional magnetometers. The system, developed by Saburo Tanaka and his group, contains three partly shielded dc SQUID magnetometers arranged along a line transverse to the direction of motion of the cheese-carrying conveyor such that the requisite detection sensitivity is assured at any point of that line above the tested object[71]. Cheese blocks pass first through a magnetization unit

[71] T. Nagaishi et al., *IEEE Trans. Appl. Supercond.* 17 (2007) 800

FIGURE 5.25:   The cheese inspection system distributed in Japan by AFT (Advanced Food Technology Co., Ltd). Tested cheese blocks, placed on the conveyer belt on the left side, pass through the magnetizing unit and subsequently under the sensing SQUIDs. The operator's console is visible on the right side (Courtesy Saburo Tanaka).

and then move under the sensing magnetometers. The external view of this system is shown in Figure 5.25. Two such systems were installed in different plants and have been operating reliably (one has been continuously operated for five years). Another such unit is supposed to be installed soon. Occasional inspection of imported cheeses indicates that they often contain inadmissible amounts of magnetic contaminants. Unfortunately, high cost still hampers broader implementation of this method[72].

5.5.6   SQUID NDE Microscopy in Semiconductor Technology

An NDE application of high-T_c SQUIDs, which might still have a bright future, is the complex failure analysis of integrated circuits (ICs). This application was first proposed in the past decade. A Joule-Thompson cryocooled scanning SQUID microscope was then developed for magnetic analysis of current distribution in multichip packages[73] and put on the market under the name MAGMA[TM]. This near-field flux imaging system permits one to image induced currents in interconnect wiring and is mainly useful for detection of shorts and high-resistance failures in IC packages held at room temperature. The close proximity of the SQUID to the tested IC, necessary for lateral resolution, is made possible by a very thin small-diameter sapphire window. Rather few such instruments have been sold for use by semiconductor industry, but they are still commercially available from Neocera, LLC., also with two complementary sensing heads–in addition to the SQUID a GMR (giant magnetoresistance) sensor provides enhanced spatial resolution[74].

The MAGMA system was perhaps developed ahead of its time because cheaper and simpler failure analysis tools were adequate in most cases. However, new opportunities might soon arise for

---

[72] A major Japanese effort in this area is planned to start in 2010–2011; it might eventually result in much broader implementation of SQUID NDE in Japanese food industry.

[73] E. F. Fleet et al., *Rev. Sci. Instr.* 72 (2001) 3281

[74] See http://www.neocera.com/semi_metrology.htm. The SQUID head lateral resolution is in the micrometer range, the GMR head resolves 250 nm. All unpublished information supplied by Antonio Orozco and Venky Venkatesan, Neocera, LLC.

the MAGMA because of the current trend towards more levels of interconnect in chips[75] and novel packaging technologies such as system-in-package, wafer-level-packaging and through-silicon-vias (TSV)[76]. These emerging multi-die 3D solutions may eventually replace today's multichip designs.

Stacking of multiple devices vertically in many opaque layers will make today's optical analysis tools useless. In contrast, magnetic field analysis appears as a natural option, provided it could localize defects both laterally and vertically with the required resolution, and identify not only shorts, but also open circuit failures. This is the performance actively pursued, but not yet fully attained, by the MAGMA development team.

A different approach to IC complex failure analysis using SQUID microscopy has been taken by Nikawa and coworkers at NEC, Japan[77]. They use laser beam excitation of photocurrents in p-n junctions, and in a two-step localization process can apparently identify and locate both shorts and open circuit failures. In the second localization step, the tested circuit is placed in the SQUID vacuum chamber to minimize the SQUID-to-circuit distance. By the time of this writing (2010), the method is apparently close to industrial implementation.

5.5.7 Concluding Remark

Of many SQUID applications, NDE is not one poised for rapid growth and major economic significance. However, in especially difficult cases, where nothing else can do, SQUID sensors are and will continue to be a valuable NDE tool.

5.6 SQUIDs — from Laboratory Devices to Commercial Products

Ronald E. Sager

5.6.1 Early SQUIDs as Laboratory Devices

I first encountered SQUIDs in 1974 when I joined Professor John Wheatley's group at the University of California San Diego as a graduate student, shortly after superfluidity in ^{3}He was discovered[78]. As part of the intense competition among research groups around the world to explore this new realm of quantum physics, John was using Zimmerman-style two-hole rf SQUIDs[79] to study ^{3}He's fascinating magnetic properties. At that time John was the unchallenged leader in using SQUIDs as tools for low-temperature research. Indeed, shortly after joining the group I was told that while most researchers were still struggling to make one SQUID work reliably, John had five SQUIDs in his cryostats and every one of them worked on every cool-down (see box). In 1971 John had published a comprehensive review article on the operation of rf SQUIDs, which became a sort of "SQUID Mechanic's Handbook" for the novice SQUID user.[80] This 78-page treatise discussed the theory of rf SQUIDs, the electronic circuits for driving them, their empirical characteristics and their operating parameters. It also included descriptions of various SQUID applications, such as

---

[75] The ITRS (International Technology Roadmap for Semiconductors) envisages, for example, 12 interconnect levels by 2014–2016, when the DRAM half-pitch (node) will decrease to 22 nm.

[76] ITRS 2009: see http://www.itrs.net/Links/2009ITRS/Home2009.htm

[77] K. Nikawa et al., *IEICE Trans. Electron.* E92-C (2009) 327

[78] D.D. Osheroff et al., *Phys. Rev. Lett.* 28 (1972) 885

[79] J.E. Zimmerman et al., *J. Appl. Phys.* 41 (1970) 1572

[80] R.P. Giffard et al., *J. Low Temp. Phys.* 6 (1971) 533

Adjusting the Critical Current in Early SQUIDs

Because the early two-hole rf SQUIDs typically used a pointed niobium screw contacting a hard flat niobium surface to create the point contact (see Section 5.2), the critical current of these devices would frequently shift rather dramatically from one cool-down to the next. Because the critical current of the SQUID needed to be within a fairly tight range for the SQUID to operate correctly, it was important to have a method for adjusting the critical current of the SQUID after the device was cold. Some researchers used a long screwdriver, extending from the top of the cryostat to the SQUID, to mechanically adjust the point contact while the SQUID was cold, but John's group developed a more innovative technique. Someone in the lab had discovered that, after the SQUID was cold, its critical current could be increased by discharging a capacitor through the rf coil—the bigger the capacitor, the more the critical current would increase. This naturally led to the practice of setting the point contacts at room temperature to have a rather low critical current and then adjusting the critical current in situ using a set of "calibrated capacitors", each of which was labeled with the expected increase in critical current.

---

pico-voltmeters and pico-ammeters, and it even described the use of SQUIDs to do thermometry at low temperatures.

5.6.2   First Commercialization of SQUIDs

In 1970 John and several of his colleagues in the low-temperature physics community invested their personal funds to found a new company in San Diego, CA, called SHE Corporation. "S", "H", and "E" were pronounced separately and represented SHE's three core competencies: superconductivity, helium and electronics. (Occasionally a reporter would telephone with a request to do an article on San Diego's "all-women company"!) SHE's business plan was to convert the dilution refrigerator and SQUID technology that already existed in John's laboratory into commercial products and sell them to the general low-temperature physics research community. The company succeeded in this endeavor from the very beginning.

Less than a year after being formed, SHE was shipping SQUIDs and dilution refrigerators on a commercial basis—the first company ever to offer a SQUID for commercial sale. These first products were essentially copies of equipment that was being used in John's lab. In particular, the first SHE rf SQUIDs used the Zimmerman two-hole design, and they were subject to the same instabilities in critical current that were inherent to this design. Consequently, each SHE SQUID was shipped with an adjustment tool so the customer could readjust the critical current of the device when it shifted.

By 1974 SHE had made two major advances which transformed the way SQUIDs were used. The first improvement was SHE's toroidal SQUID which they called the TSQ (see box). While the toroidal design itself offered some advantages, the most important improvement by far was the TSQ's permanently adjusted point contact. The new point contact design, developed by Michael Simmonds, used a small piece of electropolished niobium foil as the surface against which the pointed niobium screw was adjusted. The flexibility of the foil not only made it much easier to adjust the point contact initially, it also made the critical current very reproducible when the device was cooled down. In fact, the critical current of these devices was so stable that the TSQ was permanently sealed inside a hermetic package with just the screw terminals for the rf and input coils showing—and for the first time SHE offered a one-year warranty on their SQUIDs. This single innovation, by itself, transformed SQUIDs from temperamental devices that might fail on any cool-down into reliable tools for low-temperature research.

The Toroidal SQUID

The toroidal rf SQUID derives its name from the toroidal shape of the modulation and input coils as well as the toroidal nature of the SQUID body. The toroidal SQUID geometry has two major advantages: 1) both the coils and the point contact are shielded from external magnetic fields by the SQUID body and 2) the toroidal SQUID can be designed to have very low self-inductance which increases the SQUID's inherent sensitivity (see section 5.2.10). The toroidal geometry of the coils can be seen in Figure 5.26.

The second major innovation from SHE was a modern version of the SQUID control electronics which drove the SQUID. The new SQUID electronics, christened the Model 330 SQUID Controller, allowed the user to set up and adjust the RF drive, gain and other parameters without an oscilloscope, just by watching a meter on the front of the unit. The Model 330 was introduced about the same time as the TSQ SQUID, and together they revolutionized the use of SQUIDs by the research community.

SHE also introduced the next major improvement in commercial SQUIDs. When I joined SHE in December 1979, my first project was to assist Mike Simmonds in developing the SHE "Hybrid SQUID". This design employed true Josephson junctions formed on a silicon chip substrate using microcircuit fabrication techniques. The silicon chip, with its niobium films and Josephson junctions, was then mated to a bulk niobium SQUID body containing the toroidal modulation and input coils. The geometry of this design is shown in Figure 5.26. The design was originally developed as a dc SQUID, but since it was mechanically more stable under thermal cycling than the TSQ point contact design, we also developed an rf version of the Hybrid SQUID, which immediately replaced the TSQ. The dc Hybrid SQUID went on to become SHE's first dc SQUID, which offered a substantial improvement in sensitivity over SHE's rf SQUIDs and kept SHE at the forefront of commercial SQUID technology. However, this would be SHE's last significant contribution to that technology; the torch would soon pass to a new company about to be formed.

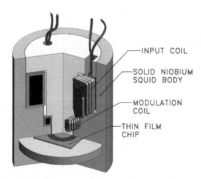

FIGURE 5.26: The SHE Hybrid SQUID showing the toroidal modulation and signal coils in a bulk niobium SQUID body. The Josephson junctions and niobium contact pads were fabricated on a thin-film chip, which was then mated to the niobium body forming superconducting contacts between it and the niobium pads on the chip.

5.6.3   SHE — Early SQUID Instrumentation

In addition to pioneering the commercialization of SQUID sensors, SHE also led the way in developing SQUID-based instruments for a variety of applications. Not surprisingly, SHE's initial

SQUID-based products were designed for applications similar to those found in John Wheatley's laboratory—a pico-ammeter, a pico-voltmeter and other equipment typically suitable for use only by highly skilled scientists. However, the company soon moved on to more user-friendly instruments. Their first major product of this type was a geophysical magnetometer system employing three SQUIDs with their respective detection loops configured as a 3-axis vector magnetometer. This instrument was constructed with the three detection loops wound around a quartz cube suspended from the end of a fiberglass cryogenic probe, with the SQUIDs mounted above in superconducting shields. The probe was then inserted into a fiberglass, nonmagnetic liquid helium cryostat. These instruments were primarily used in magnetotelluric research (see section 5.4.2).

The SHE SQUID Biomagnetometer employed a similar design with the cryogenic probe supporting the quartz substrate for the detection coil and SQUID, all within a small fiberglass cryostat. In this instrument, the cryostat had a special tail design that allowed the detection coil to be less than 1 cm from the outside surface of the cryostat tail while still being immersed in the liquid helium bath. The detection coil was wound in the form of a second-derivative gradiometer insensitive to magnetic fields which are uniform or have only a first-derivative spatial variation. Several researchers used these single channel biomagnetic gradiometer systems to perform some of the first experiments to detect the magnetic fields produced by electrical activity in biological neural systems—work which would eventually lead to their wider use in the field of biomagnetometry[81] presented in Chapter 10.

FIGURE 5.27:   The SHE Variable Temperature Susceptometer (VTS). This instrument was the first commercial SQUID sample magnetometer that provided automatic control of the magnetic field, sample temperature and measurement process. SHE sold about 45 of these instruments before they were discontinued in mid-1986.

SHE also designed and manufactured the Variable Temperature Susceptometer (VTS), shown in Figure 5.27. (I will refer to SQUID-based sample measurement systems like the VTS as sam-

---

[81] D. Cohen, *Science* 175 (1972) 664

ple magnetometers to differentiate them from SQUID vector magnetometers such as those used in magnetotelluric measurements.) This instrument employed SQUID detection coils in a gradiometric arrangement positioned at the center of a superconducting solenoid which could generate a high magnetic field. A variable temperature sample chamber located in the center of the SQUID detection coils (but thermally isolated from them) allowed the sample temperature to be controlled independently. By moving the sample vertically through the SQUID detection coils the magnetic moment of the sample could be measured with high accuracy and tremendous sensitivity over a wide range of both magnetic field and temperature. The SHE VTS was the first commercial SQUID-based sample magnetometer, but it was extremely complicated and very difficult to build.

In spite of a strong technology lead in a variety of SQUID technologies, in about 1982 SHE made a corporate decision to pursue medical applications for SQUIDs, and they chose to focus on the field of magnetoencephalography (MEG). This would entail a massive development program, and in 1983 SHE accepted its first round of investment capital to pursue this potentially lucrative business. Soon after making their initial investment, the new investors renamed the company Biomagnetic Technologies, Inc. (BTi) and hired a new president whose primary duty was to raise money to fund the development effort—tens of millions of dollars would be needed to develop the 160-channel SQUID systems and data analysis software that would be required for this demanding application. With their total commitment to the medical business, BTi's interest in laboratory equipment began to decline.

5.6.4 Quantum Design — Advanced SQUID Instruments

Just before SHE decided to pursue the MEG business, Barry Lindgren, Mike Simmonds, David Cox and I left SHE to form Quantum Design, which was officially incorporated on April 12, 1982. Our first sales were externally funded research projects, but we also began to look for products that we might develop and manufacture, and our attention soon focused on the deficiencies of the SHE VTS. Believing that we could solve many of the VTS' problems with a different design approach, we undertook the development of the Quantum Design Magnetic Property Measurement System (MPMS). The first MPMS, with a conservative 2 T magnet, was shipped to Professor Robert Shelton at Iowa State University in July of 1984 and became an immediate success. It also brought us into direct competition with SHE's VTS, which forced us to immediately redesign the MPMS to accommodate a 5.5 T magnet; this field intensity was later increased to 7 T. The first of what would become the classic Quantum Design MPMS was shipped in December 1985.[82] One of these original MPMS systems is shown in Figure 5.28. Within a few months

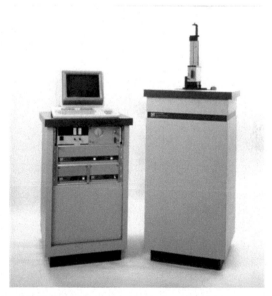

FIGURE 5.28: The classic MPMS. This instrument, introduced in 1985, would become the first and only SQUID sample magnetometer to be widely sold in a commercial market. An enhanced version of this instrument, known as the MPMS-XL, is still being sold in 2010.

---

[82] To Prof. Allen Goldman, University of Minnesota.

the MPMS had captured the entire market for SQUID sample magnetometers and in mid-1986 BTi announced that they would no longer offer the VTS for sale.

In retrospect, it is clear that this was when the industrial leadership in SQUIDs and SQUID-based laboratory instruments shifted from BTi to Quantum Design. As a result of BTi's decision to pursue the MEG medical markets, Quantum Design now had the entire market for SQUID sample magnetometers all to itself—literally within a few weeks of the publication of the discovery of the high-T_c materials[83]. The initial publication of this ground-breaking discovery received remarkably little attention until other researchers confirmed the results in late 1986, culminating in the "Woodstock of Physics" at the March 1987 APS meeting in New York City. Prior to this meeting, sales of the MPMS had been slow but steady. This changed dramatically over the next few months, particularly in the Japanese market. At the time of the "Woodstock of Physics", we had shipped only 8 MPMS systems over the previous two years. Over the next 15 months we shipped nearly 40 instruments and by mid-1988 a veritable tidal wave of orders had driven our backlog to more than 50 instruments, pushing our quoted delivery times out to 15 months. With orders still pouring in, it took us two years to fully ramp up production to meet the demand. By the time the initial excitement of the HTS discovery had subsided, Quantum Design had sold nearly 200 MPMS instruments. SQUID sample magnetometers, which had previously been specialty instruments used by only a few select researchers, had become a laboratory necessity for anyone doing materials research.

The huge surge in sales of our MPMS system also drove the next significant development in commercial SQUID technology. In 1987 we were still purchasing all of our SQUID sensors from BTi at a cost of well over $300,000/year. When our request for a volume discount was declined, under the leadership of Mike Simmonds we constructed a clean room and developed our own all-thin-film SQUID sensor, in which the modulation and input coils are fabricated right on the SQUID chip. This work led to two more significant inventions by Mike: a double-balanced SQUID design (shown in Figure 5.29), which uses a symmetrical geometry to decouple the modulation and input coils, and a technique for creating superconducting bonds to the niobium input coils on the SQUID chip which allows input coil currents exceeding 20 mA. Within 18 months Quantum Design was producing its own SQUID sensors, thereby eliminating our dependence on BTi SQUIDs. Ironically,

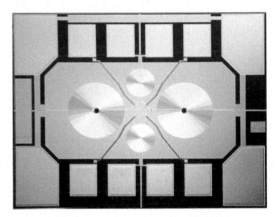

FIGURE 5.29: The Quantum Design double-balanced SQUID. The large and small circles are the input and modulation coils respectively. The SQUID itself is the octagonal structure under the coils, and the rectangular structures around the edges are the electrical contact pads. This chip is used in both rf and dc SQUIDs manufactured by Quantum Design.

Quantum Design would eventually reverse its relationship with BTi by supplying several thousand dc SQUID sensors to BTi, complete with transformer and matching network, for use in their MEG instruments. In addition, we supplied an additional 10,000 bare chips to BTi to insure their supply of SQUIDs. Although BTi no longer exists (the company declared bankruptcy in February 2009), Quantum Design still supplies SQUID sensors to several other companies that build SQUID-based instruments for special applications.

The nature of the high-T_c materials, in particular their flux-pinning behavior at very low magnetic

---

[83] J.G. Bednorz and K.A. Mueller, *Z. Phys.* B64 (1986)189

fields, had a significant impact on the further development of the MPMS technology. The need to study these behaviors in detail led to the development of the MPMS Low-Field Option with its precision magnet control and the MPMS AC Magnetization Option. Other options provided researchers with the ability to measure samples at higher temperatures and with greater sensitivity in high magnetic fields. These options made the MPMS a powerful research tool with a temperature range of 2 K to 800 K, magnetic field range of 1 μT to 7 T and the ability to measure magnetization over a dynamic range spanning 11 orders of magnitude.

The automated temperature and magnetic field control of the MPMS led to Quantum Design's next major product. In response to repeated customer requests for an "MPMS without the SQUID", the company developed the Physical Property Measurement System (PPMS), which provided the requisite temperature and magnetic field control, but was designed to offer a much broader suite of non-SQUID measurements, including ac-dc magnetization, magnetic anisotropy, heat capacity, electro-transport and thermo-transport measurements. In addition, the PPMS operating temperature range was extended up to 1000 K and down to 50 mK. The Low-Field Option was also adapted to the PPMS, and the system can now accommodate magnetic fields up to 16 T. This suite of measurement capabilities has made the PPMS an extremely versatile, multipurpose instrument.

Another major advance in the technology of SQUID sample magnetometers occurred in the mid-1990s when cryocoolers employing regenerators based on magnetic materials such as erbium-nickel and holmium-copper became available. These coolers could reliably achieve temperatures below 3 K, and in 1998 Quantum Design began to use these coolers in the EverCool option for the MPMS. After an initial transfer to cool the system down from room temperature, the MPMS EverCool completely eliminated the need for additional liquid helium. The MPMS thus became the first SQUID magnetometer to eliminate the need for constant LHe replenishment. After its introduction, the EverCool became an immediate success, particularly in Japan where liquid helium has been approximately twice as expensive as in America. By 2000 this technology was also available on PPMS systems.

During the 1990s the MPMS faced significant competition from several other manufacturers prompting us to upgrade the system in 1996 to provide faster measurements and enhanced temperature control. The upgraded instrument, designated the MPMS-XL, was still the world leader in SQUID Sample Magnetometers until it was finally eclipsed by the next major advance in SQUID instrumenta-

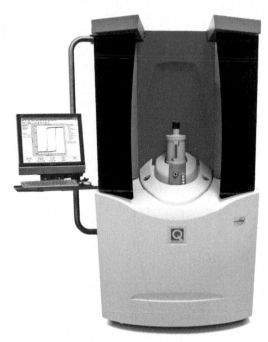

FIGURE 5.30: The Quantum Design SQUID-VSM. This instrument, which combines a vibrating sample drive with a SQUID-based detection system, provided substantial improvements over earlier SQUID sample magnetometers in sensitivity, temperature control and the speed of data collection.

tion, the Quantum Design SQUID-VSM (Vibrating Sample Magnetometer) shown in Figure 5.30.

The slow measurement process of both the MPMS and its predecessor, the SHE VTS, was a significant weakness of both instruments. The Reciprocating Sample Option introduced on the MPMS-XL improved the measurement speed, but its relatively low oscillation frequency still limited its effectiveness. To further improve the speed of data collection, the SQUID-VSM works at frequencies up to 40 Hz and employs the full power of phase-sensitive detection. When the SQUID-VSM was introduced in March 2006, its temperature control was also much faster than the MPMS-XL and a new patented superconducting magnet technology permitted high-sensitivity measurements to be made without de-energizing the magnet power supply. These features allow the SQUID-VSM to make measurements at least five times faster than the original MPMS. Furthermore, the higher operating frequency and more powerful phase detection scheme improved the sensitivity of the SQUID-VSM in high magnetic fields by nearly a factor of 100 over the original MPMS system.

5.6.5   The Market for SQUID-based Instruments Today

From the story above, it might appear that the history of commercial SQUID-based laboratory instrumentation is like a tale of two companies. There is reasonable justification for this. After developing the early commercial SQUID sensor and instrument technology, SHE Corporation chose to pursue medical applications, leaving Quantum Design as the dominant player in the SQUID instrumentation market. The extent to which this is true can be appreciated by reviewing the small group of companies that sell SQUIDs and SQUID-based instruments today. These are listed in Table 5.1, along with a brief note on the types of instruments they offer. This table is not necessarily comprehensive, and does not include most companies working in the fields of medicine, and applications discussed in other chapters of this book.

With the sole exception of Quantum Design, the companies listed in Table 5.1 have all failed to find significant commercial applications for SQUIDs; at best, some of them have found small niche markets for a particular product. For example, 2G Enterprises (presented in Section 5.4) has enjoyed a stable market for their rock magnetometers of about four systems per year for the past 25 years.

Tristan Technologies appears to offer the broadest range of products, including HTS and LTS SQUID microscopes, a rock magnetometer, a Biomagnetic Liver Susceptometer and various other SQUID-based specialty instruments. However, these instruments were all designed as specialty products for a specific customer with the hope that they will eventually find a broader market.

Other than Quantum Design, Cryogenics Ltd. (CL) is the only company in Table 5.1 that has achieved significant size or that offers a SQUID sample magnetometer. The CL Model S700X SQUID Susceptometer does compete with Quantum Design's MPMS products, but CL sells very few of these instruments, and their business is based primarily on their superconducting magnet technology. All the other companies listed in Table 5.1 offer only niche products in markets smaller than $1-$2 million dollars/year and have only a few employees. Hence, it is probably fair to say that Quantum Design is the only company in the world that has achieved significant commercial success with SQUID-based laboratory instruments.

5.6.6   Future Directions for SQUIDs in Laboratory Instrumentation

While MEG is the largest application for SQUIDs in terms of device numbers, with astronomy rapidly catching up (see Section 5.7), SQUID-based laboratory instruments will continue to be an important market. SQUID sample magnetometers have become an essential tool in materials research and this market remains robust. It is not clear, however, where SQUID enthusiasts will find other significant commercial markets. Since SHE first brought SQUIDs into the commercial market in 1971, many companies have tried to apply the devices to a variety of different problems, and most of those companies have either failed completely or remained very small.

In general, there are always two major obstacles to using SQUIDs: (1) they require a cryogenic environment to operate, and (2) their exquisite sensitivity to magnetic fields, which is their most important strength, is also their most important weakness. The first problem, the need for a cryogenic environment, is not necessarily prohibitive for many applications, but it does increase the cost and complexity of SQUID-based instruments. The second problem is equally vexing. The SQUID's incredible sensitivity to magnetic fields gives it a unique capability, but the SQUID is equally sensitive

TABLE 5.1: Companies Manufacturing SQUIDs and SQUID Instrumentation

Company	Products
2G Enterprises (USA)	LTS Rock Magnetometers
Aivon Oy (Finland)	LTS SQUIDs: SQUID Sensors & Electronics
Cryogenics Ltd (UK)	LTS Sample Magnetometer System
Cryoton (Russia)	LTS and HTS SQUIDs: 3-Axis Vector Magnetometers, MCG Systems[a]
EZ SQUID (Germany)	LTS dc/rf SQUIDS, NDE Measurement Systems
Juelicher SQUID GmbH (Germany)	HTS SQUIDs: Resonators/Phase-Locked Oscillators
Magnesensors (USA)	HTS SQUIDs: Externally Funded Research
Magnicon GmbH (Germany)	LTS and HTS SQUIDs: Vector Magnetometers/Gradiometers & Fast Electronics
Quantum Design (Worldwide)	LTS SQUIDs: SQUID Sensors & Sample Magnetometers
Star Cryoelectronics (USA)	LTS and HTS SQUIDs: SQUID Sensors & Electronics
Supracon AG (Germany)	LTS SQUIDs: MCG, JJ Standard[b], Geomagnetic Systems
Tristan Technologies (USA)	LTS/HTS Specialty SQUID Systems

[a] MCG refers to magnetocardiography, measurement of magnetic fields originating from the heart.
[b] JJ Standard refers to Josephson Junction Voltage Standard.

to background magnetic fields—and the world is full of them. There are several ways to address this problem such as counter-wound detection coils (gradiometers of various orders), magnetic shielding and the use of ancillary sensors to numerically analyze and remove the interfering signals, but these measures all increase the cost and complexity of SQUID-based instruments and, to some extent, reduce their utility. In some cases, even though the SQUID instrument is more sensitive or more capable than a competing room temperature instrument, the additional cost and complexity of the SQUID instrument makes it financially uncompetitive against its more conventional and cheaper competitors. So it seems likely that the same factors that have limited the use of SQUIDs in the past will continue to haunt them in the future.

Nonetheless, SQUIDs are a powerful tool for certain uses, and when applications are found where SQUIDs provide a unique solution to an important problem, they will be accepted in spite of their expense and complexity. In the meantime we can be sure that a small but dedicated group of researchers and companies will continue to search for promising new commercial applications for these amazing and unique sensors—and occasionally one of them may be successful.

5.7 Electromagnetic and Particle Detection and Readout

John Clarke, Kent Irwin and Peter Michelson

5.7.1 DC SQUID Amplifiers

John Clarke

The notion of using a dc SQUID as an amplifier goes back to my Cavendish days, when I coupled together two SLUGs[23]. It was not until 1979, however, after Claudia Tesche and I had worked out the noise theory of the SQUID,[27,29] that we developed the theory for the SQUID amplifier that subsequently led to the development of radiofrequency amplifiers with high gain and low noise temperature. We calculated the flux-to-voltage transfer coefficient, V_Φ, the spectral densities of the voltage noise across the SQUID, $S_V(f)$, and current noise around the loop, $S_J(f)$, and the voltage-current cross spectrum, $S_{VJ}(f)$. This partial cross correlation arises from the additional voltage noise induced across the SQUID by the circulating current noise provided $V_\Phi \neq 0$. The essential concept for a tuned amplifier is shown in Figure 5.31. An rf signal $V_i(f)$ with source impedance R_i coupled to the tuned input circuit induces a current that in turn generates a flux in the SQUID and hence a voltage $V_o(t)$ at the output. Together with the late Robin Giffard[84], we realized that the SQUID amplifier is in a sense the dual of a semiconductor amplifier, for

example, a field effect transistor (FET). In our picture, the voltage noise across the SQUID is represented as a "virtual" current noise in the input circuit, while the current noise in the SQUID loop injects an "actual" voltage noise into the same circuit. In the case of the FET, the current noise at the input is "actual" whereas the voltage noise is "virtual". Either way, by tracing the effects of the two noise sources to the output, one can calculate the amplifier noise temperature[85], T_N, which can be minimized by varying the parameters of the system. In the case of the SQUID tuned amplifier, under certain approximations, the minimum noise temperature is $T_N = (\pi f / k_B V_\Phi)(S_V S_J - S_{VJ}^2)^{1/2}$ at a frequency below that of maximum gain. The cross-term S_{VJ} reduces T_N significantly, by roughly a factor of 3,

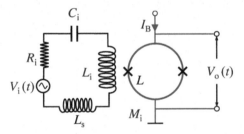

FIGURE 5.31: Configuration of SQUID tuned amplifier. V_i is the input signal with source impedance R_i, L_s is stray inductance, C_i is the tuning capacitor, L_i and M_i are the input inductance and its mutual inductance to the SQUID, which has loop inductance L and bias current I_B.

compared with its value on resonance, where S_{VJ} drops out. At the same time, off-resonance operation results in a modest reduction in gain.

Subsequently, Claude Hilbert, John Martinis and I explored SQUID tuned and untuned ampli-

---

[84] J. Clarke et al., *J. Low Temp. Phys.* 37 (1979) 405

[85] The noise temperature T_N is defined as follows. In the classical limit, a resistor R_i at the input terminals of the amplifier produces a Nyquist noise power $k_B T R_i G$ at the output, where G is the power gain. The total noise power at the output is $k_B R (T_i + T_N) G$. Thus, T_N is a simple way of representing the noise of any amplifier.

fiers in considerable detail[86],[87],[88]. In the tuned case, Claude achieved a gain of about 19 dB and a noise temperature of 1.7 K in an amplifier at 4.2 K, on resonance at 93 MHz[88]. These results were in good agreement with the theoretical model. Later, Claude, Erwin Hahn, Tycho Sleator and I used a SQUID-tuned amplifier to detect nuclear quadrupole resonance in ^{35}Cl at 30.6856 MHz. In particular, we were able to observe "spin noise"[89]. The KClO$_3$ sample was placed in a superconducting coil, replacing the resistor and voltage source of Figure 5.31. We equalized the populations of the two (doubly degenerate) nuclear spin levels by means of an rf signal at the NQR frequency, and then turned off this signal. We observed the spontaneous emission of photons emitted into the tuned circuit as the excess spin population of the upper level relaxed towards its equilibrium value. The spontaneous emission rate, given by Einstein's A coefficient, corresponds to roughly one spin flip every one million centuries! Nonetheless, there was an observable signal from the 2×10^{21} spins.

An intriguing aspect of the dc SQUID amplifier is that it can, in principle, approach the quantum limit. In 1982, the late Roger Koch, Dale van Harlingen and I revisited Claudia's SQUID equations, replacing the classical Nyquist current noise spectral density in the resistive shunts, $4k_B T/R$, with the quantum expression $(2hf/R)\coth(hf/2k_B T)$. In the limit of zero temperature, this expression reduces to $2hfR$. Roger ran the equations in the $T = 0$ limit[90], and found $T_N = hf/k_B$ for a SQUID tuned amplifier—the quantum limited value for a linear, phase-preserving amplifier. In this case, on average, one-half photon exists in the tuned circuit (the zero-point energy of a quantum harmonic oscillator) and the other one-half photon is contributed by the amplifier. The cross-spectral density makes a significant reduction in the noise when the parameters are optimized; the amplifier cannot be quantum limited on resonance.

Attaining a quantum-limited noise temperature implies a signal frequency $f > k_B T/h$; for $f = 1$ GHz, $T \approx 50$ mK. In the 1980s, we were interested in frequencies below—often far below— 100 MHz, and quantum-limited noise temperatures were not something we thought about. All this changed, however, when Leslie Rosenberg and Karl van Bibber visited me in 1994. They had recently commissioned a novel experiment at Lawrence Livermore National Laboratory to search for the putative axion (see below), and were keenly interested in an amplifier operating in the 1 GHz frequency range with a noise temperature substantially lower than that of a cold HEMT (high electron mobility transistor), typically $2 - 5$ K.

How does one do this with a SQUID? The essential problem with operating a square-washer SQUID (Figure 5.13) at 1 GHz is the parasitic capacitance between the coil and the washer that rolls off the gain at frequencies above typically 100 MHz. To tackle this problem, Marc Olivier André and I at Berkeley joined forces with Michael Mück, Jost Gail and the late Christoph Heiden at the University of Giessen in Germany. Michael made the crucial breakthrough. Instead of applying the signal between the two ends of the input coil, as in the conventional SQUID, he moved one wire and applied the signal between one end of the coil and the SQUID washer. The washer now serves as a groundplane for the coil, forming a transmission line or microstrip, thus making a virtue of the coil-washer capacitance[91] [Figure 5.32 (a)] . When the length of the microstrip corresponds to a half-wavelength of the signal, the resonance produces a substantial level of gain—typically 20 dB — as shown in Figure 5.32(b). As one progressively shortens the coil, the resonance moves to higher frequency. The actual dependence of the frequency on the coil length is somewhat complicated because the inductance of the microstrip is dominated by the inductance coupled in from the SQUID loop. Suffice it to say, the microstrip SQUID amplifier (MSA) attains useful levels of gain from 0.2 to 2 GHz.

[86] J. Martinis et al., *J. Low Temp. Phys.* 61 (1985) 227

[87] C. Hilbert and J. Clarke, *J. Low Temp. Phys.* 61 (1985) 237

[88] C. Hilbert and J. Clarke, *J. Low Temp. Phys.* 61 (1985) 263

[89] T. Sleator et al., *Phys. Rev. Lett.* 55 (1985) 1742

[90] R.H. Koch et al., *Appl. Phys. Lett.* 38 (1981) 380

[91] M. Mück et al., *Appl. Phys. Lett.* 72 (1998) 2885

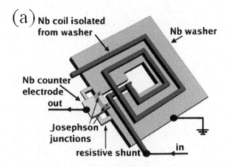 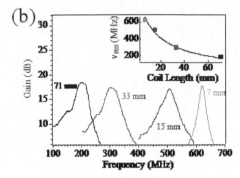

FIGURE 5.32: Microstrip SQUID amplifier. (a) Configuration. (b) Gain versus frequency for four coil lengths.

Michael, Jan Kycia and I measured the noise temperature of an MSA cooled in a dilution refrigerator. To overcome the noise of the cooled HEMT postamplifier, we used a second MSA as a postamplifier. As we cooled the experiment, we found that T_N indeed scaled with T as expected from the expression for the noise energy. Below roughly 100 mK, however, T_N flattened out. Separate experiments showed that the flattening was due to hot electrons generated in the resistive shunts by the bias current, an effect studied years earlier by Fred Wellstood, Christian Urbina and myself[92]. Nonetheless, the MSA attained a noise temperature on resonance of 47 ± 10 mK at 519 MHz, a factor of about two higher than the quantum limit[93].

Subsequently, Darin Kinion and I repeated the experiment, redesigning the MSA slightly to increase the gain so that we did not require a second MSA[94]. In addition, we added cooling fins to the shunts to lower the electron temperature. Figure 5.33 shows the gain and noise temperature versus frequency for our best device, cooled to 45 mK. The maximum gain, 20.4 dB, occurred at 620 MHz, at which frequency $T_N = 66 \pm 5$ mK. The lowest noise temperature, however, 48 ± 5 mK, occurred at 612 MHz, slightly below the resonance frequency, in qualitative agreement with our prediction 30 years earlier. The lowest achieved noise temperature was a factor of 1.6 above the quantum-limited value of 29.4 mK at 612 MHz. This value of T_N is about 30 times lower than that of the best cooled HEMT amplifiers.

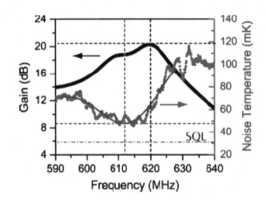

FIGURE 5.33: Performance of microstrip SQUID amplifier at 45 mK. Gain and noise versus frequency[94].

[92] F.C. Wellstood et al., *Phys. Rev. B* 49 (1994) 5942

[93] M. Mück et al., *Appl. Phys. Lett.* 78 (2001) 967

[94] D. Kinion and John Clarke, *Appl. Phys. Lett.* 98 (2011) 202503.

5.7.2 Transition Edge Sensors

Kent Irwin

The SQUID is an important detector of electromagnetic signals in its own right, but it also plays a critical role as a follow-on amplifier for other types of superconducting detectors. One of the most important of these is the superconducting transition-edge sensor (TES), which consists of a superconducting film biased in the narrow transition region between the normal and superconducting states. In this temperature region, the resistance of the film is a sensitive thermometer that can be used to detect an extraordinary range of photons and other particles. Often, measurements with a TES are more sensitive than any other technique. Arrays of SQUID-coupled TES detectors are presently deployed as detectors of photons from millimeter-wave to gamma rays, in applications ranging from astrophysics to nuclear materials analysis.

The SQUID amplifier is the natural amplifier for TES detector arrays since it can be readily noise-matched to low impedance and is obviously compatible with low-temperature operation. The implementation of SQUID amplifiers to read out the current from voltage-biased TES detectors makes it possible to implement large arrays of TESs. However, the TES was developed well before the SQUID.

The TES was invented by Donald Hatch Andrews, a professor of chemistry in Johns Hopkins University's Cryogeny Laboratory, while he relaxed on a beach in Nassau in 1938. Professor Andrews [Figure 5.34(a)] realized that, on its transition, a superconducting film had a much larger temperature coefficient of resistance than conventional resistive thermometers, and that it could be used to create a more sensitive detector. He published[95] the idea in 1938, and demonstrated[96] an infrared detector based on a TES bolometer in 1942. He even developed a real-time infrared imaging system in 1946 based on a NbN TES that used scanning optics to display live images on the screen of an oscilloscope. An infrared image of Professor Andrews is shown in Figure 5.34(b).

The low resistance of the TES made it difficult to noise-match it to a semiconductor amplifier. In recent years, this problem has been largely eliminated by the use of SQUID current amplifiers, which are readily noise-matched to low-resistance TES detectors[97]. The final barrier to the large-scale practical use of TES detectors was the difficulty of operating them within the extremely narrow superconducting transition region. When it is current-biased, Joule heating of the TES can lead to thermal runaway, and small fluctuations in bath temperature significantly degrade performance. Furthermore, variations in the transition temperature between multiple devices in an array of TES detectors can make it impossible to bias them all at the same bath temperature.

The solution to these problems was worked out[98] when I was a graduate student in Blas Cabrera's laboratory at Stanford University, developing TES detectors to search for Weakly Interacting Massive Particles (WIMPs). We realized that when TES detectors are voltage biased, they self-regulate their temperature within the transition, operate without thermal runaway, and have much less sensitivity to fluctuations in the bath temperature. Equally importantly, large arrays of TES detectors with slightly different transition temperatures can be sensitively operated at a common bath temperature when they are voltage biased. Since that time, TES detector arrays have been widely used.

The general noise theory for resistive bolometric detectors such as the TES was worked out in John Mather's classic papers[99]. However, superconducting transition-edge sensors often exhibit excess noise well beyond that predicted by Mather's theory. In 2005, I revisited Mather's calculations, and realized that the conventional noise theory of bolometers is rigorous in equilibrium systems, but

---

[95] D.H. Andrews, *American Philosophical Society Yearbook* (1938) 132

[96] D.H. Andrews et al., *Rev. Sci. Instrum.* 13 (1942) 281

[97] W. Seidel et al., *Phys. Lett. B* 236 (1990) 483

[98] K. D. Irwin, *Appl. Phys. Lett.* 66 (1995) 1998

[99] J. Mather, *Appl. Optics* 21 (1982) 1125

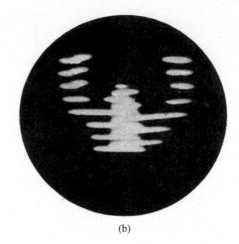

(a)   (b)

FIGURE 5.34: (a) Donald Hatch Andrews, the inventor of the superconducting transition-edge sensor. Shown in 1946 with a TES infrared imager (Photograph credit: "An Eye For Heat," *Popular Science*, July 1946, p.126.) (b) An infrared image of Donald H. Andrews with his hands raised. The image was acquired with a NbN TES in 1946 in complete darkness, and displayed in real time on an oscilloscope synchronized to the scanning optics. (Photograph credit: "An Eye For Heat," *Popular Science*, July 1946, p.126.)

not when the bolometer temperature is different from the bath temperature. Using a simple model of the electrical and thermal system of the bolometer, and applying the nonequilibrium fluctuation-dissipation relations developed by Stratonovich, I was able to derive rigorously the correlations in a nonlinear bolometer operated with a first-order deviation from equilibrium[100]. One of the predictions of this theory is that the power spectral density of the random fluctuations in the voltage across the resistive bolometer is $S_V = 4k_B TR(1 + 2\beta)$, rather than the Nyquist value $S_V = 4k_B TR$ assumed by Mather. Here $\beta \equiv [\partial log(R)/\partial log(I)]_T$, where I is the bias current. This theoretical prediction has since been strongly supported by experimental results, making it clear that many TES detectors operate with no measurable deviation from fundamental thermodynamic noise limits. The development of TES detectors with well-understood noise performance has opened up a wealth of applications in photon detection from millimeter wave through gamma rays.

5.7.3 Photon Energy Resolving Detectors

Kent Irwin

When a photon of energy E_γ is absorbed in a material with heat capacity C, the resulting temperature rise is $\Delta T = E_\gamma/C$. A microcalorimeter determines the energy of a single photon by using a sensitive thermometer to measure[101] ΔT. A TES can be used to make an excellent microcalorimeter if the heat capacity is chosen so that ΔT is a significant fraction of the superconducting transition width. The full-width-at-half-maximum (FWHM) energy resolution of this measurement of E_γ is limited by thermodynamics to $\Delta E_{\text{FWHM}} = 2.35\xi[k_B T E_\gamma(1 + 2\beta)^{1/2}]^{1/2}$, where $\xi \approx 3$ is a dimensionless parameter whose value is determined by the details of the thermal link to the heat bath, k_B is the Boltzmann constant, and T is the temperature of the TES. The $1 + 2\beta$ term is the correction

---

[100] K.D. Irwin, *Nucl. Instr. Meth.* A559 (2006) 718

[101] S. Moseley, et al., *J. Appl. Phys.* 56 (1984)1257

from non-equilibrium thermodynamics discussed earlier. As an example, putting in realistic values of $\beta = 1$ and $T = 100\,\mathrm{mK}$, we find that TES calorimeters are capable of measuring the energy of a 6 keV x-ray photon to about 1 part in 3,000. In contrast, the best commercial semiconductor detector is limited by Fano statistics of the electron-hole pair generation to an energy resolution of 1 part in 50. The startling improvement of almost two orders of magnitude available with microcalorimeters enables new applications, including x-ray astronomy and nuclear materials analysis with gamma rays.

5.7.4   X-Ray Astronomy and Materials Analysis

Kent Irwin

X-rays provide a sensitive probe of nearly all classes of astrophysical objects by either emission or absorption. Energetic sources of x-rays include plasmas in stellar winds, supernova remnants, accretion features in active galactic nuclei and intracluster gas. However, x-ray detectors that have been flown in satellite-based x-ray observatories had either extremely limited collection area (making it possible to study only the brightest sources) or relatively poor energy resolution. The TES x-ray calorimeter is an excellent candidate instrument for a satellite observatory that will provide both high collection area and excellent energy resolution.

X-ray spectral measurements are also a critical tool in materials analysis and nuclear safeguards. The excellent energy resolution provided by TES x-ray detectors can enable more precise measurement of the elemental composition of matter on a nanometer scale in a scanning electron microscope. TES x-ray calorimeters can be used to measure gamma-rays and excited x-rays from a sample of nuclear material to provide important information for nuclear safeguards and treaty verification[102].

In 1995 I joined John Martinis' group at NIST as a postdoc. John had recently developed high-bandwidth amplifiers based on series arrays of SQUIDs. We realized that the combination of these SQUID amplifiers with the voltage-biased TESs could be used to create extremely sensitive x-ray microcalorimeters. John, Gene Hilton, and I developed transition-edge sensors based on bilayers of a normal metal and a superconductor with a proximitized superconducting transition temperature of about 100 mK. Within a year, these TES calorimeters were outperforming all other superconducting x-ray detectors. Modern TES calorimeters have the highest energy resolution of all energy-dispersive detectors[103]. The first spaceborne TES x-ray calorimeters will be flown in 2012 in the Micro-X sounding rocket experiment, and TES x-ray calorimeters are now the baseline technology for the International X-Ray Observatory (IXO).

To obtain an image of an extended astrophysical object such as a supernova remnant, one requires hundreds or preferably thousands of such sensors mounted at the focal plane of an x-ray telescope. Needless to say, this is a challenging prospect. Although it is perfectly feasible to fabricate hundreds or thousands of SQUIDs, a serious concern is the thermal heat load generated by the leads when each TES requires its own readout SQUID with its own dedicated set of readout wires. In 1996, Erich Grossman at NIST suggested that we read out an array of TESs by wiring their output SQUIDs in series, and turning them on one at a time. Under appropriate conditions, the signal from each TES can be faithfully reconstructed[104]. The eventual outcome was the 32-channel time-division SQUID multiplexer (MUX) chip. In this circuit, all of the SQUIDs in a "column" are wired in series on one chip, and read out with one output wire (Figure 5.35). One SQUID is turned on at a time on each chip, using a single "row" wire going between multiple MUX chips. One feedback

[102] D.T. Chow et al., *Nucl. Instr. Meth.* A444 (2000) 196

[103] A.S. Hoover et al., *J. Radioanal. Nucl. Chem.* 282 (2009) 227

[104] J.A. Chervenak et al., *Appl. Phys. Lett.* 74 (1999) 4043

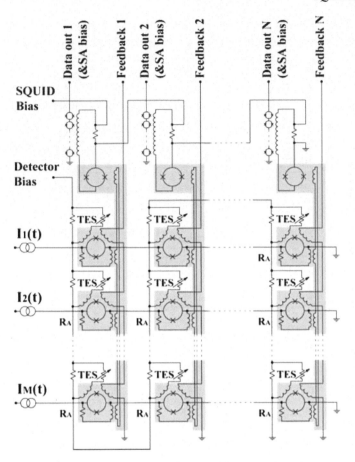

FIGURE 5.35: Circuit for a TES time-division multiplexer. Each TES (shown as a variable resistor) is coupled to a dedicated first-stage SQUID switch (on a green background). The first-stage switch SQUIDs are switched on one row at a time with common address leads $I(t)$. Each column has a readout SQUID that consists of 100 SQUIDs in series.

wire is also required for each MUX chip to linearize the SQUID response, and the value of the feedback is switched to linearize the response of whichever SQUID is on at the time. This basic chip architecture is now being used for the Micro-X TES arrays, and is baselined for IXO. A 256-channel TES calorimeter instrument using 8 SQUID MUX chips, shown in Figure 5.36(a), was co-developed by Joel Ullom at NIST and Michael Rabin at the Los Alamos National Laboratory (LANL). This instrument is now at LANL, where it is used to analyze the gamma- and x-ray spectra of complex nuclear materials for nuclear safeguards applications [Figure 5.36(b)][103].

5.7.5   Dark Energy and Cold Dark Matter

John Clarke

Measurements of the cosmic microwave background[105] (CMB) have taught us much about the origin of the universe. The CMB, which has a Planck distribution with a characteristic temperature of 2.726 K, originated 377,000 ± 3,000 years after the Big Bang and has been traveling for 13.7

---

[105] P. James et al., *Finding the Big Bang* (Cambridge University Press, Cambridge, England, 2009)

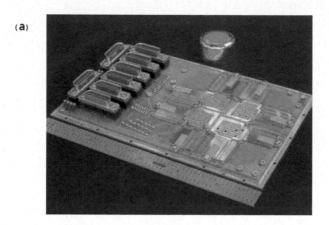

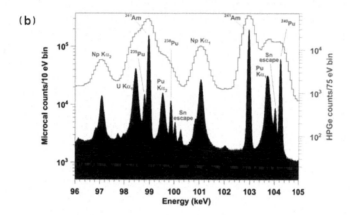

FIGURE 5.36: (a) A 256-channel TES x-ray / γ-ray detector array and a measured spectrum. A photograph of the detector array. (Photo credit: Dan Schmidt, NIST.) (b) Analysis of an isotopic mixture of plutonium and uranium. The spectrum is shown both for a state-of-the-art high-purity germanium (HPGe) detector (top line and axis on right), and for the TES calorimeter array (solid bars and axis on the left). (Figure credit: A. Hoover, LANL.)

billion years since then. We now know that the Universe consists of 0.6% neutrinos, 4.6% baryons, 73% dark energy (DE) and 22% cold dark matter (CDM). SQUID amplifiers are being used to investigate all four of these components. SQUIDs are used in direct detection experiments to search for axions and weakly interacting massive particles (WIMPs), two candidates for the cold-dark matter. Cosmic microwave background measurements using SQUID-coupled TESs are being used to constrain the properties of DE and the sum of the neutrino masses. SQUID-coupled TESs will also be used for x-ray measurements to search for the "missing" half of the baryons in the Warm-Hot Intergalactic Medium (WHIM).

5.7.6   Dark Energy: Searching for Galaxy Clusters

John Clarke

Galaxy clusters contain typically several hundred galaxies and, with a mass of 10^{14} to 10^{15} solar masses, are the largest gravitationally bound objects in the universe. Measuring the density of galaxy clusters as a function of red shift enables one to determine the parameters of the equation of state for DE. Optical surveys are being conducted to map galaxy clusters, but they are inherently biased. More distant clusters at higher red shift have weaker signals, so fewer of them will be found, leading to systematic errors in the determination of cosmological parameters.

A newly implemented search technique based on the Sunyaev-Zel'dovich Effect (SZE)[106] is now being used to survey galaxy clusters without this bias. This technique is based on the fact that the space between galaxies in a cluster contains a hot electron gas. When a microwave background photon passes through this gas, it has a small (1-2%) chance of being scattered to a higher energy, so that a fraction of the CMB spectrum is shifted to slightly higher frequencies. Thus, as the telescope scans across the sky, one expects to find regions in which there is a shifted Planck spectrum superimposed on the unshifted spectrum—the signature of a galaxy cluster. SZE surveys are based on the distortion of a signal coming from behind the cluster, and the ratio of the SZE effect to the CMB is independent of red shift. Hundreds to thousands of SZE clusters are expected to be found using this unbiased technique.

Since the CMB spectrum peaks at a frequency of about 150 GHz, one needs ultrasensitive detectors in the far infrared. The detector of choice for these measurements is the TES bolometer, which is very similar to a TES calorimeter. However, in a bolometer, the power in a photon signal, P_γ, is measured instead of the energy of a single photon. When a photon signal is absorbed in the optical element, the thermal energy is transferred to the TES, initially raising its temperature and increasing its resistance. As the resistance begins to increase, however, the dissipation generated by the voltage bias drops, so that the power dissipation in the TES remains constant. The change in the current flowing through the SQUID input coil is related to the absorbed energy. This electrothermal feedback results in a fast, linear response. Furthermore, the power dissipation is low, typically a few picowatts. The noise in an optimized sensor is limited by fluctuations in the arrival rate of photons rather than by noise in the TES bolometer itself.

To obtain an image of a galaxy cluster, one requires hundreds or preferably thousands of such sensors mounted at the focal plane of a telescope. The time-division SQUID multiplexer described earlier is used for just such measurements in the Atacama Cosmology Telescope. However, an alternative approach to multiplexing the signals from the TES arrays was developed at Berkeley for SZE measurements. In the spring of 1999, Paul Richards walked into my office to discuss the feasibility of using a single SQUID to read out an array of TESs simultaneously with a frequency-division multiplexer. I found this a very interesting problem, and the eventual outcome was the 8-channel multiplexer[107] shown in Figure 5.37. Each TES is connected in series with a capacitor and an inductor to form a resonant circuit, each with a different frequency. An oscillator provides a comb of voltages $V_i(f_i)$ at frequencies f_i corresponding to the resonant frequencies of the tuned circuit. The currents from the eight tuned circuits are summed at one end of the input coil of a SQUID (actually a 100-SQUID series array); the other end is grounded. The oscillator applies a nulling comb of voltages $-V_i(f_i)$ to the same summing point so that, in the absence of any photon flux, the total current injected into the input coil is zero. When a photon is absorbed by one of the sensors, the resulting change in current is injected into the input coil of the SQUID, which is operated in a flux-locked loop. Since the 8 frequencies are distinct, all 8 sensors can be read out simultaneously with the aid

[106] R.A. Sunyaev and Y.B. Zel'Dovich, *Comments Astrophys.* 4 (1972) 173
[107] T.M. Lanting et al., *Appl. Phys. Lett.* 86 (2005) 112511

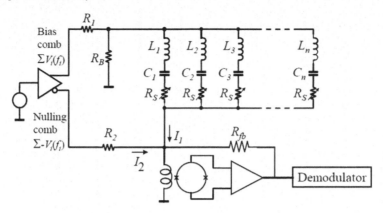

FIGURE 5.37: Circuit for a TES frequency-domain multiplexer. The readout SQUID consists of 100 SQUIDs in series. Reproduced from ref. 107, with permission.

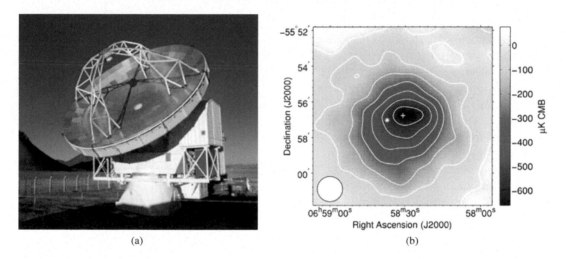

(a)   (b)

FIGURE 5.38: APEX telescope and first results. (a) Telescope at Atacama. (b) SZE image of the "Bullet Custer"; the contour interval is 100 mK$_{\text{CMB}}$. Disk at lower left represents the full width at half-maximum (FWHM) resolution. (Reproduced from Reference 108, with permission.)

of a demodulator.

The Berkeley group, in collaboration with several other groups, first implemented this scheme on APEX (Atacama Pathfinder EXperiment), a 12-m telescope at about 5100 m on the Atacama plateau in Northern Chile [Figure 5.38(a)]. The receiver contains about 280 working channels with multiplexed readout. As a first demonstration of the system operation, Figure 5.38(b) shows an SZE image of the "Bullet Cluster"—actually two merging clusters[108]. The signal-to-noise ratio is 20 in the central 1 minute of arc beam. Subsequently, the Berkeley group and others began a cluster search with SPT (South Pole Telescope), a 10-m telescope at Antarctica at about 2900 m; there are about 660 TESs with multiplexed readout. Over a two-year period, SPT will survey 4000 square degrees of sky. As of October 2011, more than 400 previously unknown galaxy clusters have been discovered and confirmed optically.

---

[108] N.W. Halverson et al., *Astrophys. J.* **701** (2009) 42

5.7.7   Cold Dark Matter: WIMPs and Axions

John Clarke

Since we do not know what CDM is, we have to postulate some kind of particle and then search for it. One of the leading candidates is the WIMP (Weakly Interacting Massive Particle)[109]. However, since WIMPs do not interact with electromagnetic forces, their direct detection is challenging. Their scattering rate in a Ge crystal is expected to be less than 0.1 per day per kilogram! A detector technology is needed that can detect the energy deposited when a WIMP scatters off a nucleus, and discriminate these events from the much more common electron-scattering events from the radioactive background.

The Cryogenic Dark Matter Search (CDMS) originated out of a marriage of the ionization detectors developed by Bernard Sadoulet's group at Berkeley and the TES detectors developed by Blas Cabrera's group at Stanford. In CDMS, large germanium crystals (10 mm thick and 76 mm in diameter) are instrumented with both types of detectors. If a WIMP scatters off a Ge nucleus, it produces both electron-hole pairs and phonons. The electron-hole pairs are collected by the ionization channel. The phonons propagate to the surface of the crystal where they break Cooper pairs in superconducting Al absorbers on the surface, producing quasiparticles. The absorbers are in electrical contact with large arrays of tungsten TESs wired in parallel. The quasiparticles diffuse into the TESs, where they are trapped, raising the temperature. Each parallel TES array is biased with a constant voltage, and its current is read out with a series array of 100 SQUIDs. Nuclear scattering events, including WIMP recoils, generate a larger fraction of phonons than electron-scattering events, so CDMS can veto the electron-scattering background. CDMS has now set the most stringent limits on WIMPs of any direct-detection experiment[110], and larger scale experiments are in development to push the limits still further.

FIGURE 5.39: Axion detector installed at Lawrence Livermore National Laboratory.

Another candidate for CDM is the axion, first proposed in 1978 to explain the absence of a measurable electric dipole moment on the neutron. The axion mass m_a is predicted to be $1\ \mu eVc^{-2}$ to $1\ meVc^{-2}$ (corresponding to 0.24 to 240 GHz). Since the measured CMB density ρ_a is $0.45\ GeVcm^{-3}$, a mass of $1\ \mu eV$ corresponds to an axion density of $4.5 \times 10^{14}\ cm^{-3}$. How does one

---

[109] G. Steigman and M.S. Turner, *Nucl. Phys.* B 253 (1985) 375

[110] Z. Ahmed et al., *Science* 327 (2010) 1619

search for this light, spinless, chargeless particle? In 1983, Pierre Sikivie[111] suggested that it could be found via Primakoff conversion: in the presence of a large magnetic field B, the axion converts to a photon (together with a virtual photon) with energy $m_a c^2$. The conversion takes place in a cooled cavity of volume V with a tunable resonance frequency. An antenna couples the signal in the cavity to a cooled, low-noise amplifier. When the cavity frequency corresponds to the frequency of photons from the putative axions, the amplifier is expected to detect a peak; off-resonance, it detects only blackbody noise. Clearly, one has to scan the cavity frequency to search for the axion. The axion-to-photon conversion power is given by $P_a \propto B^2 V g_\gamma^2 m_a \rho_a$, where g_γ is a coupling coefficient; its lowest predicted value is $|g_\gamma| = 0.36$.

The axion detector (Figure 5.39) at LLNL, known as ADMX (Axion Dark Matter eXperiment), began operation in 1996. The cavity and the surrounding 7-T persistent-current magnet were cooled to 1.5 K. The cooled HEMT amplifier had a noise temperature $T_N = 1.7$ K, giving a system noise temperature $T_S = T + T_N = 3.2$ K. For the parameters of the experiment and assuming $|g_\gamma| = 0.36$, the time to scan from $f_1 = 0.24$ GHz to $f_2 = 0.48$ GHz is $\tau(f_1, f_2) \approx 4 \times 10^{17} (T_S/1K)^2 (1/f_1 - 1/f_2)$ sec ≈ 270 years. Evidently, this experiment was not likely to find an axion very quickly!

The dilemma was solved by replacing the HEMT with a MSA. The potential impact on the scan rate of the axion detector is dramatic. As a scenario, suppose that the cavity is cooled to $T = 50$ mK with a dilution refrigerator, and that the readout amplifier has a noise temperature $T_N = 50$ mK. Thus, the system noise temperature $T_S = 100$ mK. Since the scan rate scales as T_S^2, the scan time becomes $270(0.1/3.2)^2$ years ≈ 100 days!

The very low noise temperature of the MSA spurred a first-step upgrade of the axion detector in which the HEMT was replaced with an MSA while the temperature was maintained at about 2 K. Since blackbody noise from the cavity is not reduced, the decrease in scan time is modest. Rather, the object of the upgrade was to demonstrate that the MSA could indeed operate as expected on the axion detector. In fact, the system worked extremely well at a frequency of about 842 MHz. As of this writing, 88,732, 80-sec data sets have been acquired, corresponding to a net 82 days of data[112].

5.7.8 Gravitational Radiation Detectors

Peter F. Michelson

The early ideas about cryogenic resonant-mass detectors of gravitational radiation began with discussions at Stanford University between William Fairbank and William Hamilton in the mid-1960s concerning experiments to measure gravitational effects with high precision (see Hamilton's essay[113] for more details.). Hamilton was a graduate student at the time, working with George Pake and Arthur Schawlow. Eventually Hamilton worked as a postdoc with Fairbank. In 1967 Fairbank and Hamilton visited Joseph Weber at the University of Maryland and saw what he was doing with large, resonant-mass gravity wave (GW) detectors operated at room temperature.

A resonant-mass GW detector couples to the time-dependent "tidal" force of a gravitational wave. The tidal force is transverse to the direction of propagation. A pulse of such radiation, produced by a cataclysmic cosmic event such as the coalescence of a binary neutron star system or a supernova core collapse, will excite mechanical modes of the antenna (a large, solid cylindrical bar), particularly the fundamental mode. One of the limiting sensitivity factors for such an antenna is the Brownian motion noise that is proportional to $k_B T/Q$, where T is the temperature of the antenna and Q is the mechanical quality factor of the mode[114].

[111] P. Sikivie, *Phys. Rev. Lett.* 51 (1983) 1415

[112] S.J. Asztalos et al., *Phys. Rev. Lett.* 104 (2010) 041301

[113] J.D. Fairbank et al., *Near Zero: New Frontiers of Physics* (W.H. Freeman and Company, New York, 1988)

[114] P.F. Michelson et al., *Science* 237 (1987) 150

FIGURE 5.40: The Stanford 4,800 kg low-temperature gravitational radiation detector, circa 1980. Team members shown are (left-to-right) Bill Fairbank, Mike McAshan, Steve Boughn (above), Jim Hollenhorst, Robin Giffard, and Brad Reese. Vladimir Braginsky on the far right was visiting. (Courtesy Jack Gilderoy).

It was immediately clear to Fairbank and Hamilton that an enormous gain in sensitivity could be made by designing a detector to operate at low temperature. Fairbank's initial idea was to operate a five-ton resonant-mass aluminum cylinder at 3 mK! This would reduce the thermal noise by at least a factor of 10^5. There were other benefits of low-temperature operation. In particular, the mechanical Q of the antenna material improves significantly as the temperature is reduced, thus further reducing the mechanical noise contribution. So, with much promise of major improvements in sensitivity, the group at Stanford first designed and constructed a 680 kg detector operated at 4 K, followed by a 4,800 kg detector, initially operated at 4 K. The photograph in Figure 5.40 shows some of the members of the Stanford team, circa 1980.

Aside from the cryogenic design, two of the key challenges in the design of the detector were (i) vibration isolation and (ii) the readout system for detecting the vibrations of the fundamental mode of the antenna. The readout system developed for the Stanford detector is an example of "applied superconductivity". Both Robin Giffard and Ho Jung Paik played key roles in the development of the superconducting transducer readout of the antenna, shown schematically in Figure 5.41. Oscillations of the antenna are coupled to the fundamental eigenmode of a superconducting diaphragm, which modulates the inductance of current-carrying superconducting pickup coils, causing an oscillating voltage proportional to the velocity of the diaphragm to appear at the output terminals. The output signal from the transducer is fed to the input coil of a SQUID.

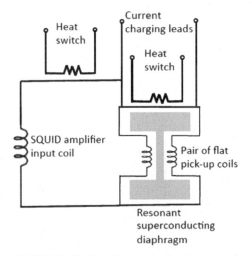

FIGURE 5.41: Superconducting motion transducer.

Giffard showed[115] that the ultimate sensitivity of a gravitational radiation detector of this type (i.e., a detector that uses a linear motion detector) is determined by the quantum-mechanical limit of linear amplifier performance, namely, the noise temperature of the system T_N obeys the limit $T_N \geq \hbar\omega/k_B$. This limit is equal to 40 nK at 850 Hz, the frequency of the Stanford detector. In operation at 4 K, this detector achieved a noise temperature of 10 mK, corresponding to a dimensionless strain sensitivity $\Delta l/l \approx 3 \cdot 10^{-18}$. The strain sensitivity of a quantum-limited detector is $\Delta l/l \approx 6 \cdot 10^{-21}$.

In recent years, long-baseline (4 km) laser interferometer detectors such as LIGO have surpassed the sensitivity of resonant-mass detectors, primarily because the baselines of interferometers are much longer (that is, 4 km versus 4 m).

5.7.9 Future Directions

John Clarke, Kent Irwin

In the spring of 2010, the axion detector (ADMX) was moved from LLNL to the University of Washington in Seattle, Washington. Over the course of the next several years a second-step upgrade will include the incorporation of a dilution refrigerator to cool the cavity and the MSA to 100 mK or lower. The combination of the cavity and the MSA cooled to this temperature will turn the axion search into an extremely viable project. As part of the upgrade, new cavities and MSAs will be designed and fabricated to extend the operating frequency to substantially higher values to increase the axion-mass range of the search. Furthermore, it has become apparent that ADMX can be used to search for other kinds of exotic particles, such as chameleons[116] and hidden sector photons[117].

Other applications of the MSA are also being pursued, notably as a near-quantum limited readout amplifier for superconducting qubits (Chapter 6).

As larger arrays of superconducting transition-edge sensors are developed, it will be necessary to increase the number of pixels that can be multiplexed in each output channel. A promising approach is to couple each TES pixel to an rf SQUID that is part of a high-Q, lithographically-patterned superconducting microwave resonator. Many thousands of these rf SQUID resonators can then be read out with a single HEMT amplifier, making it possible to develop arrays of hundreds of thousands of TES detectors.

5.8 Concluding Remarks

A. I. Braginski, John Clarke and Ronald E. Sager

This Chapter presents largely personal reminiscences on how SQUIDs and also superconducting radiation and particle detectors were conceived, understood and developed. Today, the fundamental principles of SQUIDs are well understood and the technology of low-T_c devices is relatively mature. While high quality low-T_c SQUIDs are readily available, there is an ongoing need for improved, high-yield and low-cost fabrication processes for the large SQUID arrays and related readout circuits that are now coming into use. The technology of high-T_c SQUIDs is not yet mature.

Low-T_c SQUIDs are solidly anchored in a growing number of applications at the frontiers of science—as extremely sensitive sensors of magnetic field or flux, as amplifiers, and as a variety

[115] R.P. Giffard, *Phys. Rev. D* 14 (1976) 2478

[116] G. Rybka et al., *Phys. Rev. Lett.* 105 (2010) 051801

[117] A. Wagner et al., *Phys. Rev. Lett.* 105 (2010) 171801

of readout devices. As examples, we have seen the remarkable improvements in the detection of far infrared and X-ray radiation, in the search for exotic particles, in the quest for understanding dark energy and in the detection of nuclear materials for security purposes. Some of these systems involve thousands of sensors; none of them would be remotely possible without the extremely low noise energy of the SQUID that permits one to benefit from the extraordinary sensitivity of TES detectors. Quantum Design's measurement systems using low-T_c SQUIDs have become indispensable scientific laboratory tools and a remarkable industrial and market success. Other applications do not necessarily take advantage of the ultimate noise energy of the SQUID, and are implemented because SQUIDs offer significant advantages over other techniques. An excellent example is geophysics, where the growing use of low- and high-T_c SQUIDs in geophysical mineral exploration has led to the discovery of large ore deposits. In nondestructive evaluation (NDE), low- and high-T_c SQUIDs are used to solve especially difficult problems. An example is the major success in locating impurities in niobium sheets destined for superconducting cavities for particle accelerators. In both geophysics and NDE, the use of SQUIDs resulted in a significant improvement of the methodology using conventional sensors.

Over the last decade, the supremacy of SQUIDs as the lowest noise magnetometers has been challenged by the development of optically pumped atomic (absolute) magnetometers. These operate at and above room temperature with a noise floor of the same order of magnitude as SQUIDs[118]. However, the vector nature of the SQUID response, its inherently wide bandwidth, and the fact that it is sensitive to magnetic flux (field) change rather than to its absolute value have made it hard or impossible to replace in most current applications. Nevertheless, as they are shrunk, SQUIDs achieve lower flux noise, but higher field noise. Should one need the lowest magnetic field noise in a device with extremely small area, the atomic magnetometer may be the device of choice in some future applications. The issue of SQUID cooling and the importance in many applications of replacing liquid cryogens with low-noise (and minimal maintenance) mechanical cryocoolers were emphasized in the context of laboratory devices. While in advanced and especially long-term space applications this mode of cooling might be the only possibility, the availability of suitable cryocoolers could also lower the acceptance threshold of SQUIDs in some other fields of application, especially industrial and clinical. The development of sufficiently "quiet" cryocoolers, "invisible" to the user, highly reliable, nearly maintenance-free and relatively inexpensive remains a highly desirable goal for the future of SQUID applications.

Acknowledgments

In co-authoring this chapter, John Clarke was supported by the Director, Office of Science, Office of Basic Energy Sciences, Materials Sciences and Engineering Division, of the U.S. Department of Energy under Contract No. DE-AC02-05CH11231.

---

[118] I.K. Kominis, et al., *Nature* 422 (2003) 596

5.9 Further Reading

SQUIDS and SQUID Applications

1. *The SQUID Handbook.* vol. I: *Fundamentals and Technology of SQUIDs and SQUID Systems*, eds. J. Clarke and A. I. Braginski, Wiley-VCH GmbH & Co., Weinheim, 2004.

2. *The SQUID Handbook.* vol. II: *Applications of SQUIDs and SQUID Systems*, eds. J. Clarke and A. I. Braginski, Wiley-VCH GmbH & Co., Weinheim, 2004.

3. *SQUID Sensors: Fundamentals, Fabrication and Applications*, ed. H. Weinstock, Kluwer Academic Publishers, Dordrecht, 1996.

4. Trends in SQUIDs and their applications are well covered by bi-annual proceedings of the Applied Superconductivity Conference (ASC), which appear in *IEEE Transactions on Applied Superconductivity* (TAS). The most recent are Proceedings of ASC 2008: TAS vol. 19, No. 3 (2009).

5. Another bi-annual conference covering progress in SQUIDs is the European Conference on Applied Superconductivity (EUCAS). Its proceedings appear in the *Journal of Physics: Conference Series* (JCPS). Invited papers are usually published in *Superconductor Science and Technology* (SuST). Both are Institute of Physics (IOP, UK) publications. The most recent are *Proceedings of EUCAS 2009: JCPS* vol. 234 (2010) and *SuST* vol. 23, No.3 (2010).

Superconducting Radiation and Particle Detectors

1. An excellent book reference for TES detectors is a textbook chapter: K.D. Irwin and G.C. Hilton, Transition-Edge Sensors, Ch. 3 in *Cryogenic Particle Detection*, Christian Enss, ed., *Topics Appl. Phys.* 99, 63–149 (2005), Springer-Verlag, Berlin Heidelberg 2005.

2. A representative and recent collection of articles can be found in: *AIP Conference Proceedings* Vol.1185, 13th International Workshop on Low Temperature Detectors, LTD-13. Stanford, CA, 20–24 July 2009.

6

Qubits

Editor: J. E. Mooij

6.1 Qubits
   J. E. Mooij 375
6.2 General Aspects
   J. E. Mooij 384

6.1 Qubits

J. E. Mooij

6.1.1 Introduction

Quantum mechanics and superconductivity are intimately connected on many levels. Quantum mechanics is needed to understand why a metal is a metal. The BCS condensate of Cooper pairs would not exist without quantum mechanics, nor could Cooper pairs tunnel through the barriers of Josephson junctions. The name SQUID is the abbreviation of superconducting quantum interference device. Yet another level of quantumness has been added in the last two decades. It turns out to be possible to design and fabricate superconducting objects that have macroscopic dimensions and properties, but behave as if they are single quantum particles. Amplitude and phase of the superconducting wavefunction act as conjugate quantum variables. Practical devices and circuits make use of Josephson junctions to reach the parameter values that are needed, but there is a clear distinction from the Josephson electronics circuits in the other chapters of this book. The latter may, in this context, well be referred to as being classical.

Quantum bits, or qubits, are two-level systems that can be brought into a controlled quantum superposition of their two states. Quantum entanglement of two qubits can be realized by specific gate operations on interacting qubits. Single qubits and entangling two-qubit gates are in first theory sufficient to build a universal quantum computer, which can use quantum algorithms to perform certain tasks exponentially faster than a conventional computer. However, for an algorithm-processing type of quantum computer to become relevant in a practical sense, one needs qubit numbers of more than 1 million and unrealistically high operation and measurement fidelities. No technology can provide such numbers in the foreseeable future. Quantum information processing, a completely new concept with deep fundamental roots and implications, is nevertheless likely to lead to essentially new applications that are not envisioned at this time. Superconducting circuits are attractive candidates for quantum bits and quantum circuits because of their intrinsic coherence. This coherence is lower than for the more traditional quantum particles such as atoms and photons, but the possibility to design and fabricate quantum circuits on a chip has proven to be very attractive.

Quantum computation became seriously established around 1995 when qubit manipulation was demonstrated using magnetic resonance on nuclear spins in liquids and when single trapped ions and atoms could be controlled. No superconducting quantum bits had been made at that time, but in preceding years several experiments had demonstrated remarkable quantum properties. Although no theory could guarantee that quantum mechanics would work at the level of individual control of macroscopic objects, several groups moved towards the development of intentional quantum circuits. After a number of years, breakthrough experimental results established the new area of superconducting quantum information processing. This chapter will describe some of the early experiments that preceded quantum bits, it will describe various qubit types, and it will show some of the latest developments at the time of writing. The chapter is meant to give an impression of this new aspect of superconductivity, rather than a review of the field. Several reviews of the field have been published.[1,2,3]

6.1.2   Early Experiments

Macroscopic Quantum Coherence

The potential energy of a Josephson junction that is biased with a current I is equal to $V = E_j(\cos\varphi - \varphi I/I_o)$, where φ is the phase difference, E_J the Josephson energy and I_o the critical current. This is the so-called washboard potential which has many local minima as long as $I < I_o$. Around 1985, the escape from a local well, for currents near the critical value, was addressed experimentally and theoretically. The junction has capacitance, which leads to a capacitive energy $T = CV^2/2 = C(\phi_o d\varphi/dt)^2/8\pi^2$ and a natural frequency for small oscillations in the bottom of the well, called the plasma frequency. The escape rate by thermal activation could be calculated, but at very low temperatures a higher rate was measured. This might be associated with quantum processes, but obviously a small level of spurious noise could have the same effect. The decisive experiment by the Clarke group in Berkeley showed that the escape rate could be enhanced by resonant microwave excitation of the plasma oscillations. With increasing DC current the well becomes shallower and the plasma frequency decreases. This was exactly what was observed, providing the first spectroscopic evidence for quantum behavior of the macroscopic SQUID.

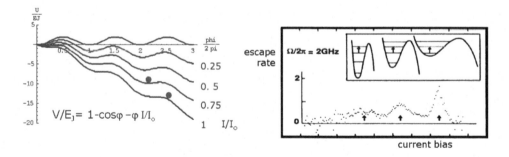

FIGURE 6.1:   Current-biased Josephson junction. Left: the washboard potential is tilted more and more with increasing current. At the critical current no metastable well remains. The oscillation frequency in the well decreases for higher current. Right: evidence for quantized oscillations, the escape rate exhibits resonances when an external RF signal is resonant with different transitions in the anharmonic well.[4]

[1]   M.H. Devoret, A. Walraff and J.M. Martinis, arXiv:cond-mat/0411174 (2004 and updates)

[2]   G. Wendin and V.S. Shumeiko, *Low Temp. Phys.* 33 (2007) 724

[3]   J. Clarke and F.K. Wilhelm, *Nature* 453 (2008) 1031

The theory for this effect was worked out by Leggett and collaborators. The equation of motion is identical to that of a particle with mass $C(\phi_0^2/4\pi^2)$ in one-dimensional space with coordinate φ. Leggett boldly quantized the equation of motion of the "particle" by replacing the "momentum" $C(\phi_0^2/4\pi^2)d\varphi/dt$ with an operator $-i\hbar\partial/\partial\varphi$. This approach gave excellent agreement with the experimental results. Caldeira and Leggett for the first time introduced damping in this quantum problem by coupling to large sets of harmonic oscillators. This method was subsequently used in many areas of physics. A good overview of experiment and theory has been presented by Clarke and collaborators[4].

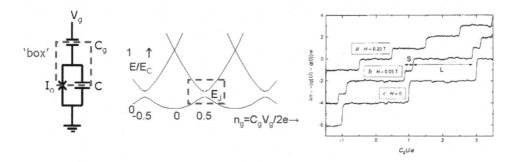

FIGURE 6.2: Cooper pair box. Left: circuit, the "box" is indicated. Middle: Energy values around a charge e on the gate capacitor. The Josephson coupling induces a superposition of the pure charge states with 0 and 1 Cooper pair on the island. Right: Measured electrical potential of the box[5]. In zero magnetic field the charge increases in units of $2e$.

Cooper Pair Box

Around 1990, several groups started to investigate the quantization of charge in a superconducting volume with small electrical capacitance. The behavior of normal metal tunnel junction circuits in the Coulomb blockade regime was by that time well understood, but it was certainly not clear what would happen in a superconducting circuit. Would the Cooper pairs be the only mobile particles, or would there always be at least a few quasiparticles? Cooper pairs are built up from many electron states in a dynamic fashion, and the range of electron states around the Fermi sphere that are involved is very vaguely defined. Could the difference between N (of order 10^7 or more) and $N + 1$ Cooper pairs have any real measurable significance? Such questions were conclusively answered in experiments on the so-called Cooper pair box. This superconducting version of the single-electron box is depicted in Figure 6.2. A superconducting "island" is a metallic volume that is separated from other metallic regions by tunnel barrier and dielectrics. There is one tunnel junction that allows transfer of particles. The charge $q = n2e$ on the box is quantized, starting from $n = 0$ where electronic charges are compensated by the positive ionic background. The charging energy is defined as $E_c = 4e^2/2C_\Sigma$, where C_Σ is the sum of all capacitance values (usually the junction capacitance C dominates). The Josephson junction creates a coupling between charge states which differ by one Cooper pair (in Figure 6.2 these are states with $n = 0$ and $n = 1$). The Hamiltonian is $H = E_c(n - n_g)^2 - E_J\cos\varphi$, where $n_g = C_gV_g/2e$. The quantum nature is introduced with the commutation relation $[\hat{n},\hat{\varphi}] = i$, and the Hamiltonian operator $\hat{H} = E_c(\hat{n} - n_g)^2 - E_J\cos\hat{\varphi}$. This leads to

---
4   J. Clarke, A.N. Cleland, M.H. Devoret, D. Esteve and J.M. Martinis, *Science* 239 (1988) 992

avoided crossings in the energy spectrum at those values of the gate voltage where charging energies for states n and $n+1$ are the same. Figure 6.2 shows the experimental evidence for clear pairing behavior of the Coulomb energy[5]. It took some years to develop the experimental conditions where the last quasiparticles could be banned, by careful filtering of leads and sometimes the addition of low-gap regions to trap the remaining single charges. Also, clear evidence for superpositions of Cooper pair states was later obtained.

Heisenberg Transistor

A peculiar experiment that nicely demonstrated the quantum nature of Josephson circuits[6] is illustrated in Figure 6.3. The experiment directly demonstrated the connection between fluctuations of the dual variables phase and charge. It is explained in the caption of Figure 6.3; squeezing of phase fluctuations leads to enhancement of charge fluctuations, visible in the transport current. The examples shown here and other experiments clearly demonstrated that Josephson circuits could be brought into the quantum regime. Almost surprisingly, the naive straightforward quantization of the macroscopic systems worked fine in explaining the new experimental findings. Now, fifteen years later, the validity of this approach is taken for granted.

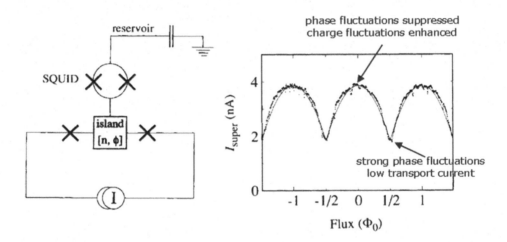

FIGURE 6.3: Heisenberg transistor[6]. On the left is the circuit with a current source connected to a Coulomb transistor without gate. The maximum supercurrent is measured. The system is in the Coulomb regime with $E_C > E_J$ and transport is possible only through fluctuations of the charge on the central island. That island is connected to a large superconducting reservoir by means of a SQUID. When the flux in the SQUID is varied, the charging energies are completely unaffected, but the supercurrent changes considerably. This is fully in agreement with quantum calculations. Strong coupling to the reservoir suppresses phase fluctuations, which in turn enhances the charge or number fluctuations. A higher supercurrent is found for zero flux in the SQUID.

5   P. Lafarge, P. Joyez, D. Estéve, C. Urbina and M.H. Devoret, *Nature* 365 (1993) 422
6   W.J. Elion, M. Matters, U Geigenmüller and J.E. Mooij, *Nature* 371 (1994) 594

6.1.3 Qubit Types

Common Features

There are many ways in which two-level states can be created in superconducting circuits. Always, number/charge and phase/current/flux act as conjugate quantum variables. Superconducting qubits make use of relatively small Josephson junctions where the ratio E_J/E_C is not too far from 1 so that quantum transitions are possible. The circuits are almost without exception made with aluminum-aluminum oxide-aluminum junctions. A high quality of the junctions is more important than a high critical temperature. Energy splittings are typically in the 5-20 GHz range; to prevent thermal occupation of levels, the experiments are performed in a dilution refrigerator at temperatures below 50 mK. Excitation and manipulation of qubit circuits is performed with resonant microwave pulses, similar to operations in magnetic resonance.

Charge Qubits

When Josephson junctions are used with charging energy E_C larger than E_J, the two levels are primarily associated with numbers of Cooper pairs on an island. The Cooper pair box, as described in the previous section, is the prototype of a charge qubit. It is used at or around a gate bias where the charge on the gate capacitor is e, half the Cooper pair charge. The first coherent manipulation of quantum states in a superconducting circuit was performed in a breakthrough experiment by Nakamura, Pashkin and Tsai[7] in 1999. In their experiment, they started by biasing the Cooper pair box at a position away from the symmetry point (Figure 6.4) where the charge is zero. Then, using

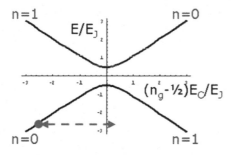
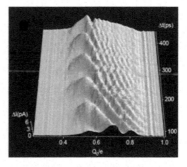

FIGURE 6.4: Left: Cooper pair box energy spectrum. The avoided crossing is shown where the Coulomb energies for zero and for one Cooper pair cross. The energies are normalized to the Josephson energy. E_j determines the splitting in the symmetry point $n_g = 1/2$. Right: Result of experiment by Nakamura et al.[7] that first demonstrated coherent dynamics of a superconducting circuit. The procedure is explained in the text. In this 3D plot x is the gate voltage normalized to e (on the scale as shown 1.0 corresponds to $n_g = 1/2$), y is the time spent in the symmetry point, z is the measurement result.

a very fast pulse generator, the bias was changed to a position close to $n_g = 1/2$ (x- axis in Figure 6.4 right) and kept there for a certain time (y-axis). The ground state is a superposition state at that bias and coming from $n = 0$ the system starts to oscillate between $n = 0$ and $n = 1$ with a frequency equal to the energy splitting. The switch back to low bias projects the system into one of the charge

[7] Y. Nakamura, Y.A. Pashkin and J.S. Tsai, *Nature* 398 (1999) 786

states. In their experiment Nakamura et al. detected the occupation of the $n = 1$ state with a special quasiparticle current detector (z-axis). A very clear oscillation pattern is visible which corresponds to the theoretical expectation. Later these experiments were expanded to direct qubit manipulation near the symmetry point and, again using ultrafast shifting techniques, to a two qubit system.

The very serious disadvantage of charge qubits is the strong noise due to variation of background charge configurations that is observed in practice. Not only the metallic gate couples to the box, but also a multitude of charged defect states. The occupation of such states varies in time with a $1/f$ spectrum that extends to high frequencies. The resulting coherence times are typically below 10 ns. So far it is impossible to fabricate circuits by means of thin film deposition and lithographic patterning without a high level of background charge noise. Using superconducting quantum bits with a ratio $E_C/E_J \gg 1$ is asking for serious trouble. Therefore the group of Devoret and Estéve in Saclay developed a hybrid charge-phase qubit called the quantronium[8] that couples the charge states to a circulating current in a junction loop. The ratio E_C/E_J is about 10 for this system; the coherence time is typically 100 ns. The emergence of a new device based on the Cooper pair box, called the transmon, with an E_C/E_J ratio significantly below 1, has now taken all priority. As the charge signal cannot be detected directly for this device, the transmon is always used in combination with a microwave resonator. The transmon will be described in Section 6.3.4 together with the resonator.

Phase Qubits

In section 6.2.1 the quantum effects in a current-biased Josephson junction were described, including the quantized states of the nonlinear resonator which is formed by the Josephson junction acting as a nonlinear inductance and the capacitance of the tunnel junction structure. The two lowest of these quantized states can be used[9] to define a quantum bit that is known as the phase qubit. This nomenclature is a bit arbitrary, as all superconducting qubits make use of phase and charge and for all of them E_J/E_C cannot be very much higher or very much lower than 1 on the penalty of losing the quantum character. The phase qubit does have the highest E_J/E_C ratio, originally around 50 but more recently about 5.

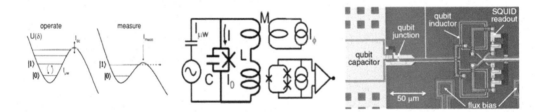

FIGURE 6.5: Phase qubit[9]. On the left the principle of operation is indicated. The bottom two levels in a metastable trap of the "washboard potential" are used as the two states of the qubit. After operations, readout is performed by increasing the bias current just enough that escape is possible from the excited but not from the ground state. The circuit is shown in the middle; a very stable current bias is created by placing the junction in a loop with large inductance. Microwaves can be applied as well as fast current shifts. If escape from the excited level has taken place, the junction phase has shifted by 2π and a fluxoid has escaped from the loop. This can be measured afterwards with the asymmetric SQUID that is at bottom right. The right picture shows the actual circuit.

---

[8]   D. Vion, A. Aassime, A. Cottet, P. Joyez, H. Pothier, C. Urbina, D, Esteve and M.H. Devoret, *Science* 296 (2002)

[9]   J.M. Martinis, S, Nam, J. Aumentado and C. Urbina, *Phys. Rev. Lett.* 89 (2002) 117901

The circuit for the phase qubit is shown in Figure 6.5. The energy splitting is tuned by the value of the bias current. For higher current the potential well becomes shallower and the oscillation frequency goes down. It is extremely important that the potential well itself is not parabolic. A harmonic oscillator would have identical distances between levels, so that selective operations on levels 0 and 1 are not possible. Here, due to the nonlinear junction inductance, the distances are different but still relatively close. Strong excitation of the qubit must be avoided to prevent leakage to the higher levels.

The phase qubit has been systematically improved, in particular by John Martinis and his group. Coherence was originally severely limited by the presence of spurious two-level defect states with energy splittings in the range of qubit energies. By reducing the junction area and by improving the materials properties, a strong reduction of their number was obtained. Coherence times are now around 500 ns. Readout is possible with a fidelity of more than 95%.

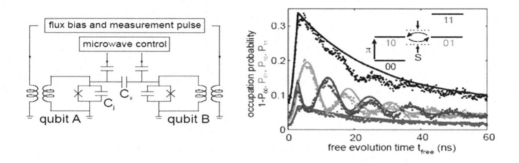

FIGURE 6.6: Swap gate for two phase qubits[10]. Left: schematic diagram showing capacitive coupling. Right: measured evolution of qubit states. The blue and the green curves indicate the states of qubits 1 and 2 respectively. They exchange their occupation periodically, for the swap gate the operation is stopped at the highest amplitude.

The Martinis group has created multi-qubit circuits by capacitive coupling of two or more qubits. Each qubit retains its own well-defined environment, driving and readout circuit. The qubits can be far apart so that cross-talk is small and qubit coherence is not reduced. Figure 6.6 shows the first SWAP gate produced in this manner[10]. Later coupling was also achieved by means of a microwave stripline, providing increased separation.

Flux Qubits

Flux qubits consist of a closed superconducting ring that is biased with a magnetic flux of about half a flux quantum.[11] Here, two degenerate or almost degenerate states are present that differ by the phase winding in the loop. The persistent currents have opposite sign. Inclusion of a small Josephson junction in the loop allows quantum tunneling of a fluxoid, at exactly half a flux quantum creating a symmetric (ground state) and an antisymmetric (excited state) of the persistent current states. The persistent currents generate a magnetic flux of about 1 $m\Phi_0$, which is measured with a SQUID. Two or more slightly bigger junctions are included to provide enough inductance for the two (meta) stable states. The tunnel coupling has a strength Δ that depends exponentially on the

---

[10] R. McDermott, R. Simmonds, M. Steffen, K.B. Cooper, K. Cicak, K.D. Osborn, S. Oh, D.P. Pappas, J.M. Martinis, *Science* 307 (2005) 1299

[11] I. Chiorescu, Y. Nakamura, C.J.P.M. Harmans and J.E. Mooij, *Science* 299 (2003) 1869

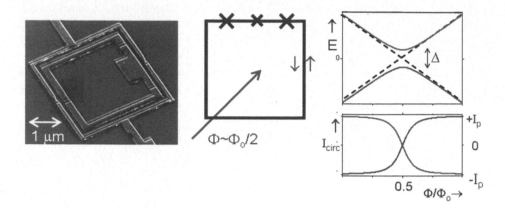

FIGURE 6.7: Flux qubit. A superconducting loop contains three or more small Josephson junctions. The flux through the loop is about $\Phi_0/2$. The classical states with lowest energy have persistent currents with opposite sign. With small Josephson junctions in the loop, quantum superpositions of these persistent current states are created as shown on the right. The ground state is blue, the excited state red. In the symmetry point 0.5, the level separation is given by the tunnel coupling Δ. The circulating currents are shown below. Qubit readout is by means of a SQUID, here shown around the qubit loop (left).

junction parameters. The flux qubit is the quantum dual to the Cooper pair box. As nature does not provide magnetic monopoles, flux noise in superconducting loops is much smaller than charge noise in capacitors. Under proper conditions the coherence time is around 1 μs at the symmetry point, where flux noise does not lead to dephasing to first order. However, at this bias point no readout signal is available as in both states the current is zero. A fast shift away from symmetry before measurement is needed. The circulating currents in the flux qubit can be used for relatively strong coupling, either to a detector, other qubits[12] or a resonator circuit. The required bias of half a flux quantum can also be provided by using a strong loop without junctions to trap one fluxoid, and attaching the qubit junction symmetrically across the middle (Figure 6.8)[13].

Coupling to Resonator, Transmon

The transmon[14] is a Cooper pair box with an E_J/E_C ratio of about 12, coupled to a resonator (figure 6.9). The energy of the charge states depends extremely weakly on the gate voltage, thereby strongly reducing the effects of charge noise. The charge states can be measured only by their dispersive effects on the resonator. The absence of other electrical circuitry helps the coherence. The system of two-level qubit plus resonator is described by means of the Jaynes-Cummings Hamiltonian

$$H_{JC} = H_r + H_a + \hbar g(a^\dagger \sigma^- + a\sigma^+) \tag{6.1}$$

which is well known in quantum optics. Here g is the coupling energy, σ indicates the qubit states and $a^\dagger$, a are the usual creation and annihilation operators for a photon. At resonance the eigenstates of the coupled system of resonator with one photon and the qubit are symmetric and antisymmetric

[12] J.H. Plantenberg, P.C. de Groot, C.J.P.M. Harmans and J.E. Mooij, *Nature* 447 (2007) 836

[13] A. Fedorov, A.K. Feofanov, P. Macha, P. Forn-Diaz, C.J.P.M. Harmans and J.E. Mooij, *Phys. Rev. Lett.* 105 (2010) 060503

[14] A. Walraff, D.I. Schuster, A. Blais. L. Frunzio, R.-S. Huang, J. Majer, S.M. Girvin and R.J. Schoelkopf, *Nature* 431 (2004) 162

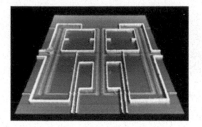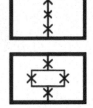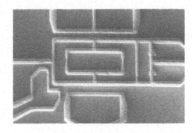

FIGURE 6.8: Flux qubits with trap loop. Left: Controlled-NOT gate[12] with two flux qubits, coupled by mutual inductance. On top are the two SQUID detectors and control lines for shifting the flux. Middle: schematic for flux qubit with trap loop. Top picture shows qubit line across the loop in which a 2π phase difference is trapped. The bias is the difference in fluxes left and right. Bottom picture shows version with tunable weakest junction. Right: flux qubit with tunable gap13. The layout is symmetric to provide decoupling of flux bias and gap tuning. SQUID is on the right.

superpositions of both elements, with energy splitting $2g$. When the frequencies differ by more than g, the apparent frequency of the resonator is shifted from its intrinsic value by $\pm g^2/\Delta$, depending on the state of the qubit. As a consequence, measuring the transmission or reflectance of the resonator gives information about the qubit state. The qubit itself can be addressed with microwave pulses that are resonant with its level splitting in the usual way. In this dispersive regime the qubit experiences shifts from two terms in the Hamiltonian. One, called the Lamb shift, is equal to the shift experienced by the resonator. The other term, called the AC Stark shift, is proportional to the number of photons in the resonator.

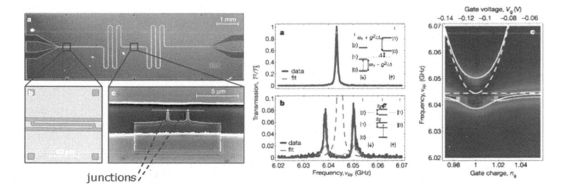

junctions

FIGURE 6.9: Transmon and resonator[14]. Left: layout of sample. The resonator is a microstripline, consisting of a center line between two ground planes. At the ends the center line is interrupted with a capacitor (bottom left) that provides partial transmission but mostly reflection. The qubit is a Cooper pair box that is capacitively coupled to the middle of the center line. The Josephson coupling strength is tuned by the magnetic flux in the junction loop. The transmission and reflection of the resonator are measured. Middle: a. transmission spectrum of resonator. At the top the qubit frequency is far away; in b. the qubit frequency is equal to the resonator frequency. Right: spectroscopy when the qubit splitting at the symmetry point is equal to the resonator frequency.

When two qubits are coupled to the same resonator and they are tuned to have the same energy splitting, they can "see" each other via the resonator, by means of the exchange of a virtual photon[15]. Figure 6.10 (left) shows the spectroscopic observation of the avoided crossing that results. For that picture the qubits are tuned with the flux through the SQUID in the Cooper pair box. The qubits can also be shifted by means of a strong microwave pulse to the cavity that increases the Stark shift. Out of mutual resonance the two qubits do not see each other. In Figure 6.10 (right) interaction is switched on for a varying time. In resonance, the two qubits exchange their quantum states (SWAP gate), as seen in the red and green traces. Independent measurement of the state of both qubits is possible through the dispersion of the cavity.

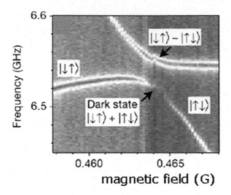 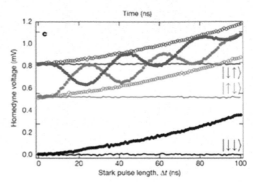

FIGURE 6.10: Coupling of two qubits via the resonator[15]. Two transmon qubits are attached to the same cavity at two different positions 4 mm apart. Left: spectroscopy on the system as a function of the magnetic field that is applied to both Cooper pair boxes, modifying the qubit frequencies. The two qubits exhibit an avoided crossing as shown. Near the crossing one qubit is not visible over a certain range, in accordance with theory. Right: First one qubit is excited with a π-pulse. Then the interaction between the qubits is switched on for varying time (Stark pulse length), resulting in counter-oscillating qubit occupations (red and green thick lines). The thin lines show behavior without coupling, the black line is the result when no π-pulse was given. The red and green open circles indicate the maximum oscillation amplitudes that should be expected.

6.2   General Aspects

J. E. Mooij

6.2.1   (De)coherence

For all superconducting quantum circuits the time of coherence for the quantum states is a severe limitation. The qubits are easy to couple to, which is often an advantage, but uncontrolled variations in the environment can have a similarly strong effect. Two different processes are distinguished,

[15] J. Majer, J.M. Chow, J.M. Gambetta, J. Koch, B.R. Johnson, J.A. Schreier, L. Frunzio, D.I. Schuster, A.A. Houck, A. Walraff, A. Blais, M.H. Devoret, S.M. Girvin and R.J. Schoelkopf, *Nature* 449 (2007) 443

characterized by a T_1 time and a T_2 time, in analogy with practice in NMR. The T_1 time indicates the rate at which the qubit exchanges energy with the environment, leading to relaxation from the excited state to the ground state. The opposite process, excitation of the qubit by a high temperature environment, is not usually important. The two main sources of decoherence are the electromagnetic coupling to the circuitry that is used for biasing, driving and readout, and coupling to materials defects that change in time.

Decohering effects from the electromagnetic environment can be approached on an engineering basis. One can calculate the spectral density of noise coming from all connections, using the real and imaginary components of the impedance as seen from the qubits over the full frequency range. For relaxation, the real component at the frequency of the qubit energy splitting is the relevant figure, as it is connected with energy transfer. For dephasing T_2 processes, it is important how the qubit energy splitting changes in time at lower frequencies. Software tools cam be used to obtain the relevant numbers and to optimize the design.

Decoherence due to materials defects can be reduced with careful optimization of fabrication processes, by operating qubits at bias values where the influence is minimal (charge or flux qubit in symmetry point), or by changing the design so that particular defects have lower effect (shunt capacitor for phase qubits). For charge and flux noise it has become clear that the noise sources are of a highly local kind and cannot be circumvented by gradiometer-type arrangements. As stated before, charge noise is very strong in the usual circuits. It is associated with electrons that are trapped and detrapped in defects. Flux noise may be connected with similar electron trapping/detrapping processes, due to the associated electron spin orientation. The choice for materials with superconducting qubits is fully dominated by the need to reduce defect noise. Aluminum yields high quality tunnel junctions, but the films are granular and have top surfaces that oxidize when exposed to air. Epitaxial fabrication of films and tunnel barriers with suitable cap layers should yield solutions. However, the small junction size that is needed to obtain the desired E_J/E_C ratio provides a severe challenge in the combination with high quality film deposition.

6.2.2 Qubit Types, General Considerations

Particular atoms all have the same quantum states with the same energies, proscribed by quantum mechanical rules. Fabricated superconducting qubits have a variation of their parameters. Josephson junction critical currents vary by several percent. It is necessary to have the ability to adjust the energy splitting. In phase qubits this is achieved by adjustment of the bias current. There is no special symmetry point where the derivative of the energy versus current is zero, so that it is essential to have a very stable current bias. In charge qubits and in the transmon the junction of the Cooper pair box is realized as a SQUID double junction, so that the enclosed magnetic flux can be used to tune the effective junction strength and the associated level splitting in the symmetry point. The magnetic flux has no effects on the Coulomb energy, which is a strong advantage. In both phase qubits and charge qubits the energy splitting is directly proportional to the junction critical current. In contrast, in flux qubits the energy splitting in the symmetry point depends exponentially on the critical current of the weakest junction in the loop. As a result, wide variations are possible after fabrication. Tunability of the weakest "alpha" junction is strongly needed, but has only recently been realized[13]. Tuning is again performed by variation of the magnetic flux in a SQUID loop, but here cross-talk to the main qubit loop has to be avoided by symmetric design or compensated.

The so-called anharmonicity is another important factor. In a harmonic oscillator all levels have the same distance, and resonant activation of the 0-1 transition has the same impact on the transitions 1-2 and higher. The phase qubit is based on oscillatory states, but due to the junction nonlinearity the oscillator has an anharmonic character (distance between 1-2 levels is 10% smaller than distance 0-1). Because these distances are still relatively close, the qubit cannot be driven strongly for fear of "quantum leakage". The transmon effectively also has an oscillator-like character and must be handled with similar caution. Qubit and gate operations have to be extended over a certain length of time. In contrast, for pure charge qubits and for flux qubits the third level is typically more than 40 GHz away and the term anharmonicity has no relevance. Qubits and gates can be driven much stronger without fear of quantum leakage.

6.2.3 Recent Results

General View

The field of superconducting qubits and quantum circuits is little more than ten years old. The first half of that time focus was on the development of qubits that were sufficiently coherent to perform the necessary single qubit operations. In the last few years emphasis has shifted to manipulation of circuits of multiple qubits, often in combination with one or more resonators. Circuit quantum electrodynamics is becoming an established field of research. Superconducting resonators in the microwave regime have a very high quality; strong coupling to the qubit two-level systems is possible. This allows experiments that are not easily paralleled in quantum electrodynamics with atoms in the optical regime. In addition, first results on mechanical oscillators that exhibit quantum behavior have been achieved. Quantum measurement, in the strongly projective as well as in the weak regime, is studied intensively.

Examples

A few examples of new developments will be given, but one should realize that even in a time as short as half a year there will be significant additional progress. Progress in the direction of performing quantum algorithms has been limited until now. The Schoelkopf group at Yale used their two-transmon system, as described in Section 6.3.5, to execute basic two-qubit algorithms[16]. In general for quantum computation one needs single-qubit gates and two-qubit gates that provide quantum entanglement. The generalized single qubit operations require accurate control of the phases of driving signals in order to reach all arbitrary superpositions of $|0>$ and $|1>$, i.e. $\sin\alpha|0> + \cos\alpha\exp(i\varphi)|1>$ with two parameters α and φ. The two-qubit gate used in their experiment is a conditional phase gate that induces phase shifts in the target qubit, depending on the state of the control qubit. After the operations, the state of the two qubits was determined by means of tomography, which consists of various combinations of single qubit phase pulses followed by readout of both qubits. The table in Figure 6.11 gives a comparison of ideal and measured values. The results are significantly better than the classical outcomes. The whole set of operations takes about 10% of the coherence time, consistent with the deviations.

Both the Yale group of Schoelkopf (transmons) and the Santa Barbara group of Martinis (phase qubits) have succeeded in creating and detecting three-qubit entanglement. Figure 6.12 illustrates the method of the latter group[17] to create two fundamentally different three-qubit entangled states, indicated as $|GHZ>$ and $|W>$. The results are not ideal, but prove genuine entanglement.

Spectacular progress has been made in circuit quantum electrodynamics with superconducting resonators coupled to qubits. Resonators can be used to manipulate and to couple qubits, as well as to read out. Conversely, qubits can be used to manipulate and couple resonator states, as well as detecting the photon states of the cavity. A beautiful example is the creation, by the Martinis group, of arbitrary Fock states of a resonator.[18] A Fock state has a distinct number of photons. With the aid of qubits it is possible not only to load photons one by one into the resonator, but also to create

---

[16] L. DiCarlo, J.M. Chow, J.M. Gambetta, L.S. Bishop, B.R. Johnson, D.I. Schuster, J. Majer, A. Blais, L. Frunzio, S.M. Girvin and R.J.Schoelkopf, *Nature* 460 (2009) 240

[17] M. Neeley, R.C. Bialczak, M. Lenander, E. Lucero, M. Mariantoni, A.D. O'Connell, D. Sank, H. Wang, M. Weides, J. Wenner, Y. Yin, T. Yamamoto, A.N. Cleland and J.M. Martinis, *Nature* 467 (2010) 570

[18] M. Hofheinz, H. Wang, M. Ansmann, R.C. Bialczak, E. Lucero, M. Neeley, A.D. O'Connell, D. Sank, J. Wenner, J.M. Martinis and A.N. Cleland, *Nature* 459 (2009) 546

Element		Grover search oracle*				Deutsch–Jozsa function†			
		f_{00}	f_{01}	f_{10}	f_{11}	f_0	f_1	f_2	f_3
$(0,0\|\rho\|0,0)$	Ideal	1	0	0	0	0	0	1	1
	Measured	0.81(1)	0.08(1)	0.07(2)	0.065(7)	0.010(3)	0.014(5)	0.909(6)	0.841(9)
$(0,1\|\rho\|0,1)$	Ideal	0	1	0	0	0	0	0	0
	Measured	0.066(7)	0.802(9)	0.05(1)	0.054(8)	0.012(4)	0.008(4)	0.031(8)	0.04(2)
$(1,0\|\rho\|1,0)$	Ideal	0	0	1	0	1	1	0	0
	Measured	0.08(1)	0.05(1)	0.82(2)	0.07(1)	0.93(1)	0.93(1)	0.05(1)	0.04(1)
$(1,1\|\rho\|1,1)$	Ideal	0	0	0	1	0	0	0	0
	Measured	0.05(2)	0.07(1)	0.06(1)	0.81(1)	0.05(1)	0.04(1)	0.012(9)	0.07(2)

Fidelity of the reconstructed output states of the Grover and Deutsch–Jozsa algorithms to their ideal outputs. These results suggest that, if combined with single-shot readout, the two algorithms executed with this processor would give the correct answer with probability far exceeding the 50% success probability of the best classical algorithms limited to single calls of the oracle' or function.

FIGURE 6.11: Quantum algorithms performed with two transmon qubits[16]. A comparison is shown between the ideal outcome values and the measured values for the Grover search and Deutsch-Jozsa algorithms for two qubits.

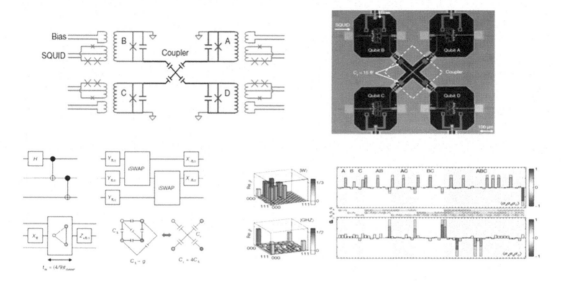

FIGURE 6.12: Experiment to create three-qubit entangled states with phase qubits[17]. Top: circuit employed, with four qubits which are capacitively coupled. Only three qubits are used in the experiment. Bottom left: protocols to create the entanglement. Bottom right: tomographic analysis of the outcome states with outcomes for the $|GHZ>$ state on top and for the $|W>$ state below. Ideal outcomes are indicated with grey, real outcomes are colored.

superpositions of these Fock states. Figure 6.13 demonstrates how close the actual result comes to the calculation.

6.2.4 Future

Can a useful quantum computer be built with superconducting qubits? Are superconducting qubits better than trapped ions when scaling up to large systems? Will spin-qubits in semiconductor quantum dots in the long run be more efficient? Such questions should be asked, but cannot be answered yet. At present, atoms and ions are still clearly ahead when it comes to the number of qubits, fidelity of operations and fidelity of readout. However, the fast rate of development for superconducting circuits will quickly reduce their advantage. Spins in quantum dots have not yet reached the coherence needed for scaling up to more than one or two qubits. Elimination of noise due to nuclear

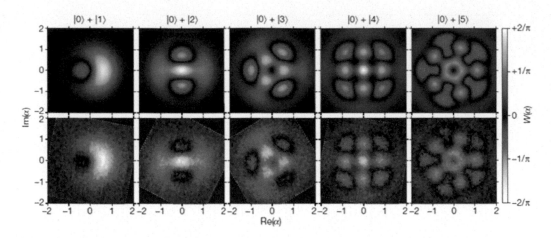

FIGURE 6.13:   Wigner tomography[18] of superpositions of the resonator state with zero photons and the resonator state with n photons, n =1-5. The top shows the result as calculated, the bottom the result as measured.

spins is needed and may be possible. Other qubit types, such as hybrid optical-electronic elements, may emerge. Both for ions and for superconducting qubits there is as yet no clear path for scaling up to very large numbers. With superconducting qubits the general notion is to use a resonator as a "bus" to couple a set of about 10 qubits. Coupling between resonators is possible using qubits.

Researchers in quantum informatics are busily developing quantum algorithms that can perform certain tasks much faster than any conventional computation method. The flagship example is Shor's algorithm for factorization of large numbers. Quantum computers will be exponentially faster for this purpose, and if they are not as reliable as Intel processors, quantum error correction may be applied. However, extreme requirements for numbers and for fidelity will have to be satisfied. Technically no hard walls against further development are visible, but it seems very unlikely that the effort of many thousands of man-years needed to obtain the materials control and the reliable operation methods will be financed by any sponsor, with no intermediate valuable goals in sights.

It is almost certain that at the time of the second centennial anniversary of superconductivity there will be important applications of quantum circuits made with superconducting elements. So far, quantum mechanics was mainly used to understand the structure of nature around us. As we learn, as engineers, to build new objects that follow quantum mechanical rules, a new world is opening. At this time full simulation of a quantum system with only 25 independent variables is not possible on the largest conventional computer system. In a few years we will be able to fabricate systems of that size and we can investigate how they behave in our experiments. We should not primarily try to imitate known substances, but we should look for completely new phenomena. Superconducting qubits and circuits are naturally suited to explore this new realm. Applications are likely to be very different from what we can imagine now.

7

Digital Electronics

Editor: Shinya Hasuo

7.1   Introduction
      Shinya Hasuo 389
7.2   Operating Principles of Digital Circuits
      Theodore Van Duzer 390
7.3   Digital Electronics in Japan
      Shinya Hasuo 397
7.4   Digital Electronics in the USA
      Fernand (Doc) Bedard 407
7.5   Digital Electronics in Europe
      Horst Rogalla 415
7.6   Integrated Circuit Fabrication Process
      Mutsuo Hidaka 424
7.7   High-Speed Digital Circuits
      Akira Fujimaki 431
7.8   History of Superconductor Analog-to-Digital Converters
      Oleg Mukhanov 440
      Acknowledgments 458

7.1   Introduction

Shinya Hasuo

Since the first electronic computer, ENIAC, was demonstrated in 1946, we human beings have been pursuing faster computers. Basic computer elements executing digital operation have been progressing continuously from the age of vacuum tubes. The basic elements have been changed to transistors, integrated circuits, and then large scale integrated circuits in this half century. During these periods, the operating speed of today's computers has increased more than one billion times that of ENIAC. Some applications, such as weather forecasting, protein design, and astronomical simulation will never be satisfied by the speed of today's computers. In the field of telecommunication, we also require very large scale routers with operation speed faster than 10 terra bits per second to deal with the huge amount of information on the Internet. Moreover, modern wireless communications requires wider and wider bandwidths in order to handle higher data rates and increase data capacity. This demands higher sample rate mixed signal circuits and higher clock rate digital signal processing circuits. In order to satisfy the need for higher speeds, researchers have been developing tiny transistors with shorter gate lengths; now they are developing transistors with twenty nanometer gate lengths. Many engineers fear, however, that the switching speed of transistors will saturate in the near future. Thus, completely new switching elements, such as superconducting devices, single electron transistors, molecular devices, spin transistors, quantum devices and so forth, are now being investigated. Among them, superconducting devices have made remarkable progress.

The most important feature of the superconducting device to be applied to the digital circuits is high-speed switching and low power consumption. The aim of development of digital circuits with superconducting devices is not only for high speed computers but also for telecommunication systems such as high-speed router systems and wireless communication systems including satellite communications. Superconducting digital circuits can also be applied to high-speed measurement systems such as real-time oscilloscopes and digital processing circuits for multichannel sensor systems and q-bit systems. Superconducting digital circuits can be applied to a wide variety of fields that require extremely-high-speed operation with low power dissipation.

The commencement of digital application of superconducting devices is historically rather old. In 1956, a superconducting device named "Cryotron" was applied to make digital circuits. It utilized switching between the superconducting state and the normal-conducting state of a thin film. Since then, superconducting devices have progressed stepwise or discontinuously. In the end of the 1960s, the Josephson junction was applied as a switching element instead of the "Cryotron". It utilized the switching between the voltage state and the superconducting state of a Josephson junction, so-called "voltage-state" switching. It was much faster than the "Cryotron". Various high speed circuits were developed using the "voltage-state" switching. From the beginning of 1990s, the circuit operation speed was drastically improved by using a different switching mode of superconducting devices. It is called the "SFQ" mode switching, which utilizes the single flux quantum as an information carrier. The "SFQ" mode operating circuits are much faster than the "voltage-state" mode circuits. Thus almost all digital circuits are now constructed with the "SFQ" mode switching of superconducting devices. Although silicon transistors, of course, have progressed steadily in this half century, superconducting devices have progressed with much faster speed because of drastic changes of the switching mode. Superconducting devices have kept an advantage of higher speed operation than that of silicon transistors of the time. These histories are described in this Chapter and the predominance of the "SFQ" over any other high-speed devices is discussed.

This chapter describes, firstly, operating principles of digital electronics, and then its brief histories in the USA, Europe, and Japan. The recent fabrication process of digital circuits with mainly niobium material is introduced. High speed digital circuits and mixed signal processing are described with the latest data, which show the advantage over other high speed systems.

7.2 Operating Principles of Digital Circuits

Theodore Van Duzer

7.2.1 The Basic Switch

There are two different and important simple circuits with embedded Josephson junctions that will act as switchable devices. Their importance lies in the very high speed at which the switching occurs as well as the very low energy dissipated in the switching. A number of families of logic and memory have been based on them.

Shortly after the demonstration of the Josephson effect in 1963, it was realized that the Josephson junction could be used as the basis of digital circuits[1,2]. Figure 7.1 shows the type of circuit for which the "0" and "1" of binary logic are represented by zero voltage and a few millivolts, respectively. Numerous families of logic were based on this so-called "voltage-state" type of switch but they were limited in speed by the need to provide clocking that would reset the switches. After the early 1990s, new logic families were almost exclusively based on a single-flux-quantum (SFQ) switching concept to be discussed below. Still, there are situations where voltage-state switching is required. One such case is the interface of high-speed SFQ Josephson logic circuits having extremely small switching energies to higher energy level semiconductor devices.

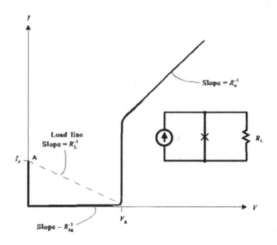

FIGURE 7.1: Basic Josephson hysteretic $I-V$ characteristic for voltage-state digital circuits and the embedding circuit (inset).

The voltage-state switch in Figure 7.1 has a resistive load, in this case matched to the normal-state resistance of the Josephson tunnel junction. Figure 7.2 shows a pattern of drive (inset) and switching response to the drive. As long as the supply current is below the "critical current" I_c, the voltage is zero. After switching, The resulting voltage across the junction builds up and contains oscillations at the Josephson frequency, $f = 2(e/h)V$, where V is the instantaneous voltage across the junction (with the oscillation amplitude determined by the circuit parameters), as long as the drive is applied. When the drive is removed, the junction resets to the zero-voltage state but with decaying oscillations.

Figure 7.3 shows the switching of a circuit representative of the switching that occurs in single-flux-quantum circuits. The input pulse causes a current to flow through the input junction that briefly exceeds its critical current. A transient voltage pulse, typically a fraction of a millivolt in amplitude and a few picoseconds in width, appears across the junction. The energy in such a pulse is typically on the order of 10^{-19} J, which is several orders of magnitude lower than the energy involved in a CMOS switching event. Some of the many logic circuits that have been based on this kind of switching will be discussed below.

---

[1] J. M. Rowell, Cryogenic Supercurrent Tunneling Devices, US Patent No. 3,281,609, October 25, 1966.

[2] J. Matisoo, *Appl. Phys. Lett.*, Vol. 9, pp. 167-168, 15 Aug.1966.

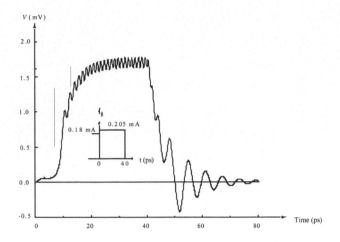

FIGURE 7.2:   Typical switching event for an overdriven Josephson junction with a hysteretic $I - V$ characteristic as in Figure 7.1. The pattern of current drive is shown in the inset.

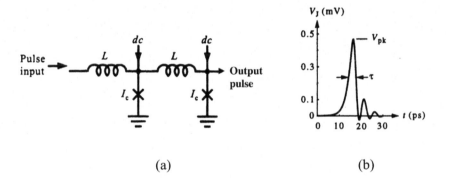

(a)   (b)

FIGURE 7.3:   (a) A common circuit situation in Rapid Single Flux Quantum (RSFQ) logic for switching a Josephson junction with a nonhysteretic I-V characteristic. Part (b) shows the voltage across the input junction when the circuit is triggered by an input pulse.

7.2.2   Logic Circuits

In the 1970s and 1980s, a variety of families of voltage-state logic circuits were developed. The largest single project was at IBM, which started just before 1970 and continued until 1983 and reached about 150 researchers at its apex. A large multi-laboratory project in Japan during the 1980s developed several voltage-state logic families. A group at Moscow State University developed a family of logic circuits based on the movement of single magnetic flux quanta[3]. These so-called Rapid Single Flux Quantum (RSFQ) circuits have been the basis of almost all subsequent research in Josephson digital circuits worldwide since then. More detail on the history of Josephson digital circuits can be found in Van Duzer and Turner[4] and Duzer[5].

---

[3]   K. K. Likharev, O. A. Mukhanov, and V. K. Semenov, in *SQUID'85*, Berlin, Germany: W. de Gruyter (1985) 1103

[4]   T. Van Duzer and C. W, Turner, *Principles of Superconductive Devices and Circuits*, 2nd Ed., Ch. 5, Prentice-Hall, Upper Saddle River, NJ, USA, 1999

[5]   T. Van Duzer, *IEICE Trans. Electron.* E91 (2008) 260

Voltage-state Circuits

Mainly in the 1970s and 1980s, there was extensive research on logic families using Josephson switches based on the principal in Figures 7.1 and 7.2. Although the logic subsequently studied in the 1990s and beyond employed single flux quantum circuits, there continues to be a need for some voltage-state circuits. An example is the interface between the SFQ logic and semiconductor devices. As mentioned above, the pulses in the single-flux-quantum circuits are of very low energy. While this is attractive for ultra-high-speed logic, it is a problem when interfacing these circuits to semiconductor circuits. Some voltage-state circuits find application in this niche. One simple compact circuit that can convert an SFQ pulse to a sustained voltage pulse with an amplitude of a few-millivolts is the 4JL gate shown in Figure 7.4 [6]. The junctions have the hysteretic I-V characteristic shown in Figure 7.1. The loop containing the junctions is small; its operation depends on the phase shifts in the junctions, not on inductance in the loop. The supplied clock sets the current through J_2 close to its critical current. An input current pulse causes J_2 to switch to the voltage state which, in turn, causes a transfer of the supply current to the right-hand branch. This switches the right branch to the voltage state thus leading to an output voltage sustained by the applied clock.

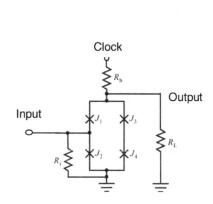

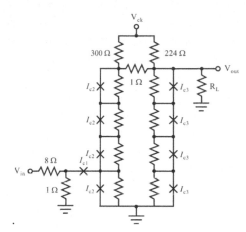

FIGURE 7.4: The basic 4JL voltage-state logic gate comprising Josephson junctions with hysteretic *I-V* characteristics.

FIGURE 7.5: Suzuki stack. Operation is similar to that of the 4JL in Figure 7.4 but with higher voltage output.

Another circuit frequently employed for interfacing millivolt Josephson circuits to volt-level semiconductor circuits employs two parallel series arrays, each with a number of Josephson junctions as shown in Figure 7.5. It is frequently succinctly referred to as a "Suzuki stack" after one of its inventors[7]. The junctions are either unshunted or somewhat resistively shunted; the voltage output is the sum of the gap voltages in the series, or somewhat less if they are shunted. The resistive shunting is used to improve the high-frequency operating margins. The input causes the bottom junction in the left branch to switch into the voltage state. This diverts the current that was flowing in the left branch into the right branch, causing, in turn, the switching of the junctions in the right branch. The voltage thus developed diverts the supply current to the left branch and switches those junctions. At that final point, all of the junctions are in the voltage state. This process takes place in

---

[6] H. Nakagawa, E. Sogawa, S. Kosaka, S. Takada, and H. Hayakawa, *Jpn. J. Appl. Phys.* 21 (1982) L198

[7] H. Suzuki, A. Inoue, T. Imamura, and S. Hasuo, *Tech. Digest*, International Electron Device Meeting San Francisco, (1988) 290. A similar double stack was used at the same time independent work: S. R. Whiteley, E. R. Hansen, G. K. G. Hohenwarter, F. Kuo, and S. M. Faris, in *Interconnection of High Speed and High Frequency Devices and Systems*, SPIE 947 (1988) 138

a few tens of picoseconds, depending on the design.

Single-Flux-Quantum Logic Circuits

Early SFQ circuits demonstrated in the 1970s and early 1980s included shift registers[8] and counters[9]. The counters comprised a series of binary dividers and were used in analog-to-digital converters. We will concentrate here on RSFQ logic circuits, which mainly came later[10].

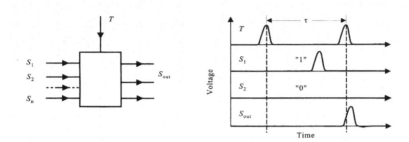

FIGURE 7.6: Method of identifying "1"s and "0"s in RSFQ logic gates shown here for an OR gate.

The basic idea of a logic block is shown in Figure 7.6 for an OR gate. The circuits are clocked by a continuous series of clock, or timing, pulses. If an input pulse arrives on one of the inputs during the interval between timing pulses, as for S_1 in the figure, it is considered to be a logical "1" input. If no pulse arrives on an input, as for S_2 in the figure, it is considered to be a logical "0". Since this example is an OR gate, an output pulse S_{out} results as shown. Other logic functions employ the same criteria for distinguishing "1"s and "0"s.

The RSFQ family contains two types of circuits, unclocked connection circuits and clocked latches. The unclocked connection circuits provide the functions of buffering the input from the output, combining inputs, and providing fan-out for latching gates. As an example of an unclocked interconnecting circuit, consider the pulse splitter shown in Figure 7.7, which is used to provide fan-out. The input at terminal A switches junction J_1, which is biased by the dc current I_{b1} close to its critical current. That switching produces a pulse that drives both output branches, causing switching of junctions J_2 and J_3 with consequent pulses at outputs B and C.

One of the common RSFQ latches is the RS flip-flop shown in Figure 7.8. In its initial state, the bias current I_b is divided between junctions J_3 and J_4 such that J_3 is biased close to its critical current and J_4 is biased at a lower level. When a pulse arrives at the input S, it drives J_3 above its critical current, which then switches briefly to the voltage state and causes a diversion of the bias current to J_4, raising its bias to a level close to its critical current. (The currents in J_3 and J_4 are equivalent to an equal current in the two junctions plus a clockwise circulating current in the J_3, L, J_4 loop corresponding to one magnetic flux quantum in the loop.) When a clock or timing pulse arrives at the reset R input, it drives J_4 beyond its critical current; a voltage pulse is generated that causes a resetting of the currents in J_3, L, and J_4 to their initial conditions and the production of an

---

[8]   T. A. Fulton, R. C. Dynes, and P. W. Anderson, *IEEE* 61 (1973) 28

[9]   J. P. Hurrell and A. H. Silver, SQUID digital electronics in *Future Trends in Superconductive Electronics*, B. S. Deaver, C. M. Falco, J. H. Harris, and S. A. Wolf (Eds.): New York, American Institute of Physics, (1978) 437 and C. A. Hamilton and F. L. Lloyd, *IEEE Electron Dev. Lett.* 3 (1982) 335

[10] K. K. Likharev and V. Semenov, *IEEE Trans. Appl. Superconduct.* 1 (1991) 3

output pulse at F. The RS flip-flop is but one of many latching logic gates that have been devised in the RSFQ family.

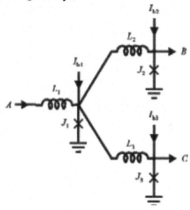

FIGURE 7.7: Splitter circuit, an un-clocked connecting circuit in the RSFQ logic family, provides fan-out.

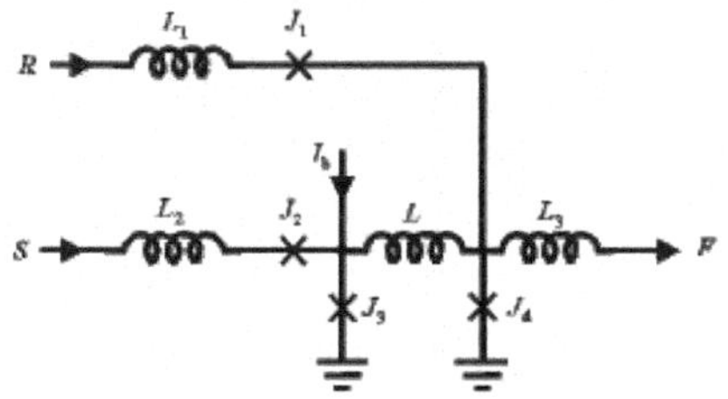

FIGURE 7.8: RS flip-flop, a latching gate in the RSFQ logic family.

7.2.3 Memory

Most of the effort to provide memory for superconducting digital systems has been directed toward random access memory (RAM), but some work has been done on first-in-first-out register-type memories[11]. Random access memories are the main interest and we focus on them here. It is well known that a circulating current in a superconducting loop will persist as long as the superconducting state is maintained. Most of the numerous concepts for random access memory cells have been based on persistent circulating currents. The earliest was reported in 1969 at IBM[12]. A later version of the IBM cell is shown in Figure 7.9. The devices shown as the write and sense gates are three-junction SQUIDs. The two states of the cell are zero circulating current to represent a logical "0" and a clockwise circulating current to represent a "1". By appropriate combinations of the currents on the I_x, I_y, and $I_{y'}$ lines, one can write either a "0" or a "1" in the cell. If a "1" is stored, the circulating current will cause the sense gate to switch when the I_{sense} is applied; the resulting voltage on the sense line signals that a "1" was stored in that cell. A complete 1 kbit memory with decoding and driving circuits was made. The difficulty of extending it to memories with larger capacity became clear and was an important part of the decision to terminate the entire IBM superconducting computer project in 1983.

Other circulating-current memory cells were developed subsequently in projects in Japan during the 1980s and complete memories were made. The project at NEC developed the vortex-transitional cell shown in Figure 7.10 and continued the development of a fully functional 4-kbit memory, which was reported in 1999 [13]. This is the largest successful memory based on circulating current memory cells that has been reported although some attempts at 16-kbit memories have been reported.

By its nature, memory employs a large number of junctions and long lines for accessing the cells. The addressing circuits are clocked at gigahertz frequencies and for practical reasons constitute an increasingly challenging design problem for memory with larger capacity. Component statistical spreads, including junction critical currents, inductances, and resistances become increasingly

---

[11] Q. P. Herr and P. Bunyk, *IEEE Trans. Appl. Superconduct.* 13 (2003) 563

[12] W. Anacker, *IEEE Trans. Magn.* MAG-5 (1969) 968

[13] S. Nasagawa, H. Numata, Y. Hashimoto, and S. Tahara, *IEEE Trans. Appl. Superconduct.* 9 (1999) 3708

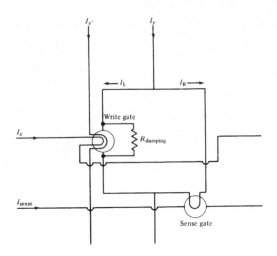

FIGURE 7.9:   The IBM circulating-current memory cell.

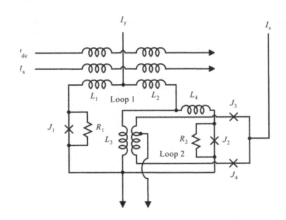

FIGURE 7.10:   NEC vortex-transitional circulating-current memory cell.

limiting as the number of components increases. An added problem for circulating-current memory cells is ambient magnetic flux and flux due to large supply currents, both of which can cause malfunctions of memory cells that function by storing magnetic flux.

An alternative approach to providing memory uses a CMOS memory which operates at volt levels and must be interfaced to the millivolt-level signals of a superconducting processor. This has the advantage over superconducting circulating-current cells in that CMOS memory structures are very compact and are highly developed. Large memory can be made with high yield. CMOS works even better at 4 K, the usual operating temperature of niobium superconducting circuits, than at 300 K, so it can be in close physical proximity to the processors.

A project at the University of California, Berkeley is focused on a hybrid of Josephson interfacing circuits and a CMOS memory as shown in Figure 7.11. The amplification of millivolt signals to volt-level signals at the input is accomplished with the combination of a Suzuki stack and a hybrid amplifier comprising a MOS device with a large series array of Josephson junctions as its load. The output current from a memory bit line is detected by a Josephson device.

An access time of 600 ps was measured for a single cell in a 64-kbit memory array using a CMOS

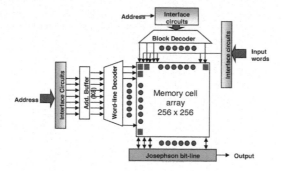

FIGURE 7.11: Architecture of a hybrid Josephson-CMOS 4 K memory.

chip made in a 180 nm process and a Josephson chip made with a 2.5 kA/cm^2 niobium process[14]. The chips were bump bonded together. Current work is directed toward writing and reading words and making improvements in access time. The use of more advanced CMOS and Josephson processes will reduce the access time considerably.

7.3 Digital Electronics in Japan

Shinya Hasuo

In this section, a brief history is described on the progress of superconducting digital electronics in Japan from the middle of 1970s to today. More than 30 years have passed since the research on digital electronics started in Japan. Many researchers have endeavored to develop advanced high speed systems during these years. But high speed conventional systems with silicon transistors have progressed steadily and continuously; the superconducting devices should have superior performance to silicon devices of the day. Thus, noteworthy changes have occurred in the history of superconducting devices for digital applications. One was the change of superconducting materials. A lead-alloy was initially used as the Josephson junction material, but it was changed to niobium (Nb) in the middle of the 1980s. In the end of the 1980s, high T_c superconductors (HTS) were introduced as junction materials, but it was difficult to make a high-quality, uniform, and reliable HTS junctions equivalent to the Nb junctions. Nb is the most reliable junction material for digital applications at present. The other change was the operating mode of the digital circuits. Until the early 1990s, "voltage-state" switches were used for the circuits, but a faster switching mode was proposed in those days. Thus, the operating mode was changed to the "SFQ" switch from the middle of the 1990s. Since then almost all digital circuits have been made with the "SFQ" mode switch, and they can be operated much faster than circuits with advanced silicon transistors. Researchers have continuously endeavored to develop superconducting high speed systems for a long time, thus a lot of remarkable and distinguished results were obtained. Typical activity and progress in these three decades are introduced here.

Many abbreviated words are used in this section for Japanese ministries and organizations, so the abbreviations used here are listed below.

---

[14] K. Fujiwara, Q. Liu, T. Van Duzer, X. Meng, and N. Yoshikawa, *IEEE Trans. Appl. Supercond.* 20 (2010) 14

AIST	National Institute of Advanced Industrial Science and Technology
CRL:	Central Research Laboratory, presently NICT
ETL:	Electrotechnical Laboratory, presently AIST
FED:	Research and Development Association for Future Electron Device
IEICE:	The Institute of Electronics, Information, and Communication Engineers
ISTEC:	International Superconductivity Technology Center
JSPS:	Japan Society for the Promotion of Science, which belongs to MESC (presently MEXT)
JST:	Japan Science and Technology Agency, which belongs to MEXT
METI:	Ministry of Economy Trade, and Industry
MESC:	Ministry of Education, Science and Culture, presently MEXT
MEXT:	Ministry of Education, Culture, Sports, Science and Technology
MITI:	Ministry of International Trade and Industry, presently METI
NICT:	National Institute of Information and Communication Technology
NTT:	Nippon Telegraph and Telephone Public Corporation of the day, presently Nippon Telegraph and Telephone Corporation
STA:	Science and Technology Agency, presently merged in MEXT

7.3.1 The Dawn of Research on Digital Applications

In around 1970, many Japanese computer companies suffered from the limitation of operating speed of bipolar transistors. With increasing integration density of the transistors, the huge power consumption caused a serious rise in temperature of integrated circuits, and it sometimes resulted in circuit breakdown. Thus, they were trying to find the next generation high speed electronic devices. They developed various kinds of high speed devices which could replace high speed bipolar transistors. Candidates were high-speed CMOS, GaAs FET (Field Effect Transistor), and many other functional devices.

It was in the middle of the 1970s that many Japanese companies and research institutes started the development of Josephson computers. IBM had been energetically developing the Josephson computer from the end of the 1960s, and IBM's results stimulated the Japanese companies. The Josephson junction device was selected as one of the candidates for the next generation high speed device. ETL, NTT and computer companies, such as Fujitsu, Hitachi, and NEC, and many other companies and universities started the research on the Josephson computer. This was the dawn of research on superconducting devices for digital application in Japan.

In these days, superconducting circuits were constructed with the "voltage-state" switching mode, which was described in Section 7.2. This mode was continually applied to digital circuits from the beginning of the research to around 1990. But thereafter a new mode of operation, the "SFQ" mode, which can construct much faster circuits than those with the "voltage-state" switching mode, was proposed by Likharev, Mukhanov and Semenov in 1985 [15]. Since then SFQ switching mode operation has become the main stream of digital circuits. But it must be noted that another operating mode, "phase mode" operation, similar to the SFQ mode operation, had been proposed by Nakajima et al. in Tohoku University in the middle of the 1970s [16].

Since Japanese researchers in companies and universities started research and development of digital applications of superconducting devices, the Japanese government strongly supported the activities. Companies increased the number of researchers for superconducting electronics, and ministries of the Japanese government, such as MITI and MESC, started national projects to support the companies and universities from the end of the 1970s. JSPS organized a specialist committee,

---

[15] K. K. Likharev, O. A. Mukhanov and V. K. Semenov, *Proc. SQUID'85*, Berlin (1985) 1103

[16] K. Nakajima, Y. Onodera, and Y. Ogawa, *J. Appl. Phys.* 47 (1976) 1620

called #146 Committee, for superconducting electronics in 1982, and it is still working. The Technical Group on Superconducting Electronics was also organized in 1982 in IEICE, and it is still active.

Since those days, almost three decades have passed. Many fruitful results have been obtained. Efforts in those days form the basis of today's research and development on digital applications.

7.3.2   Brief History of Japanese Projects on Superconducting Digital Electronics

The progress on superconducting digital electronics has been supported by the Japanese government. The progress in this field is based on the national projects. Japanese national projects in these 30 years are described here in chronological order.

Figure 7.12 shows Japanese projects on superconducting digital devices. The upper part shows the projects with LTS (low T_c superconductor) materials, and the lower shows those with HTS (high T_c superconductor) materials. The flow of research and development of Japanese superconducting digital electronics is mostly summarized in this figure, although many other small projects are omitted here. Years for the following descriptions are expressed in the Japanese fiscal year (FY), which starts on April 1 and ends on March 31.

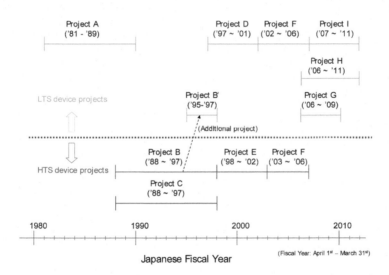

FIGURE 7.12:   Japanese projects on digital electronics 1980–2010. The name of each project is, A: Scientific Computing System, B: Superconducting Electron Devices, B': Added project to B for LTS devices as Development of Josephson Device Hybrid System Technologies, C: Superconducting Electron Materials, D: Ultimate Information Processing Function Based on a Single Flux Quantum as an Information Carrier, E: R&D of Fundamental Superconducting Application Technologies, F: Low-power Superconducting Network Device, G: Single-flux-quantum Integrated Circuits Based on Localized Electromagnetic Waves, H: Low-power, High-performance, Reconfigurable Processor Using SFQ Circuits, and I: Development of Next Generation High-efficiency Network Device Technology.

Project A

In 1981, MITI started a national project Scientific Computing System, in which 10 GFLOPS HPC (high performance computer), so-called supercomputer system, was developed. The main system was constructed with multi-CPUs (central processing units) made of high-speed CMOS. Besides the main system, three kinds of high speed devices were developed as candidates for future HPCs. These devices were Josephson junction, GaAs FET, and HEMT (high electron mobility transistor). ETL and 6 companies, Fujitsu, Hitachi, Mitsubishi Electric, NEC, Oki Electric, and Toshiba were involved in this project and they shared the technologies to be developed for the supercomputer and high speed devices. As for Josephson junction devices, ETL, Fujitsu, Hitachi and NEC were responsible for the future high speed computers.

In the beginning of the project, a lead-alloy was used for the Josephson junction material. But the lead-alloy junction was very unstable and reliable circuits could not be fabricated. From the middle of 1980s, the junction material was changed from the lead-alloy to niobium (Nb). Since then many reliable digital circuits were demonstrated. Typical distinguished results are shown later.

Project B and C

In 1988 ISTEC was established in Japan as a collaborating laboratory of many companies for the development of superconductivity technology. Many researchers joined ISTEC from not only Japanese companies and universities but also from abroad. Many Japanese companies were executing their research on HTS technology in their own laboratories and also in ISTEC by temporarily transferred researchers.

In the same year, 1988, two projects with high-T_c superconductors started. One is Project B: Superconducting Electron Devices supported by MITI through NEDO and FED, and the other is Project C: Superconducting Electron Materials supported by MITI through NEDO. They continued until 1997. Project B was carried out by ETL and eight companies, Fujitsu, Hitachi, Mitsubishi Electric, NEC, Oki Electric, Sanyo, Sumitomo Electric, and Toshiba. Nine universities collaborated with them. Project C was mainly executed by ISTEC, and ten universities collaborated. In these projects, initially various kinds of three terminal devices and YBCO thin film technology were developed. But three terminal devices were revealed to be difficult to realize with enough power gain, so the direction of the projects was slightly changed to develop YBCO thin film Josephson junctions and apply them to simple digital circuits, sampler circuits and so forth.

Project B'

LTS devices were not supported after the end of Project A (Scientific Computing System) because endeavors were concentrated on the HTS devices. But it was found to be difficult to construct large circuits with HTS devices as was done with Nb junction devices in a short term, so LTS devices were also involved in Project B from 1995 as Development of Josephson Device Hybrid System Technologies. It continued until 1997, when Superconducting Electron Devices project ended. In 1997 the LTS and HTS device projects were inherited by new projects. The LTS device was supported by STA as Project D, and the HTS device was transferred to a new MITI project (Project E).

Project D

The STA project Ultimate Information Processing Function Based on a Single Flux Quantum as an Information Carrier started in 1997 and continued to 2001. AIST, ISTEC and three companies, Fujitsu, Hitachi, and NEC, were involved and five universities, Japan Women's University, Nagoya University, Tohoku University, University of Tokyo, and Yokohama National University collaborated with them. In this project, SFQ devices and circuits were mainly developed. Interface circuits between SFQ circuits and conventional silicon devices at room temperature were also investigated.

Project E

The MITI project was funded by NEDO as R&D of Fundamental Superconducting Application Technologies. The project was a big one which included a wide area of HTS applications. It included not only HTS devices but also bulk HTS, coated conductors, and materials. In HTS devices, fabrication processes of mainly Josephson junctions and high speed circuit design were executed. Development of HTS devices was mainly carried out by ISTEC, and many companies also joined in this project such as DuPont, Fujitsu, Hitachi, Mitsubishi Electric, NEC, and Toshiba.

Project F

The LTS STA project ended in 2001, and it was inherited by a new METI project, Project F: Low-power Superconducting Network Device. In the next year, 2002, Project E ended and the HTS device was merged into Project F. Thus the project including LTS and HTS devices started as a NEDO project and continued until 2006. This project was mainly carried out by ISTEC, Nagoya University, Yokohama National University for LTS devices, and by ISTEC, Hitachi, and Advantest for HTS devices. Many remarkable results were obtained in this project both in LTS and HTS devices. These results are introduced later.

Project G

In 2006, MEXT started a new project as Grants-in Aid for Scientific Research. Six universities, University of Electro-Communication, Nagoya University, Saitama University, Tohoku University, Tokyo University of Agriculture and Technology, and Yokohama National University, joined in this project to develop basic technology for digital circuits. The project was named Single-flux-quantum Integrated Circuits Based on Localized Electromagnetic Waves. It ended in 2009.

Project H

The JST project named Low-power, High-performance, Reconfigurable Processor Using SFQ Circuits started in 2006. Kyushu University, Nagoya University, Yokohama National University, and ISTEC joined this project. The aim of this project is to develop basic technologies for a 10-TFLOPS desk-side computer based on SFQ and reconfigurable data-path (RDP) architecture. It will continue to March 2012.

Project I

After the NEDO project (Project F) for LTS and HTS devices ended, the work was inherited by new NEDO projects. The LTS device was involved in a new NEDO project, Project I: Development of Next Generation High-efficiency Network Device Technology, which started in June 2007. This project is mainly for optical communication systems. It aims at establishing enabling technologies for the next generation high-efficiency networks. In this project, the main target is to develop highly efficient large-scale edge routers, ultra high-speed local area networks and related telecommunication systems. The main technologies are developed with CMOS high-speed devices and optical I/O devices. LTS devices are also included in this project. A real-time monitoring system for optical communication will be developed using LTS SFQ devices. AIST, six IT companies and ISTEC are included in this project. Among them, only ISTEC is developing superconducting devices. The present project will continue to 2011.

The HTS device integration technologies developed in the former project (Project F) have been inherited by a new NEDO project, Technological Development of Yttrium-based Superconducting Power Equipment, so-called M-PACC Project (Materials & Power Application of Coated Conductors Project). In this project, however, HTS device technology is not applied to digital systems but to SQUID systems, so this project is not shown in Figure 7.12. By using integration technology of HTS devices, SQUID sensors are fabricated on the same chip with integrated pickup coils. The inspection system utilizing such an integrated SQUID is being used to find defects in coated conductors.

7.3.3 Progress in the 1980s

As shown in the previous section, the MITI project, Project A: Scientific Computing System, continued until 1989. ETL and three companies attained excellent results in this project. Typical achievements are summarized into two technologies. One is the fabrication technology of integrated circuits with Nb/AlOx/Nb junctions, and the other is the demonstration of high speed logic and memory circuits which are operated in the "voltage-state" mode. Fabrication technology is described in detail in Section 7.6 including recent data. High speed circuits developed in the 1980s are introduced here.

ETL demonstrated a 4-bit Josephson computer (ETL-JC1) which consisted of four chips: register and arithmetic logic unit, sequence control unit, instruction ROM (Read Only Memory) unit, and data RAM (Random Access Memory) unit[17]. The sizes of the ROM and RAM were both 1-kbit. Functional operation for 27 instructions was confirmed at low frequency. Total power consumption was 6.2 mW, and the number of Josephson junctions was 22,000. It was estimated by computer simulation that the single chip CPU including the whole circuits mentioned above could be operated at 1 GIPS (Giga-instruction per second).

Fujitsu developed two types of high-speed processors. One was a 4-bit microprocessor, and the other was an 8-bit DSP (Digital Signal Processor). The Josephson microprocessor was fabricated in 1988 for the first time[18]. It was installed in a cryostat which was connected to a refrigerator and operated in a closed cycle as shown in Figure 7.13. It functioned in the same way as the Am2901 microprocessor manufactured by Advanced Micro Devices, Inc. The Am2901 was considered the world's standard 4-bit microprocessor. A GaAs version of the Am2901 was also developed. The table in Figure 7.14 compares the performance of the Am2901 microprocessors for the three different materials. The Josephson microprocessor could be operated up to a clock frequency of 770 MHz and had a power consumption of 5 mW. Comparing these characteristics with those of the microprocessors constructed of other materials, it can be seen that the Josephson microprocessor operated at much faster speeds and with much less power than semiconductor circuits of the day.

The largest scale Josephson LSI chip of the day was an 8-bit DSP[19]. It included 23,000 Josephson junctions and it operated 100 times as fast as its CMOS logic counterpart. A photograph of the chip is shown in Figure 7.15.

The NEC group proposed a Josephson RAM with a memory cell named vortex transition memory cell. Firstly 1-kb memory circuit was published, and then it was advanced to 4-kb circuit with 21,000 Josephson junctions. Its access time was 380 ps and power dissipation was 9.5 mW[20]. This access time was the fastest of any memory chip of the day. Finally, 99.8% of the cell was confirmed to be correctly functional. A photograph of the chip is shown in Figure 7.16.

Besides the Scientific Computing System project, the Hitachi group proceeded with a new operating mode of superconducting devices, named QFP (Quantum Flux Parametron), which was originally proposed by Professor Goto of University of Tokyo. The QFP has an advantage of lower power consumption than SFQ mode operation. Hitachi demonstrated its basic function in collaboration with Professor Goto[21].

[17] H. Nakagawa, I. Kurosawa, M. Aoyagi, S. Kosaka, Y. Hamazaki, Y. Okada, and S. Takada, *IEEE Trans. Appl. Supercond.* 1 (1991) 37

[18] S.Kotani, N.Fujimaki, T.Imamura, and S.Hasuo, A Josephson 4b microprocessor, *Digest of Tech. Papers of 1988 International Solid-State Circuits Conf.* (ISSCC 1988), pp. 150

[19] S.Kotani, A.Inoue, T.Imamura, and S.Hasuo, *IEEE J. Solid-State Circuits* 25 (1990) 1518

[20] S. Nagasawa, Y. Hashimoto, H. Numata and S. Tahara, *IEEE Trans. Appl. Supercond.* 5 (1995) 2447

[21] Y. Harada, E. Goto, and N. Miyamoto, Quantum Flux Parametron, *Technical Digest International Electron Devices Meet-*

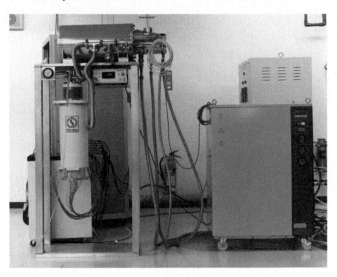

FIGURE 7.13: World's first Josephson microprocessor was installed in a cryostat, which was connected to a refrigerator and operated in a closed cycle.

Device	Si [1]	GaAs [2]	Josephson Junction
Maximum Clock Frequency (MHz)	30	72	770
Power Consumption (W)	1.4	2.2	0.005

1) AMD, 1985 Data Book
2) Vitesse, 1987 GaAs IC Symposium

FIGURE 7.14: Comparison of Am2901 type 4-bit microprocessor performance for Si, GaAs, and Josephson junction versions.

7.3.4 Progress in the 1990s

The 1990s were very busy years for researchers because HTS devices were expected to replace LTS devices. Thus projects were concentrated in the development of HTS devices. However, it was found to be difficult to make HTS Josephson junctions as reliable as the $Nb/AlO_x/Nb$ devices. In the middle of 1990s, an LTS project was added as project B', as shown in Section 7.3.2. Consequently, both HTS and LTS devices progressed to a certain extent in those days. Topical results in the Project B and B' are shown here.

A basic fabrication process for HTS junctions, i.e., a ramp-edge junction with an interface-modified barrier, was developed in 1997 by the NEC group[22]. This technology was improved in the succeeding projects (Project E and Project F) by the ISTEC group, as shown in Section 7.3.5. The NEC group also developed a sampler circuit using the ramp-edge junction[23]. This technology

---

ing (1987) 389

[22] T. Satoh, Mo. Hidaka, and S. Tahara, *IEEE Trans. Appl. Supercond.* 17 (1997) 3001

[23] M. Hidaka, S. Satoh, M. Kimishima, M. Takayama, and S. Tahara, *IEEE Trans. Appl. Supercond.* 11 (2001) 267

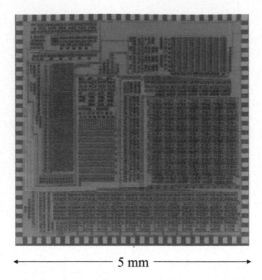

←——————— 5 mm ———————→

FIGURE 7.15:   Chip photograph of the 8-bit DSP (©1990 IEEE).

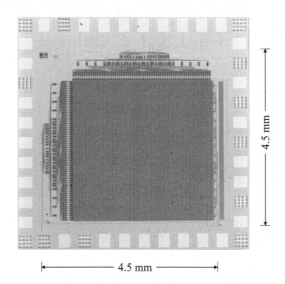

|←——————— 4.5 mm ———————→|

FIGURE 7.16:   A 4-kb memory chip (©1995 IEEE).

was inherited by the ISTEC group in Project F, and a very compact sampling oscilloscope was demonstrated as shown in Section 7.3.5. The Hitachi group demonstrated a 2:1 multiplexing circuit using QFPs with HTS bicrystal junctions[24].

---

[24] H. Hasegawa, Y. Tarutani, U. Kabasawa, N. Sugii, T. Fukazawa, and K. Takagi, *IEEE Trans. Appl. Supercond.* 7 (1997) 3446

Using LTS devices, the NEC group demonstrated a superconducting ring network system as shown in Figure 7.17 [25]. Three PCs were connected with a superconducting network switch chip. The chip has a function like a traffic rotary, i.e., data coming from one node are transferred to another node so data exchange can be performed. It was confirmed that the switch chip could be operated up to 2 GHz clock cycle. The network system was operated up to 100 MHz. This frequency was restricted by the speed of the interface ICs, not by that of the superconducting chip.

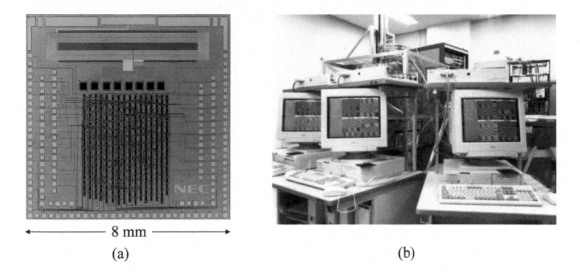

←———— 8 mm ————→

(a)   (b)

FIGURE 7.17:   Demonstration of superconducting network system. (a) shows the network switch chip and (b) shows three PCs as processor elements and a cryostat in which the switch chip and interface chips are installed (©1999 IEEE).

The interface circuit between an SFQ circuit and a room temperature electronic circuit was developed by the Fujitsu group with LTS devices[26]. They stacked eight Josephson junctions in each branch of the "Suzuki stack". They demonstrated that SFQ pulses could be amplified to 10 mV at 3 GHz and to 7 mV at 10 GHz clock. These output signals can be amplified to the order of 1 V by using semiconductor circuits at room temperature.

7.3.5   Progress in the 2000s

The integration technologies of Josephson devices with LTS and HTS materials have been advanced in Project F: Low-power Superconducting Network Device. Typical results are shown below.

LTS SFQ circuits have progressed remarkably in this project. It has become possible to integrate more than 10,000 Josephson junctions on a chip with an excellent operating chip yield as high as 10%. Josephson junctions in these SFQ circuits have been made with two different current densities. One is the critical current density J_c of 2.5 kA/cm^2 and the other is 10 kA/cm^2. The minimum line width in circuits fabricated by this process is usually 1 μm. SFQ processors and a switch for a router system have been developed. These circuits are constructed with the CONNECT cell library,

---

[25] S. Yorozu, Y. Hashimoto, H. Numata, S. Nagasawa, and S. Tahara, *IEEE Trans. Appl. Supercond.* 9 (1999) 2975

[26] N. Harada, A. Yoshida, and N. Yokoyama, *Jpn. J. Appl. Phys.* 39 (2000) L1158

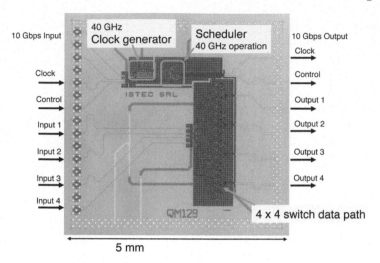

FIGURE 7.18: A 4×4 switch chip.

which includes more than 200 cells. The name "CONNECT" originally came from "Collaboration of Nagoya University, NEC, and CRL teams"[27]. But the current CONNECT group consists of IS-TEC, Nagoya University, NICT, and Yokohama National University. The cell library is the basis of the integrated circuits. By using this library, SFQ microprocessors have been developed[28]. Details will be described in Section 7.7. Another example of the CONNECT cell application is a 4×4 network switch. Its photograph is shown in Figure 7.18. This chip was operated with a clock frequency of 40 GHz; thus an effective 160 Gbps throughput was attained[29]. Using the 4×4 switch chip with a high-speed interface of 10 Gbps installed in a cryocooler, a video data transfer among four PCs was also demonstrated, as shown in Figure 7.19.

Fabrication technology for HTS integrated circuits also progressed remarkably in Project F. Typical results for the HTS fabrication technology are introduced briefly here. Figure 7.20 shows a cross-sectional view of an HTS integrated circuit. (a) is a schematic illustration and (b) shows a cross-sectional TEM (transmission electron microscope) image. The main feature of this circuit is the multilayer structure including three HTS layers, SSO ($SrSnO_3$) insulation layers, and a ramp-edge type Josephson junction with a minimum junction width of 2 μm. By optimizing the fabrication process of each layer, functional HTS circuits have been reproducibly obtained. The standard deviation $1-\sigma$ of the critical current I_c is typically 6–10% and run to run spread of I_c is ±12% [30].

This process was used to fabricate a sampler circuit. A sampling oscilloscope system has been demonstrated as shown in Figure 7.21 [31]. The system consists of a cooling unit, a control unit, and a PC. The size of the cooling unit is 140 mm$times$150 mm$times$200 mm and its weight is less than 4 kg. It is a single-stage Stirling cooler with a cooling capacity of 1 W at 77 K. A typical 50 GHz waveform has been observed with this sampler system and it has been shown that it has a potential

[27] S. Yorozu, Y. Kameda, H. Terai, A. Fujimaki, T. Yamada, and S. Tahara, *Physica* C 378-381 (2002) 1471

[28] M. Tanaka, T. Kondo, N. Nakajima, T. Kawamoto, Y. Yamanashi, Y. Kamiya, A. Akimoto, A. Fujimaki, H. Hayakawa, N. Yoshikawa, H. Terai, Y. Hashimoto, and S. Yorozu, *IEEE Trans. Appl. Supercond.* 15 (2005) 400

[29] Y. Kameda, Y. Hashimoto, and S. Yorozu, *IEICE Trans. Electron.* E91-C (2008) 333

[30] H. Wakana, S. Adachi, A. Kamitani, K.Nakayama, Y. Ishimaru, Y. Oshikubo, Y. tarutani, and K. Tanabe, *IEEE Trans. Appl. Supercond.* 15 (2005) 153

[31] H. Suzuki, M. Maruyama, T. Hato, H. Wakana, S. Adachi, K. Tanabe, T. Konno, K. Uekusa, N. Sato, and M. Kawabata, Stand alone portable HTS sampler system, EXT Abs., 11th International Superconductive Electronics Conference (2007) O-R01

to observe signals at frequencies higher than 100 GHz.

The integrated circuit technology with HTS ramp-edge Josephson junctions is now applied to fabricate a SQUID chip. The SQUID sensor with a pickup coil can be integrated on the same chip. This SQUID sensor is robust against application of AC magnetic fields up to several mT, which is four orders of magnitude higher than that for conventional HTS SQUIDs made on bicrystal substrates. This is because ramp-edge junctions and pickup coils without parasitic weak links have less probability of flux trapping. Thus, this SQUID is very useful for NDE (non-destructive examination) under fairly large excitation magnetic fields[32].

7.4 Digital Electronics in the USA

Fernand (Doc) Bedard

There have been a number of efforts to develop superconductivity into a technology for use in high performance computing.

Cryotron

Project LIGHTNING: 1956 to 1963

Josephson Junctions

Josephson Signal Processor (JSP): 1971 to 1983

Superconductive Crossbar Switch: 1991 to present

Hybrid Technology Multi-Threaded Architecture (HTMT): 1997 to 2003

Superconducting Technology Assessment (STA): 2005 to present

---

[32] T. Hato, S. Adachi, Y. Sutoh, K. Hata, Y. Oshikubo, T. Machi, K. Tanabe, *Physica* C 469 (2009) 1630

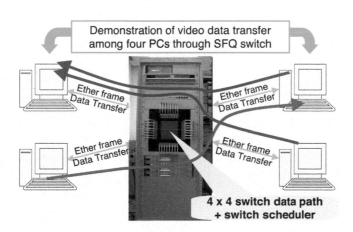

FIGURE 7.19: Video data transfer demonstration with 4×4 switch system installed in a cryocooler.

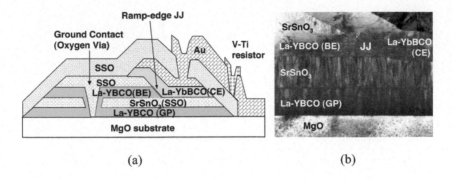

FIGURE 7.20:   Schematic cross-sectional view of an HTS integrated circuit.

7.4.1   Cryotrons

In 1954, Dudley Buck, a Department of Defense, National Security Agency (NSA) electronics engineer attending graduate school at MIT, proposed that one could make a switching device by using two features of superconductivity:

1. zero electrical resistance,

2. the suppression of superconductivity by applying a magnetic field to a superconductor.

In his doctoral thesis Buck says, The magnetic destruction of superconductivity is proposed as the basis for an electronic component having current gain and power gain. The component appears to be well adapted to a new kind of printed circuit technique wherein the entire circuit, including all active and passive elements, can be made in a single operation, such as vacuum evaporation. Circuits made in this way will make possible digital computers with greater computing capability than the present-day machines, and at the same time with smaller size, lower power dissipation, and a high

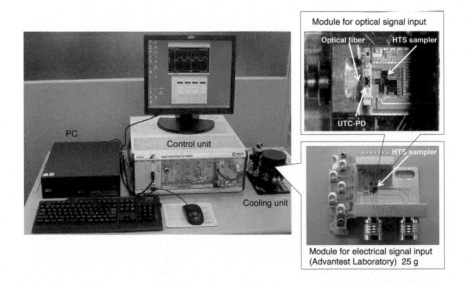

FIGURE 7.21:   Photograph of the sampling oscilloscope system with HTS integrated circuit.

degree of reliability.[33]

Buck called this device a "Cryotron". His first device used a .003 inch Niobium wire coil wound around a .009 inch tantalum wire. Current applied to the Nb coil transitioned the Ta wire to the resistive state. This demonstrated that one could use a small current to control a large current.

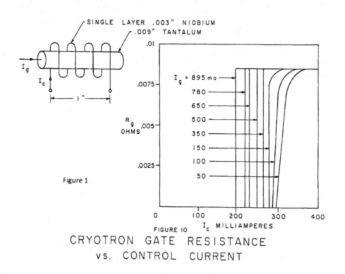

FIGURE 10
CRYOTRON GATE RESISTANCE
vs. CONTROL CURRENT

FIGURE 7.22: Cryotron gate resistance.

Buck used this "gain" to build a logic "flip-flop" which would switch the current to either of two paths. He further investigated the building of Cryotrons using crossed films which would produce smaller, faster and lower power devices in large numbers.

This prospect led to a Department of Defense (NSA) program called "LIGHTNING" with the IBM Corporation. The intent was to develop Cryotron devices for memory and logic in a high-end computing. The attraction was *very high speed at very low power*. A number of circuits were built including logic and memory, for example, a 1000 bit random access memory and some specialized logic "chips". Small systems were also fabricated. The technology was aimed toward high speed. Unfortunately, it was found that there is a fundamental limit to the time required for the Cryotron to switch states. This discovery took place at a time when semiconductor integrated circuits were also being developed and they operated at room temperature.

The program was terminated in 1963.

7.4.2 Josephson Signal Processor (JSP)

In 1971 a paper was presented by an IBM researcher, Dr. Juri Matisoo, concerning the creation of a small memory circuit using Josephson junction devices. Again, the attraction was very high speed at very low power. An NSA researcher contacted Dr. Matisoo and arranged for a meeting. The result of this exchange was the creation of a joint IBM I NSA program to investigate the applicability of this device technology, the Josephson junctions in high performance computing.

---

[33] Superconductive Electronic Components, Dudley A. Buck Dr. Science Thesis; Massachusetts Institute of Technology, June 1958

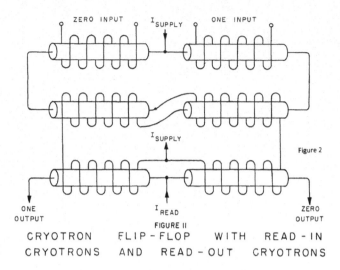

ZERO INPUT I_{SUPPLY} ONE INPUT

I_{SUPPLY}

Figure 2

ONE
OUTPUT I_{READ} ZERO
OUTPUT

FIGURE II

CRYOTRON   FLIP-FLOP   WITH   READ-IN
CRYOTRONS   AND   READ-OUT   CRYOTRONS

FIGURE 7.23:   CRYOTRON Flip Flop

The program methodically went through demonstrations of logic, memory, packaging, powering and cooling.[34,35,36,37,38] The "demonstration vehicle" chosen was a digital signal processor, the Josephson Signal Processor (JSP).

Given success, it was deemed that this would be sufficient to prove the features necessary for general purpose computing use. Fabrication materials technology moved from using lead and tin to niobium. The basic logic gate selected was "latching logic" which was found to have a switching speed limitation. (This was sufficiently high to be acceptable at that time.)

The program stopped in 1983 when it was judged that adequate memory would not be demonstrated as soon as was desired. (Also, at that time new semiconductor technologies appeared promising.)

7.4.3   Superconductive Crossbar Switch

The Microelectronics and Computer Corporation (MCC) carried out studies to address the problem of a large number of processors (~ 1000) being called upon to access a large shared memory in future high end computer (HEC) systems. Their studies investigated a number of technologie:

1. a large number of entry ports, ~1000 or more,

2. a large number of exit ports, ~1000 or more,

3. very high speed channels, multi-GHz,

4. the ability to resolve contention among requesting processors, and

---

[34] IBM Journal of Research and Development 24 (1980)

[35] W.Anacker (IBM), Computing at 4 degrees Kelvin, *IEEE SPECTRUM* 16 (1969)

[36] Wilhelm Anacker, Potential of superconductive josephson tunneling technology for ultrahigh performance memories and processors, *IEEE Trans. Mag.* MAG 5 (1969)

[37] RSFQ logic/memory family: A new Josephson junction technology for sub-terahertz digital systems, K.K.Likharev and V.K.Semenov, *IEEE Trans. on Appl. Supercond.* 1 (1991)

[38] Development of superconductor electronics technology for high-end computing, A. Silver, A. Kleinsasser, G. Kerber, Q. Herr, M. Dorojevets, P. Bunyk and L. Abelson, *Superconducting Science and Technology* 16 (2003)

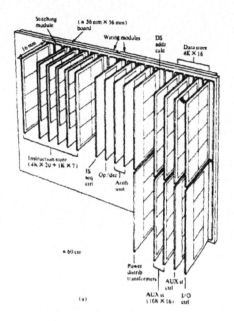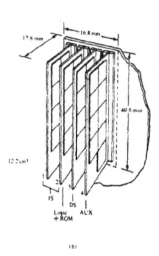

FIGURE 7.24: JSP package.

5. very high system throughput under heavy request loads.

A proposal was made by a DoD (NSA) member of MCC's Technical Advisory Board to use 11 devices in a unique architecture to solve the problem. A Crossbar switch was proposed using superconductive Josephson devices. Its architecture is based on a 32(input) × 32(output) "Switch" chip and an interface chip, "Glue", which directs and senses the data-and control-pulses coming from the room temperature processors and memory. The architecture allows expansion from the 32 × 32 basic switch to a $(N \times 32) \times (M \times 32)$ structure while maintaining the serial data rate per port and resolving contention. The basic 32 × 32 "Switch" chip contains ~ 5000 Josephson junctions.

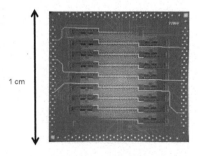

1 cm

FIGURE 7.25: 2.5 Gb/s per channel, self-routing, 16 × 16 switch based on NSA design, demonstrated to 4 Gb/s in 1996.

A development program was undertaken to evaluate performance and feasibility.

• The ability to carry high speed signals to and from the 4 K environment using copper ribbon cable was proven.

- MCC designed and built the multichip module (MCM) and the chip bonding tool.

- The Johns Hopkins University Applied Physics Laboratory carried out electromagnetic simulation and measurements of the behavior of the (MCM) and its input/output connections.

- Superconducting chips were fabricated by TRW (now Northrop Grumman), HYPRES and MIT Lincoln Laboratories.

- Semiconductor electronics, cryogenic cabling and 300 K to 4 K cabling were built by Tektronix, Inc.

- MCC performed computer simulations of the Crossbar system to evaluate its throughput and access time. These showed very high total throughput even with very high contention rates among processors.

Discrete event simulations (computer mimicking of the hardware system) were performed by David Bisant (NSA) and Marc Snir et al. (University of Illinois)[38] for a 1024×1024 Crossbar. They also showed high throughput data rates under very high processor request rates.

A 128×128 Superconductive Crossbar system was built and successfully tested by Tektronix at 2.5 Gbits per second per input port, and a 16×16 Crossbar was built and tested by TRW (now Northrop Grumman) at 10 Gb/s per input port.

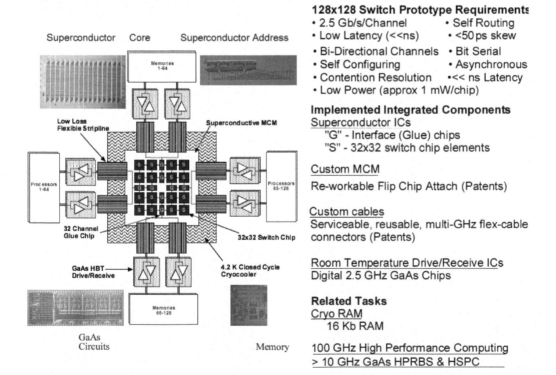

FIGURE 7.26: Superconducting crossbar switch for high performance computing: assembly and demonstration.

7.4.4 Hybrid Technology Multi-Threaded Architecture (HTMT)

A new way of using Josephson junction devices to achieve higher speed and lower power than latching logic had been developed. The concept relied on the use of single flux quanta as the individual "bits". By measuring the presence or absence of flux quanta in a time window one could carry out the required logic functions called for in a computer. The very small energy of a quantum of flux resulted in a very low power consumption even at clock speeds ~100GHz. This rapid single flux quantum (RSFQ) technology appeared to be a very promising candidate for use in very high performance computing systems.

FIGURE 7.27: HTMT facility (conceptual drawing). Cryopackage concept: 1 m^3 package, 1 kW 4 K, built with achicvable technology.

A number of studies evaluated the requirements and candidates for achieving "Petaflops" class of computing (10^{15} floating point operations per second). The studies covered computational problems, architecture, software, logic technology and memory technology. The conclusion was that the wide range of performance requirements dictated the use of a number of different technologies:

- superconductive devices,
- superconductive interconnects,
- semiconductor devices,
- metal conductors,
- magnetic media, and
- optical components.

The very high speed of Josephson junctions used in RSFQ (rapid single flux quantum) circuits and their very low power consumption would be the core of a very compact computing system having low power.

The architecture chosen was Multi-Threading. Since multiple technologies were employed in the system the effort was named HTMT hybrid technology multi threaded architecture. The program, supported by NSA and NASA (JPL/CalTech), had the goal of building a prototype petaflops system. The participants were

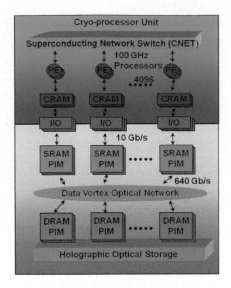

FIGURE 7.28: HTMT architecture created by Dr. Thomas Sterling (JPL).

State University of New York (SUNY)
   at Stony Brook,
Columbia University,
University of Notre Dame,
University of Delaware,

TRW (now Northrop Grumman),
Argonne National Laboratory,
California Institute of Technology, and
Jet Propulsion Laboratory (JPL).

The petaflops processor used superconducting rapid single flux quantum RSFQ circuits. The machine was partitioned into 4096 processing elements operating at 4 K. These processors interface with processor in memory (PIM) random access memory. The memories are connected to a multi-ported optical switch (DATA VORTEX) whose output was fed to a bank of conventional dynamic random access memory (DRAM). The large data store was chosen to be an optical memory.

A demonstration superconductive chip, FLUX 1, was built. It was a 5000 gate, 8 bit microprocessor RSFQ chip and used 1.75 micron 4000 A/cm^2 current density Nb-NbAlO$_x$-Nb Josephson junctions. The chip was designed to dissipate ~ 9 mW at a 20 GHz clock rate. Chip to chip communications over a 2.5 cm path through solder bumps matched transmission lines and passive MCM was demonstrated at 60 G bits per second. Due to funding restrictions the program was not continued to completion but was terminated in 2003.

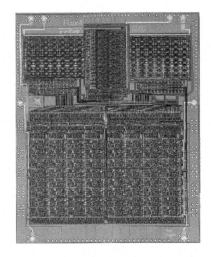

FIGURE 7.29: FLUX I.

7.4.5 Superconducting Technology Assessment (STA)

The U.S. Department of Defense (NSA) convened a team of experts to address the problem of high end computing. It was noted that:

1. "Recent industry trends clearly establish that design tradeoffs between power, clock and metrology have brought silicon to the limits of its scalability".

2. "The Semiconductor Industry Association (SIA) International Roadmap for Semiconductors (lTRS) has identified superconducting rapid single flux quantum (RSFQ) technology as the most promising technology in the continuing demand for faster processors and low power."

The panel report[39] stated: "This assessment is an in-depth examination of RSFQ technologies with the singular purpose of determining if a comprehensive roadmap for technology development is possible, aiming for industrial maturity in the 2010–2012 timeframe."

The team was composed of members from academia, industry and government. The conclusions reached were affirmative:

> "The STA concluded that there were no significant outstanding research issues for RSFQ technologies. Speed, power and Josephson junction density projections could be made reliably. Areas of risk have been identified and appropriately dealt with in the roadmap with cautionary comments on mitigation or alternatives."

The report, STA #1, was delivered in June 2005, not in time for DoD budget insertion. STA #1 was followed by a re-evaluation in 2007, STA #2, to update the original findings. STA #2 reaffirmed STA #1 and noted that, due to low funding, progress had been slow in the elapsed time.

7.5 Digital Electronics in Europe

Horst Rogalla

In this section I will give a personal view on the European situation of superconducting electronics in general and especially on digital superconducting electronics over the last 30 years. It will be neither complete nor well balanced—it just maps my personal experience in this area over the last 30+ years.

7.5.1 The Early Days

My experience with digital electronics starts about 47 years ago in the 1960s when I built a simple 2-bit arithmetic unit, as a high-school student, from old computer parts: at that time logic elements consisted of small circuit boards with individual transistors and many discrete elements soldered on top. It is no wonder, looking back, that cryotrons[40] could have been a serious speed competitor in the early days of semiconductor logic circuits. Building the arithmetic unit was a great learning experience, but it was slow, just like the semiconductors at that time. It was also the time when the future of semiconducting computing was founded — the first circuits of the 7400-family, the *de facto* industrial standard for digital circuits for a long time appeared, integrating just four NAND-gates on one chip. And with the release of the PDP-7 (which had a clock rate of about 700 kHz) the mini-computer age started. The cryotron was quickly left behind with a fundamentally limited switching speed and low complexity. It did, however, pave the way for integration in superconducting thin-film techniques.

---

[39] See http://www.nitrd.gov/pubs/nsa/sta.pdf
[40] See e.g., D. A. Buck, Memorandum 6M-3843, MIT Lincoln Laboratory (1955)

In the early 1970s, I was a student at the University of Muenster in Germany and one rainy day in November (the saying is "it rains in Muenster or the church bells ring") I stayed quite long in the library. In a popular science magazine I found an article about Josephson tunnel junctions, about their natural 0 (superconducting) and 1 (gap-voltage) state, about their incredible switching speed and how they would revolutionize the computer world. I went home wondering how you can create logic circuits with 2-terminal elements without amplification, started drawing circuits similar to the old diode-logic, and found that by far I did not know enough about superconductivity: my interest and fascination in superconductivity had caught fire. At that time few groups worked on superconducting electronics in Europe, but the prediction of the Josephson effect and its experimental verification by John Rowell et al. (see Chapter 3 Section 1) changed the picture, groups started to work with rf-SQUIDs and dc-SQUIDs, and on simple digital circuits. The development culminated in the announcement that IBM was going to build a Josephson computer. This announcement caught not only the imagination of researchers, but it was also essential for the funding situation—only national funding was actually available since there were no European Framework programs at that time (the first one (FP1) started in 1984 with only a tiny amount of funding). This was an exciting time, many new superconducting digital circuits were developed, all based on "voltage state" switching, but they were more or less copies of semiconductor logic cells and designed for standard computer architectures. I remember talks about the IBM project and about the design and properties of the superconducting digital circuits at meetings of the Low-Temperature Community of the German Physical Society—at that time still taking place in Freudenstadt in the Black Forrest in a vibrant atmosphere with the feeling that something big was happening. But at that time, I also started to get doubts, at least about the material choice for the project–lead is not a very stable material, neither mechanically nor chemically. The reports about "hillocks" growing in the planar junction area after a few thermal cycles were disturbing. In my group in Giessen, Germany in the Institute of the late Christoph Heiden, we had made a number of such junctions; initially they worked with excellent *IV*-characteristics, but after a few thermal cycles they badly degraded and mostly resulted in high sub-gap currents or even shorts. Many attempts to prevent a deterioration of the material by alloying with gold, indium, antimony, etc. improved the situation and looked promising, but did not seem to be a real solution — still the Josephson junctions were not very reliable and were sensitive to thermal cycling and humidity. The difficulty in realizing memory circuits and its scalability may have been the main reason for IBM to stop the project, but the material difficulties also must have played an important role in this decision.

In 1983 I heard about $Nb/Al/Al_2O_3/Nb$-junctions[41], and on the way to a conference I visited Michael Gurvitch at Bell Labs to see how this process worked. I had some experience with dc- and rf-magnetron sputtering of Nb in an old semiconductor evaporation system, in which we prepared $Nb/NbO_x/Nb$-junctions with bad but stable *IV*-characteristics. It became immediately clear that this was a technology with which the advantages of Nb could be combined with the excellent barrier characteristics of Al_2O_3. This technique was quickly introduced in many laboratories in Europe and we were able to use this technique on a small scale to make SQUIDs, but any attempt to enter the research on digital circuits failed—the negative fallout of the end of the IBM project hit hard: funding agencies in Europe were very reluctant to invest in a technology that industry had "proven" to be a dead end. In Europe, only a few groups were able to continue this research, leading among them Jutzi's group in Karlsruhe, Germany. Jutzi had been with IBM Rüschlikon for many years, first developing semiconductor memory circuits[42] and later becoming involved in the IBM Josephson computer project[43]. After the end of the project, Jutzi and his group continued working on digital

---

[41] M. Gurvitch, M. A. Washington. and H. A. Huggins, *Appl. Phys. Lett.* 42 (1983) 472

[42] W. Jutzi, C. H. Schuenemann, *IBM J. Res. Devel.* 16 (1972)

[43] see e.g. W. Jutzi, *Cryogenics* 16 (1976) 81

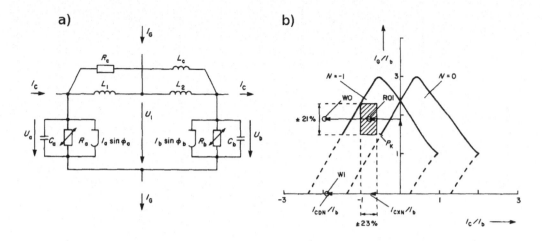

FIGURE 7.30: a) Equivalent circuit of the interferometer memory cell and b) its threshold curve. (With permission[44].)

Josephson circuits for many years, with special interest in memory cells[44]. Other important groups in Europe working on Josephson junctions and their applications were the group around Tord Claeson in Gothenburg, Sweden, in Copenhagen, Antonio Barone in Napels and Jüergen Niemeyer at the PTB in Braunschweih, Germany—each with some specialty, but no common program and no focus on digital applications. The work on Josephson voltage standards at the PTB in Braunschweig had the most advanced techniques for Josephson junction preparation, the standards required large numbers of junctions with small parameter spread and good microwave design (see Chapter 9 Section 5). The step towards complex digital circuits would have been small, but a national or a European program for a digital Josephson computer on the research laboratory or university level was in the best of circumstances "fiction".

In discussion with industry, often the argument came up that cooling Nb circuits, either with liquid helium or cryocoolers, was too painful and that one needed to use materials with an operation temperature of ~10 K—at least for small applications. This industry requirement was fulfilled when in 1985 Shoji et al. of the ETL in Japan reported the reproducible fabrication of high-quality NbN/MgO/NbN-junctions[45]. At that time Villégier at LETI in Grenoble was also working on NbN-films[46]. In the following years, he established a high-quality NbN junction fabrication in Grenoble, applying these junctions in all areas of superconducting electronics, including digital circuits. This effort still continues and this group is the most advanced in Europe in the use of NbN, either "stand-alone" or in conjunction with Nb-layers[47]. The importance of a NbN technique for industrial applications is the higher operation temperature. The A15 materials offered an even higher operation temperature and I concentrated with my group in Giessen on Nb_3Ge, which at the time had the highest T_c of up to 23 K. In this institute in Giessen, materials science work on the A15 materials had been going on for a long time: Braun had set up an ultrahigh vacuum system to co-sputter Nb_3Ge already in 1978 [48]. We used this system in 1984 and the following years to prepare Nb_3Ge-films for superconducting electronics applications. Together with Bernd David and Michael Mück we

[44] D. Drung, W. Jutzi, *Cryogenics* 24 (1984) 179

[45] A. Shoji, M. Aoyagi, S. Kosaka, F. Shinoki, H. Hayakawa, *Appl. Phys. Lett.* 46 (1985) 1098

[46] Akkermans, E., Laborde, O., Villégier, J.C., *Solid State Commun.* 56 (1985) 87

[47] E. Baggetta, M. Maignan, J.-C. Villégier, http://stinet.dtic.mil/dticrev/pdfs/ada445484.pdf (2005)

[48] H. F. Braun, E. J. Saur, *J. Low. Temp. Phys.* 33 (1978) 87

succeeded in structuring nano-bridges of less than 100 nm width and a T_c above 20 K. We also developed a multilayer technique to fabricate more complex circuits, a very challenging task because the Nb_3Ge deposition takes place at a temperature above 900 °C [49]. These circuits were tested and operated in liquid hydrogen: dc-SQUIDs and the first relaxation oscillation SQUIDs showed very nice properties. Michael Mück even successfully realized a planar Nb_3Ge Josephson junction, but the fabrication was very challenging and its properties were not great. At Stanford, the group of M. Beasley was working on a similar program with Nb_3Sn. Originally I planned to develop the Nb_3Ge technique towards more complex circuits, but neither the quality of the planar junctions was good enough for "voltage state" switching for digital circuits nor was the nano-bridge fabrication reproducible enough to apply it to the upcoming RSFQ technology. In short, interesting physics but a dead end for applications; superconducting electronics and digital applications were in general in a bad state in the mid-1980s in Europe and funding was scarce.

Then everything dramatically changed in 1986 within a very short time: the discovery of Ba-La-Cu-O by Bednorz and Müller at IBM Zürich (see Chapter 4, Section 2) and of the first material superconducting in liquid nitrogen by Paul Chu and his group in Houston (see Chapter 4, Section 4), $YBa_2Cu_3O_{7-\delta}$, started a "Cold Rush". Within a few months, groups became experts in the field that never before had been seen in superconductivity and they promised a "superconducting world" within a few years, from superconducting cables replacing the high-voltage power lines, magnetically floating cars and roller skates, to supercomputers operated in liquid nitrogen. It seemed that the more science fiction-like the proposal, the easier it got funded. Together with some colleagues from the pre-high-T_c time, we warned about overselling the high-T_c superconductivity—at that time, not even high-T_c Josephson junctions were available—but without success. Progress was made, more slowly than predicted, but steadily: the first grain boundary junctions and step-edge junctions appeared and epitaxial films were grown on single crystal substrates like $SrTiO_3$. Based on our experience with Nb_3Ge, we succeeded in setting up a multilayer technique based on $YBa_2Cu_3O_{7-\delta}$ and, using this technique, Hans Hilgenkamp prepared the first HTS dc-SQUID with an integrated multi-turn input coil.[50]

Gerrit Gerritsma and I were particularly interested in superconducting digital electronics and it was clear to us that we could surpass the performance of semiconductors, if we could make it work at sufficiently high frequencies and at temperatures easily achievable by cryocoolers, like 40 K or above. Learning from the materials science results of epitaxy on angled films, it occurred to me that by cutting a low-angle ramp in an $YBa_2Cu_3O_{7-\delta}$ film and then depositing a layer of (semiconducting) $PrBa_2Cu_3O_{7-\delta}$ on top followed by the deposition of a counter-electrode one could create a Josephson junction that had primarily a current-flow in the ab-plane of the c-axis oriented films. We tested the idea and it worked—the high-T_c ramp-type junction was born[51]. We used this Josephson junction technique together with our multilayer technique to fabricate simple digital circuits, e.g. a quasi-one-junction digitizer[52]. In Europe we were the only group with this technique at hand, later followed by a ramp-type junction technology at the Chalmers University of Technology, Sweden and a planar junction technology at DRA, Great Britain[53]. In Japan, a successful modification of this ramp-type junction, the "interface-engineered" junctions, soon appeared (see Section 3 in this chapter), followed by a similar technique at the Research Center Jülich, Germany. From the USA

---

[49] B. David, M. Muck and H. Rogalla, *Advances in Cryogenics Engineering*, A.F. Clark and R.P. Reed (eds.), Plenum Press, New York 32 (1986) 543

[50] J.W.M. Hilgenkamp, G.C.S. Brons, J.G. Soldevilla, R.P.J. Ijsselsteijn, J. Flokstra, H. Rogalla, *Appl. Phys. Lett.* 64 (1994) 3497

[51] J. Gao, W.A.M. Aarnink, G.J. Gerritsma, H. Rogalla, *Physica* C 171 (1990) 126

[52] E.M.C.M. Reuvekamp, P.A.A. Booi, M.A.J. Verhoeven, G.J. Gerritsma, H. Rogalla, *IEEE Trans. Appl. Supercond.* (1993) 2621

[53] P. J. Hirst, R. G. Humphreys, J. S. Satchell, M. J. Wooliscoft, C. L. Reeves, G. William, A. J. Pidduck, H. Willis, *IEEE Trans. Appl. Supercond.* 11 (2001) 143

exciting messages about complex digital high-T_c circuits based on high-T_c microbridges (e.g., long shift registers) appeared and vanished.

7.5.2 European Projects

Our first European program was the ESPRIT-2 project UNITED (January 1989 through 1991, coordinated by Thomson CSF)[54]. This project had the theme "HTS thin films and tunnel junction devices". Initilally we did not take part in this program because I did not believe in HTS tunnel junctions; in the final phase of the project, we joined in order to help with HTS thin films. Clearly UNITED did not reach its main goal, the HTS tunnel junction, and it was far from any application.

In June 1992 the EU ESPRIT-3 project UNITED II in the 3rd Framework Program started as a successor to the UNITED project. In contrast UNITED II was clearly dedicated to technology and application development. It addressed "...two main application areas for high-T_c superconductors: digital circuitry and magnetic field sensors ..." Denis Crete of Thomson CSF coordinated the project. In a stimulting atmosphere of industrial request, technological development and fundamental research, we developed first simple logic circuits and a 4-bit flash-type input stage for an ADC[52] (see Figure 7.31). The project was successfully completed in August 1995.

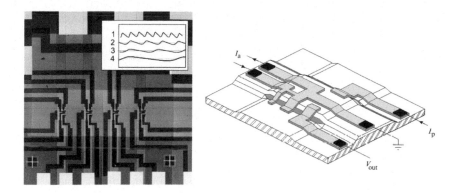

FIGURE 7.31: 4-bit flash converter with a R-2R network input and "quasi-one junction SQUID" digitizers.

Based on this success, we applied for a follow-up project, this time primarily focused on superconducting digital electronics. The program was called RSFQ-HTS and was coordinated by me. It was organized around a few groups with excellent HTS device experience: The Defense Research Agency (DRA), Great Britain brought in advanced evaporation techniques and a (planar) CAM junction process, the University of Twente–sputtering and in the ramp-type junction/multilayer process and Chalmers University of Technology/Sweden–a tri-crystal grain-boundary junction technique. The RSFQ-HTS program intended to establish a common design base for HTS digital circuits, further develop a HTS multilayer and Josephson junction technology, and demonstrate complete SFQ circuit functions. It started in January 1997 and finished at the end of 2000. In the course of the project a number of basic digital circuits were developed, among them a high-T_c T-flipflop that was operated up to 33 GHz. The main output of the project was a $\Sigma\Delta$-modulator that functioned up to an internal clock rate of 178 GHz (see Figure 7.32).

---

[54] for details of the European projects, mentioned here, search the European database CORDIS by acronym: http://cordis.europa.eu/search/index.cfm

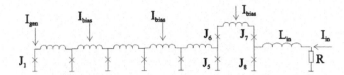

FIGURE 7.32:   1st order $\sigma\delta$-modulator.

In the meantime, the limitations of HTS Josephson junctions had become obvious: circuits with up to 10 Josephson junctions were "easy", circuits with more than 40-50 junctions were practically impossible to yield due to the wide spread of the current density in HTS Josephson junctions available at that time. So the high-T_c Josephson technology was limited to high-speed or high sensitivity frontends with a limited number of Josephson junctions.

The RSFQ-HTS project caught the attention of Erland Wikborg of Ericsson in Stockholm Sweden. Together we worked out a follow-up project, that would combine a second-order $\Sigma\Delta$-modulator at the input stage with a cold broadband amplifier interfacing with an InP-multiplexer and a CMOS decimation filter (see Figure 7.32). The low-temperature parts were all to be cooled by a specially developed cryocooler. The intention was to demonstrate the applicability of this technique to software-defined radios, which were of special interest to Ericsson. The application was successful and in January 2002 the project SUPER-ADC started.

At about middle of the project, Ericsson was forced to concentrate on its core business and sold our partner, the Ericsson Wireless Solutions to Infineon. The interest of Infineon in superconductivity was at best very small and, together with some fabrication problems of the fast electronics, the project did not reach its goal of testing all subunits together. However it succeeded in testing the major parts of the analog-to-digital converter successfully. The project finished end of June 2005.

In the early 2000s, the enthusiasm about high-T_c applications was gone—the materials science and preparation was much more difficult than many groups had anticipated and the overselling of high-T_c superconductivity now turned against the whole field: it was very difficult, nearly impossible to get funding for HTS projects. Nevertheless, after a number of attempts, we succeeded in getting funding for a project where the aim was to digitally read-out the state of a SQUID with high-T_c-electronics and then complete the data processing with semiconductor electronics at room temperature. Such a SQUID would have the advantage of a very high slew-rate and should thus

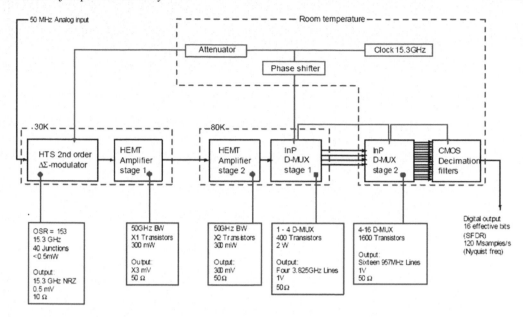

FIGURE 7.33:   The ADC concept in the SUPER-ADC project.

be able to operate without shielding in an electromagnetically noisy environment. The project had the acronym DIGI-SQUID. The project started in October 2002 and it lasted until September 2006. Many parts of the whole system were successfully fabricated and tested. But changes in the participation and personnel, especially with the industrial partners and the withdrawing of Oxxel (after going out of business) as the fabrication source of the HTS circuits, resulted in big delays and insufficient budget to finish the project in time.

In this period it was already difficult, due to decreasing funding, to keep the quite expensive and knowledge-intensive high-T_c multilayer process up and running in Twente. We nevertheless stepped in and realized at least part of the high-T_c-circuit within a short time. In principle, this project marked the end of the involvement of my group at the University of Twente in this type of research and application development—neither national nor European funding were sufficient to continue.

In the late 1980s, 1990s and early 2000s Europe concentrated its superconducting electronics research almost entirely on HTS-superconductors, primarily because of an industry requirement of operation in liquid nitrogen or with one-stage low-power cryocoolers. It forced researchers to work on applications with a not-yet mature high-T_c technology instead of further improving the technology itself and performing the necessary fundamental research (e.g., the d-wave superconductivity in high-T_c was verified in 1993 (see Chapter 3 Section 3) and the mechanism for high-T_c is still unclear today). Instead, a balanced approach for low-T_c complex electronics and high-T_c technology development would probably have been much more productive. However, such advice was unheeded, both in Brussels and by national governments in Europe.

Luckily, some low-T_c digital research proceeded quietly in some places in Europe, such as at the PTB in Braunschweig, Jena and Karlsruhe (all in Germany), in Grenoble, France, in Naples, Italy and at the Chalmers University of Technology, Sweden, where Anna Herr set up a group to design and test complex low-T_c digital circuits. Together with Ericsson she worked on circuits for telecommunication, the most complex ones designed in Europe so far.

With the rise of quantum computing and qubits the idea came up to read out qubits with short coherence time using fast low-T_c RSFQ circuits. A consortium applied successfully for a European project titled RSFQUBIT coordinated by Chalmers University of Technology, and work started in September 2004 and continued until the end of 2007.

7.5.3   Organization of Superconducting Electronics in Europe

Until 1996 there was no common body in Europe that represented applications of superconductivity at the level of the European Community or the national level. This low level of representation had impact on the political decisions of the different "Directorates" in Brussels and on the project/funding situation of superconductivity research and applications. A few attempts to realize such an infrastructure had failed until Massimo Marezio took the lead and with a number of colleagues we applied successfully for a Network of Excellence in the Information Society Technologies (ISCT) Program of the EU. This network started in October 1996 and brought together groups of superconducting electronics, superconducting materials, and large scale applications of superconductivity. It significantly strengthened the position of the superconducting electronics research in Europe. To receive funding from the European Community for digital superconducting electronics, this was an essential step: SCENET brought together research laboratories and interested industries Europe-wide, which was absolutely essential for European projects. Apart from the representation function towards the European Community, SCENET regularly organized meetings and summer schools, and supported the exchange of scientists between laboratories. This was an important step toward bringing together laboratories and people from different European countries and stimulating cooperation between scientists.

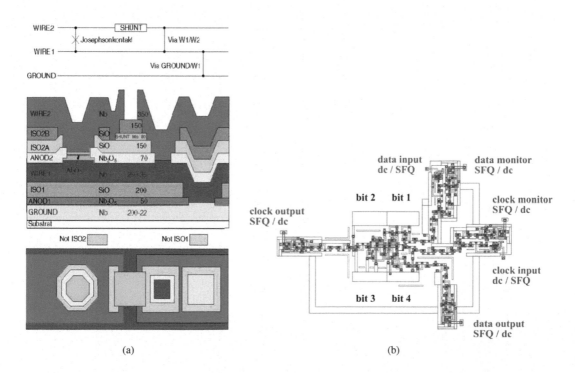

(a)   (b)

FIGURE 7.34:   a) IPHT: multilayer process b) PTB: layout of a circular RSFQ shiftregister (courtesy FLUXONICS e.V.).

In the course of time, the organizational background of SCENET in the European Community changed; the focus of SCENET became more materials and large scale oriented, but it remained an excellent platform for the applied superconductivity community in Europe. In cooperation with SCENET, I took the initiative to found, together with a number of colleagues, a European network for Superconducting Electronics, which was organized independently from the European Community as a society, according to German law, with seat at the PTB Braunschweig, Germany. The basic idea of this network was to bring together industry and research laboratories and create a "virtual" infrastructure for the design and fabrication of superconducting electronics circuits: a foundry network and a design network for analog and digital superconducting circuits.

After the formal end of SCENET in 2006, the general representation of superconductivity in Europe was taken over by the European Society for Applied Superconductivity (ESAS) and superconducting electronics by FLUXONICS. As the first director of FLUXONICS I initiated a European Roadmap for Superconducting Electronics, in which a number of colleagues wrote a common picture about the status and perspectives of superconducting electronics in Europe[55]. The second edition of this roadmap[56] was published recently as a part of the European project S-PULSE. Within FLUXONICS, superconducting digital electronics has always played an important role–the University of Ilmenau, Germany (Hannes Töpfer) has established a circuit and layout library, and regularly organized design courses for digital electronics. IPHT Jena, Germany took over the role as foundry for low-T_c circuits and a number of fabrication runs of test circuits have been made to verify its fabrication capability (see Figure 7.34). An overview of the role of FLUXONICS and Superconducting Electronics in Europe can be found in a Special Issue on Recent Progress in Superconducting Digital Electronics[57].

7.5.4 Road Ahead

Nearly all articles and talks on "alternative" digital techniques stress the fact that semiconductors "will hit a wall" soon, that Moore's law will come to an end, and that the "alternative" technique will take over. Until now the semiconductor industry has always found ways to extend Moore's law. It has to end some day for sure, not necessarily resulting in a replacement of the semiconductor technology with alternative technologies, but in complementing the existing technology with new technologies. Superconducting digital electronics is a very promising alternative technology, with the big advantage of very low switching energy and very high speed. A huge knowledge base exists for superconducting digital circuits of some complexity; design, simulation and verification techniques and tools are available. Hundreds of thousands of Josephson junctions have been successfully fabricated on a chip for voltage standard applications. High-T_c circuits of low complexity with special functions for frontends have successfully been made. New Josephson junction technologies are being investigated and especially Nb/SiNb/Nb intrinsically shunted junctions[58] show superior perfomance in terms of yield, process compatibility and electrical properties. It seems very worthwhile to invest in this technology and make it ready to complement the semiconductor technology in areas where its limit will be reached. This will require further research and development in the area of compatible memory, low bias current operation and better cryocoolers for low-T_c and high-T_c circuit operation.

---

[55] H.J.M. ter Brake et al., *Physica* C 439 (2005) 1

[56] S. Anders et al., *Physica* C 470 23-24 (2010) 2079.

[57] H. Rogalla, *IEICE Trans. Electron.* E91.C (2010) 272

[58] D. Olaya, P.D. Dresselhaus, S. P. Benz, A. Herr, Q. P. Herr, A. G. Ioannidis, D. L. Miller, and A. W. Kleinsasser, *Appl. Phys. Lett.* 96 (2010) 213510

7.6   Integrated Circuit Fabrication Process

Mutsuo Hidaka

Superconducting integrated circuits, especially single-flux-quantum (SFQ) circuits[59], are ultimate high-speed and low-power electrical devices. These features are quite attractive for use as the main elements in supercomputers and network routers. These circuits are also lately getting much attention as the peripheral circuits of quantum-bits (Q-bit) and superconducting detectors. The fabrication process to implement superconducting integrated circuits started in the 1970s using Pb as the superconductor[60] and the initial types of logic and memory circuits[61,62] were fabricated by IBM. However, the soft and unstable nature of Pb resulted in fatal flaws regarding the long-term reliability and thermal contraction on cooling to 4 K. Nb was considered an attractive superconducting material from the early days, since it was harder and more stable than Pb. Nevertheless, Pb was used at first because the fabrication of a good quality Nb Josephson junction (JJ) was a difficult process[63]. Gurvitch et al. invented Nb/AlOx/Nb JJ [64], in which a thin Al layer was used to cover the surface of the Nb base-electrode and an AlOx tunnel barrier was formed through the oxidation of the Al layer surface. This method helped to eliminate any adverse effects on the JJ quality by the Nb sub-oxide at the junction interface and to create excellent quality Nb junctions. After that, Nb circuits based on the Nb/AlOx/Nb JJs became the mainstream of superconducting electronics devices as well as digital circuits. The JJ using other superconducting materials, such as NbN, MgB_2, and high-T_c superconductors, have also been investigated. However, the controllability and reliability of these JJs have not yet reached the necessary level for use in integrated circuits, although some small circuits were demonstrated using the NbN JJs[65] and high-T_c JJs[66]. Therefore, this article focuses on the fabrication process for Nb-based integrated circuits.

7.6.1   Circuit Elements

Josephson Junction

A JJ is the active element in superconducting circuits. A Nb/AlOx/Nb tunnel-type junction is generally used as the JJ in them. Figure 7.35 shows a schematic diagram of the Nb/AlOx/Nb JJ fabrication. Thin Al, which is typically 10 nm, is deposited on the base-electrode Nb, and the surface of the Al layer is thermally oxidized to form an AlOx tunnel barrier. A counter-electrode Nb is deposited on the tunnel barrier. The JJ region is patterned by using photolithography and the counter-electrode Nb except for the JJ region is etched away using reactive ion etching (RIE) with fluorine gas. In this method, part of the Al layer remains as normal metal. The residual Al layer has little impact on the JJ quality, since the residual Al layer has superconductivity due to the proximity effect from the base-electrode Nb. A cross-section TEM photograph (a) and the current-voltage characteristics (b) of an Nb/AlOx/Nb JJ are shown in Figure 7.36. The characteristics are of an under-damp nature, which has hysteresis in it, although the SFQ circuits require an over-damp JJ.

---

[59] K. K. Likharev and V. K. Semenov, *IEEE Trans. Appl. Supercond.* 1 (1991) 3

[60] J. H. Greiner, C. J. Kircher, S. P. Klecher, S. K. Lahiri, A. J. Warncke, S. Basavaiah, E. T. Yen, J. M. Baker, P. R. Broslous, H. C. W. Huang, M. Murakami and I. Ames, *IBM J. Res. Develop.* 24 (1980) 195

[61] T. R. Gheewala, *IBM J. Res. Develop.* 24 (1980) 130

[62] S. M. Faris, W. H. Henkels, E. A. Valsamakis and H. H. Zappe, *IBM J. Res. Develop.* 24 (1980) 143

[63] R. F. Broom, R. B. Laibowitz, Th. O. Mohr and W. Walter, *IBM J. Res. Develop.* 24 (1980) 212

[64] M. Gurvitch, W. A. Washington, H. A. Huggins, *Appl. Phys. Lett.* 42 (1983) 472

[65] H. Terai and Z. Wang, *IEEE Trans. Appl. Supercond.* 11 (2001) 525

[66] M. Hidaka, T. Satoh, M. Koike and S. Tahara, *IEEE Trans. Appl. Supercond.* 9 (1999) 4081

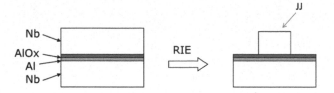

FIGURE 7.35: Schematic diagram of Nb/AlOx/Nb JJ fabrication.

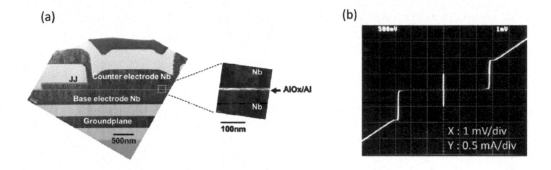

FIGURE 7.36: Cross-section TEM photograph and current-voltage characteristics of Nb/AlOx/Nb JJ.

Thus, the Nb/AlOx/Nb JJ is shunted by a resistor to change to the over-damp JJ in SFQ circuits.

The clock frequency of the SFQ circuits is proportional to $(J_c/C_0)^{1/2}$ when the McCumber coefficient of the JJ is constant, and inversely proportional to the JJ area. Here J_c is the critical current and C_0 is the characteristic capacitance of the JJ. The J_c is exponentially proportion to the tunnel barrier thickness. Figure 7.37 shows the J_c dependence on the Pt product. Here, P is the O_2 partial pressure and t is the oxidation time. The J_c can be well controlled by the self-limited logarithmic dependence on the thermal oxidation, as shown in Figure 7.37.

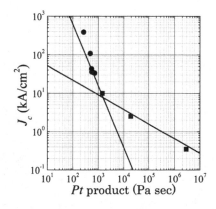

FIGURE 7.37: J_c dependence of product on O_2 partial pressure p and oxidation time t.

The basic technology for fabricating the Nb/AlOx/Nb JJs was established in the 1980s. Sustained efforts, however, have been devoted to improving the JJ quality, reliability, and controllability. For instance, Hinode et al. discovered from an I_c analysis of a huge number of JJs that the inclusion of hydrogen in the Nb electrodes affected the J_c of the JJ[67] and Noguchi et al. found that the contribution of the imaginary part of the superconducting gap energy, which was caused by incompleteness in the Nb superconductivity near the barrier interface and/or impurity in the barrier, on the JJ tunnel current[68]. Other barrier materials that can be used for a Nb JJ have been investigated. Among them, Nb_xSi_{1-x} [69] and AlN_x [70] are attractive because of their wide variety of electrical properties and excellent quality at a higher J_c, respectively. Some self-shunted JJs have been reported[71]. Some I_cR_n products from them have already been reported although they are relatively low. Here, R_n is the JJ normal resistance and the I_cR_n product represents the on-state voltage.

Wiring

Nb thin film is used as wiring material, and this wiring plays two roles. One is as the inductance element. The LI_c value is quite important in SFQ circuit, because SFQ loops are characterized by their LI_c values. Here, L is the inductance of SFQ loop and I_c is the critical current of the JJ. The wiring width, length, and insulator thickness underneath the wiring have to be well controlled to maintain a precise L value.

The other is for the interconnection between two elements. Passive transmission lines (PTL) are attractive interconnections because their SFQ pulses can propagate the speed of light. The impedance matching between the wiring and JJs at the ends of PTL has to be maintained. Microstrip line or strip line structures are adopted to define the impedance of the wiring[72].

Resistor

The typical values of resistors in integrated circuits are from a few Ω to a few tens of Ω. Mo is popular for implementing the resistors because its resistivity is appropriate for creating a 1-2 Ω sheet resistance within a 100-nm thickness. Pd is also popular because it is expandable for the Q-bit and detector peripheral circuits since it maintains a normal resistance below 1 K. Larger resistors, for instance 50 Ω in a terminate resistor for a high-speed signal line, are occasionally used. MoN_x or NbN_x, which have larger sheet resistances, are sometimes used in addition to the metal resistors to save resistor area[73].

Interlayer Insulator

SiO_2 is usually used as the interlayer insulator and dielectric material of superconducting circuits because of its high-reliability and ease of deposition and etching. Bias-sputter and PECVD (plasma enhanced chemical vapor deposition) are used for improving the step coverage of the interlayer insulator. Anodized films of Nb and/or Al are sometimes used for part of the insulator.

[67] K. Hinode, T. Satoh, S. Nagasawa and M. Hidaka, *J. Appl. Phys.* 104 (2008) 23909

[68] T. Noguchi, T. Suzuki, A. Endo and T. Tamura, *Physica* C 496 (2009) 1585

[69] D. Olaya, P. D. Dresselhaus, S. P. Benz, J. Bjarnason and E. N. Grossman, *IEEE Trans. Appl. Supercond.* 19 (2009) 144

[70] A. Endo, T. Noguchi, M. Kroug, T. Tamura and H. Inoue, *Physica* C 469 (2009) 1589

[71] M. Yu. Kupriyanov, A. Brinkman, A. A. Golbov, M. Siegel and H. Rogalla, *Physica* C 326-327 (1999) 16

[72] Y. Hashimoto, S. Yorozu, Y. Kameda and V. K. Semenov, *IEEE Trans. Appl. Supercond.* 13 (2003) 535

[73] L. A. Abelson, R. N. Elmadjian and G. L. Kerber, *IEEE Trans. Appl. Supercond.* 9 (1999) 3228

7.6.2 Integration of Circuit Elements

These circuit elements are integrated into a device structure consisting of a $Nb/AlO_x/Nb$ JJ layer, three or four Nb layers, and one or two resistor layers on a substrate with corresponding insulator layers between these metal layers. The lower three Nb layers and one resistor layer are a necessity for digital circuits and the top Nb layer and second resistor layer are options. Several institutes have developed fabrication processes for these devices[73,74,75,76]. Figure 7.38 shows a schematic cross-section diagram of an SFQ circuit fabricated by the International Superconductivity Technology Center (ISTEC) in Japan[74]. This device includes four Nb layers, such as for one groundplane and three wirings, a Mo resistor layer, and SiO_2 insulation layers between the metal ones; the device is fabricated on a 3-inch surface-oxidized Si wafer. $Nb/AlO_x/Nb$ JJs are placed between the second and third Nb layers. The minimum JJ size and wiring width are 2 μm$\times$2 μm and 1.5 μm, respectively. The target J_c is 2.5 kA/cm^2. All the metal layers were deposited using DC magnetron sputtering and SiO_2 was deposited using biased RF magnetron sputtering. An i-line stepper was used for the patterning. Each layer was etched by an RIE in which several kinds of fluorine gas were used according to the etched materials. Figure 7.39 shows a top view, cross-section, and circuit diagram for part of an SFQ circuit fabricated by using the ISTEC process. The identical parts are indicated in this figure. The top Nb layer does not appear in it.

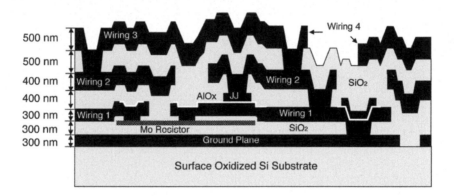

FIGURE 7.38: Schematic cross-section diagram of SFQ circuit fabricated by using ISTEC Nb four-layer process.

The circuit parameters in this process are sufficiently controlled. Run-to-run scattering of the J_c and Mo sheet resistance are both within $\pm 10\%$. Many SFQ circuits consisting of more than 10,000 JJs fabricated by this process operated correctly using several tens of GHz clock frequency[77,78].

[74] S. Nagasawa, Y. Hashimoto, H. Numata and S. Tahara, *IEEE Trans. Appl. Supercond.* 5 (1995) 2447

[75] S. K. Tolpygo, D. Yohannes, R. T. Hunt, J. A. Vivalda, D. Donnelly, D. Amparo and A. F. Kirichenko, *IEEE Trans. Appl. Supercond.* 17 (2007) 946

[76] L. Grönberg, J. Hassel, P. Helistö and M. Ylilammi, *IEEE Trans. Appl. Supercond.* 17 (2007) 952

[77] A. Fujimaki, M. Tanaka, T. Yamada, Y. Yamanashi, H. Park, N. Yoshikawa, *IEICE Trans. Electron.* E91-C (2008) 342

[78] H. Park, Y. Yamanashi, K. Taketomi, N. Yoshikawa, M. Tanaka, K. Obata, Y. Ito, A. Fujimaki, N. Takagi, K. Takagi and S. Nagasawa, *IEEE Trans. Appl. Supercond.* 19 (2009) 634

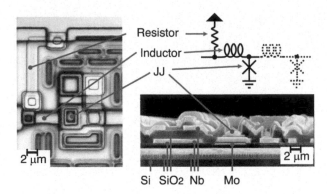

Resistor →
Inductor →
JJ →

2 μm

2 μm

Si SiO2 Nb Mo

FIGURE 7.39: Top view, cross section, and circuit diagram of part of SFQ circuit fabricated by using ISTEC Nb four-layer process.

7.6.3 Planarized Multilayer Process

In order to fabricate faster and larger integration level SFQ circuits, ISTEC also developed a new fabrication process characterized by the planarized multilayer[79]. Increasing the number of Nb layers enables not only decreasing the circuit area but also increasing the design degree of freedom. However, it is difficult to further increase the number of Nb layers past four because the vertical steps are too large to create additional layers on it without producing defects. Therefore, a new planarization technology called the caldera method was developed[80].

Figure 7.40 shows the steps used in this method for fabricating a unit wiring level. (a): After a lower Nb wiring layer has been patterned, it is covered with a SiO_2 layer using bias sputtering. The thickness of this SiO_2 layer is adjusted depending on the thickness of the Nb wiring layer, i.e. 200-300 nm. (b): A photoresist layer is then formed in the reversed pattern of the Nb wiring layer. This reversed pattern is slightly broadened (300 nm in the figure) so that it overlaps the Nb wiring layer. (c): Selective RIE is used to remove the SiO_2 on the Nb wiring layer and stops the RIE on the Nb surface resulting from the overlap, leaving narrow, 1-μm-wide at most, convex SiO_2 regions shaped like caldera volcanoes along the edges of the Nb wiring layer. This remaining caldera-shaped SiO_2 is easily flattened using a conventional planarization method such as mechanical polishing planarization (MPP). (d): A very small amount of MPP is enough to obtain a smooth surface.

The planarized surface enabled us to add more Nb layers. The critical current density J_c of the JJ was increased to 10 kA/cm^2, which was four times larger than that of the previous process shown in Figure 7.38 and is expected to have twice the operation speed. The minimum JJ size was reduced to 1 μm×1 μm to maintain I_c value and the minimum line width also decreased to 1 μm.

Figure 7.41 shows a cross-sectional view of a device structure fabricated by using the multilayer process. The device has ten Nb layers. The role of the bottom Nb layer is as a DC bias current supply. The middle five layers are devoted to the X and Y direction PTLs, which have strip-line configurations. The SFQ gates were fabricated using the top four Nb, Mo resistor, and Nb/AlO$_x$/Nb JJ layers. This part is not planarized and is the same as that of the structure in Figure 7.38 to secure the design continuity. Vertically-stacked superconductive contact holes were also adopted into this process.

The multilayer process produced high-performance SFQ circuits; for example, an 8-bit shift regis-

[79] T. Satoh, K. Hinode, H. Akaike, S. Nagasawa, Y. Kitagawa and M. Hidaka, *IEEE Trans. Appl. Supercond.* 15 (2005) 78

[80] K. Hinode, S. Nagasawa, M. Sugita, T. Satoh, H. Akaike, Y. Kitagawa and M. Hidaka, *Physica* C 412-414 (2004) 1437

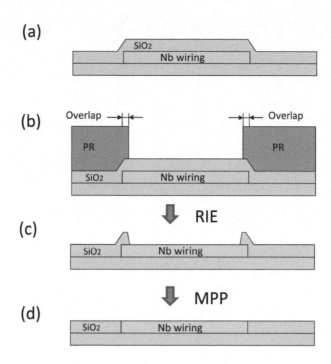

FIGURE 7.40: Steps in caldera method for fabricating unit wiring level.

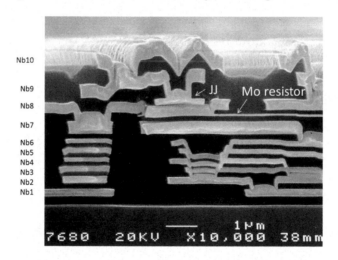

FIGURE 7.41: Cross-section SEM photograph of planarized ten Nb-layer device.

ter operated at up to 120 GHz [81] and a 16k-bit RAM consisting of 80,768 JJs at 3.8 mm×3.7 mm [82]. A 4×4 switch circuit fabricated by using the ten-layer process demonstrated a 112-GHz operation, which is more than twice that compared with the same circuit fabricated by using the four-layer

[81] H. Akaike, T. Yamada, A. Fujimaki, S. Nagasawa, K. Hinode, T. Satoh, Y. Kitagawa and M. Hidaka, *Supercond. Sci. Technol.* 19 (2006) S320

[82] S. Nagasawa, T. Satoh, K. Hinode, Y. Kitagawa and M. Hidaka, *IEEE Trans. Appl. Supercond.* 17 (2007) 177

process. Moreover, the circuit area of the ten-layer process could be reduced to 19% as shown in Figure 7.42[83].

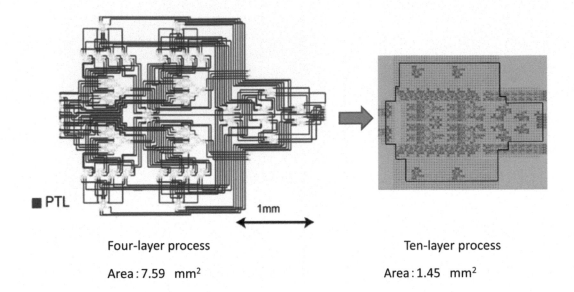

■ PTL

1mm

Four-layer process

Area : 7.59 mm^2

Ten-layer process

Area : 1.45 mm^2

FIGURE 7.42:   4×4 switch circuit fabricated by using Nb ten-layer process and schematic circuit diagram of the same 4×4 switch based on four-layer processes. The former was operated at twice the clock frequency using a 19% circuit area compared with the latter.

7.6.4   Toward Further Progress

Faster, lower-power dissipation, higher-integration, and higher-yield are the main issues for consideration towards making further progress in the field of superconducting electronics devices as well as digital circuits. These improvements are necessary for superconducting devices to become widely used and there is still room for improvement.

In order to increase the operation speed by increasing the J_c, the thickness of the JJ tunnel barrier has to be decreased without degrading the JJ quality. Figure 7.37 shows that two different tendencies were observed in the J_c dependence on the pt product. This phenomena suggest that a micro-short in tunnel barrier may occur above 10 kA/cm^2 [84]. Therefore, the other tunnel barrier needs to be investigated to get a better quality JJ with a J_c of over 10 kA/cm^2 than the AlO$_x$ barrier. The AlN$_x$ barrier is a promising candidate that was demonstrated as an excellent quality JJ with a J_c of 33 kA/cm^2 [70]. The JJ area has to be reduced to become inversely proportional to the J_c increase.

A smaller JJ is also useful for drastically reducing the SFQ circuit power dissipation. Reducing the I_c is quite effective, since the dynamic and static power dissipations in SFQ circuits are proportional to $I_c\Phi_0$ and $I_c^2 R_b$, respectively. Here, Φ_0 is an SFQ value (2.07×10^{-15} Wb) and R_b is a bias resistance. The minimum I_c in an SFQ circuit is decided by the tolerability against the ther-

---

[83] M. Ito, I. Kataeva, R. Kasagi, M. Okada, T. Koketsu, M. Tanaka, S. Nagasawa, H. Akaike and A. Fujimaki, Demonstration of a 4x4 SFQ switch fabricated with a 10kA/cm2 Nb multi-layer process, Applied Superconductivity Conference (2010) 3EY-03

[84] Sergey K. Tolpygo and Denis Amparo, *J. Appl. Phys.* 104 (2008) 63904

mal noise and the smallest makeable JJ area with a reasonable uniformity while maintaining the J_c value. The tolerability is not marginal for the current value. Thus, the I_c has the ability to be reduced by decreasing the JJ area.

The progress made when using the planarized multilayer process shows that the SFQ circuit can be improved by having a higher J_c and a larger number of Nb layers. The planarized surface shown in the multilayer process enables us to use advanced semiconductor fabrication technologies. The superconducting device fabrication process can be improved in collaboration with a deep understanding of the superconducting process and actively using advanced semiconductor fabrication technologies.

7.7 High-Speed Digital Circuits

Akira Fujimaki

In this section, recent progress in single flux quantum (SFQ) circuits where an SFQ is employed as an information carrier is described. All the SFQ circuits have the special feature that the circuits can operate in a sub-terahertz range with an energy consumption of about 1 aJ/gate (10^{-18} J/gate). In addition, signal transmission at the speed of light is possible in most of the SFQ circuits. These features become increasingly important because society's digital infrastructure depends on semiconductor large-scale integration (LSI) devices. Increased heat density and increased interconnect delays make it more and more difficult to increase the performance of LSI circuits for operation at higher frequencies. SFQ circuits with their better performance are promising devices for use in future high-speed LSI devices.

One drawback of SFQ circuits is that they require a low-temperature environment. The cooling penalty of 1000–5000 W/W (the incident power of a cryocooler required to produce 1 W of cooling power at a 4 K operation temperature) has to be paid in systems based on SFQ circuits. One of the ways to overcome this cooling penalty is to design SFQ-circuit-based systems with high energy efficiency, which is defined as performance per unit power. This will be realized in upgraded network systems including baseband digital signal processors for wireless systems[85], high-end routers in backbone networks[86], and high-end computers[87,88], in which SFQ LSI circuits and their dense packaging are essential. A third important contribution to the total power of SFQ circuits is from current biasing. Section 7.7.4 describes efforts to minimize bias power, which may further improve the performance per unit power of SFQ LSI circuits.

---

[85] A. Herr, *IEICE Trans. Electron.* E91-C (2008) 293

[86] S. Yorozu, Y. Kameda, Y. Hashimoto, H. Terai, A. Fujimaki, and N. Yoshikawa, *IEEE Trans. Appl. Supercond.* 15 (2005) 411

[87] M. Dorojevets, P. Bunyk, and D. Zinoviev, *IEEE Trans. Appl. Supercond.* 11 (2001) 326

[88] M. Tanaka, T. Kondo, N. Nakajima, T. Kawamoto, Y. Yamanashi, Y. Kamiya, A. Akimoto, A. Fujimaki, H. Hayakawa, N. Yoshikawa, H. Terai, Y. Hashimoto, and S. Yorozu, *IEEE Trans. Appl. Supercond.* 15 (2005) 400

7.7.1　Principle of the Rapid Single Flux Quantum Circuit

The rapid single flux quantum (RSFQ) circuits proposed by Likharev, Mukhanov and Se-menov[3,89] are the most notable circuits among a variety of SFQ circuits reported so far[90,91,92,93]. The coding of RSFQ circuits is defined by using clock signals. If an SFQ exists in a data path in the time interval between two adjacent clock signals, the SFQ can be interpreted as the logical "1". If there is no SFQ, it is the logical "0".

Figure 7.43 shows an equivalent circuit of a delay flip-flop (DFF), the basic element of an RSFQ circuit. The use of overdamped Josephson junctions, which have external shunt resistors, guarantees stable operation of the RSFQ circuits. In principle, the circuit is driven by a current source, and proper current distribution is achieved by bias resistor networks. The operation of the DFF begins with the arrival of an input signal at the "din" port. The data "flux quanta" is stored as a circulating current in the storage loop consisting of J_1, L, and J_2. Then the stored data is read out when a clock signal comes to the loop, and sent out at the "dout" port.

Storage loops employ the function of a latch or a memory. This function is essential for RSFQ logic gates with two inputs such as AND and XOR. Interconnects with unequal lengths are usually used to keep sufficient design flexibility, resulting in different arrival times of an SFQ at the two input ports. Thus, the input signal arriving first must wait for the other one in a storage loop.

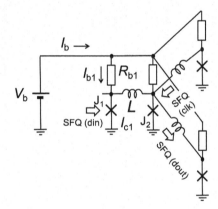

FIGURE 7.43:　Equivalent circuit of a DFF cell. Bias currents are supplied to Josephson junctions through bias resistors.

The voltage impulse called an SFQ pulse is generated only when an SFQ crosses a Josephson junction. For a typical pulse width of a few picoseconds, an energy of about $I_c\Phi_0$ is consumed in the shunt resistor, that is connected in parallel to the junction, within I_c is the critical current of the junction and Φ_0 is the flux quantum. This feature enables high-speed signal processing with very low energy or power consumption.

Passive transmission lines (PTLs) with microstrip or strip line structures are used for long inter-connects, where a voltage pulse can travel at the speed of light. For using a PTL, a transmitter is

[89]　K.K. Likharev and V. K. Semenov, *IEEE Trans. Appl. Supercond.* 1 (1991) 3

[90]　Y. Harada, E. Goto, and N. Miyamoto, *Technical Digest International Electron Devices Meeting* (1987) 389

[91]　A. Silver and Q. Herr, *IEEE Trans. Appl. Supercond.* 11 (2001) 333

[92]　Q. P. Herr and A. Y. Herr, 2008 *Applied Superconductivity Conference*, 3EY03, 2008

[93]　O. A. Mukhanov, *IEEE Trans. Appl. Supercond.*, vol. 21, no. 3, Jun. 2011

placed between a logic gate and a PTL, and a receiver is placed between a PTL and a gate. Figure 7.44 shows operating margins in bias currents provided to a receiver for PTLs having different lengths. The PTLs under test had a microstrip line structure and were made on a single chip. Bias currents are normalized by the design value. Sufficiently wide margins exceeding ±20% are obtained for PTLs with a length shorter than 50 mm, and the margin even for a 100-mm-long PTL is still within an acceptable range. This feature means that large bandwidths can be achieved not only for on-chip (gate-to-gate) communication, but also for off-chip (chip-to-chip) communication. In fact, 117 Gb/s chip-to-chip transmission has been demonstrated by using the ISTEC 10 kA/cm^2 Nb integrated-circuit technology[94].

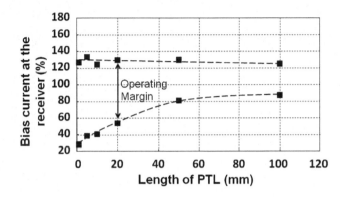

FIGURE 7.44: Length dependence of the operating margins in bias currents provided to a receiver.

7.7.2 LSI Design Technology

The sequence of arrival timing between a clock signal and a data signal has to be controlled in RSFQ circuits. In particular, timing control with pico-second order is essential for the operation under concurrent-flow clocking, which has an advantage over any other clocking method in speed of operation. Note that the delay time of a Josephson junction is sensitive to the bias current and that circuit parameters have a spread in actual chips. This makes the timing control difficult in large-scale circuits.

To overcome this situation, the introduction of computer aided design (CAD) including an analog simulator and an optimizer is necessary. The CONNECT top-down design is described here as a typical CAD for RSFQ circuits. The design is based on the CONNECT cell library[95]. Layouts of all the logic gates are designed in a square based on the 2.5-kA/cm^2 Nb/AlO$_x$/Nb LSI process called the standard process 2 (STP2)[96]. The unit length of the sides of the square is 40 μm. Input/output/clock ports are placed at designated spots inside the square. These layouts are called the logic cells. Josephson transmission lines (JTLs) and passive transmission lines (PTLs), which are used as interconnects, are also designed in a square because the standard process provides no layer dedicated to wiring.

Actual RSFQ circuits are designed by placing the logic cells and the wiring cells in the design

[94] Y. Hashimoto, S. Yorozu, T. Satoh, and T. Miyazaki, *Appl. Phys. Lett.* 87 (2005) 022502

[95] S. Yorozu, Y. Kameda, H. Terai, A. Fujimaki, T. Yamada, and S. Tahara, *Physica* C 378-381 (2002) 1471

[96] S. Nagasawa, S. Hashimoto, Y. Numata, and S. Tahara, *IEEE Trans. Appl.Supercond.* 5 (1995) 2447

area. Circuit parameters of the logic cells are optimized by using the circuit optimizer SCOPE[97] in order to exclude the interference between adjacent cells and to operate with a large margin even under parameter spread. The SCOPE optimizer can pick up the timing parameters such as setup time, hold time, delay, etc. The dependence of these parameters on the bias currents are examined and stored in the library. The timing design and verification of the RSFQ circuits are carried out in a digital domain using the stored timing parameters.

The initial CONNECT cell library was effective for circuits consisting of about 5000 Josephson junctions. However, increased bias currents in larger-scale integrated circuits induce relatively large magnetic fields themselves, and thus reduce operating margins. To suppress self fields, all the logic cells registered in the present CONNECT cell library have superconducting shields for the bias feed lines.

Recently, the 10 kA/cm^2 Nb/AlO$_x$/Nb LSI process called the advanced process 2 (ADP2) has been developed[98]. Increased critical current density enables operation at 50-100 GHz and a reduction of the width of PTLs. Moreover, the ADP2 provides 9 Nb layers. The bottom Nb layer is used for distributing bias currents, and the top 4 Nb layers are almost the same as that of the standard process. Two vertically stacked strip-line structures are formed by using the other 4 layers. The increased number of Nb layers leads to a remarkable reduction in the circuit area, because PTLs can be placed below Josephson junctions. So far, a 4×4-switch made with the ADP2 process has been demonstrated at 112 GHz using only 19% of the area of that made with the STP2[99].

Figure 7.45 (a) and (b) show a microphotograph of a DFF cell made with the STP2 and with the ADP2, respectively. In the cell with the ADP2, the unit length of the sides of the square is 30 μm. Pillars connected to the bottom layer are formed at each corner. Bias currents are provided to the circuits through these pillars. Moats are formed around the pillars.

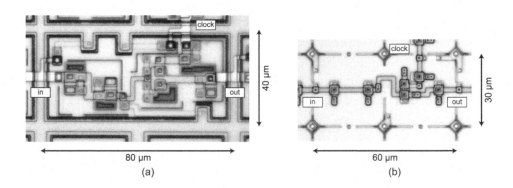

FIGURE 7.45: Microphotographs of basic cells (delay flip-flop, DFF) in the CONNECT cell library. (a) DFF cell made with the STP2. (b) DFF cell made with the ADP2.

Advanced CAD tools have been developed based on the ADP2. A new design flow and algorithms involving a PTL interconnect technique in a skewed clocking scheme were introduced. The clock frequencies of RSFQ circuits can be maximized by using these tools. The design flow is similar to that for a conventional semiconductor LSI design, and several dedicated steps are added to reinforce

[97] N. Mori, A. Akahori, T. Sato, N. Takeuchi, A. Fujimaki, and H. Hayakawa, *Physica* C 357 (2001) 1557

[98] S. Nagasawa, T. Satoh, K. Hinode, Y. Kitagawa, M. Hidaka, H. Akaike, A. Fujimaki, K. Takagi, N. Takagi, N. Yoshikawa, *Physica* C 469 (2009) 1578

[99] M. Ito, M. Tanaka, I. Kataeva, M. Okada, H. Akaike, S. Nagasawa, M.Hidaka, and A. Fujimak, 2010 Applied Superconductivity Conference, 3EY-03, 2010

it for an efficient design of RSFQ LSI circuits.

Figure 7.46 shows a design of an 8-bit carry look-ahead adder along with the new design flow[100]. In the first step, the "logic synthesis", DFFs are inserted to equalize the number of clocked gates on every data path because every RSFQ logic gate has also a latch functionality. The set of logic gates are partitioned into logic levels, where the level of the gate is defined as the number of gates on the path from the circuit input to that gate. Then, the timing of clock signals for each gate is scheduled, and the clock tree is synthesized by inserting some delay elements (JTLs) on the paths, which are required to adjust the timing[101,102]. The gate arrangement in each logic level is optimized to minimize the number of delay elements, and placement of the logic gates is roughly obtained. Figure 7.46(a) shows the schematic after the clock tree synthesis and the first rough placement.

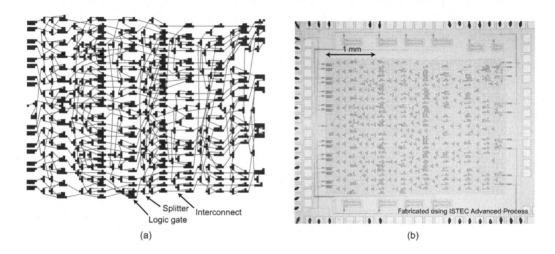

(a) (b)

FIGURE 7.46:   Design of a carry look-ahead adder with the advanced CAD tools. (a) Schematic diagram after placement of logic gates, and (b) microphotograph. Black squares and lines in the schematic represent logic gates and interconnects, respectively. PTLs are invisible in microphotograph due to planarization process. Layout size is 3.84 mm × 2.70 mm.

Using the feedback from the verification, the next steps, "place and route", are done not independently but simultaneously in order to satisfy timing requirements. After a first rough placement of logic gates and a rough routing of PTLs, a timing analysis is performed to find a critical timing path, and the detailed route of the path is determined. Repeating the analysis of timings and congestion degrees, PTLs are carefully routed one by one[103].

Figure 7.46(b) shows a microphotograph of the 8-bit adder fabricated using the ADP2. The target frequency is 50 GHz. The number of PTLs and Josephson junctions are 598 and 7,092, respectively. An actual circuit operated with a sufficient bias margin of ±20%. The routing of PTLs, which is one of the most time-consuming tasks in manual design, was done in 10 seconds.

[100] K. Takagi, N. Takagi, M. Tanaka, K. Obata, and Y. Ito, First Superconducting SFQ VLSI Workshop, A2-3, 2008

[101] K. Obata, K. Takagi, and N. Takagi, *IEICE Trans. Fundamentals* E91-A (2008) 3772

[102] K. Takagi, Y. Ito, S. Takeshima, M. Tanaka, and N. Takagi, submitted to *IEICE Trans. Electron.*

[103] S. Takeshima, K. Takagi, M. Tanaka, and N. Takagi, Second Superconducting SFQ VLSI Workshop, P5, 2009

7.7.3 Demonstration of RSFQ LSIs

So far, several RSFQ LSI circuits consisting of more than 10,000 Josephson junctions have been demonstrated. In this section, an RSFQ microprocessor based on the STP2 and an accelerator based on the ADP2 are shown as examples of RSFQ LSI circuits.

Figure 7.47 shows the microarchitecture of the RSFQ microprocessor named CORE1β. The CORE1β employed the instruction-set based on the DLX[104], a typical reduced instruction-set computer (RISC) architecture. For simplicity, the number of instructions is reduced to a minimum and only bit-serial data are handled. The lengths of data and instructions are 8 and 16 bits, respectively. The CORE1β targets almost the same performance as the semiconductor conventional one at the peak. To achieve the target, two techniques are employed. One is the introduction of a "forwarding architecture"[105], in which multiple arithmetic logic units (ALUs) are connected in cascade. As shown in Figure 7.47, the data path of the CORE1β is composed of a general register file, two bit-serial ALUs, and some buffers including a forwarding buffer (FB). The FB holds the value of the last calculation. The connected ALUs and the FB enable two calculations continuously by the execution of a series of register-to-register instructions. The latency of the cascaded ALUs is designed to approximately equal the length of the bit-serial data. As a result, the enlarged latency originating from bit-serial operations can be hidden.

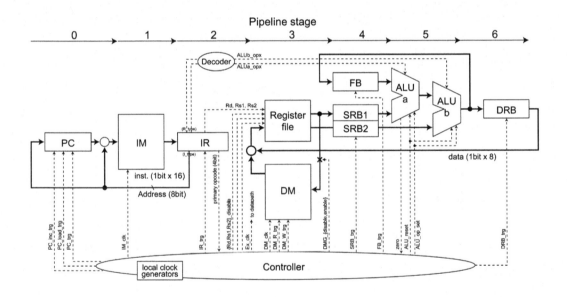

FIGURE 7.47: Microarchitecture of the CORE1β. The CORE1β is composed of a decoder, a program counter (PC), an instruction memory (IM), an instruction register (IR), a register file, two ALUs, two source register buffers (SRB1, SRB2), a destination register buffer (DRB), a forwarding buffer (FB), and a controller. The pipeline stages are shown at the top of the figure.

The other technique to obtain high performance is "pipelining". The CORE1β is made up of

[104] J. L. Hennessy and D. A. Petterson, *Computer Architecture, A Quantitative Approach*, 2nd ed. (San Francisco, CA: Morgan Kaufmann)

[105] M. Tanaka, T. Kondo, T. Kawamoto, Y. Kamiya, K. Fujiwara, Y. Yamanashi, A. Akimoto, A. Fujimaki, N. Yoshikawa, H. Terai, and S. Yorozu, *Physica* C 426 (2005) 1693

seven pipeline stages as indicated at the top of Figure 7.47. The master clock controls the pipeline execution and the local clocks. The master clock, corresponding to the so-called "clock" in a conventional semiconductor LSI, is generated in an independent clock generator with a frequency of 1–2 GHz. The local clocks with frequencies of 20–25 GHz are used for bit-serial processing. The details of the microarchitecture of the CORE1β are described in Ref. 106.

Figure 7.48 shows a microphotograph of the CORE1β. A large die with an area of 8 mm × 8 mm is used to reduce external magnetic fields generated by bonding wires or off-chip bias-feeding lines, while a typical die size is 5 mm × 5 mm. The CORE1β is composed of five main circuit blocks and consists of 10,995 Josephson junctions. Bias currents are individually supplied to each circuit block. The total bias current is 1,373 mA. The power consumption is estimated to be 3.4 mW.

8 mm

8 mm

FIGURE 7.48: Microphotographs of the CORE1β. The CORE1β was designed based on the CONNECT cell library for the STP2.

The instruction memory (IM) and the data memory (DM) are made up of 16-bit and 8-bit shift registers, respectively. The designed local clock frequency is 25 GHz for the instruction fetch and 20 GHz for the bit-serial data operation. The master clock frequency is raised up to 1.4 GHz after careful timing adjustments between components. The peak performance is calculated to be 1,400 MOPS (million operations per second) because instructions are issued every two master clocks, and two operations are executed for one instruction in the cascaded ALUs.

The operation of the CORE1β was examined in liquid helium using an on-chip high-speed test method. The test sequence included multiple ADD operations. The correct operation was obtained

---

[106]M. Tanaka, T. Kawamoto, Y. Yamanashi, Y. Kamiya, A. Akimoto, K. Fujiwara, A. Fujimaki, N. Yoshikawa, H. Terai, and S. Yorozu, *Supercond. Sci. Technol.* 19 (2006) S344

up to 37 GHz for the instruction fetch and 18 GHz for the data operation.

Another example of RSFQ LSI circuits is a reconfigurable data path (RDP) processor, which is a compact, high-performance computation engine. The architecture of the RDP processor gives a solution to the memory-wall problem. The memory-wall problem is the problem that the memory bandwidth cannot be wide enough relative to the processor performance because of the difference between the operating speed of a processor and that of the memory limits the performance of a computer.

Figure 7.49 shows a concept of a computing system based on SFQ-RDP processors. The SFQ-RDP processors are used as accelerators, while the central processing units (CPUs), based on complementary metal-oxide-semiconductor (CMOS) devices, serve as main processors. An RDP is mainly composed of a 2-dimensional array of floating point processing units (FPUs). The output of each FPU can be fed to one or more FPUs via flexible operand routing networks (ORNs). Streaming buffers are used as temporal buffers for the adjustment in the timing between SFQ RDP processors and CPUs or between SFQ RDP processors and main memories.

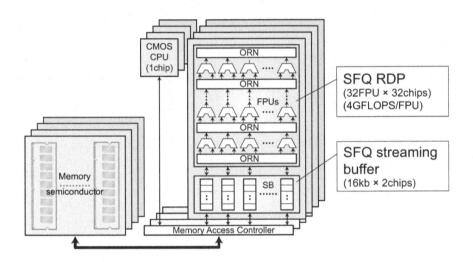

FIGURE 7.49:   Concept of a computing system based on SFQ RDP processors.

In an RDP, a data flow graph (DFG) extracted from a target application program is mapped to a 2-dimensional FPU array. To enable the mapping, an ORN consists of programmable switches, while FPUs support multiple functions such as add, sub, and multiply. By means of setting the control signals provided to FPUs and ORN switches, the function of the RDP can be changed at run time.

Since the cascaded FPUs can generate a final result without temporally memorizing intermediate data, the number of memory load/store operations corresponding to spill codes is reduced. In other words, memory bandwidth required to achieve high performance can be reduced. Moreover, since a loop-body mapped into the FPU array is executed in pipeline fashion, RDP can provide high throughput computing.

A microphotograph of a prototype of a 2×2 (double stages of an array composed of dual ALUs) RDP processor is displayed in Figure 7.50. For simplicity, ALUs are employed instead of FPUs in this design. In addition, the ALUs and ORNs are designed to handle only bit-serial data. The RDP processor prototype was made up of 11,458 Josephson junctions and occupied an area of 5.61×2.82 mm^2.

All the instructions for each ALU and reconfiguration of the 2×2 RDP processor prototype were confirmed up to a frequency of 45 GHz. The power consumption is 3.4 mW.

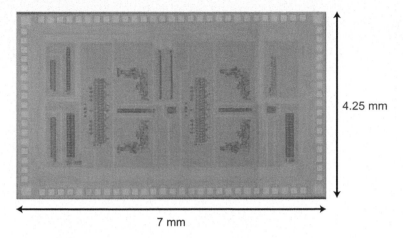

4.25 mm

7 mm

FIGURE 7.50: Microphotograph of the 2x2 SFQ RDP prototype. The prototype was designed based on the CONNECT cell library for the ADP2.

7.7.4   New Directions in SFQ Circuits

In these days, energy efficiency becomes the most important parameter in digital systems. Originally, the superconducting digital circuits, in particular the RSFQ circuits, have the special feature of low power/energy consumption. In fact, the intrinsic energy consumption of a Josephson junction within a clock period in the RSFQ circuit is roughly expressed as the product of $I_c \Phi_0$, and is estimated to be 0.3 aJ, which is small compared to semiconductor devices.

Figure 7.51(a) shows an equivalent circuit of a current-driven Josephson junction in RSFQ circuits. As shown in Figure 7.43, there are many bias resistors in an actual RSFQ circuits. As a result, Josephson junctions are considered to be driven by a voltage source with a voltage V_b through the bias resistance R_b. The typical value of R_b and V_b are 16 Ω and 2.5 mV, respectively. These values are determined to suppress interference between adjacent Josephson junctions through bias resistors in high-speed operation. However, much more energy is consumed in a bias resistor, while only a small amount of energy ($I_c \Phi_0$) is consumed in a junction's shunt resistor.

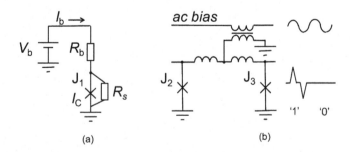

(a)   (b)

FIGURE 7.51: (a) Equivalent circuit of a current-driven Josephson junction in an actual RSFQ circuit. (b) Equivalent circuit of an RQL. No bias resistor is used.

An effective way to eliminate the energy consumption in a bias resistor is the employment of ac-biasing. The reciprocal quantum logic (RQL) is a typical model for ac-biasing[92]. An excellent summary including dc-biased energy-efficient circuits is given in Mukhanov[93].

Recently, it was found that low-V_b RSFQ circuits, RSFQ circuits driven with very low voltages such as a few tens of μV, also exhibit very good energy efficiency. In fact, at 20 GHz operation the bias resistor and the shunt resistor consumed only a total energy of about 0.1 aJ. This energy consumption is close to the Josephson coupling energy $I_c\Phi_0$, although small Josephson junctions with small critical currents were employed.

Figure 7.52 shows the energy consumption in a single switching event as a function of the clock period for different devices comprising LSI circuits such as microprocessors. Although the low-V_b RSFQ circuit described above cannot be classified as an LSI circuit, the corresponding energy consumption is marked in the figure as "low-power SFQ". The energy-delay product (EDP) is a good measure for the energy efficiency in digital circuits. The EDP values of the RSFQ circuits are 3-4 orders of magnitude smaller, and the EDP of the "low-power SFQ" circuit is 5–6 orders smaller than that of CMOS devices. This small EDP is a remarkable advantage over semiconductor devices even if the cooling penalty is taken into account.

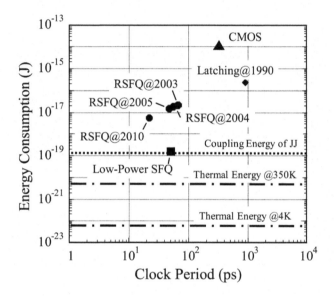

FIGURE 7.52: Energy consumption in a single switching event as a function of clock period for different devices comprising LSI circuits.

7.8 History of Superconductor Analog-to-Digital Converters

Oleg Mukhanov

An analog-to-digital converter (ADC) is a mixed-signal electronic circuit that converts an electrical signal from the analog domain to the digital domain, typically providing N binary bits at the sampling frequency f_s. The higher speed of superconducting circuits comparable to conventional circuits provided an initial interest in superconductor technology for ADC applications. One can find technical details of superconductor ADCs in earlier reviews[107,108]. In contrast here, we focus on a historical perspective of the ADC development identifying important trends and milestones as well as acknowledging main actors.

Similar to superconductor digital electronics, early superconductor mixed-signal circuits were based on cryotrons, pre-Josephson switching elements. A patent for the first superconductor analog-to-digital converter (ADC) was filed in 1960 [109], just in a few years after the initial cryotron-based digital circuit implementation. The projected speeds of the order of microseconds per ADC switching operation was rather attractive goal at that time. Several years later, the cryotrons were replaced by faster Josephson junctions in superconductor circuits including ADCs. The early Josephson-junction based ADC was a simple thermometer-code "totalizer" circuit patented in 1969 and presented an A-to-D conversion idea rather than a complete ADC circuit[110].

The first superconductor ADC circuit of a practical significance was a successive-approximation ADC invented in 1974 by M. Klein of IBM[111]. It consisted of a sample-and-hold circuit and four comparators made of serially connected Josephson junctions being switched to a voltage state at a predetermined control signal levels. This design was fabricated and tested in 1977 to show a 25 MHz bandwidth. In general, this ADC design was following semiconductor implementations using Josephson junctions. All early ADCs were focused on exploiting only the high switching speed of superconducting devices starting from cryotrons and then Josephson junctions. As in digital circuits, these first ADCs were viewed as a faster version of their semiconductor counterparts.

The realization that superconductivity offers much more than the higher speed led to innovations in ADC circuits. These new truly superconductive ADCs were exploiting fundamental features of superconductivity unavailable in other technologies: magnetic flux quantization, extremely high sensitivity, quantum accuracy, and low noise. For instance, the magnetic flux quantization provides a natural ruler which may be used to provide a large number of quantum-mechanically accurate thresholds for the ADCs.

Newer superconductor ADCs generally fall into two categories: *Nyquist-sampling parallel* ADCs and *oversampling* ADCs. An ideal Nyquist ADC samples a signal at a sampling rate $f_s = 2f_N$, where f_N is Nyquist frequency. The Nyquist ADCs are usually composed of a large number of *separate* quantizers arranged in parallel—each defining a single quantization level. The performance of such an ADC is limited by the precision of the quantization levels. These parallel-type ADCs are best for digitizing *high bandwidth* signals when a *moderate resolution* up to 8 bits is adequate.

In the oversampling ADCs, the signal is sampled at a frequency $f_s \gg 2f_N$ using a single quantizer. Then, feedback techniques and digital filtering are used to reduce or shape (move out of band) the quantization noise and enhance the effective dynamic range. Oversampling ADCs are built using a "delta" or more often a "delta-sigma" modulator (sometimes called "sigma-delta"). In semiconductor technology, more robust sigma-delta ADCs are overwhelmingly preferred to delta ADCs. In superconductor technology, the availability of a close-to-ideal integrator in the feedback loop makes delta ADCs practical. The oversampling-type ADCs are best for digitizing relatively *lower bandwidth* signals but with the *maximum possible resolution*.

---

[107] e.g., G. S. Lee and D. A. Petersen, *Proc. IEEE* 77 (1989) 1264

[108] e.g., O. Mukhanov, et al., *Proc. IEEE* 92 (2004) 1564

[109] H. T. Mann, D. G. Fladlien, Superconductive analog-to-digital converter, U.S. Patent 3 196 427, July 20, 1965 (Filed Nov. 14, 1960)

[110] M. D. Fiske, Superconductive totalizer or analog-to-digital converter, U.S. Patent 3 458 735, July 29, 1969 (Filed Jan. 24, 1966)

[111] M. Klein, Successive-approximation analog-to-digital converter using Josephson devices, U.S. Patent 3 949 395, Apr. 6, 1976 (Filed Aug. 28, 1974); M. Klein, ISSCC77, *Digest of Tech. Papers* (1977) 202)

7.8.1 Superconductor Parallel-Type ADCs

Early Superconductor Parallel ADCs

Historically, the very first true superconductor ADC was of the Nyquist type, more specifically a parallel-type Flash ADC. In 1975, H. Zappe of IBM pioneered the idea that periodic nature of superconducting quantum interference devices (SQUIDs) allows the formation of an n-bit flash ADC with only n comparators, rather than the $2n - 1$ single-threshold comparators usually required in the conventional semiconductor flash ADCs[112]. It provides a unique solution for drastic reduction of a circuit complexity and, at the same time, allows faster sampling.

This ADC design signifies the first use of a superconductor's natural magnetic flux quantization. It also exploited the extreme field sensitivity of quantum interference devices, so that a very small analog signal could be converted. This was a truly pioneering invention. This first ADC design relied on the use of SQUID comparators with different inductance values and therefore was difficult for a practical ADC implementation. These difficulties were resolved by 1979, when R. Harris, C. Hamilton and F. Lloyd of NBS (now NIST) demonstrated a 200 MS/s 4-bit Flash ADC circuit using a Pb-alloy fabrication process[113]. They improved Zappe's design by using optimally designed identical SQUID comparators with variable the coupling of signal (Figure 7.53).

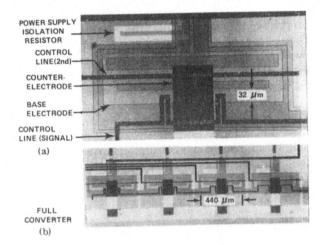

FIGURE 7.53: In 1979 the world's first implementation of superconducting flash ADC at NBS/NIST[113]: (a) microphotograph of one of 4 SQUID Flash ADC comparators. (b) microphotograph of a complete 4-bit ADC. (Reprinted with permission from 113. Copyright 1979, American Institute of Physics.)

This work inspired a significant activity aimed to improve the ADC accuracy-bandwidth performance. Hamilton and Lloyd proceeded to demonstrate a 6-bit ADC at 4 GS/s [114]. By 1988, S. Ohara, T. Imamura, and S. Hasuo of Fujitsu extended sampling rate to 5 GHz by implementing the ADC using a 3-μm Nb/AlOx/Nb process[115]. A great deal of the development took place at the University of California, Berkeley (UC Berkeley), NBS/NIST, and TRW, which aimed to develop a better SQUID comparator[107].

---

[112] H. H. Zappe, *IBM Tech. Discl. Bull.* 17 (1975) 3053

[113] R. E. Harris, C. A. Hamilton, F. L. Lloyd, *Appl. Phys. Lett.* 35 (1979) 720

[114] C. A. Hamilton and F. L. Lloyd, *IEEE Trans. Magn.* MAG-17 (1981) 3414

[115] S. Ohara, T. Imamura, and S. Hasuo, *Electron. Lett.* 24 (1988) 850

Quasi-One Junction SQUID (QOS) Comparator for Flash ADC

The most significant and influential flash ADC improvement was the 1988 invention of the Quasi-One junction SQUID (QOS) comparator by H. Ko and T. Van Duzer at UC Berkeley[116]. This new comparator had a greater than 5× improvement over the previous designs in analog bandwidth and sampling rate. The main design goal was to eliminate the hysteresis and dynamic distortions caused by the vortex-to-vortex transition in the two- and three-junction SQUID comparators. This was achieved by a low-inductance, one-junction SQUID. For signal sampling, an additional junction was inserted in the one-junction SQUID loop (Figure 7.54a). This QOS comparator set the design trend for many years. Even today, the QOS remains the basis for the latest ADC comparator designs.

In following years, the QOS-based flash ADCs were extensively developed for instrumentation applications, such as wideband transient digitizers. P. Bradley of HYPRES (originally from UC Berkeley) demonstrated a 6-bit Flash ADC with about 4 effective bits at 5 GHz and 3 effective bits at 10 GHz using a beat frequency test method[117]. The flash ADC consisted of a linear array of six active comparators with analog signal applied via an $R - 2R$ ladder and producing a Gray-coded digital output. In 1993, this Flash ADC was cryopackaged within an APD Gifford-McMahon (G-M) cryocooler and successfully operated as a rack-mounted demo unit at HYPRES (Figure 7.54b,c). In fact, this was the first superconducting ADC (or any digital or mixed-signal LTS integrated circuit) operating on a 4 K closed-cycle refrigerator.

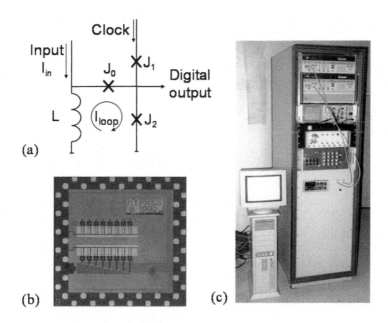

FIGURE 7.54: (a) Schematic of Quasi-Onejunction SQUID (QOS) comparator (1988)[116] (©1988 IEEE). (b) HYPRES 1992 4 GHz Flash ADC chip based on QOS comparators and fabricated using the 1993 standard HYPRES 3-μm process. (c) The demonstration setup of a cryocooled 4 GHz Flash ADC using a 3-stage APD HS-4 cryocooler with a 1 W capacity at 4.2 K. The cryocooler cold head occupied the bottom compartment while test equipment occupied the top compartment of the standard rack.

[116] H. Ko and T. Van Duzer, *IEEE J. Solid-State Cir.* 23 (1988) 1017

[117] P. Bradley, *IEEE Trans. Appl. Supercond.* 3 (1993) 2550

SFQ Flash ADC

The demonstrated flash ADC design was quite successful, although it suffered from so-called "duty cycle" problem when thresholds for the rising and falling edges appeared to be shifted for high slew rate signals. At that time, a paradigm shift from the latching type to non-latching single flux quantum (SFQ) logic influenced the designs of flash ADCs. Consequently, two new SFQ ADC comparators were proposed. One of these designs was based on Quantum Flux Parametron (QFP), which provided high accuracy although it was too complex. The alternative SFQ comparator design was introduced by S. Rylov of HYPRES in 1997 [118]. This SFQ comparator followed the general QOS basic design but was implemented using RSFQ design principles with shunted Josephson junctions and was integrated with a SQUID wheel (a phase tree). The duty cycle problem was compensated to a substantial degree by the insertion of a simple negative feedback resistor. This design became the comparator of choice for HYPRES transient digitizer.

Getting more Bits: Error-Correction and Interleaving

Like all parallel ADCs, the superconductor flash ADC is susceptible to fabrication mismatches in circuit components and conditions. This flaw can be corrected with a real-time digital error correction technique, called the look-back algorithm. This technique was developed by C. Anderson of IBM in the 1980s and then finally published in 1993 [119]. In this scheme, two comparators, offset from each other by a dc $\Phi_0/4$ flux bias, are used for each bit. This ensures that at most one comparator can be close to threshold for any input signal value. The look-back error-correction logic also converts the original Gray code output to standard binary. Furthermore, one can synthesize additional bits of lower significance. In addition to the error-correction, the bit-interleaving technique can be used to add one more bit of resolution without increasing the slew rate of a comparator. It adds an additional least significant bit (LSB) by XOR-ing outputs of 90-degree phase shifted (interleaved) additional comparators.

Flash ADC Applications: Transient Digitizer

Finally, all these techniques led to the development of a transient digitizer. There are several important areas of scientific and commercial instrumentation in which the precision, speed, and dynamic range of superconductor ADCs are of a great interest.

One of the key technical difficulties in the development of the high-speed flash ADC digitizers is the output interface bandwidth. A superconductor flash ADC produces digital output data at tens of Gbit/s which exceeds today's data link per line capabilities and speed of semiconductor electronics. The only way to connect to the conventional world is to slow down the data using two possible approaches: *on-chip memory buffering and demultiplexing*.

Figure 7.55 shows a transient digitizer developed at HYPRES by S. Kaplan et al. in 1998 [120] using an on-chip memory. It consists of a superconductor flash ADC, combined with fast RSFQ shift register memory circuits to store the digitized data for subsequent readout at much slower data rate. A prototype instrument, comprising a superconductor integrated circuit (Figure 7.55b) along with a room temperature interface and data acquisition electronics, was demonstrated for single-shot pulse capture. Each digitizer chip contained a 6-bit flash ADC coupled to a bank of 32-stage shift registers through a set of acquisition control switches. Despite the superior ADC performance compared to the semiconductor counterparts, the low capacity of the digitizer memory was the limiting factor for commercialization. As in the case of digital processors, the lack on a large capacity memory prevented the insertion of superconductor digitizers into a marketplace.

---

[118] P. Bradley and S. Rylov, *IEEE Trans. Appl. Supercond.* 7 (1997) 2677

[119] C. J. Anderson, *IEEE Trans. Appl. Supercond.* 3 (1993) 2769

[120] S. Kaplan, et al., *IEEE Trans. Appl. Supercond.* 9 (1999) 3020

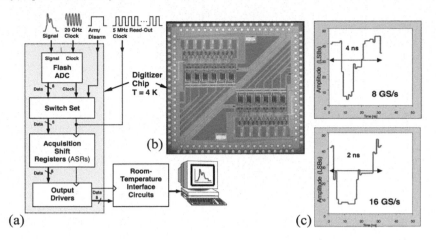

FIGURE 7.55: A 1998 HYPRES superconductor transient digitizer: (a) Block diagram. (b) A superconductor chip with two transient digitizers based on a 6-bit flash ADC and a 32-bit-deep memory. (c) A fast pulse capture: 4 ns for 8 GS/s (top) and 2 ns for 16 GS/s (bottom) (©2004 IEEE).

An alternative transient digitizer design proposed by M. Maezawa, et al. in AIST, Japan in 2001 was based on the use of demultiplexers[121]. The digitizer consisted of a 2-bit 16 GS/s flash ADC similar to the one described above. It was followed by a 1:16 RSFQ demultiplexer and output drivers. This demultiplexer-based architecture can avoid the superconductor memory bottleneck. However, the number of output lines are larger by the demultiplexing ratio (e.g., by 16), which presents a cryopackaging and cost challenge for faster and higher resolution ADCs.

Recent Flash ADCs: Complimentary QOS Comparator

Relatively fast development of parallel-type superconducting ADCs in the 1980s–1990s was considerably slowed after 2000 by the lack of dedicated government-sponsored development programs, requirements for the superfast transient measurements, and competition from conventional instruments. Similarly in the commercial marketplace, product development goals were shifted from instrumentation to wireless communications which required a different ADC type: oversampling ADCs. Fortunately, recent attention to ultra-fast optical communications has been renewed after several years of low interest caused by the "dark fiber" overcapacity problem. This has opened an attractive opportunity for flash ADCs in the optical communication systems and their measuring instruments. Sampling speeds exceeding 100 GS/s became a target for the flash ADCs. In 2008, H. Suzuki et al. from the SRL ISTEC, Japan, invented a new flash ADC comparator capable of operating with the required speed if fabricated with a 40 kA/cm2 critical current density process. The new comparator is the latest development of the famous QOS comparator, Complementary QOS (CQOS). It finally solved the long-standing "duty cycle" problem plaguing previous QOS designs. The new design is based on a differential pair of the identical SFQ QOSs connected at the decision-making part. The signal is applied to both QOSs complimentarily (Figure 7.56a) and cancels the asymmetrical distortions[122]. In 2010, a 4-bit CQOS comparator test circuit showed the 3-bit binary and 4-bit Gray-code operation at 15 GS/s in beat frequency tests. For low frequency input, correct

[121] M. Maezawa, et al., Analog-to-digital converter based on RSFQ technology for radio astronomy applications, in *Extended Abstracts of ISEC'01*, Osaka, Japan, (2001) 451

[122] M. Maruyama, et al., *IEEE Trans. Appl. Supercond.* 19 (2009) 680

high-speed sampling operation up to 50 GS/s at 10 kA/cm^2 were observed. A complete 5-bit flash ADC with the look-back error-correction and bit-interleaving circuits (Figure 7.56b) was assembled for testing on a 1 W G-M cryocooler[123].

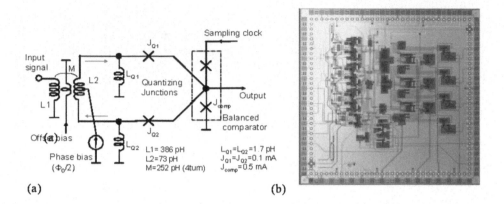

FIGURE 7.56: A 2008 Complimentary QOS comparator Flash ADC (ISTEC, Japan): (a) CQOS schematic. (b) Flash ADC chip (2010) with integrated error-correction and interleaving circuits[123] (©2009 IEEE, courtesy of H. Suzuki).

7.8.2 Superconductor Oversampling ADCs

The availability of fundamental flux quantization in superconductivity inspired several different oversampling delta and delta-sigma ADC designs. These are counting delta and sigma-delta type ADCs, a phase modulation-demodulation (PMD) delta ADC, and conventional delta and sigma-delta ADCs. While having quite different design approaches, most of them take advantage of the unique availability of *implicit feedback* and signal integration inherent to a SQUID loop, which is a result of fundamental conservation of flux in a superconducting loop. The SQUID loop automatically accumulates the total flux with opposite sign (sum of antifluxons) of all SFQ pulses (fluxons) emitted by the Josephson junction.

Counting V/F ADC

The first idea of superconductor oversampling ADC was suggested by McDonald of NBS/NIST at the 1976 Navy Summer Study on Superconductive Electronics. He proposed to use the ac Josephson effect to perform direct voltage-to-frequency (V/F) conversion. A single Josephson junction (Figure 7.57a) can act as a voltage-controlled oscillator (VCO) and produce an SFQ pulse train at a rate proportional to the applied analog voltage as $f = 2eV/h = V/\Phi_0$. The generated SFQ pulses representing threshold crossings are to be counted over a time interval. This critical ability to count low-power SFQ pulses was provided in 1978 when J. Hurrell and A. Silver invented a binary counter based on a two-junction SQUID with overdamped Josephson junctions[124,125]. The speed and simplicity of the counter circuit was a profound manifestation of the potential of superconductor circuits utilizing SFQ switching. Amazing switching speed of 100 GHz was experimentally demonstrated

[123] H. Suzuki, et al., *IEEE Trans. Appl. Supercond.*, vol. 21, no.3, Jun. 2011

[124] J. P. Hurrell and A. H. Silver, *AIP Conf. Proc.* 44 (1978) 437

[125] J. P. Hurrell, D. C. Pridmore-Brown, A. H. Silver, *IEEE Trans. Electron.* Dev. 27 (1980) 1887

by C. Hamilton and F. Lloyd of NBS/NIST in 1982 [126].

This basic design was further developed and improved to increase its performance. In order to increase resolution of quantization process, a multi-junction VCO was proposed at TRW. It is based on interleaving of several single-junction VCOs while maintaining a fixed phase shift of one junction to another[127]. In order to increase the sensitivity of the V/F ADC, one can use a SQUID with a sensitive input transformer biased into the voltage state as a pulse generator as it was proposed at HYPRES[128]. Overall, the V/F counting A/D conversion is equivalent to a low-pass first-order *sigma-delta modulation with implicit feedback*[108].

Counting Flux Quantizing (Tracking) ADCs

The fundamental linearity of flux quantization in a superconducting loop is used for constructing another counting ADC design a flux quantizing or tracking ADC which was first described by G. Lee of TRW in 1989 [129,107]. In this design, the input signal current is coupled into a SQUID loop, which generates one SFQ pulse for each Φ_0 change in flux. In contrast to the V/F ADCs, the SFQ pulses are generated in response to increments or decrements of the signal, i.e., only changes of signal are registered. Similarly to the V/F ADC, these SFQ pulses can be counted by using binary counters to reconstruct the signal. Each time the input increases the flux in the loop (Φ_{loop}) by Φ_0, the junction switches, creating a fluxon-antifluxon pair, one of which propagates as an SFQ pulse and the other decreases Φ_{loop} by Φ_0 (Figure 7.57b). This automatic subtraction of the output signal from the input makes it equivalent to low-pass first-order *delta modulation*, where the output is proportional to the signal derivative $d\Phi/dt$. Since RF signal is sensed by a SQUID transformer, this ADC has an extremely high sensitivity similar to that of a SQUID.

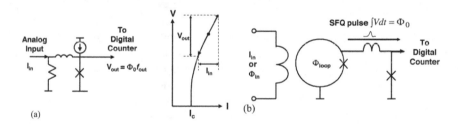

FIGURE 7.57: Basic oversampling Counting ADC modulators. (a) Voltage-to-frequency (V/F) sigma-delta ADC modulator and its operating region. (b) Flux quantizing unidirectional delta (Tracking) ADC modulator (©2004 IEEE).

This concept can be expanded to accommodate both polarities of input signal derivative. Figure 7.58a shows a scheme with a two-junction quantizer, biased such that one of them switches when the flux in the loop changes by $+\Phi_0$ and the other when it changes by $-\Phi_0$, followed by bi-directional (up and down) counting implemented using either two counters or a single bi-directional counter[107]. In 1990, the joint Moscow State University (MSU) and IREE, USSR group implemented the first RSFQ-based tracking ADC featuring bi-direction counter with non-destructive readout allowing uninterruptable signal tracking[130] (Figure 7.58b).

[126] C. A. Hamilton and F. L. Lloyd, *IEEE Electron. Dev. Lett.* 3 (1982) 335

[127] M. W. Johnson, et al., *IEEE Trans. Appl. Supercond.* 11 (2001) 607

[128] O. A. Mukhanov, et al., *Physica C* 368 (2002) 196

[129] G. S. Lee, *IEEE Trans. Magn.* 25 (1989) 830

[130] L. V. Fillipenko, et al., *IEEE Trans. Magn.* MAG-27 (1991) 2464

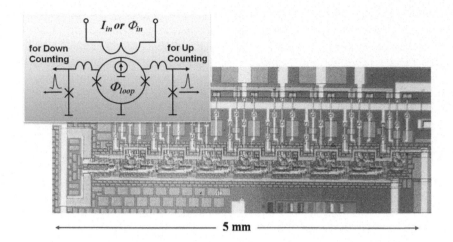

FIGURE 7.58: First RSFQ-based Tracking ADC with a bi-directional (Up/Down) counter (1990). It was fabricated with a 5-μm all-Nb process with 500 A/cm^2. Inset: Basic bi-directional ADC modulator schematic.

Phase Modulation-Demodulation (PMD) ADC

One of the problems associated with the flux-quantizing (tracking) delta ADCs is the hysteresis of the SQUID quantizer in response to changing polarity of the signal derivative. In order to solve this, a dc voltage-biased single-junction SQUID quantizer (Figure 7.59a) was introduced by S. Rylov, et al. of HYPRES in 1994 [131]. The voltage source (or phase generator) continuously pumps flux into a quantizer at a constant rate, which then leaves the quantizer via the only junction with the timing modulated by the derivative of the analog signal (Figure 7.59a). The modulated SFQ pulse train is passed for SFQ phase (time) demodulation to a synchronizer—a clocked sampling circuit generates a "1" or a '0' indicating whether or not an SFQ pulse arrived during that clock interval. Higher SFQ phase resolution can be achieved by either higher clock or by adding more channels of the synchronizer. Typical sampling clock of this PMD ADC is in tens of GHz limited by the speed of subsequent digital signal processing circuits.

Similarly to the flash ADC digitizers, the oversampling ADC produces digital output data at tens of Gbit/s which exceeds today's data link per line capabilities and speed of semiconductor electronics. There are two possible approaches to this interface problem: *on-chip digital filtering* and *demultiplexing*.

A digital decimation filter reduces the sampling rate, narrows output bandwidth, and generates additional bits. For oversampling superconductor ADCs, sinc-type digital filters were implemented using fast RSFQ logic capable of operating at the same speed as the sampling speed of the ADC. One of the key factors in successful development of the PMD ADCs was the adoption of a new design of the digital filter with programmable bandwidth developed by V. Semenov et al. in 1997 [132] and then perfected by T. Filippov et al. of SUNY Stony Brook[133]. By 2001, O. Mukhanov et al. of HYPRES

---

[131] S. V. Rylov and R. P. Robertazzi, *IEEE Trans. Appl. Supercond.* 5 (1995) 2260

[132] V. Semenov, Yu. Polyakov, and A. Ryzhikh, Decimation filters based on RSFQ logic/memory cells in *Extended Abstracts of ISEC' 97*, Berlin, Germany, (1997) 344

[133] T. V. Filippov, et al., *IEEE Trans. Appl. Supercond.*, 11 (2001) 545

and SUNY team[134] demonstrated a PMD ADC (Figure 7.59b) with over 11 effective bits or 68 dB signal-to-noise ratio (SNR) for 145 MS/s output and maximum operation speed up to 19 GS/s. This was the world's fastest operation of the most complex (6,000 Josephson junctions) superconductor digital or mixed-signal RSFQ circuit of that time.

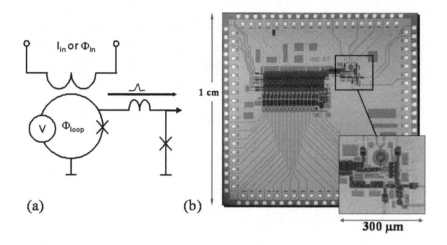

FIGURE 7.59: Phase modulation demodulation (PMD) delta ADC invented in 1994. (a) Flux quantizer for PMD ADC modulator. (b) Second-generation PMD ADC chip consisting of a low-pass PMD delta modulator (in the inset) and a 15-bit decimation digital filter operating at 19.6 GS/s. This chip contains ~6,000 Josephson junctions fabricated at 1 kA/cm^2 process in 2000 (©2004 IEEE).

A large variety of PMDs ADCs were developed including ADCs with serialized digital output, multi-rate ADC with doubled modulator sampling rate for communications and signal intelligence applications. The ADC fabrication using 4.5 kA/cm^2 allowed sampling rate up to 34 GS/s. In 2006, I. Vernik et al. of HYPRES[135] demonstrated 13.5 effective bits for 10 MHz signal with a PMD ADC chip cryopackaged onto Sumitomo G-M cryocooler. By 2008, A. Inamdar et al. of HYPRES[136] achieved 14.5 effective bits (SNR=89.2 dB) at 29 GS/s sampling clock, while the multi-rate PMD ADCs operated up to 46 GS/s. In order to increase dynamic range of the PMD ADC, a quarter-rate PMD quantizer (QRQ) was introduced by A. Inamdar and S. Rylov et al. in 2006 [137] to increase the maximum ADC slew rate and to add two more bits of resolutions.

Higher-Order Delta ADC

Improvement of ADC performance is expected with higher order ADC modulators. V. Semenov's group at SUNY developed a low-pass delta ADC modulator based on a synchronous quantizer with two feedback loops and two integrators: an implicit loop due the conservation of magnetic flux in the superconducting loop of the quantizer and an explicit loop formed with the Josephson amplification circuit and a low-pass filter. A complete ADC chip with decimation digital filter was

---

[134] O. A. Mukhanov, et al., *Supercond. Sci. Technol.* 14 (2001) 1065

[135] I. V. Vernik, et al., *IEEE Trans. Applied Superconductivity* 17 (2007) 442

[136] A. Inamdar, et al., *IEEE Trans. Applied Superconductivity* 19 (2009) 670

[137] A. Inamdar, et al., *IEEE Trans. Applied Superconductivity* 17 (2007) 446

demonstrated its operation at up to 10.5 GHz clock in 1998 [138]. Although the modulator has two loops, its performance was still similar to a first-order modulator. In order to obtain the second-order characteristics, it is necessary to increase the explicit feedback gain. Achieving higher gain is difficult in superconductor technology, since it requires the construction of SFQ amplifiers (drivers).

Sigma-Delta ADC

In 1992, J. Przybysz et al. of Westinghouse invented a low-pass sigma-delta ADC modulator based on a synchronous quantizer with an analog $L-R$ integrator (Figure 7.60a)[139]. It has an implicit feedback due to the magnetic flux conservation and demonstrates the sigma-delta "noise shaping", i.e. suppressed noise at low frequencies. By 2006, the first-order sigma-delta ADC has reached 77 dB SNR for a 10 MHz signal at 16 GS/s clock as it was demonstrated by A. Yoshida et al. of SRL ISTEC and the Hitachi group[140]. Demultiplexing was used in order to bridge the disparity in data rates between superconductor ADC modulator and semiconductor digital signal processing.

Due to the quadratically rising noise, a simple first-order decimation filter is insufficient to filter out the high-frequency noise. A second order modulator can improve SNR from 9 to 15 dB/octave of oversampling. However as for the delta ADC described above, the lack of broad-band amplification in superconductor technology makes construction of a second order sigma-delta ADC difficult. In 1994, J. Przybysz et al. proposed a two-loop modulator design employing an additional explicit feedback loop with the required substantial gain M (Figure 7.60b)[141].

There have been several attempts in the US and Japan to realize this challenging amplification task which has to deliver enough gain ($M \sim 64$) within a very short time, below the high-speed clock period. In 2003, S. Hirano et al. of SRL ISTEC, Japan demonstrated an amplifier based on a magnetically coupled Josephson transmission line (JTL) tree and serially connected SQUIDs demonstrated close to 12 dB/octave power spectrum characteristics although at a relatively low 1.2 GHz clock rate[142].

In 2004, realizing the difficulty of building a second-order sigma-delta ADC, A. Sekiya et al. of Nagoya University[143] found a way to increase first-order sigma-delta ADC SNR and sensitivity by using a multi-bit modulator following the PMD approach. The 4,000-junction ADC chip consisting of a modulator and decimation filter was demonstrated using NEC standard Nb process with 2.5 kA/cm^2.

Finally in 2008, Q. Herr et al. of Northrop Grumman[144] developed an SFQ amplifier with the required gain. A second-order sigma-delta ADC circuit featuring a high gain ($50 \, \Phi_0$) quantum-accurate feedback amplifier (Figure 7.60c) was built using HYPRES' commercial 4.5 kA/cm^2 Nb process. It achieved 81 dB SNR or 13.1 effective bits over a 10 MHz band at a 5 GS/s sampling clock. Since the ADC chip did not have an on-chip digital filter, the sampling speed was limited by an output interface. The measured performance was in agreement with the linearized model which showed that the amplifier delay does not have to be shorter than the sampling clock period.

Band-Pass Sigma-Delta ADC

Superconductor technology is particularly suitable for implementing band-pass ADC designs exhibiting a peak performance around a particular frequency. Superconductivity features two major advantages: low-loss materials allowing very high Q resonators and high clock rates allowing direct

---

[138] V. K. Semenov, Y. A. Polyakov, and T. V. Filippov, *IEEE Trans. Appl. Supercond.* 9 (1999) 3026

[139] J. X. Przybysz, et al., *IEEE Trans. Appl. Supercond.* 3 (1993) 2732

[140] A. Yoshida, et al., *IEEE Trans. Appl. Supercond.* 17 (2007) 426

[141] J. X. Przybysz, D. L. Miller, and E. H. Naviasky, *IEEE Trans. Appl. Supercond.* 5 (1995) 2248

[142] S. Hirano, et al., *Physica* C 392 (2003) 1456

[143] A. Sekiya, et al., *IEEE Trans. Appl. Supercond.* 15 (2005) 340

[144] Q. Herr, et al., *IEEE Trans. Appl. Supercond.* 19 (2009) 676

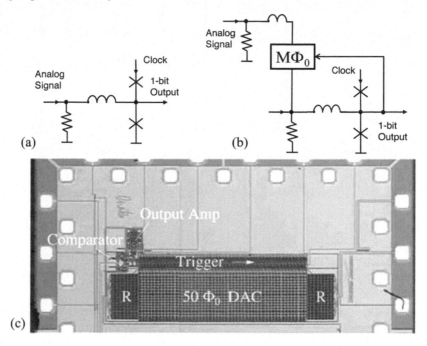

FIGURE 7.60: Superconductor low-pass sigma-delta modulator. (a) 1993 first-order modulator with implicit feedback and input $L - R$ integrator. (b) 1994 second-order modulator showing the critical feedback M Φ_0 amplifier. (c) Northrop Grumman 2-order sigma-delta ADC with a 50 Φ_0 feedback amplifier demonstrated in 2008 (©2009 IEEE).

sampling of multi-GHz RF signals. J. Przybysz and D. Miller of Westinghouse invented a first-order band-pass sigma-delta ADC modulator by replacing the front-end *LR* integrator with an *LC* resonator[145]. It suppresses the quantization noise around the resonant frequency f_{LC}, rather than at dc. The expected first-order noise shaping of sigma-delta modulator with lumped-element resonator was demonstrated by T. Hashimoto et al. of SRL ISTEC in 2001 [146]. In 2002, J. Bulzacchelli et al. of IBM[147] demonstrated a band-pass ADC with distributed microstrip-based resonators exhibiting the desired noise suppression around 2.2 GHz up to a sampling rate of 45 GHz. The demonstrated performance (SNR of 49 dB and dynamic range of 57 dB over ~20 MHz bandwidth at 2.2 GHz) exceeded that of semiconductor band-pass modulators at that time. Due to the lack of on-chip digital filters, the digitized data was stored using on-chip buffers for subsequent slow readout.

D. Kirichenko of HYPRES developed a family of continuous-time sigma-delta band-pass ADC modulators employing an implicit feedback and lamped-element resonators. Band-pass ADCs centered around various RF frequencies: 1 GHz, 4 GHz, 5 GHz, 7.5 GHz, and 20 GHz were demonstrated[148]. Similarly to the low-pass ADC, a first order sigma-delta ADC could provide sufficient performance desired for directly digitizing receiver architectures. In 2007, D. Kirichenko developed a second-order delta-sigma ADC modulator with two lumped *LC* resonators (Figure 7.61a) by introducing an explicit feedback loop using JTLs as active delay elements in addition to a D flip-flop

---

[145] J. X. Przybysz, D. L. Miller, Bandpass sigma-delta modulator for analog-to-digital converter, U.S. Patent 5 341 136, Aug. 23, 1994

[146] T. Hashimoto, et al., *Jpn. J. Appl. Phys.* 40 (2001) L1032

[147] J. F. Bulzacchelli, et al., *J. Solid State Circ.* 37 (2002) 1695

[148] D. Gupta, et al., *IEEE Trans. Applied Superconductivity* 17 (2007) 430

to control the phase of the feedback signal[149].

Typically, the clock frequency for a band-pass sigma-delta ADC is chosen to be $f_{clk} = 4f_0$, where f_0 is the center of the band of interest, e.g., 30 GHz for X-band. One can also use a lower clock frequency with some performance penalty. In this scheme, called RF undersampling, one can take advantage of the sampling process that replicates the input analog frequency band, centered at f_0, translated by multiples of the sampling frequency (f_{clk}). The world's first Ka-band ADC with a 20.362 GHz center frequency was tested in the RF undersampling mode using a 27.136 GHz clock[149].

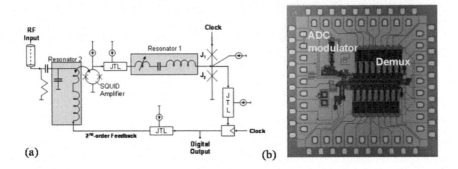

FIGURE 7.61: A 2007 HYPRES second-order band-pass sigma-delta ADC. (a) Schematics of ADC modulators with implicit and explicit feedback loops (©2009 IEEE). (b) X-band ADC chip consisting of second-order band-pass ADC modulator centered for 7.4 GHz and digital signal processor.

In 2008, D. Kirichenko invented an ADC modulator with two implicit feedback paths exhibiting a quasi-instantaneous feedback implemented by connecting two resonators directly to the comparator. The RF input was split and applied through inductive coupling to each resonator. In addition, a SQUID amplifier stage was used to connect two *LC* resonators in series to get the desired loop filter transfer function. Such a band-pass ADC equipped with a 1:16 deserializer was demonstrated with a 31.6 dB SNR in 660-915 MHz band at 10.24 GS/s clock[150].

Multi-Modulator ADC

Challenges in achieving higher performance encourage the adaptation of ADC architecture approaches known in conventional semiconductor ADC technology: time-interleaving, subranging, cascading, and others.

Time-interleaving allows the increase of effective sample rate by using several parallel comparators sampled by the same clock. The actual performance gain in the interleaved oversampling ADC depends on various factors, including feedback-loop delay in comparison to the effective clock period. In 1999, V. Semenov of SUNY[151] invented a time-interleaved delta modulator consisting of two delta modulators shifted by half a clock period, this requires interleaving both comparator and feedback functions. Since interleaving cannot be achieved with the implicit feedback of SQUID

---

[149] O. A. Mukhanov, et al., *IEICE Trans. Electron.* E91-C (2008) 306

[150] D. Kirichenko, T. Filippov, D. Gupta, Microwave receivers with direct digitization in *Proc. IMS'09*, Boston, USA, June 2009

[151] V. K. Semenov, Superconductor modulator with very high sampling rate for analog to digital converter U.S. Patent 6 608 581, Aug. 19, 2003

quantizers, separate explicit feedback loops are implemented using stacked SQUIDs.

Subranging is capable of significant increase of dynamic range by using several modulators to digitize different ranges of signal amplitude. This approach can be applicable for two-delta, two-sigma-delta, or combined delta and sigma-delta modulator combinations. A. Inamdar of HYPRES[136] showed that the two-delta subranging ADC based on the proven PMD ADC modulators is capable of a significant SNR improvement (23 dB) with a quite conventional 26 GS/s sampling clock.

Oversampling ADC Applications: Wideband Digital Receivers

Since the 1990s, rapid technological progress in wireless commercial and defense communications and related radar and electronic-warfare applications is driving the demand for much higher ADC performance. These applications can greatly benefit from the ability to directly digitize wideband RF signals. Conventional narrow-band technology implements a separate analog receiver with one or more down-conversion steps for each sub-band. A wideband software-defined radio (SDR) receiver needs mixed-signal and digital components capable of delivering extreme speed, linearity, dynamic range, noise, and sensitivity. As it was realized around 2000 in the USA, Japan, and Europe, the unique features of superconductor technology and ADC circuits, in particular, can make a true SDR possible[152,153,154].

In a *digital-RF architecture*[155], data conversion and digital processing take place at RF rather than at baseband – the analog filter and up/down-conversion stages from/to lower IF or baseband are eliminated (Figure 7.62). A wideband RF signal is applied directly to an ADC modulator producing an oversampled low-bit-width digital code at a very high data rate (tens of Gbps). This high-rate data stream is processed before down-conversion using a relatively low-complexity but very high-throughput processor, an *RF DSP*, to implement various functions such as digital signal—combining from multiple channels, true-time delay for digital beamforming, adaptive active cancellation of transmit channels, correlation-based digital filtering, etc. Finally, this high-rate data is down-converted to baseband using digital mixers and decimation filters for further processing and decoding. Back-end processing is implemented using conventional semiconductor parts and placed at ambient temperature.

The first practical implementation of the digital-RF architecture–superconducting digital-RF channelizing receivers extract different frequency bands-of-interest within the broad digitized spectrum. The single-bit oversampled data, from either a low-pass delta or band-pass delta-sigma modulator, are applied to one or more channelizers, each comprising digital in-phase (I) and quadrature (Q) mixers and decimation digital filters. On-chip digital channelization is followed by the lower-speed channelization using field programmable gate array (FPGA) chips at room temperature.

In 2004, the first digital-RF channelizing receiver chip was produced at HYPRES[155] (Figure 7.63a). It consisted of a 20 GS/s low-pass phase PMD ADC modulator. This ~11,000 junction chip was fabricated using a 1.0 kA/cm^2 process and tested up to 20 GHz clock rate. In 2005, the ADR chip was integrated onto a commercial Sumitomo SRDK 101D cryocooler capable of cooling 125 mW at 4.2 K. It was mounted into a standard 19-inch rack, which also housed a cryocooler compressor, interface and control hardware (Figure 7.63b). This world's first digital-RF receiver system was done under support and guidance from D. Van Vechten of the U.S. Office of Naval Research.

Satellite communications with high carrier frequencies and wide bandwidths (e.g., X-band: 500 MHz BW around 7.5 GHz; Ka-band: 1 GHz BW around 20.5 GHz) can benefit from wideband

[152] E. B. Wikborg, V. K. Semenov, and K. K. Likharev, *IEEE Trans. Appl. Supercond.* 9 (1999) 3615

[153] A. Fujimaki, et al., *IEEE Trans. Appl. Supercond.* 11 (2001) 318

[154] D. K. Brock, O. A. Mukhanov, and J. Rosa, *IEEE Commun. Mag.* 39 (2001) 174

[155] O. A. Mukhanov, Superconductor Digital-RF Electronics in: *Extended Abstracts ISEC'05*, Noordwijkerhout, the Netherlands, I-A.01, Sep. 2005

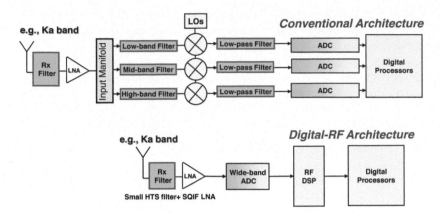

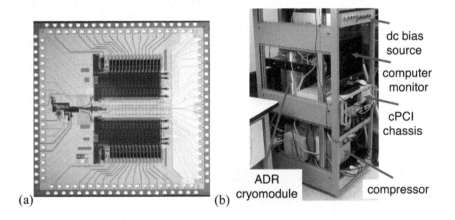

FIGURE 7.62: Comparison of conventional (top) and Digital-RF receiver (bottom) architectures. In the digital-RF architecture, data conversion is carried out directly at RF frequencies using a wideband oversampling ADC modulator. The digitized RF data stream is then processed at very high data rate in an RF DSP before being digitally down-converted and filtered to baseband for further processing.

FIGURE 7.63: World's first digital-RF system: – 2005 HYPRES channelizing receiver: (a) Microphotograph of a single-channel all-digital receiver (ADR) chip based on the first-order low-pass delta ADC, digital in-phase and quadrature (I&Q) mixer, and decimation digital filters. This 1 cm×1 cm chip consists of ~11,000 JJs and dissipates ~3.5 mW. (b) Photo of the cryopackaged ADR test setup (ADR-0) using commercial Sumitomo 2-stage GM cryocooler mounted into the lower part of standard 19-inch rack.[155]

digital-RF channelizing receivers by eliminating balky analog channelizing and downconversion stages. In 2006, D. Gupta et al. of HYPRES[148,149] assembled and delivered an X-band digital receiver system to the Joint SATCOM Engineering Center (JSEC) in Ft. Monmouth, NJ to receive wideband X-band signals from the XTAR and DSCS satellites. The receiver was integrated with a digital I&Q MODEM, demonstrating demodulation of satellite signals including a video data transmission. In 2007, the ADR-1 system was upgraded with installing a faster 30-GHz chip fabricated 4.5 kA/cm^2 process (Figure 7.64a).

In subsequent years, more ADR systems based on low-pass PMD delta ADCs and band-pass sigma-delta ADCs were assembled and delivered. In 2008, new-generation ADR systems featuring modular cryopackaging design were developed. These ADRs demonstrated the *hybrid temperature hybrid technology* (ht^2) system integration concept, where different components of the system are operated at different temperatures to optimize the overall performance. For example, such a system included a high-temperature superconductor (HTS) filter developed in University of Waterloo (Figure 7.64b) which was placed at 70 K on the first stage of the cryocooler and connected to the ADR chip located at the second (4 K) stage. In January 2009, the joint HYPRES, ViaSat, and Navy team demonstrated the world's first multi-net Link-16 data link in which analog outputs from two Link-16 terminals, operating with independent hopping patterns, were combined and applied to the superconductor chip with L-band sigma-delta ADC chip integrated with RSFQ 1:16 demux (Figure 7.64c, 7.64d)[156].

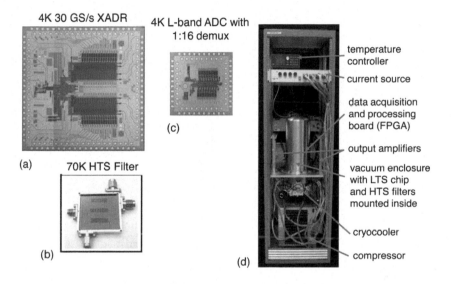

FIGURE 7.64: Digital-RF receiver and its key components: (a) World's first (2006) X-band single-chip digital-RF receiver (XADR) directly digitizes 7.5 GHz RF signal at 30 Gs/s. This 11,000-junction chip consists of a band-pass second-order continuous-time delta-sigma ADC modulator, digital I&Q mixer, and two digital filters; (b) L-band ADC integrated with a 1:16 RSFQ demux (2008); (c) HTS filter developed by University of Waterloo, Canada (2008); (d) HYPRES (2008) modular ADR system built using hybrid temperature hybrid technology (ht^2) approach.

By 2010, HYPRES ADR systems went through three generations. Third-generation ADR systems expanded to house two chip modules with 80 high-speed digital I/Os, up to four 17-channel interface amplifier assemblies and two current sources. The design is modular and allows for independent service of each chip module and quick field replacement when necessary. System reconfigurability enabled by the modular design allowed ADR-5 to be configured and used for different tasks. D. Gupta *et al.*[156] demonstrated operation with the XTAR satellite without utilizing a front-end low-noise amplifier usually required for this application. This was another step towards fully digital receivers. ADR-5 system worked up to 32 GHz clock frequency with a variety of phase and amplitude modulated waveforms at 30 Msymbol/s.

---

[156] D. Gupta, et al., Modular, multi-function Digital-RF receiver systems *IEEE Trans. Appl. Supercond.* 21 (2011)

The digital-RF channelizing receiver approach can be extended to include multiple ADC modulators and multiple channelizer units on a multi-chip module or a single chip. A chip integrating four ADC modulators (centered at 850 MHz, 4 GHz, 7.5 GHz and 12 GHz), a 1×4 digital switch matrix and a 1:16 demultiplexer was demonstrated in 2010 by S. Sarwana et al. of HYPRES[157].

ADC Applications: Sensor Readout

The inherent low noise, low power, high sensitivity, and radiation hardness of superconductor ADCs can be applied to many sensor applications especially for cooled detector arrays. Both flash-type and oversampling-type ADCs can be used for this application.

In 2001, A. Sun et al. of TRW[158] demonstrated a NbN 10 K V/F type ADC for cryocooled infrared (IR) focal plane detector arrays. In 2002, a Nb V/F type ADC with a SQUID-based VCO was used for measuring the integrated charge of a current pulse for superconductor tunnel junction (STJ) X-ray detector readout. Furthermore, this digital counter can also be used to count the number of SFQ clock pulses between successive time events to produce a time-to-digital converter (TDC) on the same chip. Such a dual-function signal and time digitizer was demonstrated in 2002 by joint US-Japanese team with 1 µA full-scale current and 30 ps time resolution[159]. The exceptionally low power of RSFQ technology allows integration of an ADC or TDC in a single cryopackage with the cooled detectors. The integration of the cooled semiconductor detector, visible light photon counter (VLPC) with Nb TDC, was demonstrated by O. Mukhanov et al. in 1998 [160]. The extreme radiation hardness of superconductor electronics was a motivation for the readout of a high-energy particle microstrip detector. A delta ADC based on a very sensitive flux-controlled comparator was demonstrated for CERN high-energy physics experiments by a joint US-Italian team[161].

7.8.3 Superconductor Materials for ADC Implementation: LTS vs HTS

Superconductor ADCs are medium-scale integrated circuits requiring a substantial number of Josephson junctions for complete systems. Even if an ADC modulator can be implemented with rather few junctions, the subsequent demultiplexer, drivers or digital filter can easily require hundreds or thousands of junctions. This requires well-controlled integrated circuit fabrication, with reproducible junction properties. Most of the complete ADC circuits have been demonstrated to date using low-temperature superconductor (LTS) niobium Josephson junctions, operating at about 4 K. Slightly higher temperate superconductors, NbN with T_c up to about 17 K, were used by TRW team to implement counting ADCs for operation at 10 K [158].

The possibility of a drastic reduction of size, weight and power (SWaP) of the cryocooler was the main motivation to implement high-temperature superconductor (HTS) ADCs. There have been major efforts to develop a reproducible technology for Josephson junctions based on $YBa_2Cu_3O_7$ (YBCO), although the high temperatures required for deposition of these materials makes a true multilayer process difficult to achieve. Several HTS ADC projects were active in Japan (Hitachi, SRL ISTEC), Europe (Twente University, Chalmers University, Karlsruhe University), and the USA (Conductus, TRW, Northrop Grumman) for a number of years. Some key components of ADCs, including a simple first-order sigma-delta oversampling ADC modulator, QOS comparator

[157] S. Sarwana, et al., Multi-band Digital-RF Receiver *IEEE Trans. Appl. Supercond.*, vol. 21, Jun. 2011.

[158] A. G. Sun, et al., *IEEE Trans. Appl. Supercond.* 11 (2001) 312

[159] S. Sarwana, et al., *Appl. Phys. Lett.* 80 (2002) 2023

[160] O. A. Mukhanov, et al., *IEEE Trans. Applied Supercond.* 9 (1999) 3619

[161] S. Pagano, et al., *IEEE Trans. Applied Supercond.* 9 (1999) 3628

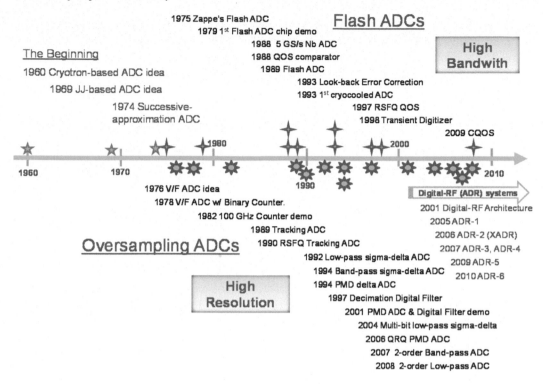

FIGURE 7.65: Superconductor ADC development timeline.

for flash ADC have been demonstrated[162,163,164,165,166]. In 1997, G. Gerritsma et al. of University of Twente[167] reported on the effort to build a 4-bit flash ADC based on QOS comparators using ramp junction technology. The functionality of QOS comparators was demonstrated at low speed. In 2004, H. Sugiyama et al. of SRL ISTEC[168] demonstrated high-speed operation of a QOS comparator based on high-T_c multilayer technology. A circuit containing 10 interface-engineered ramp-edge Josephson junctions was fabricated on a La-substituted $YBa_2Cu_3O_y$ ground plane. The output voltage as a function of the input current for the QOS indicated correct operation as a periodic comparator at clock frequency of 94 and 77 GHz at 35 and 40 K, respectively.

At this moment, it seems unlikely to have the HTS technology reach the required complexity in the near future. On the other hand, a progress in compact 4 K cryocoolers can make Nb-based ADCs useful for a wider range of applications.

7.8.4 Conclusions

We witnessed remarkable progress in superconductive ADCs over last several decades starting from early concepts to demonstrations of application systems performing satellite communication

---

[162] M. G. Forrester, et al., *Supercond. Sci. Technol.* 12 (1999) 698

[163] B. Ruck, et al., *Physica* C 326 (1999) 170

[164] A. Y. Kidiyarova-Shevchenko, et al., *Physica* C 326 (1999) 83

[165] A. H. Sonnenberg, et al., *IEEE Trans. Appl. Supercond.* 11 (2001) 200

[166] K. Saitoh, et al., *Physica* C 378-381 (2002) 1429

[167] G. Gerritsma, et al., *IEEE Trans. Appl. Supercond.* 7 (1997) 2987

[168] H. Sugiyama, et al., *Appl. Phys. Lett.* 84 (2004) 2587

tasks. Most influential ADC designs and development milestones are seen in Figure 7.65. The unique features of superconductivity, featuring the high speed of Josephson junctions and the quantum precision of magnetic flux quantization, produced a wide spectrum of superconductor ADC designs of both high-bandwidth Nyquist-sampling flash ADCs and high-resolution oversampling ADCs.

Acknowledgments

History of Superconductor Analog-to-Digital Converters

The author wishes to thank F. Bedard, D. Gupta, C. Hamilton, R. Harris, S. Hasuo, S. Kaplan, M. Ketchen, D. Kirichenko, V. Semenov, A. Silver and H. Suzuki for help and sharing their data and recollections.

8

Microwave Applications

Editor: D. E. Oates

8.1   Microwave Measurements of Fundamental Properties of Superconductors
      D.E. Oates 459
8.2   Applications of Passive Microwave Filters and Devices in Communication and Related Systems
      R. B. Hammond, N. O. Fenzi and B. A. Willemsen 471
8.3   Superconducting Quantum Electronics Enabling Astronomical Observations
      T. M. Klapwijk 484
8.4   Microwave Cooling of Superconducting Quantum Systems
      W. D. Oliver 493
8.5   Applications of Superconducting Microresonators
      Jonas Zmuidzinas 499
8.6   Further Reading 511
      Acknowledgements 513

8.1   Microwave Measurements of Fundamental Properties of Superconductors

D.E. Oates

8.1.1   Introduction

Why use microwaves to measure fundamental properties of superconductors? To understand the answer, one must consider that at microwave frequencies superconductors have finite resistance. The resistance is identically zero at dc and negligible at very low frequencies, so the superconductor does not interact with such probes. The origins of the finite resistance will be explained in Section 8.1.2. Microwaves interact strongly, and over the history of superconductivity microwave measurements have contributed considerable information to the understanding of the phenomenon. In the very early history of superconductivity, microwave technology was not developed enough to be of use in experiments of the fundamental properties. So it took the development of both the two-fluid model to illustrate the interaction of microwaves and the development of microwave technology immediately before and during World War II to make the realization of the microwave experiments possible. In this section, I use the broadest definition of microwave frequency range, from approximately 300 MHz to 300 GHz. However most of the experiments described here are in the range 1 to 20 GHz. In the microwave frequency range, one determines the resistance by measuring the quality factor Q of a microwave resonator and one determines the penetration depth by measuring the reactance of the resonator. More detail is presented below.

This section, like the whole book, is not intended to be comprehensive. It will present what I feel are the highlights of the development of microwave characterization and the findings that I think are the most important. In the more recent history, in which I have participated, the discussion will take a more personal perspective. I should add a few words on organization of Section 8.1. I will describe the two-fluid model is the basis for understanding the physics of the microwave measurements. That is followed in Section 8.1 by a brief description of the experimental techniques used in the microwave measurements. The historical highlights are then described in chronological order, Sections. 8.1.4 - 8.1.7.

8.1.2   Two-Fluid Model

To answer why microwaves, consider the existence of the finite resistance of a superconductor at microwave frequencies. The explanation invokes the two-fluid model, in which the superconductor is modeled as a mixture of normal electrons and superconducting electrons. The resistance arises because the superconducting electrons have inertia, and when the electromagnetic field reverses sign, the superconducting electrons cannot reverse direction immediately. This leads to a microwave current out of phase with the voltage, and is modeled as an inductance in an equivalent circuit model of the superconductor. An inductance generates a voltage across a sample when a time varying signal is applied. The voltage in turn will cause the normal electrons to flow and thereby generate dissipation. At dc, the superconducting electrons short out the normal electrons and no voltage or dissipation is generated. Figure 8.1 shows the resulting equivalent circuit for a piece of superconductor.

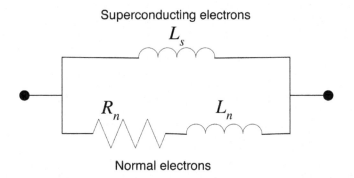

FIGURE 8.1: Equivalent circuit model for a piece of superconductor. The superconducting branch is represented by an inductance L_s, which reflects the inertia of the superconducting electrons. The normal branch is represented by a resistance R_n and inductance L_n reflecting the impedance of the normal electrons. At dc and very low frequencies the superconducting branch shorts out the normal branch, and effectively zero resistance is measured across the terminals. When a current of finite frequency is applied to the terminals a voltage develops because of the inductor, and the voltage developed induces current in the normal branch giving rise to dissipation.

An equivalent explanation of the ac losses begins with the finite penetration of the magnetic field, which is independent of frequency. Because the magnetic field is varying in time, an electric field is generated, and in the thin layer of magnetic penetration this electric field causes the normal electrons to flow and dissipate energy. The classical skin effect is similar in normal metals; for further explanation of the skin effect consult any text on classical electromagnetic theory. In superconductors the dissipation and thus resistance are very small compared with the normal resistance because the penetration depth is short compared with a normal metal.

The usual way to describe the microwave properties of a superconductor is by a complex conductivity $\sigma = \sigma_1 - i\sigma_2$, which leads to a surface impedance Z_s given by

$$Z_s = \sqrt{\frac{i\omega\mu_0}{\sigma}} \tag{8.1}$$

where

$$Z_s = R_s + iX_s = R_s + i\omega L_s, \tag{8.2}$$

where R_s is the surface resistance and X_s is the surface reactance, which is inductive. Expressions for the surface impedance can be found in many books[1] and need not be derived here. The R_s is given by

$$R_s = \frac{\omega^2 \mu_0^2 \lambda^3 \sigma_1}{2} \tag{8.3}$$

where λ is the London penetration depth, ω is the frequency in radians per second, μ_0 is the permeability of free space, and σ_1 is the real part of the conductivity given by

$$\sigma_1 = \frac{n_n \sigma_n}{n} \tag{8.4}$$

where σ_n is the normal state conductivity, and n_n/n is the ratio of normal to total electrons. The $n_n \to 0$ as $T \to 0$ and is often given by

$$n_n = \left(\frac{T}{T_c}\right)^4 n \tag{8.5}$$

where T is the temperature, and T_c is the critical temperature. The inductance is given by

$$L_s = i\omega\mu_0\lambda. \tag{8.6}$$

These are empirical formulas. A rigorous derivation of the surface impedance was presented by Mattis and Bardeen[2] using the BCS theory. Their derivation validated the expressions derived from the two-fluid model, which is more intuitive and in my opinion illustrates the physics pedagogically.

At frequencies higher than the microwave region, > 300 GHz, the two-fluid model is no longer valid. When the energy of a photon $\hbar\omega$ becomes comparable to the superconducting energy gap Δ, the excitation of quasiparticles by the photons becomes possible. In the high-frequency limit $\hbar\omega > \Delta(T)$, the resistance is equal to that of the normal state.

The two-fluid model gives a simple explanation of the microwave impedance at low levels of current. However in a superconductor, unlike in a normal conductor, the impedance is dependent on the microwave current; this is conventionally referred to as the nonlinear surface impedance. This dependence is quite complicated and cannot be given by a closed-form expression. In the simplest case, it is a matter of exceeding the critical current when the material is driven into the normal state. The dependence is also strongly influenced by the crystallinity and the nature of the superconductivity, by which we mean that the low-T_c superconductors behave differently from the high-T_c materials.

In the nonlinear case, the surface impedance can be written as

$$Z_s(I_{rf}) = R_s(I_{rf}) + iX_s(I_{rf}). \tag{8.7}$$

---

[1] For example, T. Van Duzer and C.W. Turner, *Prinicples of Superconductive Devices and Circuits* (Elsevier North Holland, New York, 1981)

[2] D.C. Mattis and J. Bardeen, *Phys. Rev.* 111 (1958) 412

The lowest-order approximation can describe many experimental results. For instance the nonlinear surface resistance is often accurately approximated by

$$R_s(I_{rf}) = R_{s0} + R_2 I_{rf}^2, \qquad (8.8)$$

where R_{s0} is the surface resistance at very low current and R_2 is a constant that can in some cases be calculated from first principles. One expects no term in Eq. 8.8 depending on the I_{rf} because the impedance should not depend on the direction of current flow in the absence of a preferred direction. Below I explain how the nonlinearity can be used to determine some important fundamental properties.

8.1.3 Microwave Measurement Techniques

Before describing the important contributions of microwave measurements a few words about the experimental techniques are needed.

Linear Measurements

Most measurements have utilized resonator techniques, either bulk cavities or transmission-line resonators for the case of thin-film measurements where the material under test is made to be some part of the cavity or the transmission line. With the exception of the first measurements of London described in Section 8.1.4, the measured Q of the resonator can be used to determine the R_s by the relation $R_s = G/Q$ where G is a geometrical factor that can be calculated from the geometry of the resonator:

$$G = \frac{\omega_0 \mu_0 \int_V H^2 dV}{\int_s H_s^2 dS}, \qquad (8.9)$$

where H is the volume magnetic field distribution in the resonator and H_s is the surface magnetic field distribution.

In general, the absolute value of λ cannot be determined from microwave measurements, but the fractional change in resonant frequency as a function of temperature $\Delta f(T)/f_0$ allows the change in penetration depth as a function of temperature to be determined by $\Delta\lambda(T)/\lambda_0 \sim \Delta f(T)/f_0$.

Nonlinear Measurements

As will be discussed in more detail below, the microwave surface impedance of superconductors shows a dependence on the microwave current I_{rf} or equivalently the microwave surface magnetic field H_{rf}. The nonlinearity is especially important in the high-T_c materials. An important consequence of the nonlinearity is the generation of higher-order harmonics, and when more than one frequency is present, intermodulation distortion (IMD) is generated. The third-order IMD is the most important and is measured in the usual way, in which two closely spaced tones of equal power at frequencies f_1 and f_2 are combined and applied to the resonator. The frequencies are centered about the resonant frequency 3-dB bandwidth. The nonlinearity of the material causes mixing products. The third-order products fall close to the fundamental frequencies and are easily observed even with narrowband devices such as high-Q resonators. The power P_{IMD} of the third-order mixing products at frequencies $2f_1 - f_2$ and $2f_2 - f_1$ is then measured in a spectrum analyzer as a function of the input power to the resonator. As discussed below the temperature dependence of this quantity yields information on the symmetry of the energy gap.

8.1.4 Early History

The first microwave measurements that I am aware of were made by Heinz London[3] and reported in 1940. The measurements were done on superconducting tin at 1.5 GHz in a split-ring resonator excited by a magnetron tube. He measured the microwave absorption and thus the resistance by a calorimetric method. The measurements were made as a function of temperature from above $T_c = 3.73$ K down to 2 K, thus also measuring the resistance in the normal state. He was able to measure the ratio of the microwave resistance in the superconducting state to that in the normal state R/R_n. This is shown in Figure 8.2. He noted that, unlike the dc resistance, R decreased gradually below the transition temperature. From this he deduced the number of the normal electrons as a function of temperature.

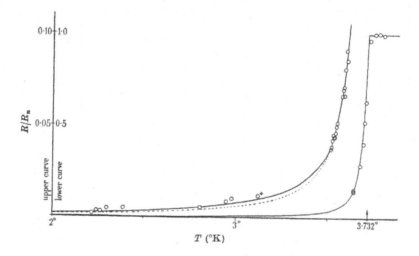

FIGURE 8.2: London's original results of microwave measurements of the resistance of a superconductor showing the gradual reduction in resistance that is the characteristic of the microwave resistance in contrast to the abrupt change in resistance for the dc case.

This experiment was the first measurement of ac resistance, provided the first verification of the two-fluid model, and supported the London theory of superconductivity. It also allowed the deduction of the temperature dependence of the number of normal electrons.

It is interesting to note that London first mentioned the possibility of measuring the microwave heating of a superconducting sample in a 1934 article, alluding to attempts that were interrupted after 1933 when he was forced to leave Germany.

8.1.5 Post World War II

The history continues after World War II. The development of microwave radar technology in the United States and Great Britain during the war transformed the experimental situation and made microwave experiments far more accessible. A. B. Pippard was one of the first to exploit the new technology following the war. Pippard had worked on radar development in England during the war, and following it was given access to much of the equipment that was then surplus[4]. In a series of

---

[3] H. London, *Proc. Roy. Soc.* A 176 (1940) 522

[4] B. Glowacki, Lectures on Superconductivity (2009) http://www.msm.cam.ac.uk/ascg/lectures/

important papers in the late 1940s and early 1950s[5], Pippard measured the microwave resistance of mercury and tin in resonators at 1.5 GHz and at 9.4 GHz. His findings agreed qualitatively with those of London mentioned above. Pippard also was able to measure the penetration depth by measuring the reactance of his resonators as a function of temperature and applied magnetic field. His values for the penetration depth agreed with those measured earlier by a completely independent method and provided validation of the numbers.

Pippard found that the measurements of penetration depth did not agree with the existing London theory, especially in impure samples and found that losses were larger than would be predicted by the London theory. He observed that the London theory is a local description of the electrodynamics where the current $\vec{j}_s$ and vector potential $\vec{A}$ are related by

$$\vec{j}_s(\vec{r}) = \frac{\vec{A}(\vec{r})}{\mu_0 \lambda^2}. \tag{8.10}$$

and his data are better fit with a nonlocal electrodynamics, where the current is that given by the vector potential averaged over a length scale which he introduced and called the coherence length ξ. The coherence length is given by

$$\frac{1}{\xi} = \frac{1}{\xi_0} + \frac{1}{\ell}, \tag{8.11}$$

where ξ_0 is an intrinsic coherence length particular to the pure material and ℓ is the mean free path of the electrons. This coherence length turns out to be identical, in the case of pure material, to the coherence length derived in the BCS theory. It is usually interpreted to be the diameter of a Cooper pair[1]. Bardeen was aware of Pippard's work when he together with Cooper and Schrieffer formulated the famous theory of superconductivity.

Pippard also discovered, as others before him had, that in the normal state, metals at low temperatures exhibit much larger losses than expected from the theory of the skin effect. This excess loss results from the mean free path being larger than the skin depth, thus making the effective skin depth larger than that calculated from the standard formulas. This effect is now known as the anomalous skin effect[2]. A similar situation arises in superconductors, as Pippard found, when the mean free path is larger than the coherence length $\ell \gg \xi_0$.

8.1.6   1960s and 1970s

Following the work of Pippard, considerable work on measuring the surface impedance of various materials continued.

Energy Gap Measurements

In the late 1950s, nearly coincident with the appearance of BCS theory, some measurements at the upper edge of what can be called microwave frequencies were reported[6] demonstrating a finite energy gap. They showed increased absorption of microwave energy when the frequency is raised above the values such that $\hbar\omega \approx kT_c$. Other data supported this and the definitive experiments of direct measurement of the energy gap were carried out in the far infrared. The development of the BCS theory gave a theoretical basis to these ideas.

Measurement in the Mixed State

Following the proposal by Abrikosov in 1958 that type II superconductors can support quantized vortices, microwave measurements played a role in the indirect verification of the proposal, before

[5]   A. B. Pippard, *Proc. Roy. Soc. Lond.* A 216 (1953) 547
[6]   M.A. Biondi et al., *Phys. Rev.* 108 (1957) 495

the later direct verification by magnetic decoration. Not only did microwave measurements support the idea of quantized vortices, but also they were able to measure the properties of the vortices, such as pinning strength and viscosity. Microwave measurements are unique in this capability, again because they interact with the superconducting state in a way that dc measurements do not. In my opinion the highlight of the effort to understand vortices is a classic series of experiments by Gittleman and Rosenblum[7], who measured the resistance from low frequencies up through the microwave region. The results are shown in Figure 8.3 where the relative power absorbed by various superconductors is plotted as a function of a reduced frequency f_0 defined as the frequency where the relative power absorbed is one half.

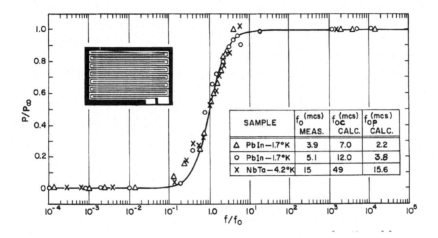

FIGURE 8.3: Relative power absorbed in the mixed state as a function of reduced frequency.

Gittleman and Rosenblum were able to understand these results using a very simple and elegant harmonic oscillator model of the vortices and pinning centers. Because the vortex-vortex interaction energy is larger than the pinning energy, the vortex lattice can be considered rigid. The vortices are driven by the microwave current, and the force constant k of the harmonic oscillator model, given by

$$k = \frac{2\pi\alpha_c\varphi_0^{1/2}}{cH_0^{1/2}}, \tag{8.12}$$

is identified as the pinning strength. In Eq. 8.12 α_c is a constant whose dependence on applied magnetic field H_0 is given explicitly and φ_0 is the flux quantum. The viscosity η of the vortex motion provides the damping term. Thus, the equation of motion for small currents and small displacements is

$$m\ddot{x} + \eta\dot{x} - kx = \frac{J\varphi_0}{c}, \tag{8.13}$$

where m is the effective mass of the vortex and J is the ac current. At low frequencies near dc the vortices are pinned, and the losses are very small. At microwave frequencies the vortices oscillate about the pinning sites freely with the losses determined by the viscosity. The vortices are essentially free at these frequencies. The transition frequency f_0 where the losses are one half of the high frequency values is given by $2\pi f_0 = k\eta$.

[7]   J.I. Gittleman and B. Rosenblum, *Phys. Rev. Lett.* 16 (1966) 734

The agreement with a number of materials is excellent when plotted as in Figure 8.3. Although more sophisticated models have been subsequently developed, this model is still used today to understand the dynamics of vortex motion and the paper[7] is quite often cited, a quite remarkable fact considering the model was proposed even before vortices had been observed directly.

8.1.7 High-T_c Era

The discovery of the high-transition temperature superconductors (HTS) in 1986 caused a large increase in the study of the microwave properties of superconductors. This was due to the potential applications in microwave devices such as filters and delay lines. In the microwave region and at 77 K, the HTS materials have a surface resistance as much as two orders of magnitude lower than that of copper at the same temperature, the most commonly used normal metal for microwave devices. The older low-T_c materials, while possessing comparably low surface resistance, had never generated the same level of interest in microwave applications because of the low operating temperatures required. Microwave characterization of HTS was of interest not only for exploration of fundamental properties but also for the relevance for applications. The use of HTS for practical microwave devices is detailed in the chapter by Hammond (see Section 8.2) and will not be discussed further in this chapter.

With the HTS materials, the primary methods used for microwave characterization of fundamental properties changed from measurements on bulk samples to measurements of thin films and single crystals. Because the HTS materials are ceramics, and bulk samples are fabricated by ceramic processes such as sintering of powders, the microwave properties of the bulk ceramic samples are dominated by the grain boundaries, which are an unavoidable consequence of the sintering process. Because of the short coherence length ξ of the HTS materials, approximately 1 nm, the grain boundaries, which are of comparable size, act like weak-link Josephson junctions, and thus dominate the microwave losses because of the significantly lower critical current of the junction compared with the bulk. When the grain-alignment angles are greater than approximately 2° the critical current is severely reduced[8,9]. In the microwave region this is manifest as a very strong power dependence in the ceramics. This power dependence can also be evident in thin films that contain large-angle grain boundaries. An example of the effects of grain boundaries is shown in Figure 8.4. This is from some of my early measurements. Shown are the results for two different films. Plotted is $1/Q$ of a stripline resonator, which is proportional to R_s, *vs* the microwave current I_{rf}, which is varied by varying the input power to the resonator. These films, one made by postannealing of a precursor film, which leaves high-angle grain boundaries in the film, and the other made by in-situ deposition of YBCO, which produces only low-angle grain boundaries, show similar values of Q at low current but differ markedly as the current is raised.

These findings were quite puzzling at first because such behavior had not been observed in the low-transition-temperature (LTS) materials. However, at the urging of several people, especially Jürgen Halbritter[10] of the Karlsruhe Research Center in Germany (now Karlsruhe Institute of Technology), I and others became convinced that the grain-boundary explanation was correct. It is now widely accepted that the grain boundaries are the cause of the strong power dependence in HTS films and ceramics. I should, however, add that as deposition methods and characterization advanced, it became clear that films with very low angle grain boundaries could be produced on lattice-matched substrates by a variety of deposition methods[11]. As mentioned above[8,9], experiments to directly measure the effects of grain boundaries on microwave surface impedance are the most convincing

---

[8]   Y.M. Habib et al., *Phys. Rev.* B 57 (1998) 13833

[9]   Y.M. Habib et al., *Appl. Phys. Lett.* 73 (1998) 2200

[10] J. Halbritter, *J. Appl. Phys.* 71 (1992) 339

[11] D.E. Oates, *J. Supercond. Novel Magn.* 20 (2007) 3

evidence in support of the grain-boundary explanation.

That films can be grown without high-angle grain boundaries is fortunate for the applications discussed in Section 8.2 by Hammond et al., because otherwise, the power-handling capability of filters would be insufficient for most applications. However the third-order intermodulation discussed in the subsection on nonlinear measurements in Section 8.1.3 is still present and the role of IMD in fundamental measurements will be discussed below in Section 8.1.7.

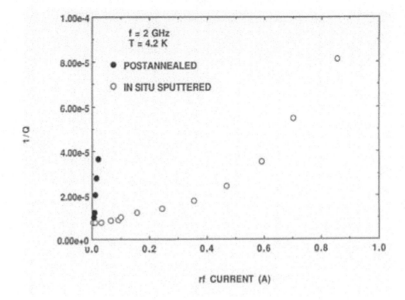

FIGURE 8.4: $1/Q$ (proportional to surface resistance) vs current for two different thin-film samples, one a postannealed film containing high-angle boundaries and an in-situ grown high-quality epitaxial film.

Microwave Determination of Penetration Depth in HTS

Early in the HTS cuprate era, it was proposed that the symmetry of the energy gap in these materials was unconventional. Whereas the symmetry of the LTS materials was s-wave, that is, spherically symmetric in momentum space, the HTS materials were, from theoretical considerations, proposed to exhibit d-wave symmetry[12]. That is,

$$\Delta(T) = \Delta_0(T)\cos(2\theta), \tag{8.14}$$

where Δ_0 is a constant and θ is the angle in momentum space. However experimental evidence for the d-wave symmetry was at best ambiguous[12]. To help clarify the issue, it was predicted that the temperature dependence of the penetration depth at low temperatures $\lambda(T)$ could give a definitive answer to the symmetry question, because the difference between s- and d-wave symmetries is distinctive. For YBCO, low temperature is between 1.3 and about 10 K. For s-wave symmetry the

---

[12] D.J. Scalapino, *Phys. Rep.* 250 (1995) 329

temperature dependence is fairly flat at low temperature, with the expression from BCS theory

$$\frac{\Delta\lambda(T)}{\lambda(0)} \cong \left(\frac{2\pi\Delta}{T}\right)^{1/2} \exp(\Delta/T),$$ (8.15)

where $\Delta\lambda(T) = \lambda(T) - \lambda(0)$. On the other hand for d-wave symmetry, a power law applies:

$$\frac{\Delta\lambda(T)}{\lambda(0)} \cong \ln(2)\frac{T}{\Delta_0}.$$ (8.16)

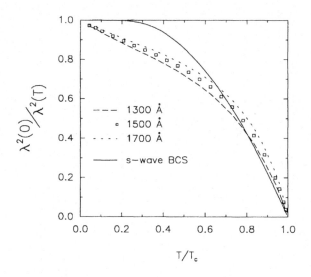

FIGURE 8.5: The quantity $(\lambda(0)/\lambda(T))^2$ *vs* T/T_c which is a measure of the superfluid density for a $YBa_2Cu_3O_{6.95}$ single crystal. The solid line is the same quantity for an s-wave superconductor. The broken lines are for different absolute values of the zero-temperature λ.

The group at the University of British Columbia led by Hardy[13] reported measurements of $\Delta\lambda$ on very high quality single crystals of $YBa_2Cu_3O_{6.95}$ free of impurities and weak links. They used a split-ring resonator at 900 MHz to make the measurements. The use of single crystals in this experiment is important because the $\Delta\lambda(T) = \lambda(T) - \lambda(0)$ changes from a linear dependence in T to a quadratic dependence $\sim T^2$ due to impurity scattering or weak links. Most measurements on thin films had shown the $1/T^2$ dependence. The results of the UBC experiments are summarized in Figure 8.5 which shows plotted vs T the quantity $\lambda^2(0)/\lambda^2(T)$ which is proportional to the fractional superfluid density n_s/n. Also shown is the predicted s-wave behavior. Clearly the d-wave nature of the YBCO sample is demonstrated. As noted above, the microwave resonator method can only measure changes in λ, not the absolute value $\lambda(0)$. The dotted and dashed lines are for different assumptions of the value of $\lambda(0)$. These results were at the time one of the most convincing pieces of evidence for the then-controversial but now widely accepted d-wave symmetry of the cuprate HTS materials.

[13] W.N. Hardy et al., *Phys. Rev. Lett.* 70 (1993) 3999

Intermodulation Distortion

Measurements

As already mentioned in the discussion of experimental methods, Section 8.1.3, IMD is a consequence of the nonlinear surface impedance. IMD is present in all superconductors but it took on new prominence in the HTS materials because of the importance in applications. As with the question of power dependence, the early measurements of IMD were puzzling, perhaps more so, and although some aspects of IMD are now well understood, as presented in the following, even now there is no comprehensive theory of IMD that explains the entirety of the measurements. One great mystery in the early development was the dependence of IMD on power. We are considering the third-order IMD, and it is expected that it arises from the quadratic dependence of the Z_s on current, Eq. 8.8. For instance if Eq. 8.8 holds then the voltage V generated by the current I_{rf} is

$$V \sim I_{rf} R_s = I_{rf} R_{s0} + I_{rf}^3 R_s. \tag{8.17}$$

The term cubic in the rf current is the source of the IMD and produces the mixing products mentioned above. The IMD voltage is proportional to the cube of the current. This implies that the IMD power P_{IMD} is proportional as well to the cube of the input rf power. On a double logarithmic plot this yields a slope of three. The early measurements produced many instances of slopes different from three. Slope 2 was observed often and a mix of slopes was frequently found. For some regions of power, slope 3 was observed and in others slope 2 or less was observed. The observation was also made that a Z_s proportional to $|I_{rf}|$ could produce the slope of two, but there was no theoretical basis. Some of these apparent anomalies have been explained by the theory presented in the next section.

Nonlinear Meissner Effect

In the midst of the efforts to understand IMD, important papers were published by Thomas Dahm and Douglas Scalapino[14,15]. This work observed that in a d-wave superconductor the nonlinear Meissner effect (NLME) would produce IMD. The NLME refers to the breaking of Cooper pairs by a magnetic field, in this case the magnetic self-field of the microwave current. For fields small compared with the critical field, this effect is negligible in s-wave materials, like the classical LTS materials, vanishing exponentially at temperatures below T_c. However, in d-wave superconductors, or any other symmetry with nodes in the energy gap, the NLME can be appreciable, and Dahm and Scalapino showed that it increases sharply at low temperatures. This is counterintuitive, but it turns out that it is easier to break pairs at low temperatures. The pair breaking manifests itself as an increase in penetration depth. Thus, as derived by Dahm and Scalapino,

$$\lambda(j,T) = \lambda(T)[1 + \frac{1}{2}b(T)\left(\frac{j}{j_{pb}}\right)^2 + \ldots], \tag{8.18}$$

where j_{pb} is the pair-breaking critical current given by $ne\Delta_0/p_F$ and p_F is the Fermi momentum.

Figure 8.6 shows the results of the calculation by Dahm and Scalapino. As is obvious in the figure the difference between s- and d-wave is large. Since the nonlinearity *increases* at low temperatures this predicts that the IMD will also increase at low temperatures. When Dahm and Scalapino published the papers all attempts to observe the NLME in static fields had been unsuccessful.

My initial impression had been that the NLME would not be observable in the IMD either, because the weak links and other defects in the films would dominate over this intrinsic effect. That

---

[14] T. Dahm and D.J. Scalapino, *Appl. Phys. Lett.* 69 (1996) 4248

[15] T. Dahm and D.J. Scalapino, *Phys. Rev.* B 60 (1999) 13125

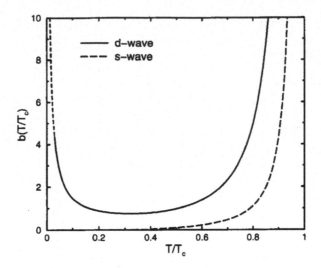

FIGURE 8.6: The leading order nonlinear coefficient $b(T/T_c)$ for the intrinsic nonlinear response of an s-wave (dashed line) and a d-wave superconductor (solid line). Here $2\Delta_0/kT_c = 6$ and currents running along the CuO bonds of the high-T_c cuprates have been assumed. The increase at low temperatures for the d-wave case will be cut off, if T/T_c falls below j/j_c .

is, the IMD was due to extrinsic effects and this intrinsic effect would be masked. Indeed many experiments had confirmed this view. A typical example is given in Willemsen et al.[16]

In spite of my doubts, with the aid of my student Sang-Hoon Park, I attempted to observe the low-temperature increase in IMD by measuring at lower temperatures than we had measured previously and using a very high quality film made by laser ablation by Gad Koren of the Technion in Israel. To my surprise we observed the effect quite clearly[17]. The results are shown in Figure 8.7. The NLME was observed this time because of the high-quality film that was used. Subsequently I have observed the same effect in many other films made by different deposition methods[11]. Although HTS films are known to contain various defects, in high-quality epitaxial films only very low angle grain boundaries are present and apparently other defects do not contribute significantly to the IMD so that the intrinsic d-wave properties dominate the IMD at low temperatures and low powers. Other researchers including the group led by Jim Booth at NIST Boulder[18] and the group of Antonello Andreone at the University of

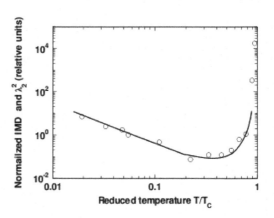

FIGURE 8.7: Comparison of data and theory. The empty circles are the IMD data from a high-quality film of $YBa_2Cu_3O_{7-\delta}$. The solid line is the calculation of d-wave theory.

[16] B.A. Willemsen et al., *Phys. Rev.* B 58 (1998) 6650

[17] D.E. Oates et al., *Phys. Rev. Lett.* 93 (2004) 197001

[18] K.T. Leong et al., *IEEE Trans. Appl. Supercond.* 15 (2005) 3608

Naples Italy[19] have observed the NLME in IMD measurements. The measurements of Andreone also included IMD of single crystals in which the NLME was observed.

Subsequently, another formulation of the theory of the NLME was developed by Dan Agassi of the Naval Surface Warfare Center in Bethesda MD. This theory was able to predict some further aspects of the NLME such as its nonlocal nature, but a full discussion is beyond the scope of this book. For detail see Agassi and Oates[20] and Oates et al.[21]

The IMD measurement of the NLME confirmed that IMD is an intrinsic effect in the cuprate HTS materials, thus leading to a better understanding of the IMD. In addition, the previous lack of success in observing the NLME was an outstanding missing piece in the d-wave hypothesis for the cuprates. The reason for the success of the IMD in observing the NLME is that the IMD is far more sensitive to changes in penetration depth than direct measurements. A value of $\Delta\lambda/\lambda$ of order 10^{-6} produces easily measurable IMD.

8.1.8 Summary and Future Prospects

I hope that I have provided some interesting examples of microwave measurement techniques that have illustrated the value of these techniques for exploring the fundamental properties of superconductors. Both historically and in the very recent past microwaves have proven to be a valuable technique in superconductivity. This was not intended to be an exhaustive listing of microwave measurements, but examples that I find particularly important.

I think that microwave techniques will continue to prove useful in characterizing superconductors. There are several very recent examples of publications dealing with the microwave properties of some of the newly discovered superconductors such as MgB_2 and the iron-pnictide materials. For example, a recent paper[22] presents measurements of MgB_2 showing evidence for unconventional symmetry as measured by microwave IMD. Also papers reporting measurements in the recently discovered pnictide superconductors are beginning to appear[23]. New explorations of the older materials will also undoubtedly continue. I cite a recent paper on the NLME in dirty s-wave superconductors, specifically niobium[24]. I expect the future of microwave characterization of superconductors to be as rich as the past has been. The situation can only improve as the technology of microwave measurements becomes more advanced and less expensive as is inevitable because of the intense interest in microwave wireless communication systems.

8.2 Applications of Passive Microwave Filters and Devices in Communication and Related Systems

R. B. Hammond, N. O. Fenzi and B. A. Willemsen

---

[19] A. Andreone et al., *IEEE Trans. Appl. Supercond.* 17 (2007) 3640

[20] D. Agassi and D.E. Oates, *Phys. Rev.* B 72 (2005) 14538

[21] D.E. Oates et al., *Phys. Rev.* B 77 (2008) 214521

[22] Y.D. Agassi et al., *Phys. Rev.* B 80 (2009) 174522

[23] T. Shibauchi et al., *Physica* C 469 (2009)590

[24] N. Groll et al., *Phys. Rev.* B 81 (2010) 020504

8.2.1 Introduction

This section focuses on the development and commercialization of HTS microwave filters for cell-phone base stations. This includes development of low-cost HTS film manufacturing and practical cryogenic systems.

In this short section we will try to address all the dimensions of a diverse subject—superconducting passive microwave devices—and we will provide suggested reading for readers interested in more detail. Most of our story will relate to the development and use of HTS filters for cellular telephone base stations. This became the primary focus of a great deal of research and technology development in HTS as well as several related fields. It also became a principal focus of commercial interest and investment in HTS during the mid-to-late 1990s. In addition, it is mostly a story that unfolded outside the public domain, mostly among the HTS companies and their wireless-industry customers. Here we attempt to reconstruct the key elements and events of that story.

Superconducting materials have been of interest for several decades to possibly replace the copper or silver used in radio frequency (RF) and microwave passive circuits where conductor loss and device size are the primary limitations to device performance and practicality. Candidate devices are primarily antennas, filters, and delay lines. Before the discovery of HTS very little development work was done. Design studies were performed for compact low-temperature-superconducting (LTS) antennas, and LTS microwave filters and delay lines were successfully demonstrated in the lab. No products resulted, however, due to the size, power requirements, reliability, and cost of the cryogenics needed to support these devices. The advent of HTS materials changed this, since operation at 77 K versus 4 K promised far more acceptable cryogenics.

Following the discovery of $YBa_2Cu_3O_7$ (YBCO) in early 1987, governments in the US, Europe, and Japan became substantial drivers in the definition of applications and funding the development of HTS materials and devices. Many large established companies became involved in the development efforts, facilitated by government funding. In addition, the excitement generated in the capital markets for the potential of HTS in the marketplace led to many new companies being formed, ultimately dozens around the world.

As early as 1987, HTS passive RF/microwave devices offered the potential for high value in a small size, since even the earliest bulk polycrystalline HTS materials showed substantially lower conductor loss than copper and silver, particularly at the lower microwave frequencies ≤ 10 GHz, and the potential reduction in size or the improvement in performance might be worth the cost of 77 K cryogenics for many applications. There were early demonstrations of high-quality-factor (Q) resonators and filters using these early materials. By April, 1990, 4 years after the discovery of HTS and 3 years after YBCO, the US Office of Technology Assessment had concluded that by ~1995 the early applications for HTS would be these devices or superconducting quantum-interference device (SQUID) sensors. Shortly thereafter it became understood that the passive microwave devices offered large existing applications and markets to justify a significant commercial investment.

Initially the market interest was for military and space applications. These both offered a wealth of diverse potential applications, where HTS passive microwave devices might provide better performance, smaller size, or both, and where those features were highly valued. By the early 1990s the explosive growth in cellular telephone networks offered another potential market.

Over about a decade from the early 1990s to the early 2000s, the cell-phone infrastructure market became the principal driver for what was ultimately the successful development of volume manufacturing of high-performance HTS materials and devices, as well as the development and volume manufacturing of reliable, maintenance-free cryogenic systems suitable for unattended operation in remote environments over many years. It is a story that grew out of the coincidence that HTS was discovered when cell-phone networks were just beginning a multi-decade period of extremely rapid growth and innovation. It involves famous people, became the focus of hundreds of millions of dollars in private sector investment, and helped some HTS start-up companies to briefly reach market capitalizations over a billion dollars in 2000.

The enormous worldwide excitement and enthusiasm generated by the discovery of the copper oxides and particularly YBCO in early 1987 led to a search for investment opportunities by large companies, investment companies, as well as private investors. New companies were started, and large companies began major R&D efforts. World-renowned scientists and engineers as well as highly successful technology entrepreneurs participated, a few directly as employees at new companies, but most as senior advisors, consultants, or board members. Two Nobel prize winners in superconductivity and the co-inventor of the integrated circuit all participated as board members. Of most relevance to this story were four new companies in the US, one in Germany, and one large company R&D effort in the US, all started within a year or two of the YBCO discovery. These attracted more than a decade of continuing investment funding, grew to efforts with ~50 or more full-time employees, with their principal focus becoming HTS passive microwave devices. Each ultimately developed and successfully trialed prototype HTS filter products in cell-phone base stations. Three of these companies established manufacturing and had their products purchased and installed into more than 100 base stations. Today there are HTS filters in approximately 10% of the 70,000 800-MHz cell-phone base stations in the US. That is about 40,000 HTS microwave filters and about 7,000 cryocoolers running continuously today in unattended locations. These HTS systems represent more than $150M in HTS product sales. They also represent the only broad application of superconducting or cryogenic systems in remote, unattended environments without requiring maintenance and with failure rates of 1% per year.

8.2.2 HTS Filters for Cellular Telephone Base Station Receivers

Background

After 1990 the search for products and markets for HTS continued unabated, particularly at the start-up companies that had to satisfy their investors to continue to receive funding. As time passed and the magnitude of the total private sector investment continued to rise, investors wanted products and markets that offered greater and shorter-term sales and profit potential.

By 1992–93, the wireless cellular telephone infrastructure original equipment manufacturers (OEMs) became important drivers for HTS-related technology development and innovation, sometimes with direct investment, but commonly and more importantly with clear, demanding application requirements: RF performance, size, reliability, manufacturing volume, and price.

The wireless opportunity for HTS narrowed fairly quickly to providing better microwave preselector filters for cell phone base-station receivers. Every cell phone base-station receiver had a preselector microwave filter at the time, which was usually a large silver-plated aluminum cavity filter, and very occasionally a dielectric-loaded cavity filter for higher performance.

It was argued that HTS filters would provide lower loss and noise, greater frequency selectivity, and/or smaller physical size than these alternatives. Thus, the geographic coverage area of the base station would be increased, adjacent band interference would be blocked, and less space would be consumed by the filters in the base station. In 1994–96, all the companies showed—in the lab—that they could make filters that did this. They used differing HTS materials and manufacturing methods, different substrates, different approaches to filter design, and different cryocoolers in their filter systems, which also began to be field tested at this time. By the late 1990s, the product focus became what was dubbed a "cryogenic receiver front-end" (CRFE) by Nippon Telegraph and Telephone (NTT) DoCoMo in Japan. NTT was a strong proponent of a CRFE solution for its planned new third-generation wireless (3G) network, which would start to roll out in 2000. The CRFE includes a high selectivity HTS filter followed immediately by a cryogenic low-noise amplifier. With this powerful combination the sensitivity of the base station receiver could be significantly improved, particularly in environments with significant adjacent-band interferors such as NTT faced with its 3G band assignment, and the US operators faced with their 800-MHz cellular band assignments.

One of the start-up companies in the US (Illinois Superconductor) became the first to consider

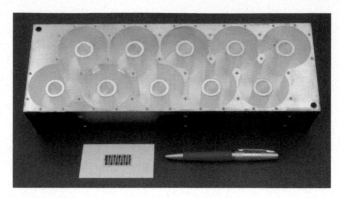

FIGURE 8.8: High performance conventional base station filter (top). Size comparison to HTS base station filter (bottom).

the cell-phone base-station market. Through ties with AT&T Bell Labs, it succeeded in getting a NIST Advanced Technology Program (ATP), funded in 1992. A second company (Superconductor Technologies Inc., STI, in Santa Barbara) at about the same time found interest from base-station OEMs Motorola and Ericsson, in this application, and Motorola funded them to develop a prototype starting in 1993. A third company (Conductus in Sunnyvale), also with ties to AT&T Bell Labs, began to look at this opportunity in 1993, exploring both HTS transmitter filters and receiver filters. A fourth company (Superconducting Core Technologies, SCT, in Colorado) began to work in this area in 1994, and focused on receiver filters primarily for increasing coverage.

Three of the US HTS start-up companies went public in 1993, mostly based on relatively broad-based stories regarding product and market opportunities for HTS; these included extremely fast fluxon-based digital electronics, SQUID sensors, magnetic resonance (MR) RF pick-up coils, RF and microwave filters for defense applications, and—most often—filters for cell-phone base stations. By about 1996, the main focus of the companies had narrowed to the cell-phone base-station opportunity, although the best funded kept some efforts going in other areas: defense/space electronics and digital electronics.

Markets

Cellular telephone systems began to appear in the early 1980s. By 1990, cell-phone networks were growing rapidly in the developed world, and continued to grow at an exponential pace throughout the 1990s and well into the 2000s. The uplink, the radio link from the cell phone to the base station, is the weaker of the two links supporting a two-way conversation, because the microwave transmit power of the cell phone is lower than the microwave transmit power of the base station. As cell phones became steadily smaller in size, cell-phone transmit power was reduced even further to maximize battery life. (Users become very annoyed when their cell phone batteries run out of juice!) This has driven a need for more sensitive base-station receivers. The CRFE provides unique benefits to cellular network performance as a simple add-on to an existing base station or as a designed-in subsystem. A CRFE is a high-selectivity, low-insertion-loss HTS filter followed by a high-linearity, cryogenic, ultra-low-noise amplifier (LNA) placed between the base-station receive antenna and the base-station receiver. The very low loss of the filter combined with the very low noise of the cryogenic amplifier combine to substantially reduce overall receiver noise level, typically by about a factor of two, thus extending uplink range and the geographic area covered by a base station. In addition, the high frequency selectivity of the HTS filter can reject large adjacent-frequency-band interfering signals that produce distortion in the base-station receiver. An added benefit is that the entire CRFE can be significantly smaller in size than conventional high-selectivity filters that might otherwise be used to provide interference protection at the expense of added loss and noise.

In the 1990s, cellular base station manufacturers and cellular network operators expressed a variety of reasons for using the CRFE, always some combination of the three benefits described above. Four are worth particular mention because of their importance to the development and sales of CRFE products. In the mid-1990s, one base station manufacturer (Motorola in the US) was interested in reducing the size of filters for its 800-MHz base stations. These were starting to dominate the base station size and were resistant to further size reduction, unlike the rest of the base station components. Also beginning in the mid-1990s and continuing right up to the present day, several US operators have been interested in a combination of uplink range extension and interference rejection in their 800-MHz base stations. In the late 1990s through early 2000s NTT DoCoMo in Japan was interested in blocking interference to the Japanese 1900-MHz 3G bands. Most recently, in the late 2000s and into the 2010s, some US carriers are interested in the CRFE to block adjacent-band interference in their new 700-MHz fourth-generation wireless (4G) bands.

OEM sales (as opposed to network-operator sales) of CRFEs are most sought after because of the opportunity offered for high-volume sales, typically tens of thousands of base stations for a single product. Despite some near misses this has not yet happened for CRFEs. (Just such an opportunity seems to be emerging in the 4G 700-MHz networks being in the US though!) Motorola's interest in reducing filter size in the mid-1990s resulted in a CRFE prototype (developed by STI) that met its most demanding requirements for frequency selectivity and would fit in one tenth the volume of its existing conventional filter product. This reduced size included the cryocooler, cryogenic package, and control electronics—the HTS filters themselves are much smaller. The CRFE prototype was qualified by Motorola for network deployment in 1996, but the HTS industry was not able to respond with sufficient manufacturing capacity to meet Motorola's needs at that time. It would take ~5 years for the HTS industry to close this gap in manufacturing capacity, but by then it was too late; that market window had closed. In the late 1990s and early 2000s NTT pushed its OEM suppliers (such as Matsushita (MCI)) to provide extremely sharp HTS filters in their 3G base stations for interference protection. Again, a product was qualified for the application and for network deployment, but this time the industry could not meet the selling price required by the OEM. Most recently a US operator has pushed its OEM suppliers to provide extremely sharp HTS filters in their new 4G base stations. Once again, a product has been qualified, and sales are expected to begin in 2012. To date, although products have been successfully developed and qualified, there have been no sales to OEMs of CRFE products besides a few dozen test and qualification units. Despite the variety of the early interests expressed by major wireless industry players in potentially using the CRFE, it has been the presence of strong adjacent-band interference in some cell-phone services that became the primary sustained driver for the development and sales of CRFE products. By far the most important, and responsible for virtually all CRFE sales to date, is the 800-MHz cellular band in the US. There are several significant sources of adjacent-band interference, as indicated in Figure 8.9.

Another important source of adjacent band interference to cell phones that drove substantial CRFE development is in the 3G bands in Japan, indicated in Figure 8.10.

In the mid-to-late 1990s when no OEM, network-designed-in, opportunities for CRFE had materialized, most of the companies had started to market their CRFE products directly to the network operators, the wireless telephone companies. It is these cellular phone service providers, close to the cell phone user's experience and thus concerned about improving service, that have proved the largest customers for HTS filter systems to date. Going this route allowed higher prices, but much smaller unit orders, lower total sales volume and higher cost of sales. Actual sales of CRFEs began with small local operators in the US, some with as few as a dozen base stations. This provided a much easier sales channel than the OEMs or even the large network operators, and the timing was right because these mostly rural carriers were experiencing the shift from car phones to handheld cell phones which had much lower transmit power. Network geographic coverage typically dropped by more than 50%. In many cases, the addition of the CRFE brought most of that back, and the chal-

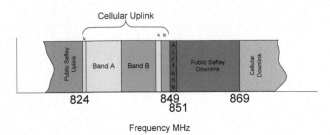

FIGURE 8.9: The 800-MHz cellular uplink bands in the US (light green A; blue B) were initially 10 MHz each. Then the A' and B' frequencies were added to make the total 12.5 MHz each and resulted in the unusual split shown in the figure. There are three sources of interference to the base-station receiver. The first is the close interleaving of the two operators' spectrum assignments which creates an unusually severe "near-far" interference. This type of interference occurs when a competing service's cell-phone user is close to your base station and far from its own base station and thus is transmitting at high power. The second is the Airfone service: high-power, always-on transmitters located near airports to provide telephone service on commercial flights (mostly not in use since ~2005). The third is high-power, always-on, public-safety transmitters that provide two-way radio voice communications for police, fire, and other emergency services.

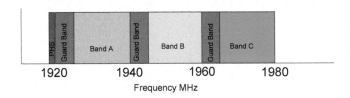

FIGURE 8.10: Personal Handyphone Services, PHS, widely used for data services in Japan, prevent the use of the first 5-MHz 3G channel in the KDDI band assignment unless an HTS filter is used. The Japanese government has prevented the other two carriers from using their first channel assignment as well, just to be fair to all.

lenge for the HTS companies was to offer a reliable product fairly quickly to capture the opportunity at the $20K to $30K per base station that the market was ready to pay.

Between 1996 and 1998, CRFE systems were sold to a number of operators in the US, totaling about 100 base stations. The first large single order was announced in 1999 (CRFE systems for 500 base stations) to a large regional cellular operator in the US (US Cellular). Another large order of 160 sites to Alltel, another large regional operator, was made by another CRFE supplier in the US at about the same time. Larger CRFE orders for 1,000 base stations by Alltel were announced during the following 3 years. Sales peaked in 2003 with CRFE systems sold and installed that year into 2,000 base stations in the US. At this point, sales dropped to under 1,000 per year, where they remain today. The early growth and the peak in 2003 was encouraged by the December, 2003, FCC-mandated change to local number portability (LNP). This would allow US cellular phone customers for the first time to retain their phone number when they changed phone companies. The change had been announced years in advance and the carriers had responded by investing in their existing network performance to give their existing customers the best possible service in advance of the deadline, hoping to reduce their motivation to switch to another carrier. In 2004, and for several years after that, the carriers focused on building out their 1900-MHz networks to expand capacity, and to utilize the spectrum bands which they had invested billions of dollars in acquiring/licensing in the late 1990s. The network build in 1900 MHz proved not to be a CRFE opportunity due to lack of adjacent band interference.

Since 2003, the US network operators have continued to rely on their 800-MHz networks as the backbone of their services for both voice and data, due to the much better propagation of the wireless signals in the atmosphere and better penetration of buildings and other physical obstacles. The CRFE has continued to be used as a tool to fix weak sites in their networks, and sales continue at several hundred sites per year.

In 2008, new spectrum was auctioned at 700 MHz in the US. Major carriers such as Verizon and AT&T invested more than $20B in the new spectrum to provide new 4G services. Some of the new bands, as was the case for the original 800-MHz bands assigned in the 1980s, suffer from strong adjacent band interference, as indicated in Figure 8.11.

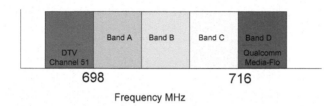

FIGURE 8.11: Broadcast services such as MediaFLO and UHF digital television provide strong interference in the lower 700-MHz bands, A thru C.

A CRFE product has been designed into the base station by one of the two major OEMs that are be building these networks. So the long-sought-after OEM opportunity for HTS filters for designed-in, initial deployment in base stations for new network roll-out may finally be on the verge of happening. Stay tuned.

HTS Film Production

Base-station RF receiver filter requirements drive the requirements on the HTS materials and filter structures utilized for HTS base-station filters. The filter microwave performance requirements are best summarized in two parameters: quality factor (Q), and third-order intermodulation-distortion intercept point (IP3). Q is a measure of the loss in the resonators in the filter and determines both the loss and frequency selectivity achievable. IP3 is a measure of the linearity of the filter response, i.e., its ability to pass signals or block interference without adding distortion to the wireless signal of interest. HTS filter resonator Q's must be ~100,000 or higher, and filter IP3 must be ~100 mW or higher for base station filter applications. These filter characteristics are determined by the surface resistance of the superconductor (Rs), any losses added by the HTS substrate material (dielectric loss tangent), the resonator structure chosen, and the microwave critical-current density of the superconductor (J_{IMD}).

The early, bulk, polycrystalline HTS materials were soon improved upon by other synthesis approaches for making practical microwave filters. Demonstrations of many approaches were successfully performed in the lab during the period 1987–90. Thick-film, melt-processed, polycrystalline YBCO films can be formed on refractory metal substrates (or ceramics) and achieve sufficient properties for filters if relatively large, 3-dimensional resonator structures are used. These also are relatively easy to put into production at low cost. Thin films of single-crystal HTS materials can be formed on single-crystal oxide wafers ($LaAlO_3$, MgO, sapphire) most often with much better microwave properties than the polycrystalline films, but the single-crystal substrates are expensive. These can often achieve sufficiently low loss for filters in small 2-dimensional microstrip-resonator structures (which require HTS films on both sides of a dielectric substrate/wafer), if the substrate has low microwave loss like LaAlO$_3$, MgO, sapphire. There are two basic approaches to forming these films. Ex-situ films are formed by depositing at room temperature a layer on a wafer with the appropriate mix of metals (Y, Ba, Cu), usually in oxide form. Then the wafer is placed in an oven in a controlled atmosphere at ~ 800 °C, for tens of minutes to grow an HTS single-crystal film from the single-crystal substrate template beneath. In-situ films are formed by depositing the appropriate mix of metals (Y, Ba, Cu) onto a single-crystal wafer heated to 800 °C in the presence of oxygen. The HTS single-crystal film grows epitaxially on the single-crystal wafer, layer by layer, while the metals are being deposited. Commonly there is a subsequent heating of the final HTS film in oxygen to complete the film synthesis.

Thick-film YBCO processes were successfully developed at a large chemical company in the UK (ICI), put into production by Illinois Superconductor, and used to develop three dimensional (3-D) HTS filters, first with fairly large cavity structures, later with more compact split-ring resonators.

Thin-film processes represent a richer story. The first successful films were formed in 1987 using ex-situ methods with YBCO. These processes are relatively easy to scale to large wafer areas, with films on both sides of a low-loss dielectric wafer (as required for HTS filters and other passive microwave devices); however, sufficient quality material could not be formed in thicknesses more than about half the penetration depth at 77 K, insufficient for devices that rely on very low microwave loss. In-situ YBCO processes received intensive development in the period 1987–92. Virtually every known technique for forming in-situ epitaxial thin films was tried. Many of these were successful in forming single-crystal films with low enough loss. However, all these techniques ultimately proved to be too expensive to scale to the manufacture of the minimum 2-inch, double-sided wafers needed for devices.

Successful in-situ growth of YBCO films requires three conditions to be met during the entire growth of the film: a) maintain a temperature of ~ 800 °C with ± 1 K stability, b) maintain specific ratios of Y, Ba, and Cu, stable to $\pm 2\%$, and c) provide an oxygen pressure at the film growth interface of $> \sim 100$ mT. This was known by the early 1990s. It was due to this new and unique combination of requirements that no pre-existing epitaxial-growth process proved scalable at low cost for 2-inch double-side wafer manufacturing. A new type of process appeared in the early 1990s that was developed specifically to meet YBCO's special epitaxial-growth requirements.

In 1992–93 a novel in-situ process was developed to form YBCO thin films which proved to be scalable for low-cost production. The basic concepts were developed at the Technical University of Munich. A company was formed, THEVA, which became the first (in 1995) successful supplier of YBCO-coated 2-inch, double-side wafers with low loss for microwave devices and other applications. Conductus adopted and adapted this approach for manufacturing YBCO wafers.

In early 1988 the thallium barium calcium copper oxide (TBCCO) family of HTS materials was discovered. For these materials ex-situ processes did prove to be successful in forming single-crystal films with low enough loss for devices. In the early 1990s DuPont Superconductivity and STI put the Tl$_2$Ba$_2$CaCu$_2$O$_8$ on LaAlO$_3$ material into production, manufacturing 2-inch, double-side HTS wafers with low microwave loss. Both companies offered these wafers for sale and also used them internally to develop devices.

In 1994 it was discovered that the existing film manufacturing processes did not meet requirements for base station filters. In 1994, Ericsson and Motorola tested early HTS base-station filters and found that they met Q requirements but not IP3 requirements. This drove the film manufacturing processes further. Beginning in 1995, a new process for ex-situ TBCCO was developed on buffered MgO wafers, first at STI then later at DuPont Superconductivity that met all the base-station filter requirements. Meanwhile, base-station filter IP3 requirements drove improvements to the in-situ YBCO wafer manufacturing that were ultimately successful first at Conductus (late 1990s) and later at THEVA at meeting filter IP3 requirements. The filter IP3 requirements also drove accurate electromagnetic analysis of superconducting microstrip structures in order to infer the limiting RF critical current density of the superconducting material from the intermodulation distortion (IMD) measurements. These studies resulted in the notation J_{IMD} for the limiting RF critical current density J_{c} in the superconductor. In general it was found that this critical current cannot be inferred from dc critical current density—different physics is responsible for these two limits. Thus, measurement of filter IMD became necessary to guide HTS material process development. As noted, this was accomplished successfully in both TBCCO and YBCO.

In December 2002, Conductus and STI merged. To achieve both the lowest manufacturing cost and the highest RF performance, in 2004 the merged company discontinued TBCCO production and changed over completely to the Conductus YBCO process, which was then scaled to high volume, semi-automated production of 2-inch double-sided YBCO wafers for filters. Interestingly, this process produces both the best RF properties for HTS films and by far the lowest manufacturing cost of any HTS film production process. The current production machine produces 22 2-inch wafers per run, and has capacity for 15,000 double-side wafers per year.

Of the 40,000 or so HTS filters in base stations today about two thirds are TBCCO and one third are YBCO.

HTS Filter Designs and Performance

The first CRFEs were built with 3-dimensional cavity resonators utilizing thick-film YBCO. These were also the first systems sold and installed into cell-phone base stations. Later the market required smaller filter systems than the multiple cubic meters required for these designs. These thick-film filters evolved to using split-ring resonators, much more compact designs, allowing systems comparable in size to that permitted by the thin film microstrip HTS filters.

Microstrip HTS filters required the development of novel resonator structures to enable the complex, high-selectivity HTS filters that were developed to be realized on compact chips. Many types of resonators were used. These evolved initially from simple half-wavelength lengths of microstrip transmission line that were edge coupled in a classic configuration for thin metal films. To reduce chip area, these structures were subsequently folded and wrapped in a variety of ways to make compact resonators that would achieve high Qs and be easily coupled with similar resonators in a controlled manner on the same chip. Figure 8.12 shows an example of an early HTS filter chip. Figures 8.13 and 8.14 show the measured insertion loss versus frequency of production HTS filters.

FIGURE 8.12: HTS filter chip compared in size to a US coin. The filter is 34 mm×18 mm×0.5 mm in size.

Other New Technologies for CRFE

Cryocooler—Long-life, no-maintenance, low-cost cryocoolers were developed.

Cryo-LNA—Low-noise, high-linearity, cryogenic, microwave semiconductor amplifiers were developed.

Cryo-cables—Low thermal conductance, low-RF-loss transmission lines were developed.

Dewars—Large, permanently sealed, long-life dewars were developed.

Automation—Automated frequency setting/tuning for manufacturing filters was developed.

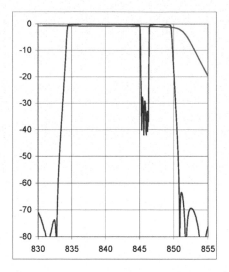

FIGURE 8.13: Measured insertion loss of a production HTS filter pair for a US 800-MHz B-Band base-station receiver. Two filters are cascaded in series: a bandpass filter followed by a bandreject filter. The red line shows the response of the standard base station receiver filter for comparison.

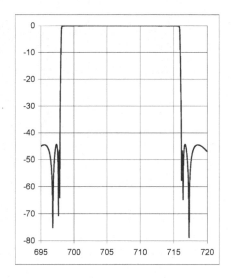

FIGURE 8.14: Measured insertion loss of a prototype HTS filter for US lower 700-MHz A,B, or C Band base station. The filter protects the base-station receiver from the digital television broadcast services at channel 51 and below and channel 55 (MediaFLO).

CRFE Products

Illinois Superconductor was the first to market with a CRFE product. It used thick film YBCO cavity resonators cooled with an off-the-shelf Gifford-McMahon (G-M) cryocooler. The filter provided very high selectivity. In the late 1990s they converted their HTS filters to a much more compact split-ring-resonator design and named the new CRFE the "ATP" (all temperature product) since

TABLE 8.1: Summary of CRFE Cryocoolers Designed Specifically for CRFE Products

Cryocooler Name	Company	Cycle	Lift, 77 K	Power	Weight	CRFE
Cryodyne M-22	CTI Cryogenics	G-M	11 W	250 W	20 kg	ClearSite
Polar SC7	Leybold	Stirling	8 W	250 W	8 kg	DE, CO, IL
Sapphire	STI	Stirling	5 W	120 W	3 kg	SuperLink

it was designed to maintain some filtering even at room temperature, in case the cryogenics failed. It utilized the Leybold cooler.

SCT was also early to market with a CRFE product. They had a system designed to be mounted on the antenna tower with an off-the-shelf G-M cryocooler. The cryocooler compressor was placed on the ground and the cold head at the top of the tower with the thin-film TBCCO filters. They transitioned this to a Stirling-cryocooler-based system at about the time the company shut down, and its technology and key staff were picked up by a new company, Spectrum Solutions Inc. (SSI) also in Colorado, which marketed the new CRFE.

TABLE 8.2: CRFE Prototypes and Products

CRFE	Company	HTS	Cooler	Power (W)	Sales (sites)
SpectrumMaster	Illinois	Thick Film Y123	G-M	2,000	200
ATP	Illinois	Thick Film Y123	Leybold	250	N/A
REACH	SCT	Thin Film Tl2212	G-M	2,000	N/A
SC200	SSI	Thin Film Tl2212	Leybold	250	~10
ClearSite	Conductus	Thin Film Y123	CTI	550	~200
SuperFilter	STI	Thin Film Tl2212	Sapphire	120	~1000
SuperLink	STI	Thin Film Tl2212/Y123	Sapphire	120	>5,000
CRFE D8	Cryoelectra	Thin Film Y123	Leybold	250	N/A

Conductus and STI both came to market with CRFE offerings in the 1996–97 timeframe. Conductus used the new G-M cryocooler from CTI designed specifically for the application. STI used an in-house developed Stirling cryocooler named "Sapphire".

Cryoelectra built several prototype CRFE systems in the mid-2000s and some were trialed in cell phone base stations in China. They also used the Leybold cryocooler.

CRFE Reliability

In the case of LTS, it had been the cost and reliability of the cryogenics that had kept superconducting passive microwave devices from being used in practical applications. With the discovery of HTS and the advent of devices operating at 77 K, the hope was that the cryogenics would be far cheaper and more practical. And indeed it was and is. However, despite that, for the companies developing CRFE products and marketing them to wireless companies, it was the poor reliability of their cryogenic systems, far more than any other factor, that became the primary stumbling block to commercial success. Several significant sources of failure had to be overcome successfully. The combination proved to be extremely challenging, and ultimately only one product line emerged successfully.

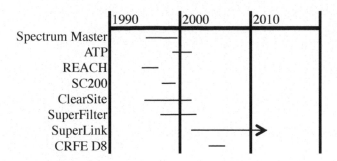

FIGURE 8.15:   CRFE product timeline.

FIGURE 8.16:   3 CRFE systems shown with one another; each servicing a single 3-sector base station with 6 base station receivers for the US 800-MHz market. ClearSite (lower right), SuperFilter (upper right), and SuperLink (upper left).

OEM requirements for base-station subsystem reliability and service (250,000-hour mean time between failures, MTBF, no maintenance) became a major challenge to CRFE companies. The CRFE is a complex system requiring a reliable cryocooler, a reliable long-life dewar, and reliable power supplies and system control electronics. Only one CRFE was qualified for OEM use in the 1990s primarily for this reason.

Operators did not qualify products rigorously like the OEMs. They were willing to try new products to see if they provided the benefits they were looking for, and then would test reliability against actual performance in the field combined with the support provided by the vendor company supplying the product. Because CRFE companies focused their attention on selling to operators, several companies made at least a few CRFE units that were tested in base stations by operators. Two com-

FIGURE 8.17: CRFE system designed into the base station of one of the two primary OEMs for the 700-MHz 4G network roll-out in the US beginning in late 2010.

panies were successful in getting orders for ~200 sites, and ultimately shipping and installing CRFE systems into those sites. None of these products though proved to be reliable enough to last more than a few months to a year or so in the field. Ultimately they were all taken out by the operators due to poor system reliability. The reasons for failure of course included the cryocoolers and the dewars, but system power supplies and control electronics also proved to be major sources of failure; these too are part of the cryogenics support system for the HTS filters. One CRFE company was successful in meeting the reliability needs of the operators. It is this company, STI, that succeeded in manufacturing reliable CRFE systems, that has sold thousands of systems and remains in this business today.

8.2.3 Other SC Passive Microwave Devices

A wide variety of HTS passive microwave devices have been successfully demonstrated in the lab over the past 2+ decades. These include antennas, filters, delay lines, multiplexers, etc. (The suggested reading provides good information on these devices.) Two are worth more detailed mention here: commercial satellite communications filters and switched-tuned filters for defense receiver applications.

COMDEV and DuPont collaborated on a program in the late 1990s, early 2000s, to develop and demonstrate HTS output-multiplexer filters for satellite-based communication transceivers. A thorough design and cost analysis was completed and lab prototypes were successfully built and tested. This is an application where there is a high premium on system weight because of the high cost of launching satellites into orbit around the earth. The microwave filters in a communications satellite comprise most of the weight of the transceiver systems and thus for decades have been an area of intense technology development and innovation. The low weight of HTS filters, even when combined with the required cryogenics systems to cool them, offers the promise of reduced total system weight and volume; additionally since the application can support relatively high cost, that is not expected to be a major barrier to broad usage. The COMDEV/DuPont joint program concluded that for the lower communications bands, e.g., C-Band (~ 4 GHz), that HTS filters could be lower weight and thus attractive compared to the best conventional solutions. At the higher frequency bands, e.g., Ku-Band (~ 12 GHz), and above, the conventional approaches would still be preferred simply on a performance/weight basis. Please see the review paper by R. Mansour in the suggested reading for a detailed review of this excellent work.

Extremely compact, high-selectivity, fixed-frequency HTS filters were demonstrated in thin-film form as early as the late 1980s. The possibility of using efficiently switched or tuned arrays of such filters offered the potential to provide extremely effective protection not otherwise possible for microwave receivers used in a wide variety of defense applications. Unlike most commercial microwave systems that utilize narrow frequency bands (generally $\ll 10\%$ total bandwidth), defense microwave systems commonly utilize frequency bands far broader in frequency ($\gg 10\%$). Because of this it is generally impractical to use high-performance microwave filters ($Q > 1,000$) in these systems due to the size and complexity of the arrays of filters that would be required to be useful. With the advent of extremely compact HTS filters, this situation changed. The volume required for a high-selectivity HTS filter at microwave frequency ($Q \gg 1,000$) could be many orders of magnitude smaller than any high-selectivity conventional filter technology. Lab demonstrations of HTS filters in the early 1990s showed more than three-orders-of-magnitude volume reduction. This sparked considerable interest and many DoD-funded development programs throughout the 1990s and well into the 2000s. An enormous variety of HTS filter switching approaches were tried. These included cryogenic semiconductor switches, photoconductors, and microelectromechanical systems (MEMS); each was implemented in a variety of circuit configurations to achieve the lowest loss and greatest linearity. In addition, a wide variety of mechanisms were tried to provide continuous frequency tuning of the HTS filters, e.g., semiconductor varactors, ferroelectric materials, magnetic materials, MEMS capacitors, micro-stepping-motor-driven HTS plates. Many impressive lab demonstrations were performed. These achieved broad tuning ranges ($> 20\%$), high Qs($\sim 100,000$), high tuning speeds (microseconds), and sometimes high linearity. No products have yet appeared, but this remains an area of great interest and potential for HTS microwave devices.

8.2.4 Summary and Looking Ahead

HTS opened the door for market entry of superconducting passive microwave devices. It simultaneously provided an impetus for the development and production of commercially viable HTS thin-film materials and reliable cryogenics systems. There is a market opportunity at this writing (October, 2010), for the emergence of the first designed-in HTS filter systems in cell phone base stations in the 700-MHz 4G roll-out in the US. In addition, in-situ YBCO manufacturing processes developed for low-cost wafers may in the future provide the foundation for low-cost HTS wire. Finally, the independent development of ion-beam-assisted-deposition (IBAD) of single-crystal MgO films on noncrystalline substrates can offer the chance to remove the last major cost element of HTS wafers — the single crystal substrate. This would open the door to the development of low cost HTS transmit filters with $Q \sim 10^6$ for high power (100s of watts or more) applications in cell phone base stations and other RF/microwave applications.

8.3 Superconducting Quantum Electronics Enabling Astronomical Observations

T. M. Klapwijk

On May 14th 2009 the Herschel Space telescope was launched from Kourou, French Guyana, to the 2nd Langrangian (L2) orbit with enough helium on board for a period of about 3 years to perform heterodyne spectrosocopy[25] of the interstellar gas with the Heterodyne Instrument for the Infrared (HIFI). While we commemorate the discovery of superconductivity 100 years ago, superconductivity plays an active role at a distance varying between 1.2 and 1.8 million kilometers from

---

[25] T. de Graauw et al., *Astronomy and Astrophysics* 518 (2010) L6

the planet Earth to unravel the evolution of the interstellar matter (Figure 8.18). When the superconductors in Herschel cease to be superconducting, expected in 2012, the Atacama Large Millimeter Array (ALMA) will be operational, and continue to carry the torch of the astronomical use of the superconducting state. It is the culmination of detector-research started around 1975, based on a phenomenon discovered in 1962 by Dayem and Martin[26] and subsequently called photon-assisted tunneling in the theoretical interpretation offered by Tien and Gordon[27]. HIFI was optimized to perform three key-tasks: (1) observations of water lines ending in the ground states, which are essential for absorption studies of cold water (557 GHz, 1.11 THz, and 1.67 THz), (2) a survey of the molecular complexity of the Universe and (3) observation of red-shifted ionized carbon[CII] at 1.9 THz. These astronomical frequencies indicate that the photon-energies are stretching the limit of the superconducting material, set by the energy gap of niobium at 700 GHz.

8.3.1 The Electrodynamics of Superconducting Films

The absorption of far-infrared radiation in thin superconducting films is one of the striking direct indications of the existence of an energy gap Δ in the electronic states of a superconductor[28,29]. The early observations even preceded the theoretical interpretation known as the BCS-theory[30] and were rapidly incorporated in the theory for the electrodynamics of superconductors by Mattis and Bardeen[31]. The real, σ_1, and imaginary, σ_2, parts of the complex impedance are given by:

$$\frac{\sigma_1}{\sigma_n} = \int_\Delta^\infty dE[f(E) - f(E + \hbar\omega)] \frac{2(E^2 + \Delta^2 + \hbar\omega E)}{\hbar\omega \sqrt{E^2 - \Delta^2} \sqrt{(E + \hbar\omega)^2 - \Delta^2}}$$
$$+ \int_\Delta^{\hbar\omega - \Delta} dE[1 - 2f(\hbar\omega - E)] \frac{\hbar\omega E - E^2 - \Delta^2}{\hbar\omega \sqrt{E^2 - \Delta^2} \sqrt{(\hbar\omega - E)^2 - \Delta^2}} \tag{8.19}$$

and

$$\frac{\sigma_2}{\sigma_n} = \int_{\Delta - \hbar\omega}^\Delta dE[1 - 2f(E + \hbar\omega)] \frac{E^2 + \Delta^2 + \hbar\omega E}{\hbar\omega \sqrt{E^2 - \Delta^2} \sqrt{(E + \hbar\omega)^2 - \Delta^2}} \tag{8.20}$$

with σ_n the normal state conductivity. In thermal equilibrium, the complex impedance is determined by the temperature through the Fermi-Dirac distribution function $f(E)$ and depends on the frequency ω. The integral runs over the energies E. For $\hbar\omega \geq 2\Delta$, where the photon energy exceeds the energy gap 2Δ, the loss rapidly increases because of the creation of quasiparticles. These expressions assume uniform superconducting properties, although in practice the superconducting films may be bilayers of niobium and aluminium or consist of material with properties that vary over the thickness of the film[32]. In both cases the energy terms will be modified and will depend on the material characteristics[33].

At low frequencies down to $\omega = 0$ the superconducting film is lossless. However, with increasing electron temperature dc resistance gradually emerges, in particular close to the critical temperature

[26] A.H. Dayem and R.J. Martin, *Phys. Rev. Lett.* 8 (1962) 246

[27] P.K.Tien and J.P.Gordon, *Phys. Rev.* 129 (1963) 647

[28] M. Tinkham, *Phys. Rev.* 104 (1956) 845

[29] R.E. Glover and M. Tinkham, *Phys. Rev.* 104 (1956) 844

[30] J. Bardeen et al., *Phys. Rev.* 108 (1957) 1175

[31] D.C. Mattis and J. Bardeen, *Phys. Rev.* 111 (1958) 412

[32] S.C. Zhu et al., *Appl. Phys. Lett.* 95 (2009) 253502

[33] S.B. Nam, *Phys. Rev.* 156 (1961) 470

T_c, due to time-dependent changes of the macroscopic quantum phase of the superconductor (See for a recent review Halperin *et al.*[34]). These processes play a stronger role in materials with a high level of disorder and they are particularly relevant for hot-electron bolometers discussed in Section 8.3.8.

8.3.2 Photon-Assisted Tunneling with Superconductors

Early on it was proposed by Burstein et al.[35] that the absorption of photons in superconducting films can be used, analogously to photo-excitation in semiconductors, to detect radiation in the submillimeter range. The created excess quasiparticles should be measured with a superconducting tunnel-junction[36], which carries a quasiparticle current I_{qp} given by

$$I_{qp} = \frac{1}{eR_N} \int_{-\infty}^{\infty} dE N_1(E) N_2(E+eV)[f(E) - f(E+eV)] \qquad (8.21)$$

with N_1 and N_2 the usual BCS density of states of the two superconductors, being 0 for $|E| < \Delta$ and equal to $E/\sqrt{E^2 - \Delta^2}$ for $|E| > \Delta$. V is the applied dc voltage and E the quasiparticle energies. Such a detection mechanism, due to absorption in one of the superconducting films, would have a low-frequency cutoff given by the energy gap and only be relevant for frequencies in the submillimeter range and higher. The tunnel current (Eq. 8.21) is merely a probe of the absorbed quasiparticles in the superconducting film and assumed to be not effected by the microwave field.

In reality, the quasiparticle tunnel current itself responds directly to the electromagnetic field. Dayem and Martin [26] discovered experimentally that the quasiparticle current of a tunnel junction increases strongly by photons with energies far below 2Δ. They find stepwise increases of current for voltages given by $eV = 2\Delta - n\hbar\omega$, with n an integer. Obviously this cannot be understood as due to photo-excitation of quasiparticles in the superconductor.

An explanation has been provided by Tien and Gordon[27], who assume that the electromagnetic field sets up a potential difference across the tunnel barrier between the superconducting electrodes given, for example, by

$$V_{rf} \cos \omega t \qquad (8.22)$$

with V_{rf} the amplitude of the voltage across the barrier and ω the radial frequency of the incoming signal. This time dependent voltage modulates the energy levels in electrode 1 with respect to electrode 2. The single-particle wave functions in electrode 1 with the microwave field are given by:

$$\psi(\mathbf{r}, t) = \psi(\mathbf{r}) e^{-\imath Et/\hbar} \qquad (8.23)$$

with the energy E referred to $V_{rf} \cos \omega t$. This leads effectively to a modification of the density-of-states to

$$N_1'(E) = \sum_{n=-\infty}^{\infty} N_1(E + n\hbar\omega) J_n^2(\alpha), \qquad (8.24)$$

with $\alpha = eV_{rf}/\hbar\omega$, J_n the n-th order Bessel-functions, and N_1 the undisturbed BCS density of states. By substituting this new expression for the density of states into Eq. 8.21, Tien and Gordon find a natural explanation for the observations of Dayem and Martin. The photon-assisted tunneling curve $I_{qp}(V)$ is given by

$$\frac{1}{eR_N} \sum_{n=-\infty}^{\infty} J_n^2(\alpha) \int_{\infty}^{\infty} dE N_1(E + n\hbar\omega) N_2(E+eV)[f(E + n\hbar\omega) - f(E+eV)]. \qquad (8.25)$$

---

[34] B.I. Halperin et al., arXiv:1005.3347v1 (2010)

[35] E. Burstein et al., Phys. Rev. Lett. 6 (1961) 92

[36] I. Giaever, *Phys. Rev. Lett.* 5 (1960) 147

Note that the theoretical approach taken by Tien and Gordon (Eqs. 8.22–8.24) is not unique to superconductive tunnel junctions. It is applicable to any quantum conductor in which the energy is conserved in a conductor coupled to two well-defined equilibrium reservoirs of electrons. However, the sharp nonlinearity resulting from the superconducting density-of-states leads to the very useful strong signatures of the photon-assisted tunneling process in the I,V curve.

This important insight into quasiparticle tunneling in the presence of microwave potentials was soon overshadowed by the subsequent developments in pair tunneling by Josephson[37].

8.3.3 Dynamic Pair Currents: Josephson Tunneling

In 1962 Josephson predicted that a tunnel barrier will also carry a zero-voltage pair current with an amplitude about equal to the quasiparticle current. The Josephson current is dependent on the differences between the macroscopic quantum phases of the two superconductors. After the initial prediction and verification it was soon realized that such a phase-dependent Josephson current is of a much more general nature than the quasiparticle current and occurs in any kind of "weak link": point contact, microbridge, normal metal, semiconductor, etc., and for an applied voltage one will get an oscillating supercurrent running in parallel with a normal current. In order to account for this complex voltage-carrying state, the resistively shunted junction (RSJ-) model was developed. In this model a Josephson element, carrying the supercurrent I_s, is described by two equations

$$I_s = I_0 \sin\phi \tag{8.26}$$

$$\frac{d\phi}{dt} = \frac{2eV}{h} \tag{8.27}$$

and in parallel a quasiparticle current I_{qp} modeled as a voltage-independent resistor

$$I_{qp} = \frac{V}{R}. \tag{8.28}$$

In this set of equations ϕ is the difference in quantum phase of the two superconductors coupled by the weak link. Both I_0 the amplitude of the Josephson current and the normal resistance R are assumed to be voltage- and hence frequency-independent. This model emphasized the unique and universal properties of the Josephson currents irrespective of the kind of weak link and came to dominate also the analysis of the response to radiation.

Heterodyne mixing is the preferred technique to obtain spectral lines revealing rotational transitions in the interstellar matter. Very high resolutions $\lambda/\Delta\lambda \approx 10^6$ are needed to measure Doppler shifts and spectral-line profiles. A signal of a local oscillator is combined with the signal to be detected in a nonlinear mixing element and the difference frequency is amplified. Several groups studied the possibility to use the Josephson current for heterodyne mixing by driving a point contact junction and studying the RSJ-model with a dc and an rf-current (for example[38,39]). Unfortunately the results were hardly any better than the best semiconductor Schottky diode devices. The noise for a driven Josephson junction turned out to be higher than expected[40,41]. This observed excess noise was subsequently[42] shown to be intrinsically due to the nonlinear and deterministic junction dynamics, exhibiting chaotic solutions associated with the appearance of strange attractors in phase space. The dynamic nature of the Josephson effect itself had shown its limitations for practical use and a return to the less universal aspects of the quasiparticle current in a tunnel junction was imminent.

---

[37] B.D. Josephson, *Phys. Lett.* 1 (1962) 251

[38] C.C. Grimes et al., *Phys. Rev. Lett.* 17 (1966) 431

[39] P.L. Richards and S.A. Sterling, *Appl. Phys. Lett.* 14 (1969) 394

[40] J.H. Claassen et al., *J. of Appl. Phys.* 49 (1978) 4117

[41] J.H. Claassen et al., *Appl. Phys. Lett.* 25 (1974) 759

[42] B.A. Huberman et al., *Appl. Phys. Lett.* 37 (1980) 250

8.3.4 Passive Nonlinear Device: Quantum Mixing

In 1973 it was recognized by McColl et al.[43] that the commonly used normal-metal-semiconductor Schottky diode could be modified in a straightforward way. By replacing the normal metal by a superconductor and by doping the semiconductor heavily, transport would be predominantly by electron tunneling, in contrast to thermionic emission. The current-voltage characteristic would be analogous to a normal-metal-insulator-superconductor (NIS) device, while maintaining, in contrast to point contacts, the stable properties of a thin-film device. In a theoretical and experimental analysis the superior properties as a video detector were demonstrated as well as in the heterodyne mode of operation[44]. Whereas in the ordinary Schottky diode the nonlinearity is due to the semiconductor, with the super-Schottky diode the nonlinearity is due to the superconducting gap and the BCS density of states. The responsivity, using classical mixing theory is given by

$$R = \frac{1}{2} \frac{d^2 I/dV^2}{dI/dV} \tag{8.29}$$

which with the appropriate expression for the nonlinearity

$$I_{\text{dc}}(V) = I_0 e^{eV/kT} \tag{8.30}$$

is predicted to lead to a responsivity given by

$$R \approx \frac{e}{2kT} \tag{8.31}$$

with e the electron charge and kT the temperature multiplied by Boltzmann's constant. Video detection at 10 GHz was reported in 1973 followed by mixing results in 1975. The super-Schottky was demonstrated to be far superior to the available alternatives, despite the fact that the semiconductor will contribute to the loss of the signal by absorbing part of the radiation[45].

The implicit consequence of Eq. 8.31 is that the responsivity will rise to infinity with decreasing temperature. This observation led Tucker[46] as early as 1975 to the insight that these resistive mixers could act as photon detectors in the millimeter spectral regions. This is first pointed out in his contribution to the Helsinki 14th Low Temperature Conference (14–20 August 1975), which closes with the sentence: "The technical problems will no doubt prove challenging, but the prospect is an extremely interesting one in our opinion". Tucker and Millea[47] proceed by analyzing the situation in which the photon energy would exceed the voltagescale over which the dc current rises rapidly (i.e., the strength of the nonlinearity). The analysis means a return to the concept of photon-assisted tunneling[27] and the development of a full quantum mixer theory. In a paper presented at the Applied Superconductivity Conference in 1978, the authors[48] applied their newly developed quantum mixer theory, as an example, to the super-Schottky diode but point out that the theory applies equally well to the superconductor-normal-metal tunnel junction (SIN) and the superconductor-insulator-superconductor junction (SIS). The full theory[49] was published in 1979, leading to the quantum responsivity in video detection of

$$R = \frac{e}{2kT} \frac{\tanh(\hbar\omega/2kT)}{\hbar\omega/2kT} \tag{8.32}$$

$$\approx \frac{e}{\hbar\omega}(\hbar\omega \gg k_{\text{B}}T) \tag{8.33}$$

[43] M. McColl et al., *Appl. Phys. Lett.* 23 (1973) 263
[44] M. McColl et al., *Appl. Phys. Lett.* 28 (1976) 159
[45] E.A. Bergin et al., *Astrongomy and Astrophysics* 521 (2010) L20
[46] J.R. Tucker, *Proc. 14th Int. Conf. Low Temp. Phys.* Vol. 4 (1975) 180
[47] J.R. Tucker and M.F. Millea, *Appl. Phys. Lett.* 33 (1978) 288
[48] J.R. Tucker and M.F. Millea, *IEEE Trans. Magn.* MAG-15 (1979) 288
[49] J.R. Tucker, *IEEE J. Quantum Electron.* QE-15 (1979) 1234

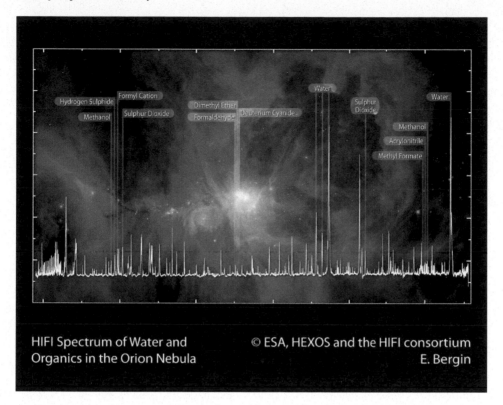

FIGURE 8.18: One of the first results of HIFI: a high resolution spectrum of water and organic molecules in the Orion nebula in the frequency range from the hitherto invisible frequency range of 1.06 to 1.12 THz. For details see Bergin et al.[45]

for a sufficiently nonlinear I,V curve. In subsequent theoretical work[50] it was demonstrated that, for heterodyne detection, the expected lower limit for the noise temperature is given by $\hbar\omega/k_B$, which implies that photon absorption is the only noise source in an ideal SIS mixer.

8.3.5 First Experimental Demonstration: Technology Base

Inspired by the experimental progress in the super-Schottky diode and taking advantage of the emerging thin-film technology for superconducting tunnel junctions, Richards et al.[51] and Dolan et al.[52] developed and studied SIS devices and demonstrated heterodyne mixing at respectively 35 and 110 GHz. After many years of focus on utilizing the Josephson effect, the field had abandoned the Josephson effect, considered it a source of unwanted noise, and returned to the Dayem-Martin effect[53] as the desired detection principle.

A very important condition was the emerging technology of making lithographically structured superconducting tunnel junctions, since those were the only superconducting weak links with a usable nonlinearity. In 1964 Matisoo[54] had proposed the use of Josephson junctions as digital switch-

---

[50] M.J. Wengler and D.P. Woody, *IEEE J. Quantum Electron.* QB-23 (1987) 613

[51] P.L. Richards et al., *Appl. Phys. Lett.* 34 (1979) 345

[52] G.J. Dolan et al., *Appl. Phys. Lett.* 34 (1979) 347

[53] T.G. Phillips et al., *IEEE Trans. Magn.* MAG-17 (1981) 684

[54] J. Matisoo, *Appl. Phys. Lett.* 9 (1966) 167

ing element with a low power consumption and fast switching speed. It led to extensive programs to develop a superconducting junction technology at IBM, Bell Laboratories, NIST and in Japan. The programs came to an end in the early 1980s but had laid the groundwork for a thin-film technology of superconducting tunnel devices, including the need to develop metallurgically stable materials such as PbBi alloys and PbInAu-alloys. A very useful invention, enabling very small scale tunnel junctions, was the Dolan-Dunkleberger stencil lift-off technique or shadow-evaporation[55,56]. The technology push towards the Josephson computer was ready to be used for astrophysics.

Just at the end of this program a major new step was provided by Gurvitch et al.[57] who found a path, based on earlier work by Rowell et al.[58], to make robust niobium tunnel junctions using aluminium-oxide barriers grown on a thin layer of aluminium on top of niobium. The superconducting proximity-effect ensured that the aluminum acted as if it had the superconducting properties of the niobium. Although the early mixer experiments relied on the soft "old materials", the subsequent developments were going to rely on this new robust niobium-based technology, which also enabled the highest possible frequency reachable with the elemental superconductors. All the submillimeter telescopes currently equipped with superconducting technology are based on this niobium technology, first realized for astrophysics by Inatani et al.[59]

8.3.6   Submillimeter-Wave Astronomy

Millimeter-wave astronomy has grown rapidly since the discovery of the cosmic microwave background radiation in 1965. Currently it has become a broad exploration of the interstellar medium. The interest in understanding this interstellar medium has increased enormously due to studies of interstellar molecule-rotation transitions. It has become possible to determine the compositions, density, temperature, velocity, and the type of the molecules (Figure 8.18). It has led to improved understanding of the galactic structure, star formation and stellar mass-loss processes. Against this background it is understandable that rather than stretching the limits of the existing radio observatories, a new generation of millimeter-wave telescopes was developed in the early 1980s. In a 1982 review by Phillips and Woody[60] the progress made in superconducting tunnel devices is communicated to the astronomical community with the unequivocal recommendation to equip the new observatories with superconducting receivers and even to consider "someday to place a large submillimeter-wave band telescope in space".

One of the first experimental papers using superconducting devices[61] identified a particular radio source (1413+135) as a far-infrared extragalactic object. New observatories were created by Caltech's Submillimeter Observatory (CSO), the UK/NL's James Clerk Maxwell telescope, the French-German IRAM 345 GHz telescope at Pico Veleta in Spain, and the Japanese Nobeyama observatory. All of them providing lots of interesting data, but limited by atmospheric transmission to a few windows at the lower frequencies and also without the possibility to enhance the spatial resolution using interferometry.

In the past decade we have witnessed the step into space with the instrument HIFI on board the Herschel Space telescope. In the next decade the step to an enhancement of the spatial resolution will be taken by the completion of Atacama Large Millimeter Array (ALMA).

---

[55] G.J. Dolan, *Appl. Phys. Lett.* 31 (1977) 337

[56] L.N. Dunkleberger, *J. Vac. Sci. Technol.* 15 (1978) 88

[57] M. Gurvitch et al. *Appl. Phys. Lett.* 42 (1983) 472

[58] J.M. Rowell et al., *Phys. Rev.* B 24 (1981) 2278

[59] J. Inatani et al., *IEEE Trans. Magn.* MAG-23 (1987) 1263

[60] T.G. Phillips and D.P. Woody, *Ann. Rev. Astron. Astrophys.* 20 (1982) 285

[61] C.A. Beichman et al., *Nature* 293 (1981) 711

8.3.7 From Proof-of-Principle to Demanding Use

In order to use superconducting tunnel junctions in these very demanding applications, two components needed careful research to exploit the ultimate possibilities. One of them, the local oscillator, is outside the scope of this chapter because it does not use superconductivity. However, the improvements in local oscillators operating up to 1.9 THz has been tremendous in the last decade. The other component, the superconducting tunnel junction, has moved forward in an equally impressive way, yet no longer being able to benefit from the Josephson computer projects (mostly terminated around 1983).

A first requirement is band coverage. The *RC* time constant of a tunnel junction should be as small as possible. Since the capacitance increases linearly as the thickness of the dielectric decreases, i.e. the tunnel barrier, and the resistance decreases exponentially with the thickness, one prefers a tunnel barrier as thin as possible (about 2 nm). In practice this means for aluminum-oxide a minimum value of 14 Ω for an area of 1 μm^2, which implies a bandwidth of about 150 GHz. A further reduction in thickness leads to a breakdown of the tunnel barrier allowing for higher order tunnel processes leading to excess shot noise[62]. Only recently, a strong improvement in bandwidth has been obtained by the introduction of much more uniformly transmissive AlN tunnel barriers[63].

A second requirement is the suppression of the Josephson current. Any tunnel junction will also carry a Josephson current. This current is suppressed with a magnetic field in order to prevent excess noise coming from the Josephson currents.

A third requirement is adequate impedance matching to the outside world. Although, in an early phase, it was customary to use mechanical tuning stubs to create optimum signal transfer, they were replaced by integrated tuning structures consisting of superconducting striplines.

A fourth requirement was to exploit as much of the frequency band as possible with a single superconducting mixing device. Since in a spacecraft the Earth's atmosphere is no longer determining the possible frequency bands, the technology can be pushed to exploit the highest possible frequencies. Since the tunnel barriers should be as thin as possible, a robust tunnel junction technology together with a high-gap superconducting material was needed. The only realistic candidate was niobium with the AlOx (or now also AlN) technology. The energy gap of niobium is at 700 GHz. Given the nature of photon-assisted tunneling this would allow the use of tunnel devices for mixing up to 1.4 THz. However, losses would occur in the striplines used for the tuning circuits. This was solved by using striplines with the high-gap superconducting material NbTiN on one side and good conducting Al on the other side.

Given these considerations the HIFI instrument (Figure 8.18) was equipped with different technologies for the different frequency bands, leading to a natural division of labor between the contributing institutes (most of them European): 450–640 GHz (the French Institutes LERMA and IRAM), 640–800 GHz (the German Institute KOSMA), 800–960 GHz (the Dutch collaboration of SRON with TUDelft), 960–1120 GHz (idem), and 1120–1250 GHz (the US collaboration of CalTech with JPL). In all cases a performance close to the quantum limit was achieved.

8.3.8 Nonequilibrium Mixers: Hot-Electron Bolometers

Beyond the upper-frequency limit of $4\Delta/\hbar$, niobium superconducting tunnel junctions no longer provide a usable signal. Unfortunately, a new tunnel barrier technology for materials with a higher superconducting gap has not yet emerged. An alternative strategy for heterodyne mixing is based on nonequilibrium hot electrons. It was first introduced by Phillips and Jefferts[64] using the temperature dependence of the resistance of the low-bandgap semiconductor InSb and used in the 100-GHz

---

[62] P. Dieleman et al., *Phys. Rev. Lett.* 79 (1997) 3486

[63] T. Zijlstra et al., *Appl. Phys. Lett.* 91 (2007) 233102

[64] T.G. Phillips and K.B. Jefferts, *Rev. Sci. Instrum.* 44, (1973) 1009

range before the emergence of superconducting mixers. The principle of hot electron bolometers was picked up again for the higher frequencies, but now based on nonequilibrium superconductivity, by Gershenson et al[65]. It uses the electrical response of a thin superconducting film to two coherent signals with a frequency higher than the energy gap. The photon energy from the local oscillator, $\hbar\omega_{LO}$, and of the signal to be detected, $\hbar\omega_s$, will break Cooper-pairs resulting in an electronic system with an enhanced electron temperature, hence the name hot-electron bolometers (HEBs). The LO-power is high enough to bring the electron temperature up to a level that the superconducting state gets close to the resistive transition, where small changes in electron temperature lead to strong changes in resistance, essentially due to the increased occurrence of phase-slip events[66]. The interesting physics of the device is contained in the time-response of the electron temperature and hence of the resistance. The hierarchy of times is given by

$$\omega_{IF} < 1/\tau_{el} < 2\Delta/\hbar < \omega_{LO} \tag{8.34}$$

with $1/\tau_{el}$ the time response of the electron system. The two oscillating fields at superTHz frequencies (ω_{LO} and ω_s) contain a component at the intermediate frequency (IF). The electron temperature at this ω_{IF} is coherently related to the signals at superTHz frequencies and leads to a modulation of the electron temperature and hence the resistance at ω_{IF}, provided $\omega_{IF} < 1/\tau_{el}$ is satisfied. In principle, one would like to have a large value for ω_{IF}, which requires a large value for $1/\tau_{el}$. For the latter condition three processes must be distinguished. First, the diffusion time $\tau_D = L^2/D$, with L the length of the device and D the diffusion constant. It is assumed that the sensitive superconducting film is connected to thermal equilibrium reservoirs. Second, the transfer of energy from the electrons to the phonon bath, characterized by τ_{e-ph}. This value is determined by the choice of materials, and materials with a strong electron-phonon interaction are preferred. Empirically it has been found that a thin film of NbN is most suitable; Third, the phonons in the film should be tightly coupled to the phonons of the substrate. Given the unavoidable acoustic mismatch between different materials this has been achieved by making the films as thin as possible (a few nanometers) in order to maximize the chances for phonon-escape, characterized by τ_{esc}.

These devices have been successfully included in HIFI of the Herschel Space telescope, providing data around the ionized carbon 1.9-THz line. Since HEB's are advantageous only at superTHz frequencies, where the atmosphere is opaque, they can be used only in experiments in the stratosphere or in space. In subsequent experiments their potential has been demonstrated up to an impressive 6.5 THz, while also the limiting factor of quantum noise has come into sight[67]. An interesting aspect is also that a very compact package has been developed by the inclusion of the newly developed quantum cascade lasers (QCL's) as a local oscillator[68]. It is very likely that this combination will within a few years be used at an observatory to be built at Dome A on Antarctica, one of the other remote places to let superconductivity function[69] for astronomy.

8.3.9   Conclusions

The unique properties of the superconducting state, as identified in the BCS-theory, have provided the extremely sharp nonlinearities in the current-voltage characteristics of superconducting tunnel junctions, which enabled quantum-limited heterodyne detection. The emerging thin-film technology, the rise of interest in millimeter and submillimeter astronomy, and the acceptance that, despite its beauty, the Josephson-effect should be suppressed to exploit fully the process of photon-assisted tunneling have conspired to the instrument Heterodyne Instrument for the Far-Infrared (HIFI) at

---

[65] E.M. Gershenson et al., *Sov. Phys. Superconductivity* 3 (1990) 1582

[66] R. Barends et al., *Appl. Phys. Lett.* 87 (2005) 263506

[67] W. Zhang et al., *Appl. Phys. Lett.* 96 (2010) 111113

[68] J.R. Gao et al., *Appl Phys. Lett.* 86 (2005) 244104

[69] Y. Ren et al., *Appl. Phys. Lett.* 97 (2010) 161105

the Herschel Space telescope and to the soon-to-be completed Atacama Large Millimeter Array in Chili.

8.4 Microwave Cooling of Superconducting Quantum Systems

W. D. Oliver

8.4.1 Introduction

Superconducting artificial atoms (qubits) are solid-state quantum systems, comprising lithographically defined Josephson tunnel junctions and superconducting interconnects. When cooled to milli-Kelvin temperatures, these qubits exhibit quantized states of charge, flux, or junction phase depending on the circuit design parameters[70]. Associated with these quantized states is a spectrum of energy levels, tunable via an external control parameter, e.g., an applied electric or magnetic field. Although generally only the lowest two energy eigenstates are utilized for quantum information science applications, the energy spectrum indeed extends to higher-energy levels corresponding to higher-excited states of the circuit. The separation between pairs of energy levels typically falls in the radio frequency and microwave regimes, and resonantly driving the artificial atom with a harmonic field can couple and induce quantum-state transitions.

There are several features that make superconducting artificial atoms excellent candidates for quantum information science applications[71]. Superconductors are intrinsically low-loss materials, enabling coherence times (both energy relaxation and dephasing) that are presently at the $1-10$ μs level, limited primarily by losses in the junction, metal surfaces, and ancillary passive components (e.g., capacitor dielectrics). Furthermore, due to their relatively large size, critical dimensions $\sim 1 - 10$μm, superconducting artificial atoms can be strongly coupled to their external microwave control fields. Consequently, gate operations with superconducting qubits can be performed on the $\sim 1 - 10$ns time scale, a factor 10^3-10^4 faster than the coherence times. These numbers are reflective of the field as of 2010. As coherence times are improved, this factor will continue to increase and ultimately reach levels required for fault-tolerant quantum error correction protocols. The states of superconducting artificial atoms can be read out efficiently and with high visibility (presently as high as $\sim 99\%$) using superconducting SQUIDs, cavity-quantum electrodynamics (C-QED) techniques, and superconducting microwave amplifiers (both phase sensitive and phase preserving).

In most QIS applications, high-fidelity state preparation is the starting point for high-fidelity gate operations. While a standard approach is simply to wait for the qubit to relax to its ground state, this imposes the constraint that state preparation will occur on the same time scale as the lifetime of the quantum system. While adequate for many of today's experiments, this approach is not conducive to future multi-qubit applications. For example, error correction protocols rely on ancillary qubits to perform syndrome measurements, and these ancillae must be rapidly refreshed each time the protocol is repeated. Another proposed technique is the use of ideal projective measurements (QND-type measurements) to put the qubit into its ground state, and this is an active area of research.

A third approach is actively cooling a quantum system using optical pumping techniques analogous to those introduced in natural atomic systems. By coupling the artificial atom to rapidly decaying energy levels, either internal levels of the atom or external levels such as as a harmonic

---

[70] J. Clarke and F.K. Wilhelm, Superconducting quantum bits, *Nature* 453 (2008) 1031; see also in this book Chapter 6 on superconducting qubits.

[71] D.P. DiVincenzo, *Fortschr. Phys. Progress of Physics* 48 (2000) 771

oscillator, energy is effectively pumped out of the quantum system of interest. In this section, we review the use of microwaves to cool superconducting artificial atoms actively to their ground state.

8.4.2 Superconducting Artificial Atoms

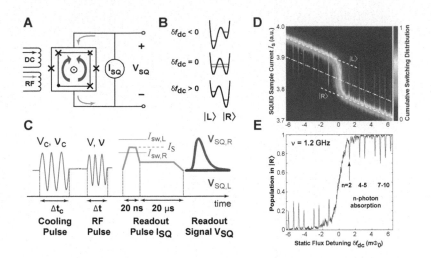

FIGURE 8.19:   Artificial atom (persistent current qubit) and measurement set-up. A schematic of the qubit and surrounding DC SQUID readout. B Double well qubit potential comprising energy levels for static magnetic flux bias δf_{dc} about $\Phi_0/2$, where Φ_0 is the superconducting flux quantum. Diabatic states of the left (right) well corresponds to a persistent current with clockwise (counterclockwise) circulation. At detuning $\delta f_{dc} = 0$, the double-well potential is symmetric and the diabatic-state energies are degenerate. Tunnel coupling opens an avoided crossing Δ. C Qubit excitation and read-out pulse sequence. The qubit is first prepared in its ground state with a harmonic cooling pulse with amplitude V_c and frequency v_c. Quantum-state transitions are induced with a subsequent harmonic RF pulse with amplitude V and frequency v. The qubit state is read-out using the DC SQUID switching response. D Qubit step. Cumulative switching current distribution of the SQUID for each δf_{dc} value following a 3-μs RF driving pulse at 1.2 GHz applied to the qubit (the cooling pulse was not used here). Resonant multiphoton transitions (of order n) are observed between states $|L\rangle$ and $|R\rangle$. The switching distribution along the dashed-dotted line discriminates between states $|L\rangle$ and $|R\rangle$ (E).

We illustrate microwave cooling of a quantum system using the specific example of a super-conducting persistent-current qubit[72,73]. The persistent-current qubit is a superconducting loop interrupted by three Josephson junctions (Figure 8.19A). When biased with a static magnetic flux $f_{dc} \sim \Phi_0/2$, where Φ_0 is the superconducting flux quantum, the system assumes a double-well potential profile (Figure 8.19B). The diabatic ground state of the left (right) well corresponds to a persistent current I_q with clockwise (counterclockwise) circulation. These two diabatic energy levels have an energy separation $\varepsilon = 2I_q \delta f_{dc}$ linear in the flux detuning $\delta f_{dc} \equiv f_{dc} - \Phi_0/2$. Higher-excited states of the double-well potential (see Figure 8.20A) will be used to cool the qubit to its ground

[72] J. E. Mooij, T. P. Orlando, L. S. Levitov, L. Tian, C. H. van der Wal, S. Lloyd, *Science* 285 (1999) 1036;

[73] T. P. Orlando, J. E. Mooij, L. Tian, C. H. van der Wal, L. S. Levitov, S. Lloyd, and J. J. Mazo, *Phys. Rev.* B 60 (1999) 15398

state.

The two-level system Hamiltonian near $\delta f_{dc} = 0$ for the lowest two energy

$$\mathcal{H} = -\frac{1}{2}(\varepsilon\sigma_z + \Delta\sigma_x), \tag{8.35}$$

where σ_z and σ_x are Pauli matrices. At detuning $\delta f_{dc} = 0$, the double-well potential is symmetric and the diabatic-state energies are degenerate. At this "degeneracy point," resonant tunneling between the diabatic states opens an avoided level crossing of energy Δ. Here, the qubit states are σ_x eigenstates, corresponding to symmetric and antisymmetric combinations of diabatic circulating-current states. Detuning the flux away from this point tilts the double well, allowing us to tune the eigenstates and eigenenergies of the artificial atom. Far from the degeneracy point the qubit states are approximately σ_z eigenstates, the diabatic states with well-defined circulating current. The qubit is read out using a hysteretic DC SQUID (superconducting quantum interference device), a sensitive magnetometer that can distinguish the flux generated by circulating current states.

In addition to the static flux biases, the artificial atom is controlled and read out using the harmonic RF/microwave pulses illustrated in Figure 8.19C. As we describe below, the qubit is first prepared in its ground state using a harmonic cooling pulse with amplitude V_c and frequency ν_c. Quantum-state transitions are then driven using a harmonic RF/microwave pulse with amplitude V and frequency ν. These fields are mutually coupled to the qubit through a small antenna. This is followed by a SQUID readout current pulse using the "sample and hold" technique. If the sample current exceeds the SQUID switching current, a voltage pulse will appear at the output during the hold phase. Threshold detection looks for the presence or absence of a SQUID voltage, and this constitutes a digital measurement of the qubit state. Alternatively, although not used in these experiments, we have incorporated the SQUID into a resonant circuit and realized qubit readout via the shift in resonance frequency and phase for both the linear and nonlinear resonance regimes.

The "qubit step," the readout of qubit in equilibrium with its environment, is shown in Figure 8.19D as a function of the SQUID sample current and the flux detuning. The diabatic states $|L\rangle$ and $|R\rangle$ correspond to different levels of sample current (dashed lines) located symmetrically about their energy degeneracy point at $\delta f_{dc}=0$. This plot constitutes a cumulative switching current distribution of the SQUID for each δf_{dc} value. Additionally, a 3-μs pulse at 1.2 GHz is applied to the qubit, and resonant transitions can be observed as fingers extending down (up) from state $|L\rangle$ ($|R\rangle$) when n×1.2 GHz becomes resonant with the energy-level separation. A best-estimator (dashed-dotted line) can be determined to provide the best statistical discrimination between states $|L\rangle$ and $|R\rangle$. The resulting qubit step with its saturated n-photon resonances along the best estimator line is shown in Figure 8.19E.

Note that in the cooling discussion in the next section, we label the multilevel energy diagram (Figure 8.20C) by the diabatic states of the left well ($|0L\rangle$ and $|1L\rangle$) and the right ($|0R\rangle$ and $|1R\rangle$). This corresponds to the energy levels of isolated wells and not the system eigenenergies. In general, the uncoupled well states have energies that are linear in flux: in Figure 8.20C energies of the states in the left well have positive slopes, while those in the right well have negative slopes. When the energy levels of these diabatic states cross, an avoided crossing opens labelled by $\Delta_{n,m}$ where n and m correspond to the level number in the left and right well respectively. For example, the state $|0L\rangle$ is the ground state at negative flux detuning in (Figure 8.20C). Following this diabatic energy as the flux is increased, it increases in energy until it becomes degenerate with the $|0R\rangle$ energy level, where an avoided crossing $\Delta_{0,0}$ opens. Continuing to increase the flux, the energy of state $|0L\rangle$ continues to increase, eventually crossing the $|1R\rangle$ energy level, where an avoided crossing $\Delta_{0,1}$ opens. Continuing further, the energy of level $|0L\rangle$ becomes becomes the second excited state on the right side of Figure 8.20C, above the states $|0R\rangle$ and $|1R\rangle$.

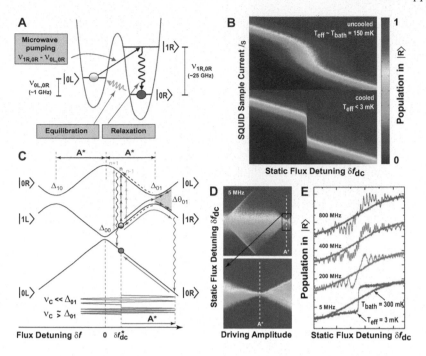

FIGURE 8.20: Cooling of an artificial atom via an ancillary excited state. A External excitation transfers the thermal population from state $|0L\rangle$ to state $|1R\rangle$ (straight line) from which it decays into the ground state $|0R\rangle$. Wavy lines represent spontaneous relaxation and absorption leading to equilibration. B Qubit step at $T_{bath} = 150$ mK in equilibrium with the bath (top) and after a 3-µs cooling pulse at 5 MHz (bottom). The average level populations exhibit a qubit step about $\delta f_{dc} = 0$, with a width proportional to T_{bath} (top) and $T_{eff} \ll T_{bath}$ (bottom). C Schematic level diagram illustrating resonant and adiabatic cooling. $|0L\rangle \rightarrow |1R\rangle$ transitions are resonant at high driving frequency ν (blue lines) and occur via adiabatic passage at low ν (red lines). Δ_{00} and Δ_{01} are the tunnel splittings between $|0R\rangle$ - $|0L\rangle$ and $|0L\rangle$ - $|1R\rangle$. D Optimal cooling parameters. State $|0R\rangle$ population *vs.* flux detuning δf_{dc} and driving amplitude A with $\nu = 5$ MHz, $\Delta t_c = 3 \mu s$, and $T_{bath} = 150$ mK. Optimal conditions for cooling are realized at $A = A^*$, where A^* is defined in C. E Cooling at driving frequencies $\nu = 800, 400, 200$ and 5 MHz. State $|0R\rangle$ population *vs.* δf_{dc} for the cooled qubit and for the qubit in thermal equilibrium with the bath (black lines, $T_{bath} = 300$ mK). Measurements for $\nu = 800, 400, 200$ and 5 MHz are displaced vertically for clarity. A cooling factor of 100, independent of detuning, is obtained in the adiabatic limit (5 MHz).

8.4.3 Microwave Cooling

Quantum operations in single qubits generally involve driving transitions within a manifold of the lowest two energy levels in the double-well potential (Figure 8.20A), which constitute the two-level qubit subsystem of a more complex energy level diagram. When higher-excited states are accessed, the driven system behavior can be markedly different from the population saturation observed when only two levels are involved (for example, as in Figure 8.19E). For example, at least three levels are required to achieve incoherent population inversion, and such a multilevel artificial atom coupled to a microwave cavity has recently been used to demonstrate masing (microwave lasing)[74]. In that work, Josephson quasi-particle states were driven to achieve inversion. Alternatively, population

---

[74] O. Astafiev, K. Inomata, A.O. Niskanen, T. Yamamoto, Yu. A. Pashkin, Y. Nakamura, J.S. Tsai, *Nature* 449 (2007) 588

inversion can be established by accessing an ancilliary excited state via direct or LZS transitions[75].

Here, by reversing the cycle that leads to population inversion, we show that one can pump population from the qubit excited state $|0L\rangle$ to the qubit ground state $|0R\rangle$ (Figure 8.20A) via an ancillary energy level $|1R\rangle$[76]. In the case where the population in $|0L\rangle$ results from thermal excitation, the transfer of population to $|0R\rangle$ effectively cools the qubit by lowering its effective temperature. This kind of active cooling represents a means to initialize and reset qubits with high fidelity, key elements for quantum information science and technology. Alternatively, in addition to the quantum system itself, the pumping mechanism can be used to refrigerate environmental degrees of freedom[77] or cool neighboring quantum systems[78,79,80].

For a qubit in equilibrium with its environment, the population in $|0L\rangle$ that is thermally excited from $|0R\rangle$ follows the Boltzmann relation

$$p_{0L}/p_{0R} = \exp[-\varepsilon/k_B T_{bath}], \tag{8.36}$$

where $p_{0L,0R}$ are the qubit populations for energy levels $\varepsilon_{0L,0R}$, $\varepsilon = \varepsilon_{0L} - \varepsilon_{0R}$, k_B is the Boltzmann constant, and T_{bath} is the bath temperature. To cool the qubit subsystem below T_{bath}, a microwave magnetic flux of amplitude A and frequency ν targets the $|0L\rangle \rightarrow |1R\rangle$ transition, driving the state-$|0L\rangle$ thermal population to state $|1R\rangle$, from which it quickly relaxes to the ground state $|0R\rangle$. Efficient cooling occurs only when the driving-induced population transfer to $|0R\rangle$ is faster than the thermal repopulation of $|0L\rangle$. The hierarchy of relaxation and absorption rates required, $\Gamma_{0R,1R} \gg \Gamma_{0L,1R}, \Gamma_{0L,0R}$, is achieved in our system owing to a relatively weak tunneling between wells (states connected quantum mechanically through a tunnel barrier), which inhibits the interwell relaxation and absorption processes $|1R\rangle \rightarrow |0L\rangle$ and $|0R\rangle \rightarrow |0L\rangle$, compared with the relatively strong intrawell relaxation process $|1R\rangle \rightarrow |0R\rangle$ (states connected within an approximate harmonic oscillator).

Figure 8.20B shows the qubit step at $T_{bath} = 150$ mK in equilibrium with the bath (top) and after a 3-μs cooling pulse at 5 MHz (bottom). Under equilibrium conditions, the average level populations exhibit a thermally-broadened qubit step about $\delta f_{dc} = 0$, with a width proportional to T_{bath}. The presence of microwave excitation targeting the $|0L\rangle \rightarrow |1R\rangle$ transition, followed by relaxation, acts to increase the ground-state population and, thereby, sharpens the qubit step. Cooling can thus be quantified in terms of an effective temperature $T_{eff} < T_{bath}$, a signature that is evident from the narrowing of the qubit steps in Figure 8.20B after cooling. Using the notation from Figure 8.20, the effective qubit temperature is obtained by fitting the temperature T_{eff} that would have been required in equilibrium to achieve the observed qubit population p_{0R},

$$p_{0R} = \frac{\varepsilon}{\sqrt{\varepsilon^2 + \Delta^2}} \left[\tanh\left(\frac{\sqrt{\varepsilon^2 + \Delta^2}}{2k_B T_{eff}} \right) + 1 \right]. \tag{8.37}$$

Universal cooling (cooling that is independent of flux detuning) occurs near an optimal driving amplitude A^* (Figure 8.20C). This is demonstrated in Figure 8.20D where we present the $|0R\rangle$ state population P_{sw} measured as a function of the microwave amplitude A and flux detuning δf_{dc} for $\nu = 5$ MHz. Cooling and the diamond feature can be understood in terms of the energy level diagram (Figure 8.20C). As the amplitude of the microwave pulse is increased from $V = 0$, population

[75] D.M. Berns, M.S. Rudner, S.O. Valenzuela, K. K. Berggren, W. D. Oliver, L. S. Levitov, T. P. Orlando, *Nature* 455 (2008) 51

[76] S. O. Valenzuela, W. D. Oliver, D. M. Berns, K. K. Berggren, L. S. Levitov, T. P. Orlando, *Science* 314 (2006) 1589

[77] A.O. Niskanen, Y. Nakamura, and J.P. Pekola, *Phys. Rev. B* 76 (2007) 174523

[78] J.Q. You, Yu-xi Liu, and Franco Nori, *Phys. Rev. Lett.* 100 (2008) 047001

[79] M. Grajcar, S.H.W. van der Ploeg, A. Izmalkov, E. Il'ichev, H.-G. Meyer, A. Fedorov, A. Shnirman, and G. Schön, *Nature Phys.* 4 (2008) 612

[80] S. Kafanov, A. Kemppinen, Yu. A. Pashkin, M. Meschke, J.S. Tsai, J.P. Pekola, *Phys. Rev. Lett.* 103 (2009) 120801

transfer first occurs when the $\Delta_{0,0}$ avoided crossing between the lowest two states is reached, i.e., $A > |\delta f_{dc}|$; this defines the front side of the observed diamond, symmetric about the qubit step. For amplitudes $A^*/2 \leq A \leq A^*$, the $\Delta_{0,1}$ ($\Delta_{1,0}$) side avoided crossing dominates the dynamics, resulting in a second pair of thresholds $A = A^* - |\delta f_{dc}|$, which define the back side of the diamond. In the region outside of the diamond's backside, the qubit is cooled. As the diamond narrows to the point $A = A^*$, the sharpest qubit step is observed. This is the universal cooling condition: only one of the two side avoided crossings ($\Delta_{0,1}$ or $\Delta_{1,0}$) is reached and, thereby, strong transitions with relaxation to the ground state result for a wide range of δf_{dc}. In contrast, for $A > A^*$, both side avoided crossings ($\Delta_{0,1}$ and $\Delta_{1,0}$) are reached simultaneously for $|\delta f_{dc}| < A - A^*$, leading once again to a large population transfer between $|0R\rangle$ and $|0L\rangle$, and opening the front side of a second diamond feature (Figure 8.20D).

The cooling exhibits a rich structure as a function of driving frequency and detuning, resulting from the manner in which state $|1R\rangle$ is accessed (Figure 8.20C). Transitions occur via a (multiphoton) resonant or adiabatic passage process when the driving frequency is high or low enough, respectively[81,82]. At high frequencies (800 and 400 MHz in Figure 8.20E) well-resolved resonances of n-photon transitions are observed and cooling is thus maximized near resonances. At intermediate frequencies (400 and 200 MHz), Mach-Zehnder interference at the side avoided crossing Δ_{01} becomes more prominent and modulates the intensity of the n-photon resonances. Below $v = 200$ MHz, individual resonances are no longer discernible, but the modulation envelope persists. At the lowest frequencies ($v < 10$ MHz), state $|1R\rangle$ is reached via adiabatic passage through the Δ_{01} crossing (Figure 8.20C), and the population transfer and cooling become conveniently independent of detuning (see $v = 5$ MHz in Figure 8.20E). As shown in Figure 8.20E, we achieve an effective qubit temperature $T_{eff} = 3$ mK, even for $T_{bath} = 300$ mK. In our qubit, our determination of T_{eff} was limited primarily by decoherence (linewidth), which limited the resolution with which we could distinguish the states $|0R\rangle$ and $|0L\rangle$ near degeneracy. Nonetheless, we can estimate the ideally resolvable cooling factor α_c for this type of cooling process using Eq. 8.36,

$$\alpha_c \equiv \frac{T_{bath}}{T_{eff}} = \frac{\varepsilon_{1R \to 0R}}{\Delta}, \tag{8.38}$$

where $\varepsilon_{1R \to 0R} \approx h \times 25$ GHz is the energy separation where the relaxation $|1R\rangle \to |0R\rangle$ occurs and $\Delta \approx h \times 0.01$ GHz for our qubit, yielding a cooling factor $\alpha_c \sim 2500$. For a bath temperature $T_{bath} = 50$ mK, this would correspond to an effective temperature $T_{eff} = 20$ μK in our qubit.

Cooling a qubit in equilibrium with the bath requires a characteristic cooling time. In turn, a cooled qubit will thermalize to the environmental bath temperature over a characteristic equilibration time. The relationship between these two times determines if it is possible to drive the qubit while it is still cold. We found in this qubit that equilibration times are at least one order of magnitude larger than cooling times at $T_{bath} < 250$ mK and up to three order of magnitudes larger at $T_{bath} < 100$ mK. This allowed us ample time to drive the qubit after cooling it. The implementation of an active cooling pulse prior to a generic driving pulse is highly advantageous. It sensibly shortens measurement times, enabling us to acquire data at repetition rates that far exceed the intrinsic equilibration rate due to interwell relaxation after each measurement trial. By adopting active cooling, we gained a factor 50 in data acquisition speed, limited by the bandwidth of our readout circuit. Furthermore, active cooling greatly reduces thermal smearing, allowing us to analyze features in the data that would have been hidden otherwise.

---

[81] W. D. Oliver, Y. Yu, J. C. Lee, K. K. Berggren, L. S. Levitov, T. P. Orlando, *Science* 310 (2005) 1653

[82] D.M. Berns, W.D. Oliver, S.O. Valenzuela, A.V. Shytov, K.K. Berggren, L.S. Levitov, T.P. Orlando, *Phys. Rev. Lett.* 97 (2006) 150502

8.4.4 Summary

We reviewed implementation of active cooling of a quantum system, a superconducting qubit. By applying microwave pulses, we pumped population from an excited state within the qubit manifold to a higher excited state, from which it quickly relaxed to the qubit qround state. This process is an example of an entropy pump, essentially, a qubit refrigerator. Using this technique, we have achieved effective qubit temperatures as low as 0.5 mK, a factor $\sim 10-100\times$ colder than the dilution refrigerator itself. The use of active cooling techniques in QIS will become more prominent as superconducting qubit coherence times increase. From a practical point of view, it reduces the amount of time one must wait between experiments to reset the qubit. Furthermore, this type of pump can be used to cool systems connected the qubit, e.g., resonators, SQUIDs, or other restricted environmental degrees of freedom. More fundamentally, active cooling has an important role in future QIS applications, enabling the rapid and high-fidelity state initialization of logical qubits as well as the ancillae required for syndrome measurements in quantum error-correction protocols.

8.5 Applications of Superconducting Microresonators

Jonas Zmuidzinas

8.5.1 Introduction

Over the past decade, interest in superconducting microresonators has risen dramatically as a result of the wide variety of emerging application possibilities. Various versions of these simple devices, produced by thin-film deposition and lithographic patterning, are now being developed for photon and dark matter detection, neutrino mass experiments, frequency-multiplexed readout of cryogenic detector arrays, quantum information experiments, coupling to nanomechanical systems, and ultra-low noise parametric amplifiers. These devices exploit a broad range of phenomena in superconductivity, including ultra-low dissipation, the kinetic inductance effect, nonlinear response, and non-equilibrium dynamics. In addition, the physics of superconducting microresonators touches on other fields of condensed matter physics, especially two-level systems in amorphous materials. To provide the necessary background, I will first briefly review the basics of superconductor electrodynamics and its historical development. I will then turn to a discussion of superconducting microresonators, with a primary focus on detector applications.

8.5.2 Linear Electrodynamics: A Brief Review

Two decades after Onnes' 1911 discovery, experiments had shown that the hallmark of superconductivity—the disappearance of the electrical resistance below the transition temperature T_c—could be seen not only using direct currents, but also with alternating radio-frequency (RF; 1-10 MHz) currents.[83] By 1940 this work had been extended into the microwave region (1.5 GHz).[84] Meanwhile, at much higher frequencies—at visible and infrared wavelengths—no significant change in the optical absorption was observed upon passing through the superconducting transition.[85] Indeed, the two effects were soon employed together to make the first superconduct-

---

[83] F. B. Silsbee et al., Phys. Rev. 39 (1932) 379; J. C. McLennan et al., *Proc. Roy. Soc.* 136 (1932) 52

[84] H. London, *Proc. Roy. Soc.* A176 (1940) 522

[85] J. G. Daunt et al., *Phil. Mag.* 23 (1937) 264

ing infrared bolometers.[86] Thus, it was apparent that the electrical conductivity $\sigma(\omega)$ must exhibit a change from superconducting behavior to normal-metal behavior at frequencies ω between the microwave and infrared bands. This transition was eventually shown to occur in the millimeter through far-infrared wavelength range,[87] consistent with a temperature-dependent electron energy gap of $2\Delta(T)$ (with $2\Delta \approx 3.5\, kT_c$ for $T << T_c$) that was a key feature of the BCS theory published shortly thereafter.[88]

Well before the electron-pairing BCS theory was proposed, Heinz London had understood a more subtle aspect of the electrodynamic behavior of superconductors.[89] London predicted that a super-conductor should in general have a small but nonzero dissipation for AC currents, in contrast to the DC case for which the resistance and associated dissipation vanish entirely at $T < T_c$. The dissipation arises because the finite inertia of the superconducting electrons allows AC electric fields to exist inside a superconductor, and therefore the presence of normal electrons–expected for temperatures above absolute zero–leads to, in London's words, "production of heat". See Section 8.1.2 for a discussion of the two-fluid model proposed by London. The subsequent application of the BCS theory to the calculation of the electrical conductivity of superconductors yielded results consistent with London's insight; the Mattis-Bardeen expressions for the complex conductivity of superconductors are given in Section 8.3.1 as Eqs. 8.19 and 8.20).

In most cases, the complex conductivity is not directly accessible experimentally; instead, the complex surface impedance $Z_s = R_s + iX_s$ is the quantity being probed. For thick films in the local limit, the surface impedance and complex conductivity $\sigma(\omega, T)$ are related by[90]

$$Z_s(\omega, T) = \sqrt{\frac{i\mu_0\omega}{\sigma(\omega, T)}} = \frac{Z_s(\omega, 0)}{\sqrt{1 + i\delta\sigma(\omega, T)/\sigma_2(\omega, 0)}} \qquad (8.39)$$

where $\delta\sigma(\omega, T) = \sigma(\omega, T) - \sigma(\omega, 0) = \sigma_1(\omega, T) - i\delta\sigma_2(\omega, T)$. The theoretical surface impedance at zero temperature is purely reactive and may be expressed in terms of the penetration depth λ

$$Z_s(\omega, 0) = i\mu_0\omega\lambda \qquad (8.40)$$

where, in the local limit, $\lambda_{local} \approx 105\,nm \times \sqrt{(\rho_n/1\,\mu\Omega\,cm)(1\,K/T_c)}$. Similar equations may be written for thick films in the "extreme anomalous" regime, and for films that are thin compared to the penetration depth.

These results may be summarized by relating the first-order fractional perturbation in the surface impedance to the fractional perturbation in the conductivity,

$$\frac{\delta Z_s(\omega, T)}{Z_s(\omega, 0)} \approx -\gamma\frac{\delta\sigma(\omega, T)}{\sigma(\omega, 0)} \qquad (8.41)$$

where $\delta Z_s(\omega, T) = Z_s(\omega, T) - Z_s(\omega, 0)$, and with $\gamma = 1$, $1/2$, and $1/3$ for the thin film, local, and extreme anomalous limits, respectively. If we introduce the conductivity quality factor as $Q_\sigma = \sigma_2(\omega, 0)/\sigma_1(\omega, T)$, and similarly the surface impedance quality factor $Q_s = X_s(\omega, 0)/R_s(\omega, T)$, Eq. (8.41) shows that they are related by $Q_s = Q_\sigma/\gamma$.

At low frequencies and temperatures, $\hbar\omega << \Delta_0$ and $k_B T << \Delta_0$, we may use the approximation[91]

---

[86] D. H. Andrews et al., *Phys. Rev.* 59 (1941) 1045

[87] G. S. Blevins et al., *Phys. Rev.* 100 (1955) 1215; R. E. Glover and M. Tinkham, *Phys. Rev.* 104 (1956) 844; *Phys. Rev.* 108 (1957) 243

[88] J. Bardeen et al., *Phys. Rev.* 106 (1957) 162; *Phys. Rev.* 108 (1957) 1175

[89] H. London, *Nature* 133 (1934) 497

[90] Accessible reviews of superconductor electrodynamics including references to authoritative and original papers may be found in J. P. Turneaure et al., *J. Supercond.* 4 (1991) 341 ; P. J. Walsh and V. P. Tomaselli, *Am. J. Phys.* 58 (1990) 644; and J. Gao, Ph.D. thesis, Caltech, Pasadena CA (2008), http://thesis.library.caltech.edu/2530.

[91] J. Gao et al., *J. Low Temp. Phys.* 151 (2008) 557

$$Q_\sigma(\omega, T) \approx \frac{\pi}{4} \frac{e^{\Delta_0/k_B T}}{\sinh(\hbar\omega/2k_B T) K_0(\hbar\omega/2k_B T)}, \qquad (8.42)$$

which explicitly[92] shows the exponential rise with decreasing temperature.

FIGURE 8.21: Left: a microstrip transmission line consists of a conducting strip of width w on the surface of a dielectric slab of thickness h backed by a conducting ground plane. The coplanar waveguide (CPW) line, illustrated on the right, has conductors only on the top surface of the dielectric, and consists of a center strip of width w separated from ground planes on either side by gaps of width g.

8.5.3   Superconducting Microresonators

Pippard[93] understood that the use of a superconductor in a parallel-strip transmission line—similar to what is today known as a microstrip line (see Figure 8.21)—would result in a change of the phase velocity due to the inductance contributed by the inductive surface impedance, $Z_s(\omega, 0) = iX_s = i\omega L_s = i\omega\mu_0\lambda_{sc}$. Indeed, the inductance per unit length of a line made using a perfect conductor ($\sigma \to \infty$) is approximately given by $\mathcal{L} = \mu_0 h/w$, where h is the dielectric thickness and w is the microstrip width, provided that $w \gg h$. The capacitance per unit length is approximately $C = \epsilon_r\epsilon_0 w/h$, where ϵ_r is the relative dielectric constant of the substrate. The phase velocity of the fundamental TEM-like mode is $\bar{c} = 1/\sqrt{\mathcal{L}C} = c/\sqrt{\epsilon_r}$. As discussed by Pippard and others,[94, 95] the use of a superconductor causes an increase in the inductance, $\mathcal{L} = \mu_0(h + 2\lambda_{sc})/w$, as if the spacing between the superconducting films had increased by $2\lambda_{sc}$. The capacitance remains unchanged. The fraction of the total inductance of the line that is contributed by the superconductor is

$$\alpha_{ms} = \frac{2\lambda_{sc}}{h + 2\lambda_{sc}}. \qquad (8.43)$$

The increased inductance causes a reduction in the phase velocity by the factor $\sqrt{1 - \alpha_{ms}}$, to $\bar{c} = (1 + 2\lambda_{sc}/h)^{-1/2} c/\sqrt{\epsilon_r}$. More accurate expressions for the properties of superconducting microstrip lines that include the effects of fringing fields may be found in the literature.[96] Alternatively, microstrip properties may be quickly calculated using modern electromagnetic simulation software.

Microstrip lines may be fabricated by depositing three film layers (superconductor, insulator, superconductor) on a substrate, and patterning the top superconductor layer. In this case, the dielectric thickness h would typically lie in the range $0.1 - 1\ \mu m$. Alternatively, the substrate may itself serve

---

[92] Note that $\sinh(x)K_0(x) \sim \sqrt{\pi/8x}$ as $x \to \infty$.

[93] A. B. Pippard, *Proc. Roy. Soc.* A191 (1947) 399

[94] J. C. Swihart, *J. Appl. Phys.* 32 (1961) 461; P. V. Mason and R. W. Gould, *J. Appl. Phys.* 40 (1969) 2039; R. L. Kautz, *J. Appl. Phys.* 49 (1978) 308

[95] J. M. Pond et al., *IEEE Trans. Magn.* 23 (1987) 903

[96] W. H. Chang, J. Appl. Phys. 50 (1979) 8129; G. Yassin and S. Withington, *J. Phys.* D28 (1995) 1983

as the microstrip dielectric, with superconducting films deposited on the top and bottom surfaces, in which case $h \sim 100 - 500\,\mu$m. In both cases h is quite small compared to the free-space wavelength $\lambda_{RF} \approx 30$ cm $\times (1 \text{ GHz}/\nu)$, so $\alpha_{ms} >> \lambda_{sc}/\lambda_{RF}$. Compared to a cavity resonator, a microstrip resonator limited by conductor losses has a considerably lower internal quality factor $Q_i = \alpha_{ms}^{-1} Q_s$ for the same value of Q_s. As Pond et al.[95] point out, the kinetic inductance fraction can even approach unity if a thin superconducting film with a high normal-state resistivity ρ_n is used.

The coplanar waveguide (CPW; Figure 8.21) is another popular superconducting transmission line structure. One advantage of CPW as compared to microstrip is that only one superconducting film layer is required. However, the kinetic inductance fraction of CPW is generally smaller than for thin-film microstrip. This can be seen from the crude estimate $\alpha_{CPW} \sim \lambda_{sc}/w$, and by noting that typical CPW center strip widths $w \sim 2 - 10\ \mu$m are larger than the $0.1 - 1\,\mu$m film thickness of deposited dielectrics used for microstrip lines. The kinetic inductance fraction of superconducting CPW lines may be accurately calculated using analytical formulae[97], conformal mapping methods,[98] or direct electromagnetic simulation, and these methods generally yield results consistent with measurements.[99]

Early Studies

Mason and Gould (1969) studied In/Ta$_2$O$_5$/Ta microstrip resonators in the 50–500 MHz range. The microstrips were made using evaporated thin In films on anodized Ta substrates. The highest resonator quality factors achieved were $Q_r \sim 1.7 \times 10^3$. Mason and Gould's results showed that a surface impedance quality factor of at least $Q_s \geq \alpha_{ms}Q_r \sim 8 \times 10^2$ was possible, a limit that was three orders of magnitude lower than contemporary cavity measurements. DiNardo et al.[100] obtained much higher quality factors, around $Q_r = 5 \times 10^5$ at 14 GHz, using thin-film lead microstrips on alumina substrates. However, the relatively thick substrates used yield low values of α_{ms}, so the inferred value of $Q_s \sim 5 \times 10^2$ is actually similar to the limit set by Mason and Gould. About a decade later, Pöpel[101] studied the losses of PbAu/SiO/Pb microstrip resonators, finding minimum losses at $T = 1.7$ K of 0.04 dB/m at 0.09 dB/m at 9.1 GHz and 27.3 GHz, corresponding to quality factors of $\sim 6 \times 10^4$. Given the the 880 nm thickness of the SiO dielectric, the kinetic inductance fraction was around $\alpha_{ms} \approx 0.12$, so a value of $Q_s \sim 7 \times 10^3$ was achieved, a significant step forward and not far from the Mattis-Bardeen value at $T_c/T = 4.2$. Pond et al.[95] studied NbN/Si:H/NbN microstrip resonators in the 0-2 GHz frequency range. Using magnetron sputtering, rather thin films (150 Å, 400 Å, 140 Å) were deposited, and the measured phase velocity $\bar{c}/c = 0.016$ was very low, indicating a kinetic inductance fraction close to unity as would be expected. The line loss was also measured, and indicated $Q_s \sim 3 - 6 \times 10^3$. As a final example, Andreone et al. (1993)[102] studied microstrip resonators made by depositing and patterning both Nb ($\rho_n = 2.6\,\mu\Omega$cm) and NbTiN ($\rho_n = 90\,\mu\Omega$cm) films on sapphire substrates. The resonator was formed by inverting the substrate and placing it on top of another sapphire substrate, with a bulk niobium foil below serving as the ground plane. The kinetic inductance fraction of this arrangement is quite low due to the 130 μm substrate thickness. The measurement fits indicate zero-temperature residual surface resistances of 1.3 $\mu\Omega$ (Nb) and 5.9 $\mu\Omega$ (NbTiN), which correspond to $Q_s \approx 1.3 \times 10^3$ for both films, comparable to or somewhat lower than earlier work.

---

[97] J. C. Booth and C. L. Holloway, *IEEE Trans. Micr. Theory Tech.* 47 (1999) 769

[98] J. Gao, Ph.D. thesis, Caltech, Pasadena CA (2008); http://thesis.library.caltech.edu/2530.

[99] J. Gao et al., Nucl. *Instrum. Meth.* A 559 (2006) 585

[100] A. J. Dinardo et al., *J. Appl. Phys.* 42 (1971) 186

[101] R. Pöpel, *IEEE Trans. Micr. Theory Tech.* 31 (1983) 600

[102] A. Andreone et al., *J. Appl. Phys.* 73 (1993) 4500

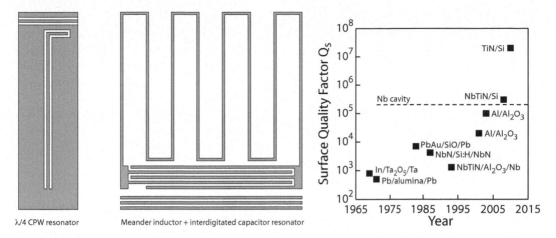

λ/4 CPW resonator   Meander inductor + interdigitated capacitor resonator

FIGURE 8.22: Far left: A $\lambda/4$ coplanar-waveguide (CPW) shunt-coupled transmission line resonator. The CPW feedline used to excite the resonator is visible at the top. Left center: A lumped-element resonator consisting of a meandered inductor and an interdigitated capacitor. The CPW feedline visible at the bottom has finite-width ground strips and couples to the interdigitated capacitor. Right: This plot shows lower limits to the surface impedance quality factor $Q_s = X_s/R_s = Q_\sigma/\gamma$ derived from measurements of superconducting microresonators. The lower limits are derived by assigning the total measured resonator loss to the superconductor. Points labeled by three materials are microstrip structures. Points labeled by two materials indicate resonators made from a single superconducting film deposited on a crystalline substrate. The dashed line indicates typical results achieved with bulk Nb cavities. The points plotted are limited to the work discussed in Section 8.5.3 and do not represent a comprehensive literature search. Nonetheless, the graph serves to illustrate the general trend.

Recent Work

Starting in 2000, the possibility of using superconducting microresonators as detectors stimulated exploratory measurements at Caltech and JPL. Initially this work was focused on thin-film microstrip resonators using materials such as Nb and Al using evaporated SiO as the dielectric, and resonator quality factors around 5×10^4 were measured (see Figure 8.24). However, these resonators did not follow the Mattis-Bardeen predictions but rather showed complex behavior as a function of temperature and microwave power. This behavior was not understood at the time; the work of Martinis et al. (2005) later showed that two-level systems in the amorphous thin-film dielectric were responsible for such effects.[103] In an effort to simplify the situation, the Caltech/JPL group switched to resonators using coplanar waveguide lines (see Figures 8.21 and 8.22), made from a single superconducting film deposited on a high-quality crystalline substrate such as silicon or sapphire. In some cases, very high quality factors were achieved,[104] around $Q_r = 2 \times 10^6$, indicating a surface impedance quality factor of at least $Q_s \sim 10^5$. These results have now been reproduced and extended by a number of other groups.[105]

In 2008, Barends et al.[106] published measurements of a CPW resonator made using NbTiN deposited on a silicon substrate. The 300 nm thick NbTiN film had $T_c = 14.8\,\mathrm{K}$ and $\rho_n = 170\,\mu\Omega\,\mathrm{cm}$,

[103] J. M. Martinis et al., *Phys. Rev. Lett.* 95 (2005) 210503

[104] B. A. Mazin et al., *AIP Conf. Proc.* 605 (2002) 309; P. K. Day et al., *Nature* 425 (2003) 817

[105] D. S. Wisbey et al., J. Appl. Phys. 108 (2010) 093918; J. M. Sage et al., ArXiV eprint 1010.6063 (2010).

[106] R. Barends et al., *Appl. Phys. Lett.* 92 (2008) 223502

giving a kinetic inductance fraction of $\alpha_{CPW} = 0.35$. The measured quality factor was $Q_r = 6 \times 10^5$, indicating a lower limit to the surface impedance quality factor of $Q_s = 3 \times 10^5$. This result was surprisingly high, over two orders of magnitude improvement compared to the measurements of Andreone et al. (1993), and comparable to or even somewhat better than the results for niobium cavities and aluminum CPW resonators. Even higher values of Q_s were established in 2010 by Leduc et al.[107] using TiN films patterned into lumped-element resonators[108] of the type shown in figure 8.22. Stoichiometric TiN films deposited on silicon substrates gave $T_c = 4.5\,\text{K}$, $\rho_n = 100\,\mu\Omega\,\text{cm}$, and a kinetic inductance fraction $\alpha = 0.74$. The measured internal quality factor $Q_i = 3 \times 10^7$ indicates a lower limit of $Q_s = 2 \times 10^7$. TiN resonators with similarly high internal quality factors have now been reproduced in other laboratories.[109]

Radiation Loss

As illustrated in Figure 8.22, the performance of superconducting microresonators has improved dramatically over the past four decades. Resonator quality factors above 10^6 are now routinely achieved using single-layer structures deposited on high-quality, low-loss crystalline substrates. Achieving high quality factors requires minimizing all potential sources of dissipation including radiation loss. Low radiation loss is possible using small, micron-scale feature sizes easily achieved with modern lithography. Indeed, a rough estimate is obtained by considering a $50\,\Omega$, half-wave CPW resonator on a semi-infinite substrate with dielectric constant $\epsilon_r = 10$. If the kinetic inductance fraction is small so that the effect on the phase velocity may be neglected, this structure has a radiation quality factor of[110] $Q_{rad} \approx 5 \times 10^{-3} \,(\lambda_0/w)^2$. Here λ_0 is the free-space wavelength at the resonance frequency and w is the center strip width. For $Q_{rad} > 10^6$, we require $w/\lambda_0 < 7 \times 10^{-5}$, or $w < 2\,\mu\text{m}$ at 10 GHz. Radiation loss may be reduced by using lower frequencies, minimizing the size of the resonator structure through the use of high kinetic inductance materials such as TiN or NbTiN, by adopting a lumped-element resonator design as shown in Figure 8.22, or all of the above.

Dissipation from Two-Level Systems

In practice, the maximum internal quality factor Q_i is often not limited by the superconducting material or radiation but instead by dissipation due to two-level systems (TLS) in amorphous dielectrics. The prevalence of TLS in amorphous materials was proposed four decades ago as a way to explain the anomalous bulk properties (e.g., heat capacity) of these materials at low temperatures.[111] TLS arise due to the random structure of amorphous materials, since occasionally it is possible for an atom or group of atoms to move between two local minima of the potential energy landscape by quantum tunneling over a barrier. The random nature of the amorphous material implies that the potential energy minima and the barrier height are also random, leading to a random, uniform distribution of TLS energy splittings.

Thermal occupation of the upper state causes the TLS loss tangent to vary as $\delta_{TLS}(\omega, T) = \delta_0 \tanh(\hbar\omega/2k_B T)$, where δ_0 is the $T \to 0$ limiting value. Therefore a simple way to reduce TLS loss is to operate at low frequencies, $\omega \ll k_B T/\hbar$, although dissipation due to TLS relaxation[112] sets in below ~ 0.3 MHz $(T/1\ \text{K})^3$. A second method to reduce loss is to use microwave fields that are strong enough to saturate the TLS.[113] Neither of these two methods is applicable if the resonator is

[107] H. G. Leduc et al., *Appl. Phys. Lett.* 97 (2010) 102509

[108] S. Doyle et al., *J. Low Temp. Phys.* 151 (2008) 530

[109] M. R. Vissers et al., *Appl. Phys. Lett.* 97 (2010) 232509

[110] A. Vayonakis and J. Zmuidzinas, unpublished (2002); B. Mazin, Ph.D. thesis, Caltech, Pasadena CA (2003), http://thesis.library.caltech.edu/3910/

[111] P. W. Anderson et al., *Phil. Mag.* 25 (1972) 1; W. A. Phillips, *J. Low Temp. Phys.* 351 (1972) 7

[112] G. Frossati et al., *J. Phys.* C. 10 (1977) L515

[113] M. Von Schickfus and S. Hunklinger, *Phys. Lett.* 64A(1977) 144

intended to operate in the quantum-mechanical regime. A third method to reduce TLS loss is to find materials that have a lower TLS density. Martinis et al.[103] showed that SiN_x ($\delta_0 \sim 2 \times 10^{-4}$) could have considerably lower loss than SiO_2 ($\delta_0 \sim 2 \times 10^{-3}$). More recent work[114] has shown that amorphous silicon and silicon-rich SiN_x films deposited using the inductively-coupled plasma chemical vapor deposition technique can achieve $\delta_0 \sim 2 \times 10^{-5}$. A final method for reducing TLS loss is to avoid use of amorphous dielectric films. Indeed, this was the route that led to high-Q CPW resonators, which are fabricated on crystalline substrates. However, even these resonators do have some TLS loss, arising from a thin (few nm) layer on the surface of the device.[115] This layer could either be surface oxides or an adsorbed layer. The dissipation caused by this surface layer may be reduced by increasing the separation of the electrodes in the capacitive portion of the resonator because this reduces the fraction of the electric field energy that is contained in the lossy TLS layer. Finally, parallel-plate capacitors with crystalline silicon dielectrics have recently been shown to have very low microwave dissipation.[116]

8.5.4 Superconducting Microresonator Detectors

The demonstration of superconducting microresonators with very high quality factors has opened up numerous new possibilities for superconducting detectors. A wide variety of schemes has been proposed and considered; the common denominator is the use of an array of microresonators spaced in frequency to allow multiplexed readout. This approach provides a simple, elegant solution to the long-standing readout problem that has impeded development of large arrays of superconducting detectors. Furthermore, wideband frequency multiplexing has become eminently practical given the advances in high-speed digital signal processing that allow a large number of carrier frequencies to be readily generated and measured by standard room-temperature electronics.[117] The simplest scheme is to use the resonator itself as the detector. The absorption of photons in the superconductor causes Cooper pair breaking and quasiparticle production, and this leads to a perturbation $\delta\sigma = \delta\sigma_1 - j\delta\sigma_2$ of the complex conductivity; to first order, $\delta\sigma$ is proportional to the change in quasiparticle density δn_{qp}. This leads to a perturbation of the complex surface impedance δZ_s, given by Eq. 8.41, which may be sensed by measuring the changes δf_r in the resonator frequency and δQ_r^{-1} in the resonator dissipation. This approach, now commonly known as the microwave kinetic inductance detector or MKID, will be covered in detail below. However, we will first review a number of important precursors that led to the development of MKIDs.

Precursors

As mentioned earlier, superconductors have been used for detection for over seven decades. The first devices were bolometers[118] operated on the resistive transition at $T = T_c$. The possibility of a superconducting detector operating at $T \ll T_c$ was first suggested by Burstein et al. in 1961[119], who proposed the use of a superconducting tunnel junction to measure the quasiparticles produced by the absorption of energy capable of breaking Cooper pairs. The development of tunneling detectors and related ideas is detailed in Section 8.3.

Another interesting suggestion for $T < T_c$ operation was McDonald's proposal to use the temperature-dependent kinetic inductance for bolometer readout.[120] This device can be understood

---

[114] A. D. O'Connell et al., *Appl. Phys. Lett.* 92 (2008) 112903; H. Paik and K. D. Osborn, *Appl. Phys. Lett.* 96 (2010) 072505

[115] J. Gao et al., *Appl. Phys. Lett.* 92 (2008) 152505

[116] S. Weber et al., arXiv:1102.2917v1 (2011)

[117] B. A. Mazin et al., *Nucl. Instrum. Meth.* A 559 (2006) 799; S. J. C. Yates et al., *Appl. Phys. Lett.* 95 (2009) 042504

[118] P. L. Richards, *J. Appl. Phys.* 76 (1994) 1

[119] E. Burstein et al., *Phys. Rev. Lett.* 6 (1961) 92

[120] D. G. McDonald, *Appl. Phys. Lett.* 50 (1987) 775

by writing

$$\frac{\sigma_2(\omega, T)}{\sigma_n} \approx \frac{\pi \Delta(T)}{\hbar \omega}[1 - 2f(\Delta(T))] \qquad (8.44)$$

which holds for $\hbar\omega \ll \Delta(T)$ and shows that the temperature dependence of the gap energy $\Delta(T)$ is responsible for the variation of inductance with temperature. McDonald proposed that the changes in kinetic inductance could be read out using a SQUID monitoring a bridge circuit excited by a ~ 100 kHz current. At temperatures far below the transition, $T \ll T_c$, the temperature variation of the kinetic inductance becomes exponentially small, $\propto e^{-\Delta_0/k_B T}$. However, the kinetic inductance continues to respond in a linear fashion to nonequilibrium changes in the quasiparticle population even as $T \to 0$. The presence of the distribution function $f(\Delta)$ in Eq. 8.44 is a clear indication of this, although the BCS gap equation shows that Δ also responds linearly to $f(E)$. This point was understood by Bluzer[121], who proposed a nonequilibrium detector that used SQUIDs to measure the nonequilibrium changes in kinetic inductance produced by pair-breaking photons. A key aspect of such a device is that the fundamental noise limit is set by the generation-recombination fluctuations of the quasiparticle population, and the noise produced by thermal quasiparticles vanishes exponentially with decreasing temperature.[122] The dissipative component of the conductivity (σ_1) also responds linearly to nonequilibrium quasiparticles, as is evident from the Mattis-Bardeen integral; the effect is analogous to the current response of a tunnel junction detector, as was pointed out by Gulian and Van Vechten.[123] They suggested that X-rays could be detected by illuminating a superconducting film with microwaves and measuring small changes in the reflected power, and discussed the possibility that this response might be enhanced through a positive feedback mechanism involving microwave amplification of the photoproduced quasiparticles.

Clearly, the detector sensitivity is controlled by the ability to measure very small changes in the conductivity $\delta\sigma$. Furthermore, minimizing the superconducting volume is desirable in order to maximize the perturbation $\delta\sigma$ that occurs in response to a fixed energy deposition δE or a change δP in the radiation power absorbed. Both of these points suggest use of superconducting microresonators, and in fact this idea was described by Michael D. Jack in a 1990 patent assigned to the Santa Barbara Research Center.[124] To quote from the patent: "It has been found that Cooper pair breaking by incident photons results in a change in the kinetic inductance and a consequent shift in the resonance frequency of the superconducting transmission line." The patent goes on to describe a readout technique involving use of Schottky-diode detectors to measure the changes in microwave power transmitted through a resonator resulting from the frequency shift, as well as a multiplexing scheme involving coupling each resonator to the readout diode sequentially by means of superconducting critical-current switches. While the patent is an early recognition of the promise of using superconducting microresonators for detection, the proposed readout scheme is complex and clearly not ideal; the electronics required for simultaneous readout of a frequency-multiplexed MKID array at noise levels approaching fundamental limits was not yet on the horizon in 1990.

Microwave Kinetic Inductance Detectors (MKIDs)

The MKID concept was developed by J. Zmuidzinas and H. G. Leduc in 1999, who were motivated by the previous work of Bluzer[121] and Sergeev and Reizer.[122] The distinguishing features included the use of superconducting microresonators, a homodyne readout method using a wideband, low-noise cryogenic amplifier and room-temperature electronics that could sense either the amplitude or phase shift of the microwave transmission, and multicarrier frequency-domain multi-

---

[121] N. Bluzer, *J. Appl. Phys.* 78 (1995) 7340

[122] A. V. Sergeev and M. Y. Reizer, Intl. *J. Mod. Phys.* B, 10 (1996) 635

[123] A. M. Gulian and D. Van Vechten, *Appl. Phys. Lett.* 67 (1995) 2560 ; D. Van Vechten et al., *Nucl. Instrum. Meth.* A 370 (1996) 34

[124] M. D. Jack, U.S. Patent 4,962,316, Oct. 9, 1990

plexing. Figure 8.23 depicts the idea as originally conceived. The method of radiation coupling is not shown; the intention was to use an antenna-fed microstrip line to couple submillimeter-wave radiation to the inductor at its center, taking advantage of the microwave virtual ground at that point.[125] Around this time, frequency-domain multiplexing of superconducting detectors was being discussed in other contexts, e.g., in conjunction with radio-frequency single electron transistor (RF-SET) amplifiers following tunnel-junction detectors,[126] or as a way of multiplexing transition-edge bolometers.[127] In all of these cases, the resonators were elements to be added to the scheme in order to allow multiplexing; in the MKID case, the resonators simultaneously served as detectors, a drastic simplification.

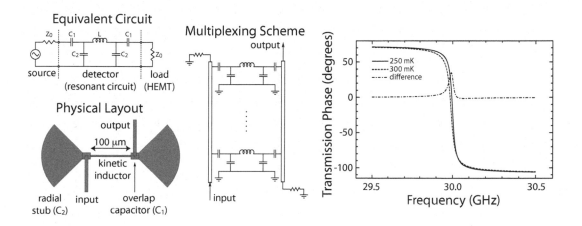

FIGURE 8.23: Description of the MKID concept as proposed by Zmuidzinas and Leduc in 1999. Left: An equivalent circuit diagram and a potential physical realization in microstrip. Center: Frequency-multiplexed readout of an array of MKIDs using two feedlines. Right: Phase response of the proposed circuit, calculated using the Mattis-Bardeen theory.

In early 2000, measurements of Nb/SiO/Nb microstrip resonators were performed in order to verify that the concept was feasible (see Figure 8.24). As mentioned earlier, the microstrip experiments displayed a number of puzzling effects, such as the variation of the resonator quality factor Q with readout power, and half a decade would pass before this behavior was connected to saturation of two-level systems in the amorphous SiO dielectric. Meanwhile, Peter Day suggested switching to CPW resonators (see Figure 8.21) in order to simplify the structure, and by mid-2001[104] this led to the demonstration of resonators with quality factors around $Q_r \sim 10^6$ (see Figure 8.25). Detection experiments using 6 keV X-rays quickly followed, leading by 2003 to the first journal paper describing and demonstrating the concept.[104] Since then, considerable effort has been expended in understanding the frequency noise of these devices, which was shown to be due to TLS-induced capacitance fluctuations.[128] Quasiparticle relaxation is another important issue that has received significant attention.[129]

---

[125] J. Zmuidzinas et al., *IEEE Trans. Microwave Theory Tech.* 42 (2004) 698

[126] R. J. Schoelkopf et al., *IEEE Trans. Appl. Supercond.* 9 (1999) 2935; T. R. Stevenson et al., *Appl. Phys. Lett.* 80 (2002) 3012

[127] J. Yoon et al., *Appl. Phys. Lett.* 78 (2001) 371

[128] J. Gao et al., *Appl. Phys. Lett.* 90 (2007) 102507; S. Kumar et al., *Appl. Phys. Lett.* 92 (2008) 123503; J. Gao et al., *Appl. Phys. Let.* 92 (2008) 212504

[129] R. Barends et al., *Phys. Rev. Lett.* 100 (2008) 257002; R. Barends et al., *Phys. Rev. B* 79 (2009) 020509

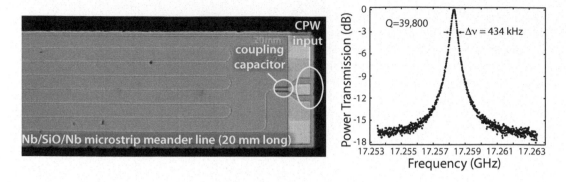

FIGURE 8.24: Left: Photograph of a half-wave Nb/SiO/Nb microstrip resonator measured in early 2000 to demonstrate MKID feasibility. Right: Measurement results, showing $Q_r = 4 \times 10^4$ at 17 GHz. Credit: A. Vayonakis and H. G. Leduc.

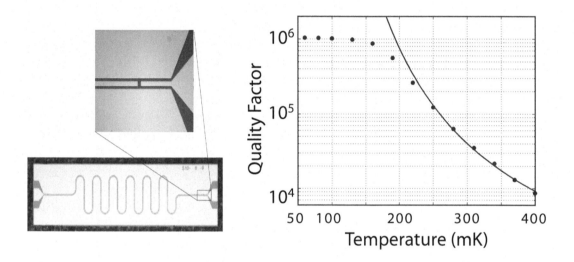

FIGURE 8.25: Left: Photograph of a JPL/Caltech half-wave CPW resonator made using an aluminum film deposited on a sapphire substrate. Right: Measurements obtained in early 2002 of the quality factor vs. temperature, showing $Q_r = 10^6$. The curve shows the Mattis-Bardeen prediction. Credit: P. K. Day, B. A. Mazin, and H. G. Leduc.

MKID Examples

One of the key issues for MKID design is is the method used to couple the photons to be detected into the resonator. Early experiments avoided this issue by flood-illuminating the entire chip with X-rays and measuring the pulse response. Figure 8.26 shows the first attempt to achieve efficient coupling in the millimeter-wave band by using slot-array antennas feeding CPW MKID resonators.[130] In later versions of these devices, the capacitive portion of the CPW resonator has been replaced by an interdigitated capacitor in order to reduce the TLS noise.[131] A multiband millimeter-wave astro-

[130] S. Kumar, Ph.D. thesis, Caltech, Pasadena CA (2008), http://thesis.library.caltech.edu/1663/; J. Schlaerth et al., *J. Low Temp. Phys.* 151 (2008) 684

[131] O. Noroozian et al., *AIP Conf. Proc.* 1185 (2009) 148

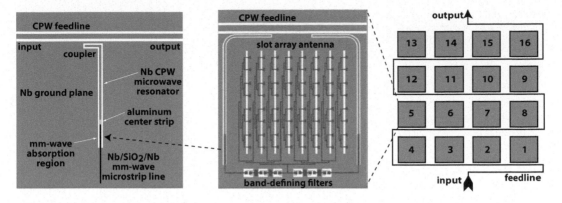

FIGURE 8.26: This figure illustrates the antenna-coupled MKID arrays developed at Caltech/JPL. Left: a $\lambda/4$ CPW resonator is shunt-coupled to a CPW feedline. Millimeter-wave radiation is brought to the resonator using a niobium microstrip line and is absorbed by the aluminum section of the CPW resonator's center strip. Center: A phased-array multislot antenna, two CPW MKIDs, and on-chip band-defining filters are used to make a dual-band pixel. Right: A 4×4 pixel array is multiplexed using a single feedline.

nomical camera using this array design is now being prepared for use on the Caltech Submillimeter Observatory.[132]

A much simpler and very clever solution to this problem was the lumped-element resonator suggested by Doyle et al.[108] and shown in Figure 8.22, in which the inductor simultaneously serves as the radiation absorber. Efficient radiation absorption requires using either very thin films of ordinary superconductors or highly resistive superconductors such as TiN; a very convenient feature of using TiN films is that T_c may be easily adjusted by changing the stoichiometry.[107] Designing for high absorption efficiency simultaneously guarantees a high kinetic-inductance fraction α; this can be understood through the Mattis-Bardeen relation between the surface inductance and surface resistance, $L_s = \hbar R_s / \pi \Delta_0$. The lumped-element pixel design is being used for the $\lambda = 2$ mm band in the "NIKA" camera built for the Institute for Millimetric Radio Astronomy (IRAM) 30-m telescope.[133] Microwave crosstalk between pixels is a serious problem with this design; however, it is possible to produce large arrays with negligible crosstalk levels by modifying the inductor design using opposite-polarity conductor pairs in close proximity to minimize the microwave dipole moment (see Figure 8.27). In addition, efficient dual-polarization absorption may be obtained using appropriate metallization patterns. A variety of lumped-element, TiN-based MKIDs are now being developed for a wide range of applications including photon detection at millimeter, submillimeter, far-IR, UV/optical, and X-ray wavelengths as well as dark matter detection.

Superconducting Microresonator Bolometers

Given the considerable progress that has been made in the performance and physical understanding of microresonators, as well as the advances in digital frequency-multiplexed readout electronics, McDonald's kinetic-inductance bolometer concept[120] is worth revisiting. Superconducting microresonators have been successfully fabricated on thermally suspended silicon nitride micromesh,[134] so a frequency-multiplexed array of bolometers is straightforward to produce. In a

[132] P. R. Maloney et al., *Proc. SPIE* 7741 (2010)

[133] A. Monfardini et al., arXiv:1102.0870v2 (2011).

[134] H. G. Leduc and P. K. Day 2010, personal communication.

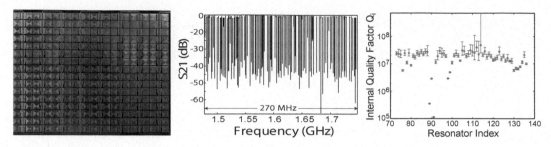

FIGURE 8.27:   Left: a 16×16 array of TiN lumped-element MKIDs using a spiral inductor and an interdigitated capacitor. The pixels are approximately 0.6 mm in size. Center: The measured resonances for half the array (128 resonators) are very deep and fall into a 270 MHz band centered at 1.6 GHz. Right: the internal quality factors are uniformly high, $Q_i > 10^7$. Credit: O. Noroozian, P. K. Day, B.-H. Eom, and H. G. Leduc.

standard MKID, quasiparticle recombination provides the bottleneck for power flow, whereas in a bolometer, the bottleneck is set by the geometry of the thermal suspension legs. Therefore, the resonator-bolometer is an interesting option when quasiparticle recombination is too rapid, e.g., at higher temperatures. In addition, the thermally suspended island provides an opportunity to spread the absorbed photon energy uniformly across the inductor, which maximizes responsivity. In contrast, this is not easy to achieve in antenna-coupled MKIDs. In addition, the quasiparticle recombination noise may be greatly reduced. Finally, the noise performance of available transistor amplifiers is considerably better than is needed to reach the fundamental sensitivity limits for bolometers (e.g., photon or phonon noise).

TES Bolometer Multiplexing

Superconducting microresonators may also be used for frequency-multiplexed readout of superconducting transition edge sensor (TES) bolometers. In this scheme, the TES current is sent through a planar coil which couples flux into a single-junction SQUID. Changes in the TES current therefore cause changes in the SQUID inductance, which is sensed by embedding the SQUID into a microresonator.[135]

8.5.5   Parametric Amplifiers

The fact that superconductors exhibit nonlinear behavior at microwave frequencies was already known in 1950 and was discussed theoretically in the context of the BCS theory by Parmenter in 1962.[136] To lowest order, the kinetic inductance varies with current as $L = L_0(1 + I^2/I_*^2)$, where I_* is related to the critical current. That reactive nonlinearities can be used for low-noise amplification has been known for at least a century; the use of the kinetic-inductance nonlinearity for this purpose was suggested and patented by Landauer.[137] Shortly thereafter, amplification was demonstrated using an ultra-thin superconducting film coupled with a waveguide-fed rutile resonator; subsequent work called into question whether the kinetic inductance or intergrain weak links provided the relevant nonlinearity.[138] This is a key issue, because low-noise parametric amplification requires that the nonlinear reactance have low dissipation. The first clear-cut demonstration of parametric

---

[135] J. A. B. Mates et al., *Appl. Phys. Lett.* 92 (2008) 023514

[136] A. B. Pippard, *Proc. Roy. Soc.* A, 203(1950) 210; R. H. Parmenter, *RCA Review,* 23 (1962) 352

[137] R. Landauer, *Proc. IRE,* 48 (1960) 1328 ; U.S. Patent Number 3,111,628, Nov 1963.

[138] A. S. Clorfeine, *Proc. IEEE* 52, 844 (1964); H. Zimmer, *Appl. Phys. Lett.* 10 (1967) 193

amplification using the kinetic inductance nonlinearity has been given only quite recently,[139] using niobium CPW resonators similar to the one shown in Figure 8.25. An alternative approach is to use a series array of Josephson junctions or SQUIDs to provide the reactive nonlinearity for a resonator.[140] One advantage of this technique is that the resonance frequency may be readily tuned through the application of a magnetic field.

8.5.6   Other Applications

In recent years, the most visible application of superconducting microresonators has been in experiments exploring the prospects for superconducting quantum computing. Superconducting microresonators are a key component for these experiments, and their dissipation must be kept very low in order to prevent loss of quantum coherence, so the use of CPW resonators has had a large impact. The first such experiment to incorporate CPW resonators was performed by the Schoelkopf group at Yale.[141] In this experiment, a Cooper pair box (CPB) acting as a two-level atom was fabricated next to a CPW resonator. In essence, this is a circuit version of a single atom in an optical cavity. The use of a superconducting microresonator allowed the strong-coupling regime to be achieved, in which the vacuum Rabi frequency is considerably larger than the relaxation rates of the CPB and resonator. This experiment received considerable attention, and led to the adoption of superconducting microresonators in a number of other low-temperature experiments. One particularly interesting example is the use of superconducting microresonators to probe the motion of nanomechanical resonators into the quantum regime.[142] These topics are discussed in move detail in Section 8.4.

8.5.7   Summary

Driven by a number of important applications, the study of superconducting microresonators has blossomed over the past decade. A deeper understanding of the physics governing microresonator dissipation and noise has led to considerably improved resonator designs. However, the performance of the best resonators is still not limited by the superconducting material but rather by extrinsic effects such as two-level systems in amorphous dielectrics. Therefore, one can expect resonator performance to continue to improve as methods for avoiding, reducing or eliminating these extrinsic effects are developed.

8.6   Further Reading

Applications of Passive Microwave Filters and Devices in Communication And Related Systems

1. Michael J. Lancaster, *Passive Microwave Device Applications of High Temperature Superconductors*, Cambridge University Press, Cambridge, 1997.

2. M. Nisenoff and H. Weinstock, eds., Microwave Superconductivity, NATO Science Series, Kluwer Dordrecht, The Netherlands, 2001.

---

[139] E. A. Tholén et al., *Appl. Phys. Lett.* 90 (2007) 253509

[140] M. A. Castellanos-Beltran et al., *Nature Physics* 4 (2008) 929

[141] A. Wallraff et al.,*Nature* 431 (2004) 162

[142] J. D. Teufel et al., *Nature Nanotechnology* 4, 820 (2009); J. B. Hertzberg et al., *Nature Physics* 6 (2010) 213

3. Raafat R. Mansour, Microwave Superconductivity, *IEEE Transactions on Microwave Theory and Techniques*, vol. 50, pp. 750-759, 2002.

4. Randy W. Simon et al., Superconducting microwave filter systems for cellular telephone Base Stations, *Proceedings of IEEE*, vol. 92, pp.1585-1596, 2004.

Microwave Cooling of Superconducting Quantum Systems

1. For a recent review of superconducting qubits:

   (a) J. Clarke and F.K. Wilhelm, Superconducting quantum bits, *Nature* 453 (2008) 1031

   (b) See also in this book Chapter 6 on superconducting qubits.

2. To read more about the cooling presented here, as well as population inversion and related works, in multilevel superconducting systems:

   (a) S. O. Valenzuela, W. D. Oliver, D. M. Berns, K. K. Berggren, L. S. Levitov, T. P. Orlando, Microwave-induced cooling of a superconducting qubit, *Science* 314 (2006) 1589.

   (b) D.M. Berns, M.S. Rudner, S.O. Valenzuela, K. K. Berggren, W. D. Oliver, L. S. Levitov, T. P. Orlando, Amplitude spectroscopy of a solid-state artificial atom, *Nature* 455 (2008) 51.

   (c) D.M. Berns, W.D. Oliver, S.O. Valenzuela, A.V. Shytov, K.K. Berggren, L.S. Levitov, T.P. Orlando, Coherent quasiclassical dynamics of a persistent current qubit, *Phys. Rev. Lett.* 97 (2006) 150502.

   (d) W. D. Oliver, Y. Yu, J. C. Lee, K. K. Berggren, L. S. Levitov, T. P. Orlando, Mach-Zehnder interferometry in a strongly driven superconducting qubit, *Science* 310 (2005) 1653.

3. To read more about cooling a superconducting qubit coupled to a superconducting resonator:

   (a) J.Q. You, Yu-xi Liu, and Franco Nori, Simultaneous cooling of an artificial atom and its neighboring quantum system, *Phys. Rev. Lett.* 100 (2008) 047001.

   (b) M. Grajcar, S.H.W. van der Ploeg, A. Izmalkov, E. Il'ichev, H.-G. Meyer, A. Fedorov, A. Shnirman, and G. Schön, Sisyphus cooling and amplification by a superconducting qubit, *Nature Phys.* 4 (2008) 612.

4. To read more about using these kinds of cooling techniques used to cool other systems:

   (a) A. Naik, O. Buu, M.D. LaHaye, A.A. Clerk, M.P. Blencowe, and K.C. Schwab, Cooling a nanomechanical resonator with quantum back-action, *Nature* 443 (2006) 193.

   (b) A.O. Niskanen, Y. Nakamura, and J.P. Pekola, Information entropic superconducting microcooler, *Phys. Rev. B* 76 (2007) 174523.

   (c) A.V. Timofeev, M. Helle, M. Meschke, M. Mottonen, J.P. Pekola, Electronic refrigeration in the quantum limit, *Phys. Rev. Lett.* 102 (2009) 200801.

   (d) S. Kafanov, A. Kemppinen, Yu. A. Pashkin, M. Meschke, J.S. Tsai, J.P. Pekola, Single-electronic radio-frequency refrigerator, *Phys. Rev. Lett.* 103 (2009) 120801.

   (e) T. Rocheleau, T. Ndukum, C. Macklin, J.B. Hertzberg, A.A. Clerk, and K.C. Schwab, Preparation and detection of a mechanical resonator near the ground state of motion, *Nature* 463 (2010) 72.

Acknowledgements

Microwave Measurements of Fundamental Properties of Superconductors

The work at Lincoln Laboratory cited in this chapter was supported by the Air Force Office of Scientific Research, the Office of Naval Research, the Department of the Air Force, and the Defense Advanced Research Projects Agency.

Superconducting Quantum Electronics Enabling Astronomical Observations

I thank Pieter de Visser, Dan Oates and Tom Phillips for a critical reading of the manuscript. I also acknowledge the collaboration on this subject with Thijs de Graauw over a period of about 25 years.

9

Quantum Metrology

Editors: Richard E. Harris and Jürgen Niemeyer

9.1 Introduction
   Richard E. Harris and Jürgen Niemeyer 515
9.2 Josephson Voltage Standard — The Ultimate Precision
   Jaw-Shen Tsai and James E. Lukens 518
9.3 The First Josephson Voltage Standards
   Thomas J. Witt 522
9.4 First Josephson Junction Array Voltage Standard
   Tadashi Endo 525
9.5 How the DC Array Standards Were Developed
   Jürgen Niemeyer 528
9.6 Making the Josephson Voltage Standard Practical
   Clark A. Hamilton 535
9.7 Programmable Josephson Voltage Standards: from DC to AC
   Johannes Kohlmann 540
9.8 Quantum-Based Voltage Waveform Synthesis
   Samuel P. Benz 546
9.9 Superconductivity and the SI (Metric) System Based on Fundamental Constants
   Edwin Williams and Ian Robinson 553
9.10 Further Reading 557

9.1 Introduction

Richard E. Harris and Jürgen Niemeyer

As thorough readers of this book have begun to discover by this chapter on metrology, the most widely used applications of superconducting electronics involve measurements of electromagnetic phenomena. Unlike most other phenomena used for measurements, superconductivity is also essential for electrical *standards*. Because it is a quantum phenomenon, it can represent voltage in terms of fundamental constants with an amazing precision of a few parts in 10^{19}. In some regards superconducting voltage standards based on arrays of Josephson junctions are the most successful application of superconductivity, considering the numbers of junctions and their worldwide use. While they are conceptually simple, it has taken almost 50 years for their underlying technology to progress from unreliable single junctions to robust, complex, three-dimensional integrated circuits having more than 300,000 junctions and producing precision ac voltages up to 10 V. Here is their story.

As described in other chapters of this book, centimeter-sized superconducting loops can carry a constant current that remains undiminished for many months or years with no sign of decline. This is a tangible manifestation of a macroscopic quantum state like only a few others in modern physics.

Within the wave function of the superconductor lies the key to using superconductors as voltage standards having incredible precision. While a macroscopic superconductor usually excludes magnetic fields, applying a large enough magnetic field causes magnetic flux to penetrate the superconductor in the form of flux bundles (Type I superconductors) or even tiny flux quanta (Type II superconductors) by destroying the superconductivity in localized regions. These normal regions co-exist with the surrounding superconductor. However as one traces a closed path through the superconductor, and around a normal region, the phase of its wave function must return to the same value plus perhaps a multiple of 2π. If the multiple is one, the normal region contains one flux quantum, Φ_0 or $h/2e$, where h is Planck's constant and $2e$ is the charge of each pair of electrons in the superconductor.

What we find is a remarkable, naturally occurring system of tiny, precisely identical magnets — the flux quanta. If we can manipulate them, we can for example induce voltages in an electrical circuit. The voltage pulse V induced by n flux quanta is

$$V = \frac{d\Phi}{dt},$$

where Φ is the flux inserted into the circuit, and the time-integral of the voltage across the circuit is

$$\int V(t)dt = n \cdot \frac{h}{2e}.$$

The flux quantum $h/2e$ has the value $2.067833667 \times 10^{-15}\mathrm{V \cdot s}$.[1]

The flux quanta can be coerced into behaving just like a microscopic electrical generator. The difficulty lies in learning how to controllably manipulate them. The method that has been used is to interrupt a superconducting circuit with a weakened region. A flux quantum can be pushed across the weakened region by putting a current, having just the right amplitude, through the weakened region.[2]

Most practical devices now use lithographed Josephson junctions, but other weakened regions have also been used, such as point contacts and lithographed weak links. Sometimes the junctions are interrupted by a normal metal barrier and sometimes by an insulating one.

Even in his original announcement of his discovery of the Josephson effect in 1962, Brian Josephson anticipated voltage standards.[3] But as early experimentalists quickly discovered, a single flux quantum produces a very small voltage pulse. Using a current to force through one flux quantum with each oscillation of a 10 GHz ac current — that is, 10 billion exactly quantized voltage pulses per second — produces a time-averaged dc voltage of only about 20 μV, a value far too small to be useful directly in calibrating typical electrical devices that operate at multiples of one volt.

In the earliest voltage standards, metrologists took advantage of constant voltage steps in the current-voltage characteristics of Josephson junctions that occurred at integral multiples n of $hf/2e$ when driven by an ac current of frequency f. The steps occur when the microwave drive pushes through n quanta in a single period. Looked at a bit more closely, a change in the bias current at the nth step and the nonlinearities in the junction generate the nth harmonic of the microwave supply. Each cycle of that harmonic drove one flux quantum through the junction, giving a total of n

[1] P. J. Mohr, B. N. Taylor, and D. B. Newell, *Rev. Mod. Phys.* 80 (2008) 633. It is illustrative of the care with which metrologists work that the uncertainty in this value is only $0.000000052 \times 10^{-15}\mathrm{V \cdot s}$. How this uncertainty may change in meaning and value over the second hundred years of superconductivity is difficult to imagine.

[2] Using flux quanta to explain the generation of quantized voltage has not been common in metrology, although Anderson used it to explain the voltage across bulk superconductors. Instead, metrologists have used the simple solution of nonlinear differential equations because they are quantitative and calculable even though perhaps a less intuitive explanation.

[3] B. D. Josephson, *Phys. Lett.* 1 (1962) 251

flux quanta in one cycle of the microwave source. The largest multiple of n that could reliably be achieved was about 250 flux quanta per oscillation of the microwaves. This produced a quantized voltage of about 5 mV. Even with this convenient multiplication of about 250, to compare this precision voltage to practical voltages often required the use of two junctions and a room-temperature bridge called a Hamon network. While single- or double-junction voltage standards of this sort were the state of the art for about a decade, the junctions often failed when cooled to liquid helium temperatures, or warmed back up, and the bridge was susceptible to errors.

It took only a high school education to feel tantalized by the prospect of directly creating higher precision voltages by putting many junctions in series. Unfortunately, that was not easy. It appeared that to create a one volt standard would require hundreds or thousands of junctions. At the time the arrays of junctions did not have sufficient uniformity to be biased by a single source of current. In an impressive effort in 1983, Endo and his colleagues in Japan succeeded in independently biasing 20 junctions in series to create 0.1 volt, as he relates in the next section of this chapter.

A major breakthrough came with lithographic fabrication and robust niobium trilayer junctions. For voltage standards, the levels of integration are still much lower than in semiconductor circuits, but have progressed from a one volt standard having 1500 junctions in 1985 to 300,000 junctions stacked three-high in 2010.

There are now three generations of standards based on arrays. The earliest produced only dc voltages but have progressed from 1 to 10 volt. For these standards, flux quanta were driven at precisely regular intervals through each junction in the array by use of a precision microwave source.

These dc standards were in routine use in standards laboratories beginning in 1985. Many are still functioning today. One that began at Hewlett-Packard's instrument division in Loveland, Colorado still functions in Indonesia in the relocated surviving company Agilent.

Because it was difficult to change the voltage of these arrays, clever people figured out how to produce arrays divided into segments having a binary distribution of numbers of junctions, so that sequentially turning on, and off, the right combination of segments could produce different voltages sufficiently fast for stepwise synthesis of frequencies as high as the kilohertz range. These "programmable" standards required new junction technologies using SINIS (superconductor-insulator-normal-insulator-superconductor junctions in Germany) and SNS (superconductor-normal-superconductor in the USA) to eliminate hysteresis and allow rapid controlled switching. Now these devices are being used for electric power standards.

Nevertheless quantum-based ac standards for synthesizing even higher frequencies required a new concept[4] that was not invented until 1995. In these standards, flux quanta are driven across each junction in a series array by the use of pulses of current. When the pulses recur rapidly, the voltage produced is high; when they recur slowly the voltage is lower. But the voltage is precisely proportional to the rate of pulsing. These standards are not yet in such widespread use as the dc and programmable ones, but as their technology advances, they offer the laboratory-demonstrated promise of a single technology for all quantum voltage standards.

All three types of standards are based on precisely manipulating the flow of flux quanta. Overcoming the technical challenges of superconducting integrated circuit fabrication, carefully designing their microwave properties, and developing easy-to-use room temperature controls has been research that spans fundamental science and practical engineering. The editors of this chapter were lucky enough to lead the two most successful laboratories in developing the technology, Harris at NIST Boulder[5] and Niemeyer at PTB Braunschweig[6]. Indeed the creation of the first one-volt stan-

---

[4] Ironically, it may be that delay in inventing ac standards had its roots in the common use of differential equations that obscured the intuitive physical ideas about flux motion that were essential to the invention of ac standards. Others believe that the delay resided in the technical difficulty of generating accurately timed pulses.

[5] NIST is the National Institute of Standards and Technology, USA. It was formerly named the National Bureau of Standards (NBS).

[6] PTB is the Physikalisch-Technische Bundesanstalt, Germany

dard was possible only because of an exceptionally successful collaboration between the two institutions that Niemeyer began by visiting NIST for six months in 1982. Unbeknownst to Niemeyer was that his determination to use NIST Boulder lithographic tools for voltage standards broke the institutional tradition that only NIST Gaithersburg worked on voltage standards. At NIST, one of the first contributions of the then-new array standard was to provide a method of verifying the accuracy of the Hamon network, and an error was found. Later NIST and PTB sometimes worked separately but collaborations recurred as new generations of voltage standards were created. Niemeyer's world travels also included ETL[7]. That laboratory also collaborated with NIST. Curiously there were striking differences among approaches to technology at NIST, PTB and ETL. NIST created fabrication technology cheaply, buying used equipment. ETL invested heavily in fabrication to participate in an attempt to construct a Josephson junction computer, although that facility was not initially available for voltage standards. PTB invested in a large clean room for fabrication in 1990. NIST will open a major fabrication facility facility in Boulder in 2012, twenty-two years later.

This section of *100 Years of Superconductivity* contains recollections of some of the pioneers in this technology. Shen Tsai and Jim Lukens tell about the incredible ultimate precision of voltage measurements. Tom Witt relates the experience of introducing the first standard measurement in national measurement institutes. Later in Japan the first arrays, independently biased, were demonstrated by Tadashi Endo. Jürgen Niemeyer describes creating the first singly biased, array as part of his close partnership between PTB and NIST in the US. Attention then turned to more widespread use of arrays as told by Clark Hamilton, a story of great international technological progress, but alas marginal commercial profitability. Subsequently programmable standards were developed as related by Johannes Kohlmann. Most recently Josephson arrays have been used in ac voltage standards and applied to a primary temperature standard that is elecctronically determined through Johnson noise thermometry, as related by Sam Benz. Finally, using large equipment one- to two-stories tall, to complement the tiny Josephson junctions, Ed Williams and Ian Robinson relate their efforts to use superconductivity and Josephson junctions to bring precision electrical measurements to the mechanical units.

This is a story of the second fifty years of superconductivity. It is a story of the technology of the ubiquitous semiconductor industry permeating the field of superconductivity and making remarkable progress possible. It leaves us barely able to fantasize what the next 50 and 100 years will bring.

9.2   Josephson Voltage Standard — The Ultimate Precision

Jaw-Shen Tsai and James E. Lukens

A voltage standard utilizing Josephson junction arrays has been adopted as an effective and convenient international standard for many years. Using specially designed cryogenic chips, laboratories around the world can reproduce voltage consistently with an accuracy of about few parts in a hundred million. The physics behind this standard is based on a very simple relation,

$$V = \frac{h}{2e} f, \tag{9.1}$$

between two macroscopic quantities: the average voltage V (electrochemical potential difference) across the junction and the frequency f of the Josephson oscillation. $h/2e$ is a quantum mechanical

---

[7]   ETL is the Electrotechnical Laboratory, Japan. It is now named AIST, National Institute of Advanced Industrial Science and Technology.

fundamental constant, where h is the Planck constant and e is the electron charge in vacuum. In other words, it equates the quantum energy hf and the classical electrostatic energy $2eV$, both associated with the Cooper-pair dynamics.

In the voltage standard, the Josephson oscillation of relation (9.1) is phase locked to an external microwave signal of frequency f, creating a Shapiro step[8] with voltage V (at the first harmonic). The approximate 10^{-8} accuracy of the voltage standard is limited by the uncertainty in the experimental determination of the quantum mechanical constant. However, there is a more fundamental question: If the fundamental constant is supposedly given with infinite accuracy, do two different Josephson junctions generate exactly the same voltage when they are driven by the same microwave frequency f, just as we expect all atoms of the same species to produce exactly the same energy eigenvalues?

To be sure, a Josephson junction is a complex macroscopic entity that consists of an enormous number of atoms, and it is possible to realize it by very different junction types (i.e., superconductor-insulator-superconductor tunnel (SIS) junctions, superconductor-normal conductor-superconductor (SNS) junctions, superconducting microbridges, and so on) and very different materials. Can such macroscopic solid state devices really behave just like microscopic quantum systems?

Theoretically, the Josephson voltage-frequency relation given above has no known corrections, which is actually quite an extraordinary statement considering the complexity of the Josephson junction. Various arguments based on different physics have pointed out the exactness of this relation, such as arguments based on the fundamental quantum mechanical consideration[9], Faraday's law[10], the exactness of gauge invariance[11], and the topological quantum number[12].

To verify experimentally the extent of the universality of the Josephson voltage-frequency relation, we have carried out a high precision comparison of Josephson voltages created in two different junctions.[13] Such a test has three-fold meaning. Scientifically, it is quite interesting to test how exactly such a macroscopic system behaves according to a quantum mechanical relation (Josephson relation). Technologically, such a test should provide a "confidence assessment" of the ultimate accuracy of the Josephson voltage standard. At the same time, in metrology, to establish a high precision measurement of a unique physical quantity is always a fascinating subject.

The experimental test was carried out by the use of a circuit set up as a dc superconducting quantum interference device (SQUID) that consists of two Josephson junctions to be compared, imbedded in a superconducting loop. Both Josephson junctions were biased by dc and ac (microwave frequency), and parked near the center of a certain harmonic of the Shapiro constant voltage step for both junctions. At Shapiro steps, the voltages across the junctions are $V = nhf/2e$, where n is the number of the harmonic.

We irradiated both junctions with the same microwave frequency f, generated from a frequency stabilized microwave source, typically at about 10 GHz. Now, suppose Eq. (9.1) is not exact, and at the two junctions, different electrochemical potential differences $V1$ and $V2$ resulted. Then the difference in these, $\Delta V = V1 - V2$, would produce a time varying flux in the superconducting loop, following Faraday's induction law:

$$\Delta V = \frac{d\Phi}{dt}.$$

By monitoring the flux, Φ, linking the SQUID, ΔV can then be detected. This method of comparison was first demonstrated by Clarke.[14] By adopting and improving this experimental scheme, we

8   S. Shapiro, *Phys. Rev. Lett.* 11 (1963) 80

9   F. Bloch, *Phys. Rev.* B2 (1970) 109

10 T. A. Fulton, *Phys. Rev.* B7 (1973) 981

11 D. G. McDonald, quoting P. W. Anderson, *Physics Today* (July 2001) 46

12 David J. Thouless, *Topological Quantum Number in Nonrelativistic Physics*, World Scientific, Singapore (1998)

13 J. S. Tsai, A. K. Jain, and J. E. Lukens, *Phys. Rev. Lett.* 51 (1983) 316

14 J. Clarke, *Phys. Rev. Lett.* 21 (1968) 1566

have succeeded in achieving an unprecedentedly high precision for Josephson voltage comparison. We monitored Φ using an rf SQUID, which was coupled to the system by a superconducting flux transformer.

The potential sensitivity of the voltage comparison scheme can be estimated as follows. When biased at a voltage step of $V = 1\ \mu V$, and assuming the integration time Δt of the flux to be 1 hour, and the flux noise of the SQUID to be $\Delta\Phi = 10^{-4}\Phi_0$ during this time interval (Φ_0 is the flux quantum), then the resolution of the voltage comparison,

$$\frac{\Delta V}{V} = \frac{\Delta\Phi}{\Delta t \cdot V},$$

will be of the order of 10^{-16}. This simplified estimation illustrates the powerful prospects for this experimental technique.

It is desirable to dc bias the two Josephson junctions at the highest possible Shapiro step (on the same harmonic for both junctions) in order to obtain the highest ΔV for a given $\Delta V/V$ ratio. To do so, it is important to make the properties of the two junctions, such as critical current and normal resistance, as symmetrical as possible. For the same reason, the inductance of two arms of the SQUID, L1 and L2, should be made as symmetrical as possible. In this way, the dc current bias would divide equally into two arms so that it is possible to bias the two junctions at the same (and as high as possible) harmonic order of the Shapiro step while reducing noise due to bias drift.

Another important factor to assure a high precision in the experiment is to secure a long integration time Δt. To do so, the most important factor is to maintain the temperature throughout the experiment as constant as possible, since temperature variations could result in variations in the I-V characteristics that would shift the dc bias point off the Shapiro step, terminating the experiment or, at a minimum, increase the flux noise. It could also cause thermoelectric effects that bring about shifts in Φ or cause phase shifts between the two microwave lines that connect the two junctions, also resulting an undesired shift in Φ.

In the experiment, we phase-locked the junctions on the microwave step, and the output of the rf SQUID was monitored for many hours with a 0.01 Hz time constant. The microwave frequency used was 18 GHz in all cases[13]. Several different types of Josephson junctions were used in the comparison experiments. In comparing two indium microbridges, we were able to bias them at the tenth harmonic step, and verify that the Josephson voltages of the two junctions were equal with a precision, $\Delta V/V$, of $\approx 5 \times 10^{-18}$. For measurements comparing an indium microbridge and a planar niobium/copper/niobium (superconducting/normal/superconducting) junction, because of the much worse symmetry in the two junctions' parameters, the bias was limited to the first step, which resulted in a slightly reduced precision limit: $\approx 2 \times 10^{-16}$. This result indicated that Eq. 9.1 does not have noticeable material[15] or junction-type dependences. Later, work by others using a similar technique demonstrated that the Josephson junction arrays that are actually used for the real voltage standards

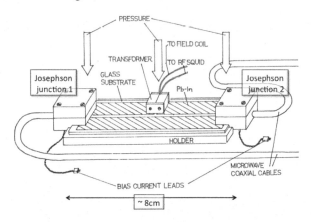

FIGURE 9.1: Schematic of the DC SQUID used for voltage comparison.

---
[15] K. Nordtvert, *Phys. Rev.* B1 (1970) 81; later theoretically refuted by D.N. Langenberg et al., *Phys. Rev.* B3 (1971) 1776, and J. B. Hurtle et al., *Phys. Rev.* B3 (1971) 1778

produce identical voltages with a precision[16] of $\approx 2 \times 10^{-17}$. Additional later investigations comparing array accuracy were carried out at PTB.[17,18]

We later compared two very similar lead tunnel junctions, further improving the precision limit[19] to $\approx 3 \times 10^{-19}$. Since the gravitational red shift in f between two junctions with a vertical separation of about $z = 3$ mm would also be $\approx 3 \times 10^{-19}$, it would be interesting to see whether one should be concerned with this effect. To test this, we made a SQUID where $z = 70$ mm. We again obtained a null result. This is expected because it is the gravito-electrochemical potential, rather than the electrochemical potential, that should be constant along a superconducting wire, just canceling the effect of the red shift. To our knowledge, this was the first confirmation of this effect.

This series of experiments established the ultimate precision of the voltage standard based on Josephson junctions. No significant dependence on material or type of the junction was found. These results show that we can reproduce the voltage within the accuracy associated with the frequency of the microwave source. It can be understood that in such Josephson junctions, most of the degrees of freedom of conducting electrons are frozen into just one quantity, the macroscopic phase difference of order parameter across the junction. Even with the surrounding quasiparticles, the resulting superconducting state is immune to dis-

FIGURE 9.2: The group at Stony Brook in about 1983: Aloke Jain, Shen Tsai, Joe Sauvageau, and Jim Lukens.

sipation and follows the relation (9.1) exactly. The equality established here for Josephson voltage is probably one of the most accurately determined equalities among all the known physical equalities. Although microwave-induced Shapiro steps are dynamical states and not energy eigenstates, the universality of the quantized energy value in this macroscopic system was established with extraordinary accuracy, much better than that confirmed in microscopic systems, such as comparison of atomic levels in two atoms. The work described in this article took place when Tsai was a student of Lukens at the State University of New York, at Stony Brook. Both comment on their personal experiences.

Lukens: I started thinking about using SQUIDs to detect gravitational effects when I was still working on my thesis in 1966. So, it was very exciting when we finally achieved the sensitivity required for such measurements while testing the universality of the Josephson frequency-voltage relation and were then able to demonstrate that, assuming the gravitational red shift of photons, it was indeed the gravito-electrochemical potential rather than the electrochemical potential that was constant in the superconducting leads of the SQUID.

Tsai: I wrote the draft of this article while trapped in Helsinki Airport for a very long time, due to the eruption of Icelandic volcano Eyjafjallajökull. As I was writing it under a rather unpleasant circumstance, all the excitements and frustrations that took place about 30 years ago came back to me. The experiment was originally designed to observe the gravitational red shift using a SQUID. For this reason, to maximize the signal, the SQUID loop had an unusually large dimension: about 8 cm in length. The idea of the "desktop" gravitational effect experiment really fascinated me in spite

---

[16] R. L. Kautz and F. L. Lloyd, *Appl. Phys. Lett.* 51 (1987) 2043

[17] J. Niemeyer, Grimm, C. A. Hamilton, R. L. Steiner, *IEEE Electr. Dev. Lett.* EDL-7 (1986) 44

[18] J. Y. Krasnopolin, R. Behr, and J. Niemeyer, *Supercond. Sci. Tech.* 15 (2002) 1034

[19] A. K. Jain, J. E. Lukens, and J. S. Tsai, *Phys. Rev. Lett.* 58 (1987) 1165

of its difficulty. From the null result obtained, we were able to demonstrate the importance of the gravito-electrochemical potential in the superconductor, in addition to the red shift. That is, in the superconductor, it is the gravito-electrochemical potential that is constant, not the electrochemical potential, as usually described in textbooks. In addition to this, as a grand bonus, we were also able to establish the universality of the Josephson frequency-voltage relation with an unprecedentedly extraordinary accuracy. All these took place in the beginning of my long low-temperature-physicist career, while working as a graduate student. Looking back now, I consider myself very fortunate to have been able to work on such challenging and rewarding topics in Jim Lukens's laboratory.

9.3   The First Josephson Voltage Standards

Thomas J. Witt

The Josephson effect is the most fruitful application of superconductivity in metrology (measurement science). In 1962, Brian Josephson, then a 22-year-old graduate student, predicted several phenomena caused by quantum mechanical tunneling of paired electrons through an insulating barrier.[20] In classical physics such tunneling effects may be compared with throwing a ball against a wall and seeing it penetrate through the wall rather than bouncing back.

The Josephson effects describe the flow of current through a Josephson junction consisting of two superconductors separated by a weak link such as a thin (order of 1 nm) insulating barrier. A bulk superconductor can often be described by a "wave function" characterized by a single phase, ϕ_1. Josephson predicted that even in the absence of an applied electromagnetic field a steady supercurrent $I = I_c \sin\phi$ would flow between the superconductors on each side of the barrier, forming a Josephson junction (the dc Josephson effect). ϕ is the difference in phase between the wave functions on the two sides of the barrier. The time dependence of ϕ is given by $\partial\phi/\partial t = (2e/\hbar)V(t)$ where e is the elementary charge, $\hbar$ is the Plank constant divided by 2π and $V(t)$ is the voltage across the junction. The application of a steady voltage V_{dc} across the junction results in the flow of an ac current of frequency $(2e/h)V_{dc}$ (ac Josephson effect). Moreover, superimposing an alternating voltage of frequency f on an applied direct voltage, the ac supercurrent is modulated. This results in constant-voltage steps at integer values $V_n = nhf/2e = nf/K_J$. n is an integer and K_J is the Josephson constant (defined as $2e/h$).

The first observation of the dc Josephson effect was reported in 1963. An account of Josephson's work and of this observation is given by Anderson.[21] Shapiro was the first to observe constant voltage steps produced by irradiating a tunnel junction with microwaves.[22] From 1965 until the early 1970s the group led by D. N. Langenberg at the University of Pennsylvania (Penn) made major contributions to the development of the use of the Josephson effect to measure $2e/h$, to access the impact of their determinations on the ensemble of fundamental constants, and to develop the Josephson effect as a voltage standard. By 1966 they had tested the Josephson frequency-to-voltage relation to 60 parts per million (ppm). In 1967 the Penn group announced a determination of $2e/h$ with an uncertainty of 6 ppm and discussed the impact of this new way to measure $2e/h$ on other constants.[23] In particular, they deduced a value of the fine structure constant α that was lower than the accepted value of the time. Using the Josephson-effect value of α resolved a 43 ppm discrepancy between the experimental value of the hyperfine splitting in atomic hydrogen and the theoretical value derived

[20] B. D. Josephson, *Phys. Lett.* 1 (1962) 251
[21] P. W. Anderson, *Phys. Today* 23 (1970) 23
[22] S. Shapiro, *Phys. Rev. Lett.* 11 (1963) 80
[23] W. H. Parker, B. N. Taylor and D. N. Langenberg, *Phys. Rev. Lett.* 18 (1967) 287

from quantum electrodynamics. Also that year the Penn group proposed to the metrology community that the ac Josephson effect could be used to maintain standards of electromotive force with an uncertainty of 1 ppm or less.

To illustrate the importance of this, I first describe the limitations of the voltage standards of the time as well as the methods and limitations of the Josephson measurements. In the Système International d'Unités (SI), the modern form of the metric system of units, the base unit for electricity is the ampere, defined by the force between idealized current-carrying wires. The volt, the unit of voltage or electrical potential difference, is derived from the ampere by assuming the equivalence of electrical and mechanical energy. Nearly all voltage measurements reduce to the determination of the ratio of an unknown voltage to some reference voltage (often incorporated in the measurement instrument itself).

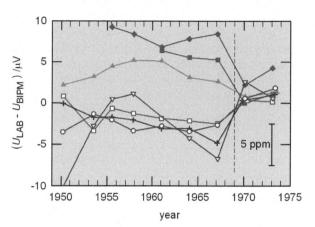

FIGURE 9.3: Comparisons of voltage standards of 7 national laboratories 1950–1973.

Calibrations of reference voltages are traceable to the volt. Determinations of the volt were, and still are, difficult to carry out. In 1967 they were limited in uncertainty to a few ppm or more. (This story is brought up to date in the article in this chapter by E. Williams and I. Robinson).

Until the early 1970s, voltage standards were maintained in national metrology institutes (NMIs) using groups of Weston standard cells (precision electrochemical cells). Each national group had a somewhat different value of voltage. International traceability was established by periodic comparisons at the Bureau International des Poids et Mesures (BIPM). Figure 9.3 shows the results of these comparisons between 1950 and 1973. Standard cells were only *representations* of the volt (as is the Josephson voltage standard today); their values in SI volts were traceable to determinations of the SI volt. The electromotive force of a standard cell (emf) is the difference in electrical potential between its terminals when no current flows. Standard cells suffer serious stability limitations: large temperature coefficients, sensitivity to shock, drift of electromotive force (emf) in time, occasional small spontaneous changes of emf and sensitivity to accidental charge or discharge. For the international comparisons, standard cells were hand-carried to the BIPM.

For all these reasons NMIs were particularly keen to explore the possibility of using the Josephson effect as a voltage standard. A major challenge in 1967 was the fabrication of adequate Josephson devices. Even good Josephson devices produced output voltages less than 800 μV. An uncertainty of 1 ppm represented only tenths of a nanovolt. To compare such voltages with the emf of a standard cell, a resistance ratio network was needed. Such networks were major sources of measurement uncertainty. The resolution of the voltage detector was of the order of 1 nV. The thermal emfs in the leads from the He cryostat to the room-temperature measurement system should be small in magnitude and stable in time. Since about 1970, little progress has been made to diminish the absolute values of the latter two limitations. Fortunately, the development of large arrays of Josephson junctions[24] (also see other contributions in this chapter) has resulted in the increase in output voltage to 10 V today, reducing an uncertainty of, say, 0.5 nV to a relative uncertainty with respect to the output of 5 parts in 10^{11}. This is roughly the present-day relative uncertainty in the direct comparisons of 10 V Josephson standards. At such output levels a resistance ratio network is not needed.

---

[24] C. A. Hamilton, *Rev. Sci. Instrum.* 71 (2000) 3600

By 1969 the Penn group achieved a 2.4 ppm uncertainty in $2e/h$ measurements. The same year, the National Physical Laboratory (NPL), UK, announced a measurement of $2e/h$ with an uncertainty of 2.2 ppm and became the first NMI to carry out high-accuracy Josephson measurements.[25] The following year the Penn group announced the first $2e/h$ measurement with sub-ppm uncertainty.[26] Shortly after, a similar uncertainty was attained by the National Standards Laboratory, Australia.[27] By 1971 the Penn apparatus was acquired by the National Bureau of Standards (NBS), USA; the claimed uncertainty was 0.05 ppm.[28] By that time the uncertainty of the NPL measurement was 0.2 ppm.[29] The Physikalisch-Technische Bundesanstalt (PTB), Germany, announced an uncertainty of 0.4 ppm.[30] The NBS organized an international comparison of voltage standards that included NPL, NSL, and PTB.[31] It was found that the differences among the participants' national voltage standards deduced from $2e/h$ results agreed with the differences found via traveling standard cells within the ranges $+0.01\ \mu V$ to $-0.11\ \mu V$ in 1971 and $-0.08\ \mu V$ to $-0.38\ \mu V$ in 1972. The drift rates of the four national standards ranged from $-0.14\ \mu V/yr$ to $-0.31\ \mu V/yr$.

On July 1, 1972, the USA abruptly redefined its voltage standard in terms of a value of the Josephson frequency-to-voltage ratio selected so that no discontinuity in value occurred on the day of the change.[32] Later in 1972 the Comité Consultatif d'Electricité (CCE), an organ of the Convention du Métre, recommended a conventional value of the Josephson frequency-to-voltage quotient that was 1.20 ppm higher than that set by the USA. Eventually most other nations redefined their voltage standards to be in agreement with the CCE's recommended conventional value, with two other exceptions: France and the USSR. This unfortunate situation lasted until January 1, 1990, when it was universally agreed to apply a new conventional value of the Josephson frequency-to-voltage quotient $K_{J-90} = 483597.9$ GHz/V with an assigned uncertainty of 0.4 ppm.[33] At the time this value was considered to be the best estimate of $2e/h$ expressed in SI units (i.e., traceable to mechanical energy). Notwithstanding progress in the determinations of e and h over the last 20 years, K_{J-90} is consistent with the present estimate of K_J in SI units. K_{J-90} is still used to assign values to the representations of the volt based on the Josephson effect.

My experience with the Josephson effect began in 1969 at the NBS. In 1971 I accepted the job of establishing a Josephson standard at BIPM. I built a voltage comparator based on that of the Penn group. In 1973 Dominique Reymann joined the BIPM Josephson volt project. Our most formidable remaining task was to use masks provided by the NBS to fabricate tunnel junctions capable of producing a 10 mV output. Figure 9.4(a) is a photograph of a 25.2 mm square substrate carrying four Pb-Pb oxide-Pb tunnel junctions connected in series to produce a 10 mV output; (b) is a sketch of the current-voltage characteristics: $I_c = 10$ mA, the voltage $E_g/2e$ gives the value of the energy gap in Pb, ≈ 2.8 mV; and (c) shows a zoomed photograph of the constant voltage steps near 4.5 mV when a junction is irradiated at 9.1 GHz.

By 1975 we had achieved an accuracy of 6 parts in 10^8 in the measurement of $2e/h$ in terms of the BIPM voltage standard. To support this uncertainty claim we transported our entire Josephson system to the PTB and compared the results of our on-site Josephson measurements of standard cells with those of the PTB. The results gave a difference of 5 parts in 10^8 with a combined standard

[25] B. W. Petley and K. Morris, *Phys. Lett.* A 29 (1969) 289

[26] T. F. Finnegan, A. Denenstein, D. N. Langenberg, *Phys. Rev. Lett.*, 24 (1970) 738

[27] I. K. Harvey, J. C. Macfarlane and R. B. Frenkel, *Phys. Rev. Lett.* 25 (1970) 853

[28] T. F. Finnegan, T. J. Witt, B. F. Field and J. Toots, in *Atomic Masses and Fundamental Constants* 4, J. H. Sanders and A. H. Wapstra, Eds., New York, Plenum (1972) 403

[29] J. C. Gallop and B. W. Petley, *Metrologia* 8 (1972) 129

[30] V. Kose, F. Melchert, H. Fack and H.-J. Schrader, *PTB Mitt.* 81 (1971) 8

[31] W. G. Eicke, B. N. Taylor, *IEEE Trans. Instrum. Meas.* IM-21 (1972) 316

[32] B. F. Field, T. F. Finnegan and J. Toots, *Metrologia* 9 (1973) 155

[33] B. N. Taylor and T. J. Witt, *Metrologia* 26 (1989) 47

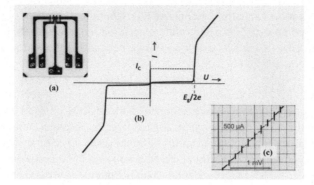

FIGURE 9.4: (a) Tunnel junctions; (b) dc characteristics; (c) induced steps.

uncertainty[34] of 6 parts in 10^8. We transported our equipment to the NPL in April 1978 where for the first time we compared Josephson voltage standards directly, bypassing the use of standard cells and found a difference between our voltage standards of (9 ± 5) parts in 10^8. This was the beginning of an ongoing series of BIPM on-site comparisons of Josephson standards that continues to this day. The later comparisons were carried out at 1 V and eventually at 10 V using arrays of Josephson junctions. Also in 1978, the ETL Japan, announced achieving outputs of 100 mV using arrays of junctions. That story will be related by Tadashi Endo in the following section.

9.4 First Josephson Junction Array Voltage Standard

Tadashi Endo

9.4.1 Introduction

In 1977 at Japan's Electrotechnical Laboratory (ETL),[35] we fabricated and began using a two-junction Josephson voltage standard. Our standard produced a maximum voltage of 12 mV using a 10 GHz microwave source. To compare this small voltage with typical calibration voltages of about 1 V, we used a precision resistance bridge between the standard and the source we wished to calibrate. As a result the voltage could be determined only to 2 parts in 10^8 in spite of the perfect accuracy of the Josephson junctions themselves.[36] Other laboratories were similarly limited.

As a research project after the Josephson voltage standard was introduced, we decided to fabricate an array of Josephson junctions. This work took place between 1977 and 1983 and was primarily done by Tadashi Endo and Masao Koyanagi at ETL. The accuracy of any step voltage developed by a Josephson junction depends only on the relatively simple accurate measurement of the frequency of the microwaves applied to the Josephson junction. We planned to use this superior feature of the

---

[34] T. J. Witt and D. Reymann, in *Atomic Masses and Fundamental Constants* 5, J. H. Sanders and A. H. Wapstra, Eds., New York, Plenum, (1976) 457

[35] The Electrotechnical Laboratory was reorganized to the present National Institute of Advanced Industrial Science and Technology, Japan (AIST) in 2001.

[36] T. Endo, M. Koyanagi, K. Shimazaki, G. Yonezaki, and A. Nakamura, AMCO-5, *Atomic Masses and Fundamental Constants* 5, Plenum Press, New York (1976) 464

Josephson junction to create an ideal potentiometer that would have, in principle, perfect linearity. The device we envisioned provided a strong motivating force to undertake this development of the first Josephson array, in spite of the fact that it had never been attempted before.

9.4.2 First Josephson Array

In order to demonstrate that a Josephson junction device could be used as a potentiometer in practical use, we had to increase the maximum voltage step to above 100 mV at least.[37] Our first task was to decide how many junctions we would use in the array. Obviously if we could make, say, 1000 junctions, the voltage achieved would be about a volt and the cumbersome bridge would not be needed at all. However each junction needed to be independently biased with a dc current. At that time, junctions differed considerably and each needed a separate bias. We were limited by the practical difficulties of creating many working junctions in series with superconducting connections between them, by supplying microwave power uniformly to the junctions, and by the formidable task of adjusting the bias supplies that were as numerous as the junctions. Because we were obtaining 5 mV from each junction in our conventional standard, we decided that a reasonable choice was an array of 20 junctions.

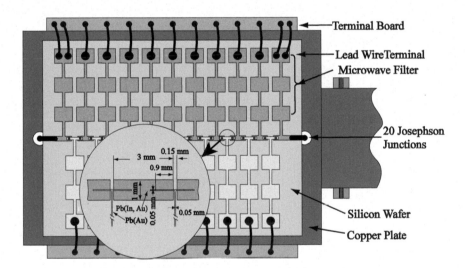

FIGURE 9.5:   The multiple Josephson junction is composed of twenty Pb(In, Au)-oxide-Pb(Au) tunnel junctions constructed on the center superconducting strip film with 1 mm width. The overall pattern of the sample is made on a 0.2 mm thick silicon wafer by vacuum-evaporating the superconducting films and patterning them using photolithography. The size and the normal resistance of each tunnel junction are 0.9 mm × 0.05 mm and about 70 mΩ, respectively.

We developed what we called, when we spoke English, the "multiple Josephson junction" as shown in Figure 9.5, and obtained a maximum voltage greater than 100 mV. The array was fabricated in our simple facility that we built based on publications from IBM Research where they were attempting to fabricate a Josephson junction computer. We worked about three years to obtain the first usable 20-junction array in 1980. Then it was necessary to learn how to use it. When using the array we found that the shape of the I-V curve changed with the level of the liquid helium in the

---

[37] T. Endo, M. Koyanagi and A. Nakamura, *IEEE Trans. Instrum. Meas.* IM-32 (1983) 267

waveguide. To solve that problem we evacuated the waveguide. Two people operated the array: one monitored the I-V curves of each junction and adjusted the bias currents; the other performed the precise voltage measurements. Eventually we were able to keep the output voltage stable for about 15 minutes. Later we understood that arrays having a single bias had similar operational instabilities. (See the next section.) Nevertheless we were able to perform real measurements.

Figure 9.6 shows that the junction array forms the upper line of a stripline with the sample holder as the ground plane and the substrate as the dielectric. This method of microwave coupling allowed us to couple uniformly the microwave power to the junctions. It also enabled us to reduce the normal resistance of each Josephson junction and to increase the current width of the voltage step. For example, the largest voltage steps having a large current width of 200 μA to 300 μA appeared in the voltage range of 3 mV to 8 mV for each junction under microwave irradiation at 9.3 GHz with a power of 120 mW.

9.4.3 Applications of the Multiple Josephson Junction

In order to demonstrate that the multiple Josephson junction could be used practically as a Josephson potentiometer, we demonstrated its use for the following two precise measurements.

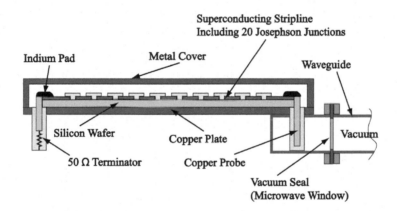

FIGURE 9.6: Cross-sectional view along the center strip film of the sample when the sample is mounted on the sample holder. A parallel plate type microwave stripline of 50 Ω is composed of the center superconducting strip film including the twenty junctions (which acts as the top electrode), the silicon wafer (which acts as the dielectric layer) and the copper plate of the sample holder (which acts as the lower electrode). A 10 GHz microwave source was coupled to the 50 Ω stripline from the waveguide through the copper probe. In order to irradiate all of the junctions by microwave as uniformly as possible, the end of the stripline was terminated by a 50 Ω resistance.

First we measured the emf (electromotive force) of a chemical standard cell using a voltage divider composed of ten 45 Ω resistors connected in series. We were able to confirm that the voltage of nominally 1 V could be measured with a relative uncertainty of 2.5×10^{-9}. Around 1982, this meant that the accuracy was increased by 1 order of magnitude compared with the 10 mV Josephson voltage standard at that time.

Second, we measured the fundamental quantity h/e^2. It was important because a resistance standard had been recently created, based on the quantum Hall effect, by defining the fundamental constant, named after von Klitzing, R_{K-90} ($\approx h/e^2$) with conventional value of $R_{K-90} = 25.812807$ kΩ since January 1990. However, h/e^2 was the quantity to be measured in 1983.[38]

We succeeded in experimentally confirming that the resistance ratio between the fourth plateau of the quantum Hall effect and a 6.45 kΩ standard resistor could be measured with a relative uncertainty of about 1.1 x 10^{-8}. This is about the level of the thermal noise of the standard resistor but not of the Josephson potentiometer.

9.4.4   Conclusion

We demonstrated the multiple Josephson junction and its use as a potentiometer having ultimate linearity by increasing the Josephson step voltage to over 100 mV. If such demonstration would have led to the present Josephson junction array, it is our great pleasure. We are also pleased to note the major advances that have come after our work and are described in the following sections of this chapter on metrology.

9.5   How the DC Array Standards Were Developed

Jürgen Niemeyer

9.5.1   Introduction

By locking a current-driven Josephson junction to an external microwave source with a precisely determined frequency f, constant voltage steps appear in the dc characteristic at $V_n = n\Phi_0 f$. As this voltage is very small (about 145 μV for f = 70 GHz and n = 1), a voltage divider was required in the early days to enhance the reference voltages to the 1 V level for practical use. But the voltage divider limited the precision, and its handling was rather complicated. For these reasons, the advantage of the quantum origin of the voltage could not be fully exploited.

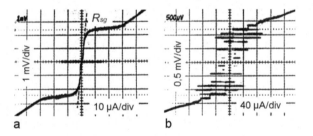

FIGURE 9.7:   Dc characteristics of a Pb-alloy tunnel junction. a: without microwaves, b: under microwave radiation of 70 GHz. R_{sg} is the subgap resistance. The rf-power is optimized for the 7th step. The steps in the back-switching part of the dc characteristic cross the zero current axis.

Therefore, the idea developed over time to increase the output voltage by connecting a certain number of Josephson junctions in series. In the 1970s, most people found the idea of preparing arrays of thousands of junctions rather ridiculous because there was no reliable Josephson circuit technology available. Early experiments were therefore restricted to series arrays with a small number of individually biased junctions.

But at the beginning of the 1980s this situation began to change because important progress was achieved in the following fields:

    1. IBM developed a Josephson circuit technology for digital computing that was based on lead

---

[38] T. Endo, Y. Murayama, M. Koyanagi, J. Kinoshita, K. Inagaki, C. Yamanouchi and K. Yoshihiro, *IEEE Instrum. Meas.* IM-34 (1985) 323

alloy tunnel junctions, the barrier of which was formed by a plasma oxidation process.[39]

2. Mogens T. Levinsen and coworkers suggested making use of zero-current voltage steps for the realization of large commonly biased arrays.[40] In this way the inevitable junction parameter spread would no longer play an important role because the array could be biased at zero current. Also voltage steps up to the limit of about 1 mV per junction could be used, which considerably reduced the number of junctions for an array with a given output voltage. See Figure 9.7.

3. Tadashi Endo at the Electrotechnical Laboratory[41] (ETL) in Japan successfully operated a series array with 20 individually biased Josephson junctions that were located as an integrated part of a superconducting stripline with the substrate serving as the dielectric.[42] (See the preceding article by T. Endo.)

4. Richard Kautz at the National Bureau of Standards[43] (NBS) in Boulder, CO was working to understand why a 5,000 junction array that NIST fabricated showed no steps at all. Kautz' work threw light on the operating conditions for a chaos-free phase lock between an external microwave source and a Josephson tunnel junction with very low damping.[44] These calculations made it possible to design correctly the single rf-driven tunnel junction.

9.5.2 Single Junction Design

A Josephson junction can be considered a driven nonlinear current oscillator with damping. To obtain zero-current voltage steps one has to use tunnel junctions with very low damping. The mechanical analog of the Josephson oscillator is a driven pendulum in the gravitational field of the earth with viscous damping. In this analog, the external microwave source characterizes the driving force; the junction capacitance, the moment of inertia; the junction conductivity, the viscous damping; and the critical current, the torque in the gravitational field. If such a "pendulum" is to be coupled to an external oscillating "force" — a microwave oscillator — the oscillations easily become chaotic or even catastrophic if the external oscillation frequency approaches the resonance frequency of the Josephson "pendulum".

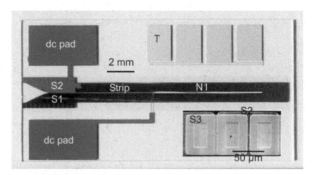

FIGURE 9.8: 102 junction array on a glass substrate. The insert shows 3 junctions of the array stripline. S1: groundplane; S2: base electrode layer; S3: top electrode layer; N1: matched load; T: test circuit for the matched load.

On the basis of the Stewart-McCumber junction model[45], Kautz found that for junctions with small damping, a chaos-free generation of constant voltage steps is possible only if the frequency of

[39] J. H. Greiner et al., *IBM J. Res. Develop.* 53 (1982) 326

[40] M. T. Levinsen, R. Y. Chiao, M. J. Feldman, B. A. Tucker, *Appl. Phys. Lett.* 31 (1977) 776

[41] The Electrotechnical Laboratory in Tsukuba, Japan, has become the National Institute of Advanced Industrial Science and Technology (AIST)

[42] T. Endo, M. Koyanagi, A. Nakamura, *IEEE Trans. Instrum. Meas.* IM-32 (1983) 267

[43] The National Bureau of Standards in the USA has become the National Institute of Standards and Technology (NIST).

[44] R. L. Kautz, *J. Appl. Phys.* 52 (1981) 35283541

[45] W. C. Stewart, *Appl. Phys. Lett.* 12, (1966) 277; D. E. McCumber, *J. Appl. Phys.* 39 (1969) 3113

the external microwave is larger than the resonance frequency of the Josephson junction oscillator. Together with the necessity of restricting the junction area for avoiding an inhomogeneous current distribution over the junction area by the magnetic field, particularly of the large rf-current and by rf-resonances inside the junction, the frequency condition determines a maximum value for the critical current. These restrictions are of great practical importance because they define a maximum current width of the constant voltage steps. The step current width should be as large as possible because the larger the steps are, the less sensitive they are to electromagnetic noise-induced spontaneous switching. At PTB in the 1970s, precision measurements with Josephson junctions often had to be performed during the night because of the reduced level of electromagnetic noise at that time. For typical lead alloy junctions with a maximum area of about 24 μm x 67 μm, the calculated optimum width of the seventh step amounts to about 220 μA, a size large enough for stable standard operation. For a driving oscillator frequency of about 70 GHz, the maximum zero current step voltage lies at the seventh voltage step, the current width of which is therefore a good criterion for the applicability of a certain junction technology for dc array standards.

In real arrays the resulting step width is much smaller than the optimum value for a single junction due to the parameter spread of the junctions and to noise-induced reduction of the step size. We had to learn that the smallest critical current of the array junctions determines the resulting step width. This and the chaos restriction of the maximum critical current made the parameter spread still play a certain role.

9.5.3 Circuit Design and First Realization

With the prerequisites described above, the NBS Cryoelectronics Section and my laboratory at PTB in Germany, headed by Johann Hinken, were encouraged to start the development that finally led to the modern Josephson voltage standards. The first NBS arrays consisted of 100 junctions, coupled capacitively to two parallel conductors that carried the rf-power of a 20 GHz microwave drive.[46] Steps were achieved up to only 27 mV, instead of the expected 80 mV. The reasons for this were seen in a noise-induced spontaneous switching between the steps and inhomogeneous microwave coupling.

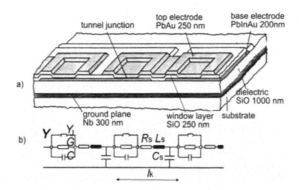

FIGURE 9.9: (a) Cross-sectional view of a portion of the 70 GHz microstripline with (b) the microwave equivalent circuit of the periodic stripline. Y is the total Josephson admittance of a stripline section with the periodic l_k, R_s, L_s, and C_s are the resistance, the inductance and the capacitance of a stripline section within the periodic length l_k.

---
[46] R. L. Kautz, G. Costabile, *IEEE Trans. Magn.* MAG-17 (1981) 780

PTB had arranged up to 166 junctions parallel to the electric field vector of a standing wave inside a 70 GHz waveguide.[47] For this array, an output Josephson voltage of 120 mV with satisfactory stability was reached. However, in this case, the space with a sufficiently homogeneous field in the standing wave pattern of the waveguide was not large enough for many more junctions. As expected from calculations for the single junction design, these results showed that a high frequency of 70 GHz is more advantageous than 20 GHz, and that a microwave distribution circuit had to be developed that would guarantee the homogeneous feeding of a large number of junctions that excluded to a certain degree standing-wave patterns. This was the reason we arranged a 70 GHz series array as part of a periodic superconducting microstripline with a matched terminating load at the end so that no standing-wave patterns could occur in the structure. The stripline impedance of a few ohms was matched to the much higher impedance of the 70 GHz waveguide by an antipodal finline taper that was inserted in a slit at the end of the waveguide. The first array contained only 102 junctions, but it could easily be extended to a higher number of junctions (Figure 9.8).

The 1.5 nm oxide barrier of the junction is extremely thin. This makes the junction fabrication a real challenge and also the junction capacitance very large so that the total rf-current in the array is mainly capacitive: the current through the capacitance is 20 times larger than that through the sub-gap resistance and 10 times larger than the inductive component of the total current. Because of the capacitive nature of the total junction admittance, the periodic stripline may be considered as a high-pass filter with a cut-off frequency of about 25 GHz. The stripline impedance of about 2.2 Ω is nearly like the impedance of the ideal stripline (Figure 9.9). The attenuation of the line is determined by the quasiparticle tunneling losses and by the resistive rf losses: The total attenuation of one stripline period is about 6×10^{-4} dB/period. If the rf-current amplitude should not be reduced by more than 5 % per stripline path, a single line should not contain more than about 700 junctions. For more details of the stripline circuit design see the review by Hinken.[48]

At that time, PTB did not have the lithographic tools to realize such a circuit. Therefore I decided to ask Richard Harris of NBS to help us with preparing the circuits. Fortunately, he invited me to stay at NBS in Boulder for six months, and with the support of the Cryoelectronics Section, the small Pb-alloy circuit (Figures 9.8 and 9.9) was realized at the end of 1983. Richard Harris wrote the following memo on November 17, 1983: "Yesterday Dr. Jürgen Niemeyer, guest worker from PTB in Western Germany, demonstrated constant voltage steps at 0.1 V using a series array of Josephson junctions. Niemeyer is working collaboratively at NBS Boulder with Dr. Richard Kautz. This result exceeds published results by a factor of about 4. It is even

FIGURE 9.10: From left: Clark Hamilton, Richard Kautz, and Jürgen Niemeyer.

more remarkable in that it was achieved with an array of only 102 junctions. Each junction produced about 1 mV, or about the maximum voltage per junction that is thought theoretically possible." This meant that we were on the right path. At that time I deposited the circuits on old-fashioned glass substrates which we had to cut to the size and shape of two-inch silicon wafers. The main difficulty with the Pb-alloy circuits was to hit the correct critical current. The problem was solved by choos-

---

[47] J. H. Hinken, J. Niemeyer, *Kleinheubacher Berichte* 28 (1985) 81

[48] J. H. Hinken, *Supraleiter-Elektronik*, Springer-Verlag Berlin Heidelberg (1988)

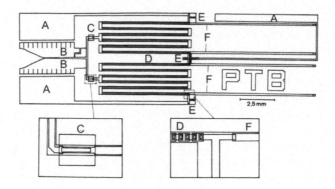

FIGURE 9.11: Series-parallel array with four microwave paths. A: dc pads, B: finline antenna, C: quarter wavelength transformers, D: array, E: bandstop filters, F: matched load.

ing the oxidation parameters in such a way that the critical currents became rather small. Suitable post-fabrication annealing enabled us to increase the small critical currents to the right level. After getting rid of the fabrication problems more or less, we finally had to increase the output voltage. By folding the array several times and simply increasing the number of junctions, step-by-step, to finally 1474 junctions, 1 V steps were generated for the first time.[49] Although the stability of 1 V output was still improvable, we enjoyed the result (Figure 9.10).

9.5.4 1 Volt Circuits

Due to inhomogeneous microwave coupling over the very long folded array stripline with a total attenuation of about 1 dB, spontaneous switching of the 1 V step occurred within a few minutes. Therefore, a new design was developed at PTB in the following year, where a long stripline with 1440 junctions was divided into four folded parallel paths, each of which coupled to the finline antenna via dc blocks and a distribution network. Each path was terminated by a lossy line (Figure 9.11). With respect to the dc bias, the paths remained series-connected. This circuit was fabricated in Pb-alloy technology at PTB and produced stable 1 V reference voltages with a standard deviation of only 3×10^{-10} V. The Cryoelectronics Section at NBS made this circuit type with $Nb/Nb_2O_5/PbInAu$ junctions at about the same time.[50] Later on, this 1 V design was adapted to be used at lower frequencies — 10 GHz at NBS and 35 GHz or 10 GHz at the Institute für Photonische Technologien (IPHT) in Jena — but finally all these versions were given up because of the better performance of the higher-frequency versions.

The main problem that remained was the susceptibility of the lead alloy circuits to damage by humidity and thermal cycling. This was the reason why IBM stopped the development of computer circuits just at the moment when we succeeded with the voltage standard circuit — although there was already a new, more robust junction technology available: M. Gurvitch and coworkers at Bell Laboratories had developed stable Nb junctions with an Al-oxide barrier.[51] On this technological basis, the Josephson computer group at ETL, headed by Hisao Hayakawa, was developing a digital circuit technology which I also found suitable for a more robust voltage standard. Besides the stability of $Nb/Al_2O_3/Nb$ junctions to destructive agents, with a current width of the seventh step

---

[49] J. Niemeyer, J. H. Hinken, R. L. Kautz, *Appl. Phys. Lett.* 45 (1984) 478

[50] J. Niemeyer, L. Grimm, W. Meier, J. H. Hinken, E. Vollmer, *Appl. Phys. Lett.* 47 (1985) 1222; C. A. Hamilton, R. L. Kautz, R. L Steiner, F. L. Lloyd, *IEEE Elect. Dev. Lett.* EDL-6 (1985) 623

[51] M. Gurvitch, M. A. Washington, H. A. Huggins, J. M. Rowell, *IEEE Trans. Magn.* 19 (1983) 79

of about 420 μA, this technology was the absolute optimum, compared to all other junction types. For the microstrip design, practically no modifications were required because at 70 GHz the surface resistance of Nb films is very similar to that of Pb films. The main difference lay in the large maximum critical current of the Nb/Al junctions, which made the size of the voltage steps nearly twice as large as for the lead alloy junctions. This is not relevant for the circuit design but facilitates the standard operation very much.

Therefore we slightly modified the Pb-alloy design, adapting it to the specific data of the Nb/Al-technology. The design looked very similar to that shown in Figure 9.11. I found support for preparing the lithographic masks at NBS and was invited by Hisao Hayakawa to realize the circuits at ETL at the beginning of 1986. Supported by ETL's Josephson computer group and by the voltage metrology group with T. Endo, M. Koyanagi, and Y. Sakamoto, we succeeded in preparing the first 1 V Nb/Al voltage standard within three months.[52] For this effort we had to modify the ETL fabrication technology because the array junction critical current density was about two orders of magnitude smaller than that for the digital circuits. In addition, the array chip size was considerably larger than that of the digital circuit chips.The NbN/MgO/NbN version was

FIGURE 9.12: ETL farewell meeting after having fabricated the first 1 V full Nb dc array standards. Clockwise from left: from the computer group Masao Koyanagi and from the metrology group Itaro Kurosawa, Masaskazu Nakanishi, Shin Kosaka, Jürgen Niemeyer, and between the young students, Akira Shoji.

not so successful. Due to the large London penetration depth of NbN, the value of the seventh voltage step, 165 μA, is much smaller than the one for the Nb/Al junctions. Together with the larger spread of the critical currents — caused by non-uniform sputtering of the MgO barrier — this makes the reference voltages unstable. This problem is irrelevant for the naturally grown aluminum oxide barriers of the Nb/Al_2O_3/Nb junctions. ETL's Akira Shoji had developed the sophisticated NbN/MgO fabrication system. The thickness of the extremely thin MgO barrier was precisely determined by counting the cycles of a rotating shutter with a small opening that uncovered the substrate only for a short time per cycle. The result became somewhat better by backing the NbN-layer with Nb, so that the junction sandwich gets the layer sequence NbN/Nb/MgO /Nb/NbN. But in this case, the advantage of the higher operating temperature of NbN — the main reason for implementing the complex technology — becomes questionable. Figure 9.12 shows members of ETL's Josephson computer group and metrology group at my farewell meeting.

9.5.5 10 Volt Circuits

After having developed the 1 V arrays, we often received requests for 10 V arrays because the direct output voltage of electronic Zener voltage references is 10 V, so that the calibration of these devices would become very simple. With the technologies in the middle of the 1980s this was, however, a real challenge — if one keeps in mind that the sum of the junction areas of a 15,000-

[52] J. Niemeyer, Y. Sakamoto, E. Vollmer, J. H. Hinken, A. Shoji, H. Nakagawa, S. Takada, S. Kosaka, Jpn. *J. Appl. Phys.* 25 (1986) L343

junction array of the described type amounts to about 15 mm^2. This required a 15 mm^2 oxide barrier with a thickness of only 1.5 nm, totally free of defects.

It was clear that with the lead-alloy technology, this was impossible to achieve. Therefore we tried to prepare 10 V Nb/Al circuits at ETL in 1988, but we failed. Although the dc characteristic of the array was good, we found that the available rf-power was not large enough to drive the circuit to 10 V. We therefore tried to replace the SiO of the dielectric layer by high quality SiO$_2$ that was deposited by means of a low temperature silane PECVD process. However, after having torched a forepump we had to stop the dangerous experiments. At that time, Susumu Takada was guiding the Josephson computer group (Figure 9.13). Unfortunately he passed away in 2008.

At NBS in 1983, some circuits Nb/Nb$_2$O$_5$/PbInAu were already being prepared from sandwiches which had a much better performance against deterioration from cycling and humidity than the pure Pb-alloy circuits. In 1987, after a long-lasting effort, the NBS group succeeded in preparing a

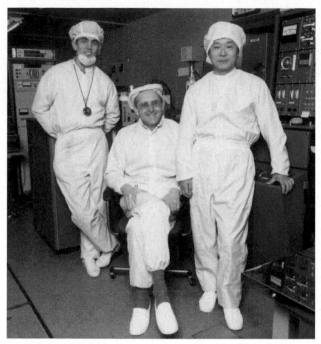

FIGURE 9.13: From left: Yasuhiko Sakamoto, Jürgen Niemeyer and Susumu Takada in the clean room facilities of the Josephson computer group at ETL. I had to sit because of an injury I had received during Kendo lessons.

14,184 junction series array with 16 parallel microwave paths, the first Josephson voltage standard with reference voltages up to 12 V[53] as shown in Figure 9.14. After having implemented a Nb/Al fabrication line, my group at PTB needed another two years until 10 V Nb/Al$_2$O$_3$/Nb arrays with up to 20,160 junctions and output voltages of up to 14.5 V could be realized.[54] Because a few problems were left with the sensitivity of the Pb-alloy top electrode layer to humidity, NIST also established Nb/Al technology and produced 10 V all-refractory metal circuits with 20,208 junctions at that time.[55]

Today, the Nb/Al$_2$O$_3$/Nb technology is commonly used by the commercial producers of dc array standards as is described in the next section by Clark Hamilton. For an overview on the development of the dc voltage standard array see especially two references.[56,57]

The final success was possible only because all collaborators were willing to share their knowledge, accept different ways of thinking and living, and avoid complex bureaucratic procedures. While working in this way, the dc array standard was introduced as the primary voltage standard

[53] F. L. Lloyd, C. A. Hamilton, J. A. Beall, D. Go, R. H. Ono, R. E. Harris, *IEEE Electron. Device Lett.* EDL-8 (1987) 449

[54] R. Pöpel, J. Niemeyer, R. Fromknecht, W. Meier, L. Grimm, *J. Appl. Phys.* 68 (1990) 4294

[55] C. A. Hamilton, C. J. Burroughs, *IEEE Trans. Instrum. Meas.* 44 (1995) 23

[56] R. L. Kautz, *Rep. Prog. Phys.* 59, 935–992, 1996

[57] J. Niemeyer, in *Handbook of Applied Superconductivity*, Vol. 2, ed. by B.Seeber, IOP Publishing, Bristol and Philadelphia (1998) 1813

FIGURE 9.14: First 10 V chip in Nb/Nb$_x$O$_y$/Pb alloy junction technology. The finline taper is inserted into the end of a 70 GHz waveguide. Dc contact is made by the two lead foils in the right part of the figure. The single junction area is 15 μm × 30 μm. (Courtesy of Clark Hamilton, NIST).

worldwide within a few years. In my opinion, the development of the dc array voltage standard is a convincing example of a fruitful international collaboration.

9.6 Making the Josephson Voltage Standard Practical

Clark A. Hamilton

A key element in the development of the first Josephson standard above one volt was the realization by Levinson[58] in 1977 that the zero crossing constant voltage steps of highly capacitive Josephson junctions provided the means to bias thousands of junctions into a quantized voltage state with a single bias current — namely zero bias current. My first observation of zero-crossing steps occurred as a graduate student of Sidney Shapiro. Shapiro was the first to confirm Josephson's prediction of rf-induced constant voltage steps in the I-V curve of a superconductive tunnel junction.[59] He suggested a thesis project to measure the detailed behavior of these steps as a function of applied microwave frequency and power. Theory predicted an oscillatory Bessel function behavior with increasing power and with a sharp deviation from Bessel behavior when the rf voltage across the junction exceeded the superconductive energy gap voltage (the Riedel peak). So, in 1969, I was making Sn-SnO-Sn

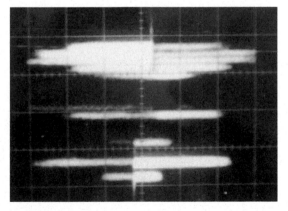

FIGURE 9.15: An early observation of zero-crossing constant voltage steps in a single tin-tin oxide-tin Josephson junction with a 10 GHz microwave bias. The steps occur at precise voltage multiples of $hf/2e \approx 21$ μV, where f is the microwave frequency, h is Planck's constant and e is the electron charge. The scales are 100 μV/div. vertical, 5 μA/div. horizontal. 18 December, 1969.

[58] M. T. Levinson, R. Y. Chiao, M. J. Feldman and B. A. Tucker, *Appl. Phys. Lett.*, 32 (1977) 776

[59] S. Shapiro, *Phys. Rev. Lett.* vol. 11 (1963) 80

junctions and observing their microwave-induced constant voltage steps. Somewhat accidentally the early results showed zero-crossing steps as in Figure 9.15.

The behavior of these steps did not match the theory because the junctions were too big and the microwave power and frequency were too small. After a year of work, I successfully suppressed those pesky zero-crossing steps and wrote my thesis confirming an obscure detail of superconductivity theory while completely ignoring the zero-crossing steps that would become, 30 years later, the basis for the most important and widespread practical use of superconductive microcircuits. So much for lost opportunities.

Sixteen years later in the early 1980s, I found myself in a group in the Cryogenics Division at NBS[60] in Boulder, Colorado. The group mission was to develop superconductor integrated circuit technology for the purpose of advancing electrical metrology. Catalyzing this research was a large program at IBM to develop superconductive logic for the next generation of ultra-fast computers. The IBM program pioneered the development of a superconductive integrated circuit process and this process was reproduced at NBS. This new technology opened the possibility to exploit Levinson's suggestion to put thousands of junctions on one chip and thus realize a Josephson standard at 1 V or more. The first attempts to do this were largely a failure owing to nonuniform distribution of microwave power and a lack of understanding of the chaotic nature of the microwave-driven Josephson junction and the consequences for stability. Eventually the microwave distribution problem was solved by Jürgen Niemeyer and Johann Hinken at PTB, and the stability problem was solved by Richard Kautz at NIST. A joint NIST/PTB effort brought these solutions together and culminated in the fabrication of the first successful 1474-junction array at NBS, in 1984.[61]

My entry into the development of the series array Josephson array voltage standard began shortly after Niemeyer and Kautz demonstrated the first 1 V chip. During the next year both PTB and NBS developed improved chips and the hardware and software required to implement the Josephson series array voltage standard.[62,63] Soon after 1985, NBS and PTB made array chips available to other national standards laboratories, and those labs developed their own measurement systems based on these chips. Early Josephson array voltage standards developed in the mid 1980s at NBS, PTB, NPL, and CSIRO made it clear that a well-funded standards lab with a highly trained research staff could reproduce the first JVS (Josephson Voltage Standard) systems built at NBS and PTB. In 1987, the NBS Boulder lab and the PTB lab in Braunschweig were the only two places in the world where these chips could be fabricated, and there was considerable interest from other labs in obtaining the new standard. So, the group in Boulder proceeded to develop a Josephson voltage standard system that could readily be reproduced and used by metrology technicians in many standards laboratories. By 1992, the NIST Boulder group had sold complete JVS systems to Lockheed, Hewlett Packard, and the US Army, Navy, and Air Force, and supplied chips and provided advice on constructing systems to many other standards laboratories throughout the world. Commercial sources were developed for every component of the system. The chip design and fabrication details were transferred to Hypres, Inc., a small company in New York that focused on practical applications of Josephson technology. Hypres became and continues today as a key supplier of JVS chips as well as complete turn-key systems. In the same way, PTB carried over its technology to the Institut für Photonische Technologien (IPHT) in Jena, Germany, and the IPHT founded a small company Supracon which delivers circuits and systems, too.

The design problem does not end with the completion of a practical and convenient voltage standard. A system like the JVS with a very small demand (at most a few systems/year) necessarily depends on components that are widely used for other purposes. As technology advances some of

---

[60] NBS became the National Institute of Standards and Technology, NIST, in 1989.

[61] J. Niemeyer, J. H. Hinken, R. L. Kautz, *Appl. Phys. Lett* 45 (1984) 478

[62] C. A. Hamilton, R. L. Kautz, R. L. Steiner, F. L. Lloyd, *IEEE Elec. Dev. Lett.* EDL-6 (1985) 623

[63] J Niemeyer, L. Grimm, W. Meier, J. H. Hinken, E. Vollmer, *Appl. Phys. Lett.* 47 (1985) 1222

these components become obsolete and unavailable. When this happens, the JVS design must be modified to use currently available hardware and software. This is an ongoing development process that is unlikely to end anytime soon. The remainder of this article focuses on the details of the JVS system and how the solutions to a variety of practical problems have evolved to make its dissemination possible. Since the author is most familiar with the NIST development, that is the story that will be told here.

Before delving into the details of the system design, one must ask what this system will be used to accomplish. The great majority of standards labs wanted a JVS system to calibrate secondary Zener-based dc reference standards[64]. Since the best Zener standards are stable at a few parts in 10^8, a JVS system with an uncertainty of a few parts in 10^9 would be perfectly adequate. A few customers wanted Josephson systems to calibrate and measure the linearity of high end digital voltmeters (DVMs), requiring about the same level of uncertainty. And then there are the metrology purists who are interested in verifying the underlying physics and uncovering system design flaws by comparing one JVS system to another with the smallest possible uncertainty. The NIST design emphasizes automation and convenience aimed at the first two classes of customers without precluding direct system-to-system comparisons at state-of-the-art accuracy.

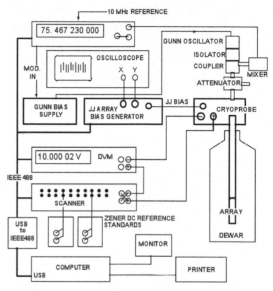

FIGURE 9.16: A typical JVS system. The figure shows a "dipper" system where the chip is cooled in a liquid helium Dewar. Liquid-helium-free cryocooled systems are also available.

The basic architecture of the system, shown in Figure 9.16, is fairly obvious, and essentially similar systems were developed at NIST and PTB. The system includes a means for cooling the chip to approximately 4.2 K in a magnetically and electrically shielded environment, supplying dc bias current for the Josephson junction (JJ) array and an accurately known microwave frequency (typically near 75 GHz), and transmitting the generated voltage to room temperature with a minimum of corruption from noise and thermal emf effects. Calibration of Zener standards is accomplished by placing the JJ array and the Zener in series opposition and measuring the difference voltage with a sensitive voltmeter. A reversing switch allows the computation of and correction for thermal voltages in the measurement loop. This reversing switch is usually a scanner that can select and reverse any one of 16 or more devices. An oscilloscope is usually required to find optimal operating conditions. Bias current for the JJ array is supplied from a variable impedance source that allows the display of I-V curves and computer control of the operating point.

The items shown in bold are specially designed for the JVS system. Over the years several small companies have made these devices, usually from designs developed at NIST or PTB. The remaining items are standard laboratory instruments that can be purchased commercially.

The computer hardware and software have seen the greatest need for updates to keep pace with advancing technology. The original system of 1989 operated under the DOS operating system using a QBasic program and communicated through the parallel port, an IEEE interface card and a special purpose digital/analog ISA card. Some of these first-generation systems are still in operation. In

---

[64] A semiconductor reversed-biased Zener diode is commonly used as the basis of commercial voltage references.

the meantime, the system has evolved to operate under Windows with much more capable control software written in Visual Basic and all communication to the DVM, microwave system, scanner and system controller through a single USB connector.

A software package (NISTVolt^{TM}) originally started at NIST has evolved over more than 20 years with contributions from many individuals. Its most used function is the automated calibration of a specified set of Zener dc reference standards. The basic algorithm of measuring the voltage difference between the Zener standard and the JJ array sounds simple enough, but when all contingencies are accounted for, this one function has about 2500 lines of code.

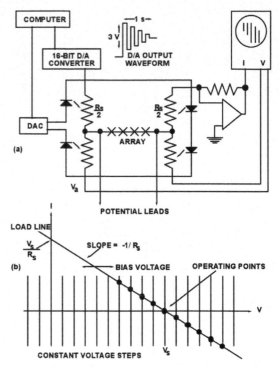

Software validation: How can a user be certain (and prove to an auditor) that the computer code is free from errors that could corrupt calibration results? NISTVolt addresses this with a simulation mode in which the JJ array, DVM, microwave system, and cable connections are all replaced by simulated devices that can be configured to include real-life defects, for example, an unstable JJ array voltage, measurement noise, thermal emfs, spurious DVM readings, DVM gain error, etc. By running the software in this simulation mode it is possible to quantify the effect of all these defects on the final calibration result and its computed uncertainty budget.

FIGURE 9.17: (a) Array bias system; (b) I-V curve and load line.

When the simulation results are combined with regular check standard measurements, systematic uncertainty evaluations, and interlaboratory comparisons, a JVS system achieves a very high level of confidence.

The array bias system shown in Figure 9.17(a) facilitates the process of setting the Josephson array on a constant-voltage step close to a specific voltage. It consists of a computer-controlled DAC and a circuit for controlling the bias impedance. Figure 9.17(b) shows the superimposed I-V characteristics of the Josephson array and the

FIGURE 9.18: A commercially available cryoprobe for a Josephson voltage system. Compliments of High Precision Devices Inc.

bias system. The vertical lines are the constant-voltage steps. The diagonal line is the bias-source load line. Its position and slope are computer controlled. The intersections of the load line and steps represent all of the possible operating points. If the load line is steep enough, then just one step will be selected. Unfortunately this solution is unstable and usually results in an oscillation among two or more steps. A better approach is to select a load line slope that intersects 10 to 40 steps and then bring the load line V-axis intercept to the desired voltage with a "ringing" waveform, as shown in the inset.

This usually results in the selection of a very stable step that is at, or very close to, the desired voltage. Once a suitable step is found, the four opto-isolators (shown as LED-controlled resistors) are turned off to establish the zero-current bias set point and to isolate the JJ array from any noise or disturbance in the bias system.

The microwave subsystem shown in Figure 9.16 is based on a Gunn oscillator and microwave source locking counter. This arrangement has been the most commonly used microwave source. As of this writing (2010) the availability and reliability of Gunn oscillators is on the wane and new sources based on a low-frequency (~ 10 GHz) synthesizer followed by a frequency multiplier/amplifier chain will likely become the dominant source in the future.

A typical cryoprobe is shown in Figure 9.18. Its function is to provide a means to cool the JJ array chip to approximately 4 K, deliver dc and microwave bias current, and to bring the precision voltage out to room temperature. The 75 GHz waveguide from room temperature to the JJ chip is a good example of how a critical component evolved over the years to improve performance and adjust to commercial availability. The initial solution was a standard bronze waveguide. About 1990, the Gore Company developed a flexible dielectric waveguide with lower attenuation and much lower thermal conductivity. With sales of only a few per year, Gore soon lost interest and priced the part beyond reach. At this point, NIST developed a rigid dielectric waveguide consisting of a 25 mm Teflon tube with a 2.5×5 mm rectangular core. About the same time PTB developed a tube waveguide that proved to be more durable. The original tube waveguide was made of German silver and later improved by using an internally silver-plated stainless steel tube. This evolution of system design is typical of many parts of the system and highlights the need for continuing technical support for users of JVS systems.

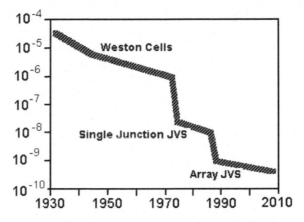

FIGURE 9.19: The approximate level of agreement in dc voltage measurements among standards laboratories through the years 1930 to 2010. The figure illustrates the major improvements that occurred as the technology evolved from Weston cells to single Josephson junction standards and finally to large arrays of Josephson junctions.

Figure 9.19 shows how the development of Josephson standards has improved the agreement among voltage standards laboratories around the world. Figure 9.20 illustrates the dissemination of these standards around the world. Although the Josephson series-array voltage standard based on zero-crossing steps has greatly advanced the state of the art in voltage metrology, it is not without its defects. First among these are the

FIGURE 9.20: The black dots show the distribution of Josephson standards around the world in 2010.

difficulty of rapidly setting a specified voltage and the tendency of noise to cause random jumps between the quantized voltage levels. As discussed in the following articles in this chapter by Johannes Kohlmann and Samuel Benz, new advances in chip design and fabrication, and very fast bias electronics now provide solutions to both of these problems.

9.7   Programmable Josephson Voltage Standards: from DC to AC

Johannes Kohlmann

9.7.1   Introduction

As described in previous sections, conventional Josephson voltage standards have been operated very successfully for dc applications since the mid 1980s. Some experts therefore already assumed in the early 1990s that the development of Josephson voltage standards would be completed. This estimation was then completely revised by new ideas. What is behind all this? Conventional Josephson array voltage standards do not enable switching rapidly and reliably between different specific voltage levels, due to their overlapping, metastable constant-voltage steps. The increasing interest in rapidly switching arrays and in highly precise ac voltages has stimulated several research activities to develop measurement tools based on Josephson arrays to meet these requirements. One approach is the programmable Josephson voltage standard (PJVS) first suggested by Hamilton et al. at NIST in 1995.[65] In principle, it is a Josephson multi-bit digital-to-analog converter based on a series array of overdamped Josephson junctions divided into a binary sequence of independently biased smaller arrays. While 2,048 junctions delivered nearly 300 mV in the first realization, modern versions contain more than 100,000 junctions for output voltages up to 10 V. Some of these fascinating developments are summarized in this contribution.

9.7.2   Principles and Fundamentals of PJVS

PJVS are intended to extend the use of high-precision Josephson voltage standards from dc to ac. This promising expectation of the suggestion by Hamilton et al. in 1995 did directly bring us to launch our own research work at PTB. PJVS are based on series arrays of overdamped Josephson junctions showing a non-hysteretic current-voltage characteristic. The current-voltage characteristic remains single-valued under microwave operation (Figure 9.21(b)). A PJVS is operated as a multi-bit digital-to-analog converter. The series array of M total junctions is divided into smaller independently biased programmable segments; see Figure 9.21(a). The number of junctions per segment often follows a binary sequence, the so-called binary-divided arrays. The output voltage $V = n \cdot M \cdot \Phi_0 \cdot f$ is given by digitally programming the junction step number n for the junctions in each segment (typically is $n = -1, 0, +1$); $\Phi_0 = h/2e \approx 2.07 \ \mu\text{V/GHz}$ denotes the flux quantum, h Planck's constant, e the elementary charge, and f the frequency of an external microwave. Each constant-voltage step between $-M$ and $+M$ can be selected by suitable programming of the different segments. A fast bias electronics enables switching times of a few tens of nanoseconds.[66] The number of junctions necessary to attain a given voltage is increased by a factor of five compared with conventional SIS[67] arrays, as PJVS are typically operated on the first-order constant-voltage step instead of the fifth one for SIS arrays.

9.7.3   Overdamped Josephson Junctions for PJVS

Overdamped Josephson junctions are needed for PJVS. Hamilton et al. demonstrated a PJVS for the first time using externally shunted SIS junctions.[65] In an array consisting of 8,192, 2048

---

[65] C. A. Hamilton, C. J. Burroughs, and R. L. Kautz, *IEEE Trans. Instrum. Meas.* 44 (1995) 223

[66] J. Lee, R. Behr, A. S. Katkov, and L. Palafox, *IEEE Trans. Instrum. Meas.* 58 (2009) 803

[67] Layered films are described by S: Superconductor, I: Insulator, N: Normal metal, I': doped semiconductor

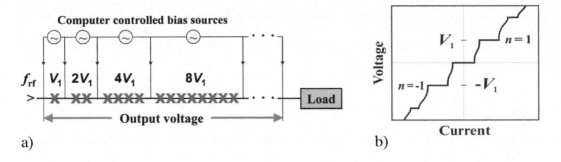

FIGURE 9.21: (a) Schematic design of a programmable Josephson voltage standard based on a binary-divided series array of Josephson junctions shown as '×'; (b) Current-voltage characteristic of an overdamped Josephson junction under microwave irradiation. The microwave power is set to equalize the widths of the zeroth and first constant-voltage steps.

junctions were operated at 75 GHz and delivered an output voltage of about 300 mV. As a design for externally shunted SIS junctions is rather complex and challenging, and the critical current and consequently the step width of these arrays are limited to a few hundred microamperes because of design restrictions, other junction types have subsequently been investigated. Calculations by Kautz gave important hints for the realization of optimized metallic-barrier Josephson junctions.[68]

Most currently fabricated series arrays are based on one of three different junction types: SNS junctions, SINIS junctions, and SI'S junctions, the barrier of which consists of a semiconductor such as Si doped with a metal and being near a metal-insulator transition. The low characteristic voltage $V_c = I_c \cdot R_n$ of SNS junctions leads to operating frequencies around 15 GHz (I_c denotes the critical current and R_n the normal state resistance). The characteristic voltage of SINIS and SI'S junctions can be tuned on the other hand over a wide range, enabling operation at frequencies either around 15 GHz or around 70 GHz. The latter is of special interest, from our point of view at PTB, as conventional Josephson voltage standards are also operated around 70 GHz and therefore the same microwave set-up can be used, and as the fabrication technology is less challenging due to the lower number of junctions necessary to reach a certain voltage level.

The first practical 1 V PJVS were realized at NIST by Benz et al. in 1997.[69] A total of 32,768 SNS junctions were embedded into the middle of a coplanar waveguide transmission line (CPW) and contained PdAu as the normal metal. The arrays were operated around 16 GHz and showed wide constant-voltage steps. The low resistivity of PdAu requires junctions of very high critical current densities around 200 kA/cm² in order to reach characteristic voltages of about 25 μV. This makes the technology very challenging, especially at the via contacts between the Nb top electrode and the Nb wiring. Different materials have been investigated at NIST in order to increase the resistivity of the normal metal and thus to make the technology more simple. An additional drawback arises from the low microwave frequency: To reach the 10 V level, more than 300,000 junctions are needed. An idea to handle this huge number is the use of stacked junctions. Silicide-based junctions were hence investigated at NIST. Voltages up to 3.9 V were generated by arrays of double- and triple-stacked junctions containing $MoSi_2$ barriers.[70] The huge number of junctions causes enormous challenges for microwave design and for the fabrication.

[68] R. L. Kautz, *J. Appl. Phys.* 78 (1995) 5811

[69] S. P. Benz, C. A. Hamilton, C. J. Burroughs, T. E. Harvey, and L. A. Christian, *Appl. Phys. Lett.* 71 (1997) 1866

[70] Y. Chong, C.J. Burroughs, P.D. Dresselhaus, N. Hadacek, H. Yamamori, and S.P. Benz, *IEEE Trans. Appl. Supercond.* 15 (2005) 461

Therefore, shortly after the first successful experiments at 16 GHz, we discussed in our group at PTB, whether practical PJVS could also be operated at frequencies around 70 GHz, which would result in a significant reduction of the number of junctions. In order to realize corresponding high values of 150 μV for the characteristic voltage, we needed materials showing a high electrical resistivity. We were still investigating etching processes for different metals, when the use of a very promising kind of junctions was presented for electronic applications at the International Superconductive Electronics Conference in 1997 in Berlin.[71] We subsequently investigated these junctions consisting of a SINIS multilayer in our group at PTB and fabricated our first small series arrays within a few months.[72] 1 V arrays were realized shortly after,[73] but nobody believed in 10 V arrays. Nevertheless, we gave 10 V a try and directly succeeded in 10 V arrays for the first time in spring 2000.[74] This success was partly caused by the use of 70 GHz junctions and by an active contribution of the junctions embedded into low-impedance microstriplines. We fabricated the SINIS arrays in a technology similar to that which is used for fabrication of conventional SIS junctions. The arrays consisted of 8,192 junctions (1 V) and 69,120 junctions (10 V), respectively.

In spite of their successful use, a serious drawback of SINIS junctions became visible in the course of time. Measurements of many arrays showed that a few junctions of SINIS series arrays are often missing, i.e., they exhibit a superconducting short (typically between 0 and 10 of 10,000 junctions). SINIS junctions seem to be very sensitive to some particular steps during fabrication, probably due to their very thin insulating oxide layers. This problem drives the search for more robust barrier materials. A very promising material was investigated at NIST consisting of an amorphous silicon layer doped with a metal such as niobium.[75] The niobium content is tuned to a value near the metal-insulator transition. This region combines a high resistivity and a sufficient conductivity needed for Josephson junctions. Nb_xSi_{1-x} barriers were initially used at NIST for pulse-driven arrays. We investigated at PTB in 2008 the use of junctions based on Nb_xSi_{1-x} barriers for operation at 70 GHz in close cooperation with NIST. The characteristic voltage was tuned to about 150 μV by adjusting the Nb content and the thickness of the barrier at NIST. 1 V and even 10 V arrays were then fabricated in our group at PTB.[76] Our first measurements confirmed that these new junctions are really more robust than SINIS junctions; fabrication is possible with a better yield, as the number of shorted junctions is significantly reduced. For the first time, programmable 10 V arrays consisting of 69,632 junctions were realized without any shorted junctions. These outstanding results were made possible by the close cooperation between our groups at NIST and PTB. This junction type presently enables the most reliable fabrication process.

Some other junction types were additionally studied recently. The advantage of array operation at higher temperatures than 4 K has been investigated at the National Institute of Industrial Science and Technology (AIST) in Japan, partly in collaboration with NIST.[77] Arrays were operated in a cryocooler at temperatures around 10 K by using NbN for the superconducting layers and TiN for the barrier. Yamamori et al. realized corresponding arrays, which consist of more than 500,000 junctions for operation at 16 GHz, generating voltages up to 17 V.[78] First steps were made toward fabrication of series arrays on a commerical basis. Hassel et al. (at the Technical Research Center of Finland (VTT)) exceeded the 1 V level using an improved design of 3315 externally shunted

[71] M. Maezawa and A. Shoji, *Appl. Phys. Lett.* 70 (1997) 3603; H. Sugiyama, A. Yanada, M. Ota, A. Fujimaki, and H. Hayakawa, *Jap. J. Appl. Physics*, 36 (1997) L1157

[72] H. Schulze, R. Behr, F. Müller, and J. Niemeyer, *Appl. Phys. Lett.* 73 (1998) 996

[73] R. Behr, H. Schulze, F. Müller, J. Kohlmann, and J. Niemeyer, *IEEE Trans. Instrum. Meas.* 48 (1999) 270

[74] H. Schulze, R. Behr, J. Kohlmann, F. Müller, and J. Niemeyer, *Supercond. Sci. Technol.* 13 (2000) 1293

[75] B. Baek, P. D. Dresselhaus, and S. P. Benz, *IEEE Trans. Appl. Supercond.* 16 (2006) 1966

[76] F. Müller, R. Behr, T. Weimann, L. Palafox, D. Olaya, P. D. Dresselhaus, and S. P. Benz, *IEEE Trans. Appl. Supercond.* 19 (2009) 981

[77] H. Yamamori, M. Ishizake, A. Shoji, P. D. Dresselhaus, and S. P. Benz, *Appl. Phys. Lett.* 88 (2006) 042503

[78] H. Yamamori, T. Yamada, H. Sasaki, and A. Shoji, *Supercond. Sci. Technol.* 21 (2008) 105007

SIS junctions operated at 70 GHz on the third order constant-voltage step.[79] Lacquaniti et al. (at the Instituto Nazionale di Recerca Metrologia (INRIM) in Italy) developed SNIS junctions based on a thick Al layer (up to 100 nm), which is slightly oxidized. 1 V SNIS arrays were subsequently fabricated within a cooperation between INRIM and PTB.[80]

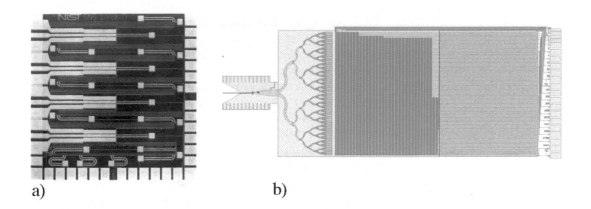

a)   b)

FIGURE 9.22:   Examples of PJVS designs: (a) 1 V design for operation at 16 GHz. 32,768 SNS junctions are arranged in the middle of 8 parallel coplanar waveguide transmission lines. The chip size is 10 mm × 10 mm. (Courtesy of S.P. Benz, NIST); (b) 10 V design for operation at 70 GHz. 69,632 SI'S junctions are embedded into 128 parallel low-impedance microstriplines. The chip size is 24 mm × 10 mm.

9.7.4   Realization of Series Arrays

The thin-film fabrication process for PJVS is based on the Nb technology of conventional Josephson voltage standards. Significant improvements at all technological levels in the last two decades enable the fabrication of series arrays containing a huge number of junctions with acceptable yield. This technological progress has been a major prerequisite for the development of PJVS. While the Josephson junctions are embedded into a low-impedance microstripline for arrays operated at 70 GHz similar to conventional SIS arrays, the junctions are integrated into the middle of a coplanar waveguide (CPW) transmission line for 16 GHz arrays. The ratio of the low junction impedance to the typical impedance of these CPWs of about 50 Ω leads to a similar situation as the microstripline for conventional SIS arrays. Attenuation of the microwave power is low because the junctions are loosely linked to the CPW. In contrast, overdamped junctions in a low-impedance microstripline attenuate microwave power at a much higher rate than for SIS arrays. The large attenuation follows from the fact that the resistive impedance of the junctions plays an important role, so that a major part of the microwave power is dissipated resistively. The high attenuation is, however, compensated in part by an active contribution of the junctions; the junctions act as oscillators.[81] Fortunately, we were unconcerned about high attenuation and active contribution when starting the development of SINIS junctions. We designed and fabricated the arrays, and they did work very well. Only later did

[79] J. Hassel, P. Helistö, L. Grönberg, H. Seppä, J. Nissilä, and A. Kemppinen, *IEEE Trans. Instrum. Meas.* 54 (2005) 632

[80] V. Lacquaniti, N. De Leo, M. Fretto, A. Sosso, F. Müller, and J. Kohlmann, *Precision Electromagn. Meas. Conf. Digest* (2010) 145

[81] H. Schulze, F. Müller, R. Behr, J. Kohlmann, J. Niemeyer, and D. Balashov, *IEEE Trans. Appl. Supercond.* 9 (1999) 4241

we realize the microwave behavior of overdamped Josephson junctions.

The main challenge for the design of PJVS is the homogeneous distribution of microwave power among the junctions. While this request is reasonably possible for 1 V arrays typically consisting of 8,192 junctions (70 GHz) and around 33,000 junctions (16 GHz), respectively, the requirements significantly increase for 10 V arrays containing nearly 70,000 junctions (70 GHz) and even more than 300,000 (16 GHz), respectively. The series arrays are therefore arranged in several parallel microwave lines. The CPW of 1 V arrays operated at 16 GHz is typically split into 8 parallel lines containing roughly 4,000 junctions each, as shown in Figure 9.22(a). Further microwave design improvements are needed for 10 V arrays. Tapered CPWs and improved power dividers have been introduced.[82] The arrays developed at AIST in Japan contain 524,288 SNS junctions, which are arranged in 16 parallel lines.[78]

The 8,192 junctions for operation at 70 GHz are integrated into 64 parallel microstriplines. The attenuation of microwave power is rather low, due to the small number of junctions (only 128) in each line. The width of the first-order constant-voltage step reaches the value of the critical current I_c, which enables the same width of both steps (Figure 9.23(a)), and is therefore near its maximum value of about $1.095 \cdot I_c$.[68,83] The number of junctions in each line has been increased for 10 V arrays up to 582 junctions for a single line; 69,632 junctions have then been arranged within 128 parallel lines as shown in Figure 9.22(b)[76] and Figure 9.23(b).

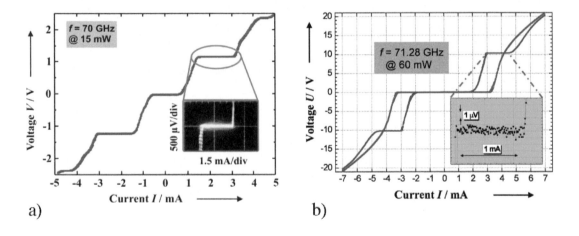

FIGURE 9.23: Current-voltage characteristics (IVC) of PJVS under 70 GHz microwave irradiation showing constant-voltage steps. The insets show the steps at high resolution: (a) 1 V array consisting of 8192 SINIS junctions. Zeroth and first-order steps reach the same width above 1 mA; (b) 10 V array consisting of 69,632 SI'S junctions. The graph also shows the IVC without microwave irradiation.

The dimensions of the junctions should be at the order of the Josephson penetration depth at most, as calculations have shown.[68] As the Josephson penetration depth is determined mainly by the critical current density of the junctions, their lateral size therefore ranges from a few micrometers for SNS junctions with critical current density above $100 \, kA/cm^2$ to about $20 \, \mu m \times 50 \, \mu m$ for SINIS junctions with a current density below $1 \, kA/cm^2$. As the step width of the constant-voltage step at

---

[82] M. M. Elsbury, P. D. Dresselhaus, N. F. Bergren, C. J. Burroughs, S. P. Benz, and Z. Popovic, *IEEE Trans. Microwave Theory Tech.* 57 (2009) 2055

[83] O. Kieler, R. Behr, F. Müller, H. Schulze, J. Kohlmann, and J. Niemeyer, *Physica C* 372 (2002) 309

1 V or 10 V should be about at least 1 mA for applications, the junction size is chosen such that the critical current ranges between 1 mA and 10 mA at most.

9.7.5 Applications

Good and reliable 1 V PJVS arrays have been available since the late 1990s; the availability of 10 V arrays has been considerably improved within the last few years. The number of applications has significantly increased in parallel, with 1 V arrays being initially used. An important first test for different junction types is the validation of the constant-voltage steps by a direct comparison with a conventional Josephson voltage standard. Several dc comparisons at 1 V and 10 V showed that PJVS enable the same very low uncertainty level of a few parts in 10^{11} as using conventional Josephson voltage standards.[73,76]

Since PJVS allow rapid and reliable switching between different voltage levels, they are now used in many different high-precision voltage applications. As the description of all applications is not possible within this brief survey, the interested reader is referred to many references.[84] Applications include, among others, calibrations, linearity measurements, potentiometers, watt balance experiments, metrological triangle experiments, and quantum voltmeters.

The synthesis of stepwise-approximated ac waveforms was an early promise connected with the development of PJVS. But initial precision measurements already revealed the difficulties in generating ac waveforms with low uncertainty due to the transients occurring during the switching between constant-voltage steps. Besides these errors, the accuracy of the rms voltage is limited by higher harmonics resulting from the digital nature of the generated waveform. The synthesis of waveforms is therefore restricted to low frequencies, such as power applications at 50 Hz or 60 Hz.[66,85] The use of sampling voltmeters additionally improves measurements at low frequencies by discarding those data points acquired in the transient regions. These investigations are partly discussed in more detail in the next article by Sam Benz.

9.7.6 Conclusions and Outlook

Programmable Josephson voltage standards have been the next step in the exciting story of applications of the Josephson effect in metrology. PJVS have now achieved a high level of maturity after roughly one and a half decades of development. The significant progress of the fabrication technology within this period has been a major prerequisite for the development of series arrays for PJVS. While 1 V arrays are routinely fabricated, first 10 V arrays containing tens or even hundreds of thousands of Josephson junctions are available by now. These arrays belong to the most complex circuits in superconducting electronics. The circuits have been used for several different applications demanding rapid and reliable switching between different specific voltage levels or highly precise ac voltages, respectively. Some of these exciting results have been achieved within the framework of fruitful cooperations and projects. PJVS will significantly improve applications like conventional Josephson voltage standards did for dc applications. I expect that PJVS will therefore be commercially available soon. They will then replace more and more conventional Josephson voltage standards, as PJVS are easier to operate and provide exciting additional possibilities and applications.

---

[84] For example, B. Jeanneret and S.P. Benz, *Eur. Phys. J. Special Topics* 172 (2009) 181, or in the special issues of the biennial Conference on Precision Electromagnetic Measurements (CPEM) in *IEEE Trans. Instrum. Meas.*

[85] C. J. Burroughs, A. Rüfenacht, S. P. Benz, and P. D. Dresselhaus, *IEEE Trans. Instrum. Meas.* 58 (2009) 761

9.8   Quantum-Based Voltage Waveform Synthesis

Samuel P. Benz

More than a decade of research and development was required to practically exploit the quantum behavior of superconducting Josephson junctions for ac applications. Sine waves and arbitrary waveforms had to be synthesized with sufficiently large voltage amplitudes; and measurement techniques with appropriate accuracy had to be developed. Two waveform synthesis methods, pulse-driven and stepwise-approximation, have now been implemented in a number of practical systems that are presently being used to calibrate audio-frequency ac voltages and power meters, characterize the stability and linearity of analog components, and as an arbitrary waveform source at the heart of an electronic primary thermometer. Some of the important milestones from these efforts, from my perspective, portray the importance of collaboration as well as the challenges that are typical of technology development. Most importantly, these events have shown how important it is to ensure the accuracy of electrical measurements, even when using quantum-based systems.

The first method that successfully synthesized quantum-based voltage waveforms was the programmable Josephson voltage standard (PJVS), which was conceived by Clark Hamilton in 1991. He proposed synthesizing step-wise approximated sine waves, similar to a multi-bit digital-to-analog converter (DAC), by independently current biasing series-connected arrays on different quantized voltage steps (Shapiro steps). During my postdoc at NIST in September of that year, Clark asked me to research prototype circuit designs. I based the circuit designs on aluminum-oxide-barrier tunnel junctions with external shunt resistors. At the time, many labs, including NIST, were using this junction technology for single-flux-quantum logic circuits, because the junctions were the most reproducible and they were non-hysteretic. PJVS operation required single-valued, stable, constant-voltage steps that were possible with such shunted junctions, which had electrical characteristics very different from those used in the conventional Josephson voltage standard (JVS). I simulated a number of prototype circuits for Clark's PJVS, including circuits in which the microwave bias was inductively coupled to series junctions or series SQUIDs. I chose not to fabricate series arrays with these inductively coupled circuits, since they would have required multilayered wiring, significant chip area per junction, and complicated microwave designs.

The first PJVS prototype circuit that I fabricated used direct-driven series arrays of junctions. In order to further simplify the layout, I chose to shunt each pair of junctions with a single shunt resistor. Although this circuit appeared to work correctly in simulation, the measured current-voltage characteristics (I-V curves) of the prototype circuit showed only half the expected voltage, indicating that only one junction in each shunted pair entered the voltage state. This result taught me the importance of including stability analysis in simulations and the importance of measuring real circuits! Three years later, and after much more research, Clark, Charlie Burroughs and Richard Kautz made and measured the first PJVS circuit with multiple arrays and with all (individually-shunted) junctions contributing to the output voltage. In research and development there usually are many false starts and lessons learned before the correct approach is apparent, as will be seen throughout this chapter.

During the early 1990s, a number of laboratories around the world were working on intrinsically shunted junctions with high-temperature superconductors (HTS). It was clear that the PJVS circuit would be much simpler and have higher junction density if the junctions were intrinsically rather than externally shunted. At that time, and it remains so today, it was difficult in HTS technology to fabricate a large number of junctions with sufficient uniformity to produce practical voltages. During my graduate research with Chris Lobb and Mike Tinkham, I had gained experience with superconductor-normal-superconductor (SNS) junctions. With the help of Martin Forrester and Horst Rogalla, I fabricated two-dimensional (2D) arrays of SNS in-line Josephson junctions with thin-film niobium islands patterned on top of a copper film. The Nb was deposited with a magnetron sputtering head that Horst built in Giessen. The Nb islands were patterned using a custom-made reactive-ion etching chamber. Having observed interesting quantum effects in these 2D circuits with planar SNS junctions, which were not particularly uniform, I was optimistic that planar trilayer SNS junctions might produce sufficiently uniform series arrays for the PJVS because the barrier would be defined by the normal-metal film thickness instead of a lithographically defined Nb gap having length of about 1 μm. However, most of my more experienced colleagues, including Richard Harris, were much less optimistic, and one expert insisted that SNS junctions simply would not work. This pessimism was well founded because the current density of SNS junctions depends exponentially on the barrier thickness.

Richard Kautz calculated in April, 1994, that SNS junctions would be feasible with regard to appropriate electrical characteristics. He showed that high $I_c R_n$ products would be difficult to achieve with AuPd and noted that higher resistivity materials would be needed for operation at frequencies of at least 70 GHz, the frequency typically used for the conventional JVS. I used AuPd as the barrier material in my first Nb-based SNS junctions, because I had recently implemented it as the resistor material for our single-flux-quantum (SFQ) digital circuits, because it had higher resistivity than that of InAu. Two challenges with producing these junctions were (a) modifying the wet aqua regia[86] gold etch to adequately remove the Pd, and (b) increasing the current density of the Nb wiring contacts to the junction's Nb counter electrode. By November, 1994, I successfully fabricated a wafer of test chips that had series arrays of junctions

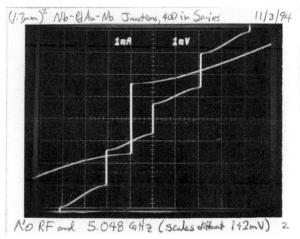

FIGURE 9.24: 1994 photo of the oscilloscope I-V curves of the first "uniform" 400-junction SNS array of Nb-AuPd-Nb trilayer junctions with (2 mV/div) and without (1 mV/div) microwave bias. The I-V curve without microwaves was exposed after moving the origin and changing voltage scale (1 mV/div). $I_c = 1.5$ mA, $R = 5.5$ mΩ/JJ, $(1.7$ μm$)^2$, and 43 nm thick AuPd.

with different areas and uniform electrical characteristics. Figure 9.24 shows I-V curves of the first uniform arrays that produced constant voltage steps with an applied microwave bias. The Shapiro steps are greater than 1 mA and are surprisingly flat[87], even though the junctions were not embedded within a microwave circuit and the microwaves were radiatively coupled to the test circuit with a wire coil in close proximity. Sometimes research and development produces unexpectedly favorable results.

The first 1 V PJVS circuit based on SNS junctions was demonstrated in 1995. It contained 32,768 junctions and implemented coplanar waveguide microwave designs, including filters and impedance-matching elements, and custom DAC bias electronics. The intrinsically stable steps were a unique feature, which allowed us to conceive, for the first time, of a turn-key Josephson voltage standard that could be fully automated to produce arbitrary voltages. For the next ten years at NIST,

---

[86] Aqua regia is a highly corrosive mixture of nitric and hydrochloric acids.

[87] Flat means constant in voltage over a large current range, which is called the "operating margin".

we applied the PJVS primarily to applications requiring stable and accurate dc voltages, such as supporting the watt balance experiment for measuring Planck's constant (see the next section) and helping Yi-Hua Tang build a new dc voltage dissemination chain, which allowed NIST to retire its traditional electrochemical standard cells.

With regard to ac synthesis with the PJVS, continued research on the SNS and the resistively shunted junction circuits showed that the slow ~300 ns rise times of the transitions between the quantized voltage steps limited the uncertainty of root-mean-square (rms) voltage measurements with the PJVS multi-bit stepwise-approximated waveforms, especially for the target frequencies above 1 kHz that were of primary interest for ac-metrology. However, interesting and useful rms measurements were performed at lower frequencies by Hitoshi Sasaki and Burroughs and separately by PTB researchers, who used SINIS-junction PJVS circuits[88] (see the preceeding section by Kohlmann). Both experiments characterized the frequency response of a thermal converter[89] with the fast-reversed dc method that used simple waveforms with only three voltages, which minimized the transient effects.

On the morning of July 26, 1995, John Przybysz and Hodge Worsham, superconducting electronics colleagues from Westinghouse Research and Development Center in Pittsburgh, visited NIST, Boulder. They presented to Clark and me their interests in developing DACs and analog-to-digital converters (ADCs) for U.S. Navy applications. I remember that the discussion began with the concept of waveform synthesis by switching between two Josephson voltage levels, similar to what we had demonstrated with the PJVS system, but with much faster sampling rates. We already knew that perfect quantization was compromised in stepwise waveforms, due to transients. However, we were all involved in research on SFQ digital circuits, which controlled the movement of SFQ through superconducting integrated circuits and changed their quantum states with properly timed junction pulses. Clark pointed out that waveform synthesis would be intrinsically accurate if the quantized voltage pulses of the junctions were exactly controlled and the resulting voltage would be proportional to the pulse spacing. It was typical to use "average voltage" measurements to characterize the time-dependent pulsed-voltage outputs of SFQ circuits. Semiconductor pattern generators were suggested as a possible bias signal because they could produce pulse waveforms by programming either of two voltages, depending on a digital pattern stored in memory. We realized that Josephson arrays could produce quantum-accurate waveforms by biasing them with a digitally controlled pulse waveform.

I was thrilled to have participated in this creative process, which required complementary expertise from all parties, and generated a potentially useful and important new idea. That afternoon I hunted for and found an HP8082A pulse generator[90] with an appropriately short, 2 to 10 ns pulse width that was capable of 10 to 250 MHz pulse repetition rates. The next morning I successfully produced quantum-accurate dc voltages with 100 μA margins by pulse biasing an SNS array circuit from a prototype PJVS design. It was very exciting to have experimentally realized the pulse-driven bias technique within 24 hours of conception. This was a critical first step toward demonstrating accurate ac waveform synthesis. I was also able to show that the step voltage changed with pulse period and that the current range of the step decreased with wider pulses, due to the frequency response of the junctions. Only the smallest array (the PJVS least significant bit) with 512 junctions (each having $I_c = 1.9$ mA, $R = 4.4$ mΩ, and $(2 \ \mu m)^2$ area) showed steps because of insufficient uniformity of the pulse signal in the longer arrays.

---

[88] A SINIS junction is formed with layers of superconductor-insulator-normal metal-insulator-superconductor.

[89] Thermal converters are rms detectors that determine the equivalence between alternating current (ac) signals and direct current (dc) signals by measuring the difference in their heating values.

[90] Commercial instruments are identified in this paper only to adequately specify the experimental procedure. Such identification does not imply recommendation or endorsement by the NIST, nor does it imply that the equipment identified is necessarily the best available for the purpose.

Figure 9.25 shows the first 265 μV voltage step that was produced with the pulse-driven array at the fastest 250 MHz pulse rate. The pulse amplitude was adjusted to center the first constant-voltage step on the zero-current axis. The voltage was correct for the number of junctions and the pulse frequency, and corresponded to exactly one quantized Josephson voltage pulse for every input pulse. The higher voltage steps are for higher order quantization (n = 2 and 3 Josephson pulses for every input pulse).

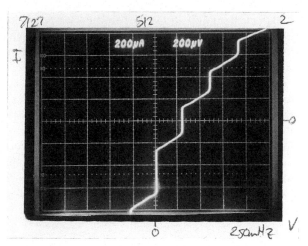

FIGURE 9.25: First pulse-biased SNS array I-V curve of 512 series-connected junctions showing flat steps that indicated reasonable pulse uniformity.

Flat steps were larger than expected, for two reasons. First, the 250 MHz pulse frequency was much slower than the 4 GHz junction characteristic frequency, so that the current ranges of the steps were inherently small, due to the junction response. Second, the pulse waveform reaching each junction was probably not uniform because the PJVS circuit had multiple bias taps with filters designed for 11 GHz. This PJVS circuit was intentionally chosen because it didn't contain on-chip blocking capacitors, which would have blocked the low-speed pulse drive. Sometimes your best available resources (such as the circuits and pulse equipment, in this case), can still produce useful results.

A few days later, I discovered the 1996 paper by Maggi that showed his simulations of pulse-driven junctions. For many months, we were unaware of a related 1990 paper by Monaco. Both of these researchers were interested in "step-width" enhancement of zero-crossing steps, not waveform synthesis, for increasing the current range of dc voltages. Maggi's simulations gave us confidence that our experimental results were reasonable. I began my own simulation investigation of pulse-driven junction dynamics by modifying one of Kautz' FORTRAN programs[91] and characterizing the junction response with different pulse-drive and junction parameters, including different pulse shapes. Research then shifted to understanding digital waveform synthesis (with Richard Schreier's delta-sigma modulator Matlab programs) and oversampling techniques, and characterizing the performance of state-of-the-art commercial pattern generators. This research gave me an acute appreciation of the broadband nature of our pulse-drive waveforms and how significantly more challenging it would be to ensure that all junctions receive the same pulse bias as compared with the single-frequency continuous-wave (CW) biases of both the PJVS and conventional JVS. It became necessary to develop lumped-element filters and other microwave circuit techniques that were appropriate for broadband waveforms and our specific applications.

Increasing the rms voltage to at least 100 mV was the greatest challenge to making the ACJVS a useful system. At lunch one day in the NIST cafeteria, Charlie Burroughs and I were discussing different schemes to produce bipolar pulses. On a napkin we sketched I-V curves and pulse waveforms. We considered combining the two-level pattern generator signal with a CW microwave signal and speculated that pulses of both polarities might be possible if the CW frequency were 3/2 that of the pattern generator's clock. Charlie arrived the next morning with calculated combined waveforms that showed bipolar pulses. Experimentally combining the two bias waveforms required creativity, such as minimizing attenuation of the two-level pattern generator signal by using a microwave coupler (in reverse). Developing techniques to optimize the phase and amplitude of the two bias signals

---

[91] FORTRAN is a programming language widely used for scientific computing, especially in the previous century.

was also challenging. After many failed attempts to achieve operating margins, we finally synthesized the first bipolar pulse-driven sine wave with quantum-based accuracy, as shown in Figure 9.26. We found that the −52.7 dBc second harmonic and lower-amplitude third harmonic were due to nonlinearities of the differential preamplifier and spectrum analyzer measurement instruments. Every junction in this array produced exactly one quantized voltage pulse for every input pulse, so that the first perfect bipolar sine wave was synthesized with no measureable distortion above the noise floor (more than 90 dB below the fundamental tone).

It took many more years of technology development in fabrication, microwave circuit design, and bias techniques before circuits were able to produce useful 100 mV sine waves. Automating the ACJVS for ac-dc difference measurements enabled the first quantum-based ac voltage calibrations in 2007. Also during these many years of development Paul Dresselhaus, Yonuk Chong, Nicolas Hadacek, and many other collaborators helped improve the fabrication process and experimented with new barrier materials. The barriers we found that produced the most uniform and reproducible junctions were high-resistivity metal-silicides. These barriers enabled us to increase output voltage for both the ACJVS and PJVS circuits. The maximum rms output voltage of the ACJVS is currently 275 mV. We also worked with a number of collaborators and com-

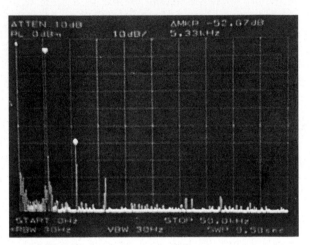

FIGURE 9.26: First spectrum (2/25/1998) that demonstrated operating margins of an ACJVS-synthesized bipolar sine wave and with the largest amplitude to date of 13.4 mV rms — 95 % of peak amplitude waveform, 1000 junctions biased with a 6.44 GHz clock and 9.66 GHz microwave drive.

panies to produce custom pattern generators that had more memory and performance optimized for the ACJVS. A German company, Sympuls LLC, now makes a ternary pulse pattern generator that directly generates bipolar pulse waveforms, which they developed in collaboration with our metrology colleagues at the NMi Van Swinden Laboratory in The Netherlands.

Understanding the limitations (and especially the systematic errors) of Josephson voltage standard systems is extremely important, and influences the measurement techniques used for a particular application. The most important step in ensuring that a Josephson system is producing an accurate voltage is demonstrating that it produces the same voltage for a range of values for every bias parameter. Even if a measured voltage appears repeatable and can attain a low uncertainty, it can still be inaccurate! The accuracy also depends on systematic errors, such as thermal voltages (for dc signals) and ac signals from other sources (electromagnetic interference or bias-related signals) or unexpected circuit paths (especially for frequencies greater than 100 kHz). Other effects that compromise the accuracy of measured voltage waveforms include the frequency response of the output transmission line, voltages induced from bias signals driving the inductance of the superconducting wiring between the junctions, nonlinearities from wiring connections (that produce distortion), and signals produced by the digitization process, which are particularly detrimental in rms measurements.

Some of the above challenges to measurement accuracy were reinvestigated beginning in 2005 as researchers again attempted to make rms measurements of PJVS stepwise waveforms at frequencies up to 2 kHz. It was hoped that a bias source with a faster 100 ns rise time that enabled shorter transients would reduce rms measurement uncertainties. Burroughs and other NIST colleagues showed

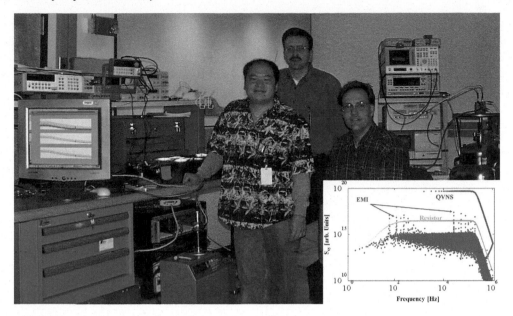

FIGURE 9.27:   First (November, 2001) cross-correlation measurement (left monitor display) with NIST Johnson noise thermometry electronics (on bench). Pictured are Sae Woo Nam, Wes Tew, and Sam Benz. The sense resistor probe is in the water cell inside the thermoelectric cooler (on the floor). The ACJVS system is in the upper right. Inset shows the first published spectra of the resistor and QVNS signals and challenges with EMI.

that the uncertainty of rms measurements of stepwise waveforms, even with rise times as short as 20 ns, was still limited by bias offsets and also by other bias parameters that influence the shape of an array's I-V curve, such as the applied microwave power. These "flat spot" measurement results convinced most researchers to focus on sampling measurement techniques, which can eliminate transient and bias-related errors when stepwise waveforms are measured.

In parallel with the rms investigations, Waldemar Kürten Ihlenfeld and Luis Palafox and colleagues used stepwise waveforms and direct sampling to calibrate the digital voltmeter and demonstrated the technique for power calibration. In 2005, Ralf Behr et al. proposed a differential sampling method with a PJVS voltage reference. Alain Rüfenacht and collaborators optimized and demonstrated this differential sampling technique by characterizing ac sources. At frequencies up to 200 Hz, the lowest uncertainty for a commonly used commercial voltage source was a few parts in 10^7, which was limited by the stability of that source. An order of magnitude lower uncertainty was found when comparing the multilevel stepwise waveforms of two PJVS signals (one source and one reference). In 2008, Waltrip et al. developed a new power calibration system for NIST that used differential sampling and an integrated PJVS as the voltage reference. The most challenging research aspects for this system were developing the stable semiconductor synthesized source, the permuting dividers, and the differential measurement technique. Frequently, when implementing a quantum system in a new application, the challenges may lie in integration or other instruments or components required for the measurement.

Finally, I would like to mention my favorite application of quantum-based waveform synthesis. It is my favorite because it has been and continues to be challenging, has required many smart and capable collaborators, and exploits quantum-accurate waveform synthesis in a completely new area for me, namely thermometry. In 2000, John Martinis conceived of improving Johnson noise thermometry (JNT) with quantum-based waveforms because they could provide a calculable, accurate, pseudo-noise voltage reference. Rod White (Measurement Standards Laboratory, New Zealand)

and Wes Tew (NIST) provided the thermometry expertise for our collaboration. Sae Woo Nam, with Martinis, constructed the cross-correlation electronics. The initial experiments were designed to demonstrate the new approach to JNT and to investigate differences between thermodynamic temperature and the ITS-90 temperature scale, which has been characterized primarily by gas and radiation thermometry. In Figure 9.27, Sae Woo Nam, Wes Tew and I pose with the JNT apparatus after the first measurements. More recently, we focused our JNT research toward producing an electronic measurement of Boltzmann's constant k_B at parts in 10^6 uncertainty.

The unique aspect of our JNT measurement technique is the use of a low-voltage version (only eight junctions) of the ACJVS, called a "quantized voltage noise source" (QVNS), which is used to synthesize a waveform constructed from harmonic tones of identical amplitudes and random relative phases. This quantum-accurate "pseudo-noise" waveform is used to characterize the amplitude-frequency response of the JNT electronics over its 1 MHz Nyquist measurement bandwidth. We have also used two-tone and multi-tone waveforms synthesized with both the QVNS and the ACJVS to uncover, characterize, and reduce nonlinearities by revealing the distortion in both active and passive electronics components, the input transmission lines and wiring, and the sampling ADCs. As with the power calibration system, the most difficult aspects of the JNT experiment are the non-Josephson components, namely, improving the low-noise measurement electronics. Horst Rogalla (University of Twente, the Netherlands), Jifeng Qu (National Institute of Metrology, China), Alessio Polarollo (Istituto Nazionale di Ricerca Metrologica, iNRiM) and Chiharu Urano (AIST/National Metrology Institute of Japan) are recent guest researchers working on the NIST JNT experiment and their improvements are now yielding a measurement of Boltzmann's constant with an uncertainty less than 12 μK/K.

In conclusion, there are two different quantum-based ac voltage sources that produce either step-wise approximated waveforms (PJVS) or perfect digital-to-analog conversion (ACJVS). New measurement techniques that use these systems have been developed and demonstrated for ac voltage and power applications. These systems and measurement techniques have the potential to shift the paradigm for measuring ac signals from one based on rms detectors to one based on intrinsically accurate voltage sources. The present region of impact in ac voltage metrology for each of the two systems is indicated in Figure 9.28. The uncertainties displayed in this plot, which are based primarily on rms detection with thermal converters, are those offered in 2010 by the NIST ac volt-

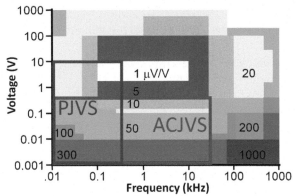

FIGURE 9.28: Voltage vs. frequency plot showing the uncertainty boundaries (in units of μV/V) for ac voltage calibrations at NIST (courtesy of T. Lipe and J. Kinard, NIST). Boxes indicate regions presently impacted by the PJVS and ACJVS quantum-based voltage sources.

age calibration service, which is used for calibrating thermal converters and voltage sources. The Josephson systems should be able to improve upon these uncertainties by at least an order of magnitude, and this has already been demonstrated at a number of frequencies and voltages. In order to fully realize the paradigm shift to quantum-based voltage sources, additional technology development will be required to improve the Josephson systems and measurement techniques, particularly in bias electronics and system automation. I am optimistic that the remaining technical challenges will be overcome, as usual, through collaborative interactions between the metrology and superconducting electronics researchers working in this field.

9.9 Superconductivity and the SI (Metric) System Based on Fundamental Constants

Edwin Williams and Ian Robinson

Superconducting measurements that help connect the unit of mass, the kilogram, to the Planck constant h are bringing the time when the scale for all units in the International System of Units (SI) can be defined directly through fundamental and atomic constants. The metric system, now known as the SI, has served science, engineering and commerce very well since it was adopted internationally at the signing of the Treaty of the Meter in 1875. One reason for its success is that the system is adaptive to change. New science and technologies, such as those associated with superconductivity, have provided scientists with a connection between the atomic world that is governed by quantum mechanics and the more familiar world of macroscopic measurements of voltage and current. We need to extend this link to the kilogram, which is defined as the macroscopic mass of a metal cylinder made of platinum-iridium that, since 1889, has been kept in a vault at the Bureau International des Poids et Mesures, BIPM, in Sèvres, a suburb of Paris. In 1967 developments in atomic time-keeping allowed the General Conference on Weights and Measures (CGPM), an intergovernmental organization which oversees the SI, to set the scale for the second by fixing the value of the ground-state hyperfine transition frequency of the cesium 133 atom. Likewise in 1983, advancements in optical frequency measurements allowed the CGPM to fix the value of the speed of light, c, which, in conjunction with the previously defined second, sets the scale for the meter. In this section we describe experimental progress of "watt balance determinations" that measure, in terms of the current definition of the meter, kilogram and second, accurate values of the Planck constant, h; from which can be derived both the electron charge, e; and the Avogadro constant, N_A. Sufficiently accurate measurements of h, e, and N_A will allow the scale of the units kilogram, ampere, and mole to be set by fixing the value of these three constants.[92,93] Then, the watt balance experiment, or similar ones, will be used to realize or maintain the unit of mass, much like atomic clocks presently maintain the unit of time. Of course, as in the past, we will ensure that the values chosen for these constants will minimize any discontinuity with the existing scale of the units. One clear advantage of this new SI is that the constants controlling the voltage and resistance scales will have fixed values.

Worldwide voltage measurements can be traced back to calibrations made using the Josephson effect, which produces a voltage dependent only on the Josephson constant $2e/h$ and frequency, as has been discussed earlier in this book. Likewise, accurate resistance measurements are calibrated against a quantum of conductance e^2/h (the inverse of the von Klitzing constant h/e^2), using the quantum Hall effect[94]. A current can be measured as the number of elementary charges, e, flowing per second via the voltage drop across a resistance measured in terms of the values of the Josephson and von Klitzing constants, respectively. However, in the SI, the unit of current is defined by the force between two wires and therefore must be related to the macroscopic kilogram mass standard. Experiments that presently measure the values of h or N_A in terms of the kilogram, if sufficiently accurate, can easily be reversed to measure mass in terms of fixed values of these constants. As the constants h and N_A are related by an expression that is more accurate than the present uncertainties of either h or N_A, one set of experiments can be used to provide a check on the other. At present, the two techniques display similar accuracies with the promise of even better accuracy in the future. A number of laboratories in the world are building, and will be operating, independent watt balance experiments; this will provide the significant advantage that results can be compared, and

[92] I. M. Mills, P. J. Mohr, T. J. Quinn, B. N. Taylor, and E. R. Williams, *Metrologia* 43 (2006) 227.

[93] E. R. Williams, *IEEE Trans. Instrum. Meas.* 56 (2007) 646

[94] K. von Klitzing, G. Dorda, and M. Pepper, *Phys. Rev. Lett.* 45 (1980) 494

then combined, to maintain the global mass scale with greater accuracy than any of the individual experiments.

The watt balance uses the quantum Hall resistance and the Josephson volt to measure virtual electrical power in an experiment that equates virtual mechanical (SI) power to virtual electrical power. This is accomplished in a two-stage process, one stage using a balance that compares the force on a coil in a magnetic field to the force of gravity on a kilogram mass. The second stage is a clever technique suggested by Bryan Kibble[95], which allows us to measure the quantity that relates the force and the current in the first part of the experiment, which depends on the magnetic field and geometry of the coil, by moving the coil in the field at a measured velocity and measuring the induced voltage. The heart of the apparatus is a coil of wire of effective length l, which is suspended in a strong magnetic flux density B that is arranged to produce a vertical force $F = Bli$ when a current i is passed through it. This force is compared to the weight, mg, of a mass m by suspending the coil and mass from one arm of a precise balance. The product Bl is measured by moving the coil ver-

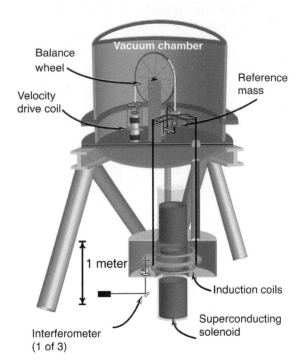

FIGURE 9.29: This two-story apparatus, the NIST watt balance, is used to measure h and calculate the electron mass and other atomic masses and constants.

tically with a velocity u and measuring the generated voltage $v_m = Blu$. By combining the two measurements, the virtual electrical power $v_m i$ can be equated to the virtual mechanical power mgu. These measurements can be used to calculate the Planck constant (see Eq. (5) in Williams et al.[96]), or the Introduction in Robinson[97]) and thereby reduce the uncertainties of many other fundamental constants, including the values of the SI elementary charge and the mass of the electron. The NIST watt balance apparatus will be used to illustrate a few details.

Figure 9.29 shows the configuration of the experiment. The axial force on a loop of wire in a purely radial field ($\mathbf{B} = [B_a(z)/r]\hat{\mathbf{r}}$, where $B_a(z)$ is nearly constant with vertical displacement, z, and time) is independent of the wire shape. A superconducting magnet consisting of two solenoid sections (shown in blue) wound in opposition produces a 0.1 T radial field outside the magnet Dewar. $B_a(z)$ varies by $+150\ \mu T/T$ from the center over a ± 40 mm vertical displacement, maintaining this variation over days. Because the field, and not the flux, must be constant, the magnet is operated in a constant-current mode with 5.6 A flowing in the series-connected solenoid sections.

The magnet has 200,000 turns and an inductance of 5000 H. Three induction coils, the center coil with 2376 turns and the upper and lower coil each having half that number, are located in the radial

[95] B.P. Kibble, in *Atomic Masses and Fundamental Constants*, J. H. Sanders and A. H. Wapstra, Eds., vol. 5. New York: Plenum (1976) 541

[96] E. R. Williams, R. L. Steiner, D. B. Newell, and P. T. Olsen, *Phys. Rev. Lett.* 81 (1998) 2404

[97] I. A. Robinson, *IEEE Trans. Instrum. Meas.* 58 (2009) 942

field. The lower and upper induction coils are fixed to the support structure of the balance, and the center moving induction coil is attached to a wheel balance located above the Dewar. The wheel balance, induction coils and their connecting rods are all in one vacuum system that is separate from the Dewar vacuum. Aligning the magnetic field perpendicular to gravity, and the inductive coils to the field, is essential, so that all forces and velocities measured are vertical. Thus, the balance is a 31 cm radius wheel that operates like a pulley, where the inductive coils, mass standards, and countermass hang from flat bands of 50 strands of wire rolling on the wheel, allowing the coil to move strictly vertically for 100 mm as the wheel rotates ± 10 degrees. The five degrees of freedom for the induction coil, other than vertical, are monitored, and excitations are damped. This monitoring of coil motion plus some mutual inductance techniques are used to align the experiment and to estimate the alignment errors.

For the velocity phase of the measurement, no mass standards are on the pans, and a small force on the countermass side is applied, via an auxiliary coil and permanent magnet, to produce a velocity u of about 2 mm/s that generates a constant voltage v_m (1.018 ± 0.001) V across the moving induction coil. We synchronously measure the time, the voltage difference between the moving and fixed induction coils, and the distance between these coils, eliminating voltage and motion common to both. Three interferometers, spaced equally apart on the coils, record the coil center-of-mass position, while three digital voltmeters integrate voltage against a Josephson volt standard between successive position readings with less than 200 ns dead time. The interferometry is performed in vacuum. The resulting v_m/u ratio has a vibration related noise of about 0.002 % and must be extensively averaged. The $B_a(z)/B_a(z = 0)$ variation is measured with 650 v_m/u measurements timed uniformly over 85 mm travel. The field's z-dependence is modeled with an eighth order orthogonal polynomial from hundreds of curves measured daily, which is then used in calculating the temporal changes $(\pm 0.1 \; (\mu B_a/B_a)/h)$ for each v_m/u ratio at the position where the weighings are made. A set of 10 up and 10 down velocities takes 30 minutes.

In the balance phase, a tare weight of 500 g placed on the countermass pan is balanced by a -10.18 mA servo current in the induction coil. As the 1 kg PtIr standard mass is placed on and off the pan, the moving induction coil current is reversed whilst causing minimal rotation of the balance. The reversal of the flow of the current through the 100 Ω standard resistor causes the voltage across its potential terminals to change between $\pm$ 1.018 V. The change is measured by comparison to voltages generated by a Josephson junction array. The four-terminal value of the resistor is known via calibration against a quantum Hall resistance standard. Five mass on-off sequences take 30 minutes. The result of each weighing phase is combined with the results of the before and after moving phases to produce a single measurement of the Planck constant or equivalently SI power. Both phases of the measurement have equivalent relative statistical uncertainties of 0.02 μW/W.

There are eight active experiments in progress in the world today that are planned to measure h, e, and N_A accurately. Six are watt balance type experiments: five at national measurement institutes, NMIs, of Canada (transferred from Great Britain in 2009), US, Switzerland, France and New Zealand, and one is at the BIPM. These six watt balance experiments each involve significant differences in design from the others, and we refer the reader to a summary report.[98] The most accurate watt balance measurement to date was reported by NIST,[99] but others may soon report even better results. The NMI of China is designing an energy balance experiment that measures the magnetic field and geometry by measuring mutual inductance gradients rather than measuring the voltage and velocity of a moving coil.[100] The eighth approach does not involve superconductivity or accurate electrical measurements. The AVOGADRO Project, an international collaboration involving research laboratories in Germany, Italy, Australia, Japan, Belgium, and the US, is measuring the

[98] I. A. Robinson, *IEEE Trans. Instrum. Meas.* 58 (2009) 942

[99] R. L. Steiner, E. R. Williams, R. Liu, and D. B. Newell, *IEEE Trans. Instrum. Meas.* 56 (2007) 592

[100] Z. Zhang, Q. He, and Z. Li, in *Proc. Conf. Precision Electromagn. Meas. Dig.* (2006) 126

number of atoms in a kilogram sphere made from an isotopically pure ^{28}Si crystal.[101,102] Agreement between the AVOGADRO Project and the seven experiments that involve superconductivity will demonstrate the universality of the measurements. At present, the AVOGADO project and the watt balance method display similar accuracies with the promise of even better accuracy in the future. A number of laboratories in the world are building, and will be operating, independent watt balance experiments; this will provide the significant advantage that results can be compared, and then combined, to maintain the global mass scale with greater accuracy than any of the individual experiments.

Figure 9.30 shows the results of a number of measurements, which have been transformed to yield the measured value of the Planck constant normalized to the value of the Planck constant h_{90} derived from the conventional values of the Josephson and von Klitzing constants K_{J-90} and R_{K-90}. There have been inconsistencies between the NIST watt balance, the NPL (National Physical Laboratory, Great Britain) watt balance, and the Avogadro results, but results presented at conferences but as yet unpublished (shown with a * in Figure 9.30) indicate that better agreement is being reached as possible sources of error are investigated in each of these ex-

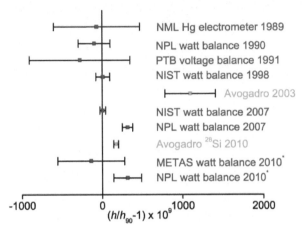

FIGURE 9.30: Measurements of the Planck constant h.

periments. Further results are expected and consistency will be the key to when a redefinition will occur. You will notice that the h_{90} value used to maintain the volt will not need a large adjustment when a redefinition occurs. The 1990 value h_{90} was a weighted average, and the breakthrough measurement at NPL[103] had the dominant weight. Figure 9.31 compares the changes in the sister copies of the International Prototype of the Kilogram, IPK, with a few hundred measures of the Planck constant made over four years, using the NIST watt balance. The IPK has been compared with its sister copies just twice in the last century. The Planck constant data uses a one kilogram Pt-Ir mass standard that was made and measured at the BIPM in 2004 and is periodically measured against the NIST Pt-Ir mass standard to provide a link back to the IPK. Tests are continuing at NIST to detect any systematic errors in the results. The vertical scale has the same sensitivity for both, but the IPK data cover 100 years, while the NIST data cover 4 years. The IPK data are not statistical noise, but are a measure of real differences between the IPK and its sisters. These differences, plus the fact that we must wait a half-century between calibrations of the world's mass standards against the IPK, are a driving force behind the move to an SI that includes fixed values of the three constants h, e, and N_A.

The connection between the quantum phenomena and the electrical measurements was first discovered 100 years ago. We are still learning new ways of using these phenomena to make our scientific world a better, more accurate community. The CGPM has the final authority to change the

[101] K. Fujii, A. Waseda, N. Kuramoto, S. Mizushima, P. Becker, H. Bettin, A. Nicolaus, U. Kuetgens, S. Valkiers, P. Taylor, P. De Biévre, G. Mana, E. Massa, R. Matyi, E. G. Kessler, Jr., and M. Hanke, *IEEE Trans. Instrum. Meas.* 54 (2005) 854

[102] B. Andreas, Y. Azuma, G. Bartl, P. Becker, H. Bettin, M. Borys, I. Busch, M. Gray, P. Fuchs, K. Fujii, H. Fujimoto, E. Kessler, M. Krumrey, U. Kuetgens, N. Kuramoto, G. Mana, P. Manson, E. Massa, S. Mizushima, A. Nicolaus, A. Picard, A. Pramann, O. Rienitz, D. Schiel, S. Valkiers, and A. Waseda,"An accurate determination of the Avogadro constant by counting the atoms in a ^{28}Si crystal," (2010) published on the web at http://arXiv.org/abs/1010.2317v1

[103] B. P. Kibble, I. A. Robinson and J. H. Belliss, *Metrologia*, vol. 27, pp. 173-192, 1990.

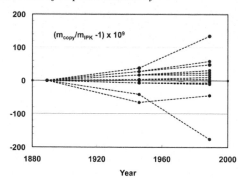 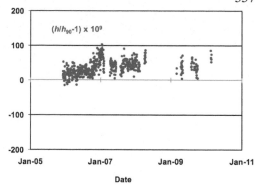

FIGURE 9.31: Historic comparisons of Pt-Ir kilograms (left) and a series of measurements of the Planck constant at NIST (right).

SI, and the scientific committees that advise them are recommending that e, h, and N_A be used as the constants to be fixed, but they have not yet recommended a timetable.[104] The CGPM meets every four years, and late 2011 will be the next one. The advantages of an SI based only on the constants of nature were foreseen by J. C. Maxwell in 1870, and, with the help of the quantum properties of superconductors, we are only now on the brink of achieving this goal.

9.10   Further Reading

For readers interested in more detailed discussion of metrology using quantum effects in superconductors, the following review papers may be helpful. They also contain many additional references to original literature.

1. R. A. Kamper, Superconducting devices for metrology and standards, in *Superconductor Applications: SQUIDs and Machines*, edited by Brian B. Schwartz and Simon Foner, Plenum Press (1976) 189.

2. J. Niemeyer, Josephson voltage standards, in *Handbook of Applied Superconductivity*, edited by B. Seeber, vol. 2, Institute of Physics, Philadelphia, PA (1998) 1813.

3. C. A. Hamilton, Josephson voltage standards, *Rev. Sci. Instrum.* 71 (2000) 3611.

4. R. L. Kautz, Design and operation of series-array Josephson voltage standards, in *Metrology at the Frontier of Physics and Technology*, edited by L. Crovini, T.J. Quinn, North-Holland, Amsterdam (1992) 259.

5. S. P. Benz and C. A. Hamilton, Application of the Josephson effect to voltage metrology, *Proc. of the IEEE* 92 (2004) 1617.

6. B. Jeanneret and S. P. Benz, Application of the Josephson effect in electrical metrology, in *Proceedings of the International School on Quantum Metrology and Fundamental Constants*, Les Houches, France (2007), edited by F. Piquemal and B. Jeckelmann, published jointly by EDP Sciences and Springer Verlag in *The European Physical Journal Special Topics* 172 (2009) 181.

---

[104] Andrew Wallard, *Metrologia* 47 (2009) 103

10

Medical Applications

Editor: Harold Weinstock

10.1 Introduction
   Harold Weinstock 559
10.2 Medical Applications of Magnetoencephalography
   *Cosimo Del Gratta, Stefania Della Penna, Vittorio Pizzella and Gian-Luca
   Romani* 562
10.3 MCG Instrumentation and Applications
   Riccardo Fenici, Donatella Brisinda, Anna Rita Sorbo and Angela Venuti 582
10.4 MRI (Magnetic Resonance Imaging) Instrumentation and Applications
   Jim Bray and Kathleen Amm 602
10.5 Ultralow Field NMR and MRI
   John Clarke 610
10.6 Superconductivity in Medical Accelerators for Cancer Therapy
   Peter A. Zavodszky 619
10.7 Further Reading 625
   Acknowledgements 625

10.1   Introduction

Harold Weinstock

Of the many applications of superconductivity that are found in this volume, there is only one with which the general population in developed countries has any major contact. I refer, of course, to magnetic resonance imaging, or as it has become generally known, MRI. By some estimates, this is a greater than $5B per year enterprise with companies in Europe, Asia and the US involved in marketing their MRI systems to the medical community. There are other biomedical applications of superconductivity that will be covered in this section, some that have been around for as long as (or even longer than) MRI has. However, for reasons that are mostly nontechnical, these other applications have had rather limited success in developing a significant market. In some instances, these markets may never develop because of continuing improvements in other technologies that are less expensive and/or that require less training of medical technicians. Yet there is one application, namely magnetocardiography, that surely will find widespread use as a diagnostic tool for routine screening of otherwise undetected heart ailments and for triage in hospital emergency rooms when patients appear with chest pain. The only caveat to this prediction is that the development of other non-cryogenic, less expensive and equally sensitive magnetometers could replace SQUID-based magnetocardiography. Nevertheless, it is the use of SQUID magnetometry that has opened the door to magnetocardiography and that has served to prove its effectiveness in numerous clinical studies around the world. Furthermore, the introduction of closed-cycle refrigeration should make SQUID-based magnetometry more user friendly. In the end, reliable operation and overall cost (including shielded enclosures for the non-cryogenic systems) will determine which form of magnetometry becomes universally adopted. There have been many previous attempts to preempt SQUID technology with non-cryogenic forms of magnetometry, but, thus far, SQUID systems have maintained a clear advantage for sensitivity and reliability.

For the most part, biomedical applications can be divided into 2 major categories, those that are based upon (1) superconducting magnets, e.g., MRI and particle accelerators for cancer therapy, and (2) SQUID-based magnetic gradiometers, e.g., magnetoencephalography (MEG) and magnetocardiography (MCG). There is yet another application that uses superconducting pick-up coils to produce sharper MRI images when applied to human limbs. Finally, in this introduction, it is worth noting that there is a new form of MRI that operates at low magnetic fields (not requiring superconducting magnets), but that utilizes SQUID magnetometry to yield images that are competitive in some instances with those produced by conventional high-field MRI.

In this introduction there is focus on some extraordinary individuals whose vision gave rise to these biomedical applications of superconductivity. First, I would like to acknowledge 2 gentlemen who were in large measure responsible for the development of the superconducting magnet industry. I doubt whether either of them would have considered that one day the companies they founded would be characterized as "big business." I refer to Sir Martin Wood, who, along with his wife Audrey, founded Oxford Instruments (OI) in 1959 in their garden shed, and Carl Rosner, who with Paul Swartz, was "spun off" by General Electric (GE) to form Intermagnetics General Corporation (IGC) in 1971 to work primarily on construction of superconducting magnets. At the time of formation of these companies, the main customer base for superconducting magnets was university and government laboratories engaging in mostly basic and some applied research. The advent of MRI in the late 1970s changed all that dramatically. You can read about the development of MRI in the section of this chapter written by Jim Bray and Kathleen Amm. You will note their mention of the skepticism of some GE scientists regarding the penetration of the magnetic field into the human body. It was assumed that inside the body the field would be less than at the surface and perhaps somewhat distorted. Carl Rosner has recently confirmed that IGC's first order in 1979 for a superconducting magnet that was large enough for a human body to fit into its bore, came from the University of Wisconsin Medical School. IGC ultimately delivered the requested magnet, but Carl stated that at a price of $50,000 IGC probably lost money on the deal. In June of 2006 IGC was sold to Royal Philips Electronics for $1,300,000,000. In 1982 OI spun off its large-magnet manufacturing division into Oxford Magnet Technology (OMT), and in 1989 Siemens purchased 51% of OMT from OI. In November 2003 Siemens purchased the remaining 49% of OMT, and renamed the company Siemens Magnet Technology in June 2004. The first whole-body magnet built by OI in 1979 was a conventional resistive electromagnet, and just a year later it manufactured its first superconducting whole-body magnet. I recall a plenary talk that Sir Martin Wood gave at the 1994 Applied Superconductivity Conference in Boston. He recounted a profile of him and OI by a financial newspaper. The article he quoted referred to him as a rather odd CEO who wore sandals on his feet and drove an 11-year-old automobile. He then said he still wears sandals and that the car he drives was then 18 years old. He concluded by listing about 5 or 6 well-known international companies and concluded by saying that the previous year OI had declared a profit that was greater than that of all of the companies listed combined.

In citing people who have made a difference, I would like to mention Ed Edelsack, an Office of Naval Research (ONR) program manager who in the late 1960s had initiated programs in superconducting electronics and biophysics. In doing so, he was the catalyst that brought Jim Zimmerman together with David Cohen. At that time Jim was a scientist employed by the Philco-Ford Corporation, although he left shortly thereafter to help form a company, SHE (standing for Superconducting Helium Electronics), and he then moved to the National Bureau of Standards (now NIST) in 1970. Ed was funding David Cohen at MIT's Francis Bitter National Magnet Laboratory under his biophysics program, and was funding Jim Zimmerman under his superconducting electronics program. The son of a medical doctor, Ed had a keen interest in trying to employ the then-new SQUID sensors to detect signals emanating from the human body. In late December of 1969, Cohen, Zimmerman and Edelsack came together to test the ability of Zimmerman's RF SQUID (co-invented with Arnold Silver) to detect signals from the human heart and the human brain under the favorable conditions created inside the magnetically-shielded room recently constructed by Cohen. The ability to detect a magnetic signal from the human heart using a SQUID was accomplished on December 31, 1969. Later, other signals from the brain were observed using either the same shielded room at MIT by Cohen or a SQUID-based magnetic gradiometer as initially developed by Jim Zimmerman and N. V. Frederick.

Ed Edelsack continued his interest in exploring medical applications of SQUID gradiometry and arranged for the purchase of a commercial (single-channel) second-derivative SQUID-based gradiometer by the Naval Research Laboratory (NRL) in the hope that he and Marty Nisenoff at NRL could take it to some nearby medical center. Somehow they could not make the right connections with the medical community. Meanwhile, I had started to work for ONR part-time from Chicago in late 1979, and after some time, Ed provided funding that helped me take sabbatical leave at NRL from my academic position in Chicago. While the primary goal of the planned research with the SQUID gradiometer was in other areas, Ed encouraged me to seek interest from the National Institutes of Health. After a couple of abortive attempts I made contact with the Director of the Epilepsy Branch of the National Institute for Neurological Diseases and Stroke (NINDS). He saw the potential of the SQUID to move forward in locating "interictal spikes" noninvasively, as opposed to removing part of the skull and inserting numerous electrodes for several days. Over a period of a few months I worked with NINDS scientist/physicians and support engineers to help establish the potential for evaluation of epileptic patients. As my sabbatical year was ending, NINDS placed an order with BTi (the new name for SHE) for the first 7-channel system to be built. It will become clear from reading the section of this chapter on magnetoencephalography (MEG) that working with a 7-channel system unshielded, one reduces the time required to produce a magnetic map of the brain, but it is not at all close to what can be done with systems with over 200 channels operating in a shielded room. Nevertheless, an employee of BTi told me the order from NINDS brought new life to that company and to the field of MEG more generally.

Finally, I wish to cite an individual who made an enormous contribution to the field of MEG, namely Sam Williamson of New York University (NYU). Sam had a remarkable career in a variety of areas of physics, but his greatest influence was in MEG. He pioneered this field somewhat by accident, as his training was in low temperature condensed matter physics. At NYU at some faculty social event he met Lloyd Kaufman, a cognitive psychologist. From this chance meeting, the field of MEG emerged. Working with a single-channel system in a building sitting above an underground train line and without shielding, Williamson and Kaufman gave birth to MEG. One of Sam's favorite lines was that if they could obtain valid measurements at NYU, then it could be done anywhere. As their results became known, visiting scholars from around the globe came to engage in research at NYU and then returned to their home institutions to set-up their own MEG systems. Within a short time Sam had a joint appointment in the NYU Medical School. Sam looked for other neurological SQUID applications and mentioned the possibility of using MEG to learn more about Alzheimer's Disease. Sadly he was diagnosed with this disease when he was approaching 60 years of age. He had enough foresight to take early retirement. His collaborator, Lloyd Kaufman, himself recently retired then, arranged a 60th birthday, retirement event that drew dozens of senior scientists and former students from everywhere that MEG was being done. It was a truly stirring tribute. One of his former students spoke of how much Sam had meant to him, and not knowing of Sam's affliction, pleaded for him not to retire.

To end on a happier note, I wish to mention one of Sam Williamson's more amusing achievements. It seems that an acupuncturist realized that one of his needles had broken and that part of it was in his patient's body, although he didn't know exactly where. X-rays and MRI scans were of no help. Finally, someone recalled the work being done at NYU. Surveying the patient's body with the SQUID gradiometer, the embedded needle was located. It had to be removed by a surgeon. Afterwards a short article was prepared for publication in which the patient was referred to as Mr. Haystack.

I offer one final observation. In editing the various sections of this chapter, I realized that a considerable amount of the text involved medical terminology, some of which was foreign to me, even though I've had significant contact with the medical community. Since most of the readers of this book will be physical scientists and engineers, I either removed some of the more esoteric sections or provided some explanatory addenda. I somewhat suspect that this does not endear me to the authors, nor will it endear me to the readers without a strong connection to the biomedical community. Such is the nature of compromises. Despite this caveat, I invite the readers to enjoy the wonderful biomedical applications of superconductivity. Some have yet to reach their full potential, but all of them are quite interesting in my humble opinion.

10.2   Medical Applications of Magnetoencephalography

Cosimo Del Gratta, Stefania Della Penna, Vittorio Pizzella and Gian-Luca Romani

10.2.1   Introduction: The Origin of Magnetoencephalography

The first successful attempts to reveal magnetic signals generated by neural activity in the human brain were performed in the mid 1960s, and were carried out using a non-superconducting detector. The quality of those signals was, however, very poor and certainly inadequate for any possible medical use. The first measurement of a good-quality magnetic signal associated with human cerebral spontaneous activity (alpha rhythm) was carried out only at the very end of 1969. What had happened in that four-year time span? Indeed, in 1969 there were two important developments: i) Jim Zimmerman, a leading physicist in the field of low temperature physics and superconductivity, had invented (with Arnold Silver) the radiofrequency (RF) biased SQUID[1], creating the most sensitive magnetometer of its time; ii) David Cohen had built an efficient 5-layer magnetically-shielded room at MIT, thus providing a dedicated and necessary environment to detect the weak magnetic signals associated with bioelectric currents in the human heart and eventually in the brain. Cohen and Zimmerman were brought together by Ed Edelsack, a program manager at the Office of Naval Research with programs in superconductivity and biophysics. As a result, in late December 1969, the first superconductivity-based human magnetocardiogram was obtained, and a related paper was published a few months later[2] (see Figure 10.1).

At that time few people other than the 3 authors of that article would have imagined that this measurement had marked a milestone, opening a new avenue to the study of the brain. This, however, became more evident less than 2 years later, when David Cohen recorded the first good quality magnetoencephalogram (MEG) inside the same shielded room (see Figure 10.1c) using a commercial SQUID[3]. After this second paper, the interest of several scientists — mostly physicists — was aroused, and a number of laboratories started studying the brain with this new tool.

After this brief historical overview, it is appropriate to dwell a little on the reasons that made the detection of cerebral magnetic signals so challenging in view of the fact that its electric counterpart,

---

[1]   J.E. Zimmerman et al., *J. Appl. Phys.* 41 (1970) 1572.

[2]   D. Cohen, E.A. Edelsack, J.E. Zimmerman, *Appl. Phys. Lett.* 16(1970) 278.

[3]   D. Cohen, *Science* 175 (1972) 664

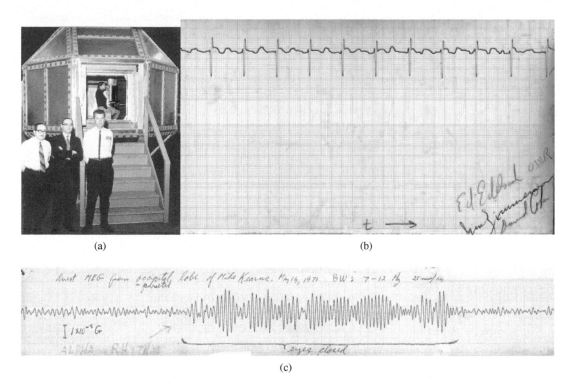

FIGURE 10.1: (a) Ed Edelsack, David Cohen and Jim Zimmerman in front of the MIT shielded room in December 1969. The first low-noise magnetocardiogram had been recorded therein using Zimmerman's SQUID. The long-tail glass dewar which contained the SQUID, is visible inside the room. (b) The first magnetocardiogram recorded in the MIT shielded room. It is signed by Edelsack, Zimmerman and Cohen. (Courtesy of David Cohen) (c) First MEG measured with a commercial SQUID, in the MIT shielded room (May, 1971). The subject's eyes were open at the beginning of the trace, then closed, resulting in the high amplitude alpha rhythm, then open again. (Courtesy of David Cohen)

the electroencephalogram (EEG), had first been recorded forty years earlier. As we shall see in the following paragraph, intracellular currents flowing inside neurons, the primary source of the magnetic signals, are indeed quite small, and the intensity of the generated fields strongly decrease with distance. Since detection occurs outside the scalp, the effective intensity of the field generated by a single neuron is extremely small, being of the order of a few attoteslas. Fortunately, the actual signals that are measured are generated by coherent activity of many thousands (or more) of neurons, and the overall field intensity reaches values of tens to hundreds of femtoteslas. An extremely sensitive detector is required to reveal such tiny fields, and no room-temperature device can provide such performance. The SQUID, however, immediately demonstrated an unrivaled sensitivity (as detailed elsewhere in this book) and, equally important, a wide operational bandwidth extending down to quite low frequencies, thus permitting satisfactory detection of the magnetic signals associated with cerebral activity, the frequency bandwidth of which typically spans from 0.1 to 100 Hz.

The existence of a sufficiently sensitive detector was only a partial solution to a two-fold problem, since one typically is required to measure this extremely small MEG signal in an extremely noisy environment. As will be seen, the amplitude of ambient magnetic noise is relatively quite large in a typical laboratory or hospital. For example, the earth's magnetic field, which is of the order of 50 μT, undergoes fluctuations in the low frequency range that can be several orders of magnitude larger than MEG signals. Therefore, this noise must be reduced by several orders of magnitude to permit successful detection. The most effective way to achieve such a reduction is to build a magnetically-shielded environment that, by using several layers of high-permeability magnetic material alternating with thick layers of a high-conductivity material, reduces the external unwanted signal down to an acceptable level.

10.2.2   Basics of MEG

Generation of Neuromagnetic Fields

Analysis of neuromagnetic signals recorded above the scalp requires an understanding of how these signals are generated by underlying neuronal currents. Since the measured magnetic signals are of low frequency (typically below 100 Hz) the Biot-Savart law is sufficient to describe the generation of neuromagnetic fields. In particular, it expresses the magnetic field generated by an infinitesimal current element.

$$dB = \frac{\mu_0}{4\pi} I dL \frac{r - r_0}{|r - r_0|^3}. \tag{10.1}$$

In the above expression, dB is the infinitesimal magnetic field generated at point r by the infinitesimal element of wire at point r_0, of length dL, and carrying a current of intensity I. The Biot-Savart law is usually integrated along the wire in order to obtain the total magnetic field.

What is the actual shape of electric current circuits in the brain? Current does not flow in a wire, but rather in a large conductive medium, the head tissue. This medium may be considered (for simplicity) to have homogeneous and isotropic conductivity, and, if the current is not too deep inside the brain, we may consider the medium boundary, i.e., the scalp, where the magnetic field is measured, to be plane and infinite, while more accurate models of the conductive medium will be described later. The current carriers are, on one hand, ions driven by neural activity in the intracellular space or across the cell membrane, and, on the other hand, free ions moving in the extracellular space in response to an electric field. We view the first type of ion flow as an active current, directly related to neural signaling. This current is usually called the impressed current. The other type of ion flow is a passive current, generated by the electric field resulting from charge distribution over the cell membrane. This current is usually called the volume current. Together, the impressed and volume currents close the loop to satisfy the principle of conservation of charge.

Given this complex current pattern, finding an expression for the neuromagnetic field appears to be a difficult task. Fortunately, to a good approximation, the magnetic field generated by volume currents is tangential to the scalp surface, while the magnetic field generated by impressed currents is orthogonal to the scalp surface. This property is rigorously true if the conductive medium is plane or spherical, as will be seen later, but it only approximates a real head. The contribution of volume currents can be neglected if one measures only the magnetic field component orthogonal to the scalp. Then one only needs to calculate the field of the impressed current. In the simplest case, that of a post-synaptic potential (PSP), the impressed current is the minute current flowing in a neuron dendrite due to membrane depolarization (or hyperpolarization) after neurotransmitter intake at the synapse. The PSP may be modeled as a small current element similar to the infinitesimal current element in the Biot-Savart law. Such a current element is called a current dipole and is represented by a vector, the current dipole moment, the direction of which is the current direction, and the magnitude of which is the product of length and current intensity (usually expressed as Am):

$$Q = IL. \tag{10.2}$$

According to the Biot-Savart law, the magnetic field generated by a current dipole is then

$$dB = \frac{\mu_0}{4\pi} \frac{Q \times (r - r_0)}{|r - r_0|^3}. \tag{10.3}$$

Is it possible to measure the magnetic field of a PSP? First, one must estimate the intensity of the corresponding current dipole moment. From Ohm's law, $Q = \sigma A \Delta\Phi$, where σ is the conductivity of the intracellular medium, A is the cross-section of the dendrite, and $\Delta\Phi$ is the potential difference along the dendrite. Inserting typical values (1 μm = dendrite diameter, $\sigma = 0.25\ \Omega^{-1}m^{-1}$, $\Delta\Phi = 10\ mV$) yields a current dipole moment intensity of 2×10^{-15} Am. According to Eq. 10.3, this current dipole generates a field of 3×10^{-19} T at a distance of five centimeters (where the field sensors are located). This field intensity is about five orders of magnitude lower than that typically measured in recordings of evoked activity. It also is four orders of magnitude lower than the sensitivity of SQUID magnetometers. This indicates that when one records neuromagnetic fields, one receives signals from a large number (about 50,000) of synapses activated simultaneously (or in synchrony), the fields of which add via the principle of superposition to yield a measurable value. Although it is not possible to record the magnetic field generated by the activity of a single synapse, the current dipole is still useful as a model of an assembly of synapses, corresponding to a small activated cortical patch, since 1 mm^2 of cortex may contain as many as 10^8 synapses. In this case, the current dipole is usually called an equivalent current dipole (ECD) to emphasize that it is a macroscopic current dipole, equivalent, by the principle of superposition, to a large number of microscopic current dipoles.

Another type of neural electric activity that generates magnetic fields is the action potential (AP). This is a depolarization wave flowing along the axon of a neuron, followed by a repolarization wave restoring the rest state transmembrane potential. The two waves are associated with intracellular currents in opposite directions. The AP may be modeled as a pair of current dipoles with opposite directions and equal intensity, i.e., a current quadrupole. Due to partial cancellation of the magnetic fields generated by the two current dipoles, the current quadrupole generates a total magnetic field an order of magnitude smaller than that of the current dipole. Although it is possible to detect the magnetic field of an AP from a single exposed axon in an animal model, as well as of APs from nerve fibers in the human noninvasively (e.g., above the arm), magnetic fields of APs from cortico-cortical or cortico-spinal connections have never been observed, likely due to weak signals and a lack of synchronicity. In summary, with MEG one is mostly sensitive to synaptic activity, and it requires several tens of thousands of synapses to be active synchronously to detect them. This large number of synapses may be modeled as an ECD. Therefore, the ECD is the basic current pattern in the brain with regard to MEG measurements.

What does the magnetic field of a current dipole look like? One may use the above equation for the magnetic field of a current dipole to calculate the orthogonal component of the magnetic field over the plane boundary of the conductive medium described earlier. Note that, if the current dipole is orthogonal to the boundary, its magnetic field will be parallel to the boundary and therefore will have no orthogonal component. In fact, for an orthogonal current dipole, the total magnetic field, including the field of volume currents, vanishes, as may be shown using Ampère's theorem applied to a circular loop centered over the current dipole and parallel to the boundary. This shows that MEG is mostly sensitive to currents flowing parallel to the scalp, while it is weakly sensitive to currents flowing orthogonally to the scalp. We shall comment on this in a later section, but for now we calculate the orthogonal component of the magnetic field of a current dipole parallel to the plane boundary. We use an orthonormal coordinate system with its origin on the boundary, the z axis orthogonal to the boundary, pointing outwards, and the x and y axes on the boundary. Let the current dipole Q lie on the z axis at a depth d below the boundary, and oriented parallel to the x axis. The 2D map of the orthogonal component of the magnetic field over the boundary is obtained from Eq. (10.3):

$$B_\perp = \frac{\mu_0}{4\pi} \frac{Qy}{(x^2 + y^2 + d^2)^{3/2}} \tag{10.4}$$

where x and y are the coordinates of a point on the boundary, and the current dipole lies at a depth d below the origin. This map is shown in Figure 10.2. Note that the magnetic field map has a maximum and a minimum for $x = 0$ and $y = d/\sqrt{2}$, and for $x = 0$ and $y = -d/\sqrt{2}$, respectively. The distance between these extrema is $d\sqrt{2}$, so it is proportional to current dipole depth, and is independent of the current dipole moment. Both extrema lie on the y axis, so that the current dipole is at right angles to the straight line joining the extrema, and lies halfway between the extrema. This shows that, given the position of the extrema, one can readily find the location of the current dipole. The intensity of the current dipole may then be calculated from the value of the field at the extrema, which is proportional to Q/d^2. Note also from Eq. (10.4) that the magnetic field above the current dipole vanishes, while its spatial derivative along the line joining the extrema is maximal. In summary, from an analysis of the magnetic field map over the scalp, one can infer the location of the currents in the brain.

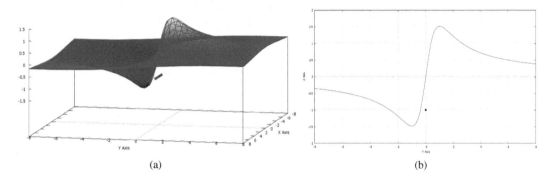

(a)   (b)

FIGURE 10.2:   Map of the orthogonal component of the magnetic field of a current dipole in a homogeneously conducting half-space. The boundary is in the xy-plane at z = 0, and is not shown on the figure. Units are arbitrary. The coordinates of the dipole are (0,0,-1). Left: 3D representation of the graph of function (2.4). Right: the same function along the y-axis.

Although the procedure is more complex than indicated in this simple account, as will be described later, the above equations provide the principle of 3D localization of neural activity using MEG.

Instrumentation

Detection Coils

Usually, the MEG field is measured by means of a superconducting detection coil (made of Nb) coupled to a low T_c SQUID. The main reasons for not using SQUIDs alone are:

i) The SQUID inductance should be as small as possible since its energy resolution is proportional to it — the energy resolution is $\epsilon \approx 9kTL_{\text{SQUID}}/R$ for a typical low-T_c SQUID.

ii) The geometry of the detection coil can define the spatial sensitivity of the instrument.

The first issue suggests that the SQUID loop should be as small as possible (≈ 100 pH), but this strategy is not adequate to measure MEG signals since the sensitivity to sources in the human brain would be dramatically reduced. Indeed, if the magnetic field generated by a current dipole is measured, the flux sensitivity of a loop is $\Phi = CA_1Q/r^2$, where Q is the source strength, r is the distance of the source from the loop, A_1 is the loop area and C is a constant. For typical values, loops with a diameter of the order of a centimeter should be used to increase the device effective area and

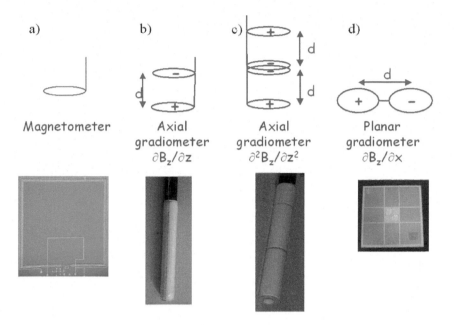

FIGURE 10.3: Schematic drawing and related pictures of a) magnetometer integrated on the same chip as a SQUID; b) wire-wound, first-order axial gradiometer; c) wire-wound, second-order axial gradiometer; d) two orthogonal planar gradiometers integrated on the same chip as the SQUID.

thus obtain a signal above device noise. The second issue suggests that the geometry of a MEG channel can be arranged to be sensitive to brain sources while rejecting background noise at the same time. Typical coils used to detect MEG signals are simple magnetometers, axial gradiometers and planar gradiometers, as shown in Figure 10.3. The magnetometer is a single or multi-turn loop whose flux sensitivity (as discussed above) indicates that this detection coil measures the signal generated by near and far sources, including environmental noise. Gradiometers are made of two or more magnetometers wound in opposite directions and displaced by a baseline d. Specifically, the loops of axial gradiometers are displaced along the loop axis, whereas in planar gradiometers the displacement is coplanar with the loops. Gradiometers provide reduced sensitivity to faraway sources, such as environmental and subject's noise (i.e., the heart signal). If we consider a first-order gradiometer as an example (Figure 10.3 b), the total signal is the difference between the flux detected by the lower coil (closer to the MEG source than to the noise source) and the flux detected by the upper one (far from both the MEG source and noise sources). The gradiometer principle is that noise has virtually the same value in both coils, with opposite signs, and therefore cancels. On the contrary, the source signal is much lower in the farther-away coil, due to the decay of signal with distance, and therefore does not cancel. How does one select the proper geometry for the MEG channels? The answer depends on the specific environment, since magnetometers pick up the signal of shallow and deep sources in the brain together with environmental noise, whereas gradiometers reject ambient noise, but suffer from a reduction of sensitivity to deep brain sources[4]. Magnetometers and planar gradiometers can be integrated on the same chip containing the SQUID, whereas axial gradiometers are made of Nb wire wound on a non-magnetic support (usually MACOR).

---

[4] Romani et al., 1982, NATO ASI series, for an extensive review.

Cryogenics

The MEG detection coil coupled to the SQUID should be cooled below the critical temperature of Nb (9.3 K), and the usual strategy is to place them in a bath of liquid helium at 4.2 K. The main feature of a cryostat for MEG is that it should be made of non-magnetic material, e.g., fiberglass. In addition, in order to place the sensors as close as possible to the scalp, the bottom of the cryostat must be much thinner (about 1 to 2 cm) than for typical cryostats. The cryostat consists of one reservoir and an outer container connected by a very thin neck (a few millimeters thick), to reduce thermal input by heat conduction. The chamber between the two shells is vacuum-pumped to reduce convection. In this chamber one or more thermal metallic shields and superinsulation made of blankets of reflective material are connected to the cryostat neck and reduce input by radiation. The boil-off gas cools the neck, the thermal shields and the superinsulation. Since the sensing channel should be as near as possible to the brain source, the distance between the inner and outer shells is smaller at the cryostat bottom than elsewhere, implying a larger thermal input. Additionally, thermal shielding is designed to produce low eddy-current noise, thus the shields are made of metallic stripes instead of a continuous sheet of metal.

Shielding

Since environmental noise can be up to 6 orders of magnitude higher than the brain's magnetic field, the use of magnetometers would be impaired and the operation of gradiometers would be troublesome. Additionally, RF interference could prevent the SQUID from being stable. To overcome this issue a MEG channel can be operated inside a magnetic shield. RF interference can be locally shielded by copper sheets wrapped around individual channels or around the cryostat itself. Alternately, since local shields can introduce eddy-current noise, a channel may be placed inside an RF-shielded room made of aluminum walls placed far from the sensor. Low-frequency shielding is based on high-permeability magnetic sheets, such as μ-metal, forming a room containing the MEG sensor and the subject to be investigated. The shape of the room determines the shielding factor, which is better for spherical rooms than for cubic ones, the latter being easier to fabricate. Additionally, the larger the number of concentric layers comprising the room, the better is the shielding. In summary, a shielding factor greater than 10^5 above 100 Hz and 10^2 at 0.01 Hz can be constructed using concentric layers of high-permeability materials and one layer for RF shielding. Magnetometers can be effective only with the use of high-quality shielding, e.g., a shielding factor greater than 200 at 0.01 Hz.

Read-out Electronics

The common scheme to drive the SQUID of a MEG channel is the flux locked loop (FLL), used to obtain a linear output from the SQUID. Negative feedback is used to stabilize the operating point of the SQUID. Many variants of the FLL scheme have been adopted, each aimed at improving the noise, dynamic range and/or bandwidth. Flux modulation has been used extensively in MEG sensors. Generally, a modulating flux with an amplitude about $\Phi_0/4$ is applied to the SQUID, the output of which is amplified and processed by means of a lock-in detector, utilizing the flux modulation frequency as the reference frequency. The resulting signal, which is the feedback signal, is a linear function of the flux difference $\Phi_{err} = \Phi_e - n\Phi_0/2$. The bandwidth required for MEG recordings is from DC up to a few kHz, whereas environmental background determines dynamic range, which may be as large as tens of nanoteslas.

Acquisition system

The read-out electronic unit is interfaced to an A/D converter (ADC) driven by a personal computer. Usually the ADC input range is matched to the sensor output, and ADCs with 20 to 24 bits are used to provide quantization noise[5] that is less than SQUID noise. The ADC includes a low-pass filter applied to sensor output to avoid aliasing, i.e., the passband of the sensor should be at most one half of the sampling frequency. Additionally, signal preconditioning, such as bandpass filtering, decimation and rejection of power-line noise, are available and can be applied by the operator.

Measurement procedure

One critical factor for practical MEG recording is subject positioning. It is necessary to know the position of the subject with respect to the sensor to co-register a reference system relative to the subject's head. This allows the operator to locate the physiological source of the magnetic field in the subject's brain with the aid of MRI. A typical set-up for subject positioning consists of a set of small coils—at least 3—placed on the subject's head at known positions, typically corresponding to anatomical landmarks. The coils are energized by an AC current before and after the recording session or even during it, provided that the frequency of the driving current is well above the band of cerebral signals. During the recording of the coil signals and the related brain signal, the subject is cautioned to keep his/her head stationary.

10.2.3   First MEG Studies

Following the initial results described in Section 1, many groups worldwide began to study the brain using these newly developed SQUID-based detectors. Foremost among them was the neuromagnetism group at New York University (NYU) led by Sam Williamson, a physicist, and Lloyd Kaufman, a cognitive psychologist. That group produced several important publications, not only in basic research, but also in the area of modeling. Indeed, in 1975—almost contemporaneous with Cohen's group[6]—the first visually-evoked fields were detected by Brenner and coworkers[7] at NYU, using a 10-Hz flickering grid as a stimulus. Some years later the same authors measured somatotopically-evoked fields under

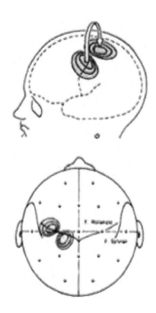

FIGURE 10.4: First attempt to measure a somatosensory magnetic field, evoked by electric stimulation of the thumb of the right hand. Note the spatial field distribution and the field direction over the somatosensory cortex. (modified from Brenner et al. [8])

steady-state stimulation of the thumb and of the little finger, providing a first hint of a possible somatotopy of the corresponding sources in the contralateral primary somatosensory cortex[8]—see Figure 10.4.

[5]   A.V. Oppenheim and R.W. Schafer. *Digital Signal Processing*, Prentice-Hall, Inc.: Englewood Cliffs, NJ, 1975.

[6]   D. Cohen D, *IEEE Trans. Magnetics*. MAG-11 (1975) 694

[7]   D. Brenner et al., *Science* 190 (1975) 480

[8]   D. Brenner et al., Science 199 (1978) 81

In 1978 the first auditory-evoked magnetic field was measured by Martin Reite and coworkers[9], and in the two following years some progress in the field was achieved by Farrell and coworkers in Detroit[10], and by Elberling and coworkers in Denmark[11].

It is worth stressing that this first decade of MEG measurements was characterized by studies that substantially replicated, from a magnetic point of view, studies that had been carried out earlier using EEG, thus demonstrating the existence of the magnetic "counterpart" of an electric cerebral signal associated with neuronal activity. This last statement is not meant to diminish the value of those first MEG findings; rather it is meant to indicate that the potential of the magnetic approach was not yet fully understood. Advances were made shortly thereafter by Williamson and Kaufman[12], who provided a theoretical analysis of the origin of neuromagnetic fields, their properties, and how they affect source localization.

It was only after this theoretical advance that in 1982 Romani, Williamson and Kaufman were able to demonstrate the existence of tonotopic organization of the human auditory cortex. It is worth noting that this research yielded the first demonstration of the capability of the neuromagnetic method to identify 3D sources in the human brain, and therefore it was considered a milestone in source localization, as well as in basic neuroscience. Romani and coworkers[13] presented the first evidence that the source of the evoked field elicited by auditory stimulation by tones with increasing frequency are systematically—and logarithmically—located at increasing cortical depth (see Figure 10.5). This result, which is similar to that found in animals, demonstrated that an equal number of neurons are devoted to each octave of sound, a result never previously obtained with EEG. Now researchers are accustomed to whole-head multichannel systems. Back in the early 1980s, in order to obtain data with a single-channel instrument in an unshielded environment and produce a signal map over the entire scalp with sufficient detail to allow source localization, it required many hours of data acquisition (often at night to reduce environmental noise). Imagine the stress for the volunteer subjects! Another major paper was published by the NYU group in 1982: an extensive review of biomagnetic instrumentation and modeling that in the years since has been considered a useful tool for both experts and

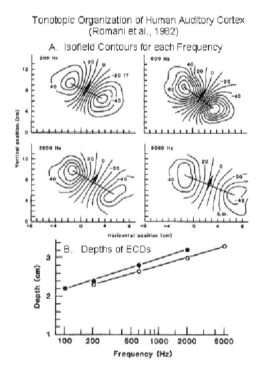

FIGURE 10.5: Upper: Isofield contours recorded over right hemisphere of subject SW. Arrows reflect position and orientation of sources for 200, 600, 2000, and 5000 Hz tones. The origin is at the ear canal. Lower: Depths of equivalent current dipoles (ECDs) for 2 subjects as a function of the logarithm of the frequency. Modified from Romani et al., 1982[9].

[9]   M. Reite, J. Edrich, J. T. Zimmerman, J. E. Zimmerman, *Electroenceph. Clin. Neurophysiol.* 45 (1978) 114

[10]   D. E. Farrell, J. H. Tripp, R. Nogren, T. J. Teyler, *Electroenceph. Clin. Neurophysiol.* 49 (1980) 31

[11]   C. Elberling, C. Bak, B. Kofoed, J. Lebech, K. Soermark, *Acta Neurol Scandinav* 65 (1982) 553

[12]   S.J. Williamson, L. Kaufman, *Mag. and Mag. Mat.* 22, (1981) 129

[13]   G.L. Romani et al., *Science* 216 (1982) 1339

newcomers to the field[14].

Other studies performed in the ensuing years using similar localization procedures yielded new results in complementary fields: Maclin and coworkers[15] observed — albeit qualitatively — retinotopic organization of the visual cortex; Okada and coworkers[16] revealed somatotopic organization of the somatosensory cortex, namely the somatosensory homunculus, and Hari and coworkers[17] observed signals originating from the (human) second somatosensory cortex. All of these new observations were later confirmed and extended using commercial instruments as well as self-made sensors, as the number of groups becoming involved in this "new" field of research was growing. New results were generated around the world, and with international collaboration. For example, an investigation of the spontaneous cerebral alpha rhythm was carried out in Rome in 1983 by Chapman and coworkers, providing the first tentative identification of widespread generators inside the occipital cortex.

In Figure 10.6 we see most of the experimenters participating in that study depicted around the self-constructed system featuring a DC SQUID for sensing.

Probably the first use of clinically relevant MEG was in the study of epilepsy. The possibility of studying the relatively strong magnetic signals originating from interictal "spikes" was intriguing, although the major difficulty experimenters faced was that quite often this activity is not localized. It involves a relatively large area in a cerebral lobe, then rapidly spreads to the rest of the brain. Identifying and following this activity with a single-channel system was definitely a challenge, even with the relatively large amplitude of epileptic signals. Nevertheless in 1982, Barth and coworkers[18] initiated a procedure that provided a solution to this problem. The "trick" was to record just the very

FIGURE 10.6: Some of the researchers working in the biomagnetics group at the Solid State Institute of the National Research Council in Rome in 1983. (second and third from left: Cosimo Del Gratta and Vittorio Pizzella; sixth and seventh: Bob Chapman and Gian Luca Romani)

first epileptic spike that initiates the more complex activity, and average it with similar initial spikes. When averaged data produced an acceptable S/N ratio, the sensor was moved to an adjacent position, with the same process repeated at that location. In this manner, the field distribution over the scalp was obtained, as well as the location of one source in one hemisphere. The authors also could identify a mirror focus in the contralateral hemisphere. In a subsequent paper the same authors were able to observe the spread of activity from one source to another in the same hemisphere.

During this pioneering period, enthusiasm for this new technique encouraged scientists to initiate a series of periodic meetings. The inaugural International Conference on Biomagnetism took place in Boston in 1976 and was repeated biennially around the world in subsequent years in such places as Berlin, Rome and Vancouver, and it continues to the present time, with an average of about 700 attendees. Last, but not least, a NATO Advanced Study Institute on Biomagnetism was organized by Williamson and Romani in Grottaferrata (Italy) in 1982, the proceedings of which have been used to

[14] G.L. Romani et al., *Rev. Sci. Instrum.* 53 (1982) 1815

[15] Maclin et al., *Il Nuovo Cimento D*, 2 (1983) 410

[16] YC Okada et al., *Exp. Brain Res.*, 56 (1984) 197

[17] R Hari et al., Acta *Neurologica Scandinavica*, 68 (1983) 207

[18] D.S. Barth et al., *Science* 218 (1982) 891

help train numerous students in biomagnetism in succeeding years. Additionally, the International Conference on Functional Source Imaging was held in 1986 at Villa Gualino, Torino (Italy). It attracted leading experts in the field and marked the first attempt to establish standards for MEG source identification.

10.2.4   A Step Ahead: Technological Developments

Modeling

As shown in a previous section, the magnetic field pattern generated by a current dipole over the scalp features a maximum and a minimum, with the distance separating them proportional to the current dipole depth. Conversely, in order to localize a small area of activated cortex, it is necessary to map the magnetic field over a surface including both the maximum and minimum. Therefore, the deeper the activation, the larger is the area that should be mapped. This is a simple example of the principle that, in order to increase the information content, it is necessary not only to collect more data points, but also to enlarge the area to be mapped. As one maps over a larger area, the assumption of a plane boundary for the conductive medium, which may be valid locally, is no longer valid globally, so one must resort to volume conductor models bounded in space. In particular, a spherical model is a better approximation of the head as a volume conductor. More generally, for the practical application of MEG, there is a need for more flexible models for data interpretation that, on the one hand, remove simplifying assumptions that may be unjustified in practice, and on the other hand, represent the biological medium more realistically. For example, the question arises as to whether the skull, as a source of inhomogeneous conductivity, has a strong influence on the magnitude and direction of the magnetic field at each point. Another example is the difference between the actual head shape and a sphere: Is a spherical conductor model accurate if the activity is in the temporal area, which is an area that clearly deviates from a spherical shape? Does one need a more realistic conductor model? In that case, if many sensors are located on a rigid surface (in the cryostat), it is not possible to place them all orthogonal to the scalp, so that the formula for the magnetic field of only primary currents is no longer accurate, and a more general formula including the magnetic field of volume currents is needed. Finally, it is possible that complex patterns of a magnetic field may be measured, likely due to the simultaneous activation of multiple areas. In some cases the resulting magnetic field map may not be interpreted in terms of a single ECD, and more complex current patterns must be envisaged, such as multiple ECDs, or higher-order terms of the multipole expansion of the current. We now present a few examples of volume conductor and neural activation models. A simple way to describe the inhomogeneity of the head as a volume conductor is to segment it into three compartments: brain, skull, and scalp. Although gross, this subdivision allows one to account for the presence of the skull which can influence volume currents due to its low conductivity with respect to the other tissues. A general equation for the magnetic field of currents inside a piecewise homogeneous conductor was given by Geselowitz[19]:

$$\mathbf{B}(\mathbf{r}) = \mathbf{B}_0(\mathbf{r}) + \frac{\mu_0}{4\pi} \sum_{j=0}^{k} (\sigma_{j+1} - \sigma_j) \int_{S_j} \frac{\Phi \mathbf{n}_j' \times (\mathbf{r} - \mathbf{r}')}{|\mathbf{r} - \mathbf{r}'|^3} da'. \tag{10.5}$$

In this relation, the homogeneous compartments are assumed to be nested, with the impressed currents flowing into the innermost compartment only. The field is given at a point $\mathbf{x}$ outside the outermost compartment. The total magnetic field $\mathbf{B}(\mathbf{r})$ at point $\mathbf{r}$ is the sum of the magnetic field of the impressed currents $\mathbf{B}_0(\mathbf{r})$. For the case of an ECD, this is equal to equation 10.3 and to the magnetic field due to volume currents. The latter term is expressed as a sum of surface integrals, one over each of the interfaces S_j between adjacent homogeneous compartments (where $\mathbf{n}_j'$ is the unit vector lo-

---

[19]   D.B. Geselowitz, *IEEE Trans. Magn.* 6(1970) 346

cally orthogonal to the surface and pointing outward). The sum is obtained from the innermost to the outermost surface. A comparison of the integrand with Equation 10.1 shows that each integral represents the total magnetic field of an array of current dipoles, orthogonal to the surface, of intensity $\Delta\sigma\Phi$ per unit area, where $\Delta\sigma$ is the conductivity difference across the surface, and Φ is the electric potential over the surface. Therefore, the calculation of the magnetic field of volume currents requires knowledge of the electric potential over the interfaces between compartments with different conductivity. The electric potential may be obtained by solving an integral equation[20]. Note that the current dipole distribution over the interfaces is fictitious; it is merely a practical means for calculation. The only real currents are the impressed and volume currents. Although seemingly complex, Equation 10.5 leads to a great simplification: First, it replaces a volume integral (the integral of the Biot-Savart law over the entire conductor), with a sum of surface integrals; second, it replaces the problem of solving for the volume current distribution inside the conductor with the problem of solving for the electric potential above the interfaces.

If the piecewise homogeneous conductor is spherical, the contribution of the volume currents is a vector tangential to the sphere, as may be understood by noting that the projection of the numerator of the integrand onto a radial vector $\mathbf{e}_r$ at point $\mathbf{r}$, $\Phi\mathbf{n}_j' \times (\mathbf{r} - \mathbf{r}') \bullet \mathbf{e}_r$ vanishes because these three vectors lie in a plane (with the origin taken as the center of the sphere). The radial component of the magnetic field depends only on the impressed currents. In the spherical case, it is possible to circumvent the calculation of the electric potential, and to obtain the magnetic field of a current dipole in closed form[21] by deriving it from the scalar magnetic potential, which in turn may be calculated by integrating the radial component of the field from a sphere's surface to infinity. Since the latter does not depend on conductivity, neither does the total field depend on conductivity. It is remarkable that, in this case, the difference in conductivity between compartments does affect the volume current distribution, but not the magnetic field outside the sphere

$$\mathbf{B}(\mathbf{r}) = \frac{\mu_0}{4\pi F^2}(F\mathbf{Q} \times \mathbf{r}_0 - \mathbf{Q} \times \mathbf{r}_0 \bullet \mathbf{r}\, \mathrm{grad}\, F) \tag{10.6}$$

where F depends only on $\mathbf{r}$ and $\mathbf{r}_0$ (not on $\mathbf{Q}$).

Equation 10.6 gives the total magnetic field (i.e., the sum of impressed and volume current contributions) at point $\mathbf{r}$, generated outside the sphere by a current dipole at point $\mathbf{r}_0$ with current dipole moment $\mathbf{Q}$. Note that, as already observed for a plane conductor, a radially-oriented current dipole does not generate a field outside the sphere, since in that case $\mathbf{Q} \times \mathbf{r}_0$ vanishes. Although this feature is true for a spherical conductor, it gives a hint of what can happen in a real head: Currents that are oriented "radially" will generate a very low field, and will be undetectable by MEG due to noise. Conversely, MEG will be most sensitive to currents flowing orthogonal to the radial direction, i.e., tangentially. As a consequence, since most PSP currents flow orthogonal to the cortex surface in the apical dendrites of neurons, MEG will be most sensitive to activations located inside the cortical fissures. On one hand, this limits the sensitivity of MEG because part of the activation is silent; on the other hand, this very limit, by reducing the detectable area of cortical activation, leads to more focused magnetic field patterns and makes the currents easier to localize. Stated informally, with MEG "we see a little less, but what we do see, we see more clearly than with EEG". It should be noted that the reduced sensitivity of MEG has not yet hindered its development, probably because the largest part of the cortex is located in fissures that include the primary sensory regions.

In summary, the spherical volume conductor model is a bounded conductor model that provides an analytical relation for the total magnetic field due to impressed and volume currents. As a consequence, the exact geometry (position and orientation) of the sensors may be incorporated into the model. Despite these advantages, the spherical model may still be inaccurate in some cases. Indeed, the human head does not have a spherical shape, and while a sphere will approximate the

[20] D.B. Geselowitz, *Biophys. J.* 7 (1967) 1

[21] J. Sarvas, *Phys. Med. Biol.* 32 (1987) 11

head surface locally, it will fail globally. Since the deviation from the real head shape may be a source of localization errors, researchers have chosen more realistic models based on numerical representations of an individual subject head.

Geselowitz's Equation 10.5 becomes quite practical for the implementation of realistic head models. In such models the surface integrals must be evaluated on numerically-represented surfaces, and are thus discretized. This is a particular implementation of a more general method called the boundary element method (BEM). From an MRI of the subject undergoing MEG, scalp, skull and brain compartments are segmented based on voxel gray values, and their interfaces reconstructed numerically, then tessellated with a large number of small plane triangular panels. Each triangle vertex is called a node and lies on one of the interfaces; its coordinates are obtained from MRI. Tessellated models of the head typically include several thousand panels, and field computations are costly. The above mentioned integral equation for the electric potential over the interfaces is also discretized into a linear system of equations that yields the electric potential either at the centroid of each panel of the tessellation, or at each node, depending on the particular implementation. The advantage of using nodes over triangles is that on each surface, the former are about half as many as the latter, thereby reducing the matrix size. At each triangle centroid, or at each node, a current dipole of appropriate direction and intensity is placed to represent the fictitious current density orthogonal to the interface. For example, for the case of the centroid-based implementation, the current dipole is orthogonal to the panel, with intensity $A\Delta\sigma\Phi$, where A is the panel area, $\Delta\sigma$ is the conductivity difference across the interface, and Φ is the potential at the centroid. An estimate of the tissue conductivity values also is required. Conductivity values are usually taken from the literature, and are based on post mortem measured values. In this respect, in vivo measurements of conductivity are of great value for improved accuracy. The scalp and brain have similar conductivities, while the skull has a conductivity about one hundred times smaller. Finally, the total magnetic field may be calculated from a discretized version of Equation 10.5[22]. An important issue is the accuracy of the calculation with the BEM. Due to the low conductivity of the skull, the potential values on the two outer surfaces (air-scalp and scalp-skull interfaces) are much lower than the potential on the inner one (brain-skull interface). In the linear system for the electric potential, the potential at one point depends on the potential at all other points from all surfaces. Therefore, inaccuracies in the potential on the brain-skull interface propagate to the potential on the lower-valued surfaces where they are larger in relative value, leading to poor accuracy. To improve the accuracy, one possibility is to increase the number of panels, thus increasing computational cost. Another possibility is to find the solution by a two-step method: First the potential on the brain-skull interface is approximated assuming the skull conductivity is equal to zero (with this assumption the potential on the two outer interfaces is zero), then a correction term is found using the actual skull conductivity and added to the approximate solution[23]. One finds that the correction is small and that the approximate solution with zero skull conductivity is already a good one. For this reason, the model often used in MEG data analysis is a single-compartment homogeneous model, where the boundary is the brain surface. This represents an excellent compromise between numerical accuracy and ease of computation.

In summary, with relatively simple analytical or numerical calculations, it is possible to take into account the gross inhomogeneities of the head as a volume conductor, as well as the complex shape of the head. An important feature of the model is the position and orientation of the magnetic field sensors. These must be recorded together with magnetic field data during the recording session. This is a distinctive feature of MEG — and also of EEG — compared to other brain imaging techniques. The most popular models for MEG data analysis are the homogeneous sphere, and the realistic, single compartment, head model.

To interpret the measured magnetic field data in terms of underlying cortical activity and its

[22] Barnard et al., *Biophys.* J. 7 (1967) 463

[23] M.S. Hämäläinen, J. Sarvas, *IEEE Trans. Biomed.* Eng. 36 (1989) 165

source, the data are usually fitted to the selected model. In addition to a model of the volume conductor (and sensors), a model of the source must be adopted. As described earlier, the basic source model is the ECD, but this may be too specific. For example, if two brain areas are active simultaneously or close to it, the magnetic field pattern will be, by the principle of superposition, the sum of two dipolar patterns. In such a case, a source model consisting of two ECDs would be more appropriate. A multiple ECD model is a useful generalization of the ECD model, with more ECDs used as more areas are activated. However, there are 6 parameters for each ECD: 3 for the position and 3 for the moment (or 5 if the spherical model is used); so that the number of ECDs that can be fitted in a single time sample in the case of 100 noisy data channels is limited. It is usually assumed that the ECD locations remain fixed during the time period of activation, which may last a few tens of milliseconds, while their moments are allowed to vary at each time data point. This reduces the number of position parameters to be estimated, thus increasing the robustness of the fit. Another widely-used source model consists of an array of a large number of fixed dipoles, usually covering the entire cortical surface. In this case the parameters to be estimated are only the ECD moments, so that the relationship between source and field is linear. The problem of source reconstruction is then a problem of linear estimation of an underdetermined system. Such a problem requires the application of a regularization procedure that introduces constraints on the possible solutions[24]. Typical constraints are, for example, the need for the current distribution to have some degree of smoothness[25], or that some dipoles have a higher probability of being activated[26]. Other types of constraints may incorporate information from other imaging techniques such as, for example, fMRI.

Instrumentation

The very first MEG instruments were single-channel ones, usually operated in an open environment using cryostats equipped with local RF shields. The gradiometric detection coils were coupled to RF SQUIDs. A picture of a typical single channel system during set-up and in the measurement position is shown in Figure 10.7. Since mapping of the magnetic field over the entire scalp with

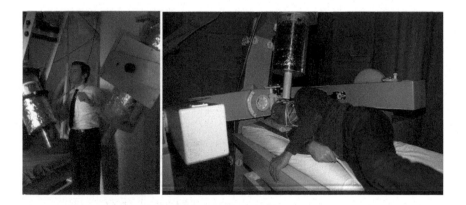

FIGURE 10.7: Left: Gian-Luca Romani setting up two rf-SQUID single-channel MEG systems. Right: Measurements being made by placing the sensor in different positions during repeated auditory stimulation.

[24] A.N. Tikhonov, V.Y. Arsenin, *Solution of Ill-Posed Problems* (1977), Washington, DC: Winston
[25] R.D. Pascual-Marqui et al., *Int. J. Psychophysiol.* 18 (1995) 49
[26] A.A. Iaonnides et al., *Inverse Probl.* 6 (1990) 523

dense sampling is needed to estimate the cerebral sources, usually the tip of the cryostat was placed at each of the sampling points of a regular grid where the evoked magnetic field was recorded. The stimulation sequence was repeated for every sample point; thus it took several hours to complete a set of measurements. Additionally, the recording conditions, such as the subject's attention and environmental noise, might change during measurement. Due to these instrumental limits, MEG in its early stages did not attract the interest of the clinical and neuropsychological communities. Fortunately, advances in SQUID design and fabrication technology produced DC SQUIDs with uniform properties and improved sensitivity ($\sim 20 \times 10^{-15}$ THz$^{-1/2}$), leading in the early 1980s to the first multiple sensor systems with on the order of ten channels. These early multi-channel systems covered part of the scalp so that either single-shot measurements were possible when focal sources were under study, or a reduced number of repositionings were necessary for more extended sampling. At this stage the first commercial systems (i.e., the twin 7-channel systems by BTi) were produced together with self-made systems, each representing a step forward in the development of whole-head systems. Even though these first multi-channel units featured only a small number of channels, they required unique cryostat design, read-out electronics and data acquisition systems. The sensing array, made of wire-wound gradiometers coupled to DC SQUIDs, was arranged on a curved surface, thus requiring larger cryostats with small curved bottoms. Suitable FLL schemes were initiated to reduce crosstalk among channels, for example, coupling of the feedback coil to the gradiometer instead of to the SQUID loop in order to null the supercurrent in the detection coil, or using separate shielded read-out electronics based on flux modulation. The parameters of each SQUID were manually set, and the output stored via the first multichannel ADC boards interfaced with a personal computer.

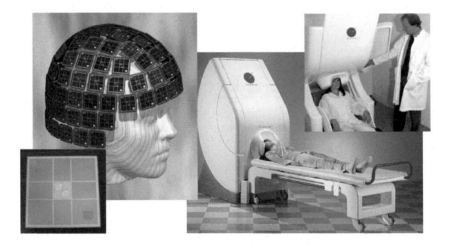

FIGURE 10.8: The Elekta Neuromag systems consisting of 306 channels arranged on 102 sampling positions on a helmet surface. Each measurement module consists of two orthogonal planar gradiometers and one co-planar magnetometer all integrated onto a single chip.

Medium-scale systems were implemented in the early 1990s, comprising about 20 to 40 channels, such as the 28-channel hybrid system in Rome, the 37-channel "Magnes" from BTi, and the 24-channel system in Helsinki. In each, the measurement surface was enlarged yielding a larger number of protocols for which single-shot measurements were possible. Additionally, these larger systems were placed in medium-quality shielded rooms that were made of one to two layers of high magnetic permeability materials and one layer of aluminum. At this stage, the number of MEG systems in the world increased thanks to their more stable performance. The first whole-helmet MEG systems, the Neuromag 122 and the 64-channel CTF system, entered the market. They represented the next step toward the large-scale systems operating today, one example of which is shown in Figure 10.8. With these systems featuring a few hundred channels, a single-shot measurement of the brain's magnetic field over the entire scalp is possible.

From the 1990s until now, advances in thin-film fabrication led to magnetometers and planar gradiometers integrated on the same chip as the SQUID, suitable to be densely packed over the measurement surface. Additionally, alternate read-out schemes better suited to a large number of channels, e.g., direct read-out with applied positive feedback and voltage bias with noise cancellation, were adopted in place of the standard flux modulation of the FLL. These schemes are simpler, with a smaller number of wires per channel fed through the cryostat and with reduced crosstalk among channels. The read-out electronics is set remotely by a digital interface, integrated in a user-friendly acquisition system. The use of whole-head systems required new cryostat designs to host the extensive number of sensors at the bottom, and as near as possible to the room-temperature surface. In some cryostats the helmet is placed with its axis at 45° to the vessel axis, allowing for measurement in either the supine or seated position. With these systems installed in high-quality shielded rooms, clinical studies are feasible due to advanced engineering and the resultant ease of operation, both from the point of view of the clinical investigator, and from that of the patient.

10.2.5 Brain Mapping and Clinical Studies

The development of multichannel MEG devices improved the quality of MEG studies and allowed scientists to investigate cross-modal sensory interaction or cognitive brain functions such as language or attention. An example of interaction of auditory and visual sensations is described by Sams and coworkers[27]. In that paper the magnetic field generated by the activity of the auditory sensory cortex was recorded while the subject was hearing a Finnish syllable "pa" repeated every second. At the same time the subject was looking at a videotaped face articulating the same syllable. However, in 16% of the cases the face articulated a different syllable: "ka". This discordance caused a clear difference in the measured field about 200 ms after the sound, thus supporting the idea that cross-modal effects can be observed even in primary areas.

One of the features of the human brain is its plasticity, i.e., the ability to change its functional configuration not only during development and maturation, but also whenever deviations from homeostasis are transiently or permanently imposed. This long-term plasticity involves changes in the synaptic configuration of large brain segments. This type of plasticity occurs over weeks or months during which the brain's wiring is profoundly reorganized. However, there must be different and more rapid plasticity. Short-term plasticity takes minutes to become effective and involves a type of software alteration of the brain's functional configuration. The first demonstrations of this form of plasticity using MEG were provided by Rossini and coworkers[28]. In this work the authors induced an anesthetic block of the sensory information from all but one finger of the right hand. Somatosensory-evoked fields (SEFs) were recorded during electrical stimulation of the unanesthetized finger, this finger being the 1st, 3rd or 5th finger. MEG data showed that only 30 minutes after the anesthetic block, the primary cortical area activated by stimulating the unanesthetized finger enlarged to include nearby areas usually responding to the anesthetized fingers (Figure 10.9). A similar study, but involving a long-term plasticity effect, was performed by Elbert and colleagues[29]. In it the somatosensory cortical organization of the hand was measured in nine musicians playing string instruments. Since string players use their left hand differently from the right one, the dif-

[27] M. Sams et al., *Neurosci. Lett.* 127 (1991) 141

[28] P.M. Rossini et al., *Brain Res.* 642 (1994) 169

[29] T. Elbert et al., *Science* 270 (1995) 305

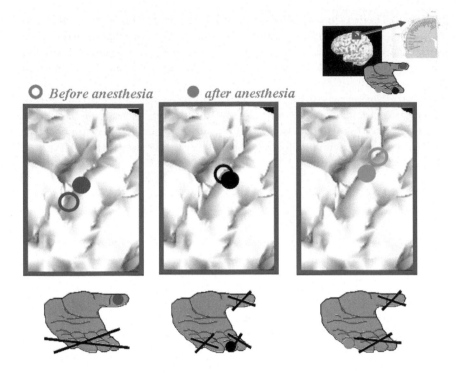

O *Before anesthesia*   ● *after anesthesia*

FIGURE 10.9:   Top right inset: "homuncular" distribution of the hand representation in the primary sensory cortex is observed by MEG. Bottom: transient rearrangements of finger representation in primary somatosensory cortex induced by an anesthetic block of the sensory inflow from adjacent fingers: the topography of the source responsible for the cortical responses from the unanesthetized finger changed significantly, invading the region devoted to the cortical representation of the other anesthetized fingers.

ference should be found in the brain's somatosensory areas. In fact, MEG measurements showed that the cortical representation of the digits of the left hand was larger than that in a control group, while no differences were observed for the representations of the right-hand digits. Moreover, the amount of cortical reorganization in the representation of the fingering digits was correlated with the age at which the person had begun to play. During the second half of the 1990s the development of whole-head MEG systems gave new vitality to the use of MEG in neuroscience, as well as in clinical studies. The spread of MEG beyond research laboratories matches the marketing of whole-head systems featuring over 100 SQUID sensors. The reason for this is simple: the human brain is very complex, and to understand its organization, one must detect the spatial distribution of the magnetic field associated with its operation as accurately as possible. Moreover, many diseases originate or result from incorrect interplay of different brain areas. Simultaneous detection of the activity of different neuronal pools is therefore necessary to highlight changes in neuronal processing or conduction that may result in dyslexia or more severe forms of cognitive impairment.

Basic Neuroscience

The human brain is comprised of two distinct hemispheres. Each hemisphere mainly collects sensory information from the contralateral side of the body, but there are important exceptions for relevant stimuli, such as those involving pain. Whole-head MEG systems have proven to be very ef-

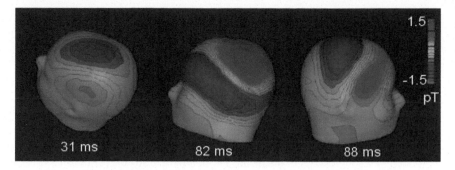

FIGURE 10.10: Isofield contour maps of the magnetic response recorded by magnetometers to galvanic right thumb stimulation at 31 ms, 82 ms and 88 ms after the stimulus. Positive values indicate the magnetic field emerging from the head. The dipolar field distribution suggests the presence of a cortical source in the contralateral primary somatosensory area (left), contralateral (middle) and ipsilateral (right) secondary somatosensory area with different latencies.

fective in detecting cortical activation due to somatosensory stimulation, and most importantly, they serve to quantify the possible difference in latency between the response of the two hemispheres. An example of this ability is provided in one of the simplest MEG protocols: the detection of the magnetic field generated by brain activity after electric stimulation of the thumb. When looking at the brain's response after stimulation of the thumb of the right hand, the following observations are made (Figure 10.10). Complete coverage of the head enables detection of all three major neuronal pools synchronously activated by the stimulus and the quantification of the latency difference between the first response of the left hemisphere (contralateral to the stimulus) and the later response occurring in the so-called "second somatosensory area" of both hemispheres, but with the ipsilateral — meaning same side - responding a few milliseconds later, since information must travel farther to reach the target area. The second somatosensory area response changes with different stimulus intensity. If the stimulating current exceeds the pain threshold, the cortical activation is larger and new areas activate synchronously, thus suggesting a different parallel pathway.

Language studies also benefited from the development of whole-head MEG systems. One of the first studies performed using the Neuromag 122-channel system concerns picture naming. During measurement, the subject looks at simple black and white figures and names them overtly (aloud) or covertly (silently). Through MEG, it was possible not only to identify several brain areas involved in the task, but also to establish precise arrival times. During this naming process, the translation from visual to symbolic representation progressed bilaterally from the occipital cortex, where the visual areas are located, to temporal and frontal lobes. In addition to language areas being involved during overt and covert naming, these areas also exhibited a response to simple passive viewing, although more weakly or after a longer delay[30].

Clinical Studies

One of the first clinical studies made possible by whole-head MEG systems concerns dyslexia. Dyslexia is a learning disability that impairs reading ability, yet it cannot be attributed to diminished intellectual capacity. Dyslexia is distinct from reading difficulties resulting from poor or inadequate reading instruction. MEG measurements during visual presentation of words (passive viewing of concrete/abstract words, pseudowords and letter strings) revealed differences between control and

---

[30] R. Salmelin et al., *Nature* 368 (1994) 463

dyslexic adults 0–200 ms after word presentation in the left inferior temporo-occipital cortex (word form area) followed by differences of 200–400 ms in the left temporal lobe. These results suggest that dyslexia is related more to the timing of signal processing rather than to a difference in sensory performance[31].

Epilepsy is the brain disease where magnetoencephalography proved to be most effective. The main use of MEG in epilepsy studies is the localization of pathological activity in patients with focal epilepsy. The goal of epilepsy surgery is to remove the epileptogenic tissue while sparing healthy brain areas. Knowing the exact position of important brain regions (such as the primary sensorimotor cortex and areas involved in speech production and comprehension) helps to avoid surgically-induced neurological deficits. Because epileptic activity is not stationary, it is important to record activity of both hemispheres simultaneously. One of the first studies performed with a whole-head MEG system on an epileptic patient proved the importance of this technique. In fact, mirror foci in homotopic areas of the hemispheres were detected. The precise time resolution of MEG led to detection of which area was the primary focus and which was the mirror one[32].

Language function may reside in both or one hemisphere in patients with epilepsy. Determination of laterality is important to preserve as much language and memory function as possible during resective surgery. The intracarotid amobarbital (Wada) test has long been used for language and memory localization. It has both merits and shortcomings when compared to newer forms of testing. It is invasive, uncomfortable and carries some risk. MEG may be used as a substitute or supplement to the Wada test to identify the eloquent cortex, i.e., a part of the cortex that controls various senses or actions, for removal of brain tumors or arteriovenous malformations. This was evaluated in a review by Frye and coworkers[33]. The authors concluded that MEG seems well suited for efficiently mapping language in neurologically-intact individuals. Furthermore, MEG-derived activation profiles appear to be reliable and valid indices of neural activity associated with receptive language function. While certain limitations still preclude MEG from completely replacing invasive diagnostic methods in clinical practice, there is little doubt that the information obtained using this technique is an effective adjunct to existing pre-surgical routines. Today magnetoencephalography or magnetic source imaging (MSI) is considered medically necessary for presurgical evaluation in patients with intractable focal epilepsy, in order to identify and localize areas of epileptiform activity when discordance or continuing questions arise from other techniques designed to localize a focus. In line with this view, the American Medical Association has identified three CPT (Current Procedural Terminology) codes:

95965 Magnetoencephalography (MEG), recording and analysis; for spontaneous brain magnetic activity (e.g., epileptic cerebral cortex localization)

95966 Magnetoencephalography (MEG), recording and analysis; for evoked magnetic fields, single modality (e.g., sensory, motor, language, or visual cortex localization)

95967 Magnetoencephalography (MEG), recording and analysis; for evoked magnetic fields, each additional modality (e.g., sensory, motor, language, or visual cortex localization)

10.2.6   Recent Developments and Perspectives

In the past ten years most of the work performed in the field of magnetoencephalography dealt with data analysis. In fact, the toughest issue for MEG is the identification of multiple magnetic field generators from measurements of the resulting field. Among the various algorithms that have been

[31] R. Salmelin et al., *Ann. Neurol.* 40 (1996) 157

[32] R. Hari et al., *Neuroreport* 5 (1993) 45

[33] R. Frye et al., *Physics of Life Rev.* 6 (2009) 1

developed and successfully used to process MEG data, independent component analysis (ICA) is the most successful. The basic idea of the ICA algorithm is the following: different sources, especially if related to external disturbances, are often independent of each other. In MEG data analysis, ICA can isolate disturbances due to heart activity or an eye blink. Moreover, data regularization is required for a linear estimate of the sources of the magnetic field, and this regularization is easier to achieve if the time series of the different channels are separated into statistically independent components.

A different, but successful, approach to identify the activity of a specific brain region is the so-called "beamforming" approach. This method is borrowed from radar technology, where different "beams" are used to build an image. The beamforming approach falls into the category of adaptive spatial filters. Therefore, it is a data-driven method, whereas the more traditional least squares fit is model-driven.

As a result of new multichannel systems and new data analysis techniques, challenges are being met more successfully. Perhaps the biggest challenge is related to brain functional connectivity. Recently, more importance has been given to the idea that the bases of peculiarity of the human brain should not be sought in its size or cell number, but rather in the specificity and abundance with which neurons are connected. The study of maps of functional cerebral connections has become the basis for understanding how the brain works. Recently, the concept of resting state networks (RSNs) has been introduced for investigating functional connectivity in studies that do not imply external stimuli or task execution. RSNs are defined as neuronal pools activated in spatial patterns that are active, even in the absence of a task. RSNs are thought to reflect the efficacy of anatomical connections present between brain areas. It is surmised that their change during development or aging might reveal changes in brain functional organization. RSNs were first observed (using fMRI) as correlated fluctuations of the transit time from brain areas that are known to belong to the same functional network. These correlated fluctuations show that different nodes of a particular network have similar activations even if they are not engaged in a task, suggesting that some form of ongoing communication is constantly present in the network. fMRI signals have poor temporal resolution: the hemodynamic response is slow and delayed with respect to neuronal activity, and acts as a low-pass filter on the neuronal time interval of an activity. The observed RSN correlations have a one-second temporal resolution, which is quite large compared to the time scale of neuronal events. Recently RSN correlations of electric activity have been observed with MEG[34], taking advantage of its superior temporal resolution, its comparatively high spatial resolution, and by applying advanced data analysis methods. MEG patterns of RSN correlations have shown richer variability over different time scales. It was found in particular that correlation in a single RSN across hemispheres, spans much shorter time intervals (about 10 s) than correlation within a single (stationary) hemisphere, suggesting that RSNs are not really resting, but rather are somewhat restless.

In conclusion, MEG has now reached a level of maturity that ranks it as one of the most important neuroimaging technologies. Although less versatile than MRI, and dependent on MRI for anatomical imaging and model construction, it produces direct detection of neural activity with unrivalled temporal resolution. With better spatial resolution than EEG, comparatively low influence of volume conduction in tissues, and complete noninvasiveness, it is a unique tool for large-scale electrophysiological imaging. Recent results in studies of brain connectivity demonstrate its potential for neurological investigation. The recognition that impaired connectivity is a biomarker of many neurological diseases, such as stroke and degenerative diseases, suggests an important future role for MEG in neurological diagnostics.

---

[34] F. de Pasquale et al., *PNAS* 107 (2010) 6040

10.3   MCG Instrumentation and Applications

Riccardo Fenici, Donatella Brisinda, Anna Rita Sorbo and Angela Venuti

An invitation to write about the history of something is usually an acknowledgment of competence in the field. However, it creates at least two problems. First, you must accept the fact that you have become old enough to be considered part of that history, and that is usually not a good feeling. Second, it is highly likely that someone else has already written that history and in a better form. So it may be difficult to provide information that cannot be found elsewhere.

The concept of magnetocardiography (MCG) was born in 1963, when Baule and McFee[35] recorded the first magnetic heart signals with a huge set of copper coils with millions of windings arranged in a gradiometer configuration. However, systematic MCG recordings became possible only after the invention of the superconducting quantum interference device (SQUID). Indeed, I missed the beginning of this history, as well as the first decade of superconductive magnetocardiography in the early seventies when a few scientists pioneered basic research in MCG. Their work created the basis of cardiac magnetic field interpretation and explored several potential applications of MCG as a method to improve noninvasive diagnosis of cardiac abnormalities. In addition to many excellent papers, there are many reviews covering the history of the first 10 years of MCG[36,37,38], so it would be difficult for me to find something to add here. Instead, being a small part of MCG history starting in the 1980s, I shall focus on the last three decades. However, since Gerhard Stroink[39] has provided an elegant review of the history of magnetocardiography at the International Conference on Biomagnetism 2010, I shall try to integrate his exhaustive work from the point of view of a clinical cardiologist, whose destiny was to became "gray-haired" expending energy in an attempt to bring MCG to the patient's bedside and to use it daily as a diagnostic tool.

10.3.1   A New Boost in the 1980s

The first time I was introduced to the concept of superconductivity was in late November 1979, thus almost ten years after the first high-quality magnetocardiogram had been recorded in David Cohen's magnetically-shielded room (MSR) at MIT using Jim Zimmerman's SQUID. I received a call from Ivo Modena, head of the Istituto di Elettronica dello Stato Solido (IESS) of the Consiglio Nazionale delle Ricerche (CNR) in Rome, who invited me to visit his laboratory and see the first SQUID-based Italian instrumentation for biomagnetic measurements. At that time I was mostly involved in invasive cardiology with special interest in cardiac arrhythmias and advanced clinical electrophysiology (i.e., monophasic action potential recording, cardiac pacing). I had never heard of biomagnetism or magnetocardiography. At IESS I met Gian Luca Romani, a young physicist who had constructed the Italian SQUID. He gave me a quick introduction to the fundamentals of superconductivity. I learned the SQUID was not only a sea creature, but also a superconducting quantum interference device, invented by Jim Zimmerman and capable of detecting the very weak magnetic field (MF) produced by the electrical activity of biological sources. With Zimmerman's SQUID installed in MIT's 5-layer MSR, Cohen had recorded the first (Zimmerman's) MCG, which was as clear as a conventional ECG. The combination of the SQUID and MIT's MSR provided a high signal-to-noise ratio. However, a MSR was expensive and not easy to install everywhere,

---

[35] G.M. Baule, R. McFee, *Am. Heart.* J. 55 (1963) 95

[36] D. Cohen, *Neurology and Clinical Neurophysiology* 30 (2004)

[37] P. Siltanen, in: *Comprehensive Electrocardiology,* Pergamon Press 2 (1989) 1405

[38] J.P. Wikswo, Society of Photo-Optical Instrumentation Engineers 167 (1979) 181

[39] G. Stroink, BIOMAG 2010, *IFMBE Proceedings* 28 (2010) 1

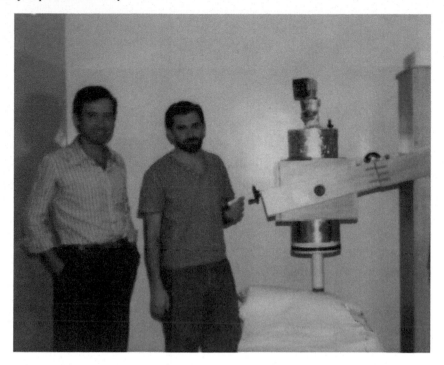

FIGURE 10.11:   The original single-channel SQUID-based instrumentation constructed at IESS by Gian Luca Romani (left) and installed, in 1980, by Riccardo Fenici (right) at the Gemelli Hospital of Catholic University, Rome, where the first clinical high-resolution MCG measurements were carried out.

especially in clinical environments. Thus, it was evident that, seeking clinical applications of bio-magnetism, a cheaper and simpler solution was needed to damp electromagnetic noise. The first tentative solution to record biomagnetic signals without electromagnetic shielding derived from Zimmerman's invention of the first superconducting gradiometer connected to a SQUID. The un-shielded approach offered the potential to involve more scientists in biomagnetism, even those with modest research budgets. This method was used at the Helsinki University of Technology, where MCGs were obtained with a SQUID-based gradiometer in a wooden cottage, as well as below ground at the Helsinki Hospital[40,41] and at Dalhousie University[42], where there was considerable experience and technology available to combine cardiac magnetic recordings with body surface potential mapping (BSPM).

At that time, with only a single channel available, the normal component of the cardiac MF had to be measured sequentially at different locations, typically in a rectangular grid. An alternate approach was developed at Stanford, where the three orthogonal components of cardiac MF (vector magnetocardiography) were measured at a single position over the heart, on the assumption that, as in vector electrocardiography, the 3D motion of the magnetic heart vector represents the overall activity of the heart.

In Rome, Gian Luca Romani had constructed a more sophisticated gradiometer capable of improving the S/N ratio of magnetic signals by accurately balancing the gradient. That solution was

[40] M. Saarinen et al., *Cardiovascular Res.* 8 (1974) 820

[41] M. Saarinen et al., *Ann. Clin. Res.* 10 (1978) 1

[42] G. Stroink et al., *Med. Biol. Eng. Comp.* 23 (1985) 61

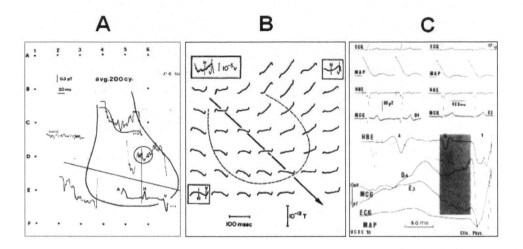

FIGURE 10.12:   Three historical high-resolution MCG recordings. (A) The first validation of High Resolution (HR)-MCG of His-Purkinje System by comparison with catheter recording of His-Bundle Electrogram in a patient with first degree infra-hisian block [modified from R. Fenici et al., *G. Ital. Cardiol.* 10 (1980) 1366]. (B) First complete mapping of High Resolution (HR)-MCG of the P-R segment of a normal subject. The inversion of polarity and asymmetry of the "ramp" pattern from the upper right to the lower left is evident [modified from D.E. Farrell et al., SQUID 80: Proc. Of the Second International Conf. on Superconducting Quantum Devices, Berlin 1980 (de Gruyter, NY, 1980) eds: H. D. Hahlböhm + H. Lübbig 273]. (C) First simultaneous recording of MCG and invasive electrophysiological signals with amagnetic catheters [modified from R. Fenici et al., *Med. Biol. Eng. Comp.* 23 (1985) 1475].

very useful in adapting the instrumentation to various noisy environments.

Nevertheless, the appearance of the IESS system (Figure 10.11) was not appealing to a clinical cardiologist! However, that afternoon I was very impressed to see my own cardiac signal, similar to an ECG trace, coming through my shirt and in a noisy physics laboratory. That was my first magnetocardiogram. Ivo and Gian Luca asked me to suggest what could be done with "that thing" in a clinic. As stated earlier, I was unaware that scientists (mostly physicists) elsewhere had followed David Cohen's lead — his pioneering research has been of great value to the field of MCG — and had already explored many potential applications of MCG as an alternative to or as complementary to ECG for noninvasive investigation of cardiac electric activity, nor did I know of the debate on the potential of MCG to provide "unique information" not attainable via ECG. I returned home fascinated by the fact that MCG was a "contactless" technique, thus potentially an easier method to perform cardiac surface mapping without sanding the skin and attaching tens of electrodes. In this way it was theoretically possible to use MCG to map the dynamics of cardiac electrical activity even through the sterile dressing covering the patient during interventional electrophysiological procedures. Another interesting fact I learned from the physicists was that the cardiac MF, directly induced by ionic action currents, reached the sensor unaffected, to a first approximation, by the varied conductivity of interposed tissue, thus facilitating the solution of the inverse problem for noninvasive 3D localization of intracardiac sources.

This seemed almost the perfect solution to the problems that clinical electrocardiologists had in finding a link between intracardiac (and cellular) electrophysiology and BSPM. The Italian instrumentation was potentially functional in an electromagnetically noisy environment such as a cardiac catheterization laboratory. Thus, in principle it was possible to install it in my hospital on short notice.

I had a "fever attack" during the night (dreams or nightmares?) and I awoke in the morning with proposals to the IESS's physicists to use MCG to visualize the dynamics of cardiac electrophysiology more easily than with BSPM, and to attempt improved noninvasive detection of the His-Bundle electrogram, as it was performed with high-resolution (HR) signal-averaged (SA) ECG, to evaluate noninvasively the infranodal conduction time (H-V interval). I was discouraged about the amount of time it took for MF mapping: rather time-consuming with a single channel and no software available for isofield maps reconstruction, but we quickly moved the single-channel instrument to Gemelli Hospital of the Catholic University. Using signal averaging we attempted the first "clinical" MCG to detect HR signals during the P-R interval. We found a magnetic spike as a function of time correlated with the His-Bundle electrogram recorded via cardiac catheterization[43] (Figure 10.12A). Then we learned that, independently, David Farrell et al. from Case Western Reserve University, had performed a preliminary HR recording of P-R activity with MCG and had described notches in the last 40 ms prior to the QRS segment of the heart cycle that were similar to those recorded with SA-ECG[44]. This was the harbinger of a new era in MCG history, leading to much future research and hot discussions that are still ongoing. In addition to spikes, a "ramp-like" pattern was observed, changing in polarity along the longitudinal diameter of the heart. This was interpreted as the field generated by the His-Purkinje system (HPS)[45] based upon the contemporary findings of Berbari et al. using SA-ECG[46].

10.3.2 The History of High Resolution MCG

At the Third International Conference on Biomagnetism in Berlin in the spring of 1980, Barbanera et al.[47] and Farrell et al.[48] presented the first data on HR-MCG. Farrell et al. showed a complete magnetic mapping of the "ramp-pattern" during the terminal part of the P-R interval (Figure 10.12B) and supported the interpretation of its electrogenesis from HPS activation on the basis of their model of the HPS[49]. We had doubts about the interpretation of the "ramp" as a field generated by the Purkinje system alone because during the terminal part of the P-R interval atrial repolarization, it widely overlaps the activation of the HPS tissue [as demonstrated by simultaneous monophasic action potential recording, (Figure 10.12C)]. However, Tripp's theory was convincing and the "ramp pattern" was accepted as related to HPS activity (Figure 10.13A). I had the opportunity to continue that research also in the Berlin MSR, having been invited to the Physikalisch-Technische Bundesanstalt (PTB) by Sergio Erné — another key person in my personal history of MCG — to work on HR-MCG/ECG (Figure 10.14). At that time the focus was still on HR waveform analysis to give a physiological explanation of the "spikes" and "ramps". In September 1982, at the Fourth Biomagnetism Conference in Rome, theories, models and experimental results were presented[50].

---

[43] R. Fenici et al., G. *Ital. Cardiol.* 10 (1980) 1366

[44] D.E. Farrell et al., *Pro. Soc. Photo-Opt. Instrum. Eng.* 167 (1978) 173

[45] D.E. Farrel et al., *IEEE Trans. Biomed. Eng.* BME-27 (1980) 345

[46] E.J. Berbari et al., *IEEE Trans. Biomed. Eng.* 26 (1979) 82

[47] S. Barbanera et al., W. de Gruyter & Co, Berlin (1981) 283

[48] D.E. Farrel et al., W. de Gruyter & Co, Berlin (1981) 273

[49] J.H. Tripp, D.E. Farrell, W. de Gruyter & Co, Berlin (1981) 259

[50] J.L. Patrick et al., *Il Nuovo Cimento* 2D 2 (1983) 255

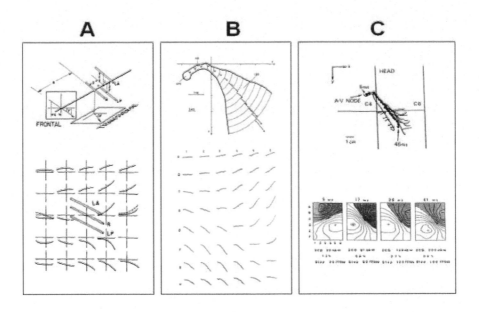

FIGURE 10.13:   Mathematical models of the His-Purkinje conduction system. (A) [modified from J.H. Tripp, D.E. Farrell, SQUID 80: Proc. Of the Second International Conf. on Superconducting Quantum Devices, Berlin 1980 (de Gruyter, NY, 1980) eds: H. D. Hahlböhm + H. Lübbig 259]; (B) [modified from R. Leoni, G.L. Romani, *Il Nuovo Cimento* D 1 (1982) 737]; (C) [modified from Y. Uchikawa, S.N. Erné, I.S.I. *Word Scientific* (1989) 13].

Mark Leifer from Stanford came with an experimental demonstration that the "ramp" pattern was due to atrial repolarization[51]. Roberto Leoni, at IESS, had developed another model of the HPS[52] that generated both "spikes" and "ramps", in good agreement with experimental findings (Figure 10.13B). We reported the first "manual" reconstruction of "isofield maps" of the P-R interval and calculated 3D localization of equivalent dipoles generated by the HPS[53]. I was sure we needed to adopt that type of analysis to improve the diagnostic potential of magnetic cardiac mapping. After days of discussion a final consensus was reached that additional systematic research was needed to clearly identify and separate the contribution to the MF distribution of the P-R interval arising from the atrial and the HPS fibers.

More advanced and realistic mathematical modeling[54,55,56,57,58] as well as basic experimental

[51] M. Leifer et al., *Il Nuovo Cimento* 2D 2 (1983) 266

[52] R. Leoni, G.L. Romani, *Il Nuovo Cimento* D 1 (1982) 737

[53] R. Fenici et al., *Il Nuovo Cimento* 2D 2 (1983) 280

[54] J. Nenonen, T. Katila, *Int. J. of Cardiovascular Imaging* 7 (1991) 177

[55] A. van Oesterom, G.J. Huiskamp, *Int. J. of Cardiovascular Imaging* 7 (1991) 169

[56] R.S. Gonnelli, M. Agnello, *Phys. Med. Biol.* 32 (1987) 133

[57] Y. Uchikawa, S.N. Erné, in: *Biomagnetism* 87 Tokyo Denki University Press (1987) 322

[58] M. De Melis, Y. Uchikawa, BIOMAG2010, *IFMBE Proceedings* 28 (2010) 420

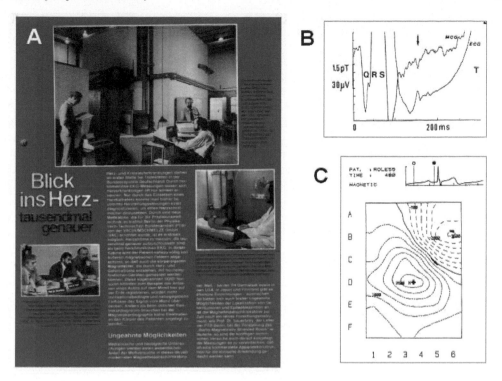

FIGURE 10.14: High resolution MCG at the Physikalisch Technische Bundesanstalt, in 1981. (A) The Berlin MSR with Sergio Erné (center), Z. Trontelj (left) and Hans Lehmann (right). (B) Simultaneous HR-MCG and ECG recordings of the P-R interval in the MSR. Both "spikes" and "ramps" were appreciable [modified from S. Erné et al., *Journal of Electrocardiology*. 16 (1983) 355]. (C) First automatic isofield map reconstruction and ECD localization at an instant of the P-R interval.

research[59] were required to understand the fundamentals of biomagnetic localization of cardiac sources. Although there were hot tempers and sleepless nights in Berlin, Erné agreed that simple HR waveform analysis was a limitation and accepted the task of writing software to image quickly and automatically HR isomagnetic maps of the P-R interval and to localize three-dimensionally the corresponding cardiac sources using the equivalent current dipole (ECD) model. When we finally were able to examine the dynamics of MF maps along the P-R interval, it became evident that the "ramp-like" MF pattern was mostly due to atrial repolarization. In fact, that MF was opposite in polarity to the one generated during atrial depolarization (the P wave). I then convinced Erné to write the software to subtract the atrial repolarization field component from the P-R segment. We found that after this subtraction, it was possible to identify a weaker (remaining) MF generated by sources (localized with the ECD inverse solution) that moved from the A-V junction downward along the interventricular septum[60]. That MF was consistent with the activation of the HPS. The correctness of this interpretation was validated by comparison with the MF pattern generated by a more realistic mathematical model of the HPS, developed at the PTB by Yoshinori Uchikawa[61,62]

---

[59] J.P. Wikswo et al., *Il Nuovo Cimento* 2D 2 (1983) 368

[60] S.N. Erné et al., in: *Biomagnetism Application and Theory*. Pergamon Press NY (1985) 132

[61] S.N. Erné et al., in: *Biomagnetism: Application and Theory*. Pergamon Press NY (1985) 123

[62] Y. Uchikawa, S.N. Erné, I.S.I. *World Scientific* (1989) 13

in collaboration with my coworker Mariella Masselli (Figure 10.13C). Later it was confirmed by other investigators[63,64,65,66].

The debate on the nature of magnetic "signals and ramps" recorded during the P-R interval served to stimulate a change of focus from standard and HR time-domain waveform analysis to HR analysis of MF mapping and to motivate the development of software tools for MF imaging[67]. Meanwhile HR MCG had been applied to investigate other parts of the cardiac cycle, first of all the terminal portion of the QRS, to identify the magnetic counterpart of the so-called "electric late-potentials", widely accepted as non-invasive markers for arrhythmogenic risk in post-myocardial infarction patients[68,69]. Mariella Masselli moved to Berlin and made a significant contribution to that work. The first results on magnetic detection of "late potentials activity" were published in 1982 and stimulated interesting discussions at the 1982 International Biomagnetism Conference, held after a NATO Advanced Study Institute in Grottaferrata. There, accompanied by much guitar and ocarina music, the first ad hoc Committee for Biomagnetic Standardization was established. It included Toivo Katila, Mark Leifer, Jim Zimmerman and me (R.F.), with the very difficult task "to find agreement between physicists and cardiologists about the polarity of the coordinate system to be used in clinical MCG and to standardize the colors to identify MF polarity". After hours of discussion in a cafeteria, the committee decided that the Frank lead coordinate system polarity—a convention for ECG coordinates suggested by Frank[70]—be accepted as a standard for cardiomagnetism as well. This means that a negative MF (i.e., the field leaving the anterior chest wall) had to be indicated by blue isofield lines, and that a positive MF (i.e., the field entering the chest wall) had to be designated as red. After the Biomagnetism Conference in Rome, there was much enthusiasm for the future of MCG!

10.3.3   From Magnetic Field Mapping to 3D Cardiac Source Localization

As stated earlier, one of the most appealing aspects of MCG was that it could, at least in theory, provide accurate 3D localization of intracardiac sources. This meant for a cardiologist that it could be used to localize arrhythmogenic substrates noninvasively. This was particularly important at a time when experience with antiarrhythmic surgery had shown that the outcome of antiarrhythmic interventions depended very much on preoperative knowledge about the localization and stability of the arrhythmogenic substrate to treat. It is obviously more important now, with catheter ablation proposed as the first line of therapy for several forms of arrhythmias. Although single-channel mapping was quite time-consuming, we thought that it was worth investigating. After successful localization of the His-Bundle activity using ECD[71] analysis, and validation by comparison with invasive recording of the His bundle electrogram[72], we attempted the first noninvasive localization of accessory pathways in patients with the Wolff-Parkinson-White (WPW) syndrome and paroxysmal tachycardias with MCG[73]. Also the localization of accessory pathways was validated with catheter recording of the Kent Bundle's electrogram, invasive cardiac mapping and pacing[74]. That work created the basis for future application of MCG for pre-interventional localization of arrhyth-

---

[63] C. MacAulay et al., *Med. Biol. Eng. and Com.* 23 (1985) 1479

[64] S. Yamada et al., *Circ.* J. 67 (2003) 622

[65] B.J. Ten Voorde et al., *Med. Biol. Eng. and Com.* 26 (1988) 130

[66] H.E. Lorenzana et al., *Biomagnetism: Applications and Theory.* Pergamon Press NY (1984)

[67] S.N. Erné et al., *Il Nuovo Cimento* 2D 2 (1983) 291

[68] S.N. Erné et al., *Jap. Heart* J. 23 (1982) 703

[69] S.N. Erné et al., *Il Nuovo Cimento* 2D 2 (1983) 340

[70] E. Frank, *Circulation* 13 (1956) 737

[71] R. Fenici et al., *Il Nuovo Cimento* 2D 2 (1983) 280

[72] R. Fenici et al., *Med. Biol. Eng. Comp.* 23 (1985) 1483

[73] R. Fenici et al., *Med. Biol. Eng. Comp.* 23 (1985) 1475

[74] R. Fenici et al., *New Trends in Arrhythmias* 3 (1985) 455

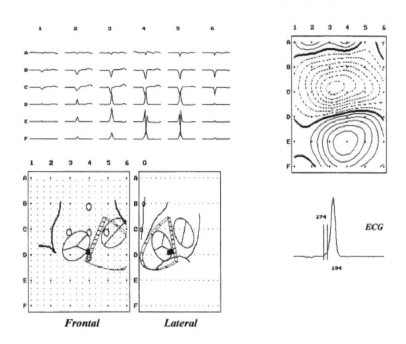

FIGURE 10.15:   Typical output of the software written by Martin Burghoff in 1989, providing MF distribution pattern and 3D localization of a right paraseptal accessory pathway, in a patient with WPW syndrome. All kinds of elaboration were analyzable in an interactive and dynamic way directly on the PC screen. The heart silhouette, interventricular septum and atrioventricular plane drawings were derived from fluoroscopy and ultrasound imaging of the individual patient. At that time it was a significant step forward to speed-up clinical evaluation of MCG data.

mogenic substrates susceptible to interventional ablation[75,76]. Other authors reproduced this type of measurement, validating the accuracy of MCG by comparison with the results of invasive electrophysiological procedures or surgery[77,78,79].

Source localization was a promising application, but a major problem in the 1980s was the lack of software adequate to visualize MF maps and localization data quickly and directly on the computer screen. Post-processing of MCG data was terribly slow, time consuming and definitely not adequate for "clinical use". Unfortunately, at least in Italy, the growing success of neuromagnetism had absorbed the major part of manpower and research funding. Thus, while in other labs physicists had difficulty finding cardiologists interested in MCG, a cardiologist (me) who seriously believed in MCG and wished to bring it to a patient's bedside was unable to find physicists interested in developing software appropriate for clinical work. Luckily, in 1987 I was invited by Hannes Novak to give a lecture on MCG in Jena. There I met a young physicist, Martin Burghoff. I was surprised when he showed me that he was able to visualize isofield maps on the screen of a "homemade"

[75] R. Fenici et al., *J. Interv. Cardiol.* 8 (1995) 825

[76] R. Fenici et al., I.S.I. *World Scientific* (1989) 102

[77] J. Nenonen et al., *Eur. Heart J.* 14 (1993) 168

[78] M. Makijarvi et al., *Eur. Heart J.* 14 (1993) 46

[79] M. Nomura, *Clin. Cardiol.* 17 (1994) 239

PC-like computer. I suggested that he come to Rome. It took us almost two years to get all necessary permissions, but in the end Martin arrived in Rome on the last week of September 1989, and he stayed in my laboratory for several months. In less than three weeks he was able to adapt his original software to almost all of my multiple requests. We were soon able to upload our averaged MCG data to a PC and perform interactively all analytic procedures (filtering, high-resolution analysis and butterfly presentation of signals, movies of MF maps and localization results obtained with different models, etc.) almost in real time. This was a significant step forward to evaluate MCG clinically. I was enthusiastic and already dreaming about the next step: the development of an automatic procedure for data fusion with anatomical cardiac images inferred from fluoroscopy and echocardiography, when, on November 9, 1989, the Berlin Wall came down. Martin had to rush home to Jena, and my dreams of new developments for MCG once again had to be put on hold. We did on our own have at least a form of a scaled 3D anatomical outline of the heart superimposed on the MF maps (Figure 10.15).

Meanwhile, in order to objectively validate the accuracy of 3D localization of intracardiac sources with MCG, I had refined the nonmagnetic catheter technique[80], constructed different types of nonmagnetic catheters for electrophysiology studies and collected a number of maps during nonmagnetic pacing with different types of dipoles in tanks filled with saline and in patients during electrophysiology studies[81]. In our unshielded catheterization laboratory the average 3D uncertainty of the catheter source localizations in patients was on the order of 12 mm — not bad for an experiment to link noninvasive MCG to invasive clinical electrophysiology. At the end of 1989, a patent application was made based upon using MCG in combination with a nonmagnetic catheter, as a minimally invasive method for magnetically-guided electrophysiology studies, endomyocardial biopsy and ablation, directly aimed at an arrhythmogenic substrate[82]. Experiments with a dog and in vitro had provided data about the MF maps generated by a current dipole implanted at different distances from the sensor[83].

On December 1990 a workshop was organized in Rome by the European Concerted Action on Biomagnetism, on "Non-pharmacological treatment of cardiac arrhythmias: present and future" to inform interventional cardiologists and cardiac surgeons about the potential use of MCG as a method to guide ablation of arrhythmias. The results of several research groups were in good agreement and substantially confirmed that localization of cardiac arrhythmias was a major application of MCG with good prospects. However, the consensus was that simultaneous mapping from multiple locations was mandatory for clinical application and that MCG had to be available in the hospital and utilized routinely by clinicians. Preliminary clinical results, independently reported by Werner Moshage[84] and Vinzenz Hombach[85], obtained with the first multichannel system of 37 first-order axial gradiometers arranged on a flat hexagonal grid (with an instrument named KRENIKON®) were outstanding. This system produced accurate localization of arrhythmogenic substrates and provided data fusion for MRI images of patients with WPW syndrome and with ventricular tachycardia[86]. The accuracy of localization was validated by comparison with ISPECT-radionuclide ventriculography, intracardiac mapping/pacing and with successful catheter ablation. We saw for the first time that, with additional refinements, MCG could yield a new method for 3D imaging during interventional electrophysiology. Later, we had the opportunity to compare our localization results with those obtained with the KRENIKON® on the same patient, who was afflicted with WPW syndrome (Figure 10.16).

---

[80] R. Fenici et al., *New Trends in Arrhythmias.* 2 (1986) 357

[81] R. Fenici et al., Plenum Press New York, (1989) 361

[82] R. Fenici, CNR, US Patent 5,056,517 (1991)

[83] E. Costa Monteiro et al., *Phys. Med. Biol.* 32 (1987) 65

[84] W. Moshage et al., *Int. J. Cardiac Imaging* 7 (1991) 217

[85] V. Hombach et al., *Int. J. Cardiac Imaging* 7 (1991) 225

[86] P. Weissmüller et al., *Eur. Heart J.* 14 (1993) 61

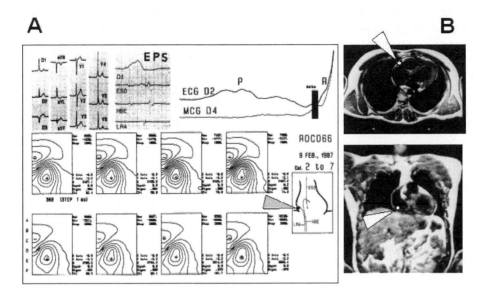

FIGURE 10.16: Patient with WPW syndrome, studied in Rome with a single-channel unshielded MCG on February 9, 1987. (A) MF maps and ECD localization were obtained with Erné's software installed in Rome. The accuracy of localization of the accessory right lateral pathway was validated by catheter mapping (EPS) and marked with the grey arrow onto the heart silhouette inferred by fluoroscopy. In 1992 the same patient was restudied in Erlangen with the 37-channel KRENIKON® system. (B) The localization results, automatically transferred into the patient MRI, marked with white arrows, confirmed the accuracy achievable even with single-channel in an unshielded catheterization laboratory.

10.3.4   From 3D localization to Magnetic Source Imaging

Since the beginning of the 1990s, an increasing number of multichannel systems for MCG mapping were constructed. Some were developed in academic institutions, others as commercial products[87]. Unfortunately, all multichannel systems at that time required electromagnetic shielding. That was the most efficient strategy to reduce ambient electromagnetic noise. For quite some time this had been an expensive drawback to installation of MCG in hospitals for routine daily investigation of a large number of patients. Furthermore, complexity related to cooling via a liquid helium transfer system required coupling to physicists in practically all biomagnetic laboratories, including those in hospitals. This probably created ambivalence toward MCG even among the most interested clinicians because it was not considered something that they could handle on their own, and that at best it was considered as research to be assigned to Ph.D. students. In spite of those drawbacks, since the second half of the 1990s, the increasing availability of high-performance multichannel systems in hospitals has significantly changed the scenario, providing a tool for advanced research in clinical MCG. One typical example of optimal multidisciplinary integration was the set-up of

---

[87] J. Nenonen, *SQUID Handbook.* Berlin Whiley-VCH, 2 (2005)

the BioMag Laboratory at the Helsinki University Hospital by scientists of the Helsinki University of Technology. In 1995 I had the opportunity to join the Helsinki Group, in the framework of the BIRCH Cardiomagnetism Program, with a research project to validate the accuracy of the improved nonmagnetic catheter technique for magnetically-guided interventional electrophysiology[88], profiting by the high performance of their new multichannel Neuromag system for MCG, installed at the BioMag Center. In Helsinki I enjoyed the friendship and professionalism of top scientists such as Jukka Nenonen (who developed almost in real-time all software interfaces that we needed) and who taught me much about modeling and signal analysis. Markku Makijarvi, in addition to performing interventional procedures in the catheterization laboratory, was a fundamental clinical partner. Katja Pesola was fully dedicated and did part of her excellent Ph.D. thesis working on that project. After hundreds of measurements to test the accuracy of catheter imaging in patients and in a patient-specific boundary element torso model constructed by Uwe Tenner (from Jena), we were able to determine that the average 3D localization precision of the catheter tip was 6–7 mm in patients and 2 mm in phantoms[89,90]. It also was demonstrated that the use of a realistic and patient-tailored torso model is one of the most important factors in determining the accuracy of 3D localization[91,92].

The obvious limitation of the simple ECD model is that it cannot be applied when multiple current sources are simultaneously activated. Under this condition, it is more convenient to solve the inverse problem for a current distribution in terms of the lead fields with a minimum-norm estimate and appropriate techniques to stabilize the solution. Those experiments initiated the transition from the concept of magnetic localization of cardiac sources to that of magnetic source imaging (or MSI) with current density reconstruction. Current density reconstruction from MCG was used to localize the origin of ventricular late fields, atrial reentry circuits, accessory pathways and ischemic regions induced by exercise in patients with ischemic heart disease (IHD)[93,94]. Further refinement of source modeling was the implementation of the uniform double-layer source model to define the sequence of cardiac excitation, which was validated with intra-operative epicardial recordings in patients undergoing open-chest surgery[95].

High-performance multichannel systems and progressive development in mathematical techniques by combining magnetocardiographic functional imaging with patients' cardiac 3D models obtained with MRI or CT scans, allowed for higher level validation of previous findings on localization of arrhythmias with MCG[96].

10.3.5   To Shield or Not to Shield? The History of Unshielded MCG

One of the most debated issues among investors and people involved in industrial development has been the perennial question of why clinicians did not trust MCG and what should be the best approach to gain their favor? The choice was between highly reliable large-scale MCG systems providing the best signal quality in MSRs and the compromise of less sensitive but budget-priced unshielded MCG-systems. In the 1990s the majority of companies favored the former approach, but costs and space needed for the installation of shielded systems have been major drawbacks for the acceptance of MCG in clinical environments. Indeed, no system was readily available at a patient's bedside, nor was MCG user friendly in the view of clinicians. Furthermore, the sophisticated tech-

---

[88] R. Fenici, CNR US Patent 6,527,724, (2003)

[89] R. Fenici et al., *Pacing Clin. Electrophysiol.* 21 (1998) 2492

[90] R. Fenici et al., *Pacing Clin. Electrophysiol.* 21 (1998) 2485

[91] R. Fenici et al., *Pacing Clin. Electrophysiol.* 21 (1998) 2485

[92] K. Pesola et al., *Phys. Med. Biol.* 44 (1999) 2565

[93] U. Leder et al., *Int. J. Cardiol.* 64 (1998) 83

[94] J. Nenonen et al., *J. Electrocardiol.* 34 (2001) 37

[95] T. Oostendorp, K. Pesola, *Proc. 12th Int. Conf. Biomagn.* (2001)

[96] W. Moshage et al., *Radiology* 180 (1991) 685

nology and the complexity of system management (cooling, liquid helium transfer, maintenance, etc.) implied that MCG systems were driven by physicists, since even when a system was installed in a hospital, medical personnel were uncomfortable and skeptical. The lack of "clinical" equipment and of evidence-based clinical benefits in using MCG impeded the creation of a market thus leading to a vicious cycle: No market ⇒ No investments ⇒ No development of "clinical" equipment ⇒ No market.

Only a few people, including myself, were convinced that the only way to break the vicious cycle and promote MCG as a clinical tool was to favor the development of user-friendly multichannel instrumentation capable of working routinely in unshielded hospital cardiology departments and to place this instrument directly in the hands of clinicians. My conviction was based on the Italian experience in which neuromagnetism had absorbed most of the research funding, and there was no way to get institutional support to install an expensive MSR mandatory for commercial multichannel MCG systems available at that time. On the other hand, based on our 10 years' experience with the IESS's single channel, I was sure that it was possible to construct a gradiometer-based multichannel system capable of working even in an unshielded catheterization laboratory. Having a goal of using MCG as an ambulatory diagnostic tool and as a real-time method to assist interventional electrophysiology, I continued to publicize my vision, but with little success. It took almost ten years to find someone willing to provide resources to construct an unshielded multichannel MCG system. Meanwhile, thanks to the Helmholtz reciprocity theorem, someone else invented and commercialized the first "magnetic" method for 3D catheter-based electro-anatomical imaging[97], which became a "best seller", and these days is used in all catheterization laboratories for interventional electrophysiology.

Unexpectedly, in 1997, I received an invitation to speak about MCG to a small group of experts in cardiac electrophysiology, at a meeting organized by Cryogenic Electronic Systems (CES), in connection with a meeting of the American College of Cardiology in Anaheim, to stimulate the interest of American cardiologists for MCG. As always, I faced more skepticism than real interest. However, I also had the opportunity to visit Alexander Bakharev's company in Springfield, MA and to play with his 7-channel prototype for unshielded MCG. With some manipulation of the system and much discussion, we considered scientific cooperation to develop "my" multichannel system, which became a reality a few years later after receiving a grant from the National Ministry of Research. This provided me the opportunity to co-finance the project thanks to the trust of Carl Rosner, CEO of the newborn CardioMag Imaging Inc., the successor company to CES. We had to proceed step-by-step. First a 9-channel prototype was installed in my lab as a test system. The 9-channel prototype was ready to record clinical MCG at the beginning of 2001[98]. Based on that experience, the unique unshielded 36-channel system was constructed and installed in our catheterization laboratory in 2002[99]. With that instrumentation it was finally possible to study many patients a day on an ambulatory basis, with immediate diagnostic feedback (the time required for a standard ambulatory mapping was only 90 seconds and post-processing for 3D imaging took less than a minute). We soon confirmed that imaging of the cardiac MF was possible, even on a beat-to-beat basis, during interventional electrophysiology. Preoperative localization of arrhythmogenic substrates was successful to guide ablation in patients with WPW syndrome, atrial fibrillation and ventricular arrhythmias. It also was proven that biomagnetic driving of a nonmagnetic catheter for action potential recording of an arrhythmogenic target was possible, and useful to define arrhythmogenic mechanisms[100] (Figure 10.16). These features also were confirmed by extensive experimental research on small animals. They underwent multiple magnetic mappings with simultaneous invasive electrophysiologic

---

[97] S.A. Ben-Haim et al., *Nature Medicine* 2 (1996) 1396

[98] R. Fenici et al., *Biomedizinische Technik* 2 (2001) 219

[99] R. Fenici R et al., *Int. J. Bioelectromagnetism* 5 (2003) 80

[100] R. Fenici, D. Brisinda, *J Electrocard.* 40 (2007) S47

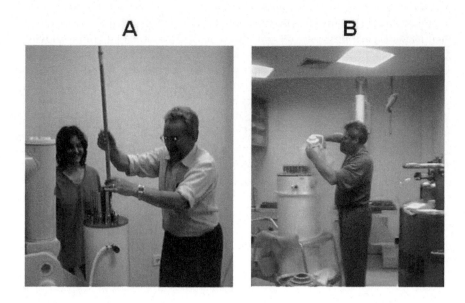

FIGURE 10.17:   History of unshielded multichannel MCG in the catheterization laboratory of the Biomagnetism Center, at the Catholic University of Rome. (A) Alexander Braginski and Donatella Brisinda working at the installation of the 9-channel "scout" system in 2001. (B) Alexander Bakharev is starting the cooling procedure of the 36-channel system in 2002.

procedures, carried out with nonmagnetic catheters 3D and localized via MCG[101,102,103]. The next mandatory step was the development of appropriate software for real-time 3D and 4D biomagnetic imaging of the nonmagnetic catheter during interventions, but this was unfortunately considered a "low priority" by venture capitalists, who focused investment on development on automatic diagnosis of myocardial ischemia[104]. A (9-channel) scaled-down version for unshielded MCG, received the approval of the US Food and Drug Administration (FDA) as a tool for clinical analysis of cardiac health. The 9-channel system was installed in leading hospitals and research centers of the US, Germany and China, especially in intensive care units or emergency rooms, where it was successfully tested for the triage of patients with chest pain, with or without myocardial ischemia (see below). At the same time, similarly downscaled systems with different numbers of channels were used in Germany and Russia[105,106]. Those systems for unshielded MCG have been mostly disregarded by the supporters of huge supersensitive shielded installations, but these unshielded systems have made a significant contribution to bringing MCG closer to the patient's bedside, to increase the number of patients investigated in clinical environments, and to enhance the interest of cardiologists for this

---

[101] D. Brisinda et al., *Basic Res. Cardiol.* 99 (2004) 193

[102] D. Brisinda et al., *Am. J. Physiol. Heart Circ. Physiol.* 291 (2006) H368

[103] D. Brisinda et al., *Physiological Measurement* 28 (207) 773

[104] A.A. Bakharev, PCT Application U.S. Prov. Appl. No.: 60/228,640 (2001)

[105] A. Gapelyuk et al., *J. Electrocardiol.* 40 (2007) 401

[106] R. Fischer et al., *Am. J. Physiol. Heart Cir., Physiol.* 293 (2007) H1242

diagnostic tool, at least for its use for early detection of ischemic heart disease[107,108,109]. That is a brief history of unshielded MCG. However, now the debate "to shield or not" is becoming obsolete. In fact, technological progress can provide more flexible shielding solutions which are more easily adaptable in almost all clinical environments at an affordable cost. Thus, the answer to the title question of this section is no longer "yes" or "no", but rather what kind of shielding might be necessary to guarantee optimal reliability and sensitivity in various clinical environments.

10.3.6 Clinical Perspectives

Although from an analysis of the literature the number of patients so far investigated with MCG is on the order of thousands, clinical validation of MCG is still far from the standard required on the basis of the AHA/ACC/ESC recommendations for the acceptance of new technologies. In fact, with few recent exceptions, reported clinical experience is based on studies that have rarely enrolled more than 250 patients. Furthermore, randomized controlled multicenter trials of MCG haven't been done. On the other hand, the safety and reproducibility of human recordings have been successfully demonstrated, and the FDA has approved MCG as a clinical method to measure the cardiac MF since 2004. Meanwhile, thanks to the increasing number of trials, clinical experience in MCG is progressively growing worldwide and new insights have been provided[110,111]. Successful results have been reported in several clinically-relevant applications that are summarized in Table 10.1.

TABLE 10.1: Most Relevant Clinical Applications of Magnetocardiography

Detection of myocardial ischemia and viability (rest and stress MCG)
Arrhythmogenic risk assessment
Noninvasive electroanatomical imaging of arrhythmogenic substrates
- WPW syndrome
- Ventricular arrhythmias
- Atrial fibrillation
- Brugada syndrome
Preoperative programming of robotic ablation systems
Fetal magnetocardiography

10.3.7 Detection of Myocardial Ischemia and Viability (Rest and Stress MCG)

By far, the major focus of research in clinical MCG has been the definition of its predictive value for early diagnosis of IHD and myocardial viability. This is understandable since IHD is the leading cause of death in western countries. Moreover, noninvasive early detection of myocardial ischemia is still a major clinical problem, because early ischemia induces changes in the electrophysiological properties of the myocardium that often are not visible in at-rest ECG, but which are detectable

---

[107] K. Tolstrup et al., *Cardiology* 106 (2006) 270

[108] J.W. Park et al., *Critical Path. Cardiology* 1 (2002) 253

[109] W.W. Quan et al., *Chin. Med. J.* 121 (2008) 22

[110] R. Fenici et al., *Expert Review of Molecular Diagnostics* 5 (2005) 291

[111] I. Tavarozzi et al., *Ital. Heart J.* 3 (2002) 151

by MCG. In fact, compared with ECG, MCG is more sensitive to tangential currents, curl currents and transmural current flow; and it can detect a closed-looped current that is invisible to ECG[112]. A first demonstration of the unique potential of MCG for the study of acute myocardial ischemia was provided in the 1970s by Cohen et al., who used experimental measurements to reveal near-DC changes related to ischemic injury currents, and which were not detected by ECG[113]. Further studies have subsequently confirmed the reproducibility of MCG for longitudinal investigation of acute and chronic ischemia-induced alteration of myocardial electrophysiological properties[114,115,116]. Saarinen et al. and later Savard et al. were the first to explore S-T shifts in MCG due to exercise tests[117]. Unfortunately, although the method was feasible in patients, it had no follow-up until the end of the 1990s when Brockmeier et al. demonstrated that pharmacological stress in normal subjects showed more distinct repolarization changes in multichannel MCG than in simultaneously recorded 32-lead ECG[118].

In 2000, Hänninen et al. described, in IHD patients, a rotation of the maximum spatial gradient of the MF pattern at the S-T-segment (α angle) induced by exercise, which slowly reverts to the basal pattern during the recovery phase. Such exercise-induced rotation of the "α angle" parameter was useful to differentiate patients with coronary artery lesions.

In patients with multi-vessel coronary artery disease, equivalent current-density estimation, computed from MCG after exercise-induced ischemia, showed good agreement between segments of high current density and viable myocardium in PET, as well as between low current density and scars[119]. It also was reported that percutaneous transluminal coronary angioplasty (PTCA) reperfusion is characterized by a more homogeneous current density distribution during repolarization as early as one month after the procedure[120]. In 2003, Kanzaki et al. reported 83% diagnostic accuracy for exercise MCG and concluded that the MF in the depolarization process can detect the subtle myocardial ischemia induced by exercise[121]. Brisinda et al. confirmed that the same exercise-induced ischemic alteration of MCG parameters described by Hänninen was detectable also with unshielded MCG[122]. Many other studies, carried out in both shielded and unshielded laboratories, have confirmed that several parameters, calculated along the ST interval and T-wave, were typically abnormal in at-rest MCG for patients with IHD, and they seem to be sensitive diagnostic markers of early ischemia. On this basis MCG recently has been proposed as the front-line method for non-invasive triage of patients with chest pain, but who exhibit normal ECG and cardiac enzyme screening[123,124,125,126,127,128,129,130]. Hailer et al. studied 177 patients with no prior myocardial infarction and with stable angina and coronary artery disease (CAD) documented by coronary an-

[112] J.P. Wikswo, J. Barach, *J. Theor. Biol.* 95 (1982) 721

[113] D. Cohen et al., *Science* 172 (1970) 1329

[114] E.C. Monteiro et al., *Physiol. Meas.* 18 (1997)191

[115] R. Fischer et al., *Int. Cong. Series* 1300 (2007) 484

[116] A. Brazdeikis et al., *Neurol. Clin. Neurophysiol.* 30 (2004) 16

[117] M. Saarinen et al., *Cardiovascular Res.* 8 (1974) 820

[118] K. Brockmeier et al., *J. Cardiovasc. Electrophysiol.* 18 (1997) 615

[119] J. Nenonen et al., *J. Electrocardiol.* 34 Suppl (2001) 37

[120] T. Hecker et al., *Proc. 12th Int. Conf. Biomagn.* (2001) 572

[121] H. Kanzaki et al., *Basic Res. Cardiol.* 98 (2003) 124

[122] D. Brisinda et al., *Biomed. Tech.* 48 (2004) 137

[123] K. Tsukada et al., *Int. J. Card. Imaging.* 16 (2000) 55

[124] J. Shiono et al., *Pacing Clin. Electrophysiol.* 25 (2002) 915

[125] . Van Leeuwen et al., *Pacing Clin Electrophysiol.* 26 (2003) 1706

[126] H. Kanzaki et al., *Basic Res. Cardiol.* 98 (2003) 124

[127] D. Brisinda et al., *Lect. Notes Comput. Sc.* 2674 (2003) 122

[128] B. Hailer et al., *Clin. Cardiol.* 26 (2003) 465

[129] B.G. Schless et al., *J. Med. Eng. Technol.* 28(2004) 56

[130] J.W. Park et al., *ANE* 3 (2005) 312

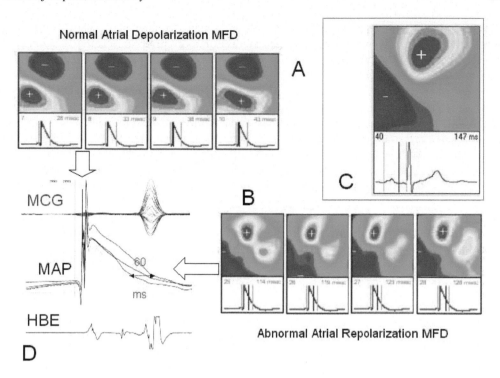

FIGURE 10.18: Example of MCG detection of atrial electrophysiological abnormality in a patient with paroxysmal atrial flutter. High-resolution MCG of atrial activation evidences a normal MF distribution during atrial depolarization (A), but abnormal MF distribution during atrial repolarization (B); for comparison an example of normal atrial repolarization MF is shown in the upper right square (C). The simultaneous recording of multiple monophasic action potentials with an amagnetic catheter (MAP in D) validates the noninvasive MCG finding as related to an abnormal (60 ms) local dispersion of atrial repolarization [modified from R. Fenici, D. Brisinda, *J. Electrocardiol.* 40 (Suppl. 1) (2007) 47].

giography. They also studied 123 patients admitted to hospital with typical chest pain, but without hemodynamic relevant stenosis of the coronary arteries. They concluded that, contrary to ECG, unshielded MCG identified CAD patients at rest with satisfactory sensitivity (73.3%) and specificity (70.1%)[131]. Tolstrup et al. reported two multicenter studies from 2 sites in the US, one in Germany and one in Italy, evaluating ventricular repolarization by MCG, in more than 100 subjects with chest-pain syndrome. All had a normal or nonspecific 12-lead ECG and normal troponin level. They concluded that at-rest MCG accurately detects myocardial ischemia in chest pain patients when 12-lead ECG and troponin appear normal or nonspecific[132,133]. Similar results were cited by other authors[134], even investigating non-STEMI patients[135], i.e., patients with partially blocked arteries.

Several attempts have been reported aimed at reliable methods for automatic computing of MCG

[131] B. Hailer et al., *Pacing Clin. Electrophysiol.* 28 (2005) 8

[132] K. K. Tolstrup et al., *Circulation* suppl 110 (2004) 743

[133] K. Tolstrup et al., *Cardiology* 106 (2006) 270

[134] K. On et al., *Circ.* J. 71 (2007) 1586

[135] H.K. Lim et al., *Ann. Med.* 39 (2007) 617

ventricular repolarization parameters[136]. The Machine Learning method, evaluated in preliminary clinical MCG studies, provided 75% sensitivity, 85% specificity, 83% positive predictive value, 78% negative predictive value, and 80% predictive accuracy for automatic MCG detection of IHD. More recently, a study with a sensitivity, specificity and accuracy of 86.2%, 72.7% and 80.4%, respectively, was reported[137].

MCG's ability to assess the viability of heart muscle in patients with previous myocardial infarction (MI) also has been investigated. The results were compared with PET and SPECT studies. The amplitudes of the R and T waves were identified as parameters with the highest selectivity for determining myocardial viability. They correctly classified, in retrospect, all patients with regard to the extension of myocardial scarring within the viable tissues[138]. In a subsequent study, current density reconstruction from MCG provided an accurate imaging of viable myocardium compared with PET results[139]. Park et al., in 100 patients with an intermediate pretest probability for IHD, have recently found that in dobutamine stress tests, MCG compared to ECG showed a significantly higher accuracy for the detection of relevant coronary artery stenosis[140]. Unfortunately, although recent studies unequivocally confirm that compared to at-rest ECG, at-rest MCG has a much higher diagnostic predictability of early myocardial ischemia, still major cardiology journals give "low priority" to this relevant information, which remains poorly exploited.

10.3.8   3D Electro-anatomical Imaging of Arrhythmogenic Substrates and Pre-interventional Virtual Programming of Ablation Procedures

Invasive dynamic 3D electro-anatomical imaging is widely used for precise ablation of arrhythmogenic substrates[141] and to visualize robotic navigation of electro-catheters from a remote radiation-free location[142]. Noninvasive MCG 3D imaging of the patient's heart obtained from cardiac MDCT (Multi-Detector Computed Tomography) or MRI is envisioned as a means to provide a virtual model to plan the ablation procedure non-invasively and to test non-invasively its efficacy prior to the intervention. This already has been proven effective for the targeted ablation of accessory pathways and ventricular arrhythmias[143]. The new challenge is to develop pre-interventional classification of patients with AF, an arrhythmia that impairs quality of life, leading to increased risk of cardiovascular complications in millions of patients. Although the HRS/ECAS/EHRA expert Task Force agrees that catheter ablation of AF in general should not be considered as the first choice for therapy[144], there is a consensus that symptomatic AF is an indication for catheter ablation. Catheter ablation of AF is a demanding intervention that may result in complications, thus it should be planned only after carefully weighing the risks and benefits of the procedure.

A first tentative MCG recording of paroxysmal supraventricular tachyarrhythmias was reported in 1994[145], but real imaging of atrial re-entry arrhythmias with MCG appeared only after 2000. Since then several authors have shown that magneto-cardiographic mapping is able to identify different activation patterns in the sinus rhythm and during atrial arrhythmias, with accuracy good enough to improve clinical understanding of AF pathogenesis and to select subgroups for patient-tailored therapy. Yamada et al. reported that the time-motion of the tangential component of the atrial MF

---

[136] F.E. Smith et al., *Pacing Clin. Electrophysiol.* 26 (2003) 2096

[137] T. Tantimongcolwat et al., *Computers in Biology and Medicine* 38 (2008) 817

[138] A. J. Morguet et al., *Coronary Artery Dis.* 15 (2004) 155

[139] M. Goernig et al., *Comput. Med. Imaging Graph.* 33 (2008) 1

[140] J.W. Park et al., *Clin. Hemorheol. Microcirc.* 39 (2008) 21

[141] C. Blomstrom-Lundqvist et al., *Circulation* 108 (2003) 1871

[142] S. Ernst et al., *Circulation* 109 (2004) 1472

[143] D. Brisinda, R. Fenici, *Pacing Clin. Electrophysiol.* 30 (Suppl 1) (2007) S151

[144] H. Calkins et al., *Heart Rhythm* 4 (2007) 816

[145] R. Fenici et al., in: *High-Resolution Electrocardiography Update*. Moduzzi Edizioni (1994) 73

was useful to noninvasively identify the mechanisms responsible for different atrial tachyarrhythmias[146]. In 2005 Nakai et al. used a 64-channel MCG and current density imaging to generate a 3D heart outline in patients with AF and atrial flutter and demonstrated that the conduction pattern generated within the heart had a counterclockwise rotation in patients with atrial flutter and random micro-reentry in the case of AF[147]. Another demonstration that MCG is an efficient method for mapping and imaging of atrial currents has been provided by Kim at al., who used the results to guide minimal AF surgery in patients undergoing mitral valve replacement[148]. MCG was useful to distinguish clinical subclasses of patients with paroxysmal lone AF with distinct signal profiles of atrial depolarization and to demonstrate that susceptibility to paroxysmal lone AF is associated with altered intra-atrial conduction modalities[149].

Numerous other studies suggest that, although needing further technological development, pre-interventional noninvasive characterization of patients with different kinds of AF seems to be one of the most promising clinical applications of MCG in the near future, especially since magnetic imaging of AF activity is possible in an unshielded catheterization laboratory[150].

10.3.9   Arrhythmogenic Risk Assessment

Noninvasive identification of patients at risk of sudden arrhythmic death is typically based on the demonstration of myocardial activation abnormality during ventricular depolarization, on abnormal inhomogeneity of ventricular repolarization, and on abnormal heart rate variability parameters. Each of the above-mentioned parameters can be noninvasively studied with ECG. However, non-contact cardiac mapping with MCG is faster, provides evaluation of all parameters with a single recording session, and might exhibit 3D imaging of arrhythmogenic mechanisms. Moreover, in comparison to ECG, MCG is more sensitive to tangential arrhythmogenic currents predominantly induced by structural alteration occurring after a myocardial infarction[151] and can detect electrically silent sources such as "vortex currents"[152,153,154]. Arrhythmogenic ventricular depolarization abnormalities can be detected with high-resolution MCG as low-amplitude "late magnetic fields" at the end of QRS[155,156], or with intra-QRS fragmentation[157].

Intra-QRS fragmentation analysis, obtained by applying binomial filtering to detect polarity changes inside QRS, was efficient in identifying patients prone to ventricular tachycardia (VT) after a MI[158]. Increased intra-QRS fragmentation also can be an early sign of diabetes-associated cardiomyopathy[159]. MCG identifies abnormal inhomogeneity in ventricular repolarization[160]. An abnormal dispersion of the late part of the T wave interval ($T_{apex} - T_{end}$ interval), which is an index of transmural repolarization inhomogeneity, was found in patients with sustained VT, due to previous MI and with idiopathic dilated cardiomyopathy. Interestingly, MCG simultaneously iden-

[146] S. Yamada et al., *Europace* 5 (2003) 343

[147] K. Nakai et al., *Int. J. Cardiovasc. Imaging* 21 (2005) 555

[148] D. Kim et al., *Interactive Cardiovascular and Thoracic Surgery* 6 (2007) 274

[149] R. Jurkko et al., *Europace* 11 (2009) 169

[150] R. Fenici, D. Brisinda, *The Anatolian J. Card.* 7 (Suppl. 1) (2007) 23

[151] D.R. Bolick et al., *Circulation* 74 (1986) 1266

[152] K. Brockmeier et al., *J. Cardiovasc. Electrophysiol.* 18 (1997) 615

[153] F.J. van Capelle, D. Durrer, *Circ. Res.* 47(1980) 454

[154] M. Liehr et al., *Ann. Biomed. Eng.* 33 (2005) 240

[155] P. Korhonen et al., *J. Cardiovasc. Electrophysiol.* 11 (2000) 413

[156] P. Korhonen et al., *Pacing Clin. Electrophysiol.* 25 (2002) 1339

[157] M.K. Das, E. Masry, *Curr. Opin. Cardiol.* 25 (2010) 59

[158] P. Korhonen et al., *Pacing Clin. Electrophysiol.* 24 (2001) 1179

[159] K. Brockmeier et al., *J. Electrocardiol.* 30 (1997) 239

[160] K. Nakai et al., *Int. Heart J.* 48 (2007) 701

tified late fields and $T_{apex} - T_{end}$ alteration only in post-MI patients. In post-MI patients, both MCG late fields and intra-QRS fragmentation parameters differed significantly between patients with and without a propensity to sustained VT, but the abolition of the arrhythmic substrate with antiarrhythmic surgery rendered both late field and intra-QRS fragmentation parameters almost similar to those of the post-infarction patients without a VT propensity[161,162]. MCG has been used to localize late fields and to investigate their relation to the location of the arrhythmogenic substrate[163]. The prolongation of the QT interval (i.e., the intervals from the QRS onset to the offset of the T-wave) and its inhomogeneity is another indicator of propensity to life-threatening ventricular arrhythmias and sudden death. Automatic calculation of QT interval duration at multiple locations above the heart and the reconstruction of the spatial distribution of QT dispersion from MCG has proven more sensitive than 12-lead ECG recordings in screening post-MI patients at risk to develop sustained VT[164].

In addition to morphological HR analysis of MCG signals, the extraction of the spatial features of MF distribution provides additional information about vulnerability to arrhythmia. Another example of a potential clinical application of non-invasive imaging of myocardial arrhythmogenic currents with MCG is the study of patients with the Brugada syndrome (BS), an inherited arrhythmogenic disease characterized by a typical electrocardiographic pattern with ST-segment elevation in leads V1 through V3, incomplete right bundle-branch block (RBBB) and an enhanced risk of sudden cardiac death because of ventricular fibrillation (VF) or rapid polymorphic VT[165].

10.3.10 Fetal Magnetocardiography

Fetal monitoring is another unquestionable clinical application of MCG, despite recent improvements in detecting fetal ECG, not only during labor,[166,167] but also with surface electrodes during pregnancy[168]. A recent review is available that covers the history of fetal magnetocardiography (FMCG)[169]. Thus we summarize only some key points which are relevant to clinical application. It is well known that FMCG was initiated in Finland in 1974 with an unshielded SQUID system[170]. Thereafter, most FMCG studies have been carried out in MSRs with different MCG systems, including a 151-channel system specifically designed for magnetic study of fetal cerebral, cardiac and uterine activities[171]. Now FMCG is recognized as a reliable method for noninvasive surveillance of fetal cardiac electrophysiology, especially during the second half of pregnancy, when the presence of electrically insulating vernix cascosa impairs the quality of fetal ECG recording with abdominal leads[172].

Shielded FMCG has proven reliable to measure standard cardiac intervals[173,174], heart rate and

[161] L. Oikarinen et al., *J. Cardiovasc. Electrophysiol.* 12 (2001) 1115

[162] P. Korhonen et al., *J. Cardiovasc. Electrophysiol.* 12 (2001) 772

[163] P. Weismüller et al., *Eur. Heart J.* 14 (1993) 61

[164] P. Van Leeuwen P et al., *Pacing Clin. Electrophysiol.* 19 (1996) 1894

[165] P. Brugada et al., *Circulation* 97 (1998) 457

[166] K.G. Rosen, Curr. *Opin. Obstet. Gynecol.* 17 (2005) 147

[167] I. Amer-Wahlin et al., *B.J..O.G.* 112 (2005) 160

[168] F. Mochimaru et al., *Jpn. J. Physiol.* 54 (2004) 457

[169] S. Srinivasan, J. Strasburger, *J. Curr. Opin. Pediatr.* 20 (2008) 522

[170] V. Kariniemi et al., *J. Perinat. Med.* 2 (1974) 214

[171] C.L. Lowery et al., *Am. J. Obstet. Gynecol.* 188 (2003) 1491

[172] S. Comani et al., *Physiol. Meas.* 26 (2005) 193

[173] S. Comani, G. Alleva, *Physiol. Meas.* 28 (2007) 49

[174] P. van Leeuwen et al., *Physiol. Meas.* 25 (2004) 539

its variability (HRV)[175,176], to monitor fetal neurodevelopment[177], to detect and classify fetal ar-rhythmias[178], to identify and treat arrhythmogenic risk factors, such as heart block, ventricular pre-excitation[179,180] or long QT syndromes[181] and other pathological conditions which can jeopardize the fetus or the survival of the newborn after the delivery, such as intrauterine growth retardation, fetal distress and congenital heart defects. Although MSRs provide ideal conditions to study the FMCG on a beat-to-beat basis according to established standards, more widespread routine clinical use of this innovative method is expected by bringing FMCG to the patient's bedside and by mini-mizing its cost. For this reason, several attempts have been reported (since 1982) to record FMCG without employing MSRs[182]. The quality of unshielded FMCG signals was good enough to detect and analyze the FMCG waveform, to measure cardiac intervals and to analyze fetal heart rate vari-ability.[183] More recent instruments provided real time FMCG at the 31th week of gestation, with an acceptable signal-to-noise ratio, even using HTS-SQUID technology in MSR[184]. Additionally, multichannel mapping of FMCG has been successfully performed in unshielded clinical environ-ments on an ambulatory basis. In this clinical arrangement, real-time FMCG signals were noisy, but visually perceptible from the 28th week of gestation. However, after averaging, signal resolution was good enough to measure cardiac intervals, to reconstruct MF maps and to localize fetal ven-tricular activity[185]. Further improvement of the signal quality was achieved with the Independent Component Analysis approach[186], making it possible to measure cardiac intervals beat-to-beat and to evaluate the HRV parameters.

10.3.11   Conclusions

As stated in the introduction, this review of MCG has been written focusing on personal memories linked to achievements in the field, rather than as a dry scientific monograph. It is worth noting that the number of indexed papers in clinical MCG has increased almost exponentially through the years, and that there are many others that have not appeared in PubMed. Thus, we apologize to those scientists whose work might have been unintentionally forgotten or quoted superficially.

Concerning the future of MCG, we believe that the most important achievements have yet to come. Certainly SQUID development and improvements had a key role for the progress of biomag-netism in general and cardiomagnetism in particular. However, although it might seem unfair to close a review on clinical applications of superconductivity with this statement, we believe that the widespread clinical use of MCG will occur only with the advent of a new generation of multichan-nel instruments based on non-cryogenic sensors such as, for example, laser-based optically-pumped magnetometers, which should be less expensive, more compact, practically maintenance-free[187] and hopefully available soon at the patient's bedside. Preliminary validation of single-channel optic MCG mapping showed that the method is promising[188]. Since then significant progress has been

[175] S. Lange et al., *Early Hum. Dev.* 85 (2009) 131

[176] U. Schneider et al., *Early Hum. Dev.* 86 (2010) 319

[177] R.B Govindan et al., *Am. J. Obstet. Gynecol.* 196 (2007) 572

[178] A. Fukushima et al., *Heart Vessels* 25 (2010) 270

[179] L.K. Hornberger, K. Collins, *J. Am. Coll. Cardiol.* 51 (2008) 85

[180] H. Zhao et al., *J. Am. Coll. Cardiol.* 2008 51 (2008) 77

[181] T. Menendez et al., *Pacing Clin. Electrophysiol.* 23 (2000) 1305

[182] I. Awano et al., *J. Exp. Med.* 138 (1982) 367

[183] J.A. Crowe et al., *Physiol. Meas.* 16 (1995) 43

[184] Y. Zhang et al., *Supercond. Sci. Technol.* 19 (2006) 266

[185] D. Brisinda et al., *Prenat. Diagn.* 25 (2005) 376

[186] S. Comani et al., *Phys. Med. Biol.* 52 (2007) N87

[187] G. Bison et al., *Appl. Phys.* 76 (2003) 325

[188] R. Fenici et al., *Biomedizinische Technik* 48 (2004) 192

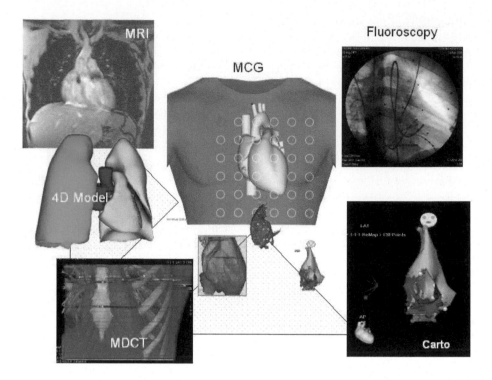

FIGURE 10.19: Future imaging for interventional electrophysiology is foreseen as the multi-modal integration between preinterventional MCG-based electro-anatomical imaging into patient's 4D model of beating heart in a breathing chest obtained with cardiac magnetic resonance imaging (MRI) and interventional fluoroscopic and non-fluoroscopic catheter navigation [modified from R. Fenici, D. Brisinda, *The Anatolian Journal of Cardiology* 7 (Suppl. 1) (2007) 23].

made with the preliminary clinical testing of a laser-pumped 19-channel second-order magnetometer[189] and the development of new technology for the construction of a compact sensor module for a multichannel sensor head with 57 primary channels to be installed in hospitals[190]. The new sensor head should allow the recording of cardiac MF maps with a spatial resolution comparable to existing SQUID-based systems and with sensitivity adequate to record diagnostically relevant information at least in the QRS-complex and the T-wave. At present optically-based magnetometry still needs some shielding, but the reduction of cost of such systems might favor the acceptance of MCG in clinical environments in the near future.

10.4   MRI (Magnetic Resonance Imaging) Instrumentation and Applications

Jim Bray and Kathleen Amm

MRI provides the largest commercial market using superconductors today, at a size currently of about $3.6 billion/year and growing. Here we discuss how this market arose and how superconductors are employed within it.

---

[189] A. Weis et al., BIOMAG 2010, *IFMBE Proceedings* 28 (2010) 58

[190] G. Lembke et al., BIOMAG 2010, *IFMBE Proceedings* 28 (2010) 62

10.4.1   Basic Principles of MRI

Since being discovered by Paul Lauterbur[191] in 1973, MRI took only a little over a decade to become a major medical imaging technology. The MR signal has its origin in the spin of atomic nuclei with an odd number of protons or neutrons.

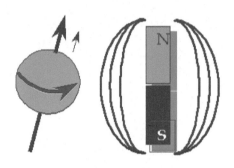

Spin is a basic property of elementary particles and produces a magnetic dipole in the particle (see Figure 10.20) with a magnetic moment given by the equation

$$\mu = \gamma \hbar \mathbf{I}$$

where γ is the gyromagnetic ratio (the ratio of dipole moment to the angular momentum), $\hbar$ is Planck's constant divided by 2π, $\mathbf{I}$ is the spin and μ is the magnetic moment. When the magnetic moments are placed in a magnetic field B_0, they become partially magnetized. In a 2-state system such as spin = 1/2, the distribution of the magnetic moments will be

FIGURE 10.20:   Dipole magnetic moment.

determined by a Boltzmann distribution:

$$\frac{N^-}{N^+} = e^{\Delta E/k_\mathrm{B} T} \qquad \Delta E = \gamma \hbar B_0$$

$N^-/N^+ \approx 0.999993$ for hydrogen at $B_0 = 1.5$ tesla, where N^+ and N^- are the spins aligned parallel to the field and anti-parallel to the field, respectively, T is the absolute temperature, and k_B is Boltzmann's constant. The energy difference between the up and down spins, ΔE, corresponds to a frequency ω of RF radiation by the equation $\Delta E = \hbar\omega$, which specifies a photon energy. Precisely, this energy causes a resonant transition between the two spin states, and this transition is the basis for NMR. The higher energy spin state can be populated by absorption of resonant photons. It will then decay back to its equilibrium population governed by the temperature T and magnetic field B_0, and the emitted photons from this decay process can be detected with appropriate receiving antennas in the vicinity. These decay times are labeled by names such as T_1 and T_2 and are defined by the detailed interactions among the spins and magnetic fields in the system. This absorption and emission of precise photons forms the basis for NMR.

MRI adds to the NMR process the spatial localization of the spins. Since the NMR RF frequency is dictated by the magnetic field at the nucleus, we can add a gradient magnetic field to B_0 to produce a different field at each location in the imaged body. Every location will then have its own resonant NMR frequency that can identify it when it emits from an excited state. Since a certain number of spins are necessary to achieve a practical signal-to-noise ratio (SNR) for each frequency, the imaged body is typically divided into voxels, which are the minimum volume elements required to achieve an acceptable SNR. To date, a plethora of methods have been developed for applying the gradient magnetic fields and RF spin-excitation fields in time and space to achieve optimal desired results in the image. Different features in the human body can be emphasized by various combinations of gradients and RF, which may emphasize spin density or different decay processes of the spins (e.g., T_1, T_2).

In an MRI machine, only the main, large, static field, B_0, is produced by superconductors (SC)[192]. B_0 is produced axially inside the bore of a superconducting solenoidal coil in most MRI machines,

[191] P. C. Lauterbur, *Nature* 242 (1973) 190

[192] Y. Lvovsky and P. Jarvis, *IEEE Trans. Appl. Supercon.* 15 (2005) 1317

although some machines ("open MRI") produce this field with a pair of Helmholtz coils. The open MRI machines allow larger access to a less tunnel-like volume, sometimes relieving patient claustrophobia, but cannot support as large a magnitude of B_0. From the above discussion, it should be apparent that a strong requirement on B_0 is that it be very uniform in the imaging space, since spatial localization of the nuclei depends on knowing precisely the field value at each voxel location. Inhomogeneities in B_0 will also cause loss of SNR from faster decay of excited nuclear spins. Imaging and SNR improve as B_0 increases, and this provides another motivation for using superconducting coils, since they can produce higher fields than copper, although there are limits to this process[193]. The ever-changing gradient and RF fields are produced by copper wires driven by appropriate electronics since superconductors are not lossless under ac fields or currents. The emitted photons from the excited nuclei are also generally received by copper antennas, although HTS (high temperature superconducting) materials have occasionally been employed for this to enhance SNR[194].

The future of HTS in MRI is an interesting question. At present, all commercial MRI systems use LTS (low temperature superconductors), and almost all of these employ NbTi wires, the workhorse of superconductivity. The only uses of HTS in MRI so far have been in experimental RF receiver coils[194] and as current leads for a few low-volume systems. Yet, it would seem obvious that operating at the higher temperatures afforded by HTS would lead to simpler construction of the cryogenic subsystems. The answer lies in the economics: even with the benefit of the cost reduction of the cryogenics, HTS wires are at present too expensive to compete with NbTi, which is easily available in many forms at a cost of around $2/kA-m. HTS wire prices are at present around 100× higher, and this rules out their use. Additional disadvantages of HTS are availability of HTS in limited geometric forms (mostly tapes) and lengths, less manipulability than NbTi, and pinning (current carrying capability in large magnetic fields) which is sub-optimal around 77 K. All of these issues are being addressed in research labs, and it is possible that HTS will one day supersede NbTi in MRI systems.

The human body is mostly fat and water, both containing hydrogen, which comprises $\sim 63\%$ of the atoms in the body. Therefore, hydrogen nuclei (protons), which produce a strong NMR signal, are the primary nuclei imaged in MRI.

10.4.2   History of MRI — A General Electric (GE) Perspective

That protons possess a magnetic dipole moment was proposed by Pauli in 1924. The possibility of detecting these moments by NMR was first suggested by Gorter[195] in 1936, and the first successful NMR experiments were done on particle beams moving through a vacuum by Rabi[196] in 1939. In 1946 Purcell[197] at Harvard and Bloch[198] at Stanford were the first to perform NMR in condensed matter (solids or liquids), for which they received the Nobel Prize in Physics in 1952. The possibility of whole-body NMR was suggested by Jackson[199] in 1968. The first images based on proton NMR were obtained by Damadian (using a technique called field-focusing) and by Lauterbur[191] (using the back projection technique used in CT X-ray reconstruction). Lauterbur's introduction of gradient field coils to produce position-dependence in the NMR signal was a crucial innovation. In 1975, Ernst[200] proposed a two-dimensional Fourier transform reconstruction technique using phase and frequency encoding that has become the standard acquisition technique for MRI. In the late 1970s, initial human images using either electromagnets or resistive air core magnets were produced by two

[193] T.W. Redpath, *Br. Jour. Radiology* 71 (1998) 704

[194] http://www.time-medical.com/products_technology_hts.htm

[195] C. J. Gorter, *Physica* 3 (1936) 995

[196] I.I. Rabi et al., *Phys Rev.* 53 (1938) 318

[197] E. M. Prucell et al., *Phys Rev.* 69 (1946) 37

[198] F. Bloch et al., *Phys Rev.* 69 (1946) 127

[199] J. A. Jackson et al., *Rev. Sc. Instrum.* 39 (1968) 510

[200] A. Kumar et al., *J. of Mag. Resonance* 18 (1969) 69

groups at the University of Nottingham, one under Andrew[201], the other under Mansfield[202], and by a group under Mallard[193] at the University of Aberdeen. In 2003, Lauterbur of the University of Illinois and Mansfield of the University of Nottingham were awarded the Nobel Prize in Medicine for their development of MRI.

The first whole body MRI systems were large air-core resistive magnets that required large power supplies and water cooling. Most importantly, these thermal constraints limited field strengths to about 0.2 T. This limitation at first did not appear to be particularly significant. In the late 1970s it was commonly believed that human proton MRI would be effective only at frequencies below about 10 MHz due to screening by the human body of the RF fields at higher frequencies[202]. During this period the total worldwide effort on MRI was dwarfed by the development of CT scanning. Near the end of this decade, CT had grown rapidly and was a highly competitive enterprise, as it revolutionized the practice of radiology. After a late start, GE become a major force in the CT marketplace after a 2-year period of R&D at its Central Research Lab in Niskayuna NY. The CT research effort, which had been led by physicist Dr. Rowland Redington, was reduced rapidly as the technology was transferred to GE's imaging business headquartered in Milwaukee, WI. A small group of investigators, including Redington, another physicist, Howard Hart, and a computer scientist, Bill Leue, began to follow MRI developments elsewhere and to formulate some new approaches, but initially there was little or no experimental imaging being done. As part of this work, Richard Likes, who was working at the GE labs while preparing for graduate school, developed some advanced MR imaging concepts, including the introduction of k-space.

In 1978, a physicist, John Schenck, who had temporarily left the company to complete a medical degree, returned to GE and was asked by Redington to investigate the desirability of an expanded MRI effort. Within a few weeks of his return, he attended a 1978 NMR Gordon Conference where a session was devoted to recent work on NMR imaging. This lively session generated intense interest and skepticism about the practicality of MRI. By this time, CT had become a major business facing intense competition, generating significant revenue and numerous technological innovations. In this climate, MRI, with numerous impediments, was viewed as a distraction. One comment was that "MRI is the technique of the future — and probably always will be." Nonetheless, it was decided to make a significant investment to investigate improving the resolution of proton imaging and the possibility of metabolic studies using phosphorus NMR. Permission was obtained to commission a high-field superconducting magnet and to hire two additional staff members. Several possible magnets were considered. A resistive magnet with a (barely) human-size bore and operating at 0.12 T was ordered from the Walker magnet company for initial imaging. It was known that more powerful magnets could be built using superconducting technology, but it was expensive, there were uncertainties about RF field penetration at higher frequencies, and much of the work that had been done on them involved nuclear weapons research and was classified. On the other hand, high fields offered the possibility of improved signal-to-noise ratio, leading to improved imaging resolution, and the ability to detect low concentration metabolites.

In the UK, Oxford Instruments (OI) was in the process of constructing two whole-body MRI superconducting magnets operating at about 0.3 T. These systems were eventually installed at Hammersmith Hospital in London and at University of California, San Francisco (UCSF)/Diasonics facility in San Francisco. The options considered at GE included 0.3-0.5 T systems similar to the Hammersmith and UCSF systems, and medium-bore, high-field magnets that could be used for animal imaging studies. In the midst of a huge snowstorm in the winter of 1980, OI executives visited GE. They said they were confident it was feasible to make a whole-body magnet operating at a field strength on the order of 1.5 to 2 T. Although this increased the cost and risk involved in the magnet acquisition, it also offered the possibility of achieving a leadership position in this new field.

---

[201] R. Andrew et al., *Phys. Med. Biol.* 22 (1977) 971

[202] P. Mansfield, *J. Phys. C: Solid State Physics* 10 (1977) L55

After considering bids from OI and another company, Intermagnetics General Corporation (IGC), OI was judged to have the edge in technical magnet performance and experience, and in July 1980, OI was awarded the contract for a one-meter clear-bore system to be delivered in 18 months. The agreement was that the magnet would be designed to reach 2 T but only contractually required to operate at 1.5 T. Offers were made to researchers who had been active in MRI developments. Bill Edelstein had been a leading member of Mallard's team at Aberdeen and Paul Bottomley had been a leader in Andrew's team at Nottingham. They joined Redington's team in late summer and early fall of 1980. The group constructed an initial imaging system with a 0.12 T Walker magnet, which was eventually transferred to the University of Pennsylvania for early clinical evaluation. The OI magnet arrived in 1982, and Redington's team assembled an imaging system to operate with it. However, when the magnet was ramped to about 1.6 T, a significant downward drift was evident. The magnet drifted so fast at higher fields that it never reached 2.0 T. However, the drift rate decreased as the field was lowered, and Hart calculated that at 1.5 T he could ramp the shim coil enough to balance the drift for about a day. To prepare for imaging every few days, Hart had to ramp the field to just above 1.5 T, a tedious and scary process that involved inserting the leads and heating up the superconductive switches with the potential risk of quenching the magnet. Late one night, between 3 a.m. and 4 a.m., with John Schenck lying in the magnet on a wooden plank and Hart operating the scanner, a 1.5 T brain image was acquired at 63.4 MHz. To Hart's surprise, he noted that the entire brain was present in the image, and there was no evidence for the hypothesized shielding of the deep tissues by skin-effect processes. Thus, the team at the GE labs became the first in the world to take an MRI image in a 1.5 T MRI whole-body magnet and experimentally dispel the concerns that proton imaging of humans would be severely limited above 10 MHz.

John Schenck presented these imaging results at a conference in Colorado Springs in February 1983 to a group that was a precursor to today's International Society for Medical Imaging (ISMRM) in a talk which received the best paper award. This presentation was somewhat disruptive to the plans of several groups who had pioneered and were planning to commercialize MRI at lower field strengths using resistive magnets. One of the investigators from another group told Schenck that customers would be confused with regard to what field strength and type of magnet would be relevant clinically. Subsequently, the GE imaging business in Milwaukee decided to introduce a clinical 1.5 T scanner which had the effect of freezing the MRI market for about two years. The first GE 1.5 T products, based on magnets purchased from OI, were introduced in 1984, two years after the first images in 1982. The first systems were placed at the University of Pennsylvania, Duke, and UCSF, and these groups rapidly established the clinical utility of MRI using 1.5 T superconducting magnets. Subsequent magnets easily achieved stable operation at 2 T, but by about 1985, 1.5 T scanners had become a standard product in radiology and remained so even in 2010 when on the order of 10,000 of these magnets are in use worldwide. It is ironic that the quirks of a single early magnet, and the ability of its users to obtain sufficient short-term stability to conduct initial imaging studies, played a major role in establishing this standard field strength. The issue of the penetration of RF fields into human subjects is substantially more complex than at first appreciated. It is relatively unimportant at or below 1.5 T. However, whole-body magnets of 3 T (128 MHz), 7 T (300 MHz) and higher, have been introduced into clinical medicine. Often these high-field systems show stronger, rather than attenuated signals from deep-lying tissues.

GE made the decision that magnet technology was a critical technology for MRI, as 80% of the system cost was from the OI magnet. So in 1982, GE developed its first 0.8 T magnet to demonstrate proof of concept with a superconducting magnet team led by Evangelos (Trifon) Laskaris, who had been working on superconducting generators for GE Energy. GE developed the magnet, and CVI developed the cryostat. Once the business decision was made to go to 1.5 T after the successful first images, a program was launched to develop an MRI magnet at 1.5 T in 1983 and was completed in 1984. During the development of the first cryostats, engineers had many challenges with regard to the suspension systems, which could be damaged easily during transport. GE convened a meeting of experts from its research center and its aerospace business, to design a suspension system, allowing

GE to launch a robust MRI product.

Once it had launched its 1.5 T MRI product, GE began to design a variety of MRI products to address a broad and rapidly developing MRI marketplace. It developed technologies for a low-cost MR system called MR Max, a passively-shielded system launched at a meeting of the Radiological Society of North America (RSNA) with a fully-powered MRI on display. The MR Max sold 25 systems at its launch. With the advent of high performance cryocoolers operating at 10 K, the team began looking at conduction-cooled Nb_3Sn magnets.

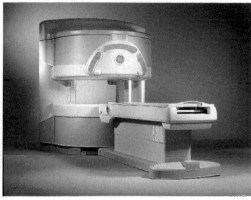

FIGURE 10.21:   MRT system magnet.   **FIGURE 10.22:**   Openspeed 0.7T MRI System.

The head of MR marketing in the late 1980s asked the team to look at interventional systems. The team then began looking at different MR configurations that could enable surgery and therapies to be done in an MR system. This led to the MRT system, a conduction-cooled Nb_3Sn MR system that consisted of a "double donut" vertical Helmholtz pair — see Figure 10.21. These systems enabled surgery and other therapies to take place in MR systems and led to the development of interventional MR. The installation of the first system took place at the Brigham and Women's (B&W) hospital in Boston in the middle of a huge snowstorm with -20 F temperatures. During the installation, Trifon and Bijan Dorri, who were in Boston to help with the installation, heard a large explosion and were concerned about the magnet. The magnet had been transported from GE's Global Research Center to Boston on an inflatable suspension system. What they soon realized was that the magnet was not damaged, but due to the cold temperatures, the inflatable transport system had become brittle and failed. Despite the transport system issues and the snowstorm, they were able to install the magnet in B&W Hospital, where the system was used for 13 years to save many lives with advanced, image-guided MR surgery. Advances continued in MR technology into the 1990s. Bill Chen identified an emerging cryocooler technology being developed by Sumitomo. He then fostered a partnership between GE and Sumitomo Heavy Industries to develop the first zero boil-off system for MR imaging. This has greatly expanded the MR market into markets where the cost of helium service previously made MR impossible. The technology that enabled MR Max to be a low-cost, passively-shielded system was extended to develop a low-cost actively-shielded 1.5 T MR system. GE introduced the highest-field open MRI system with the introduction of the 0.7 T Openspeed system in 1999 — see Figure 10.22.

In the late 1990s and early 2000s, GE acquired key MR magnet companies Elscint and portions of Magnex. In the early 2000s, GE developed the first whole body 3 T MR system with the acquisition of the technology from Magnex. The mid-2000s saw the introduction of a new type of open system with a 1.5 T 70 cm wide-bore system by Siemens called the Espree and later a 3 T wide-bore system.

GE introduced the first full field-of-view 1.5 T wide-bore system in 2009 with the introduction of the MR450W. GE acquired ONI in 2009, a specialty MR company founded by Peter Roemer, a GE alumnus, that specializes in limb MRI imaging — see Figure 10.23.

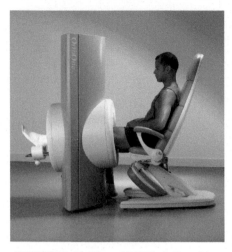

FIGURE 10.23: ONI limb imaging system.

FIGURE 10.24: GE MR350 Brivo MR value system.

Innovations continue in the development of MRI magnets — see Figure 10.24. GE, Siemens, Philips, Toshiba and other MRI companies have begun introducing value systems that address the challenges of making MR cost attractive for emerging markets. Companies around the world are focused on developing MR that can address the clinical needs of both emerging and developed markets and that can help to enable MR become more accessible to thus improve patient care globally.

The Expansion of MRI and Its Effect on Superconductors

MRI has proven to be a rich tool for biological research and medical applications, and its utilization has expanded. Examples are flow imaging (MRI angiography)[203,204], magnetic transfer contrast[205], magnetic resonance elastography (MRE)[206], functional magnetic resonance imaging (fMRI)[207], and magnetic resonance spectroscopy (MRS)[208]. Some of these have led to increased demand on the superconductors used to construct MRI machines and will be discussed further here.

fMRI is used for brain function analysis in humans and animals. The basic premise is that neural activity will cause increased associated metabolic activity and blood flow in the adjacent vessels to feed the activity, and this metabolic activity will increase the use of oxygen from the blood hemoglobin by active neurons. Unoxygenated hemoglobin (deoxyhemoglobin), which is paramagnetic, results from the metabolic loss of oxygen from the hemoglobin, which is diamagnetic. The magnetic fields from the paramagnetic deoxyhemoglobin will affect the MRI signal (especially,

[203] D.G. Nishimura, *Magn. Reson. Med.* 14 (1990) 194.

[204] C.L. Dumoulin et al., *Magn. Reson. Med.* 9 (1989) 139

[205] J. Eng et al., *Magn. Reson. Med.* 17 (1991) 304

[206] T. Wu et al., *Magn. Reson. Med.* 43 (2000) 111

[207] Scott A. Huettel et al., *Functional Magnetic Resonance Imaging*, Sinauer Associates, 2004

[208] G.B. Matson and M.W. Weiner, *Magnetic Resonance Imaging*, D.D. Stark & W.G. Bradley, Jr. Mosby Year Book, St. Louis, (1992) 438

spin-echo T_2 or gradient echo T_2^*), thereby labeling the location of the neural activity in the MRI scan[209]. The study of the location of brain activity using this method (often called BOLD for blood oxygen level dependence fMRI) is now a significant research area and has various medical applications to identify brain pathologies. The association of neural activity with BOLD has been established recently[210]. The MRI signal change is small, however, and this requires MRI base fields (B_0) of 4 T or higher, and the imaging voxel size is usually larger than in a standard image (e.g., 3 mm as opposed to 1 mm) to help compensate for this small signal and get good signal-to-noise (SNR) ratios.

In normal MRI procedures, the ^{1}H nuclei (protons) have their spins polarized in thermal equilibrium in the base field B_0 of the MRI system, usually 1.5 or 3 T. A very small fraction of the spins become polarized in this situation, and the polarization P is given by

$$P = \frac{N^+ - N^-}{N^+ + N^-} = \tanh \frac{\gamma \hbar B_0}{2k_B T}$$

where $N^{+,-}$ is the number of "up" and "down" spins (relative to the applied magnetic field B_0,), γ is the gyromagnetic ratio, $\hbar$ is Planck's constant divided by 2π, k_B is Boltzmann's constant, and T is the absolute temperature. This leads to $P = 5 \times 10^{-6}$ for protons in $B_0 = 1.5$ T at room temperature, and this is why protons are usually the only imaged nuclei, since their abundance in living tissue (water, lipids, proteins, etc.) compensates for the low value of P. A problem arises when one wishes to image other interesting nuclei which have polarizable moments and are found in living tissue, such as ^{13}C, ^{19}F, ^{23}Na, ^{31}P, and ^{15}N, but which are far less dense than ^{1}H. One recently popular MRI technique invented to deal with this problem and requiring additional superconducting magnets is hyperpolarization.It is possible with hyperpolarization[211] to overcome the problems of dealing with some low-volumetric-density nuclei like ^{13}C. At room temperature and 1.5 T, $P(^{13}C) = 1 \times 10^{-6}$. Hyperpolarization can increase P to near 1, and this hyperpolarization can be maintained at room temperature within useful (in-vivo compatible) molecules (such as pyruvic acid) long enough to make useful MRI images of ^{13}C, giving new insight into its role in various living tissues and in-vivo processes. Hyperpolarization also can be used with some inert gases, ^{3}He and ^{129}Xe, which are of course not normally found in the body, but can be safely breathed into the lungs to make detailed MRI images of the interior of the lungs, which normally are invisible to MRI since they just contain air. All of these hyperpolarization techniques require a polarizing magnetic field from 0.3 to over 3 T, and these are best made by superconducting magnets quite apart from the MRI imaging system.

When hyperpolarization is not available to enhance signals from the nuclei other than ^{1}H, one is left with 2 methods to make the signal-to noise (SNR) acceptable for imaging: increasing the voxel volume (which makes the image more grainy and less localized), and increasing B_0. Imaging chemical-shift aspects of the various NMR-accessible nuclei in living tissue is often called MRS. One encounters this need for enhanced SNR when one wishes to perform MRS which examines the NMR chemical shifts of ^{1}H in order to distinguish and image the different tissue environments in which ^{1}H is bonded. Dividing the available ^{1}H this way not only lowers the signal available for each chemical shift in a voxel, but the chemical shifts are small and proportional to B_0, again calling for high-field magnets. This need for higher fields has led to creating human-size MRI magnets up to the 10 T range. At this field range, the preferred LTS superconductor of choice switches from NbTi to Nb$_3$Sn, despite less favorable mechanical properties, in order to increase field capability. Since full-body magnets with this field are expensive and difficult to construct, only a few have been built, and only for research purposes.

[209] S. Ogawa et al., *Proc. Natl. Acad. Sci.* USA 89 (1992) 5951

[210] J. H. Lee et al., *Nat. Lett.* 465 (2010) 788

[211] K Golman et al., *British J. Radiology* 76 (2003) S118

10.5   Ultralow Field NMR and MRI

John Clarke

10.5.1   Introduction

Nuclear magnetic resonance (NMR) and magnetic resonance imaging (MRI) are wonderfully successful technologies. High-field NMR machines are widely used, for example, to determine the structure of new pharmaceutical molecules with exquisite accuracy; Bruker has recently introduced a system operating at 1 GHz in a magnetic field of about 23.5 T. There are perhaps 30,000 MRI machines deployed worldwide[212], capable of imaging most parts of the human body with a spatial resolution of about 1 mm. Probably few patients are aware that they are surrounded by roughly 100 km of NbTi wire wound in a persistent-current solenoid and immersed in about 1000 L of liquid helium! The majority of MRI machines operate at a magnetic field of 1.5 T but there is an increasing use of 3-T machines. MRI machines are expensive, however, roughly $2 M for a 1.5-T system, have a large footprint, and are heavy: it is not unknown for the cost of reinforcing the building to exceed the cost of the machine.

Despite the trend to higher fields, in this section I take a contrarian approach and discuss what would happen if we tried to image at much lower fields. Would it be possible to image in fields, say, four orders of magnitude lower? If so, would the images be better in some way?

10.5.2   Background

MRI[213] is based on NMR[214,215], almost always of protons, which have magnetic moment μ_p and spin 1/2. In the presence of an applied magnetic field B_0, the protons align either parallel or antiparallel to B_0, the energy level splitting being given (in angular frequency units) by the Larmor frequency $\omega = \gamma B_0$; γ is the gyromagnetic ratio. The NMR frequency is 42.58 MHz/T. The magnetic moment of N protons is $M_0 = N\mu_p^2 B_0/k_B T$ in the limit $\mu_p B_0 \ll k_B T$ (Curie Law). At room temperature and in achievable magnetic fields, the magnetic moment is very small. For example, for $B_0 = 1$ T and $T = 300$ K, $M_0/N\mu_p = \mu_p B_0/k_B T \simeq 3.6 \times 10^{-6}$.

In NMR, M_0 is initially in thermal equilibrium, aligned with B_0 along the z-axis. The application of a "$\pi/2$ pulse" along the x-axis at the Larmor frequency tips M_0 into the x-y plane in which it precesses about the z-axis at frequency $\omega_0/2\pi$. During this precession, M_0 undergoes two relaxation processes. First, it relaxes its direction towards the z-axis, where it eventually regains thermal equilibrium in a characteristic longitudinal (spin-lattice) relaxation time T_1. Second, individual spins dephase via local field fluctuations produced at each site by neighboring spins in the transverse (spin-spin) relaxation time T_2. In water at room temperature, T_1 and T_2 are about 1 s; $T_1 \geq T_2$ always. The full-width at half-maximum (FWHM) linewidth of the NMR line is $\Delta f = 1/\pi T_2$ ("homogeneous broadening"). Inhomogeneities in the magnetic field, however, can substantially broaden the linewidth ("inhomogeneous broadening"), and reduce T_2 to a value T_2^* given by $1/T_2^* = 1/T_2 + 1/T_2'$; T_2' is the inhomogeneous lifetime. The very important spin echo

---

[212] See *Section* 10.4

[213] E.M. Haacke, et al., *Magnetic Resonance Imaging: Physical Principles and Sequence Design.* Wiley & Sons, New York, 1999

[214] A. Abragam, *The Principles of Nuclear Magnetism.* Clarendon Press, Oxford, 1961

[215] C.P. Slichter, *Principles of Magnetic Resonance*, 3rd ed. (Springer, New York, 1989)

technique invented by Erwin Hahn[216] — used in virtually all NMR and MRI pulse sequences — eliminates inhomogeneous broadening but not homogeneous broadening. In MRI, one uses NMR to determine spatial structure by means of three orthogonal magnetic field gradients that define the magnetic field B_0 in a small volume or "voxel". Gradient switching translates the voxel through the patient to construct the magnetic resonance image of a specified region.

10.5.3 Why ULF NMR and MRI?

Given the success of high-field MRI, why would one consider ultralow-field (ULF) imaging? Part of the answer relies on the very low magnetic field noise of detectors based on the SQUID (Superconducting QUantum Interference Device)[217] and the fact that it responds to magnetic flux rather than the rate of change of magnetic flux. As a result, the response of the SQUID to an oscillating magnetic field is frequency independent. In conventional NMR the oscillating magnetic signal induces a voltage across an inductor shunted with a capacitor to form a resonant circuit. The voltage across the tank circuit, by Faraday's Law, scales as $\omega_0 M_0$, that is, as B_0^2. Thus, at first sight, reducing B_0 would seem to be exactly the wrong thing to do. Two factors counter this thinking. First, consider replacing the tank circuit with a SQUID-based detector. Its frequency independent sensitivity eliminates one factor of B_0 in the response, so that the output voltage scales as B_0 rather than B_0^2. Second, one can prepolarize[218] the spins in a magnetic field B_p much greater than B_0. After B_p has been turned off, the spins retain a corresponding magnetic moment $M_\mathrm{p} \gg M_0$ that decays in a time T_1, so that, although the spins precess at frequency $\gamma B_0/2\pi$, they produce a signal amplitude proportional to B_p. Thus, the amplitude of the SQUID-detected NMR signal becomes independent of B_0; one can choose B_0 at will provided $B_0 \ll B_\mathrm{p}$.

There is an immediate advantage of low-field NMR and MRI. For a given inhomogeneity ΔB_0 and a fixed relative inhomogeneity $\Delta B_0/B_0$, the inhomogeneous linewidth $\Delta f'$ scales as $(\Delta B_0/B_0)B_0$, that is, as B_0. Thus, narrow linewidths—and high spatial resolution—can be achieved in relatively inhomogeneous fields. For example, to obtain a 1-Hz linewidth in a 1-GHz NMR system, it is necessary to shim the magnetic field homogeneity to 1 part in 10^9 over the volume of the sample. Although achievable, this is challenging. Furthermore, spatial variations in magnetic susceptibility within a sample produce linewidth broadening—and a loss of spatial resolution in MRI—that cannot be compensated. In contrast, at an NMR frequency of 2 kHz (for protons, corresponding roughly to the Earth's field), the field homogeneity required for a 1-Hz linewidth is only 1 part in 2,000, which is easily obtainable; in addition, the effects of susceptibility variation are negligible.

10.5.4 Ultralow Field NMR

In 2001, in a collaboration with Erwin Hahn and Alex Pines, Robert McDermott—then a graduate student in my group—built a SQUID-based system for experiments on low-field NMR and NQR (nuclear quadrupole resonance). The essence of the system is shown in Figure 10.25. The magnetic flux signal is detected by a dc-SQUID—two Josephson junctions connected in parallel on a superconducting loop[217]. The SQUID—cooled in liquid helium to 4.2 K—is current-biased in the voltage state, and the voltage is periodic in the applied magnetic flux with a period of one flux quantum, Φ_0. The flux in the SQUID is kept constant by means of a current fed back into a coil by a flux-locked loop (not shown). Typically, the flux noise is on the order of $10^{-6}\Phi_0 \, \mathrm{Hz}^{-1/2}$. The superconducting input coil on the SQUID washer[219] is coupled to a superconducting flux transformer configured as

---
[216] E.L. Hahn, *Phys. Rev.* 80 (1950) 580

[217] See Chapter 5 "SQUIDs and Detectors"

[218] M. Packard and R. Varian, *Phys. Rev.* 93 (1954) 941

[219] See Chapter 5, Figure 5.13(b)

a first-derivative, axial gradiometer. When a magnetic flux is applied to the lower pickup loop of the gradiometer, a current is induced in the transformer that in turn induces a magnetic flux in the SQUID. Both the transformer and SQUID respond at arbitrarily low frequencies. Ideally, the gradiometer is insensitive to uniform magnetic field fluctuations, and thus attenuates external magnetic field noise. As indicated in Figure 10.25, room-temperature access enables one to maintain samples at room temperature. Michael Mück at the University of Giessen fabricated the low-noise SQUID.

Our thinking of what to do with our new system, however, was diverted by the arrival of Andreas Trabesinger as a postdoctoral scholar with Alex Pines. Andreas became intrigued with the notion of NMR in liquids at very low fields, and we set up an experiment to study the NMR of protons in mineral oil[220]. We placed the sample, maintained near room temperature, in the lower loop of the first-derivative gradiometer. Separate coils supplied B_p, an oscillating field to tip the spins through $\pi/2$, and B_0—which was deliberately made relatively inhomogeneous. Figure 10.26(a) shows the results of a conventional spin echo measurement with $B_0 = 1.8$ mT. The signal-to-noise ratio is poor, despite the 10,000 averages, and the linewidth is broad, about 1 kHz. In contrast, the spectrum in Figure 10.26(b) was obtained at 1.8 µT fol-

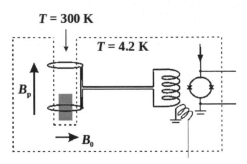

FIGURE 10.25: Configuration of ULF NMR system. The sample, which is nominally at room temperature, is lowered into the lower loop of a superconducting, first-derivative gradiometer inductively coupled to a dc SQUID operated in a flux-locked loop. The directions of the polarizing field B_p and Larmor field B_0 are indicated.

lowing prepolarization of the spins at 1.8 mT. To obtain this spectrum, B_p was turned off nonadiabatically (in a time short compared with the NMR period) so that the spins precessed about B_0. After a time τ, we reversed the direction of B_0, producing a spin echo at 2τ. As expected, the linewidth was reduced by a factor of 1,000, and, even with only 100 averages, the signal-to-noise ratio of the spectrum was greatly increased. This result vividly illustrates the increase in signal amplitude and reduction in linewidth one can achieve by reducing the magnetic field for a given relative inhomogeneity.

10.5.5 Ultralow Field MRI

Our next step was to apply our ULF NMR methodology to the rather more complicated technique of ULF MRI. Robert constructed an MRI system[221,222,223] made from wood with copper wire coils and assembled the required electronics. The wooden cube, 1.8 m on a side, supports two pairs of coils that cancel the Earth's magnetic field in the vertical (x) direction and one horizontal (y) direction. A Helmholtz pair provides a horizontal measurement field in the z-direction that adds to the z-component of the Earth's field to produce a field of 132 µT at the center of the cube. The corresponding Larmor frequency for protons is 5.6 kHz. The cube also supports three sets of coils that produce the gradients $G_z = \partial B_z/\partial z$, $G_x = \partial B_z/\partial x$ and $G_y = \partial B_z/\partial y$, typically 200 µT/m. A copper-wire coil, placed immediately below the sample, produces a polarizing field in the x-direction, and a Helmholz pair produces the $\pi/2$ and π pulses required to create a spin echo. The NMR signal is

[220] R. McDermott et al., *Science* 295 (2002) 2247

[221] R. McDermott et al., *Proc. Natl. Acad. Sci* 101 (2004) 7857

[222] R. McDermott et al., *J. Low Temp. Phys.* 135 (2004) 793

[223] J. Clarke et al., *Annu. Rev. Biomed. Eng.* 9 (2007) 389

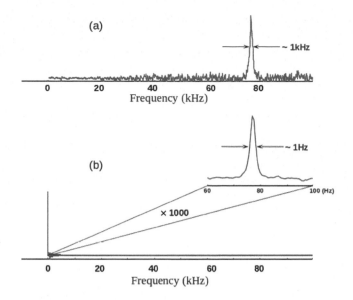

FIGURE 10.26: NMR spectra of 5 mL of mineral oil obtained with the circuit shown in Figure 10.25. (a) Spectrum obtained using a conventional Hahn spin echo with $B_0 = 1.8$ mT and 10,000 acquisitions. The linewidth is about 1 kHz. (b) Spectrum obtained with $B_p = 1.8$ mT and $B_0 = 1.8$ μT using 100 acquisitions. The linewidth is about 1 Hz, and the signal-to-noise ratio is greatly improved compared to that in (a).

detected by a second-derivative gradiometer contained in a fiberglass dewar mounted immediately above the sample. The dewar, built by Robert, has very low magnetic noise, following the design of Seton et al.[224] Instead of the traditional aluminized mylar, the superinsulation consists of aluminized polyester cloth. The finely divided aluminum coating reflects well in the infrared, thus giving a low liquid helium boil-off, while producing much lower magnetic noise generated by Nyquist noise currents. An important feature is an array of 25 Josephson tunnel junctions in series with the gradiometer and the input coil of the SQUID. These junctions switch rapidly to the voltage state to protect the SQUID from the large currents that would otherwise be induced when the various fields are pulsed. Once the fields stabilize, the junctions rapidly revert to the superconducting state, enabling us to acquire the data. The magnetic field noise referred to the lowest loop of the gradiometer is typically 1 fTHz$^{-1/2}$. Later, Robert and Bennie ten Haken, who was on leave from the University of Twente in The Netherlands, surrounded the entire system with a 3-mm thick aluminum shield to reduce both radiofrequency interference and 5.6-kHz magnetic field noise.

To obtain 3-D images, we proceed as follows. The measurement field, $B_0 = 132$ μT, is applied in the z-direction throughout the image acquisition. The pulse sequence begins with the application of the polarizing field along the x-axis for a time several times longer than T_1, so that the proton spins are fully polarized. This field is turned off adiabatically (in a time longer than the NMR period) so that the spins reorient to align with B_0, retaining their initial polarization. Subsequently, a $\pi/2$ pulse at the 5.6-kHz Larmor frequency is applied along the x-axis, causing the spins to precess about the z-axis. Shortly after, the frequency encoding and phase encoding gradients are switched on; these persist until the π pulse is applied to form the spin echo. In the x-direction, the image is frequency encoded, that is, a gradient G_x produces a frequency variation $\omega = \gamma(B_0 + xG_x)$. In the

---

[224] H.C. Seton et al., *Cryogenics* 45 (2005) 348

y- and z-directions, the image is phase encoded: gradient pulses G_y and G_z produce phase changes $\Delta\phi(z) = \gamma z G_y \tau$ and $\Delta\phi(y) = \gamma y G_z \tau$, respectively. Each complete pulse sequence involves one value of G_x and one value of G_y. The SQUID electronics is enabled shortly after the π pulse is complete, so that it is not saturated when the magnetic fields are switched. The data are subsequently decoded to reconstruct the image. The entire sequence is repeated as necessary to improve the signal-to-noise ratio.

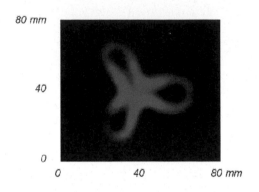

FIGURE 10.27: Image of red pepper slice obtained at 132 μT.

Our first 2-D image was of a pepper slice, shown in Figure 10.27. Notice that the outer edges of the pepper fade out—this is because of the somewhat restricted field of view set by the diameter of the superconducting pickup loop. Subsequent images of "phantoms" demonstrated an in-plane resolution of better than 1 mm in an acquisition time of a few minutes. In one set of experiments we placed a titanium rod on top of the phantom[225]. In high field, the magnetic susceptibility of the rod introduced a substantial distortion of the local magnetic field, and hence of the image of the phantom. In ULF, however, the distortion was unobservable. This result is entirely expected from the earlier discussion showing that linewidth broadening scales as B_0 for a given relative inhomogeneity $\Delta B_0/B_0$. Thus, another potential advantage of ULF MRI is its ability to obtain undistorted images in the vicinity of, say, a titanium screw.

Michael Mößle, a postdoctoral scholar from Tübingen, Germany, joined the group, and we set about obtaining 3-D images in vivo. Figure 10.28a shows Michael's arm being imaged[226], and the resulting images are shown in Figure 10.28b. Each of the six slices is 10 mm thick, the in-plane resolution is about 2 mm and the total acquisition time was 6 minutes. The two small white dots in each image represent bone marrow, and the black annuli around them are bone. The narrow white region on the underside of each slice arises from a fat layer. The white regions are prominent because T_1 is relatively long in fatty tissue. The arm muscle is poorly defined. This is because its T_1 is short, typically 40 ms, so that the signal amplitude diminishes significantly during the imaging sequence. As we shall see, differences in T_1 turn out to be a major strength of ULF MRI.

10.5.6 T_1-weighted Contrast Imaging

An important issue in MRI is distinguishing different tissue types, e.g., healthy and cancerous tissues, even though the proton densities may be identical. In high-field MRI this distinction can, in principle, be made using the fact that T_1 differs in different tissue types so that one can resolve them by weighting the image with T_1. In practice, this technique is not always successful. For example, to image a breast tumor it is necessary to inject a contrast agent—usually a Gd salt–into the bloodstream. The blood tends to flow preferentially to the tumor, where the paramagnetic Gd ions increase the NMR relaxation rate, shortening T_1. However, this technique tends to produce a high rate of false positives. On the other hand, it is known that T_1 contrast can be significantly

[225] M. Mößle et al., *J. Magn. Reson.* 179 (2006) 146

[226] Our protocol was approved by both the environmental health and safety and the human subjects committees.

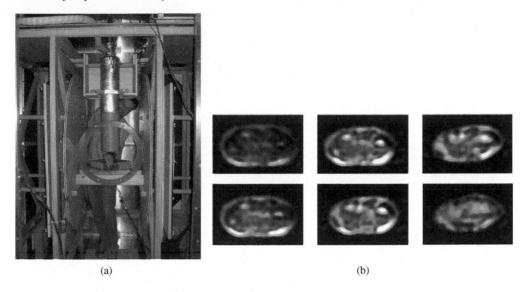

(a)   (b)

FIGURE 10.28: (a) Imaging Michael Mößle's arm. Fiberglass dewar containing the SQUID and gradiometer is suspended immediately above his arm. Polarizing coil is below his arm, blue coil is one of the Helmholtz pair of coils that provides $\pi/2$ and π pulses. The innermost pair of larger coils with axis in the z-plane provides B_0; outside these are the biplanar coils that provide $\partial B_z/\partial x$ and $\partial B_z/\partial y$ and the Maxwell pair that provides $\partial B_z/\partial z$. (b) Six arm slices acquired with the system shown in (a). Slice thickness 10 mm, $B_0 = 132\ \mu$T, field gradients 150 μT/m, in-plane resolution ~ 2 mm, acquisition time 6 min.

enhanced in low fields[227], and ULF MRI is readily adapted to this modality.

SeungKyun Lee , Whit Myers and Nathan Kelso joined our project to develop ULF T_1-weighted contrast imaging[228]. The essence of the technique is as follows. Suppose two regions, A and B, have relaxation times T_{1A} and T_{1B}, with $T_{1A} > T_{1B}$. After polarizing the proton spins fully in a polarizing field B_p, we reduce B_p adiabatically to an intermediate value B_{pi}. Following a delay of a few tens of milliseconds during which the spins decay toward equilibrium at different rates in the field $(B_{pi}^2 + B_0^2)^{1/2}$, we reduce B_{pi} adiabatically to zero and apply the imaging sequence in the field B_0. The image of region A will be brighter than that of region B since the magnetization at the beginning of the imaging sequence is higher. To illustrate this principle, we studied samples with 0.25% and 0.5% concentrations of agarose gel in water; this jelly-like substance is often used to replicate tissue in MRI phantoms. Figure 10.29 shows the variation of the relaxation rate $1/T_1$ versus the intermediate magnetic field and Larmor frequency in which the spins were allowed to relax; for reference, we have included the relaxation rate of protons in water. We see immediately that, while $1/T_1$ in water remains constant, in the two gel solutions the rates are almost identical at high field (above about 30 mT), whereas below about 500 μT the T_1 value for the 0.25% concentration is a factor of two longer than for the 0.50% concentration. This plot shows graphically the enhancement in T_1 contrast at ultralow fields.

As a further illustration of the enhanced T_1-weighted contrast achievable at low fields, Figure 10.30(a) shows a phantom consisting of a tube of 0.5% agarose gel containing nine plastic straws filled with water. The inner diameter of the tubes ranges from 1 mm to 6 mm. Figure 10.30(b) shows

[227] S.H. Koenig and R.D. Brown, III in R.K. Gupta (ed.) *NMR Spectroscopy of Cells and Organisms* Vol. II (CRC Press, Boca Raton, Florida, 1987) p. 75

[228] S-K. Lee et al., *Magn. Reson. Med.* 53 (2005) 9

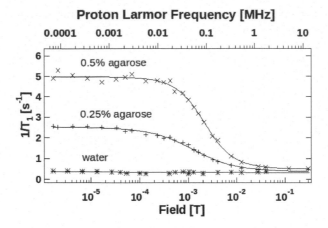

Proton Larmor Frequency [MHz]

FIGURE 10.29: T_1 dispersion of water agarose gel. Relaxation rate $1/T_1$ is plotted versus B_0 (lower abscissa) and Lamor frequency (upper abscissa).

an image of the phantom acquired with an intermediate field of 100 mT where the T_1 values of water and gel are approximately 1.6 s and 1.5 s, respectively. The largest of the water columns is barely distinguished. In contrast, in the image acquired—see Figure 10.30(c)—with an intermediate field of 132 μT where T_1 is reduced to 0.3 s for gel and is unchanged for water, the contrast is stark: all nine straws are clearly visible. These results indicate the potential of T_1-weighted contrast imaging at low fields for distinguishing different tissue types in the human body.

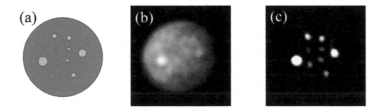

FIGURE 10.30: MRI of phantom at different fields. (a) Phantom consists of column of 0.5% agarose gel in water with 9 plastic straws containing water with diameters ranging from 1-6 mm. (b) T_1-weighted contrast image at 100 mT. (c) T_1-weighted contrast image at 132 μT.

10.5.7 T_1 Contrast in Ex Vivo Prostate Tissue

Prostate cancer is generally diagnosed by means of a PSA (prostate specific antigen) test or a digital rectal examination. Imaging the extent of the cancer is, unfortunately, extremely difficult: in particular, high-field MRI—with or without a contrast agent—is unable to distinguish cancerous and healthy tissues by means of T_1 contrast. Treatment of the disease poses a dilemma because of the wide range of malignancy, and is usually based on biopsy. Treatments include "active surveillance", surgery, radiation therapy and cryosurgery. Thus, reliable, cost-effective imaging would greatly benefit the diagnosis of prostate cancer. In fact, there are MRI-based techniques that can image prostate cancer[229], but these techniques require a 3-T system, and their complexity unfortunately results in a cost that is too high for routine clinical use.

For these reasons, and encouraged by our ULF MRI results on T_1-contrast, Sarah Busch and Michael Hatridge embarked on a study of ex vivo prostate tissue[230]. This study is in collaboration with Dr. Jeff Simko at the Comprehensive Cancer Care Center at the University of California, San Francisco. After a cancerous prostate is removed surgically, it is taken immediately to the pathology laboratory where two small specimens are excised. One is judged (by expert visual inspection) to be nominally cancerous and the other nominally healthy. The specimens are enclosed in a biohazard bag, placed on ice, rushed to Berkeley and placed in our ULF MRI machine. We measure their T_1 values simultaneously in a gradient field which separates the NMR frequencies of the two specimens. The specimens are maintained at 4 °C to slow their degradation. Subsequently, we freeze the specimens on dry ice and return them to UCSF, where they are formalin fixed, sectioned and stained. Jeff examines the thin slices under a microscope to determine the fraction of tumor cells.

Although the values of T_1 in healthy tissue and in 100% tumor tissue vary from patient to patient, for a given patient we find that T_1 is approximately 50% higher in healthy tissue than in tumor tissue. Furthermore, we find that the "image contrast" $(T_{1A} - T_{1B})/T_{1A}$ scales approximately linearly with $(\%\text{tumor})_B - (\%\text{tumor})_A$; T_{1A} and T_{1B} ($T_{1A} > T_{1B}$) are the relaxation times of two tissue specimens A and B from the same patient, where A contains less tumor than B.

In separate experiments, we obtained "T_1 maps" of several larger specimens on which Jeff subsequently performed the histology. Not surprisingly, these T_1 maps showed a similar range of T_1 values as those observed in the small specimens.

These ex vivo results suggest that microtesla MRI with T_1 contrast may be a viable technique for in vivo imaging of prostate cancer. Only in vivo studies, however, can determine the spatial resolution, sensitivity and specificity. We simply do not know how measurements of T_1 made on tissue specimens a few hours after surgery and cooled to 4 °C relate to in vivo values of T_1. If, indeed, the method proves successful in imaging prostate cancer with reasonable resolution—say 2 mm—what might its role be? One application would be in assessing the severity of the disease following the initial diagnosis, e.g., as an intermediate step between initial diagnosis and biopsy. ULF MRI could provide a reference image to guide subsequent biopsy. Because of its potentially low cost per image, it could be used routinely to monitor the progression of prostate cancer during active surveillance or radiation therapy. Since radiation therapy involves the accurate placement of metallic radioactive seeds, ULF MRI—with its insensitivity to the presence of metallic objects—could be used to monitor the insertion of the seeds with the aid of a previously acquired reference image. A broader issue concerns the applicability of ULF MRI to imaging other cancers, for example, breast or brain tumors.

10.5.8   Research in Other Institutions

Numerous other groups have established research programs on microtesla NMR and MRI, and I give a few examples. Qiu et al.[231] measured a proton linewidth of 0.034 Hz in benzene in the Earth's field using a high-T_c SQUID, and Bernarding et al.[232] a linewidth of 0.34 Hz in a J-coupling spectrum. Volegov and coworkers[233] made simultaneous measurements of a somatosensory response and microtesla NMR from the brain — no claim was made that the two are correlated. Zotev et al.[234]

[229] J. Kurhanewicz et al., *Curr. Opin. Urol.* 18 (2008) 71

[230] S. Busch, M. Hatridge, M. Mößle, W. Myers, T. Wong, M. Mück, K. Chew, J. Simko, A. Pines and J. Clarke *(unpublished)

[231] L. Qiu et al., *Appl. Phys. Lett.* 91 (2007) 072505

[232] J. Bernarding et al., *J. Am. Chem. Soc.* 128 (2006) 714

[233] P. Volegov et al., *Magn. Reson. Med.* 52 (2004) 467

[234] V.S. Zotev et al., *J. Magn. Reson.* 194 (2008) 115

used a 7-channel SQUID system to acquire 3-D images of the brain at 46 µT using a polarizing field of 30 mT.

On the technical front, a European-Union consortium involving a dozen institutions and led by Aalto University in Helsinki, Finland, is focused on combining microtesla MRI with a 300-channel system for magnetoencephalography (MEG). Such a system would allow one to perform MRI and MEG with the same system, eliminating co-registration problems and potentially reducing costs. The use of 300-SQUID gradiometers to acquire the same NMR signal could greatly increase the signal-to-noise ratio. Inseob Hahn and coworkers[235] at the Jet Propulsion Laboratory/Caltech and the University of California, Los Angeles have developed a SQUID MRI system involving a cryocooler that runs continuously and achieves a temperature of 3.5 K. In preliminary experiments on a phantom, using an imaging frequency of 5.7 kHz and a polarizing field of 10 mT, they achieved an in-plane resolution of 3.5 mm × 3.5 mm. This is a significant step towards implementing clinical systems. Alternative magnetic field sensors are being investigated. Savukov et al.[236] at Los Alamos National Laboratory used an atomic magnetometer with a magnetic field noise of 12 fTHz$^{-1/2}$ to image a phantom. Myriam Pannetier, Claude Fermon and coworkers[237] at Saclay, France are developing a "mixed sensor"—a thin-film loop of YBCO coupled to a giant magnetoresistance (GMR) detector. They have achieved a magnetic field noise of a few fTHz$^{-1/2}$ at frequencies f above the $1/f$ noise, and imaged water phantoms.

10.5.9 A Perspective

We have seen that it is possible to obtain MR images in magnetic fields four orders of magnitude lower than in clinical MRI machines, and yes, ULF MRI does have a distinct advantage in terms of enhanced T_1 contrast. But what would it take to develop a ULF MRI machine that would have some impact on clinically-important imaging?

First of all, it should be realized that ULF MRI will never compete with high-field MRI in terms of the combination of speed and spatial resolution. Furthermore, high-field MRI has the great advantage that one can place an appropriately-shaped, room-temperature pickup coil over or around any part of the human body to obtain an image. In this respect, the use of a cryogenic detector is restrictive. In addition, the field of view of a single detector is limited to a few centimeters, although the use of multiple detectors—already very well developed for 300-channel, whole-head systems for MEG—would overcome this problem.

The major hurdle in developing a useful ULF MRI system is to improve the signal-to-noise-ratio. To achieve a given spatial resolution, the required imaging time scales as $S_B(f_0)/B_p^2$, where $S_B(f_0)$ is the spectral density of the magnetic field noise of the detector at the Larmor frequency. This scaling factor has tremendous leverage. Currently, the best SQUID-based detectors used in our ULF MRI system have a noise of about 0.5 fTHz$^{-1/2}$. However, in our experience, this noise level is generally increased by radiofrequency interference coupled in via various leads and by noise at the Larmor frequency resulting from an imperfectly balanced gradiometer. With improved rejection of these parasitic noise sources and the use of a lower noise SQUID, I believe it is possible to achieve a noise as low as 0.2 fTHz$^{-1/2}$ or conceivably 0.1 fTHz$^{-1/2}$—roughly the noise produced by Nyquist noise currents in the human body at about 5 kHz. Evidently, an order-of-magnitude reduction in the noise would decrease the imaging time by two orders of magnitude.

Similarly, one must accept the fact that higher polarizing fields are going to be a way of life. We have recently implemented a 150-mT water-cooled polarizing coil, kindly donated by Steve Conolly at UC Berkeley. The coil, consisting of 240 turns of 4 mm × 4 mm copper tube, has an

[235] I. Hahn et al. (unpublished)
[236] L.M. Savukov et al., *J. Magn. Reson.* 199 (2009) 188
[237] M. Pannetier et al., *Science* 304 (2004) 1648

outer diameter of 400 mm and a mass of 31 kg. A 20-kW power supply is required to supply the 200-A current pulse that produces the 150-mT field in the center of the coil. The use of such a coil leads to some challenges, the most serious of which are the eddy currents produced in the aluminum shield when the field is ramped down and the need to make the coil utterly safe for human subjects. Both difficulties have recently been overcome.

I believe there is significant potential to develop a practical, relatively inexpensive, clinically useful ULF MRI system with an initial focus on imaging cancer. As always with new technologies, realizing such a system will require major investments of time and money in science and engineering.

10.6 Superconductivity in Medical Accelerators for Cancer Therapy

Peter A. Zavodszky

In the developed world cancer is the second leading cause of death. During 1990–2006 the cancer mortality rate in the USA decreased from the peak value of 215 to 175 (per 100,000 population), mainly due to earlier detection and better treatment methods. About 50% of cancer patients receive some form of radiotherapy. The success rate of this method of treatment is second only to that for surgery in the case of localized tumors.

Ideally, radiation should kill only the cancerous cells and spare the surrounding healthy tissue. In reality, there is no such radiation treatment, and the total maximum dose deliverable to the tumor is determined by the radiation tolerance of the surrounding area. Over the past 100 years great effort was made to increase the precision of the irradiation to minimize the chance of inducing negative effects (secondary cancers) in the surrounding area of the tumor. Today 90% of radiation therapy involves high energy Bremsstrahlung radiation from electron linear accelerators. This has replaced low energy X-ray sources and ^{60}Co gamma ray sources because of a better depth-dose distribution. Utilizing multi-leaf collimators, intensity modulation and multiple irradiation directions, modern treatment facilities can achieve significantly better dose distribution, and, consequently, better clinical results.

Particle accelerators played an important role in fighting cancer since their invention. John Lawrence and Robert Stone began clinical trials in 1938 to treat cancer with neutrons produced using protons accelerated with the 37-inch cyclotron at the Radiation Laboratory at the University of California, Berkeley[238]. This was the origin of a different mode to use radiation with increased biological effectiveness. Photons have low linear energy transfer, primarily breaking one strand of DNA, which can be repaired by the damaged cell. In contrast, neutrons interact primarily through nuclear reactions, breaking both strands in the DNA helix, which can not be repaired.

It was in 1946 that Robert R. Wilson realized that charged heavy particles, such as protons or carbon ions, have superior characteristics, which make them more suitable than photons or neutrons to kill deep-seated tumors. In his seminal paper[239] he showed that protons and ions lose 85% of their energy near the end of their range, making them suitable to spare healthy tissue surrounding the tumor and killing only cells that are located at the end of that range in the human body. This increase in ionization density had been measured in 1903 for alpha particles by Sir William Bragg and is known as the Bragg profile. Figure 10.31 compares the relative dose at different depths of water—the human body contains mainly water—deposited by 18 MeV photons, 135 MeV protons,

---

[238] J. L. Heilbron and R. W. Seidel, *Lawrence and His Laboratory: A History of the Lawrence Berkeley Laboratory*, Volume I. University of California Press, Berkeley, (1989)

[239] R. R. Wilson, *Radiology* 47 (1946) 487

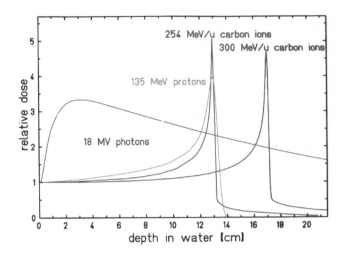

FIGURE 10.31: Comparison of the depth-dose distribution of (conventionally used) photons, protons and carbon ions. With photons the dose decreases exponentially with increasing depth, i.e., the dose in the target volume of deep-seated tumors is smaller than the dose delivered to the healthy tissue above it. Protons and carbon ions exhibit an inverse dose profile, i.e., the dose increases with increasing penetration depth. This profile can be shifted by energy variation over the target volume, leading to a much higher dose deposition inside the tumor than outside in the healthy tissue (see G. Kraft, *Prog. Part. Nucl. Phys.* 45 (2000) S473).

254 and 300 MeV/u ^{12}C ions. By adjusting the energy of the ions, the location of the damage caused by the energy deposition can be controlled with high precision.

The first direct use of ions in cancer treatment took place at the Radiation Laboratory at Berkeley in 1952. Cornelius Tobias and John Lawrence used the 184-inch synchrocyclotron to accelerate deuterium and helium ions to several hundred MeV/u energies, beginning a new era in fighting cancer[240].

The completion of the Bevalac, an accelerator complex formed from a 6 GeV synchrotron (Bevatron), and the SuperHILAC (Heavy Ion Linac) led to a major program of clinical trials to treat human cancer using light ion beams (^{4}He to ^{28}Si with ^{20}Ne most commonly used). From 1977 to 1992 (when the Bevalac became inactive, about 700 patients were treated with ^{4}He ions and about 300 patients, with ^{20}Ne ion beams[241].

Soon after the invention of the cyclotron by Ernest O. Lawrence[242], these accelerators played a crucial role in a variety of biological, medical and industrial applications. The final energy (T) of a given ion with charge $Q = Ze$ and mass Am_0 accelerated in a cyclotron of radius (r) and magnetic field (B) can be calculated with the following simple formula:

$$T = \frac{Q^2}{A} \frac{e^2}{2m_0} r^2 B^2.$$

This shows that for a particular type of charged particle and final energy, a cyclotron can be more compact if the operating magnetic field is stronger. However, the maximum field attainable with room temperature magnets is typically less than 2 tesla. Fraser and colleagues at Chalk River Nuclear Laboratories proposed that by using superconducting magnets to exceed 3 T, one can reduce

[240] C.A. Tobias, H.O. Anger, J.H. Lawrence, *Am. J. Roentgenol., Radium Ther., and Nucl. Med.* 67(1) (1952) 1

[241] W.T. Chu, *Proc. of the 1st Int. Conf. on Particle Accelerators*, Kyoto (2010) 21

[242] E.O Lawrence and N.E. Edlefsen, *Science* 72 (1930) 376

the size and the cost of the cyclotron significantly[243].

Although the initial treatments using neutrons produced poor results, later Mary Catterall[244] obtained encouraging results treating cancer patients with fast neutrons produced in the Medical Research Council Cyclotron Facility at Hammersmith Hospital in London, UK. The National Cancer Institute in the USA in 1979 initiated a 10-year program to design, develop and build hospital-based neutron therapy machines, and to conduct phase III clinical trials[245]. These machines were all room-temperature, magnet-based cyclotrons producing neutrons via the bombardment of a beryllium target with high energy protons ($E_p > 42$ MeV). The use of isocentric gantries allowed irradiations from several directions, spreading the unwanted dose to a larger tissue volume. Neutrons could be produced more easily with deuterium projectiles, due to the larger production cross section, but the size of the cyclotron and gantry capable of producing sufficiently energetic deuterons was deemed excessively large and expensive.

After constructing the world's first superconducting cyclotron in 1982 (the K500, still in service for research as an injector for the larger K1200 superconducting cyclotron completed in 1988) at Michigan State University's National Superconducting Cyclotron Laboratory (NSCL), Henry Blosser and his team designed a more compact cyclotron equipped with superconducting magnets and capable of operating on a rotating gantry[246]. This K100 superconducting cyclotron was built jointly by NSCL and the MedCyc Corporation, commissioned at NSCL in April 1989, and installed the following year at Harper Grace Hospital in Detroit, Michigan[247]. The superconducting magnet produced a field of 4.6 T in the center of the cyclotron and of 5.4 T in the "hill region", with 206 A current in its coils. The coils were potted windings located in a specially designed cryostat that contained liquid He even when the cyclotron was rotated 360° by the gantry. The liquid He consumption was 2.5 l/h with no beam and 6 l/h with a maximum deuterium beam of 15 µA, up to 50 MeV final energy. It was not necessary to extract the beam from the cyclotron, greatly simplifying the design. The neutrons were produced bombarding an internal beryllium target, being collimated and shaped with a specially designed multi-leaf collimator. This cyclotron is shown during the assembly on the gantry in Figure 10.32. Neutron radiotherapy never became a mainstream oncology therapy, but this machine was in clinical operation until 2008, when the neutron radiotherapy program at Harper Grace was discontinued, mainly due to a lack of funding.

As particle therapy slowly gained acceptance as an effective tool fighting cancer, it was realized that treatment can be performed most efficiently in a dedicated, hospital-based facility. In the United States the first such facility was inaugurated at the Loma Linda University Medical Center (LLUMC) in 1990. The accelerator is a synchrotron, designed and built at the Fermi National Accelerator Laboratory, whose first director was R.R. Wilson. This center treats 140–160 patients every day in five treatment rooms, three of them equipped with isocentric gantries. It was generally believed that synchrotrons had certain advantages in particle therapy.

In 1991 IBA, a Belgian company specializing in cyclotron design and fabrication, entered the particle therapy business. They introduced an effective cyclotron design, the C230, and today most particle therapy centers use cyclotron technology. Some of the advantages of cyclotrons are simplicity, reliability, lower cost, smaller size, and the ability to modulate the particle beam current rapidly and accurately, which is required for intensity-modulated particle therapy (IMPT). Since commissioning the first C230 at Massachusetts General Hospital in 2002, IBA has installed a total of 13 additional facilities worldwide, 60% of the total number of commercial proton therapy facilities (30) installed or under construction as of 2009.

[243] C. B. Bingham, J.S. Fraser and H. R. Schneider, Chalk River Nuclear Laboratories Report AECL-4654 (1973)

[244] Mary Catterall and D.D. Vonberg, *Brit. Med. J.* 3 (1974) 137

[245] S. Zink, J. Antoine and F. J. Mahoney, *Am. J. Clin. Oncol.* 12 (1989) 277

[246] H.G. Blosser, et al., *IEEE Trans. Nucl. Sci.* NS-32 (1985) 3287

[247] R.L. Maugan, W.E Powers and H. G. Blosser, *Med. Phys.* 21 (1994) 779

FIGURE 10.32: The K100 superconducting cyclotron during assembly on the rotating gantry at Harper Grace Hospital, Detroit, Michigan. Prof. Henry Blosser, the inventor of the superconducting cyclotrons, is on the left; on the right is Prof. William Powers, the chief physician of the neutron treatment program.

After the successful completion of the Harper Grace Hospital neutron therapy system, the NSCL team led by Prof. Blosser performed a detailed design study of a 250 MeV superconducting cyclotron for proton therapy[248]. This energy is required to reach deep-seated tumors in the human body. This machine was ultimately built by a commercial company, ACCEL, for the Paul Scherrer Institute in Villigen, Switzerland. In parallel, a second unit with gantries and a complete treatment clinic was constructed for the first European commercial proton therapy facility, RPTC, in Munich, Germany. This facility was completed by Varian Medical Systems, which acquired ACCEL in January 2007. The use of superconducting technology in the fabrication of the COMET cyclotron by ACCEL produced an accelerator with considerably less weight and diameter (90 tons, d = 3 m) than the competing IBA room temperature C230 cyclotron (220 tons, d = 4 m), and with decreased electric power consumption[249]. The four cryocoolers used to keep the magnets at liquid He temperature consume 40 kW, compared to the 200 kW needed to operate a similar room-temperature cyclotron. The magnetic field at the center of this cyclotron is 2.4 T, with a 3 T field at the extraction region. The accelerated proton extraction efficiency exceeds 85%, due to a larger pole gap that is enabled by the stronger superconducting magnet. This reduces the unwanted activation of the cyclotron's internal components, allowing for easier maintenance and simpler decommissioning at the end of the useful lifetime of the facility (25–30 years). The machine stays powered overnight, greatly reducing the switch-on time in the morning before patients are treated. This cyclotron is shown in Figure 10.33.

Building on the success of the K100 superconducting cyclotron mounted on a gantry, Henry Blosser began a design study for a 250 MeV proton synchrocyclotron small enough to be mounted on a gantry[250]. An 8 T test magnet was built[251] and demonstrated short–term operation, with plans to transform this magnet to a medical accelerator. This effort was discontinued due to lack of funding, but served as a precursor for a recent project initiated by Timothy Antaya at the Massachusetts Institute of Technology.

In 2003–2004, Blosser introduced a superconducting synchrocyclotron design with a 9 T field for

---

[248] H.G. Blosser et al. *MSUCL*-874, 1993

[249] D. Krischel, et al., *IEEE Trans. Appl. Supercond.* 17 (2007) 2307

[250] H. G. Blosser, *Nucl. Inst. and Meth.* B40/41 (1989) 1326

[251] J. Kim, et al., *IEEE Trans.Appl. Supercond.* 3 (1993) 266

FIGURE 10.33: The COMET superconducting 250 MeV proton cyclotron during the commissioning at the ACCEL facility.

proton beam radiotherapy[252]. This very high field magnet allowed shrinking the mass of the accelerator to less than 25 tons, with an overall diameter of about 1.8 m. This permits mounting the entire synchrocyclotron on a gantry, just as for the K100 neutron therapy machine. A new company, Still Rivers Systems (Littleton, MA.), was formed to commercialize this design. Barnes Jewish Hospital in St. Louis, MO. and the M.D. Anderson Cancer Center in Orlando, FL. have made preliminary commitments to purchase this system, which is anticipated to cost $20M.

A conceptual drawing of the gantry-mounted, high-field 250 MeV proton synchrocyclotron is shown in Figure 10.34. The magnet was successfully tested in February 2009 and the installation was begun in 2010 in the two above-mentioned hospitals. The large cost reduction and single-treatment-room concept will enable many hospitals with existing radiation treatment facilities to procure proton therapy installations. So far the high cost (about $100M–150M for a 3–4 room treatment facility) was the major reason this treatment modality didn't gain wider acceptance in radiation oncology.

Even in the early days of particle therapy it was clear that heavier ions than protons would be more suitable for cancer treatment. Carbon ions were found to be the best compromise between increased relative biological efficiency (RBE) and fragmentation due to nuclear reactions. These ions have a better ratio of the inside/outside of the tumor dose. Due to the larger RBE, they have a smaller lateral scattering in the human tissue compared to protons, making them more suitable to treat deep-seated tumors near sensitive organs, and online dose verification is possible using a positron emission tomography (PET) camera around the patient because a fraction of the ^{12}C projectiles are transformed into ^{11}C, a positron-emitter isotope.

There is considerable evidence for the effectiveness of carbon-ion therapy, mainly accumulated since 1994 at the National Institute of Radiological Science in Chiba, Japan, and the Heavy Ion

---

[252] T.A. Antaya, US patent no. 7,541,905 B2 (June 2, 2009)

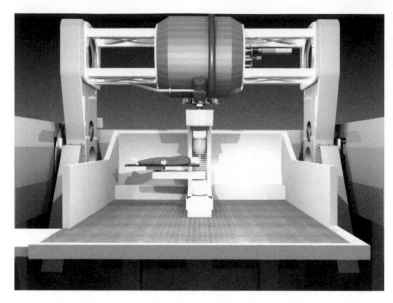

FIGURE 10.34:   The conceptual design for the Still Rivers Systems single-treatment-room Proton Beam Radiation Treatment system showing the Monarch250TM mounted on a gantry within the treatment room.

Research Institute, GSI, in Darmstadt, Germany. In all carbon therapy facilities to this day, ions are accelerated by synchrotrons. In order to reach deep-seated tumors, the energy of the carbon ions should be 400 MeV/u, a value not reachable with present medical cyclotrons.

IBA in Belgium has begun the design of a 400 MeV/u superconducting cyclotron in collaboration with physicists from the Joint Institute for Nuclear Research, Dubna, Russia[253]. The main motivation for a cyclotron to produce carbon ions is to reduce the size and the cost of the accelerator, to be able to modulate rapidly the intensity of the ion beam and to have one accelerator instead of three in series, as is the case for synchrotrons, which require a radiofrequency quadrupole (RFQ) magnet and a linear accelerator (LINAC) as injectors. It is expected that in the space and for the cost of a carbon synchrotron, it will be possible to install a cyclotron and a compact carbon gantry, equipped with superconducting dipole magnets. This machine will accelerate ions with $Q/A = 1/2$ (H^{2+}, He^{2+}, Li^{3+}, B^{5+}, C^{6+}, N^{7+}, O^{8+}, and Ne^{10+}) to 400 MeV/u. The ions will be produced by three external ion sources and will be injected in the cyclotron using an electrostatic spiral inflector. The design of this C400 cyclotron is similar to the IBA C230 proton cyclotron, but with higher magnetic field thanks to superconducting coils, and increased diameter (6.3 m vs. 4.7 m). The C400 cyclotron will operate at fixed energy (400 MeV/u, except for protons at 265 MeV). The energy will be changed in a graphite degrader followed by an electromagnetic energy selector (EES). The nuclear fragments produced during the energy degradation will be eliminated by the EES. Preliminary estimates show that the current on the degrader will be large enough to produce clinically usable beams of ^{11}C, and this would serve for direct, online monitoring with a PET camera of the irradiation received by the patient. The C400 cyclotron will have a magnet with 4 strongly spiraled sectors, 45° at extraction, capable of producing a 4.5 T field on hills and 2.45 T in valleys, and 1.29×10^6 amp-turns produced by superconducting (Nb-Ti) coils. The superconducting coils will have low current density (31.5 Amm^{-2}), with a total stored energy of 60 MJ. The first of these accelerators will be installed in Caen, France, for the ARCHADE Consortium.

---

[253] Y. Jongen et al. Cyclotrons and Their Applications 2007, Eighteenth International Conference (2007) 151

As Denis Friesel and Timothy Antaya concluded in their excellent review paper on medical cyclotrons[254]: "The development of the cyclotron into more powerful and compact machines is continuing at a rapid pace to provide smaller, more powerful, more flexible, and less expensive isotope production and hadron therapy facilities for medical applications". These advances would not have been possible without the extensive use of superconductors.

10.7 Further Reading

Section 2: Medical Applications of Magnetoencephalography

Several excellent review papers and books have been published on the topic of the present chapter. Some of them are historical landmarks in the field. In some cases, their scope is not restricted to magnetoencephalography but covers the whole field of biomagnetism. In other cases, MEG is presented as an application of SQUID magnetometry. The most recent developments of MEG may be found in the proceedings of the BIOMAG conference published biannually.

1. S.J. Williamson L. Kaufman, Biomagnetism, *J. Magnetism and Magnetic Materials*, 1981, 22, 129-201.

2. Romani G.-L., Williamson S.J., and Kaufman L., Biomagnetic instrumentation, *Rev. Sci. Instr.* 1982, 53, 1815-1845.

3. Hämäläinen M., Hari R., Ilmoniemi R., Knuutila J., and Lounasmaa O., Magnetoencephalography — Theory, instrumentation, and applications to noninvasive studies of the working human brain, *Rev. Mod. Phys.*, 1993, 65, 413-497.

4. Biomagnetism: An Interdisciplinary Approach, NATO ASI series, Williamson S.J., Romani G.-L., Kaufman L., and Modena I. eds., Plenum Press, New York (USA), 1983.

5. SQUID sensors: *Fundamentals, Fabrication and Applications*, NATO ASI series, Weinstock H. ed., Kluwer Academic Publisher, Boston (USA), 1996.

6. *The SQUID Handbook — Fundamentals and Technology of SQUIDs and SQUID Systems*, Clarke J. and Braginski A.I. eds., Wiley-VHC Verlag, Weinheim (Germany), 2004.

Acknowledgements

Section 2: Medical Applications of Magnetoencephalography

The authors are grateful to Laura Marzetti, Filippo Zappasodi and Roberto Guidotti for their assistance in the preparation of this manuscript.

---

[254] D. L. Friesel and T.A Antaya, *Reviews of Accelerator Science and Technology* 2 (2009) 133

Section 5: Ultralow Field NMR and MRI

I thank Sarah Busch, Erwin Hahn, Bennie ten Haken, Michael Hatridge, SeungKyun Lee, Nathan Kelso, Robert McDermott, Michael Mössle, Michael Mück, Whit Myers, Alex Pines, Jeff Simko, Daniel Slichter, Andreas Trabesinger and Travis Wong for their many contributions. Development of the original NMR and MRI systems was supported by the Director, Office of Science, Office of Basic Energy Sciences, Materials Sciences and Engineering Division, of the U.S. DOE under Contract No. DE-AC02-05CH11231. Acquisition of the ex vivo prostate tissue data was supported by NIH grant R21CA133338.

11

Wires and Tapes

Editor: David Larbalestier

11.1 The Long Road to High Current Density Superconducting Conductors
David Larbalestier 627
11.2 Nb-Ti — from Beginnings to Perfection
Peter J. Lee and Bruce Strauss 643
11.3 History of Nb₃Sn and Related A15 Wires
Kyoji Tachikawa and Peter J. Lee 661
11.4 Bi-Sr-Ca-Cu-O HTS Wire
Martin W. Rupich and Eric E. Hellstrom 671
11.5 Coated Conductor: Second Generation HTS Wire
A. P. Malozemoff and Y. Yamada 689
11.6 The MgB₂ Conductor Story
René Flükiger and Hiroaki Kumakura 702
11.7 Further Reading 711
Acknowledgments 712

11.1   The Long Road to High Current Density Superconducting Conductors

David Larbalestier

11.1.1   Overview

Multi-tesla, high-field superconducting magnets are possible only when a superconductor can be made into a wire or tape conductor form hundreds of meters or even kilometers long with high current density, acceptable strength, high critical field and affordable price. Thousands of materials are superconducting, yet only six, Nb47wt.%Ti (Section 11.2), Nb_3Sn (11.3), $Bi_2Sr_2CaCu_2O_{8-x}$ and $(Bi,Pb)_2Sr_2Ca_2Cu_3O_{10-x}$ (11.4), $YBa_2Cu_3O_{7-x}$, or more generally any rare-earth (RE) variant of $YBa_2Cu_3O_{7-x}$ (11.5) and MgB_2 (11.6), are available commercially as conductors. Chapter 11 describes many aspects of the development and property evolution in these six conductors, this opening Section 11.1 concentrating on the reasons why the well-developed vision of superconducting magnets of 1913 actually took almost 50 years to implement. It is in many ways a truly remarkable story that intersects with some of the most heartening and terrible aspects of 20th century science and politics. We can define several eras that have governed the development of conductor technology:

- Onnes's vision for superconducting magnets — 1913

- Alloyed superconductors, the type I and type II transition and the collective failure to understand it — 1930–1960

- Theoretical developments and their disconnect from experiment — 1930–1957

- High current density, high field superconductors — 1960 on

- Coexistence of high temperature and low temperature superconductors — 1987 on

- Superconducting conductors beyond the Centennial — 2011 on.

11.1.2   Onnes's Vision for Superconducting Conductors and 10 T Magnets

In 1913, two years after his discovery of superconductivity in mercury, Kamerlingh Onnes prepared a paper for the International Institute of Refrigeration in Chicago about his latest investigations with liquid helium. One part of his paper[1] presents an amazing breadth of view about the potential for superconductors in comparison to the drawbacks of using Cu cooled by liquid air for generating powerful fields well beyond the ~2.5 T saturation of Fe alloys. Only extensive quotations from this paper can properly recognize the breadth of Onnes' vision.

> Mercury has passed into a new state, which on account of its extraordinary electrical properties may be called the superconductive state... The behavior of metals in this state gives rise to new fundamental questions as to the mechanism of electrical conductivity.

> It is therefore of great importance that tin and lead were found to become superconductive also. Tin has its step down point at 3.8 K, a somewhat lower temperature than the vanishing point of mercury. The vanishing point of lead may be put at 6 K. Tin and lead being easily workable metals, we can now contemplate all kinds of electrical experiments with apparatus without resistance.... .

> The extraordinary character of the state can be well elucidated by its bearing on the property of producing intense magnetic fields with the aid of coils without iron cores. Theoretically it will be possible to obtain a field as intense as we wish by arranging a sufficient number of ampere windings round the space where the field has to be established. This is the idea of Perrin, who made the suggestion of a field of 100,000 gauss being produced over a fairly large space in this way. He pointed out that by cooling the coil by liquid air the resistance of the coil ... could be diminished ... To get a field of 100,000 gauss in the coil with an internal space of 1 cm radius, with copper cooled by liquid air, 100 kW would be necessary ... The electric supply, as Fabry remarks, would give no real difficulty, but it would arise from the development of Joule-heat in the small volume of coil ... to the amount of 25 kg–calories per second, which in order to be carried off by evaporation of liquid air would require ... about 1500 l of liquid air per hour ...

> But the greatest difficulty, as Fabry points out, resides in the impossibility of making the small coil give off the relatively enormous quantities of Joule heat to the liquefied gas. The dimensions of the coil to make the cooling possible must be much larger, by which at the same time the electric work in the amount of liquefied gas required becomes

---

[1]   H. Kamerlingh Onnes, *Commun. Physical Lab.*, Univ. of Leiden, Suppl. 34b to 133-144 (1913) 37

greater in the same proportion. The cost of carrying out Perrin's plan even with liquid air might be about comparable to that of building a cruiser...

We should no more get a solution by cooling with liquid helium as long as the coil does not become superconductive.

The problem which seems hopeless in this way enters a quite new phase when a superconductive wire can be used. Joule heat comes not more into play, not even at very high current densities, and an exceedingly great number of ampere windings can be located in a very small space without in such a coil heat being developed. A current of 1000 A/mm^2 was sent through a mercury wire and 460 A/mm^2 through a lead wire, without appreciable heat being developed in either...

There remains of course the possibility that a resistance is developed in the superconductive by the magnetic field. If this were the case, the Joule heat ... would have to be withdrawn. One of the first things to be investigated... at helium-temperatures... will be this magnetic resistance. We shall see that it plays no role for fields below say 1000 gauss.

The insulation of the wire was obtained by putting silk between the windings, which being soaked by the liquid helium brought the windings as much as possible into contact with the bath. The coil proved to bear a current of 0.8 ampere without losing its superconductivity. There may have been bad places in the wire, where heat was developed which could not be withdrawn and which locally warmed the wire above the vanishing point of resistance...

I think it will be possible to come to a higher current density ... if we secure better heat conduction from the bad places in the wire to the liquid helium ... in a coil of bare lead wire wound on a copper tube the current will take its way when the whole is cooled to 1.5 K practically exclusively through the windings of the superconductor. If the projected contrivance succeeds and the current through the coil can be brought to 8 amperes ... we shall approach to a field of 10,000 gauss. The solution of the problem of obtaining a field of 100,000 gauss could then be obtained by a coil of say 30 cm in diameter and the cooling with helium would require a plant which could be realized in Leiden with a relatively modest financial support... When all outstanding questions will have been studied and all difficulties overcome, the miniature coil referred to may prove to be the prototype of magnetic coils without iron, by which in future much stronger and ... more extensive fields may be realized than are at present reached in the interferrum of the strongest electromagnets. As we may trust in an accelerated development of experimental science this future ought not to be far away.

What a description! So many of the essential aspects of superconducting magnet technology were sketched out by Onnes already in 1913. The conception of powerful magnets, the problem of heat removal from compact windings, the economic feasibility of superconducting, as opposed to resistive, magnet operation at current densities of 1000 A/mm^2 and temperatures down to 1.5 K—all of these are crucial aspects of our present superconducting magnet technology. Elsewhere in the same article, Onnes describes the melting of superconducting wires following an abrupt transition from the superconducting to the normal state, and prefigures modern composite filamentary conductors by describing a composite wire made by coating a strong constantan wire with a superconducting tin layer.

It is perhaps only the last sentence ("As we may trust in an accelerated development of experimental science this future ought not to be far away.") that strikes a false note. Sadly the false note became clear to Onnes later in 1913, when he measured the influence of the magnetic field on the properties of lead[2], finding that superconductivity was destroyed by only 600 gauss. More than 20 years would have to go by before the impossibility of constructing magnets from pure metals would come to be dimly understood. And when in the 1935–1937 time frame the way forward to make multi-tesla superconducting magnets using Pb alloys appeared, everything closed down due to the imprisonment and state murder of Lev Shubnikov, even though he showed—and published[3]—the way that alloying a pure metal separated the magnetic transition at a field H_{k1} where flux partially enters the superconductor while still allowing a superconducting transport state up to a significantly higher field H_{k2}, as opposed to a pure metal at which the two transitions occur at a single field. We now know that the pure metals are type I superconductors which lose superconductivity at a critical field H_c which is always less than 1000 gauss, while so-called mixed or Shubnikov-state superconductors can allow superconductivity up to fields H_{c2} well over 10^6 gauss in the cuprate superconductors. What is utterly amazing is that Shubnikov's breakthrough, carried out in a well-known laboratory doing experiments being paralleled in Leiden, Oxford and elsewhere at a time when Landau too was in Kharkov, was completely forgotten until Abrikosov recalled it in 1957[4] but even then not generally known until after Kunzler's quite unexpected demonstration of very high critical current densities in crude wires of the intermetallic compound Nb_3Sn in 1961[5].

The excitement of the pre-world war I era was slow to rebuild and only in the late 1920s did work relevant to Onnes's vision restart, when, in Leiden, the properties of alloys of Pb and In were taken up. An interesting and very important review of this period has been given recently by A. G. Shepelev[6]. The son of Shubnikov's first Ph. D. student, G. D. Shepelev, he tells a riveting story of the discovery of type II superconductivity and its tragic neglect until well after World War II. The tragedy is broad, for Shubnikov had done beautiful experiments that clearly showed how to make superconductors useful for Onnes's vision, had published them, and then was thrown into jail and very quickly executed. His excellent links to the world scientific community were undoubtedly one of the reasons for his execution, making the failure to exploit the discoveries by his scientific colleagues even more mystifying. An abbreviated version of the story[6] follows in the next section, contributed by A. G. Shepelev.

11.1.3 Alloyed Superconductors, Type I and type II Transition and the Collective Failure to Understand It.

(contributed by Anatoly Shepelev, National Science Center Kharkov Institute of Physics & Technology, Ukraine)

"The phenomenon (type II superconductivity) was discovered experimentally by the Russian physicist, Shubnikov[3], around 1937", as reported by twice Nobel Laureate in Physics, J. Bardeen[7] at the Superconductivity in Science and Technology Conference (Chicago, 1966).

On April 11, 1936 the editors of the *Physikalische Zeitschrift der Sowjetunion* received an article from staff scientists of the Ukrainian Physical-Technical Institute (Kharkov) L. V. Shubnikov,

---

[2] H. Kamerlingh Onnes, *Commun. Physical Lab.*, Univ. of Leiden 13 to 139f (1913) 65

[3] L.V. Shubnikov et al., *Zh. Exper. Teor. Fiz.* (USSR) 7 (1937) 221; *Ukrainian J. Phys.* 53, Special Issue (2008) 42 (Reprinted in English)

[4] A.A. Abrikosov, *Sov. Phys. JETP.* 5 (1957) 1174

[5] J. E. Kunzler, E. Buehler, F. S. L. Hsu, and J. H. Wernick, *Phys. Rev. Lett.* 6 (1961) 89

[6] A.G. Shepelev, In: *Superconductor* (ed. A.M. Luiz), Sciyo, Rijeka (2010), p.17; http://www.intechopen.com/books/show/title/superconductor

[7] J. Bardeen, *Superconductivity in Science and Technology* (ed. M.H.Cohen), U. Chicago Press, Chicago & London (1968), p.4.

V. I. Khotkevich, G. D. Shepelev, Yu. N. Rjabinin[8] in which they reported on their research into the effects of magnetic field on the magnetic properties of single crystals of pure metals and alloys. Now, this is known the world over as the experimental discovery of type II superconductors. These superconductors play a very major role now in modern science and technology (for example for the LHC[9] and ITER[10]).

Type II superconductors represent the predominant number of superconducting alloys and compounds discovered over the past 50 years and are characterized by the Ginzburg-Landau parameter κ:

$$\kappa = \lambda/\xi > 1/\sqrt{2},$$

where λ is the magnetic field penetration depth and ξ the coherence length of electrons in the Cooper pair.

Studies on the influence of magnetic field on the *electric* properties of polycrystals of the superconducting alloys were begun in 1928[11] at the Kamerlingh Onnes Laboratory. Then, it was found that, as opposed to pure superconductors (where the superconductivity is destroyed immediately at the critical field H_c), alloys had a broad applied field range in which the superconductivity went gradually away. The experimental studies on the *magnetic* properties of superconducting alloys (regrettably, mainly inhomogeneous two-phase polycrystals) were made in 1934-1935 by four out of the five cryogenic labs that had then the liquid helium capabilities in Canada[12], Britain[13], Holland[14] and the Soviet Union[15]. De Haas and Casimir-Jonker[14] found late in 1934 that, quite different from pure superconductors (whose magnetic properties were well known to change with a leap at H_c), superconducting alloys had a region of gradual penetration of magnetic field into the alloy starting from the weaker magnetic field $H < H_c$. This observation was borne out by the British[13] and Soviet[15] authors, the latter of whom introduced the designations H_{c1} for the incipient penetration field and H_{c2} for the total superconductivity destruction field in the alloy.

These differences in the electric and magnetic properties between alloys and pure superconductors were attributed by all the authors of that time to sample inhomogeneities. The famous "Mendelssohn sponge" hypothesis[13] pre-supposed the existence in the alloys of non-uniformities of the composition, structure and internal stress that were deemed to lead to multiply connected thin structures of anomalously high fields that allowed active high field supercurrent paths. This hypothesis was predominant in the scientific literature for 20–25 years, although in the same year of 1935 the Gorter[16] and London[17] theories were brought out indicating that the superconducting alloys without inhomogeneities could be broken down into alternating thin superconducting and normal layers that ran parallel to the magnetic field. When the superconducting layer thicknesses were smaller than the magnetic field penetration depth λ, superconductivity could exist under large applied magnetic fields. However, neither those theories nor the Mendelssohn sponge could account for the magnetic field penetration into the alloy at $H < H_c$.

The clarification that could have cleared all this up right then was then undertaken by Shubnikov, Khotkevich, Shepelev, Ryabinin[3,8,15], who researched the magnetic properties of carefully prepared

[8] L. W. Schubnikow et al., *Sondernummer Phys. Z. Sowiet.* Arbeiten auf dem Gebiete tiefer Temperaturen, Juni 1936, 39; Phys.Z.Sowiet. 10 (1936) 165

[9] L. Rossi, *Supercond. Sci. and Technol.* **23**, 034001 (2010).

[10] E. Salpietro, *Supercond. Sci. Technol.* **19**, S84 (2006).

[11] W.J. De Haas, J. Voogd, *Comm. Phys. Lab. Univ. Leiden* 199c (1929) 31; ibid 208b (1930) 9; ibid 214b (1931) 9

[12] F.G.A. Tarr, J.O. Wilhelm, *Canad. J. Research* 12 (1935) 265

[13] T.C. Keeley, K. Mendelssohn, J.R. Moore, *Nature* 134 (1934) 773 ; K. Mendelssohn, J.R. Moore. ibid, 135 (1935) 826; K. Mendelssohn, Proc. Roy. Soc. (London) 152A (1935) 34

[14] W.J. De Haas, J.M. Casimir-Jonker, *Proc. Roy. Acad. Amsterdam, Proc. Sec. Sci.* 38 (1935) 2; *Nature* 135 (1935) 30; *Comm. Phys. Lab. Univ. Leiden* 233c (1935) 1

[15] J. N. Rjabinin, L. W. Schubnikow, *Phys. Z. Sowjet.* 7 (1935) 122; Nature 135 (1935) 581

[16] C.J. Gorter, *Physica* 2 (1935) 449

[17] H. London, *Proc. Roy. Soc.* (London) 152A (1935) 650

single crystals of single-phase alloys of Pb-Tl and Pb-In. They made the landmark discoveries that clearly showed that alloying produced a new type of superconductor:

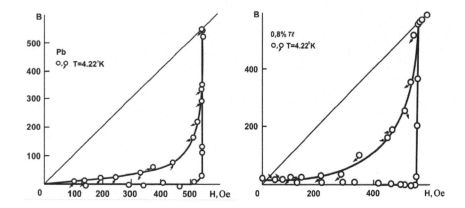

FIGURE 11.1: The magnetic induction curves of long cylinders of pure single-crystal Pb and single crystal alloy of Pb+0.8wt%Tl in longitudinal magnetic field. Both graphs show type I behavior.

1. They found a critical impurity concentration below which the magnetic properties resembled those of pure superconductors—the total Meissner Effect at fields that were smaller than critical and a sudden disruption of the superconductivity upon further magnetic field increasing (Figure 11.1).

2. Upon increasing the impurity concentration beyond the critical concentration (as we understand today, this is when the Ginzburg-Landau parameter $\kappa > 1/\sqrt{2}$) the magnetic properties of the alloys start to differ drastically from those of pure superconductors: The Meissner effect now exists only as far as the magnetic field H_{c1}, and upon further increase of field the alloys remain superconducting to a field H_{c2} much greater than H_c with the magnetic field gradually penetrating into the alloy (Figure 11.2) between H_{c1} and H_{c2}.

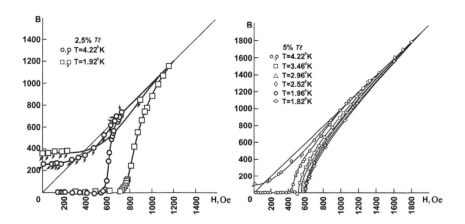

FIGURE 11.2: The induction curve of long cylinders of single-crystals of the alloys Pb+2,5wt%Tl and Pb+5wt%Tl which show type II behavior.

3. With increasing impurity concentration and growing Ginzburg-Landau parameter κ the interval between H_{c1} and H_{c2} broadened, *i.e.* H_{c1} became smaller, while H_{c2} grew (Figure 11.3).

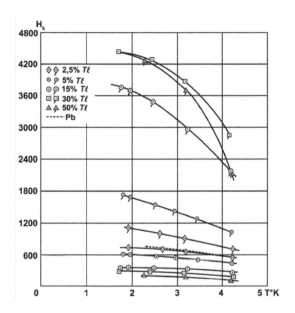

FIGURE 11.3: Temperature dependence of H_{c1} and H_{c2} for single-crystal alloys of Pb-Tl with Tl content varying from 5 to 50% in comparison to the behavior of the single critical field H_c for pure lead and Pb2.5%Tl. The units of the vertical axis are Oe.

4. The unusual properties found in the superconducting alloys could not be attributed to hysteretic phenomena, since the phenomenon was almost reversible.

5. The difference in free energy of the superconducting and normal states was given by the area of the magnetization curve:

$$\Delta F = \int M dH,$$

where M is the magnetization. The entropy difference is given by the derivative:

$$\Delta S = -(\partial F / \partial T)_B.$$

The computation of the entropy difference made in reference[3,8] for these alloys indicated that the magnitudes were of the same order and had similar temperature dependence as for pure superconductors. For this reason, the jump in the heat capacity during the superconducting transition in zero magnetic field for the alloy was comparable to that of a pure superconductor.

6. The X-ray studies made on the superconducting alloys indicated that there was no disintegration of the solid solution and that the alloys were single-phase.

It was exactly in those pioneering works by Shubnikov, Khotkevich, Shepelev, Ryabinin[3,8] that the well-grounded and correct conclusion was made about the existence of a new superconductor type, which conflicted with the earlier experimental research that attributed the results produced until that time to the sample compositional and structural inhomogeneities.

Shubnikov and his coworkers' results[3,8] were in fact instantly known to the scientific community abroad[18,19,20,21,22,23], since copies of *Sondernummer Phys. Z. Sowjet.* that ran the article[8] were handed out to participants of the 6th International Conference on Refrigeration held in the Hague (1936) by M. Ruhemann, who presented Shubnikov's results on the alloyed magnetic properties because Shubnikov had not been allowed out of the country by the Soviet authorities. It was exactly the article[8] published by *Sondernummer* that was referred to later by Jackson[20], Burton, Smith, Wilhelm[21] and Mendelssohn. Over and above, Journal Phys. Z. Sowjet. itself came out in German and English with the circulation of 1700 copies and was distributed to all of the biggest physics research centers across the globe.

Sadly, the research[3,8] was far ahead of its time and its triumphal recognition was made only in 1963 at the International Conference on the Science of Superconductivity (Hamilton, NY)[24] after Abrikosov[4] had constructed the theory of this phenomenon based on this experimental research[3,8] and the Ginzburg-Landau theory[25]. The concept of Type II Superconductors (Shubnikov phase) has found its place into the *Golden Treasury of World Science*, and is expounded in all current monographs on superconductivity. It goes absolutely without saying that the Type II superconductors will enrich the humankind with more very useful applications. In more detail, information on this subject can be found in the book and on-line[6].

11.1.4   Theory and the Long Disconnect from Experiment

It is extraordinary reading academician Shepelev's story in the previous section that the report of the Shubnikov results by Ruhemann in the Hague in 1936 where all the major players were represented did not lead to their duplication, confirmation and extension. But they did not. Even in 1950 when Ginzburg and Landau (G-L) developed their phenomenological model of the superconducting state[25] and explicitly considered that solutions of the G-L equations existed not just for type I superconductors with $\kappa < 1/\sqrt{2}$, but also for $\kappa > 1/\sqrt{2}$—but they considered the latter unphysical. What makes this even more surprising is that Landau had been in Kharkov with Shubnikov in the 1930s and was close to him. Abrikosov's difficulties with getting Landau to accept his vortex solutions of the G-L equations for very large κ are well known[26], as is the delay of several years that ensued. One grace note is that when Abrikosov did prepare his breakthrough 1957 paper on the vortex state in high κ superconductors, he recalled seeing the original results of Shubnikov and refers to Shubnikov's 1937[3] paper. Abrikosov comments that many experimentalists did not accept vortices until the decoration experiments of Essmann and Träuble[27] were published in 1967. This may be a theorist's viewpoint. By 1967, I was midway through my experimental Ph. D. studies and all around me were experimentalists happily exploiting the benefits of high-J_c, high-field superconductivity.

[18] M. and B.Ruhemann, *Low Temperature Physics*, U. Press, Cambridge (1937) 313

[19] D.Shoenberg, *Superconductivity*, U. Press, Cambridge (1938) 82

[20] L.C. Jackson, *Repts. Progr. Phys.* 6 (1940) 338

[21] E.F. Burton, H.G. Smith, J.O. Wilhelm, *Phenomena at the temperature of liquid helium*, Reinhold Publ. Corp., N.Y. (1940) 319

[22] K. Mendelssohn, *Repts. Progr. Phys.* 10 (1946) 362

[23] D. Shoenberg, *Superconductivity* (2nd ed.), U. Press., Cambridge (1952) 41

[24] J. Bardeen, R.W. Schmitt, *Rev. Mod. Phys.* 36 (1964) 2; C.J. Gorter, ibid, 6; K. Mendelssohn, ibid, 10; B.B. Goodman, ibid, 15; T.G. Berlincourt, ibid, 20.

[25] V.L. Ginzburg, L.D. Landau, *Zh. Exper. Teor. Fiz.* (USSR) 20 (1950) 1064

[26] A. A. Abrikosov, *Physics Today*, pp 56-60, January 1973.

[27] U. Essmann and H. Träuble, *Phys. Letts.* 24A (1967) 526

11.1.5 The 1960s: The Age of High Critical Current Densities and, Finally, High Field Magnets

An interesting view of the transition from the 1950s to the 1960s was given at the 1982 Applied Superconductivity Conference by John Hulm, then Director of the Westinghouse Research Laboratory in Pittsburg[28]. At the University of Chicago, in 1951 he and his graduate student George Hardy had discovered superconductivity at 17 K in the A15 compound V_3Si (T_c of 17 K). This A15 structure was long to be that which guaranteed the highest T_c until the discovery of superconductivity in the cuprates. Matthias, his former Chicago colleague, then at Bell Labs, quickly picked up this A15 discovery and discovered more than 30 superconducting A15 compounds, most importantly Nb_3Sn with T_c of 18 K. A curiosity that perhaps appealed to Hulm more in industry than in academia was the finding of Yntema in 1955[29] that cold-drawn Nb could achieve a critical current density of more than 10^5 A/cm^2 in fields of a few kgauss and moreover that this Nb wire could be used to magnetize an iron-cored magnet. Autler[30] and then Kunzler[31] made similar wires out of cold-drawn Nb and Mo-Re alloy, finding that quite high current density was retained in fields that seemed incompatible with the old ideas of the "sponge" and its sparse and very small filaments. All was thus set for matters to come to a head which they did in December 1960 when Kunzler's group[5] put primitive wires of Nb_3Sn into their 88 kG magnet and found them not just still to superconduct at 88 kG, but to do this at a current density J_c exceeding 10^5 A/cm^2.

A second fascinating synoptic review of the very long unification of theory and the practice of superconducting applications was given by Ted Berlincourt, another of the small band of 1950s superconductivity researchers, at the Applied Superconductivity Conference in Baltimore in 1986 for celebrations of the 75th anniversary of the discovery of superconductivity[32]. Berlincourt's paper opens with two key sentences (the italics are mine): "The discoveries of Yntema and of Kunzler, Buehler, Hsu and Wernick demolished a myth. Despite earlier contrary indications, it was suddenly found that *large critical current densities* could, after all, be supported in superconductors at high magnetic fields." As someone who has spent almost his whole superconducting life concerned with the critical current density J_c, it is remarkable for this chapter to have gotten so far with so little mention of the critical current density, the property essential for magnet applications. For if a new high T_c discovery is reason enough for at least a nomination for a Nobel prize, no step towards Onnes's magnet dream was ever going to occur without high J_c at high fields. Reflecting on the curious disconnects that occurred between Shubnikov's homogeneous single crystals with almost no J_c (they had no hysteresis), the Mendelsohn sponge idea which rationalized low J_c by assuming just a trivial volume of residual superconducting filaments, and the general idea of fundamental scientists, expressed most clearly by Ginzburg[33] that "alloys are an unsavory business", one can see how fertile the field was in the late 1950s. All the key discoveries had been made—but the connections between them were absent and their "so what?" unappreciated. Berlincourt's paper pulls these strands together from the viewpoint of an experimentalist trying to understand alloying effects in the refractory-metal superconductors like Nb, itself a marginally type II elemental superconductor, not the much lower κ type I materials based on Pb, In and Tl with which the 1930s community was playing. When Kunzler placed Nb_3Sn into the strongest DC field of 88 kgauss available at Bell Labs in December 1960 and found J_c values greater than 10^5 A/cm^2, the incipient, unappreciated potential era of superconductivity crystallized into an almost immediate reality and uncertainty evaporated

[28] J. Hulm, IEEE Trans. *On Magnetics* 19 (1983) 161

[29] G. B. Yntema, *Phys. Rev.* 98 (1955) 1197

[30] S. H. Autler, *Bull. Amer. Phys. Soc.* 4 (1959) 413; *Rev. Sci. Instr.* 31 (1959) 369

[31] J. E. Kunzler, E. Buehler, F. S. L. Hsu, B. T. Matthias and C. Wahl. *J. of Appl. Phys.* 32 (1961) 325

[32] T. G. Berlincourt, *IEEE Trans. on Magnetics* 20 (1987) 403

[33] V. L. Ginzburg, On superconductivity and superfluidity, in T. Fraengsnyr (Ed.), *Le Prix Nobel*. The Nobel Prizes 2003 (Nobel Foundation, Stokholm), p103.

almost overnight. Their paper[5] was submitted to *Physical Review Letters* on January 9, 1961 and appeared in the February 1, 1961 issue (has publication in *PRL* ever been so rapid?). By the fall of 1961, small solenoids made from Nb_3Sn and Nb-Zr wire had been used by Bell Labs, Westinghouse and Atomics International to generate fields of more than 60 kG [28]. Now the superconductor was generating the ampere turns and not being used as a low dissipation helper to saturate iron. Superconductivity as a technology was at hand!

The story of how this breakthrough generated a technology of superconducting magnets is the subject of the rest of Chapter 11 and Chapter 12. Given the 50 years that had occurred between the discovery of superconductivity in Hg and in Nb_3Sn, the speed with which the technology progressed in the 1960s was quite as impressive as it would have been unexpected at the beginning of 1960. At first monofilament conductors of cold-drawn Nb-Zr wire were used but stable charging of current into the magnet and protection of the magnet from burn up during quench of the superconducting state both required a parallel, high conductivity normal metal path. However, it was hard to bond copper to Nb-Zr and attention soon turned to Nb-Ti, both because it had higher H_{c2} and because Cu bonded well to it, as Section 11.2 describes. Nb_3Sn too made it to conductor status and to high field magnets, first as a tape (Section 11.3.2) and then in the 1970s as a filamentary conductor. By the end of the 1960s all of the essential understanding of how to make superconducting materials into conductors useful for magnets had been developed. The need to subdivide the superconductor transverse to the field to avoid large shielding currents which would initiate flux jumps and quench the superconducting state led to so-called multifilament conductors (Figures 11.17 and 11.23) with tens, hundreds or even thousands of filaments < 50 μm in diameter. Minimizing coupling of the filaments across the normal Cu matrix by the changing external field of the magnet was controlled by twisting the filaments, while controlling the current capacity of the magnet conductors was addressed by cabling strands of typical diameter 0.5-1 mm into transposed cables, the most widespread of which is the 20–30 strand Rutherford cable (see box in Section 11.2.11) or the multi-hundred strand cables used for cable-in-conduit conductors like that for ITER (Figure 11.25), which enable stable operation of conductors with currents as high as 100 kA. With these two materials, Nb47wt.%Ti and Nb_3Sn, was born the superconducting wire industry. Nb-Ti supplied the particle accelerators at Fermilab (the Tevatron), at DESY (HERA), the most immense accelerator at CERN, the Large Hadron Collider (LHC), and the truly commercial application of superconductivity in magnetic resonance imaging (MRI). Nb_3Sn was pulled by the need for very high field magnets above the approximately 10 T capability of Nb-Ti, above all the application to very high field NMR above proton frequencies of about 500 MHz, which have recently culminated in the 1 GHz Nb-Ti/Nb_3Sn system installed by Bruker in Lyons in 2010 (Figure 11.26). Application of more than 600 tonnes of Nb_3Sn conductor for the magnet coils of ITER starting in about 2009 shows that the Nb-base conductors retain their grip on the high field magnet market even 50 years after Kunzler's discovery and 25 years after the high temperature superconductor discovery.

11.1.6 Coexistence: High Temperature and Low Temperature Superconductors — 1987 to Present

Displacing an established technology by an "improved" one is always hard and so it has proved for the new higher T_c superconductors discovered in 1986–1988. When Bednorz and Muller showed a T_c onset at about 35 K in a multiphase La-Ba-Cu-O[34], few in the applied community paid much attention, but by early Spring 1987 when T_c rose above 77 K [35], it seemed that everyone (*Time, Popular Science, The New York Times,* ...) was paying attention. One reason was that an immense new superconducting machine, the Superconducting Super Collider (SSC), was in advanced design and

---

[34] J. G. Bednorz and K. A. Muller, *Z. fur Physik* B64, 189 (1986).

[35] M. K. Wu et al., *Phys. Rev. Letts.* 58 (1987) 405

preconstruction status. Some advocates of small-scale science saw an opportunity to advocate for the new superconductors, almost as soon as the T_c of 90 K in $YBa_2Cu_3O_{7-x}$ was announced at the "Woodstock of Physics" in New York City where the March 1987 American Physical Society Meeting convened. The struggle soon seemed to default to the too-simple question of whether to invest in a big new high energy physics machine or in smaller scale science—rather than to advocate for both. This led to the truly ambitious, if temporary, advocacy that the SSC should be built not with Nb-Ti, but with one or more of the new cuprate superconductors, which seemed to appear with ever higher T_c almost every month. With hindsight, this optimism about the new cuprate superconductors, now generically called high temperature superconductors (HTS), seems rather naïve. If it had taken 25 years (from about 1963 to 1988) to optimize the "simple", binary Nb47wt.%Ti solid-solution on the industrial scale, how could conductors be made of complex oxides like $YBa_2Cu_3O_{7-x}$ or $Bi_2Sr_2CaCu_2O_{8-x}$ in 5 years or less? And so it has proved, as Section 11.4 on the development of $(Bi,Pb)_2Sr_2Ca_2Cu_3O_{10-x}$ and $Bi_2Sr_2CaCu_2O_{8-x}$ conductors and Section 11.5 on conductors of $YBa_2Cu_3O_{7-x}$, show.

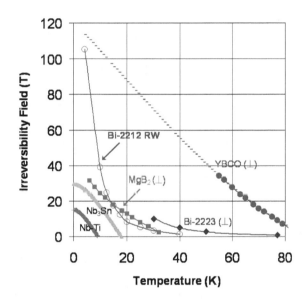

FIGURE 11.4: Irreversibility fields for representative wires of MgB_2, Bi-2212, Nb-Ti, Nb_3Sn and YBCO tape for H parallel to the c axis. Data on round wire Bi-2212 provided by J. Jiang and on SuperPower YBCO tape by Z. Chen, both unpublished. Bi-2212 and YBCO have H_{irr} exceeding 100 T, more than 3 times Nb_3Sn.

At first sight the advantages of the higher T_c superconductors are truly formidable, as can be seen in Figure 11.4 by comparing the immensely larger domain of superconductivity of HTS as compared to LTS materials. Nb_3Sn with twice the T_c of Nb-Ti also has twice the H_{c2}. By steady innovation, Nb_3Sn now allows construction of magnets up to about the 23.5 T needed for 1 GHz proton NMR. But the cuprates YBCO and Bi-2212 and Bi-2223 have H_{irr} values at 4 K which are at least 3 times that of Nb_3Sn. In principle their huge H_{irr} advantages can be taken either for the construction of much higher field magnets than with Nb-base conductors or for the construction of lower-field magnets at much higher temperatures and thus with much more economical refrigerants than the liquid helium used for most Nb-base magnets. But we now must come back to the delicate

Wires and Tapes

question of the attainable critical current density J_c in wires because there is a threshold of J_c below which applications are not viable. Crudely speaking this occurs at about 1000 A/mm^2 in the desired field and temperature domain. This threshold was passed almost immediately for Nb-Ti and Nb$_3$Sn, but for the cuprate superconductors a huge problem almost immediately appeared: superconductivity was strongly depressed at arbitrarily misoriented grain boundaries (GBs) (see also Chapter 3.6), greatly inhibiting long-range current flow in the polycrystalline forms needed for conductors. This GB impediment is the single most important reason why HTS conductors have not yet driven out LTS conductors. HTS conductors have multiple challenges. First is the need for high J_c within the grains without sacrifice of the required crystallographic texture that allows it over long lengths. Second is to transition from the "superconductivity-at-77-K-is-so-marvellous-that-it-must-be practical" push-phase to a broad market-pull phase, above all by the electric utility industry. Third is to do this with conductors that have technical and cost properties needed by applications. As Chapter 12 shows, HTS conductors can enable huge segments of the electro-technology market, but sadly the electric utility industry has never been as active an application-pull agent for HTS as the accelerator community has been for Nb-Ti and Nb$_3$Sn.

11.1.7 Superconducting Conductors beyond the 2011 Centennial

In fact the accelerator community does have a long-term interest in the very high fields that can be generated by HTS conductors, now well over 30 T in small prototype coils[36]. Two such applications might be a Muon Collider that would use fields of 30–50 T to shrink the beam to small enough size to get useful particle densities and an LHC tripler that might use 20–24 T dipole magnets to steer the beam within the existing LHC tunnel. Moreover, high magnetic field laboratories that presently supply 30+ T fields using 10–20 MW into resistive magnets now are exploring the use of HTS conductors for the generation of such fields with only a few kW of refrigeration. The 32 T user-magnet now being built at the National High Magnetic Field Laboratory in Tallahassee will combine a 15 T Nb-Ti/Nb$_3$Sn outsert with a 17 T REBCO insert as the first such example. To get an idea of what conductors might in 2011 support such projects, Figure 11.5 is useful, comparing both LTS and HTS conductors at 4.2 K, where the highest field magnets are possible. Here is plotted not the J_c in the superconductor cross-section but the overall conductor current density, J_E. The dominance of Nb-Ti and Nb$_3$Sn in their respective J_c domains below about 8 T for Nb-Ti and below about 17-18 T for Nb$_3$Sn is evident. What does not appear in the plot is that the costs for LTS conductors are well known and reasonable and that their conductor architectures are flexible and the wires readily made into cables that allow arbitrary amperage conductors to be made. The superconductor cross-sections in such conductors are variable but typically 25–40%. A similar statement is true for the round wire Bi-2212 and for MgB$_2$, though the hard nature of MgB$_2$ tends to restrict superconductor cross-sections to the 10-20% range. It is interesting that round wire Bi-2212 is getting to the stage at which it is competitive in J_c with Nb$_3$Sn below 20 T, especially because ongoing work within the Very High Field Superconducting Magnet Collaboration (VHFSMC)[37] is gradually "unpeeling its onion". It is clearly near to the critical J_c take-off point. The truly remarkable properties of REBCO coated conductors do not jump out quite so immediately from the graph because with REBCO layers of 1 μm and a total tape thickness of 100 μm, actually the superconductor is 1% or even less of the total conductor. Even so REBCO tape in its inferior orientation (H perpendicular to the tape plane), which generally defines the performance limit of a magnet, is almost identical to round wire Bi-2212 with about 25% of superconductor. Moreover some variants are manufactured on very strong

[36] H.W. Weijers, U.P. Trociewitz, W.D. Markiewicz, J. Jiang, D. Myers, E. E. Hellstrom, A. Xu, J. Jaroszynski, P. Noyes, Y. Viouchkov, and D. C. Larbalestier, *IEEE Transactions on Applied Superconductivity* 20 (2010) 576

[37] Very High Field Superconducting Magnet Collaboration, a collaboration of groups at Brookhaven National Laboratory, Fermilab, Lawrence Berkeley Laboratory, Los Alamos National Laboratory, the National High Magnetic Field Laboratory, North Carolina State University, Texas A & M University seeking to understand and apply round wire Bi-2212.

substrates like Hastelloy (Section 11.5) that yield great resistance to the very high stresses found in high field magnets. Thus, HTS applications to magnets, perhaps even some invasion of the domain of LTS conductors, can be foreseen.

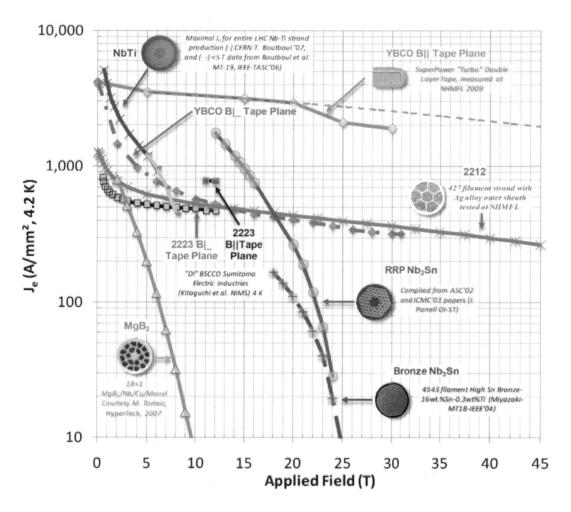

FIGURE 11.5: Engineering critical current density values J_E for recent representative wires of MgB_2, Bi-2212, Nb-Ti, Nb_3Sn wires where anisotropy is absent. The REBCO tape data is for H parallel to the c axis and at 90°, where H lies in the tape plane. Plot maintained by Peter Lee at the National High Magnetic Field Laboratory. (http://magnet.fsu.edu/~/ee/plot/plot.htm)

How rapidly might such a changeover from LTS to HTS occur? The key underpinning of any superconducting magnet technology is the conductor. As might be expected from an author of a centennial article, I believe that history tells us that this will take some time. Conductor technology development occurs slowly, mostly because of the increasing complexity of the superconductor in the conductor. It took 25 years from about 1962 to 1987 to perfect an understanding of the basic metallurgy, the development of optimized two-phase nanostructures, and the processing and integration into large-scale manufacture of Nb-Ti (Section 11.2). But the final 10 years of progress was accelerated hugely by the decision of the US High Energy Physics Community to build the Tevatron and then the SSC. Out of this came a clear demand-pull and in response a highly integrated magnet-builder, industrial fabricator and basic research culture. The US still has the biggest, the most capable and the most diverse superconducting wire industry, both for LTS and for HTS. So the future ought to be bright. This multi-sector culture has been particularly effective because it responded to the whole problem of developing conductors, as the peeling the onion cartoon in Figure 11.6 of Hem Kanithi, himself a former Berkeley superconducting materials scientist Ph. D. gone industrial (and still in the business at Luvata in Waterbury, CT), shows.

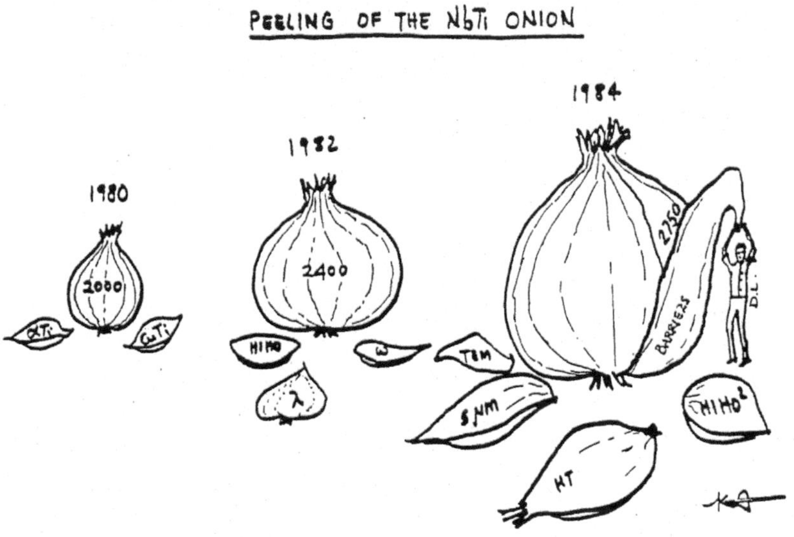

FIGURE 11.6: Peeling the Nb-Ti onion. The increase in J_c measured in A/mm^2 at 5 T, 4.2 K corresponded to a continuous process of understanding on the one hand how to increase vortex pinning and the local J_c by homogenizing the starting alloy (HiHo and HiHo^2) by enhancing precipitation of α-Ti, while on the other suppressing degrading factors like filament sausaging that gated the locally high J_c by periodic reductions in filament cross-section. Sketch courtesy of Hem Kanithi. Although the detail is specific to Nb-Ti, this process occurs for all conductors.

The essential lesson of the "onion" is that while manufacture of small samples with compelling properties in highly sophisticated laboratories can provide the rationale to go forward with a conductor technology, actual technology progress is often about going early to manufacture, using the product, understanding the process defects and the difference between tolerable and intolerable ones, continuously innovating and keeping the wire-device product cycle short and ever-evolving. Judicious activities in the basic science can then work on avoiding defects that degrade J_c, and continued enhancement of vortex pinning and GB properties. Obstructive grain boundaries, inherent structural anisotropy, and chemical complexity continue to hold back HTS conductors, as does the restricted market demand of a few, if now quite large, electric-utility, demonstration projects. Smaller, but still very significant magnet applications are, in my view, a great way to broaden HTS conductor use, especially by integrating YBCO with cryocoolers which Figure 11.4 shows could enable the 10–20 T field range of Nb$_3$Sn magnets, not just at 2–4 K, but up to 55 K. 20 years of support by the US Department of Energy – Office of Electricity and its predecessors cultivated a rich, synergistic culture of industry-national laboratory-university collaborations that hugely helped HTS technologies. The 2010 US-DOE decision to phase out support for conductor development has

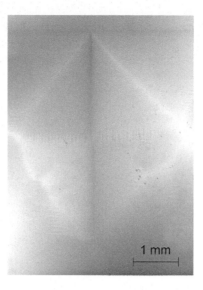

FIGURE 11.7: Magneto-optical image of the perpendicular field above a $Ba(Fe_{1-x}Co_x)_2As_2$ thin film grown on a $9°[001]$ tilt $SrTiO_3$ substrate. The GB runs vertically upward in the center of the images and the roof pattern due to supercurrent flow in each grains is strongly perturbed by the GB which only allows about 10% of the intragrain current to pass. Image courtesy of A. Polyanski (NHMFL)[40].

been justified by the belief that HTS conductors are now commercial within the US. The experience of LTS conductor development suggests that this decision is premature.

In fact the grain boundary problem in cuprates is still not understood sufficiently well to really define the long-term capabilities of the technology, largely because grain boundaries in real polycrystalline forms are significantly more complex than in the thin-film bicrystals preferred for doing science. It is an interesting point that the critical angle θ_c at which the GB of epitaxial bicrystals starts to fall below the grain is only $3°$ (Section 3.6). But for the meandered, real GBs found for example in the ex situ processes used on RABiTS conductors, it is about twice as large. While this enhancement may appear small, it is utterly vital for the RABiTS approach to coated conductors, where template texture is harder to develop than in the IBAD-MgO approach, as section 11.5 tells. Without this enhancement, the IBAD coated conductors would have a 2–3 times advantage in J_c, as compared to the RABiTS. In this case then, greater complexity is also better![38] Figure 11.7 shows a visually appealing way of showing this obstruction using magneto-optical imaging. Applied in detail much earlier to the study of YBCO bicrystals[39], here it was applied to epitaxial thin films of the recently discovered $Ba(Fe_{1-x}Co_x)_2As_2$ ferropnictide superconductor. Although we conclude that the GB suppression of J_c is proportionately not quite as large as in YBCO[40], it is still significant, ensuring that bulk samples of these compounds tend to have low transport critical current density[41]. In spite of the attractive low anisotropy of the compounds, in general lower than any cuprate, they are at least for now weak-linked, meaning that conductors will need to borrow some of the same texturing strategies that have been so vital for HTS conductors.

Another major problem occurs when anisotropy is present. As the electronic anisotropy in-

[38] D. M. Feldmann et al., *J. Appl. Physics* 102 (2007) 083192

[39] A. A. Polyanskii et al., *Phys. Rev.* B53 (1996) 8687

[40] S. Lee et al. *App. Phys. Lett.* 95 (2009) 212505

[41] A Yamamoto et al., *Supercond. Science and Techn.* 21 (2008) 095508

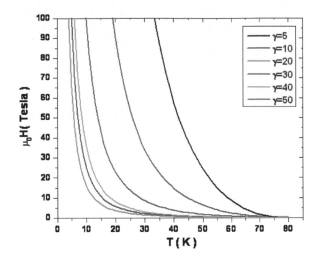

FIGURE 11.8: The scaled c-axis irreversibility field for a 90 K superconductor with H_{c2} anisotropy appropriate for YBCO (5) and Bi-2212 (50). Note the very strong suppression of the superconducting domain where J_c is greater than zero as the anisotropy increases.

creases, vortex stiffness declines (see Chapter 3.9), thermal fluctuations play a larger role and the field and temperature domain in which finite J_c is possible diminishes. This is conveniently measured by the irreversibility field H_{irr} at which $J_c = 0$. As Figure 11.4 shows the $H_{irr}(T)$ curves for the much more anisotropic Bi-2212 and Bi-2223 cuprates are depressed well below that of YBCO. In this respect the 123 crystal structure of the REBCO compounds is much more favorable for applications than any other cuprate, in spite of the fact that T_c does not exceed 95 K, while the highest T_c cuprate Hg-1223 superconducts until about 135 K. The anisotropy can be expressed by the parameter γ, which represents the ratio of H_{c2} with field parallel and perpendicular to the superconducting planes[42]. The huge influence of large γ (Bi-2212 is ~50, YBCO ~5) is quite evident in both the calculation of Figure 11.8 and the real data in Figure 11.4. The key point of the figure is that anisotropic-layered superconductors are likely to have a limited ability to develop high J_c in strong fields, a concern for some preferred models of the search for higher T_c superconductors. In any case, it is clear that there is no slam dunk for applications just because a higher T_c superconductor is found. Many factors control the technology and T_c is only one of them. It is worth recalling that more than 90% of all superconductors made are Nb-Ti, the lowest T_c superconductor discussed here. They are strong, tough, understood, readily fabricated, easily wound and predictable in performance. By far the biggest use is in the MRI industry and few patients in MRI machines understand that the heart of the machine is sitting only 4 K above absolute zero. But HTS conductors do offer us opportunities that are impossible with Nb-Ti and Nb_3Sn, especially YBCO with its low anisotropy and Bi-2212 with its highly desirable round wire form. If new discoveries can expand the range of conductors too, then perhaps Onnes's manifest excitement of a magnet technology 1913 might be manifested by another of equal importance in 2013—or even sooner!

---

[42] M. Tinkham, *Phys. Rev. Letts.* 61 (1988) 1658

11.2 Nb-Ti — from Beginnings to Perfection

Peter J. Lee and Bruce Strauss

11.2.1 The High Field Revolution in Retrospect

Towards the end of a cryogenics session at the January 1955 Annual Meeting of the American Physical Society in New York, George Yntema of Cornell reported that he had achieved a field of 0.71 T at 4.2 K in a 3 mm gap using a cold-worked superconducting Nb wire produced by Fansteel. The 4296 turns of 0.05 mm diameter Formvar insulated cold-drawn Nb strand around a soft Fe core carried a reported 1.8 A at 888 A/mm^2.[43] In Ted Berlincourt's review of the early history of Nb-Ti, he pondered what might have been if Yntema (who had shared laboratory space with him while they were graduate students in Cecil T. Lane's group at Yale) had his additional finding, that the Nb wire (0.05 mm in diameter) had carried 1.5 A in an externally applied field of 0.5 T (a field well above the H_{c2} of annealed Nb), reported in terms of the substantial current density of 740 A/mm^2.[44] Even though Yntema helpfully pointed out that "such windings should be useful in various cryogenic experiments," not much attention was paid to this first high current density superconducting magnet. In 1986 Yntema recalled that the only person he knew who had noticed his abstract was John Hulm[45]. Hulm had gone to work at Westinghouse Research Laboratories in Pittsburgh in 1954 after leaving the University of Chicago, where, at the instigation of Enrico Fermi, he and Bernd Matthias had been looking for new superconducting compounds (see Section 11.3)[46]. Indeed Hulm followed up his reading of Yntema's abstract by constructing a solenoid using enameled cold-worked Nb wire which achieved 0.6 T in a 19 mm bore in 1955[47].

11.2.2 The High Field Revolution—The One that Was Noticed

Unaware of Yntema's work, Stan Autler at Lincoln Laboratories published his first paper on Nb solenoids in 1959[48] and oversaw the construction of a 0.43 T solenoid using silk-insulated Nb wire (127 μm diameter), which reached a field of 1.4 T with an iron core[49]. He also suggested that even greater fields could be reached using materials such as Ti-Mo and Nb$_3$Sn. Learning of this work while visiting Lincoln Labs, Rudolf Kompfner, at that time the Director of Electronics and Radio Research at Bell Telephone Laboratories (BTL), asked Ted Geballe and Bernd Matthias if this technology could be used for a superconducting shield for masers; Geballe and Matthias indicated that they had been unable to get metallurgists to make the required alloys, and so Kompfner encouraged Earle Schumacher and Morris Tanenbaum of the metallurgy group to help out[50]. Working under Tanenbaum at the time was J. E. Kunzler, who was diverted from his studies of transport in metals to lead the new superconductor effort in the metallurgy group[51]. Kunzler first started work on Mo-Re (discovered by Hulm in 1955[52]). An ingot was float-zone melted and the eventual 178 μm

---

[43] G. B. Yntema, *Phys. Rev.* 98 (1955) 1197 Abstract W8

[44] T. G. Berlincourt, *Cryogenics* 27 (1987) 283

[45] G. Yntema, *IEEE Transactions on Magnetics*: 23:2, (1987) 390

[46] J. Hulm, *IEEE Trans. on Magnetics* 19 (1984) 161

[47] Private communication to G.B. Yntema reported in G. B. Yntema(1955)

[48] S.H. Autler, *Bull. Am. Phys. Soc.* 11 (1959) 413

[49] S. H. Autler, *Rev. Sci. Inst.* 31 (1960) 369

[50] J. R. Pierce, in "Rudolph Kompfner 1909-1977, A Biographical Memoir, National Academy of Sciences (1983)

[51] J. Kunzler, *Rev. Mod. Phys.* 33 (1961) 501

[52] J. K. Hulm, *Phys. Rev.* 98 (1955) 1539

diameter monofilamentary wire was plated with Au for turn-to-turn insulation at the suggestion of Ted Geballe. As outlined in the patent, [53] the advantage of using a good-conducting normal metal as "insulation" was that the coil windings could be decoupled from each other while the filament was superconducting but would couple and protect the magnet from destruction when the wire entered the normal state and started to dissipate I^2R heat. Furthermore, metallic "insulators" were much more space-efficient than organic insulation. The resulting 30,000 turn solenoid attained 1.5 T, to their delight matching closely the field predicted by their short sample tests[54]. A resistive solenoid capable of 8.8 T was available at Bell Labs and the next available time was booked to test the highest fields that could be sustained by superconducting strand. To make things more interesting, a wager was initiated between Tanenbaum and Kunzler; Tanenbaum would provide a bottle of Scotch for every 0.3 T achieved over 2.5 T while Kunzler would provide one Beefeater martini for each week a much overdue paper on magneto-thermal oscillations was delayed[55]. Because Nb_3Sn had the highest critical temperature, it was also expected to have the highest critical field, but the intermetallic compound was too brittle to wind a solenoid. However, Ernie Buehler, who had fabricated the Mo-Re ingot, developed a method of making Nb_3Sn precursor strand using a mixture of Sn and Nb powders in a Nb tube, providing a route to wind-and-react magnet. To their great surprise[55], a rectangular rod of bulk Nb_3Sn, that had been sintered and then melted at 2400 °C, was still superconducting at the maximum field of 8.8 T on the very first day of testing, December 14, 1960. Not only did this represent 21 bottles of Scotch (the wager with Tanenbaum would eventually be halted at 10 T) but it exceeded the ~7 T estimate for critical field by another group at BTL earlier in the year[56]. Perhaps because of the prevalence of the sponge theory of superconductivity at the time (which did not predict useful high field transport currents in superconductors) Bozorth et al.[56] did not note the significance of the current in the tested material (Berlincourt later estimated a critical current density of ~60 A/mm^2 at 7 T from their magnetization data)[45,57] but to magnet builders the new strand data represented a revolution in the potential of superconductors for magnet application. They next tested the strand and it carried 50 times more J_c than the bulk sample[58]. The best powder-in-tube (PIT) strand with 10% excess Sn (over that required to fully reacted the Nb powder) reached almost 1500 A/mm^2 at 8.8 T (4.2 K) and was given the lowest heat treatment temperature (970 °C). They remarked that "still higher current densities might be obtained at still lower reaction temperatures". Indeed this was a prescient comment!

The era of high current high field superconductivity had truly begun and the race was on for the highest magnetic field solenoid and the best materials from which to make it.

11.2.3   First International Conference on High Magnetic Fields

James Wong had been experimenting with alloying Nb with Zr at the Wah Chang Corporation to increase the strength of reactor shielding for a nuclear powered plane but he was encouraged by John Hulm and Stan Autler (who had known him when he was a student at MIT)[59] to provide them with Nb-Zr wire, optimized for ductility at 25% Zr. Unlike the BTL PIT Nb_3Sn strand, which required the magnet to be wound in the not-yet superconducting state and then be heat treated at ~1000 °C to make it superconducting, Nb-Zr (and Nb-Ti) was superconducting as made, making it much more useful for early experimentation. By September 1961 Wah Chang was producing long lengths of Nb-Zr to support the increasing demand for laboratory scale superconducting mag-

[53] Theodore H. Geballe, US Patent number: 3109963, Filing date: Aug 29, 1960:date: Nov 5, 1963

[54] J. E. Kunzler et al., *J. Appl. Phys.* 32:2 (1961) 325

[55] J. Kunzler, *IEEE Trans. MAG* 23:2 (1987) 396

[56] R. M. Bozorth, A. J. Williams, and D. D. Davis, *Phys. Rev. Lett.* 5:4 (1960) 148

[57] T.G. Berlincourt, IEEE Trans. MAG 23:6 (1987) 403

[58] J.E. Kunzler et al., *Phys. Rev. Lett.* 6 (1961) 89

[59] A. Wood, in *Magnetic Venture: The Story of Oxford Instruments*, Oxford University Press, Oxford, UK (2001)

nets[57]. The First International Conference on High Magnetic Fields attracted some 500 attendees to MIT from November 1 to 4, 1961 and a major source of interest were the latest reports of high field superconducting coils from Bell Telephone Laboratories (6.9 T, 1.5 K, Nb_3Sn[60], Westinghouse (5.6 T using Nb-25Zr)[61], Atomics International (5.9 T using Nb-25Zr)[62]. In *Magnetic Venture* Lady Audrey Wood, cofounder with Martin Wood of Oxford Instruments (OI), captured the excitement: "Through the next three days of papers and discussions on current ways of making and testing high magnetic fields, the concourse buzzed with talk of the recent breakthrough in superconductivity." This culminated in a special session on the Saturday afternoon that was added to allow the presentation of the latest results. "It was at this Saturday afternoon session that superconductivity really 'arrived' on the magnet-making scene."[59] The competition to report for the highest fields that afternoon was intense and yet the technology was still very primitive; Ted Berlincourt of Atomics International (AI) was surprised to learn later that both Bell Telephone Laboratories and Westinghouse used batteries to power their magnets (in the case of Bell Telephone Laboratories a simple car battery) and the AI team was winding its magnets with bare Nb-Zr monofilaments with bare Cu foil between the layers, having been unable to apply an insulator successfully[63]. The antics at AI inspired a limerick by W. J. Tomasch[57]:

> *A group of young men so frenetic*
> *Struggle with matters magnetic,*
> *Each day they conspire,*
> *To wind super wire,*
> *A pastime which some deem pathetic.*

As John Hulm recalled, "Those tiny, primitive magnets were, of course, terribly unstable and tended to damage themselves on normalization, for reasons that are now well understood. One had to have faith to believe that these erratic toys of the low temperature physicist would ever be of any consequence as large engineered devices"[64]. New to the faith, Martin and Audrey Wood decided that evening that their fledgling company would build superconducting magnets and, on their return to the UK, they ordered one pound of Nb-25Zr from Jimmy Wong at Wah Chang (who would found Supercon as an independent commercial superconducting wire manufacturer). In March 1962 OI had built the Europe's first magnet (4 T, 4.2 K, 18 mm bore) using the new superconducting wire. In the end, Nb-Zr proved to be a dead end because the magnets produced by OI using Cu-plated Nb-Zr strand deteriorated over time and they switched to multifilamentary Nb-Ti made by Imperial Metal Industries (IMI) as soon as it became available in March 1967 [59].

11.2.4   God Save the Queen

Hulm and Matthias had started a study of solid solution alloys of the transition metals, such as Nb-Ti and Nb-Zr (alloys that would fit well with the empirical electron to atom ratio rules established by Matthias in 1955)[65] while in Chicago their attempts to sinter samples from powder in a poor vacuum system resulted in "a lot of nitride and oxidized samples"[66]. At Westinghouse, however, Hulm had excellent facilities for alloy melting and, with Richard Blaugher, would eventually measure the critical temperatures of all the nearest-neighbor body-centered cubic binary solid-solution alloys.

[60] J. Kunzler, *Rev. Mod. Phys.* 33:4 (1961) 501

[61] J. K. Hulm et al., in *High Magnetic Fields: Proc. Conf. November* 1-4, 1961, at the Massachusetts Institute of Technology," Cambridge, MA (1962) 332

[62] R. R Hake, T. G Berlincourt, and D. H Leslie (1962), ibid, 341

[63] Ted Berlincourt, personal communication for this article (2010)

[64] J. Hulm, *IEEE Trans. MAG* 19:3 (1983) 161

[65] B. T. Matthias, *Phys. Rev.* 97:1 (1955) 74

[66] J. K. Hulm, J. E. Kunzler, and B. T. Matthias, *Phys. Today* 34, (1981) 34

Richard Blaugher, who had been hired by John Hulm in 1957, recalled the inception of Nb-Ti for the memorial session for John Hulm at the 2004 Applied Superconductivity Conference (Richard Blaugher recalls it occurred just before Christmas 1959)[67]:

> *One day I was sitting at my desk at Westinghouse R&D and John Hulm literally burst into the lab. "Blaugher" he says, "I have been thinking." John started to sketch out on the blackboard the periodic chart for groups four, five and six. "We have studied V-Nb, Nb-Ta, V-Cr, V-Mo, Nb-Cr, Nb-Mo, Nb-W, Nb-Hf, and Nb-Zr" (which he emphasized by drawing straight lines linking the various binaries)". We MUST do Nb-Ti!" Why I asked? John replied, because, "God save the Queen, we'll have the Union Jack." Needless to say with this overwhelming logic I proceeded to make up some samples to study this binary. Were it not for John's English background we would have omitted the most important binary for our study.*

Hulm and Blaugher published their wide-ranging study of transition element alloy superconductors in September 1961[68].

11.2.5   The Slow Emergence of Nb-Ti

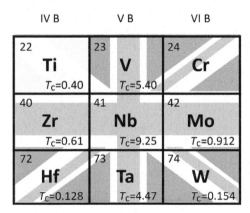

FIGURE 11.9:   The transition elements making up the ductile BCC superconducting alloys and their relationship to Hulm and Blaugher's investigation of Nb-Ti. The first row of transition elements is notably missing from Matthias's table of transition temperatures.

The Atomics International (AI) group pursued critical field and current (key to application), measuring its first Nb-Ti samples in April 1961[69]. A paper by B. B. Goodman at the IBM Conference on Fundamental Research on Superconductivity in June 1961[70] caused great interest at AI because it accurately predicted the upper critical field of U-Mo from the phenomenological Ginzburg-Landau-Abrikosov-Gorkov (GLAG) theory utilizing Berlincourt's experimental data on the transition temperature and normal state resistivity and Goodman's own measurement of the normal state specific heat. Berlincourt and Richard Hake proceeded to test a wide variety of U-Mo, Ti-Mo and Ti-V alloys and not only confirmed the GLAG predictions of a critical field that was independent of cold work but also showed that Nb-Ti possessed the highest critical field (~14.5 T) of all the ductile superconductors[71].

Early manufacturers of Nb-Ti included Atomics International (Ti-22 at.%Nb = Nb-65wt.%Ti), Westinghouse and Mitsubishi of Japan (Nb-Zr-Ti). Dating back to the days of Nb strands the Nb and Nb-alloys strands were typically finely drawn monofilaments ~ 250 μm in diameter in order to get the high levels of cold work that gave useful critical current densities for magnet building.

Very large successful bubble chamber magnets were made at this time, starting with the 12 ft Argonne bubble chamber magnet in 1966[72] which, as the current density of 7.75 A/mm² indicates, was

---

[67]   Richard Blaugher, personal communication for this article (2010)

[68]   J. K. Hulm and R. D. Blaugher, *Phys. Rev.* 123:5 (1961) 1569

[69]   T. G. Berlincourt et al., Atomics International Laboratory Notebook B063551, April 17th 1961

[70]   B. B. Goodman, *IBM J. Res. Develop.*, 6:1 (1962) 63

[71]   T. G. Berlincourt and R. R. Hake, *Phys. Rev. Lett.* 9:7 (1962) 293

[72]   J. R. Purcell, Proc. *4th Int. Conf. on Magnet Technology*, 19-22 September 1972, Upton, NY, 201

so heavily protected with Cu that it was essentially a helium cooled copper magnet that happened to have several, large, superconducting strands embedded within its copper matrix (Nb-Ti is much more easily bonded to a Cu or Cu-Ni stabilizer matrix than Nb-Zr). The remarkable co-extruded conductor, developed by Jimmy Wong at Supercon, had six 4 mm wide filaments embedded in a 50 mm by 2.5 mm OFHC Cu matrix. The filaments were arranged in 3 separated pairs (2 pairs at each edge and 1 pair in the center) so that the conductor could be riveted at the joints without hitting the filaments[73]. At Fermilab (then the National Accelerator Laboratory (NAL)) John Purcell, builder of the 12 foot ANL bubble chamber magnet, was commissioned to make the magnet for the 15 foot bubble chamber. It too used large, superconducting filaments embedded within the copper matrix. The time constant for the eddy currents between the untwisted filaments to die down was on the order of a day. The magnet which was very conservatively designed had a long successful run at NAL.

A group under Ron Fast was responsible for design of some early beam line and detector magnets at NAL. These magnets had fairly open windings to maximize the wetted surface of the conductor and all used the principle of cryogenic stability developed by John Stekly[74]. Conductors used by Fast's group were small rectangular monoliths insulated by Formvar. There was a learning curve in the use of Formvar. Designers at the laboratory had to learn that there were dimensional standards for the base conductor, especially corner radii, that had to be met in order to ensure even coating of the insulating varnish. The magnets were so-called superferric magnets which used the superconductor to magnetize Fe cores and which did not require significant winding current densities. Many of these magnets were superferric versions of more conventional copper and iron magnets.

An important step towards the high J_c strand of today was the discovery by Atomics International in 1965 that they could get useful critical currents (sufficient for a 3.5 T coil) from their Nb-78 at.%Ti (65wt.%Ti) alloy using a single final size heat treatment for a few hours at 400 °C.[75,76] By using heat treatments to introduce precipitation of a non-superconducting phase, an additional method of raising critical current density was now available. The filaments were still "insulated" by electroplating with a thin ($\sim$ 20 μm thick) layer of Cu which was susceptible to oxidation[77]. A major improvement at AI was the in-house fabrication of Cu clad monofilament using Nb Ti rods inserted in Cu tubes before wire drawing[78]. This was also compatible with a desire for thicker Cu layers for improved magnet stability. Larger currents could be carried by cabling multiple strands together and different combinations of Cu-clad monofilaments could be combined for different coil sections. Al McInturff and his colleagues at AI found that that similar critical current densities could be produced for a wide range of Nb-Ti diameters using their final size heat treatment of Nb-78 at.%Ti (65wt.%Ti) and they speculated that "it is now possible to replace many small conductors with one large conductor. For example, stabilized cables can be constructed in which one large Ti (22 at.% Nb) wire replaces many smaller wires, without loss in current capacity" (1967) [79]. They soon put the idea into practice for a multisectioned large bore solenoid for Brookhaven, making one section of the magnet with one fairly large copper clad superconductor wire, with copper strands around it. To their great disappointment, this section was totally unstable and was replaced by a multi-strand conductor cable[78]. Not only was the dependence of large filaments to be electromagnetically unstable unknown at the time but the importance of the twist inherent in the cabling process was not realized outside the Rutherford Laboratory/IMI collaboration, a key insight that would burst on the scene one year later at the 1968 Brookhaven Summer School.

---

[73] J. Purcell, *IEEE Trans. MAG* 23:2 (1987) 413

[74] Z. J. J. Stekly and J. L. Zar, *IEEE Trans. Nuc. Sci.*, 12:3 (1965) 367

[75] J. B. Vetrano and R. W. Boom, *J. Appl. Phys.* 36:3 (1965) 1179

[76] A. D. McInturff et al., *J. Appl. Phys.* 38:2 (1967) 524

[77] C. Laverick and G. Lobell, *Rev. Sci. Inst.* 36:6 (1965) 825.

[78] G. G. Chase, C. N. Whetstone, A. D. McInturff, personal written history for this article (2010)

[79] A. D. McInturff *et al.*, *J. Appl. Phys.* 38 (1967) 524

AI closed its superconducting strand operation in 1966 and the members of the AI team formed an influential diaspora across the US with Roger Boom and Bob Remsbottom moving to the University of Wisconsin, Al McInturff to Brookhaven National Laboratory and Clay Whetstone joined Eric Gregory and Bruce Zeitlin at Airco working on scale up techniques using extrusion of Nb-Ti and Cu strand.

Another step toward the multifilamentary strand of today came from an unlikely source; F. P. Levi, an associate director of Rola, a loudspeaker manufacturer in Melbourne, Australia was looking for a way to create powerful permanent magnets using a microscopic array of magnetically insulated ferromagnetic filaments and came up with the idea that this could be achieved by "repeated drawings of assemblies of iron wire encased in non-magnetic tubes. Len [*Rola founder A. L. C. Webb*] agreed enthusiastically, I was relieved of all other duties and within eighteen months and at a cost of about $20,000 ... we discovered a new method of making permanent magnets"[80]. Filaments as small as 20 nm were produced[81]. The idea was picked up by Rose and Strauss at MIT, where, the same technique was used to produce Nb filaments as small as 10 nm in diameter in a Cu matrix[82]. Cline, Rose and Wulff were also successful in producing a Nb-Ti in Nb matrix stack[83], shown in Figure 11.10, that would look very familiar to modern manufacturers but their attempt at the fabrication of a copper-clad Nb-Ti composite broke after the first bundling (which they suggested was most likely due to the limited ductility of the Nb-Ti alloy).

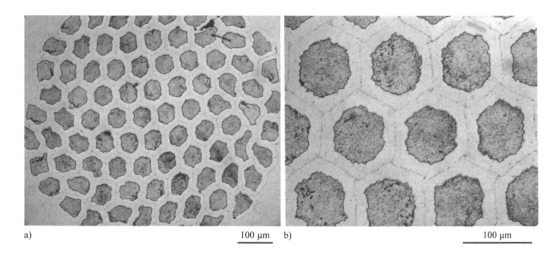

a)   100 μm   b)   100 μm

FIGURE 11.10:   Nb60%-Ti40% in niobium matrix. This figure is taken from NASA CR 54103[83] and used with permission of NASA.

11.2.6   Rutherford CEGB/IMI Strand and the 1968 Brookhaven Summer School

The first commercial multifilamentary composite, in the Cu-stabilized form we understand it today, was the result of a collaboration in the UK between, first, the Central Electricity Generating Board (CEGB) under Peter Chester and a little later, Rutherford Laboratory, under Peter Smith (after earlier experimental filamentary wire made for Rutherford by IRD Newcastle in late 66–early 67)[85],

---

[80] Letter from Dr. F. P. Levi, formerly Associate Director, Rola (Australia) Pty. Ltd., March 6, 1976 quoted in R. J. Clarke, Innovation in Australian high technology industries: Two case studies. *Australian Economic Papers* 18:32 (1979) 89

[81] F. P. Levi, *J. Appl. Phys.* 31:8 (1960) 1469

[82] H. E. Cline et al., *J. Appl. Phys.* 37:1 (1966) 5

[83] H.E. Cline, R.M. Rose, and J. Wulff, NASA Report CR 54103 (1964)

and Imperial Metal Industries, IMI. The "Niomax M" commercial strand contained a hexagonal array of 61 filaments of Nb-60 at.% Ti (~44 wt.% Ti) in a Cu matrix[84]. The Rutherford group needed a wire with which they could build a synchrotron[85]. Production quantities also made their way to Oxford Instruments for magnet production in March 1967 [59]. The concepts behind the development of the CEGB/IMI strand were first presented at the 2nd Magnet Technology Conference[86] but its major impact came from the Summer School held at Brookhaven National Laboratory in the USA in 1968 where the success of the new twisted multifilamentary strand in reducing flux jumps [87] was discussed with the leading large scale magnet designers,[88] much to the chagrin of IMI.[85]

11.2.7 Filamentary Superconductors

Contributed by *Martin Wilson*

In the early 1960s, not long after the discovery of high field type 2 superconductors, several manufacturers started to produce long lengths of good quality niobium zirconium and niobium tin wire, closely followed by niobium titanium. Although these wires performed well when tested as short samples immersed in a background magnetic field, they did not perform at all well when wound into magnet coils. Upon investigation, it emerged that one of the problems was a phenomenon known as "flux jumping". When a field is applied to any type 2 superconductor, it responds by setting up persistent screening currents which try to impede the movement of flux into the center of the wire—a bit like conventional eddy currents only, being superconducting, these current do not decay. If flux does move through a type 2 superconductor, it dissipates heat. Furthermore, the amplitude of the screening currents decreases if the temperature is raised. It is the combination of these two properties which gives rise to an electromagnetic thermal instability known as a flux jump, whereby the screening currents can suddenly collapse, releasing energy which takes the conductor above its critical temperature. The cure for flux jumping is very simple - subdivide the wire into fine filaments; for NbTi, these filaments must be less than ~ 50 µm in diameter. For ease of handling and also to ensure stability of current flow, many filaments are embedded in a matrix of normally conducting metal such as copper to make a wire of diameter 0.5 mm to 1.0 mm. To avoid magnetic coupling between the filaments, the wire must be twisted like a rope. The first filamentary superconductors were developed in the UK by the Rutherford Laboratory, in collaboration with Imperial Metal Industries using Nb-44wt.%Ti.

With flux jumping cured, it became possible to build magnets operating at very high current density and producing high magnetic fields. Careful design is still needed, however, to avoid the possibility of mechanical movement under the enormous electromagnetic forces, which can release frictional energy and cause unreliable performance. This problem remains with us right to the present day.

Fine subdivision also brings another advantage: it reduces ac losses. Although superconductors have zero loss under dc conditions, changing magnetic fields cause flux to move within the superconductor which dissipates heat. Reducing the filament size reduces this flux motion and hence the losses, in much the same way that laminating a transformer core can reduce eddy current loss. Keeping the ac losses within reasonable bounds has made it possible to build large particle accelerators, such as the Tevatron, RHIC and LHC, in which the superconducting magnets must

---

[84] P F Chester, *Rep. Prog. Phys.* 30:2 (1967) 561

[85] Martin Wilson, personal communication for this article (2010)

[86] P.F. Smith, in *Proc. 2nd Magnet Technology Conference*, Oxford (1967) 543

[87] P. F. Smith, M. N. Wilson, C. R. Walters and J. D. Lewin, in *Proc. 1968 Summer Study of Superconducting Devices and Accelerators*, BNL, 913

[88] *The Proceedings of the 1968 Summer Study on Superconducting Devices and Accelerators*, June 10 – July 19, 1968, Brookhaven National Laboratory, Upton, NY, has been archived at:
http://www.bnl.gov/magnets/Staff/Gupta/Summer1968/index.htm

be ramped up from low field to high field so that they keep pace with the increasing energy of the particle beam. Finer filaments are needed for particle accelerators, typically in the range 5–10 μm. Even finer filaments, in the sub-micron range, have been tried with a view to building 50–60 Hz superconducting machinery for use in power generation and distribution, but unfortunately the refrigeration power needed to remove the residual ac losses at the very low temperatures involved is still too high for economic operation. There are, however, good reasons for hoping that, if filamentary conductors can be made with the newer high temperature superconductors, they will revolutionize the way that we generate and distribute electrical power.

11.2.8   After the 1968 Summer School

The impact of the 1968 Summer School was immediate; Bill Hassenzahl was working on a high intensity proton accelerator at the Los Alamos Meson Physics facility, using existing monocore strand to make quadrupole magnets of sufficient field (3.4 T)[89]. At less than half the 550 A design current, however, their prototype quadrupole quenched with a large "whoosh" of helium.

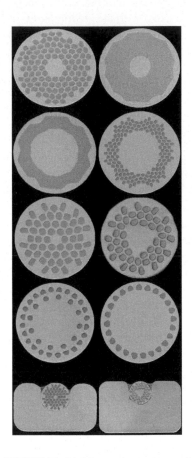

> "During the tests we learned of the new multifilamentary superconducting wire and decided to immediately build and test a racetrack with this advanced conductor. It reached about 320 A, and the quench had a bigger 'whosh', but there was another sound just prior to the opening of the relief valve. I heard a very distinct ping. Neither John Rogers nor Henry Laquer heard the sound. During the course of the design and testing of these magnets, the two of them had been my mentors in cryogenics and superconductivity. Henry Laquer was arguably the first person to observe flux creep in superconductors, and had the earliest patents on flux pumps[90,91], though he did not call them by that name. Probably my only advantage was younger ears. As usual on experiments, this was late on Friday, and several more quenches yielded a small increase in current, and the same ping that only I could hear. On Monday morning, we opened the cryostat and found that there had been significant mechanical motion in the coils. The source of the pings was readily determined to be from sheared rods that had been used to hold a separator in place. This phenomenon was not present in the other coils. When the rod had sheared, a significant fraction

FIGURE 11.11:   A variety of Nb-Ti strand and monolith cross-sections from just one manufacturer's current production. Images courtesy of Manfred Thoener of Bruker EST (BEST). The top left strand is a restacked composite for the LHC dipoles and has 7 μm diameter filaments. The large filament strands and strand in channel conductors towards the bottom are for MRI application.

---

[89] J. D. Rogers et al., *J. Appl. Phys.*, 1971

[90] Henry Laquer, US Patent: 3145284, August 18, 1964

[91] Henry Laquer, US Patent: 3150291, September 22, 1964

of the coil had moved by nearly 7 mm. Prior to shearing the rod it had moved by nearly 3 mm. We decided to remove other rods in a similar configuration and retest the coils.[92]

During this test, the coils exceeded the design current of 500 A without quenching, even though there was internal motion of about 1 cm at maximum current. The 'Whoosh' was louder during several quenches in the 515 to 550 A range than it had been in earlier tests.

Finally, two quadrupoles were constructed of the twisted multifilamentary wire. They were tested up to the design current, and were eventually used on one of the LAMPF beam lines."

11.2.9   Making Multifilamentary Strand

Bill Marancik recalls his work with Eric Gregory
in developing early multifilamentary Nb-Ti at the
Air Reduction Central Research Laboratory
(later AIRCO, Oxford AIRCO and eventually OST):

"We used various techniques using NbTi rods inserted between concentric copper tubes; each rod was also separated by copper rods. This composite was swaged from about four inches to several millimeters, insulated with tape and wound into a coil. Because of the gross swaging the copper surface was covered with flakes and the magnet was completely shorted. Using the techniques for the manufacture of flux core welding wire we wrapped copper foil around both NbTi powder and also around NbTi rods for stacking into copper billets. We then turned to inserting NbTi rods into copper tubes, drawing these to various shapes such as squares, rectangles and finely into hexes. With the hex stacking a more compact configuration was formed in which we were able to produce the first fine multifilament conductor consisting of 98 filaments (we thought this was a huge number). The billet was four inches in diameter by a foot long and was extruded at the DuPont Metal Center and drawn to final size"[93].

> After his brief stay at Airco, Whetstone formed his own company, Cryomagnetics in 1967, to produce Nb-Ti strand and magnets and he was soon joined there by his former AI colleagues, Gordon Chase and Al McInturff. Wanting to produce multifilamentary Nb-Ti in a Cu extrusion can, they found that gun drilling holes for the Nb-Ti rods into 24 inch Cu billets was not economical at the time, so they assembled billets by stacking 4 inch interlocking Cu pancakes, which were easy and cheap to drill one at a time[78]. They were now using a Nb-45 wt%Ti (62 at.%Ti) that had been identified as the optimum alloy by McInturff while still at AI (it was not published until much later)[94]. Optimized processing of this low-Ti strand required drawing strains after precipitation heat treatment, which meant that final wire drawing was occurring for filaments with flow stress much increased by the introduction of precipitates. Bonding the Cu pancakes required high extrusion temperatures which caused excessive formation of brittle Cu-Ti intermetallic at the Cu/Nb-Ti interface.

Airco was having similar problems, which were addressed using solutions that are used to this day:

> "All our development billets were two, three or four inches in diameter and weighing a few pounds. Several problems became apparent: a Cu-Ti reaction caused hard inclu-

---

[92] W. V. Hassenzahl, personal communication for this article, 2010

[93] W. Marancik, personal communication for this article, 2010

[94] A. D. McInturff and G. G. Chase, *J. Appl. Phys.* 44:5 (1973) 2378

sions to be formed, which was solved by inserting Nb foil between the copper and the NbTi rods. The second problem was the ID of the copper tubes turned out to present a contamination problem in the ID of the tubing. This could be eliminated only by forming the single strand composite by extrusion of an Nb ingot in a Cu can. This was drawn, hexed and stacked into an extrusion can. As with any composite where the yield strengths of the components are widely different, center burst is likely to occur and did with the initial billets. This was eliminated using the criteria developed by Betzalel Avitzur. Control of both drawing die angle and percent reduction as well as location of the filaments within the array eliminated this difficulty. This has been a key component of our processing in all subsequent conductor developments."[93]

The arrangement of filaments included an optimum filament spacing for mechanical stability[95].

11.2.10   The First Strain Measurements

contributed by *Jack Ekin*

Sometimes discovery is the result of the need for survival. A solid state physicist specializing in low temperature electrical measurements, I found myself a postdoc thrown into the midst of a group of mostly metallurgists at the National Bureau of Standards specializing in fracture, deformation, and stress. Not much use for an electrical type. This was 1974 at the bottom of the great recession of the time. With few job prospects and one year left on my postdoc appointment, I had a two-minute "bathroom conversation" with sponsor Mike Superczynski about looking for a possible connection (hopefully) between stress and electrical properties in practical superconductors. The literature from a decade earlier showed there was a very small stress effect on critical temperature, but no one had observed (or looked for, to my knowledge) any effect on the really important parameter, critical current. Fortunately, I got "funded" for $10 k, but never mind, that was enough to justify myself to the administration to at least have a first look.

I kluged together a stress rig made of a stainless steel cable running up out of the measurement cryostat, up over two pulleys, and down to a pan, which I progressively loaded with lead-shot-filled bags (yes real buckshot), 5 lbs each. I measured critical current while progressively loading the pan, until eventually the wire sample broke and the pan came crashing down onto the floor with a bang that made you jump. Also, I had no protective circuit to "kill" the power supply if there was a thermal runaway quench, so I ran the experiments with my trigger finger constantly poised on the power supply switch, ready to snap it off at the slightest sign of a rapid voltage rise across the sample. If I was too slow by a fraction of a second, my precious sample would melt into a shapeless blob. After many days of such tension between the expectation of the pan crashing down and vigilance at the power button, my nerves were frazzled. But there it was, a reversible stress effect on the critical current (and potentially a job)!

Thus was the first observation of stress/strain effects on the critical current of NbTi, and later Nb_3Sn. An unassuming beginning, but the elegant scaling laws that grew out of this work have for decades served the important design work needed for large (*i.e.* high stress) superconducting magnets.

11.2.11   The Birth of Nb-46.5wt%Ti

The Vietnam War and the ensuing recession also took its toll on the strand manufacturing community and Cryomagnetics closed its doors in 1971, leaving only three US wire vendors, AIRCO, MCA

---

[95] E. Gregory, T. S. Kreilick, J. Wong, A. K. Ghosh, and W. B. Sampson, *Cryogenics*, 27, (1987) 178

and Supercon. The poor state of the commercial superconducting strand business caused concern at Fermilab, which was planning a 1000 GeV superconducting upgrade to its recently completed Cu and Fe 400 GeV main ring. It knew that it would need both a competitive market for the large quantities of strand that would be required, as well as a consistent and high performance product. This incited a program to directly develop production methods for accelerator quality strand[96]. The upgrade to the accelerator, a project known at various times as the "Energy Doubler", the "Energy Saver" and the "Energy Doubler/Saver" and finally the Tevatron, required 774 dipole magnets to steer the beam and 216 focusing quadrupoles. The peak field of 5 T would mean that high-strand current densities would be required. Cryostatic stability was not an option under these conditions. Adiabatic stability, ably covered in the Brookhaven Summer Study, required small (<50 μm diameter) twisted filaments, while the high strand current density implied a low copper to superconductor ratio (<2:1). Fermilab chose to make a 23-strand rectangular cable using 0.68 mm diameter strands. The cable configuration, known as a Rutherford Cable[97,98], is basically a flattened helix and is now the standard method for making accelerator magnets (see inset).

Rutherford Cables contributed by *Martin Wilson*

A large synchrotron accelerator comprises a ring of some hundreds of magnets which focus the beam of particles and constrain it to follow a circular orbit. To make sure that the orbit is exactly circular and in the right place, it is essential for each bending magnet to produce exactly the same field, i.e., to carry exactly the same current. The easiest way of ensuring that all magnets carry the same current is to connect them in series. A power supply must apply voltage across the magnet terminals to overcome the self inductance and ramp the magnets up to follow the beam energy. When the magnets are connected in series, this voltage adds up around the ring. To keep the voltage down to a manageable level, the magnet inductance must be minimized, which requires high operating currents of ~ 10 kA. Such high currents require 30–50 wires to be connected in parallel and, because they have no resistance, it is difficult to ensure that the current distributes equally between the wires. To achieve such equal distribution, the wires must be perfectly transposed, i.e., they must all change places with each other along the length of the ca-

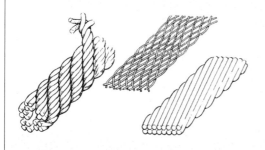

FIGURE 11.12: Three types of transposed cable: rope, braid and Rutherford[99].

ble. Figure 11.12 sketches three examples of transposed cables: a rope, a braid and a flat twisted cable similar to the Roebel bar. Prototype accelerator magnets were built from each type of cable but the flat twisted type gave the best performance, probably because it can be compacted to high density without damaging the strands. Because it was developed at Rutherford Laboratory, this cable has come to be known as Rutherford cable and has been used in all large superconducting accelerators built to date.

---

[96] B. Strauss et al., *IEEE Trans. MAG* 13:1 (1977) 487

[97] G. E. Gallagher-Daggitt, Rutherford Laboratory Memorandum No. RHEL/M/A25 (1975) unpublished

[98] C. Walters, *IEEE. Trans.MAG* 11:2 (1975) 328

Each of the three domestic strand suppliers for the Tevatron had a proprietary alloy composition. In order to broaden competition and not be beholden to a vendor with a proprietary alloy, Fermilab decided to supply raw materials to the individual vendors. This also enabled a savings of time, as the laboratory had kits of materials on hand and delivery schedules from the fabricators were significantly reduced. The Tevatron would eventually consume 90% of the Nb-Ti that had ever been produced and the selection of alloy composition would have a major long term impact. Bill McDonald of Wah Chang recalls:

> "Wire suppliers were buying 45Ti and 48Ti. Wah Chang was the primary (essentially the only) supplier. We had a meeting in the cafeteria at Wah Chang, including Paul Reardon FNAL, Bruce Strauss FNAL, Bob Marsh (Wah Chang salesman), and me (Bill McDonald). We had met to generate the spec for the alloy to be used in the accelerator magnets. The technical review was finally settled by splitting the difference between 45Ti and 48Ti, so that no customer would be given a process advantage. I remember it as being proposed by me, and Bruce remembers it as being proposed by him. No matter. We all agreed and the official alloy was established as Nb46.5Ti for the Fermi Lab magnets."

The origin of the Fermilab billet filament stacking arrangement is recalled by Bruce Zeitlin:

> "It was in the early to mid seventies while I was working for Dr. Eric Gregory at Airco (acquired by BOC) as a young supervisor/engineer of the superconducting wire pilot plant that a challenge came to us from Westinghouse's Mike Walker. Their superconducting generator program required several thousand fine filaments in a monolithic rectangular conductor. At that time only several hundred filaments were possible due to the difficulties of stacking the hexagonal elements into the extrusion billet. NbTi was typically clad with copper and drawn through a hexagonal die. The resulting rod could not be straightened to the required precision to stack the billet. It occurred to me that if we could obtain straight hexagonal tubing and precision straightened Nb-Ti rod, then assembly would be much easier. We did such. The first 1000 to 2000 filament conductors were assembled by stacking the hexagonal copper tubes and then inserting the precision straightened NbTi rods. This was a quick and scalable process."

After the acquisition of the technology and people by Magnetic Corporation of America (MCA), Z.J.J. Stekly's company, this technique was applied to the Fermi accelerator conductor. This is one of the few incidents in which non-accelerator requirements led to improvements in superconductor for accelerators.

For what became known as the Fermi-kit, Fermilab supplied copper in hexagonally shaped OD and round ID tubes, as well as extrusion cans. The copper supplied was from Phelps-Dodge and had the highest residual resistivity ratio available at that time thanks to the use of virgin Cu rather than remelt (which inevitably includes Fe from recycling).

Bruce Zeitlin also recalled a less fortunate event that led to a standard improvement in the processing of the extrusion billets:

Los Alamos in the later part of the seventies required a mixed matrix conductor for John Roger's energy storage program. An ambitious conductor was designed with 2100 filaments of NbTi clad with copper and 90/10 copper nickel alloy assembled in a 254 mm diameter copper nickel billet. Extrusion at RMI was a disaster. The billet stalled. The extrusion temperature was too low for the tough copper nickel. Inspection of the extrusion revealed that we had some major folds and creases in the billet. I entered Dr. Stekly's office at Magnetic Corporation of America in trepidation, knowing that we had just cost the company and program 20 to 30 thousand dollars and delayed

---

[99] M. Wilson, *Superconducting Magnets* (Monographs on Cryogenics, 2), Oxford University Press (1983)

follow-on contracts for the device. John calmly just asked what was the length-to-diameter ratio of the individual rods. He promptly stated that the internal component buckled, as there was enough free space within the billet such that the rods were unstable under the columnar load. This led to the standard practice of cold isostatic compaction of any billet with a large number of filaments. It was also important to the Fermi conductors as the void space introduced by the slide fit in 2000 tubes led to instances of internal buckling as well. It was a measure of the man that John almost instantly understood the problem and focused on the solution as opposed to chastising me.

The Fermilab Tevatron strands were eventually manufactured by Intermagnetics General Corporation (now Luvata) and Magnetic Corporation of America (now out of business) using a FNAL recipe that used a single long precipitation heat treatment followed by a cold work final drawing strain used by Airco[100]. Cabling was accomplished by New England Electric Wire Corporation in Lisbon, New Hampshire. Art Green of NEEC contributes this history of the FNAL cable:

11.2.12 Manufacturing of Cable for the Fermilab Tevatron

Contributed by *Art Greene*[101]

A decision was made on 23 as the number of wires in the Tevatron cable because the conventional cablers at New England Wire had 24 payoffs. Additionally, it was understood then that the geometry of the cross-section of a Rutherford cable required an odd number of strands. During the final manufacturing phase four cablers were utilized, the "Jumbo" cabler where most of the development work was done, two other planetary-type cablers and a tubular strander which ran at a much higher rpm.

New England Wire worked with Fenn Manufacturing Co. to develop Turks-Head rolls which were used to shape the originally round 23-strand cable into the desired keystoned filament cross section. All of the Fermilab cable was rolled twice, first into a rectangular cross section and subsequently into the final keystoned shape. Special mandrels were designed with nose pieces to prevent crossovers of strands just prior to rolling. As magnet fabrication and testing progressed at Fermilab, it was realized by the magnet designers that some type of coating was required on the individual superconductor strands, so that the magnetic field quality would not be adversely affected by eddy currents generated in the cable during ramping of the magnets. The initial choice was to coat all strands with Stabrite, a 95wt % tin with 5wt % silver eutectic alloy. In late-1978 additional magnet testing showed that another type of coating system was required to further reduce eddy current effects. It was suggested that a copper oxide coating be created on the surface of the wire. Further investigation revealed the existence of an Ebonol coating process which was at that time commonly used to blacken the outside of caskets covered in copper foil. Finally, after extensive studies, it was decided by Fermilab that the best results for both good magnet stability and reduced eddy current effects came from cable with both Stabrite and Ebonol coated strands (11 and 12 respectively) then given the name Zebra cable. This decision somewhat complicated things, because all of the strand then available for making cable had already been coated with Stabrite. Next came the request to Robert Meserve to develop a method to remove the Stabrite from almost 10,000,000 feet of wire! A process to de-plate the Stabrite was developed and two Ebonol coating lines were also constructed at New England Wire in addition to the original Stabrite lines. The first Zebra cable for magnet production was shipped to Fermilab on August 1, 1979.

---

[100] P. R. Critchlow, E. Gregory, and B. Zeitlin, *Cryogenics* 11:1 (1971) 3

[101] Information contained in this summary of the fabrication of Rutherford-type cable for the Fermilab Tevatron comes partially from meticulous log books of daily events kept by Robert Meserve at New England Wire in Lisbon, New Hampshire. Robert, now partially retired as Vice President for Engineering, managed all development work and manufacturing of the superconducting cable for Fermilab magnets.

FIGURE 11.13:   Manufacturing of Tevatron cable at New England Wire on one of the three large planetary cablers. Photograph courtesy of FNAL.

In late-1979 and 1980 a regular production process developed for manufacturing the required large quantities of Tevatron cable. Strand received from Supercon, MCA and IGC was inspected, cleaned and coated with either Stabrite or Ebonol, some after de-plating. Then the strand material was appropriately spooled and fabricated into cable. A final step was to process the cable through a vapor degreaser to remove any oil picked up in cabling and then to apply Kapton tape with a 50% overlap followed by space-wrap of glass tape which had been impregnated with B-stage epoxy. The epoxy was activated by heat at Fermilab after the magnet coils were wound to create a solid superconducting coil package.

One of the complexities of the process was created by the use of the B-stage epoxy which needed to be maintained cold and in low humidity. Spools of glass tape were shipped by Fermilab to New England Wire, stored under refrigeration and later used for insulating the cable. Then a system of refrigerated delivery trucks was created to ship completed spools of cable from New Hampshire to Fermilab. Over 5,000,000 feet of 23-strand cable was manufactured using these methods to create the world's first superconducting magnet accelerator.

11.2.13   Toward a Complete Description of Nb-Ti

A US DOE program to better understand the development of high J_c in Nb-Ti conductor was initiated under David Larbalestier, who had joined Roger Boom at the University Wisconsin-Madison in 1976. Boom needed the cheapest possible conductor, using the metric of $/kilo-ampere-meter, to make his huge diurnal superconducting magnetic energy storage dreams feasible. Running at 1.8 K using Nb-Ti was central. Larbalestier had been at Rutherford Laboratory in Martin Wilson's group developing the first filamentary Nb$_3$Sn conductors in collaboration with Jimmy Lee's group at Harwell. When he came to Madison, he got some lengths of Fermilab strand from Bruce Strauss and started to heat treat them according to the published recipes. But the results were quite non-systematic. Over the next 3 or 4 years as the Fermilab and Brookhaven programs proceeded towards large wire production for the Isabelle and Tevatron, there was intense competition in the industry between the remaining US vendors. Larbalestier was very struck by the fact that Airco was approved for Isabelle strand but not for Fermilab strand. Although much more excited about Nb$_3$Sn developments, he was pushed into Nb-Ti studies both by the need to support the SMES program and by his sheer frustration that logically planned experiments on Nb-Ti produced random outcomes. Finally in 1979–1980, he concluded that the processing of Nb-Ti was sufficiently unpredictable that he wrote a proposal to understand Nb-Ti processing, and finally Dave Sutter in DOE-High Energy Physics decided to fund it. According to Dave the reviews were very mixed. Some people said it was all known anyway, while others said that more understanding would make it impossible to run a commercial business. Sutter imposed one simple condition: hold an annual workshop at which the industry and the magnet builders would come to hear what Larbalestier's group was finding. This review, christened the Nb-Ti Workshop, was first held in July 1983 in Madison. It has evolved into a remarkably successful meeting of the users, fabricators and understanders of conductors, and is now

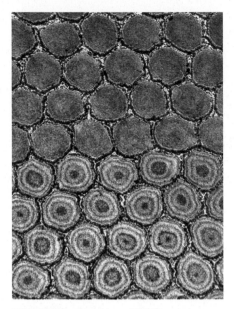

FIGURE 11.14: Highly inhomogeneous tree-ringed Nb-Ti filaments mixed with Homogeneous Nb-Ti in a prototype SSC strand.

called the Low Temperature/High Field Superconductor Workshop. Over 30 have been held, the latest in Monterey, CA, in November 2010. At the first workshop, formal presentations were made by UW researchers only; however, it was clear from strong discussion by the attendees that such a researcher-application user-industrial supplier forum was highly appreciated. And indeed, using this workshop-discussion format, there would be a strong contribution from the entire community in future workshops. An important focus of the first workshop was on the fact that US Nb-Ti wire was not competitive with the best R&D wires from Europe and that systematic experiments with US industrial Nb-Ti generally did not produce systematic responses, and that studies of the precipitate structures seen in modern wires were not similar to those reported in the literature. The prospect of doubling the regular US wire production critical current density, J_c, values from $\sim$1600-1800 A/mm^2 at 5 T in the Isabelle and Tevatron wires to the values over 3500 A/mm^2, 5 T, 4.2 K, 10^{-14} $\Omega \cdot$m reported by German[102] and Chinese[103] groups attracted both strong applications and industrial interest. Of particular note, the Chinese group at the Baoji Institute for Non-ferrous Metal Research, led by Zhou Lian, had made a wire with J_c at 5 T well above 3000 A/mm^2. What was decisive was that he gave a length to David Larbalestier at the 1982 Applied Superconductivity Conference and after careful evaluation of all relevant parameters, it became clear that Baoji had indeed made a major J_c breakthrough using their own Nb-Ti ingots and in-house composite wire fabrication. Zhou Lian invited Larbalestier to visit China in 1983. Out of this came the visit of the young Baoji researcher, Li Chengren, to Madison, where he would push J_c to 3680 A/mm^2 (5 T, 4.2 K) using standard Wah Chang Nb-46.5Ti monofilament strand [104]. What made this possible was the development of an earlier understanding in the period 1980–1982 of the hidden variable that was making systematic experiments on Tevatron and Isabelle wires so unpredictable.

The Tevatron made many huge contributions to the superconductor industry (2011 marks the 29th and final year of the Tevatron's operation). Of crucial importance for this story was the establishment of rigorous specifications for all of the components needed to make multifilamentary composite wires. Starting in about 1978, Larbalestier's first PhD student David Hawksworth, later the leader of the MRI-fabrication company Oxford Magnet Technology, was trying to raise the upper critical field of Nb-Ti by alloying so that better fusion devices could be built. After many studies, an alloy of Nb43Ti25Ta was found to have the highest critical field and a special melt was procured from Wah Chang for a fusion magnet coil to be built at General Atomics by John Purcell. But the J_c developed by standard processing was very disappointing, well below the Tevatron values. Examination of the microstructure by graduate student Bill Warnes showed an enormous range of local composition and α-Ti precipitate size that made a hugely non-uniform microstructure.

This observation was a decisive demonstration for Larbalestier that experiments with bad outcomes are often those with the biggest lessons to teach. When the Tevatron program had fixed the

[102] Juergen Willbrand and Wolfgang Schlump, *Zeitschrift fuer Metallkunde* 66:12 (1975) 714

[103] Li Cheng-ren, Wu Xiao-zu, and Zhou Nong, *IEEE Trans. MAG* 19:3 (1983) 284

[104] Li Chengren and D.C Larbalestier, *Cryogenics*s 27:4 (1987) 171

specification of the Nb-Ti alloy, they had fixed the composition, the impurity contents, the grain size and a host of other variables. But they had not fixed the permissible composition variation at the local scale. Because there is a 300 °C difference between the liquidus and solidus of the Nb-Ti alloy, extensive coring is possible in the cast ingots. What the Nb-Ti-Ta alloy showed was local composition variations of 9 at%, the variation mainly being in the Ti content, and on looking at binary Nb-Ti billets, it immediately became clear that these too were inhomogeneous, the local composition varying by as much as ±5 wt% [105]. It was immediately obvious that systematic studies failed so often because the local Ti content, which controlled the α-Ti precipitation, was not under control. Meetings with Wah Chang were immediately set in place and a long-running collaboration between Bill McDonald at Wah Chang and Peter Lee and David Larbalestier in Wisconsin lead to a 5 year program of more and more homogeneous alloy development. McDonald was immediately able to greatly reduce the coring by turning on electromagnetic stirrers. This first alloy, christened 906 for the last 3 digits of its heat number, was ready for Li Chengren when he came to Madison to continue his high J_c experiments.

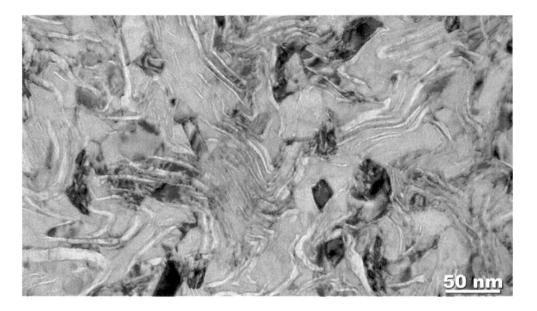

FIGURE 11.15:   TEM image of the microstructure (transverse cross-section) of the first 3700 A/mm^2 (5 T, 4.2 K) multifilamentary strand from a US manufacturer (OST). This previously unpublished image taken on September 5, 1986, shows the dense array of folded α-Ti ribbons (lighter contrast) that create the strong vortex pinning.

Solidified by the DOE-HEP support for the Madison group and all of the strong interactions that were developing in the superconducting wire industry and with the HEP magnet builders, Wah Chang was asked to make a more homogeneous alloy that was free of large Ti-rich regions called freckles, a product that became known as "Hi-Ho"[106].With a tight composition range, a complete description of the optimization of the Nb-Ti microstructure could be reached because systematic experiments became possible[107]. A linear relationship between precipitate volume and critical current

[105] A. West et al., *IEEE Trans. MAG* 19:3 (1983) 749

[106] D. Larbalestier et al., *IEEE Trans. MAG* 21:2 (1985) 269

[107] P.J. Lee and D.C. Larbalestier, *Acta Metallurgica* 35:10 (1987) 2523

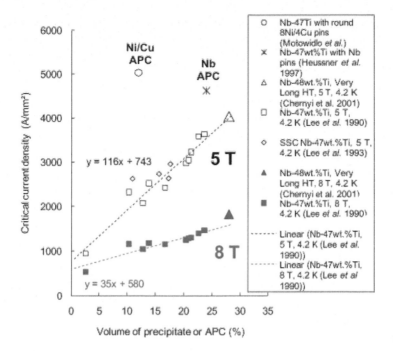

FIGURE 11.16: The critical current density of conventionally heat treated Nb-47wt.%Ti has been shown to be linearly dependent on the volume % of the α-Ti pinning center[108] and Oleg Chernyi and his colleagues extended the 5 T J_c beyond 4000 A/mm^2 by using 48 wt% and very long heat treatments to obtain 28 vol%-Ti [110]. Using Nb pinning centers[111] the 5 T J_c has been extended to 4600 A/mm^2 and using ferromagnetic pinning centers (Ni/Cu) the 5 T, 4.2 K J_c has been extended to over 5000 A/m^2 [112].

density, indicating full summation of the individual vortex-precipitate interactions was soon seen[108]. The extension of a high homogeneity alloy supply over a wide range of compositions[109] enabled a unified a view of the impact of composition on precipitate morphology, sensitivity to strain and precipitation rate the explained previous inconsistencies.

By the Sixth Workshop on Nb-Ti Superconductors (now jointly organized with Ron Scanlan at Lawrence Berkeley Laboratory) all the US wire manufacturers were able to report critical current densities exceeding 3000 A/mm^2 (5 T, 4.2 K) in large filament trial billets in preparation for Super-conducting Super Collider, which was intended to be the Tevatron's successor, and by the Eighth Nb-Ti workshop, held at Asilomar, CA, in March 1988, the technology for obtaining SSC current requirements and beyond had been developed in both 6 and 9 μm diameter filamentary conductors.

In 1993 the SSC was cancelled, but high critical current Nb-Ti had reached maturity and would require little further technical development for the LHC (strand example shown in Figure 11.17), where the highest J_c strand had a critical current density at 5 T, 4.2 K of 3194 A/mm^2.[113] Neverthe-

[108] P. J. Lee, J. C. McKinnell, and D. C. Larbalestier, *Advances in Cryogenic Engineering* (Materials) 36 (1990) 287

[109] P.J. Lee, J.C. McKinnell, and D.C. Larbalestier, *IEEE Trans. MAG* 25:2 (1989) 1918

[110] O. V. Chernyi et al., *Advances in Cryogenic Engineering*, 48B (2002) 883

[111] R. W. Heussner et al., *Applied Phys. Lett.* 70:7 (1997) 901

[112] L.R. Motowidlo, M.K. Rudziak, and T. Wong, *IEEE Trans. Applied Superconductivity* 13:2 (2003) 3351

[113] Private communication from Thierry Boutboul, LHC, September 2006.

less, the performance range of ductile Nb-Ti superconductors has been shown to extend well beyond that used by LHC by increasing the volume of α-Ti or changing the pinning centers completely as shown in Figure 11.16; Oleg Chernyi and his colleagues extended the 5 T J_c beyond 4000 A/mm^2 by using a higher Ti content alloy (48 wt%) and extremely long heat treatments to obtain 28 vol% α-Ti[114]. By further alloy modification it seems likely that further improvements could be made. Mechanically introducing pinning centers and multiple restacks, "Artificial Pinning Center" (APC) composites can be created that are not restricted in pinning center composition or geometry[115]; using Nb pinning centers the 5 T J_c has been extended to 4600 A/mm^2 with only 24 vol% of pinning centers[116] and by using ferromagnetic pinning centers (Ni/Cu) the 5 T, 4.2 K J_c has been extended to over 5000 A/mm^2.[117]

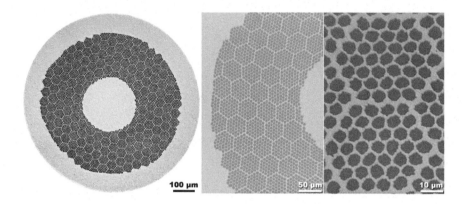

FIGURE 11.17:   Outokumpu AS (now Luvata) LHC inner strand cross-section showing double stacked filament array.

11.2.14   Nb-Ti as a Commodity

"There is no base business for the applied superconductivity industry without MRI"[118].

Although much of the push for advancing the technology of Nb-Ti superconductors came from the needs of the synchrotron and accelerator communities, the business of Nb-Ti is dominated by wires for MRI imaging. What can be said is that the initial intense developments that took place from 1965 to about 1985 were driven technically by the need to make the highest possible J_c values for fine filament (5–10 μm) conductors that could be made into Rutherford cables. Technical, rather than cost, considerations were dominant. The success of the wire industry was built, however, on the emerging market for MRI conductors—much simpler, largely monolithic ones, that needed to be made very reliably at the lowest possible cost. The wonderful synergy that developed between the technical understandings emerging from the Nb-Ti workshop–the optimum ribbon nanostructure, and the demonstration that Nb47Ti was indeed the optimum composition supported the largely hidden technology developments of MRI. Nb-Ti is the tonnage conductor of superconducting applications because it is thoroughly understood, manufacturable and the workhorse for more than 90% of all superconducting magnets. But as subsequent chapters show, Nb$_3$Sn, MgB$_2$, Bi$_2$Sr$_2$CaCu$_2$O$_x$,

[114] O. V. Chernyi et al., *Adv. Cryog. Eng.*, 48B (2002) 883

[115] G. L. Dorofejev, et al., *Proc. 9th Int. Conf. on Magnet Technol.*, (1985) 564

[116] R. W. Heussner et al., *Appl. Phys. Lett.* 70:7 (1997) 901

[117] L.R. Motowidlo, M.K. Rudziak, and T. Wong, *IEEE Trans. Appl. Supercond.* 13:2 (2003) 3351

[118] Seung Hong, Oxford Superconducting Technologies, private communication for this article (2010)

$(Bi,Pb)_2Sr_2Ca_2Cu_3O_x$, and now $YBa_2Cu_3O_{7-\delta}$ are all vying to take part or all of its place. History will report in another 10–20 years whether or not the huge industrial and scientific base of Nb-Ti could outweigh the advantages of transition temperatures more than 10 times higher than the 9 K of Nb-Ti.

11.3   History of Nb_3Sn and Related A15 Wires

Kyoji Tachikawa and Peter J. Lee

11.3.1   Introduction

Nb_3Sn occupies a special place in any history of superconductivity as the first material to show, 50 years ago in 1961, that superconductivity actually could exist in very high fields. It is both fitting and also perhaps remarkable that it so quickly made the transition from a *Physical Review Letter*[119] to high field magnets of 10 T or more that fulfilled Onnes's 50-year-delayed dreams. Initially made as a few µm thick layers in tape forms on strong substrates, so as to allow application to magnets without performance being compromised by the hard and brittle nature of its A15 crystal structure. However, the electromagnetic instabilities of superconducting tapes were soon recognized. Following the development of filamentary forms of Nb-Ti in about 1965, the urge to develop a filamentary technology for A15 compounds occurred too. It was soon discovered that the presence of a small amount of Cu enabled formation of the relevant A15 compound V_3Ga or Nb_3Sn at temperatures of 600–700 °C, some 200–300 °C below the temperatures needed in the binary mixtures. Cu-Sn or Cu-Ga bronzes fulfilled this condition and allowed fabrication of wires containing many hundreds or even thousands of Nb or V filaments in the relevant bronze by conventional extrusion and drawing processes. This bronze route was the breakthrough that enabled Nb_3Sn multifilamentary conductor forms capable of producing stable 10–23 T field magnets required for fusion, NMR, and all manner of other systems. Due to these widespread applications, Nb_3Sn can thus be regarded as one of the key materials in science and technology. In this article the chronological progress of Nb_3Sn wires will be surveyed, together with some discussion of its brother A15 compound, V_3Ga, which played an important role too.

The A15 compounds like Nb_3Sn have the nominal composition ratio of A_3B, but, although the stoichiometric A_3B composition is obtained by diffusion, actually all possible compositions that are stable also form, thus producing a range of superconducting properties. For technology it is vital to work with at least ternary systems, since the formation temperature of Nb_3Sn is lowered from ~950 °C to ~700 °C by the addition of Cu. The constituent elements, Nb, Sn and Cu, have a favorable workability at room temperature, allowing multifilamentary composites to be assembled at larger size, co-drawn to final wire dimensions, and then reacted to form the A15 filament structure. Because optimum reaction conditions do not allow reaction to completion[120], the measured transition temperature T_c and upper critical field B_{c2} of Nb_3Sn wires have a width that depends on the variable stoichiometry of the filaments. A central compromise of all multifilamentary conductors is that grain boundaries provide the primary flux-pinning centers requiring a low reaction temperature, while the highest T_c and B_{c2} distribution requires a somewhat higher reaction temperature that enlarges the grain size and reduces the flux pinning. Third elements can also improve the performance of Nb_3Sn wires and commercial strand normally incorporates at least one of these. Most commonly,

[119] J. E. Kunzler et al., *Phys. Rev. Let.* 6:3 (1961) 89
[120] P. J. Lee and D. C. Larbalestier, *Cryogenics*, 48:7-8 (2008) 283

small amounts of Ti and Ta addition are used to enhance the B_{c2} by increasing the normal state resistivity of the A15 phase which results in higher critical current density at fields above 12 T (4.2 K). Moreover the addition of Ti appreciably accelerates the formation of Nb_3Sn.

11.3.2   Chronological Progress in the Fabrication of Nb_3Sn and V_3Ga Wires

At the 1982 Applied Superconductivity Conference, John Hulm recalled the circumstances surrounding the discovery of superconductivity in the A15 compounds:[121]

In the spring of 1952 I was working with a graduate student, George Hardy. We decided that the carbides and nitrides were more or less exhausted, so we moved down to silicides and germanides in the second and third periods. We also began arc-melting our samples. These were two very fortunate moves. Not only was the general quality of the samples improved over our earlier sintered materials, but we soon discovered a new high T_c superconductor, V_3Si, at 17 K.[122] It belonged to what was then known, erroneously, as the beta-tungsten structure; of course this was subsequently changed to A15. I told this news to Bernd Matthias almost immediately.

By then Bernd Matthias had teamed up with Ted Geballe, Ernie Corenzwit, and Seymour Geller at the Bell Laboratories. These investigators proceeded to execute a tour-de-force in creative synthesis by discovering about 30 new A15s, including several new high T_c materials, most prominently Nb_3Sn at 18 K, in 1954.[123]

On December 15, 1960, a group at Bell Laboratories led by J. E. Kunzler tested the high field properties of a rectangular rod of bulk Nb_3Sn, that had been sintered and then melted at 2400 °C and to their "complete surprise" [124], found that it was still superconducting at their maximum field of 8.8 T. The details of the road to this discovery are covered in more detail in the ductile superconductor history Section 11.2 but it should be repeated here that this achievement represented a prize of 21 bottles of Scotch whisky courtesy of a wager with Morris Tanenbaum. The 2400 °C bulk sample was impractical as the magnet material they had hoped to use for Masers. However, "super technician"[125] Ernie Buehler (Tanenbaum and Buehler formed the team that made the first Si transistor), devised the powder-in-tube (PIT) route to make wire, placing Sn and Nb powders in a Nb tube, a ductile combination that could be drawn to final size before the reaction heat treatment that would create the brittle Nb_3Sn. The first strand carried 50 times higher J_c than the bulk sample,[126] and their best PIT strand with 10% excess Sn than the A15 composition almost 1500 A/mm^2 at 8.8 T (4.2 K) using a much more manageable heat treatment of 970 °C. Kunzler suggested that Nb_3Sn wires might enable a convenient and economical high-field magnet without the consumption of huge electric power and cooling water required in a water-cooled Cu magnet.[127]

The next major milestone was the development of a vapor deposition technique to produce Nb_3Sn by J. J. Hanak (who became the father of combinatorial chemistry) at RCA Laboratories.[128] Chemical vapor deposition produced a thin enough layer of Nb_3Sn on a wire or tape that was flexible enough to wind magnets. Furthermore, the use of substrates with thermal expansions greater than

[121] J. Hulm, *IEEE Trans. on Magnets*, 19:3 (1983) 161

[122] George F. Hardy and John K. Hulm, *Phys. Rev.* 93:5 (1954) 1004

[123] B. T. Matthias et al., *Phys. Rev.* 95:6 (1954) 1435

[124] J. Kunzler, *IEEE Trans. on Magnetics*, 23:2 (1987) 396

[125] Tanenbaum's description of Ernie Buehler in "Oral-History: Goldey, Hittinger and Tanenbaum" interview conducted by Michael Riordan and Sheldon Hochheiser, Murray Hill, New Jersey, September 25, 2008, Interview #480 for the IEEE History Center, The Institute of Electrical and Electronics Engineers, Inc.

[126] J. E. Kunzler et al., *Phys. Rev. Lett.* 6:3 (1961) 89

[127] J. Kunzler, *Rev. Mod. Phys.* 33:4 (1961) 501

[128] J. J. Hanak, Vapor deposition of Nb_3Sn, in *metallurgy of Advanced Electronic Materials*, (AIME) 19, Ed. G. E. Brock. (New York: Interscience) (1963) 161

FIGURE 11.18: Production facilities for Nb$_3$Sn wires using the continuous CVD process were established at RCA already in 1966 [129].

Nb$_3$Sn placed the brittle A15 layer under beneficial compressive strain. The technique involved production of gaseous NbCl$_4$ and SnCl$_4$ by direct reaction of Nb, Sn and Cl$_2$. The chlorides, together with H$_2$ and HCl were passed into a reaction chamber in which a continuously fed metallic substrate tape or wire was resistively heated between two carbon electrodes to ~1000 °C.[129] After growth of the Nb$_3$Sn, a layer of Ag was applied as the stabilizer.

Figure 11.18 illustrates the industrial CVD facilities developed for the winding of 10 T-class superconducting magnets. A major effort was also underway at General Electric, and in competition with RCA, GE developed the Nb$_3$Sn diffusion process in which

a Nb substrate tape was passed through a molten Sn bath.[130] Nb$_3$Sn layers were formed on both sides of the tape after the reaction heat treatment at 950 °C. Stabilizing Cu layers were soldered to the tape using the residual Sn left behind after reaction, as shown in Figure 11.19. Many 15 T-class superconducting magnets, mostly used for solid state physics research, were wound from these tapes by Intermagnetics General Co. (IGC), the spin-off company that emerged from GE. Thus the vision of Kunzler[127] was realized rather quickly.

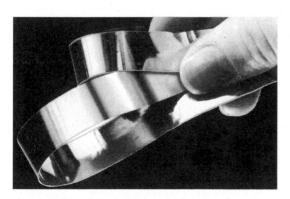

FIGURE 11.19: Nb$_3$Sn tape produced at General Electric using diffusion between a liquid Sn bath and a Nb foil, later the basis of magnets made by Intermagnetics General Corp.[130]

Nb$_3$Sn diffusion tape was also used for a 110 m-long power transmission model cable operated at 7 K that was the forerunner of the many electric utility application dreams that flowered after the discovery of superconductivity in the cuprates in 1987.[131]

However all tape magnets suffered from flux jumps that made them slow to charge and sometimes irregular in performance. As monocore wires of Cu-clad Nb-Zr gave way to better-bonded Cu-Nb-Ti (see Section 11.2) and then to metallurgically bonded, twisted multifilamentary Nb-Ti conductors, superconducting magnets became stable, capable of fast ramps and generally much more predictable. The drive to develop multifilamentary A15 wires was on!

A key study in this transition from tapes to wires was made on V$_3$Ga wires. A small Cu addition to molten Ga was found to change the diffusion mode for growth of V$_3$Ga from a grain boundary one to a bulk one, without the Cu impairing the properties of the growing V$_3$Ga layer. Cu accumulated at the diffusion boundary, decreasing appreciably the formation temperature of V$_3$Ga as illustrated in Figure 11.20.[132] The area fraction of V$_3$Ga with fine grains is

[129] J. J Hanak, K. Strater, and G.W. Cullen, *RCA Review*, September 1964

[130] M.G. Benz, G.E. Research. & Development Center Report No. 66-C-044 (1966)

[131] E. Forsyth and G. Morgan, *IEEE Trans. on Magnetics*, 19:3 (1983) 652

[132] K. Tachikawa and Y. Tanaka Japanese *J. Applied Physics* 6:6 (1967) 782

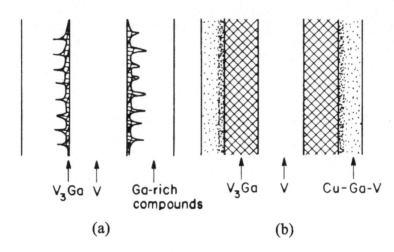

FIGURE 11.20: Formation of V_3Ga layer by a diffusion reaction between V and Ga. (a) diffusion at 800 °C without Cu addition, and (b) at 700 °C with Cu addition. The small dots represent the Cu in the Cu-Ga-V ternary alloy[132].

much increased by the lower temperature reaction enabled by the Cu addition. The patent describing the effect of Cu on the promotion of V_3Ga and Nb_3Sn synthesis was filed in 1966.[133]

The V_3Ga tape exhibits an appreciably better performance than pure Nb_3Sn tape in fields above 15 T. The 17.5 T superconducting magnet system shown in Figure 11.21, which held the field record for an all-superconducting magnet for ~10 years after 1975, was made by using a Nb_3Sn outer and a V_3Ga inner section.[134] Only in the 1980s was it possible to exceed this performance with multifilamentary conductors of Nb_3Sn with Ti addition.

The route to multifilamentary A15 conductors was greatly simplified by the work with Cu additions to V-Ga tapes, and multifilamentary V_3Ga wires were soon fabricated using a Cu-Ga bronze matrix and V cores.[135] The first commercial multifilamentary V_3Ga wire was produced by Furukawa Electric in

FIGURE 11.21: 17.5 T superconducting magnet system operated at the National Research Institute for Metals[134].

1972 which was wound into a stable 10.4 T magnet, and by this time groups in Europe[136] and the US[137] had also developed the bronze route to make multifilamentary Nb_3Sn wires. Because bronze wires required many anneals to allow wire drawing, interest rapidly grew in modifications of the

[133] K. Tachikawa, Y. Tanaka and S. Fukuda, Japan Pat. 0670619 (Filed June 25, 1966)

[134] K. Tachikawa et al., *J. Cryogenic Soc. of Japan* 11:6 (1976) 252

[135] K. Tachikawa, Proc. ICEC-3, *Illife Sci. & Technol. Pub.* (1970) 339

[136] Brian Howlett, Great Britain Pat. 52,623/69 (Filed Oct. 27, 1969). US Pat. 3,728,165 (Filed Oct. 19, 1970)

[137] A. R. Kaufman and J. J. Pickett, *Bull. Ame. Phys. Soc.* 15 (1970) 833

bronze process, in which the more ductile pure metal components of Sn, Cu and Nb were used. The greatest longevity has come from variants of the internal tin (IT) process, in which a composite wire constructed from a Cu matrix, Sn and Nb cores (see Figure 11.22), which was initiated by Mitsubishi Electric in 1974.[138]

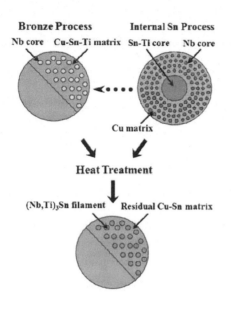

Bronze Process **Internal Sn Process**
Nb core Cu-Sn-Ti matrix Sn-Ti core Nb core

Cu matrix

Heat Treatment

(Nb,Ti)$_3$Sn filament Residual Cu-Sn matrix

FIGURE 11.22: Schematic illustration of bronze and internal Sn (IT) process.

Some of the earliest work on multifilamentary conductors using the bronze process was performed at the UK Atomic Energy Research Establishment at Harwell, where Nb$_3$Sn tapes were also found to produce too much electromagnetic instability when in magnets. One of the young researchers in the Harwell group, Phil Charlesworth, recalls:[139]

It also soon became clear that the critical problem in constructing superconducting magnets is ensuring stability against catastrophic quenching caused by small movements of flux vortices rearranging themselves against pinning sites during field ramping. For this purpose, a wire containing many thin filaments of superconductor embedded in a normal conducting matrix of low resistivity, providing dynamic stabilisation, is much more attractive than a tape, as the filaments, if fine enough, are adiabatically stable and the matrix can help to damp out propagation of fluctuations to neighbouring filaments. Work along these lines was started in Brian Howlett's group at Harwell in 1969 with the progressive development of the so-called bronze-route method of fabrication using niobium rods embedded in copper-tin bronze, the Nb$_3$Sn created by heat-treatment to create a reaction layer between the tin and the niobium. In early 1971, the group moved into the Chemistry Division with Jim Lee as group leader.

I was involved in this program to the extent of measuring the properties of the material produced but the bulk of the effort went into development of a production line for ever more sophisticated composites. The fabrication route involved drilling a matrix of 37 holes in a cylindrical bronze billet, using a gun drill, inserting niobium rods and then swaging and drawing the material to bring the bronze into firm contact with the niobium and to reduce the overall diameter. The process could then be repeated, one or more times, using the previous stage composites as inserts. Many people worked on this program but Derek Armstrong, who toured the country acquiring second-hand machine tools to build up the industrial-scale fabrication shop we needed, threw himself with enthusiasm into developing the fabrication route.

All along this work had been in collaboration with a team at the neighboring Rutherford Laboratory interested in producing practical solenoids with possible particle accelerator applications in mind. Again there were several people involved but our liaison was chiefly through David Larbalestier and Chris Scott. A joint paper in 1974[140] describes composites with filament counts up to 42,439 and including regions of pure copper protected by a diffusion barrier to add extra stabilisation. Another paper of the same date describes the suc-

[138] Y. Hashimoto, K. Yoshizaki and M. Tanaka, *Proc. ICEC-5*, K. Mendelssohn, Editor, (1974)332

[139] Phil Charlesworth, personal communication for this article, October 17, 2010

[140] D. Larbalestier et al., *IEEE Trans. on Magnetics* 11:2 (1975) 247

cessful performance of several small solenoids made from our bronze-route material by winding the material as-fabricated and then heat-treating the whole coil to produce the Nb_3Sn.

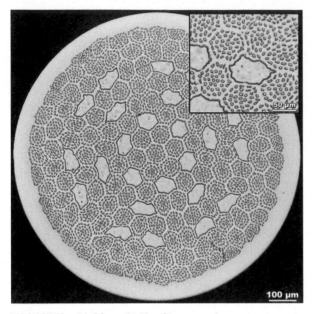

FIGURE 11.23: 5143 filament bronze process strand fabricated by Harwell/Rutherford group in early 1974. Application of a diffusion barrier is shown protecting internal stabilizer elements of Cu[140].

It is interesting that Howlett's bronze route patent was filed in 1969, and the work at Brookhaven National Laboratory in the USA was clearly ongoing at this time too. Having efforts in Japan, the UK and the USA led to competition that enormously enhanced progress. In particular, the joint Harwell-Rutherford program had led to intrinsically stable multifilamentary conductors (shown in Figure 11.23) which had been wound into solenoids generating more than 12 T in 1974.

11.3.3 Bronze-Processed Nb_3Sn Wires

Figure 11.22 schematically illustrates the bronze and internal tin (IT) processes, while Figure 11.24 shows typical bronze ingots prepared on the industrial scale. Diameter, length and weight of the ingots are ~200 mm, ~800 mm and ~250 kg, respectively. Typically 19, 37 or some other hexagonal-array number of holes are drilled in the bronze ingot, into which Nb cores are inserted. The composite stack is then fabricated to hexagonal rods by extrusion and drawing. These hexagons are then restacked in a Cu cylinder which is protected from the bronze by a Ta or Nb diffusion barrier, whose purpose is to prevent poisoning of the high quality stabilization of the Cu sheath by any Sn leakage. Bronze billets are normally hot extruded, and then drawn into multifilamentary wire. Wires are typically heat treated at ~675 °C for ~200 h to form the Nb_3Sn layers around each original Nb filament.

It was soon found that a small addition of Ti to either the bronze or Nb cores was favorable to J_c, as also was raising the amount of Sn in the bronze.[141] Tachikawa had originally intended the Ti addition to the bronze to improve the mechanical strength of the Sn-poor bronze left behind after the diffusion reaction was ended. Unfortunately the Ti was entirely incorporated into the Nb_3Sn layer or accumulated at the Nb_3Sn/bronze interfaces, and was not effective for reinforcement. Fortunately Ti incorporation into the A15 layer appreciably improved the high-field performance of Nb_3Sn by raising B_{c2}. A small amount of Ta added to the Nb filaments also enhances the high field performance. Both additions are used still today. Figure 11.25 shows bronze-processed Nb_3Sn strands used in a prototype ITER cable-in-conduit conductor (CICC).[142]

ITER requires 502 tonnes of multifilamentary Nb_3Sn wires to provide the high magnetic fields (as high as 13 T) required to confine the plasma (Toroidal Field System) and induce the main plasma current (Central Solenoid). Both IT and bronze route wires are being manufactured for this program in a worldwide effort.

---

[141] K. Tachikawa, H. Sekine and Y. Iijima, *J. Appl. Phys.* 53:7 (1982) 5354

[142] P. Bruzzone, R. Wesche, and F. Cau, *IEEE Trans. on Appl. Supercond.* 20:3 (2010)470

As the ITER use suggests, the bronze route to Nb$_3$Sn has had remarkable longevity. However, almost every part of the process has been subject to continuous development, resulting in a better, cheaper and longer piece-length product. A higher Sn:Cu ratio was recognized very early to be valuable for raising J_c but high-Sn bronzes were difficult to cast without inhomogeneity and difficult to draw without excessive bronze work-hardening. By such processes of continuous improvement and production, the Sn content of the bronze has been increased from 13wt% to 16wt% in the 1990s, allowing a doubling of the J_c at 4.2 K and 20 T. Such bronze wires have found wide application in NMR magnets, including the most recent delivery of a 1 GHz magnet by Bruker, as shown in Figure 11.26.[143] However, the specification of the wire used in this magnet is not reported. A variety of refrigerator-cooled magnets, vertical, horizontal and split pair magnets, have also been developed commercially using bronze-processed Nb$_3$Sn wires. The magnet shown in Figure 11.27 generates 15 T in a 171 mm room temperature bore.[144] In addition Nb$_3$Sn wires for the AC use have been fabricated by the bronze process. A 50 mm-bore 2 T Nb$_3$Sn magnet successfully operated at 53 Hz.[145] The diameter of Nb filaments in the wire has been reduced to as fine as 0.2 μm.

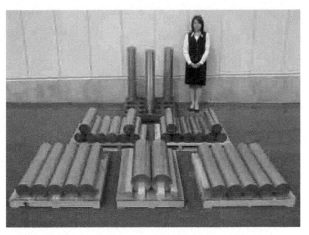

FIGURE 11.24: Typical bronze ingots for Nb$_3$Sn wires (courtesy of Osaka Alloying Works Co., Ltd.).

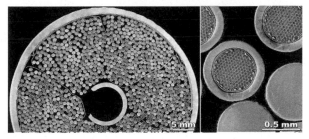

FIGURE 11.25: Partial cross-section (left) of a prototype ITER Toroidal Field Cable-in-Conduit conductor (CICC) EUTF5 (ENEA) containing a mixture of bronze-process Nb$_3$Sn strands and Cu strands (right) within 6 cable petals (29.6% void fraction). Cable supplied to FSU after testing in the Sultan facility[142].

11.3.4 Internal Sn-Processed Nb$_3$Sn Wires

The principle of the internal tin or IT process has been shown in Figure 11.22. In a bronze alloy the Sn solubility limit is ~15.8 wt.%, while a larger Sn fraction may be possible in the IT process. Since the IT processed wire is composed of three components, a variety of cross-sectional designs are possible. Figure 11.28 is an example of an IT-processed wire using 19 filamentary subelements protected by a single diffusion barrier.[146]

Figure 11.29 shows a part of the Rutherford cable (see box in Section 11.2.11) for the record-breaking D20 dipole at LBNL[147] using IT-processed strands like the one shown in Figure 11.28.

[143] Bruker Biospin : http://www.bruker-biospin.com/pro091202.htmL?&L=11, March1 (2010)

[144] R. Hirose et al., *IEEE Trans. on Appl. Supercond.* 16:2 (2006) 953

[145] K. Tachikawa et al., *Proc. ICMC* 17, D. Dew-Hughes, ed., IOP Press (1998) 439

[146] M. B. Field et al., *Adv. in Cryogenic Eng. Materials* 54 (2008) 237

[147] R. Benjegerdes et al., *Proc 1999 Particle Accelerator Conference*, 5 (1999) 3233

FIGURE 11.26: The 1 GHz NMR magnet constructed by Bruker now installed in Lyon, France[143].

FIGURE 11.27: Refrigerator-cooled superconducting magnet[144].

Wires composed of as many as 100–200 subelements, each covered by the diffusion barrier (distributed barrier) have been also developed. Design of IT wires is still rapidly evolving as the trade-offs between high J_c (favored by a small Cu fraction in the Cu-Sn-Nb composite needed to increase Nb$_3$Sn fraction) and a small effective filament diameter (favored by a large fraction of Cu to keep the A15 filaments separated) are explored. The requirements for high field magnets for NMR, for ITER and for high energy physics applications are still driving this not-yet mature technology.

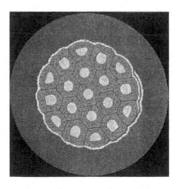

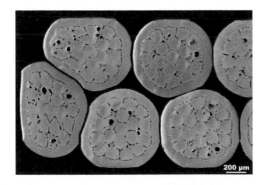

FIGURE 11.28: Typical cross-section of 0.78 mm diameter 19 sub-element IT wire before heat treatment. The thin white ring is the diffusion barrier[146].

FIGURE 11.29: IT strands cabled for the record-breaking 13.5 T D20 dipole magnet at LBNL (cable supplied to FSU by Dan Dietderich—LBNL) imaged in transverse cross-section after heat treatment[147].

In the heat treatment of IT wires, the Cu and Sn elements are first converted into high-Sn bronze phases at temperatures up to about 500 °C, and only after this is the wire heated to 600–700 °C to react with the Nb filaments. The IT reaction thus requires a multistep heat treatment in which the filaments move away from the Sn cores during heat treatment, often resulting in bridging of the filaments, increasing the effective filament diameter and the hysteresis loss. The arrangement of Nb filaments can be designed to account for this change as in the example.[148]

Figure 11.30 shows the non-Cu J_c versus magnetic field curves of the distributed barrier IT-processed wire[146] in which ~3000 A/mm^2 is achieved at 12 T, as part of the US DOE HEP Conductor Development Program, nearly 3 times larger than that in bronze-processed wires. However, increasing the separation of Nb filaments by enhancing the Cu:Sn ratio to avoid filament bridging reduces the J_c to the level of bronze-processed wires. High-field accelerator magnets are being developed using IT-processed Nb$_3$Sn wires due to their large non-Cu J_c. IT-processed Nb$_3$Sn wire is used also for fusion magnets, e.g., ITER and KSTAR (Korea Superconducting Tokamak Advanced Research) facilities.

Comparing the bronze and IT process, a clear advantage for the IT process is in the wire fabrication, since the bronze process requires frequent intermediate anneals to soften the work-hardened bronze. However, it is not possible to extrude composites containing Sn cores because of the danger of melting the Sn, so temporary cores must be used which can be replaced by Sn at a later stage. The heat treatment profile is also simpler for bronze wires. The J_c is larger in IT-processed wires, while AC loss is smaller in bronze-processed wires. The *n*-value and the irreversible strain tend to be better in bronze-processed wires. Thus there are advantages and disadvantages to both processes. Figure 11.31 illustrates the Nb$_3$Sn layer J_c, reflecting the intrinsic quality of the Nb$_3$Sn, versus magnetic increase in Sn supply available in bronze processed wires[149]. Furthermore the strain effect in Nb$_3$Sn wires which is practically an important topic has been extensively studied.[150]

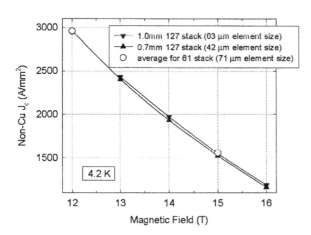

FIGURE 11.30: Non-Cu J_c of distributed barrier IT wires as a function of magnetic field. The $J_c - B$ curves are nearly identical for strands containing either 61 or 127 subelement rods[146].

11.3.5 Conclusions and Future Outlook

We have described a remarkable development of the A15 compound Nb$_3$Sn from its "accidental" discovery as a high-field superconductor 50 years ago in 1961 through rapid implementation as a tape conductor made either by CVD or by diffusion-processed Nb$_3$Sn tapes and then to its replacement by multifilamentary conductors, starting in the early 1970s. The discovery that Cu could allow synthesis of V$_3$Ga and Nb$_3$Sn at much lower temperatures than was possible for binary mixtures enabled this move to a multifilamentary technology. Additions such as Ti, which enhanced B_{c2}(4.2 K) of Nb$_3$Sn from ~21 T to ~26 T further enhanced the high-field capability of Nb$_3$Sn wires. Industrial

[148] Y. Kubo et al., *IEEE Trans. on Applied Superconductivity* 16:2 (2006) 1232

[149] P. J. Lee, A. A. Squitieri and D. C. Larbalestier, *IEEE Trans. on Appl. Supercond.* 10:1 (2000) 979

[150] J.W. Ekin, *Adv. Cryogenic Eng: Mat.* 30 (1984), 823

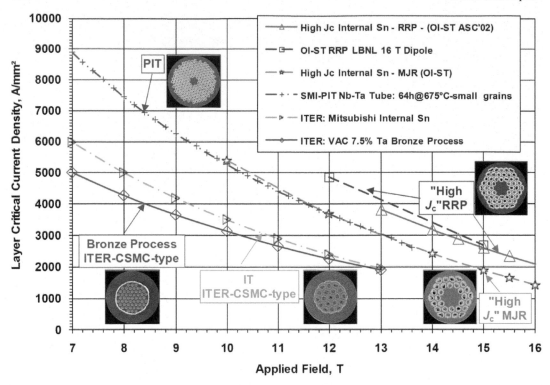

FIGURE 11.31: The layer J_c of the Nb_3Sn in a variety of Nb_3Sn strands. The higher Sn contents available in PIT and high-J_c type IT strands can double the intrinsic J_c of the Nb_3Sn[149], RRP: Restacked Rod Process[146] (IT), MJR: Modified Jelly Roll[151] (IT), PIT: Powder in Tube[152].

fabrication of Nb_3Sn wires was well established in the 1970s on so that it was not disrupted by the discovery of high-T_c cuprates in 1987 [151]. In the 1990s and the 2000s marked improvements in the performance of Nb_3Sn wire continued to be achieved using the usual techniques of gaining a better understanding of manufacturing improvement and better understanding of how to best package the mixture of Sn, Cu, Nb and Ti or Ta needed to produce the superconducting A15 phase.

Because Nb-Ti cannot provide fields greater than 10-12 T, now there is a clear market for high-field superconducting coils in the range up to ~23 T that Nb_3Sn can address. In fact alternative processes to high performance conductors are still being explored. Besides the bronze and IT processes, the PIT process, using powders of Sn-rich compounds such as $NbSn_2$ or Nb_6Sn_5[152] and the Nb tube process[153], yields Nb_3Sn wires with excellent J_c performance and a well-defined and small filament diameter. Such innovations show clearly that Nb_3Sn has not yet lost its place as the high-field superconductor, a rather remarkable state for the "original" high-field superconductor 50 years on. Fresh water may still come out from an old spring in the future.

[151] J. A. Parrell et al., *Adv. in Cryogenic Eng. Mat.* 48 (2002) 968

[152] A. Godeke et al., *Cryogenics* 48 (2008) 308

[153] E. Gregory et al., *Adv. Cryogenic Eng.: Mat.* 54 (2008) 252

11.4   Bi-Sr-Ca-Cu-O HTS Wire

Martin W. Rupich and Eric E. Hellstrom

11.4.1   Introduction

The years following the discovery of HTS materials saw many attempts to develop them into practical wires. However, these efforts were frequently stymied by the inability to transform the brittle ceramics into strong, flexible wires that carried high critical currents. The first wire, based on $YBa_2Cu_3O_7$ (YBCO), was prepared by Jin et al.[154] in 1987 using an oxide-powder-in-tube (OPIT) approach. However, the maximum current density (J_c) of 1.75 A/mm^2 (77 K, self-field) was limited by the inability to sinter the grains together and the presence of numerous high-angle grain boundaries[155]. In 1989, Heine et al.[156] at Vacuumschmelze in Germany packed some of the recently discovered Bi-Sr-Ca-Cu-O material[157] inside a Ag tube and found that melting the powder resulted in wires with J_c higher than 100 A/mm^2 (4.2 K, 25 T). Such a "high" J_c in a polycrystalline $Bi_2Sr_2CaCu_2O_8$ (2212) wire was the first genuine proof-of-principle that HTS wires could be fabricated in a form useful for the envisioned applications. This successful demonstration revitalized the HTS community at a time when the tremendous excitement of 1987 about visions of HTS applications was waning.

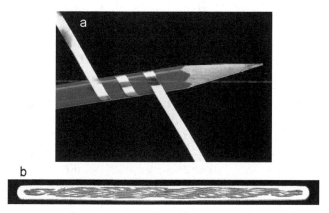

FIGURE 11.32:   1G 2223 wire is produced in continuous lengths (>1 km) as a flexible, strong composite tape consisting of multiple 2223 filaments embedded in a Ag matrix. (a) The composite architecture of 1G wire enables winding to small diameter. (b) Optical micrograph showing cross-section of the Ag/2223 composite wire. The wire is ~4.3 mm × 0.2 mm. (Courtesy American Superconductor Corp.)

Two decades later, 2223 [$(Bi,Pb)_2Sr_2Ca_2Cu_3O_{10}$] based HTS wires are being manufactured in kilometer lengths with critical currents approaching 200 A at 77 K, self-field in a robust, multifilamentary composite wire. The 2223 conductor, also called First Generation (1G) wire, is produced in

[154] S. Jin, R. C. Sherwood, R. B. van Dover, T. H. Tiefel, and D. W. Johnson, *Jr., App. Phys. Lett.* 51 (1987) 203

[155] D. Dimos, P. Chaudhari, J. Mannhart, and F.K. LeGoues, *Phys. Rev. Let.* 61 (1988) 219

[156] K. Heine, J. Tenbrink, and M. Thoner, *App. Phys. Lett.* 55 (1989) 2441

[157] H. Maeda, Y. Tanaka, M. Fukutomi, and T. Asano, *Jap. J. App. Phys.* 27 (1988) 209

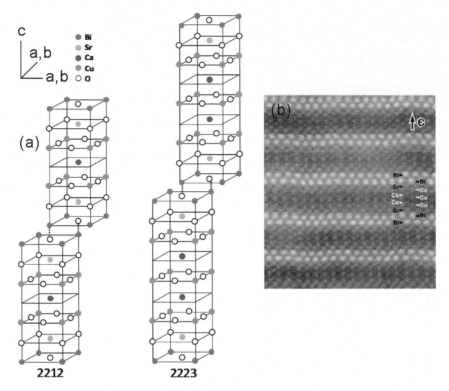

FIGURE 11.33: (a) Crystal structures of 2212 and 2223 and (b) TEM image (z-contrast along [100] projection) of a 2223 grain in a 1G 2223 wire. (TEM image courtesy American Superconductor Corp. and Nigel Browning.)

a flexible tape shape, containing multiple, fine ceramic HTS filaments embedded in a silver matrix as shown in Figure 11.32. The multifilamentary composite architecture results in very consistent properties along the length of the wire and imparts adequate mechanical properties that can be enhanced by lamination with metal stabilizing strips. Development of these wires was a major technical triumph, overcoming challenges from the complex chemistry and inherent brittleness of the ceramic material, and establishing the remarkable manufacturing control needed for defect-free, high-yield production. The 2223 wires have been successfully used in a variety of cable, rotating machine, current lead and magnet prototypes and products. These have established, beyond doubt, the technical credibility of HTS wire technology for power applications. Although Second Generation (2G) wires based on the YBCO materials, described in Section 11.5 by Malozemoff and Yamada, are the focus of most current efforts to develop commercial HTS wire, 2223 wire sets the standard for performance, particularly in terms of critical current and reproducibility.

The Bi-Sr-Ca-Cu-O (BSCCO) system contains multiple phases, $Bi_2Sr_2Ca_{n-1}Cu_nO_{2n+4}$, which differ primarily in the number of Ca and CuO_2 layers in the crystal structure[157,158]. The two phases most relevant for wire applications are $Bi_2Sr_2CaCu_2O_8$ (2212) and $(Bi,Pb)_2Sr_2Ca_2Cu_3O_{10}$ (2223) with a T_c of 85 and 110 K, respectively. The 2212-based HTS wires, which have high J_c at low temperatures and fields up to at least 45 T[159], are of particular interest for their potential to create high magnetic fields well beyond the capabilities of NbTi and Nb_3Sn wires.

---

[158] C. Michel, M. Hervieu, M. M. Borel, A. Grandin, F. Deslandes, J. Provost, and B. Raveau, *Z. Phys.* B 68 (1987) 421

[159] U.P. Trociewitz, J. Schwartz, K. Marken, H. Miao, M. Meinesz, and B. Czabaj, National High Magnetic Field Laboratory, Tallahassee, FL USA Report #297 (2005)

The 2212 and 2223 crystal structures, shown in Figure 11.33a, consist of superconducting CuO_2 layers interleaved with Ca, BiO and SrO layers. The individual layers are clearly seen in the TEM image of a 2223 sample in Figure 11.33b. Preparing single-phase compositions is difficult since the energies of formation of the different phases are very similar. Thus the materials are generally contaminated with intergrowths of the other phases, which appear as stacking faults lying in the *ab* plane. The formation and stability of the 2223 phase is enhanced by doping with Pb.

Unlike the YBCO family of materials, which require a high degree of biaxial alignment for current to pass from grain to grain, the BSCCO materials carry reasonably high current with only alignment of the c-axis[160,161,162,163]. Fortunately, the BSCCO materials have a micaceous-like structure with weak bonding between the BiO layers, allowing the CuO_2 planes to be aligned through simple deformation texturing. This feature is essential to the fabrication of 2223 wire.

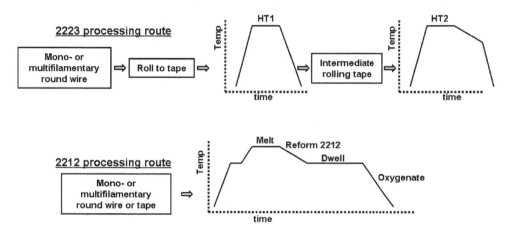

FIGURE 11.34: Processing routes developed for fabricating 2223 and 2212 composites. The 2223 wire requires two heat treatments with an intermediate rolling step. The 2212 wire only requires one heat treatment.

Although both 2212 and 2223 wires are produced as a multifilamentary composite with silver, differences in their chemistries have led to significantly different processing paths, as shown in Figure 11.34. The 2212 wire, either round or tape-shaped, is processed with a single heat treatment, during which the 2212 phase melts and reforms upon cooling into well-connected, c-axis textured grains. In contrast, the 2223 phase does not crystallize from a melt but rather forms by a liquid-mediated reaction of 2212 and alkaline earth cuprates. The *ab*-plane texture, necessary for high critical current density, is introduced by mechanically aligning the 2212 phase prior to formation of 2223, restricting the wire to a flat, tape-like architecture.

The grain structure of BSCCO wire consists of highly aspected, overlapping platelet-like grains with the high conductivity *ab*-planes textured in the plane of the wire. Both high- and low-angle grain boundaries exist between grains in the *ab*-plane, as is shown in Figure 11.35. The historic work by Dimos et al.[155] showed that grain boundary misalignments greater than a few degrees drastically reduce current density. In order to understand current flow in BSCCO wires with high grain bound-

[160] L.N. Bulaevskii, J.R. Clem, L.I. Glazman, and A.P. Malozemoff, *Phys. Rev.* B 45 (1992) 2545

[161] L.N. Bulaevskii, L.L. Daemen, M.P. Maley, and J.Y. Coulter, *Phys. Rev.* B 48 (1993) 13798

[162] B. Hensel, J.C. Grivel, A. Jeremie, A. Perin, A. Pollini, and R. Flukiger, *Physica* C 205 (1993) 329

[163] G.N. Riley, Jr., A.P. Malozemoff, Q. Li, S. Fleshler, and T.G. Holesinger, *J. Metals* 49 (1997) 24

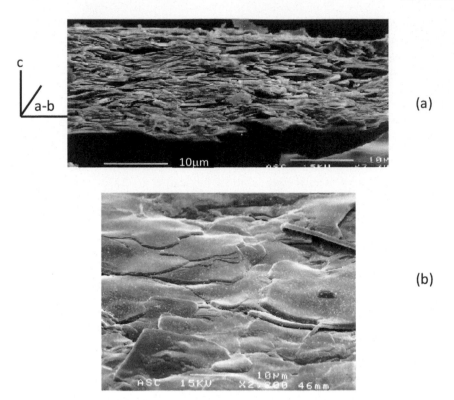

FIGURE 11.35: SEM images of (a) fractured short transverse cross-section of a 2223 HTS filament and (b) surface of a 2223 HTS filament (Ag removed) showing the overlapping arrangement of
platelet-like 2223 grains. Grain boundaries along the ab-plane consist of both low- and high-angle
grain boundaries. (Courtesy American Superconductors Corp.)

ary misalignments, various models were proposed.[160,161,162,163] In the "brick-wall" model[160,161],
currents encountering a high-angle grain boundary along the *ab*-plane turn and flow along the *c*-axis
into an adjacent BSCCO grain where the current resumes travel along the *ab*-plane. Thus high-angle
grain boundaries in the *ab*-plane are bypassed.

The following sections provide a high-level review of a few key advances that enabled the development of high-performance 2223 and 2212 wires.

11.4.2   2223 Wires

In the early years of 2223 wire development, it was naively thought that the materials could
be easily transitioned from the research and development stage to long-length wires that would
find rapid adoption in a wide range of commercial applications. However, it was soon realized
that the complexity of the materials presented significant challenges to the scientists and engineers
developing HTS wire.

One early approach to 2223 wire development was the deformation processing of a metallic
precursor consisting of a composite architecture of Bi, Pb, Sr, Ca, and Cu filaments in a silver
matrix. The precursor filaments were internally oxidized to form the desired HTS compound[164].
Nevertheless, the most successful approach for developing 2223 wires was the oxide-powder-in-

---

[164] A. Otto, C. Craven, D. Daly, E.R. Podtburg, I. Schreiber, and L.J. Masur, *J. Metals* 45 (1993) 48

tube or OPIT process, first used for 2223 by Sumitomo Electric Industries (SEI)[165]. Over time, this approach was adopted by all companies manufacturing 2223 wire.

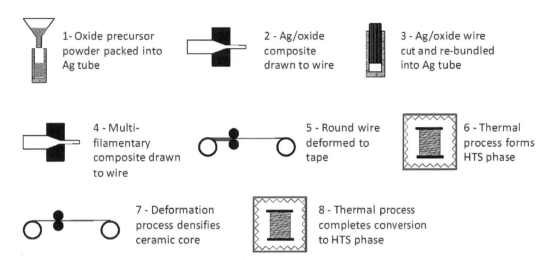

FIGURE 11.36: Oxide-powder-in-tube (OPIT) process used for fabricating 2223 wire. Ag-sheathed 2212 conductor is fabricated similarly, but 2212 requires only one heat treatment, so it ends after step 6. Round 2212 wire skips step 5.

The basic OPIT process is schematically illustrated in Figure 11.36. A silver tube is filled with an oxide precursor powder and then drawn into a hexagonally shaped rod that is 1-2 mm across the flats. The rod is then cut into short lengths and arranged as bundles of 7–85 filaments in a second silver tube. This composite bundle is drawn into a wire that is rolled to a tape-like architecture with typical dimensions of ~ 0.2 mm × 4 mm. The tape is heated in an atmosphere of ~7.5% O_2 to 830–850 °C to partially convert the precursor powder to the 2223 phase. This process results in de-densification of the ceramic cores, leading to significant porosity in the center of the filaments. Thus a second deformation and heat treatment process is used to partially remove the porosity and complete formation of the 2223 phase.

The achievement of fine filaments in a composite structure was a key breakthrough and solved a fundamental problem for HTS wire: how to achieve adequate mechanical flexibility and strength with a brittle ceramic. Early efforts at wire development using solid rods of HTS material rather than powder suffered from the complete lack of flexibility and high risk of fracture. By contrast, present day 2223 tapes can tolerate a mechanical bending strain of up to 0.5%, which, given their thin (200 µm thickness) cross-section, is enough to enable helical stranding onto several centimeter diameter flexible formers for power cables, or pancake and layer stranding into magnets with minimum radii of curvature of a centimeter. The bare Ag-composite wire is capable of withstanding tensile stresses of up to 100 MPa, while conductors laminated with stainless steel strips can tolerate tensile stresses of several hundred MPa.

The use of silver as the sheath material is critical to the success of the OPIT process for both 2223 and 2212, as it is the only material that is chemically compatible with BSCCO and permeable to oxygen at the processing temperature. It also lowers the 2223 formation temperature. At the same time, Ag is also the Achilles heel of 1G wire, comprising approximately 60% of the cross-section and thereby contributing to a significant fraction of the 2223 wire cost.

---

[165] T. Hikata, K. Sato, and H. Hitotsuyanagi, *Jap. J. App. Phys.* 28 (1989) 82

Although the basic fabrication process shown in Figure 11.36 was followed by all manufacturers, individual manufacturers implemented subtle differences that led to different degrees of success in developing high-current, long-length wires. Throughout the 1990s numerous organizations including American Superconductor Corp. (AMSC), Intermagnetics General Corp., Sumitomo Electric Industries (SEI), Siemens, Showa Electric Wire and Cable Co., Australian Superconductor, NKT and others worked on the development of the 2223 wires. During this time, SEI and AMSC became leaders in 2223 wire and demonstrated the first commercial manufacturing capability for HTS wires.

The 2223 Wire Development Group

In 1990, the Wire Development Group (WDG) was organized by AMSC in an effort to bring together a diverse group of physicists, material scientists and chemists working as a cohesive group exploring the fundamental material science of the 2223 system. The WDG (Figure 11.37) was organized around a core group of scientists (Eric Hellstrom, Terry Holesinger, David Larbalestier, Alex Malozemoff, Vic Maroni, Bart Riley and Marty Rupich) although numerous other outstanding individuals played key roles over the years. This group was recognized around the world for driving the 2223 material science and was a major contributor to the advances achieved at AMSC. During this time, the performance of 2223 wire produced at AMSC increased at a rate commonly referred to as the "Malozemoff Law" shown in Figure 11.38.

FIGURE 11.37: Photograph of the Wire Development Group in 1993. From left to right: Vic Maroni, Dingan Yu, Alex Otto, David Larbalestier, Yibing Huang, Don Kroger, Qi Li, Bart Riley, Eric Hellstrom, Dominic Lee, Terry Holesinger, Jeff Willis, Marty Rupich and Ron Parrella.

As the performance of 2223 wire improved throughout the 1990s, both AMSC and SEI began producing longer-than-kilometer lengths of robust, high performance wire. These 1G wires were successfully used in multiple prototype demonstrations validating the HTS technology for applications in the electric power grid. However, in the early 2000s, AMSC and most other organizations switched development and manufacturing from 1G to 2G wires. One reason for this switch was concern about the projected 1G wire cost stemming from the cost of silver, the inefficiency of batch manufacturing and the lower J_c of 2223 wire compared to that of biaxially aligned YBCO wire. However, SEI continued to develop 1G wire and now leads the world in performance and manufacturing capacity.

When I (Rupich) began working at AMSC in the early 1990s, there was tremendous excitement within the company following a significant advance in performance resulting from a new processing route referred to internally as the "magic heat treatment". The "magic heat treatment", which originated from a furnace controller that was misprogrammed late one Friday evening, provided the first understanding that residual liquid affected the grain connectivity and second-phase formation during the cool-down step. Over the next decade, a detailed study of the 2223 chemistry and materials properties led to major improvements in the properties of 2223 wire. A few key aspects responsible for the successful development of the 2223 technology at AMSC and SEI are discussed below.

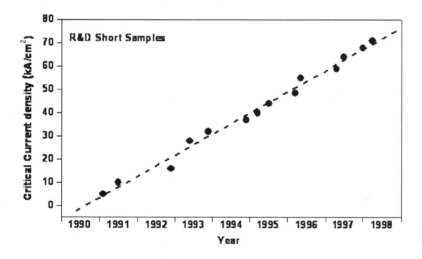

FIGURE 11.38: Time line, referred to as the "Malozemoff Law," showing the increase in J_c of 2223 wires produced at AMSC during the 1990s. Measurements made at 77 K, self-field. (Courtesy American Superconductor Corp.)

Precursor Powder

Early in the development of the 2223 process, it was realized that controlling the properties of the precursor powder was critical to the 1G manufacturing process and that any changes in the particle size distribution, composition or starting phase content of the powder affected the later deformation and thermal treatment steps. A broad range of routes for preparing the precursor powder with various compositions and phase contents was investigated. A first basic insight was that a starting powder consisting of the superconducting 2223 phase could not be used since 2223 grains could not be effectively sintered together, limiting their ability to carry significant current over length. Instead, the strategy was to start with a mixture of oxide precursor powders that could be reacted

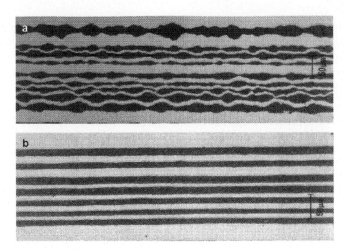

FIGURE 11.39: Longitudinal cross-sectional images showing 2223 precursor filament structure after (a) multi-pass rolling (MPR) deformation and (b) single-pass rolling (SPR) deformation. (Courtesy American Superconductor Corp.)

to the superconducting 2223 phase once the wire was in near-final form, thus establishing electrical connectivity over length. AMSC settled on a precursor powder with an overall composition of $Bi_{1.8}Pb_{0.34}Sr_{1.9}Ca_{2.1}Cu_{3.04}O_x$, prepared from nitrate salts using either a spray drying or aerosol process. The starting phase content of the precursor powder consists primarily of 2212, Ca_2PbO_4, CuO and $(Sr,Ca)_{14}Cu_{24}O_x$. As mentioned earlier, doping the 2223 phase with Pb is critical to enhancing the formation and stability of the 2223 phase. In addition, it was realized that both the Pb content and the specific Pb phases present throughout the thermal-mechanical processing are critical to controlling the formation of the highly-textured 2223 phase.

Contaminants, especially carbonaceous materials that oxidize to CO_2 during the thermal processing, are especially deleterious because they lead to the formation of balloon-type defects in the composite wire that can destroy the filament structure or even rupture the Ag sheath.

Deformation Processing

Texturing of the 2223 precursor powder occurs during the mechanical deformation, which transforms the round composite wire into a tape (step 5 in Figure 11.36). The easy slip plane of the 2212 component in the precursor powder is critical to improve deformability. The conventional approach to deform the composite, based on modeling studies, used a multi-pass rolling procedure with small reductions during each pass[166]. However, multi-pass deformation of the ceramic-Ag composite results in an irregular interface between the Ag and the precursor filaments producing what is commonly called a "sausaged" cross-section, as shown in Figure 11.39a. Contrary to initial expectations, work at AMSC found that just the opposite deformation procedure is needed to prevent sausaging. The best approach is to carry out the deformation from the round wire to tape in a single step in which the total reduction is greater than 85%[167,168,169]. This single-pass reduction (SPR) was a major breakthrough that produced the very uniform Ag/filament interface shown in Figure 11.39b.

---

[166] D.A. Korzekwa, J.F. Bingert, E.J. Podtburg, and P. Miles, *App. Supercond.* 2 (1994) 261

[167] Q. Li, K. Brodersen, H.A. Hjuler, and T. Freltoft, *Physica* C 217 (1993) 360

[168] Q. Li, E.R. Podtburg, P.J. Walsh, W.L. Carter, G.N. Riley, Jr., M.W. Rupich, E. Thompson, and A. Otto, U.S. Patent 6, 247, 224 (2001)

[169] H. Mukai, EP Patent 435286 (1997)

A second key advance was the realization that the phase content of the precursor powder not only affects the chemical conversion process but also the deformation process. Studies of the deformation process by the WDG found that wire drawing is best done with the starting precursor phase content given above, but that the SPR deformation is much more effective in aligning the 2212 component when the Pb is in the 2212 phase. Thus a key process step is to heat the round composite in a low pO_2 atmosphere to reduce the Ca_2PbO_4 and form the Pb-doped 2212 phase in the precursor powder just prior to the SPR process. This thermal process, combined with SPR, results in significant enhancement of both the through-thickness filament density and c-axis texturing of the 2212 phase, compared to that obtained from Pb-free 2212 and Ca_2PbO_4 phases deformed by a multi-pass rolling (MPR) process.

Thermal Treatment

Another crucial advance in the development of the 2223 process was careful optimization of the heat treatment and phase content of the precursor after the SPR process. Studies had shown that maintaining Pb in the 2^+ oxidation state of the Pb-doped 2212 enabled better control of the liquid phase and subsequent 2223 phase formation[170]. However, since thermal processing of the wire is carried out in an atmosphere with ~7.5% O_2, the Pb^{2+} oxidizes to Pb^{4+}, and Ca_2PbO_4 reforms during the slow temperature ramp to the processing temperature[171]. The simple solution to prevent re-oxidation of the Pb^{2+} to Pb^{4+} is to use a fast ramp rate through the temperature range where oxidation occurs[172].

At the end of the final heat treatment, controlling the cool-down rate is critical to controlling the residual liquid and preventing the formation of 2212 intergrowths and Pb-rich phases that disrupt the current flow between 2223 grains. Understanding and optimizing this process, which evolved from the previously described "magic heat treatment," was a major focus of the WDG throughout the 1990s.

These process innovations resulted in long-length 2223 wires with a critical current of ~150 A in a wire cross-section of 4.3×0.2 mm^2. Examination of the microstructure showed that the supercurrent in these wires was still limited by poor alignment of the 2223 grains, cracks, porosity, 2212 intergrowths and numerous non-superconducting phases in the interior of the filaments[173]. Most of these defects, particularly the porosity and poor grain alignment, were a consequence of the de-densification that occurred during the thermal-mechanical process.

Efforts within the WDG to improve the texture and density of the 2223 core focused on carrying out the second heat treatment at pressures above 0.1 MPa, called overpressure (OP) processing, in an effort to prevent de-densification of the 2223 filament cores[174,175,176]. In another major breakthrough, short research and development samples processed at pressures around 15 MPa showed higher density in the 2223 filament cores, fewer 2212 intergrowths, fewer cracks and higher J_c compared to samples processed at 0.1 MPa.

In spite of the technical success of the 1G wire development, AMSC, for reasons mentioned

[170] J. C. Grivel, A. Jeremie, B. Hensel, and R. Flukiger, *Supercond. Sci. Tech.* 6 (1993) 725

[171] M. W. Rupich, W. L. Carter, Q. Li, A. Otto, and G. Riley, Jr., U.S. Patent 6,295,716 (2001)

[172] Q. Li, E. R. Podtburg, P. J. Walsh, W. L. Carter, G. N. Riley, Jr., M. W. Rupich, E. Thompson, and A. Otto, U.S. Patent 6, 311, 386 (2001)

[173] J. Jiang, X.Y. Cai, A.A. Polyanskii, L.A. Schwartzkopf, D.C. Larbalestier, R.D. Parrella, Q. Li, M.W. Rupich, and G.N. Riley, Jr., *Supercond. Sci. Tech.* 14 (2001) 548

[174] Y. Yuan, R.K. Williams, J. Jiang, D.C. Larbalestier, X.Y. Cai, M.O. Rikel, K.L. DeMoranville, Y. Huang, Q. Li, E. Thompson, G.N. Riley, Jr., and E.E. Hellstrom, *Physica* C 372-376 (2002) 883

[175] M.O. Rikel, R.K. Williams, X.Y. Cai, A.A. Polyanskii, J. Jiang, D. Wesolowski, E.E. Hellstrom, D.C. Larbalestier, K. DeMoranville, and G.N. Riley, Jr., *IEEE Trans. App. Supercond.* 11 (2001) 3026

[176] Y. Yuan, J. Jiang, X.Y. Cai, S. Patnaik, A.A. Polyanskii, E.E. Hellstrom, D.C. Larbalestier, R.K. Williams, and Y. Huang, *IEEE Trans. App. Supercond.* 13 (2003) 2921

earlier, ended its 2223 wire effort in the early 2000s and shifted its development efforts to 2G YBCO wire as described in Section 11.5.

Starting in the early 2000s, SEI began using the OP process to manufacture 2223 wire[177,178]. The SEI process, called ConTrolled Over-Pressure (CT-OP^{TM}), utilizes high pressure (~30 MPa) to densify the 2223 filaments during the final heat treatment. The development of a full-scale, high-pressure oxidation furnace for processing long lengths of wire is a major technical achievement. In addition, careful control of the 2223 formation results in a significant reduction of non-superconducting phases, particularly alkaline earth cuprates, leading to higher density and better alignment of the 2223 grains (as seen in Figure 11.40) and ultimately higher J_c. The CT-OP^{TM} process continues to improve the performance of both research and development scale and production length 2223 wires, as seen in Figure 11.41. The process now yields the highest performance 2223 wire ever produced.

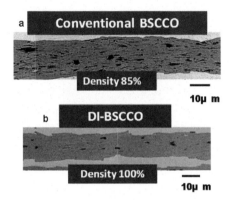

FIGURE 11.40: Cross-sectional micrographs showing the density and grain alignment of 2223 filaments prepared using (a) the conventional OPIT process at 1 bar total pressure and (b) the over-pressure (CT-OP) process. (Courtesy Sumitomo Electric Industries.)

2223 Wire Status in 2011

After nearly two decades of development and a series of remarkable technical breakthroughs, 2223 wire is now manufactured in km lengths with critical currents in 4.3 mm wide wires approaching 200 A at 77 K, self-field. The 1G wire has been used successfully in numerous demonstration projects including AC transmission and distribution cables, motors, generators and transformers. 1G-based HTS cable projects have been successfully operated in the US, China, Korea, Mexico, Russia, and Spain. This includes the world's first permanently operating transmission cable installed in the commercial power grid at the Holbrook substation on Long Island, New York. The 600-meter long, 138 kV cable was energized in April 2008. (See Hassenzahl and Tsukamoto in Chapter 12 section 5.)

The performance of today's 1G wire, over long length, still exceeds that of the best 2G wire. As the cumulative knowledge of 2223 grows, it is anticipated that the critical current will continue to increase. The real challenge to the ultimate commercial success of 1G wire is still overcoming

[177] S. Kobayashi, K. Yamazaki, T. Kato, K. Ohkura, E. Ueno, K. Fujino, J. Fujikami, N. Ayai, M. Kikuchi, K. Hayashi, K. Sato, and R. Hata, *Physica* C 426-431 (2005) 1132

[178] N. Ayai, S. Kobayashi, M. Kikuchi, T. Ishida, J. Fujikami, K. Yamazaki, S. Yamade, K. Tatamidani, K. Hayashi, K. Sato, H. Kitaguchi, H. Kumakura, K. Osamura, J. Shimoyama, H. Kamijyo, and Y. Fukumoto, *Physica* C 468 (2008) 1747

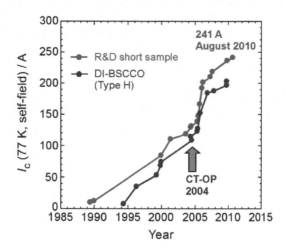

FIGURE 11.41: Time line showing the increase in the critical current of 2223 composite wires produced at Sumitomo Electric Industries. DI-BSCCO Type H is their high-critical-current density wire. (Courtesy Sumitomo Electric Industries.)

the cost disadvantages of batch processing and the large-volume fraction of silver required for the composite architecture, compared to the rapidly emerging 2G wire technology.

11.4.3   2212 Conductors

The development of 2212 conductors has been distinctly different than 2223 conductors for several reasons. A prime reason is that 2212 can usefully operate only up to ~ 20 K, while 2223 can operate in self-field at 77 K, making it clear that large, new electric utility markets could not be addressed by 2212. This created a very different dynamic in 2212 development compared to 2223 because there was less commercial pressure to develop proprietary understanding. The complex processing of both materials and the fact that the thermal processing for 2212 is significantly different than that for 2223 tended to separate the respective efforts. More recently development of 2212 to extend the 4.2 K high-field domain of superconducting magnets has focused on a high J_c conductor that can be made as a round wire and then reacted to its high J_c condition with one heat treatment *after* winding into a coil. By contrast, 2223 needs two heat treatments and must be wound when it is in its final, somewhat fragile, textured, flat-tape form. The following part of this review traces the development of 2212 from a historical perspective, starting with the initial discovery that melt processing 2212 leads to high J_c, progressing through the various wire architectures that have been studied over the years to maximize J_c, and ending with remaining challenges to develop viable 2212 conductors for very-high-field magnets.

Heine et al.[156] fabricated the first 2212 wire in 1989. Their OPIT, monocore, round wire was not the first HTS wire but, as Figure 11.42 shows, it was the first that carried significant supercurrent, attaining a $J_c = 5.5 \times 10^2$ A/mm^2 at 4.2 K, SF, which only decreased by about a factor of 4 to 1.5×10^2 A/mm^2 at 4.2 K, 26 T. Never before had a superconducting material carried such large supercurrent density in such a high magnetic field. This performance inspired magnet builders to begin thinking about developing all-superconducting, very-high-field magnets, but in fact the difficulty of developing a reliable and high J_c conductor kept programs small for many years. The first HTS coil to generate 24 T, which is beyond the capabilities of Nb$_3$Sn conductors, was made with

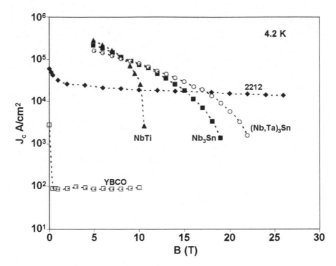

FIGURE 11.42: J_c as a function of applied field for Ag-sheathed 2212 and Ag-sheathed YBCO wires. This figure is redrawn from Heine et al.[156] so it shows data for commercial NbTi, Nb$_3$Sn, and (Nb,Ta)$_3$Sn multifilamentary wires Vacuumschmelze produced in 1989.

2223[179]. In the past few years, tests at the National High Magnetic Field Laboratory have shown HTS coils can generate over 30 T with a 2212 coil producing 1.1 T in a background field of 31 T for a total of 32.1 T and a YBCO coil generating 2.8 T for a total of 33.8 T[180]. The US DOE, Office of High Energy Physics has recently renewed interest in 2212 for future accelerator magnets in the 30–50 T range. The 2212 conductor is of particular interest since it is the only HTS material that can be made as a round wire, the form that magnet builders prefer for winding both small and large magnets, especially because of its ability to be made into Rutherford cables.

Melt Processing of 2212

The heat treatment schedule for 2212 conductors, shown in Figure 11.34, consists of a single heat treatment in which the discrete 2212 phase particles are melted at 880–900 °C and connected grains of the 2212 phase form during cooling. This development of a high connectivity is vital to achieving high J_c since solid-state sintering does not by itself develop a well-connected structure as was clearly recognized very early by Tenbrink et al.[181] They showed that the J_c of sintered 2212 that had not melted was about 100 times lower than melt-processed conductor. In contrast to 2212, 2223 cannot be melt-processed, since cooling from the melt produces 2212 rather than 2223; however, reaction sintering of 2223 allows a connected 2223 grain structure to develop. Melt processing is also key to understanding why 2212 can form round wire. As shown by Reaction 11.1 the 2212 phase melts incongruently forming a mixture of liquid and crystalline phases. Again, in contrast to 2223, the starting 2212 precursor powder usually consists of the 2212 phase.

$$2212 \; precursor \; powder \xrightarrow{heating} liquid + AEC + CF + O_2 \xrightarrow{cooling} 2212 \qquad (11.1)$$

In Reaction 11.1 AEC is an alkaline earth cuprate phase, CF is a copper-free phase, and O_2 is oxygen that is released from 2212 during the melting process. Oxygen is released on melting and

[179] K. Ohkura, K. Sato, M. Ueyama, J. Fujikami, and Y. Iwasa, *App. Phys. Lett.* 67 (1995) 1923

[180] H.W. Weijers, U.P. Trociewitz, W.D. Markiewicz, J. Jiang, D. Myers, E.E. Hellstrom, A. Xu, J. Jaroszynski, P. Noyes, Y. Viouchkov, and D.C. Larbalestier, *IEEE Trans. App. Supercond.* 20 (2010) 576

[181] J. Tenbrink, M. Wilhelm, K. Heine, and H. Krauth, *IEEE Trans. Mag.* 27 (1991) 1239

must be taken up by the 2212 phase as it reforms from the melt on cooling. This makes the oxygen permeability of the Ag sheath critical for forming 2212. The composition of the AEC and CF phases varies with the ambient oxygen partial pressure[182] but melt processing is now almost universally done in 0.1 MPal oxygen, where AEC = $(Sr,Ca)_{14}Cu_{24}O_x$ and CF = $Bi_9(Sr,Ca)_{16}O_x$. The melting temperature of 2212 depends on the oxygen partial pressure; thus isothermal processing methods, in which the 2212 is melted under low oxygen pressure and crystallized at higher oxygen pressure, are also possible[183].

On paper, Reaction 11.1 is simple; however, no one has yet succeeded in driving it to completion. Thus the final conductor is always contaminated with AEC, CF and 2201 ($Bi_2Sr_2CuO_6$) phases that can disrupt grain texture and grain connectivity, limiting J_c. In fact there are so many variables that full understanding of how best to optimize the process to make a high-J_c 2212 conductor has not yet been achieved.

2212 Conductor Architectures

The basic process used today for fabricating 2212 round wire is still similar to that used by Heine et al. [156] in 1989. Wires are made by the OPIT technique shown in Figure 11.36. 2212 require only one heat treatment, so its processing ends after step 6 in Figure 11.36.

Between the first work on monocore 2212 round wire[156] and today's multifilamentary 2212 round wire, much work was done on flat films that are a few tens of μm thick, and flat-tape conductors, since J_c in the first round monocore wire[156] and flat tape[184] were 1.6×10^2 A/mm^2 and 4.5×10^2 A/mm^2 at 4.2 K, 20 T, respectively. As Figure 11.33 shows, the 2212 phase is a two-dimensional superconductor that, like 2223 and YBCO, requires at least c-axis texture to carry high supercurrent across grain boundaries. Early on it was discovered that crystallizing 2212 from a melt produced a high J_c material[156,184] because it forms highly aspected lath-like grains. This 2D morphology suggested that processing 2212 on flat substrates would promote alignment of the grains on cooling. This led to 2212 films on Ag foil and flat, Ag-sheathed, 2212 tape.

Publication Date Determines Group Credited with First 2212 Wire

An interesting historical tidbit discovered while reviewing the 2212 literature is that although Heine et al.[156] are credited with having made the first 2212 wire, Enomoto et al.[184] submitted a manuscript on melt-processed flat, Ag-sheathed tape on August 21, 1989, eight days before the manuscript by Heine et al. was received on August 29, 1989. However, the Heine manuscript was accepted for publication on October 2, 1989, four months before the Enomoto manuscript (Febuary 8, 1990). Thus Heine et al. are credited with making the first 2212 wire. This historical detail shows how the idiosyncrasies of the review and publication process can affect the later historical perceptions of "discovery" when there is a very rapid and competitive science and applications community in place.

Compared to Ag-sheathed 2212 conductors 2212 films with high J_c are relatively easy to fabricate. Long-length 2212 films are made by a coating slurry of 2212 precursor powder suspended in

---

[182] W. Zhang, and E.E. Hellstrom, *Supercond. Sci. Tech.* 8 (1995) 430

[183] T.G. Holesinger, J.M. Johnson, J.Y. Coulter, H. Safar, D.S. Phillips, J.F. Bingert, B.L. Bingham, M.P. Maley, J.L. Smith, and D.E. Peterson, *Physica* C 253 (1995) 182

[184] N. Enomoto, H. Kikuchi, N. Uno, H. Kumakura, K. Togano, and K. Watanabe, *Jap. J. App. Phys.* 29 (1990) 447

a mixture of organic solvents onto a thin Ag foil. The slurry layer must be thick enough to form a 10-30 μm thick 2212 film. After coating, the organics are carefully burned out in a separate step before the heat treatment shown in Figure 11.34. Wei Zhang, a post doc who worked in Hellstrom's group, within a few weeks after joining the group fabricated doctor-bladed 2212 films that had high J_c . He then spent several years working to achieve comparable J_c values in Ag-sheathed 2212 tape. To date 2212 films made by the PAIR process, which is discussed below, have the highest J_c yet attained and it is not yet understood why Ag-sheathed 2212 conductors have lower J_c.

Researchers at NRIM (National Research Institute for Metals, now the National Institute for Materials Science, in Tsukuba, Japan), carried out extensive studies on 2212 films and tape[185]. Their goal was to make 2212 pancake coil magnets. Although they successfully developed a process to heat treat 2212 films for high J_c, they had recurring difficulties heat treating long-length 2212 tapes. Their process used thin, wide Ag foil—in one study it was 80 μm thick, 3 cm wide, and 3.6 m long[186]. After coating the Ag foil with slurry, it was wound in a loose spiral so adjacent layers in the coil did not touch and placed on its side in a box furnace for the heat treatment. During the heat treatment, such tapes often buckled because of the low mechanical properties of the thin Ag foil at the high processing temperature.

The film technique was improved using the PAIR (preanneal, intermediate rolling) process[187]. The initial PAIR work was done with a single layer of dip-coated foil in which a preanneal (PA) step was used to burn out the organics just below the melting point of 2212. An intermediate rolling (IR) was done after the preanneal, reducing the film thickness by ~20–30%, and densifying the precursor powder. The foil was then melt processed to produce a dense, well-aligned, high J_c tape. NRIM and Showa Electric Wire and Cable Co. quickly modified this single-layer PAIR process to produce a stack of PAIR-processed 2212 films encased in a Ag sheath[188]. In this case, the preanneal step was done on the individual foils to insure complete removal of residual carbon from the organics before the stack of 2212 foils was wrapped in a Ag foil and rolled. It was critical to remove all carbon from the precursor mixture so CO_2 did not form during melt processing, leading to the ballooning described earlier for 2223. Complete carbon removal and densification were key to achieving a J_c of 5×10^3 A/mm^2 at 4.2 K and 10 T[189], which is the highest J_c achieved to date for a 2212 conductor.

Ag-sheathed 2212 tapes, made by rolling round OPIT wires into tapes (see Figure 11.36), were also developed. Multifilamentary 2212 tapes were made by restacking round 2212 wire, drawing and rolling. The tape architecture imparts a 2D geometry on the filaments, promoting 2212 grain alignment during cooling. However, J_c in Ag-sheathed tapes was usually lower than in films.

J_c in the 2D tapes and films is strongly anisotropic, as shown by Enomoto et al.[184], meaning that J_c varies with the angle between the applied magnetic field and the tape plane. As with 2223 and YBCO tapes, J_c is highest with the magnetic field parallel to the ab-plane.

To minimize this anisotropy and fabricate the 2212 wire in the preferred round wire geometry, Hitachi developed a ROSAT (Rotation-Symmetric Arranged Tape-in-Tube) architecture[190]. ROSAT wires shown in Figure 11.43 were built by stacking monocore or multifilamentary 2212 tapes in a radially symmetric pattern. This assembly of 2212 filaments was drawn to a small diameter and heat treated. The radial symmetry of the 2212 tapes in ROSAT wire reduced the J_c anisotropy to

[185] J. Kase, N. Irisawa, T. Morimoto, K. Togano, H. Kumakura, D.R. Dietderich, and H. Maeda, *App. Phys. Lett.* 56 (1990) 970

[186] N. Tomita, M. Arai, E. Yanagisawa, T. Morimoto, H. Fujii, H. Kitaguchi, H. Kumakura, K. Inoue, K. Togano, H. Maeda, and K. Nomura, *App. Phys. Lett.* 65 (1994) 898

[187] H. Miao, H. Kitaguchi, H. Kumakura, and K. Togano, *Cryogenics* 38 (1998) 257

[188] T. Hasegawa, T. Koizumi, Y. Aoki, H. Kitaguchi, H. Miao, H. Kumakura, and K. Togano, *IEEE Trans. App. Supercond.* 9 (1999) 1884

[189] H. Miao, H. Kitaguchi, H. Kumakura, K. Togano, T. Hasegawa, and T. Koizumi, *Physica* C 303 (1998) 81

[190] M. Okada, K. Tanaka, T. Wakuda, K. Ohata, J. Sato, H. Kumakura, T. Kiyoshi, H. Kitaguchi, K. Togano, and H. Wada, *IEEE Trans. App. Supercond.* 9 (1999) 1904

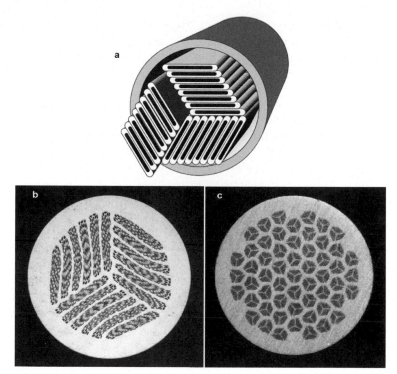

FIGURE 11.43: 2212 ROSAT wire[191]. (a) A schematic diagram showing how Ag-sheathed monocore or multifilamentary tapes are packed to make ROSAT wire. (b) A 1.62 mm diameter wire made from multifilamentary flat tape. (c) A 1.89 mm diameter wire made by double restacking flat, monocore tape, like that in (a).

less than 5% and J_c was within $\pm 10\%$ of J_c (800 A/mm^2, 4.2 K, 10 T) in the long 2212 tapes used for magnets[191]. We might wonder why Hitachi developed this complicated technology rather than pursuing round 2212 wire with round filaments. The answer is that they knew that tapes had higher J_c than round wire, thus using flat tapes to build the round ROSAT wire was a reasonable approach to maximize J_c and eliminate J_c anisotropy in the preferred wire geometry.

The 2212 was initially heat treated in air. Endo et al.[192] at Sumitomo Metals were first to report heat treating 2212 in pure oxygen, which increased J_c by a factor of roughly 16 giving 1.2×10^3 A/mm^2 (4.2 K, 8 T) in a flat tape. Heat treating in pure oxygen is almost universally practiced now. But it was not a direct path from heat treating in air to pure oxygen. Just after their work was published, Hellstrom visited these researchers at Sumitomo Metals in Japan and found they had switched to heat treating in 0.1% oxygen[193] in spite of the high J_c achieved processing in pure oxygen. They had run into a technical problem heat treating in pure oxygen when going from their original 10 cm long samples to 20 cm and longer samples: the longer samples ballooned, similar to the ballooning in 2223 described above. They thought the oxygen released during melting (see Reaction 11.1) caused the ballooning and found experimentally that reducing the oxygen content prevented ballooning. They believed this was due to reduced oxygen release on melting, but it turned out that the ballooning was not caused by oxygen release but rather by residual carbon in the 2212 precursor powder. Reducing the C content in the 2212 precursor powder is crucial to prevent

[191] K. Ohata, J. Sato, H. Okada, K. Tanaka, H. Kumakura, T. Kiyoshi, K. Togano, H. Wada, *Hitachi Cable Rev.* 18 (1999) 81

[192] A. Endo and S. Nishikida, *IEEE Trans. App. Supercond.* 3 (1993) 931

[193] M. Yoshida, and A. Endo, *Jap. J. App. Phys.* 32 (1993) L1509

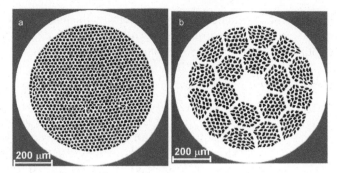

FIGURE 11.44: 2212 round wire. (a) 1025 filament single restack made by Supercon, Inc. (b) 37×18 double restack made by Oxford Superconducting Technologies.

ballooning[194].

Motowidlo et al.[195] were first to fabricate and heat treat double-restack round wire. Figure 11.44 shows the cross-section architecture of single and double-restack round wire. Their work was also done in pure oxygen and they found that heat treating in oxygen produced a more complex filament structure than in air, which was characterized by significant bridging of 2212 filaments across the Ag matrix[196]. Only in the last two years, through detailed quench studies, has the formation of the complicated round-wire microstructure been understood[197].

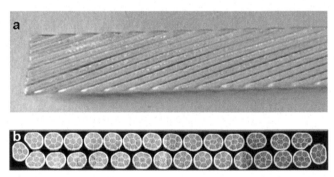

FIGURE 11.45: 2212 Rutherford cable. (b) Cross section of a Rutherford cable made without a center core before heat treatment. The individual wires in (a) and (b) are 0.8 mm in diameter. (Courtesy A. Godeke, LANL.)

Ron Scanlan had eyed the possibility of using round wire 2212 for Rutherford cables for many years and made extensive studies of them. In an early summary of their work reported in 1998[198], various architectures and metallic cores were explored in order to cut down the face-face coupling currents[199]. The study showed that 2212 round wire survived the cabling process and could be

[194] W. Zhang and E.E. Hellstrom, *Physica* C 234 (1994) 137

[195] L.R. Motowidlo, G. Galinski, G. Ozeryansky, W. Zhang, and E. E. Hellstrom, *App. Phys. Lett.* 65 (1994) 2731

[196] L. R. Motowidlo, G. Galinski, G. Ozeryansky, W. Zhang, E. E. Hellstrom, M. Sumption, and T. Collings, *IEEE Trans. App. Supercond.* 5 (1995) 1162

[197] T. Shen, J. Jiang, F. Kametani, U.P. Trociewitz, D.C. Larbalestier, J. Schwartz, and E.E. Hellstrom, *Supercond. Sci. Tech.* 23 (2010) 025009

[198] M.D. Sumption, R.M. Scanlan, and E.W. Collings, *Physica* C 310 (1998) 291

[199] E. W. Collings, M. D. Sumption, R. M. Scanlan, D. R. Dietderich, L. R. Motowidlo, R. S. Sokolowski, Y. Aoki, and T.

successfully processed, but that the metallic cores could react with the filaments, degrading the J_c of the 2212. Many Rutherford cables are now made without a center core[200] as shown in Figure 11.45, but the need for cores both for strength and to reduce coupling losses remains.

The Microstructure of 2212 Conductors

The texture evolution and grain alignment in melt-processed 2212 conductors has been investigated through quench studies with both 2212 tapes[201] and round wires[197]. Studies have shown that 2212 grains grow fastest along the *b*-axis, slightly slower along the *a*-axis, and very slowly in the *c*-axis direction. In films and tapes, this leads to a 2D grain structure with the *c*-axis aligned perpendicular to the plane of the tape and the *a*- and *b*-axes randomly oriented in the tape plane, similar to that in 2223 wire, as discussed above.

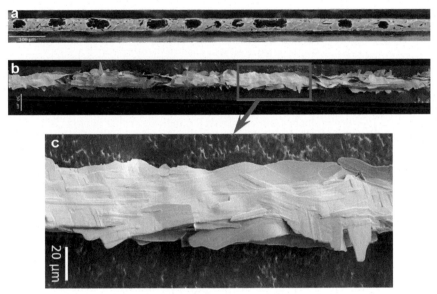

FIGURE 11.46: Sections from individual 2212 filaments extracted from a wire with widely separated filaments made by Oxford Superconducting Technologies. (a) Filament from wire quenched from the melt state before 2212 began to form. The elongated black regions are bubbles formed from agglomerated pores. They have almost the same diameter as the filament. (b) Filament from a fully processed wire. There are local regions along the length of the fully processed filament, such as the one shown in the boxed region in (b), that are highly aligned. These regions are randomly rotated along the length of the filament. (c) Detail of a local, highly-aligned region showing the lath-like structure of the individual 2212 grains. The 2212 *b*-axis is parallel to the long axis of the lath-like grains. (Courtesy F. Kametani Applied Superconductivity Center, FSU.)

In round wire, 2212 grains grow with their *b*-axes aligned parallel to the wire axis and their *c*-axes randomly oriented in the radial direction[202]. The SEM image (Figure 11.46) of a fully-processed 2212 filament extracted from a round wire shows localized regions comprised of well-aligned stacks

---

Hasegawa, *Supercond. Sci. Tech.* 12 (1999) 87

[200] T. Hasegawa, J. Nishioka, N. Ohtani, Y. Hikichi, R. Scanlan, R. Gupta, N. Hirano, and S. Nagaya, *IEEE Trans. App. Supercond.* 14 (2004) 1066

[201] R.D. Ray, II, and E.E. Hellstrom, *Physica* C 175 (1991) 255

[202] F. Kametani, Presented at the Applied Superconductivity Conference, Washington D.C., Aug. 2010

of 2212 platelets. These local regions consist of 2212 grains with the fast-growing *b*-axes aligned with the wire axis and the *c*-axes aligned parallel with each other. These locally aligned regions repeat along the filament length with the *c*-axis orientation of each region randomly rotated around the filament axis. The averaged orientation of these randomly-rotated, locally aligned regions over the several hundred filaments in a wire results in an isotropic J_c behavior in round wire.

The Future of 2212

There are still many questions that must be addressed before 2212 round wire can be considered a technically viable conductor.

First, the filament J_c and the overall conductor J_E[203] of 2212 round wire are still lower than desired for high-field applications, particularly in long-length wires used in coils. Godeke et al.[204], who are interested in dipole magnets for future accelerators, have proposed that the overall current density (J_E) for 2212 wire should be ~600 A/mm^2 at 4.2 K in 20 T, which is about twice that of the best Nb$_3$Sn wire. Currently J_E in the best long samples of 2212 is 200-300 A/mm^2 (4.2 K, 20 T), while in short (a few cm long) samples it is ~400-500 A/mm^2. The constraints on insert solenoid coils for high-field magnets are less severe. In any case, the clearest way to higher J_E is to increase the 2212 grain connectivity, a conclusion already reached by Tenbrink et al. in 1991[181].

Recent attempts to define specific defects in the microstructure responsible for degrading the filament connectivity have focused on the bubbles that consume a large volume fraction of each filament[205]. Studies of wires quenched directly from the melt show that bubbles form as soon as the 2212 melts. These bubbles do not shrink after formation and, although partially bridged by 2212 grains, they occupy as much as 40 vol% of the filament and are major obstacles to current flow even in fully-processed wire[202,205]. Understanding and reducing these bubbles is an active area of 2212 technology development and many of the hopes for significant J_c improvement depend on the outcome of techniques for controlling them.

Optimizing the 2212 composition is also thought to be important to increase J_c. The current 2212 powder used in round wire was developed in a compositional study by Nexans and Oxford Superconductor Technologies[206]. The current composition of Bi$_{2.14}$Sr$_{1.66}$Ca$_{1.24}$Cu$_{1.96}$ results in J_cs up to four times higher than compositions with lower Sr/Ca ratios. Future studies may allow better control of the 2212 grain growth and minimize non-superconducting phases leading to better grain connectivity and higher J_c. Alternatively a better understanding of how to drive Reaction 11.1 towards completion would avoid current blocking by non-superconducting phases in each filament. Numerous efforts to increase J_c in 2212 round wire for high-field magnet use are ongoing and we are optimistic the J_E of round 2212 wire will show the steady and significant progress needed to make 2212 round wire the preferred conductor for high-field magnets.

11.4.4   Conclusions

The development of 2212 and 2223 wires has been a tour-de-force, bringing together deep insights into their complex chemistry and composite deformation properties enabled by clever engineering, to achieve long-length, robust and high-quality wires. Development continues, driving ever higher performance of 2223 for electric power applications and of 2212 round wire for high-field

---

[203] J_c is the critical current I_c divided by the area of the superconductor; J_E (engineering critical current density) is I_c divided by the total area of the conductor.

[204] A. Godeke, D. Cheng, D.R. Dietderich, P. Ferracin, S.O. Prestemon, G. Sabbi, and R.M. Scanlan, *IEEE Trans. App. Supercond.* 17 (2007) 1149. To find J_E needed for the next generation of magnets, divide the non-Cu current density for the oval in Figure 11.33 in the reference by a factor of 2. This gives J_E = 600 A/mm^2 at 4.2 K, 20 T.

[205] F. Kametani et al., subm. to *Super. Sci. and Tech.* March 2011.

[206] M.O. Rikel, S. Arsac, E. Soileux, J. Ehrenberg, J. Bock, K. Marken, H. Miao, C.E. Bruzek, S. Pavard, A. Matsumoto, E.E. Hellstrom, and L. Motowidlo, *J. Phys.: Conf. Ser.* 43 (2006) 51

magnet applications. While competition from the newer 2G wires is increasing, BSCCO HTS wire is likely to find important, if specialized, commercial roles in the future. They illustrate the best of what can be achieved when scientists and engineers pool their insights and work closely together to create a remarkable new technology of great future value for society.

11.5 Coated Conductor: Second Generation HTS Wire

A. P. Malozemoff and Y. Yamada

Second generation (2G) high temperature superconductor (HTS) wire, also called coated conductor, is rapidly becoming the most promising type of HTS wire for bulk power and magnet applications. It is comprised of a flexible textured tape-shaped template on which is deposited an epitaxial film of the rare-earth/barium/copper/oxide $REBa_2Cu_3O_7$ ("REBCO"), along with a surrounding normal metal stabilizer. Its development is a triumph of scientific insight, sophisticated processing and determined scale-up efforts over almost twenty years, and it is now being manufactured commercially and coming into use in a variety of applications. In this brief review, we summarize the basic concept of 2G HTS wire, describe its many alternative processing routes and their status, and briefly mention some of the 2G-wire-based prototypes which have already been demonstrated.

The basic concept of 2G wire originates from an innovative solution to the fundamental problem in achieving high superconductor current density in long HTS wire. The problem stems from grain boundaries which act as weak links in these materials, drastically limiting current flow. This phenomenon is described in more detail in a companion article by Mannhart and Dimos[207]. The grain boundary problem arises from a fundamental property of HTS material, namely its short coherence length, which, is given by $\xi = a\hbar v_F/kT_c$ (where a is a constant of order unity, $\hbar$ is Planck's constant divided by 2π, v_F is the Fermi velocity, k is Boltzmann's constant and T_c is the superconductor transition temperature). Because ξ depends inversely on T_c, and because T_c is high, ξ is very small, on an Angstrom scale. This has important consequences for current flow in HTS materials: Any disturbance of lattice periodicity can severely perturb the superconductor gap and create an obstacle to current flow[208]. In particular, grain boundaries, being planar defects, form the dominant obstacle to current flow once physical defects like cracks and secondary phases are eliminated. And indeed, polycrystalline high temperature superconductors, filled with grain boundaries, have very low current densities, while single crystal films of $YBa_2Cu_3O_7$ achieve MA/cm^2 levels even at 77 K. So to achieve a km-length wire, it appeared necessary to produce a km-length single crystal, a challenge which seemed for many years beyond the realm of possibility.

The first hint to solve this grain boundary problem came from the ground-breaking work of Dimos, Chaudhari and Mannhart[209], who measured the critical current of thin YBCO films deposited epitaxially on bicrystals with controlled grain boundary misorientation. As shown in Figure 3.44 of Section 3.6 by Dimos and Mannhart, they found that critical current density J_c across the grain boundary falls off exponentially with increasing grain boundary misorientation angle. Only at very small angles, of order 4 degrees or less, does the intergrain J_c flatten out and approach the intragranular, crystalline value of J_c. This can be understood from the fact that at small angles, the grain boundary separates into localized regions of coherent lattice between dislocations.

Dimos *et al.*[209] also demonstrated that the reduction in grain boundary current density occurs irrespective of the type of grain boundary misorientation: twist, tilt, etc. Thus, to avoid significant

---

[207] J. Mannhart and D. Dimos, Chapter 3 Section 6

[208] K. A. Mueller and G. Deutscher, *Phys. Rev. Lett.* 59 (1987) 1745

[209] D. Dimos, P. Chaudhari and J. Mannhart, *Phys. Rev.* B 41 (1990) 4038

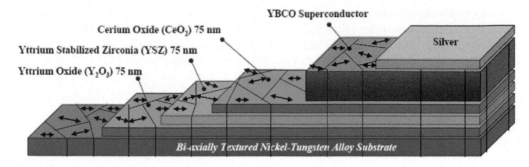

FIGURE 11.47: Schematic of epitaxial configuration of a substrate, buffers and superconductor, for the specific case of AMSC's 2G HTS wire stack. Double-headed arrows indicate alignment of grains; preferred alignment along the wire axis illustrates biaxial texture[211].

grain boundary obstacles to current flow, the material must be textured "biaxially," that is, not just with one axis of texture but two (once two are textured, the third axis also becomes textured). Their remarkable results suggested an alternative path to long-length wire: texturing the HTS material biaxially so as to reduce the grain boundary angles to the range of a few degrees.

But how to do this consistently over long lengths? A partial step came using the HTS BiSr-CaCuO family, which has a mica-like weak slip plane within its BiO double layers and which can therefore be *uniaxially* textured by deformation processing to align the CuO_2 *ab*-planes (i.e., c-axis grain alignment); this is the basis of first generation (1G) HTS wire. However, in spite of many attempts, deformation processing has failed to yield "biaxial" texture, with all three axes—*a*, *b* and *c*—aligned. Other methods to directly texture HTS material have relied on single crystal growth and are limited to stubby bulk pieces; it has not been possible to extend such methods to lengths of a meter—not to speak of a kilometer.

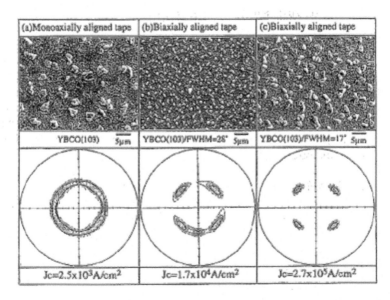

FIGURE 11.48: Correlation of critical current density J_c (77 K) and [103] x-ray pole figure texture in historic original Fujikura work.[212] Later work improved texture to an x-ray full width half maximum of order 6 degrees, and J_c to as high as 3.5 MA/cm^2 at 77 K.

The breakthrough came with the innovative work of Iijima et al.[210] of Fujikura Ltd., announced in 1991. They recognized that while difficult to directly texture HTS material over long lengths, it was possible to texture other materials, like yttria-stabilized zirconia (YSZ), using processes such as applying an argon-ion beam at an angle to a substrate during YSZ deposition. This method, called ion beam assisted deposition (IBAD) was compatible with reel-to-reel processing of long lengths of material on a flexible substrate. The HTS material could then be deposited epitaxially onto the textured template to achieve highly textured HTS. A schematic of epitaxial YBCO on a textured substrate is shown in Figure 11.47[211], and original results from Fujikura[210,212] showing increasing current density with improving texture are shown in Figure 11.48.

This fundamental process concept of using a textured template with an epitaxial HTS layer gave birth to 2G HTS wire. It was quickly recognized by the international technical community that materials on long-length flexible substrates could be textured in multiple ways; in fact not only buffer layers but also the substrates themselves could be textured. And the HTS layer could be deposited epitaxially on the textured template in multiple ways. This led to a tremendous variety of work around the world in developing and scaling up 2G HTS wire.

11.5.1 Texturing the Template

The coated conductor process starts with a long and flexible tape-shaped metallic substrate, typically 50–100 μm thick. The substrate is coated with a multifunctional oxide barrier or buffer layer, typically less than 1 μm thick, usually consisting of multiple layers and providing a textured template surface for epitaxial deposition of the 1–4 μm thick REBCO HTS layer. A passivating Ag layer of a few μm, a thicker metallic stabilization layer and possibly insulation complete the conductor (e.g., see Figure 11.47). The textured template is created by either texturing the buffer layer by IBAD[210] or inclined substrate deposition[213] (ISD—now largely abandoned and not discussed further here), or by deformation-texturing the metal substrate with the rolling assisted biaxially textured substrate approach[214] and applying epitaxial oxide buffer layers (trademarked RABiTS[TM] by Oak Ridge National Laboratory).

IBAD Template

Figure 11.49 shows the schematic arrangement for the IBAD process and a photograph of an actual deposition. An argon ion gun sputters an oxide target such as yttria-stabilized zirconia, YSZ, gadolinium zirconate $Gd_2Zr_2O_7$ (GZO) or magnesium oxide MgO onto a polycrystalline substrate such as Hastelloy. An additional argon ion gun irradiates the growing oxide at the incident angle of 55 degrees for YSZ and GZO or 45 degrees for MgO, generating biaxial texture. This method using YSZ and GZO was initially developed for long-length wire by Fujikura, ISTEC-SRL, the University of Göttingen (now transferred to Bruker Energy and Supercon Technologies, BEST), and SuperPower (subsidiary of Royal Philips Electronics), but the ~1 m/h production rate with the ~1 μm thick IBAD-YSZ or GZO layers is considered by many to be too slow for mass-production.

A breakthrough came from the Stanford group[215] which found that magnesium oxide MgO only 10 nm thick could be textured by IBAD, thus enormously accelerating the process rate. The optimum incident angle of the assisting ion beam in this case is 45 degrees rather than 55 degrees, a difference related to the different crystal structures of MgO (cubic) and YSZ or GZO (fluorite), although the

[210] I. Iijima et al., *Appl. Phys. Lett.* 60 (1992) 769; *IEEE Trans. Appl. Supercond.* 11 (2001) 2816

[211] Figure courtesy of AMSC; X. Li et al., *IEEE Trans. Appl. Supercond.* 19 (2009) 3231

[212] K. Onabe et al., *Adv. Supercond.* XI (Springer, Tokyo, 1999) 781

[213] B. Ma et al., *Physica* C 403 (2004) 183; W. Prusseit et al., *Physica* C 426-431(2005) 866

[214] A. Goyal et al., *Appl. Phys. Lett.* 69 (1996) 1795; Physica C 357-360 (2001) 903

[215] C. P. Wang et al., *Appl. Phys. Lett.* 71 (1997) 2955

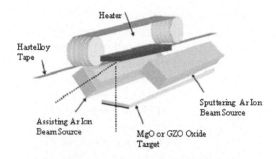

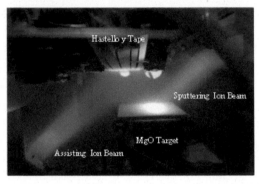

FIGURE 11.49: (top) Schematic ion-beam-assisted deposition (IBAD) system. Two sets of ion sources are used: for sputtering target material and for angled ion bombardment of growing films to optimize texture. The optimum IBAD angle between assisting ion beam and the substrate tape normal is 45 degrees for MgO or 55 degrees for YSZ or GZO. (bottom) IBAD-MgO in operation. (Courtesy of ISTEC[220].)

IBAD mechanism is still not fully understood. Low metal surface roughness, less than the IBAD-MgO layer thickness, is required and has been achieved by electropolishing the substrates. The in-plane texture of MgO-IBAD, as low as 3–4 degrees FWHM, is unusually sharp.

Los Alamos National Laboratory (LANL) demonstrated the high production rate of IBAD-MgO in a reel-to-reel system for the first time[216] and afterward worked with SuperPower to optimize and scale up the process. The SuperPower tape architecture[217] is shown in the cross-sectional TEM of Figure 11.50. Each layer is so thin that, in the latest results, the production rate reached 300 to 760 m/h of 12 mm wide tape, and for the first time, lengths greater than 1 km were achieved[218]. The buffer layers in Figure 11.50 consist of Al_2O_3 for a diffusion barrier, Y_2O_3 for a nucleation layer, the biaxially textured IBAD-MgO, epitaxial MgO for developing and stabilizing the epitaxy and $LaMnO_3$ (LMO) for lattice matching to the REBCO layer. The process has also been successfully applied to a long-length template in Korea by the SuNAM company; working with Seoul University and KERI[219]; they achieved a 1-km long template with a similar production rate to SuperPower. Subsequently, in Japan, ISTEC-SRL Nagoya[220] also succeeded in fabricating long IBAD-MgO substrate with a modified architecture using GZO instead of Al_2O_3, and Y_2O_3, and CeO_2 instead of

[216] P. N. Arendt and S. R. Foltyn, *MRS Bulletin* 29 (2004) 543

[217] V. Selvamanickam, US DOE Annual Peer Review, Aug., 2009; http://www.htspeerreview.com/2009/ agenda.html

[218] X. Xiong et al., *IEEE Trans. Appl. Supercond.* 19 (2009) 3319

[219] K. P. Ko et al., *Physica* C 463-465 (2007) 564

[220] Y. Yamada et al., *IEEE Trans. Appl. Supercond.* 19 (2009) 3236

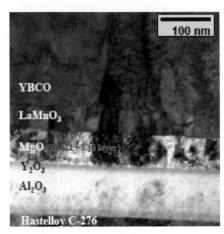

FIGURE 11.50: Cross-sectional TEM image of a 5 layer IBAD-MgO buffer stack, Al_2O_3/Y_2O_3/IBAD-MgO/epitaxial MgO/$LaMnO_3$ (LMO), on Hastelloy with YBCO on top. (Courtesy of SuperPower[217].)

FIGURE 11.51: Reel-to-reel IBAD production system with the world's largest IBAD ion guns (350x500cm). The production speed of IBAD-MgO is 1 km/h. (Courtesy of Fujikura Ltd.[222].)

$LaMnO_3$. The PLD-CeO_2 layer at the top of this buffer layer stack, also developed by Fujikura, provides a "self-epitaxy" effect[221] enhancing the texture during deposition. Recently Fujikura introduced the world's largest ion guns with dimensions 350 cm x 500 cm and reel-to-reel equipment shown in Figure 11.51[222], processing 1 km of IBAD template at 1000 m/h[223]. Development continues worldwide to further simplify and lower IBAD production costs by finding architectures with fewer layers.

RABiTS Template

In the RABiTS approach, [100] cube texture is created in the substrate by conventional rolling and recrystallization of a metal slab. RABiTS was initially developed with pure nickel[214]; however alloys with stronger mechanical properties and reduced magnetism, notably nickel with 5 atomic%

---

[221] Y. Yamada et al., *Physica* C 392-396 (2003) 777

[222] H. Fuji et al., *Physica* C 468 (2008) 1510

[223] S. Hanyu et al., *Physica* C 470 (2010) 1227

tungsten, have now replaced[224] pure Ni and achieved x-ray FWHM below 5 degrees. High surface smoothness requires regular repolishing of the finishing rolls. RABiTS has a cost advantage in processing large amounts of material rapidly and simply and is the basis of AMSC's commercial wire. An even higher strength and completely non-magnetic textured alloy with 9 at% W is also being developed.[224]

The buffer layer on the textured substrate is multifunctional: it provides a strongly adhering metal-oxide interface, a metal and oxygen diffusion barrier, and cap layer lattice-matching to REBCO and promoting its epitaxial growth. As a result, the buffer is today always a stack of different materials, although research continues to probe for the ideal single buffer layer. AMSC's RABiTS buffer stack is shown schematically in Figure 11.47 with three ~75 nm thick layers of Y_2O_3, YSZ and CeO_2, all deposited with DC reactive sputtering[211]. A remarkable phenomenon, not yet understood, is that subsequent buffer layers can significantly improve in-plane texture over that of the underlying layer or substrate (the "self-epitaxy" effect). Thus, a RABiTS template with 6.5 degrees FWHM of in-plane texture in the NiW substrate can attain 5 degrees FWHM on the CeO_2 template surface[211].

11.5.2   REBCO HTS Layer Deposition

The epitaxial REBCO layer has been deposited successfully using both in-situ and ex-situ methods. With in-situ methods, the superconductor forms directly during the deposition process. Examples include pulsed laser deposition (PLD) and metal-organic chemical-vapor-deposition (MOCVD). With ex-situ methods, a precursor is deposited and subsequently reacted to form RE-BCO, as in the metal-organic deposition (MOD) process.

PLD Process

This in-situ process has been widely used as a convenient research tool for HTS film synthesis because it preserves REBCO stoichiometry in during film formation from a bulk target. An excimer laser beam from KrF or XeCl irradiated on the target in a vacuum chamber generates a plume of REBCO constituents which deposit on the facing substrate. To fabricate a long conductor with this process, a reel-to-reel system was developed by Fujikura[225], Göttingen University[226], and ISTEC-SRL Nagoya[227].

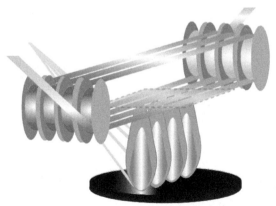

FIGURE 11.52:   Schematic diagram of multi-plume/multi-turn pulsed laser deposition system[227].

[224] D. Verebelyi et al., *Supercond. Sci. Technol.* 16 (2003) L19; J. Eickemeyer et al., ibid. 23 (2010) 085012

[225] Y. Iijima et al., *MRS Bulletin* 29 (2004) 564

[226] A. Usoskin et al., *MRS Bulletin* 29 (2004) 583

[227] T. Watanabe et al. *IEEE Trans. Appl. Supercond.* 150 (2005) 2566

FIGURE 11.53: 212 m long MPMT-PLD-processed YBCO coated conductor after silver layer deposition. (Courtesy of ISTEC[228].)

To enhance deposition rate, a multi-plume and multi-turn (MPMT) process was proposed[227], in which the laser beam is rastered over the large REBCO target area. To increase the deposition area as shown in Figure 11.52, the template tape was wound a few times above the laser plume, which was divided into several plumes via mirror scanning. Figure 11.53 shows a 200-m long 1-cm wide YBCO-coated conductor fabricated with the MPMT-PLD process[228] after silver layer deposition on the YBCO layer; it achieved I_c at 77 K and 0 T of 245 A. Furthermore, in Japan, wires based on GdBCO were developed instead of the conventional YBCO because of its higher I_c and J_c and higher deposition rate[229]. Fujikura and Sumitomo Electric Industry[230] now use MPMT and GdBCO for long coated conductor wires. Fujikura produces[231] lengths over 500 m with self-field 77 K I_c of 350 A/cm-width and production rate now reaching a few tens of m/h. I_c was also increased to above 600 A/cm-width for a 100 m class length. In 2011, Fujikura announced the new world record of $I_c \times L$, 466,981 A.m (572 A x 816.4 m). Bruker (BEST) in Germany has also developed a high rate PLD system for long conductor fabrication, successfully using a cylindrical form for a large deposition area of the tape substrate[226]. They supplied 3 km of wire using this method for the European Super3C cable project in March 2009 [232]. R&D on short samples has achieved substantially higher I_c: Los Alamos National Laboratory achieved the world highest I_c of 1400 A/cm-width[233] (at 75 K) using YBCO/CeO$_2$ multilayers, and Fujikura obtained a thick GdBCO film conductor with 1040 A/cm-width[234] in a reel-to-reel system.

MOCVD Process

Initially, metal-organic chemical vapor deposition was difficult to apply to REBCO because lack of gas- or liquid-phase metal-organic precursors limited the process to solid precursors, where RE-

[228] A. Ibi et al., *Physica* C 445-448 (2006) 525

[229] Y. Yamada et al., *IEEE Trans Appl. Supercond.* 17 (2007) 3371

[230] K. Abiru et al., *IEEE Trans. Appl. Supercond.* 21 (2011) 2941.

[231] S. Hanyu et al., *IEEE Trans. Appl. Supercond.* 19(2009) 3240

[232] Super3C Project news release, March 17, 2009: http://www.bruker-est.com/super3c.html

[233] S. R. Foltyn et al., *Appl. Phys. Lett.* 87 (2005) 162505

[234] K. Kakimoto et al., *Physica* C 469 (2009) 1294

BCO composition is difficult to control because of vapor phase composition instability. After much R&D, Superpower has now succeeded with a single source liquid delivery system[235]. Precursors of M (tmhd, Tetramethyl HeptaneDionate) (M=RE, Ba, Cu) are dissolved in an organic solvent and delivered by a micro-pump or liquid mass-flow controller to a flash-vaporizer. The vaporized precursor is transported to the CVD reactor by an inert carrier gas. SuperPower[236] and Chubu Electric Power Company[237] have been leading the development and scale-up. Combined with its IBAD-MgO process, SuperPower attained production length above 1 km as shown in Figure 11.54. Their $I_c x L$ (length) value of 300,330 Am was until recently the world record for coated conductor[217].

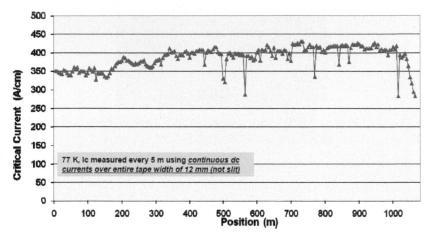

FIGURE 11.54: I_c distribution of 1 km long, 12 mm wide coated conductor fabricated by SuperPower using the IBAD-MgO/MOCVD process. Minimum current (I_c) = 282 A/cm over 1065 m gives I_c x length = 300,330 A-m (Courtesy of SuperPower[217]).

Electron-beam (EB) Deposition

The Technical University of München and Theva deposited REBCO by an electron-beam process, using an innovative two-zone deposition technique, with a low pressure zone for RE, Ba and Cu metal deposition, and a high oxygen-pressure zone for the reaction to REBCO[238]. Rapid cycling of deposition and oxygen reaction, initially achieved through a rotating chamber, solved critical issues of temperature uniformity via a black-body-type heater, and oxygen pressure in a high vacuum deposition environment via a differentially high oxygen pressure enclosed in the heater. The process has been used for long coated conductors in Germany by Prusseit and coworkers at Theva[239], and in Korea by Youm and coworkers at KAIST[240] and Oh et al. at KERI[241]. A schematic of the production system using a cylindrical drum for the wire winding by KERI, applied to the SmBCO composition, is shown in Figure 11.55. Very recently SuNAM in Korea applied this technique to a reel-to-reel

[235] V. Selvamanickam et al., *IEEE Trans. Appl. Supercond.* 11 (2001) 3379;

[236] V. Selvamanickam et al., *IEEE Trans. Appl. Supercond.* 19 (2009) 3225

[237] T. Watanabe et al., *IEEE Trans. Appl. Supercond.* 17 (2007) 3386

[238] P. Berberich et al., *Physica* C 219 (1994) 497

[239] W. Prusseit et al., *Supercond. Sci. Technol.* 13 (2000) 519

[240] B. S. Lee et al., *Supercond. Sci. Technol.* 17 (2004) 580

[241] S. S. Oh et al., *Supercond. Sci. Technol.* 21 (2008) 034003; Korea-Japan Superconductivity Workshop, Changwon, South Korea, April 15, 2010

system and obtained 200 m long 4 mm wide GdBCO tape with 220 A/cm-width[242] at a production rate of 600 m/h, which is promising for mass production.

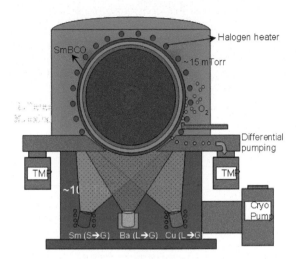

FIGURE 11.55: Schematic co-evaporation deposition system for long SmBCO coated conductor. The vacuum chamber is separated into a low pressure part (below) for the EB deposition and a high pressure part for reaction and oxygenation of SmBCO. The drum carrying the tape template rotates to alternate deposition and reaction. (Courtesy of KERI[241].)

MOD Process

The ex-situ methods start by depositing a precursor typically containing fluorine, which is subsequently reacted to form REBCO, although non-fluorine based processes have also been developed. While EB deposition of metal constituents and BaF_2 has been demonstrated[243], the leading ex-situ method, which has the potential for lowest cost, uses a liquid-phase chemical route[244] called metal-organic-deposition (MOD) or chemical solution deposition (CSD). Original work used trifluoroacetates as carriers for all the cations, but many other chemical carriers have since been explored.

Let us consider the case of YBCO, although the process can equally well be practiced with a mixture of rare earths and yttrium. After a slot-die coating or dip-coating and initial drying, the precursor undergoes a "calcination" or decomposition process forming an yttrium-barium oxyfluoride (YBaOF) phase in conjunction with CuO. In this critical process step, a large amount of material is burned off, collapsing the initially coated thickness by as much as a factor of ten. Cracking and other defects in thicker deposits at this stage have so far limited the final reacted YBCO thickness in a single coating to around 1 micron, although ongoing research is steadily pushing this limit upward and so increasing the wire's current-carrying capacity[245]. Multiple coatings have also been demonstrated to further increase thickness, and using as many as ten coatings, each containing material for 0.2 microns of final REBCO thickness, the ISTEC group has achieved performance as high as 735 A/cm-width in films totaling 2 microns or more[246].

Finally, reaction in water vapor at ~700 °C starts by converting YBaOF to yttria and BaO, releas-

[242] S. H. Moon, presentation at The Korean Superconductivity Society Meeting 2010, Yongpyong Resort. Gangwondo, Korea. July 7, 2010

[243] M. Suenaga, *Physica C* 378-381 (2002) 1045

[244] A. P. Malozemoff et al., *Supercond. Sci. Technol.* 13 (2000) 473

[245] M. W. Rupich et al., *Supercond. Sci. Technol.* 23 (2010) 014015; M. Miura et al., Physica C 469 (2009) 1336

[246] T. Izumi et al., *Physica C* 469 (2009) 1322

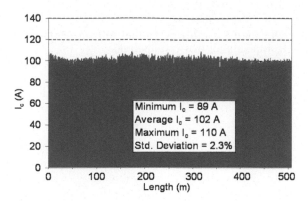

FIGURE 11.56: Critical current of RABiTS/MOD 2G HTS wire at 77 K and self-field, measured in 1-m gauge length. (Courtesy of AMSC[211].)

ing HF, and then

$$3\,CuO + 2\,BaO + 0.5\,Y_2O_3 \;\rightarrow\; YBa_2Cu_3O_{6.5}\,.$$

Oxygenation at the end of the process, usually after deposition of a thin silver passivation layer, converts the YBCO phase to the fully superconducting $YBa_2Cu_3O_7$.

AMSC's version of this process achieves J_c of 3-3.5 MA/cm^2 in 0.8 micron thick REBCO layers, enabling 100 A in a 4 mm-wide conductor, corresponding to 250 A/cm-width at self field and 77 K. This process has been scaled up to 500-m lengths with high I_c uniformity, even in 4-mm wide wires measured on a 1-m scale, as shown in Figure 11.56[211].

FIGURE 11.57: Cross-section of laminated 4.4 mm wide HTS wire, showing two 50 μm thick copper stabilizer strips sandwiching the "insert", which consists mostly of NiW substrate 75 μm thick, carrying the (invisible at this magnification) 225 nm thick buffer stack and ~1 μm thick REBCO layer. Solder forms the fillets at the edges. (Courtesy of AMSC[211].)

AMSC's substrate, buffer and REBCO reaction processes are carried out on >40 mm wide strips which are then roll-slit into eight 4-mm wide "insert" tapes (discarding the edges)[211,247]. The ability to process so many inserts simultaneously provides a major cost advantage. Further scale-up is planned to 100-mm wide strips, which will yield as many as 23 4-mm wide insert tapes in one pass through the process equipment. After slitting, the inserts are solder-laminated to surrounding metallic stabilizer strips, which can be copper, brass, or stainless steel, providing flexibility to optimize the wire for the given application. A cross-section is shown in Figure 11.57. High conductivity copper is desirable for conduction-cooled lower temperature magnet applications. Brass is a good compromise for some cable applications in liquid nitrogen, while stainless steel, with its high resistivity, is most favorable for resistive current limiters. The slit width can also be adjusted, depending on application needs. For example, 12-mm wide tapes have been slit from the original 46-mm wide strip and used in a fault current limiter demonstration with Siemens[247]. Wider tapes reduce cost in

---

[247] W. Schmidt et al., *IEEE Trans. Appl. Supercond.* 17 (2007) 3471

magnet applications by reducing the number of parallel coils needing to be wound. Multiple insert tapes can also be solder-laminated together to double the critical current. AMSC has delivered 75 km of the 4 mm wide, brass-laminated tape to LS Cable of Korea for a 500 m 22.9 kV in-grid cable at the Icheon substation in the KEPCO grid.

11.5.3 Artificial Pinning

Enhancing I_c, particularly in an applied magnetic field, is an important issue for 2G HTS wire. High in-field critical current is controlled by a phenomenon known as "pinning," whereby quanta of magnetic flux, called flux lines or vortices, are immobilized in the superconductor by defects, which are often called artificial pinning centers (APC). If the Lorentz force $J \times B$ on the flux lines exceeds the defect pinning force, the flux lines move, generating resistive losses, thus limiting the critical current. Although a quantitative prediction of experimentally observed I_c is still lacking, clear experimental evidence correlates different defects with enhanced I_c, as shown in Figure 11.58 for AMSC's MOD HTS layer. A powerful way to identify different pinning centers is through measurements of I_c as a function of field angle as it is rotated in the plane perpendicular to the current flow. Peaks, usually along the principal axes, can be correlated with pinning centers such as stacking faults or randomly located second-phase nanoparticles[248].

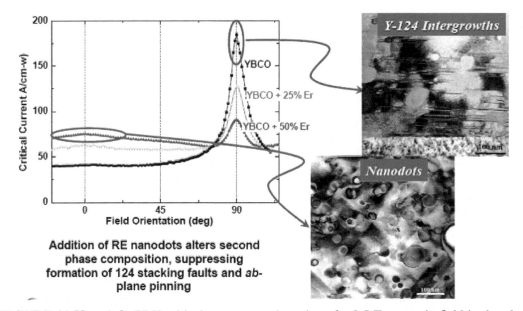

Addition of RE nanodots alters second phase composition, suppressing formation of 124 stacking faults and *ab*-plane pinning

FIGURE 11.58: (left) 75 K critical current *vs* orientation of a 0.5 T magnetic field in the plane perpendicular to current flow (0 degrees corresponds to field perpendicular to the tape), in a series of MOD wires with different concentrations of Er_2O_3 nanoprecipitates called nanodots. (right) TEM micrographs illustrate different kinds of defects correlating with peaks in the I_c plot. (Courtesy of T. Holesinger and L. Civale, Los Alamos National Laboratory[248].)

Perhaps the most remarkable pinning center has been found in in-situ deposited films by several researchers[249,251] starting with the work of Haugan et al.[249] using second-phase barium zirconate

[248] T. Holesinger et al., *Adv. Mat.* 20 (2008) 391; W. Zhang et al., *IEEE Trans. Appl. Supercond.* 17 (2007) 3347

[249] T. Haugan et al., *Nature* 430 (2004) 867; J. L. M. Driscoll, et al., *Nat. Mat.* 3 (2004) 439 ; A. Goyal et al., *Supercond. Sci.*

BaZrO$_3$ (BZO). The growth process leads, in a mechanism still poorly understood, to nanoparticles of BZO lining up one on top of the other, to form a nanorod structure, sometimes called a bamboo structure[250], oriented perpendicular to the tape plane, as shown in Figure 11.59. This pinning defect is one of the strongest known, enhancing J_c or I_c several times above the values in conventional REBCO films in fields applied perpendicular to the tape plane. In the PLD process, the target can be, for example, a mixture of YBCO with BZO pinning center material sintered together. This target can be ablated with no other significant process changes. Nanorod pinning has also been successfully introduced in the MOCVD process using Gd and Zr addition[251]. I_c is increased as high as a few tens of A/cm-width at 3 Tesla and 77K (without the nanorods, I_c is negligible under these conditions). However, I_c at high fields and high temperatures must be increased further, as several hundred A/cm-width is a practical operational goal for most magnet applications, and so research in this area continues actively. Several hundred A/cm-width performance in several tesla is achieved in existing wires at lower temperature (below about 40 K).

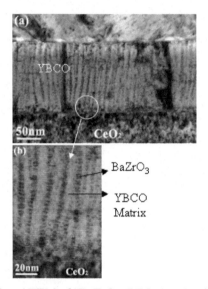

FIGURE 11.59:   Cross-sectional TEM of "BaZrO$_3$ (BZO) bamboo" structure, an artificial pinning center, in YBCO IBAD/PLD coated conductors. (Courtesy of ISTEC[250].)

11.5.4   Drivers for 2G HTS Wire

As reviewed in Section 11.4 of this book[252], first generation (1G) HTS wire using the bismuth-strontium-calcium-copper-oxide family ("BSCCO") has been very successful technically, with scaled-up manufacturing producing over kilometer lengths, high currents and adequate mechanical properties, when laminated with stainless steel or other strong stabilizing metals. The 1G wire has been the vehicle for most of the cable, rotating machine and magnet demonstrations to date. So why the interest in 2G wire?

A major driver is cost[244], though this remains controversial. Supporters of 2G HTS wire point to its much reduced use of silver compared with 1G, which is a major issue considering today's silver

Technol. 18 (2005) 1533

[250] Y.Yamada et al., *Appl. Phys. Lett.* 87 (2005) 13502

[251] V. Selvamanickam et al., *Supercond. Sci. Technol.* 23 (2010) 014014

[252] M. Rupich and E. Hellstrom, Section 11.4

prices, and to the ability to process many wires in tandem using wide strips, which reduces labor costs. Cost of the elaborate vacuum systems required for physical vapor deposition, used in most 2G buffer processing, is minimized by keeping buffer layers as thin as possible. An all-chemical liquid phase deposition process for buffers and HTS layers is particularly cost-advantageous[253]. All in all, cost models for coated conductors, particularly those made by nonvacuum methods, predict long-term values below \$50/kAm (77K, 0T), a range which is competitive with today's effective price-performance of copper in power cables[244]. Manufacturers do not reveal their manufacturing costs, but particularly as economies of scale are achieved, and rising silver prices impact 1G wire cost, 2G wire is expected to be the cost winner, which is critical for its widespread commercial use.

There are other drivers for 2G HTS wire. Because of the significantly higher irreversibility contour, in the temperature-field plot of Figure 11.4, of REBCO compared to BSCCO[254], second generation HTS wire offers the possibility of higher temperature operation in a given magnetic field. This could be important for a host of magnet and rotating machine applications. However an approximate minimum engineering critical current density for practical rotating machinery coils is 200 A/mm^2, corresponding to 400 A/cm-width in typical wires, a level achieved so far at Tesla-level fields only well below 77 K. Enhancing REBCO in-field critical currents is therefore the focus of intense ongoing work.

A third advantage of 2G wire is its flexibility in choice of stabilizer. Because the wire contains a minimum amount of low resistivity (0.3 $\mu\Omega$cm at 77 K) silver, use of a high electrical or thermal resistivity stabilizer can enable resistive fault current limiter and current lead applications. Mechanical properties also meet application needs. And the wires are chemically stable and avoid toxic elements.

Interest in 2G HTS wire has never been higher. Active R&D is underway at Los Alamos, Oak Ridge and Argonne National Laboratories, at the International Superconductivity Technology Center (ISTEC) in Japan, at KERI and other laboratories in Korea, and at a variety of European laboratories and universities. Industrial programs are underway in the US at AMSC, MetOx, Royal Philips Electronics subsidiary SuperPower Inc. and Superconductor Technologies Inc., in Japan at Fujikura Ltd., Furukawa Electric Co. Ltd., SWCC Showa Holdings Co. Ltd., Chubu Electric Power Co. Inc. and Sumitomo Electric Industries Ltd., in Korea at SuNam Co. and in Europe at Zenergy Power, Bruker-BEST and Nexans S. A.

Several short-length cable demonstrations have been made with 2G wire: by SEI and Super-Power[255] in the US, by a consortium headed by Nexans in Europe (the Super3C project)[232], and by SEI and Furukawa in Japan[256]. A project is underway to replace one phase of the 600-m Long Island Power Authority 1G superconductor transmission cable, developed by Nexans and AMSC, by a 2G-based cable[257]. A fault-current-limiting cable technology has been demonstrated by AMSC, Southwire, Con Edison and Oak Ridge National Laboratory[257]. Laboratory demonstrations of distribution-level resistive current limiters using 2G wire have been made by Siemens and AMSC[258], by Toshiba[259], and by Hyundai[260]. And a 2G superconducting insert coil generating 2.8 T in a 31 T resistive background field, has been demonstrated by SuperPower and the National High Magnetic Field Laboratory[261], opening the possibility of a new generation of ultra-high-field superconductor magnets.

---

[253] M. Baecker and T. Schneller, *Nachrichten aus der Chemie* 55 (2007) 1202; V. Cloet et al., *Adv. Sci. Technol.* 47 (2006)153

[254] D. Larbalestier, Section 11.1.6

[255] H. Yumura, *IEEE Trans. Appl. Supercond.* 19 (2009) 1698

[256] M. Ohya et al., *IEEE Trans. Appl. Supercond.* 19 (2009) 1766; S. Mukoyama et al., *Physica C* 469 (2009) 1688

[257] J. F. Maguire et al., *IEEE Trans. Appl. Supercond.* 19 (2009) 1692 and 1740

[258] H.-W. Neumueller et al., *IEEE Trans. Appl. Supercond.* 19 (2009) 1950

[259] T. Yazawa et al., *Physica C* 469 (2009) 1740

[260] J.B. Na et al., *Physica C* 469 (2009) 1754

[261] H. W. Weijers et al., *IEEE Trans. Appl. Supercond.* 20 (2010) 576

The recently announced order by LS Cable, the leading South Korean cable manufacturer, for 3 million meters of AMSC wire, to be used in cables for KEPCO's South Korean grid and internationally, is a major benchmark in commercial demand for 2G HTS wire and in utility interest in HTS power technology. As manufacturers scale up production capacity, cost/performance improves and commercial power equipment is installed in the grid, 2G HTS wire finally appears set to fulfill the dream from the first exciting years of the discovery of HTS: to facilitate a dramatic revolution in electric power and magnet applications, bringing broad benefits for society.

11.6   The MgB$_2$ Conductor Story

René Flükiger and Hiroaki Kumakura

11.6.1   Reasons for Excitement at the Unexpected Discovery of Superconductivity in MgB$_2$

The compound MgB$_2$ is an "old" material, known since the early 1950's, but found to be a superconductor only in 2001[262]. The discovery of superconductivity at 39 K in MgB$_2$ caused an intense interest in the field of fundamental and applied research[262]. Almost immediately there were strong signs that grain boundaries were not intrinsic weak links[263]. Indeed, MgB$_2$ was a "simple" superconductor, and showed properties of an electron phonon BCS superconductor. For the many who had wrestled very hard with the great complexities of HTS compounds, there was a tremendous hope that medium temperature superconductivity, even at half the T_c of the HTS materials, would still make a very practical conductor. Very soon after, researchers of the Genoa group[264] were able to show that a Powder-in-Tube conductor process could produce reasonable J_c values, even without heat treatment, giving additional practical confirmation to the weak-link free nature of grain boundaries. As first made, MgB$_2$ was clean and had low upper critical field, but Eom[265] showed quickly that it could be shifted towards the dirty limit, resulting in strongly increased upper critical fields. Not very long after the HPCVD process showed that films of great purity could be grown[266]. Another important factor was that industry jumped in to address this opportunity. Columbus in Genoa and Hyper Tech in Columbus USA both made manufacture of MgB$_2$ a high priority. Since both Mg and B are abundant and not intrinsically expensive, the way was open for operating superconducting devices cooled by liquid hydrogen or by closed-circuit refrigerators. A review article published in the same year[267] contained already 270 citations, and by 2010 several thousand articles had been published. In the following, the development of MgB$_2$ conductors will be followed from the discovery up to the present day, based on a choice of representative articles.

MgB$_2$ is a clean superconductor, the normal state electrical resistivity at 40 K, ρ_o, being 0.38 $\mu\Omega$cm[268]. Its coherence length ($\xi_o = 50$ Å) is similar to that of Nb$_3$Sn, and the supercurrents flow weak-link-free across the grain boundaries[263,264]. A first prototype MgB$_2$ wire was prepared by diffusion of Mg vapor into B fibers reinforced by a W core[268]. The transport critical current den-

---

[262] J. Nagamatsu et al., *Nature* 410 (2001) 63

[263] D.C. Larbalestier et al., *Nature* 410 (2001) 186

[264] G. Grasso et al., *Appl. Phys. Lett.* 79 (2001) 230

[265] C.B. Eom et al., *Nature* 411 (2001) 558

[266] X. H. Zeng et al., *Nat. Mat.* 1 (2002) 35

[267] C. Buzea and T. Yamashita, *Supercond. Sci. Technol.* 14 (2001) 1

[268] P.C. Canfield et al., *Phys. Rev. Lett.* 86 (2001) 2423

sities, J_c, were already quite high: of the order of 10^5 A/cm^2 at 4.2 K. Remarkably high values of the upper critical field, B_{c2}, were reported in MgB$_2$ films grown on single-crystal substrates by Eom[265]. Later on, B_{c2} values close to 70 T perpendicular to the c direction were reported by Braccini et al.[269] In bulk or filamentary MgB$_2$ samples, the highest reported B_{c2} values are close to 40 T[270].

Gurevich[271] quickly established a two-band model for describing the variation of B_{c2} versus T, which showed why anomalously high values of H_{c2} were attainable in two-band superconductors. MgB$_2$ crystallizes at the stoichiometric composition, and no noticeable width of the phase field has yet been reliably established. The phase diagram[272] of the system Mg-B contains the phases MgB$_2$, MgB$_4$ and MgB$_7$. Bulk pinning was found to be weak in MgB$_2$ single crystals, but grain boundaries were found to be the dominant pinning mechanism in filamentary MgB$_2$ wires, due to grain sizes as small as 10-50 nm. When describing the transport properties of MgB$_2$ conductors, an important aspect is the anisotropy of B_{c2}. The extrapolated $B_{c2}(0)$ values for a binary single crystal were reported as 23 and 3.1 T for field directions // and $\perp$ to the ab plane, respectively, while an anisotropy ratio $\gamma = H_{c2}^{//ab}/H_{c2}^{\perp ab}$ of 6 at low

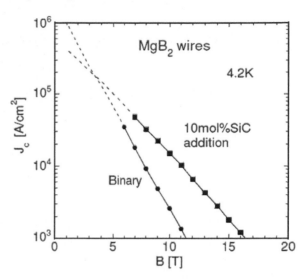

FIGURE 11.60: Comparison between the variation of J_c vs applied field B for binary and SiC additions to in situ MgB$_2$ wires.

temperature was reported[273]. As a consequence of the intrinsic anisotropy of the hexagonal MgB$_2$ phase, the current path in bulk or filamentary samples is non-uniform. The current transport occurs by percolation, thus depending on grain connectivity. Rowell[274] proposed a simple correlation between electric resistivity and grain connectivity, which has been verified by numerous authors. The importance of the mixed state parameters in defining the limiting conditions for loss-free currents and thus the critical current densities in MgB$_2$ has been discussed in detail in a review by Eisterer[275].

11.6.2 The Effect of Carbon or Carbon-based Additives on the Transport Properties

Important progress was achieved by Dou[276], who reported a strong enhancement of J_c after adding SiC nanopowders to the initial Mg+B powder mixtures. The effect of SiC additives on the transport J_c of MgB$_2$ wires at 4.2 K is shown in Figure 11.60: J_c increases at high fields, but decreases at low fields. The effect of additives on J_c of MgB$_2$ has been the subject of numerous investigations; more than 35 additives comprising elements, compounds, oxides and organic solutions were mentioned by Collings[277] in a review article. All these data show that a sizeable and

[269] V. Braccini et al., *Phys. Rev.* B 71 (2005) 012504.

[270] B.J. Senkowicz, *Supercond. Sci. Technol.* 21 (2008) 035009

[271] A. Gurevich, *Physica* C 456 2007 160

[272] Z. K. Liu, *Appl. Phys. Lett.* 78 (2001) 3678

[273] M. Angst, *Phys. Rev. Lett.* 88 (2002) 167004

[274] J. M. Rowell, *Supercond. Sci. Technol.* 16 (2003) R17

[275] M. Eisterer, *Supercond. Sci. Technol.* 20 (2007) R47

[276] S. X. Dou et al., *Appl. Phys. Lett.* 81 (2002) 3419; S. Soltanian et al., *Physica* C 361 (2001) 84

[277] E. W. Collings et al., *Supercond. Sci. Technol.* 21 (2008) 103001

reproducible enhancement of J_c of MgB_2 is only observed for elemental carbon[278] and for C-based additives, e.g., SiC[15], $C_9H_{12}+SiC$[279], and malic acid $(C_4H_6O_5)$[280]. In the case of C-based additives, it was found that the free carbon arising from their decomposition during the final heat treatment reacts with the precursor powders and substitutes for B in the MgB_2 lattice. Single crystal observations show that the substitution of C on the B lattice sites leads to an enhanced residual resistivity ρ_o[274], which is correlated to the observed enhancement of the upper critical field, B_{c2}, and the irreversibility field, B_{irr}, in MgB_2. The enhancement of ρ_o with lattice disorder in alloyed MgB_2 wires was clearly recognized as the dominant effect enhancing the critical current density, J_c, at high magnetic fields. The vortex pinning behavior of MgB_2 is affected only a little by the presence of C on the B sites. This is confirmed by relaxation measurements on bulk samples which show that the pinning energy U_o of SiC alloyed MgB_2 is unchanged with respect to that of the binary compound[281]. No changes of pinning or of the connectivity have to be assumed to describe the data on alloyed MgB_2 wires. The value of T_c in MgB_2 decreases almost linearly with the C content: for $x = 0.18$ in the formula $MgB_{2-x}C_x$ a T_c value of 30 K is obtained. The addition of C also reduces the anisotropy factor $H_{c2}^{//ab}/H_{c2}^{\perp ab}$, as shown on single crystals, where γ at T=0 decreases from 5.3 for binary MgB_2 to a value ~ 1 for 10 % C[282]. The Carbon content for the maximum H_{c2} in MgB_2 wires is of the order of $x = 0.05$; at this composition, alloyed MgB_2 tapes exhibit a considerably reduced anisotropy.

11.6.3   Factors Influencing the Transport Properties of MgB_2 Conductors

The Initial Powders

Since B and MgB_2 are hard materials and cannot be plastically deformed, MgB_2 conductors have to be prepared by powder metallurgical methods. From the wealth of published data it follows that the main requirements for optimized transport properties in MgB_2 wires are high purity and a small size of the initial powder particles, as well as of the C-based additives. The desired final MgB_2 grain sizes being of the order of 10–50 nm [270], the appropriate initial particle size should be ideally in the nanometer range. Since the total amount of oxygen present at the large surface of nanopowders reaches several at% and cannot be removed, it will thus be present in the reacted filaments, mainly as MgO. However, Mg particles with sizes < 1 μm are very expensive and have to be treated under inert gas due to fire hazard, while amorphous, nanosized boron powders with sufficiently high purity are very difficult to obtain. For these practical reasons, industrial MgB_2 wires may always exhibit somewhat lower J_c values than would be possible with ideal powders.

The Metal Matrix and the Reaction Barrier

It was found that there is only a moderate reaction between MgB_2 and the Fe or Ni sheath during the heat treatment[283]. Meanwhile, even less reaction was found for Nb, Ta and Ti, which thus constitute excellent barrier materials. The commonly used matrix materials for industrial MgB_2 wires are Ni or Monel. The transport properties of MgB_2 wires are sensitive to the uniaxial applied stress and thus also to the precompression induced by the matrix material. The variation of J_c *vs* the applied uniaxial strain ε for multifilamentary wires shows an almost linear increase up to $\varepsilon \sim 0.4\%$, after which degradation occurs[284].

---

[278] R. H. T. Wilke et al., *Phys. Rev. Lett.* 92 (2004) 217003

[279] H. Yamada et al., *Supercond. Sci. Technol.* 19 (2006) 175

[280] J.H. Kim, *Appl. Phys. Lett.* 9 (2006) 142505

[281] C. Senatore et al., *IEEE Trans. Appl. Supercond.* 2941 (2007)

[282] T. Masui et al., *Physica C* 412-414 (2004) 303

[283] S. Jin et al., *Nature* 411 (2001) 563

[284] H. Kitaguchi et al., *Physica C* 401 (2004) 246

Texturing

MgB$_2$ tapes exhibit a marked texture as a consequence of the rolling process, as studied by Kovac[285]. For in situ tapes, synchrotron radiation measurements by Abrahamsen[286] have shown that the initial rolling of the Mg + B powder mixture inside a Fe sheathed tape causes an elongation of the Mg grains, as shown in Figure 11.61. This elongation leads to local texturing, which is subsequently transferred to the MgB$_2$ grains during reaction. This is possible since the reaction between Mg vapor and B occurs already at T < 600 °C, i.e., well below the melting temperature of Mg, 649 °C. As suggested by Figure 11.61, the same local texturing of the grains can be assumed in MgB$_2$ wires, too. However, diffraction analysis over a cross section cannot detect it, local effects being averaged out.

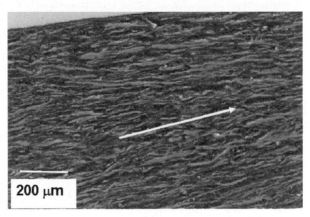

FIGURE 11.61: Elongation of Mg particles in the Mg+B mixture of an in-situ filament after drawing. The arrow indicates the drawing direction. (SEM picture, after polishing.)

11.6.4 The Fabrication Methods of MgB$_2$ Wires and Tapes

Important efforts have led to three main routes for the fabrication of MgB$_2$ wires, the fabrication steps being schematically shown in Figure 11.62. The powder-in-tube (PIT) method can be applied industrially for both the ex-situ and in-situ processes to fabricate wires in km length. The internal Mg diffusion process (IMD), which combines B powder and Mg rods, has the potential for large scale production, too. In the following, the three mentioned techniques for fabricating multifilamentary MgB$_2$ wires are described.

The *ex situ* Technique

The first ex-situ MgB$_2$ wires were produced by filling Fe tubes with reacted MgB$_2$ powder and deforming them to fine wires, followed by a recrystallization heat treatment above 950 °C [264]. The reaction time being quite short (< 15 minutes), the thickness of the reacted zone at the MgB$_2$/Fe bound-

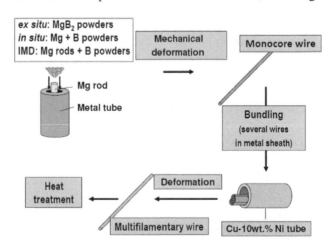

FIGURE 11.62: Fabrication of MgB$_2$ wires by metallurgical powder. The main difference between ex-situ, in-situ and IMD processing, resides in the initial powder mixture.

[285] P. Kovac et al., *Supercond. Sci. Technol.* 18 (2005) L45

[286] A.B. Abrahamsen et al., *IEEE Trans. Appl. Supercond.* 17 (2007) 2757

ary was limited to only a few μm. Remarkably high J_c values were obtained after introducing C in ex situ MgB_2 tapes: $J_c(4.2K) = 10^4$ A/cm^2 at 13 T [287], but the homogeneous introduction of C required crystallization heat treatments at very high temperatures, which leads to thicker reaction layers.

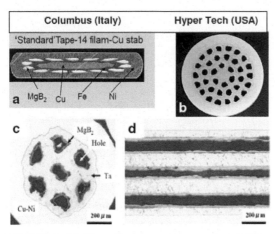

FIGURE 11.63: Cross sections of multifilamentary MgB_2 wires prepared by different techniques: a: *ex situ*, 3.6.× 0.65 mm, b: *in situ*, ø = 0.83 mm, c: IMD, cross section, ø = 1.00 mm, d: IMD, longitudinal cross section, after reaction at 640 °C for 1 hr.

Binary ex situ MgB_2 wires are presently industrially produced in km lengths by Columbus for applications at low magnetic fields, in particular for MRI magnets and current leads. A typical cross section of an ex situ tape is shown in Figure 11.63 a. Details about the preparation of industrial binary ex situ wires are described in the following[288]: Prereacted MgB_2 powders (99.99% Mg and amorphous, 99% B) with a molar ratio of 1:2 were preheated at 760 °C for 1 h in an Ar - 5%H$_2$ atmosphere to avoid oxidation of Mg during reaction and minimize the amorphous MgO layer around the MgB_2 grains. The MgB_2 powders were inserted into a Nb tube, surrounded by a Monel tube and drawn to a wire of about 3.5 mm diameter. Several pieces of wire were then inserted in a Monel tube, together with OFHC Cu for thermal stabilization. The so prepared composite conductor was groove rolled and drawn to a round wire with a diameter of 2 mm, which was twisted

before rolled to a tape of 3.6 × 0.65 mm. In Figure 11.63a, the Cu is placed at the center of the wire, but other configurations are also possible where the Cu is located externally. The heat treatment occurred at 965°C for 4 min in an Ar-5%H$_2$ gas flow. Typical critical current densities of ex situ tapes[289] are shown in Figure 11.64, the corresponding values for in situ wires[290] are also shown for comparison.

The *in situ* technique

The in situ technique[291] was developed simultaneously with the ex situ process, the initial MgB_2 powder mixture in Figure 11.63 being replaced by Mg+B powder mixtures. Industrial wires using this process are fabricated by Hyper Tech Research (US), a cross section[292] being shown in Figure 11.63b. With respect to the ex situ route, this alternative has the advantage that the reaction to MgB_2 occurs at considerably lower temperatures (600–700 °C), thus reducing the reaction layer at the metallic sheath. Even at these temperatures, C can be substituted into the B plane of the MgB_2 lattice. The highest J_c reported so far for *in situ* monofilamentary MgB_2 wires was obtained by Susner[293], who used the following procedure: The starting powders used were B of 100 nm size, pre-doped with various C contents by plasma spray synthesis and 99% Mg

---

[287] V. Braccini et al., *IEEE Trans. Appl. Supercond.* 17 (2007) 2766

[288] A. Malagoli et al., *Supercond. Sci. Technol.* 22 (2009) 105017

[289] V. Braccini, *Physica* C 456 2007 209

[290] M. Tomsic, *Physica* C 456 (2007) 203

[291] B. A. Glowacki et al., *Supercond. Sci. Technol.* 14 (2001) 193

[292] M. D. Sumption et al., *Supercond. Sci. Technol.* 19 (2006) 155

[293] M. A. Susner et al., *Supercond. Sci. Technol.* 24 (2011) 012001

of 25 mm size. After thorough mixing and ball-milling in Ar atmosphere, monofilament wires of 0.83 mm diameter with a Nb barrier and a Monel outer sheath (MgB$_2$/Nb/Monel) were manufactured. The reaction was performed at 700 °C for 40 min, the final composition in the final MgB$_2$ filament being 2.54 mol% C. This wire exhibited J_c(4.2K) = 1 × 10^4 A/cm^2 at 13.2 T.

The Internal Mg Diffusion (IMD) Process

The in situ and ex situ PIT methods lead to low relative densities of MgB$_2$ in the filaments, typically between ~45 and ~70% of full density, producing a low connectivity between MgB$_2$ grains which strongly limits J_c. A method yielding considerably higher MgB$_2$ mass densities is the internal Mg diffusion (IMD) process, a modification of the infiltration process[294]. The fabrication of multifilamentary MgB$_2$ wires by the IMD Process follows the scheme in Figure 11.63. According to Hur [295], a pure Mg rod with a diameter of 2.0 mm is placed at the center of a metal tube and the hollow space between the metal sheath inner wall and the Mg rod is filled with a boron + SiC powder mixture.

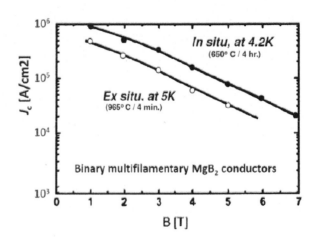

FIGURE 11.64: Comparison between J_c of industrial binary, multifilamentary MgB$_2$ wires. In situ: Hyper Tech (USA), 19 filaments, Monel/Nb sheath Ex situ: Columbus (It), 14 filaments, Ni sheath.

The composite is cold worked into a wire of 1.3 mm diameter by groove rolling, followed by drawing. Either 7 or 19 pieces of the monocore wires are bundled and inserted into a Cu-Ni tube and then reacted. During the heat treatment, Mg diffused into the B layer and reacted with B to form MgB$_2$. Figure 11.63 (c) and (d) show optical micrographs of the transverse and longitudinal cross sections of Ta-sheathed 7-core wires after 645 °C for 1 hour[296]. After the heat treatment a reacted layer with a thickness of 10–30 μm forms along the inner wall of the Ta sheath, and holes are formed at the center of each filament, where the Mg core was located before reaction. The longitudinal cross section of heat-treated wire shows that the reacted layer forms quite uniformly along the length of the wire, provided that the reaction temperature remains below 649 °C, the melt-

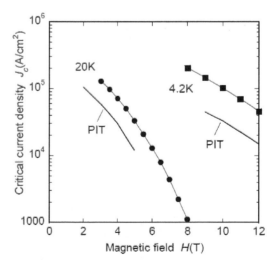

FIGURE 11.65: J_c vs B for an IMD processed, multifilamentary MgB$_2$ wire with SiC additives.

[294] G. Giunchi et al., *Supercond. Sci. Technol.* 16 (2003) 285.

[295] J. M. Hur et al., *Supercond. Sci. Technol.* 21 (2008) 032001

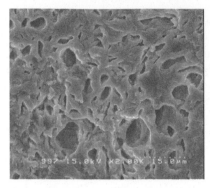

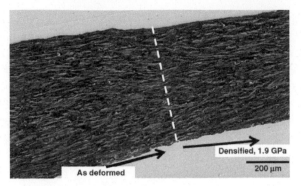

FIGURE 11.66: SEM image of an in situ MgB_2 filament after reaction[297]. Mass density: 50% .

FIGURE 11.67: SEM image of an in situ MgB_2 densified filament after reaction at 1.9 GPa.

ing temperature of Mg. X-ray diffraction analysis shows that the major phase in the reacted layer is MgB_2, with small amounts of impurity phases, mainly MgB_4, particularly in the region near the Ta sheath. From the lattice parameter a = 0.3082 nm, a C content of $x = 0.1$ in the formula $Mg(B_{1-x}C_x)_2$ was deduced, indicating that C atoms arising from the decomposition of the SiC additive substitute for B in the MgB_2 phase formed in the same manner as in the PIT processes. In case of the single filament wire, a thick B+SiC layer remained unreacted between the reacted layer and the Ta wall. The presence of unreacted B and C is strongly reduced in the 7-filament wire, and almost disappears in the 19-filament wire heat treated at 640 °C C, the penetration of Mg into the hot pressed MgB_2. As a consequence, the values of J_c are higher than for wires prepared by PIT processes. The value of J_c at 4.2 K increases with reaction temperature and reaches a maximum at around 640 °C. Figure 11.65 shows J_c *vs* B curves at 4.2 K and 20 K for a 7-filament wire heat treated at 640 °C for 1 hr. The values of J_c, calculated for the reacted area, are 1.3×10^3 A/cm^2 and 5×10^5 A/cm^2 at 20 K/3 T and 4.2 K/10 T, respectively. These J_c values are the highest ones reported so far for MgB_2 wires.

11.6.5 Densification Effects in MgB_2 Wires

As already mentioned, the maximization of J_c in MgB_2 wires requires a high mass density of the MgB_2 filaments. This is so far achieved only by the IMD method[295,296]: additional means have to be applied to PIT wires to enhance the mass density and thus the critical current density J_c in PIT processed wires. The microstructure of an in situ filament with a mass density of ~ 50% is illustrated in Figure 11.66[297]. Recently, the relative mass density of MgB_2 of *in situ* wires was enhanced from ~ 45 to ~ 58 % by a new technique developed in Geneva, the Cold High Pressure Densification or CHPD[298]. The densification was obtained by pressing simultaneously on the 4 sides of a wire, thus yielding square wire cross sections. The data obtained so far on binary and alloyed in situ wires, for both, mono- and multifilamentary configurations show a substantial increase of J_c after CHPD processing[299]. As shown in Figure 11.68, an increase of J_c by a factor 2 at 4.2 K, by a factor 5 at 20 K and by a factor 6.5 at 25 K was obtained after densification at $p = 1.5$ GPa on monofilamentary wires alloyed with $C_4H_6O_5$ additives and reacted at 600 °C for 4 hours.

---

[296] K. Togano et al., *Supercond. Sci. Technol.* 22 (2009) 015003

[297] J.H. Kim et al., *Physica C* 470 (2010) 1426

[298] R. Flükiger et al, *Supercond. Sci. Technol.* 22 (2009) 095004

[299] R. Flükiger et al., *IEEE Trans. Appl. Superconductivity* 21 (2011) 2649; M.S.A. Hossain et al., *Supercond. Sci. Technol.* 22 (2009) 095004

For fields parallel to the *ab* plane, the value of $J_c = 1 \times 10^4$ A/cm^2 at 4.2 and 20 K was reached at 13.8 and 6.4 T, respectively[299], the anisotropy ratio being 1.05. B_{irr} was enhanced by almost 2 T, due to the enhanced C content in the filaments, an unexpected consequence of the shorter reaction path in CHPD treated wires. A higher connectivity is observed, the superconducting volume fraction for percolation decreasing from 25.3 to 12.5%[300]. The CHPD process has been also successfully applied to longer wire lengths using a newly developed machine, which is promising in view of industrial applications[299].

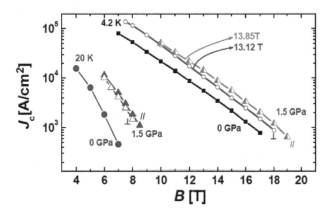

FIGURE 11.68: Comparison between the variation of J_c *vs* B at 4.2 and 20 K for an in situ MgB$_2$ wire, unpressed and CHPD treated at 1.5 GPa[299] (monofilamentary, C$_4$H$_6$O$_5$ additives, 4 h/600 °C).

11.6.6 Perspectives for Further Development of MgB$_2$

The progress achieved in the field of MgB$_2$ conductors in the last 10 years is impressive, from the theoretical, as well as from the experimental point of view. The three known methods used for the fabrication of industrial wires have been presented, each one having individual advantages. Ex situ wires show a high homogeneity and have their best performance at low fields, while the advantage of in situ and IMD processing consists of an easier introduction of C-based additives and thus in higher J_c values at high fields. The potential of MgB$_2$ wires in a round or practically isotropic square geometry can be estimated from Figure 11.69, which shows the variation of J_c *vs* B for thin films[301], MgB$_2$ tapes[299] and for wires produced by in situ[290,292] and IMD[295,296] processing.

The present limits of J_c can be estimated from the data for highly textured thin films[301], which exhibit very high J_c values for fields parallel to *ab*, the pinning force F_p being comparable to that of Nb$_3$Sn.

As we write in 2011, MgB$_2$ is a "niche" superconductor, but the niche is apparently quite large. Why? First, it is cheap: going by the projections of Columbus, at least 10 times cheaper than HTS conductors. Second, it can be made by the PIT process in round wire or in tape shape, thus showing a flexible architecture. Third, although it cannot challenge either Nb-Ti or Nb$_3$Sn at 4 K, it can challenge them at 20 to 30 K. Because of its low cost, it clearly challenges any HTS conductor for low field applications in this temperature domain too where MRI and high current busbar applications

[300] C. Senatore et al., *IEEE Trans. Appl. Supercond.*, 21 (2011) 2680

[301] A. Matsumoto et al., *Appl. Physics Express* 1 (2008) 021702

need an affordable and capable superconductor. MgB_2 is viable here, where the cost of HTS rules it out for such applications. Especially in planned upgrades at the LHC at CERN, there seems to be a near term large market that MgB_2 can address.

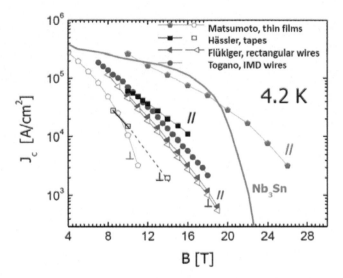

FIGURE 11.69: Comparison between J_c *vs* B for various configurations of MgB_2. Strongly anisotropic J_c behavior: thin films: Matsumoto[301]; in situ tapes: Häßler[302]; almost isotropic J_c behavior: in situ wires: Flükiger[299]; IMD wires: Togano[296]. The values for Nb_3Sn have been added for comparison.

What are the perils for MgB_2? The largest is perhaps the fact that the promise of low cost in manufactured form will not occur. But there are at least two companies worldwide competing with each other trying to demonstrate that low cost is feasible, so one can be plausibly optimistic on this front. A second concern is that the low fill factor of the superconductor in the total wire cross-section, typically less than 20%, is not showing significant increase at the present time. This means that there is much expensive metal around the superconductor, much of which is alloyed and therefore not very effective as a protection or stabilizing element in the conductor. The third concern is that the critical current density of the MgB_2 itself is not showing much sign of improving except by working on controlling known defects of MgB_2 such as its low relative density.

What are the future opportunities for MgB_2? First of all is the possibility that the critical current density will be raised by working effectively on vortex pinning by additions. So far vortex pinning occurs mainly by making very fine grains, something which is not too difficult when one starts with ball-milled MgB_2, but which is more difficult when starting with the in situ process. The second really big opportunity would come in greatly expanding the niche within which MgB_2 exists if it were possible in fact to enhance the upper critical field B_{c2} by introducing very strong scattering into the grains, without at the same time decreasing the grain to grain connectivity. Given the long period over which all other superconductors have been developed to their present state, it is reasonable be optimistic about further significant advances in the conductor technology of MgB_2.

───────────
[302]W. Häßler et al., *Supercond. Sci. Technol.* 23 (2010) 065011

Summary

All in all, MgB_2 is at a promising stage in its progression from an interesting material to a practical conductor. Its cost structure is attractive and it has some significant and potentially large-scale applications that really can use it. These advantages should keep it relevant for still a few years yet.

11.7 Further Reading

Nb-Ti—From Beginnings to Perfection

1. In 1986 the Applied Superconductivity Conference celebrated the 75th anniversary of the discovery of superconductivity with a symposium on the history of superconductivity. The symposium is published in full in *IEEE Trans. Magn.*, 23, pp. 354-415, 1986.

2. The most detailed account of the events surrounding the discovery of Type II Superconductivity can be found in A. G. Shepelev's the The Discovery of Type II Superconductors (Shubnikov Phase) in *Superconductor*, Edited by: Adir Moyses Luiz, ISBN 978-953-307-107-7, Publisher: Sciyo, August 2010.

   http://www.intechopen.com/books/show/title/superconductor.

3. The development of the Nb-Ti nanostructure is covered in more detail in: Conductor Processing of Low-T_c materials: The Alloy Nb-Ti, L. D. Cooley, P. J. Lee, and D. C. Larbalestier, in *Handbook of Superconducting Materials*, ed. David A Cardwell and David S Ginley (Institute of Physics Publishing, Ltd, Bristol 2003), Volume I: Superconductivity, Materials, and Processes, Chapter B3.3.2, pp 603–637. Also see P. J. Lee, "Abridged metallurgy of ductile alloy superconductors," in J. G. Webster, ed., *Wiley Encyclopedia of Electrical and Electronics Engineering*, Vol. 21, New York: Wiley, pp. 75–87, 1999.

History of Nb$_3$Sn and Related A15 Wires

1. In 1986 the Applied Superconductivity Conference celebrated the 75th anniversary of the discovery of superconductivity with a symposium on the history of superconductivity. The symposium is published in full in *IEEE Trans. Magn.*, 23, pp. 354–415, 1986.

2. In 2008 a special edition of the journal *Cryogenics* brought together articles on the current state of the art of low temperature superconductor science and technology: *Cryogenics* 48 (2008).

3. A history of technical superconductors in Russia, including Nb$_3$Sn, is covered in A. K. Shikov et al., The History of Technical Superconductors Development in Russia, *IEEE Transactions on Applied Superconductivity* 17, no. 2 (2007): 2550–2555.

Acknowledgments

Section 1: The Long Road to High Current Density Superconducting Conductors

David Larbalestier thanks Peter Lee for many discussions and Alex Malozemoff for a critical manuscript review. The vital help in final chapter preparation by Dmytro Abraimov is gratefully acknowledged.

Section 1.2: Alloyed Superconductors, The Type I and Type II Transition and the Collective Failure to Understand It.

A. G. Shepelev acknowledges with gratitude the discussions with academician N.F. Shul'ga, Director of NSC KIPT/Akhiezer Theoretical Physics Institute and with members of the Scientific Council.

Section 4: Bi-Ca-Sr-Cu-O HTS Wire

MWR thanks Alex Malozemoff, David Larbalestier and Bill Carter for critical review of this manuscript and acknowledges the contribution of the WDG members to advancing the understanding of BSCCO material science.

EEH acknowledges support from the DOE-HEP, NSF-DMR, and the State of Florida, as well as useful discussion with colleagues at the Applied Superconductivity Center at the National High Magnetic Field Laboratory at Florida State University.

Section 5: Coated Conductors: Section Generation HTS Wire

The authors thank David Larbalestier for comments on the manuscript.

12

Large Scale Applications

Editors: Peter Komarek, Bruce Strauss, and Steve St. Lorant

12.1 Introduction
   Steve St. Lorant 713
12.2 The History of Superconductivity in High Energy Physics
   Steve A. Gourlay and Lucio Rossi 716
12.3 Magnet Engineering—Study in Stability and Quench Protection
   Luca Bottura and Al McInturff 737
12.4 The History of Fusion Magnet Development
   Jean-Luc Duchateau, Peter Komarek and Bernard Turck 753
12.5 Electric Power Applications of Superconductivity
   William Hassenzahl and Osami Tsukamoto 769
12.6 Magnetic Separation
   Christopher Rey 797
12.7 Superconducting Induction Heating of Nonferrous Metals
   Niklas Magnusson and Larry Masur 811
12.8 Superconducting Magnets for NMR
   Gerhard Roth 817

12.1 Introduction

Steve St. Lorant

Rarely in the history of science has so great and useful a discovery as superconductivity had so long a period of gestation before acceptance by the technical world. From the very beginning the phenomenon was burdened by the stigma of arcane science, metals such as mercury, tin and lead that are unsuitable for any practical application, nonexistent theoretical underpinnings, the very low and difficult-to-achieve temperatures, and even early doubts by the discoverer as to the practical value of zero electrical resistance. In the following two years attitudes changed so much so that by the Third International Congress of Refrigeration held in Chicago in 1913 Kammerlingh Onnes was able to speculate: "...When all outstanding questions will have been studied and all difficulties overcome, the miniature coil referred to may prove to be the prototype of magnetic coils without iron, by which in future much stronger and ... more extensive fields may be realized than are at present reached in the interferrum of the strongest electromagnets. As we may trust in an accelerated development of experimental science this future ought not to be far away ..."

But the phenomenon brooked no rapid penetration into its mysteries. In 1914 Onnes discovered that resistance was restored by a magnetic field leading to the realization that one consequence of a critical magnetic field leads to a limiting current strength supported by a superconductor. Two years later Silsbee pointed out that the restoration of resistance in a wire was simply due to the magnetic field produced by the flowing current and not caused by any inherent properties of the metal.

The ensuing forty or so years were devoted to intense practical and theoretical investigations into the phenomenon, highlighted by the discovery of the Meissner-Ochsenfeld effect and obscured by confusing nomenclature defining superconductors as Type I, Type II, hard, dirty, sponge-like, mixed, intermediate, etc., thereby demonstrating its complexity. This confusion reached a zenith at the 1963 International Conference on the Science of Superconductivity where the participants unanimously petitioned the IUPAP to rule on the use of the symbols H_{c1} and H_{c2}! And that happened two years after the discovery of the properties of niobium-tin. At the same conference, in his closing remarks Pippard said: "... we are in the process of handing over to engineers the problem of superconductivity. If engineers have got to be trained to the point where they are competent to understand a time-dependent Ginzburg-Landau equation, or perhaps BCS theory and Green function formalism, I suspect they'll have precious little brains left for the useful arts and will become bad scientists and worse engineers."

How wrong was he! Another dynamic was at work—the needs of physicists and engineers in other fields, in energy, in particle physics, in transport, in industries not usually associated with state of the art engineering. Mere months after the discovery of the high-field properties of niobium-tin, the particle physics community (they were not yet HEP in those days) met to discuss the potential use of this new material for the construction of magnets, for detectors initially. For accelerators conventional magnets were still de rigueur. Meanwhile in the industrial sector an intense search for suitable materials followed, with the result that niobium-zirconium became what we may call the first superconductor produced on an industrial scale, Avco Everett's NbZr Supergenic Strip, SG 700, consisting of nine wires embedded, in parallel, in a copper substrate, boasting of a current density of 2.78×10^4 A/cm^2.

We should not forget the simultaneous search for high T_c compounds, Nb$_3$N, V$_3$Si, V$_3$Ga, Nb$_3$Ge and ultimately, at ~21 K, the heroic Nb$_{79}$(Al$_{73}$Ge$_{27}$)$_{21}$! Of the alloys, NbZr and NbTi remained and of these only the latter survived, but not before a NbZr magnet became for a short while the first and largest superconducting magnet, built by Avco, a 3 meter, 4 tesla device, a working model of a type that would be used in a large-scale commercial MHD generator. The saddle-shaped magnet was constructed of five concentric modules, each containing two concentric winding layers and used 20,700 meters of conductor. In announcing this event, the Avco Everett Public Relations Department release casually noted that "... should the windings develop a resistive "hot" spot, the Avco design transfers the current out of that area and into the copper with no change in field strength. In effect, the copper acts as a shunt, allowing the current to bypass the "hot" spot until the spot cools and resumes superconductivity". And thus was the concept of cryostatic stability formally announced to the world.

Shortly thereafter Avco abandoned NbZr and reformulated its "Supergenic" strip with NbTi, in company with a number of other manufacturers. In parallel, the RCA and GE companies began manufacturing niobium-tin ribbon, marketed by the former as "Vapodep", the latter as plain "GE Superconductive Tape", together with a range of tape-wound high field superconducting laboratory solenoids, so much so that by the end of 1966 glossy product catalog literature was being distributed. The industrialization of superconductivity had begun.

This advertising literature interspersed with technical papers makes fascinating reading. One is struck immediately by the obvious lack of understanding of the principle of cryostability; somehow the importance of the triad of field, current and temperature was not fully appreciated, and the presence of an "excess" copper was regarded more as a trick of the trade rather than something which had a sound physical basis. But this changed in the summer of 1968 when at the Brookhaven Summer Study where the theoretical underpinnings of stability in cooled superconducting magnets were discussed at the same time as the first tests of the Argonne 12-foot bubble chamber magnet were presented, a magnet which was designed with this principle in mind. Also at this meeting the participants were treated to another, quite unanticipated, criterion, that of "intrinsic" stability. Taken together, this gathering of scientific and engineering talent in the name of superconductivity quite negated Pippard's earlier concerns, a solution of GLAG equations was obviously not required to build useful superconducting magnets!

Every science needs a forum. Initially progress in the technical field of superconductivity was announced in the pages of the proceedings of the Cryogenic Engineering Conference. Then in November 1966 a euphemistically called National Superconductivity Information Meeting was hosted by the Brookhaven National Laboratory, sponsored by the US Atomic Energy Commission, which quickly morphed into the Applied Superconductivity Conference. For several years this gathering sought a home for the publisher of its proceedings, in 1974 it finally found refuge in the IEEE and proudly displayed its logo and that year's papers on the front cover of an issue of *IEEE Transactions on Magnetics*. ASC papers are now published in an issue of *IEEE Transactions on Applied Superconductivity*. Of course in any nascent science or technology there is a growing demand for publication space, our science being no exception, so that in the succeeding years the number of journals and publishers demanding our emanations proliferated.

Let us return now to hardware. In the four years following Brookhaven 1968, we saw intense activity in the DC magnet construction business as well as the first tentative entries into pulsed magnets, catalyzed by the work at the Rutherford Laboratory, to which we owe the eponymous conductor design we all, or almost all, use today. During this time also numerous proposals for the use of superconducting magnets in fusion research, levitation, medicine, power generation and transport were made. Who can forget the daring superconducting electromagnetic guns intended to hurl equipment into space, the high speed levitated trains and the promise of new medical applications for intravascular navigation or tumor therapy? Unfortunately this period also saw the demise of a number of promising conductor fabricators whose products, while conceptually well designed, were often far too expensive for the meager demand. Yes, the applied superconductivity community was replete with wonderful ideas, enthusiastically presented at the ASC meetings, but by and large they were paper studies, prompting the late Paul Reardon to remark on one occasion that he was "...but a part of the theoretical applied superconductivity community".

This state of affairs did not last long; various governments with deep pockets began to understand the importance of superconductivity and began to support projects in which the phenomenon played an important part. In the following pages we will read about the history of fusion magnets, about the efforts the power generation industry made to introduce novel concepts into established and very conservative utilities and about successes in industries which once seemed to be at the most remote end of the interest spectrum: ore beneficiation, and metal treatment. High energy physics due to its high demand for magnets quickly subsumed the major conductor development in an effort to satisfy the growing need for larger and more powerful accelerators and detectors; not only were improvements in the conductors sought, but industrialization of the entire facility construction process was invented. We will be reminded by our contributors that the road to the Large Hadron Collider, HEP's most ambitious undertaking was often littered both with niggling failures and monumental mistakes, and we expect that the next major project involving superconductivity, ITER, will encounter similar bumps in its road to a successful demonstration of its potential.

Several articles will be conspicuous by their absence, magnetohydrodynamic power generation, levitated transport and some facets of medical applications. Magnetic resonance imaging systems will be found elsewhere in this book; and while the design and fabrication of the superconducting magnets for this application is by now a major, worldwide industrial endeavor, the initial attempts to produce a patient-friendly diagnostic system make fascinating reading.

Magnetohydrodynamic power generation seems to have receded into the background, obscured no doubt by the extensive attention focused on fusion. Nevertheless some impressive superconducting magnets for MHD were designed and built, none more notable than the Argonne production of the magnetic channel for use in the bypass loop of the U-25 MHD facility in Moscow, a part of the 1976 US-USSR magnetohydrodynamics information exchange. Regrettably this program never came to fruition and whatever data were obtained disappeared into forgotten archives. A comparably sized dipole magnet was designed and built by the same team for the UTSI Coal Fired Flow MHD Research Facility in 1981; channel corrosion problems as well as NO_x formation and control led to the termination of the program in 1995.

Magnetohydrodynamic drives were popularized by the *Star Trek* universe and were featured in a number of science fiction stories, the best known being the film adaptation of the *Hunt for Red October* which described the drive as a "caterpillar drive" for submarines, a stealthy, silent propulsive system. In the novel of the same name the drive was a "pumpjet", a variant of which appeared in reality in Japan, in 1992. The Ship and Ocean Foundation set up a research organization in 1985 to demonstrate that a ship can actually be propelled by MHD thrusters: the *Yamato 1* was the result. This 185 ton ship had a drive consisting of six dipole superconducting magnets of 4 tesla each, linked together in a hexagonal pattern around the seawater ducts and a small but complete cryogenic recondensing facility on board. The magnets were operated in the persistent mode. Its design speed was about 8 knots. On the basis of this program, the participants concluded that to make superconducting MHD ship drives a practical proposition, very much larger magnets with higher fields were required, but, even more importantly, a higher electrical conductivity of the sea water was essential!

Turning to levitation, particularly of people movers, we find intense initial interest in the superconductor community, an interest which inevitably always seemed to be on the verge of a major breakthrough. The Magneplane of 1972 epitomized this thinking and for virtually the first time addressed issues not directly connected with magnets, superconducting or otherwise: passenger comfort, the needs of the infrastructure, rights of way and the like. Subsequently a number of short track demonstration projects were launched in Europe and Asia culminating in the opening for public service in 2004 of the Shanghai Maglev Train, a 30.5 km long track covered at a commercial maximum speed of 431 km per hour; its EMS system however powered by resistive magnets.

As we will learn in more detail in the following pages, the phenomenon of superconductivity has earned its place in the world of magnet engineering, power management and varied industrial applications. It is not a universal panacea for things magnetic, but it is a very benign phenomenon when approached with understanding and treated with respect.

12.2   The History of Superconductivity in High Energy Physics

Steve A. Gourlay and Lucio Rossi

12.2.1   Introduction

Scientists spent the last half of the century putting together what is called the Standard Model of particle physics. The Standard Model, which explains the basic interactions of fundamental particles that make up everything we see, is the most complete physical theory in history, yet it leaves 95% of the universe unexplained! Particle physicists use accelerators to recreate the conditions of the early universe in an attempt to piece together the complex puzzle of how we got to where we are today. These huge machines are used to accelerate particles like electrons, protons and ions of various masses to high energies where they are brought together in collisions that generate particles that only existed a few moments after the Big Bang that created the universe 15 billion years ago.

An electron or proton accelerated by a potential difference of 1 V acquires an energy of one electron-volt or 1 eV. Typical accelerators today operate in the range from millions of electron-volts (MeV) to Tera electron-volts (TeV). Accelerators enable particle physicists to explore the fundamental nature of matter and gain insight into the rules that govern their interactions.

The Large Hadron Collider, or LHC, is located near Geneva, Switzerland and is the largest and most powerful particle accelerator in the world with a circumference of 27 km. Protons with energies of 7 TeV will be brought to collision inside giant detectors used to reconstruct the complex collisions that consist of hundreds of particles. This gargantuan "time machine" will generate conditions that existed approximately 20 billionths of a second after the Big Bang, and, if nature is kind, will uncover phenomena never seen before.

Accelerators were invented at the beginning of the 1930s and soon became key instruments for nuclear and particle physics. Ernest O. Lawrence, who won the Nobel Prize in 1939 for the invention of the cyclotron, a circular accelerator that opened the door to high-energy physics, was the first to use the new technology for medical applications.

In this history of accelerators, of high-energy physics, (HEP), seen from the point of superconductivity we would be remiss if we failed to mention the first purpose built accelerator, the Bevatron. When operations started in 1954, this weak focusing proton synchrotron with resistive magnets had initially but one mission: to demonstrate the existence of the antiproton. It achieved that goal signally and in the years following gave the nascent HEP community an incredibly useful tool with which to explore the nuclear realm. Figure 12.1 is a picture of this historic particle accelerator during decommissioning, a sight guaranteed to evoke many a memory in the readers of this book, and at the same time highlight the remarkable progress which the phenomenon of superconductivity has brought to HEP.

FIGURE 12.1:   The Bevatron. (Courtesy LBNL.)

Figure 12.2, the so-called Livingston Plot, illustrates the increase in accelerator energy over time; it charts the evolution from Lawrence's cyclotron, capable of 0.03 to 0.1 GeV, in the early 1930s, through the Cosmotron and Bevatron to today's LHC and on to possible future projects. As can be seen from the plot, starting in the 1980s, superconducting magnets have been a major enabling technology in the development of particle accelerators.

The adventure started in the mid-1960s, thanks to the pioneering work of W.B. Sampson at Brookhaven National Laboratory (BNL), who built a 76-mm-aperture, 85-T/m quadrupole magnet model wound from Nb_3Sn ribbons and cold tested in January 1966. The feasibility and reliability of large superconducting magnet systems was demonstrated by the Tevatron at FNAL, which was commissioned in 1983. The Tevatron paved the way to commercial applications of applied superconductivity (such as Magnetic Resonance Imaging or MRI systems) and to a series of ever more ambitious projects (HERA, SSC, UNK, RHIC and now LHC), which have continuously pushed the technology forward.

A wide variety of acceleration methods have been developed over the years, but they all require a means of generating a time-varying electric field to accelerate the charged particles. Modern accelerators come in two main configurations. Linear accelerators or "linacs" built from a series of radio frequency (RF) structures that generate the accelerating field and circular machines (cyclotrons and synchrotrons) that contain the particles using magnetic fields, sending the particles through the accelerating structure again and again until they have reached their final energy. Each can operate with a single beam, as pure accelerators, or as a collider, where two oppositely directed beams are focused and brought into collision inside the particle detectors placed around the ring. In the collider mode the center of mass energy is maximized, but the beam stability and accuracy of the trajectory must be one to two orders better than that in a pure accelerator, and the size and position of the beam at the collision must be controlled from the micro- to the nanometer level.

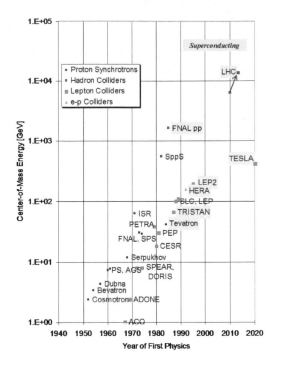

FIGURE 12.2: The Livingston Plot for accelerators and colliders.

It was quickly realized that in these machines, superconductivity could be used to obtain much higher energies. Resistive magnets have a maximum field of < 2 T and, in principle, if a ring of magnets were large enough, could be used to achieve extremely high energies, but the power consumption would be enormous, not to mention the capital costs of building the tunnel to contain them. Without superconductivity, the field of high energy physics (HEP) would not be what it is today. However, the advantages of superconducting magnets did not come without considerable challenges. Superconductivity, that is the loss of electrical resistance in matter, is a phenomenon largely restricted to the very low temperature regime. Thus superconducting magnets must be kept "cold", very cold indeed, and the operation of small and large cryogenic systems to do so is one of the major challenges. The materials which are superconducting themselves are another; at the beginning of the era of practical applications there were a number of candidates, of these various alloys of niobium and titanium still dominate conductor species used for HEP magnets. For applications requiring magnetic fields beyond the properties of niobium-titanium alloys, the so-called A15 compounds, true chemical compounds of niobium and tin, Nb_3Sn, are being considered, as well as the so-called high temperature superconductors (HTS). The latter materials, discovered about twenty five years ago have unique properties but unfortunately all high field materials are inherently brittle and strain sensitive, a particularly regrettable situation given that higher fields lead to higher forces and thus a higher stress and hence strain on the conductors.

12.2.2 Accelerators

Manipulation of particle beams is based on electric and magnetic fields. The basics of cyclic accelerators, by far the most common accelerators in the physical sciences and in medicine, are depicted in Figure 12.3. Electromagnetic fields are shaped to have the electric components parallel to the particle trajectory and in phase such that when the particle crosses the gap between the electrodes it receives a positive kick of energy. Magnetic fields perpendicular to the particle trajectory are shaped in such a manner as to bend the particles and recirculate them into the gap repeatedly, thereby adding continuously to the energy kick.

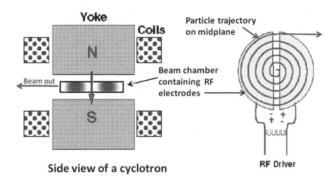

FIGURE 12.3: The principle of the cyclic accelerator.

Obviously the voltage between the electrodes must vary in sign; whatever dc voltage is impressed initially, it cannot always be positive, in order that the particle is accelerated at each crossing of the gap. This means that the field polarity must reverse with a frequency ranging from tens of MHz to a few GHz, according to the type of particle, energy and size of the accelerator. In addition to guiding the particle beam, usually accomplished by a uniform transverse dipole field, magnetic fields have to provide the necessary focusing elements to stabilize the beam against inevitable perturbation of the trajectory, of the main field or against intra-beam and inter-beam effects.

This restoring or focusing force is given by a field gradient, usually in the form of a quadrupole field. Figure 12.4 illustrates this action. "Higher order" magnetic fields are required to nudge the beams and keep the particles in a stable orbit in the machine throughout the acceleration phase and

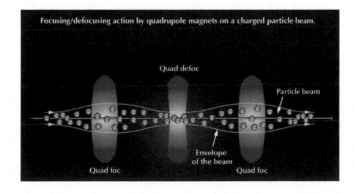

FIGURE 12.4: The properties of a quadrupole magnet. (Courtesy CERN.)

during storage. The focusing of a quadrupole depends on the energy of each particle and particle beams have a spread in energy and momentum. This leads to an effect called "chromaticity" and requires the use of a sextupole field to compensate. An octupole field provides special features, damping of certain beam instabilities for example, but it needs to be carefully controlled. In colliders, where the beam maintains its top energy for hours and millions of turns, higher order multipole magnetic fields need to be carefully controlled to avoid resonant behavior and instabilities. For example, in the LHC the magnetic fields are carefully controlled as far as the dodecapole component, a requirement which sets particularly stringent specifications on the magnet builders, with concomitant demands on manufacturing quality control.

In the case of a circular accelerator the main parameter is the bending strength, that is the product of the field strength, B, and length, L, of each dipole magnet. For relativistic accelerators one can show that the top energy scales as $E \sim 0.3BR$, where B (in Tesla) is the maximum field in the dipoles and R is the bending radius (in km). With this choice of units, E is measured in TeV. No wonder that the history of high energy physics has been driven by size, nurtured by increasing beam energy. Despite that, the high magnetic fields provided by superconductivity help mitigate this rush toward larger size, the current winner in the accelerator race, the LHC, is a ring of 27 km in diameter. Not surprisingly, in the name of science, even 200-300 km diameter ring structures have been proposed in recent years, the Eloisatron and the VLHC, where the meaning of the name or of the letters is left to the discerning reader. The main components of a collider are illustrated in Figure 12.5, which is a schematic representation of the LHC collider.

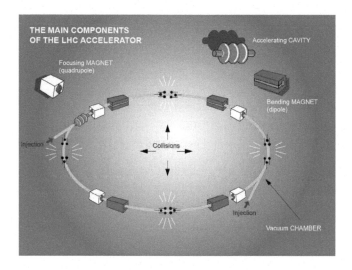

FIGURE 12.5:   A picture of the LHC: the combination of an accelerator and a collider.

An impressive number of accelerators have been built for various uses, from basic research to medicine and industry, many of them not even recognizable as such. Even the ubiquitous X-ray machine is a miniature accelerator and the detector a film or a semiconducting screen. Industrial accelerators serve a large range of applications, as simple irradiation tools in the food industry for preservation, to sophisticated ion implantation devices in the semiconductor industry. As Figure 12.6 shows, the end "product" of an accelerator is not necessarily only the advancement of our understanding of the constitution of matter on the nuclear or elementary particle level, but it is also a rich field of applied techniques for the benefits of society. We note further that a detector is not necessarily an essential component except in fundamental research where accelerators provide

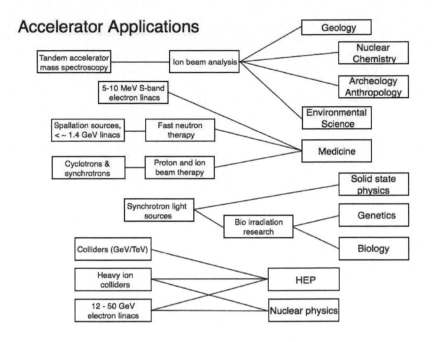

FIGURE 12.6: The world of accelerators.

new particles and new conditions that must be detected by special devices, to be described in the following section.

12.2.3 Detectors

Detectors have a part that is pure calorimetry that is the measurement of the particle energy and a part that we call spectrometry, which measures the charge and the momenta of the reaction products. While the total deflection of a particle is of course proportional to the product of the field and the length of a magnet, $B \times L$, the main parameters of the accelerator as mentioned above, HEP detectors usually measure the amount of bend (proportional to the particles' momentum), with

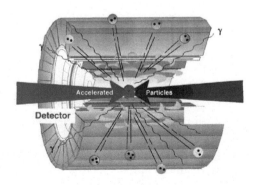

FIGURE 12.7: The detector at the collision point of two opposing beams of particles. (Courtesy CERN.)

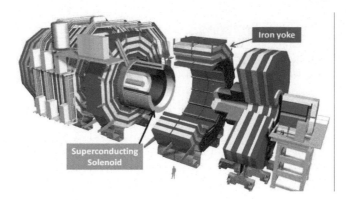

FIGURE 12.8: The CMS detector at the LHC. The superconducting magnet is usually a relatively small, less complex, though very important, component of a typical detector. (Courtesy CERN.)

detectors positioned inside the magnetic field to reconstruct the particle trajectories. In such a case the resolution of a magnetic spectrometer is proportional to $B \times L^2$: therefore large volumes pay off more than high fields. That is why the magnet size in detectors is huge: the largest magnetic system, the one for the ATLAS detector at the LHC, is a cylinder 20 m in diameter and 25 m long!

The principle of a detector at a typical collider is illustrated in Figure 12.7. Two opposing beams of accelerated particles, they can be electrons or protons or ions, or even a mix for example, are made to interact, collide head on in effect, inside an evacuated beam pipe. The products of the collision are then analyzed with a collection of specialized instrumentation chosen to maximize the information yield.

Given the magnetic field and spatial requirements, it is not surprising that despite the medium field level, 1 T average, 4 T peak, the construction challenges are enormous. Another detector magnet for the LHC, the one for the CMS experiment, Figure 12.8, is a detector with a classical solenoid.

FIGURE 12.9: The ATLAS detector. The superconducting magnet is contained in the eight toroidal cryostats. (Courtesy CERN.)

However its size with an internal diameter of 6 m, a length greater than 12 m and field, 3.8 T, gives it a world record 2.5 GJ of stored energy in a single superconducting magnet circuit. The other challenge inherent in detector magnets is that they are intimately integrated with the various particle subdetectors, electronics, wiring and other subsystems that constitute the main components of the detector. Often the magnet becomes the supporting structure around which the detector is built. Figure 12.9 shows the Barrel Toroid coils of ATLAS after their assembly and of the immensity of the detector as a whole.

One has to imagine that all the empty space must then be filled by a variety of subdetector electronics and other very complex systems, among which a larger liquid argon chamber, end cap toroids for spectrometry in the forward direction, and an inner thin superconducting solenoid must all be integrated into the detector.

12.2.4 Main Characteristics of Accelerator Magnets

The coils of the dipole magnets used in HEP accelerators may be arranged in different configurations to generate the desired uniform transverse field. These basic shapes are illustrated in the top half of Figure 12.10. A uniform field may be generated either by a constant current density with a geometry given by oppositely intersecting ellipses, or by a shell in which the current density is maximum at the midplane and vanishes toward the pole region with a current distribution given by $J = J_0 cos\theta$. In practice, real magnets are composed of shells of constant current to fit the cosine distribution. Quadrupoles need a $cos2\theta$ current distribution and so on. The lower half of Figure 12.10 illustrates the coil cross-section of the actual LHC dipole.

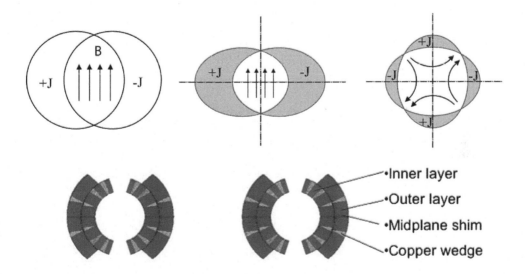

FIGURE 12.10: Top: Ideal current distributions to generate dipoles and quadrupole fields. Bottom: The LHC twin dipole configuration.

12.2.5 The Benefits of Superconductivity

The central field produced by a cosine distribution scales as $B \times J_0 \times a$, where J_0 is the current density and a is the thickness of the coil package. Obviously, the higher the current density, the higher the field. In addition, the coil volume increases faster than the field level. In other words the current density is by far the most important engineering parameter of accelerator magnets. Table 12.1 lists the ranges of values of the most important characteristics of magnet systems, superconducting as well as resistive. We note that while the tokamak systems have higher field and stored energy values, accelerator magnets require the materials capable of providing the highest possible current densities.

TABLE 12.1: The Typical Overall Current Densities of Various Magnet Systems

Magnet System (DC only)	Current Density (J_{overall} A/mm^2)	Operating Current (kA)	Typical Field Range, (T)	Stored Energy in the System (MJ)
Resistive—air cooled	1–5	1–2	1	0.01
Resistive—water cooled	10–15	1–10	2	0.05
Large S.C. coils for detectors	20–50	2–20	2–6	5–2,500
S.C. MRI Magnets	20–50	1	1–10	1–40
S.C. Tokamaks for Fusion*	25–50	5–70	8–13	5–40,000
S.C Laboratory Solenoids	100–200	0.1–2	5–20	1–20
S.C. Accelerators	200–400	1–12	4–10	1–10,000

The benefit of superconductivity is evident from the previous considerations and the LHC can be used to quantify this benefit and to emphasize the virtues and advantages of the technology of superconductivity.

The LHC ring is 26.7 km long and requires some 126 t of liquid helium inventory to operate its 8.3 T superconducting magnets. The power required by the refrigeration system is about 40 MW. If the LHC had been built with classical resistive magnets having fields of 1.8 T, the circumference would have been more than 100 km and have required about 900 MW of installed power. This would have led to prohibitive construction and operating costs; 900 MW is the output of a fairly good sized nuclear power plant. In addition it would have had an unacceptable impact on the environment since the 900 MW would be rejected as virtually unusable, warm water.

Using the LHC as a specific example, the main characteristics for accelerator magnets are:

- High overall current density is the first key parameter. This is obtained through a high value of the critical current density, J_c where by J_c we mean the critical current (I_c) at a particular field divided by the area of the superconductor (I_c/A_{Sc}); this is of course of prime importance. Not only must the intrinsic J_c performance be high, but clearly we need to operate the system at a small fraction of the critical current. For example, the LHC dipoles run at 85% of the load line intersecting the critical surface, but they have been designed, and many of them tested, to

---

* High numbers refer to ITER under construction.

operate up to 93%. Therefore even a small variation in the J_c performance, usually negligible in other systems, may have a direct effect on the magnet performance.

- Low stabilizer content. Copper is of course necessary but we need to keep it to a minimum, compatible with stabilization and protection requirements. Usually the copper to superconductor ratio, or Cu/Sc (or $A_{\text{stabilizer}}/A_{\text{nonstabilizer}}$ in a more complex system) ranges between 1.5 to 2 for NbTi-based systems and 0.9 to 1.5 for Nb_3Sn magnets. The stored energy can be as high as 8 MJ for each of these magnets.

- High compaction cable. Following initial attempts to use fully transposed braids, flat double faced cable, of which the so-called Rutherford cable is the prime example, is the invariable choice. It has a 90% compaction factor, which can increase to 93–94% after coil curing. The interstices between the strands are such as to allow helium to percolate inside to increase stability, particularly in NbTi magnets. Such a high compaction is not trivial considering the necessity of multi-kilo-Ampere cables as we will see later.

- Thin insulation. Total inter-turn thickness is less than 250 μm in modern accelerator magnets, usually consisting of polyimide tapes, to maximize the current density of the coil while withstanding discharge voltages in the kilovolt range.

- Collaring systems to withstand the forces. Dipoles, like toroids, are not self-supporting systems with respect to the forces generated by the magnetic field, as Figure 12.11 indicates. For accelerator magnets the restraining system must be outside the coils, to avoid reducing the current density. This is provided in the form of collars tightly surrounding the coils to prevent any motion which could lead to quenches (transition from the superconducting to resistive state caused by motion and subsequent heat generation) and provide the precise coil geometry for the required field quality.

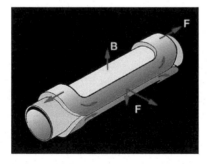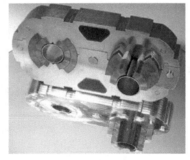

FIGURE 12.11: Forces on a dipole, left; a section showing the LHC dipole retaining structure, right.

- Very low stability margin. This is a consequence of the high Joverall, and translates into a very small stability margin, of the order of 10 to 1000 μJ. When we compare this to the 8 MJ of stored energy in the magnet, it is not surprising that accelerator magnets are designed to train and to withstand multiple quenches. Training is the process by which a magnet approaches the theoretical maximum operating current through a series of quenches, relieving stress in the coil and converging on a more stable mechanical configuration.

- It is important that the magnet has the ability to maintain memory of the process, and once trained to operating field, it can reach this field level with little or no further training. Since energy release is mostly generated by movement of the coil, the very low stability margin dictates not only a good system of force restraint but also a very accurate mechanical construction, which as we will see later, affects field quality.

- Protection through active heaters and bypass diodes to reduce the current in 100 to 500 ms to avoid dangerous hot spot temperatures. Magnets quench when a small volume of the coil, usually due to a microscopic movement of the conductor, transitions to the normal state. The current, formerly carried by the superconductor, shunts into the copper matrix generating a current density in the copper (J_{Cu}) of 1000 A/mm^2 which is not sustainable without damaging the magnet. Once such an event is detected, heater strips, placed within the coil are fired, causing the entire magnet to become normal and thus dissipating the stored energy throughout the entire mass of the coil, while at the same time the current (13 kA in the case of the LHC dipoles) is bypassed around the magnet via cold diodes. Development of 13 kA cold diodes and heaters near the coil, which are exposed to the same forces and stresses, has been a major endeavor of the LHC.

- Field quality. First of all, the magnets in the ring must all have the same field within a few units, a unit being 100 ppm of the main field, or about 1 mT for the LHC magnets at full field. Then, all higher order harmonics must be controlled at a level better than one unit, in some at the level of 0.1 units. Such a goal is relatively easy to achieve with electromagnets where the iron pole plays the primary role in shaping the field, but this is a real challenge in superconducting magnets as it requires the control of the conductor position to ±50 μm. Coil geometry dominates the field quality at the collision energy and thus requires control of the persistent currents, snapback effects and the coupling currents in the cable, that is governed by the interstrand resistance, to a very high degree. These superconducting effects are very important at low field during injection, and during ramp up.

12.2.6   Early History

Probably the first truly large scale application of superconductivity was in particle physics. As a technology eminently suited to the purpose, it allowed physicists initially to build detector magnets of a prodigious size and later to reach energies not accessible by any other means. Even so, it was not always taken for granted: as a former DOE Program Manager, David F. Sutter frequently remarked, rather acidly, "Never use superconductivity unless you absolutely have to!" Superconductivity was not fully embraced by the High Energy Physics community either, an experimentalist once remarked that, "superconductivity will be the death of High Energy Physics!"

The long relationship between superconductivity and high energy physics began with the discovery of Nb$_3$Sn in 1961, 50 years after Onnes' discovery. Shortly after the publication of this latest discovery appeared, Lawrence Berkeley Laboratory (LBL) laid out plans for development of bubble chamber magnets with fields of 10 T or more. This was a rather ambitious goal given that the initial conductor samples were microscopic and far from the "engineering" material required for magnet fabrication and the magnets were never realized. This sort of bold extrapolation has been repeated with the discovery of every new superconductor since, with mixed success. There is always a huge gap between the appearance of a new superconducting material and the subsequent actual engineering application. Despite the visionary goals of the LBL team, the first viable superconducting device used in high energy physics, a 25 cm bubble chamber magnet, was built at Argonne National Laboratory (ANL) in 1965. At this time there was a growing realization among high energy physicists that future experiments would require superconducting technology. In the early 1970s, three groups in Europe, three in the U.S. and one at KEK in Japan were studying the high energy superconducting accelerator. This flurry of activity was eventually reduced to two main projects in the late 1970s: ISABELLE and the Fermilab Energy Doubler/Energy Saver/Tevatron. The Tevatron was the machine that was ultimately built and it has been in operation since 1983.

A pivotal event in the application of superconductivity in high energy physics came in 1968, at a summer study held at Brookhaven National Laboratory (BNL). During this six week meeting, international experts examined many application-related challenges and laid the groundwork for the future of superconductivity in high energy physics.

In 1972 a team of physicists and engineers at the Rutherford Laboratory in the UK, under the leadership of Martin Wilson, figured out that superconducting cables could be made to work for accelerator magnets (and a host of other applications!) if the wire were made of a filamentary composite, in which very fine filaments of superconductor are embedded in a matrix of copper. The theory associated with the invention of this "Rutherford Cable" was written up by Wilson as Rutherford Memo A26, but never published! Magnets powered with superconducting coils can produce much higher fields, enabling higher energies to be reached for a given machine size. This key development ignited a new wave of HEP magnet applications. Government support has played, and continues to play, a large role in the development of superconductivity.

In the 1970s, the leaders in high energy physics considered superconductivity a necessity for growth of the field. The main focus at that time was filling in the missing pieces of the Standard Model of particle physics.

During that period Fermilab Main Ring was the largest operating proton synchrotron with a radius of 1 km, a peak field of 2 T, a power consumption of more than 50 MW and an operating energy of 400 GeV. Type II superconductors were becoming available that offered the prospect of operating at fields higher than 4 T with no resistive losses. Superconductivity would allow Robert R. Wilson, the founding director of the laboratory to double the energy while reducing power consumption. Hence the original name Energy Saver/Doubler, which was later changed to Tevatron. The opportunity was too good to pass up: Wilson brought the initial accelerator project in considerably under cost and planned to use the left over funds to build a second superconducting ring with an energy of 1 TeV. The approach in designing the machine can be seen in a paragraph from the The Energy Doubler Design Study: "The design process, and if carried out, the construction of the Doubler, builds upon our experience at NAL. We have not proceeded on the basis of deciding what is readily practicable, designing to that, adding up the cost and attempting the result. Instead, we have set a cost goal and keep designing, redesigning, haggling and improving until we have done what we set out to do. Occasionally, we are forced to admit that we are not clever enough to achieve our cost goal and admit defeat, but not without a struggle." This was a fruitful time in accelerator development with several other applications being considered by the community. A 4 GeV experimental ring, ESCAR, was under construction at LBL, Brookhaven was developing a 400×400 GeV proton-proton collider called ISABELLE, and the Rutherford Laboratory undertook some crucial development of superconducting cable while studying the possibility of building the Super Proton Synchrotron at CERN with superconducting magnets, an effort that was eventually discarded in favor of conventional technology.

The construction of the Tevatron initiated the development approach and basic engineering that is still used today. This would be the first time that industrial-scale production of a superconductor became a necessity. Over 1,000 superconducting magnets were required: a model magnet program, based on short model magnets, was used to rapidly produce magnets with varying parameters. This program was the basis for development of fabrication techniques such as cable and insulation and ways to consistently produce magnets with the required field quality. In parallel, a long magnet production program was taking input from the short model program to develop full length magnets. Over 200 model magnets were produced in the two programs. Much of the technology developed for the Tevatron has been used on subsequent accelerators with minor evolutionary improvements.

HERA at DESY began construction in 1984, soon after the Tevatron came on and pushed the technology further. The HERA magnets were longer at 8.824 m as opposed to 6.4 m and worked at a higher field, 4.7 T versus 4 T at the Tevatron. The HERA magnet had an adequate amount of margin that was used later on to run the magnets at 5.5 T, boosting the proton ring energy from 800 to 920 GeV.

Among several innovations resulting from these projects, HERA was the first to have the magnets assembled industrially. This involved transferring the design to industry and carefully monitoring the product through the production cycle. The construction started in 1984 and the machine became operational in 1990.

During the period from 1985 to 1995 there was a great deal of intense work on superconducting magnets at other locations. The Relativistic Heavy Ion Collider (RHIC), at Brookhaven National Laboratory developed a very simple single shell coil that used the iron collars directly to contain the winding (Figure 12.12). Another project of the early nineteen eighties was the Superconducting Super Collider (SSC), the most ambitious application of superconductivity ever conceived. This machine involved all of the US National laboratories and much of the technology we have today was developed for it. Alas, a victim of politics and poor management, it died a sad death in 1993.

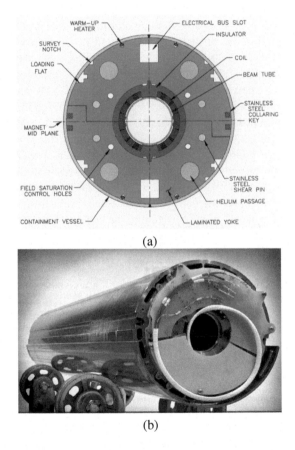

(a)

(b)

FIGURE 12.12:   (a) Cross section of a RHIC arc dipole. (b) The cold mass of a RHIC helical dipole. (Courtesy of Brookhaven National Laboratory-RHIC Collaboration.)

12.2.7   The Large Hadron Collider

The demise of the SSC proved to be a boon to the Europeans, who were in the process of developing the Large Hadron Collider at CERN in the ring space then occupied by LEP. Building on the technology of the SSC, the LHC contributed some key developments in accelerator technology: solving the problem of putting two magnets in one yoke and cryostat, operation in superfluid helium and large-scale industrial production. Beset with teething problems, the initial expectations were not fully met, but with an operating field of 8.33 T it will have the highest field value to date once the idiosyncrasies of this very complex machine are fully understood. Situations like this, where technology is pushed to the limit and fallen short of initial expectations, are common and have been repeated many times in the past. Establishing the parameters of such a complex system, trying to squeeze the best possible performance while trying to reduce cost, is a complex process only understood in hindsight. Let us look at the history in a little more detail. The original specification was for nominal 10 T, 50 mm bore, 10 m long, twin aperture with a separation of 180 mm, using a 17 mm wide, 15 kA cable. By the time of the EPAC'90 conference it was in the range 8 to 10 T. The work had started around 1983 with some winding tests at CERN, but the 1 m models were made in industry. At the time it was thought that it would be a good idea to whet the appetite of industry in order to drum up enthusiasm and increase the chance of being funded. It is not so clear now that that was a good idea: progress would probably have been much faster if the models had been made at CERN. Industry is more interested in large scale manufacturing than model work. In the early 1990s, when the SSC was canceled, things got serious and it was realized that the aperture was too tight; it was increased to 56 mm and at the same time the beam separation was increased to 194 mm. The 17 mm cables were hard to wind, so were reduced to 15 mm, and the magnet length was increased to 15 m in order to reduce the total dead space between magnets so as to maximize the integrated dipole field around the ring. It was then that 8.33 T (7 TeV) was decided, with 9 T "ultimate"—i.e., ancillary equipment designed so that if 9 T could be reached in the dipoles there wouldn't be a bottleneck elsewhere.

The LHC represents the summit of 20 years of development of Nb-Ti based accelerator magnets. The main characteristic is the two-in-one structure, i.e., the fact that two magnets are contained in the same cold mass, to accommodate the counter-circulating particle beams.

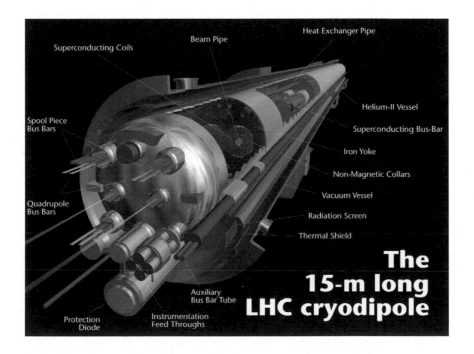

FIGURE 12.13: The LHC two-in-one dipole magnet. (Courtesy CERN.)

Figure 12.13, an artist's impression of the main LHC dipoles, shows this unique concept. The magnets are cooled by superfluid helium pressurized to 1 bar to boost the performance of NbTi. The collider uses the 27 km long tunnel of the previous CERN accelerator, LEP, and given the fact that the tunnel length was fixed, increasing the field as much as possible was instrumental in getting approval for the project. Moreover, the transverse tunnel dimensions were fixed as well, so a compact structure became essential, dictating the two-in-one design. In addition, it turned out to be a cost-saving factor: the twin aperture design fully magnetically and mechanically coupled, giving maximum compactness. Subsequent analysis has shown that this design feature saved at least 15% of the cost, or 200 MCHF.

Ultimately the LHC magnet R&D, design, construction and installation lasted more than 20 years. Figure 12.14 illustrates the LHC magnet roadmap with the main milestones. A few of the most interesting highlights in this process are presented here in order to illustrate the challenges in large-scale production of extremely complex scientific devices.

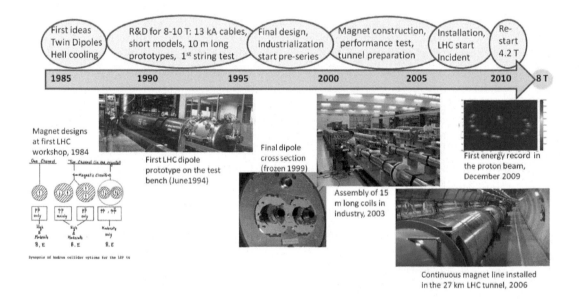

FIGURE 12.14: The path to success: a glimpse at the milestones and accomplishments of the LHC project.

There are more than 1200 dipole magnets in the collider which in principle must all be identical, in practice this is impossible. As there will always be manufacturing deviations from the ideal, the fabrication process must include all the lessons learned from the extensive prototyping process in which the various spatial and prestress correlations are established and elaborate quality assurance controls established. Considering that more than 5,000 dipole coils were produced, the coil size variation along its 15 meter length of better than 30 μm is a superb achievement. The field quality, the most essential parameter of all, is even better than originally desired which taken together with the superconductor fabrication, is one of the most striking achievements of the whole project.

The success of the LHC superconducting magnet construction, within schedule and with a negligible extra cost of less than 2%, meeting or exceeding all specifications shows the maturity of this technology. The entire installation with its huge detector magnets working very satisfactorily is yet another step in the quest of knowledge, a step that has been made possible by superconductivity.

12.2.8 Applications in Japan

Superconducting magnet systems play a key role in the Japan Proton Accelerator Research Complex (J-PARC). The experimental facilities utilize three superconducting magnet systems; a muon transport solenoid, a superconducting spectrometer and a 150 m superconducting beam line in the Neutrino Experimental Facility.

An excellent example illustrating that there is still room for creativity and improvement is a superconducting combined-function magnet developed by the Japan High Energy Research Organization, KEK, to optimize the magnet performance with a strong emphasis on cost effectiveness. This magnet, which features a coil radius of 170 mm, provides dipole (2.6 T) and quadrupole (19 T/m) fields superimposed in the same bore. The optics in the bending section of the beam line is simplified by using 14 pairs of these combined-function magnets. Also, it offers the advantage of requiring only a single component to design instead of separate dipoles and quadrupoles. The magnet design has been greatly simplified by the choice of a single layer coil surrounded by plastic collars and an iron, flux-return yoke.

The KEK-B BELLE detector is of course well known for its extensive use of superconducting elements in its B meson factory experimental program. The original detector, now in the process of being dismantled in preparation for a proposed upgrade, is shown in Figure 12.15. The beam pipes for the two counter-rotating electron beams are clearly visible in the foreground.

FIGURE 12.15:   The KEK-B BELLE detector. (Courtesy KEK, Japan.)

Another Japanese innovation pushing the limits of existing technology is an extremely thin superconducting solenoid magnet launched via balloon to high altitudes for more than 2 weeks, used to investigate cosmic-ray antiparticles. Called the Balloon-borne Experiment with a Superconducting Solenoid or BESS, the design goals were extreme: maximize the volume, minimize the mass and heat load, and be able to survive up to 10 g. The magnet has a coil thickness of only 3.4 mm to minimize particle interactions and is wound with advanced aluminum stabilized superconductor to achieve the required strength to weight ratio. The central field is 1.2 T in a volume of 0.9 m and a length of 1.4 m. Figure 12.16 is a photograph of the preparations for launch in the Antarctic of this imaginative experiment.

FIGURE 12.16:   BESS on the launch gantry in the Antarctic. (Courtesy Dr. J. W. Mitchell, NASA.)

12.2.9   RF Superconductivity

In our initial discussion of accelerators we stated that the manipulation of particle beams is based on electric and magnetic fields and thus far our attention has been focused on the magnetic field requirements for this manipulation. But particle beams once produced by a source need to be accelerated and this is where another component of the modern accelerator plays an essential role, the electromagnetic cavity resonator. The resonant frequency depends on the application, usually lying between 100 MHz and 3000 MHz, the type of cavity used in turn depends on the velocity of the particles to be accelerated. Initially cavities made of copper were universally used, but as our understanding of superconductivity progressed, the advantages of superconducting structures became increasingly obvious. One of the attractions is that the dissipation in the walls of the superconducting resonator can be many orders of magnitude less than that in a copper unit of the same species which is particularly attractive for accelerators which operate in the continuous wave (CW) mode. In other words, as the ohmic loss in the cavity walls increases as the square of the accelerating voltage, resistive cavities become increasingly uneconomical. In such a case as the surface resistance of a superconducting cavity is generally five orders of magnitude less than that of copper, even when the refrigeration needs are taken into account, there is a net gain of many hundreds. Figure 12.17 illustrates a collection of accelerator cavities. Clockwise from the left a 345 MHz double spoke resonator (ANL), an 800 MHz, 6 unit elliptical cell (JLAB/MSU), the pioneering Stony Brook Heavy Ion Accelerator resonator, and a variety of superconducting cavities. In the left foreground the DESY 9 cell TESLA cavity, on the right a CESR 500 MHz cavity, a small-Sscale version of which is on the left. The large Nb-Cu cavity in the background is a 200 MHz resonator developed by Cornell and CERN for future muon accelerators and neutrino factories.

The application of the technology of RF superconductivity to particle accelerators traces its beginnings to about 1965 with the acceleration of electrons in a lead plated resonator at Stanford. Contem-

FIGURE 12.17:   A selection of superconducting accelerator cavities for different applications. (Courtesy Dr. H. Padamsee, Cornell University.)

poraneously slow-wave structures were beginning to be developed at Argonne National Laboratory which culminated in the heavy ion linac known as ATLAS. By 1989 this machine had accumulated more than 40,000 hours of beam-on-target operating time, in the company of numerous other superconducting linacs around the world, accelerators designed not only for heavy ions but for proton and electrons as well. As a result of this collective experience it could be demonstrated that the dramatic reduction of the RF surface resistance could be achieved even in complex resonator geometries.

In the 1 to 10 GeV range, superconducting cavities have special advantages for electron accelerators intended for nuclear physics: a continuous beam, high average current with excellent beam quality. Such a machine would be used for precision measurement of electromagnetic cross sections and for coincidence detection of reaction products in electron-nuclear collisions. CEBAF at the Thomas Jefferson National Accelerator Facility in the USA and LEP-II at CERN in Europe became the two largest superconducting RF installations: at CEBAF, originally designed for 4 GeV, a beam energy of 6.5 GeV was achieved in five recirculating passes with a CW beam current of 200 μA. LEP-II installed a total of 465 meters of superconducting RF cavities. The accelerating gradients in

FIGURE 12.18:   CEBAF 5-cell cavities, 50 cm in length. (Courtesy Jefferson Laboratory.)

FIGURE 12.19:   350 MHz Nb/Cu cavities for LEP-II. (Courtesy CERN.)

the cavities rose commensurately, at CEBAF from the initial 5 MV/m to more than 7 MV/m with major improvements readied for the planned upgrade. Figure 12.18 illustrates the original assembly of 5-cell cavities operating at 1497 MHz.

In contrast, the LEP-II cavities consisted of a thin film of niobium sputtered onto a copper cavity shell which was initially capable of 6 MV/m, to be later upgraded to 7 MV/m. By mid 2000 the average sustainable gradients approached 7.2 MV/m. An assembly of these 350 MHz Nb-Cu cavities is shown in Figure 12.19.

The history of the remarkable progress achieved at continuous electron beam accelerator facility (CEBAF) is particularly interesting. When the decision was made to make the recirculating electron machine superconducting, it was based on a Cornell single cell design developed for storage rings. This cavity had routinely achieved a gradient of 5 MV/m and was proven in a beam test at cornell electron storage ring (CESR). The accelerator was built with major input from industry which provided the essential high purity niobium in unheard-of quantities, enough to make more than 400 cavities, establish mass production protocols for manufacture and assembly of these devices under exacting conditions and execute them.

The Spallation Neutron Source (SNS) at Oak Ridge National Laboratory is driven by a high intensity proton linac: the 1.4 MW machine is based on eighty-one 804 MHz superconducting cavities and accelerates protons from 200 MeV to 1000 MeV. Los Alamos and Jefferson Lab were partners in the development effort; Figure 12.20 illustrates the assembly of a 1 m long cavity string.

The LHC with 14 TeV in the CM is certainly in keeping with the historical energy growth rate. It too has the necessary complement of superconducting RF cavities: at 400 MHz, 16 Nb-Cu cavities in four cryomodules will provide 16 MW per beam and deliver about 180 kW of beam power. Figure 12.21 is a composite view of such an assembly.

So far we have concentrated our attention on the needs of high and very high energy machines but the world is full of other machines which over the years

FIGURE 12.20:   A 1 m long SNS RF cavity during assembly. (Courtesy Jefferson Laboratory.)

FIGURE 12.21: A superconducting RF LHC cryomodule. (Courtesy CERN.)

have become dependent on superconducting RF structures. High luminosity machines so equipped, noteworthy among others is CESR in the US and KEK-B in Japan. At the latter the two beams intersect at a finite angle to reduce the background at the interaction point and to simplify the local beam optics, but this reduces the luminosity and has other undesirable influences on the circulating beams. So called "crab" RF cavities were invented to deflect the bunches and thus position them for head-on collisions. Four 500 MHz cavities are required for this beam manipulation.

Another application is storage ring light sources whose numbers grew from just a few machines in the 1960s to about 70 machines worldwide either in various stages of development or built and fully operational. Not included in these numbers are the free electron lasers and SASE FELs such as the TESLA Test Facility at DESY in Germany, all of which depend on one or more superconducting RF components. In other words, RF superconductivity has also come of age.

12.2.10   New Applications

Wigglers and Undulators

The demand for more sources of synchrotron radiation is growing, leading to the construction of new laboratories and upgrades of existing facilities in order to provide higher performance beams. Light sources depend on undulators to provide high brilliance beams of tunable radiation. Currently short period undulators based on permanent magnet technology represent the state-of-the art but performance beyond the limits imposed by permanent magnets will require superconducting technology. Superconducting magnets facilitate larger gaps, shorter periods and higher fields, but aside from the obvious issues associated with field quality, a major challenge will be the development of a cryogenic system that supports small gap operation in the presence of beam heating. In addition, there will be a premium on cryogen-free operation using cryocoolers since most existing light sources do not have a cryogenic infrastructure.

Fast-Cycling Magnets

Until recently, there have been few applications of fast pulsed superconducting accelerator magnets. One example is the dipole design used for the Nuclotron at JINR in Dubna. Despite the complications of field quality degradation and cryogenic losses caused by persistent currents in the superconductor and eddy currents in the cable, iron and surrounding structures, the increasing demand for high intensity beams requires superconducting magnets with high ramp-rate-capability. Note that all the designs considered for this purpose rely on NbTi technology. The international Facility for Antiproton and Ion Research (FAIR), currently under construction at the Helmholtzzentrum für Schwerionenforschung (GSI), consists of two synchrotrons in one tunnel, the SIS 100 (100 T-m rigidity) and SIS 300 (300 T-m rigidity). The SIS 100 is the heart of the machine, accelerating protons and ions at a high repetition rate and distributing them to other parts of the complex. The required dipole ramp rate is 4 T/s up to a field of about 2 T for SIS 100 with a 1-Hz duty cycle and 1 T/s up to 6 T for SIS 300, with a duty cycle of 50%.

Future Applications

Superconducting magnet technology is continually evolving in order to meet the demanding needs of new accelerators and to provide necessary upgrades for existing machines. A variety of designs are now under development, including high fields and gradients, rapid cycling and novel coil configurations. The performance, or physics "reach" of a colliding beam accelerator is determined by the energy of the particles and the luminosity, where, for a given energy, the luminosity is a measure of the production rate for a particular process. The particles are brought to collision in the interaction regions by sets of quadrupoles placed on either side of the collision point. As machines go to higher energies, there is a trend toward higher gradients. In many cases, it is aperture that dominates the requirements, leading to higher fields on the conductor. Also, the high radiation/high heat load environments in which these magnets are expected to operate, make them very challenging.

Recent progress in the development of Nb_3Sn has encouraged the prospects for use in accelerator magnets but the application for the LHC upgrade adds additional issues to the already formidable list. The magnets will operate in a high radiation environment, subject to unprecedented beam induced heat loads, that will require development of radiation hard materials for coil construction and understanding the heat transfer characteristics of composite coils. The design process will necessitate working closely with accelerator physicists to understand trade-offs between magnet performance limits and the interaction region upgrade options.

Magnet applications for future colliding beam machines can be separated into two categories; upgrades of existing facilities such as the LHC, described below, and new facilities. High fields are the obvious choice for an upgrade scenario, while other factors such as tunnel cost, cryogenics, logistics and location are factors that have to be considered along with magnet performance and cost at a green-field site.

The second phase of the LHC upgrade proposal is to increase the energy in the existing LHC tunnel. This will require major changes to the LHC arcs, new dipoles and quadrupoles, as well as the injector chain. Increase in energy a factor by two implies an operating dipole field of at least 16 T, quadrupoles with a gradient of 450 T/m and apertures of ~50 mm. A program to develop these magnets may take 15 to 20 years and will necessarily focus on cost as well as technology.

So far, superconductivity has not been the death of high energy physics, but rather a crucial enabling technology. New materials tempt HEP to continue its relationship with this fickle partner. It remains to be seen how it ends.

12.3 Magnet Engineering—Study in Stability and Quench Protection

Luca Bottura and Al McInturff

12.3.1 Outline

This chapter documents the historical perspective of stability and the closely related subject of protection in superconducting magnet engineering. We will follow the historical path, showing how the parallel development of materials suitable for building magnets and the achievement of increased current density, was related to conductor stability. The interplay of high current density and stability features drove the increasing sophistication of the magnet/device protection systems.

12.3.2 The Infancy of Superconducting Magnet Technology

Soon after his discovery of superconductivity, H. K. Onnes realized the potential for generating high magnetic fields using negligible electrical power. The concept was rapidly tested using Sn and Pb wires. But his announcement at the Third International Congress of Refrigeration, held in Chicago in 1913, that a 10 T solenoid was close at hand[1], was rapidly *quenched* by the realization that any of the material used would not carry any significant current at magnetic fields exceeding 50 mT [2]. Onnes himself concluded that superconducting magnets were not practical, and this can be regarded as the first and certainly not the last superconducting magnet project canceled. It was not until 1954 that Yntema at the University of Illinois wound superconducting coils with cold-worked niobium and produced magnetic fields close to 1 T [3], a range of practical application. The results were published in a short abstract[4] that was nearly ignored, but for a few scientists like John Hulm at Westinghouse who is credited with making the second superconducting magnet in 1955, an air-core solenoid that achieved 0,6 T.

The pace increased in the 1960s. Autler of the Lincoln Laboratory at M.I.T is reported to have made in 1959 iron-cored niobium magnets reaching a field of 1.4 T, to provide the magnetic field required by a solid state maser. It was then that Kunzler, at Bell Labs, filed the first patent for a 1.5 T superconducting magnet wound in layers with a molybdenum-rhenium wire and cooled in a cryostat by a bath of liquid He [5]. His patent on "Superconducting Magnet Configuration" was filed on September 19, 1960, beating a similar patent request by Autler by 2 weeks. This patent[6], issued 4 years later, in April 1964, can be regarded as the actual start of superconducting magnet technology. Early in 1961, Kunzler's group reported the discovery of superconductivity in Nb and Sn compounds up to field of 8.8 T [7], and the race for the original dream of Onnes was on again.

The groups at the respective laboratories of Bell, Westinghouse, Atomics International, Lincoln Laboratory, and Oak Ridge were competing for the highest field achieved by the time of the 1961 International Conference on High Magnetic Fields at MIT[8]. Field records were broken daily throughout that week. Kunzler's group, reportedly spurred by a bet at Bell that would result in a bottle of

[1]   H.K. Onnes, *Commun. Phys. Lab. Univ. Leiden* 34b (1913) 55

[2]   H.K. Onnes, *Commun. Phys. Lab. Univ. Leiden* 139f (1913)

[3]   G.B. Yntema, *IEEE Trans. Magn.* 23 (1987) 390

[4]   G.B. Yntema, *Phys. Rev.* 98 (1955) 1197

[5]   J.E. Kunzler et al., *J. Appl. Phys.* 32 (1961) 325

[6]   J. E. Kunzler, US Patent 3,129,359, filed September 19, 1960, received April 14, 1964

[7]   J. E. Kunzler et al., *Phys. Rev. Lett.* 6 (1961) 89

[8]   H. Kolm et al., *High Magnetic Rields*,Cambridge, MA, Nov. 1-4, 1961, New York, NY, Wiley, 1962

Scotch for every 0.3 T above 2.5 T [9], was the one achieving the highest field at the time, 6.8 T using a wire made of Nb tubes filled with crushed powder of Nb and Sn, and heat treated at 1000 °C to chemically react the precursors into Nb_3Sn. This barely surpassed the two main competitors, Hulm from Westinghouse and Hake and Berlincourt from Atomics International, who each reported fields of 6.6 T using Nb-Zr in their windings. Using the Nb_3Sn technique, Kunzler's group later achieved a record field of 10 T, finally reaching the goal of H. K. Onnes, 60 years after it was speculated. Superconducting magnet engineering was born.

12.3.3   The Appeal of Nb_3Sn and the First Stumbling Steps

The high level of interest for Type II superconductors resumed because the new form of Nb_3Sn conductor produced by Kunzler and colleagues had excellent $J_c(H,T)$ characteristics. Indeed, superconducting magnets need current density and a large part of the race for the first 10 T magnet was the search for a superconducting material with a sufficiently high current density. The cold-worked Nb wires used by Yntema for his magnet had critical current density J_c just below 1 kA/mm^2 in an applied field of 0.5 T and at 1.7 K. The Mo_3Re wires in Kunzler's initial magnets had a J_c of 500 A/mm^2 at the operating field of 1.5 T. The first samples of Nb_3Sn barely achieved 40 A/mm^2 at 9 T and 1.5 K, but that was soon raised to values of 1000 A/mm^2 at 9 T [7], an impressive achievement for the time. There was limited interest in superconducting magnet construction prior to the 1950s primarily because most of the research into superconductivity concentrated on an understanding of the theoretical underpinnings of the phenomenon. Particle physics was still dependent on resistive components for its accelerators and beam lines. Superconductivity in niobium tin was discovered in 1950, yet for a decade aroused no interest.

From the magnet construction point of view the engineering current density J_E, the prime requirement for magnet design, is directly related to the critical current density $J_c(H,T)$; the ability to modulate the amount of superconductor in the winding area was far too low for most practical applications. Superconducting devices become economically viable and efficient when the engineering current density is in the range of few hundred A/mm^2. Although this was met by the 0.5 mm diameter, Nb_3Sn conductors produced by Kunzler and coworkers, it was industry, especially RCA and GE, that produced tape and filamentary material on a substantial scale, opening a whole new range of magnetic fields for engineering applications and devices, and the concomitant surge in the drive to exploit them. (See Appendix 1 of this section) Once again, though, the sudden and enthusiastic rush to wind and test the new materials was somewhat reduced by the issues of flux-jumps, stability and training. Indeed, early coils failed to reach the same current level of short samples of conductor. This was eventually traced to a problem in a conductor where the heat generated by the movement of the magnetic flux was too large to be conducted out of the superconductor before it reached the critical temperature, i.e. an unstable flux-jump. We will elaborate further on this issue. The result was a serious limitation on the use of the early Nb_3Sn wires, made in large filaments and with high current density. However, it was also discovered in magnetization studies in the mid to late 1960s that a wire sample powered in a high background magnetic field produced in a different flux arrangement in the conductor and much improved stability. An ingenious solution to the problem of unstable flux-jumps was to subdivide the coil in multiple winding sections that were independently powered. For example in a solenoid the outer windings would be wound with a wire of lower current density conductor (more stable against flux jumps) than the inner sections (where high current density was needed). Powering the outer coil before the inner one would lower the magnetization of the inner section at a given current, and thus allow reaching higher fields. (See Appendix 2 of this section) The technical literature as well as the glossy advertising produced by industry was full of examples of this type of "research magnet." These magnets could now be found in every properly

---

[9]   J. E. Kunzler, *IEEE Trans. Magn.* 23 (1987) 396

equipped low temperature laboratory as well as in various experiments at accelerator establishments such as Brookhaven or the Bevatron at Berkeley.

12.3.4 Malleable Nb-Based Alloys: Technology Goes to Industry

The supremacy of Nb_3Sn, and other more exotic and brittle intermetallic compounds such as Mo_3Re, did not last long. Already in the first half of the 1960s, Type II malleable alloys were commercially available. The first wire offered commercially, in 1962, was Nb-Zr at Avco Everett Research Laboratory[10], soon after followed by Nb-Ti at Westinghouse[11]. These two alloys were mechanically very tough, as they were originally developed for high strength rivets. Among the two, Nb-Ti was easier to manufacture, had superior mechanical strength, and had a 2-Tesla advantage in the upper critical field. For these reasons Nb-Ti became the dominant conductor used in device designs and construction[12]. This domination can be ascribed to two factors: ease of fabrication and the ability to sustain the Lorentz forces without loss of current carrying capacity. Fabrication of NbTi became an industrial-type process with the advent of the multifilamentary conductor first commercially produced by Imperial Metal Industries Ltd. The price to be paid, when compared to the intermetallic, brittle A15 structure superconductors, Nb_3Sn, Nb_3Al, and V_3Ga, was only an approximate factor of two reduction in critical temperature and field. At that time this was a minor issue when compared to the degradation of coil performance with respect to the theoretical current carrying capability of short samples.

12.3.5 Training and Degradation, the Discovery of Stability

The first superconducting magnets built in the decade around 1960 had their first transition to the normal state, that is quenched, much before reaching the expected critical current, largely disappointing the constructors. At the first powering only the early magnets reached a small fraction of the critical current of the single wire after which they quenched. The following attempt to power the magnet resulted in a higher current before a quench. The process continued, more or less regularly, at each attempt, and the maximum current that could be reached increased quench after quench, slowly approaching a plateau. This behavior became known as training,[13] and the curve of quench current vs. powering attempt became known as the training curve. The plateau current reached, however, was still below the expected maximum current carrying limit of the wire. The current limitation observed was originally thought due to bad spots in the wires or cables, and thus contributed to poor homogeneity in the quality of the superconductor. This idea produced the concept of degradation of the conductor performance. Although training clearly showed that a physical degradation could not be responsible for the bad performance, the misleading name remained as an inheritance of the misty understanding. Particularly puzzling was the fact that the degradation depended on the coil construction and on its geometry. A principle not yet fully understood at the time was that of stability of the cable with respect to external and internal disturbances. Insufficient stability and large disturbances were the key issues to the failure of the early experiments on superconducting magnets. (See Appendix 3 at the end of this section)

The pieces started to come together in the early 1960s when it was recognized that the perturbations that were limiting the achievable currents in small solenoids were originating from magnetic instabilities called flux-jumps. Magnetization studies of these conductors showed that under certain circumstances, the magnetization of the wire in a varying field could collapse suddenly and catas-

---

[10] Avco Supergenic Strip SG 700
[11] J. K. Hulm and R. D. Blaugher, *Phys. Rev.* 123 (1961) 1596
[12] T. G. Berlincourt, *Cryogenics* 27 (1987) 283
[13] M.A.R. LeBlanc, *Phys. Rev.* 124 (1961) 1423

trophically. The flux linked with the filaments could change nearly instantaneously, jump, and cause local energy dissipation in the wire. The source was identified and was shown to be a function not only of field and field change, but was also dependent on the conductor geometry with respect to the field. Numerous workers began to consider ways in which these instabilities could be controlled and their reports and conjectures appeared in the literature of the day. The devices that were being designed and constructed in this mid-to-late 1960s were based on one or more forms of stability enhancement.

In the midst of this bubbling atmosphere, Brookhaven National Laboratory sponsored in 1968 a Summer Study on Superconducting Devices and Accelerators, which did an excellent job of bringing the various contributors to this state of the art and represents a veritable historical milestone. The proceedings of the workshop[14], as well as the summaries and highlights[15], provide an excellent record which traces well the understanding and progress made.

12.3.6 Stabilization Strategies

With improved understanding of the causes of instability, it was finally possible to design a coil so that it would be stable. In the following sections we will review some of the main stability strategies developed in a selection of the previously mentioned references and authors. Each strategy added to the change in understanding and in some cases effected a change in the direction of design or construction of a device even after their start.

12.3.7 Cryogenic Stability

The first and earliest form of stabilization was used in a large MHD generator magnet model at Avco Everett. This magnet with windings in a saddle-shaped configuration produced 4 T over a region 3 m long and 0.3 m in diameter. Constructed of Avco SG 700 9-strand NbZr conductor embedded in a copper strip, the dipole had a stored energy of 5 MJ at a current of 785 A [16]. Although the device never contributed to the advancement of MHD technology, it gave the superconducting community a new principle for the construction of stabilized superconducting coils[17], a principle which became known as *cryogenic stability* and the conductors dubbed as *cryostable*. Simply stated, a cryogenically stable conductor has a sufficient amount of low electrical resistivity material in parallel, and a sufficiently large surface in contact with an adequate volume of liquid or pressurized helium, so that even if the superconductor is driven into the normal state, whatever the cause, it is cooled down and can recover the superconducting condition below the current sharing temperature T_{CS}. Taking as a simple example, a single strand of cross section A operating at a current I, the maximum joule heating per unit length is given by $I^2 \rho_n / A$ where ρ_n is the normal state resistivity, averaged over the whole strand cross section. The minimum heat flux per unit length to the helium bath can be written as $hP(T_c - T_{OP})$, where h is the heat transfer coefficient to the helium, P is the wetted perimeter, T_c the critical temperature and T_{OP} the temperature of the helium. Stekly[18,19] was the first to formulate the condition of cryogenic stability as:

$$\frac{\rho_n I^2}{A} \leq hP(T_c - T_{OP}) \tag{12.1}$$

[14] *Proceedings of the 1968 Summer Study on Superconducting Devices and Accelerators*, Brookhaven National Laboratory, June10–July 19, (1968)

[15] H. R. Hart, *Proceedings of the 1968 Summer Study on Superconducting Devices and Accelerators*, Brookhaven National Laboratory (1968) 571

[16] Avco-Everett Research Laboratory press release, *Cryogenic Engineering News* (1966)

[17] A. R. Kantrowitz and Z. J. J. Stekly, *Appl. Phys. Lett.* 6 (1965) 56

[18] Z. J. J. Stekly and J. L. Zar, *IEEE Trans. Nucl. Sci.* 12 (1965) 367

[19] Z. J. J. Stekly et al., *J. Appl. Phys.* 40 (1969) 2238

or the equivalent conditions $\alpha_{\text{Stekly}} \leq 1$ in terms of the parameter α_{Stekly}:

$$\alpha_{\text{Stekly}} = \frac{\rho_{\text{n}} I^2}{hPA(T_{\text{c}} - T_{\text{OP}})}. \tag{12.2}$$

The above criterion provides a limit for the maximum current up to which the conductor operates in cryo-stable conditions. In a strand of radius R, cross section $A = \pi R^2$ and wetted perimeter $P = 2\pi R$, the maximum cryo-stable engineering current density $J_{\text{cryostable}}$ is:

$$J_{\text{cryostable}} = \frac{2q''}{\rho_{\text{n}} R}. \tag{12.3}$$

Once the strand size, R, the materials and their fractions are chosen, T_{c} and ρ_{n}, we see that the heat flux q'' is key to the stability of the winding. An average value of heat flux normally found for a 50% wetted surface in a pool of liquid helium is $q'' \approx 0.4\,\text{Wcm}^{-2}$. Minimum values are obtained for very small channels and less than 10% wetted surface, for which $q'' \approx 0.1\,\text{Wcm}^{-2}$. The upper extreme is the case of completely open bath and more than 60% wetted surface being $q'' \approx 0.8\,\text{Wcm}^{-2}$. Yet higher values can be obtained in transient or super-critical helium, with the heat flux reaching values $q'' > 1.0\,\text{Wcm}^{-2}$.

The corresponding engineering current density $J_{\text{cryostable}}$ is in the range of 100 to 200 A/mm^2. This form of stability, which is also the most conservative form of magnet winding protection, is normally adopted in large coils where the size/and or bulk of the winding is not an issue. It is still used in modern devices and systems on the buswork and interconnects, and to guarantee stable operation of large-scale magnets subjected to a large energy spectrum such as the ITER coils. On the other hand, when the current density needs to be high, as close to the magnet bore as possible, and the winding is tightly packed with little helium (i.e., in accelerator magnets), cryostability is not applicable.

A dramatic example of a cryo-stable magnet was the Big European Bubble Chamber (BEBC) at CERN[20,21]. Illustrated in Figure 12.22, BEBC was a 4.7 m bore split solenoid with a 0.5 m gap producing a maximum field in its center of 3.5 T, corresponding to a maximum field at the conductor of 5.1 T, and a stored energy of 800 MJ. At nominal operating conditions (5700 A) the conductor was operating at a Stekly parameter α_{Stekly} of approximately 0.5, i.e., largely in the cryo-stable regime. (See Appendix 4 at end of this section)

12.3.8 Adiabatic Stability

The concept of enthalpy stabilization in adiabatic winding is at the other extreme of the strategies of stabilization. The idea is that the conductor has the enthalpy to absorb the heat locally, without exceeding the current sharing temperature, preventing the instability to initiate. This concept applies to tightly packed, impregnated windings that are not permeated by the helium coolant (adiabatic windings) and operate at high current density, a bonus for the economics of the magnet system. The price to pay is that the enthalpy of a strand between typical operating temperatures of 1.9 to 4.2 K, and current sharing conditions is small, a few mJ/cm^3. Correspondingly, the superconductor can survive events depositing only very small energy, and great care must be put to reducing all sources of magnetic and mechanical perturbations. By contrast, a cryo-stable conductor can in principle survive any perturbation.

This form was presented in early papers by Hancox[22], Chester[23] and Smith[24] and is normally

[20] E. U. Haebel and F. Wittgenstein, *Proc. 3rd Int. Conf. Magnet Technol.* (1970) 874

[21] F. Wittgenstein et al., *Cryogenics* 9 (1969) 158

[22] R. Hancox, *Phys. Lett.* 16 (1965) 208

[23] P. F. Chester, *Rep. Prog. Phys.* XXX (1967) 561

[24] P. F. Smith, *Proc. 2nd. Int. Conf. Magnet Technol.* (1967) 543

FIGURE 12.22: The solenoid of the Big European Bubble Chamber during winding, and a cross section of one of the conductors used, consisting of straight Nb-Ti filaments in a large Copper matrix.

used to evaluate the upper bound on the size of a superconducting filament that can sustain a flux-jump. Their analysis considered the effect of a small temperature increase δT on the magnetization of a superconductor in an external field. The decrease of J_c at increasing temperature, i.e., a negative derivative dJ_c/dT, causes a reduction of the magnetization and an energy dissipation δE in the superconductor. This energy in turn leads to a further temperature increase $\delta T'$. An instability is initiated when $\delta T' > \delta T$, a flux jump. This condition was presented as an upper limit on the superconductor characteristic size d:

$$d \leq \frac{1}{J_c} \frac{3CT_0}{m_0} \tag{12.4}$$

where T_0 is given by $J_c/(dJ_c/dT)$ and falls in the range of $T_c/2$ and the difference between the critical and operating temperature, i.e., $(T_c - T_{op})$. The above criterion was derived for the case of a superconducting slab, but can be used as an approximation of the diameter of a round filament. For the highest current density NbTi the maximum diameter compatible with flux jumps and adiabatic stabilization is approximately 25 μm. Wires made with single filaments of this dimension are clearly not practical for applications requiring ampere-turns at the level of few hundreds of kA-turns that is necessary, as for example in accelerator applications.

12.3.9 Dynamic Stability

The successful experience on magnets built with ribbon or tape superconductors, tapes obtained by vapor deposition of Nb_3Sn on Hastelloy or silver, the "Vapodep" ribbon from RCA, or the diffusion processed niobium tin tape from GE, of dimensions exceeding the flux jump limit but sandwiched with a copper tape, demonstrated that the adiabatic stability condition given above is too conservative. An important realization was that while the magnetic flux diffusion time in a superconductor is a few orders of magnitude faster than the thermal diffusion time, for very pure conductors such as Al and Cu the situation is the opposite. Therefore, if the correct combination of pure metal foils and thin layers of superconductor are combined in a winding, it is possible to "freeze" the magnetic flux for time sufficiently long to allow cooling. Such a package is hence stable thanks to a dynamic heat balance, whence the name of dynamic stabilization given to this strategy. A report by Dahl[25] lists several early Nb_3Sn magnets with on this form of stability. In the dynamic case the stability limit on the superconductor size d (the tape thickness or filament diameter) was modified as follows:

$$d \leq \frac{1}{J_c} \frac{6k(T)T_0}{\rho_n} \frac{1-\lambda}{\lambda} \tag{12.5}$$

where λ is the fraction of superconductor in the strand, and $k(T)$ is its effective thermal conductivity and all other quantities were defined earlier. The factor $(1-\lambda)/\lambda$ for a strand with 50% superconductor is equal to 1. The above limit is much less restrictive than the adiabatic limit, resulting in maximum filament sizes larger by one order of magnitude (i.e. around 200 μm). It is evident that distributing the superconductor in a low resistance matrix metal has multiple advantages. Small filaments in one multifilamentary strand had lesser tendency to adiabatic flux jumps, and even if they did they could be dynamically stabilized.

12.3.10 MultiFilamentary Wires and Twisting

Most of the work so far described originated in the early to mid 1960s. By this time multifilamentary copper/NbTi wires with good steady state stability properties began to be available, and new issues were found related to field ramps. (See Appendix 5 at end of this section) The use of small filaments embedded in a matrix of high conductivity material was sufficient to prevent flux-jumping to trigger an instability, but it was clear that this was not the end of the story. (See Appendix 6 at end of this section) We return once more to that seminal event, the "Woodstock of Superconductivity", the 1968 Summer Study on Superconducting Devices and Accelerators, and in particular to the work of the Superconducting Applications Group at the Rutherford Laboratory in England[26].

12.3.11 Cable Stability: The Invention of Rutherford Cable and CICC

Wires and tapes can carry currents in the range of few hundreds of amperes and are appropriate to wind small magnets, when the magnet inductance and stored energy are not an issue as we will see in a later discussion on protection. On the other hand, the large-scale magnets of interest for accelerator or fusion applications store energies that can reach hundreds of MJ. It was soon realized that it was important to decrease the magnet inductance, so to limit the operating voltage. This calls for the use of cables made of several wires in parallel, capable of carrying much larger currents, typically in the range of 10 kA. Similar to a multifilamentary strand, the strands in the cable need to be electromagnetic de-coupled through geometric transposition[26]. Early efforts at effecting this transposition of filament elements in superconducting cables in the manner of high frequency litz cables proved unsatisfactory, but eventually the eponymous cable geometry developed at the Rutherford Appleton Laboratory[27] became the standard. A Rutherford cable is composed of two dimensionally transposed, twisted wires, which ensures good current distribution. The main advantage of this geometry is the possibility of achieving a tight twist and a high compaction ratio without degrading the strand performance. In practice, 80 to 90% of the cable cross section is filled

---

[25] P. F. Dahl, BNL Report BNL50498 (1976)

[26] P. F. Smith et al., *Proceedings of the 1968 Summer Study on Superconducting Devices and Accelerators*, Brookhaven National Laboratory (1968) 913

[27] G. E. Gallagher-Daggitt, Rutherford Laboratory Memorandum No. RHEL/M/A25 (1973)

with strands, with minimal loss of engineering current density. In addition, the cable can be shaped to a precisely defined and tightly controlled geometry. Rutherford-type cables soon dominated the design of accelerator magnets. All superconducting accelerators to date have been built using this type of cable. Figure 12.23 shows the example of an LHC Rutherford cable used to wind the main dipoles.

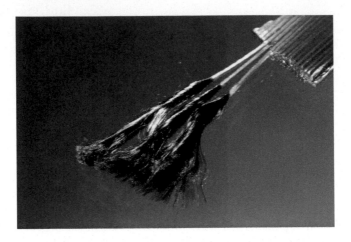

FIGURE 12.23:   A recent example of Rutherford-type cable, used for the construction of the LHC magnets. A few strands have been etched to expose the 7 μm thick Nb-Ti filaments.

Compacted cables such as those of the Rutherford type rely on the same stability concepts as single tapes and wires. A shift in paradigm was achieved with the concept of a cable-in-conduit conductor (CICC), which evolved from the internally cooled superconductors (ICS) that had found applications in magnets of considerable size between the late 1960s and early 1970s in particular the work of Morpurgo[28]. In ICSs the helium is in a cooling pipe that either includes or is in intimate contact with a superconductor, very much like standard water-cooled copper conductors. This conductor can be wound and insulated using standard technology, a considerable advantage for large systems requiring high dielectric strength. A major drawback of this concept, however, was the fact that in order to achieve good heat transfer and thus stability and high operating current density the helium would either have to be in the supercritical state or have to flow at very high flow rates. The advantage of the increase of the wetted perimeter obtained by subdivision of the strands was clear already at the beginning of the development of ICS[23]. In 1974 Hoenig, Iwasa and Montgomery proposed the first CICC concept, whose properties were successively elaborated by Hoenig and collaborators[29,30,31,32,33].

[28] M. Morpurgo, *Proc. 1968 Summer Study on Supercond. Devices and Accelerators*, Brookhaven National Laboratory (1968) 953

[29] M. O. Hoenig and D. B. Montgomery, *IEEE Trans. Magn.* 11 (1975) 569

[30] M. O. Hoenig et al., *Proc. 5th Int. Conf. Magnet Technology* (1975) 519

[31] M. O. Hoenig et al., *Proc. 6th Int. Cryo Eng. Conf.* (1976) 310

[32] L. Dresner, *IEEE Trans. Magn.* 13 (1977) 670

[33] L. Dresner and J. W. Lue, *Proc. 7th Symp. Eng. Problems* (1977) 703

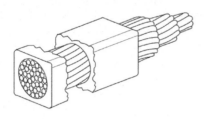

ROUND BUNDLE WITH 37 STRANDS ENCLOSED IN RECTANGULAR
CONDUIT - SHOWING TRANSPOSITION OF STRANDS

FIGURE 12.24: The first proposal of a CICC configuration, by Hoenig (left), contrasted to a prototype ITER TF CICC with central cooling channel. (Courtesy of ITER.)

A bundled conductor is obtained cabling superconducting strands, with a typical diameter in the mm range, in several stages. The bundle is then jacketed into a helium-tight conduit that provides structural support. Helium flows in the conduit within the interstitial spaces of the cable. The small hydraulic diameter insures a high turbulence, while the large wetted surface achieves high heat transfer, so that their combination gives the known excellent heat transfer properties. CICCs were initially designed to satisfy the Stekly criterion. Strictly speaking, a CICC cannot be considered to be cryostable because the amount of helium available for its stabilization representing the dominant heat capacity is limited in any case to the volume in the local cross section. A large enough energy input will always cause a quench, so the question is the magnitude of the minimum energy input producing a quench in a particular operating condition. This parameter is the stability margin[34]. A typical behavior of the stability margin was found in measurements of Nb-Ti CICCs as a function of the operating current[35]. For a sufficiently low operating current a region with high stability margin, named by Schultz and Minervini[36] the *well-cooled* regime of operation, is observed. In this regime the stability margin is given by the total enthalpy available in the cross section of the CICC up to current-sharing, including both strands material and helium. As the current increases a fall in the stability margin to low values, the *ill-cooled* regime, appears. The proper naming of this regime received much attention by many native language reviewers at conferences and international journals, proposing various, more grammatically correct alternatives, such as *poorly-cooled* or *badly-cooled*, demonstrating yet again that the smallest perturbations to a system arouse the widest consequences, just as in a superconducting circuit! The fundamentals are that in this regime the stability margin is one to two orders of magnitude lower than in the well-cooled regime and approaches the enthalpy of the superconducting strands. A good CICC design is hence optimum from the stability point of view when operating just below the well-cooled/ill-cooled transition.

Stability was the main argument that won the conductor competition in the European Fusion Program of the late 1980s. The CICC concept that was prototyped in Europe by Brown-Boveri Switzerland lived through a few minor changes during the first years of the ITER Conceptual Design Activities. It was modified to the present form which includes a central cooling channel during the ITER interregnum of P. H. Rebut, the official motivation being the fear that the cable would act as a filter and eventually be clogged by particles or impurities in the helium coolant. Although the overall current density was reduced, and the fear of impurities may not be justified, the present ITER cables

---

[34] M. O. Hoenig and D. B. Montgomery, *Proc. 7th Symp. Eng. Problems Fusion* (1977) 780

[35] L. Dresner, *Cryogenics* 20 (1980) 558

[36] H. Schultz and J. V. Minervini, *Proc. 9th Int. Conf. Mag. Technology* (1985) 643

are extremely robust with regard to their stability properties. This is also due to the fact that the cable can sustain a significant resistive voltage without quenching because of the fairly low resistive transition index with common values in the range of 5 to 10, a feature that makes magnet engineers refer to the ITER coils as "hyperconducting."

12.3.12   Stability in the New Millennium

As we write, we are profiting from the spectacular development of Nb_3Sn multifilamentary conductor for High Energy Physics that took place at the turn of the millennium: the J_c of this material has increased by a factor of three within the last decade The implication of this development can be appreciated considering the record setting dipoles D20 and HD1, built at the Lawrence Berkeley National Laboratory in the 1990s and 2000s, respectively. In 1997, the D20, a 4 layer cos-θ coil, produced a central field of 13.5 T using a Nb_3Sn conductor with J_c of approximately 750 A/mm^2 at 12 T and 4.2 K [37]. In 2004, the HD1 dipole reached a bore field of 16.1 T by virtue of a conductor with J_c of approximately 3000 A/mm^2 at 12 T and 4.2 K [38]. If D20 had used the conductor available for HD1, and ignoring issues of electromagnetic forces, it would have been a two-layer coil, and would have attained significantly higher field.

Although beneficial, this boost of J_c has also led to the rediscovery of the importance of stability, its relation to filament and strand diameter, and the electrical properties of the stabilizer. Specifically, magnet engineers are being reminded of the early finding that we have described in this chapter. Filament sizes in high-J_c internal tin wires are in excess of 100 μm, i.e., on the edge of flux-jump stability, and caution is in order. A modest residual resistivity ratio (RRR) is in itself not a problem if it is uniform in the wires, and in general magnets built with low-RRR wires, say in the range of 15, are easier to protect as they are characterized by a faster quench detection, thus easier to spread the energy and a faster dump. However, it was found that the presence of local spots with low RRR caused by tin leakage from the subelements damaged during cabling can lead to poor stability and magnet performance limitations in the intermediate field region of 8 to 12 T.

A few more lessons have been learned and are common practice in superconducting magnet engineering. It is important to hold the winding pack in compression in all directions, with a force that is greater than the Lorentz load, including an appropriate safety factor. This is done to reduce the energy inputs of mechanical origins that can trigger instabilities. Good conditions are attained when the winding does not have freedom to move and is always in contact with the force-bearing surface. An initial load is applied on this surface, sometimes just a few MPa sufficient to remove the fluff. Caution must be used to avoid cracking or tearing at the interfaces, and in particular those bonded or glued during an impregnation process. In some cases it may be of advantage to intentionally remove the bond between the windings and the surfaces that are not supporting the Lorentz load, especially at the surfaces that tend to separate. In this case the coil would be allowed to move as much as required, e.g., by field quality considerations. These principles may be difficult to apply in cases where the force distribution is complex, e.g., the large coil systems of future fusion experiments, or in high order field configurations such as the multipole corrector magnets of particle accelerators. Such magnets are then designed with a larger operating margin to cope with the increased perturbation energy spectrum. High-temperature superconductors are just at the beginning of this process of learning again from the past, and will eventually have to deal with stability at liquid Helium temperature. They are ideal perspective candidates for high field inserts in magnets where the background field is generated by coils wound from A-15 or Nb-Ti conductors. For reasons of magnet economy, HTS conductors still need an increase of J_c by a factor of 5. Assuming that this is successfully achieved, and if the present effective filament size does not decrease, HTS should reach the known

---

[37] R. Scanlan et al., *Inst. Phys. Conf. Ser.* 158 (1997) 1503

[38] A. Lietzke et al., *IEEE Trans. on Appl. Supercond.* 14 (2004) 345

stability boundary conditions. Stability as it concerns filaments in a matrix via dynamic criteria á la Chester or an enhanced cryostability scenario á la Stekly will result again in a criterion on the maximum dimension of the filament that we quoted earlier. History repeats itself.

12.3.13 Quench and Protection—a Burning Issue

A quench is an inherent part of the life of a superconducting magnet system. Quenches happen either accidentally, as we have described at length earlier, or intentionally, e.g., during a magnet test. A normal zone starts suddenly or gradually, the resistance and Joule heating grow, the temperature, initially barely above the superconducting transition, starts rising more and more rapidly, and a normal voltage appears at the coil terminal. Whatever the initiation, the final result of a quench is always a coil dump. The burning question is how rapidly the current is reduced to zero. As early as 1963, that is right after the superconducting rush for high field applications, Smith documented the concern that: "... the dissipation of the stored energy of the coil gives rise to local heating which, for large coils, may be sufficiently great to damage the insulation or permanently alter the properties of the wire ..." [39]. Indeed, recall that a superconducting wire carrying a current of 100 A can have a typical diameter of 0.5 mm, i.e., no larger than the wire in a household fuse that breaks at 15 A in fractions of a second. This comparison was inspiring to those that were finding criteria for magnet design. Conductors were considered adiabatic, and the temperature increase was computed equating the power produced by Joule heating $I^2 \rho_n / A$ to the rate of enthalpy change $AC\frac{dT}{dt}$, where C is the volumetric heat capacity. With minimal manipulation, the same design used for fuse wires, it is possible to relate the peak conductor temperature at the end of the current dump, also called hot-spot temperature T_{hot}, to the integral in time of the square of the current:

$$\int_{T_{op}}^{T_{hot}} \frac{C}{\rho_n} dt = \int_0^\infty I^2 dt. \qquad (12.6)$$

The interest in the equation above is the integral on the left. It is a property of the conductor geometry and materials only, while the integral on the right depends only on the current waveform during the dump. The equation thus defines the relation between the hot-spot temperature and the current rating of a specific dump. The units of the above integrals are those of squared current times seconds. A new, convenient definition of this unit was found at Fermilab in the 1970s: the MIITS, which measures the current integral in 10^6 (A^2s), a practical number for conductors that carry a few kA and need to be dumped in fraction of a second to avoid over-heating. (See Appendix 7 at end of this section.) The use of the MIITS curve caught-on very rapidly. This is a relatively easy criterion to apply, and somewhat conservative as a conductor is not adiabatic inside a coil. One difficulty remains, though, namely that to predict temperature one has to know the current waveform during a coil quench. This, in turn, depends on the coil resistance, that is on the size of the normal zone and the temperature in the coil which we are seeking to determine, on the coil inductance, a geometrical parameter that is affected by the cable size through the number of turns, as well as on actions external to the coil such as power supply or switch response. This difficulty shows the typical iterative process associated with the choice of a suitable protection strategy for a magnet system. In broad terms, a good protection strategy is a balance among a safe temperature increase (material integrity), limited temperature gradients (thermal stresses), coil voltage level (electrical standoff) and the complexity of the quench detection and protection circuitry. We can distinguish two extremes when considering different protection strategy. At one extreme the magnetic energy inventory is fully dissipated within the magnet cross section. This is appropriate for small magnets that store little energy compared to the material damage level. The main issue in this case is how

---

[39] P. F. Smith, *Rev. Sci. Inst.* 34 (1963) 368

to spread evenly the energy, and to increase the resistance rapidly so that the coil discharges fast. The best situation is that of quick quench propagation in the coil. At the opposite end of protection strategies we have the case in which the magnetic energy is nearly completely drawn from the coil and dumped, e.g., in extraction resistors or diodes switched into the circuit. This is mandatory in large systems that store quantities of energy far exceeding the local damage limit. The issue in this case is whether the external circuit can be fast enough, i.e., essentially a question of allowable MIITS and terminal voltage. It is now clear why in large coil systems it is beneficial to reduce the magnet inductance by winding the coil with large conductors. Of course, the two extreme strategies are not mutually exclusive, and can be combined as appropriate. These ideas developed fast in the early days. It is remarkable that the earlier work of Smith[39] already reviewed protection strategies ranging from self-protection, subdivision, coupling to additional circuits and fast insertion of resistors, covering most conditions of relevance, and including considerations of magnets in persistent mode, and concern on the maximum voltage developed internally and externally to the coil by the various methods. Apart for the use of quench heaters embedded in the coil, an expedient to speed-up quench propagation in the high current density and high field magnets connected in the kilometer-long strings of modern accelerators such as the LHC, most of the fundamental work had been completed.

12.3.14   Hot Helium Bubbles

Electrical and thermal considerations are not the only issues during a quench. In most cases the energy deposited to the helium is so large that trying to contain the ensuing pressure increase would neither be practical nor safe. A closed volume of liquid helium would pressurize to about 400 bar upon warming up to 100 K, which is still a modest temperature for a typical quench. This explains why all cryogen-cooled superconducting magnets are duly equipped with quench pressure relief lines. While in the early days these would blow cold helium into the atmosphere of the laboratory, they presently vent to a helium recovery system, a much more environmentally and economically correct solution. Internally cooled cables, and especially CICCs represent a special case for the process of helium heating, pressurization and expulsion. If a quench happens within the coil, the helium in the normal zone is heated and pressurized. The quench relief line in this case is the cable itself that sees a rush of warm helium trying to escape the normal zone. In doing so, the helium actually propagates the quench with a speed that is comparable to the flow velocity. This is in fact the dominating quench propagation mechanism in coils built from ICSs and CICCs. The understanding of this phenomenon was strongly motivated by the work on the CICC coils for fusion, and in particular the Westinghouse coil built within the framework of the Large Coil Task Program, a toroidal assembly of 6 coils built at major world laboratories and industries and tested at Oak Ridge National Laboratory[40]. It was Dresner who proposed the "hot helium bubble" mechanism to explain quench propagation in CICCs[41]. Miller and Lue, systematically tested Dresner's theories in specially constructed sub-scale conductor experiments[42,43,44]. In the years around 1980, this team addressed most of the issues in CICC design, stability, quench and protection, thus providing a well-founded background to the design of the large coil system that will be at the core of ITER. An interesting sidelight to this work is the thermo-hydraulic quench-back. A process such as quench in a CICC, involving heat transfer in the cable and compressible fluid-dynamics, is not a trivial analytical task, and it is natural to resort to computer modeling for this type of work. Computing quench

---

[40]   D. S. Beard et al., *Fus. Eng. Des.* 7 (1988)

[41]   L.Dresner, *Proc. 10th Symp. Fusion. Eng.* (1983) 2040

[42]   J. R.Miller et al., *Proc.8th Int. Cryo. Eng. Conf.* (1980) 321

[43]   L.Dresner, *Proc. 9th Symp Eng. Prob. Fusion Research.* (1981) 618

[44]   L.Dresner, *Proc. 11th Symp Fusion Eng.* (1985) 218

evolution as a function of operating conditions, Luongo[45] found that under certain circumstances the quench propagation speed could increase from the few m/s typical of heating-induced flow to tens and hundreds of meters per second. The first obvious reaction was skepticism in the results, initially attributed to an issue in numerical convergence. As understanding improved, it was again Dresner[46], and two mathematicians, who identified analytically conditions upon which the adiabatic compression and frictional heat could generate a thermo-hydraulic quench-back wave traveling in the coil up to the sound of speed in helium[47]. Lue reproduced this condition experimentally in 1993 [48], thus demonstrating that computer simulations can indeed be useful.

Helium expulsion can be an effective means to identify a quench. Flow conditions in a cryogenic circuit depend strongly on local heating, and already modest power cause flow locking and thermally induced flow. These effects can be measured easily and used as a backup quench detection system. (See Appendix ix at end of section)

12.3.15   Coda

One hundred years since discovery, the promise of exceedingly large magnetic fields at little cost still inspires the imagination. Invariably, though, among the many facets that are vital to the successful engineering of a superconducting magnet system, stability and protection are two key issues that determine performance, reliability and safe operation. In our first sentence here we indulge in a little hyperbole: exceedingly large magnetic fields are definitely not available at little cost, projects involving them have become international ventures involving participating specialists numbering in the thousands. The results are spectacular, for they all rest on the shoulders of the small band of dreamers, thinkers and tinkers of yesteryear.

12.3.16   Appendix: Recollections and Reminiscences

A.M. = Al McInturff, L.B. = Luca Bottura

1. On the Discovery of High Field Superconductivity, A.M.

I remember my thesis advisor Professor Charles Roos discussing the possibility of obtaining a measurement of H_{c2} of Kunzler type Nb_3Sn in a 200 kGauss pulsed solenoid by converting the Michigan State University old surplus RF capacitor bank into a pulsed power supply with another Vanderbilt graduate student Clay Whetstone and I, using "war surplus" radar parts. This pulsed power supply we would in turn use to power the spare "Bitter Type" solenoid (15–25T 30 millisecond pulse) I was using to keep the photo-produced electron pairs out of my emulsion stack exposed at the Cal Tech Synchrotron that were to be scanned for photo produced sigma particles at Vanderbilt's Scanning Laboratory. We would turn the magnet on its end, so that the axis was vertical and then mount a double liquid volume glass cryostat (Dewar) in the bore with the reservoirs of liquid nitrogen and liquid helium above the magnet with the glass Dewar resting on a cork washer sitting on the upper magnet end plate. The Nb_3Sn sample mounted on a couple of current leads held by a G10 clamp supported on a stainless steel threaded rod. It was a four-terminal measurement to determine H_{c2}. The plate at the top was clamped to an external support to hold the sample in the right spot during the magnet pulse to measure their H_{c2} resistively. Our "Kunzler" type samples were fabricated by our ORNL colleagues Drs. G. Kniepe, J. Betterton and R. Boom. Of course we were unaware that the pulsed field was going to increase the temperature of our sample. As I remember,

[45] C. Luongo et al., *IEEE Trans. Magn.* 25 (1989) 1589

[46] L.Dresner, *Cryogenics* 31 (1991) 557

[47] A. Shajii and J. P. Freidberg, *Int. J. Heat Mass Transfer* 39 (1996) 491

[48] J. W. Lue et al., *IEEE Trans. Appl. Supercond.* 3 (1993) 338

the samples were a cross between dried black spaghetti and hard rubber and to add to the challenge about as easy to solder! Our first LHe level gauge was an Ever-Ready flash light and a ruler. First rotate the non-silvered strip of the outer cryostat to align with the inner one and put the light behind to reflect off the helium surface and then measure. The liquid helium was obtained from ORNL in a rental truck driven by Clay or myself to Vanderbilt in Nashville from Oak Ridge. Our measured value was a few Tesla low but still served to give the highest early H_{c2} for the A15 compound. Such a high value only increased the interest in applications.

2. On the AGS K Meson Beam Line, A. M., in Connection with Early High Field Magnets:

"One such example was the solenoid at the BNL AGS K-beam line designed and constructed by William Sampson. This device was used in an experiment (1967–1968) at the BNL AGS K-beam line to measure the magnetic moment of the short life time Ξ^- hyperon. Due to the life time the Ξ^- traveled only 10 cm on the average. The split pair solenoid magnet would be located right behind a solid H_2 target which together would be located inside a "Cerenkov" counter. A spark chamber followed behind the magnet centered on the beam line. This magnet was a four split-coil array stabilized by anodized high purity Al foils separating the layers for two sections and a Mylar/cu-foil/Mylar for the other two sections and protected against high voltages by copper shorting foil strips across each layer. The conductor was a vapor deposited Nb_3Sn on a Hastelloy ribbon. The stability of the higher J_c inner section was enhanced by the field of the lower J_c outer sections. The magnet produced 12.5 T bore field and had a persistent switch across the input leads during HEP operations, a very early feature which has become common place nowadays. The magnet had an eight degree conical inner bore (3.8 to 6.6 cm id), the production angle of the Ξ^-, 0.89 million pictures were taken during the HEP data run and 1.7×10^4 possible Ξ^- events were found.

3. Training and Degradation, Recollections of J. Hulm:

"Those tiny, primitive magnets were, of course, terribly unstable [...] One had to have faith to believe that these erratic toys of the low temperature physicist would ever be of any consequence as large engineered devices."

J. K. Hulm, "Superconductivity research in the good old days," Talk given at the banquet of the Applied Superconductivity Conference, Knoxville, TN, Dec. 1, 1982. The talk is available for download at : http//www.sainc.com/asc04/johnHulm.htm

4. On the BEBC at CERN:

A dramatic example of a cryo-stable magnet was the Big European Bubble Chamber (BEBC) built and operated at CERN (Haebel, 1970). BEBC was a 4.7 m bore split solenoid with a 0.5 m gap producing a maximum field in its center of 3.5 T, corresponding to a maximum field at the conductor of 5.1 T, and storing an energy of 800 MJ. The magnet was cooled in a bath of 17,000 l of liquid helium, and the bubble chamber sitting in the bore contained 35,000 l of liquid hydrogen. These dimensions are spectacular even on the scale of modern HEP detector magnets, BEBC was the largest European superconducting magnet at that time, and long after. Each coil was wound in 20 pancakes out of a flat monolithic conductor with 3 mm thickness and 61 mm width. This conductor was itself a composite containing 200 untwisted NbTi filaments with a diameter of about 200 μm in an OFHC copper matrix. At nominal operating conditions (5700 A) the conductor was operating at a Stekly parameter α_{Stekly} of approximately 0.5, i.e., largely in the cryo-stable regime. The construction of the magnet started in the late 1960s, only a few months before the P. Smith and the Rutherford Group discussed the necessity of twisting the filaments to decouple them (see later). In the recollection of the BEBC magnet project engineer, Francois Wittgenstein, by that time the conductor production in Germany (Siemens) and France(CGE) was too advanced to introduce a major change in the manufacturing sequence, such as twisting of the filaments. Besides, earlier prototype conductor testing in a small-scale solenoid model built for this purpose did not show evidence of performance issues (Wittgenstein, 1969). Hence, the production commitment could not be stopped, especially when such a minor issue such as the threat of flux-jumps was to be compared to the strikes organized by the French Unions in May 1968, right in the middle of the extrusion of 2 km conductor lengths required for winding. The group managed to overcome these very practical difficulties, and wound the solenoid that reached nominal operating conditions, as expected. Indeed, thanks to cryostability, the BEBC coil could be energized up to the operating current in spite of the fact that the conductor with large untwisted NbTi filaments had very large magnetization associated with the currents that flowed in the electromagnetically coupled filaments. A feature that was built in the coil design was a set of heaters that could be fired to suppress the large magnetization produced by these coupling currents. In practice, however, the heaters were never used in routine operation. But there was another effect associated with the large magnetic moment of the conductor, namely the force distribution in the solenoid was not as expected. As a result, at locations the solenoid shrank in diameter, instead of expanding. This caused mechanical interference with the cryogenic pipe work and the appearance of shorts that, strangely enough in the eye of the operators, would disappear after the ramp was stopped (and the induced currents in the conductor would decay). The problem was cured by adding mechanical support and reinforcing electrical insulation during a technical stop, when finally, in light of the evidence, the influence of the induced currents was recognized.

5. On Magnetization Studies, A.M.

I was doing magnetization studies on various conductors that were available and had already figured out that the field/current path geometry was driving the instabilities that I saw. I was primarily using tapes (even fabricated NbTi tapes for comparison). The first MF samples that I tested were very stable until I reached very high rates of field change.

6. On Impressions of Smith's Exposition of Filament Twist in Relation to Stability of Wires, A.M.

As I listened to him explain the twisting arguments canceling the filament-to-filament current coupling, a light went on and I understood the problem with the MF mini-coils I had made for the next stage of magnetization studies. I went back to the lab and took one sample coil apart and put it on a lathe with a brake and pulled it through a 10% reduction die and back on the sample holder. I then tested the small coil and the magnetization was about a magnitude lower and at twice the rate of change of background field. I was convinced and I called my old colleague Whetstone to tell him to twist the multifilament conductor that Cryomagnetics, Inc. was making for the outer NbTi winding of a bubble chamber coil in partnership with IGC who would provide the inner Nb_3Sn tape coil. The NbTi outer coil would provide a 7 T field to be added to the inner Nb_3Sn coils 4 to 5 T resulting in an 11 T field for the bubble chamber data. IGC had used the "dynamic" scheme for stabilizing the 30 cm id inner winding plus of course the background field of the outer NbTi coils contributing to the stability. The magnet during test after a quench or two reached a bore field of 11.5 T and it was decided to mount the bubble chamber inside and run at 11 T in the CERN K-beam line. The contract was with the Max Plank Institute in Munich and was in collaboration with our former Professor Charles Roos. This bubble chamber and high field magnet system designated "HYBUC" ran in the CERN PS Kaon beam line for over a year resulting in a statistically accurate determination of the Σ^+ and Σ^- magnetic moments. I had set the twist rate high in my magnetization check and used that same value for the HYBUC magnet, but another colleague (G. Morgan) in our group at BNL actually worked out a model and published a result for the critical pitch length. In order to decouple the filaments in a composite, the twist pitch "$l_c(max)^2$" $\leq (10^8 \lambda \rho a J_c d)/[(dH/dt)_c(a+d)]$ where λ is a geometry factor which equals 1 for Cu, "a" is the filament separation, and ρ is the resistivity of the matrix, according to G. H. Morgan, *J. Appl. Phy.*, 41, p. 3673 (1970). Actually the

FIGURE 12.25: A remarkable little bubble chamber, called HYBUC, was built to measure magnetic moments of hyperons. It was one of the first rapid cycling chambers, and, using a superconducting magnet achieved very high fields of 11 Tesla. (Courtesy CERN.)

magnetization data indicated about a third of that pitch length for a normal charge rate is ok.

7. On MIITS, A. M.

Early in the prototyping of the Energy/Saver Doubler (Tevatron) work, Moyses Kuchnir and Joe Tague measured this type curve for the E/S cable. Moyses and Joe mounted an E/S cable on a long sample mount, which they instrumented with thermometry, voltage taps, and a heater in the middle of the cable. After covering the cable and associated instrumentation with insulating silicone grease, they placed the rig in a cryostat which subsequently was filled with LHe. A power supply was connected to the conductor and a DAQ unit continuously monitored and recorded cable current, voltage between taps, cable temperature, and the pulsed heater's current and voltage versus time. The heater was manually pulsed until the cable quenches, this quench voltage appearance causes the DAQ to speed up and after a few seconds to stop taking data. The data stored is a few seconds before the quench and a few after the quench. The data logged yields the peak temperature versus MIITS curve. I do not recall if they actually degraded or burnt any of the cables, but I would assume so. When I arrived at Fermilab after accepting a position to work on the Energy/Saver Doubler, I computed the MIITS curve for the E/S cable using Martin Wilson's "Quench" code and compared it to Moyses' data and found the computed values to be slightly conservative especially at the start (which is what one might have expected for an adiabatic calculation).

8. On Helium Expulsion Effects During a Quench of a CICC Conductor, L.B.

We were testing the ITER PF Conductor Insert at the CS Model Coil Test Facility in Naka, and one important item in the test preparation, before any serious powering, was to make sure that a quench could be detected timely. To this aim the Insert was equipped with a rather involved quench circuitry that included cowound wires to provide a reliable cancellation of inductive voltages. These instrumentation wires are embedded in the winding, and can fail under thermal stresses or electromagnetic forces. One such issue at the start of the test was solved by rewiring the instrumentation, modifying the balanced bridges, adjusting them slowly and carefully until the team had confidence enough that a quench would be suitably protected. The whole balancing procedure took very many hours, under the pressure of those who wanted to have a quick answer on the cable performance limits. Finally, with enough confidence in the adjusted quench detection, the cable quench current was measured, right at the expected value, 45 kA at 6.3 K and 6 T. The quench was detected by an interlock on the inlet massflow, much earlier than the voltage threshold. In fact, all but one of the quenches were detected by flow interlocks, not by the voltage signals!

12.4 The History of Fusion Magnet Development

Jean-Luc Duchateau, Peter Komarek and Bernard Turck

12.4.1 General Introduction

The need for superconducting magnets in large fusion devices was already recognized in the middle of the 1970s and the initial major development programs were started based on the first definitions for the required magnet parameters. These programs followed two lines. First was the development of conductor and magnet systems for small and medium sized plasma devices such as T-7 and T-15 in the Soviet Union, TRIAM in Japan and most prominently TORE SUPRA at CEA in France. Second, in the form of an international project, was the development of a conductor and magnet arrangement with parameters pertinent to large fusion devices and their test in a dedicated special facility, the Large Coil Task (LCT) project, within the framework of an agreement by the International Energy Agency (IEA). These projects were followed in Europe, Japan and the US by the basic development of poloidal field coils with principal emphasis on conductor development.

Following the successful execution of these projects, the design and construction of larger fusion devices with superconducting confinement magnets were started. Some are already in operation—tokamaks EAST in China and KSTAR in Korea, the stellarator LHD in Japan—or are still in the final construction stage, the SST1 in India, tokamak and the stellarator Wendelstein W7-X in Germany. Finally, the international fusion community felt brave enough to start the development, in a worldwide effort, of ITER, a tokamak-type reactor accompanied by a somewhat smaller tokamak, JT-60SA, in Japan, with major contributions from Europe. This chapter gives a brief survey of these developments and the present status of superconductivity for thermonuclear fusion.

12.4.2 The Large Coil Task (LCT)

As the result of an invitation from US Department of Energy, EURATOM and Switzerland, at that time not yet a member of EURATOM, joined in 1976 the so-called Large Coil Task (LCT)[49].

---

[49] D.S. Beard et al., *Fusion Engineering and Design* 7 (1998) 1

This project involved the development of a torus with six superconducting D-shaped coils with a bore of about 2.5 m × 3.5 m. Three coils were to be developed in the US, one in Japan, invited as an additional partner, one by EURATOM, and one by Switzerland. The work was formally organized through an Implementing Agreement of the International Energy Agency (IEA). All but one of the LCT magnets were constructed with a NbTi conductor meeting quite exacting specifications; one US coil was designed with Nb_3Sn conductor, a technology still unproven at that time. In contrast to the almost standard design at that time, namely pool boiling cooling for large coils, the advantages of a forced flow cooling concept within a strong conduit were emphasized. Its advantages are optimal force transmission to the coil case, well defined heat transfer and fully predictable voltage properties. This was a big challenge for the project groups involved and for their industrial partners. Budget and the time table became very critical, but all groups stayed on schedule and all 6 coils were operated together very successfully in 1986–87, Figure 12.26, in a test facility constructed for that purpose at the Oak Ridge National Laboratory. During extended tests some coils reached up to 9.1 T at 3.6 K before transitioning smoothly to normal conductivity, a record magnetic field value at that time.

FIGURE 12.26: View of the toroidal arrangement of the LCT coils in the test cryostat at ORNL. (Courtesy ORNL.)

One major finding of this project was that for large fusion magnets, forced flow cooling of the conductor was the right choice to cope with the huge mechanical forces and the losses, and all later fusion projects followed this path. Another important experience was that this project was the beginning of very well working, formalized, international collaboration in the area of fusion magnets, greatly needed for resource sharing in projects which were growing larger and larger.

In connection with the LCT project, a smaller test facility TOSKA, a cryostat with a 4.5 m bore and 8 m in height, was built at FZ-Karlsruhe, now Karlsruhe Institut für Technologie, for the first tests of the EURATOM LCT coil as a single coil prior to shipment to the USA. This facility was also planned for future upgrades and future European fusion coil projects, a very wise decision as we will see later.

After its operation at ORNL, the EURATOM LCT coil was shipped back safely to Karlsruhe. As an extension of the existing program, the FZ-Karlsruhe team proposed to investigate the coil performance at reduced temperatures, down to 1.8 K, to demonstrate the feasibility of higher field operation of such coils with a NbTi superconductor. Following some hard negotiations, the EU-RATOM authorities finally approved this experiment. For this purpose the coil had to be reinforced with external thick steel bands to withstand the huge mechanical forces existing in a single coil operation. The tests were performed in the TOSKA facility and proved to be very successful. While the highest operation point at ORNL had been 16 kA at 3.6 K, achieving 9.1 T at the winding, the coil was now operated up to 19 kA at 1.8 K, achieving 11 T. The coil quenched only slightly above that value, close to the critical current load line. With these results in 1998 everybody in Europe was now happy that this coil could be used later as a background coil for the testing of an ITER TF Model Coil and the Wendelstein W7- X demonstration coil, as will be described later in the appropriate sections.

12.4.3 Early Work on Poloidal Field Coils for Tokamaks

The challenge of this coil system is the need for pulsed operation in the same mode as the plasma current. Shortly after the LCT, the international collaboration began to establish initially a number of medium sized programs to meet these needs, mainly concentrating on conductor problems, with particular attention on the reduction of AC losses. Due to the fact that the magnetic field here is much smaller than in the toroidal field coils, NbTi can be used for the conductor, but sophisticated cabling of the conductor strands and strand coating with less conductive material is essential, requiring a special conductor development program. Even today, including the ITER design, several approaches are under investigation resulting in complicated and more expensive conductors. One of the largest test coils, 3 m in diameter, was developed at FZK earlier and tested in 1996 in the TOSKA facility. Currently a cost and effort intensive program is devoted to the ITER project as we will see in a later section.

12.4.4 *Tore Supra*

Introduction and Context

Until the beginning of the 1980s, all the fusion magnet systems were resistive with silver alloyed copper conductors to improve their mechanical properties and to resist the large electromagnetic forces. This kind of solution was possible due to the small size of the machines and to their pulsed mode of operation. The largest machine of this type was the Joint European Torus (JET) with a major radius of 2.98 m. Here the power required to energize the system was more than 1 GW, produced by flywheel generators, a solution possible only because of the short duration, 10 to 30 s, of the JET plasma discharges.

The first significant introduction of superconducting magnets into magnetic fusion systems was the construction of the 2.4 m diameter Tore Supra (TS) tokamak in France and of the T-15 in the Soviet Union. In both machines the magnetic field on the conductor was nominally around 9 T, making the classical use of NbTi conductors at 4.2 K impossible.

This led to a debate at the end of the 1970s centered around two possible choices:

- the use of forced flow Nb_3Sn conductors at a temperature in the range of 4 K,

- the use of bath cooled NbTi conductors with pressurized helium at 1.8 K through a new technique developed at CEA in France.

The inadequate industrial maturity of Nb_3Sn technology was clearly seen during the acceptance tests of T-15 where resistive parts in the magnets prevented steady state operation of the tokamak, which ultimately stopped operation in 1991.

The main design goal for the toroidal field system of Tore Supra at the beginning of the eighties was to extend the application of the NbTi superconductor, a cheap material, insensitive to mechanical strain, to a magnetic field of at least 9 T using superfluid helium, at 1.8 K and 1 bar, as a coolant.

This industrialization of refrigeration at 1.8 K was really a breakthrough, furthering the production of higher field NbTi as well as Nb$_3$Sn magnets. Only the toroidal field coil system was chosen to be superconducting, with the cryostat walls as close to the coils as possible All the other pulsed coils were conventional copper coils operating at ambient temperatures. This choice proved to be the correct one; as operational experience with the plasma control systems increased, it was possible to run with plasma discharges as long as 6 minutes.

Tore Supra and the Invention of Pressurized Superfluid Helium as a Cooling Technique for Superconductivity

Tore Supra was the first large sized superconducting machine using a substantial volume of superfluid helium, 5 m^3 at 1.2 bar of He II, mainly contained within the 18 toroidal field coils. The cooling of the pressurized baths is achieved through heat exchangers feeding the saturated baths. The design and sizing of all the 1.8 K refrigeration circuits were the result of studies at CEA during the 1970s, particularly the low temperature pumping of the saturated baths by two stage centrifugal compressors. This technique, used here for the first time in the world, was selected to obviate the intensive development of the low-pressure exchangers which necessarily must have a very small pressure drop. The choice of warm oil ring pumping was also the result of comparative studies from the technical, investment and operating cost points of view.

This novel cooling technique[50], developed at CEA, Grenoble, not only provided a lower operating temperature and thus a higher critical current density but the superfluid helium under atmospheric pressure has outstanding heat conduction properties. Also the total enthalpy between 1.8 K and 2.16 K, approximately 3×10^5 J/m^3, is available to ensure the stability of the conductor against disturbances.

In order to make a compact winding with good overall mechanical characteristics while reducing ac losses and increasing the wetted perimeter, a rectangular monolithic conductor wound edgewise in double pancakes was chosen. The superconducting filaments were extruded in a mixed matrix of copper and cupronickel in order to limit coupling losses by cupronickel barriers. The copper amount was also reduced as a result of the superfluid helium cooling resulting in a current density in the conductor of 89.2 A/mm^2.

The superconducting winding, Figure 12.27, was made of 26 double pancakes of conductor edge wound on the narrow side and made as rigid as possible, especially in the radial direction, to take the large void associated with superfluid helium into account. Each turn was electrically insulated with a pre-impregnated glass-fiber epoxy tape co-wound with the conductor. The interconnects were located on the outer part of the coil, in a low field and low mechanical stress region.

The pancakes were separated by insulating spacers, providing additional mechanical cohesion of the winding while ensuring a very large volume of helium in direct contact with the conductor. After completion of the assembly process, the winding was heated to polymerize all pre-impregnated parts and to ensure a rigid coil.

Tests of a Prototype Coil and Difficulties Linked to the Acceptance Tests of the Assembled System

Once the superfluid helium cooling at 1.8 K and 0.12 MPa process was established, and a half scale model coil had been fabricated and its performance checked, a prototype coil was fabricated and tested at CEA, Saclay. These tests constituted the essential step to confirm the performance of the coils and to finalize the process for operation, with particular attention devoted to protection.

A stability test showed that, apart from a more elevated temperature zone around the initiating heated point, the coil was entirely quenched in 2 or 3 seconds after the opening of the circuit breaker,

---

50 G. Claudet et al., *Cryogenics* 26 (1986) 443

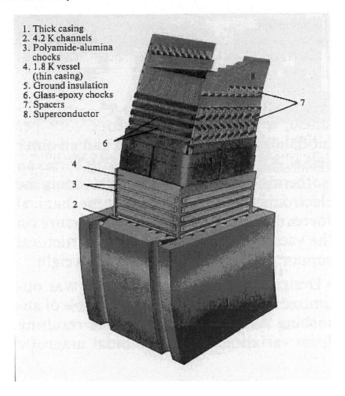

1. Thick casing
2. 4.2 K channels
3. Polyamide-alumina chocks
4. 1.8 K vessel (thin casing)
5. Ground insulation
6. Glass-epoxy chocks
7. Spacers
8. Superconductor

FIGURE 12.27: Sketch of one coil of Tore Supra showing the bare conductors immersed in superfluid helium confined in a thin casing. (Courtesy CEA.)

at a temperature of about 25 K. This proved that the quench propagation was very fast thanks to strong helium convection throughout the magnet. This process was driven by the increase of pressure and the escape of the helium through a safety valve located at the bottom of the coil in the low field zone. The final average temperature of the coil was 45 K, while the highest temperature located on the second pancake at the opposite of the gas outlet was estimated to be 130 K. Similar results were confirmed later in 1989, when during operation at 1400 A the coil BT4 suddenly quenched and triggered the magnet safety discharge system. This quench incidentally was the only quench observed during the 20 years of the Tore Supra operation. It was caused by the heat deposited by a fast electron beam from a severe plasma current disruption.

It may be recalled that in 1988 the acceptance tests of the entire system at CEA, Cadarache, were interrupted at 600 A by the appearance of a short circuit in coil BT17 during a test of the safety discharge system. This fault produced constant power dissipation in the coil. Pulsed operation of the coil was initiated during the next few months with a dedicated power supply to begin the first plasma runs in the machine. Then BT17 was replaced with a spare coil, the former prototype coil BT19, a process which took some 6 months. The damaged coil was removed, inspected, repaired and tested at CEA Saclay to be used as a possible spare coil. Only a limited part of it was damaged: 4 out of a total of 52 pancakes. In addition a more sophisticated analysis of the stability and quench test results of the prototype coil was undertaken, based on the experience with the toroidal field coil protection system. The purpose was to find a way to protect the magnet effectively while reducing the discharge voltages to a lower level. After analysis and additional testing, a complete revision of the protection scheme and a modification of the electrical circuit was instituted; it was demonstrated that the voltages could be divided by a factor of 10 in cases of fast safety discharges. This conclusion was reached after an in-depth examination of the specific behavior of the coil in He II after a quench, shown in the prototype tests and in the supplementary tests.

Commissioning the Tore Supra Superconducting Toroidal Field System at Nominal Current

After the coil replacement and revision of the protection scheme, in November 1989 the current was progressively increased to 1455 A, corresponding to a magnetic field on the strand of 9.3 T and to a stored energy of 650 MJ [51].

During these acceptance tests, several discharges were triggered to test the safety system at increasing current levels. In the course of this phase, the coil current was reduced by extracting the magnetic energy through a dump resistor with a typical time constant of 120s.

One Day of Operation

On December 4, 2003, a day specifically chosen to study the impact of plasma operation on the cryogenics, two 4-minute plasma discharge runs were made with 1 GJ energy injected in the plasma.

Figure 12.28 represents the impact of a plasma operation on the temperature of coil BT15, this coil being representative of the behavior of the 18 coils of Tore Supra.

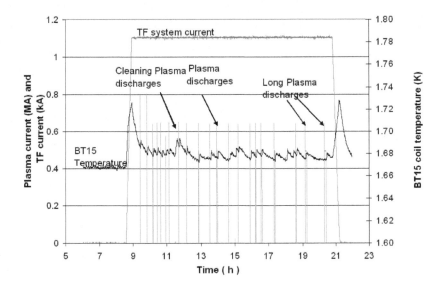

FIGURE 12.28:   BT15 temperature during one day of operation.

On a normal day of operation the current in the toroidal field system was raised at the beginning of the day and kept constant during the whole day until reduced at the end of the day. The changes in the current levels had but a small impact on the coil temperature, the changes being about 0.06 K, recovery from which took place within half an hour.

The effect of plasma operation on the coil temperature is due to hysteretic and coupling losses, induced in the toroidal field conductors by the associated field variations. These field variations also induce eddy currents and heat dissipation in the thick coil casings. During the day, a phase of cleaning plasma discharges to recondition the vacuum vessel after a disruption produces a typical temperature increase of 0.25 K in the thick casings, needing about half an hour for the system to

---

[51] B. Turck, *IEEE Trans.* Mag. 25 (1989) 1473

recover. The rest of the operation has little impact on the cryogenics. Heat dissipation is always associated with plasma current initiation but the related temperature increases are small, less than 0.01 K for the coil temperature and approximately 0.08 K for the thick casing. It should be noted that the two last plasma discharges of the day, the 1 GJ discharges, had no particular impact on the cryogenics.

Summary

After more than 20 years of operation and 45000 plasma discharges without a major failure, this first marriage between fusion and superconductivity proved to be very positive[52]. It is also worth noting that while only one quench in a coil had been experienced during the whole life of the system, about 100 fast safety discharges of the magnet system occurred, resulting each time in a shutdown of the system for about two hours. These fast safety discharges were always triggered for reasons not associated with the magnet but rather by interference events in the sensors or in the electronics. Minimizing this type of disturbance must be an objective for the operation of all large devices including ITER.

The Tore Supra experience has demonstrated that superconducting magnets can safely and successfully be operated in long term plasma physics research. Far from being a burden, the steady state operation of the toroidal field system leads to a simplification in the preparation of the plasma discharges. No significant heat load is apparently associated with long plasma discharges but the circular shape of the Tore Supra plasma may attenuate the thermal load associated with plasma control, a phenomenon which must be confirmed on a machine with an elongated plasma. Future long pulse operation of the later tokamaks such as EAST and KSTAR will help to clarify this important point. The steady state operation of a large magnet system is also an advantage with regard to the mechanical integrity, fatigue life and aging of the coil. In Tore Supra no sign of mechanical aging was apparent judging from the mechanical measurements.

Clearly, the path is really open for a long and extensive program for superconducting machines for fusion. Certainly the superconducting magnet system of Tore Supra, with its 20 years of safe operation, has demonstrated and emphasized the way for a necessary and fruitful relationship between superconductivity and fusion.

12.4.5   Tokamaks with Superconducting Magnet Systems in Recent Time

Tore Supra is no longer the only tokamak with superconducting magnets. In 2006, the first fully superconducting tokamak with NbTi magnets, EAST[53] with a radius of 1.8 m, produced its first plasmas in China then, in 2008 the Korean tokamak KSTAR with a radius of 1.8 m with both NbTi and Nb_3Sn magnets with forced flow conductors in the manner of ITER was put into operation. It was a big success for industrial fabrication. It is remarkable that KSTAR[54] is so far the only machine where no cold testing of the magnets took place before onsite installation, Figure 12.29 Another tokamak, SST1 in India, with NbTi superconducting coils is expected to be commissioned in the near future.

[52] J.L. Duchateau et al., *Fusion Sci. and Technol.* 56 (2009) 1092

[53] S. Wu and EAST Team, *Fusion Engineering and Design* 82 (2007) 463

[54] Y.C. Oh et al., *Fusion Engineering and Design* 84 (2009) 334

FIGURE 12.29:   The KSTAR tokamak at KAERI in Korea[54]. (Courtesy KAERI, Korea.)

12.4.6   ITER as a Worldwide Collaboration Project in Thermonuclear Fusion

General

The ITER adventure was initiated in November 1985 when Ronald Reagan and Mikhail Gorbachev met at Geneva and encouraged the formation of an international collaboration with the goal of finally mastering fusion energy. However, it was only in 1991 that four entities, Europe, Russia, Japan and the US finally began a 6-year funded program with a dedicated project team: the ITER project was born. The ITER project has now been extended to 3 other countries, India, Korea and China; the construction of the machine was officially started in 2006 on the Cadarache site in France. To prove the feasibility of thermonuclear fusion as a potential source of energy for humanity, ITER is departing in a major way from the currently active fusion machines, the JT60 in Japan and the JET in Europe. JET has a major radius of 3 meters, while for ITER in its selected configuration a major radius of 6.2 meters is necessary, the major radius being the radius of the plasma torus. This geometry is necessary to sustain stable plasma discharges for as long as 500s resulting in a fusion power of 500 MW with an amplification of energy Q of 10, the ratio of the output power to the input power.

If the extrapolation from JET, based on plasma physics scaling laws, is considered to be adequate, ITER will also have to handle additional numerous technological challenges. These challenges include, for example, the plasma facing components, that is the inner blanket, the high power plasma heating sources and the very large size of the components.

Of these, the superconducting magnet system is probably the most remarkable one, as it is the real backbone of the machine. In practice, deeply buried in the very heart of the tokamak, repairs are hardly feasible except for a few protruding components like joints for instance, leading to a rigorous no-fault strategy in design, construction, assembly and test. The magnet system represents about 30% of the investment cost of the device.

In the case of ITER, for which a 500 second plasma discharge duration is planned, and probably longer in the later operational stages, a solution with resistive magnets is no longer possible and the magnet system must be superconducting. The electrical power associated with the refrigerator to compensate for all losses at cryogenic temperatures is estimated to be about 20 MW, a value to be compared to the 2 GW that would be needed for an equivalent resistive system. In machine design, the factor of merit $\xi = R^2 \times B_t^3$ is a key driver of the machine performance with respect to the fusion power and the amplification factor Q. Here R is the major radius of the tokamak and B_t is the magnetic field at the plasma center. The selection of the two parameters R and B_t has to be made in close consideration of questions involving cost, available technology and accessibility to the plasma through the ports. To satisfy the objectives of ITER, the two parameters were set at $R = 6.2$ m and $B_t = 5.3$ T. Due to the toroidal shape, the magnetic field from the center of the plasma to the conductor on the magnet system is increased by a factor of more than 2, which clearly mandates the use of Nb$_3$Sn for the superconducting material of the toroidal field magnet system system.

The Cable-in-Conduit-Conductor

For the ITER coils, the requirements for high currents in the 70–80 kA range and for very high voltages in operation, 10 to 20 kV to ground for the poloidal field and central solenoid systems inherent in the size of the magnetic systems, led to the selection of the cable-in-conduit-conductor as the best choice for the conductors in the present state of the superconducting technology. Moreover this type of conductor is well adapted to support fast heat deposition. The principle of the CICC is not recent: M. Hoenig at MIT (US) introduced it in 1975 and a number of coils have been made with this type material, mainly in the US and in Japan as we have noted before. The coil with the largest size to use this concept to date was the Westinghouse coil in the Large Coil Task described above, where Nb$_3$Sn was the superconductor. The maximum performance of that magnet was unfortunately limited by some spreading out of a resistive phase in the magnet.

A modern CICC is basically made in several stages by cabling superconducting and copper strands and then by compacting the cable inside a conduit generally made of stainless steel. A CICC such as the one for ITER is composed of several components, superconducting strands, copper strands, steel bandages, one or more helium channels and the steel encasing conduit, as shown in Figure 12.30. In a project like ITER, the optimum composition of the conductor components is defined by the system design criteria.

FIGURE 12.30: Conductors of the two ITER model coils. On the left CS model coil (51 mm × 51 mm, 40 kA) on the right an exploded view of the TF model coil (diameter 40.7 mm, 80 kA). (Courtesy ITER.)

The CICC was invented to benefit from the very high volumetric heat capacity of helium, about 500 times the volumetric heat capacity of metallic materials, thereby limiting the temperature excursions in the case of fast energy deposition. This occurs in tokamaks after a very fast decrease to zero of the plasma current when plasma disruption occurs. In this case, a fast magnetic field variation of the order of 100 ms affects the whole coil over lengths well over several meters, creating losses in the superconducting strands. This is rather similar to the kind of event which can occur in high field test facilities where the outer superconducting magnet of a hybrid magnet is affected when the central copper magnet disrupts. The cable in conduit offers an adequate solution to this problem by providing:

- a local helium reservoir,

- a very long wetted perimeter. The diameter of the ITER TF cable is 39.7 mm. It is made of 900 superconducting strands 0.82 mm in diameter with a void fraction in the cable of 30%. This fine subdivision of the strands can be translated into a total of 2.3 meters of wetted perimeter in the cable section facilitating a large heat transfer to the helium reservoir,

- small ac losses for the conductor by controlling its time constant through the contact resistance between strands.

Thanks to the high wetted perimeter, a large heat flux from the conductor to helium takes place during the transients, characterized by the thermal time constant of the CICC which is in the range of 50 ms, of the same order of magnitude as the time constant of the magnetic field variation caused by the disruption. The temperature excursion in the conductor is therefore limited and the conductor ability to withstand the transient is enhanced. During operation, the helium mass flow circulating in the conductor limits the temperature increase due to the residual nuclear heating and to the ac losses generated by the varying magnetic fields during a plasma discharge. The central channel helps to keep the pressure drop at an acceptable level. To date, the community of CICC users is rather small but the number of applications is increasing, each, with somewhat limited associated practical experience as an operational system. There are only four big systems currently in operation in the world: the poloidal field system of the 1999 Japanese Large Helical Device, the winding of the hybrid system at the National High Magnetic Field Laboratory in the US, and the conductors of two superconducting tokamaks: EAST and KSTAR.

The ITER Magnet System

The ITER magnetic field system[55] is composed of three major components shown in Figure 12.31.

- Toroidal Field system (TF)

- Central Solenoid (CS)

- Poloidal Field system (PF) and some correction coils (CC)

The TF system is made up of 18 coils. It provides the main magnetic field to confine the plasma charged particles. The 6 CS coils provide the inductive flux to ramp the plasma current and shape the plasma current.

The main characteristics of these systems are presented in Table 12.2.

The PF system is responsible for the stabilization and positioning of the plasma current. The TF system is a dc system while the PF and the CS systems are pulsed coils. Due to the moderate magnetic field of the PF system, the superconducting material selected for this system is NbTi. The

---

[55] N. Mitchell et al., *Fusion Engineering and Design* 84 (2009) 113

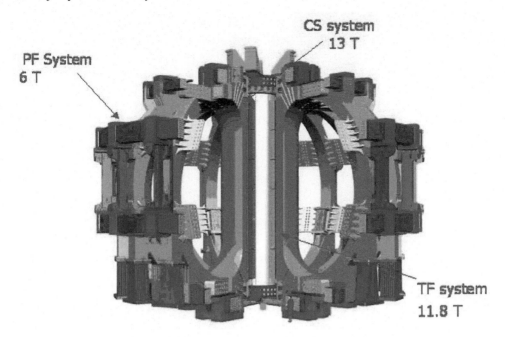

FIGURE 12.31:   ITER superconducting systems. (Courtesy ITER.)

TABLE 12.2:   Main Characteristics of the ITER Superconducting Systems

System	Energy (GJ)	Peak Field (T)	Total Ampere-Turns (MA)	Conductor Length (km)	Total Weight (t) (strand)
Toroidal Field, TF	41	11.8	164	82.2 Nb_3Sn	6540 (396)
Central Solenoid, CS	6.4	13	147	35.6 Nb_3Sn	974 (118)
Poloidal Field, PF	4	6	58.2	61.4 NbTi	2163 (224)
Correction Coils, CC		4.2	3.6	8.2 NbTi	85

multifilamentary composite is made of very fine filaments, 6 μm in diameter. Due to the high magnetic field (12-13 T), Nb_3Sn has been selected for the CS and the TF system. The total production of 514 tons of Nb_3Sn for ITER will mark the beginning of the industrialization of this superconducting material, the present production being in the range of a few tons per year. It should be emphasized that a very important internationally led project resulted in two model coils during the preparation phase of ITER (1997-2002). The two coils were designed, manufactured and tested as follows:

- A model coil of the central solenoid which was manufactured by the US and Japan and tested at the JAERI facility in Japan. This is illustrated in Figure 12.32. Japan, US and Europe shared the fabrication of the conductor.

- A model coil of the toroidal field system manufactured in Europe and tested at the FZK facility is shown in Figure 12.33.

FIGURE 12.32:   The outer module CS model coil of ITER at JAERI. (Courtesy ITER and JAERI, Japan.)

These model coil experiments during 2000 through 2002 were crucial in testing the behavior of large Nb_3Sn CICCs with actual dimensions and with the relevant lengths. Some unexpected degradation in the critical performance was found to be due to the great strain sensitivity of the Nb_3Sn strands. The strain is mainly due to the differential thermal contraction between the steel conduit and Nb_3Sn caused by the extreme temperature range between the reaction temperature of 650 °C at which Nb_3Sn is formed, and the cryogenic operating temperature. Some extra strain including bending strain was also indicated during the model coil experiments. This extra strain can be related to the Lorenz force loading the strands at the nominal current, very important findings leading to extensive discussions in the international community. The design of the ITER CICCs was later corrected to take these effects into account.

The CS model coil with a stored energy of 600 MJ was the first very large coil to be tested in a fast varying field related to the ITER operational conditions by discharging the magnet at the rate of 0.6 T/s without quenching. This high rate was made possible by the limited ac losses of the conductor. The successful series of tests effectively constitute a real milestone in applied superconductivity, confirming that the CICC is an appropriate choice for the ITER magnets.

The Construction of ITER

The organization responsible for the construction of ITER is objective and task specific. While the specifications of the various components are prepared by the central ITER organization, the individual components are ordered and constructed under the supervisions of the so-called domestic agencies, which are the seven ITER participants. The budgetary responsibilities of the seven parties to the project are in the form of in-kind contribution. Thus Europe is responsible for bringing contributions representing about 50% of the ITER cost investment. Six partners with the exception of India are involved in the delivery of conductors and magnets where the CICCs represent almost 50% of the magnet investment cost. The challenge is now in the hands of the companies and of the project teams: for example most contracts involving the conductors have already been placed. There is no doubt that quality assurance with all the controls and the acceptance tests will play a major role, it will have to be carried out at a global level, which in itself is another challenge. Again, this task is coordinated by the ITER organization.

FIGURE 12.33: The TF model coil at the FZK test facility. (Courtesy FZK Karlsruhe.)

Introducing DEMO

The next step after ITER will be DEMO[56], a fusion reactor that should provide 1000 MW of dc electrical power. Once ITER has delivered significant results, its anticipated construction could start, 20 years hence. Whether the DEMO magnet system will be an extrapolation of the ITER magnet system or whether a technological revolution will be needed, for example will the conventional NbTi and Nb_3Sn CICCs of ITER, operating at 5 K, be able to satisfy the technological specifications of the DEMO magnet systems or will it be necessary to use new, emerging superconducting materials operating at higher temperatures in the 20 to 50 K range are questions which will have to be answered through development programs yet to be defined. Necessarily the answers are not restricted to the applied superconductivity community only. The dimensioning of such a reactor and its optimization has to be found, in a manner similar to that of ITER, through a collaborative action involving scientists from all the disciplines. For an industrial plant on the scale of DEMO, the utilities will have to play a very strong participatory role in defining the constraints pertinent to dimensioning, construction, operation and maintenance.

JT-60SA Tokamak

Apart from the decisions related to the construction of ITER, an additional agreement between Europe and Japan was established, the so-called Broader Approach for Fusion Energy. This consists of 3 satellite facilities: IFERC, a computation center, a materials research facility, the IFMIF–EVEDA; and a superconducting tokamak JT-60SA[57], replacing the existing Japanese copper coil

---

[56] P. Komarek, *Fusion Engineering and Design* 81 (2006) 2287

[57] M. Mutsukawa et al., *Fusion Engineering and Design* 83 (2008) 795

tokamak operating in the short pulse, 5 s mode. This new machine will be installed in the existing tokamak hall at Naka in Japan. With a major radius of 2.95 m and a plasma current of 5.5 MA, its dimensions and plasma current are similar to those of JET but JT-60SA will not be a nuclear machine, that is it will not allow the injection of deuterium or tritium. By virtue of the superconducting magnets, the machine will be the largest superconducting tokamak capable of providing plasma discharges as long as 100 s. It is scheduled to operate in 2016. The JT-60SA program aims at supporting the ITER program and to complement ITER towards DEMO, the steady state demonstration reactor mentioned above. Europe will contribute to its construction to the tune of about 50% of the investment cost. Europe will supply an important part of the magnet system: the cryostat, the TF magnets, the power supplies, the HTS current leads and the cryogenic system for the whole magnet system. This activity is coordinated by Fusion For Energy, the cognizant European agency.

A comparison of the plasma dimensions of tokamaks under construction with those of existing installations is shown in Figure 12.34.

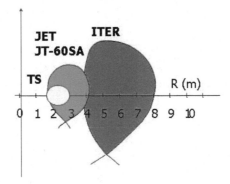

FIGURE 12.34:   Plasma dimensions of superconducting tokamaks under construction, ITER, and JT-60SA compared with JET and Tore Supra (TS). (Courtesy ITER.)

Since 2006, much work has been devoted to the optimization of the toroidal field system, which is as always, the most important magnet component of the tokamak. This effort has resulted in important changes from the initial design of the conductor and of the magnet as well. The main characteristics of the JT-60SA TF system are summarized in Table 12.3.

The conductor of the TF system is also a CICC with no central channel and a moderate copper to NbTi ratio of about 2 in the superconducting strands. The winding pack is made of 6 double pancakes, each with 6 turns. A wedge-shaped inboard structure was adopted to allow space for segmentation of the central solenoid into several modules for better control of the plasma shape.

The expected behavior of the TF system has been analyzed in detail assuming normal operation taking into account the nuclear losses and the conductor ac losses. A substantial temperature margin of the conductor, 1.4 K at the end of burn, is a guarantee for safe operation and is adequate to prevent a disruption of the plasma at full current from quenching the coils. Before being shipped to Japan, the 18 coils of the TF magnet system will be tested in a cryogenic test facility, the construction of which has started at CEA, Cadarache.

TABLE 12.3: Main Characteristics of the JT-60SA TF Superconducting System

System	Energy (GJ)	Peak Field (T)	Conductor Current (kA)	Conductor Length (km)	Total Weight (t) (strand)
Toroidal Field TF	1.06	5.65	25.7	24.4 NbTi	280 (33.4)

12.4.7 Superconducting Magnets for Stellarator-Type Fusion Devices

The Large Helical Device

The National Institute for Fusion Science (NIFS) in Toki, Japan, has been working for many years on torsatron-type stellarators. In 1998 a sophisticated large machine, the Large Helical Device with a radius of 3.7 m and superconducting NbTi-coils became operational[58]. While the magnetic field strength is moderate, the construction of the helical coil was a major challenge for industry. It had to be done on site with the plasma vessel already in position, which required a specially designed winding machine. The different ring coils (several meters in diameter) were made with a forced flow cooled CCIC conductor. A view of the LHD during installation is shown in Figure 12.35.

FIGURE 12.35: View of the LHD helical coil during installation at NIFS in Toki, Japan. (Courtesy of the National Institute for Fusion Science.)

---

[58] T. Satow et al., *IEEE Trans. Appl. Supercond.* 10 (2000) 600

The Modular Stellarator W7-X

A large modular stellarator with superconducting coils, W7-X, is under construction at the Greifswald branch of the Max Planck Institute for Plasmaphysics (IPP)[59]. This modular stellarator experiment of about JET size is a follow-up to the first modular system with copper coils, W7-AS, successfully operated at IPP in Garching in the 1980s. A set of nonplanar coils provide both the toroidal and poloidal field components to achieve the necessary twist of the magnetic field lines. The size of the machine and the potential of the stellarator in general for steady-state operation called for a superconducting confinement magnet system. In an extended development effort, the appropriate superconductor, the coil manufacture and the complicated cryostat were designed and engineered in the 1990s.

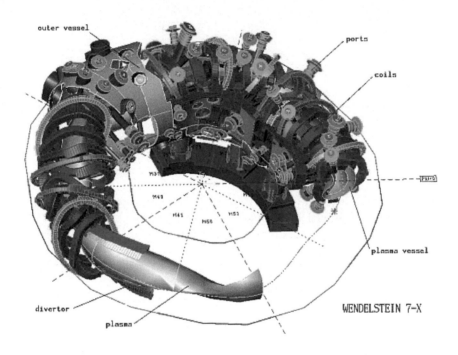

FIGURE 12.36: A representation of the W7-X coil system. (Courtesy of the Max Planck Institute.)

The ultimate goal of this program was the construction and testing of a full size demonstration coil, the completion of which in 1999 was marked by its successful testing in the TOSKA facility at KIT. The main parameters of the W7-X magnet system, illustrated in Figure 12.36, are a plasma major radius of 5.5 m and a magnetic field up to 6.7 T at the winding yielding a central field of 3 T on the plasma axis. This means that NbTi as superconductor is adequate. A set of five different types of nonplanar coils are periodically arranged in the torus. In addition, 20 superconducting planar coils are placed around the torus to follow the variation of the rotational transform of the plasma between the values 5/6 and 5/4 by modifying the main B field. Each nonplanar coil has a height of about 3.3 m, a width of 2.5 m and a depth of 1.5 m.

The superconductor is of the cable in conduit type with NbTi-strands. Due to the need to bend the superconductor through two axes the conduit material is a specially developed 6063 aluminum

---

[59] L. Wegener, *Fusion Engineering and Design* 84 (2009) 106

FIGURE 12.37: The W7-X coil system during installation at IPP in Greifswald. (Courtesy of the Max Planck Institute.)

alloy which is soft enough after co-extrusion with the bundle of strands to facilitate the winding work, but will be hardened afterward during the curing of the winding pack with its resin at about 1000 C and thereby achieve the necessary stiffness. Its dimensions are 16 x 16mm with 243 strands 0.58 mm in diameter with helium flowing through the voids between the strands. The rated current is 17.6 kA. The construction status in 2010 is well advanced, all coils have been manufactured and pre-tested at CEA Saclay by subcontract and installation in Greifswald is proceeding, as shown in Figure 12.37.

12.5 Electric Power Applications of Superconductivity

William Hassenzahl and Osami Tsukamoto

12.5.1 Introduction

The development of superconducting applications for the electric grid is driven by their promise of improved efficiency, smaller size, and reduced weight as compared to existing technologies, and by the possibility of new applications based on the unique characteristics of superconducting materials. Superconducting power components can also contribute to improved power quality and increased system reliability. This chapter addresses historical developments and technology status of five superconducting power applications: fault-current limiters, superconducting magnetic energy storage (SMES), rotating machinery, power cables, and transformers.

It is instructive to provide a simple description of an electric power system before exploring how superconductivity might contribute to its performance. Though there are several important exceptions, electricity is produced by a "generator", converted to an appropriately high voltage by a transformer, carried by a transmission line over long distances, transformed to a voltage that is appropriate for local distribution systems, carried to a local load by a distribution line or cable, and finally used for a variety of purposes. Along the path, there are various elements, controls, and feedback systems that ensure the near-continuous operation of the power grid, even under upset conditions that may be short- or long-lived. Note: not all power applications involve a major utility grid: communities in remote areas operate power systems without ties to an extended power grid, large ships have significant electric power systems and may be driven by electric motors that connect directly to the propulsion system.

As mentioned above, there are exceptions that need mentioning. The "generator" is taken here to be a device that converts energy of some form into electric power. It can be a windmill, a gas or steam turbine, a solar cell, etc. These generators can produce ac or dc electricity. In addition, power may be transmitted either as ac or dc electricity, depending on a variety of factors. Transformers only operate on ac power and thus the use of dc requires modification of the process of taking power from the source to the user. Usually the choice of form of electricity used in any situation and any part of the grid is based on ease and efficiency of use and, ultimately, cost.

In principle, most of the conventional components of an electric power system could be replaced by a superconducting equivalent. However, the tremendous developments of conventional technology that have occurred over the past century—as a large fraction of the world has been electrified– have led to standard components that are effective, economical, and simple. These attributes combined with large-scale production deliver capabilities at low costs and provide a significant challenge to the introduction of any new technology.

As with most new concepts, when superconductivity was discovered in 1911, there was a rush by scientists to understand it—and an even greater rush by the power industry, by practical engineers, and by hucksters to capitalize on it. The early euphoria for practical applications was to be repeated over and over with each new breakthrough as the phenomenon was better understood and as materials with properties that seemed better or at least more practical were discovered. Though newspaper headlines at various times from 1911 through the 1950s suggested that superconductivity might have practical applications in the electric power arena, there was no real progress until the early 1960s. That was when practical Type II superconducting materials such as Nb-Ti and Nb_3Sn were discovered. Power applications such as motors, generators, and power cables were suggested early on. New concepts with no conventional equivalent, such as flux pumps for converting power and charging high current magnets, superconducting magnets for energy storage, and fault current limiters were soon to follow. However, developing superconducting systems with the same capability as existing conventional systems turned out to be a significant challenge. In addition, it was not until the last decade, that superconducting materials that could be used in practical devices had characteristics that allowed them to operate effectively above the temperature of liquid helium.

We mentioned above that the ability to carry large currents in the presence of magnetic fields is important for superconductors to be effective in most power applications. In fact, there are several characteristics that can make a superconductor an effective material. Most power applications have large alternating currents, the ability to carry these currents with low losses is critical because the losses occur at the operating temperature and must be removed by a refrigeration system that has a net efficiency of 1 to 10%. That is, the effective room temperature power loss is 10 to 100 times higher than the loss within the conductor itself. Perhaps the most important characteristic of large-scale power equipment is the need to operate at voltages of 100 kV or higher. These two enabling technologies, cryogenic refrigeration and cryogenic dielectrics have also seen rapid development in recent years.

However, considerable development and many full scale demonstrations will be needed before superconducting devices can be integrated into the utility environment. To date, most of these efforts are directly funded by government entities or have a great deal of government support. Initial commercial installations will be for niche applications and will be where space is limited, where power demands are increasing, and/or where initial development costs can be offset by enhanced performance.

Electricity as we use it is not available in a natural form. It must be converted from some other source. By tracing the path of the energy, it is clear that only a fraction of the initial energy consumed is converted to electricity and that an even smaller fraction is delivered to the consumer. Most electricity is produced by converting heat to rotary mechanical motion that powers a three-phase electrical generator. Generators are typically 95% efficient or better in carrying out this conversion. However, the systems that convert thermal energy to mechanical energy are subject to thermodynamic limits, referred to as Carnot efficiency[60], and which is determined by the upper and lower temperatures of the process. Generally the efficiency of gas and steam turbines is in the range of 30 to 50%. Hydroelectric power generation is 85 to 90% efficient, so it is no surprise that Europe and the US use almost all the water resources available to produce electricity. In addition, many developing countries are building dams for power production. Wind and solar will eventually become important sources of electricity but, in the near term, they continue to have their own sets of barriers to large-scale implementation.

Once electrical energy is produced, it is converted to high voltage for transmission and then converted back to a voltage appropriate for the end user. About 10% of the electrical energy from the generator is lost in transformers and the transmission and distribution systems. Though this may seem to be a small fraction of the total, it is an enormous quantity of energy. Future technical improvements, including superconductivity, can hope to save only a portion of this 10%. There are, however, other drivers for new technologies. One is the ever increasing electricity use in urban settings, which, when combined with environmental and other requirements, requires greater power flow through existing right of ways. Another is the increased electrification of the planet earth. As we begin the 21st century, there are 6 billion people on the earth of which fully one-third do not have electric power. Electric systems are being expanded and new infrastructure will be developed to deliver power to a much larger fraction of humanity as the century progresses. Perhaps the most intriguing aspect of electric power systems vis-á-vis new technologies is the need for the assured delivery of high-quality power. For example, industrial processes such as the manufacture of silicon chips take several days to complete. Loss of power for a fraction of a second can result in the loss of one or more days of preparation and work.

The applications in which superconductivity has the potential to be effective in an electric power system can be separated into two general classes. The first class includes cables, motors, generators and transformers where superconductors replace resistive conductors. The second class includes technologies that will be enabled by superconductivity and that have little or at most limited capability if conventional materials are used. Examples are superconducting magnetic energy storage (SMES), fault current limiters (FCLs), and fault current controllers (FCCs). These technologies are addressed here, but other technologies, such as a superconducting substation[61,62] and an integrated power system in which liquid hydrogen is piped as a fuel and simultaneously functions as a coolant for superconducting cables,[63] are not discussed.

Rather than going into the rationale for use of superconducting applications in the power grid, we note that several earlier documents addressed the potential advantages of superconducting power equipment. Two articles[64,65] in the October, 2004 issue of the *Proceedings of the IEEE* described several power technologies. In 2000, several articles on power applications of superconductors were published in the May, June and August issues of the *Power Engineering Review* (*PER*), at the time the official journal of the IEEE Power Engineering Society. Prior to that time, an issue of the *IEEE Spectrum* (July, 1997) was dedicated to developments of HTS applications for electric power systems.

---

[60] M. J. Moran and H. J. Shapiro, *Fundamentals of Engineering Thermodynamics* 5th ed., Wiley Text Books, (2003)

[61] GuoMin Zhang et al., (Accepted for publication) *Proc. 2010 Appl. Supercond. Conf.*

[62] W. V. Hassenzahl, EPRI Technical Report TR-1000915 (2000)

[63] J. R. Bartlit et al., *Proc. Int. Cryogenic Eng. Conf.* (1972) 176

[64] W. V. Hassenzahl et al., *Proc. IEEE* 92 (2004) 1655

[65] S.S. Kalsi et al., ibid.

12.5.2   Fault-Current Limiters

Faults in transmission and distribution networks cause short-circuit currents that typically reach 10 to 50 times the rated current. These faults may be a temporary short between phases of a three-phase system, a short between one phase and ground, a lightning strike induced short, etc. They may be very short in duration, up to a few cycles, or of much longer duration. No matter the source or duration of the fault, each component of the grid must be designed to withstand the mechanical and thermal loads of the high current during the fault condition. As electrification expands, the number of power stations, other grid components, and interconnections on the existing system increase. The result is a reduction of the effective system impedance and generally greater power flow. Since the voltage is fixed, however, the lower impedance induces larger fault currents. It is not reasonable (nor economical) to replace older network elements just to deal with higher fault currents; thus, some method of limiting short-circuit currents is needed. In principle, an increase of the internal impedance of each element or inserting impedance in series with each element might accomplish this end. There are, however, two undesirable side effects of this type of solution. First, higher impedances usually cause increased energy loss, and second, fluctuating loads would cause greater feedback to control systems, which would decrease overall grid stability.

Today, the electric utilities either limit fault currents, typically by the use of inductors in low-voltage systems, or interrupt the fault current with special switchgear in high-voltage, high power systems. Nevertheless, all grid components are designed to accommodate the forces produced by the maximum current reached. The switchgear that is available on the high-voltage transmission grid is effective today. However, ever-increasing fault currents are pushing the limits of existing switchgear, which are near their theoretical maximum. A device having a small impedance during normal operation and the ability to develop a considerably higher impedance within a very short time (less than a quarter of an AC cycle) when a fault occurs could limit the short-circuit current to a reasonable value. Ideally, such a fault current limiter (FCL) would operate passively and would return to its normal operational mode immediately after the fault clears.

Several FCL development projects are underway, some of which are based on superconductivity, and others use silicon based switches. A discussion of the various locations where an FCL can be installed on a utility are discussed below, and then some of the existing programs are covered. It is of note that of all the superconducting power technologies, only one device has been purchased directly by a utility from a vendor with no government support. That device is a superconducting fault current limiter (SCFCL) fabricated by Nexans and installed at a substation in Boxberg, Germany.

Applications of Fault Current Limiters

In general, FCLs can be used at all voltage levels in utility grids, industrial systems, and isolated power systems such as large ships. The electric utilities have determined that there are several locations within the power grid where fault current limiters could be effective[64]. Eleven of these are shown in Figure 12.38. Specifications such as voltage, current, power, and response time, etc., depend on the location of the device in the network.

Some of the technical and economic benefits are listed in Table 12.4. They depend very much on the application itself, the structure of the FCL and the specific situation in the power system. As a general rule, the installation of SCFCLs seems most attractive in grids and power systems with rapidly increasing power demand, in heavy load areas, in highly meshed systems, and in locations where new generator units have to be introduced. Irrespective of the voltage level, the two main applications may be classified as coupling (3, 4, 5 and 11 in Figure 12.38) and feeder locations (1, 2, 6, 7, 8 and 10 in Figure 12.38). The main advantage of a coupling location is that the network impedance and the required short-circuit capability are reduced. The feeder location enables the reduction of short-circuit currents in the sources.

TABLE 12.4: Summary of Technical and Economic Benefits of SCFCLs

Technical benefits
- Reduction of high short-circuit currents during fault conditions
- Greater short-circuit capacity during normal operation
- Use of lower impedance (lower loss) transformers
- Improved power quality
- Increases steady-state and transient stability
- Greater flexibility in locating generation
Economic benefits
- Use of more compact devices of installations and devices
- Postponement of component upgrades
- Reduced network losses
- Increased availability
- Replacement of conventional devices

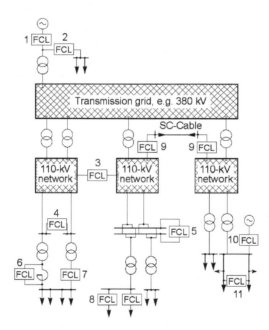

FIGURE 12.38: Locations for FCLs in power systems. (Figure from M. Noe, and P. Komarek.)

Types of SFCLs

Electric utilities, entrepreneurs, and research scientists are investigating a number of technologies in their search for a functional fault current limiter. Perhaps the most promising involve superconductivity because it can offer such a function by changing from "zero" resistivity to a high normal resistivity when either critical current, critical temperature, and/or critical field are exceeded. Several types of Superconducting Fault Current Limiters were proposed long before HTS materials were discovered[66,67,68]. Some of those concepts have been developed further with HTS materials, while other concepts have come forward and appear to be effective at higher operating temperatures. There are several conceptually different types of FCLs[69,70,71,72,73,74,75]. Four concepts are summarized in Table 12.5 and are described briefly below. In addition, hundreds of patents exist for the technology.

TABLE 12.5: General Characteristics of Superconducting Fault Current Limiter Technologies

Technology	Losses	Triggering	Recovery	Size/Weight	Distortion
Resistive SFCL	Hysteretic (depends on HTS materials)	Passive	HTS conductor must be re-cooled	Potential to be small, because HTS performs limiting action	Only during first cycle
Hybrid Resistive SFCL	Hysteretic (depends on HTS materials)	Passive or Active	Faster than the resistive FCL because less energy is deposited in the HTS	Potential to be small, but additional components may increase size	Only during first cycle
Shielded-Core SFCL	Hysteretic (amount depends on HTS materials)	Passive	Faster than resistive, but re-cooling required	Large and heavy due to iron core and windings	Only during first cycle
Saturable-Core SFCL	Power to DC that saturates the iron core, Joule heating in copper coils	Passive	Immediate	Large and heavy due to iron core and conventional windings	Some; cause by nonlinear magnetic characteristics

Resistive and Hybrid SCFCLs

Figure 12.39 shows the principle of the resistive SCFCL and defines the quantities used here. The current limiting element has an appropriate quantity of superconducting wire, tape or bulk material, which is in parallel with a normal resistor R_p (e.g., a thin conductive sheet on the surface of the superconductor). The superconductor must be rapidly driven normal over its whole length by the fault current to avoid hot spots and for stability reasons. When the superconductor is normal, the

[66] R. T. Kampwirth and K. E. Gray, *IEEE Trans. Magn.* 17 (1981) 565

[67] J. D. Rogers et al., *IEEE Trans. Magn.* 19 (1983) 1051

[68] H. J. Boenig and D. A. Paice, *IEEE Trans. Magn.* 19 (1983) 1054

[69] W. V. Hassenzahl et al., *Proc. IEEE* 92 (2004) 1655

[70] L. Salasoo et al., *IEEE Trans. Appl. Supercond.* 5 (1995) 1079

[71] M. Noe and K.P. Jüngst, *Nachrichten des Forschungszentrums Karlsruhe* 31 (1999) 309

[72] M. Noe and B.R. Oswald, IEEE Trans. Appl. Supercond. 9 (1999) 1347

[73] M. Noe and M. Steurer, *Supercond. Sci. Technol.* 20 (2007) R15

[74] K. Hongesombut et al., *IEEE Trans Appl. Supercond.* 13 (2003) 1828

[75] Masaki Nagata et al., *IEEE Trans Appl. Supercond.* 11 (2001) 2489

parallel normal resistance causes a reduction of the superconductor current $i_{SC}(t)$ as

$$i_{sc}(t) = i_{initial}(t)\frac{R_p}{R_p + R_{SC}(t)} \qquad (12.7)$$

The passive transition and current sharing with the normal resistor protects the power grid. Opening the conventional switch, S, after a time t is required to protect the SCFCL itself and to eventually restore grid operation.

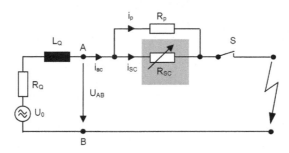

FIGURE 12.39: Schematic of a resistive SFFCL showing the superconductor as a variable resistance. There are several variations on this design that have been used by various manufacturers.

The hybrid resistive SFCLs include a separate, fast switch in series with the superconducting element. This switch quickly isolates the superconductor after most of the current has transitioned to the shunt element, thereby allowing the superconducting element to begin the recovery cycle while the limiting action is sustained by the shunt.

Inductive SCFCL

There are several types of inductive SCFCLs. The two most often considered are the "shielded core" design and the "saturable core" design. In the "shielded core" FCL, a superconducting coil and a conventional copper coil are wrapped around an iron yoke. During normal operation, transformer action causes the current in the superconductor to oppose the current in the copper coil. The flux from these two coils cancel and the iron yoke is essentially out of the circuit. When a fault occurs, the current in the superconductor increases and may exceed the critical current. The resistance of the superconductor causes its current to decrease so that it no longer shields the iron core. The impedance of the copper coil is now determined by the iron yoke. To be effective, this transition must occur in about a millisecond and, once the fault has cleared, the superconductor must recover while the FCL is under normal load.

Unlike resistive and shielded-core SFCLs, which rely on the quenching of superconductors to achieve increased impedance, saturable-core SFCLs utilize the dynamic behavior of the magnetic properties of iron to change the inductive reactance on the AC line. The concept utilizes an iron core, that is normally saturated by a DC superconducting coil, and an AC winding made of a conventional conductor. The conventional conductors are wrapped around the core and together form an impedance in series with the AC line. The system is designed so that at normal current levels, the iron is completely saturated and the conventional coil appears as a low-impedance air core reactor. When the current in the AC coils increase during a fault, the iron goes out of saturation, which increases the FCL's impedance. This change occurs in microseconds so that the core goes in and out of saturation twice during each cycle. The superconductor in this type of FCL does not quench so that there is no issue of recovery after a fault.

State of the Art SCFCL

There is ongoing development of several types of SCFCL. Tables 12.6 and 12.7 give a snapshot of many of the existing SCFCL projects.

TABLE 12.6:   HTS SFCL Projects in the United States and Asia

Project	Avanti	AEP-TIDD	Puji	Korea
Location	Los Angeles, CA	Ohio	Kunming, China	Gochang, Junbuk Province
Site	Shandin Substation	TIDD Subtation	Puji Substation	Gochang Power Testing Center
Status	Operating at 13 kV	To be installed May 2011	Operating	Operation tests
Developer	Zenergy Power	Zenergy Power	Innopower	Consortium[76]
Utility/ Host	Southern California Edison (SCE)	AEP	Yunnan Electric Power Grid	KEPCO
Start Date	January 2009	July 2011	December 2007	TBD[3]
End Date	October 2010	October 2012	TBD	TBD
Type	Saturable Core	Saturable Core	Saturable Core	Hybrid Resistive
Phases	3	3	3	3

TABLE 12.7:   SFCL Projects in Europe

Project	Nexans	Nexans	ATA
Location	Lancashire, UK	Boxberg, Germany	North Italy
Site	Bamber Bridge	Local Power Plant	San Dionigi Substation (MI)
Status	Operating	Operating	Fabrication of first prototype
Developer	Nexans	Nexans	ERSE Spa
Utility/ Host	Consortium[1]	Vattenfall Europe Generation AG	A2A Reti Elettriche Spa Group
Start Date	Fall 2009	Fall 2009	Early 2010
End Date	Mid 2010	Late 2010	End of 2011
Type	Resistive	Resistive	Resistive
Phases	3	3	3

Several SCFCLs have been installed on utility systems. Perhaps the most impressive is the resistive SFCL shown in Figure 12.40. It was constructed by Nexans and installed at a substation in Boxberg, Germany. Another FCL that saw nearly two years of operation on a utility grid is shown in Figure 12.41. It was constructed by Zenergy Corporation and was installed at the Avanti substation in California.

---

[76]Gochang Power Consortium

FIGURE 12.40: A resistive SCFCL constructed by Nexans and installed at the Boxberg substation. (Courtesy Nexans-Boxberg.)

FIGURE 12.41: A saturated core SCFCL constructed by Zenergy and installed at the Avanti substation. (Courtesy Zenergy Corporation.)

12.5.3 SMES

Efficient and reliable electric energy storage technologies have become more and more important in the past few decades. Superconducting magnetic energy storage (SMES) has the potential of becoming the most efficient of all energy storage systems, including the various kinds of batteries, flywheel energy storage, pumped-hydro, and capacitors. Furthermore, SMES has long calendar life, long cycle life, and a rapid response to power demands. Therefore, from a technical sense, SMES is an ideal storage system for electric utilities. SMES may be used by a utility in several application areas.

- Load leveling or arbitrage: for efficient use of power generation systems coping with diurnal demand fluctuations and also the non-dispatchable power generation of renewable energy systems such as solar and wind. Today, the most often used large-scale storage technology is pumped hydro. However, its efficiency is only 70–80%, whereas the efficiency of SMES is expected to be more than 90%. In addition, most of the effective sites for pumped hydro are already in use.

- Power quality (PQ): bridging the voltage dips and sags in the power grid. A voltage dip of only a few ac cycles can cause multimillion dollar losses to a modern factory, such as those that make semiconductors and electronic systems. PQ systems are very effective and often have short payback periods.

- System stability and spinning reserve: large-and extended power grids occasionally exhibit low frequency oscillations, which may lead to instabilities. In addition, loss of critical generation can induce widespread failure and lead to black outs. A storage system such as SMES can help stabilize the power grid to control oscillations and rapidly provide a reserve of energy "spinning reserve" to prevent a black out. Most large grids require power generators to maintain a certain percentage of capacity in the form of spinning reserve.

SMES System Design

Figure 12.42 is a simplified schematic of a SMES system. Key components are a superconducting magnet, a switch, which maybe superconducting, and AC/DC inverter/converter.

The magnet and the switch are maintained at an appropriate operating temperature. The energy stored in a magnet is given by:

$$W = \frac{LI^2}{2} \tag{12.8}$$

$$= \oint_V \frac{B^2}{2\mu_0} dV \tag{12.9}$$

where the energy W is in Joules, the inductance L is in Henry s, current I is in amperes, the field B is in Tesla, and μ_0 is the permeability of free space. One can look at the first of these two formulations and conclude that there is a trade-off between inductance (or number of turns) and the current carried by the SMES coil. Looking at the second formulation, it is clear that for a given size and geometry, the higher the magnetic field, the greater the stored energy.

The magnet is charged and discharged from and to the AC power line, respectively, through the AC/DC inverter/converter, which is often referred to as a power conditioning system (PCS). If the switch is superconducting rather than a normal conductor, the device is more efficient. However, the response time of the system is much longer than is acceptable for either PQ or stability applications.

In other energy storage systems such as batteries, flywheels, and pumped-hydro, the electric energy is converted to chemical, kinetic and potential energy, respectively, and the conversion of the energy inevitably involves losses. In SMES, electric energy is stored directly in the form of electricity in the superconducting magnet, which has very low loss. Intrinsically, SMES is quite efficient and has the ability to respond within less than an AC cycle to demands from a power controller.

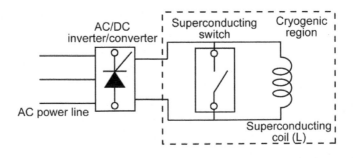

FIGURE 12.42: Simplified SMES schematic.

History of SMES

The concept of using SMES on a utility grid to level diurnal power demand variations was proposed in France in 1969 [77]. Feasibility studies of a large-scale SMES as an alternative to pumped-hydro were conducted in the early 1970s by several institutions[78,79]. These designs were based on large-scale superconducting magnet and semiconductor switching device technologies that saw great strides in 1960–1970.

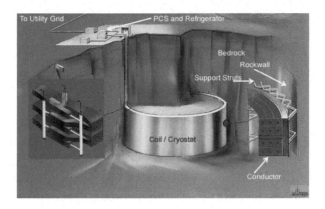

FIGURE 12.43: Large-scale SMES plant constructed underground. (Courtesy Los Alamos National Laboratory.)

Figure 12.43 shows one concept of a large SMES plant constructed underground. This device stores several thousand MWh of energy. The diameter of the superconducting solenoid magnet depends on the operating field and is ≈200 m. Internal structures to support the electromagnetic force in the magnet are unacceptably costly in a magnet of this size. Therefore, it was determined that bedrock structure was required. The magnet is installed in an underground tunnel and the magnetic force is transmitted to the surface of the rock wall. The magnet is placed more than 100 m below the earth's surface, to use the structural capabilities of hard rock and to reduce the magnetic field at the surface. The technical feasibility of that type of SMES was demonstrated, but, at the time, its cost was determined to be too high for practical development.

A different and more realistic application of SMES was proposed by Los Alamos and the Bonneville Power Administration in 1976 [80]. Its purpose was to stabilize a specific instability that occurred in the 900 mile long Pacific Power Corridor from near Seattle to Southern California. A 30 MJ coil, Figure 12.44, coupled to the grid with a 10 MW PCS, was constructed and installed at the Tacoma substation south of Seattle in the early 1980s [81]. Following this project, some other projects were conducted, mostly in the US and Japan. However, those projects showed that, except in specific niche markets, the economic merits of even medium-scale SMES were very limited. A summary of the status of SMES late in the 1980s is described in Hassenzahl[82].

Near the end of the 1980s, business development and research efforts were started on small or mi-

---

[77] M. Ferrier, *Low Temperature and Electric Power* (1970) 425

[78] W.V Hassenzahl, *IEEE Trans. Magn.* 11 (1975) 482

[79] R. W. Boom et al., ibid. 475

[80] L Cresap (BPA), W. Hassenzahl (LASL), private communication

[81] E. Hoffman et al., *IEEE Trans. Magn.*, 17 (1981) 521

[82] W. V. Hassenzahl, *IEEE Trans. Magn.* 25 (1989) 750

cro SMES devices that stored up to several MJ for a variety of applications in the PQ area, including compensation of load fluctuations of factories and military applications. Superconductor Inc. (later purchased by American Superconductor Corporation: AMSC) developed micro SMES storing about 1 MJ and coupled to the power system by modern motor drive electronics with a power capacity of 430 kVA. These devices were functional in terms of reducing voltage sags in the power line, particularly those associated with overload power conditions. The complete microSMES system including refrigerator and converter was installed in a transportable semitrailer. The effectiveness of this micro SMES was verified at several sites.

FIGURE 12.44:   The 30 MJ coil constructed for installation on the BPA power grid. (Courtesy by Los Alamos National Laboratory.)

Also in the mid1980s, utilities in Japan became interested in SMES because the diurnal load variation was increasing as overall demand grew rapidly. SMES was expected also in Japan to be promising as a technology to shave the peak demand and stabilize the utility grid. In this situation, the national project of SMES was established by the Ministry of International Trade and Industry (MITI) and New Energy Development Organization (NEDO) in 1991. Much of the recent work on SMES has been carried out as Phase I-IV SMES projects. In addition, several small SMES systems have been constructed in Korea, China, Australia, and Europe.

In phase III of the METI and NEDO SMES project (Japanese fiscal years 2004–2008), a multi-purpose 20 MVA SMES for utility grid use was developed using Nb-Ti conductors cooled by liquid helium at 4.2 K based on the R&D results of the phase I and II projects. The SMES was installed in a copper refinery that experienced considerable power fluctuations. It interfaced with the refinery and the power grid. In this project, the load fluctuation caused by the refinery was suppressed (note that the 30 MJ BPA SMES was placed adjacent to an aluminum refinery for the same reason, but was never used for that purpose). Chubu Electric Power Company and Toshiba developed and installed a 5 MVA (later increased to 10 MVA) SMES in a modern LED panel factory to compensate voltage dips in the utility grid in 2003[83] (Figure 12.45). The four-pole configuration was used for the SMES coil to reduce stray magnetic flux. This device is still in operation.

The biggest issue at present in SMES technology has to do with the potential impact of HTS materials. The advantage will be operating at temperatures above those available with liquid helium.

---

[83] S. Nagaya et al., *IEEE Trans. Appl. Supercond.* 14 (2004) 699

FIGURE 12.45: SMES installed in the LED factory. (Courtesy of Chubu Electric Power Company.)

HTS coated conductors (CCs) technology has progressed remarkably and these materials are now becoming commercially available. In the near future, fabricating a SMES coil wound of CCs will be possible. Critical current density of CCs at 20 K is much higher at fields above 10 T than that of low temperature superconductor wires (LTS wires) made of Nb-Ti and Nb$_3$Sn at 4.2 K. Because of the backing material, the mechanical strength of CCs is greater than those of LTS wires. As a result, a high field and thus compact coil operated at $\approx$ 20 K can be made using CCs. The disadvantage of CCs today is cost. To be economically competitive in a SMES device, the conductor cost must drop to 10 or 20 \$/kAm at 20 k and 10 T. The likelihood of reaching this target in the near future is small. However, the potential merit of a compact CC based coil is obvious as shown in Figure 12.46. Some effort is underway to build SMES devices with HTS materials.

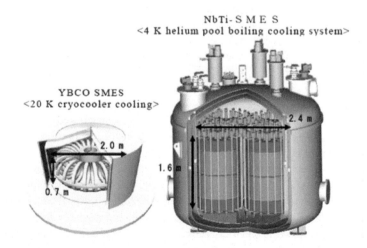

FIGURE 12.46: Comparison of HTS CC coil and NbTi LTS coil for 20 MJ class SMES. (Courtesy Los Alamos National Laboratory.)

12.5.4   Superconducting Rotating Machines

Three types of rotating machines are used in power applications. The two most generally used are generators and motors. In addition, synchronous condensers, which are much like generators in design, are used for a few special applications. The key to the effective implementation of superconducting machines is the very high magnetic flux density delivered by the high current density available in superconductors. Today superconducting motors and generators are progressing rapidly to commercial scale in a variety of areas using HTS materials.

Because superconductors have losses when carrying alternating currents or when exposed to varying magnetic fields, the preponderance of superconducting machines use superconductors in configurations where both current and magnetic field are nearly constant. There are various ways to build generators and motors. By far, the geometry used most often, for both conventional and superconducting machines, has a coil on the rotor that produces a magnetic field. However, this geometry requires cryogenic cooling on the rotor, which presents thermodynamic challenges to the designer.

It was clear early on that the higher magnetic field would, for a given size, increase the available torque between the rotor and stator, and thus allow superconducting machines to be smaller and lighter than conventional machines. Thus, work on superconducting machines began soon after the discovery of Nb-Ti and Nb3$_S$n in the early 1960s. All of these early machines were limited by the need to use liquid helium as a coolant. However, the early work on these machines and the lessons learned have contributed greatly to recent machine development using HTS materials. An early observation, which remains true today, was that to be effective and economically competitive a superconducting machine must operate near practical design limits in many areas, which encompass, though not necessarily exclusively, the parameters listed in Table 12.8.

TABLE 12.8:   Design Considerations for Superconducting Rotating Machines

Design Requirements	
Magnetic field	Efficiency
Mechanical stress	Electric stress
Internal impedance	Mechanical stiffness
Cryogenic cooling	Normal cooling
Superconductor	Stabilizer

Several different machine types were chosen as likely to be most successful with superconductors for electric power applications. Details of the various early designs and construction of different machines have been published in[84,85,86,87,88]. However, the most effective designs, then and now are the synchronous rotating machines and the homopolar motor. Homopolar generators can also be effective, but rectified ac power in many ways provides a more controllable source of direct current. Some of the early machines that were built and tested are shown in Table 12.9.

During the 1970s the main approach for superconducting generators was to build very large machines that would match the capabilities of anticipated very large nuclear reactors. These would be

---
[84] H. H. Woodson et al., *IEEE Trans. Power App. Syst.* PAS-85 (1966)

[85] A. Atherton, *Proc 1972 Appl. Supercond. Conf.* (1972) 16

[86] J. L. Smith et al., *IEEE Transa. Magn.*, 11 (1975) 129

[87] H. O. Stevens et al., *IEEE Trans. Magn.* 13 (1977) 269

[88] C. E. Oberly, ibid. 260

TABLE 12.9: Early, Low Temperature Rotating Machines

Description	Power	Group	Date
Homopolar disk solid brushes	50 hp	IRD England	1966
Homopolar liquid brushes	1000 hp	NSRDC USA	1972
Homopolar disk solid brushes	3 MW	Toshiba Japan	1973
AC Synchronous Rotating field	2 MVA	MIT USA	1973
AC Synchronous Rotating field	5 MVA	Westinghouse	1973
AC Synchronous Rotating field	1 MVA	Leningrad USSR	1973

in the 1GW range, which, if conventional, would often exceed the capacity of rail and road transportation systems in the US. The apparent advantages of superconducting generators in the GW range were lower weight and smaller footprint, which allowed for easier transportation and a slight increase in efficiency. Several programs addressed the design of these large generators, the most interesting of these in the US were carried out by Westinghouse, General Electric, and MIT, all under the support of the Electric Power Research Institute[89]. These programs continued for several years and led to the development of conceptual designs and prototypes. However, the effort on these generators ended when large nuclear power plants went out of favor. The development of homopolar motors and generators continued through the early 1980s [90]. In 1980, the US Navy installed a 300 kW superconducting homopolar generator and a 400 hp superconducting homopolar motor on the naval ship Jupiter II, making it the first naval vessel to be propelled completely by superconductivity. In 1983 the Jupiter II drive system was upgraded by installing a conventional ac generator and a rectifier, which powered a 3000 hp homopolar motor[91]. This ship was stationed at the Baltimore harbor for viewing during the 1986 Applied Superconductivity Conference.

In 1987 Japan began a program to build a superconducting generator in the 200 MW range with a possible extension to 600 MW. The intention was to have modular generators that would in combination exceed 1 GVA, but which could also be used with smaller nuclear plants, and possibly with conventional thermal plants. Besides the advantages discussed earlier, superconducting generators intrinsically have a small internal synchronous reactance, which is effective for control of system voltages. They can also be designed to have more rapid response to demands than conventional generators because of the reduced inertia. The Japanese program was called "Super-GM project", and was reorganized by NEDO in 1991 and supported by MITI as a part of Japan's New Sunshine Project. Several rotors were constructed with LTS field windings and were tested in a grid environment. A 70 MW class model generator was operated for more than 1500 h in the Osaka power station. A thorough description of the results of the Super-GM project is described in 9 papers in the 2002 issue of Cryogenics. Nitta[92] is the first paper of the set. Figure 12.47 shows the field test installation of one of the Super-GM generators.

With the discovery of high-temperature superconductors in 1986, several programs were initiated to use these new materials in rotating machines. A great deal of effort has gone into HTS materials development over the past 25 years. One result of that work is that today these materials have the capabilities that will allow the development of commercially viable superconducting rotating machines. However, the key breakthrough is the ability to design and build high power density, highly efficient machines that can operate at 30–40 K. These machines use conduction cooled coils

---

[89] Mario Rabinowitz, ibid. 255

[90] A. Arkkio et al., *IEEE Trans. Magn.* 17 (1981) 900

[91] M. Superczynski, *IEEE Tran. Magn.* 23 (1987) 348

[92] Tanzo Nitta, *Cryogenics* 42 (2002) 151

FIGURE 12.47: A Super-GM 70 MVA generator under field test at the Osaka Power Station. (Courtesy OEPCO.)

and can operate with standard Gifford-McMahon cryocoolers. A summary of the state of the art of rotating electrical machines in 2004 and some of the details and history noted above are available in Kalsi et al.[93]

One of the most widely used superconducting rotating machine topologies is an air-core topology illustrated schematically in Figure 12.48. The field winding is mounted on a metallic rotor that is nonmagnetic (hence the term "air core") and consists of several coils (poles) that are conduction-cooled through the support structure.

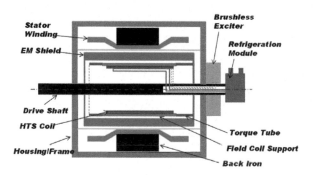

FIGURE 12.48: Air-core electric machine topology.

The primary components of the rotor assembly are as follows:

- HTS field winding operating at 30–40 K

- Rotor support structure

- Cooling loop

- External cryocooler module connected to the motor's cooling loop at the non-drive end of the shaft through a rotating coupling

- Room-temperature electromagnetic shield

---

[93] S.S. Kalsi et al., *Proce. IEEE* 92 (2004) 1675

- Torque tube for transferring torque from the "cold" (cryogenically-cooled) environment to the "warm" shaft ends.

The superconducting field winding is an assembly containing multiple polesets, each fabricated using wire that is designed to withstand the magnetic and mechanical forces experienced in the rotor. The polesets are attached to a metallic support structure on the rotor which provides support against centrifugal and torsional loading. The flux pattern is shown in Figure 12.49.

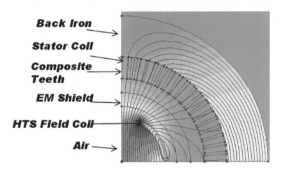

FIGURE 12.49: Flux distribution in an "air-core" electric machine topology; note the use of non-magnetic teeth in the stator.

The polesets and support structure are enclosed in a vacuum-sealed cryostat that minimizes radiant heat input and provides the insulated operating environment required by the HTS field coils. An electromagnetic (EM) shield, which is located at the outside surface of the cryostat, reduces losses in the field winding by attenuating field variations caused by the stator winding. It also carries a high transient torque during a fault, and provides damping for low-frequency torsional oscillations, negative sequence currents and any other harmonic currents generated by a variable-frequency drive.

An alternative design, Figure 12.50, uses an iron core in the rotor and magnetic teeth in the stator. Though fully saturated, the iron core nevertheless increases the flux density at the stator, and thus contributes to some additional reduction in footprint.

In addition to the synchronous machines, work continues on homopolar motors. They have somewhat limited interest because of need for brush contacts that must carry very high currents.

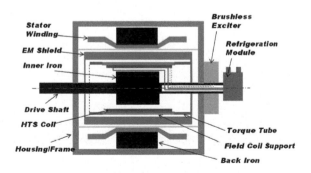

FIGURE 12.50: Machine topology with rotor iron.

An impressive array of superconducting machines has been constructed and demonstrated during the last decade, as summarized in Table 12.10. As a result, they appear to be one of the most promising applications of superconductivity.

TABLE 12.10:   Recent HTS Machine Developments

Manufacturer	Machine	Timeline
AMSC (US)	5 MW demonstration motor	2004
	8 MVA, 12 MVA synchronous condenser	2005/2006
	40 MVA generator design study	2006
	36 MW ship propulsion motor	2008
	8 MW wind generator design study	2010
Converteam (UK)	200 kW demonstrator	2006
	1.25 MVA hydro generator	2008
	500 kW demo-generator	2008
	8 MW wind generator design study	2010
Dooson, KERI (Korea)	1 MVA demonstration generator	2007
	5 MW motor (homopolar)	2010
GE (US)	100 MVA utility generator	2006 (discontinued)
	5 MVA homopolar induction motor	2008
LEI (US)	5 MVA high speed generator	2006
Reliance Electric (US)	10.5 MVA generator design study	2008
IHI Marine, SEI (JP)	265 kW ship propeller motor	2010
	2.5 MW ship propeller motor	2010
Siemens (Germany)	400 kW demonstration motor	2001
	4 MVA industrial generator	2005
	4 MW ship propeller motor	2000

Several programs around the world are addressing HTS motors. A 1 MW HTS motor by KERI and Doosan Heavy Industries has been developed under the Korean DAPAS program. In addition to radial flux type HTS machines, a Japanese team including Sumitomo Electric and IHI finished the development of a 365 kW axial flux type HTS motor cooled by liquid nitrogen.

Perhaps the greatest progress in rotating machines has been in the area of ship propulsion motors using HTS based, air-core synchronous AC technology. These motors offer significant benefits for both naval and commercial shipping applications[93,94,95,96], including high power density, high efficiency and low structure-borne noise. Significant advances have been supported by the US Navy in the area of propulsion motors with 120–150 revolutions/min and ratings above 30 MW. HTS technology can be 2.5 to 5 times lighter and more compact than conventional technology in this size range. After a series of early motor development by American Superconductor (AMSC) and Reliance Electric (later Rockwell Automation), the U.S. Navy's Office of Naval Research (ONR)

---

[94] Kalsi S et al., Naval Symposium on Electric Machines, Philadelphia, PA (2000)

[95] Eckels P and Snitcher G, Naval Symposium on Electric Machines, Philadelphia, PA (2004)

[96] Weeber K K et al., *IEEE 2003 Power Engineering Society Annual Meeting*, Emerging Technologies Panel Session, Toronto, ON, (2003)

FIGURE 12.51:   A 36 MVA, HTS synchronous motor under test. (Courtesy US Navy, ONR.)

funded AMSC and Alstom (now Converteam) to build a 5 MW output power, 230 r/min propulsion motor to validate technologies for larger ship propulsion motors.

Based on this success, the ONR funded a full-scale 36 MVA advanced technology demonstrator, a successful program now representing the state-of-the-art in HTS rotating machinery; see Figure 12.51. The motor has an output shaft speed of 120 rpm and generates over 2.9 million Newton-meters (2.2 million ft-lbs) of torque. The development team was led by American Superconductor (AMSC) as prime contractor, with Northrop Grumman, Electric Machinery, Inc., BMT-Syntek, and CAPS. The motor's design characteristics were achieved or exceeded as shown in Table 12.11.

TABLE 12.11:   Characteristics of 36 MVA Motor Developed for the US Navy

Parameter	Design	Measured
Design Power	37.2 MW	> 37.2 MW
Voltage	5.8 kV	6.6 kV
Current	1.275 kA	1.28 kA
Efficiency	97 %	97 %

Siemens has designed, manufactured and tested a series of HTS synchronous machines, including, a generator for marine application rated 4 MVA at 60 Hz and 3600 rpm. The company carried out the first-ever synchronization of an HTS generator onto the grid in 2005. The rotor consisted of HTS pancake coils manufactured from 1G HTS wire from what is now Bruker HTS. Zenergy Power and Converteam have announced a project to build a 1.25 MW hydrogenerator.

Another significant development is the recognition that compact, lightweight HTS generators can enable 10 MW-class generators for offshore wind turbines. Programs towards this goal have been launched by Converteam, AMSC in collaboration with TECO-Westinghouse, and by the Danish Technical University in collaboration with Denmark's Riso National Laboratory.

A different design concept is used for a development carried out by General Electric for a 5 MW high rotational speed generator supported by the US Air Force. Here, a homopolar design concept uses a stationary HTS winding that excites the machine's rotor iron. KERI and Doosan Heavy Industries have also announced plans to explore this idea.

As mentioned early in this section, another application of rotating machine technology is the synchronous condenser, which is essentially a generator without a power source. It provides reactive, out-of-phase power (MVAR) rather than real, in-phase power (MVA). AMSC and Tennessee Valley

Authority designed, built, installed, and operated an 8 MVAR 13.8 kV dynamic synchronous condenser[97]. The machine employed coils of BSCCO wire, an inner iron topology, and a liquid neon cooling loop. The unit was installed near an arc furnace of the Hoeganaes steel plant in Gallatin, Tennessee. It began grid operation on October 10, 2004, was brought into regular operation in January 2005 and operated until November 2005. In practice, it reduced voltage flicker in the area that was caused by the arc furnace.

At the Hoeganaes site, transient disturbance+489s occurred during the 30 minutes to 1 hour melt cycles in the arc furnace at 1 hour intervals. With active changes in reactance during each melt cycle, the unit experienced an enormous number of reactive events—of order 5 million—during the year. While certain peripheral equipment required maintenance during this period, the rotating machine itself performed very successfully in what must be considered one of the most rigorous in-site tests of HTS equipment to date. In spite of this success, utility needs are for much larger, 50 MVAR machines; so further development is required for a commercial entry point. In summary, although superconductor rotating machinery has not yet entered the commercial realm, the technology based on HTS wire is well developed, and the prospects look bright for commercialization in a variety of applications.

12.5.5 Superconducting Cables

Power transmission and distribution by cables as compared to overhead lines has developed for a variety of reasons. The most important are right-of-way and environmental considerations. Power use per square meter in high population areas where there is limited space, such as city centers, and where there is need for vehicle and pedestrian access is providing an ever-increasing demand for underground power cables. The key to the effectiveness of cables compared to lines in air is that the voltage breakdown capability (V/m) of the insulators used for cables is 100 to 1000 times better than air. Thus, the power flow along underground cables in a restricted corridor can be considerably greater than is possible with overhead lines. In addition, as designed, the stray electric field from an underground cable is negligible. The rationale for various conventional cable designs and the development thereof are described in detail in the EPRI publication referred to as the Green Book[98]. Superconducting cables appear to provide an even greater power density than conventional cables. Thus, just as conventional cables have been used to increase power flow in specific corridors and right-of-ways, in the not-too-distant future, superconducting cables will replace both overhead lines and conventional cables.

As with other applications, superconducting power cables were considered for development with each new development. At present, both ac and dc superconducting power cables appear promising for future installations. Below, after a brief description of the history of the various efforts on superconducting cables, some of the existing programs will be described.

Serious development on superconducting cables began in the late 1960s after Nb-Ti and Nb_3Sn became available on a near commercial scale. Early work using low temperature superconductors began in Europe[99,100,101] and later in the US[102,103,104] and Russia[105]. Significant development programs existed for ac and dc superconducting cables. These programs addressed the significant issues

---

[97] S. S. Kalsi et al., *IEEE Trans. Appl. Supercond.* 15 (2005) 2146

[98] *Underground Transmission Systems Reference Book*, EPRI product number 1012334 (2006)

[99] P. Klaudy, *Adv. Cryogenic Eng.* 11 (1967) 684

[100] P. Klaudy et al., *IEEE Trans. Magn.* 17 (1981) 153

[101] F. Moisson and J. M. Leroux, *J. Appl. Phys.* 42 (1971) 154

[102] H. M. Long and J. Nottaro, ibid. 155

[103] E.B. Forsyth and R.A. Thomas, *Cryogenics* 26 (1986) 599

[104] H. L. Laquer et al., *IEEE Trans. Magn. Mag*-17 (1977) 182

[105] P. I. Dolgosheyev et al., *IEEE Trans. Magn. Mag*-15 (1979) 150

of superconducting cable technology and the research in the 1970s and early 1980s is generally applicable to many of the technical issues associated with HTS power cables today.

Perhaps the largest LTS power cable effort was in the US, which included the ac cable program at BNL[103] and the dc cable program at LASL[104]. These programs addressed very different issues associated with electric power. The ac cable effort explored a variety of issues associated with relatively short, multi-kilometer systems, while the dc cable effort was directed to very high power, long distance transmission systems. The latter had been proposed by Garwin and Matisoo[106] in 1967 as a means to transmit power from collections of massive electricity generators, e.g., nuclear power farms or hydroelectric dams to distant load centers.

These LTS cable projects were abandoned, for several reasons.

- In 1973, OPEC established an oil embargo that led to greater conservation efforts and the growth rate of electric power decreased to almost zero for several years.

- Nuclear plants went into disfavor and their costs and construction time, with environmental add-ons, made their profitability less certain.

- The operating temperature of LTS systems is so low that refrigerator size and efficiency impact the overall economics, limiting their attractiveness to only the very highest power capacities.

- The diameter of the cryostat for LTS cables is significantly larger than the existing ducts in use for many conventional cables, so they could not be simple retrofits that upgraded power flow at many sites.

The discovery of high-temperature superconductors in 1986 sparked renewed interest in superconducting power cables. HTS based power cables have significantly lower refrigeration requirements than LTS based cables. Thus, capital and operating costs for this component are less. However, today, and for the foreseeable future, the cost for HTS conductors is significantly higher than that of LTS conductors, which offsets the refrigeration advantages. In addition, today, HTS materials are more difficult to make in a form suitable for power cables. On the other hand, since the energy required to power the refrigeration equipment affects the overall efficiency of the system, HTS cables, both ac and dc, are more efficient than either an LTS or a conventional cable. In the near term, it appears that HTS ac cables can be effective at relatively low power levels, 10 to 100s of megawatts that are needed in urban and suburban environments. HTS dc cables appear to be more likely for very high power, long distance power transmission.

Power cables and transmission lines are designed to operate at a fixed voltage while instantaneous power flow is determined by the dc or rms-ac current. Because superconductors can carry orders of magnitude more current than conventional conductors of the same cross section, in a given envelope (or existing duct) HTS superconducting cables can carry several times more power than a conventional cable.

A number of HTS cable designs have been developed to take advantage of the benefits of superconductivity, while minimizing the additional capital and operating costs that result from the material refrigeration and vacuum insulation requirements. Different cable architectures have important effects in terms of efficiency, radiated stray electromagnetic fields, and reactive power requirements. There are several types of HTS power cable under development. One ac cable design is based on a single conductor, with HTS wires stranded around a flexible core in a channel filled with liquid nitrogen coolant as depicted in Figure 12.52.

This design is referred to as a "warm dielectric design" because it employs an outer dielectric insulation layer at room temperature. Of the various superconducting ac cables, it uses the least

---

[106] R. L. Garwin and J. Matisoo, *Proc. IEEE.* 55 (1967) 538

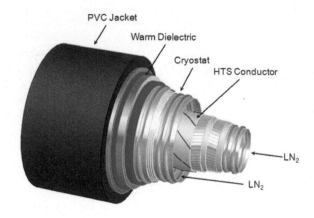

FIGURE 12.52:   Single-phase warm dielectric cable. (Courtesy Nexans Corporation.)

amount of HTS wire for a given level of power transfer, but it has a high inductance and has an external magnetic field. Because the three phases of the system are close together, the magnetic field produced by one phase induces losses in the other phases. An advantage of this cable design is its use of insulation developed for conventional cables. The second ac cable design employs two concentric layers of HTS wire separated by a cold dielectric as shown in Figure 12.53.

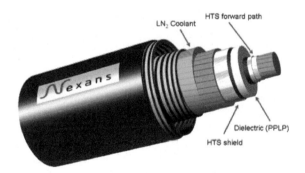

FIGURE 12.53:   Single phase cold dielectric cable. (Courtesy Nexans Corporation.)

It is commonly referred to as a "cold dielectric design". Liquid nitrogen coolant may contact avoiding both cooling and dielectric insulation between the center conductor layer and the outer shield layer. Compared to the warm dielectric design, it has a lower inductance, a higher current carrying capacity, reduced ac losses, and very low stray magnetic fields. It also uses more superconductor than either of the other designs.

The third ac cable design is a concentric triplex, or three-conductor, design as shown in Figure 12.54. This design appears to be optimum at low to medium voltage levels. It has a very low inductance, and it uses less superconducting material than the coaxial design.

HTS dc cables are being considered for long distance, high power transmission, *vis-á-vis* [106] and the present interest in renewable energy sources such as wind and solar. They are also under consideration for shorter power transmission for the interconnection of large ac power systems that do not operate synchronously. Most of the HTS ac cables built to date use multiple cryostats that

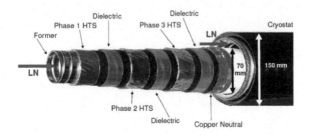

FIGURE 12.54: Triaxial three phase cable installed by Southwire at AEP's Bixby substation. (Courtesy Southwire Corporation.)

have lengths of about 100 m and are permanently evacuated. For power transmission of 1000s of kilometers, the risk associated with failure of one or a few of these cryostats requires a different solution. Figure 12.55 shows a concept of a superconducting dc cable that can carry up to 10 GW [107].

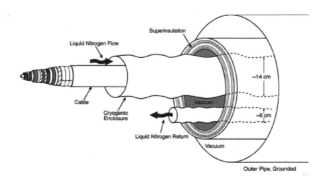

FIGURE 12.55: A superconducting dc cable designed to carry 10 GW over distances greater than 500 km. (Courtesy EPRI.)

This design has a long outer pipe, such as those used for gas pipelines, which contains two cryostat pipes, one for the power cable and the other for coolant return. The size of the outer pipe is determined by the vacuum requirements including the installation of pumps every kilometer or so and refrigeration stations separated by 10 to 20 kilometers. Besides the obvious increase of power density, the cold dielectric designs have inductances that are roughly about one-sixth those of conventional cables and 1/20th of that of overhead lines in the same voltage class[108]. Just as electrical current flows through the path of least resistance, power flow selects the lowest impedance path. As a result, coupled with the high current capacities of HTS cables, this quality provides superconducting cables with the potential to relieve transmission bottlenecks. Several evaluations of their lower impedance in a utility environment support this conclusion. Assessments of the Italian power grid[109] and of specific bottlenecks in the Chicago area[110] showed significant improvements

---

[107] W. Hassenzahl et al., *EPRI report* 1020458 (2009)

[108] N. Kelley et al., *Proc. IEEE/PES, Trans. and Dist. Conf.* 2 (2001) 871

[109] A. Mansoldo et al., *Proc. IEEE/PES Winter Meeting*, New York (2002) 142

[110] R. Silberglitt et al., RAND Corporation Science and Technology Policy Institute Report (2002)

TABLE 12.12: Superconducting Cables that Have been Installed and Operated at Utility Voltages

Lead Manufacturer	Use / Location	Voltage	Current	Installed	Removed
Southwire	In Plant, Grid Connected, USA	12.4 kV	1.25 kA	2000	2005
Sumitomo	In Plant, Japan	67 kV	1.0 kA	2001	2002
nkt Cables	On Grid, Sweden	30 kV	2.0 kA	2001	2003
Sumitomo/ Superpower*	National Grid, USA NY	34.5 kV	0.8 kA	2006	2008
Southwire	On Grid AEP Ohio USA	13.2 kV	3.0 kA	2006	Present
AMSC/Nexans	LIPA/ Long Island, USA NY	138 kV	2.5 kA	2007	Present
Keri/LS-Cable	South Korea Test Center	22.9 kV	1.5 kA	2006	2009
Changtong Cable	Baiyin, China	10.5 kV	1.5 kA	2004	2007
* After operating for a year with BSCCO tape, a 30 m section was removed and replaced with a coated conductor based section. Thus it was the first superconducting cable on a power grid with Gen II HTS material.					

in power flow.

The downside of the lower impedance of superconductive cables is the potential for increased fault current. During a fault, a cable carrying 5 kA may see as much as 100 kA for short periods. Superconducting cables can be protected against such currents by using a conventional conductor such as copper in intimate contact with the superconductor. The HTS conductors would transition to the normal state when their critical current is exceeded and the stabilizer would carry the bulk of the current until the fault is cleared. On the flip side, when designed properly, the superconducting cables nonlinear impedance can be an advantage as it can assist in limiting the fault current[111].

Several HTS power cables have been installed and operated for extended periods of time. Some of these installations are listed in Table 12.12. Other projects that are under development are also listed in the table. Rather than describe the details of some of these projects, we include Figures 12.56 to 12.58, which are representative of the various installations. Descriptions of the various figures explain significant details

12.5.6 Superconducting Transformers

Issues associated with superconducting transformers were initially addressed in the mid 1960s[112], with the conclusion that, to be effective, 50 to 60 Hz power transformers would require a warm steel yoke and special low-loss superconductors. It was realized that cryogenics was a critical enabling technology and the need for thermal insulation, which added several centimeters to the coil dimensions, meant that at a 4 K operating temperature, only transformers having power ratings above 100 MW had a chance of becoming economically competitive. An additional barrier to LTS based transformers was the drive to improve conventional transformers. This effort increased the overall efficiency of power transformers from $\approx 99\%$ to $\approx 99.5\%$ from 1960 to 2000 [113].

There were some efforts to design low-loss Nb-Ti superconductors with fine filaments (< 0.1 micron) for transformer use, but the complication and expense of refrigeration at 4 K were of tantamount importance and, through the mid-1990s, only a few small experimental LTS devices were

---

[111] Pat Duggan, Consolidated Edison, private communication (2009)

[112] S. H. Minnich, *Proc. 1996 Appl. Supercond. Conf.* (1967) 32

[113] B. W. McConnell, *IEEE Power Eng.Rev.* 18 (1998) 8

FIGURE 12.56: Pulling Southwire's triaxial cable into a cable duct at AEP's Bixby substation near Columbus, Ohio. (Courtesy Southwire Corporation.)

FIGURE 12.57: Installation of the Changtong warm dielectric HTS cable at the Changtong cable plant in Baiyin, China. (Courtesy Changtong Inc., China.)

ever made. The major work in this area was in Europe by Alsthom and ABB, in Japan by Toshiba, and at several universities as described in a review article in 2004 [69].

Although transformers for ac power applications had not advanced to a practical stage with LTS materials, special "dc transformers" use LTS materials and are effective elements in test facilities for certifying very high current conductors. In a typical dc transformer, the secondary and the specimen to be measured are both maintained at the operating temperature. Thus, the only resistance in the circuit is due to the joints between the conductors of the transformer secondary and the specimen. This allows high currents to be maintained for periods of minutes. Several of these dc transformers were designed and built by various laboratories around the world from the 1970s to the present. The most spectacular of these systems was built for tests of a 300 kA conductor[114] for a 5000 MWh SMES system that was under development in the US in the late 1980s. The special transformer for this test consisted of a single-strand, 200-turn primary and a two-turn, 30-strand secondary. The system operated in a bath of superfluid helium at 1.8 K and included a superconducting dipole

---

[114] D. L. Walker et al., *IEEE Trans. Magn.* 25 (1989) 1596

FIGURE 12.58:   Terminations for the three phases of the LIPA cable produced by Nexans and AMSC. (Courtesy Nexans Corporation)

that could produce a background field up to 7.5 T. The highest current reached in the tests was an impressive 280 kA at a field of 5.8 T [115]. More recently this technique has been applied to tests of 50 kA conductors for ITER[116].

12.5.7   Conventional Transformer Characteristics

Large, conventional power transformers use oil as a coolant and as a component of the electrical insulation. The specialized practices of designing and building these transformers are described in many articles and several books[117]. The electric power grid uses several types of conventional transformers. Each generator has step-up transformer to convert its high-current low-voltage (typically 20 kV to 30 kV) electric power to an appropriate voltage (129 kV to 500 kV) for the transmission grid. These units have limited flexibility and operate at near capacity most of the time. They are designed to have efficiencies of 99.5 to 99.8%, which is considerably greater than the other transformers in the grid.

The step-down transformers at substations and in distribution centers are designed to be more flexible and to allow different voltages as the load changes. About half of them have tap changers, which operate either automatically under load or manually during a brief out-of-service period. The addition of this feature to a transformer makes it heavier, more expensive, and considerable more complicated.

Transformers and cables must meet several insulation tests. The first is the "basic insulation level" or "basic impulse level" referred to as "bil". The bil requirement of conventional transformers and cables is based on the anticipated voltage that they may experience during a lightning strike or other short period fault event. The bil levels for 69 kV and 345 kV transformers are 350 kV and 1300 kV respectively. The bil impulse test has a specific waveform and lasts about 50 μs [98,117].

Another dielectric issue for transformers and cables is a phenomenon referred to as "partial discharge". Even if the insulation is capable of withstanding the bil tests, all insulation has occasional imperfections, e.g., voids or particles with significantly different dielectric constants. These lead to local increases in electric fields at the operating voltage and may cause internal discharges with cur-

[115] J. Zeigler et al., *IEEE Trans. Magn.* 27 (1991) 2395

[116] Y. Shi et al., *IEEE Trans. Appl. Supercond.* 20 (2010) 1155

[117] J. Harlow, ed. *Electric Power Transformer Engineering*, 2nd Edition, 2007 CRC Press Florida

rents measured in picocoulombs. The energy associated with these partial discharges can damage insulation and decrease transformer and cable life[98].

To be effective in the power grid, HTS based transformers must mimic conventional transformers in performance and durability. However, operation at cryogenic temperatures imposes several major constraints on the design and operation. Although the first thought might be that the superconductor and its losses are the critical issue, in fact, a combination of cooling and insulation provides the real challenge. There are two design approaches to accommodate this combination in the transformer. The first is to replicate conventional transformer design by using liquid nitrogen as a replacement for the oil. The second is to use a solid insulation and indirect cooling. Both approaches have advantages and disadvantages. The use of liquid nitrogen and near conventional design provides a ready solution to the insulation issue, but conductor performance is so low at temperatures available with this solution, i.e., above 65 K, that a great deal of material is required. Existing HTS materials operate best below 50 K or so, at which temperature they can better meet the demands of field and current in a transformer. However, solid, low-temperature insulation that can be effective in physically large devices is not proven. Several programs are in process to develop some of these materials[118,119]. In the future, transformers using HTS materials may offer a variety of benefits to the electric utility. These include:

1. enhanced efficiency (because losses in superconductors are less than losses in conventional conductors at 300 K, and because losses at 50 K to 70 K are not so important as at 4 K),

2. the ability to run in an overloaded condition without impacting insulation life,

3. a reduced footprint (both weight and size are less),

4. the potential for lower leakage reactance,

5. ease of siting because oil is not used, and

6. the ability to limit fault currents.

These advantages and a discussion of the economic benefits and many significant developments of HTS transformers are described in several background articles[120,121,122,123,69].

The advantages of superconducting transformers are generally derived from the higher current density available and, in those that are intended as fault current limiters, the ability of the conductor to transition to a resistive state. The higher current density leads to a reduced dimension for the windings. In this regard, an additional advantage of the fact that the conduction cooled superconducting transformers do not use oil is that they are much lighter and smaller than any other type of transformer. There are other impacts of the use of superconductors. For example, there is a lower leakage reactance due to the higher magnetic fields within the HTS windings. Detailed analyses looking at the various tradeoffs are required to determine the optimum coil geometry.

Roughly half of the losses in a conventional transformer are in the iron core (hysteresis) and half in the conductor. However, if the iron losses were to occur at the cryogenic operating temperature, it would cause the efficiency to be so low that the transformer would not be competitive. Thus, the iron core must be thermally isolated from the coils and the cryogenic refrigeration system so that none of its hysteresis losses appear as a cryogenic load. Though there are no resistive losses in the

[118] M. Hara and H. Okubo, *Cryogenics* 38 (1998) 1083

[119] J. H. Choi et al., *IEEE Trans. Appl. Supercond* 19 (2009) 1972

[120] S. Mehta et al., *IEEE Spectr.* (1997) 44

[121] J. Laumond, *Handbook of Applied Superconductivity* 2 Institute of Physics Pub. (1998) 1613

[122] V. R. Ramanan et al., *47th Int. Wire and Cable Symp.* (1998)

[123] G. Donnier-Valentin et al., *IEEE Trans. Appl. Supercond.* 11 (2001) 1498

superconducting coils, there are, however, ac losses and heat flow from the surroundings that must be removed by a cryogenic refrigerator.

Fault currents are one of the most significant issues with all transformers. In the superconducting transformer, these currents can be accommodated by the use of co-wound normal conductors that carry the fault current for a few cycles. A powerful feature of an HTS transformer can be operational characteristics under long-duration overload conditions. By designing the HTS windings and the refrigeration system so that the transformer can carry two or more times the design current without transitioning to the normal state, one can avoid redundant installations and thus reduce system costs. The ac losses, some of which are proportional to the square of the current, will be considerably greater at higher currents so the refrigeration requirements during overload become a critical part of transformer design. Conventional transformers are designed to operate under such overload conditions, but high temperatures damage the insulation and result in loss of transformer life.

For widespread adoption, HTS transformers must meet or surpass the performance characteristics of today's conventional transformers, provide some of the added benefits mentioned above, and be economically competitive. These requirements provide a huge challenge to future development. Nevertheless, today, there are several HTS transformer development programs underway worldwide.

Many programs groups began developing superconducting transformers in the late 1990s. A few small model devices were constructed early on and a few near utility scale transformers were constructed and tested. The early work at universities and laboratories has continued on and off over the past decade, but most of the larger projects reached a test phase and ended either with successful tests or a variety of failures. No superconducting power transformers have been operated for long periods. In addition, because of lack of availability of appropriate superconducting wires, none of the existing devices use materials that will be needed for a commercially competitive transformer in the future. Nevertheless, there have been several successful tests of superconducting transformers. A group in Japan[124,125]—including Fuji Electric, Kyushu Transformer, Taiyo Toyo Sanso Co. Kyushu University, and Kyushu Electric Power—developed, tested, and connected to the grid a 1MVA, 22/6.9kV single-phase, liquid nitrogen bath-cooled transformer. The transformer withstood a 100 kV bil test. The group has carried out some development work on a similar 3MVA 3-phase system. A group in the US—including Waukesha Electric Systems, SuperPower, Oak Ridge National Lab, and Rochester Gas & Electric—built and tested a cryo-cooled 1MVA, 13.8/6.9kV single-phase transformer. The same team has completed construction and some preliminary testing of a 5/10 MVA, 24.9/4.6kV 3 phase prototype transformer. The testing was ended by the observation of considerable partial discharge. A group, including ABB, American Superconductor, Los Alamos National Laboratory, and Southern California Edison, worked on the design of a 10 MVA, three-phase liquid nitrogen bath cooled transformer until it became clear that the cost of BSCCO material would not allow the development of a competitive device. They did test a 630 kVA, 18.7 kV/420 V three-phase liquid nitrogen bath cooled transformer with Electricitè de France. This transformer was successfully connected to the grid in Geneva, Switzerland[126]. More recently[127], there have been several studies as to the incorporation of tap changing systems into a superconducting transformer. In slightly different arena, Siemens[128] built and tested a 1 MVA liquid nitrogen bath cooled HTS transformer for railway applications, where reduction in weight is dominant and the system must operate under a somewhat different set of performance and design criteria. More recently, a group in Japan[129] tested a 4 MVA transformer designed for use on the Shinkansen.

---

[124] K. Funaki et al., *Cryogenics* 38 (1998) 211

[125] K. Funaki et al., *IEEE Trans. Appl. Supercond.* 11 (2001) 1578

[126] H. Zueger, *Cryogenics* 38 (1998) 1169

[127] S. W. Kim et al., *IEEE Trans. Appl. Supercond.* 17 (2007) 1939

[128] R. Schlosser, *IEEE Trans. Appl. Supercond.* 13 (2003) 2325

[129] H. Kamijo et al., *IEEE Trans. Appl. Supercond.* 17 (2007) 1927

12.6 Magnetic Separation

Christopher Rey

12.6.1 Introduction

Commercial applications using the principles of magnetic separation have been around for more than one hundred years. Magnetic separators have been used since the time of Joseph Henry, when electromagnets were used to remove nails from horses' feed[130,131]. Principles of magnetic separation are widely used in commercial applications today. Typical uses range from the simple removal of coarse tramp iron and steel from garbage, to the more sophisticated separations, such as the removal of weakly magnetic mineral contaminants from paper-coating clays. Technical advancements in magnetic separator design have led to the commercial use of high gradient magnetic separators (HGMS). These devices are capable of removing weakly magnetic particles and process tons of material per hour. This continued progress has greatly broadened and enhanced the commercial magnetic separations market. For example, it is estimated that the introduction of HGMS into the purification of kaolin clay has nearly doubled the worldwide useful reserves[132] by making lower grade ores economically attractive. Until the development of HGMS, magnetic separation techniques had been confined to manipulating mixtures that contained one or more of the three strongly magnetic (ferromagnetic) elements: iron, nickel, and cobalt. HGMS is potentially applicable to many more elements, mixtures, and compounds. For example, there are over 56 weakly magnetic elements (diamagnetic and paramagnetic) contained within the periodic table. Perhaps of even greater potential benefit is the possibility of manipulating *non-magnetic* substances (the term *non-magnetic* is typically reserved for materials that display extremely weak paramagnetic or diamagnetic properties), e.g. pollutants in water, using appropriate magnetic "seeding" techniques. Recently HGMS, fabricated with superconducting coils, have been introduced into the kaolin industry. Superconducting HGMS devices can produce even higher magnetic fields and operate with less than one fifth the total power consumption of conventional resistive units. Higher magnetic fields translate to better particulate selectivity and increased productivity by processing more material per unit canister volume. For a more complete description of the history and development of high gradient magnetic separation, see reference[133].

12.6.2 Principles of Magnetic Separation

Magnetic Phenomena

All materials possess magnetic properties to some extent. The relative strength of these magnetic properties varies widely among different materials. Materials are typically classified into four categories based upon the strength of their magnetic properties: (1) ferromagnetic, (2) strongly magnetic, (3) weakly magnetic and (4) non-magnetic. With the exception of ferromagnetic, the boundary defining the difference between strongly magnetic, weakly magnetic, and non-magnetic is arbitrary and typically application dependent. Any particle introduced to a magnetic field will

---

[130] F. Knoll, *Perry's Chemical Engineers' Handbook*, 17th ed., ed. by D. Green and J. Maloney, New York: Mcgraw-Hill, (1997) 19–40

[131] C. M. Rey, *Engineering Superconductivity*, ed. By P. Lee, New York, John Wiley & Sons, (2001) 448

[132] J. Iannicelli, *Ultrafine Grinding and Separation of Industrial Minerals*, ed. by S. Malghan, New York: American Institute of Mining, Metallurgical and Petroleum Engineers Inc. (1983) 105

[133] H. Kolm, J. Oberteuffer, and D. Kelland, *Scientific American* 233 (1975) 46

become magnetized. The magnetic moment of an atom results from the electron spins, their orbital angular momentum and the change induced in the orbital angular momentum[134].

Materials that exhibit a positive magnetization when placed in an external magnetic field are described as paramagnetic. Materials that exhibit a negative magnetization are described as diamagnetic. All materials will display some degree of diamagnetism due to the moment induced by an applied magnetic field; however, this type of diamagnetism is relatively weak and can be negligible if other forms of magnetic properties are present, e.g., paramagnetism or ferromagnetism. Paramagnetism results from the electron spins and their orbital angular momentum. Atoms with unpaired electrons typically exhibit paramagnetism. Paramagnetic materials are further classified as strongly or weakly paramagnetic depending upon the strength of the magnetization, the magnetic moment per unit volume, when placed in an external magnetic field. A large number of elements and their compounds exhibit paramagnetism. Some common examples are hematite (Fe_2O_3) and pyrite (FeS_2). Table 12.13 lists the magnetic susceptibility of some common elements and minerals. When domains of paramagnetism are created in some materials such that long-range order is established, the magnetization can be quite large and the materials are described as ferromagnetic. These elements include iron, nickel, and cobalt and a relatively small number of compounds of these elements. An example of a ferromagnetic ore is magnetite (Fe_3O_4).

TABLE 12.13:   Magnetic Susceptibility of Some Common Elements and Minerals

Substance	Susceptibility (10^{-6} cgs)	Substance	Susceptibility (10^{-6} cgs)
Aluminum	+10.5	Ferberite	+39.3
Al_2O_3	-37	Galena	-0.4
Apatite	+1.0 to +18.0	Garnierite	+30.7
Aragonite	-0.4	Gold	-28.0
Asbolan	+150.0	Ilmenite	+15.45 to +70.0
Azurite	+12.2 to +19.0	Lead	-23
Anatase	+0.96 to +5.60	Malachite	+10.5 to +14.5
Beryl	+0.4	Millerite	+0.21 to +3.9
Braunite	+35.0 to +150.0	Molybdenite	+4.9 to +7.1
Biotite	+40.0	Molybdenum	+89.0
Barite (pure)	-71.3	Rutile	+0.85 to +4.78
Brannerite	+3.5	Scheelite	+0.13 to +0.27
Chromium	+180.0	Siderite	+65.2 to +103.8
Chromite	+125.6 to +450.0	Titanium	+150.0
Cobalt	Ferromagnetic	Tungsten	+59.0
Cobaltine	+2.0	Uranium	+395.0
Cobaltite	+0.34 to +0.64	Vanadium	+255.0
Columbite	+32.6 to +37.2	Vanadinite	-0.2 to +0.27
Copper	-0.1	Wolframite	+42.2

---

[134] R. M. Bozorth, *Ferromagnetism*, New York: IEEE Press, re-issued 1993

12.6.3 Magnetic Separation Dynamics

Magnetic Forces

Several magnetic separation concepts have been proposed throughout the years, but they all rely on the same electromagnetic principle: a particle exposed to an external spatially varying magnetic field, i.e., a magnetic field gradient, will experience a force in newtons (N) equal to:

$$\mathbf{F}_M = V\, \mathbf{M(B)} \cdot \text{grad} \mathbf{B} \tag{12.10}$$

where V is the volume of the particle in cubic meters (m^3), $\mathbf{M(B)}$ is the magnetic field-dependent magnetization (in amperes/meter) of the particle and grad$\mathbf{B}$ is the gradient of the magnetic induction (in tesla/meter)[134]. The magnetic force acting on a particle can be rewritten in terms of the magnetic susceptibility where $\mathbf{M(B)} = \chi_m(\mathbf{B})\mathbf{H}$ such that:

$$\mathbf{F}_M = \chi_m(\mathbf{B}) V\, \mathbf{H} \cdot \text{grad} \mathbf{B}. \tag{12.11}$$

The implications of Equation (12.11) are that in order to have large magnetic separation forces, it is not only the particles magnetic susceptibility that is important, but also a combination of both high magnetic field intensity and magnetic field gradient that determine the magnitude of the magnetic force.

Fluid Dynamic Effects

For many magnetic separation processing techniques (i.e., wet processing), the particles are often mixed within an aqueous fluid (slurry) which moves past the high gradient collector. Typically, the high gradient collector is a mesh fabricated from thin ribbons or wires of highly permeable stainless steel, which concentrates the magnetic field lines. In general magnetic field gradients are large near sharp edges or corners. Thus, the magnetic force, given in Equations (12.10) and (12.11), acting on the particle can be quite large. The mesh acts as a filtering mechanism by trapping magnetic particles attracted to the wire and allowing non-magnetic particle to pass freely. While traveling within this fluid, however, there are several other competing forces acting on each mixture of particle, which include

F_M = magnetic force, $\qquad$ F_D = fluid viscous drag force,
F_G = gravitational force, $\qquad$ F_B = fluid buoyant force.

Neglecting the buoyant force of the particle, Newton's law for the equation of motion for this particle in a fluid is given by Equation (12.12):

$$\rho V (d^2 r / dt^2) = F_M + F_D + F_G, \tag{12.12}$$

where ρ is the particle density in kg/m^3, V is the particle volume in m^3, r is the position coordinate in meters, t is the time in seconds, and the magnetic force has been defined previously in Equation (12.10) and (12.11). To a first order approximation, the viscous drag force is given by

$$F_D = 3\pi \eta v a, \tag{12.13}$$

where ρ = particle density (kg/m^3), v = fluid velocity (m/s), a = particle diameter (m), and η = fluid viscosity ($N\text{-}s/m^2$). Figure 12.59 illustrates the various magnetic forces acting on a magnetic particle described in Equations (12.10) and (12.11). For a magnetic particle to be collected at the surface of the wire within the magnetized volume of the separator, the following condition must be met:

$$F_M \geq F_D + F_G. \tag{12.14}$$

In the analysis above, both the density and particle size (diameter) are extremely important in the magnitude of the magnetic and viscous drag force. The magnetic force varies as the cube of

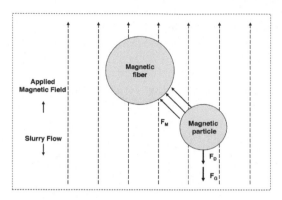

FIGURE 12.59:   Schematic representation of a magnetic fiber in a high gradient magnetic separator and the forces acting on a magnetic particle.

the particle diameter, a^3. The viscous drag force varies linearly with the particle diameter. This implies that separations between particles with similar magnetic susceptibilities are possible only when significant size and density difference exists.

12.6.4   Open Gradient Magnet Systems

The large magnetic field gradients used in separation devices are typically created using two different methods. The first method is known as the open gradient magnet system (OGMS). The second method is the matrix/filter. Open gradient devices separate particles with different magnetic susceptibilities by preferentially directing them in open spaces using magnetic forces. In the open gradient technique, the conductor geometry or suitably designed pole pieces generate magnetic field gradients and hence the magnetic forces on the particles. The magnetic forces are often in competition with other forces such as gravity, electrostatics, drag, and buoyancy. A significant shortcoming of open gradient systems is that the magnetic field gradients that can be generated are typically less than 2 T/cm. Typically, these low magnetic field gradients limit the use of OGMS devices to the separation of strongly magnetic minerals and metal scrap.

Open gradient systems have the advantage in high-volume processing. No magnetic matrix or filter is required to trap the magnetic particle mechanically. In standard operation, the particle stream is passed down the bore and magnetic particles are either pushed radially inward or outward depending on whether they are diamagnetic or paramagnetic. The particles can then be physically separated further down stream.

12.6.5   Matrix/Filter Systems

The second method most often used to generate magnetic field gradients in separation devices is the use of a magnetic matrix or filter. Frantz patented the conceptual basis for the modern ferromagnetic filament-type collector matrix[135] in 1937. The Frantz magnetic filter (Frantz Ferrofilter®) utilized a matrix of ferromagnetic type 430 stainless steel screens fabricated from thin sharp ribbons. In general, magnetic field gradients are large near sharp edges or corners. The introduction of a large number of spheroids, wires, or a mesh of wires will similarly create regions of high gradients near the wires. The magnetic field gradient near the vicinity of the wire can be as high as 10,000 T/cm. A magnetic particle traveling past the wire will experience two magnetic fields: one from the

---

[135] S.G. Frantz, U.S. Patent 2,074,085, March 16, 1937

background ambient field B_0 of the magnet and the other from the collector wire B_c, where the total magnetic field is the vector sum of the two contributions, namely, $B_t = B_0 + B_c$. A more detailed analysis on the shape effects of the matrix on the capture cross section of the particle can be found elsewhere[136].

A significant disadvantage of the matrix/filter system is the problem of build-up of magnetic particles. A magnetic separation device that utilizes a matrix/filter to mechanically trap magnetic particles must be periodically cleaned in order to prevent clogging. This means that during the cleaning of the matrix/filter, the magnetic field must be reduced to zero (or near zero) in order to minimize the magnetic trapping force. Periodically cycling the magnetic field in order to clean the matrix/filter reduces processing efficiency.

12.6.6 Characteristics of High Gradient Magnetic Separation

The efficiency of a magnetic separation device is typically expressed in two ways[137]. First, it may be characterized by the so-called "Grade". The Grade is defined as the percentage ratio of the desired component in mass relative to the total mass of the magnetic fraction (Mags). Second, the separation efficiency can be characterized by the so-called "Recovery". The Recovery is defined as the percentage ratio of the amount of magnetic material recovered relative to the total amount of magnetic material in the "Feed". The two measures are independent quantities and together determine the efficiency of the separation. For most separation devices, there is a relation between the grade and the recovery of the processed material. By adjusting the operating parameters of the separator, the grade can be increased at the expense of the recovery, and vice versa. For example, Figure 12.60 shows processing data for the magnetically trapped product of a taconite ore.

The percent Recovery and Grade is plotted as a function of the slurry flow rate at a constant applied field of 2 T. As the Recovery of the magnetically trapped product decreases with increasing flow rate, the Grade correspondingly increases.

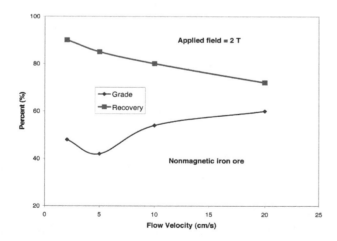

FIGURE 12.60: Grade and Recovery of a non-magnetic iron ore (non-mags) vs. slurry flow velocity at a constant applied field of 2 T.

[136] Z. J. Stekly and J. V. Minervini, *IEEE Trans. Magn.* 12 (1976)

[137] S. Foner and B. Schwartz (ed.), *Superconducting Machines and Devices*, New York, Plenum Press, (1974) 581

12.6.7 Magnetic Separation Equipment

There are several terms that are often used to characterize magnetic separation devices, such as permanent magnet, electromagnet, high gradient, open gradient, etc., but the devices themselves basically fall into two separate categories: (1) batch-type and (2) continuous-type. Table 12.14 summarizes some common types of magnetic separators and their ranges of potential applications. The following information concerning commercial magnetic separations units can be found in more detail in Knoll[130] and Rey[131].

Batch-type

Batch-type separators are most useful when the unprocessed feed materials contain a relatively small amount of magnetic material to be trapped in the collection volume. This allows for a convenient duty cycle for the device. Batch-type separation devices are most often fabricated using iron bound solenoid electromagnets surrounding a cylindrical canister. The cross section of the canister is circular in order to minimize the amount of conductor used to magnetize the collection volume. Utilizing a long coil where the height of the coil exceeds the canister height further maximizes the efficiency of a batch-type solenoid electromagnet. Iron pole pieces of the magnet are then designed to extend into the top and bottom of the solenoid. Batch-type separators operate on the principle of cycling the magnetic field from a maximum value down to zero applied field. It is important to have a magnetically soft material with a low remnant field as the collection matrix, so that when the applied field is reduced to zero the force acting on the magnetic particles is minimized. In normal operation, typical duty cycles for mineral separations vary between 10% and 80%.

Resistive Solenoid Electromagnets

In March 1973, the first large-scale HGMS was installed at the Freeport Kaolin Co. in Gordon, Georgia[130,131]. The separator was a 2.1 m diameter resistive electromagnet fabricated by Pacific Electric Motor Co. The separator could generate magnetic fields up to 2 T and process nearly 3800 liters of slurry per minute or 60 metric tons per hour (dry basis). The central magnetic field was 2 T, with an electrical power consumption of nearly 500 kW. The conductor consisted of 16 hollow water-cooled copper coils surrounded by a vault like steel enclosure that was 3.65 m × 3.65 m × 2.44 m [138,139]. For its day, this magnet represented a thirteen-fold increase in process capacity over the largest previous commercial magnetic separator. Continued improvements in magnet design have led to similarly sized units operating at 2 T with less than 300 kW of electrical power consumption. In 1982, the first 3 m diameter resistive electromagnetic separator went into operation[140]. This enormous unit could process 130 tons per hour of kaolin with about 400 kW of electrical power consumption. To date, there have been twenty-nine resistive electromagnets, 2.1 m and 3 m in diameter, installed in the US and eight others in the rest of the world for the purification of kaolin clay[140].

Low Temperature Superconducting Solenoid Electromagnets

In a low temperature superconducting (LTS) magnet, the magnetic field is generated in exactly the same way as in a conventional resistive electromagnet. The only real difference between the two is that the conductor in the LTS magnet must be maintained at a suitably low temperature in order to remain in its superconducting state. The key benefits offered by superconducting magnets

[138] J. Iannicelli, paper presented at the AIME/SME Meeting in Atlanta, GA, March 6-10, 1983. Reprinted by Aquafine Corporation Brunswick, GA.

[139] J. Iannicelli, Clays and *Clay Minerals*, 20, Great Britain: Pergamon Press, (1976) 64

[140] Aquafine Corporation, Products Catalog, Brunswick, GA, 1996

are (1) very low power consumption, resulting from zero resistance in the conductor windings (see section below) and (2) much higher magnetic fields resulting in better selectivity of particles and higher separation efficiency. However, the zero resistance property of a superconductor is only true for direct current (dc). For applications where the electromagnetic field is changing in time (ac), a superconducting material no longer operates with zero electrical resistance. Therefore, for batch-type magnetic separators, which periodically cycle the magnetic field, there is a practical limitation on the maximum allowable magnetic field ramp rate for these devices.

Several LTS materials have been studied, but the most prominent in terms of conductor fabrication and commercial implementation is an alloy of niobium and titanium (NbTi). In order to enhance electric and thermal stability of the superconductor, the NbTi is typically embedded in a normal metal matrix of copper or aluminum. This allows for greater heat transfer through thermal conduction and also provides a low resistance electrical path in the event that the superconductor comes out of its superconducting state. One disadvantage of the normal metal matrix is in ac applications. In ac applications, additional Joule heating is caused by the generation of induced currents in the normal metal matrix. To fabricate these composite superconductors, fine filaments of NbTi are either drawn or extruded with the aluminum or copper. The filament size of the NbTi can vary between about 5 and 30 microns depending upon the application. Copper and aluminum clad superconducting windings using NbTi conductor are now common place in the research community and have been used in applications such as motors, generators, transformers, and magnets.

An important aspect of the design of an LTS magnet is the choice of refrigeration or cryogenic cooling system that will be used to cool the conductor windings. There are two basic refrigeration routes that have been used successfully. The first and most prevalent route is the use of liquid or gaseous cryogens. The NbTi conductor windings are typically cooled by either immersion in a bath of liquid helium or by forcing cold, two-phase, helium gas around the conductor. Helium gas will liquefy at 4.2 K at atmospheric pressure. The helium gas that is boiled-off or heated by ac-loss is recirculated through a liquefier, cooled and recondensed into liquid or a two-phase mixture. The second method used to cool superconducting windings is through indirect cooling using a refrigerator. These refrigerators are commonly referred to as cryocoolers. Cryocoolers, based on various thermodynamic cycles, allow cooling to extremely low temperatures. One such device commonly used in the cryogenic industry is the Gifford-McMahon cryocooler. Recent advances in Gifford-McMahon cryocooler technology permit these units to generate temperatures of 4 K or lower. The units enable superconducting windings to be cooled indirectly and without the presence of liquid or gaseous cryogens and are particularly advantageous in small-scale systems where the cost of a liquefier cannot be justified. The two major drawbacks to these systems are (1) the constant supply of electrical power that is essential for reliable operation and (2) the relatively small cooling capacity of cryocoolers, typically a few watts, at 4 K. In 1996, the first successful demonstration of kaolin benefaction using a high temperature superconducting (HTS) magnetic separator was reported[141]. In this report, five different types of kaolin clays representing major worldwide deposits were processed in a 5 cm diameter warm bore HTS magnet in fields up to 2.5 T. However, because of the higher cost of the HTS wire per ampere-meter compared to its LTS counterpart, it is unclear at this time if the economic benefits from increased refrigeration efficiency will outweigh the additional capital cost of the HTS wire.

Energy Efficiency

One of the major drawbacks of conventional resistive type magnetic separators using water cooled copper conductor windings, is the high operating cost due to the large electrical power consumption. The power consumption in a conventional copper magnetic separator can be as high as 400 kW in a 3 m diameter unit. In a superconducting unit, the primary source of power consumption is the

---

[141] Eriez Magnetics, Products Catalog, Erie, PA, 1997

refrigeration unit. An equivalent 3 m diameter LTS unit is rated at about 50 kW. In production environments these units operate approximately 8000 hours per year with an expected lifetime between five and ten years. In the southern US, where most of the processing plants are located, the present average price of electricity is approximately $0.05/kWh. This translates to an annual cost saving of about $140,000 (dollars) per year with a lifetime saving of $1.4 million (dollars) for a superconducting magnet over its resistive counterpart. For processing plants located in non-industrialzed countries, the savings in annual operating cost is even more substantial as price and availability of electrical power is at a premium.

Installed Systems

The first large-scale LTS magnetic separator went into operation in 1986. The device had a 2.1 m diameter bore and was fabricated by Eriez Magnetics of Erie, Pennsylvania[142]. The device was installed at the J. M. Huber Company in Georgia and used for the benefaction of kaolin clay. The installed cost of this device was around 2 million dollars. To date, twelve batch-type LTS magnetic separation systems have been installed in the US. Eight of the twelve LTS systems installed have been retrofits, where the existing resistive electromagnets have been replaced with superconducting windings. Worldwide, five other batch-type LTS magnetic separation systems have been installed for the benefaction of kaolin clay in Australia, Brazil, China, England, and Germany. LTS magnetic separation systems appear to be displacing their conventional resistive counterparts.

Continuous-type

One of the primary shortcomings of the batch-type separator is processing efficiency. Batch-type separators operate on a duty cycle, so there is a period of time when material is not being processed. Continuous-type separation devices are designed to maximize the duty cycle and minimize the processing time. Continuous-type separators are advantageous when the magnetic fraction of the unprocessed feed material is relatively high.

Drum and Pulley Magnets

In the earliest magnetic separators, and in many that are still applicable for attracting strongly magnetic materials, permanent magnets were used in an open single surface device. These devices consisted of suspended magnets, pulleys, conveyors, or drums. These devices produced fields in the neighborhood 0.06 T and gradients of the order of 0.05 T/cm. Drum and pulley magnetic separation equipment has been used since the time of Thomas Edison, where he used a magnetic pulley for the concentration of nickel ore. These are among the most common types of magnetic separators in the world today. They can be made from either permanent or electromagnets and process either wet or dry feeds. Dry magnetic drums can be designed to perform as lifting magnets or pulleys. Magnetic drum devices have stationary magnets while pulley drums rotate. Schematics for a drum-type separator are shown in Figure 12.61.

Wet Drum Magnetic Separators

Wet drum separators are used exclusively for the processing of wet feed material for the separation of strongly magnetic coarse particles. The key processing variable that determines the size and processing capacity of the device are slurry volume, percent magnets and solids in the slurry, and

---

[142] Z. J. Stekly, *IEEE Trans. Magn.* 11 (1975) 1594

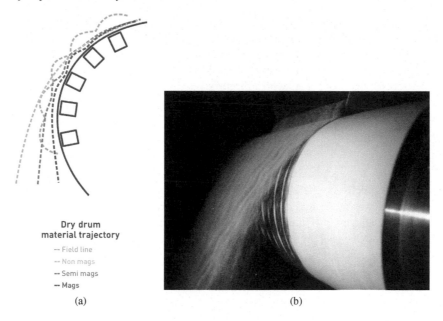

**Dry drum
material trajectory**

-- Field line
-- Non mags
-- Semi mags
-- Mags

(a)

(b)

FIGURE 12.61: (a) The principle behind a rare-earth permanent magnet drum-type separator. (b) Rare-earth drum-type separator in operation. (Courtesy Outotec Inc.)

the required recovery and concentration of magnetic particles. Several vendors manufacture these devices. Typical drum sizes vary from 0.023 m to 1.2 m in diameter with heights up to 3 m. An example of a wet drum magnetic separator is shown in Figure 12.62. Unprocessed feed material is introduced into the first drum, which is used to remove the strongly magnetic coarse particles. The processed material is then fed into the second drum for the removal of finer magnetic particles. Units with single drums can process slurry with magnetics up to 20% by weight and units with two drums can process slurry up to 45% magnetics by weight. Recommended maximum particle size is 6 mm. Concurrent devices can process finer particles down to about 0.8 mm (20 mesh) with an optimum solids content of about 30% by weight. Wet drum separators have the advantage of being able to process material with a wide variation in particle size and throughput. The installed cost of a single wet drum-type separator can vary between $25,000 and $75,000 per meter of magnet width. Multiple drum cost increase in direct proportion to the number of drums required.

LTS Reciprocating Magnet System

As early as 1975, studies were carried out on a new type of magnetic separator design that benefits from the zero resistance property of superconducting coils[143]. In 1989, Carpco (now Outotec) introduced the first commercial reciprocating magnetic separator. Unlike its LTS batch-type predecessor that cycles its magnetic field, this magnet maintains the field at a constant level and instead cycles the matrix/filter canister in and out of the active magnetic field region. This design allows for semicontinuous processing and reduces ac losses by not cycling the magnetic field. The magnet operates in what is known as the "persistent mode". Below the superconducting transition temperature, a superconducting magnet can be energized and then disconnected from the power supply, and current will continue to flow without additional power input. The basic processing cycle for the reciprocating magnetic separator is shown in Figure 12.63. The key feature is that while one matrix

---

[143] Z. J. Stekly, *IEEE Trans. Magn.* 11 (1975) 1594

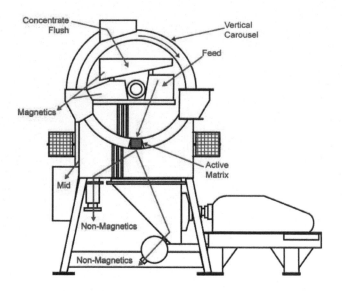

FIGURE 12.62:   Wet drum magnetic separator arrangements. (Courtesy Outotec Inc.)

canister is processing material in the central magnetic field region, the other matrix canister is being cleaned/flushed in the low field region[144].

Installed System

In 1989, the first LTS reciprocating magnetic separations unit used in the purification of kaolin clay went into operation in Cornwall in the United Kingdom. This unit consists of a NbTi conductor winding operating in a bath of liquid helium. It has a warm bore diameter of about 0.28 m with a maximum central field of 4 T and can process between 2 and 5 tons of kaolin per hour. In 1992, the second LTS reciprocating magnet system was installed in southern Germany for the purification of kaolin clay. This unit has a 0.26 m warm bore diameter with a maximum central field of 5 T and can process up to 5 tons of kaolin per hour. To date (July 2010), twenty-six more industrial scale LTS reciprocating magnetic separators have been installed worldwide. Reciprocating systems with warm bore diameters up to 1 meter and central fields of 5 T are presently in operation, see Figure 12.64. Ten smaller diameter LTS reciprocating units, operating in research laboratories and pilot scale production lines, have also been installed. One might expect that the complexity of superconducting technology would restrict the commercial viability of these units to developed and industrialized areas. In reality, the simplicity and reliability of the low-loss cryogen technology coupled with the reciprocating canister principle has enabled operation of these units in Munguba and Rio Caprim[144], which are remote areas of the Amazon rain forest.

12.6.8   Applications of Magnetic Separation

There are several commercially available mineral separation technologies utilizing specific gravity, magnetic separation, electrostatic separation, column flotation, etc. All of these techniques exploit various discernible properties among mixtures of minerals. A comparison of different separa-

---

[144]Carpco (now Outotec), Products Catalog, Jacksonville, FL, 1998 or see www.Outotec.com

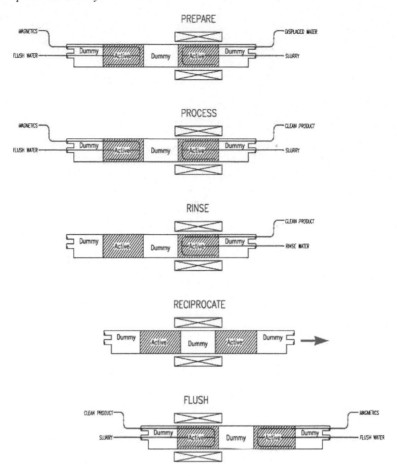

FIGURE 12.63: A schematic representation of a Reciprocating-type operating cycle. (Courtesy Outotec Inc.)

tion methods as a function of particle size range is shown in Figure 12.65.

Each separation technology has its own particular strength; however, it is typically a combination of techniques that provides the best industrial minerals separation processes. Commercial magnetic separators come in a variety of shapes and sizes depending upon the required application. To select the most appropriate magnetic separator for a specific application requires an evaluation of several variables including type of material being processed, wet or dry processing, particle-size, magnetic characteristics, and processing rate. Table 12.14. summarizes some of these defining parameters.

There are several commercial/industrial applications which utilize magnetic separation or in which magnetic separation could play a role in the future e.g., kaolin clay purification, titanium dioxide purification, waste water remediation, waste soil remediation[145], coal purification[146] chem-

---

[145] L. A. World, F. C. Prenger, D. D. Hill, D. D. Padilla, to be published in *Emerging Technologies in Hazardous Waste Management*, Proceedings of the 16th American Chemical Society Conference, Boston MA, August, 1998

[146] J. Iannicelli and T. Webster, Electric Power Research Institute Grant No. NP- 3273, research project S106-1, 1983

FIGURE 12.64: A 5 T low temperature superconducting reciprocating magnet with a 0.5 m bore, installed in Georgia. This magnet can process 20 tons of kaolin clay per hour. (Courtesy Outotec Inc.)

TABLE 12.14: Potential Applications of Magnetic Separators

Device	Magnet	Maximum Field (T)	Matrix	Maximum Gradient (T/cm)	Required Susceptibility	Particle Size (mm)
Grate	Permanent	0.05	Rods	0.05	Ferromagnetic	< 12
Pulley	Permanent	0.02	–	0.01 - 0.1	Ferromagnetic	< 50
Drum	Permanent	0.05 - 0.1	–	0.05 - 0.1	Ferromagnetic	0.02 - 20
Belt	Electromagnet	0.01 - 0.1	–	0.01 - 0.1	Ferromagnetic	0.15 - 30
Induced roll	Electromagnet	2	–	20	Paramagnetic	0.03 - 3
Carpco	Electromagnet	2	Steel balls	4.5	Paramagnetic	0.01 - 1
C-frame; Jones	Electromagnet	2	Grooved Plates	20	Paramagnetic	0.01 - 2
Marston Sala	Electromagnet superconducting	2 - 5 T	Steel wool	2500	Paramagnetic weak	0.0001 - 2

ical processing[147,148,149], and solid-liquid separation[150]. For brevity purposes, only kaolin clay and titanium dioxide are presented in more detail.

Kaolin Processing

Kaolin is a naturally occurring white clay consisting of microscopic platelets of aluminum silicate. The U.S. is the largest producer and exporter of kaolin in the world, with over 10 million tons valued at over 1.3 billion dollars produced in 1993. Georgia generates 80% of the tonnage and 90% of the value of kaolin in the U.S[138,139]. Kaolin is used in the manufacture of fine porcelain and as a base filler in the manufacture of high-grade paper. One of the most widely recognized industrial

[147] H. J. Schneider-Muntau (ed.), *High Magnetic Fields*, New Jersey: World Scientific, (1997) 31

[148] S. Dale, S. Wolf, and T. Schnieder, *Energy Applications of High-Temperature Superconductivity*, 2, 1976

[149] R. Mitchell and D. Allen, *Industrial Applications of Magnetic Separation*, IEEE Catalog No. 78CH1447-2, p.142, 1979

[150] C. M. Rey, K. Keller, and B. Fuchs, *Magnetic Processing in Magnetic Fields*, NewJersey: World Scientific, 2004

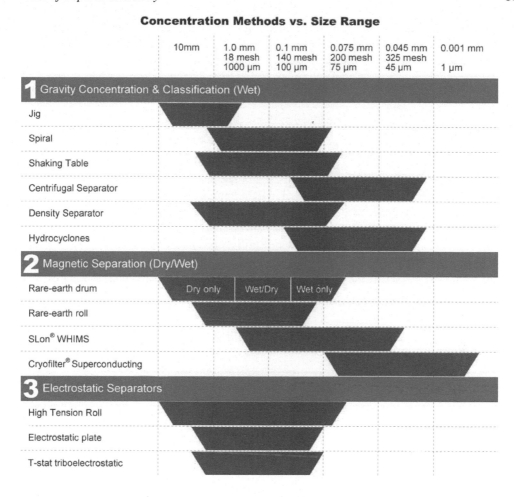

Concentration Methods vs. Size Range

	10mm	1.0 mm 18 mesh 1000 μm	0.1 mm 140 mesh 100 μm	0.075 mm 200 mesh 75 μm	0.045 mm 325 mesh 45 μm	0.001 mm 1 μm

1 Gravity Concentration & Classification (Wet)

Jig

Spiral

Shaking Table

Centrifugal Separator

Density Separator

Hydrocyclones

2 Magnetic Separation (Dry/Wet)

Rare-earth drum — Dry only | Wet/Dry | Wet only

Rare-earth roll

SLon® WHIMS

Cryofilter® Superconducting

3 Electrostatic Separators

High Tension Roll

Electrostatic plate

T-stat triboelectrostatic

FIGURE 12.65: A comparison of separation methods as a function of particle size. (Courtesy Outotec Inc.)

applications of high gradient magnetic separation is in the purification of kaolin clay. Magnetic separation has been used in the kaolin industry for over 25 years. The primary benefit is increased whiteness or brightness of the kaolin product for paper or ceramic applications. Magnetic separation has offered additional benefits to kaolin processing such as improved viscosity or rheology. Magnetic separation is used primarily to remove small paramagnetic impurities of titanium and iron compounds that discolor the clay. The impurities' content typically comprise between 2 and 5% of the weight of kaolin. The quality of the processed clay is determined by the resulting brightness. The industrial association that establishes the brightness standards for the kaolin industry is the Technical Association of the Pulp and Paper Industry (TAPPI). Brightness is typically measured using a reflectance technique and results compared with industry standards, e.g., TAPPI T 646 om-94 for pulverized material with 45°/0° geometry[151]. The improvement in brightness after magnetic separation depends upon many processing variables such as the initial quality of the unprocessed clay, magnetic field strength, mesh size and density, retention time, etc. Typical improvement between 1 and five brightness units can be expected as a result of processing kaolin in a magnetic field. Es-

---

[151] *TAPPI Test Methods*, Atlanta: TAPPI Press, 1997

sentially all kaolin is processed through magnetic separators. The most common separator size in current use can process kaolin slurry at typically 70 to 120 m^3/h, which translates to a production rate of approximately 23 to 45 metric tons per hour (dry basis). Most of the magnetic separation devices used in the purification of kaolin clay are resistive magnets that operate in batch mode. However, both batch-type and reciprocating-type superconducting magnets are quickly displacing their resistive counterparts.

Titanium Dioxide

Rutile or anatase is a naturally occurring ore containing microscopic particles of titanium dioxide (TiO_2). It is the most effective white pigment used in the paint, paper, and plastics industries. It is widely used because it efficiently refracts visible light imparting whiteness, brightness, and opacity when incorporated in a huge variety of fabricated products. Titanium dioxide is chemically inert, insoluble, and thermally stable under the harshest processing conditions. TiO_e is produced commercially in two crystal forms—anatase and rutile. Rutile based pigments are preferred because they refract light more efficiently, are more stable, and less photoreactive. Worldwide, over 2.2 million metric tons of TiO_2 are produced annually. A typical method for the separation of titanium dioxide is shown in Figure 12.66.

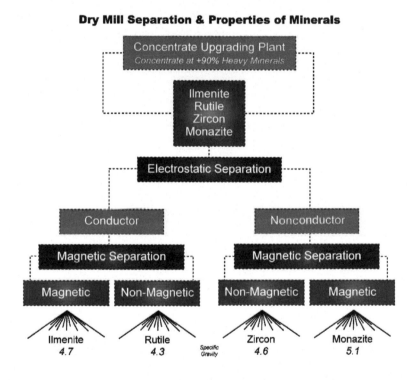

FIGURE 12.66: Dry mill separation process of rutile and ilmenite. (Courtesy Outotec Inc.)

Magnetic separation is used in both the front end screening process as well as the final processing of the "fines" of the product. As shown in Figure 12.66, both ilmenite ($FeTiO_3$) and rutile (TiO_2) are electrically conductive, but, successful separation is possible because ilmenite is strongly paramagnetic and rutile is very weakly paramagnetic.

12.6.9 Summary

Magnetic separation has been used in the processing of materials for more than one hundred years. Magnetic separation has a variety of modern uses ranging from the simple removal of tramp iron to the highly sophisticated removal of weakly paramagnetic minerals from clays. There are several different types of magnetic separation devices available. The choice and type of magnetic separator best suited for a particular application depends upon several variables such as wet or dry processing, magnetic susceptibility of the materials present, magnetic fraction of the feed materials, particle-size, and processing rate. The introduction of High Gradient Magnetic Separation greatly expanded the role that these devices play in the mining of minerals. It has been estimated that the introduction of HGMS has nearly doubled the worldwide useful reserves of kaolin clay, by allowing the mining of lower grade material. It is clear that in large-scale high gradient magnetic separators, low temperature superconducting technology is displacing conventional water-cooled copper magnets. With the recent discovery of high temperature superconductivity, it remains to be seen if these new ceramic-oxide superconductors will replace the traditional intermetallic low temperature superconductors.

Magnetic separation techniques are being explored in many non-traditional applications such as wastewater cleanup, the removal of pyritic sulfur and ash from pulverized coal, chemical processing, and the removal of uranium oxide compounds from contaminated soil. The use of "magnetic seeding" techniques on non magnetic materials may open an entirely new area of benefits and applications. As more research and development is being performed, magnetic separation techniques continue to find new areas of potential environmental and commercial benefit.

12.7 Superconducting Induction Heating of Nonferrous Metals

Niklas Magnusson and Larry Masur

12.7.1 Conventional Aluminum, Copper and Brass Induction Heating

Electromagnetic induction is widely applied for industrial heating of metals. The process is fast, clean, and in most cases very energy efficient[152]. However, when heating non-magnetic materials with low electrical resistivity such as aluminum, copper and brass, the efficiency becomes low. In aluminum extrusion plants, 50/60 Hz induction heaters are used to preheat billets, large cylinders, about 0.2 m in diameter and up to or above 1 m in length, from room temperature to around 500 °C to soften the metal before the billet is pressed through the extruder.

In conventional billet heaters, AC currents are passed through copper coils to generate a strong time-varying magnetic field. The aluminum billet is placed in this field, and the resistive losses due to the electric currents being induced in the billet generate heat, Figure 12.67. The efficiency of the process depends to a large extent on the losses in the induction coil. The current density of the windings of the induction coil is extremely high, typically around 20 A/mm^2, and, consequently, modern aluminum billet heaters operate with efficiencies in the 50% range, being one of the large-scale electrotechnical components with absolutely the poorest energy efficiency.

Large heaters have power ratings exceeding 1 MW of which 500 kW are losses dissipated in the hollow, water-cooled copper conductors and then converted into useless cooling water at temperatures around 40 °C. With 4000 operating hours per year and an energy cost of $0.05/kWh, the

---
[152]N. Magnusson and M.Runde, Proc. Eighth Int. Extrusion. Tech. Seminar, Orlando, FL, USA, (2004) 389

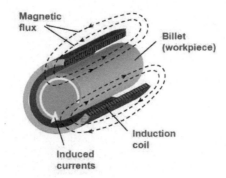

FIGURE 12.67: Principle of conventional induction heating[153].

annual value of the energy losses becomes $ 100,000 for one unit.

12.7.2 AC Superconducting Induction Heating

To increase the efficiency, the losses in the induction coil have to be reduced. One approach is to introduce windings consisting of superconducting material. In 2002 a small-scale (10 kW) superconducting AC induction heater shown in Figure 12.68, was designed, built and tested. The aluminum billet to be heated was placed in the center and was enclosed by high-temperature thermal insulation. The BSCCO/Ag HTS coil was built up of 24 double pancake coils. At the coil ends flux-diverters were inserted in the form of transformer sheets to straighten the magnetic field and hence to reduce the radial magnetic field, which otherwise would result in unacceptably high losses. The HTS coil was immersed in liquid nitrogen.

The energy efficiency of the small-scale AC superconducting induction heater was 47%, which is about the same as for conventional heaters. However, studies showed that scale up advantages and the use of an AC optimized superconductor could increase the efficiency to 80-90%. Hence, by replacing the copper windings of the AC induction coil with windings of HTS tapes, there is a potential for increasing the efficiency beyond 80%. However, the investment costs for such a solution (especially for the cooling device) are rather high and it requires the development of an AC optimized HTS tape.

12.7.3 DC Superconducting Induction Heating

The AC superconducting induction heater was, although clearly energy saving, not economically viable. At this point it was necessary to go back to the electromagnetic basics and reconsider the concept[154].

From Lenz's law we know that currents are induced in a loop when the magnetic flux enclosed by the loop is subjected to a change. For the conventional induction heater, this change is obtained by applying an AC magnetic field to the material to be heated. An alternative solution to obtain the change in magnetic flux is to rotate the material in a static magnetic field. When the loop is rotated with constant speed in the DC magnetic field, the enclosed flux varies sinusoidally in the same way as a static loop in a sinusoidal magnetic field, Figure 12.69.

To accomplish the change in enclosed flux in the aluminum billet, the topology of the rotating

---

[153]Reprinted with permission from the *Proceedings of the Eighth International Aluminum Extrusion Technology Seminar*, published by the Extrusion Technology for Aluminum Profiles Foundation.

[154]N. Magnusson, *Proc. Int. Symp. Heat Electromag. Sourc.*, Padua, Italy, (June 2007) 497

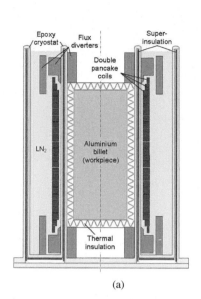

(a)   (b)

FIGURE 12.68: Schematic drawing of the AC induction heater (a) and the superconducting induction coil (b)[153].

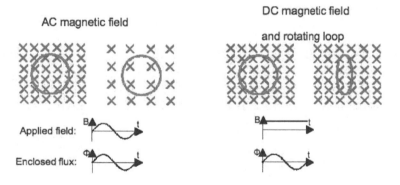

FIGURE 12.69: Applied magnetic field and enclosed magnetic flux in a loop for a time-varying field and a static loop (left) and a static field and rotating loop (right)[155].

billet heater becomes different from the one of a conventional heater. With the magnetic field oriented in parallel with the billet axis, the enclosed flux is not changed as the billet is rotated around its axis. Instead, the magnetic field should be applied perpendicular to the billet axis. Consider the schematic drawing in Figure 12.70. A DC current in a coil generates a DC magnetic field oriented perpendicular to the axis of the aluminum billet. The aluminum billet is rotated around its axis. For any given loop in a plane parallel to the rotational axis, the magnetic flux enclosed by the loop will change as the billet is rotated and hence currents are induced and these induced currents cause resistive heating in the billet. As the billet is rotated by a motor, mechanical energy from the motor is converted into heat in the billet.

The DC currents generating the DC magnetic field are passed losslessly through the coil by the

---

[155]Reprinted with permission from the *Proceedings of the 2007 Heating by Electromagnetic Sources Symposium.*

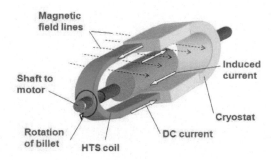

FIGURE 12.70: Principle of DC superconducting induction heating[153].

use of superconducting wires or tapes making the cooling easy and inexpensive. The efficiency of the system becomes determined by the efficiency of the motor used for rotating the billet and by the need of cooling power and the cooling efficiency. Asynchronous electric motors in the 500 kW class typically operate with efficiencies well above 90% yielding an overall efficiency of up to 90% when also the efficiency of the cooling system is considered.

12.7.4   Commercial Deployment

The first magnetic billet heater was put into commercial operation at the German aluminum extrusion works Weseralu GmbH & Co KG, in July 2008. Weseralu extrudes approximately 14,000 t of aluminum profiles per year for an annual turnover of about €50 million. Weseralu's market focus is on highly demanding profiles for the automotive and engineering sectors; hence high technical achievement in billet heating is an important part of the manufacturing strategy.

FIGURE 12.71: Photograph of magnetic billet heater operating at Weseralu aluminum profile extrusion plant[156].

―――――――――――
[156]Reprinted with permission from *Light Metal Age Magazine* 67:2 (2009)

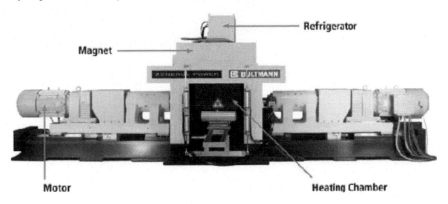

FIGURE 12.72: The four main components of the magnetic induction heater for aluminum billets[156].

This magnetic heater which has been in commercial operation at Weseralu is equipped with a DC-powered superconductive magnetic coil. Due to the uninhibited flow of electricity the coil delivers a sufficiently strong magnetic field for induction heating with a power input of merely 10 W. As described above, the rotation of the billet induces eddy currents, which work to oppose the rotation and create a strong braking torque. This is overcome by industrial electric motors of a size of 100 kW-500 kW. The energy consumed by the motors is directly converted into heat within the rotating billet. The power supply of the motors involves standard frequency converters which cause electric losses on the order of 2–3% of the total power consumption of the magnetic heater. The cooling system and the power supply of the magnet consume about 13 kW. The total energy efficiency of a magnetic heater including all losses caused by peripheral technical devices is greater than 80%. A photograph of the plant is shown in Figure 12.71.

12.7.5 Simple Technical Design

As shown in Figure 12.72, the central component of a magnetic heater is a superconducting magnet. A refrigeration system consisting of commercial off-the-shelf components keeps it at its operating temperature. The magnet generates a magnetic field which penetrates into two thermally insulated heating chambers in which the billets rotate. Electric motors on either side provide the rotational energy. The motors can slide in and out to accommodate different billet lengths and a hydraulic system locks the billets to the drive systems. The machine focuses the heating effect exclusively on the rotating billets. In contrast to AC induction heaters, no critical component in the magnetic heater is subjected to significant temperature increase, vibrations or any other mechanical stress factors.

12.7.6 Low Frequency Billet Heating

A magnetic heater reduces the power consumption of heating aluminum billets to an average 150 kWh/t and at the same time improves the quality of the heating process. AC induction heaters operating at normal power grid frequencies of 50–60 Hz induce eddy currents mostly close to the surface of billets. In contrast, lower frequencies and a more powerful magnetic field lead to a deeper penetration of heating energy. For magnetic heating it has proved favorable to rotate billets within the magnetic field at speeds between 240 rpm and 750 rpm. This corresponds to frequencies of 4–12.5 Hz. Figure 12.73 illustrates the difference in penetration depth between an AC induction heater operating at 60 Hz and a magnetic heater operating at 4 Hz. The x-axis shows the position along the

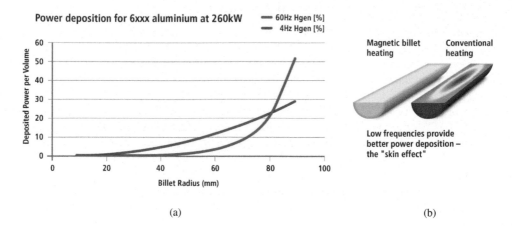

(a)   (b)

FIGURE 12.73: Comparison of deposited heat per volume as a function of distance from billet center between 60 Hz and 4 Hz magnetic induction frequencies for aluminum[156].

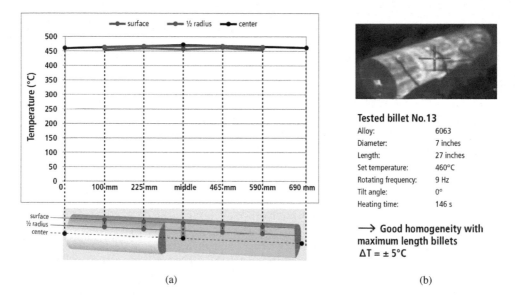

(a)   (b)

FIGURE 12.74: Comparison of temperature along the length and at the center and surface of an aluminum billet after heating. Note the negligible temperature gradient[156].

radius of the billet while the y-axis gives the heating power. Even though the total power delivered to the billet is the same in both cases, the 60 Hz curve is delivering higher power at the surface of the billet and then falling to less than 20% within about 15 mm of the surface. On the other hand, at 4 Hz the power input is much more uniform, penetrating about 50 mm before falling to 20%. With a magnetic heater the energy penetration is three times as deep as with an ac induction heater. This leads to a much more uniform heating, Figure 12.74, which provides better preconditions for the subsequent extrusion. Moreover, the heating process can be run faster and does not involve a risk of locally melting the material.

12.7.7 A Field Report: Experiences with Operation

This magnetic heater is optimized for 152-177 mm diameter billets with a length of 690 mm. Its electric drives have a power output of 360 kW and the machine has a capacity of 2.2 t/hr when heating aluminum. In commercial production the magnetic billet heater has been found to be advantageous both economically and technically. Heating 152 mm×690 mm aluminum billets takes only 140 s per billet; heating two billets simultaneously allows a heated billet to be delivered to the extrusion press every 70 s. Furthermore, the target temperature of the billet could be lowered by more than 30 °C due to the enhanced temperature homogeneity throughout the billet. The lower billet temperature in turn enabled improvements in extrusion, especially with complex and highly precise aluminum profiles. Moreover the quality of surface finishes has been improved. At Weser-alu GmbH & Co KG, the increase of productivity directly attributable to the deployment of the magnetic billet heater has been found to amount to an average of 25% across a variety of profiles. Achieving these results did not involve any significant additional expense; the results occurred as a consequence of optimizing the billet temperature and extrusion speed. At the same time the cost of heating aluminum billets was reduced by 50% compared with conventional induction heating. The combined economic effect of energy savings and productivity improvements resulted in a payback period of less than two years for the magnetic billet heater.

12.8 Superconducting Magnets for NMR

Gerhard Roth

12.8.1 Introduction

When Edward Purcell in 1945 and Felix Bloch in 1946 independently recorded the first Nuclear Magnetic Resonance (NMR) signal[157,158], for many years no one thought that NMR would ever develop into such a large field of diverse applications, a field of research which today is giving us deep insight into the chemical structure of molecules, biomolecules and since 1978 even into living bodies. Very much like the phenomenon of superconductivity, NMR started as a physics discovery leading to the Nobel Prize for a number of researchers who first described the new effect[159,160,161,162]; the phenomenon then disappeared for many years amid the vast amount of new science, but once the time was propitious, the background technology appeared and industrialization led to its widespread practical use. Even particle physics became involved in the late 1960s when experiments involving dynamic polarization of various nuclei became an essential tool in the study of the structure of matter using the then rapidly increasing access to electron and proton beams from high energy accelerators[163,164].

Today, NMR methods are continuously developing as an advanced analytical tool, expanding

---

[157] F. Bloch, W. W. Hansen and M. Packard, *Phys. Rev.* 70 (1946) 460

[158] E. M. Purcell, H. C. Torrey and R. V. Pound, *Phys. Rev.* 69, (1945) 37

[159] P. C. Lauterbur, *Nature* 242 (1973) 190

[160] M. K. Stehling, R. Turner, and P. Mansfield, *Science* 254 (1991) 43

[161] A. Kumar, D. Welti and R. R. Ernst, *J. Magn. Res* 18 (1975) 69

[162] G. Otting, E. Liepinsh, B. T. Farmer and K. Wuethrich, *J. Biomol.* NMR 1 (1991) 209

[163] M. Borghini and A. Abragam, *Compt. Rend.* 248 (1959) 1803

[164] H. Desportes and B. Tsai, *Proc. Int. Symp. Magn. Techn.* 1965, 509, Stanford Linear Accelerator Center, Stanford University, Stanford, California, Sept. 1965

into new fields of applications in industrial process control, as an analysis and screening tool in food industry, in medicine for the screening of newborns for certain inherited deceases. The growth of NMR is intimately connected with the development of superconducting magnets for the simple reason that the signal-to-noise ratio of the NMR signal increases with the square of the magnetic field strength and the spatial resolution between individual signals increases linearly with the magnetic field strength. This was recognized as early as 1965, and became especially important once the chemical shift was discovered and the relation of this shift to the chemical neighborhood of the observed nucleus.

The limitations of permanent and electromagnets were appreciated in the early 1960s; further progress in NMR techniques became closely connected to progress in superconducting magnet technology, which in turn became the most important and also the most expensive part of an NMR spectrometer[165]. We tend to forget the role which the other, equally important enabling technology, played in the development of the spectroscopy industry, namely the development of the digital computer. Both technologies were responsible for the unprecedented growth of a relatively arcane phenomenon and its intrusion into everyday life, specifically MRI, the offshoot of NMR, both of which today account for the only truly large scale commercial application of superconductivity.

NMR is an important method for structure determination in analytical chemistry and biology as well as in various specialized applications such as, for example, its younger brother, MRI, where it is widely used for diagnostics in medicine. Many building blocks from different areas needed to come together to form today's complex instruments and one key component of these is quite naturally the magnet, the superconducting magnet. In its early days NMR had to do with a permanent or a resistive magnet, limited in performance to the registration of simple signals from elements or proton signals in a few simple molecules. However, for today's largely diversified and demanding applications the use of high field magnets, that is superconducting magnets, is essential.

As an illustration what an early NMR spectrometer looked like, Figure 12.75 is an example of the first commercial NMR spectrometer manufactured in 1962 by the Trüb-Täuber company of Switzerland, equipped with a 25 MHz (0.59 T) permanent magnet.

We should comment here that, as in other branches of science, the terminology of NMR usage had assumed its own course: the magnetic field strength is always expressed in MHz, corresponding to the resonance frequency of protons at that particular field. The "frequency to field" ratio is fixed for each individual nucleus. For hydrogen, this ratio is 42.576 MHz per tesla, thus the magnetic field strength of this first magnet would have been 0.59 T. The necessary spectrometer electronics needed three large cabinets, filled with advanced RF-tube technology of that time. Experimentally it was hard work to get all parameters adjusted in the right and stable way to be able to record a resonance signal. Contrast this with Figure 12.76 which shows a modern 300 MHz (7.05 T) spectrometer with experimental capabilities exceeding by far those implemented in the first 25 MHz system and a 1000 MHz magnet, the latter being the dominant part of a very high field spectrometer. The lowest frequency superconducting NMR system available today is based on a 7 T magnet.

The 1000 MHz spectrometer installation has a 23.49 T magnet, reasonably compact electronics and is a system which offers a huge variety of experimental possibilities. The magnet weighs about 15 t and stores an energy of some 50 MJ. It represents the zenith of both the current thinking in advanced magnet design and the latest superconductor wire technology.

Superconducting magnets for high resolution NMR have certain features, which although not unique to this application do necessitate a somewhat different approach in magnet design than standard precision laboratory magnets. The requirements are demanding and restrictive:

- Stability: very low drift over time with a field decay in the range of $< 1 \times 10^{-8}$ Hz/h.

---

[165] P. Grivet and M. Sauzade, *Proc. Int. Symp. Magn. Techn.*1965, 517, Stanford Linear Accelerator Center, Stanford University, Stanford, California, Sept. 1965

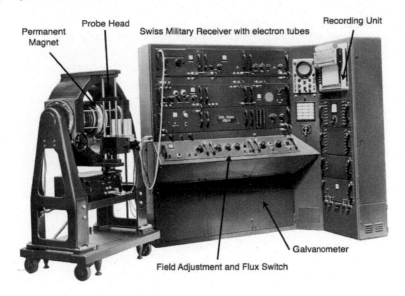

FIGURE 12.75: First commercial 25 MHz (0.57 T) NMR spectrometer in 1962 by Trüb-Täuber. (Courtesy T. Keller, Bruker Biospin.)

FIGURE 12.76: 300 MHz NMR spectrometer, left, 1000 MHz NMR magnet, right. (Courtesy Bruker-Biospin.)

- Homogeneity: in the range of $\sim 1 \times 10^{-10}$ over a sample volume 10 mm in diameter by 30 mm in length.

- Helium consumption: A cryostat system with very low helium consumption to ensure good long term stability and requiring little or no attention.

12.8.2 Stability

To meet the required criterion of very high stability over time the NMR magnet must be operated in the persistent mode, a continuous feed from a power supply simply will not do. In this mode, once the magnet has been ramped up to full field, the magnet windings are connected to a superconducting short circuit which allows the current to circulate in what amounts to a resistance free loop. Once the closed loop is superconducting, ramping down the current in the current leads to zero will not affect the current in the magnet and the magnetic field becomes persistent.

This is effected with a superconducting switch which is a length of superconducting wire surrounded by an electrical heater. When the heater is activated, the wire in the switch becomes resistive and, when ramping up the magnet, the current will flow through the superconducting path. The resistive voltage drop along this normal conducting wire length corresponds to the inductive voltage drop across the superconducting coil generated by ramping up the magnet. Once full field is reached, the voltage drop across the magnet and the switch will level off after some time. Then the heater of the superconducting switch is switched off and the wire becomes superconducting again. When the current of the power supply is reduced again, the current change will take place only along the resistance-free path and not along the inductive path of the coil. At this point the power supply current can be reduced to zero and the current leads and the power supply can be removed from the magnet. The process is illustrated in Figure 12.77. The magnet is now in the persistent mode, a "persistent" magnet which will stay at field as long as it remains immersed in liquid helium. In this simple electrical network one needs to provide a truly superconducting path for the loop current; these superconducting connections required to establish such a resistance-free path were and still are one of the biggest challenges in NMR magnet technology, especially for very high field magnets. In modern superconducting magnets such switches can be manufactured in an almost perfect way, so that the resulting magnet drift frequently falls below 2 Hz/h which corresponds to a decay time of 30,000 years. Obviously such a decay time would meet the most demanding requirements of any NMR spectroscopist.

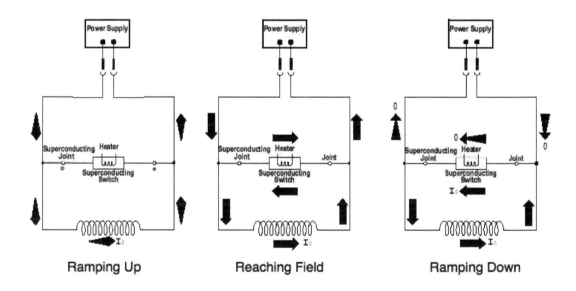

FIGURE 12.77:   Making a magnet persistent.

12.8.3   Homogeneity

A magnetic field precision of 10^{-10} implies a frequency resolution of 0.1 Hz at 1 GHz, which corresponds to a high resolution line width. Typically the goal for the superconducting magnet is to reach homogeneity of $\sim 10^{-5}$ over an axial length of ~ 5 cm, which can be achieved by designing the magnet as a solenoid with correction coils, which compensate for the second, fourth and eventually higher order field errors. In addition, such a solenoid is equipped with a set of further small correction coils, the superconducting shim coils which can be adjusted individually. Normally the magnet as manufactured will not reach the calculated homogeneity, as there are always winding errors and imperfect mechanical tolerances, which will introduce unacceptable errors in the field, usually irregular gradients. The usual procedure after fabrication is to survey the existing field gradients by measuring the magnetic field irregularities with NMR probes and then correcting the errors by energizing the compensating superconducting shim coils.

At the outset when the magnetic field strengths were still low, it was sufficient to apply a first order correction in *x*, *y* and *z* only (Figure 12.78).

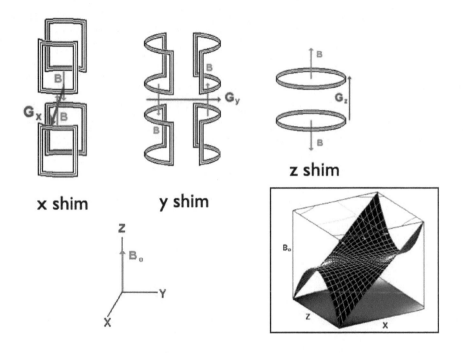

FIGURE 12.78:   Three first order *x*, *y*, *z* shims. Insert: xz^2 shim field correction plot.

With increasing field strengths, z^2 needed to be added and for magnets containing Nb_3Sn it became common practice to add second order shim coils as well to correct *zx*, *zy*, *xy* and $x^2 - y^2$ field errors. As the error correction contribution of these shims decreases rapidly with their distance from the magnet center, the correction coils require a very large number of turns in large magnets. This large number of turns cannot be wound perfectly which introduces another effect here: as the influence of the first order shims decreases less rapidly with the distance from the magnetic center, winding errors in the second and higher order shim coils lead to a considerable change in the first order shim settings when the second order shims are applied. When a third order shim is applied the influence on the lower order shims dramatically increases so that for large superconducting shims this can no longer be done.

With an appropriate set of first and second order superconducting shims the basic homogeneity requirements can be satisfied. Once this is achieved, a third set of higher order correction coils is added, consisting of resistive room-temperature shims inserted into the room temperature bore of the complete magnet. In low field magnets these additional room-temperature shim systems usually contained a total of 12 different shims. For 500 MHz devices, this was extended to 20 and later to 32 different shims correcting up to the fifth order. In current high field magnets, the room-temperature shim systems typically contain around 40 different shims, which can be addressed separately and are able to correct gradients up to the eighth order.

12.8.4 Cryostat

The cryostat housing the NMR magnet system is an equally important component defining the long-term stability. NMR devices are sufficiently sensitive that they can monitor changes in the thermal equilibrium conditions in the outer cryostat system, that is changes in cryostat temperature through ambient temperature changes. Ambient pressure changes can also be observed, even the bubbling of evaporating liquid helium and liquid nitrogen when evaporation rates are too high are registered. At lower fields this was not an issue, fortunately cryostat technology improved in step with increasing field strength and increasing sensitivity of the spectrometers. In the early days typical helium hold times were only a few days, but they were soon extended to a month and more. Today hold times are approaching values in the range of up to 1 year, depending on magnet type and even for small cryostats.

Progress was not achieved by simply increasing the storage volume of the helium and nitrogen vessels, but by reducing the helium evaporation rate down to incredibly low levels, to ~ 5 ml/h for the small magnets, so that the annual consumption dropped well below 50 l per year, corresponding to helium costs of typically less than € 500 per year.

Until the early 1990s all magnet cryostats were operated at 4.2 K. By the time NMR frequencies approached 500 MHz, which of necessity would involve Nb_3Sn technology, several attempts were made to improve the performance by taking advantage of the field shift in the critical field of NbTi wires by ~3 T, corresponding to ~130 MHz, by lowering the temperature of the helium bath[166].

In an early attempt to establish a long-term stable helium cooling technique suitable for NMR magnets, a 500 MHz (11.7 T) magnet was successfully designed and built for operation at 2.3 K at the Francis Bitter National Magnet Laboratory at MIT[167] (Figure 12.79). The method to generate temperatures below 4.2 K was not exactly unique as the magnet bath was directly pumped in the usual manner. The helium vessel was in turn surrounded by another 4.2 K vessel, isolated from the magnet bath. Helium was transferred from the 4.2 K vessel through a needle valve into the 2.3 K magnet bath in a "flash evaporation" process while pumping. Although this system was operational for a while, it turned out that this method was not really suitable for a long-term stable NMR magnet with little or no need for customer interference to operate the magnet. Also, with a line width of 1.5 Hz, this was not really a high resolution system, one which would normally require a line-width typically is in the range of 0.2 Hz.

The transition to a long-term stable and disturbance free high resolution NMR magnet took more than another decade, until the first fully sub-cooled magnet using pressurized He II technology was introduced in 1992 to push the highest available field strength to 750 MHz (17.6 T).

A similar direct method of directly pumping on a helium bath was extensively used in high energy physics (HEP) experiments at various accelerator facilities when dynamically oriented and polarized targets were used in numerous nuclear structure investigations[168]. In this NMR application the magnetic fields employed were relatively low, rarely exceeding 6 tesla, but the magnets themselves were unusually complex Helmholtz systems, with sixth to eighth order corrections. Access for detectors was the reason for the open magnet structures, and helium consumption was irrelevant in the presence of large helium recirculating liquefaction plant.

[166] W. D. Coles, G. V. Brown, E. H. Meyn and E. R. Schrader, *J. Appl. Phys.* 40 (1969)

[167] J. E. C. Williams, L. J. Neuringer, E. Bobrov, R. Weggel, D. J. Ruben and W. G. Harrison, *Rev. Sci. Instrum.* 52 (1981) 649

[168] T. Powell, M. Borghini, O. Chamberlain et al., *Phys. Rev. Lett.* (1970) 753

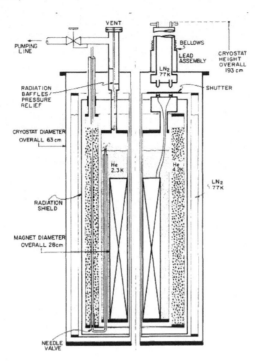

FIGURE 12.79: First NMR magnet operated at ~2.2 K, which was designed and built at MIT in 1979.

12.8.5 Major NMR Magnet Development Steps

The history of this science is relatively long. On occasion, years were needed to solve certain issues before progress could be made, progress which depended not only on the continuous development of superconducting strands but also on new magnet technology; see Figure 12.80.

A good example of this is the long time delay between the 500 MHz and 600 MHz magnets where no new superconducting materials were introduced when compared with the relatively short period for the 750 MHz, 800 MHz and 900 MHz magnets which occurred after the introduction of the new 2 K sub-cooling technology. However NMR did not develop in a technological vacuum; other large scale research applications instigated the intense and detailed research into superconducting materials, as we have observed before, thereby greatly benefiting the advances in related fields using superconducting materials and superconducting magnets. Modern technology is only too familiar with such multi-parameter relationships, not necessarily linear, the growth in size of particle accelerators being an example of technical progress driven by multidisciplinary science needs.

As mentioned before, early NMR spectrometers used permanent or resistive electromagnets which ultimately covered frequencies up to 90 MHz. The first major system involving superconductors was the introduction in 1968 of a 180 MHz (4.2 T) magnet, made with single-core or mono-filamentary NbTi conductor. This frequency was chosen because it was a multiple of 90 MHz, a frequency common to a number of installed spectrometers. The frequency also had the advantage that it simply could be reached by just doubling the existing 90 MHz spectrometer frequency.

Soon a new generation of 200 to 300 MHz (4.7 to 7 T) magnets appeared, which was quickly extended to 360 MHz (8.5 T), the reason being that at this field/frequency the limit for magnets built from mono-filamentary NbTi wire was reached. The widely observed occurrence of flux jumps led to numerous magnet quenches and a number of magnet failures in the production. The problem, as

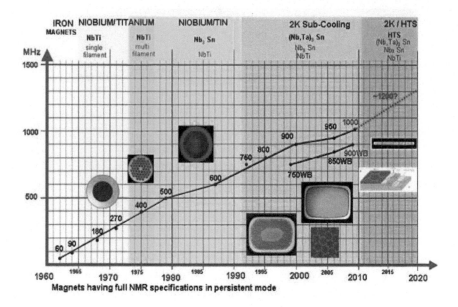

FIGURE 12.80:   Progress of the NMR frequencies as a function of time.

always, was caused by too large a filament diameter, somehow violating the then understood stability criteria. During this period the underlying theory of stability was being introduced, conductor manufacturers were beginning to understand the problem and were quickly introducing new multi-filamentary conductors with many small filaments to satisfy demand. In spite of new conductors, the magic threshold of 8.5 T, and thus the magnet with the highest field, remained uncrossed for a few more years.

To reach 400 MHz (9.4 T) required more NbTi wire development and NMR spectroscopy at that time but fortunately an unrelated branch of physics, HEP, needed more advanced materials. Not only were large particle detectors being built with superconducting magnets, but the accelerators themselves were assembled from many hundreds of superconducting magnets. This activity demanded not only industrial production of the materials, but also major qualitative improvements. As soon as there was production experience, the necessary process adjustments could be made and reliable materials with substantial critical currents at ~9.5 T could be manufactured reliably. The composition of NbTi was adapted for high field use so that from the early 1980 on, 9.4 T magnets could be built routinely. Ultimately though, in spite of significant improvements in the high field properties of NbTi, the limit was reached at this field.

12.8.6 Nb_3Sn Technology

The next step to reach even higher fields required the use of superconductors, with higher upper critical fields, so that the field range could be extended. There were several promising superconducting compounds; among these were Nb_3Zr, Nb_3Al, Nb_3Sn and Nb_3V. Initially the properties of these diverse compounds were similar, but the industrial production of Nb_3Sn in tape form quickly assured the dominance of this material. However the detrimental properties of tape were soon recognized so that attention reverted to the production of filamentary conductor. Again the demand subsided and such developments which took place focused on understanding the mechanical properties of the material rather than on improving the electrical and magnetic behavior. At this stage, at the beginning of the 1970s, another outside interest began to influence the thinking behind niobium-tin: plasma physics and its demands in connection with fusion reactors. These devices required high magnetic

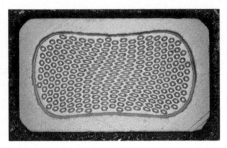

FIGURE 12.81: Examples of Nb$_3$Sn conductor with inner copper stabilization, (left) and with outer copper stabilization, (right).

fields in large volumes spurring the industry to re-examine its programs vis-à-vis niobium-tin.

In the Large Coil Task Demonstration Program at the Oak Ridge National Laboratory in the US, a sextuplet of large FD-shaped coils manufactured in the US, Europe and Japan, demonstrated that niobium-tin has potential; the project rekindled interest in this material. In the ensuing decade interest in other niobium compounds faded, Nb$_3$Sn became the material of choice for high field DC applications and its production followed in the footsteps of NbTi in the form of multi-filamentary copper clad conductor. This conductor was produced in every variant demanded by the application, strand, wire, cable, monolith, just like NbTi, but with one very important difference: the magnet with which it was wound needed to be heat treated to form the compound. This heat treatment requirement resulted in a number of different manufacturing processes, each appropriate for its intended application.

The bronze process described below is suitable for DC applications such as NMR where magnetization issues at low fields are less important than for example, in pulsed accelerator magnets. Other manufacturing processes are used for such applications, but we must emphasize that in every application the triad of current, field and temperature must be adjusted to maximize the effectiveness of the device. The left side of Figure 12.81 shows an earlier round type of Nb$_3$Sn conductor with approximately 6,000 filaments with a diameter of ~4 μm each. The filaments are pure Nb embedded in a 13% CuSn Bronze tube. The boundary of the CuSn bronze is separated by a Ta barrier from the stabilizing Cu in the center. This is necessary as the material needs to be heat treated at 700 °C, to form the superconducting Nb$_3$Sn phase, a process in which tin diffuses into the niobium filaments in a solid state diffusion process. Without this barrier, tin inevitably would also diffuse into the surrounding stabilizing Cu, this way forming bronze, which would destroy the electrical and stabilizing properties of Cu at low temperatures. The conductor shown on the right in the figure is a modern high performance rectangular Nb$_3$Sn conductor with an inverted architecture. The stabilizing copper is on the outside of the conductor and the bronze is inside, again separated by a tantalum barrier. Embedded in the bronze are ~50,000 filaments, which may be of pure niobium or may in addition contain some tantalum. The bronze may in addition to Cu and Sn also contain some small amounts of Ti. After heat treatment, this structure will result in a (NbTaTi)$_3$Sn conductor having the best high-field superconducting properties for bronze type conductors available today.

The new wire types faced some interesting problems in high-field NMR magnets caused by the complexity of the material. The material needed to be heat treated which reduced the ductility of the wire; the conductor became too brittle to be bent into the small radii of ~30 mm or so which would be needed for an NMR magnet. This led to the "wind and react" method mentioned earlier; that is the magnet coil was wound completely with the ductile, unreacted wire and then heat treated. Not only does this process have a rather long learning curve, new joining technologies had to be developed as well for connecting a small number of NbTi filaments with a large number of Nb$_3$Sn filaments ($\sim$5,000 – $\sim$50,000)) using a superconducting solder[169].

The early niobium-tin NMR magnets were prone to drifting which was initially attributed to malfunctioning joints. However, it turned out that most of the observed problems were due to the superconducting wire itself. Its brittleness, the complicated manufacturing process and the thermal heat treatment, included too many variable parameters to define a consistent manufacturing process. These problems existed for quite a while until the parameters governing bronze quality, filament size, drawing process, heat treatment conditions, magnet design itself and winding details were were fully understood. Early "good" magnets were considerably distant from what we expect good drift performance to be today.

Once the bronze route became established, many 500 MHz (11.7 T) magnets were routinely manufactured beginning in the early 1980s, followed later by 600 MHz (14 T) magnets. The long development time between these two systems was an unwelcome indicator that with existing technology progress towards even higher fields would be very tedious and slow, unless new methods were discovered thereby again speeding up high field development.

In 1985 the Research Center in Karlsruhe (now KIT) and Bruker entered into a basic development program aimed at introducing new technologies and extending the existing 11.7 T field strengths by at least 50%. This was a challenging project and the goal could be reached only by either obtaining superconductors with much higher critical currents at high fields or by increasing the critical current of existing superconductors by lowering the helium bath temperature below the lambda point.

12.8.7 Subcooling Technology

Initially it was believed that the extreme sensitivity of NMR experiments made the operation of high resolution magnets at helium temperatures below 4.2 K unlikely. Pumping on the helium bath was expected to introduce vibrations which would be detrimental to the resolution and the refilling of a low pressure bath would complicate matters. Fortunately in the development cooperation between the Research Center and Bruker in Karlsruhe alluded to before, the pressurized He II cooling method, invented at CEA in Grenoble, proved to be the route through which these magnets could be operated at about 2 K, thereby exploiting the properties of the superconductors used to the fullest.

The Research Center in Karlsruhe had existing experience with pressurized He II technology at that time which had been transferred successfully from the ideas developed and demonstrated at CEA during the TOSKA and Tore Supra programs[170]. The "HOMER" facility in Karlsruhe housed a readily available 19.3 T laboratory magnet[171], the field of which was extended to 20.3 T in 1988, at that time a world record for a pure superconducting solenoid[172]. This magnet had a large helium consumption, somewhere in the range of $\sim$15 l/h, nor was it persistent and and its homogeneity was certainly not in a range that one could think of an NMR experiment.

At Bruker, also located in Karlsruhe, there was lot of experience with low-loss cryostats, with persistent magnets, with high homogeneity NMR magnets, with additional shim coils. Bruker also had the other components, which make an NMR spectrometer, such as probe heads, filters, highfrequency electronics, pulse programs and the necessary spectrometer software. The close vicinity of these two facilities was an excellent opportunity to combine the expertise in high field magnets and the expertise in NMR and start a collaboration with the goal of raising the available spectrometer frequency.

The "HOMER" installation also facilitated the study of the properties of new advanced superconductors at high fields, so that there was no necessity to extrapolate measured data to the desired

---

[169] C. A. Swenson and W. D. Markiewicz, *IEEE Trans. Appl. Supercond.* 9.2 (1999) 185

[170] G. Claudet et al., *Cryogenics* 26 (1986) 443

[171] P. Turowski and T. Schneider, *Cryogenics.* 27 (1987) 403

[172] P. Turowski and T Schneider, *Physica* B 155 (1989) 87

field strength for the planned magnets. This provided a degree of confidence in magnet design, in addition to the existing high fields expertise, so that the first project was able to reach its goal successfully after only seven years of development. The progress in wire technology did not stop either and with $(NbTa)_3Sn$ capable of higher current densities at high fields, it became possible to design 18 T magnets at 4.2 K. Also the new higher current densities allowed the rapid advance to 800 HHz (18.8 T), a frequency which was previously accessible reliably with sub-cooled magnets only.

The sub-cooling technology for high resolution NMR magnets proved to be one of the more intricate components of the collaboration, as for once the problem was not technological but cultural: how to adapt the physics laboratory technology to NMR customer, who is not interested in operating a complex superconducting magnet, but only in performing analytical experiments. The difficult part here was to make the system reliable and convenient enough, with limited attention by the user. The helium consumption had to be reduced as far as possible and the system was required to exhibit extreme long term stability without supervision. Ultimately the combination of Bruker's low-loss cryostat expertise and the experience of the Research Center Karlsruhe in the design and construction of pressurized He II cryostats resulted in a "100 mW-cryostat". This was a big step forward, as now the helium consumption was reduced by a factor of ~100 in comparison with the performance of other laboratory magnets of similar size. The helium usage of the first systems was 140 ml/h with a liquid hold-time of two months. The sub-cooling method required for long term stability is quite simple: the helium vessel is divided into two parts, isolated thermally but connected hydrostatically. Helium from the upper reservoir is expanded through a J T valve at about 10 mbar, then passes through a heat exchanger and leaves the cryostat via a narrow bore pumping line. Such a system has the necessary long term thermal stability and requires the minimum of customer attention once it has been properly set up.

The other big issue was to suppress any vibration, which could arise from the pumping system required for the low-temperature operation. Thus from the very beginning the design preconditions required that vibration sources be reduced to the lowest possible level; the pumping line from the magnet to the pumping station was equipped with vibration stops so that no pump vibration effects were observed in the NMR signal.

It is only natural that when vibration issues occurred, as happened from time to time due to external sources such as traffic, building vibrations and so on, it became common practice of the service personnel to blame the pumps and to switch the pumping system off as a first test. It took quite a long time before everybody realized, that the origin of the vibrations was never the pumping system, but invariably some other external source. Once the peripheral issues were settled, the truly sub-cooled high field magnet using pressurized He II technology became a fully usable, commercially viable, high resolution NMR magnet. Figure 12.82 illustrates the new 750 MHz (17.6 T) magnet as well as the sub-cooling technology described.

No sooner was the basic design methodology for very high field magnets established, than it became increasingly obvious that it would be difficult to extend this field to even higher values as the known superconductors were approaching their upper critical field values with rapidly decreasing current densities. Of course the available current densities also improved slowly, and ultimately the NMR frequency could be pushed to the current 1 GHz [173]. In 1996 an actively shielded 600 MHz magnet was introduced as a result of an earlier parallel magnet technology development program for 4.2 K lower field magnets. In this program the focus was shifted from purely increasing the field strength to increasing the customer convenience. The introduction of the first actively shielded 800 MHz magnet set the starting point for the replacement of all high field magnet types with actively shielded magnets. With better and even more advanced superconductors it became possible, by 2004, to actively shield 900 MHz (21 T) systems [174]. Two years later the maximum frequency for

---

[173] G. Roth, *Bruker SpinReport* 152/153 p. 14

[174] G. Roth, *Bruker SpinReport* 156, p. 33

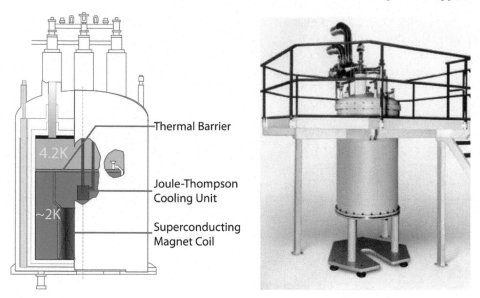

FIGURE 12.82:   On the left, the sub-cooling scheme; on the right, the first 17.6 T magnet using this technology (1992), installed in Japan and in continuous operation since 1994. (Courtesy Bruker-Biospin.)

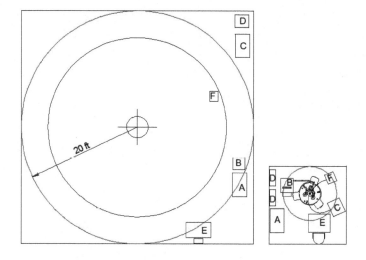

FIGURE 12.83:   Space comparison of an 800 MHz unshielded magnet NMR spectrometer (1995, left); with an 800 MHz UltraShield Plus spectrometer in the same floor space (2006, right).

actively-shielded magnets was pushed up to 950 MHz (22.3 T), while in the same year it became possible to shrink the size of actively-shielded 18.8 T magnets to a much smaller format requiring only about 1/3 of the winding volume of the earlier models.

This second generation of actively shielded magnets clearly demonstrates how far reaching the small improvements in wire technology over time really are when applied simultaneously to developments in magnet technology. Concurrently these developments resulted in significant customer benefits with the result that today the installation of an 18 or 20 T magnet in an NMR lab is as simple as the installation of a 7 T magnet was in former years. Figure 12.83 shows the space needed to set up an unshielded 800 MHz (18.8 T) system compared with the space necessary for today's actively shielded 800/850 MHz (18.8/20 T) systems. The 5 Gauss stray field line for an unshielded 18.8 T magnet needed an area of 12 m by 12 m with a ceiling height requiring two floors, or 4.9 m. Today in the same space one could set up eight 800 MHz spectrometers, and the 5 Gauss line would demand a clearance area of only 3 m by 3 m.

As the second generation actively shielded magnet fits into a single story floor, today one could in principle install sixteen 18.8 T magnets in the volume required by one first generation 800 MHz system. It would appear that at 950 MHz (22.3 T) the maximum field for actively shielded magnets using conventional LTS wire technology had been reached. The following Table 12.15 summarizes some of the characteristics of superconducting NMR magnets: the weight and the stored energy as a function of field strength. Note how the stored energy increases with increasing field strength.

TABLE 12.15: Stored Energies and Weights for High Resolution NMR Magnets from 7 to 23.5 T

Field Strength	Frequency	Weight	Stored Energy
7.05 T	300 MHz	~0.25 t	0.25 MJ
14.09 T	600 MHz	~1.2 t	0.8 MJ
19.97 T	850 MHz	~3.4 t	3.6 MJ
22.32 T	950 MHz	~6.7 t	14.6 MJ
23.49 T	1000 MHz	~15 t	50 MJ

It is technically possible to push the maximum field up to 1 GHz, as long as non-actively shielded versions of the systems are concerned because once the critical triad limit, current, temperature and magnetic field has been reached, a magnet with active shielding would become even more demanding and even more expensive. The table also illustrates the rate at which the mass of high resolution magnets increases as a function of field strength: raising the field from 20 T to 23.5 T increases the mass by a factor of more than 4; at the same time with the same advanced conductor technology in both cases, the superconductor mass increases by a factor of 8 and the amount of Nb_3Sn required increases by a factor of over 20! This dramatic increase is illustrated in Figure 12.84.

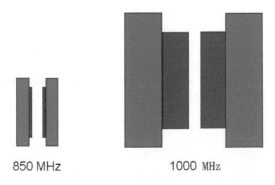

850 MHz   1000 MHz

FIGURE 12.84: Illustrative comparison of the masses of a 20 T and a 23.5 T superconducting magnet using the same advanced wire technology.

Plans to break the 1 GHz barrier already exist; if one may speak of a barrier in this case, plans for which have been around for quite a while. The discovery of high temperature superconductors in 1986 raised expectations that the magnet world would change rapidly and that high fields substantially exceeding 1 GHz would become readily available. The high critical transition temperatures and the relatively high currents at high fields, publicized at that time, were very tempting and the task appeared to be easy.

From the point of the user unfortunately, the progress in handling HTS materials and in making their properties suitable for the desired applications proved to be very slow due to their complexity. The first generation of high temperature superconductors of the BSCCO type were ruled out as an unsuitable high field magnet material. The second generation of YBCO conductors on the other hand seems to be a very promising candidate. As the production of this material ramps up and becomes more of a routine manufacturing operation, our understanding of the stability requirements and mechanical properties will increase to the extent that the material may be ready for use in a practical application of the type that we have been reviewing. Optimistically we expect to be able to reach 1.2 GHz in about 5 to 7 years from now, a frequency region which promises to be of particular interest to the researchers in the life sciences.